SUPRAMOLECULAR POLYMERS

Second Edition

Edited By
Alberto Ciferri

Duke University
Durham, NC

University of Genoa
Genoa-Italy

 CRC Press
Taylor & Francis Group
Boca Raton London New York

CRC Press is an imprint of the
Taylor & Francis Group, an **informa** business
A TAYLOR & FRANCIS BOOK

CRC Press
Taylor & Francis Group
6000 Broken Sound Parkway NW, Suite 300
Boca Raton, FL 33487-2742

First issued in paperback 2019

© 2005 by Taylor & Francis Group, LLC
CRC Press is an imprint of Taylor & Francis Group, an Informa business

No claim to original U.S. Government works

ISBN-13: 978-0-367-39295-6

Library of Congress Cataloging-in-Publication Data

Catalog record is available from the Library of Congress

**Visit the Taylor & Francis Web site at
http://www.taylorandfrancis.com**

**and the CRC Press Web site at
http://www.crcpress.com**

Preface

The five years that have elapsed from the presentation of the first edition of the book have witnessed an unprecedented expansion of the area of supramolecular polymers (SPs). The organization of the present edition in two sections (Theory/Structure and Properties/Functions) highlights the directions of said expansion. The first section describes developments that have occurred in the synthesis of complex structures, arising from an understanding of chemical design principles that underline self-assembly (Chapters 1, 6, 7, 8, 10, and 11). These developments have been accompanied by the elaboration of theoretical models of growth processes, allowing predictions of the field of stability, degree of polymerization, and shapes of a variety of assemblies (Chapters 2, 3, 4, and 9).

The self-assembly of natural and synthetic multifunctional unimers occurs, in fact, by a combination of classical molecular recognition and growth mechanisms. Self-assembled, reversible structures include linear, helical, columnar, and tubular polymers; micelles; monolayers; and three-dimensional phases and networks. Self-assembled systems based on an interplay between covalent chains and supramolecular interactions are also important. Typical examples are side-chain SPs (Chapter 5), host-guest polymeric composites and dendrimers (Chapters 2 and 7), polymers with mechanical bonds (Chapter 8), and block copolymers (Chapters 9–11).

The second section describes properties and proposed applications of synthetic SPs of both the self-assembled and engineered types. Naturally occurring SPs are included, because they demonstrate the rich potential that may eventually be matched by synthetic systems. An impressive amount of work has indeed been reported. Chapter 19, for instance, documents the impressive threefold increase of references on layered polyelectrolytes that has occurred during the five years between the two editions.

Systems included represent a selection adequate to show both the potential applications of SPs and the relationship with the basic principles outlined in the first section. Bottom-up nanotechnology applications based on DNA templates (Chapter 12) include the self-assembly of electronic components, DNA actuators, and even molecular computing. Chapter 13 describes supramolecular amphiphiles forming membranes, hydrogels, and organic/inorganic nanocomposites. Opto-electronic devices (for instance, columnar π-stacks and molecular wires) are extensively described in Chapter 14. Linear SPs are receiving attention for tunable, adaptive features expressed by their rheological and self-healing properties (Chapter 15), while tubular SPs may host polymer molecules or act as ion-selective channels (Chapter 16). Planar SPs are applied as templates for immobilization and patterning in molecular biosensors and biocompatible surfaces (Chapters 17, 18, and 19). Molecular imprinting (Chapter 20) is a recognized approach to the preparation of tailor-made receptors, artificial enzymes, and catalytic antibodies. Helical SPs coupled to chemical stimuli allow motion or force generation (Chapters 21 and 22).

Several chapters from the first edition have been updated or rewritten, and an equal number of new chapters have been added. Rather than a collection of scattered chapters, the book is an attempt to provide a unified overview of the field, with an emphasis on fundamental principles, chemical design, and applications. This was made possible by the commitment and patience of the authors and their participation in minimeetings.

Supramolecular organization can be controlled at the nanometer level but can also be programmed to attain meso- and macroscopic levels through a hierarchical ordering sequence. Thus, the general principles outlined in the book, and their further elaboration, will be guidelines for the currently emphasized nanofabrication, for mimicking functional biological structures, and for bottom-up technologies where the "bottom" limit is based on molecules but the true extent of the "up" limit is wide open to future exploration.

About the Editor

Alberto Ciferri received a D.Sc. degree in physical chemistry from the University of Rome and performed postdoctoral work with H. Benoit in Strasbourg, Germany, and P.J. Flory in Pittsburgh, Pennsylvania. He held the positions of scientist at the Monsanto Company and director of research at the National Research Council. He is currently a professor of chemistry at the University of Genoa (from 1982 until 2005) and has been a visiting professor at Duke University since 1975. He received numerous visiting appointments and has been involved in the promotion of advanced education in developing countries through his JEPA Foundation. He has published over 200 papers as well as several books and patents in the areas of polymer mechanics, biopolymers, salt interaction, liquid crystals, and supramolecular assemblies.

Contributors

Volker Abetz
Institut für Chemie
GKSS-Forschungszentrum Geeshachhht
  GmbH
Geesthacht, Germany

Xavier Arys
Department of Materials and Processing
  Science
Université Catholique de Louvain
Louvain-La Neuve, Belgium

J. Benjamin Beck
Department of Macromolecular Science
Case Western Reserve University
Cleveland, Ohio

Anton W. Bosman
SupraPolix BV
Eindhoven, The Netherlands

Alberto Ciferri
Department of Chemistry
Duke University
Durham, North Carolina

Davide Comoretto
Università di Genova, DCCI
Genova, Italy

Perry S. Corbin
Department of Chemistry
Ashland University
Ashland, Ohio

Stephen L. Craig
Department of Chemistry
Duke University
Durham, North Carolina

Bernd Fodi
Fachbereich Physik
Bergische Universitaet
Wuppertal, Germany

Reinhard Hentschke
Fachbereich Physik
Bergische Universitaet
Wuppertal, Germany

Wilhem T. S. Huck
Department of Chemistry
University of Cambridge
Cambridge, United Kingdom

Alain M. Jonas
Department of Materials and Processing
  Science
Université Catholique de Louvain
Louvain-La Neuve, Belgium

Takashi Kato
Department of Chemistry and
  Biotechnology
School of Engineering
The University of Tokyo
Tokyo, Japan

Nobuo Kimizuka
Department of Chemistry and
  Biochemistry
Graduate School of Engineering
Kyushu University
Fukuoka, Japan

Makoto Komiyama
Department of Chemistry and Biotechnology
University of Tokyo
Tokyo, Japan

Thomas H. Labean
Department of Computer Science
Duke University
Durham, North Carolina

Andre Laschewsky
Institute für Chemistry
Universität Potsdam
Potsdam, Germany

Roger Legras
Department of Materials and Processing
  Science
Université Catholique de Louvain
Louvain-La Neuve, Belgium

Jean-Marie Lehn
Laboratoire de Chimie Supramoleculaire
ISIS-ULP-CNRS-UMR-7006
Strasbourg, France

Peter Lenz
Fachbereich Physik
Bergische Universitaet
Wuppertal, Germany

Katja Loos
Department of Polymer Chemistry
University of Groningen
Groningen, The Netherlands

F. Mallwitz
Institute für Chemistry
Universität Potsdam
Potsdam, Germany

Alex Mogilner
Department of Mathematics
University of California
Berkeley, California

Sebastian Muñoz-Guerra
Departamento de Ingenieria Quimica ETSIIB
Universidad Politecnica de Cataluña
Barcelona, Spain

Colin Nuckolls
Department of Chemistry
Columbia University
New York, New York

Fumio Oosawa
Aichi Institute of Technology
Toyota-Shi, Japan

George Oster
Department of Molecular and Cell Biology
University of California
Berkeley, California

Dietmar Pum
Center for Ultrastructure Research
  and Ludwig Boltzann Institute for Molecular
  Nanotechnology
University of Agricultural Sciences
Vienna, Austria

Stuart S. Rowan
Department of Macromolecular Science
Case Western Reserve University
Cleveland, Ohio

Margit Sara
Center for Ultrastructure Research
  and Ludwig Boltzann Institute for Molecular
  Nanotechnology
University of Agricultural Sciences
Vienna, Austria

Bernhard Schuster
Center for Ultrastructure Research
  and Ludwig Boltzann Institute for Molecular
  Nanotechnology
University of Agricultural Sciences
Vienna, Austria

Rint P. Sijbesma
Laboratory of Macromolecular and
  Organic Chemistry
Eindhoven University of Technology
Eindhoven, The Netherlands

Uwe B. Sleytr
Center for Ultrastructure Research
  and Ludwig Boltzann Institute for Molecular
  Nanotechnology
University of Agricultural Sciences
Vienna, Austria

Paul van der Schoot
Polymer Physics Group
Department of Applied Physics
Eindhoven University of Technology
Eindhoven, The Netherlands

Donald A. Tomalia
Dentritic Nanotechnologies Limited
Central Michigan University
Mt. Pleasant, Michigan

Mark Whitmore
Department of Physics and Astronomy
University of Manitoba
Manitoba, Canada

George M. Whitesides
Department of Chemistry
Harvard University
Cambridge, Massachusetts

Jun Xu
Department of Chemistry
Duke University
Durham, North Carolina

Lin Yan
Pharmaceutical Research Institute
Bristol-Meyers Squibb
Princeton, New Jersey

Wei Zang
Department of Chemistry
Columbia University
New York, New York

Stephen C. Zimmerman
Department of Chemistry
University of Illinois
Urbana, Illinois

Contents

Part I: Theory and Structure

Part II: Properties and Functions

Part I

Theory and Structure

Chapter 1

Supramolecular Polymer Chemistry — Scope and Perspectives

Jean-Marie Lehn

CONTENTS

I. INTRODUCTION

Beyond molecular chemistry based on the covalent bond, lies supramolecular chemistry, the chemistry of the entities generated via intermolecular noncovalent interactions [1–3]. The objects of supramolecular chemistry are thus defined on one hand by the nature of the molecular components and on the other by the type of interactions that hold them together (hydrogen bonding, electrostatic and donor–acceptor interactions, metal–ion coordination, etc.). They may be divided into two broad, partially overlapping classes: (1) *supermolecules*, well-defined oligomolecular species resulting from the specific intermolecular association of a few components; (2) *polymolecular assemblies*, formed by the spontaneous association of a large number of components into a large supramolecular architecture or a specific phase having more or less well-defined microscopic organization and macroscopic characteristics depending on its nature (films, layers, membranes, vesicles, micelles, mesophases, surfaces, solids, etc.).

The extension of the concepts of supramolecular chemistry [1–4] from supermolecules to polymolecular entities leads in particular, to the implementation of *molecular recognition* as a means for controlling the evolution and the architecture of polymolecular species as they spontaneously buildup from their components through *self-organization* [1–4]. Such recognition-directed self-assembly is of major interest in supramolecular design and engineering. In particular, its combination with the chemistry of macromolecules and of organized assemblies led to the emergence of the areas of supramolecular polymers and of supramolecular phases such as liquid crystals [1,3–17].

A very rich and active field of research thus developed involving the designed manipulation of molecular interactions and information through recognition processes to generate, in a spontaneous but controlled fashion, supramolecular polymers and phases by the self-assembly of complementary monomeric components, bearing two or more interaction/recognition groups. These systems belong to the realm of *programmed supramolecular systems* that generate organized entities following a defined plan based on molecular recognition events [1,3]. Three main steps may be distinguished in the process: (1) *selective binding* of complementary components via molecular *recognition*; (2) *growth* through sequential binding of the components in the correct relative *orientation*; (3) *termination* requiring a built-in feature, a *stop signal*, which specifies the end point and signifies that the process has reached completion.

In addition, supramolecular chemistry is a *constitutional dynamic chemistry* [4] due to the *reversibility* of the connecting events, that is, their kinetic lability allows the exploration of the energy hypersurface of the system. It confers to self-assembling systems the ability to undergo *annealing* and *self-healing* of defects and to manifest tunable degree of polymerization and cohesive properties (rheology). In contrast, covalently linked, nonlabile species cannot heal spontaneously and defects are permanent.

Since this book provides a wide selection of relevant topics presented by some of the major actors in the domain, we shall emphasize here the conceptual and prospective aspects, illustrated by a brief retrospective of our own work. It started with the exploration of the concept of supramolecular polymers, introduced in 1990 [3a], through the implementation of the principles of supramolecular chemistry to generate polymers and liquid crystals of supramolecular nature from molecular components interacting through specific hydrogen-bonding patterns. The chemistry of supramolecular polymeric entities based on these as well as on other types of noncovalent interactions has since then actively developed [3–17].

II. GENERATION OF HYDROGEN-BONDED SUPRAMOLECULAR POLYMERS AND LIQUID CRYSTALS

Intermolecular processes occurring in a material may markedly affect its properties. Thus, supramolecular polymerization could be expected to induce changes in phase organization, viscosity, optical features, etc. For instance, the interaction between molecular units that by themselves would not be mesogenic could lead to the formation of a supramolecular species presenting liquid-crystalline behavior. It might then be possible to take advantage of selective interactions so that the mesogenic supermolecule would form only from complementary components. This would amount to a *macroscopic expression of molecular recognition*, since recognition processes occurring at the molecular level would be displayed at the macroscopic level of the material by the induction of a mesomorphic phase.

A. Formation of Mesogenic Supermolecules by Association of Complementary Molecular Components

The most common type of molecular species that form thermotropic liquid crystals possess an axial rigid core fitted with flexible chains at each end. One may then imagine splitting the central core into two complementary halves, whose association would generate the mesogenic supermolecule, as schematically represented in Figure 1. This was realized with the derivatives P_1 and U_1 of the heterocyclic groups 2,6-diamino-pyridine P and uracil U presenting complementary arrays, DAD and ADA, respectively, of hydrogen-bonding acceptor (A) and donor (D) sites. Whereas the pure compounds did not show liquid-crystalline behavior, 1:1 mixtures presented a metastable mesophase of columnar hexagonal type as indicated by x-ray diffraction data. Its existence may be attributed to the formation of a supermolecule via molecular recognition-directed association of the complementary components U_1 and P_1, followed by the self-organization into columns formed by stacks of disk-like plates each containing two units of the supermolecule, arranged side by side (Figure 2) [12].

Supramolecular discotic liquid crystals may be generated via the initial formation of disk-like supermolecules. Thus, the tautomerism-induced self-assembly of three units of the lactam–lactim form of disubstituted derivatives of phthalhydrazide yields a disk-like trimeric super-molecule (Figure 3). Thereafter, these disks self-organize into a thermotropic, columnar discotic mesophase [13].

Related processes are the formation of discotic mesogens from hydrogen-bonded phenanthridinone derivatives [14a] and especially of helical mesophases based on tetrameric cyclic arrangements (G-quartets) of guanine-related molecules [14b].

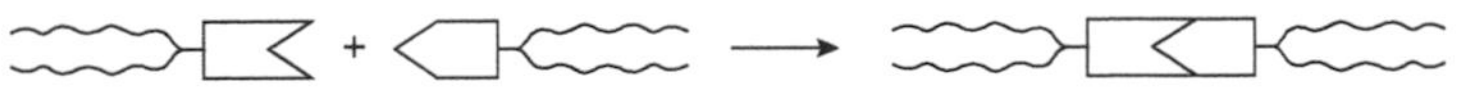

Figure 1 Formation of a mesogenic supermolecule from two complementary components.

$$k, l = 10; m = 11; n = 16 \Rightarrow$$
$$k, l = 16; m = 15; n = 16 \Rightarrow$$

37.0 Å
40.4 Å
3.5 Å

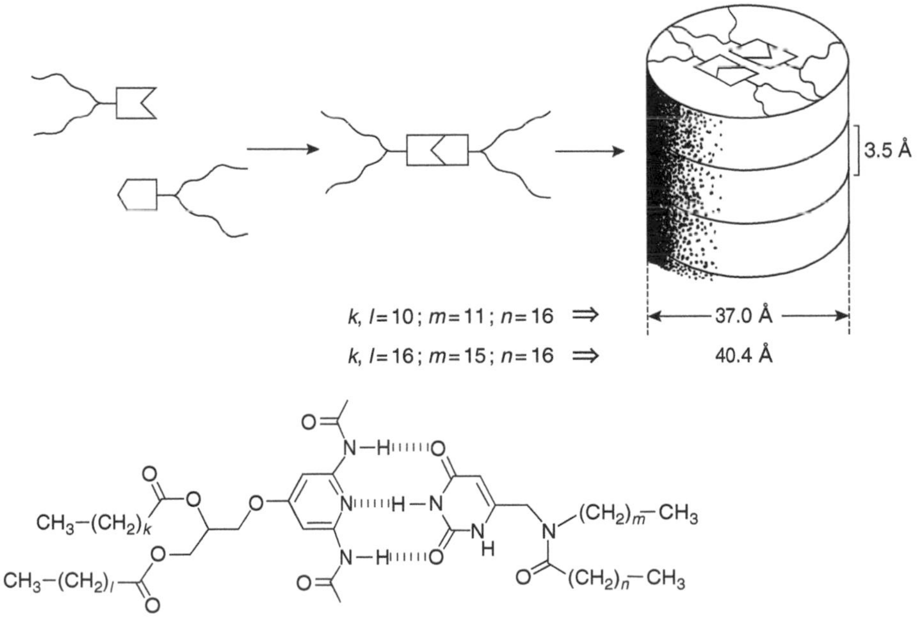

Figure 2 Formation of columnar mesophases (top) from disks constituted by the side by side arrangement of two units of a supermolecule (bottom) resulting from hydrogen bonding between the complementary components P_1 (bottom left) and U_1 (bottom right).

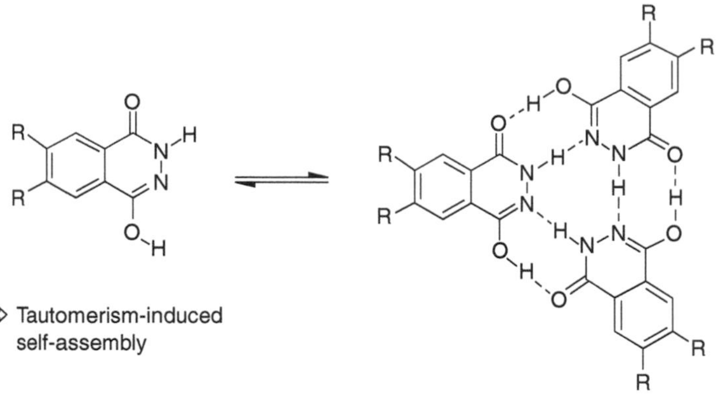

⟹ Tautomerism-induced
self-assembly

Figure 3 Tautomerism-induced self-assembly of a supramolecular cyclic trimer from the lactam–lactim form of phthalhydrazide derivatives.

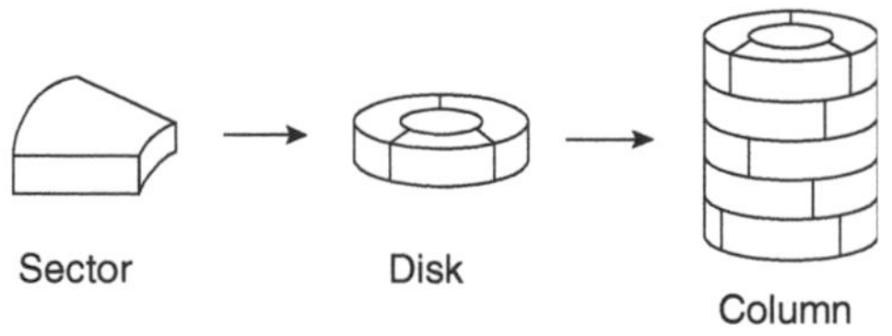

Sector   Disk

Column

Figure 4 Hierarchical self-assembly. Self-assembly of sector components into a disk is a prerequisite for the subsequent self-assembly of the disks into a discotic columnar architecture; the case illustrated is that of a trimeric mesogenic supermolecule (see Figure 3).

All these cases represent overall examples of *hierarchical self-assembly*, a conditional process where the initial assembly of molecular components into a disk-like mesogenic supermolecule is a prerequisite for the subsequent formation of a discotic columnar architecture by stacking of the disks (Figure 4; also see Section II.B).

B.   Generation of Supramolecular Liquid-Crystalline Polymers and Fibers

Mixing molecular monomers bearing two identical hydrogen-bonding subunits should lead to the self-assembly of a linear "polymeric" supramolecular species via molecular recognition-directed association. Figure 5 schematically represents such a process which served as the basis for the concept of supramolecular polymers [3c]. The resulting supramolecular polymeric material may be expected to present novel features resulting from its polyassociated nature, for instance, liquid-crystalline properties if suitable chains are introduced on the components.

Condensation of the complementary groups P and U with long-chain derivatives of tartaric acid T (T = L, D, or M) gave substances TP_2 and TU_2 each bearing two identical units capable of undergoing supramolecular polymerization via triple hydrogen bonding [3c]. Whereas the individual species LP_2, LU_2, DP_2, DU_2, MP_2, and MU_2 are solids, the mixtures ($LP_2 + LU_2$), ($DP_2 + LU_2$), and ($MP_2 + MU_2$) display *thermotropic mesophases* presenting an exceptionally wide domain of liquid crystallinity (from <25 to 220–250°C) and a hexagonal columnar superstructure, with a total column diameter of about 37 to 38 Å. The materials have the aspect of a highly birefringent glue that forms fibers upon spreading. The overall process may be described as the molecular recognition-induced self-assembly of a supramolecular liquid-crystalline polymer $(TP_2, TU_2)_n$ **1** (Figure 6). Supramolecular telechelic polymers and block copolymers based on related hydrogen-bonding units have been described recently [15].

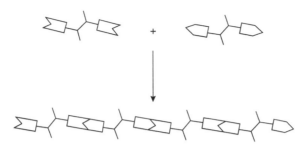

Figure 5 Formation of a polymeric supramolecular species by association of two complementary ditopic molecular components.

The x-ray patterns for $(LP_2, LU_2)_n$ are consistent with columns formed by three polymeric strands having a *triple helix superstructure*, whereas those for the $(MP_2, MU_2)_n$ mixture fit a model built on the three strands in a zig-zag conformation (Figure 7).

Electron microscopy studies revealed the successive states of self-assembly of **1** from nuclei to filaments and then to very long helical fibers of opposite chirality for the (L, L) **1** and (D, D) **1**, whereas the achiral (M, M) **1** material showed no helicity (Figure 8) [16]. Thus, molecular chirality is transduced into supramolecular helicity which is expressed at the level of the material on nanometric and micrometric scales.

The *racemic mixture* of all four components LP_2, LU_2, DP_2, and DU_2 yielded long superhelices of opposite handedness that coexisted in the same sample, pointing to the occurrence of spontaneous *resolution* through *chiral selection* in molecular recognition-directed self-assembly of supramolecular liquid-crystalline polymers. Such chiral selection features of self-organized entities are of general significance in connection with the questions of spontaneous resolution and of chirality amplification. This was confirmed by subsequent studies on a variety of helical supramolecular polymers (see Chapters 2, 3, and 6).

Supramolecular polymers have been obtained with other types of interaction patterns between monomers, from a single hydrogen bond between a carboxylic acid and a pyridine unit [9] to four hydrogen bonds between self-complementary heterocyclic groups. The latter case is represented, for instance, by the supramolecular unit **2**, that reaches degrees of polymerization up to 1000 in isotropic (nonliquid-crystalline) solutions. It has led to especially broad and fruitful developments, yielding a variety of extended entities, which display a number of interesting physico-chemical properties [11,17]. Modifications of the liquid-crystalline properties have been induced in ternary mixtures by means of chiral additives [18a] and nanofibers are formed from ditopic complementary nucleose monomers [18b]. DNA-based supramolecular polymers are generated by duplex formation between complementary bis-oligonucleotide modules [18c].

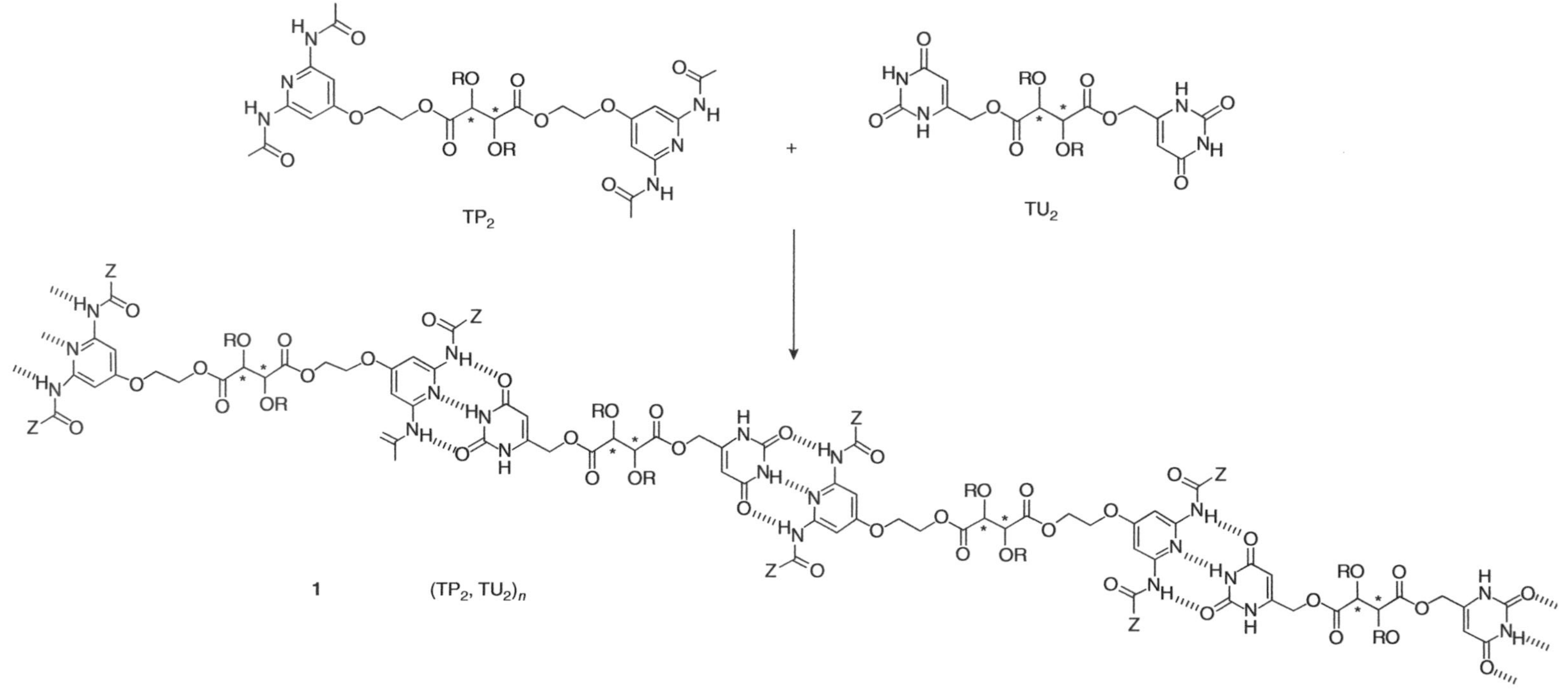

Figure 6 Self-assembly of the polymolecular supramolecular species (TP$_2$, TU$_2$)$_n$ **1** from the complementary chiral components TP$_2$ and TU$_2$ via hydrogen bonding; T represents L-, D-, or M-tartaric acid; R = C$_{12}$H$_{25}$.

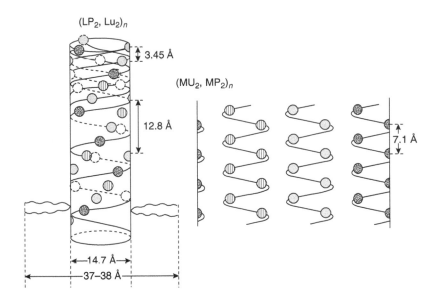

Figure 7 Schematic representation of the columnar superstructures suggested by the x-ray data for $(LP_2, LU_2)_n$, (left) and $(MP_2, MU_2)_n$ (right); each spot represents a PU base pair; spots of the same type belong to the same supramolecular strand; the dimensions are compatible with an arrangement of the PTP and UTU components along the strands indicated; the aliphatic chains stick out of the cylinder, more or less perpendicularly to its axis. For $(LP_2, LU_2)_n$, a single helical strand and the full triple helix are respectively represented at the bottom and at the top of the column. For $(MP_2, MU_2)_n$, the representation shown corresponds to the column cut parallel to its axis and flattened out.

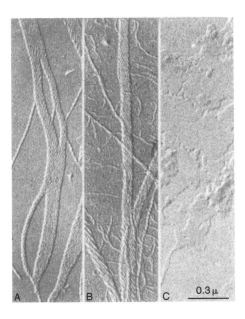

Figure 8 Helical textures observed by electron microscopy for the materials formed by the mixtures (A) $LP_2 + LU_2$, (B) $DP_2 + DU_2$, and (C) $MP_2 + MU_2$.

Supramolecular polymeric chains **3** have been obtained [19a] from the association of two homoditopic heterocomplementary monomers through sextuple hydrogen bonding [19]. They form fibers and a variety of different materials depending on conditions. Fiber formation was shown to be strongly influenced by stoichiometry, as well as by the addition of end-capping agents and of the tritopic cross-linking unit **4**.

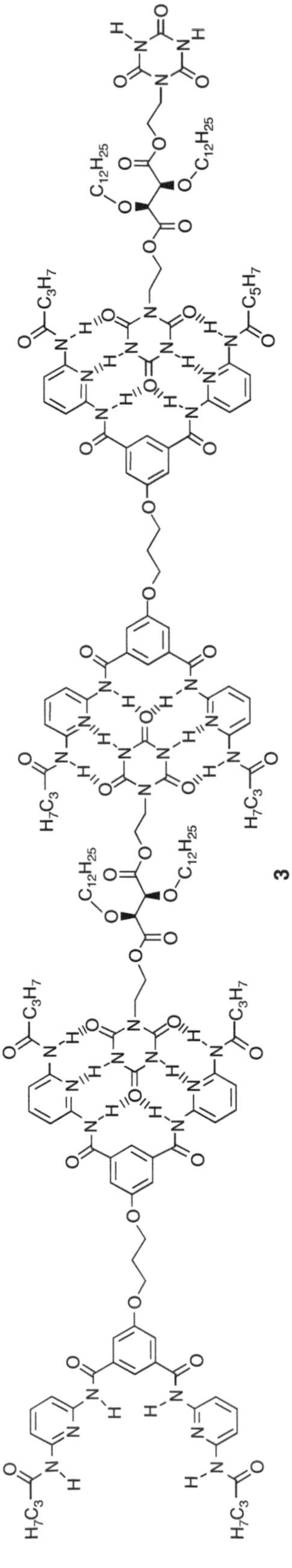

3

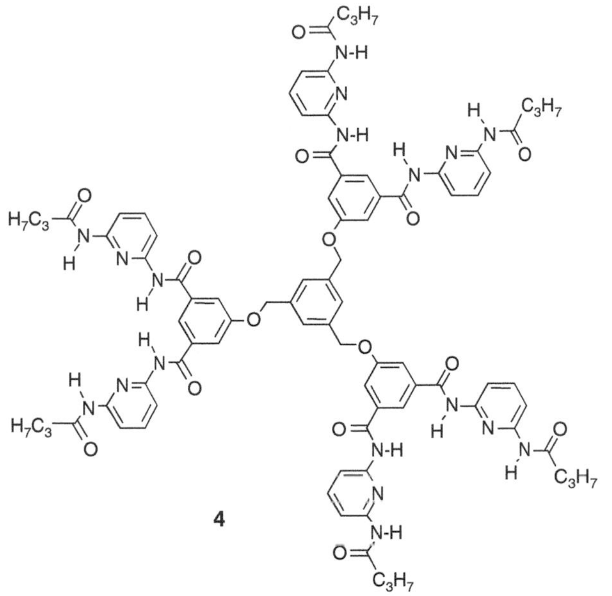

4

Multiple hydrogen bonding between calixarene-derived groups yields polymeric capsules [20].

C. Mesophases from Combination of Monotopic and Ditopic Complementary Components

In line with the processes described earlier, it may be possible to obtain liquid-crystalline materials from 2:1 mixtures of species containing, respectively, one and two recognition sites as represented schematically in Figure 9.

Indeed when an uracil component U_1 is combined with the complementary LP_2 unit in 2:1 ratio a mesophase is obtained; its occurrence may be attributed to the formation of mixed 2:1 supermolecules such as **5**, which forms a mesophase having a columnar structure of rectangular section and a very wide domain of liquid crystallinity (from <20 to 111–116°C, for R = $C_{12}H_{25}$, $m = C_{11}H_{22}$, $n = C_{16}H_{32}$ in **5**). Similar observations were made for the 2:1 combination of a diaminopyridine unit P_1 with the complementary LU_2 component [21].

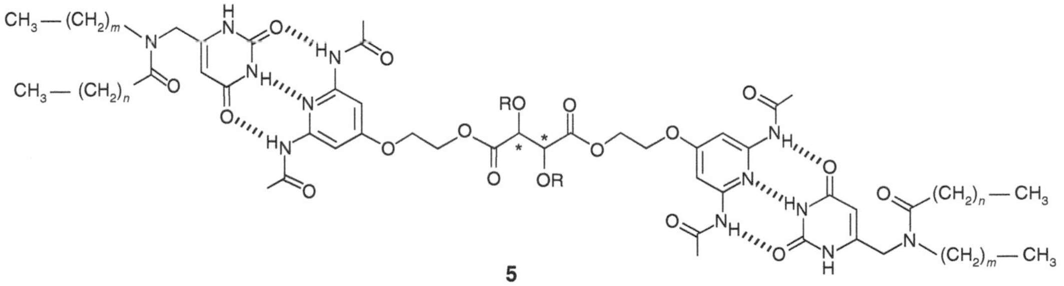

5

One may note that the U_1 and P_1 components (Figure 2) represent *chain termination* groups inducing a control of chain length via end-capping on addition to the polymeric entities **1** [19a,22]. Related effects are induced when nonstoichiometric amounts of the two homo-ditopic monomeric components in **3** are used, the component in excess acting as chain-capping agent [19a].

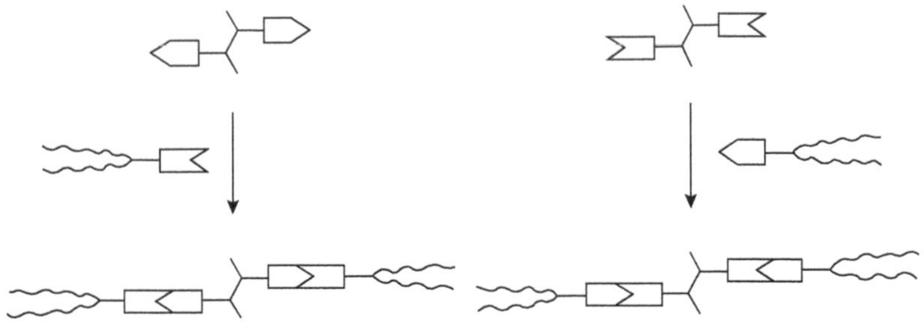

Figure 9 Formation of mesogenic species by 2:1 association of complementary monotopic and ditopic components.

Figure 10 Schematic representation of: (top) a self-assembled rigid rod supramolecular system from two rigid complementary components, and (bottom) a self-assembled mixed system from a rigid unit and a complementary flexible one.

D. Rigid Rod Supramolecular Polymers

The introduction of rigid molecular units into macromolecular species has been extensively pursued in view of the novel physico-chemical that the resulting rigid rods may present. Self-assembling rigid components may be designed by attaching recognition groups to a rigid core. Combination of two such complementary components may result in the formation of rigid rod supramolecular systems (Figure 10). Mixed materials would be formed by combining a rigid unit with a complementary flexible one, such as the LP_2 or LU_2 species described in Section II.C. The two complementary rigid components AP_2 and AU_2, each containing two identical recognition groups linked to an anthracenyl core, self-assemble to yield the rigid rod supramolecular polymeric entity **6** which was found to present a lyotropic mesophase whereas AP_2 and AU_2 themselves are solids [23].

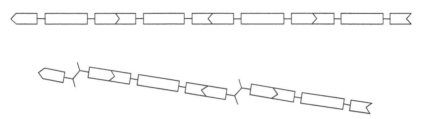

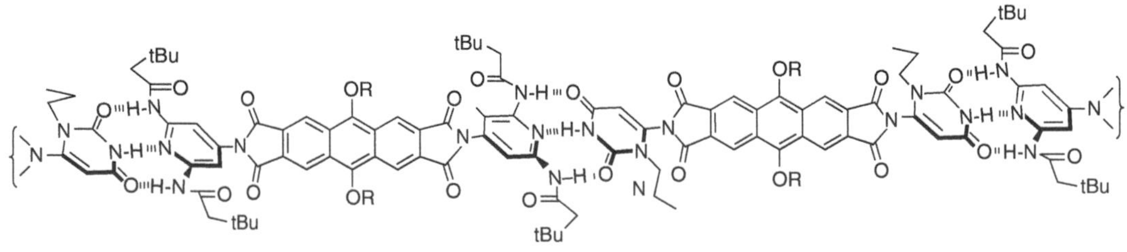

6 $(AP_2, AU_2)_n$

Hairy rigid rod polymers in which flexible side chains are attached to a rigid core, present attractive features [24]. A supramolecular version of such materials may be devised on the basis of components containing two recognition sites and on the capability of forming "hairy" ribbon-like structures. Thus, hydrogen-bonding recognition between double-faced Janus-type recognition units, such as barbituric acid and triamino-pyrimidine derivatives, leads to ordered molecular solids through formation of polyassociated supramolecular strand bearing lateral chains

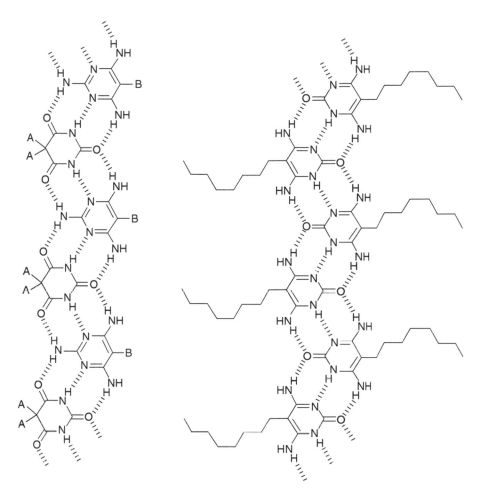

Figure 11 Self-assembled supramolecular "hairy" strands derived from components bearing side chains and containing heterocomplementary (left) or self-complementary (right) recognition sites; A and B: aliphatic chains; for the corresponding crystal structures [24a,b].

(Figure 11, left) [25a]. In the process molecular sorting out and left–right differentiation occurs, so that it is possible to obtain extended structures carrying identical or different chains of various lengths on each side.

A similar type of species bearing identical chains on each side is formed from components containing self-complementary recognition arrays of diamino-pyrimidone type (Figure 11, right) [25b]. However, whereas the (barbituric acid, triamino-pyrimidine) combination may in principle yield either a linear strand or a cyclic entity [25b,c], this latter recognition group enforces the formation of a strand only. Furthermore, one may also point out that, in this case, the arrangement of interaction sites presents some relationship with the Oosawa model of chain growth (see Figure 6 in Chapter 2, this volume).

One may note that the triple helical supramolecular species described above (Figure 7, left) displays the features of a hairy cylinder. The self-assembled species **6** bearing long R chains represent supramolecular hairy rigid rods. Mesoscopic supramolecular assemblies of cylindrical rigid rod-type, of about 150 Å diameter and several tens of micrometers length, have been obtained from a tricyclic bis-imide Janus molecule and a long-chain triaminotriazine derivative [26]. Numerous variations in the nature of the side chains of these compounds may be envisaged, giving in principle access to a variety of materials.

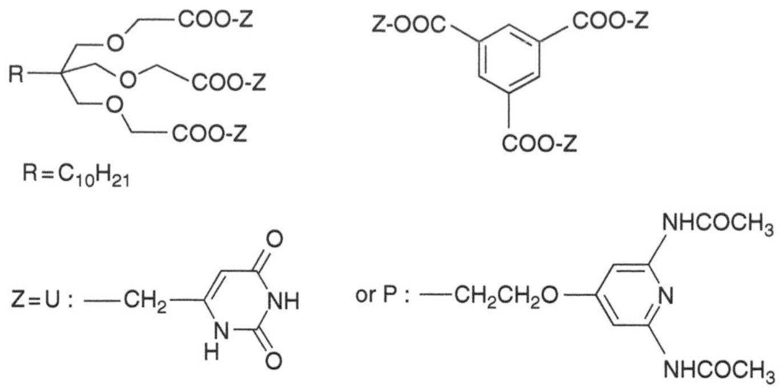

Figure 12   Ternary recognition components for the cross-linking of supramolecular polymeric species.

E.   Cross-Linking of Supramolecular Polymers

Extending further the procedures of polymer chemistry to supramolecular species, one may envisage to devise supramolecular cross-linking agents. Thus, tritopic components containing three equivalent recognition subunits may be expected to establish two-dimensional networks when mixed with linear polyassociated species such as those described in Section II.D, yielding cross-linked supramolecular polymers (Figure 12) [27a]. For instance, addition of **4** to **3** leads to marked changes in the material obtained, from initial long fibers to much shorter ones and to loss of structure, as observed by electron microscopy [19]. Of course, one may also imagine corresponding tetravalent components bearing four interaction groups which would then yield formally three-dimensionally cross-linked entities. Such polytopic monomers may also generate supramolecular branched species, in particular, of dendrimeric type (see Section II.F) [28].

F.   "Ladder" and Two-Dimensional Supramolecular Polymers from Monomers Containing Janus-Type Recognition Groups

Janus-type recognition groups, such as barbituric or cyanuric acid and triamino-pyrimidine or triazine derivatives (see also Figure 11), represent a special type of cross-linking units by virtue of their ability to interact through their two hydrogen-bonding faces. The incorporation of two such groups into ditopic molecular monomers provides entries toward the generation of "ladder" or double-ribbon polyassociations (when only a single ditopic component is used) or of two-dimensional supramolecular polymeric networks (from two complementary ditopic components). This is schematically illustrated in Figure 13.

Indeed, monomers in which two such groups are grafted onto tartaric acid units (e.g., replacing P and U units in TP_2 and TU_2, see Figure 6) generate very high molecular weight aggregates that may be characterized by various physical methods [27b].

G.   Supramolecular Coordination Polymers

When the monomeric components carry metal–ion binding subunits, polyassociation occurs on addition of a suitable ion, yielding a supramolecular coordination polymer [29]. This is the case for the L-tartaric acid derivative **7** bearing two methylated bipyridine groups. The binding of Cu (I) ions may *inter alia* generate chains where the components **7** are connected through [Cu (I) (bipy)$_2$] centers of tetrahedral coordination. Indeed, addition of $Cu(CH_3CN)_4PF_6$ to **7** resulted in the formation of organized phases, in particular *self-assembled inorganic nanotubes* of very regular structure, resulting presumably from the helicoidal winding of a large tape-like entity, as revealed by electron microscopy (Figure 14) [29a].

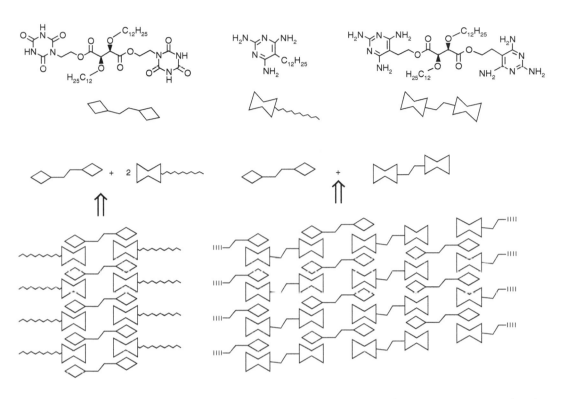

Figure 13 Schematic representation of the H-bond-mediated self-assembly of monomer components bearing Janus-type complementary recognition groups (e.g., cyanuric acid and 2,4,6-triaminopyrimidine); forming supramolecular polymers: (left) of "ladder" or double-ribbon type from a double-Janus and a single-Janus component, (right) of two-dimensional cross-linked nature from two double-Janus components.

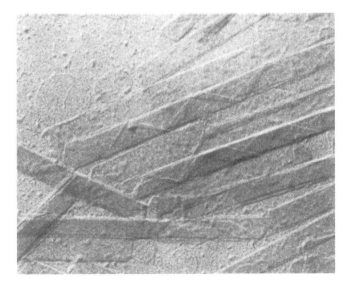

Figure 14 Inorganic nanotubes formed by the ditopic ligand **5** with Cu(CH$_3$CN)$_4$PF$_6$ observed by electron microscopy [29a].

Soluble coordination polymers of variable degrees of polymerization have been obtained from bis-bipyridylketone and Cu (II) ions in different solvents; they have been characterized by electrospray mass spectrometry revealing molecular weights reaching >60,000 Da [29b]. More recent examples of coordination polymers are discussed in Chapter 6 (see also Refs. 29c,d).

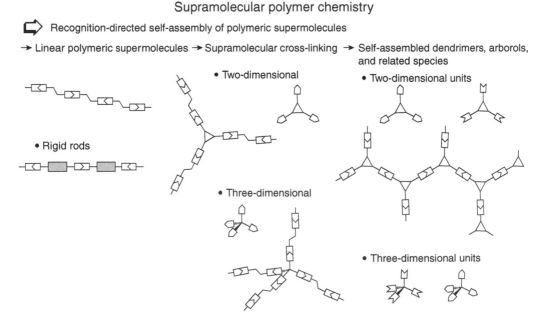

III. BASIC FEATURES OF SUPRAMOLECULAR POLYMERS

The results presented earlier illustrate the rich domain that emerges from the combination of polymer chemistry with supramolecular chemistry. It involves the generation of polymeric superstructures by the designed use and manipulation of molecular interactions and information through molecular recognition processes. Figure 15 presents some of the different types of such

Figure 15 An aspect of the panorama of supramolecular polymer chemistry; formation of different types of polymeric supermolecular entities by recognition-directed self-assembly of monomers through noncovalent interactions, such as hydrogen bonds.

"informed" supramolecular polymers that may be generated by recognition-directed self-assembly of complementary monomer species.

Broadening the scope, we may briefly consider a nonexhaustive panorama of various types and features of supramolecular polymers depending on their constitution, characterized by three main parameters: the nature of the core/framework of the monomers, the type of noncovalent interaction(s), and the eventual incorporation of functional subunits. The interactions may involve complementary arrays of hydrogen-bonding sites, electrostatic forces, electronic donor–acceptor interactions, metal–ion coordination, etc. The polyassociated structure itself may be of main-chain, side-chain, or branched, dendritic type, depending on the number and disposition of the interaction subunits. The central question is that of the size and the polydispersity of the polymeric supramolecular species formed. Of course their size is expected to increase with concentration and the polydispersity depends on the stability constants for successive associations. The dependence of the molecular weight distribution on these parameters may be simulated by a mathematical model [19]. These features are detailed in Chapters 2, 3, and 6 for various growth mechanisms.

The basic characteristics of *main-chain supramolecular polymers* are presented schematically in Figure 16. Designating the monomer core residues by R_i, monomers bearing two identical interaction/recognition groups (homoditopic), may yield either *homopolymers*, when $R_i = R_j$ or regularly alternating *copolymers* when $R_i \neq R_j$.

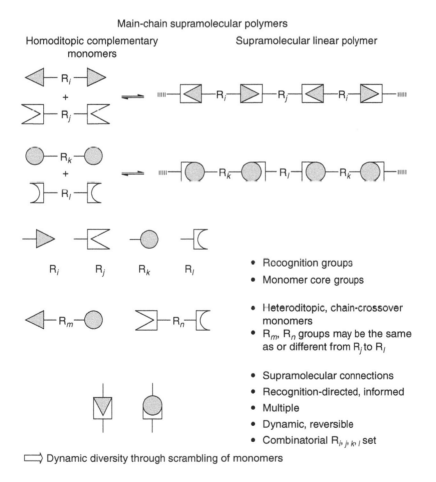

Main-chain supramolecular polymers

Homoditopic complementary monomers   Supramolecular linear polymer

R_i R_j R_k R_l

- Recognition groups
- Monomer core groups

- Heteroditopic, chain-crossover monomers
- R_m, R_n groups may be the same as or different from R_j to R_l

- Supramolecular connections
- Recognition-directed, informed
- Multiple
- Dynamic, reversible
- Combinatorial $R_{i, j, k, l}$ set

⟹ Dynamic diversity through scrambling of monomers

Figure 16 Schematic representation of the formation of linear main-chain supramolecular polymers from complementary homoditopic monomers, and of the constituting subunits. Dynamic diversity may be generated by scrambling of monomers containing different core groups.

When several different core residues are used, a large number of polymeric objects may be generated. Thus, for a chain $(R_i\,R_j)_k$ of k pairs length, formed from m different monomers R_i and n different monomers R_j, the total number of different sequences is $(m \times n)^k$. Since chains of any length ($k = 1$ to p) can be formed, the total number of different objects that can be present, comprising all possible sequences of all lengths (i.e., the full virtual combinatorial diversity [see Section IV.C]) is $mn[(mn^p - 1)/(mn - 1)]$. The fractions of the species as a function of chain length follow the size distribution curve of the system considered.

Initiation and chain growth occur on mixing of the complementary monomers. Chains of different compositions may be formed side by side, in principle without crossover, when several different pairs of complementary recognition groups are put to use (see Figure 16, top). Heteroditopic monomers combining recognition groups from different pairs may act as chain-crossover components, allowing the combination of two different chains.

Self-complementary monomers where the R units bear two complementary recognition groups yield homopolymers (only a single R_i) or random copolymers (two or more different core units R_i, R_j, R_k, . . .). Initiation and growth occur immediately after generation of the interaction groups.

In all cases, chain growth can be initiated by setting free one of the recognition groups by an external stimulus (e.g., light) from a derivative bearing a protecting group. Chain length may be altered/reduced by the addition of substances bearing a single recognition group, which act as *chain termination component* by end-capping [22].

Side-chain supramolecular polymers result from the binding of residues bearing recognition groups to complementary groups attached to the main chain of a covalent polymer (Figure 17) [9,11,30]. Of course, such main-chain covalent polymers may be cross-linked in a supramolecular fashion by means of double-headed complementary additives establishing bridges between the side groups of two different chains.

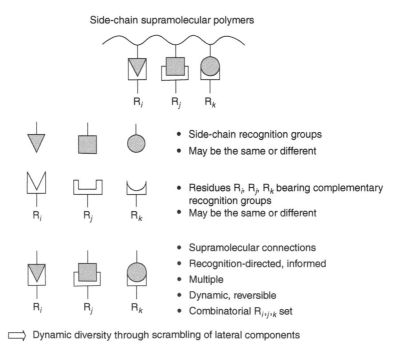

Figure 17 Schematic representation of the formation of side-chain supramolecular polymers from a covalent polymer bearing recognition groups, that bind complementary components, and of the constituting subunits. Dynamic diversity may be generated by scrambling of lateral components containing different residues R.

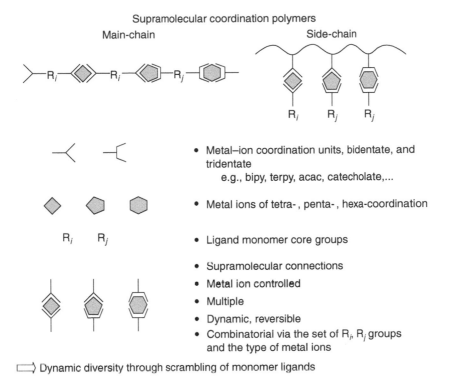

Figure 18 Schematic representation of the formation of supramolecular coordination polymers of main-chain or side-chain type from ligands containing bidentate and/or tridentate complexation subunits (such as bipyridine and terpyridine) binding metal ions of tetra-, penta-, or hexa-coordination, and of the constituting subunits. Dynamic diversity may be generated by scrambling of different ligand monomers.

Supramolecular coordination polymers represent a special class where the monomers are ditopic ligand molecules possessing two metal binding groups and where the connection is provided by metal–ion coordination (Figure 18). The metal ions play the role of association mediators enabling one to select the ligand components and to direct the polyassociation according to the combination (metal–ion/ligand binding site) in operation. One may take advantage of the vast set of metal binding units available and of their more or less selective coordination with specific ions. Initiation and growth occur only on addition of given metal ions and preferentially between specific ligands. Thus, in the presence of a mixture of ditopic ligands bearing different metal binding subunits, the nature of the "monomers" participating in the formation of main-chain coordination polymers may be determined/directed by the choice of the ion introduced. For instance, with bidentate (B) tridentate (T) metal binding sites, metal ions (M) of tetra-, penta-, and hexa-coordination are expected to yield (BMB), (BMT), and (TMT) connections, respectively (Figure 18). Of course, side-chain coordination polymers can also be obtained, as well as discotic-type coordination assemblies, such as the columnar mesophases formed from trimeric gold complexes of pyrazole derivatives [31] in a fashion similar to the processes shown in Figure 3 and Figure 4. Components combining organic (e.g., hydrogen bonding, etc.) and metal–ion binding sites may be expected to yield mixed organic/inorganic composite polymers. (For metallosupramolecular polymers combining hydrogen bonding and metal–ion binding sites see Ref. 32).

Similar considerations hold for *supramolecular organometallic polymeric* entities [33]. *Supramolecular cross-linking* may be achieved by introducing molecular monomers bearing more than two organic or metal–ion binding subunits. Such compounds provide links between chains in a way directed by the nature of the interacting groups. Figure 19 schematically represents cross-linking

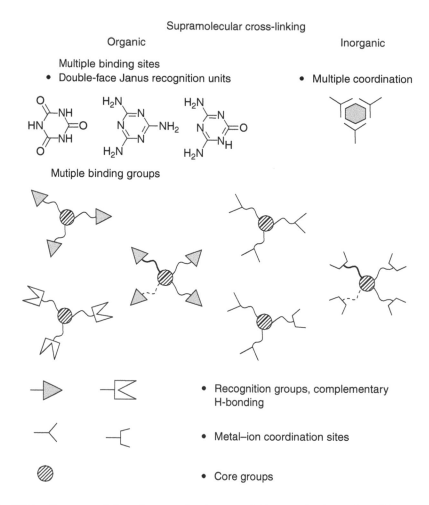

Figure 19   Schematic representation of supramolecular cross-linking agents of organic and inorganic types, and of the constituting subunits.

components of organic and inorganic types. An intriguing case is that of diamino-triazinone which possesses three different H-bonding faces DAA, ADD, and DAD, a sort of triple-Janus! Of course one may also envisage combining organic and inorganic interaction sites, which would allow the cross-linking of organic and inorganic supramolecular polymeric chains.

The use of suitable components containing multiple recognition groups should allow the designed generation of supramolecular species possessing a desired architecture. On the other hand, such components also greatly increase the dynamic combinatorial diversity of the system (see Section IV.C).

As in the case of covalent macromolecules, the supramolecular associations may also present internal (intrasupramolecular) interactions between sites located either in the main-chain or in side-chain appendages, thus leading to chain folding and structuration of the supramolecular entity.

Branched supramolecular polymers are obtained by means of cross-linking monomers. Of special significance is the fact that equimolar mixtures of complementary ternary or quaternary components lead in principle to the spontaneous generation of tree-like species that represent recognition-directed, *self-assembling supramolecular dendrimers* of the usual dichotomic as well as of trichotomic types (see also Figure 15). This holds both for organic and inorganic dendritic entities [28]. Mixed organic/inorganic dendrimers would be accessible from tri- and tetratopic monomers bearing both organic and metal–ion binding sites.

Hierarchical self-assembly takes place when several self-assembling events occur in sequence via a conditional process, a given step being a prerequisite for the subsequent one. This is, for instance, the case in the formation of discotic-type columnar liquid crystals from self-assembled supramolecular disks (see Figure 4). One may distinguish three steps in the generation of the liquid-crystalline entities described in Section II.B: formation of a supramolecular strand (**1**), assembly of three strands into columnar superstructures (Figure 7), and finally generation of fibers (Figure 8) by association of several columnar entities. The control of hierarchical organization at different scales in supramolecular polymeric materials has been described [34].

A basic common feature of all these types of supramolecular polyassociations is that they are *reversible polymers* due to the lability of the noncovalent connections, and as a result they also possess the ability to generate dynamic diversity through scrambling of the monomer components (see Section IV). Furthermore, since only correct recognition-directed complementary pairing is expected to occur, *self-selection* [35] between compatible units should take place and control of the self-assembly process is in principle possible.

In view of the lability of the associations, supramolecular polymers present features of "living" polymers capable of growing or shortening, of rearranging their interaction patterns, of exchanging components, and of undergoing annealing, healing, and adaptation processes.

Growth control and *regulation of structure and composition* may be achievable by means of external effectors (temperature, pH, metal ions, competing ligands, end-capping units, etc.). A relevant case is that of a molecular strand which undergoes a structural reorganization upon recognition-directed binding of a complementary effector to give a coiled disk-like object which thereafter self-assembles into extended fibers (Figure 20) [36].

All these most interesting features displayed by supramolecular polymers, however, raise important but difficult questions of *characterization* concerning composition, size, persistence length,

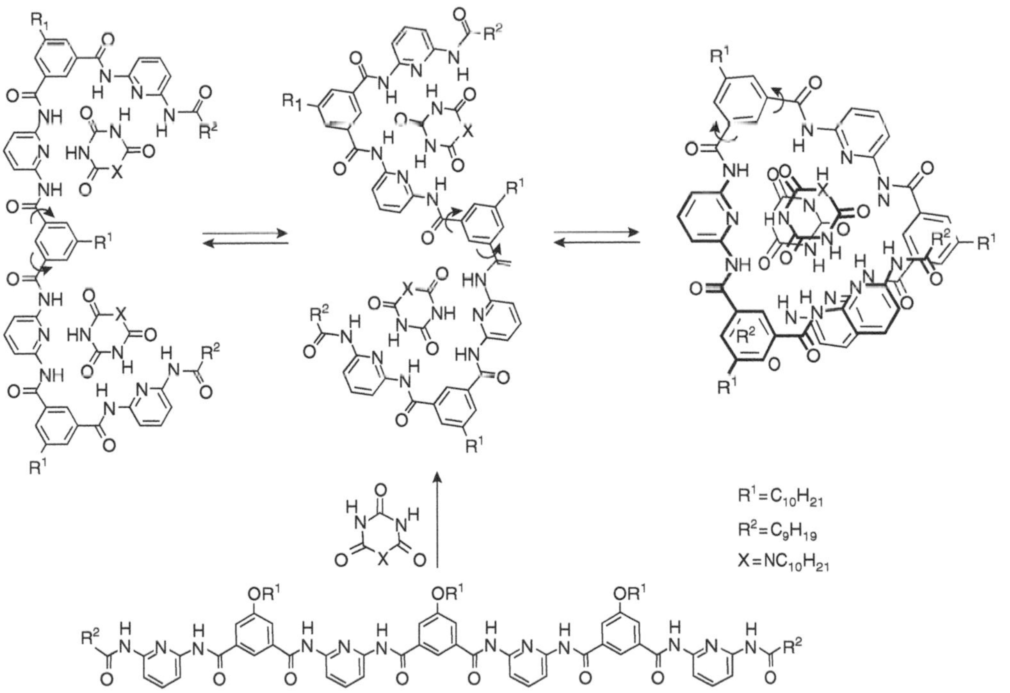

Figure 20 Folding of a molecular strand (bottom) induced by the binding of a cyanuric acid-derived template, generates different possible conformers (top) and leads to the formation of fibers from the helical conformer (top right).

shape, structure, etc., of the entities formed. To this end, an array of physico-chemical methods is required and must be put to use, such as vapor-phase osmometry, differential scanning calorimetry, electrospray mass spectrometry, NMR spectroscopy, gel permeation chromatography light scattering, electron microscopy, near field microscopies, etc. (For a relevant case, see Refs. 17 and 37.)

Supramolecular versions of the various species and procedures of molecular polymer chemistry may be imagined and implemented providing a wide field of future investigation that offers a wealth of novel entities and functionalities. Possible extensions concern, for instance, the introduction of various central cores, in particular those already known to yield molecular liquid crystals, the incorporation of photo-, electro-, or iono-active functional units, the potential use for detection devices as well as the extension to various other recognition components particularly those of biological nature (see, for instance, the case of the guanine-type quartets [14b]). Recently suggested applications are presented in Chapter 14.

IV. SUPRAMOLECULAR POLYMERS AS SUPRAMOLECULAR MATERIALS

A. Supramolecular Materials

The properties of a material depend both on the nature of the constituents and on the interactions between them. Supramolecular chemistry may thus be expected to have a strong impact on materials science via the explicit manipulation of the noncovalent forces that hold the constituents together. These interactions and the recognition processes that they underlie, allow the design of materials and the control of their buildup from suitable units by self-assembly.

Through recognition-directed association, self-assembly, and self-organization processes, supramolecular chemistry opens new perspectives in materials science toward an area of supramolecular materials whose features depend on molecular information and which involve "smart" materials, network engineering, polymolecular patterning, etc. As shown in Section III, liquid-crystalline polymers of supramolecular nature presenting various supramolecular textures may be obtained by the self-assembly of complementary subunits. This amounts to a macroscopic expression of molecular information via a phase change which, being a highly cooperative process, also corresponds to an amplification of molecular recognition and information from the microscopic to the macroscopic level.

Supramolecular engineering gives access to the molecular information-controlled generation of nanostructures [38,39] and of polymolecular architectures and patterns in molecular assemblies, layers, films, membranes, micelles, gels, colloids, mesophases, and solids as well as in large inorganic entities, polymetallic coordination architectures, and coordination polymers.

Molecular recognition processes may be used to induce and control processes between polymolecular assemblies such as organization of, and binding to, molecular layers and membranes [10,40], selective interaction of lipidic vesicles with molecular films [41], aggregation and fusion of *recosomes*, vesicles bearing complementary recognition groups [42,43], etc.

The buildup of supramolecular architectures and materials may involve several steps and proceed in particular via hierarchical self-assembly processes (see also above). On the one hand such sequential conditional processes enable the progressive buildup of more and more complex systems in a directed ordered fashion; on the other hand, they offer the intriguing possibility to intervene at each step so as to either suppress the following ones or to reorient the subsequent evolution of the system into another direction, toward another output entity.

Molecular recognition-directed processes also provide a powerful entry into *supramolecular solid state chemistry* and *crystal engineering*. The ability to control the way in which molecules associate may allow the designed generation of supramolecular architectures in the solid state. Modification of surfaces with recognition units leads to extended exo-receptors [1,3] that could display selective surface binding on the microscopic level, and to recognition-controlled adhesion on the macroscopic scale [41,44].

B. Supramolecular Chemistry: Constitutionally Dynamic

Supramolecular chemistry is intrinsically a dynamic chemistry in view of the lability of the interactions connecting the molecular components of a supramolecular entity. Moreover, and most significantly, the reversibility of the associations allows a continuous reorganization by both modification of the connections between the constituents and incorporation or extrusion of components by exchange with the surroundings, conferring therefore *combinatorial* features to the system.

Supramolecular chemistry thus has a direct relationship with the highly active area of combinatorial chemistry, however in a very specific fashion. Indeed, reversibility being a basic and crucial feature of supramolecular systems, the dynamic generation of supramolecular diversity from the reversible combination of noncovalently linked building blocks falls within the realm of the emerging area of *dynamic combinatorial chemistry* (DCC) which involves dynamic combinatorial libraries, of either virtual (VCL) or real nature depending on the system and the conditions [45,46]. The concepts and perspectives of the DCC/VCL approach have been outlined, *inter alia* with respect to supramolecular polymers [45].

Consequently, *supramolecular materials* are by nature *dynamic materials*, defined as materials whose constituents are linked through *reversible* connections (covalent or noncovalent) and undergo spontaneous and continuous assembly/disassembly processes in a given set of conditions [47]. Because of their intrinsic ability to exchange their constituents, they also have combinatorial character so that they may be considered as *dynamic combinatorial materials* (DCMs). Supramolecular materials thus are *instructed*, *dynamic*, and *combinatorial*; they may in principle select their constituents in response to external stimuli or environmental factors and therefore behave as *adaptive materials* [45].

In an other vein, DCMs may be considered as *five-dimensional materials* with three dimensions of space, one dimension of time/dynamics, and one dimension of constitution, representing the different constitutional combinations.

C. Supramolecular Polymers as Constitutional Dynamic Materials, Dynamers

It follows from the previous considerations that supramolecular polymer chemistry is both dynamic and constitutionally diverse. Supramolecular polymers present dynamic constitutional diversity and are therefore *constitutional dynamic materials* (CDMs) belonging to the realm of *constitutional dynamic chemistry* (CDC) [4]. They are based on *dynamic polymer libraries* whose constituents have a constitutional diversity determined by the number of different monomers (see Section III). Similar views apply to supramolecular liquid crystals.

The components effectively incorporated into the polyassociation depend in particular on the nature of the core groups and on the interactions with the environment, so that supramolecular polymers possess the possibility of *adaptation* by association/growth/dissociation sequences. The selection of components may occur on the basis of size commensurability [18], of compatibility in chemical properties, in charge, in rigidity/flexibility, etc. An example is given by the formation of homochiral helical fibers with chiral selection from a racemic mixture of monomeric tartaric acid derivatives: $LU_2 = LP_2 + DU_2 + DP_2(LU_2LP_2)_n + (DU_2DP_2)_n$ (see Section II.C) [16].

On the other hand, the selective incorporation of components presenting specific functional properties (energy transfer, electron transfer, ion binding, etc.) may be brought about in a recognition-controlled fashion. In particular, the use of suitable monomers allows to envisage applications for such diverse purposes as drug delivery, gene transfer, mechanical action (e.g., triggered changes in shape or size), viscosity adjustment, hydrophilicity/hydrophobicity modulation, optical and electronic effects, etc.

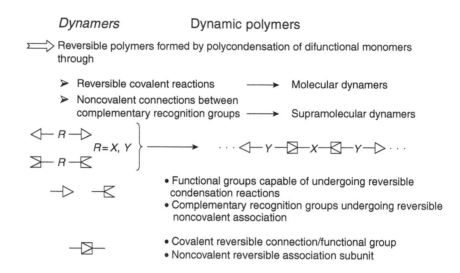

Figure 21 Dynamers: dynamic polymers defined as reversible polymeric entities of either molecular or supramolecular nature.

Broadening the scope, one may also consider covalent polymers formed from difunctional monomers that may undergo polycondensation through reversible chemical reactions (for instance, imine formation or others see Ref. 45). Such entities, where the reversible connections are covalent, are also dynamic and generate constitutional diversity [48–50]. Taken together, supramolecular polymers, dynamic by nature, and reversible molecular polymers, dynamic by design, define the class of dynamic polymers or "dynamers" [49a] (Figure 21), which may display a range of novel properties, merging molecular and supramolecular features both in main-chain and side-chain processes [51].

Depending on the nature and variety of core/interaction/functional groups in mixtures of several different monomeric components, the dynamic features give access to higher levels of behavior such as healing, adaptability, and response to external stimuli (heat, light, pressure, shear, additives, etc.).

V. CONCLUSION

Molecular information-based recognition events represent a means of performing programmed materials engineering and processing of biomimetic or abiotic type and may lead to *self-assembling nanostructures*, organized and functional species of nanometric dimensions that define a *supramolecular nanochemistry*, an area to which supramolecular polymer chemistry is particularly well suited and able to make important contributions.

Nanoscience and *nanotechnology* have become and will remain very active areas of investigation, in view of both their basic interest and their potential applications. Here again, supramolecular chemistry may have a deep impact. Indeed, the spontaneous but controlled generation of well-defined functional supramolecular architectures of nanometric size through self-organization offers a very powerful alternative to nanofabrication and to nanomanipulation, providing a chemical approach to nanoscience and technology [1]. One may surmise that rather than having to stepwise construct bottom-up or top-down prefabricated nanostructures, it will become possible to devise more and more powerful *self-fabrication* methodologies resorting to self-organization from instructed components. The results described above give an aspect of possible routes toward self-organized nanostructures. The dynamic features and constitutional diversity of such supramolecular architectures confers to them the potential to undergo healing and adaptation, processes of great value for the development of "*smart*" nanomaterials.

Widening the perspectives, one may consider that the science of five-dimensional supramolecular materials in general and supramolecular polymer chemistry in particular, as part of dynamic constitutional chemistry [4], will strongly contribute to the emergence and development of *adaptive chemistry* [47] on the way toward complex matter.

REFERENCES

1. J.-M. Lehn, *Supramolecular Chemistry — Concepts and Perspectives*, VCH, Weinheim, 1995.
2. *Comprehensive Supramolecular Chemistry*, J.L. Atwood, J.E.D. Davies, D.M. MacNicol, F. Vögtle, and J.-M. Lehn, Eds., Pergamon, Oxford, 1996, Vol. 9; D. Philp and J.F. Stoddart, *Angew. Chem. Int. Ed. Engl.* 1996, *35*, 1155.
3. (a) J.-M. Lehn, *Angew. Chem. Int. Ed. Engl.* 1990, *29*, 1304; (b) see also Ref. 1, chapter 9; (c) C. Fouquey, J.-M. Lehn, and A.-M. Levelut, *Adv. Mater.* 1990, *2*, 254.
4. J.-M. Lehn, *Proc. Natl. Acad. Sci. USA* 2002, *99*, 4763.
5. J.-M. Lehn, *Makromol. Chem., Macromol. Symp.* 1993, *69*, 1.
6. V. Percec, H. Jonsson, and D. Tomazos, in *Polymerization in Organized Media*, C.M. Paleos, Ed., Gordon and Breach, Philadelphia, 1992, p. 1.
7. C.T. Imrie, *Trends Polym. Sci.* 1995, *3*, 22; A. Ciferri, *Trends Polym. Sci.* 1997, *5*, 142; N. Zimmerman, J.S. Moore, and S.C. Zimmerman, *Chem. Ind.* 1998, 606; R.P. Sijbesma and E.W. Meijer, *Curr. Opin. Colloid Interf. Sci.* 1999, *4*, 24; J.S. Moore, *Curr. Opin. Colloid Interf. Sci.* 1999, *4*, 108; R.F.M. Lange, M. Van Gurp, and E.W. Meijer, *J. Polym. Sci., Polym. Chem. Ed.* 1999, *37*, 3657.
8. C.M. Paleos and D. Tsiourvas, *Angew. Chem. Int. Ed. Engl.* 1995, *34*, 1696.
9. T. Kato and J.M.J. Fréchet, *Macromol. Symp.* 1995, *98*, 311; T. Kato, M. Fujimasa, and J.M.J. Fréchet, *Chem. Mater.* 1995, *7*, 368; T. Kato, *Structure Bonding*, 2000, *96*, 95; P. Zhang and J.S. Moore, *J. Polym. Sci., Part A, Polym. Chem.* 2000, *38*, 207.
10. A. Reichert, H. Ringsdorf, P. Schuhmacher, W. Baumeister, and T. Scheybani, in Ref. 2, Vol. 9, p. 313.
11. L. Brunsveld, J.B. Folmer, E.W. Meijer, and R.P. Sijbesma, *Chem. Rev.* 2001, *101*, 4071; R.P. Sijbesma and E.W. Meijer, *Chem. Commun.* 2003, 5.
12. M.-J. Brienne, J. Gabard, J.-M. Lehn, and I. Stibor, *J. Chem. Soc., Chem. Commun.* 1989, 1868.
13. M. Suarez, J.-M. Lehn, S.C. Zimmerman, A. Skoulios, and B. Heinrich, *J. Am. Chem. Soc.* 1998, *37*, 9526.
14. (a) R. Kleppinger, C.P. Lillya, and C. Yang, *Angew. Chem. Int. Ed. Engl.* 1995, *34*, 1637; (b) G. Gottarelli, G.P. Spada, and A. Garbesi, in Ref. 2, Vol. 9, p. 483.
15. W.H. Binder, M.J. Kunz, and E. Ingolic, *J. Polym. Sci., Part A* 2004, *42*, 162; M.J. Kunz, G. Hayn, R. Saf, and W.H. Binder, *J. Polym. Sci., Part A* 2004, *42*, 661.
16. T. Gulik-Krczywicki, C. Fouquey, and J.-M. Lehn, *Proc. Natl. Acad. Sci. USA* 1993, *90*, 163.
17. R.P. Sijbesma, F.H. Beijer, L. Brunsveld, B.J.B. Folmer, J.H.K.K. Hirschberg, R.F. Lange, J.K.L. Lowe, and E.W. Meijer, *Science* 1997, *278*, 1601; A. El-ghayoury, E. Peters, A.P.H.J. Schenning, and E.W. Meijer, *Chem. Commun.* 2000, 1969; B.J.B. Folmer, R.P. Sibesma, and E.W. Meijer, *J. Am. Chem. Soc.* 2001, *123*, 2093.
18. (a) C. He, C.-M. Lee, A.C. Griffin, L. Bouteiller, N. Lacoudre, S. Boileau, C. Fouquey, and J.-M. Lehn, *Mol. Cryst. Liq. Cryst.* 1999, *332*, 251/2761; (b) T. Shimizu, R. Iwaura, M. Masuda, T. Hanada, and K. Yase, *J. Am. Chem. Soc.* 2001, *123*, 5947; (c) J. Xu, E.A. Fogleman, and S.L. Craig, *Macromolecules* 2004, *37*, 1863.
19. (a) V. Berl, M. Schmutz, M.J. Krische, R.G. Khoury, and J.-M. Lehn, *Chem. Eur. J.* 2002, *8*, 1227; (b) For the introduction of the same binding motif into supramolecular pseudo-block copolymers see: W.H. Binder, M.J. Kunz, C. Kluger, G. Hayn, and R. Saf, *Macromolecules* 2004, *37*, 1749.
20. R.K. Castellano, R. Clark, S.L. Craig, C. Nuckolls, and J. Rebek, *Proc. Natl. Acad. Sci. USA* 2000, *97*, 12418.
21. M.-J. Brienne, C. Fouquey, A.-M. Levelut, and J.-M. Lehn, unpublished work.
22. B.J.B. Folmer, E. Cavini, R.P. Sijbesma, and E.W. Meijer, *Chem. Commun.* 1998, 1847.
23. M. Kotera, J.-M. Lehn, and J.-P. Vigneron, *J. Chem. Soc., Chem. Commun.* 1994, 197.
24. G. Wegner, *Thin Solid Films* 1992, *216*, 105 and references therein.
25. (a) J.-M. Lehn, M. Mascal, A. Decian, and F. Fischer, *J. Chem. Soc., Chem. Commun.* 1990, 479; (b) J.-M. Lehn, M. Mascal, A. Decian, and J. Fischer, *J. Chem. Soc., Perkin Trans.* 1992, 461;

(c) For the enforced formation of supramolecular macrocycles see: A. Marsh, M. Silvestri, and J.-M. Lehn, *Chem. Commun.* 1996, 1527; M. Mascal, N.M. Hext, R. Warmuth, M.H. Moore, and J.T. Turkenburg, *Angew. Chem. Int. Ed. Engl.* 1996, *35*, 2204; P.S. Corbin, L.J. Lawless, Z. Li, Y. Ma, M.J. Witmer, and S.C. Zimmerman, *Proc. Natl. Acad. Sci. USA* 2002, *99*, 5099.

26. N. Kimizuko, S. Fujikawa, H. Kuwahara, T. Kunitake, A. Marsh, and J.-M. Lehn, *J. Chem. Soc., Chem. Commun.* 1995, 2103.

27. (a) C. Fouquey and J.-M. Lehn, unpublished work; (b) M. Krische, A. Petitjean, E. Pitsinos, D. Sarazin, C. Picot, and J.-M. Lehn, to be published.

28. (a) See Chapter 6, this volume; (b) F. Zeng, S.C. Zimmerman, S.V. Kolotuchin, D.E.C. Reichert, and Y. Ma, *Tetrahedron* 2002, *58*, 825; (c) D.A. Tomalia chap. 7; (d) G.R. Newkome, C.N. Moorefield, and F. Vögtle, *Dendritic Molecules*, VCH, Weinheim, 1996; (e) for inorganic, metallodendritic structures see for instance: C. Gorman, *Adv. Mater.* 1998, *10*, 295; D. Astruc, *Top. Cur. Chem.* 1991, *160*, 47; S. Campagna, G. Denti, S. Serroni, A. Juris, M. Venturi, V. Ricevuto, and V. Balzani, *Chem. Eur. J.* 1995, *1*, 211 and references therein.

29. (a) C. Fouquey and T. Gulik-Krczywicki, see *Annuaire Collège de France* 1992–93, 305; (b) H. Nierengarten, J. Rojo, E. Leize, J.-M. Lehn, and A. Van Dorsselaer, *Eur. J. Inorg. Chem.* 2002, 573; (c) U. S. Schubert and C. Eschenbaumer, *Angew. Chem. Int. Ed.* 2002, *41*, 2892; (d) for recent examples of solid state coordination polymers, see for instance: L. Carlucci, G. Ciani, and D. M. Proserpio, *Chem. Commun.* 1999, 449; T. Ezuhara, K. Endo, and Y. Aoyama, *J. Am. Chem. Soc.* 1999, *121*, 3279; A. Mayr and J. Guo, *Inorg. Chem.* 1999, *38*, 921.

30. See for instance: C. Bamford and K. Al-Lame, *J. Chem. Soc., Chem. Commun.* 1993, 1580; H.A. Asanuma, T. Hishiya, T. Bau, S. Gotoh, and M. Komiyama, *J. Chem. Soc., Perkin Trans.* 1998, *2*, 1915.

31. J. Barberá, A. Elduque, R. Giménez, L.A. Oro, and J.L. Serrano, *Angew. Chem. Int. Ed. Engl.* 1996, *35*, 2832.

32. (a) A.D. Burrows, C.-W. Chan, M.M. Chowdry, J.E. McGrady, and D.M.P. Mingos, *Chem. Soc. Rev.* 1996, *25*, 329; (b) Z. Qin, H.A. Jenkins, S.J. Coles, K.W. Muir, and R.J. Puddephatt, *Can. J. Chem.* 1999, *77*, 155; (c) H. Hofmeier, A. El-ghayoury, A.P.H.J. Schenning, and U.S. Schubert, *Chem. Commun.* 2004, 318.

33. F.T. Edelmann and I. Haiduc, *Supramolecular Organometallic Chemistry*, Wiley-VCH, Weinheim, 1999.

34. J. Ruokolainen, R. Mäkinen, M. Torkkeli, T. Mäkelä, R. Serimaa, G. ten Brinke, and O. Ikkala, *Science* 1998, *280*, 557.

35. R. Krämer, J.-M. Lehn, and A. Marquis, *Proc. Natl. Acad. Sci. USA* 1993, *90*, 5394.

36. V. Berl, M.J. Krische, I. Huc, J.-M. Lehn, and M. Schmutz, *Chem. Eur. J.* 2000, *6*, 1938.

37. E.E. Simanek, X. Li, I.S. Choi, and G.M. Whitesides, in Ref. 2, Vol. 9, p. 595.

38. For the assembly of triblock copolymers into nanostructures see: S.I. Stupp, V. LeBonheur, K. Walker, L.S. Li, K.E. Huggins, M. Keser, and A. Amstutz, *Science* 1997, *276*, 384.

39. See for instance the self-assembly of rod-like hydrogen-bonded nanostructures: H.A. Klok, K.A. Jolliffe, C.L. Schauer, L.J. Prins, J.P. Spatz, M. Möller, P. Timmerman, and D.N. Reinhoudt, *J. Am. Chem. Soc.* 1999, *121*, 7154.

40. T. Kunitake, in Ref. 2, Vol. 9, p. 351; K. Ariga and T. Kunitake, *Acc. Chem. Res.* 1998, *31*, 371.

41. V. Marchi-Artzner, F. Artzner, O. Karthaus, M. Shimomura, K. Ariga, T. Kunitake, and J.-M. Lehn, *Langmuir* 1998, *14*, 5164; V. Marchi-Artzner, J.-M. Lehn, and T. Kunitake, *Langmuir* 1998, *14*, 6470.

42. V. Marchi-Artzner, L. Jullien, T. Gulik-Krzywicki, and J.-M. Lehn, *Chem. Commun.* 1997, 117; S. Chiruvolu, S. Walker, J. Israelachvili, F.-J. Schmitt, D. Leckband, and J.A. Zasadinski, *Science* 1994, *264*, 1753.

43. C.M. Paleos, Z. Sideratou, and D. Tsiourvas, *Chembiochem* 2001, *2*, 305.

44. For the interaction between a molecule and a modified nanocrystalline solid bearing a complementary recognition group, see for instance: L. Cusack, S.N. Rao, and D. Fitzmaurice, *Chem. Eur. J.* 1997, *3*, 202.

45. J.-M. Lehn, *Chem. Eur. J.* 1999, *5*, 2455 and references therein.

46. A.V. Eliseev, *Curr. Opin. Drug Discov. Develop.* 1998, *1*, 106; A.V. Eliseev and J.-M. Lehn, *Curr. Top. Microbiol. Immun.* 1999, *243*, 159; G.R.L. Cousins, S.A. Poulsen, and J.K.M. Sanders, *Curr. Opin. Chem. Biol.* 2000, *4*, 270.

47. J.-M. Lehn, in *Supramolecular Science: Where It Is and Where It Is Going*, R. Ungaro and E. Dalcanale, Eds., Kluwer, Dordrecht, 1999, p. 287.

48. For a review including reversible covalent polymers, see: S.J. Rowan, S.J. Cantrill, G.R.L. Cousins, J.K.M. Sanders, and J.F. Stoddart, *Angew. Chem. Int. Ed. Engl.* 2002, *41*, 899.

49. (a) W.G. Skene and J.-M Lehn, *Proc. Natl. Acad. Sci. USA* 2004, **101**, 8270; (b) T.Nobori, T. Ono, N. Giuseppone, and J.-M. Lehn, to be published.

50. (a) Based on the formation of helical molecular strands by imine condensation [50b], extended helical polymers may be obtained from suitable difunctional monomers; J.-L. Schmitt and J.-M. Lehn, *Helv. Chim. Acta*, 2003, *86*, 3417; (b) K.M. Gardinier, R.G. Khoury, and J.-M. Lehn, *Chem. Eur. J.* 2000, *6*, 4124; (c) For the stacking induced self-assembly of helical monomers into extended fibers, see: L.A. Cuccia, J.-M. Lehn, J.-C. Homo, and M. Schmutz, *Angew. Chem. Int. Ed. Engl.* 2000, *39*, 233; L.A. Cuccia, E. Ruiz, J-M. Lehn, and J.-C. Homo, *Chem. Eur. J.* 2002, *8*, 3448.

51. Both dynamic processes, reversible molecular (covalent) and supramolecular (non-covalent), may be incorporated into the same polymeric entity; E. Kolomietz and J.-M. Lehn, submitted.

Chapter 2

Growth of Supramolecular Structures

Alberto Ciferri

Contents

A. Definitions

The broadest definition of a supramolecular polymer (SP) is that of a system based on the association of many unimers through supramolecular (noncovalent) interaction. The unimers may be covalent molecules and macromolecules, and also supramolecular entities such as supermolecules (cf. Chapter 1) and complex micellar structures. Although such a broad definition includes all the systems that have been described as SPs, it does not readily illustrate structural features and potential applications of this exciting class of new materials. For instance, even an organic crystal might be regarded as a SP [1] and supramolecular interactions play a controlling role on the intramolecular conformation and the intermolecular ordering of covalent polymers.

Attempts to restrict the definition of SPs have been made. The most restrictive definition was proposed by Meijer and coworkers [2] to regard SPs as only those systems exhibiting chainlike behavior in diluted solutions. Such a definition is of interest since linear SPs do have peculiar properties (e.g., rheological ones) that widen possible applications of conventional polymers. However, a significant degree of supramolecular polymerization (DP) may not occur in dilute solutions while novel properties may appear in concentrated phases or in the bulk. Moreover, a less restrictive definition of SPs may highlight important features of supramolecular polymer chemistry better, in particular the use of approaches that cut across traditional boundaries between colloid, polymer, and solid-state science.

A definition that may be proposed (not as broad, nor as restrictive as the two discussed above) is based on the distinction between *self-assembling* and *engineered* SPs. The former type of assemblies is the main focus of the present chapter and is characterized by thermodynamically controlled structures. By contrast engineered SPs, important for a variety of applications, are materials assembled by controlled deposition or synthesis. A more detailed definition of all systems that have been described as SPs (including self-assembling and engineered ones) is presented in Section I.B [3].

B. Classification

Figure 1 schematizes several classes of systems described as SPs. The reference model at the top of the figure represents the classical covalent chain resulting from molecular polymerization of small bifunctional monomers. The covalent chain is an open one, meaning that it could in principle grow to a large DP distribution, irreversible in solution and under a wide range of external variables.

Class A. The major components of this class are *equilibrium polymers* based on processes that can appropriately be regarded as *supramolecular polymerizations* [4,5]. Uni- and multidimensional growth mechanisms are coupled to molecular recognition processes typical of supramolecular chemistry. The linear chains are self-assembled, open, growing to a distribution of DP, and in a state of thermodynamic equilibrium sensitive to solvent type, concentration, and external variables. The geometrical shapes of the unimers in class A (Figure 1) bring to bear that repeating units in supramolecular polymerization can be of several forms and sizes. Specific chemical structures will be discussed in Section III. At variance with molecular polymers all interaction, including those stabilizing the main chain, are of supramolecular nature. Therefore we include in class A polymers based on unimers with functionality ≥2, when a variety of multidimensional assemblies (linear, planar, three-dimensional) becomes possible. Examples of linear systems are hydrogen-bonded polymers, coordination polymers, and micelles (cf. Chapter 6). Examples of more complex geometries are helical, helical–columnar, and tubular structures (cf. Chapters 2, 16), protein layers (cf. Chapter 17), and composite systems such as block copolymers (cf. Chapters 10, 11). Random networks and blends stabilized by multifunctional supramolecular linkages (cf. Chapter 6) may also be included.

Class B. This class includes self-assembled structures based on supramolecular binding of monofunctional unimers. These unimers cannot undergo open supramolecular polymerization, but can form closed assemblies involving low- or high-MW species. Classical host–guest complexes, base pairing of simple nucleosides, and supermolecules are low-MW examples. Polymeric examples include side-chain binding of

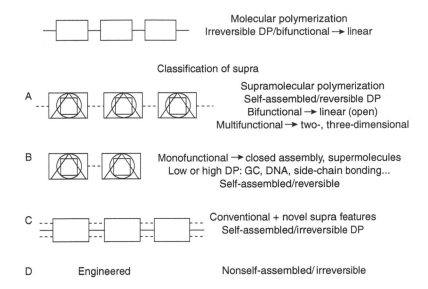

Figure 1 Classification of supramolecular polymers. Class A (reversible polymers obtained by supramolecular polymerization) is the main topic of this chapter. (From A. Ciferri. *Macromol. Rapid Commun.* 23:511, 2002a. Copyright Wiley-VCH 2002.)

a monofunctional unimer to a covalent chain (e.g., binding of low-MW mesogens to a covalent polymer, cf. Chapter 5, counterion binding), double- and triple-chain assemblies, and globular structures unable to further growth when complementary monofunctional sites are internally saturated.

Class C. SPs displaying novel supramolecular features have been obtained by a combination of covalent and supramolecular bonds. Such systems are self-assembling but the covalent component will not show reversible association–dissociation equilibria. The supramolecular organization may either precede, be simultaneous to, or follow the formation of covalent bonds. Examples of the first type are the rotaxane and catenane polymers described by Stoddart and coworkers (cf. Chapter 8), the growth of dendrimers through successive generations, and other attempts to stabilize a supramolecular assembly by subsequent formation of covalent bonds (cf. Chapters 7, 12). The final covalent system may retain supramolecular features. Alternatively, the precursor organization may just be a step of a supramolecularly assisted synthesis of a complex structure. Examples in which the supramolecular and the molecular order are simultaneously established include the synthesis of dendrons possessing polymerizable functionality at their focal points, noticeably reported by Percec and Schlüter (described in Chapter 7). These assemblies display most interesting composite architectures such as columns of disks based on dendrons with a main covalent chain running in the center of each column. Cases in which the covalent structure precedes the formation of the supramolecular one include the dendronization of a covalent polymer, reported for instance by Tomalia and Majoros (cf. Chapter 7), and the self-assembled monolayers (SAMs, cf. Chapter 18) regarded as supramolecular assemblies of short hydrocarbon chains covalently grafted to a gold surface.

Class D. The class of engineered assemblies includes systems that do not spontaneously form ordered structures under normal conditions. Their classification as SPs can be justified since elements of supramolecular interaction still assist the final organization. Some examples are layered assembly of complementary polyelectrolytes obtained by stepwise deposition under kinetic control (cf. Chapter 19), and polymer brushes prepared by grafting a polymer chain over a SAM of an initiator [6]. Both approaches allow a fine-tuning of surface properties and patterning possibilities. Tailored performance in applications, such as biocompatibility, biocatalysis, integrated optics and electronics have been considered. Additional differences between self-assembled and engineered SPs are discussed in Section I.C.

C. Self Versus Engineered Assemblies: Reversibility and Stability

Self-assembled systems may undergo association ↔ dissociation equilibria under the influence of state variables, composition, and external conservative fields. Note that structural reversibility applies only to the component assembled via secondary interaction. Reversible DP is shown by systems in class A, but not by systems exhibiting covalent main chains in class B and C.

In engineered systems the supramolecular interaction component is still important but the shape of the assembly or the distribution of components may be controlled by specific deposition techniques. For instance, random aggregates or double-helical structures may be produced by molecular recognition of complementary polyelectrolytes in solution [7]. The same recognition process allowed the engineered formation of spherical skins or layered films in the following engineered examples. Larez et al. [8] reported the formation of spherical assemblies (diameters up to 500 μm) by letting drops of a chitosan solution free-fall into a solution of oxidized scleroglucan. The skin of the assembly, separating the external polyanion from the internal polycation solution, acted as a semipermeable membrane. Laschewski and coworkers (cf. Chapter 19) reported a stepwise, layer-by-layer adsorption of oppositely charged elements enabling the engineered growth of these assemblies on suitable substrates. Due to the kinetically controlled deposition, interfacial complexation gives rise to stratified supramolecular structures that strongly differ from the ones obtained by direct polyelectrolyte complexation.

Deposition techniques may also allow the fabrication of functional assemblies of incompatible components such as rod and coiling polymers. This is the case of high-performance composites and natural systems such as connective tissue or the vitreous body of the eye [9]. Engineered systems may exhibit even greater stability than self-assembled ones since use of compatibilizers, fast quenching techniques, or covalent cross-linking may prevent or retard the dissipation of their organization.

D. Open Versus Closed Systems

All the examples cited for class A (Figure 1(A)) refer to supramolecular systems for which the addition of successive repeating units exposes sites at the end of the growing chain to which additional units can bind, a situation typical of the polymerization of molecular polymers. These systems are classified as *open assemblies*. Planar and three-dimensional assemblies growing by a mechanism similar to a phase transition may also be classified as open assemblies (cf. Section II.A.3). The examples cited for class B represent instead *closed assemblies* for which the binding sites are internally compensated and the complexes have a definite stoichiometry. A too strict classification of open or closed assemblies should nevertheless be avoided since some closed assemblies may have residual sites allowing growth to occur. For instance, although the stoichiometry of the collagen triple helix is strictly defined, it is difficult to observe a stable solution of tropocollagen at neutral pH. Side-by-side or head-to-tail aggregation continue until phase separation occurs.

On the other hand, some supramolecular assemblies may have a large number of repeating units but their growth is strictly limited. These systems are not oligomeric ones and it is convenient to classify them as open systems undergoing supramolecular polymerization once a proper termination step is recognized. Several situations may occur, for instance, a size limitation and a size distribution may be the natural result of the stochastic nature (cf. Section II.B.2) of the assembly mechanism. Size limitation may also result from termination by a monofunctional unit unable to grow further (Figure 3(c)). Situations of this type are encountered even in biological polymerizations (e.g., termination of actin growth by gelsolin; cf. Section III.B). The growth of cylindrical micelles slows down when the length of the assembly attains the value of the persistence length (cf. Section II.C.2). Giant mesoscopic vesicles based on either simple surfactants [10] or block copolymers [11] (cf. Chapter 11) are structures that may be conceived as arising from the closing up of extended bilayers exhibiting randomly fluctuating local curvature [12,13]. Spherical surfactant micelles have the typical aggregation number 100, their growth to infinite size being prevented by the peculiar nature of amphiphilic molecules (cf. Section II.C.2). The size of virus capsides may be controlled by the polynucleotide guest [3,14,15] (cf. Section III.B).

II. PRINCIPLES OF SUPRAMOLECULAR ORGANIZATION

A. Unimer Functionality and Assembly Dimensionality

1. Site Distribution

The number of complementary sites per unimer (S) and their distribution control the functionality of the unimer (F) and the dimensionality of the assembly [3,15]. Figure 2 illustrates the above concept in the case of a unimer schematized by a small square. Monofunctional unimers are those with one or more sites pointing in just one direction. Such unimers cannot linearly polymerize and form closed host–guest complexes, a subject of classical supramolecular chemistry. Bifunctional unimers, those with two or more sites pointing toward opposite (North and South) directions, can form linear polymers or closed rings (Figure 3(a)). Tetrafunctional unimers with four or more sites pointing toward azimuthal directions (N, S, E, W) form planar polymers. However, helical polymers are expected even in the case $S = 4$, $F = 4$ when two sites are located on the same surface (pointing toward NE and SE, Figure 3(b)). Three-dimensional ordered networks are expected if N–S sites longitudinally increase the functionality of the planar system. It appears that the dimensionality of the final assembly is crucially dependent upon site distribution.

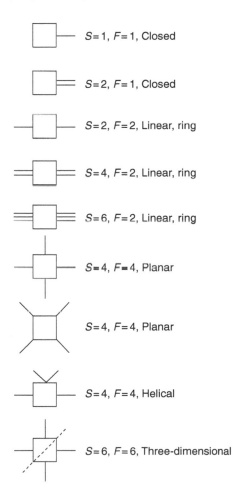

Figure 2 Schematic representation of unimer functionality (F), site number (S), and distribution. The geometry of the resulting assembly is indicated.

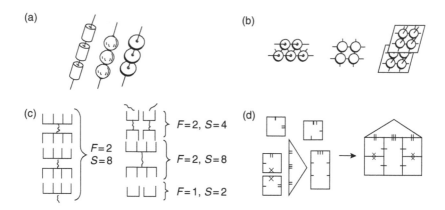

Figure 3 Shapes and functionality: (a) linear assemblies of bifunctional rodlike, spherical, disklike unimers ($F = 2$, $S = 2$); (b) helical, planar, three-dimensional assemblies ($F > 2$, $S > 2$); (c) linear polymers, branching, and termination for various F and S; (d) assembly of unimers with different shapes and functionalities (note two square unimers not consistent with the final structure). (From A. Ciferri. Encyclopedia of Supramolecular Chemistry, J.L. Atwood and J.W. Steed Eds., New York: M. Dekker, 2004. With permission www. tandf.co.uk/journals)

Supramolecular unimers can be large in size, and often more than one binding site occurs over each complementary surface (Figure 3(c)). If full compensation occurs, there will be no alteration of the functionality of the unimer, or the dimensionality of the assembly. Multiple sites on properly designed complementary surfaces have a high probability of being mutually compensated. Should a mismatch occur, or unimers with different functionality or number of sites be mixed, alterations in the above assembling patterns are expected, that is, termination or random network formation, as shown schematically in Figure 3(c). More complex patterns are expected upon mixing unimers differing in shape, functionality, and bond type (Figure 3(d)). These situations are not yet under active investigation. Computational and simulation methods might be used for describing assemblies of increasing complexities (cf. Chapter 12). The algorithmic feature of the assembling process schematized in Figure 3(d) is evidenced by the two unimers indicated as an improper choice. These unimers might be included in the growing assembly, but should eventually be rejected by the final structure.

2. Shape Effects

The process of molecular recognition, underlining the interaction of complementary sites discussed above, implies both chemical and shape recognition. Although molecular and shape recognition are closely related, it is often convenient to distinguish them. With supramolecular polymers it is further convenient to distinguish two types of shape effects. The first type (shape I effect) pertains to the shape of the molecules or of the assembly forming the repeating unit. Shape II effects pertain instead to the assembled polymer. The overall assembly process may start as a molecular and shape I recognition. Eventually this process produces well-defined shapes. Thereafter, the evolution toward more complex structures may be simply described by shape II recognition and geometrical shape parameters (cf. Section II.C.2).

Shape I effects are important at the molecular recognition level. *Endo* recognition occurs when binding sites are oriented into a molecular concavity. A well-known example is the enzyme–substrate catalysis when the specificity toward the hydrolysis of a particular peptide bond is controlled by the binding of the adjacent side chain inside a pocket of the enzyme [16]. The great selectivity and complex stability observed in closed cavities is attributed to an enhancement of the strength of pairwise attraction (particularly of the dispersive and hydrophobic type) with respect to binding over flat surfaces [17]. Site distribution within a cleft may favor the convergence of groups capable of directional binding, thus enhancing the strength of H-bonds [18]. On the other hand, *exo* recognition occurs when binding sites are directed outward to flat surfaces. Similarity of size of the surfaces and multiple pairwise interactions enhance the binding free energy.

Shape II effects are particularly relevant to excluded volume and liquid crystallinity, and also suggest intriguing correlations between macroscopic and molecular design concepts. The dendrimers (cf. Chapter 7) may be regarded as a growing sequence of similar chemical steps exhibiting fractal shapes around an initiator core [19]. Dendrimers in the nanometric range may assembly further through a repetition and composition of the basic geometrical design. Macroscopic analogies are the growth of corals, branching of trees, nesting of spheres into a large sphere, and so on [20]. Intracellular networks reflect a supramolecular assembly design of protein chains that display occasional pentagonal meshes within a hexagonal topology [21]. The pentagons determine the curvature of the network and stabilize the discocythe shape of the erythrocytes. Similar design strategies are found in both molecular systems such as fullerenes [22], and in macroscopic objects such as bamboo vases and football spheres [23].

3. Dimensionality and Growth

Israelachvili [13] summarized general thermodynamic considerations with regard to the growth of supramolecular aggregates. The growth process is strongly determined by the dimensionality of the assembly. Dimensionality indexes $p = 1$, $p = \frac{1}{2}$, $p = \frac{1}{3}$ describe, respectively, unidimensional, planar, and three-dimensional growth. Following the formation of a critical aggregate nucleus at a critical aggregate concentration (CAC), further growth upon increasing unimer concentration produces macroscopic aggregates (aggregation number $\to \infty$) whenever $p < 1$. Finite-size distributions of assemblies are instead produced whenever $p = 1$. Spherical micelles share the latter feature of the $p = 1$ system due to the amphiphilic nature of constituent molecules [13]. The predicted phase diagrams (aggregation numbers versus free unimer concentration) include a critical concentration above which the concentration of aggregates increases and that of the free unimer remains constant.

On the above basis, supramolecular polymerization can be discussed within a general context that cuts across traditional boundaries between colloid, polymer, and solid-state science. The aggregation of two- and three-dimensional systems can be described as a none or all (crystallization) process, whereas the formation of unidimensional assemblies (linear, columnar, helical, cf. Figure 3) may display large cooperativity, but not a true phase transition. For unidimensional systems, different growth mechanisms (cf. seq.) have been identified within the above framework.

B. Localized Interactions

1. The Supramolecular Bond

For all types of SPs, the stabilization of well-defined structures is due to combinations of interactions that may be localized at specific atomic groups of the unimers, or more uniformly distributed over the assembly surface. In all cases, shape complementarity is an integral part of the process of molecular recognition. The present section discusses the role of localized interaction; smoothed-out interaction will be discussed in Section II.C. Detailed quantitative assessment of the role of classical localized interactions is described in the literature of low-MW host–guest complexes [17]. In the case of SPs, most important types of localized interactions are based on H-bonds, π–π stacking, charged groups, or metal–ion coordination. Several combinations of these interactions may additively contribute to the overall contact energy [17]. Solvents may compete or enhance the formation of localized bonds through localized effects. For instance, H-bonds are stronger in apolar not exchanging solvents and ionizable groups are sensitive to pH. Solvents also display nonlocalized effects (solvophobic/solvophilic interaction) on the formation of supramolecular bonds (to be described in Section II.C). Localized interactions are described by the respective sets of potential functions involving point charges, dipolar interaction, and separation distances.

(a) Hydrogen bonds

These occupy a major role in the assemblies of both synthetic and biological molecules due to their strength and directionality. The H-bond involves a proton donor (C–H, O–H, N–H, F–H) and a basic acceptor (O, N) atom with distance of separation of ~3 Å and a strong directionality that primarily reflects the anisotropy of charge distribution (lone pair) of the acceptor atom. The bond results from an interplay of van der Waals and Coulombic interactions with the latter playing a predominant role. Potential functions have been given using either the point charge or the dipole interaction. The expression [24]

$$V = C[(r_0/r)^{12} - (r_0/r)^6] - (\mu_1\mu_2/r^3)g(\theta_1\theta_2, \phi_1\phi_2) \tag{1}$$

includes the steric or dispersion interaction terms of the Lennard–Jones potential where r_0 and r are, respectively, the van der Waals and the actual $NH \cdots O$ separation distances. The second term in Eq. (1) represents the Coulombic interaction for the dipoles attached to the NH and O=C bonds, as illustrated in Figure 4(a). The $NH \cdots O$ distance is thus a predictable function of the $NH \cdots O=C$ angle and azimuthal orientation. The strength of a H-bond can be determined from

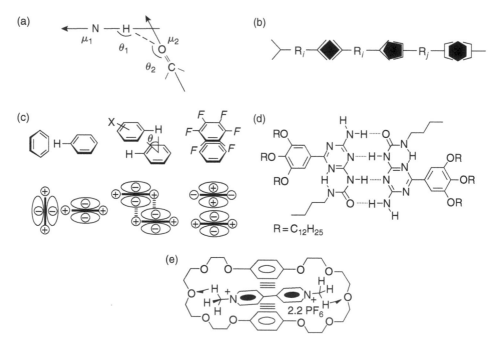

Figure 4 (a) Dipole–dipole interaction parameters for H-bonds; (b) scheme for coordination of SPs based on bidentate/tridendate complexation unit with metal ion and tetra- to hexa-coordination. (From J.-M. Lehn. *Supramolecular Polymers*. A. Ciferri, Ed., New York: Marcel Dekker, 2000. With permission.); (c) Geo-metries of aromatic interactions: edge-to-face, offset stacking, face-to-face stacked. (From M.L. Waters. *Curr. Opin. Chem. Biol.* 6:736, 2002. With permission.); (d) Monofunctional ureidotriazines assemble via quadruple hydrogen-bonding in a disk capable of columnar stacking. (From L. Brunsveld, B.J.B. Folmer, E.W. Meijer, and R.P. Sijbesma. *Chem. Rev.* 101:4071, 2001. Copyright 2001 ACS.); (e) π–π stacking interaction plus H-bonds contribute to the stabilization of a complex between paraquat [PQT]$^{2+}$ and a cyclophane-like macrocyclic polyether with hydroquinol rings. (From M.C.T. Fyfe and J.F. Stoddart. *Acc. Chem. Res.* 30:393, 1997. Copyright 1997 ACS.)

simple dimers forming a single bond, or by the analysis of data for complexation of compounds forming multiple H-bonds. The observed linear increase of ΔF for the complexation of amide-type complexes in CDCl$_3$ with the number of H-bonds is a verification of the important principle of additive binding increments mentioned above. The strength of a single H-bond in a particular solvent was evaluated to be 7.9 kJ/mol ($\sim$2 kcal/mol). The parallel or antiparallel arrangement of multiple H-bonds (e.g., AAA-DDD, ADA-DAD) in a given complex may increase or reduce the product of single-bonding constants due to secondary electrostatic interaction (cf. Chapter 6 and also Ref. 25). Other data quoted by Schneider [17] suggest a broader range of strength between $\sim$2 and $\sim$20 kJ/mol per amide link. It is important to note that polar solvents compete and destabilize H-bonds of supramolecular structures.

(b) π–π or arene–arene stacking

These attractions are also introduced by proper chemical design in SPs. The geometries of π–π interactions are schematized in Figure 4(c) [26b]. Quadruple moments arise due to the uneven electron density on the face and on the edge of rings. The nature and size of rings and substituents, and also the solvent interactions, are the variables. The offset stacked geometry (typical of DNA base stacking) maximizes buried surface area and thus dispersive and solvophobic interactions. Face-to-face stacking is favored by different substituents (opposite quadruple moments) and also by solvents or divalent cations. π–π interaction between two π-poor aromatic systems could be even more favorable than in the case of two π-rich, or one poor and one π-rich systems [26]. Stacking interaction in planar aromatic systems (discotics) occurring in polar solvents appears strengthened by solvophobic interaction [27] (cf. also Chapters 14, 16). However, the strength of stacking is not as strong nor as directional as for the H-bond in apolar solvents. Therefore, combinations of arene–arene and H-bonding have been exploited to produce columnar assemblies of discotic unimers characterized by strong contact energies (and thus large DP) in a variety of solvents [2]. The supermolecule in Figure 4(d) illustrates a coupling of several interactions. Dimerization by H-bonding stabilizes an extended core discotic supermolecule favoring π–π longitudinal stacking interaction and producing an apolar environment for stabilizing the H-bonds. Electrostatic components modulate the strength of the DDDA sequence of H-bonds. Interaction with solvents (solubility, solvophobic/solvophilic) can be modulated by the proper use of polar or apolar side chains R.

Figure 4(e) illustrates an example of superimposition of H-bonds and charge transfer in a 1:1 complex between the crown ether (host) bisparaphenylene-34-crown-10 having electron-rich hydroquinol rings, and the bipyridinium derivative paraquat [PQT]$^{2+}$ (guest) with electron-poor rings [28]. The complex is a good example of stabilization due to a variety of interactions such as charge transfer, hydrogen bonds (involving a hydrogen atom of [PQT]$^{2+}$ and the polyether oxygen atom of the crown ether), ion-dipole, and dispersive forces. The occurrence of charge transfer bands in the electronic spectra should be verified for a conclusive proof of electron transfer from high-level occupied to low-level unoccupied molecular orbitals for a given complex in a given solvent.

(c) Other electrostatic interactions

The electrostatic interactions between fixed and complementary ionizable groups are also frequently exploited (binding energies up to 10 kJ/mol) for both host–guest complexes and SPs. In line with the general features of the Debye–Huckel theory, a decrease of the association constants with increasing charge separation, solvent polarity, and ionic strength was frequently observed [29–31]. Stabilization of columnar SPs (Section III.B) by formation of salt bridges between discotic unimers was also exploited [32,33]. Triphenylenes form alternating donor–acceptor SPs in solution when doped with equimolecular amount of electron donors favoring unidirectional charge transfer along the columnar axis [34]. Metal–ion coordination favors linear association of unimers having two metal binding sites. Figure 4(b) illustrates the coordination scheme between metal ions with tetra- to hexa-coordination and bidentate or tridentate binding terminals of a covalent segment.

(d) Dispersive forces

The dispersive forces in host–guest supramolecular chemistry do not have the same all-important significance manifested by the conformation of molecular polymers. This type of interaction lacks the selectivity of other attractive interactions such as the H-bond. Nevertheless, dispersive interactions can be rather strong. The gas-phase calculated interaction between two C–H bonds amounts to ~0.2 kcal/mol, but the cumulative effect of the bonds occurring in two n-hexane molecules amounts to ~6 kcal/mol [17]. The interaction is reduced in polarizable organic solvents but is not much affected by water. Dispersive interactions can therefore be expected to play an essential role in situations (cf. seq.) in which long aliphatic segments undergo a molecular recognition that is not as specific, or pointlike directed, as other types of interactions.

The complex balance of localized interactions is the main component of binding constants and contact forces that promote supramolecular polymerization. Accordingly, the two main assembly mechanisms based on localized interactions will be presented below. Assembly mechanisms prevalently based on nonspecific (smoothed-out) interactions will instead be presented in Section II.C within the context of the latter type of interaction.

2. Assembling via MSOA (Isodesmic Polymerization)

The scheme of supramolecular association of n monomeric units (unimers) M_1 into a linear sequence (multistage open association (MSOA)) is [35–41]

$$M_1 + M_1 \iff M_2, \quad K = \frac{|M_2|}{|M_1|\,|M_1|}, \qquad C_2 = K(C_1)^2$$

$$M_2 + M_1 \iff M_3, \quad K = \frac{|M_3|}{|M_2|\,|M_1|}, \qquad C_3 = KC_2C_1 = K^2(C_1)^3 \tag{2}$$

$$M_3 + M_1 \iff M_4, \quad K = \frac{|M_4|}{|M_3|\,|M_1|}, \qquad C_4 = K^3(C_1)^4$$

$$\cdots \qquad\qquad \cdots \qquad\qquad \cdots$$

$$C_n = K^{-1}(KC_1)^n \tag{3}$$

where C_n is the concentration of the n-mer and the identical equilibrium constant K for each step (no cooperation) is assumed. The total concentration (C_p) of all species coexisting at a given polymerization degree (unimers, oligomers, polymers) and the total initial concentration of unimers (C_0) are

$$C_p = \sum C_n = \sum K^{-1}(KC_1)^n = \frac{C_1}{1 - KC_1}$$

$$C_0 = \sum nC_n = \sum nK^{-1}(KC_1)^n = \frac{C_1}{(1 - KC_1)^2} \tag{4}$$

The above equations show that $C_0 > C_p$ and $KC_1 \leqslant 1$. The extent of growth can thus be expressed as

$$DP_n = \frac{C_0}{C_p} = \frac{1}{1 - KC_1} \tag{5}$$

showing that the number average degree of polymerization, $DP_n \rightarrow \infty$ when $KC_1 \rightarrow 1$. Note that the dominant variables are C_0 and K. $C_0 = C_1$ only at the beginning of the polymerization and the various $C_1, C_2, \ldots$ could only be assessed by fractionation. It is easy to show that the corresponding weight average can be expressed as

$$DP_w = \frac{1 + KC_1}{1 - KC_1} \tag{6}$$

and the width of the length distribution $DP_w/DP_n = 1 + KC_1$ widens to the limit of 2 when $KC_1 \rightarrow 1$ or C_1 approaches K^{-1}.

It is important to note that the above polymerization scheme applies to supramolecular polymerization when bond formation occurs *without byproducts*, in contrast to the case usually observed with molecular polycondensation [35,41]. The most significant difference is that DP increases with the initial unimer concentration (C_0 or C_1) in the supramolecular case, while it is independent of concentration for conventional isodesmic polycondensation. In the latter case, using the extent of reaction p (<1) and $C_p = C_0(1 - p)$ the Carothers equation yields $DP_n = 1/(1 - p)$. In both polymerizations, DP increases with increasing K. Plots illustrating the role of unimer concentration and equilibrium constants in supramolecular polymerization are shown in Figure 5(a) [41] and Figure 5(b) [2]. In the bulk phase (volume fraction = 1) DP is simply related to K by the approximate relationship (valid for $K \gg 1$)

$$DP \approx K^{1/2} \tag{7}$$

3. Assembling via HG (Cooperative Helical Growth)

A new situation is expected when binding of one unit promotes binding of successive units along a helical pattern. This type of cooperation was introduced by Oosawa [4,37] to interpret the occurrence

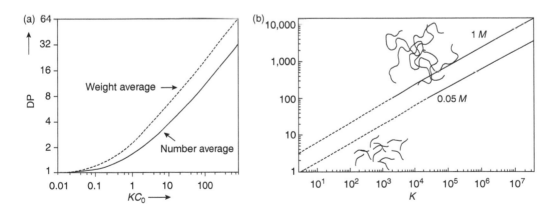

Figure 5   (a) Plots of DP_n and DP_w as functions of KC_0 for isodesmic polymerization without byproduct. (From D. Zhao and J.S. Moore. *Org. Biomol. Chem.* 2003. Copyright 2003 RCS.); (b) Theoretical relationship between the association constant K and DP according to the multistage open association model. (From L. Brunsveld, B.J.B. Folmer, E.W. Meijer, and R.P. Sijbesma. *Chem. Rev.* 101:4071, 2001. Copyright 2001 ACS.)

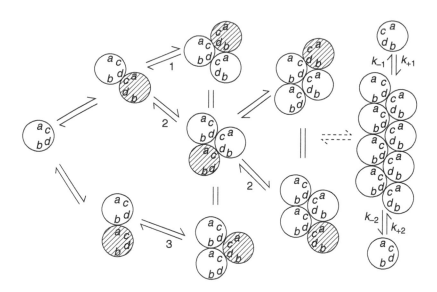

Figure 6 Scheme for the polymerization of actin–ADP. Two dimers (*cd* and *ab* bonds) originate identical trimers that may elongate by addition of unimers at either end through the simultaneous formation of *ab* and *cd* bonds. A steady-state situation is eventually reached with each end independently at equilibrium with unimers. (From E.D. Korn. *Physiol. Rev.* 62:672, 1982. With permission.)

of extremely large linear assemblies during the G $\leftrightarrow$ F (globular $\leftrightarrow$ fibrous) transformation in proteins. The approach is a simple thermodynamic treatment inspired by the statistical mechanical treatment of Zimm and Bragg [42] for the coil $\rightarrow$ helix transformation in linear polypeptides. In the α-helix, each amino acid residue is bound to two residues by main-chain covalent bonds and to two additional residues by H-bonds. In analogy, Oosawa imagines a helical assembly in which each unit can make two kinds of supramolecular bonds with four neighboring units: two along a linear sequence and two along a helical pattern [43]. In this case the shortest oligomer having a helical sequence is composed of four units (Figure 6 and Figure 3(b)). Only linear aggregates (or a transformation from helical to linear assemblies) are expected under conditions disfavoring the formation of one type of bonds.

The distinctive features of the Oosawa mechanism, in particular a large DP of the supramolecular polymer, emerge upon detailed consideration of the model. The formation and growth of the basic helical nucleus of four units (Figure 6) is characterized by the fact that the addition of the fourth (and successive) unit involves a larger number of bonds per unit, and a larger binding constant (K_h) relative to the constant K of the linear sequence. Thus, in analogy with Eq. (1),

$$M_1 + M_1 \iff M_2 \quad \rightarrow \quad K = \frac{|M_2|}{|M_1||M_1|} \quad \rightarrow \quad C_2 = K(C_1)^2$$

$$M_2 + M_1 \iff M_3 \quad \rightarrow \quad K = \frac{|M_3|}{|M_2||M_1|} \quad \rightarrow \quad C_3 = KC_2C_1 = K^2(C_1)^3$$

(8)

$$M_3 + M_1 \iff M_4 \quad \rightarrow \quad K = \frac{|M_4|}{|M_3||M_1|} \quad \rightarrow \quad C_4 = K_hC_3C_1 = K_hK^2(C_1)^4$$

$$\cdots \qquad\qquad \cdots \qquad\qquad \cdots$$

$$C_n = \sigma K_h^{-1}(K_hC_1)^n \quad n \geq 4$$

(9)

where

$$\sigma = (K/K_h)^2 \tag{10}$$

is the key parameter of the theory accounting for the low probability of initiating the helical sequence and the cooperativity of the multiple equilibria process. The total initial concentration C_0 can be written as

$$C_0 = C_1 + \sum n\sigma K_h^{-1}(K_h C_1)^n \approx C_1 + \frac{\sigma C_1}{(1 - K_h C_1)^2} \tag{11}$$

and a plot of $K_h C_1$ versus C_0 is given in Figure 7(a) [41].

A schematic representation of the overall trend is given in Figure 7(b). At low C_0, almost all units occur as dispersed unimers and short linear polymers ($C_0 = C_1$) since the second terms on the right of Eq. (11) can be neglected on account of the small value of σ. Upon further increase of C_0, a critical concentration C^* is reached when

$$C^* = K_h^{-1} \tag{12}$$

For $C_0 > C^*$ the second term in Eq. (11) increases with C_0 and all excess units form helical supramolecular polymers coexisting with monomers and short linear sequences having constant concentration C_1. The average DP of the helical polymer

$$DP_n = \sum DP_n C_{nh} / \sum C_{nh} = 1/(1 - K_h C_1) = (C_h/C^*)^{1/2}\sigma^{-1/2} \tag{13}$$

becomes very large near the critical concentration. For instance if $C_h/C^* \sim 1$ and $\sigma = 10^{-6}$, $DP_n = 10^3$. Comparison of Figure 7(a) and Figure 5(a) reveals the role of $\sigma \ll 1$ in increasing cooperativity ($K_h C_1$ reflects DP_n according to Eq. [13]). Cooperation is lost when $\sigma \rightarrow 1$ and the MSOA mechanism is recovered. Thus, the helical supramolecular polymerization exhibits

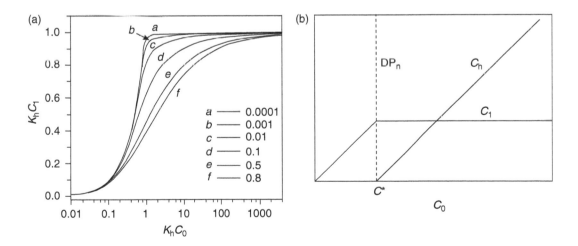

Figure 7   (a) Plot of $K_h C_1$ as functions of $K_h C_0$ for the indicated values of σ. (From D. Zhao and J.S. Moore. *Org. Biomol. Chem.* 2003. Copyright 2003 RCS.); (b) Features of helical supramolecular polymerization: ordinate: unimer + linear oligomer concentration C_1; helical polymer concentration C_h; average degree of polymerization DP_n. Abscissa: total initial unit concentration. (From F. Oosawa and S. Asakura. *Thermodynamics of the Polymerization of Protein.* London: Academic Press, p. 25, 1975. With permission.)

a high degree of cooperativity and essentially involves two extremes: dissociated units (G) and very long assemblies (F). The enhancement of DP_n over the value predicted by isodesmic polymerization (Figure 5, compare Eqs. [4] and [11]) is determined by the cooperativity parameter σ. A broad equilibrium distribution can still be expected with an exponential decrease of C_n with DP and $DP_w/DP_n \approx 2$. The critical concentration is expected to depend upon temperature and solvent type through the value of K_h in Eq. (12). The G $\rightarrow$ F transformation is thus predicted to occur in isotropic solution before the liquid-crystalline phase is formed.

Oosawa emphasized the nucleation of the helix with $n = 4$ that appeared to best describe the experimental data for actin (cf. Section III.B). However, helix nucleation by critical nuclei having n smaller or larger than 4 is also described by Oosawa's theory. Recent work has considered the model with $n = 2$ for the G $\rightarrow$ F transformation [44] (see also [41]). In fact, the scheme in Figure 7(b) has general validity for a host of nucleation processes in solution including, for instance, the nucleation of spherical micelles and their sphere to rod transition (cf. Section II.A.3) [13].

Recently, a true statistical mechanical treatment of the nucleation of all supramolecular helices was presented by van der Schoot and coworkers (45, 46) and successfully applied to experimental data for columnar assemblies of chiral discotic molecules (cf. Section III.B). They elaborated a complete partition function for the nucleation process without an *a priori* specification of a molecular model or a critical nucleus, but included the role of chain-end conformation: confined helical (H), nonhelical (N), or free (F) [46b]. A typical phase diagram for the case in which both ends are unrestricted (FF) is reproduced in Figure 8. The diagram shows the field of stability of monomer, weakly aggregated species, and helical polymers as a function of the excess helical bond energy $(-P)$ and the difference in the chemical potential of a monomer in free solution and in the aggregate $(\Delta\mu)$. A transition to helical polymer is predicted for strong helical bonds, and nonhelical aggregates are predicted in a wide region of the diagram. Significant alteration, including a reentrance of the nonhelical aggregate (cf. Ref. 44) are predicted for other assignments of chain-end boundary. The partition function evaluated for the Oosawa model appeared to agree with the van der Schoot model only for the FF boundary and high cooperativity. A significant difference between the two approaches is that Oosawa considers a specific model for site interaction requiring an *a priori* specification of the nucleus size, whereas such a restriction is unnecessary with the generalized

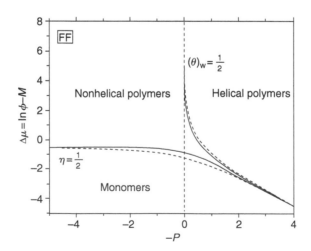

Figure 8 Theoretical state diagram under free boundary conditions (FF) for $\sigma = 1$ (dashed line) and $\sigma = 1.5 \times 10^{-3}$ (drawn line). Vertical axis: $\Delta\mu = -M + \ln\phi$; horizontal axis: $-P = \ln s$. The top line is the helical transition line, where half the bonds are helical and $(\theta)_w = \frac{1}{2}$. The bottom line is the polymerization line, where half the material is in the aggregated state and $\eta = \frac{1}{2}$. (From J. van Gestel, P. van der Schoot, M.A.J. Michels Langmuir 19:1375, 2003. Copyright 2003 ACS.)

theory. Consequently, the latter approach needs to be complemented by an *a posteriori* identification of a structural model.

C. Nonspecific Interactions

1. *Assembling by Incompatibility Effects*

The solute/solvent interaction is an important driving interaction in supramolecular assembling. Chemical compatibility or incompatibility between solution components affects the solubility, the solvation, and the conformation of polymers and oligomers. The well-known hydrophobic effect is due to the poor affinity of water for nonpolar molecules. Transfer of hydrocarbons from water to hydrocarbon solvents is accompanied by a large entropy increase, an often negligible enthalpy contribution, and a decrease of heat capacity [12,17]. The entropy gain has been attributed to a fluidification of the water shell surrounding the dispersed component although this interpretation disregards the role of attractive dispersion interaction within each component [47]. Macroscopic phase separation and ordering is eventually observed on increasing temperature.

The interaction of a particular segment with a poor solvent may be generalized to the interaction occurring when two incompatible segments are present. If the incompatible segments are chemically or supramolecularly connected, a selective solvent may promote solvation and exposure of the solvophilic component and association of the solvophobic one, a situation that leads to cases of globular proteins and micelles characterized by hydrophobic cores and hydrophylic shells. The solvophilic component actually prevents macroscopic phase separation of the solvophobic one. Even in the absence of the solvent, the segmental incompatibility will generate a supramolecular structure based on the *microsegregation* of segments in domains separated by the surface containing the intersegmental bonds.

In spite of their weakness, nonspecific interactions play a fundamental role in determining supramolecular architecture and properties. A relevant example is the possibility of carrying in solution even classical solid-state complexes by supramolecular association to anionic lipids (cf. Figure 12 in Chapter 13) thereby producing processable molecular wires. In fact, incompatible segments have been widely used in the design of amphiphilic molecules and supermolecules forming supramolecular structures in selective solvents. Relevant thermodynamic parameters deriving from the theory of polymer solution [48] describe the affinity of any pair of components down to the ultimate phase separation, also to be regarded as a self-assembly process. Therefore, we analyze the basis of assembly processes due to generic solvophobic interaction, phase separation, micellization, and microsegregation.

(a) *Solvophobic interaction and macroscopic phase separation*

The solubility of SPs in a given solvent is often controlled (cf. Section III.A) by the use of side chains compatible with the particular solvent. For instance, columnar assemblies of *m*-phenylene ethynylene rings having apolar aliphatic substituents are (poorly) soluble only in nonpolar solvents. More polar substituents result in increased solubility in polar solvents. While the side chain experiences a solvophilic interaction with the solvent, the core of the molecule experiences a solvophobic environment that may actually reinforce the contact forces.

This intricate scheme of interaction may be described in terms of sets of compatibility parameters easily calculated (or measured) for relevant binary systems (e.g., side chain/solvent, core/solvent, core/side chain) [49,50]. The knowledge of pairwise parameters is also necessary for a quantitative assessment of the temperature variation of solubility and demixing (two liquid phases or crystallization). According to an approximate treatment of binary solutions originally developed for mixtures of poorly interacting apolar polymers, a liquid–liquid phase separation is expected to occur at a critical temperature $T = \theta > 0$ at which a balance of the enthalpy (κ) and entropy (ψ) components of dilution is achieved [48].

$$\theta/T = \kappa/\psi \tag{14}$$

The condition $\theta > 0$ is fulfilled provided both the heat and entropy of dilution parameters are of the same algebraic sign. Under normal situations the phases separate on cooling and an upper critical consolute temperature (UCST) corresponds to $\psi = \kappa$ with $\psi > 0$ and $\kappa > 0$. However, under the so-called *inverted* situations [51,52] the phases separate on heating and the lower critical consolute temperature (LCST) requires $\psi < 0$ and $\kappa < 0$. Inverted transitions are typical of hydrophobic interaction when more order may be said to occur in the solution than in the phase-separated system. Some degree of aggregation must therefore have occurred at least for one of the solution components. Moreover, the entropy gain due to the breaking of the aggregation must prevail over the entropy loss resulting from demixing. Liquid–liquid phase separation involving an isotropic and a liquid-crystalline phase can also be of the normal or inverted type [49] and conform to the above general principles [53]. For instance, the solvophobic effect in the binary system (hydroxypropyl)cellulose in H_2O produces the ultimate effect of an inverted transition leading to the formation of a liquid-crystalline phase at the smallest concentration (~0.4% v/v, Figure 9) at which cholesteric order was ever detected [54,55]. Note that solution demixing into two coexisting phases (a diluted and a concentrated one), which cannot entail any DP alteration for a covalent polymer, will instead cause significant DP alterations for SPs. Phase transitions can thus be included among the mechanisms of supramolecular polymerization.

Inverted melting transition in the presence of a diluent has also been described [51]. It is characterized by an *increase* of the melting temperature with the amount of diluent, in contrast to the normal case of a depression of the melting temperature by a diluent. Even in this case entropy and enthalpy changes at the transition must each be described by contributions due to two components. The total enthalpy exchange is the sum [51]

$$\Delta H_{tot} = \Delta H^\circ + \Delta H_{dil} \tag{15}$$

where $\Delta H^\circ > 0$ is the melting enthalpy of a pure component, and ΔH_{dil} is its dilution enthalpy. The condition for normal melting requires $\Delta H_{tot} > 0$, whereas the condition for inverted melting is $\Delta H_{tot} < 0$. Therefore the inverted transition requires $\Delta H_{dil} < 0$ (or $\kappa < 0$) and also $|\Delta H_{dil}| > \Delta H^\circ$.

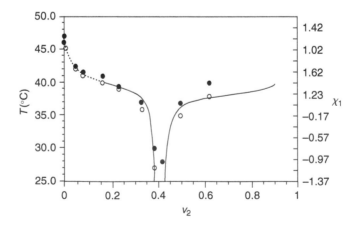

Figure 9 Macroscopic separation: inverted phase transition for (hydroxypropyl)cellulose in H_2O. An isotropic solution with polymer volume fraction $\gtrsim 0.05$ produces a biphasic (isotropic + liquid-crystalline) mixture at $T > 42°C$. The condition $T = \Theta$ ($\chi \sim 0.5$) occurs at $T = 41°C$. (From C.V. Larez, V. Crescenzi, and A. Ciferri. *Macromolecules* 28:5280, 1995. Copyright 1995 ACS.)

The mutual solubility, or compatibility, of two components 1 and 2 (polymer/solvent or polymer/polymer) may also be described in terms of cohesive energy density (CED) using the interaction parameter defined as [48,49]

$$\chi = z\Delta\omega/kT = V_{seg}(\delta_1 - \delta_2)^2/RT \tag{16}$$

where $\Delta\omega = \omega_{12} - \frac{1}{2}(\omega_{11} + \omega_{22})$ is the difference in energy for the formation of 1, 2 contacts out of 1, 1 and 2, 2 contacts, z is a coordination number, k is the Boltzmann constant, $\delta^2 =$ CED is the Hildebrand solubility parameter (calculated with QSPR methods), and V_{seg} the segment volume.

When large differences in the CEDs of the components occur, χ exceeds the critical value and the two components segregate into two macroscopic phases. Values of $\chi < \frac{1}{2}$ characterize instead thermodynamically good solvents in which polymer segments experience chain expansion and solvation. The term $\frac{1}{2} - \chi$ describes the excess mixing (dilution) free energy, which attains pseudo ideal values when [48]

$$T = \Theta, \qquad \chi = \frac{1}{2}, \qquad \psi = \kappa \tag{17}$$

Demixing for a ternary system is predicted to occur when $\chi > \frac{1}{2}(1/DP_2^{1/2} + 1/DP_3^{1/2})$. Solubility parameters for complex mixtures such as two block copolymers poly(A-co-B) + poly(C-co-D) x,y being volume fractions of A and C are evaluated according to [49]

$$\chi = xy\chi_{AC} + x(1-y)\chi_{AD} + (1-x)y\chi_{BC} + (1-x)(1-y)\chi_{BD} - x(1-x)\chi_{AB} - y(1-y)\chi_{CD} \tag{18}$$

(b) Micellization

The poor affinity of nonpolar hydrocarbons with water is manifested in a migration to the air–water interface and in a macroscopic phase separation equivalent to the formation of aggregates of infinite size, as discussed above. However, if a hydrophilic head group is attached to the hydrocarbon molecule the apolar tails can avoid phase separation by forming, above a critical concentration, micellarlike supramolecular structures stabilized by the exposure of the head group to water. Similar structures are expected if two incompatible segments A and B are chemically connected in an amphiphilic block copolymer and dissolved in a selective solvent for either A or B. Self-assembled supramolecular structures formed by surfactants, lipids [12,13,56], and block copolymers [57] in selective solvents have been extensively investigated. Structures involve spherical, cylindrical, and inverted micelles as well as bilayers and vesicles (Figure 10). In all cases the solvophilic groups point toward the solvent and in the case of vesicles there is also a solvent-filled cavity. The simplest structures in Figure 10 can be regarded as the repeating building blocks of larger assemblies. Suitable amphiphilic ABC triblock copolymers may form liposomic, vesicular structures and functional channels within copolymer membranes [58]. The relevance of these structures to the formation of biological membranes has been extensively discussed [12,13,56–58], and their relevance to supramolecular polymerization has been pointed out [5]. Supramolecular amphiphiles, when the solvophobic and solvophilic components are linked by a noncovalent bond, are also known. Kimizuka [59] has described hydrogen-bond-mediated bilayer membranes (cf. Figure 3 in Chapter 13).

The distinctive feature of block copolymer micelles is the occurrence of a polymer segment as the head group protruding from the core. Figure 11(a) illustrates cases in which the relative length of the two flexible blocks determines a large core and a thin corona, or vice versa. Cases in which the selective solvent was a homopolymer of the A or B type have been described [60]. In the case of micelles formed by a rod/coil copolymer (Figure 11[b]), a spherical core formed by the rigid block is not favored. Cylindrical cores or bilayers should be favored instead [57,60–62]. A lamellar sheet typical of the microsegregation occurring with an ABA undiluted triblock copolymer is schematized in Figure 11(c). The formation of bilayers and giant vesicles has also been reported [10,11]. Micellar and microdomain structures are detailed in Chapters 9, 11, and 13.

The formation of spherical micelles occurs with a significant degree of cooperativity at very low amphiphile concentrations and should be regarded as a true phase transition only in the case of an infinite micelle [12,13]. The critical micelle concentration (CMC) marks the limit at which micellar aggregates are formed. Further addition of surfactant does not cause a large increase of free amphiphile molecules but rather an increase of number and average size of micelles (cf. figure 3

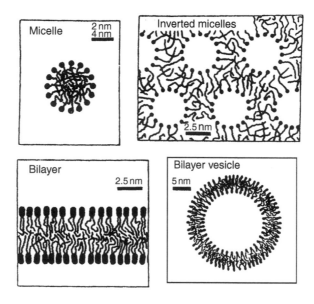

Figure 10   Micellization: micelles, bilayers, and vesicles formed by single- or double-chained surfactants in water. (From J.N. Israelachvili. *Intermolecular and Surface Forces*. London: Academic Press, 1992. With permission.)

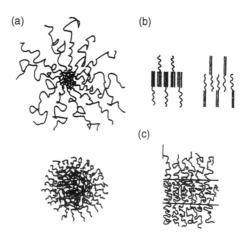

Figure 11   Micellization: (a) micelles in solutions of coil–coil diblock copolymers having different lengths of the solvophilic and solvophobic block. (From A. Halperin, M. Tirrel, and T.P. Lodge. *Adv. Polym. Sci.* 100:31, 1991.); (b) bilayers of rod–coil diblock copolymers in solvents affine for either block. (From A. Gabellini, M. Novi, A. Ciferri, and C. Dell'Erba. *Acta Polimerica* 50:127, 1999. With permission.); (c) microdomain structurization for a triblock copolymer in absence of a diluent. (From A. Halperin, M. Tirrel, and T.P. Lodge. *Adv. Polym. Sci.* 100:31, 1991. Copyright 1991 Springer-Verlag.)

and figure 4 in Chapter 4). A second CMC at which spherical micelles assume a cylindrical shape has often been discussed [63]. Spherical micelles thus exhibit a size distribution corresponding to variable numbers (n) of constituent amphiphilic molecules, and broadening with total surfactant concentration. Minimum $\langle n \rangle$ values range from 50 to 100 for typical ionic surfactants, average sizes are in the order of nm, and the CMC are in the order of mM, decreasing with charge screening and increasing with the size of the apolar tail. It is important to note that the essential features of micellar formation are similar to those illustrated in Figure 7(b) describing helical supramolecular polymerization. In fact, both phenomena follow the general thermodynamic considerations regarding growth of supramolecular aggregates discussed in Section II.A.3 [13]. Micellar parameters can be deduced from a balance between the attraction of apolar tails and the repulsion of charges at the rim, respectively decreasing and increasing with separation distance (64). The approach was recently expanded by Kegel and van der Schoot to describe the somewhat similar assembly of the hepatitis B virus capside [14,65,66]. The stabilization of the core of micelles (or capsides) is driven by a cooperative association due to weak hydrophobic interactions (expectedly characterized by rather small equilibrium constants) counteracted by repulsive electrostatic interactions concentrated at the rim. The cooperative character of assembly formation is attributed to the high rim energy for conformations intermediate between the dispersed unimers and the fully formed micelle, coupled to the large translational entropy loss upon aggregation.

An alternative attempt in using geometrical parameters of the amphiphile to explain micellar shapes (without a detailed knowledge of specific interactions) is due to Israelachvili [13]. He describes the geometrical constraints that affect the interfacial surface area in terms of the area of the solvophilic head group (a_0), the volume (v), and extended length (l) of the aliphatic tail (Figure 12). The parameter $v/a_0 l$ controls the critical packing shape and the most stable structure for a given amphiphile in a given solvent environment. For instance, large head group areas (e.g., ionic amphiphiles in low salt) favor conical packing shape ($v/a_0 l < \frac{1}{3}$) and spherical micelles (Figure 12). On the other hand, cylindrical packing shapes (i.e., double-chained lipids with small head group, $v/a_0 l_c \sim 1$) favor planar bilayer, while a truncated cone ($v/a_0 l \sim \frac{1}{2} - 1$) favors a vesicle.

Upon increasing the amphiphile concentration an evolution toward more asymmetric shapes (rodlike or disklike) and decreasing surface/volume ratio is observed. Eventually cylindrical (capped) micelles, bilayers (extended open sheet with rounded edges), and closed vesicles are formed.

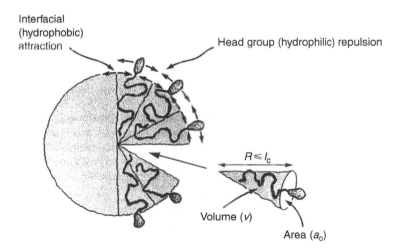

Figure 12 Micellization: the geometrical shape of the amphiphile, expressed by the ratio $v/a_0 l$, determines the stability of micelles and bilayers. Spherical micelles are stabilized by conical shapes: $v/a_0 l < \frac{1}{3}$ (single-chained, large head group area). (From J.N. Israelachvili. *Intermolecular and Surface Forces*. London: Academic Press, 1992. With permission.)

The prediction of the stability of the more complex geometrical shapes has been one of the outstanding goals of micellar studies. A simple treatment of the solvophobic core as a structureless continuum does provide a justification for all micellar shapes, including vesicles [12]. More complex, however, is the detailed description of how amphiphilic molecules can pack within the micellar structures. For instance, the external surface area of a vesicle is larger than the internal one, requiring a larger number of molecules in the section of the curved bilayer pointing outward.

Intermicellar forces are generally of a repulsive nature (i.e., charged amphiphiles) and a reduction of such repulsion accompanies the transformation from spherical to cylindrical micelles. Further increase of concentration results in the formation of linear assemblies and lyotropic mesophases (cf. Section II.C.2). Not only nematic (N_c and N_d for rodlike or disklike shapes, respectively), hexagonal, and smectic phases, but also biaxial (mixtures of N_c and N_d) and complex cubic phases (bicontinuous networks or plastics crystals) were reported by Israelachvili [67,68]. For block copolymers with long segments protruding from the core, interlocking may instead occur upon increasing concentration (cf. Chapter 9).

(c) Microsegregation

In the preceding section, AB block copolymers in solution were shown to produce micellar structures. If the same AB block copolymers are studied in the absence of a solvent, the chemical bond prevents the macroscopic phase separation expected for unconnected A and B. Supramolecular structures will instead occur in which all A-type and B-type segments microsegregate in domains separated by a surface that contains the intersegmental bonds.

A detailed mean-field theory [69,70] (namely, self-consistent field theory — SCFT) describes the supramolecular organization of block copolymers in terms of the favorable attraction of similar blocks (controlled by χ) counteracted by the conformational entropy loss by the other blocks confined in neighboring domains (controlled by chain flexibility and the total number of units N). Figure 13(a) illustrates the predicted range of stability of cubic, hexagonal, lamellar, and other phases in terms of the product $N\chi$ versus the fraction of A units in an A–B uncrystallizable block copolymer [71]. Instability modes generated in undiluted melt direct the formation of the various phases [71]. A review of the most recent elaborations of the SCMF theory that unifies the weak [72] and strong [73] segregation regimes was presented by Matsen [70]. Extension of the theory to the calculation of the relative stabilities of liquid-crystalline phases occurring in copolymer solutions was also attempted [74] and it is described in more detail in Chapter 9. A simple, qualitative description

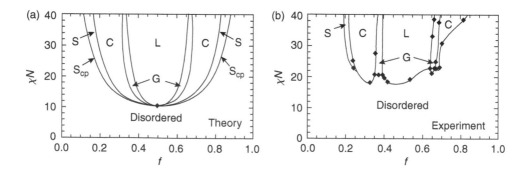

Figure 13   Microsegregation: (a) theoretical and (b) experimental equilibrium phase diagram for amorphous diblock copolymers $(A)_n-(B)_m$ calculated using SCFT and measured using polystyrene–polyisoprene diblock copolymers. f is the fraction of A segments. L = lamellar, C = hexagonal cylindrical, S = spheres, G = gyroid phases. (From M.W. Matsen. *J. Phys. Condens. Matter* 14:R21, 2002. With permission.)

[3,15] of hexagonal and lamellar mesophases in solution and in the bulk, based on the supramolecular polymerization of suitable building blocks, will be described in Section III.D.

2. Assembling by Orientational Fields

The formation of liquid-crystalline phases by covalent rigid, wormlike, and segmented chains has been extensively described [50]. Anisotropy and orientation are characterized at the molecular level by the order parameter and at the mesoscopic level by director orientation. In the case of supramolecular polymers orientation and growth may occur according to the following mechanisms:

1. SPs pregrown in the isotropic phase by MSOA or HG subsequently forming a mesophase.
2. SPs growing simultaneously with the formation of their own nematic alignment.
3. SPs growing within a preexisting liquid-crystalline phase of another compound.
4. SPs growing under the action of external fields.

The difference between mechanisms 1 and 2 is illustrated in Figure 14 schematizing the transition from isotropic to nematic phase for molecularly dispersed rodlike polymers (Figure 14(a)), for closed (Figure 14(b)), and open (Figure 14(c)) supramolecular assemblies. Whereas molecular and closed supramolecular polymers are just oriented in the nematic phase, in the case of open SPs development of orientation is simultaneous with an enhancement of polymerization [5].

(a) Liquid crystallinity of molecular and closed SPs

The following interactions assist the formation of mesophases: *soft anisotropic attraction* that is the prevailing orienting component for low-MW mesogens and for segmented polymers in thermotropic melts (a segmented polymer is based on low-MW mesogens connected by flexible spacers along the main chain) [50,75]; *hard repulsion* (shape II recognition), which is the prevailing orienting component for rigid polymers in lyotropic solutions; *soft isotropic interaction* representing the role of solvent as expressed by χ parameters already discussed in connection with Figure 9. Soft attraction results from the orientation-dependent intermolecular energy $\varepsilon(\theta)$ of nonspherical molecules related

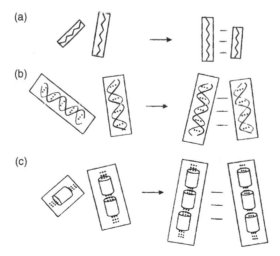

Figure 14   (a) Schematization of the isotropic → nematic transition for molecularly dispersed polymers; (b) closed supramolecular polymers; (c) open (linear) supramolecular assemblies. Coupling of contact interactions (···) with hard and soft interactions (—) causes growth simultaneous to orientation for case (c). (From A. Ciferri. *Liq. Cryst.* 26:489, 1999. Copyright 1999 Taylor & Francis.)

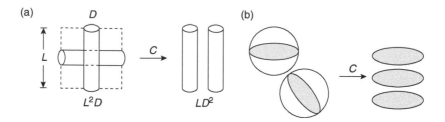

Figure 15 Excluded volume effects: a decrease of free volume favors the formation of (a) parallel assemblies of rods and (b) columnar assemblies of disks upon increasing unimer concentration. (From A. Ciferri, *Liq. Cryst.* 31:1487, 2004. Copyright 2004 Taylor & Francis. www.tandf.co.uk/journals)

to their anisotropy of polarization $\Delta\alpha$ [76,77]

$$\varepsilon(\theta) = C(\Delta\alpha/\bar{\alpha})^2 \varepsilon_{\mathrm{iso}} \bar{V} S\left(1 - \tfrac{3}{2}\sin^2\theta\right) \tag{19}$$

where C is a constant, θ is the angle between the molecular and the domain axis, $\bar{\alpha}$ is the mean polarizability, $\varepsilon_{\mathrm{iso}}$ the isotropic intermolecular energy, $\bar{V}$ the ratio of actual to hard core volume, and S is the order parameter. Hard interactions reflect instead the shape-dependent geometrical anisotropy of the molecules causing (Figure 15) a decrease of volume exclusion from random to parallel assembly of rods or columnar assembly of disks. The tendency to reduce excluded volume leads to a driving force for the orientation of rods or the stacking of disks. The effect begins to be detectable even in isotropic solutions (cf. Chapter 4) and eventually leads to a transition to a nematic mesophase at a critical concentration. The particle anisotropy is expressed by its axial ratio X (length/diameter > 1 for rods, thickness/diameter < 1 for disks).

Extensive theoretical and experimental investigation [50,75] has shown the limits under which the experimental behavior of low- and high-MW mesogens is described by the corresponding theoretical approaches. The following important conclusions are relevant to the present discussion.

1. A critical value of the axial ratio (X^i), varying from ∼4 to ∼8 for different theories of rigid chains [78], determines the limit at which an undiluted mesophase becomes "absolutely stable" [76], implying that when $X > X^i$ the mesophase is primarily stabilized by hard interaction and compositional changes (lyotropic systems). In this case a critical solute volume fraction (v^i) can be defined, decreasing with X according to

$$v^i \approx X^i/X \qquad (3 < X^i < 8) \tag{20}$$

    Critical values of axial ratio for hard rods have been obtained from simulation [79]. For the above system, the nematic → isotropic transition is not influenced by temperature changes meaning that $T_{\mathrm{NI}} \to \infty$, unless a large temperature coefficient dX/dT of rigid conformation does occur [50]. The theory is well-developed for large rods ($L/D \gg 1$), but nor for large disks ($L/D \ll 1$). However, simulation studies support the formation of nematic and columnar phases in solutions of thin disks characterized by $0 < L/D < 0.1$ [80]. Note from Figure 15 that disks exhibit a larger excluded volume than rods, if similar values of relevant dimensions are considered. On the other hand, when $X < X^i$ soft interaction prevails, the low-MW mesogens or the segmented polymers are thermotropic ($T_{\mathrm{NI}} > 0$) and may admit only a small amount of isotropic diluent. For fully covalent low-MW mesogens, the T_{NI} temperature can be predicted from first principles (i.e., anisotropy of polarizability, cf. Eq. [19]). It is important to note that for mesogens including supramolecular bonds, a pronounced temperature dependence of the bond strength will play a significant role on T_{NI}.

2. The partial rigidity of long chain polymers is characterized by the persistence length P, which assumes the role of the limiting rigid segment stabilizing the mesophase. This implies that v^i will decrease with L to an asymptotic value. For the model of the freely jointed chain, the axial ratio in Eq. [20] is expressed in terms of the Kuhn segment [78]

$$X = 2P/D \tag{21}$$

Persistence lengths for molecular polymers are in the range 10–200 nm corresponding to rather large values [50] of the critical volume fractions (in the range 0.02–0.2 taking $X^i = 6$ and $D = 5$–10 Å). Within the mesophase, a wormlike chain may be forced to assume a more extended profile than in the isotropic phase due to the restriction imposed on the director by the order parameter [78,81]. Semirigidity in the nematic state is characterized by the deflection length λ

$$\lambda = P/\alpha \tag{22}$$

where α is a parameter larger than unity, increasing with concentration and inversely related to the width of the angular distribution of the chain tangent vectors.

In the case of the closed supramolecular liquid crystal (SLC) (Figure 14(b)), when no further association $\rightarrow$ dissociation equilibria accompany the formation of the mesophase, their liquid-crystalline behavior is undistinguishable from that of a molecular liquid crystal (LC). The relevant axial ratio is determined by the geometry of the assembly with no need for account of contact interaction.

(b) Open SLC

Quite different is the case of open assemblies for which a coupling may occur between the contact forces that stabilize the assembly and the hard/soft interactions that stabilize the mesophase. Formally, it is necessary to add a term F_{intra} accounting for the stabilization of the assembly through the contact energy to the molecular LC terms describing hard interactions, soft interactions, and any conformational rearrangement of semirigid mesogens within the nematic field. The free energy of the open SLC becomes

$$F' = F'_{ster} + F'_\varepsilon + F'_{el} + F_{intra} \tag{23}$$

The result is an enhancement of growth of the assembly occurring simultaneously with the formation of the ordered mesophase (cf. scheme in Figure 14(c)). The detailed theory for *growth coupled to nematic orientation* was proposed by Herzfeld and Briehl [82] and by Gelbart et al. [83] to describe the assembly of micelles into linear particles. Odijk [84,85] revised the mechanism by recognizing that catastrophic growth in the nematic state is prevented by the flexibility of the linear assembly, resulting in a decoupling of growth in correspondence to the persistence or deflection length of the assembly. The theoretical expectations [5], schematized in Figure 16, show the encroachment of stepwise association to nematic ordering at the critical concentration C^i (generally larger than C^* in Figure 7(b)). The extent of increase of DP at C^i is related to the rigidity of the formed assembly, as expressed by its persistence or deflection length. The DP attained at C^i may be approximated by the ratio P/L_0 where L_0 is the length of the unimer.

The original theory [84], developed for linear assemblies of cylindrical micelles in nematic solutions, was later extended to discotic molecules displaying hard interactions and showing nematic, hexagonal, and higher order phases [86]. The theory predicts that the nematic phase may be skipped for particular combinations of contact energy and rigidity [87]. Extension to systems displaying soft interactions (thermotropic melts) was also considered [88]. A more detailed account of the theory is given in Chapter 4.

More recent considerations relating to the shape of the unimer and to the structure of SPs formed via the open SLC were presented by the author [89]. In particular, growth-coupled-to orientation needs not to be restricted to linear and discotic rigid assemblies. Helical SPs could also assemble by the open SLC mechanism. In fact, the stabilization of supramolecular helices (shown in Section II.B.2 to be due to a nucleation process) could be favored by the SLC even in the absence of nucleation. The process has some analogies with the selection of allowed ordered conformations occurring during crystallization. A related coupling between orientational and supramolecular order was discussed in

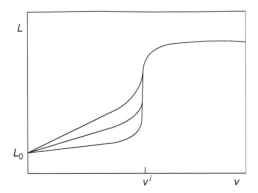

Figure 16   Schematic variation of the length L of a supramolecular polymer with the volume fraction of unimers having length L_0. At the critical volume fraction v^i sudden growth is simultaneous with the formation of the nematic phase. At $v < v^i$ growth occurs according to the MSOA mechanism — three curves for increasing values of the contact energy being represented. (The expected biphasic gap for a first-order transition in lyotropic systems is omitted for clarity.) (From A. Ciferri. *Lig. Cryst.* 26:489, 1999. Copyright 1999 Taylor & Francis.)

the case of the H-bond scheme of the α-helical conformation when the coil $\rightarrow$ helix transformation was shown to be enhanced by the simultaneous formation of the mesophase [53,90]. Another interesting feature needing additional investigation is the possible coalescence of unimers to form a continuous filament. This possibility is supported by simulation studies on the growth of end-cupped micelles of block copolymers [91]. Another recent suggestion [3,15] is the possibility of assembling extremely complex, composite, and functional structures by supramolecular polymerization of specifically designed *building blocks* as described in more detail in Section III.D.

(c)   Spherical particles in a nematic field

Theoretical description of the assembly of small spherical particles of diameter σ within a nematic solution of rods $(D \ll \sigma \ll L)$ has recently been presented [92]. The coupling of excluded volume of rods and spheres drives the spheres in a zone that is significantly depleted of rods. The depletion zone, oriented parallel to the nematic director, has the approximate shape of a cylinder of length L and diameter related to the average excursion of the rod tips from their aligned position. For low volume fraction of rods, an effective attraction between the spheres is induced, and the formation of their chainlike aggregates oriented along the nematic director is predicted (Figure 17). For a smectic A-type layered structure, the globules concentrate in the interlayer space. Larger spheres tend to demix (note that demixing is invariably predicted and observed for mixtures of rods and coiling polymers) [50]. Experimental results for mixtures of fd virus and PS spheres $(\sigma = 1000$ nm) [93] were consistent with the above theory.

(d)   External fields

Orientation induced by elongational flow field has been shown to promote the formation of nematic order at concentration below the critical value in the absence of flow [94]. For covalent systems, the effect of local ordering (described by the order parameters) was shown to be not as dramatic as that occurring over a mesoscopic scale due to director orientation. It is known that application of flow, electrical, and magnetic fields may lead to perfectly ordered single LCs. Theoretical analysis for supramolecular polymers has so far been restricted to the relatively simpler case of helices in magnetic and electrical fields [46]. For polarizable helices in a quadruple field, and for helices with a permanent dipole in a dipolar field, growth enhancement was predicted as more pronounced

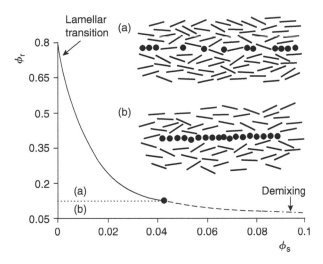

Figure 17 Calculated state diagram for hard globules with diameter σ dispersed in nematic rods with axial ratio $L/D = 100$. ϕ_s is the volume fraction of spheres, and ϕ_r that of the rods. $L/\sigma = 10$. The solid line marks the spinodal instability to the lamellar phase, and the dash–dotted line that to macroscopically demixed phases. The dotted line separates the region where self-assembled chains are of (a) the "open" type from that where they are of (b) the "dense" type. (From P. van der Schoot. *J. Chem. Phys.* 117:3537, 2002. With permission.)

the smaller the angle between chain axis and field direction (cf. Chapter 3). For linear micelles, a decrease of the critical concentration with flow gradient was experimentally observed [95,96]. Even in the absence of a nematic transition, an increase of DP under elongational flow can be related to an increased correlation between growing segments at an effectively larger local concentration.

An interesting example of growth promoted by the application of an electric field was reported [97,98]. A solution of (1,2-dimethoxybenzene) subjected to anodic oxidation (1.6 V) produced a discotic cation radical (1,2,5,6,9,10-hexamethoxytriphenylene). The presence of tetrafluoroborate counterions favored the next and subsequent additions of disklike units to the growing end, resulting in the formation of polyveratrole. Fibers emanating from the electrode surface could be isolated with no loss of properties (including paramagnetism) after several years.

III. SELF-ASSEMBLED POLYMERS

In this section selected data for synthetic and natural polymers in class A are analyzed making a systematic comparison between experimental behavior and theoretical mechanisms presented in Section II. Systems are grouped according to assembling mechanisms rather than chemical structure.

A. Linear Chains and Columnar Stacks: MSOA

Linear and columnar assemblies that were assembled by a noncooperative growth mechanism have been reported. Among earlier linear systems we find the association of glutamate dehydrogenase [38], tropomysin [99], and synthetic SPs stabilized by a single main chain H-bond [100]. For glutamate dehydrogenase in diluted isotropic solutions an association constant $K = 9 \times 10^5 /M$ was reported, and the experimental $M_w - C$ dependence in the oligomeric range (DP $\rightarrow \sim 15$) was well represented by the theoretical prediction illustrated in Figure 5(a). In the case of the SPs reported by Hilger and Stadler [100], undiluted (non LC) systems were considered and DP in the order of 15 was reported.

This result can be attributed to a much smaller association constant than in the case of glutamate dehydrogenase and is consistent with the theoretical plot in Figure 5(b), assuming $K \sim 500/M$ as determined for the pyridine/benzoic acid association [101].

The goal of obtaining large DP occurring in isotropic solution can be realized by the use of multiple H-bonds. A main chain link based on three H-bonds in the AAA-DDD configuration is expected to generate binding constants in the order of $500^3 \approx 10 \times 10^7$ adequate to attain DP in the order of 1000 according to Figure 5(b). This expectation was amply verified by Meijer and coworkers ([2,102], cf. Chapter 15) who studied systems (Table 1, polymer 1) based on the dimerization of ureidopyrimidone characterized by $K = 5 \times 10^7/M$ in CDCl$_3$. A covalent segment (low- or high-MW linkers) was terminated by ureidopyrimidone forming a bifunctional unimer allowing four H-bonds on each terminal surface ($F = 2, S = 8$, cf. Figure 3(c)). The configuration AADD-DDAA reduced, by a factor in the order of 10^3, the much larger value of K expected for the regular AAAA-DDDD configuration. These linear systems did attain DP in the order of 1000 in diluted isotropic solution, when contributions from the HG and SLC mechanisms were ruled out.

Evidence for large growth was also reported for coordination polymers [104,108]. One polymer based on functionalized porphyrin (Table 1, polymer 2) attained DP in the order of 100 in a 7×10^{-3} M CHCl$_3$ solution. Preliminary reports have appeared describing SPs stabilized by DNA base pairing interaction. Rowan and coworkers [109] used only one nucleobase (A^p, C^p, G^p, T) as a terminal group of a short poly(tetrahydrofuran) segment. Solid-state properties were considerably altered but insignificant polymerization was detected in solution. Craig and coworkers [105] used oligonucleotides with partly complementary sequences involving seven or eight base pairs (Table 1, polymer 3). Nucleotide recognition promoted supramolecular polymerization rather than a closed double-helical system of class B type. Polymers formed in solution exhibited viscosity behavior typical of double-stranded DNA while scaling as expected for SPs (cf. Chapter 12).

It is noteworthy that DP obtained by the MSOA mechanism may reach values even larger than those obtained by ordinary polycondensation (DP $\sim$ 100 can only be obtained using irreversible conditions for aliphatic polyamides) [35]. SPs may thus exhibit strong growth in spite of relatively weak bonds, allowing readjustment of donor/acceptor patterns and DP alteration in response to concentration, temperature, and other external variables. Novel properties are thus expected for this class of dynamic, adaptive, smart, self-healing, combinatorial materials (cf. Chapters 1,15). Meijer and coworkers [102] described the peculiar rheological features of networks cross-linked through a four H-bond scheme. Groups displaying supramolecular interaction have also been used to improve mutual compatibility in polymer blends, or in chain extension [103,110]. Supramolecular block copolymers have also been investigated [111] with supramolecular joints based on either metal complexes, e.g. bis(2,2′:6,2″-terpyridine)ruthenium(II), [111a] or multiple H-bonds, e.g. UPy or base pairing, [111b-e]. Corresponding micellar, tubular structures were characterized. Electrostatic binding of copolymers having complementary charged block were also described [111f].

Polycaps based on H-bonded calixarene units functionalized with urea and hosting a small guest molecule were also reported. Isotropic solutions in o-dichlorobenzene revealed polymer like properties, notably strong normal forces supporting the permanence of a significant DP even under a flow field [112].

Columnar stacks are exemplified by the following systems. Triphenylenes with varying substitution (Table 1, polymer 4) form columnar stacks in isotropic solution of hexadecane or heptane ($<10^{-3}$ M) due to arene–arene interaction. DP was determined using small-angle neutron scattering (SANS) by Sheu et al. [106], but no data are included in their paper. From the reported association constant, a maximum DP in the bulk phase is in the order of 300 according to Eq. (7). In very diluted solutions a loose stack of the unimer along the columnar axis was reported, and no helicity was detected for a chiral triphenylene in n-heptane [2]. Computer simulation [113] confirms that association in isotropic solution follows the MSOA mechanisms and is modulated by concentration and temperature changes [114]. Upon increasing concentration, a liquid-crystalline phase is formed causing a large DP increase due to encroachment with the SLC mechanism. This behavior will be

Table 1   Linear and Columnar SPs in Isotropic Solutions (MSOA)

	Polymer	DP	Phase	References
1		→ 1000	I	2,103
2		→ 100	I	104
3		→ 100	I	105
4		→ 300[a]	I (→T)	2,106
5		→ 120[a]	I (→T)	27
6		→ 50[a]	I (→T)	107
7		→ 20	I (→A)	33

Notes: [a]: Theoretical in bulk, I: Isotropic solution, A: LC solution, T: Thermotropic melt.

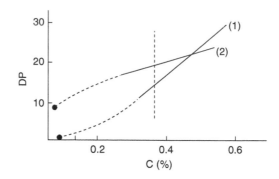

Figure 18 Variation of the number of stacked tetrameric disks (cf. Table 1, polymer 7) with folate concentration in pure H_2O (1) and 1 M NaCl (2) at 30°C. The vertical broken line indicates the I → H transition. (Plotted using data taken from G. Gottarelli, G.P. Spada, and A. Garbesi. *Crystallography of Supramolecular Compounds*. G. Tsoucaris, J.L. Atwood, and J. Lipkowski, Eds., Kluwer Academic Publishers), From A. Ciferri. *Macromol. Rapid Commun.* 23:511, 2002. Copyright Wiley_VCH 2002.

discussed in Section III.C using data for a triphenylene with polar substitution. These systems are attracting attention as molecular-scale wires endowed with charge carrier mobility that may be of use in electronic and optical devices (cf. Chapter 16).

Columnar stacks assembled by the MSOA mechanism include soluble *m*-phenylene ethynylene cycles adopting a completely flat conformation (Table 1, polymer 5). These SPs are stabilized by $\pi-\pi$ stacking of the core and form isotropic solutions and thermotropic melts. A study by Moore and coworkers [27] clarifies the role of hydrophobic interaction and the nature of the substituent on the strength of $\pi-\pi$ interaction and extent of association. Polar side chains such as tri(ethylene glycol) connected by an ester linkage to the macrocycle promoted solubility in a variety of solvents. Measured association constants varied from 50 to 15,000/M, respectively in chloroform and in acetone. In terms of Eq. (7) these values suggest corresponding theoretical DP from 10 to 120 in the undiluted phase. Solubility and association constants were affected by the polarity of the substituent and even by its linkage to the macrocycle.

An additional example of ring-stacking motifs is based on cyclic peptides with alternating D- and L-amino acids giving rise to flat conformations stacking in nanotubes via antiparallel β-sheet H-bonding [107,115,116]. Due to their insolubility in non-H-bond breaking solvents, nanotubes were characterized in the solid state and in lipid bilayers where they formed transmembrane channels. To derive information on the assembling mechanism, Ghadiri and coworkers investigated soluble cylindrical dimers obtained by proper substitution at alternating residues. The dimer association constant for cyclo[(-L-Phe-D-^{NE}N-Ala)$_4$-] (Table 1, polymer 6) was 2540/M in chloroform (a somewhat larger value would have been expected in view of the large number of H-bonds). On the basis of this constant, we can predict values of DP in the order of 50 in the undiluted system according to Eq. (7).

The tetrameric H-bonded supermolecules of folic acid (Table 1, polymer 7) are thin disks of diameter $D \sim 30$ Å and thickness $L \sim 2.35$ Å ($L/D \sim 0.13$) forming columnar stacks of low DP. The plot in Figure 18 was constructed [3] selecting data for the stacked folate disks determined by SANS in both isotropic and liquid-crystalline aqueous solutions [33]. It is evident that no jump in the DP versus folate concentration occurs upon entering the mesophase region confirming that columnar growth is occurring by the MSOA mechanism. Contribution from excluded volume effects and the origin of mesophase will be discussed in Section III.C.

B. Helices and Helical Columns: HG

Several reports have illustrated the formation of helical SPs in isotropic solutions. Cases in which the helical structure nucleates from linear or from columnar sequences can be successfully described in terms of the HG mechanism. Table 2, polymer 8 schematizes the growth of actin F-filaments.

Table 2 Helices and Helical Columns in Isotropic Solutions (HG)

	Polymer	DP	Phase	Chirality	References
8		→ 4000	I	+	117
9	54a, R = C₈H₁₃ X R = C₁₂H₂₅ X R = C₁₂H₂₅ 54b, R = 56a, R = 56b, R =	→ 1000	I	+,−,!	118–121
10	50a, R = C₁₂H₂₅ 50b, R = 52a, R = 52b, R =	→100	I	+,−	2,122
11	X=H, R= ...OCH₃ X=Me, R= ...OCH₃ X=H, R= (S) ...OCH₃ X=H, R= (S) ...CH₃	....	I(A → T)	+,−,!	123,124
12	R = ...OBn	......	I(A → T)	+	125
13		→140	I	+	126

Notes: I: Isotropic soln, A: LC solution, T: Thermotropic melt. + = Chiral, − = Achiral, ! = Chiral amplification.

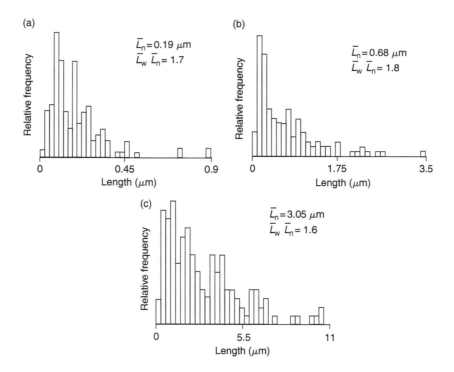

Figure 19 Length distribution determined from electron micrographs for filaments polymerized from solutions having actin/gelsolin mole ratio (a) 61:1; (b) 256:1; and (c) 2048:1. (From P.A. Janmey, J. Peetermans, K.S. Zanert, T.P. Stossel, and T. Tanaka. *J. Biol. Chem.* 261:8357, 1986. With permission.)

Verification of the theoretical HG mechanism (cf. Figure 7) was best performed on actin–ADP systems *in vitro* to avoid the complicating effect of the dephosphorilization reaction of ATP usually bound to the protein. In line with Oosawa's theory, the experimental phase diagram [117] reveals the occurrence of a critical concentration C^* at which HG begins and the concentration of unimers and oligomers attains a constant value. The double-helical structure conforms to the site distribution in Figure 6 [43,127]. In the few reports in which HG was not halted by a chain stopper (gelsolin), filaments in excess of 11 μm were reported in isotropic solutions at $C < 0.04$ mg/ml [127]. The latter value corresponds to a DP of ~4000 and is comparable to the persistence length of F-actin (cf. Section III.C). Note the much larger value of the critical concentration for appearance of the liquid-crystalline phase (~2 mg/ml) [128]. Figure 19 illustrates the length distribution of F-actin filaments polymerized at different actin/gelsolin ratios [127]. The ratio $L_w/L_n \sim 1.7$ and the theoretical exponential distribution was verified. $DP_n (=L_n/27 \text{ Å})$ is usually close to the actin/gelsolin ratio. Sharp, Poisson type distributions can nevertheless be observed under nonequilibrium conditions in nucleation-controlled polymerization or *in vivo*. In the former case, the uniform length is controlled by the ratio nuclei/monomers and by a slow reverse (depolymerization) reaction that prevents a length redistribution [129]. In the *in vivo* case special controlling mechanisms may be involved [4].

The complexity of the growth process *in vivo* is related to a coupling between the polymerization and the ATP → ADP reaction resulting in a cycling of G unimers from one end to the other of the growing filaments [130]. As a result the polymer translates (treadmilling effect). A related dynamic instability controls the *in vivo* functioning of microtubules [131]. These effects are at the basis of molecular engines described in Chapters 21 and 22.

The disklike C_3-symmetrical molecules synthesized by Palmans et al. [118] (Table 2, polymer 9) stack due to both arene–arene and H-bonding interaction and exemplify the stabilization of a helical structure of a columnar assembly. Side chains had either achiral or chiral character and varying polarity allowing study in either nonpolar or polar solvents. In isotropic solutions of *n*-hexane (10^{-6} M),

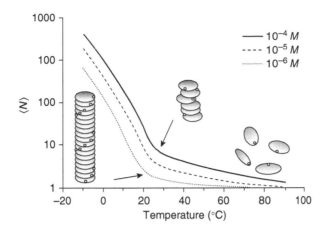

Figure 20 Helical columns: theoretical variation of the average DP ($\langle N \rangle$) of chiral C_3 symmetrical molecules (cf. Table 2, polymer 9) with temperature. Formation of achiral assemblies and transition to chiral helices occurs upon lowering T. (From J. van Gestel, P. van der Schoot, and M.A.J. Michels. *J. Phys. Chem. B* 105:10691, 2001. Copyright 2001 ACS.)

long chiral, helical assemblies with large binding constants ($10^8/M$) were detected [118]. Chirality is expected to favor helicoization but not necessarily a cooperative transition. The cooperative effect leading to helix formation was attributed to a conformational transition from flat to propeller shape of the arms of each disk, allowing a maximization of site interaction [45].

In the case of more polar homochiral C_3 molecules, studied in isotropic solutions of butanol (10^{-8} to 10^{-4} M), the apolar core is shielded from the polar solvent by the polar side chains. These molecules showed a sequence of two assembly steps upon decreasing temperature (Figure 20). Low DP achiral stacks were stable in the ~85°C to ~25°C range, but were transformed below 25°C into large DP helical assemblies (DP → 1000) with a simultaneous induction of the unimer chirality to the chirality of the whole assembly [119–121]. The chirality amplification extended to over 400 molecules before helix inversion was detected by sergeant and soldier [132] experiments. Helical order was strongly dependent upon solvent type. In water, helix inversion was detected after about 12 disks. The temperature variation of DP is reproduced in Figure 20. The theoretical lines represent the fitting of the van der Schoot and coworkers theory [45,46]. The two assembling steps may be described as the encroachment of MSOA to HG growth. Corresponding binding constants ($K < K_h$) and cooperativity parameters σ were derived from the theoretical fit. The role of boundary conditions was also discussed (cf. Section II.B.3 and Figure 8). Oosawa's theory does afford a good fitting of the data for actin (for a more detailed discussion, cf. Chapter 3).

The above results allow the definition of one important feature of the assembly of discotic molecules in isotropic solution. There seem to be conditions (controlled by temperature, concentration, and solvent type) in which contact forces are weak and loose binding of the unimers produces short columns with low DP and little or no chiral amplification. Cooperative growth ensues even though a detailed mechanism is often unclear. The critical nucleus size is not readily identified from theory (cf. Section II.B.3) but might be associated with the number of disks included in the pitch of the highly correlated helix forming when contact forces increase at low temperature.

The next case in Table 2 adds additional insight on helix formation by discotic components. Polymer 10 based on the bifunctional ureidotriazine unimer ($F = 2, S = 8$) is stabilized by quadruple H-bonds. At variance with the somewhat similar polymer 1 (Table 1), a very short spacer connects the two hydrogen-bonding terminals of each unimer. A coiling chain might be expected with H-bonded extended core discotics interconnected by the spacers. However, the arene–arene interaction should favor a columnar stacking of the disks externally connected by the spacers. In fact, SANS data for decane solutions (0.2% to 1.0%) of the achiral unimer (R = $C_{12}H_{25}$) revealed the occurrence of cylinders with radii corresponding to the monofunctional ureidotriazine dimer and length between 100 and 190 Å (corresponding to DP → 100 for a 2 Å disk thickness). The helical nature of the assembly was confirmed by the Cotton effect in decane when chiral alkyl chains were used for the R substituent. The increased correlation between unimeric disks due to the short linker favors cooperative helix formation. Simple columnar stacking was instead observed with the related monofunctional ureidotriazine units illustrated in Figure 4(d) [122].

The spontaneous formation of a columnar–helical structure based on disks interconnected by short covalent segments is significant for analogies with the new family of molecular foldamers and with complex biological assemblies. *Foldamers* are oligomeric segments able to fold into a programmed conformation in solution [133,134]. Their role is better understood in terms of the tertiary structure of proteins. The distribution of polar and apolar substituents along the polypeptide sequence does ultimately result in a globular structure with a solvophobic core and a solvophilic skin. For any specific structure, chain folds must be programmed to occur at particular locations along the sequence. Oligomeric foldamers have been studied mostly in connection with synthetic covalent sequences. However, polymers based on both supramolecular and covalent main chain sequences may also be induced to assume programmable structures, as indicated by the case of polymer 10.

An additional example of a conformationally programmed structure is offered by polymer 11 in Table 2. The unimers are short segments of m-phenylene ethynylene ($n = 8 \rightarrow 18$). The m-substitution of rigid subunits favors a change from the coiled to helical conformational for oligomers that are long enough ($n > 8$) to allow π–π stacking of aromatic rings [123]. The effect occurs in isotropic solutions of polar solvents inducing cooperative solvophobic packing of phenyl rings. Helical folding induces the simultaneous piling up of oligomers into helical SPs, evolving to lyotropic phases and hexagonal packing in the solid state [124]. The helical nature of the polymer was demonstrated by the Cotton effect revealing chiral amplification: a chiral oligomer amplified its chirality even to achiral foldamers supramolecularly following the helical pattern. The polymerization of helicenes is included in Table 2 (polymer 12) for comparison with polymer 11. Due to its fixed folded sequence, helicene is not considered a foldamer. Thus, the folding cooperative contribution is absent in the polymerization process. Its discotic shape allows the formation of liquid-crystalline phases exhibiting large DP and chiral effects [125]. However, the role of the mesophase might not be essential since also nonliquid-crystalline helicenes were able to self-assemble in very diluted (0.0005 M) dodecane solutions [125].

Chromatin is an interesting foldameric assembly based on a sequence of discotic nucleosomes (histone protein octamers) wounded up and interconnected by a long DNA superhelix (Figure 21) [135]. The final folded structure, often described by a solenoid-type model [136] is the result of histone–histone and histone–DNA interactions. The former can be studied by separating the histones from DNA either in 2 M NaCl or by enzymatic cleavage (the reassociation is not completely reversible [137]. Results showed that the discotic octamers formed loose stacks in isotropic solutions, eventually evolving to columnar and finally lamellar organizations [138]. The latter appears stabilized by interactions of electrostatic nature, and by nonspecific interactions in the transversal direction. The interaction by which DNA folds and wraps around the nucleosome is primarily of an electrostatic nature, as supported by the dissociation at high ionic strength. Note that the spacing between the nucleosomes (about 15 nm) is smaller than the persistence length of DNA. It is unlikely that rigid chains can wrap around spherical particles [139], and therefore strong interactions and possible local conformational alterations may be involved. A more detailed analysis of the folding mechanism and the final tertiary structure is still being investigated [140].

The last system included in Table 2, the tobacco mosaic virus (TMV), illustrates the induction of helicity in host–guest polymeric assemblies [3,126]. Columnar, helical, and helical–columnar assemblies often have a cavity in which guest molecules can be hosted. A cavity of only 6 Å diameter can host a polymer chain: for instance, stacks of α-cyclodextrin rings can host a poly(ethylene oxide) chain [141] without induction of helicity. Columnar stacks have also been observed with

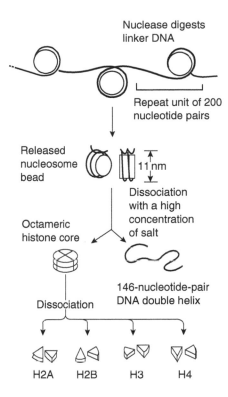

Nuclease digests
linker DNA

Repeat unit of 200
nucleotide pairs

Released
nucleosome
bead

11 nm

Dissociation
with a high
concentration
of salt

Octameric
histone core

Dissociation

146-nucleotide-pair
DNA double helix

H2A   H2B   H3   H4

Figure 21   The chromatin assembly based on discotic supermolecules (histone octamers) linked by a DNA super-helix. (From B. Alberts, D. Bray, M. Raff, K. Roberts and J.D. Watson. Molecular Biology of the cell. 3rd Edition. New York: Garland Publishing 1955 page 343. With permission www.tandf.co.uk/journals)

covalent chains having dentritic side chains that self-assemble into disks [142]. No helicity is manifested by the above systems. Other known cases are those in which the polymeric guest induces helicization of tubular stacks of disks. In the case of TMV, the guest is a RNA molecule and the host is a helical–columnar assembly composed of identical tapered protein molecules. The structure can be disassembled and reassembled by pH changes in isotropic solutions with or without RNA. Without RNA, a population of dimeric disks, helical columns, and columnar stacks is observed, each disk comprising 17 protein units. However, the native helical structure based on 2310 proteins and 16.3 units per turn is reassembled when RNA is present. The complex role of RNA for the whole structure is evident. RNA acts like a crankshaft that stabilizes the proteins bound to it into a helical pattern and simultaneously provides the information for the proper length and DP of the host. The assembly mechanism of the overall TMV structure can thus be described in terms of a supramolecular polymerization of the external columnar assembly, coupled to the formation of monofunctional side chain bonds between host and guest [3]. A quantitative approach along the above lines was recently reported by van der Schoot and coworkers [14] for the spherical capside of the hepatitis B virus.

C.   Liquid Crystallinity in Supramolecular Polymers

Following the emphasis on the assembling power of the open SLC (Figure 14(c), [5,143]) several authors [2,144,145] have suggested an enhancement of growth associated with the formation of a mesophase for a large variety of SPs. It is therefore essential to critically distinguish cases in which growth is effectively coupled to orientation from cases in which the supramolecular polymerization is uncoupled to the occurrence of liquid crystallinity [89]. There is a need for more studies of the open SLC and several aspects of its theoretical mechanism have not yet been clarified.

1. Liquid Crystallinity for Molecular and Closed SPs

The closed SLC is exemplified in Figure 14(b) by a class B system when sites are internally compensated and no further growth accompanies the formation of the mesophase. The behavior of the closed SLC is thus indistinguishable from that of a molecular LC (Figure 14(a)). Relevant cases are DNA [146], adequately described by the theory of the molecular LC (Section II.C.2), and poly(p-benzamide) (PBA) in N,N-dimethylacetamide/LiCl solutions. An assembly of seven PBA molecules with a side-by-side shift of one fourth the molecular length was detected in both isotropic and lyotropic solutions. Even the axial ratio of the assembly (~ 104) was undistinguishable from the axial ratio (~ 100) of molecularly dispersed PBA [147].

2. Liquid Crystallinity Uncoupled to Growth

If strong growth, in isotropic solutions has produced at $C < C^i$ a wormlike chain with length comparable or exceeding the persistence length, no sudden growth due to the open SLC occurs at the critical concentration even though growth continues at $C > C^i$ driven by the MSOA or HG mechanism. In the case of actin (Table 2, polymer 8) the lowest critical concentration for appearance of the mesophase reported by Furukawa et al. [128] was ~ 2 mg/ml for a gelsolin terminated filament having DP ~ 1780 corresponding to a length of ~ 5 μm. Janmey et al. [127] were however able to grow filaments with a larger length at concentration of ~ 0.04 mg/ml suggesting that actin grows to a length comparable to the persistence length in isotropic solutions ($C^* < C^i$).

Figure 22 illustrates dynamic association–dissociation cycles of α- and β-tubulin into microtubules growing to a size visible under the polarizing microscope. The data by Hitt et al. [148] suggest a synchronous occurrence of growth and liquid crystallinity. The polymerization $\rightarrow$ depolymerization cycles are modulated by the reversible GTP $\leftrightarrow$ GDP reaction that uncovers sites on β-tubulin (cf. Chapter 21). A direct coupling between growth and alignment cannot be confirmed by these data (and similar cases in the literature) since the tubulin concentration (15 mg/ml) was larger than either the critical concentration at which the helix nucleates or the mesophase appears.

In the case of actin, and other systems characterized by strong binding constants, growth and mesophase are "uncoupled" but "hierarchically related" since liquid crystallinity arises as a consequence of a preassembling step in isotropic solution. Cases of growth totally uncoupled to liquid crystallinity have also been reported. One example is the tetrameric H-bonded supermolecule of folic

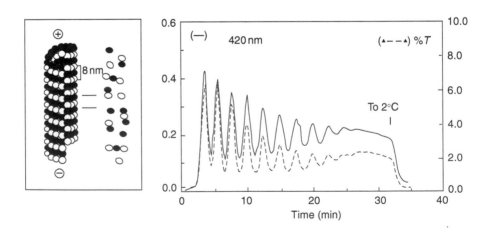

Figure 22 Dynamic assembly and disassembling of microtubules. Time variation of turbidity (right) and birefringence (left) at 420 nm for solutions of tubulin (15 mg/ml) in pH 6.9 buffer $+12$ mM MgSO$_4$ $+2$ mM GPT. Measurements at 37°C followed by quenching at 2°C. (From A.L. Hitt, A.R. Cross, and C.R. Williams Jr. *J. Biol. Chem.* 265:1639, 1990. With permission.)

acid (Table 1, polymer 7) for which the DP was determined by SANS. A continuous decrease of the number of stacked tetramers with concentration, undisturbed by the occurrence of the mesophase, is evidenced by the data collected in Figure 18. The intensity of contact forces or rigidity do not allow cooperative growth due to either the SLC or the HG mechanism. The occurrence of the mesophase can therefore be attributed to the large excluded volume of disks (cf. Figure 15). Similar behavior was reported by Ben-Shaul and Gelbart [149].

3. Liquid Crystallinity Coupled to Growth

Verification of the open SLC model is based on a sudden increase of polymerization when the nematic phase appears (Figure 16). Data regarding the rigidity (persistence length) of the assembly are desirable. Systems for which growth-coupled-to-orientation was adequately documented are described below [85,150–152,156]. In the case of micelles (Table 3, polymer 14), Odijk [85] has critically

Table 3 SPs in Lyotropic and Thermotropic Phases

	Polymer/complex	DP	Phase	References
14		$\longrightarrow$ 1000	A	84
15		$\longrightarrow$ —	A	152
16		$\longrightarrow$ 20[a]	T	88
17		—	T	88
18		$\longrightarrow$ 100[a]	T	153,154
19		—	T	155

Notes: A: LC solution, T: Thermotropic melt.
[a] Theoretical in bulk.

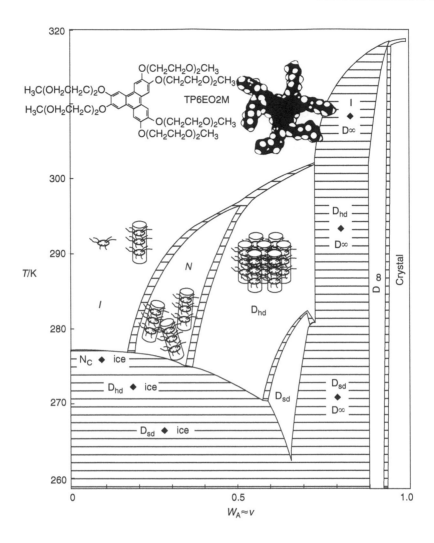

Figure 23   The discotic amphiphile 2,3,6,7,10,11-hexa-(trioxoacetyl) triphenylene (cf. Table 3, polymer 15) is the repeating unit in a columnar stack stabilized by solvophobic interactions in D_2O. The temperature versus volume fraction phase diagram shows single columns occurring in the nematic phase (*N*) at volume fraction as low as 0.2. Hexagonal columnar and eventually crystalline phases evolve upon increasing concentration. (From R. Hentschke, P.J.B. Edwards, N. Boden, and R. Bushby. *Macromol. Symp.* 81:361, 1994. With permission.)

reviewed data supporting their linear polymerization. Upon increasing surfactant concentration the hierarchical sequence

$$\text{dispersed molecule} \rightarrow \text{spherical micelles} \rightarrow \text{end-cupped micelles} \rightarrow \text{growth} + \text{nematic} \quad (24a)$$

is observed. The broad features of the predicted phase diagram were verified [150,151]. Persistence length data for several surfactants support a large linear growth [5,84] (cf. Table 4).

   The phase diagram of the discotic amphiphile based on triphenylene with polar side chains (2,3,6,7,10,11-hexa-(1,4,7-trioxoacetyl) triphenylene in D_2O, Table 3, polymer 15) [152,157] is reported in Figure 23. Large columnar stacks formed at a volume fraction $v \sim 20\%$ (RT) simultaneously with the isotropic $\rightarrow$ nematic transition. The nematic phase was stable up to $v \sim 40\%$, when

Table 4 Persistence length for supramolecular and molecular polymers.

System	Shape	P, μm	D μm	P/D	DP*	References
Poly(p-benzamide)/ DMAc/3%LiCl	Linear	7.5×10^{-2}	5.10^{-4}	150	110	161
Dimethyleylamine Oxide/H_2O + 10^{-2}M NaCl	Linear	$(0, 2\rightarrow)1.7$	6.10^{-3}	280	20.000	162
Actin/ Phalloidin stabilized +Buffer	Helical	$(6\rightarrow)17.7$	5.10^{-3}	3.500	6.500	163–165
Microtubules/Taxol stabilized + buffer	Tubular	$(79\rightarrow)5.200$	3.10^{-2}	170.000	17.10^6	163, 166

* Degree of Polymerization

the hexagonal columnar phase appeared, followed at $v \gtrsim 60\%$ by higher order phases. The data are consistent with predictions of the growth-coupled-to-orientation theory as elaborated by Hentschke [86], but no persistence length data are available. The hexagonal phase, evolves due to its better packing efficiency from the nematic one. The sequence of phases reveals a hierarchical evolution of the assembly process from nanoscale to mesoscale dimension.

$$\text{dispersed disks } (I) \rightarrow \text{growth + nematic } (N) \rightarrow \text{hexagonal } (H) \rightarrow \text{higher order phases} \qquad (24b)$$

Comparison of osmotic pressure versus concentration data for nonaggregating and sickle cell hemoglobin reported by Hentschke and Herzfeld [156] supports the simultaneous onset of linear growth and alignment at ~20% concentration. Assemblies attained length in the μm range.

Among features of the SLC mechanisms that await more detailed investigations there is a need for a quantitative assessment of the range of contact forces and flexural rigidity that allow stabilization of the nematic phase. Systems with varying number of interacting sites and persistence length need to be investigated. In fact, for some surfactants and block copolymer micelles [158,159] the nematic phase is not observed and it is not clear if the effect is due to the delicate balance of the two parameters required by theory [87].

Persistence lengths were determined only for a few supramolecular assemblies and reveal a much larger rigidity than reported for molecular polymers or closed SPs [160,161]. Table 4 compares data for elongated assemblies having different shapes (linear, helical, tubular) with data for poly(p-benzamide) a typical rigid polymer. The scatter of determinations performed with various techniques is indicated in the table. For micelles, light scattering data for the nonionic surfactant dimethyloleylamine oxide are reported [162]. Alternative determinations for other types of micelles provided values of P ranging from 0.02 to 10 μm [84]. The P values for cytoskeleton assemblies were evaluated from their flexural rigidity measured from thermally driven fluctuations in shape [163], or from fluctuations of the end-to-end distance [167]. The persistence length of microtubules, larger than the length of the samples used for its determination and of microtubules found in cells, reflects the larger cross-section with respect to actin. The tubular shape, based on a lateral association of 13 to 16 protofilaments with a linear sequences of α- and β-tubulin molecules (cf. Figure 22), maximizes the bending stiffness.

4. Discotic SPs

The author has recently pointed out some peculiarities that need to be considered when the liquid-crystalline behavior expected for discotics interacting by excluded volume effects is coupled to the occurrence of contact forces and supramolecular polymerization [89]. The L/D ratio of self-assembling discotics such as those in Tables 1 to 3 is in the order of 0.1. Simulation studies [80] show that such a large asymmetry should lead to the formation of mesophases (N_D and columnar) even in the absence of contact forces between their surfaces.

The superimposition of contact forces reduces the asymmetry of particles. For large association constants, elongated columnar assemblies ($L/D > 1$) will be formed with individual disks separated by short separation distances and no longer freely rotating. The N_D phase is disfavored with an evolution toward cubatic phases ($L/D = 1$) [168] and rodlike ($L/D > 1$) behavior. For weak equilibrium constants only short oligomers are expected, and the N_D phase will be simply shifted to larger critical concentrations due to the modest increase of particle thickness.

Inspection of the data in Tables 1 to 3 support the subtle way in which the geometrical asymmetry of disks couples with contact forces for the formation of LC phases. In cases of strong association constants, discotic columns have been shown to form via the HG mechanism in isotropic solutions (Table 2, polymers 9, 10). Liquid-crystalline behavior eventually develops at larger concentration promoted by the excluded volume of long rigid particles, uncoupled but hierarchically related to growth as discussed in Section III.B.

Cases in which equilibria constants are rather weak (exemplified by folic acid tetramers, Table 1, polymer 7, cf. Section III.C.2) show complete uncoupling between growth and liquid crystallinity. Data in Figure 18 evidence no cooperative growth producing large (>1) axial ratios. The mesophase appears at $\sim$40% concentration and DP is $\sim$10 when the axial ratio is still <1.

In the case of a favorable combination of moderate association constants and flexural rigidity (e.g., Table 3, polymer 15), the coupling of growth and liquid crystallinity again produces a suppression of the N_D phase.

5. Thermotropic SPs

The formation of supramolecular dimers, or low DP oligomers between similar or dissimilar components, is often accompanied by the formation of LC phases. Hydroxypyridine dimers [169] or H-bonded complexes between adenine and thymine [170] are capable of forming LCs. Nonmesogenic pyridine and carboxylic acid derivatives also develop liquid crystallinity upon complexation [171]. Complexation at the side chain of a nonmesogenic flexible polymer, described in Chapter 5 [172,173] and theoretically by Tanaka and coworkers [174], may also result in the formation of mesophases (side chain SPs are not considered in this chapter since they belong to class B systems, unable to grow). The linear segmented assemblies between a dipyridyl and a diacid originally reported by Griffin and coworkers [88] also exhibit a thermotropic nematic phase. In terms of the discussion in Section II.C.2, liquid crystallinity in segmented chains is primarily a reflection of the soft anisotropic interaction occurring for the low-MW mesogens incorporated in the main chain [76]. For covalent segmented polymers occurrence of polymerization could be evidenced by larger transition temperatures with respect to the corresponding low-MW mesogen [175, 176]. However, both polymer 16 and dipyridyl complexed with two monofunctional acids (supermolecule 17) exhibit comparable T_{NI} temperatures (respectively 180° and 178.5℃). When supramolecular bonds occur, an additional complication in interpreting T_{NI} is due to the lability of the secondary bond.

In principle, bonding in thermotropic systems should be enhanced by the formation of the nematic phase, just as in the case of the lyotropic systems, polymers 14 and 15 in Table 3 (cf. Section II.C.2). Bladon and Griffin [88] have indeed theoretically shown that growth is expected for a wormlike chain forming a nematic phase stabilized by soft interactions of the Maier–Saupe type. However, polymer 16 conforms to the model of a segmented chain rather than to the model of a wormlike chain. The flexible joint in segmented chains prevents the attainment of large persistent lengths [175]. Therefore, growth due to the open SLC mechanism is precluded likely, and the liquid crystallinity can be simply due to the soft anisotropic interaction of the low-MW mesogen in the main chain. The actual DP of polymer 16 may be evaluated using Eq. (7) derived for the MSOA mechanism which, on the basis of a binding constant in the order of $500/M$ (cf. discussion in Section III.A), yields a value in the order of 20. A recent review [145], describing other examples of thermotropic SPs based on a single H-bond scheme, does not support the occurrence of considerable DP or offer evidence for a correlation between liquid crystallinity and growth.

Thermotropic SPs based on unimers with multiple H-bonding confirm the difficulty in assessing DP and growth mechanism. In a truly pioneering work, Lehn and coworkers ([153,154], cf. Chapter 1) prepared polymer 18 from bifunctional tartaric acid derivatives (D, L, or meso M) terminated with either two 2,6-diamino-pyridine (P) or two uracil (U) derivatives. The P–U bond is based upon a triple H-bond scheme with the DAD-ADA arrangement and binding constant that ought to be in the order of $10^4/M$. All polymers exhibited thermotropic mesophases in the 25°C to 250°C range. X-ray and TEM data revealed a triple-helical structure and fiber-forming properties for the chiral polymer (LP_2, LU_2) in contrast to the achiral sample (MP_2, MU_2) that exhibited individual chains in a zig-zag conformation and other features typical of a flexible polymer. The interpretation currently favored is that of growth developing simultaneously with liquid crystallinity in the undiluted state (cf. Section III.C.3). No independent support for this interpretation does however exist.

In view of the flexibility of a single chain of polymer 18 it is unlikely that the open SLC mechanism might have been operative. On the other hand, the circumstantial evidence reported by Lehn and coworkers supports the possibility that the triple helix of the chiral polymer formed even in the presence of the $CHCl_3$ solvent. Therefore, it is possible that the triple helix is nucleated by the HG mechanism whereas liquid crystallinity reflects the excluded volume of the helical assembly (a hierarchical relationship between growth and liquid crystallinity was discussed in Section III.C.3). The expected DP of the chiral polymer should be larger than predicted by the MSOA mechanism (DP ~ 100 according to Eq. [7] when $K \sim 10^4/M$).

In the case of the achiral polymer (MP_2, MU_2), by the open SLC mechanism is again precluded by single chain flexibility. Moreover, even growth due to the HG mechanism is ruled out since no cooperative helix formation was observed. DP should therefore be close to the value predicted above by the MSOA mechanism and liquid crystallinity should be related to intrinsic characteristics of the P–U assembly (soft anisotropic interactions or the ability to form disklike assemblies such as those documented for the supermolecule 19 described below).

The monofunctional P–U complex (supermolecule 19, Table 3) has been shown to form disklike dimers that can self-assemble into columns displaying thermotropic behavior. Each disk has a thickness/diameter ratio of ~ 0.1 [155]. The formation of columnar mesophases by similar discotic supermolecules is described in the previous section when cases in which liquid crystallinity and growth are hierarchically related, coupled, or totally uncoupled were considered. Lack of data on the equilibrium constants, or DP, prevents a definite assessment of the assembling mechanism of discotics based on supermolecule 19.

D. Multidimensional Assemblies

1. Planar Assemblies

Planar assemblies conforming to the site distributions of the $S = 4$, $F = 4$ type (cf. Figure 2) were reported. A typical example is that of S-layers (cf. Chapter 17) forming a protective monolayer for the external surface of bacterial cells. The constituent proteins have a quasi-spherical shape with an equatorial distribution of H-bonding sites. An additional single site on the south pole allows weaker electrostatic anchoring to the proteins of the cell membrane. Proteins can be isolated and reassembled *in vitro* allowing the preparation of purely H-bonded monolayers standing over inert surfaces [177]. The assembly $\leftrightarrow$ disassembly process was described as a two-dimensional crystallization, which is consistent with the general thermodynamic considerations presented by Israelachvili (cf. Section II.A.3) for $p = \frac{1}{2}$ systems. Crystalline morphologies corresponding to oblique, square, hexagonal lattice symmetries were evidenced by TEM with unimer center to center distance from 3 to 30 nm. Bottom-up nanotechnology applications and patterning have been suggested. Of particular interest is the behavior of the *in vivo* assembly when features of a "dynamic closed surface crystal" are exhibited. The curvature and finite size effects are controlled by the curvature of the bilayer cell membrane to which the protective layer is anchored. A unique feature is the way in which the assembly ↔ disassembly process of the crystalline layer responds to cell growth and division. A continuous synthesis of unimers and their incorporation at particular dislocation sites is believed to cause "intussusceptive" growth of this dynamic crystal [178].

Another example of planar systems conforming to the $S = 4$, $F = 4$ scheme of Figure 2 is based on the assembly of tetrafunctional branched DNA [179,180] (cf. Chapter 12). Consider a double-helical DNA strand terminated with two single-chain complementary segments. This molecule ($S = 2$, $F = 2$) can linearly polymerize by recognition of complementary single-chain ends (cf. Table 1, polymer 3). Suppose now that two phosphoesther bonds are cleaved at suitable locations near the center of the double helix, followed by ligation at the $3'$ and $5'$ positions with two double helical DNA side chains. When the new arms are terminated with two complementary single-chain segments, a $S = 4$, $F = 4$ unimer is produced that is able to self-assemble into planar systems. The approach allows the design of a variety of DNA nanoconstructions and templates (cf. Chapter 12).

Planar synthetic systems include the self-assembled monolayers (SAMs, cf. Chapter 18) that result from physisorption or chemisorption of small molecules such as $CH_3(CH_2)_nX$ over a planar substrate. Acid-functionalized *n*-alkanes [181] or alkanethiolates [182,183] can form bonds of increasing strength between the head group X and alumina or gold surfaces, respectively. The case most extensively studied is that of decanethiol over the 111 face of gold single crystals, when a covalent bond actually occurs. With increasing coverage, the assemblage of monolayers evolves from an initial phase with molecular axes of decanethiol parallel to the gold surface (Figure 24) to a final phase characterized by bound axes nearly perpendicular to the surface. The lateral intermolecular interaction between grafted alkyl segments drives the supramolecular organization in this

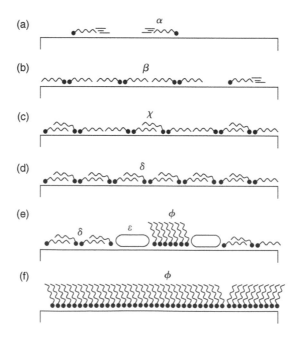

Figure 24 Sequence of monolayer phases with increasing coverage of decanethiol on Au(111). (From G.E. Poirier. *Langmuir* 15:1167, 1999. Copyright 1999 ACS.)

two-dimensional assembly including covalent bonds. Alignment of liquid-crystalline layers over a surface may also be described [184]. Spontaneous growth of highly organized films of amphiphiles over various surfaces was described by Shimomura [185].

2. Three-Dimensional Assemblies

Three-dimensional assemblies of spheres with a symmetrical distribution of equivalent sites (cf. Figure 2, $S = 6$, $F = 6$) should conform to the expectation for the $p = \frac{1}{3}$ systems. More interesting are situations in which either a geometrical anisotropy of the unimer or differences in the recognition pattern of different components exist. In these cases it might be possible to identify the modes of growth along longitudinal and transversal directions [139].

A most interesting example was described by Muñoz-Guerra and coworkers [186], (cf. Chapter 11). They synthesized poly(α-n-alkyl-β-aspartate)s, a family of comblike polyamides characterized by a rigid (helical) backbone and side chains of flexible aliphatic segments having up to 22 methylene units. The assembly produced a solid-state composite structure, characterized by a regular distribution of rigid and flexible components possessing similarities to the structure of keratin fibers. They observed the occurrence of one crystalline and two liquid-crystalline phases upon increasing temperature (cf. Figure 25). In the crystalline phase of the sample with 18 methylene units, the side chains crystallized in a separate hexagonal lattice with an interlayer spacing of 3.1 nm. In the high temperature phase the side chains are completely molten (spacing = 2.3 nm) but the features of a nematic phase attributed to the molecular helices diluted by the flexible components are retained. The intermediate phase may be described as a SLC in which the rigid helices are correlated by interdigitation with partly molten side chains. It is thus possible to suggest that a supramolecular unimer, based on an elementary cluster of helices correlated by interdigitating side chains, was a *building block* of the solid-state structure. The latter could then be described in terms of the planar and longitudinal repetition of building blocks correlated by interdigitation along the lateral dimension and by recognition of the exposed alkyl segment along the helical axes.

The above building blocks could not be isolated even in dilute solutions due to the all-or-none character of the phase transition. In fact, planar and longitudinal growth can be operationally separated only in special cases. For instance, if longitudinal growth could be prevented (i.e., by end-capping north and south surfaces), a planar assembly having a thickness corresponding to the length of the helices might be produced. A system in which longitudinal versus lateral growth could be controlled is based on rigid DNA to which dodecylpyridinium cations are electrostatically bound [139]. In this case, linear micellarlike building blocks might be isolated in solution by controlling the ratio of the two components. Solution and solid-state morphologies of systems based on charged polypeptides having long alkyl side chains electrostatically bound are described in Chapter 11.

The above approach led the author [3,15] to suggest the possibility of describing solid systems showing complex and ordered structurization in terms of the self-assembling of specifically

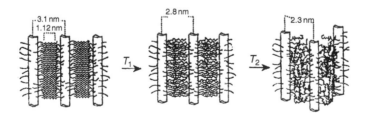

Figure 25 Three-dimensional assemblies: reversible transitions for comblike polymer Pα AA-18. The aliphatic segments interdigitate loosely in the liquid-crystalline phase at $T_1 < T < T_2$ forming supramolecular clusters. (From F. Lopez-Carrasquero, S. Monserrat, A. Martinez de Harduya, and S. Muñoz-Guerra. *Macromolecules* 28:5535, 1995. Copyright 1995 ACS.)

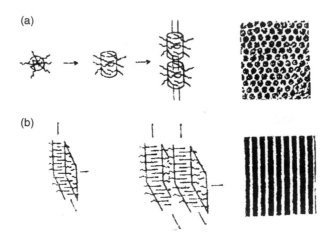

Figure 26 Building-block assemblies: (a) supramolecular polymerization of micelles of coil–coil block copolymer in a selective solvent. Hexagonally packed morphology by TEM; (b) bilayers of rod–coil copolymer in a solvent selective for the coil block, growing within the plane of the layer and mutually orienting in the perpendicular direction. Lamellar morphology by TEM. (From A. Ciferri. *Recent Res. Dev. Macromol.* 6:129, 2002b.)

designed building blocks. A similar concept has been used to describe the solid-state organization of interpenetrating nets [187]. To characterize the solid structure resulting from an assembly process started in solution, the specification of *basic building blocks and their supramolecular polymerization mechanism* is required. The approach is exemplified by a description of solid-state hexagonal-cylindrical, and lamellar phases observed for several crystalline and amorphous materials. The formation of the hexagonal phase is illustrated in Figure 26(a). A spherical micelle undergoes the series of hierarchical events schematized in the sequences (24a) and (24b). The hexagonal phase may be regarded as an embryo of the final morphology in the condensed state. The process is basically a unidimensional growth of micellar unimers according to the mechanism of growth-coupled-to-orientation. Elongated micelles, formed from spherical ones, have been suggested to coalesce [91] (cf. Chapter 9) transferring end-cupping to the head and tail of a continuous filament. No lateral attraction between filaments needs to be invoked: the order is simply produced by the growth-coupled-to-orientation process.

The formation of silkworm fibers may be related to formation of supramolecular elongated structures starting from micellar structures [188,189]. Spherical micelles (100 to 200 nm diameter) were observed in aqueous solution of reconstituted silkworm silk fibroin. Aggregation of these micelles into larger structures upon increasing fibroin concentration was observed. Shearing of these solutions produced a fibrillar structure with morphological features typical of silkworm fibers.

The expectation that linear growth controls the formation of hexagonal columnar mesophases is not limited to unimers with long molecular axes normal to the growth direction. Bifunctional unimers unable to grow along the lateral dimensions should in general be candidates for linear growth. The keratin fibers is characterized by a hexagonal arrangement of microfibrils ($L \sim 1\ \mu m, D \sim 8$ nm) each composed of eight photofibrils which are left-hand cables of two strands, each composed of two right-handed α-helices [190,191]. The microfilaments are imbedded in a conformationally disorder protein matrix rich in cystine residues and –S–S– cross-linkages. Considering that the length of the microfibrils by far exceeds the length of constituent chains and falls in the range of the persistence length reported for similar systems, it is plausible that the assembly of the fibril is directed by the growth-coupled-to-orientation mechanism. A mechanism for the assembly of keratin fibers *in vivo* might include the following sequence: extrusion of low-sulfur protein into extracellular fluids and protofibril assembly → growth to microfibrils with simultaneous orientation in the mesophase → stabilization of microfibrils by internal –S–S– bridges and two non-α-helical terminals → cross-linking of the sulfur-rich matrix in the narrow interfibrillar space [3].

The formation of lamellar structures could also be described by the identification of basic building blocks and corresponding assembly mechanisms. In Figure 26(b), the basic unimer is a bilayer growing along two perpendicular in-plane directions by a crystallization mechanism and assembling along the other direction through the formation of a nematic–discotic phase (cf. Section III.C.4). The segments protruding from the lamellae may dangle over the surface, as discussed for the cylindrical assemblies [3,15].

The qualitative character of the approach based on preformed building blocks should be emphasized. The approach is of interest within the general context of polymerization mechanisms and when simple and well-defined structures are involved. In the case of block copolymers, the SCFT approach (cf. Section II.C.1 and Chapter 9) affords a more detailed quantitative description. Figure 13 compares the theoretical SCFT prediction with the experimental behavior of polystyrene–polyisoprene amorphous block copolymers in the bulk phase. At variance with the building block approach, the SCFT includes a specific account for the role of the length and flexibility of the A and B blocks. The lamellar phase is predicted (Figure 13(a)) to be more stable than the hexagonal one for comparable size of A and B segments ($f \sim 0.5$), the hexagonal phase is instead preferred when there is an imbalance in segment size. These basic features appear well supported by the experimental data shown in Figure 13(b). The predicted sequence of phases in copolymer solutions also appear in line with experimental data [74,192]. (cf. also Chapters 9 and 10).

The foregoing discussion reveals the existence of seemingly different approaches to describe the assembly of different types of amphiphilic molecules. In particular, conventional surfactants and discotic molecules are described by the growth-coupled-to-orientation theory, while diblock copolymer molecules are described by the mean-field theory in terms of a balance between chain stretching and compatibility parameters. Hopefully, these differences ought to be rationalized in terms of a unified treatment.

Acknowledgment

The author expresses his appreciation to Prof. P. van der Schoot for many stimulating and clarifying discussions.

References

1. G.R. Desiraju, Ed. *The Crystal as a Supramolecular Entity*. New York: Wiley, 1996.
2. L. Brunsveld, B.J.B. Folmer, E.W. Meijer, and R.P. Sijbesma. *Chem. Rev.* 101:4071, 2001.
3. A. Ciferri. a) *Macromol. Rapid Commun.* 23:511, 2002. b) Encyclopedia of Supramolecular Chemistry. New York: M. Dekker, 2004.
4. F. Oosawa and S. Asakura. *Thermodynamics of the Polymerization of Protein*. London: Academic Press, p. 25, 1975.
5. A. Ciferri. *Liq. Cryst.* 26:489, 1999.
6. J. Rühe and W. Knoll. *J. Macromol. Sci. Polym. Rev.* 42, 91: 2003.
7. S.E. Kudaibergenov. *Polyampholytes: Synthesis, Characterization and Application*. Dordrecht Kluwer Academic Publishers, 2002.
8. C.V. Larez, V. Crescenzi, M. Dentini, and A. Ciferri. *Supramol. Sci.* 2:141, 1995.
9. A. Ciferri. *Prog. Polym. Sci.* 20:1081, 1995.
10. R. Wick, P. Walde, and P.L. Luisi. *J. Am. Chem. Soc.* 117:1435, 1995.
11. S.A. Jenecke and X.L. Chen. *Science* 279:1903, 1998.
12. C. Tanford. *Formation of Micelles and Biological Membranes*. New York: Wiley, 1980.
13. J.N. Israelachvili. *Intermolecular and Surface Forces*. London: Academic Press, 1992.

14. W.K. Kegel and P. van der Schoot. *Biophys. J.* 86:3905, 2004.
15. A. Ciferri. *Recent Res. Dev. Macromol.* 6:129, 2002.
16. J. Rebeck. *Angew. Chem. Int. Ed. Engl.* 29:245, 1990.
17. H.-J. Schneider. *Angew. Chem. Int. Ed. Engl.* 30:1417, 1991.
18. J.C. Adrian and C.S. Wilcox. *J. Am. Chem. Soc.* 113:678, 1991.
19. D.A. Tomalia, A.M. Naylor, and W.A. Goddart III. *Angew. Chem. Int. Ed. Engl.* 29:138, 1990.
20. B.B. Mandelbrot. *The Fractal Geometry of Nature.* New York: Freeman, 1983.
21. E. Sackman. *Macromol. Chem. Phys.* 195:7, 1994.
22. J. Baggott. *Perfect Symmetry: The Accidental Discovery of a New Form of Carbon.* London: Oxford University Press, 1994.
23. E. Osawa, M. Yoshida, and M. Fujita. *MRS Bull.* 11:23, 1994.
24. P. De Santis. *Nature* 206:456, 1965.
25. J. Sartorius and H.-J. Schneider. *Chem. Eur. J.* 2:1446, 1996.
26. (a) C.A. Hunter and J.K.M. Sanders. *J. Am. Chem. Soc.* 112:5525, 1990; (b) M.L. Waters. *Curr. Opin. Chem. Biol.* 6:736, 2002.
27. S. Lahiri, J.L. Thompson, and J.S. Moore. *J. Am. Chem. Soc.* 122:11315, 2000.
28. M.C.T. Fyfe and J.F. Stoddart. *Acc. Chem. Res.* 30:393, 1997.
29. A. Marcus. *Ion Solvation.* New York: Wiley, 1986.
30. H.-J. Schneider, D. Güttes, and V. Schneider. *J. Am. Chem. Soc.* 110:6449, 1988.
31. H.-J. Schneider, R. Kramers, S. Simova, and V. Schneider. *J. Am. Chem. Soc.* 110:6642, 1988.
32. V. Percec, J. Heck, G. Johansen, G. Tomazos, and M. Kawagumi. *J. Macromol. Sci. Pure Appl. Chem. A* 11:1031, 1994.
33. G. Gottarelli, G.P. Spada, and A. Garbesi. *Crystallography of Supramolecular Compounds.* G. Tsoucaris, J.L. Atwood, and J. Lipkowski, Eds., Kluwer Academic Publishers, p. 307, 1996.
34. D. Markovitsi, H. Bengs, N. Pfeffer, F. Charra, J.-M. Nunzi, and H. Ringsdorf. *J. Chem. Soc. Faraday Trans.* 89:37, 1993.
35. G. Odian. *Principles of Polymerization.* New York: McGraw-Hill, 1991.
36. K. Markau, J. Schneider, and A. Sund. *Eur. J. Biochem.* 24:293, 1972.
37. F. Oosawa and M. Kasai. *J. Mol. Biol.* 4:10, 1962.
38. H. Sund and K. Markau. *Int. J. Polym. Mat.* 4:251, 1976.
39. W. Burchard. *Trends Polym. Sci.* 1:192, 1993.
40. R. Bruce Martin. *Chem. Rev.* 96:3043, 1996.
41. D. Zhao and J.S. Moore. *Org. Biomol. Chem.* 1:3471, 2003.
42. B. Zimm and J.K. Bragg. *J. Chem. Phys.* 31:526, 1959.
43. E.D. Korn. *Physiol. Rev.* 62:672, 1982.
44. P.S. Niranjan, J.G. Forbes, S.C. Greer, J. Dudowicz, K.F. Freed, and J.F. Douglas. *J. Chem. Phys.* 114:10573, 2001.
45. J. van Gestel, P. van der Schoot, and M.A.J. Michels. *J. Phys. Chem. B* 105:10691, 2001.
46. a) J. van Gestel. Theory of helical supramolecular polymers. Thesis. Eindhoven: Technische Universiteit Eindhoven, Proefschrift. 2003. b) J. van Gestel, P. van der Schoot, M.A.J. Michels Langmuir 19:1375, 2003.
47. M.H. Abraham. *J. Am. Chem. Soc.* 104:2085, 1982.
48. P.J. Flory. *Principles of Polymer Chemistry.* New York: Cornell University Press, 1953.
49. F.H. Case and J.D. Honeycutt. *Trends Polym. Sci.* 2:259, 1994.
50. A. Ciferri. *Liquid Crystallinity in Polymers.* A. Ciferri, Ed., New York: VCH, 1991.
51. T.A. Orofino, A. Ciferri, and J.J. Hermans. *Biopolymers* 5:773, 1967.
52. C. Balbi, E. Bianchi, A. Ciferri, A. Tealdi, and W.R. Krigbaum. *J. Polym. Sci. Polym. Phys.* 18:2037, 1980.
53. P.J. Flory. *Adv. Polym. Sci.* 59:1, 1984.
54. C.V. Larez, V. Crescenzi, and A. Ciferri. *Macromolecules* 28:5280, 1995.
55. R.S. Werbowyj and D.G. Gray. *Macromolecules* 13:69, 1980.
56. H. Ringsdorf, B. Schlarb, and J. Venzmer. *Angew. Chem.* 100:117, 1988.
57. A. Halperin, M. Tirrel, and T.P. Lodge. *Adv. Polym. Sci.* 100:31, 1991.
58. W. Meijer, C. Nardin, and M. Winterhafter. *Angew. Chem. Intl Ed.* 39:4599, 2000.
59. N. Kimizuka. *Curr. Opin. Chem. Biol.* 7:702, 2003.
60. D. Whitmore and J. Noolandi. *Macromolecules* 18:657, 1985.
61. A. Gabellini, M. Novi, A. Ciferri, and C. Dell'Erba. *Acta Polimerica* 50:127, 1999.

62. L.H. Radzilowski and S.I. Stupp. *Macromolecules* 27:7747, 1994.

63. F. Mackintosh. *Europhys. Lett.* 12:697, 1990.

64. R. Nagarajan and K. Ganesh. *J. Chem. Phys.* 90:5843, 1989.

65. P. Ceres and A. Zlotnick. *Biochemistry* 41:11525, 2002.

66. M.E. Cates and S.J. Candau. *J. Phys. Condens. Matter* 2:6869, 1990.

67. J.N. Israelachvili, D.J. Mitchell, and B.W. Niham. *J. Chem. Soc. Faraday Trans. 2* 72:1525, 1976.

68. A. Stroobants and H.N.W. Lekkerkerker. *J. Phys. Chem.* 88:3699, 1984.

69. M.W. Matsen and F.S. Bates. *Macromolecules* 29:1091, 1996.

70. M.W. Matsen. *J. Phys. Condens. Matter* 14:R21, 2002.

71. E. Helfand. *J. Chem. Phys.* 62:999, 1975.

72. L. Leibler. *Macromolecules* 13:1602, 1980.

73. N. Semenov. *Sov. Phys. JETP* 61:733, 1985.

74. J. Noolandi, A.C. Shi, and P. Linse. *Macromolecules* 29:5907, 1996.

75. A. Ciferri, W.R. Krigbaum, and R.B. Meyer, Eds. *Polymer Liquid Crystals.* New York: Academic Press, 1982.

76. P.J. Flory and G. Ronca. *Mol. Cryst. Liq. Cryst.* 54:311, 1979.

77. W. Maier and A. Saupe. *Z. Naturforsch* 15:287, 1960.

78. A.R. Khokhlov. *Liquid Crystallinity in Polymers.* A. Ciferri, Ed., New York: VCH, 1991.

79. P. Bolhuis and D. Frenkel. *J. Chem. Phys.* 106:666, 1997.

80. M.A. Bates and D. Frenkel. *Phys. Rev. E* 57:4824, 1998.

81. G.J. Vroege and T. Odijk. *Macromolecules* 21:2848, 1988.

82. J. Herzfeld and R.W. Briehl. *Macromolecules* 14:1209, 1981.

83. W.M. Gelbart, W.E. McMullen, and A. Ben-Shaul. *J. Phys.* 46:1137, 1985.

84. T. Odijk. *J. Phys.* 48:125, 1987.

85. T. Odijk. *Curr. Opin. Coll. Interface Sci.* 1:337, 1996.

86. R. Hentschke. *Liq. Cryst.* 10:691, 1991.

87. P. van der Schoot. *J. Phys. II* 5:243, 1995.

88. P. Bladon and A.C. Griffin. *Macromolecules* 26:6004, 1993.

89. A. Ciferri. *Liq. Cryst.* 31:1487, 2004.

90. A. Abe and M. Ballauff. *Liquid Crystallinity in Polymers.* A Ciferri, Ed., New York: VCH, 1991.

91. M. Kenward and M.D. Whitmore. *J. Chem. Phys.* 116:3455, 2002.

92. P. van der Schoot. *J. Chem. Phys.* 117:3537, 2002.

93. Z. Dogic, D. Frenkel, and S. Fraden. *Phys. Rev. E* 62:3925, 2000.

94. G. Marrucci and A. Ciferri. *J. Polym. Sci. Polym. Lett.* 15:643, 1977.

95. L. Walker et al. *Langmuir* 12:6309, 1996.

96. G. Massiere et al. *Europhys. Lett.* 57:127, 2002.

97. O.P. Marquez, B. Fontal, J. Marquez, R. Ortiz, R. Castillo, M. Choi, and C. Larez. *J. Electrochem. Soc.* 144:707, 1995.

98. C. Boraz, E. Weinhold, W. Cabrera, O.P. Martinez, and R.O. Lenza. *J. Electrochem. Soc.* 144:3871, 1997.

99. T. Ooi, K. Mihashi, and H. Kobayashi. *Arch. Biochem. Biophys.* 98:1, 1962.

100. C. Hilger and R. Stadler. *Makromol. Chem.* 192:805, 1991.

101. J.Y. Lee, P.C. Painter, and M.M. Coleman. *Macromolecules* 21:954, 1988.

102. R.P. Sijbesma, F.H. Beijer, L. Brunsveld, B.J.B. Folmer, J.H.K.K. Hirschberg, R.F.M. Lange, J.K.L. Lowe, and E.W. Meijer. *Science* 278:1601, 1997.

103. B.J.B. Folmer, R.P. Sijbesma, R.M. Verstgeen, J.A. van der Rijt, and E.W. Meijer. *Adv. Mater.* 12:874, 2000.

104. U. Michelsen and C.A. Hunter. *Angew. Chem. Int. Ed.* 29:764, 2000.

105. E.A. Fogleman, W.C. Yount, J. Xu, and S.L. Craig. *Angew. Chem. Int. Ed.* 41:4026, 2002.

106. E.Y. Sheu, K.S. Liang, and L.Y. Chiang. *J. Phys. (Paris)*, 50:1279, 1989.

107. T.D. Clark, J.M. Buriak, K. Kobyashi, M.P. Isler, D.E. McRee, and M.R. Ghadiri. *J. Am. Chem. Soc.* 120:8949, 1998.

108. H. Nierengarten, J. Rojo, E. Leize, J.-M. Lehn, and A. van Dorsselaer. *Eur. J. Inorg. Chem.* 3:573, 2002.

109. a) S.J. Rowan, P. Suwanmala and S. Sivakova. *J. Polym. Sci. Polym. Chem.* 41, 3589, 2003. b) S. Sivakova and S.J. Rowan. *Chem. Soc. Rev.* 9: 21, 2005.

110. C. Hilger, M. Drager, and R. Stadler. *Macromolecules* 25:2498, 1992b.

111. a) J.F. Gohy, B.B.G. Lohmeijer, S.K. Varshney, U.S. Schubert. *Macromolecules* 35:7427, 2002.
     b) W.H. Binder, M.T. Kunz, C. Kluger, R. Saf, G. Hayn. *Macromolecules* 37:1749, 2004. c) H.S. Bazzi,
     H.F. Sleiman. *Macromolecules* 35:9617, 2002. d) K. Yamauchi, A. Kanomata, T. Inue, T.E. Long. *Mac-romolecules* 37:3519, 2004. e) Z. Li, C.A. Mirkin. *Nanoletters* 4:1055, 2004. f) J.-F. Berret, B. Vigolo,
     R. Eng, P. Herve, I. Grillo, I. Yang. *Macromolecules* 37:4922, 2004.

112. a) R.K. Castellano and J. Rebek Jr. *J. Am. Chem. Soc.* 120:3657, 1998. b) R.K. Castellano, R. Clark,
     S.L. Craig, C. Nuckolls, and J. Rebek Jr. *Proc. Natl. Acad. Sci. USA.* 97:12418, 2000.

113. R.G. Edwards, J.R. Henderson, and R.L. Pinning. *Mol. Phys.* 86:567, 1995.

114. S.U. Vallerien, M. Werth, F. Kremer, and H.W. Spiess. *Liq. Cryst.* 8:889, 1990.

115. P. De Santis, S. Morosetti, and R. Ricco. *Macromolecules* 7:52, 1974.

116. D.T. Bong, T.D. Clark, J.R. Granja, and M.R. Ghadiri. *Angew. Chem. Int. Ed.* 40:988, 2001.

117. F. Oosawa. *Biophys. Chem.* 47:1010, 1993.

118. R.A. Palmans, J.A.J.M. Vekemans, E.E. Havinga, and E.W. Meijer. *Angew. Chem. Int. Ed. Engl.* 36:2648,
     1997.

119. L. Brunsveld, H. Zhang, M. Glasbeek, J.A.J.M. Vekemans, and E.W. Meijer. *J. Am. Chem. Soc.* 122:6175,
     2000.

120. L. Brunsveld, B.G.G. Lohmeijer, J.A.J.M. Vekemans, and E.W. Meijer. *Chem. Commun.* 23:2305, 2000.

121. P. van der Schoot, M.A.J. Michels, L. Brunsveld, R.P. Sijbesma, and A. Ramzi. *Langmuir* 16:10076, 2000.

122. J.H.K.K. Hirschberg, L. Brunsveld, A. Ramzi, J.A.J.M. Vekemans, R.P. Sijbesma, and E.W. Meijer.
     Nature 407:167, 2000.

123. R.B. Prince, J.G. Saven, P.G. Wolynes, and J.S. Moore. *J. Am. Chem. Soc.* 121:3114, 1999.

124. C. Kubel, M.J. Mio, J.S. Moore, and D.C. Martin. *J. Am. Chem. Soc.* 124:8605, 2002.

125. C. Nuckolls and T.J. Katz. *J. Am. Chem. Soc.* 120:9541, 1988; *J. Am. Chem. Soc.* 121:79, 1999.

126. A. Klug. *Angew. Chem. Int. Ed. Engl.* 22:565, 1983.

127. P.A. Janmey, J. Peetermans, K.S. Zanert, T.P. Stossel, and T. Tanaka. *J. Biol. Chem.* 261:8357, 1986.

128. R. Furukawa, R. Kundra, and M. Fechheimer. *Biochemistry* 32:12346, 1993.

129. S. Asakura, G. Eguchi, and T. Iino. *J. Mol. Biol.* 10:42, 1964.

130. M.O. Steinmetz, D. Stoffler, A. Hoenger, A. Bremer, and U. Aebi. *J. Struct. Biol.* 119:295, 1997.

131. T. Horio and H. Hotani. *Nature* 321:605, 1986.

132. M.M. Green, N.C. Peterson, T. Sato, A. Teramoto, and S. Lifson. *Science* 268:1869, 1995.

133. S.H. Gellman. *Acc. Chem. Res.* 31:173, 1998.

134. D.J. Hill, M.J. Mio, R.B. Prince, T.S. Hughes, and J.S. Moore. *Chem. Rev.* 101:3893, 2001.

135. G. Fensenfeld and M. Groudine. *Nature* 421:448, 2003.

136. G. Wedemann and J. Langowski. *Biophys. J.* 82:2847, 2002.

137. A. Ruiz Carrillo, J.L. Jorcano, G. Elder, and R. Lurz. *Proc. Natl. Acad. Sci. USA.* 76:3284, 1979.

138. A. Leforestier, J. Douchet, and F. Livolant. *Biophys. J.* 81:2414, 2001.

139. A. Ciferri. *Macromol. Chem. Phys.* 195:457, 1994.

140. a) G.S. Manning. *J. Am. Chem. Soc.* 125:15087, 2003. b) A. Perico, G. LaPenna and L. Arcesi. Submitted
     to Macromolecules.

141. L. Huang and A.E. Tonelli. *J.M.S. Rev. Macromol. Chem. Phys. C* 38:781, 1998.

142. V. Percec and D. Schlueter. *Macromolecules* 30:5783, 1997.

143. A. Ciferri. *Trends. Polym. Sci.* 5:142, 1997.

144. J.-M. Lehn. *Supramolecular Polymers.* A. Ciferri, Ed., New York: Marcel Dekker, 2000.

145. C.M. Paleos and D. Tsiourvas. *Liq. Cryst.* 28:1127, 2001.

146. K. Merchant and R.L. Rill. *Macromolecules* 27:2365, 1994.

147. P. Cavalleri, A. Ciferri, C. Dell'Erba, M. Novi, and B. Purevsuren. *Macromolecules* 30:3513, 1997.

148. a) A.L. Hitt, A.R. Cross, and C.R. Williams Jr. *J. Biol. Chem.* 265:1639, 1990. b) E. Mandelkov,
     E.M. Mandelkov. *Curr. Op. Struct. Biol.* 4:71, 1994.

149. A. Ben-Shaul and W.M. Gelbart. *Micelles, Membranes, Microemulsions and Monolayers.* W.M. Gelbart,
     A. Ben Shaul, and D. Roux, Eds., New York: Springer-Verlag, 1994.

150. P. van der Schoot and M.E. Cates. *Europhys. Lett.* 25:515, 1994.

151. T. Shikata and D.S. Pearson. *Langmuir* 10:4027, 1994.

152. R. Hentschke, P.J.B. Edwards, N. Boden, and R. Bushby. *Macromol. Symp.* 81:361, 1994.

153. T. Gulik-Krzywicki, C. Fouquey, and J.-M. Lehn. *Proc. Natl. Acad. Sci. USA.* 90:163, 1993.

154. C. Fouquey, J.-M. Lehn, and A.M. Levelut. *Adv. Mater.* 2:254, 1990.

155. M.-J. Brienne, J. Gabard, J.-M. Lehn, and I. Stibor. *J. Chem. Soc. Chem. Commun.* 24:1869, 1989.

156. R. Hentschke and T. Herzfeld. *Phys. Rev. A* 43:7019, 1991.
157. N. Boden. *Micelles, Membranes, Microemulsions and Monolayers.* W.M. Gelbart, A. Ben Shaul, and D. Roux, Eds., New York: Springer-Verlag, 1994.
158. D.C. Morse and S.T. Milner. *Phys. Rev.* 47:1119, 1993.
159. P. Linse. *J. Phys. Chem.* 97:13896, 1993.
160. Q. Ying and B. Chu. *Macromolecules* 20:871, 1987.
161. G.L. Brelsford and W.R. Krigbaum. *Liquid Crystallinity in Polymers.* A. Ciferri, Ed., New York: VCH, 1991.
162. T. Imae and S. Ikeda. *Coll. Polym. Sci.* 262:497, 1984.
163. F. Gittes, B. Mickey, J. Nettleton, and J. Howard. *J. Cell. Biol.* 120:923, 1993.
164. S. Burlacu, P.A. Janmey, and J. Borejdo. *Ann. J. Physiol.* 262:C569, 1992.
165. T. Yanagida, T. Nakase, K. Nishiyama, and F. Oosawa. *Nature* 307:58, 1984.
166. J. Mizushima-Sangano, T. Maeda, and T. Miki-Noumura. *Biochim. Biophys. Acta* 755:257, 1983.
167. L.D. Landau and E.M. Lifshitz. *Statistical Physics.* Tarrytown, NY: Pergamon Press, 1980.
168. R. Blaak, D. Frenkel, and B.M. Mulder. *J. Chem. Phys.* 110:11652, 1999.
169. S. Hoffman, W. Witkowski, G. Borrman, and H. Shubert. *WZ Weissflog Chem.* 18:403, 1978.
170. C.M. Paleos and J. Michas. *Liq. Cryst.* 11:773, 1992.
171. T. Kato, H. Adachi, A. Fujishima, and J.M.J. Fréchet. *Chem. Lett.* 21:265, 1992.
172. U. Kumar, J.M.J. Fréchet, T. Kato, S. Ujiie, and K. Timura. *Angew. Chem. Int. Ed. Engl.* 31:1531, 1992.
173. C.T. Imrie. *Trends Polym. Sci.* 3:22, 1995.
174. M. Shoji and F. Tanaka. *Macromolecules* 35:7460, 2002.
175. A. Sirigu. *Liquid Crystallinity in Polymers.* A. Ciferri, Ed., New York: VCH, 1991.
176. G. Ronca and A. Ten Bosch. *Liquid Crystallinity in Polymers.* A. Ciferri, Ed., New York: VCH, 1991.
177. U.B. Sleytr and M. Sára. *Trends Biotechnol.* 15:5, 1997.
178. U.B. Sleytr and P. Messner. *Electron Microscopy of Subcellular Dynamics.* H. Plattner, Ed., Boca Raton, FL: CRC Press, 1989.
179. N.C. Seeman. *Synletters* X:1536, 2000.
180. N.C. Seeman. *Nature* 421:427, 2003.
181. Y.-T. Tao, G.D. Hietpas, and D.L. Allara. *J. Am. Chem. Soc.* 118:6724, 1996.
182. N. Camillone III, T.Y.B. Leung, and G. Scoles. *Surf. Sci.* 373:333, 1996.
183. G.E. Poirier. *Langmuir* 15:1167, 1999.
184. W.J. Miller, V.K. Gupta, N.L. Abbott, H. Johnson, M.W. Taso, and J. Rabolt. *Liq. Cryst.* 23:175, 1997.
185. M. Shimomura. *Supramolecular Polymers.* A. Ciferri, Ed., New York: Marcel Dekker, 2000.
186. F. Lopez-Carrasquero, S. Monserrat, A. Martinez de Harduya, and S. Muñoz-Guerra. *Macromolecules* 28.5535, 1995.
187. S.R. Batten and R. Robson. *Angew. Chem. Int. Ed.* 37:1460, 1998.
188. F.N. Braun and C. Viney. *Int. J. Biol. Macromol.* 32:59, 2003.
189. H.-J. Jin and D.L. Kaplan. *Nature* 424:1057, 2003.
190. M.K. Hartzer, Y.-Ys. Pang, and R.M. Robson. *J. Mol. Biol.* 365:376, 1985.
191. K. Arai, F. Hirata, S. Nishimura, M. Hirano, and S. Naito. *J. Appl. Polym. Sci.* 47:1973, 1993.
192. M. Svensson, P. Alexandridis, and P. Linse. *Macromolecules* 32:637, 1999.

Chapter 3

Theory of Supramolecular Polymerization

Paul van der Schoot

Contents

I. Introduction

The term *supramolecular polymer* applies to any type of polymer-like assembly that spontaneously forms by the reversible linear aggregation of one or more type of molecule in solution or in the melt. The crucial factor discriminating supramolecular from conventional or so-called "dead" polymers, is that for the former the monomeric and the polymeric states are in thermal equilibrium with each other, while for the latter this is not so (on the relevant experimental timescale). Examples of supramolecular polymers include the so-called giant surfactant micelles [1], peptide β-sheet ribbons [2], self-assembled stacks of discotic molecules [3], protein fibers such as those formed by sickle cell hemoglobin [4], and so on. Chains of colloidal particles found in quite diverse contexts [5–8] and living polymers of chemically reactive species [9] also belong to the class of supramolecular polymers, if only in principle.

The types of intermolecular interaction that give rise to supramolecular polymers are almost as diverse as the supramolecular polymers themselves [10]. The bonds that link the molecular building blocks together are typically (but not exclusively) of noncovalent, physical origin. They can be specific, such as is the case when hydrogen bonds are involved, but may also be the result

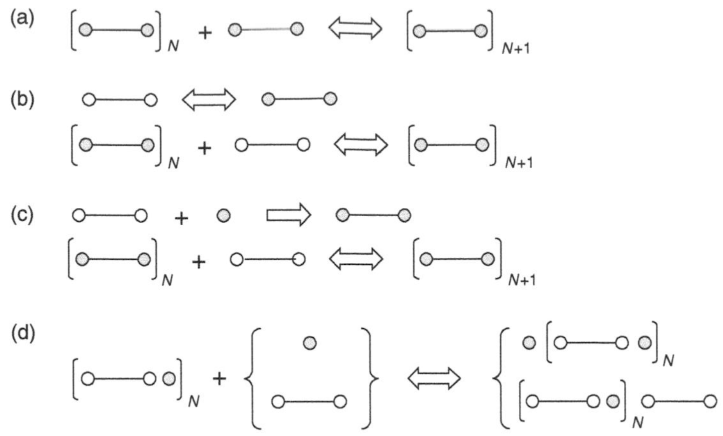

Figure 1 Classification scheme of supramolecular polymerizations adhered to in the main text: (a) EP; (b) TAEP; (c) CAEP; (d) CP. Active (inactive) monomers are indicated by the filled (open) symbols.

of a competition between fairly generic solvophobic and solvophilic interactions [11]. The principle driving forces of supramolecular polymerization are discussed in considerable detail in Chapter 2, to which the reader is referred for more information. Here, we take a phenomenological point of view and presume that supramolecular polymerization occurs as a result of some system-specific driving force that we need not consider in detail.

Not surprisingly, supramolecular polymers cannot all be lumped into a single group. They constitute such an overwhelmingly broad range of materials that it is in fact quite remarkable that a relatively small number of classes may, at least theoretically, be distinguished. Adapting the classification scheme put forward by Tobolsky and Eisenstein [12], we distinguish the following main groups of linearly self-assembling systems that share many important features. This list is not exhaustive, but should cover most materials of interest from Class A of Chapter 2 (see also Figure 1):

1. *Equilibrium polymers (EPs)*. Apparently the most common type of supramolecular polymer, in which all the monomers present in the system are active, that is, are able to bond with each other and form polymeric assemblies. Examples are given in Figure 2. In single-component EPs all bonded interactions are plausibly equivalent, a case sometimes also referred to as the isodesmic equilibrium polymerization, or, in Chapter 2, as the multistage open association (MSOA).

2. *Thermally activated equilibrium polymers (TAEPs)*. For these, only a small portion of the monomers are strongly active and polymerize into active polymers. The weakly active (or inactive) state and the strongly active state of the monomers are subjected to a thermal equilibrium that favors the inactive state. Polymerization of these materials occur via thermal activation and may be highly cooperative. Any equilibrium polymerization with a highly unfavorable intermediate stage such as the helical polymerization of actin [13,14], or the fibril formation in peptide solutions [15], behaves like a TAEP (see also Chapters 2 and 21). Sometimes ring-opening polymerization is viewed and modeled as a TAEP [16,17].

3. *Chemically activated equilibrium polymers (CAEPs)*. If the monomers are inactive unless activated through chemical reaction with initiator molecules, it obeys a chemically activated equilibrium polymerization. The amount of initiator added determines the number of activated monomers that in turn fixes the number of EPs in the system. The living polymerization of poly(α-methylstyrene) is an example of a CAEP, provided the polymerization reaction is not (chemically) terminated [18].

4. *Coordination polymers (CPs)*. Bifunctional molecules that rely on the complexation of ligands with metal ions or other components are, strictly speaking, copolymeric, and give rise to a special group of EPs referred to as CPs or as chelating supramolecular polymers. A discussion of this class of materials can, for example, be found in Chapter 1. In a way, CPs may be seen as two-component EPs but they behave somewhat differently from them because of the prescribed stoichiometry [19].

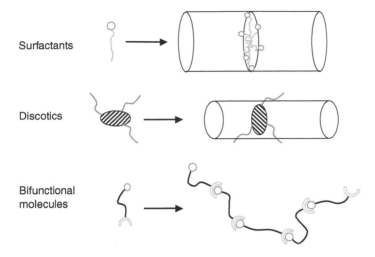

Figure 2 Examples of EPs to which the theory of Section II applies: cylindrical micelles of surfactant molecules, self-assembled stacks of discotic compounds, and chains of bifunctional molecules. The chains grafted to the core of the discotics enhance their solubility.

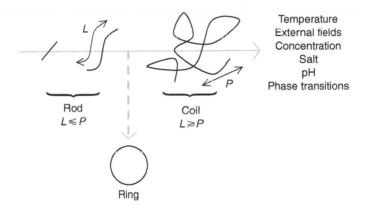

Figure 3 Supramolecular polymers respond to changes in the external conditions by changing their mean degree of polymerization. Hence, the rod and coil limits of linear supramolecular polymers may be bridged by varying the temperature, imposing external fields, and so on. However, for conditions such that the mean length L of the assemblies is of the order of their persistence length P, rings may spontaneously form and predominate the population of assemblies (see Section VI). The persistence length measures the mean distance along the contour of a chain over which it loses its angular memory as a result of thermal agitation. Assemblies shorter than this length are rodlike, while longer ones resemble coils [20].

Despite the apparent differences between these four classes of supramolecular polymer, they have in common a mean degree of polymerization that is a function of the external conditions. Indeed, in clear contrast to conventional polymers, supramolecular polymers respond not only to changes in the temperature, the concentration of dissolved material, the solvent quality, the presence of external fields, but also to phase transitions.

The sensitivity of the degree of polymerization to the external conditions implies that the assemblies may take on the shape of a rod or that of a coil, depending on their mean length relative to the intrinsic stiffness of the polymer backbone (both of which are influenced by the external conditions) (see Figure 3). If the intrinsic stiffness of the backbone is large enough, the self-assembled chains may spontaneously organize into liquid-crystalline phases [21,22]. (Supramolecular liquid crystals are discussed in Chapter 4.) The chains may in fact change their topology too, for the chains need not be linear but could close back onto themselves and form rings [23]. Under what conditions the rings form is not well understood [24], and presumably depends on the rigidity of the self-assembled polymer backbone, on the nature of the monomers, and on the type of supramolecular polymerization [25,26].

In the remainder of this chapter, a brief and quite incomplete overview of the theory of supramolecular polymerization is presented. The aim is to highlight the role of interactions, in particular the impact of bonded interactions, nonbonded interactions, and a coupling to external orienting fields. Theoretically the simplest case to deal with is that of *ideal* supramolecular polymers, for which the nonbonded interactions are unimportant. As we shall see, the coupling between linear aggregation and nonbonded interactions is subtle, and can often be ignored altogether. In Section II, EPs are described in some detail, and in Sections III to V in slightly less detail the other three cases of TAEPs, CAEPs, and CPs. The impact of ring closure is discussed in Section VI. Recent work on helical aggregation and the breakdown of the so-called ground-state approximation is described in Section VII. Sections VIII and IX deal with the impact of excluded-volume interactions and of external orienting fields upon the growth of supramolecular polymers, showing the important role of chain flexibility. Concluding remarks and an outlook is presented in Section X.

II. EQUILIBRIUM POLYMERS

Equilibrium polymerization is often described by a MSOA scheme such as given in Chapter 2, also known as the ladder model [27], the isodesmic model [28], or the free association model [29], where the equilibrium constants of the sequential back and forward reaction steps are presumed to be equal. (For a discussion of variants of this assumption, e.g., see [30].) Here, we give a statistical–mechanical prescription that is equivalent to it but that allows for a more straightforward generalization.

If we suppress the formation of rings, the grand potential Ω of an assembly of noninteracting (linear) EPs may be written as the sum of an ideal entropy of mixing and a free energy of the individual species that drives the self-assembly [1]

$$\frac{\beta\Omega}{V} = \sum_{N=1}^{\infty} \rho(N)[\ln \rho(N)\upsilon - 1 - \ln Z(N) - \beta\mu N] \tag{1}$$

where as usual $\beta \equiv 1/k_B T$, with k_B Boltzmann's constant and T the absolute temperature. In Eq. (1), υ denotes the volume of a monomer, V the volume of the system, $\rho(N)$ the number density of chains of degree of polymerization N, and μ the chemical potential of the monomers. The reader is referred to Ref. 31 for a discussion of the choice of length scale implicit in the identification of υ with the monomer volume. The partition function $Z(N)$ counts the configurational states of the chains according to their Boltzmann weight, and includes the free energy gain of association. The solvent is treated as a continuum that somehow influences the bending rigidity of the chains, as well as the strength of the (net) bonded or contact interaction between two neighboring monomers in a chain.

The equilibrium size distribution optimizes the free energy Eq. (1). Setting $\delta\Omega/\delta\rho(N) = 0$ gives

$$\rho(N) = \upsilon^{-1}Z(N)\exp(\beta\mu N) \tag{2}$$

where $\delta^2\Omega/\delta\rho(N)^2 > 0$ irrespective of the functional form of $Z(N) > 0$, so Eq. (2) indeed minimizes the free energy. Because $\rho(N) \propto Z(N)$, the size distribution must be a sensitive function of the configurational state of the polymers. From established polymer theory we expect

$$Z(N) = Z(1)\lambda^{N-1} \exp(-\beta\varepsilon(N-1)) \tag{3}$$

to hold all the way from the rod to the flexible coil limit with $\varepsilon < 0$ the free energy of the bonded interaction (the "bare" contact energy — see below). $Z(1)$ and λ are nonuniversal, model-dependent parameters that are a function of the chemical details of the monomers and of the bending stiffness of the supramolecular chains [32,33].

Equations (2) and (3) force the conclusion upon us that all extensive terms can be absorbed into an effective chemical potential, $\beta\tilde{\mu} \equiv -\beta\mu + \beta\varepsilon - \ln\lambda$, and all intensive ones into an effective free energy gain of binding, $\beta G \equiv -\beta\varepsilon - \ln Z(1)/\lambda$. Hence, the size distribution simplifies to $\rho(N) = \exp(-\beta\tilde{\mu}N - \beta G)$ with G a phenomenological parameter that is some function of the bending rigidity of the chains, and of the thermodynamics of the bonding between two monomers in a chain that obviously also depends on the solvent type [28,34]. Typical values for the binding free energy G are 10–40 $k_B T$ equivalent to $\sim$25–100 kJ mol^{-1} [1,3,27,35,36].

If ϕ denotes the volume fraction of self-assembling material present in the solution, we must have $\sum_{N=1}^{\infty} N\rho(N)\upsilon = \phi$ by the conservation of mass. For a given concentration, it is straightforward to eliminate the chemical potential from Eq. (2), to give

$$\rho(N) = \upsilon^{-1}\left(1 - \frac{1}{\overline{N}_n}\right)^N \exp(-\beta G) \tag{4}$$

with $\overline{N}_n$ the number-averaged degree of polymerization,

$$\overline{N}_n = \frac{\sum_{N=1}^{\infty} N\rho(N)}{\sum_{N=1}^{\infty} \rho(N)} = \frac{1}{2} + \frac{1}{2}\sqrt{1 + 4\phi\exp(\beta G)} \tag{5}$$

Both $\overline{N}_n$ and $\rho(N)$ are a function of the molecular architecture and the solvent type only through the parameter G, which in the field of giant micelles is usually referred to as the end-cap energy [1].

It turns out to be useful to express the strength of the mass action implicit in Eq. (5) in terms of a single parameter $X \equiv \phi\exp(\beta G)$. For $X \ll 1, \overline{N}_n \sim 1 + X$, while for $X \gg 1, \overline{N}_n \sim \sqrt{X}$. The latter growth law is the well-known square-root law valid for large degrees of polymerization [1]. In the long-chain limit $\overline{N}_n \gg 1$, the distribution approaches an exponential one, $\rho(N)\upsilon \sim \exp(-N/\overline{N}_n - \beta G) = \phi\overline{N}_n^{-2}\exp(-N/\overline{N}_n)$.

The fraction material absorbed into polymers, η, is related to the mean aggregation number, $\overline{N}_n$, through the equality

$$\eta = 1 - \overline{N}_n^{-2} \tag{6}$$

that is, in principle, a measurable quantity, accessible by means of spectroscopic techniques [37]. The weight-average degree of polymerization $\overline{N}_w = \sum_{N=1}^{\infty} N^2\rho(N)/\sum_{N=1}^{\infty} N\rho(N)$ is also obtainable experimentally, for example, from radiation scattering experiments. It obeys the relation $\overline{N}_w = 2\overline{N}_n - 1$, allowing us to immediately connect between our approach and that of the MSOA of Chapter 2. We conclude that the dimensionless strength of the mass action X is related to the dimension-bearing concentration of monomers C_0 [M] and equilibrium constant K [M^{-1}] to $X = \phi\exp(\beta G) = C_0 K$.

It is interesting to note that although the crossover from the monomer-dominated regime to the polymer-dominated regime is gradual, we can formally (and somewhat arbitrarily) define a polymerization "transition" for those conditions where $\eta = \frac{1}{2}$, that is, where half the dispersed

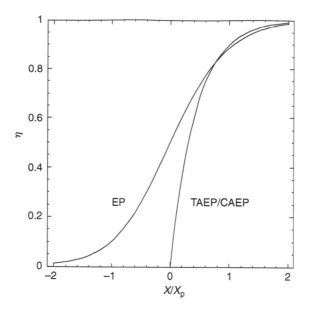

Figure 4 Universal polymerization curves of EPs, TAEPs, and CAEPs. Plotted is the fraction of polymerized material, η, as a function of the ratio X/X_p. The quantity X is the strength of the mass action as defined in the main text, proportional to the overall concentration of dissolved material. X_p denotes its value at the polymerization transition, which for EPs is defined as the point where half the material is absorbed into polymers. The curve for CAEPs and TAEPs is the limiting one for conditions where the polymerization transition becomes a true phase transition.

material is absorbed into polymers (see Figure 4). At constant temperature, we find from Eq. (5) that the number-averaged degree of polymerization is a universal function of the ratio ϕ/ϕ_p,

$$\overline{N}_n = \frac{1}{2} + \frac{1}{2}\sqrt{1 + \hat{\gamma}\frac{\phi}{\phi_p}} \tag{7}$$

with $\hat{\gamma} \equiv (2\sqrt{2} - 1)^2 - 1 \approx 2.34$ a numerical constant and ϕ_p the (temperature dependent) crossover concentration. The latter is uniquely linked to the equilibrium constant through the equality $4X_p = 4\phi_p \exp(\beta G) = \hat{\gamma}$. This implies that η is a universal function of ϕ/ϕ_p, and that it does not contain any information on the energetics of the polymerization except through the value of ϕ_p, which is also a measurable quantity [38]. That this is so, is shown in Figure 5 where experimental data on five chemically distinct supramolecular building blocks are collected — all data collapse onto the universal curve predicted by Eqs. (6) and (7). Note that $\overline{N}_n = \sqrt{2}$ if $\eta = \frac{1}{2}$, so at the polymerization transition the material is still largely held up in monomers and in dimers.

Alternatively, one may obtain information on the thermodynamics of the polymerization at fixed concentration by probing the temperature dependence of η or, equivalently, of $\overline{N}_n$ since

$$\overline{N}_n \simeq \frac{1}{2} + \frac{1}{2}\sqrt{1 + \hat{\gamma}\exp\left(-\frac{h_p}{k_B T_p^2}(T - T_p)\right)} \tag{8}$$

to leading order in $T - T_p$, where T_p is the temperature at which $\eta = \frac{1}{2}$ for the given concentration of material. The quantity $h_p \equiv h(T_p)$ is the enthalpy gain of a single link at $T = T_p$. A consequence of Eq. (8) is that (i) η is a universal function of $h_p(T - T_p)/k_B T_p^2$, at least near T_p, and (ii) the larger

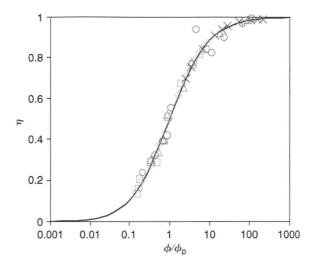

Figure 5 Universal polymerization curve of EPs. Plotted is the fraction polymerized material η as function of the concentration ratio ϕ/ϕ_p, where ϕ_p is the concentration at the half-way or "polymerization" point. The line gives the theoretical prediction of the isodesmic model. The symbols indicate experimental data on five chemically distinct oligo(phenylene vinyl)s in the solvent methyl cyclohexane taken at the temperature of 293 K [38]. These compounds form helical assemblies in solution [37]. From the measured half-way concentrations ϕ_p corresponding to 10^{-8}–10^{-5} M link free energies G in the range from 12 to 18 $k_B T$ or 30 to 44 kJ mol^{-1} are obtained.

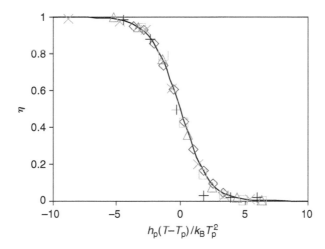

Figure 6 Universal polymerization curve of EPs. Plotted is the fraction polymerized material η as a function of the dimensionless ratio $h_p(T - T_p)/k_B T_p^2$ with h_p the net enthalpy gain of the formation of a single link, T_p the concentration-dependent "polymerization" temperature, and k_B Boltzmann's constant. The line gives the theoretical prediction of the isodesmic model. The symbols indicate experimental data on five chemically different oligo(phenylene vinyl)s in the solvent methyl cyclohexane at a concentration of 10^{-5} M [38]. By fitting to the data, values of h_p are obtained from 24 to 70 $k_B T$ equivalent to 60 to 170 kJ mol^{-1}.

the value of h_p, the stronger the growth of the EPs is with temperature and, hence, the steeper the fraction polymerized material $\eta(T)$ (see Figure 6 and Figure 7). In fact, it is possible to extract $h_p/k_B T_p^2$ from the polymerization curve without a full curve-fitting procedure, for at the polymerization transition $(\partial\eta/\partial T)_{T=T_p} \approx -0.227 h_p/k_B T_p^2$ [3]. Alternatively, h_p may also be determined from

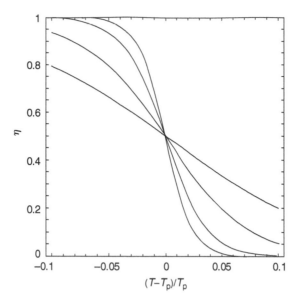

Figure 7 Polymerization curves of EPs as a function of the temperature T relative to the reference temperature T_p at the crossover from the monomer to the polymer-dominated regimes. Different curves give the fraction polymerized material η for different values of the dimensionless link enthalpy $h_p/k_B T_p = 15$, 30, 60, 90 (from bottom to top).

a Van't Hoff plot of the logarithm of the equilibrium constant, K, although care must be taken to render this quantity dimensionless.

III. THERMALLY ACTIVATED EQUILIBRIUM POLYMERS

For TAEPs a highly active state is in equilibrium with a weakly active state of the same material. The theory of TAEPs may be set up in a very similar fashion as was done for EPs in Section II, and is equivalent to that of the cooperative helical growth mechanism described in Chapter 2 [39].

Let $\rho_1(1)$ denote the volume fraction of weakly active material, which we presume to be so weakly active as to be essentially inactive and to remain monomeric. (This is not a true restriction, but simplifies the analysis considerably.) The grand potential of the (noninteracting) inactive monomers can be written as $\beta\Omega_1/V = \rho_1(1)[\ln \rho_1(1)\upsilon - 1 - \ln Z_1(1) - \beta\mu]$, where β again denotes the reciprocal of the thermal energy, V the volume of the system, μ the chemical potential of the monomers, and υ their volume. Next, let $\rho_2(N)$ denote the dimensionless concentration of N-mers of the active species that may be described by the grand potential Ω_2 given in Eq. (1), where we replace $\rho(N)$ by $\rho_2(N)$. The grand potential of the entire system of active and inactive species is given by the sum $\Omega = \Omega_1 + \Omega_2$.

The optimal distribution of material over the active and inactive states follows from a free energy minimization, requiring that $\delta\Omega/\delta\rho_1(1) = \delta\Omega/\delta\rho_2(N) = 0$. This gives expressions for $\rho_i(1) = \upsilon^{-1}Z_i(1)\exp(\beta\mu N)$ for $i = 1, 2$, where only $Z_2(N)$ is presumed to be of the form of Eq. (3). Following the recipe of Section II, the chemical potential μ of the monomers in the solution can be eliminated from the description by invoking the conservation of mass, which now reads $\phi = \rho_1(1)\upsilon + \sum_{N=1}^{\infty} N\rho_2(N)\upsilon$ with ϕ being the overall volume fraction of active and inactive monomers.

The equilibrium constant of the reaction from the inactive to the activated state is given by $K_a = Z_2(1)/Z_1(1)$, where we presume $K_a \to 0$ so that the activated state becomes highly unfavorable.

In this limit, we find a sharp polymerization transition reminiscent of a phase transition (see also below). The polymerization transition takes place for conditions where the quantity $X \equiv \phi \exp(\beta G)$ (the "mass action") is equal to unity, with G again the free energy gain associated with the formation of a bond, defined in a similar fashion as in Section II.

For $X \leqslant X_p \equiv 1$ all the material remains in inactive form if $K_a \to 0$, with $\rho_1(1)\upsilon = \phi$ and $\rho_2(N)\upsilon = 0$. On the other hand, if $X > X_p$ we obtain for the concentration of inactive species

$$\rho_1(1) = \upsilon^{-1}\left(1 - \frac{1}{\overline{N}_2}\right)\exp(-\beta G) \tag{9}$$

while for the active species

$$\rho_2(N) = \upsilon^{-1}K_a\left(1 - \frac{1}{\overline{N}_2}\right)^N \exp(-\beta G) \tag{10}$$

where $\overline{N}_2$ denotes the number-averaged degree of polymerization averaged over the active species only. It obeys the algebraic equality

$$1 - \overline{N}_2^{-1} + K_a\overline{N}_2(\overline{N}_2 - 1) = X \tag{11}$$

which may be obtained from the requirement of the conservation of mass, giving $\overline{N}_2 \sim 1 + X$ in the monomer-dominated regime with $X \ll X_p$, $\overline{N}_2 \sim K_a^{-1/3}$ at the polymerization transition with $X = X_p$, and $\overline{N}_2 \sim \sqrt{X/K_a}$ in the polymerized regime with $X \gg X_p$.

The connection between the present approach and that of the cooperative helical aggregation model of Oosawa and Kasai [13,39] described in Chapter 2 is now easily established. Our dimensionless mass action X is related to the dimension-bearing concentration of monomers C_0 [M] and equilibrium constant K_h [M^{-1}] through $X = \phi \exp(\beta G) = C_0 K_h$. Hence, our polymerization condition $\phi_p \exp(\beta G) = 1$ corresponds to that of $C_* K_h = 1$ given in Chapter 2, where C_* denotes the "critical" polymerization concentration. It also follows that our dimensionless equilibrium constant K_a takes on the role of the cooperativity parameter σ in the helical aggregation model [39].

The number-averaged degree of polymerization $\overline{N}_n$ averaged over all active and inactive species is given by

$$\overline{N}_n = \frac{1 + K_a\overline{N}_2^2}{1 + K_a\overline{N}_2} \tag{12}$$

Hence, the aggregation number grows very sharply beyond the polymerization transition at $X = X_p = 1$, and much more so than is the case for EPs if $K_a \to 0$. Even fairly close to the polymerization transition, in the regime $K_a\overline{N}_2 \ll 1 \ll K_a\overline{N}_2^2$, the activation step leads to a much stronger growth with $\overline{N}_n \sim 1 + K_a\overline{N}_2^2 \sim 1 + X$. This deviates from square-root law found for EPs if $X \gtrsim 1$. More deeply into the polymerized regime where $K_a\overline{N}_2 \gg 1$ does produce a square-root growth law, $\overline{N}_n \sim \overline{N}_2 \sim \sqrt{X/K_a} \gg 1$, although the growth remains strongly enhanced by the activation step that essentially provides a cooperative growth mechanism [13]. In this regime we also retrieve our previous result for the size distribution $\rho_2(N)\upsilon \sim \phi\overline{N}_n^{-2}\exp(-N/\overline{N}_n)$.

The fraction of active (polymerized) material is given by

$$\eta = K_a\overline{N}_2(\overline{N}_2 - 1)/X \tag{13}$$

In the limit $K_a \to 0$, we get $\eta = 0$ if $X \leqslant 1$ and $\eta = 1 - X^{-1}$ if $X > 1$. At constant temperature, we may define the polymerization concentration $\phi_p \equiv \exp(-\beta G)$, so that for $\phi > \phi_p$ we obtain a universal polymerization curve

$$\eta = 1 - \frac{\phi_p}{\phi} \tag{14}$$

irrespective of the free energy of bonding that fixes ϕ_p. Plotted in Figure 4, Eq. (14) exemplifies the distinct difference between EPs and TAEPs (and in fact CAEPs, see Section IV). We note that the larger the value of K_a, the less abrupt and the more gradual ("rounded") the polymerization transition at $\phi = \phi_p$ becomes [29].

At fixed concentration ϕ, we can alternatively define a polymerization temperature $k_B T_p \equiv -G_p / \ln \phi$ with $G_p = G(T_p)$ the free energy gain of a single bond at the temperature $T = T_p$, so that

$$\eta \simeq 1 - \exp\left(\frac{h_p}{k_B T_p^2}(T - T_p)\right) \tag{15}$$

to leading order in the temperature relative to the transition temperature T_p, with $h_p = h(T_p)$ the enthalpy gain of a single link at $T = T_p$. Note that for enthalpy-driven aggregation $h_p > 0$, so that Eq. (15) is only defined for $T < T_p$ and $\eta \simeq 0$ for $T > T_p$. T_p is then sometimes called a "ceiling" temperature. On the other hand, for entropy-driven aggregation $h_p < 0$ and Eq. (15) applies only for $T > T_p$ and $\eta \simeq 0$ for $T < T_p$. T_p is then a so-called "floor" temperature. Note that higher order corrections in the temperature may lead to a nonmonotonic temperature dependence of the fraction aggregated material [14].

In line with the results for EPs, we find that the steepness of the growth of the amount of polymerized material depends on the ratio $h_p/k_B T_p^2$, which is thus amenable to experimental determination (see Figure 8). The quantity $h_p/k_B T_p^2$ can be extracted from the polymerization curve without a full curve-fitting procedure, for at the polymerization transition $(\partial\eta/\partial T)_{T=T_p} = -h_p/k_B T_p^2$.

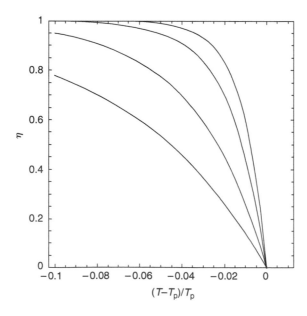

Figure 8 Polymerization curve of TAEPs and CAEPs as a function of the temperature T relative to the polymerization temperature T_p. Different curves give the fraction polymerized material η for different values of the dimensionless link enthalpy $h_p/k_B T_p = 15, 30, 60, 90$ (from bottom to top). Results are shown for the limiting case where the polymerization transition becomes a true phase transition.

IV. CHEMICALLY ACTIVATED EQUILIBRIUM POLYMERS

In CAEPs a fixed number of monomers is activated through chemical reaction with initiator molecules [9]. Activated monomers react with inactive monomers to form activated polymers, the total number of which is conserved. Despite the obvious differences between chemical and thermal activation of equilibrium polymerization, the theory describing the self-assembly of CAEPs is closely related to that of TAEPs [40].

Let $\rho_1(1)\upsilon$ be the volume fraction of inactive material and let $\rho_2(N)\upsilon$ denote the dimensionless concentration of activated N-mers (υ is the monomer volume). The grand potential of the inactive monomers is given by $\beta\Omega_1/V = \rho_1(1)[\ln\rho_1(1)\upsilon - 1 - \ln Z_1(1) - \beta\mu]$, while that of the active ones, Ω_2, is given by Eq. (1), where as before we replace $\rho(N)$ by $\rho_2(N)$. The equilibrium distribution of material over the active and inactive states is determined by the set of equations $\delta\Omega/\delta\rho_1(1) = 0$ and $\delta\Omega/\delta\rho_2(N) = \psi$, where $\Omega = \Omega_1 + \Omega_2$. Here, ψ acts as a Lagrange multiplier (or chemical potential) that keeps the number of active chains constant.

The free energy minimization gives $\rho_1(1) = \upsilon^{-1}Z_1(1)\exp(\beta\mu N)$ and $\rho_2(N) = \upsilon^{-1}Z_2(N)$ $\times \exp(\beta\mu N + \psi)$. Again, we presume that $Z_2(N)$ is given by an expression of the form of Eq. (3). Without loss of generality we may put the single-particle partition function of the active and inactive monomers to be equal, $Z_2(1) = Z_1(1)$, because this only renormalizes the Lagrange multiplier ψ.

The chemical potential μ of the material is eliminated from the description by invoking the conservation of mass $\phi = \rho_1(1)\upsilon + \sum_{N=1}^{\infty} N\rho_2(N)\upsilon$, where ϕ denotes as before the overall volume fraction of dispersed material. The Lagrange parameter ψ follows from the conservation of the number of active chains, which we express as $r \equiv \sum_{N=1}^{\infty} \rho_2(N)\upsilon/\phi$. Here, the dimensionless parameter r accounts for the fraction initiator added to the solution. The stoichiometry of the initiation reaction is absorbed into the definition of r, so our definition is somewhat different from what is conventionally done in the literature [40].

In analogy to the thermally activated case, the polymerization transition of chemically activated polymers becomes sharp in the limit $r \to 0$ and takes place for conditions where $X = X_p \equiv 1$. As before, we have $X \equiv \phi\exp(\beta G)$ and G denotes the free energy gain associated with the formation of a bond. In that case almost all of the material remains in inactive form provided $X < 1$, with $\rho_1(1)\upsilon = \phi$ and $\rho_2(N)\upsilon = 0$. On the other hand, if $X > 1$ we obtain a concentration of inactive species

$$\rho_1(1)\upsilon = (1 - \overline{N_2}^{-1})\exp(-\beta G) \tag{16}$$

while for the active species we have

$$\rho_2(N) = \upsilon^{-1}\frac{r\phi}{\overline{N_2} - 1}\left(1 - \frac{1}{\overline{N_2}}\right)^N \tag{17}$$

In Eqs. (16) and (17), the mean degree of polymerization $\overline{N_2}$ averaged over the active species only obeys

$$\overline{N_2}^{-1} = \frac{1}{2}(1 - X) + \frac{1}{2}\sqrt{(1 - X)^2 + 4rX} \tag{18}$$

This gives $\overline{N_2} \sim 1 + X$ in the monomer-dominated regime $X \ll X_p = 1$, $\overline{N_2} \sim 1/\sqrt{r}$ at the polymerization transition with $X = X_p$, and $\overline{N_2} \sim 1/r$ in the strongly polymerized regime for $X \gg X_p$. Note that in the vicinity of the polymerization transition the active chains can in the mean be very long if $r \ll 1$. In fact, $\overline{N_2}$ saturates for values of X not very much larger than unity. Provided $X \geqslant 1$ and $r \to 0$, Eq. (17) reduces to an exponential distribution $\rho_2(N)\upsilon^{-1} \sim r\phi\overline{N}_2^{-1} \exp(-N/\overline{N}_2)$, not dissimilar to those obtained for EPs and TAEPs.

The degree of polymerization $\overline{N}_n$ averaged over all species now obeys

$$\overline{N}_n = \frac{1 - \overline{N}_2^{-1} + rX\overline{N}_2}{1 - \overline{N}_2^{-1} + rX} \tag{19}$$

which also grows sharply beyond the polymerization transition if $r \to 0$. Close to the polymerization transition where $rX \ll 1$, $\overline{N}_n \sim 1 + X$ grows strongly with concentration since $X > 1$. Much more deeply into the polymerized regime, when $Xr \gg 1$, we find that the mean degree of polymerization saturates to $\overline{N}_n \sim \overline{N}_2 \sim 1/r$.

The fraction activated material is given by $\eta = r\overline{N}_2$. In the limit $r \to 0$, we have $\eta = 0$ for $X \leqslant 1$ and $\eta = 1 - X^{-1}$ if $X > 1$. Hence, CAEPs and TAEPs behave in quite similar ways (see Figure 4 and Figure 8). Indeed, Eqs. (14) and (15) also apply to CAEPs, signifying that CAEPs and TAEPs belong to a different "universality class" than EPs do. In fact, the formal limits $r \to 0$ and $K_a \to 0$ are equivalent in that the polymerization transition becomes a true (continuous) phase transition. That this must be so, can be inferred from the behavior of the chemical potential of the monomers near $X_p = 1$. For both systems the (dimensionless) chemical potential $\ln(1 - \overline{N}_2^{-1}) \sim -\overline{N}_2^{-1}$ exhibits a discontinuity at the critical point $X_p = 1$. In addition, the heat capacities calculated within the given theoretical framework are typical of mean-field theories near a critical point, that is, their values jump at the critical point [29].

The mean-field critical behavior is a result of the ideal entropy of mixing implicit in Eq. (1) that presumes Gaussian density fluctuations. Strong fluctuations in the distribution over the inactive and active states invalidate the mean-field approximation, and one has to resort to more advanced treatments to be able to deal with those accurately [22,41,42]. However, K_a and r are typically not all that close to zero, implying that in reality the polymerization transition is not quite critical. As a result, a mean-field treatment of the polymerization transition of CAEPs and TAEPs is often remarkably accurate [9].

V. Coordination Polymers

If the monomeric building blocks in supramolecular polymers are held together through bicomplexation with an ion or some other small molecule, we deal with the special class of CPs. Although a large number of examples of this type of supramolecular polymer are known, the topic has not received a great deal of attention from the theoretical community. Recently, an early poly-condensation theory of Jacobson and Stockmayer [43] allowing for ring closure has been reanalyzed in the context of CPs [44]. Here, we focus on those cases where rings are suppressed either for entropy or for enthalpy reasons [45]. In Section VI, we discuss under what conditions rings can be ignored.

Because CPs are, strictly speaking, copolymers, we define a number density $\rho(N_1, N_2)$ of copolymers containing N_1 linking agents and N_2 bifunctional monomers, where, as a result of the prescribed stoichiometry of the copolymerization, $|N_1 - N_2| \leqslant 1$. For the grand potential we can then write

$$\frac{\beta\Omega}{V} = \sum_{N_1=0}^{\infty} \sum_{N_2=N_1-1}^{N_1+1} \rho(N_1, N_2)[\ln \rho(N_1, N_2)\upsilon - 1 - \ln Z(N_1, N_2) - \beta\mu_1 N_1 - \beta\mu_2 N_2] \tag{20}$$

where $\rho(N_1, N_2) \equiv 0$ if $N_1 = 0$ and $N_2 \leqslant 0$. We take υ to be the monomer volume — the linking agents are presumed to be pointlike. $Z(N_1, N_2)$ is the partition function of a chain of N_1 linker molecules and N_2 monomers, and μ_1 and μ_2 the chemical potentials of the two components.

For every bond between a ligand and a linker molecule, we assume that this yields constant free energy $\varepsilon < 0$, except when the linker is at the end of a chain and bonds to one monomer. In that case, a free energy $\varepsilon + \Delta\varepsilon$ is attributed to the link, where $\Delta\varepsilon$ can be positive or negative. If negative a free end is favored, and if positive not so. In other words, if $\Delta\varepsilon < 0$ two ligands linking to a single cofactor interfere with each other.

This allows us to put forward a generalization of Eq. (3) for the partition function of CP chain,

$$Z(N_1, N_2) = Z_2(1)\lambda^{N_1+N_2-1} \exp(-\beta\varepsilon(N_1 + N_2 - 1)) \exp(-\beta\Delta\varepsilon(N_1 - N_2 + 1)) \quad (21)$$

provided $N_1 \geqslant 1$ and $N_2 \geqslant 1$. Here, $Z_2(1)$ and λ are again nonuniversal constants that we need not specify, as they can be absorbed in the chemical potentials and in the free energies of binding. For the individual species we put $Z(1,0) \equiv 1$ and $Z(0,1) \equiv Z_2(1)$, since we presume the linker molecules to be pointlike on the scale of the monomers.

The equilibrium distribution of the chains follows upon minimizing Eq. (20), and setting $\delta\Omega/\delta\rho(N_1, N_2) = 0$. This gives, as expected, $\rho(N_1, N_2)\upsilon = Z(N_1, N_2) \times \exp(\beta\mu_1 N_1 + \beta\mu_2 N_2)$. The chemical potentials can be eliminated by invoking the conservation of mass, $\phi_i = \sum_{N_1=0}^{\infty} \sum_{N_2=0}^{\infty} N_i \rho(N_1, N_2)\upsilon$ with ϕ_i a dimensionless concentration for components $i = 1, 2$. These concentrations can be varied independently. It is useful to introduce the dimensionless concentrations or mass actions $X_1 \equiv \phi_1 \exp(\beta G)$ and $X_2 \equiv \phi_2 \exp(\beta G)$, where $\beta G \equiv -\beta\varepsilon - \beta\Delta\varepsilon + \ln\lambda - \ln Z_2(1)$ is the free energy gain of a link between a ligand and a linker molecule. There is one additional energetic parameter, H, that measures the free energy penalty if the bond between a ligand and a linker is at the end of the chain, $\beta H = -2\beta\Delta\varepsilon - \ln Z_2(1)$. Both parameters G and H should be treated as phenomenological ones.

The coordination polymerization is within the present description as gradual as that of EPs, albeit that two instead of one dimensionless concentrations or mass actions should be large. It is only if $X_1 \gg 1$ and $X_2 \gg 1$, and then only if $|X_1 - X_2| \ll 1$, that is, at nearly equal concentrations of both components, that long chains can form, with $\overline{N}_2 \sim \sqrt{X_2}/(1 + \exp(\beta H))$. Note that long chains cannot form if $H \gtrsim 1$, because the linkers then strongly favor binding to as many free ends as possible thereby strongly reducing the mean degree of polymerization. If $H \lesssim -1$ the linkers avoid binding to free ends, so, at fixed X_2, there is no opposite effect of appreciable growth enhancement.

Even if the mass actions of both components are large, $X_1 \gg 1$ and $X_2 \gg 1$, long-chains cannot grow if $|X_1 - X_2| \gg 1$. The reason is that in that case the mass action of the dominant component favors binding as much material as is possible. As a consequence, if we increase the concentration of the linker species X_1 at a fixed monomer concentration $X_2 \gg 1$, this initially leads to the growth of the chains until equal molarities of both species is reached (presuming H is small). Beyond this compensation (or equivalence) point, the chains shrink with increasing linker concentration. Clearly, if either X_1 or X_2 is small, or both, most of the material remains in the monomeric state.

VI. Rods, Rings, and Worms

Free ends are removed from the solution by chain growth if we let the strength of the bonded interaction βG between monomers along a supramolecular chain become large. However, chain ends can also be removed from the solution by ring closure, a possibility we have so far suppressed. Rings should, in principle, always form as this generates an entropy of mixing in addition to the extra bond energy gained per chain [24]. Naïvely, all types of supramolecular polymer should therefore be expected to behave in a way that is similar to that of TAEPs, with a ring-opening polymerization that is not gradual but becomes sharper the larger βG [23,46].

There is strong evidence for a ring-opening type of polymerization in some supramolecular systems [44,47–49], and rings might also be involved in the living polymerization of poly(α-methylstyrene) [18,26]. However, it appears that very often either rings do not form,

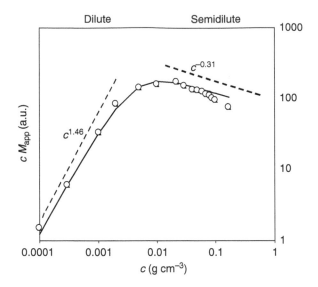

Figure 9   Dimensionless apparent molecular weight M_{app} times the concentration c of giant surfactant micelles of hexadecyl trimethyl ammonium bromide in brine, obtained by means of light scattering measurements at a temperature of 306 K. Circles: experimental data of Buhler and coworkers [50]. Line: renormalization group theory for self-assembled flexible chains fitted to the data [51,52]. Also indicated by the dashed lines are the predictions of scaling theory in the dilute and semidilute regimes [1,53], which agree well with results of Monte Carlo simulation [52].

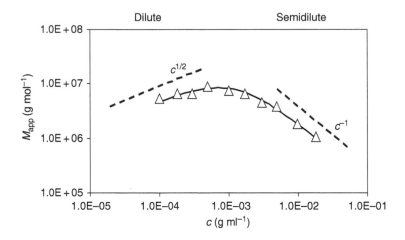

Figure 10   Apparent molecular weight M_{app} as a function of the concentration c of giant surfactant micelles of sodium sulfopropyl octadecyl maleate in brine, obtained by means of light scattering at a temperature of 323 K. Triangles: experimental data of Berlepsch and coworkers [54]. Line: theory for self-assembled rigid rods fitted to the data [55]. The link free energy obtained from the fit amounts to $G \approx 26 \, k_B T$, which is equivalent to about 65 kJ mol^{-1}. The micelles have an estimated persistence length of 120 nm and a diameter of 6 nm. Also indicated by the dashed lines are the predictions of mean-field theory in the dilute and semidilute regimes [56].

or, if they do, their influence is minor. Indeed, the growth of supramolecular polymers can very often be described accurately by ignoring the generation of rings altogether, as shown in Figures 5, 6, 9, and 10. In some cases the absence of ring closure can be rationalized when in a living anionic polymerization charges are generated on the (growing) free ends. On the other hand, if the assemblies would in principle allow for ring closure but have a high bending rigidity, tight rings are energetically suppressed leading to an important reduction of the fraction material absorbable into rings (especially if βG is not all that large) [25]. An entropic ring-suppression mechanism ensues if the monomers are not small but of flexible polymeric nature themselves [45].

To illustrate the impact of chain flexibility on ring closure in EPs, let $\rho_1(N)$ be the number density of rings of degree of polymerization N, and $\rho_2(N)$ that of the linear chains. The grand potential of rings plus linear chains is given by $\Omega = \Omega_1 + \Omega_2$ where the contributions of each species is given by Eq. (1). The usual minimization of the grand potential gives $\rho_i(N) = \upsilon^{-1} Z_i(N) \exp(\beta \mu N)$ for $i = 1, 2$, with υ the volume of the monomers and μ their chemical potential. Here, $Z_i(N)$ denotes the partition function of a ring if $i = 1$ and that of a linear chain if $i = 2$, where we presume that the latter is accurately described by Eq. (3). We relate the latter to the overall volume fraction of aggregating material ϕ through the conservation of mass $\sum_{N=1}^{\infty} N(\rho_1(N) + \rho_2(N))\upsilon = \phi$. This allows us to immediately write down $\rho_2(N) = \upsilon^{-1} \exp(-\beta \tilde{\mu} N - \beta G)$ for the linear chains, with $\tilde{\mu}$ the renormalized chemical potential and G the free energy gain of the formation of a link.

An accurate analytical expression for the partition function $Z_1(N)$ of closed loops is not known for arbitrary degrees of polymerization and bending stiffness. However, simple scaling arguments can be invoked to produce limiting relations for $Z_1(N)$ from $Z_2(N)$, the reason being that the ratio $Z_2(N)/Z_1(N)$ must be equal to the probability of opening a loop [25]. The ring-opening probability is proportional to the number of places a ring can break, N, and a Boltzmann weight accounting for the loss of the bond energy $\exp(-\beta G)$. It must also be proportional to the increased volume two monomers can explore after their bond has been severed, and, for tight rings, a Boltzmann weight stemming from the release of the elastic energy upon ring opening.

If the contour length L of a ring is smaller than its persistence length P, the free energy of this ring is dominated by an elastic energy equal to $2\pi^2 P/L$, and less so by its (entropic) bending modes [57]. This elastic energy is released upon ring opening. After the ring opening, the mean lateral excursion the tips can make relative to the intrinsic length $l = L/N \ll P$ given by the length a single monomer contributes to the length of a chain must be of order L^3/Pl^2 [58]. Hence, we deduce that for $L \lesssim P$ the ring-opening probability must scale as $Z_2(N)/Z_1(N) \sim N \exp(-\beta G)(L^3/Pl^2) \exp(2\pi^2 P/L)$.

The leading-order exponential $\exp(2\pi^2 P/L)$ agrees with a more accurate calculation [57], and it is this term that so very strongly suppresses the emergence of tight rings. This has recently been confirmed by direct electron microscopic observation of giant surfactant micelles [59]. Note that the subdominant power law L^3 that we naïvely get from considering the entropy gain of the free ends is in fact an underestimation. There are additional contributions stemming from bending fluctuations ignored in our analysis, giving rise to an additional L^3 in front of the exponential [57] as well as a subdominant term linear in $L/P \ll 1$ in the exponential [58].

It follows that we need to only consider rings longer than about a persistence length. These are entropy dominated, and behave to a first approximation like Gaussian rings. The volume accessible to the two ends following a ring-opening scales as the radius of gyration of the chain to the third power, $(LP)^{3/2}$, which has to be compared to much smaller phase volume of order l^3 available before the scission of the bond. Hence, for $L \gtrsim P$ we have

$$Z_1(N) = \zeta N_P^{-3/2} N^{-5/2} Z_2(N) \exp(\beta G) \qquad (22)$$

for $N \geqslant N_P$ where ζ is an unknown constant of order unity and $N_P \equiv P/l \gg 1$ the degree of polymerization of a chain of one persistence length. The weight $N_P^{-3/2}$ can be quite small if the persistence length is large enough, and reduces the concentration of floppy rings, for Eq. (22) gives a distribution $\rho_1(N \geqslant N_P) = \zeta \upsilon^{-1} N_P^{-3/2} N^{-5/2} \exp(-\beta \tilde{\mu} N)$. The scaling of the size distribution with $N^{-5/2}$ is in agreement with computer simulations [26] and with the electron microscopic study of [59]. It is useful to mention that even for flexible supramolecular polymers we expect a value for this quantity of about 5 to 10, rising to many hundreds or thousands for stiff assemblies [25].

From the theory of Section II, it is easy to see that $\exp(-\beta\tilde{\mu}) = 1 - \overline{N}_2^{-1}$ where $\overline{N}_2$ is the mean degree of polymerization averaged over the linear chains only. Conservation of mass lets us fix this quantity in the usual way, giving

$$X = \overline{N}_2^2 - \overline{N}_2 + \zeta N_P^{-3/2}\exp(\beta G)\sum_{N=N_P}^{\infty} N^{-3/2}\left(1 - \overline{N}_2^{-1}\right)^N \tag{23}$$

with $X = \phi\exp(\beta G)$ as before the mass action. Here, we set $\rho_1(N < N_P)v = 0$, which seems quite a reasonable approximation given that, tight rings are exponentially suppressed.

Equation (23) teaches us that if $\zeta N_P^{-4}\exp(\beta G)$ is sufficiently small, that is, if the chains are sufficiently rigid, rings never constitute a significant portion of the assemblies. We have plotted the fraction linear chains as a function of the control parameters $\zeta N_P^{-4}\exp(\beta G)$ and $\overline{N}_2/N_P$ for the case $N_P \gg 1$ (see Figure 11). Even if $\zeta N_P^{-4}\exp(\beta G)$ is not small, linear chains dominate provided $\overline{N}_2 \ll N_P$. If $\overline{N}_2 \gg N_P$, the mass absorbed into rings becomes a constant $\phi_1^{\max} \approx 2\zeta N_P^{-2}$ that can be very small if $N_P \gg 1$. In the limit $\overline{N}_2 \gg N_P$ we retrieve the square-root law we found in Section II albeit slightly modified, with $\overline{N}_2 \sim \sqrt{(\phi - \phi_1^{\max})}\exp(\beta G)$. In this regime the rings just deplete a fixed amount of material from the solution [24]. Note that the mean degree of polymerization of the rings, $\overline{N}_1$, has a value which is between N_P and a few times N_P for all $\overline{N}_2 > 1$.

The theory needs to be amended if $P/l \ll 1$ and the monomers themselves are of a polymeric nature [45]. In that case all rings formed are loose ones, including those formed by monomers folding back onto themselves. This means that for all degrees supramolecular polymerization $N \geqslant 1$ we must have $Z_2(N)/Z_1(N) \sim N\exp(-\beta G)(LP/P^2)^{3/2}$ because for Gaussian chains the smallest relevant length is the persistence length P. The only other difference with the previous case is that the contour length of the chain is now given by $L = NMP$, where M is the degree of polymerization of the monomers. Hence, we obtain $Z_1(N) = \zeta M^{-3/2}N^{-5/2}Z_2(N)\exp(\beta G)$ for all $N \geqslant 1$ showing

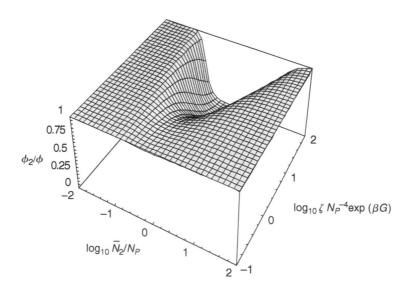

Figure 11 Fraction of material absorbed into linear chains, ϕ_2/ϕ, as a function of the quantities $\zeta N_P^{-4}\exp(\beta G)$ and $\overline{N}_2/N_P$, for the case where ring closure is allowed. N_P denotes the degree of polymerization of a chain with a contour length equal to its persistence length, βG the dimensionless free energy associated with a link, ζ a numerical coefficient of order unity, and $\overline{N}_2$ the mean degree of polymerization of the linear chains. Rings deplete the amount of mass absorbed in chains only if $\overline{N}_2/N_P \approx 1$ and then only for large enough link energies $\zeta N_P^{-4}\exp(\beta G) \gtrsim 1$.

that we only need to replace N_P by M in the above theory for the case $P/l \gg 1$, and set the lower bound in the summation over the rings to unity, for example, in Eq. (23). The conclusions we can draw for $P/l \ll 1$ remain the same as those for the opposite case $P/l \gg 1$, that is, rings are strongly suppressed if $\zeta M^{-4} \exp(\beta G) \ll 1$ albeit that now the reasons are entropic not energetic. If rings do form, the maximum amount of material absorbed in rings is very small with $\phi_1^{\max} \approx 2\zeta M^{-3/2} \ll 1$ if $M \gg 1$.

VII. Helical Aggregation

The entire theory we have so far set up hinges on the assumption that the partition function of a linear chain that can be written as Eq. (1). For most cases this is indeed so, producing a mass action $X = \phi \exp(\beta G)$ that depends on a single effective free energy gain of binding G (the end or link energy). This link energy, and therefore also the associated dimensionless equilibrium constant of the polymerization, $\exp(\beta G)$, is somehow connected with chemical details of the aggregates but also of the solvent (if present) [28,34]. Clearly, one of the great challenges in the theory of supramolecular polymers is to predict the link energy, in particular how it depends on the chain stiffness [32,33], how it is impacted upon by the presence of charges on the assemblies [60–65], and so on [66–68].

Most work in this context has been done in the field of giant micelles, albeit typically within a fairly coarse-grained description of the assemblies. In spite of this, quite reasonable agreement with experimental data have been obtained, for example, for the ionic strength dependence of G of highly charged cylindrical micelles [35]. However, the issue remains fraught with difficulty, not least because all intensive contributions to the model free energy of a chain end up in G, contributions which in the context of conventional polymers are usually considered unimportant and hence ignored. This makes G quite sensitive to the type of model description and even to the mathematical approximations invoked [69]. In view of this, the most useful strategy is to treat G as a phenomenological parameter, as we have done so far, and interpret experimentally found values *a posteriori*.

A fact not widely recognized is that the concept of a phenomenological link energy may under certain conditions lose its usefulness, even if the supramolecular chains behave ideally. The reason is that if there is more than one bonded state between sequential monomers, these states may become correlated along a single chain and give rise to long-range conformational fluctuations. Indeed, Eq. (3) only holds if the chains are sufficiently long relative to some decay length that describes the range of the correlations between the bond fluctuations. These are particularly important for a class of supramolecular polymer that exhibit a configurational transformation similar to the helix-coil transition observed in biopolymers such as polypeptides, polysaccharides, and polynucleotides [3,37,70,71] (see Figure 12).

Helix-coil type transitions in conventional as well as supramolecular polymers are well represented by a model description that can be mapped onto the Ising chain and that is known as the Zimm–Bragg theory in the field of biopolymers [72]. Within this theory, sequential chain units can switch between a weakly bonded (nonhelical or disordered) state and a strongly bonded (helical or ordered) state. This switching may be induced by the formation or breaking of hydrogen bonds, but also by a combination of solvophobic interactions and the geometry of the monomers, such as seems to be the case in some propeller-shaped molecules shown in Figure 13 [3,71].

The statistical weights of the two conformational states are described by a phenomenological free energy ε_1 for the disordered (nonhelical) bonded conformation and $\varepsilon_2 \equiv \varepsilon_1 + \Delta\varepsilon$ for the ordered (helical) bonded conformation. Crucially, the theory introduces a free energy penalty $\varepsilon_3 > 0$, assigned whenever a disordered region along the backbone of the chain switches into an ordered one and vice versa because this could lead to a conformational frustration of monomers at the switching point.

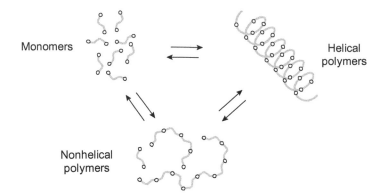

Figure 12 Schematic of helical supramolecular polymerization. Monomeric, nonhelical polymeric and helical polymeric states are in thermal equilibrium as indicated by the arrows. The equilibrium constants between the various species may differ greatly in magnitude. If helical polymers cannot form directly from the monomers, they need to be nucleated through the formation of nonhelical chains that may be energetically unfavorable.

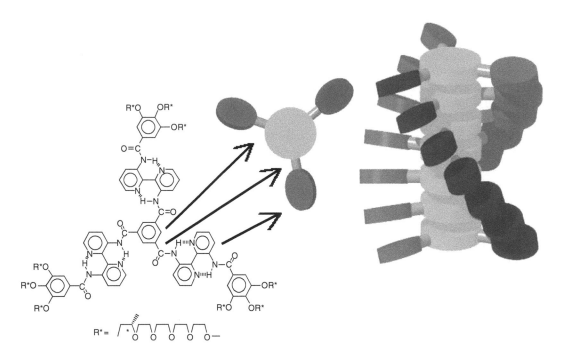

Figure 13 Chemical and coarse-grained structure of propeller-shaped, so-called C_3 discotics that in solution self-assemble into stacks [71]. The "blades" of the propeller can rotate freely about their bond with the central core of the molecule. At high temperatures these stacks are nonhelical but they undergo a configurational transition and become helical at a sufficiently low temperature as evidenced by circular dichroism spectroscopy [71]. (Schematic taken from P. van der Schoot et al. Langmuir 16:10076–10083, 2000. With permission.)

Since the model is effectively one-dimensional, the partition function of an assembly of N monomers can be worked out exactly by standard methodology, to give [73,74]

$$Z(N) = (x\lambda_+^{N-2} + y\lambda_-^{N-2}) \exp(-\beta\varepsilon_1(N-1)) \tag{24}$$

for $N > 2$, where

$$\lambda_\pm = \tfrac{1}{2}[1 + s \pm \sqrt{(1 - s)^2 + 4\sigma s}] \tag{25}$$

as well as the quantities x and y are functions of the so-called Zimm–Bragg parameters $s \equiv \exp(-\beta\Delta\varepsilon) \geqslant 0$ and $\sigma \equiv \exp(-2\beta\varepsilon_3) < 1$ [72]. The precise form of x and y, and that of the partition function of the dimers $Z(2)$, depends on the binding properties of the monomers at the end of the chains. The binding of the monomers at the ends may differ substantially from the ones in the centre of the assemblies, suggesting that the conformational properties of the dimers may be different from those of trimers, tetramers, etc. [74]. For the present purpose, we need not specify these quantities.

According to Eq. (25), λ_+ is always quite larger than λ_- except for $s = 1$ because then $\lambda_\pm = 1 \pm \sqrt{\sigma}$. This implies that for large enough chains, that is, for chains longer than a bare correlation length that is of the order $1/\sqrt{\sigma}$ [72], we may in fact neglect the term associated with the smaller λ_- for all values of the parameter s, retrieving what is essentially the same result as Eq. (3). This is called the ground-state approximation [20].

Within the ground-state approximation, the mean fraction of helical bonds $\bar{\theta}(N)$ in a chain can easily be calculated [3]

$$\bar{\theta}(N) = \frac{1}{N - 2}\frac{\partial \ln Z(N)}{\partial \ln s} \sim \frac{1}{2} + \frac{1}{2}\frac{s - 1}{\sqrt{(1 - s)^2 + 4\sigma s}} \tag{26}$$

and does not depend on the specifics of the chain ends nor the degree of polymerization. Equation (26) should hold for supramolecular chains if their mean degree of polymerization is sufficiently large, and indeed describes the experimental data in that regime very well as may be inferred from Figure 14.

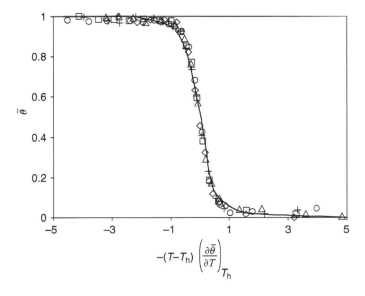

Figure 14 Master curve for the mean fraction helical bonds $\bar{\theta}$ as a function of a dimensionless temperature scale. T is the temperature and T_h the temperature at the helical transition temperature. The symbols give the experimental results obtained from a solution of the C_3 discotic molecules shown in Figure 13 [3,71]. Different symbols represent data taken at different concentration in the range from 10^{-6} to 10^{-2} M in the solvent n-butanol. The drawn curve is obtained from the theory of [73] that fits the data reasonably well [73].

The crossover from the nonhelical to the helical polymeric state occurs for $s = 1$, and becomes sharper the smaller the value of the parameter σ. It is for this reason that σ is often viewed as a parameter that describes the cooperativity of the helical transition [72]. This quantity is accessible experimentally by means of circular dichroism spectroscopy because to leading order in the temperature we have $\ln s = \Delta h_h (T - T_h)/k_B T_h^2$ and $(\partial \bar{\theta}/\partial T)_{T=T_h} = \Delta h_h/4\sqrt{\sigma} k_B T_h^2$. Here, T_h denotes the helical transition temperature where $\bar{\theta} = \frac{1}{2}$, and Δh_h the excess enthalpy associated with the formation of a helical bond [3]. The steepness of the measured helicity versus temperature curve (obtained, for example, by means of circular dichroism spectroscopy) depends on the ratio $\Delta h_h/\sqrt{\sigma}$, where Δh_h is obtainable independently from microcalorimetry [71]. Values of $\sigma \approx 0.01 - 0.001$ and $\Delta h_h \approx 50$ kJ mol^{-1} have been obtained for the discotic of Figure 13, suggesting that the helical transition in supramolecular polymers can indeed be highly cooperative [3,73].

As advertised, it is possible at the level of the ground-state approximation to define a link energy $\beta G \equiv -\beta \varepsilon_1 - \ln x + 2 \ln \lambda_+$ that inserted into Eq. (5) produces a mean degree of polymerization. If we do this, we find that the chains exhibit a growth spurt at the helical transition temperature in particular if σ is small, because λ_+ then increases sharply from unity to s at the transition. Indeed, for $\sigma \rightarrow 0$ the link energy becomes $\beta G = -\beta \varepsilon_1 - \ln x$ if $\varepsilon_2 > \varepsilon_1$, and $\beta G = -\beta \varepsilon_2 - \ln x$ if $\varepsilon_2 < \varepsilon_1$, where we note that both ε_1 and ε_2 are negative or otherwise there would be no chains. This effect may further be strengthened by the contribution from the ends through the $\ln x$ term. For free ends, however, this term is close to zero for all temperatures, and the helical aggregation behaves like an EP [74] (see Figure 15).

The situation becomes complex when ε_1 is not large and negative, because then the ground-state approximation no longer holds and the exact partition function Eq. (24) has to be implemented into the theory. The state of the ends can change the polymerization from an EP to a TAEP, in which case the helical polymerized state has got to be nucleated in a similar way as in the Oosawa–Kasai theory [13] or variants thereof [14]. For instance, if the ends are for some reason not able to attain a helical conformation, nucleation of the helical aggregates can only occur via highly unfavorable nonhelical dimer and trimer states [75]. Application of the full theory shows that two regimes emerge, one where helical transition occurs deeply into the polymerized state of the dissolved material, and one where the material polymerizes directly into helical polymers. In the former regime, the ground-state approximation is accurate, in the latter it is not.

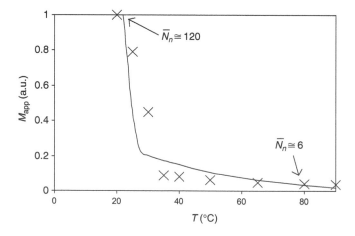

Figure 15   Temperature dependence of the apparent molecular weight (in arbitrary units) of the assemblies formed by the C_3 discotic of Figure 13 in the solvent n-butanol at a concentration of 2.39×10^{-3} M. The crosses indicate the small-angle neutron scattering data of [3]. The drawn line was obtained by fitting the theory to ultraviolet and circular dichroism spectroscopy data [73]. The onset of the growth spurt at low temperatures coincides with the helical transition temperature. Indicated at two temperatures are the calculated number-averaged mean degrees of polymerization $\bar{N}_n$.

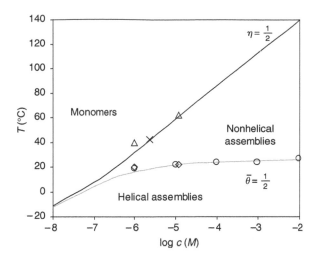

Figure 16 Diagram of states of the C_3 discotic of Figure 13 in the solvent n-butanol. The symbols represent data obtained by various experimental methods, and the lines the theoretical fits to the experimental data [75]. At low concentrations of material the monomers polymerize directly into helical polymers, while at higher concentrations a nonhelical polymerized regime intervenes. The demarcation of the monomeric and the polymeric regimes is set at 50% fraction material in polymers, $\eta = \frac{1}{2}$, and that between the helical and nonhelical states by a 50% fraction of helical bonds, $\bar{\theta} = \frac{1}{2}$.

There is some experimental support for the existence of two helical polymerization regimes, as in fact is shown in Figure 16. For the particular system the data of which are shown in the figure, these regimes can be found at different concentrations of dissolved material. Since the regimes are the result of a combination of mass action and the relative as well as the absolute magnitude of two free energies of binding, the crossover from one to the other regime may also be induced by, say, the solvent quality, small changes in the architecture of the monomers, and so on.

VIII. INTERACTING SUPRAMOLECULAR POLYMERS

Neighboring monomers along a supramolecular chain interact through a net attraction at contact that we have termed the bonded interaction. If two monomers are not direct neighbors they may interact too, even if they are separated by a considerable distance along the chain's backbone. This happens if the chain is sufficiently flexible so that thermal agitation lets the chain fold back onto itself. In addition, there may be interactions between monomers on different chains, if the concentration of chains is not dilute. Such (nonbonded) intra- and inter-chain interactions can be attractive or repulsive. Typically, if the solvent quality is good for the polymeric species, the monomers repel each other. If not, then the net interactions can become attractive, ultimately leading to demixing. If the chains are charged, they interact via repulsive Coulomb interactions that may couple quite strongly to the supramolecular polymerization [61,63].

Within a mean-field theory, the impact of nonideality on (uncharged) supramolecular polymers is readily evaluated. For instance, a Flory–Huggins type of theory for any of the four classes of supramolecular polymers discussed can be constructed straightforwardly by adding to the grand potential an excess free energy [16,29,76]

$$\frac{\beta \upsilon \Omega_{exc}}{V} = (1 - \phi) \ln(1 - \phi) + \chi \phi (1 - \phi) \tag{27}$$

with χ the familiar Flory–Huggins interaction parameter [20]. It is immediately clear that the solvent quality does not *directly* influence the supramolecular polymerization, because the mean-field excess free energy depends only on the overall volume fraction of dispersed material and not on how this material is distributed over the various degrees of polymerization. It is only through the effects of mass action that interactions modify the mean degree of polymerization, for example, in coexisting demixed phases. The location of the critical point does couple to the self-assembly, as it is a sensitive function of the mean degree of polymerization of the chains. For EPs it becomes a function of the link energy G, with a critical volume fraction $\phi_c \sim (\frac{3}{4})^{2/5}(\exp \beta G)^{-1/5}$ and a critical Flory–Huggins parameter $\chi_c \sim \frac{1}{2}$ to leading order in $\exp \beta G \gg 1$ [76]. For TAEPs and CAEPs additional dependencies arise on the equilibrium constant K_a and the initiator fraction r [29].

In a good solvent χ is so small that demixing cannot occur, allowing for arbitrary concentrations of material to be dissolved. In this case the mean-field treatment predicts the theory of Sections II to V to hold all the way into the melt regime where $\phi \to 1$. Here, we have implicitly assumed that the ends of the supramolecular chains interact with the other monomers in the same way as the monomers in the central parts of the chains do. This, however, need not be so. Within a mean-field approximation, we expect corrections to the excess free energy of the order of $\phi^2 \overline{N}_n^{-1}$ if the mean degree of polymerization, $\overline{N}_n$, is large and the concentration of free ends, $\phi \overline{N}_n^{-1}$, is low. It is easy to show that, for example, for EPs this leaves the theory of Section II intact, except for a renormalization of the link energy βG that is replaced by an effective, concentration-dependent one $\beta G \to \beta G + \kappa \phi$ with κ a constant of order unity. Growth is then enhanced by the interactions albeit not enormously so. Computer simulations bear out this conclusion [26]. For very stiff supramolecular polymers, modeled as smooth cylinders interacting through a hard-core repulsion, more elaborate estimates have been derived [55,77]. These are discussed in Chapter 4.

A mean-field description of the configurational states of interacting polymeric chains is accurate if they are in the mean not much longer than about $(P/D)^2$ with D the diameter and P the persistence length of the chains [20] (see, e.g., Figure 10). The reason is that self-interactions are strongly attenuated in semiflexible chains. For flexible chains, however, P/D is not a great deal larger than unity and mean-field theory must be expected to break down even at fairly low degrees of polymerization. In this case at least three regimes may be distinguished, termed the dilute, the semidilute, and the concentrated regime. In the concentrated or melt regime, interactions are screened and mean-field theory becomes accurate again [78]. (The very small fraction of chains that are much longer than the average chain length may be swollen by the smaller ones, however [51,79].)

In dilute solution the chains do not overlap but the self-interactions that cause the chains to swell have to be accounted for. As is well-known, the partition function $Z_D(N)$ of a self-avoiding linear chain obeys the scaling relation $Z_D(N) \sim N^{\gamma-1}Z(N)$ for large N, where $Z(N)$ is the ideal chain partition function Eq. (3), and where γ is a critical exponent associated with the "end enhancement" [78]. In three spatial dimensions $\gamma \simeq 1.165$ [80]. Inserting this in Eq. (1) gives for EPs an equilibrium size distribution of the Zimm–Schulz type if we extrapolate the scaling law down to $N = 1$, which is reasonable so long as $\overline{N}_n \gg 1$. The growth law deviates from the square-root law discussed in Section II albeit only slightly, with $\overline{N}_n \sim \phi^{1/(\gamma+1)} \approx \phi^{0.46}$ in three-dimensional space. This modified scaling law has been confirmed by renormalization group treatments [51,79], by Monte Carlo simulations [26,80], and is consistent with results from light scattering experiments such as the ones shown in Figure 9. Sometimes considerably higher exponents are found [81], irreconcilable with current theory and with results available from computer simulations.

In the semidilute regime, supramolecular chains larger than the so-called blob size ξ overlap and strongly interact. Deeply into the polymerized regime we have $\xi \approx P\phi^{-\nu/(\nu d-1)}$ with ν the critical exponent of the end-to-end distance of self-avoiding chains, and d the dimensionality of space [78]. In three-dimensional space, $\nu \simeq 0.588$ [26]. At the scale above the blob size interactions are screened, implying that the free energy Eq. (1) may still be used to calculate the size distributions provided the polymers are treated as chains of blobs [78,80]. Within this blob picture, we have for the partition function of a single chain in the semidilute regime $Z_{SD}(N) \sim (\xi/P)^{(\gamma-1)/\nu}Z(N)$ for $N^\nu P \gtrsim \xi$, and $Z_{SD}(N) \sim N^{\gamma-1}Z(N)$ for $N^\nu P \lesssim \xi$ because chains smaller than the blob size remain swollen.

Using this in Eq. (1), we retrieve in the semidilute regime an exponential distribution for the EPs but with a growth exponent that is now slightly larger than one half, $\overline{N}_n \sim \phi^{(vd+\gamma-2)/(vd-1)2} \approx \phi^{0.60}$ for $d = 3$ [24]. Again, good agreement is found with renormalization group calculations [51,79] and with results from Monte Carlo simulation [80]. Note that the crossover value of the mass action to the semidilute regime, X_*, does indeed occur deeply into the polymerized regime because $\ln X_* \approx \beta G(\gamma + 1)/(\gamma + vd) \gg \ln X_p$.

Chemically activated equilibrium polymerization of material in a good solvent was recently investigated by means of a scaling theory similar to that described for EPs [40]. For this type of supramolecular polymer, the effect of excluded-volume interactions on the growth seems to be even more subtle than for EPs, and to be the most prominent very close to the polymerization transition. The location of the polymerization transition remains unaffected by the interactions but it becomes slightly more rounded because the initiator fraction r in Eq. (18) is replaced by $\gamma r \simeq 1.2r$. For CAEPs the polymerization transition (at temperature T_p) and the crossover to the semidilute regime (at temperature T_*) turn out to be quite close to each other, quite unlike the situation for EPs, because of their strong growth beyond T_p. Provided $\phi \gg r^{vd-1} \approx r^{0.76}$ for $d = 3$, the temperature difference between the two obeys [40]

$$T_* - T_p \simeq \frac{k_B T_p^2}{h_p} r^{(vd-1)/vd} \phi^{1/vd} \tag{28}$$

where h_p is the (effective) enthalpy gain of a link. Equation (28) is in quite good agreement with living polymerization experiments on poly(α-methyl stryrene), showing that in practical situations T_p and T_* need not differ by more than a few Kelvin.

In the limit of vanishing initiator fraction $r \to 0$, the polymerization transition and the crossover to the semidilute regime merge, that is, one either has a solution consisting mainly of monomers if $X < X_p$, or a polymerized solution of long and strongly interpenetrating chains if $X > X_p$. The absence of a dilute polymerized regime has nontrivial consequences for properties such as the osmotic compressibility of the solution, which becomes strongly enhanced (and nonclassical) near X_p if r is sufficiently small [42]. Plausibly, it should also lead to a sharp increase in the solution viscosity.

A similar kind of behavior has to be expected for thermally activated equilibrium polymerization, as in fact for the ring-opening transition of EPs. (For a discussion of the latter, see the work of Petschek et al. [23] and of Cates [24].) Applying the scaling analysis for CAEPs [40] to TAEPs, one finds that volume exclusion should not shift the polymerization transition, only to make it slightly *less* rounded. Straightforward algebra gives

$$T_* - T_p \simeq \frac{k_B T_p^2}{h_p} K_a^{(vd-1)/(vd+\gamma)} \phi^{-(\gamma+1)/(vd+\gamma)} \tag{29}$$

for the difference between the crossover temperature to the semidilute regime, T_*, and polymerization temperature, T_p. This difference vanishes as $K_a \to 0$. Focusing on the concentration dependence, the growth law of the TAEPs in the semidilute regime is similar to that of EPs with $\overline{N}_2 \sim \phi^{0.6}(1 - X^{-1})^{0.1}$ for $X > 1$ and $d = 3$. Note that for CAEPs in a good solvent, the growth law remains that of ideal, noninteracting ones.

IX. COUPLING TO EXTERNAL FIELDS

The size distribution of supramolecular polymers should be sensitive not only to changes in the solvent conditions, but also to externally applied fields including electric and magnetic fields [82,83], flow fields [84–86], gravitational and centrifugal fields [87,88], confining walls [89,90], and so on.

A vast amount of theoretical work has been done in this context, although almost exclusively on EPs. Here we restrict ourselves to a discussion of the coupling of the self-assembly of rodlike EPs to a quadrupole ordering field in order to highlight the important role that a supramolecular bending flexibility plays in suppressing a runaway growth instability, often interpreted in terms of a gelation transition [85,86]. Our conclusions extend to *molecular* ordering fields that drive the isotropic–nematic phase transition [21,69,91,92], discussed in more detail in Chapter 4.

In order to describe self-assembled, noninteracting rods in a static orienting field, we need only to slightly modify Eq. (1) [91],

$$
\frac{\beta \Omega}{V} = \sum_{N=1}^{\infty} \int d^2 \vec{u} \rho(N, \vec{u}) [\ln \rho(N, \vec{u}) \upsilon - 1 - \ln Z(N, \vec{u}) - \beta \mu N] \tag{30}
$$

where $\rho(N, \vec{u})$ is the number density and $Z(N, \vec{u})$ the partition function of the rods of degree of polymerization N, pointing in the direction along the unit vector $\vec{u}$. The integration $\int d^2 \vec{u} \equiv \int_0^{\pi} d\theta \sin \theta \int_0^{2\pi} d\phi$ is over the polar angles θ and ϕ.

If the rods are subjected to an external orienting field $U(N, \vec{u})$, we have $Z(N, \vec{u}) = Z(N) \exp(-\beta U(N, \vec{u}))$ with $Z(N)$ the partition function in the absence of the field, presumed to be of the form of Eq. (3). The equilibrium distribution minimizes Eq. (30), so by putting $\delta \Omega / \delta \rho = 0$ we obtain similarly to Eq. (2), $\rho(N, \vec{u}) = \upsilon^{-1} Z(N) \exp(\beta \mu N - \beta U(N, \vec{u}))$. The size distribution is a function of the orientation of the rods, unless $\beta U \to 0$. Restricting ourselves to quadrupole fields that couple linearly to the length of the rods, such as experienced by induced electric and magnetic dipoles on the assemblies or by a coupling to a nematic solvent, we write $\beta U(N, \vec{u}) = KN(\cos \theta)^2$ with K the dimensionless field strength and θ the angle between the rod and the field direction (defined along the z-axis). Clearly, if $K > 0$ the field favors rod to align perpendicular to the field direction, representing a disorienting field, while if $K < 0$ it favors alignment, representing an orienting field.

Following the now standard prescription, we eliminate the chemical potential by invoking the conservation of mass, $\phi = \sum_{N=1}^{\infty} \int d^2 \vec{u} N \rho(N, \vec{u}) \upsilon$. By defining a dimensionless number density of rods $\rho = \sum_{N=1}^{\infty} \int d^2 \vec{u} \rho(N, \vec{u}) \upsilon$, the number-averaged mean degree of polymerization can be calculated from $\overline{N}_n(K) = \phi / \rho$. It is an explicit function of the field strength K. For $K > 0$ we find [75]

$$
\frac{\overline{N}_n(K)}{\overline{N}_n(0)} = \frac{1}{2} \left[1 + \frac{\sqrt{K \overline{N}_n(0)}}{1 + K \overline{N}_n(0)} \cdot \frac{1}{\arctan \sqrt{K \overline{N}_n(0)}} \right] \tag{31}
$$

provided the degree of polymerization $\overline{N}_n(0)$ in the absence of the field is sufficiently large. According to Eq. (31), the rods shrink in the presence of a disorienting field albeit not hugely so and then only if $K \overline{N}_n(0)$ is not exceedingly small (see also Figure 17). In the limit $K \overline{N}_n(0) \to 0$ the rods remain of the same size [82], but do order perpendicularly to the field direction.

If $K < 0$, we get [75]

$$
\frac{\overline{N}_n(K)}{\overline{N}_n(0)} = \frac{1}{2} \left[1 + \frac{\sqrt{-K \overline{N}_n(0)}}{1 + K \overline{N}_n(0)} \cdot 2 \left(\ln \frac{1 + \sqrt{-K \overline{N}_n(0)}}{1 - \sqrt{-K \overline{N}_n(0)}} \right)^{-1} \right] \tag{32}
$$

which diverges as $-K \overline{N}_n(0) \to 1$ (see Figure 17). In the limit $K \overline{N}_n(0) \to 0$, the rods do order along the field direction but remain of the same size.

The divergence for $-K \overline{N}_n(0) \to 1$ is caused by a positive feedback mechanism in which the mean length of the rods increases due to the ordering, which in turn leads to a stronger coupling to the external field and hence to an increased ordering, and so on (see Figure 18). One way to deal

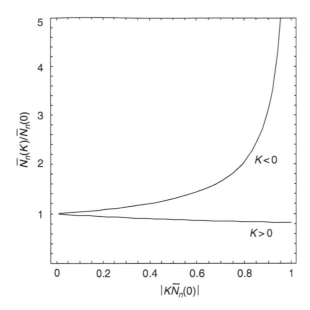

Figure 17 Growth of rodlike equilibrium polymers in response to a quadrupole external field. Plotted is the number-average aggregation number $\overline{N}_n(K)$ as a function of the dimensionless field strength $|K\overline{N}_n(0)|$, relative to number-average aggregation number $\overline{N}_n(0)$ in the absence of the field. $K < 0$ represents an ordering field and $K > 0$ a disordering field.

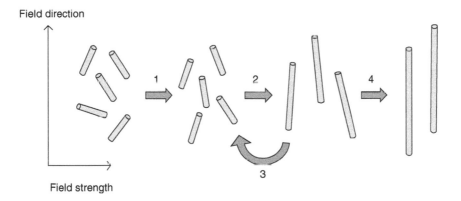

Figure 18 Response of self-assembled rods to an orienting field. (1) At low field strengths the self-assembled rods align along the direction of the ordering field. (2) At higher strengths the rods grow, (3) in turn leading to an enhanced ordering. (4) The positive feedback leads at sufficiently high fields to a runaway instability producing perfectly ordered, infinitely long rods. (Schematic adapted from M.S. Turner and M.E. Cates. *J. Phys. Condens. Matter* 4: 3719–3741, 1992. With permission.)

with the divergence is to account for any tension that may build up in the chains, such as has been done in the context of elongational flow fields, and by allowing for chain scission to occur if the local stretching energy increases beyond the link energy [84]. There is indeed evidence for flow-induced chain scission [93].

A similar runaway instability occurs for interacting self-assembled rods as a result of a *molecular* orienting field that drives the nematic phase ordering, both in a Maier–Saupe and in an Onsager type of description [55,92,94]. In nematic liquid crystals of self-assembled rods this divergence can be suppressed by incorporating in the description a finite bending rigidity that gives rise to a nontrivial correction to Eq. (30) of the form [91]

$$\frac{\beta \Omega_{\text{flex}}}{V} = -\frac{1}{3}\frac{l}{P}\sum_{N=1}^{\infty} N \int d^2\vec{u}\rho^{1/2}(N,\vec{u})\Delta_{\vec{u}}\rho^{1/2}(N,\vec{u}) \tag{33}$$

where l is again the length a monomer contributes to the total length of the assembly, and $\Delta_{\vec{u}}$ is the Laplacian on the unit sphere. Equation (33) is valid provided the assemblies are in the mean much smaller then one persistence length P, that is, provided $\overline{N}_n(K)l \ll P$.

An analysis along the lines of [91] shows that Eq. (32) remains valid for slightly flexible rodlike assemblies, provided we replace the actual field strength K by an effective one K_{eff}. The effective field strength K_{eff} is a function of the mean length of the assemblies in the absence of the field, $\overline{N}_n(0)l$, of the actual field strength K and of the persistence length P of the assemblies. To a good approximation we find

$$K = K_{\text{eff}}\left[1 + \frac{1}{4}\frac{\overline{N}_n(0)l}{P}\cdot\frac{1 - K_{\text{eff}}\overline{N}_n(0)}{1 + K_{\text{eff}}\overline{N}_n(0)}\right] \tag{34}$$

for $K < 0$, showing that in order for the mean aggregate size to diverge as $K_{\text{eff}}\overline{N}_n(0) \to -1$, the actual field strength must diverge too, that is, $-K\overline{N}_n(0) \to \infty$ (see Figure 19). Hence, the runaway instability is completely suppressed by the effects of supramolecular flexibility, however small, provided $\overline{N}_n(0)l/P > 0$. In the rigid-rod limit $\overline{N}_n(0)l/P = 0$ this ceases to be the case. This implies that the rigid-rod limit must be singular. Note that although the runaway instability is suppressed, long aggregates can still be formed at high enough field strength as is shown in Figure 20.

In conclusion, the growth and alignment coupling of rodlike EPs is attenuated by the effects of supramolecular flexibility, which, as we have seen, also strongly impacts upon the ring closure and

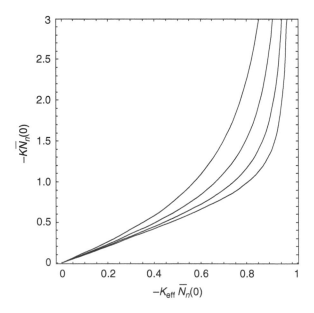

Figure 19 Relation between the actual coupling constant, K, to external quadrupole ordering fields and the effective coupling constant, K_{eff}, for semiflexible equilibrium polymers. Indicated from top to bottom are results for increasingly stiff chains with a mean length in the absence of the field of $\overline{N}_n(0)l/P = 0.8, 0.4, 0.2, 0.1$, measured in units of the persistence length. Here, $\overline{N}_n(0)$ denotes the number-average aggregation number of the assemblies in the absence of the field, l the length each monomer contributes to the total length of an aggregate and P its persistence length.

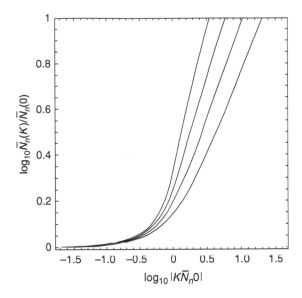

Figure 20 Growth of semiflexible equilibrium polymers in response to a quadrupole field. Plotted is the number-average aggregation number $\overline{N}_n(K)$ as a function of the dimensionless field strength $|K\overline{N}_n(0)|$, relative to number-average aggregation number $\overline{N}_n(0)$ in the absence of the field. Indicated from bottom to top are the results for increasingly stiff chains with a mean length in the absence of the field of $\overline{N}_n(0)l/P = 0.8, 0.4, 0.2, 0.1$ measured in units of the persistence length. Here, l denotes the length each monomer contributes to the total length of an aggregate and P its persistence length.

the nonideal behavior of EPs. An additional attenuation may find its source in contributions of the ordering field that cause the chains to stretch, an effect ignored here. Clearly, it would be interesting to investigate if and how external ordering fields impact upon the polymerization transitions of CAEPs and TAEPs.

X. CONCLUSIONS

In this contribution, a flexible statistical mechanical treatment of various types of supramolecular polymerization has been described based on the free energy functional Eq. (1). The treatment deals with the distribution of material over the monomeric and polymeric states at the mean-field level, but does allow for a nonmean-field analysis of the configurational statistics of the chains. It should be accurate in most practical situations away from a critical point.

There are essentially only two general types of equilibrium polymerization, that is, those that require nucleation and those that do not. Chain growth is almost explosive for the former but much more gradual for the latter. A central role is played by a link (free) energy that measures the net free energy gain of connecting two chain ends. The larger the link energy, the larger the mean degree of polymerization irrespective of the type of equilibrium polymerization.

The link energy is easier to measure and to interpret than to calculate from molecular models, because it is a function of the chemical details of the monomeric units and of the solvent and of the inherent bending stiffness of the polymeric species. Although that fairly generic features such as the influence of (screened) charges or of the bending stiffness are amenable to theoretical analysis, the link energy is inherently very sensitive to model approximations. Computer simulation may prove useful albeit that the issue of model sensitivity remains.

Important but underappreciated and very incompletely understood is the impact of the chain flexibility (or the lack thereof) on supramolecular polymerization. Indeed, chain rigidity appears to be instrumental in reducing ring-closure effects, in attenuating the influence of spontaneous fluctuations, and in suppressing the runaway growth of EPs in ordering fields, both external and molecular. Only a concerted effort of chemists and physicists working in the field of supramolecular science is likely to lead to an improvement of the current state of affairs.

ACKNOWLEDGMENTS

I am grateful to Prof. Alberto Ciferri for stimulating discussions, for a critical reading of the manuscript and for his hospitality. I am also grateful to Albert Schenning for discussions and for providing the experimental data for Figure 5 and Figure 6 prior to publication.

REFERENCES

1. M.E. Cates and S.J. Candau. *J. Phys. Condens. Matter* 2:6869–6892, 1990.
2. I.A. Nyrkova, A.N. Semenov, A. Aggelli, and N. Boden. *Eur. Phys. J. B* 17:481–497, 2000.
3. P. van der Schoot, M.A.J. Michels, L. Brunsveld, R. Sijbesma, and A. Ramzi. *Langmuir* 16:10076–10083, 2000.
4. R. Hentschke and J. Herzfeld. *Phys. Rev. A* 43:7019–7030, 1991.
5. P.G. de Gennes and P.A. Pincus. *Phys. Kondes. Materie II* 11:189–198, 1970.
6. S. Fraden, A.J. Hurd, and R.B. Meyer. *Phys. Rev. Lett.* 63:2373–2376, 1989.
7. P. van der Schoot. *J. Chem. Phys.* 117:3537–3540, 2002.
8. J.M. Tavares, J.J. Weis, and M.M. Telo da Gama. *Phys. Rev. E* 59:4288–4395, 1999.
9. S. Greer. *Annu. Rev. Phys. Chem.* 53:173–200, 2002.
10. J.S. Moore. *Curr. Opin. Colloid Interface Sci.* 4:108–116, 1999.
11. A. Ciferri. *Macromol. Rapid Commun.* 23:511–529, 2002.
12. A.V. Tobolsky and A. Eisenstein. *J. Am. Chem. Soc.* 82:289–293, 1960.
13. F. Oosawa and M. Kasai. *J. Mol. Biol.* 4:10–21, 1962.
14. P.S. Niranjan, J.G. Forbes, S.C. Greer, J. Dudowicz, K.F. Freed, and J.F. Douglas. *J. Chem. Phys.* 114:10573–10576, 2001.
15. I.A. Nyrkova, A.N. Semenov, A. Agelli, M. Bell, N. Boden, and T.C.B. McLeish. *Eur. Phys. J. E* 17:499–513, 2000.
16. R.L. Scott. *J. Phys. Chem.* 69:261–270, 1965.
17. A.V. Tobolsky and A. Eisenberg. *J. Am. Chem. Soc.* 780:780–782, 1959.
18. S. Sarkar Das, J. Zhuang, A. Ploplis Andrews, S. Greer, C.M. Guttman, and W. Blair. *J. Chem. Phys.* 111:9406–9417, 1999.
19. A.V. Tobolsky and G.D.T. Owen. *J. Polym. Sci.* 59:329–337, 1962.
20. A.Yu. Grosberg and A.R. Khokhlov. *Statistical Physics of Macromolecules*. AIP Press: New York, 1994.
21. M.P. Taylor and J. Herzfeld. *J. Phys. Condens. Matter* 5:2651–2678, 1993.
22. A. Ciferri. *Liq. Cryst.* 26:489–494, 1999.
23. R.G. Petschek, P. Pfeuty, and J.C. Wheeler. *Phys. Rev. A* 34:2391–2421, 1986.
24. M.E. Cates. *J. Phys. Fr.* 49:1593–1600, 1988.
25. G. Porte. *J. Phys. Chem.* 87:3541–3550, 1983.
26. J. Wittmer, P. van der Schoot, A. Milchev, and J.L. Barrat. *J. Chem. Phys.* 113:6992–7005, 2000.
27. P.J. Missel, N.A. Mazer, G.B. Benedek, and C.Y. Young. *J. Phys. Chem.* 84:1044–1057, 1980.
28. J.R. Henderson. *Phys. Rev. E* 55:5731–5742, 1997.
29. J. Dudowicz, K.F. Freed, and J.F. Douglas. *J. Chem. Phys.* 119:12645–12666, 2003.
30. R.B. Martin. *Chem. Rev.* 96:3043–3064, 1996.
31. H. Reiss, W.K. Kegel, and J. Groenewold. *Ber. Bunsenges Phys. Chem.* 100:279–295, 1996.
32. D.C. Morse and S.T. Milner. *Phys. Rev. E* 52:5918–5945, 1995.
33. W. Carl and Y. Rouault. *Macromol. Theory Simul.* 7:497–500, 1998.
34. P. Attard. *Mol. Phys.* 89:691–709, 1996.
35. A. Duyndam and T. Odijk. *Langmuir* 12:4718–4722, 1996.

36. J. Narayanan, W. Urbach, D. Langevin, C. Manohar, and R. Zana. *Phys. Rev. Lett.* 81:228–231, 1998.

37. A.P.H.J. Schenning, P. Jonkheijn, E. Peeters, and E.W. Meijer. *J. Am. Chem. Soc.* 123:409–416, 2001.

38. F. Würther, Z. Chen, F.J.M. Hoeben, P. Osswald, C.-C. You, P. Jonkheijm, J. van Herrikhuyzen, A.P.H.J. Schenning, P. van der Schoot, E.W. Meijer, E.H.A. Beckers, S.C.J. Meskers and R.A.J. Janssen.

39. F. Oosawa and S. Asakura. *Thermodynamics of the Polymerization of Protein.* Academic Press: London, 1975.

40. P. van der Schoot. *Macromolecules* 35:2845–2850, 2002.

41. J.C. Wheeler, S.J. Kennedy, and P. Pfeuty. *Phys. Rev. Lett.* 45:1748–1752, 1980.

42. J.C. Wheeler and P.M. Pfeuty. *Phys. Rev. Lett.* 71:1653–1656, 1993.

43. H. Jacobson and W.H. Stockmayer. *J. Chem. Phys.* 18:1600–1606, 1950.

44. T. Versmonden, J. van der Gucht, P. de Waard, A.T.M. Marcelis, N.A.M. Besseling, E.J.R. Sudhölter, G.J. Fleer, and M.A. Cohen Stuart. *Macromolecules* 36:7035–7044, 2003.

45. P. van der Schoot, S. Schmatloch, A.M. van den Bergh, and U.S. Schubert. *Polymer Preprints* 45:466–467, 2004.

46. R. Cordery. *Phys. Rev. Lett.* 47:457–459, 1981.

47. G. Ercolani, L. Mandolini, P. Mencarelli, and S. Roelens. *J. Am. Chem. Soc.* 115:3901–3908, 1993.

48. B.J.B. Folmer, R.P. Sijbesma, and E.W. Meijer. *J. Am. Chem. Soc.* 123:2093–2094, 2001.

49. S.H.M. Söntjens, R.P. Sijbesma, M.H.P. van Genderen, and E.W. Meijer. *Macromolecules* 34:3815–3818, 2001.

50. E. Buhler, J.P. Munch, and S.J. Candau. *J. Phys. II Fr.* 5:765–787, 1991.

51. P. van der Schoot. *Europhys. Lett.* 39:25–30, 1997.

52. A. Milchev, J.P. Wittmer, P. van der Schoot, and D. Landau. *Europhys. Lett.* 54:58–69, 2001.

53. J. Appell and G. Porte. *Europhys. Lett.* 12:186–190, 1990.

54. H. von Berlepsch, H. Dautzenberg, G. Rother, and J. Jäger. *Langmuir* 12:3613–3625, 1996.

55. P. van der Schoot and M.E. Cates. *Langmuir* 10:670–679, 1994.

56. Z.G. Wang, M.E. Costas, and W.M. Gelbart. *J. Phys. Chem.* 97:1237–1242, 1993.

57. H. Yamakawa. *Helical Wormlike Chains in Polymer Solutions.* Springer: Berlin, 1997.

58. T. Odijk. *Macromolecules* 26:6897–6902, 1993.

59. M. In, O. Aguerre-Chariol, and R. Zana. *J. Phys. Chem. B* 103:7747–7750, 1999.

60. T. Odijk. *J. Phys. Chem.* 93:3888–3889, 1989.

61. F.C. MacKintosch, S.A. Safran, and P.A. Pincus. *Europhys. Lett.* 12:697–702, 1990.

62. J.C. Eriksson and S. Ljunggren. *Langmuir* 6:895–904, 1990.

63. T. Odijk. *Biophys. Chem.* 41:23–29, 1991.

64. A. Heindl and H.H. Kohler. *Langmuir* 12:2464–2477, 1996.

65. P. van der Schoot. *Langmuir* 13:4926–4928, 1997.

66. S. May, Y. Bohbot, and A. Ben-Shaul. *J. Phys. Chem. B* 101:8648–8657, 1997.

67. S.A. Safran, L.A. Turkevich, and P. Pincus. *J. Phys. Fr.* 45:L69–L74, 1984.

68. A.L. Frischknecht and S.T. Milner. *J. Chem. Phys.* 114:1032–1050, 2001.

69. T. Odijk. *Curr. Opin. Colloid Interface Sci.* 1:337–340, 1996.

70. J.H.K.K. Hirschberg, L. Brunsveld, A. Ramzi, J.A.J.M. Vekemans, R.P. Sijbesma, and E.W. Meijer. *Nature* 407:167–170, 2000.

71. L. Brunsveld, H. Zhang, M. Glasbeek, J.A.J.M. Vekemans, and E.W. Meijer. *J. Am. Chem. Soc.* 122:6175–6182, 2000.

72. B.H. Zimm and J.K. Bragg. *J. Chem. Phys.* 31:476–535, 1959.

73. J. van Gestel, P. van der Schoot, and M.A.J. Michels. *J. Phys. Chem. B* 105:10691–10699, 2001.

74. J. van Gestel, P. van der Schoot, and M.A.J. Michels. *Langmuir* 19:1375–1382, 2003.

75. J. van Gestel. *Theory of Helical Supramolecular Polymers.* PhD Thesis, Eindhoven University of Technology, 2003.

76. D. Blankenstein, G.M. Thurston, and G.B. Benedek. *Phys. Rev. Lett.* 54:955–958, 1985.

77. W.M. Gelbart, A. Ben-Shaul, W.E. McMullen, and A. Masters. *J. Phys. Chem.* 88:861–866, 1984.

78. P.G. de Gennes. *Scaling Concepts in Polymer Physics.* Cornell University Press: Ithaca, 1979.

79. L. Schäfer. *Phys. Rev. B* 46:6061–6070, 1992.

80. J. Wittmer, A. Milchev, and M.E. Cates. *J. Chem. Phys.* 109:834–845, 1998.

81. P. Schurtenberger, C. Cavaco, F. Triberg, and O. Regev. *Langmuir* 12:2894–2899, 1996.

82. T.J. Drye and M.E. Cates. *J. Chem. Phys.* 98:9790–9797, 1993.

83. P. van der Schoot and M.E. Cates. *J. Chem. Phys.* 101:5040–5446, 1994.

84. W.M. Gelbart, A. Ben-Shaul, S. Wang, and R. Bruinsma. *Surfactants in Solution* 11:113–126, Eds. K.L. Mittal and D.O. Shah, plenum press: 1991.

85. R. Bruinsma, W.M. Gelbart, and A. Ben-Shaul. *J. Chem. Phys.* 96:7710–7727, 1992.

86. M.S. Turner and M.E. Cates. *J. Phys. Condens. Matter* 4:3719–3741, 1992.

87. A. Duyndam and T. Odijk. *J. Chem. Phys.* 100:4569–4574, 1994.

88. V. Baulin. *J. Chem. Phys.* 119:2874–2885, 2003.

89. V. Schmitt, F. Lequeux, and C.M. Marques. *J. Phys. II Fr.* 3:891–902, 1993.

90. J. van der Gucht, N.A.M. Besseling, and G.J. Fleer. *J. Chem. Phys.* 119:8175–8188, 2003.

91. P. van der Schoot. *J. Phys. II Fr.* 5:243–248, 1995.

92. T. Odijk. *J. Phys. Fr.* 48:125–129, 1987.

93. C. Chen and G.G. Warr. *Langmuir* 13:1374–1376, 1997.

94. E.M. Kramer and J. Herzfeld. *Phys. Rev. E* 58:5934–5947, 1998.

Chapter 4

Supramolecular Liquid Crystals: Simulation

Reinhard Hentschke, Peter Lenz, and Bernd Fodi

CONTENTS

I. INTRODUCTION

Generating specific functionality of supramolecular assemblies based on tailor-made molecular building blocks has been quite successful in the past [1,2] and certainly will continue to be of increasing importance in the future. Thus, for the underlying design, concepts are largely based on intuition and on simple theoretical concepts [3–5]. More detailed theories of supramolecular assembly face a principle challenge: predicting the structural and dynamic properties of complex macroscopic or mesoscopic systems based on the microscopic interactions of their constituent molecular units. Most likely, predictions of this type will not come from analytic theory but in the future will become possible using computer modeling.

The purpose of this chapter is twofold. First, we want to introduce the reader to the basic principles of the most frequently used simulation techniques to study reversible molecular assembly and (liquid-crystalline) structuring. Second, we survey the recent literature and likewise briefly discuss typical applications. This should provide the reader with a feeling for the present possibilities as well as limitations of computer simulations.

In Section II of this chapter we present an overview of the common computer modeling methods. Section II.A familiarizes the reader with the principles of two of the most important approaches, molecular dynamics (MD) and Monte Carlo (MC). The various models describing the system of interest are presented in Sections II.B to II.D. We discuss different approximations ranging from atomic scale modeling of single aggregates (or fractions of single aggregates) to rather abstract lattice models for self-assembling polymers. In Section III we go over some of the thermodynamic concepts underlying the description of reversible assembly at low concentrations (Section III.A), and we discuss simple theories for structural ordering (Section III.B). Section IV combines the two previous sections. It contains a discussion of typical computer simulation applications to structuring in reversibly assembling systems. We also illustrate the current possibilities and limitations of molecular modeling. The focus, hereby, is on force field approaches in Section IV.A, allowing a detailed description of molecular interaction, and on lattice models in Section IV.B, allowing to extend the limited time and spatial scales accessible by the former.

II. COMPUTER MODELING OF REVERSIBLY ASSEMBLING BULK SYSTEMS

A. Simulation Methods

Computer studies on molecular systems most commonly are based on either of the two methods, MD or MC. MD stands for the numerical solution of Newton's equations of motion,

$$m_i \frac{d^2 r_i}{dt^2} = F_i \tag{1}$$

where r_i is the position of either a real or an effective atom i with mass m_i subject to a net force F_i exerted by all other atoms or effective atoms in the system. Subtracting the Taylor series expansions of $r_i(t \pm \Delta t)$ yields a numerical solution:

$$r_i(t + \Delta t) = 2r_i(t) - r_i(t - \Delta t) + \frac{1}{m_i} F_i \Delta t^2 + O(\Delta t^4) \tag{2}$$

where Δt is the integration time step, which must be significantly smaller than v_{max}^{-1}, the inverse of the highest frequency in the system. Thermal averages obtained on the basis of Eqs. (1) and (2) for a fixed number of particles, N, confined to a constant volume V are microcanonical or NVE-ensemble averages. The volume is kept constant via the minimum image convention according to which all interparticle distances r_{ij} are calculated via $r_{\alpha,ij}^{min} = |r_{\alpha,ij} - L_\alpha \cdot \text{ANINT}(r_{\alpha,ij}/L_\alpha)|$. Here $r_{\alpha,ij}$ is the α-component of r_{ij}, and ANINT is a FORTRAN function rounding the argument to the nearest integer. Using the minimum image convention the original N-particle system is replaced by an N-particle system in a box of volume $V = L_x L_y L_z$ embedded in an infinite 3D lattice of its own images. The same minimum image convention is applied in MC simulations as well. In addition it can be adopted to other rectangular geometries.

Usually it is desirable to control the temperature, T, in a system, rather than its total energy, E. For this purpose Eq. (1) may be modified to

$$\ddot{r}_i = \frac{1}{m_i} F_i - \zeta \dot{r}_i \qquad \text{with} \quad \zeta = \frac{1 - \lambda}{\Delta t} \tag{3}$$

where the term $-\zeta\dot{r}_i$ corresponds to an effective friction force. The form of this equation becomes more clear if we realize that the $\dot{r}_i \to \lambda\dot{r}_i$ formally corresponds to an additional acceleration, $\ddot{r}_i \to \ddot{r}_i + [(\lambda - 1)/\Delta t]\dot{r}_i$. According to the two frequently used thermostats we may write

$$\zeta \approx \frac{1}{2\tau_T}\left(1 - \frac{T_B}{T^{(\text{inst})}}\right) \tag{4}$$

(Berendsen thermostat) or alternatively

$$\dot{\zeta} \approx \frac{3Nk_B T^{(\text{inst})}}{Q_T}\left(1 - \frac{T_B}{T^{(\text{inst})}}\right) \tag{5}$$

(Nosé–Hoover thermostat). Here $T^{(\text{inst})}$ is the instantaneous temperature, and T_B is the temperature of the bath. The quantities τ_T and Q_T are parameters controlling the time dependence of the temperature relaxation. In order to integrate Eqs. (3) through (5) one often employs the "leap-frog" version of Eq. (2):

$$\dot{r}_i\left(t + \frac{\Delta t}{2}\right) = \dot{r}_i\left(t - \frac{\Delta t}{2}\right)(1 - \zeta_i \Delta t) + \frac{F_i}{m_i}\Delta t + O(\Delta t^2) \tag{6}$$

$$r_i(t + \Delta t) = r_i(t) + \dot{r}_i\left(t + \frac{\Delta t}{2}\right)\Delta t + O(\Delta t^3) \tag{7}$$

For brevity we have included terms of the leading order in Δt only.

Equation (4) generates no particular ensemble except in the limit of large τ_T, where the microcanonical ensemble is approached. Equation (5) on the other hand can be shown to generate canonical averages. For sufficiently large N this is of little importance unless fluctuation correlation functions involving δE are to be calculated. Equation (4) has the advantage of excellent numerical stability. Similarly, it is possible to control the pressure in the system, that is, it is possible to carry out NPT-simulations (this does not necessarily mean that one computes NPT-ensemble averages, however!).

A popular alternative to simple MD is Metropolis MC. Here a new system configuration k' characterized by a potential energy $U_{k'}$ is created at random from the previous configuration k characterized by a potential energy U_k. The potential energy difference $\Delta U_{k'k} = U_{k'} - U_k$ is calculated, and the new configuration is accepted if

$$\min\{1, \exp[-\beta U_{k'k}]\} \geq \xi \tag{8}$$

where ξ is a random number between 0 and 1, and $\beta^{-1} = k_B T$. If the new configuration is rejected, then the old configuration is accepted. Acceptance means that this configuration is added to a list of configurations used to compute the desired thermal averages. The Metropolis algorithm, if applied to a system with constant number of particles, N, in a fixed volume, V, will yield canonical or NVT-ensemble averages. An additional requirement, however, is that the "moves" creating new configurations from previous ones do not violate detailed balance. The criterion (8) is straightforwardly extended to other common ensembles. Notice that $\exp[-\beta U_{k'k}]$ is the ratio of the canonic probability densities $p_{k'} = c\exp[-\beta U_{k'}]$ and $p_k = c\exp[-\beta U_k]$, where c is a constant. In the grand canonical case of a one-component system consisting of N single-interaction-site particles, for instance, the corresponding densities are computed via $p_{\mu VT}(\{r\}, N) \propto \exp[-\beta U(\{r\}) + N \ln aV - \ln N!]$, where $a = \Lambda_T^{-3}\exp[\beta\mu]$. μ is the chemical potential of the particles, and Λ_T is their thermal wavelength.

There are advantages as well as disadvantages associated with MD and MC. MC in its above form does not yield quantities which require to correlate information at different times. However,

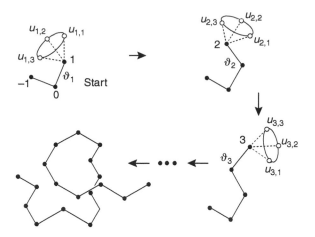

Figure 1   Chain construction as part of the Rosenbluth method.

the lack of an explicit timescale offers the advantage that configuration space may be sampled far more efficiently by introducing unphysical moves.

An example is the frequently used Rosenbluth construction of polymer conformations. Figure 1 illustrates the principle for a polymer chain consisting of "beads," whose configurations are determined via torsion angles $\vartheta_1, \vartheta_2, \ldots, \vartheta_{n-3}$. The latter may assume a finite number (k) of values. Here we use $k = 3$ for simplicity. The potential energy of the polymer chain therefore is $U(\{r\}) = U(\{\vartheta\}, r_{-1}, r_0, r_1)$, where r_i are the positions of the first three beads.

The Rosenbluth algorithm for "growing" a polymer chain is:

1. Chose r_{-1}, r_0, r_1 with probability

$$p_1 = \frac{1}{\omega_1} e^{-\beta u_1} \tag{9}$$

   where $\omega_1 \equiv k \exp[-\beta u_1]$. Here u_1 is the potential energy of the first three "beads."

2. Select the position of the next bead according to the possible k torsion angles with probability

$$p_i = \frac{1}{\omega_i} e^{-\beta u_i} \quad \text{where} \quad \omega_i = \sum_{j=1}^{k} e^{-\beta u_{i,j}} \tag{10}$$

3. Continue at 2.

Here $u_{i,j}$ denotes the potential energy of bead i at the rotational position j, and u_i denotes the potential energy of the actually chosen position (e.g., $u_i = u_{i,2}$ if $j = 2$ is the actually chosen k-value). Note that both u_i and $u_{i,j}$ include the interaction of bead i with the beads added in all previous steps as well as all other interaction sites in the system. It is important to note that this algorithm does not generate conformations according to a Boltzmann distribution of the total potential energy. However, the proper canonical averages of a quantity A may be computed according to

$$\langle A \rangle = \frac{\langle A(\{r\}) W(\{r\}) \rangle_R}{\langle W(\{r\}) \rangle_R} \tag{11}$$

where the subscript R indicates averages based on the above Rosenbluth configurations, and

$$W(\{r\}) = \prod_{i=1}^{n} \frac{\omega_i}{k} \tag{12}$$

is the so-called Rosenbluth factor. Moreover, the chemical potential of the polymer chain may be obtained via

$$\beta\mu = \beta\mu_{ideal} - \ln\langle W \rangle \tag{13}$$

In practice this only works for short chains and additional "tricks" must be used in order to grow longer chains.

Another advantage of MC is that it may be applied to systems for which $F_i = -\nabla U$ cannot be calculated or is not meaningful. This is of particular importance for modeling self-assembling systems on a lattice. Attaining chemical equilibrium everywhere in the system requires frequent exchange of monomers between aggregates, which is possible for extremely simplified lattice models only.

According to the level of detail, we may distinguish three basic types of simulation used to investigate self-assembling systems. Computationally the most expensive are simulations based on empirical potential energy expressions including chemical detail on the atomic level. Both MD and MC may be used in this case to study a single small aggregate or micelle interacting with explicit solvent. However, MD is preferred because of the additional dynamic information. On the next level, molecular detail is abandoned and large groups of atoms are lumped together into one effective interaction site. Neighboring sites usually are connected with harmonic springs. Nonbonding interactions are often simplified to Lennard–Jones like interactions possibly including the leading Coulomb-contribution (multipole expansion). Again both MD as well as MC may be used to study somewhat larger systems, for example, a few hundred monomers assembling into a labile supramolecular structure. However, MD simulations most likely are difficult to equilibrate on this stage. To study the behavior of bulk systems one has to resort to lattice models, and MC becomes the method of choice. In the following sections we briefly discuss the three types of simulation.

Above we have listed only very basic MD and MC algorithms, and the interested reader is referred to the literature for more information. Introductory texts on computer simulation methods are abundant. Somewhat older but still one of the best is Ref. 6. Also highly recommendable is Ref. 7. A special text on the details of MC is Ref. 8. Simulation techniques especially for polymers are discussed in Ref. 9.

B. Chemically Realistic Systems

In order to construct a numerically useful MD force field or a corresponding potential energy for a molecular system, one has to map the quantum mechanical interactions onto a reasonably simple classical expression. Here, the Hellmann–Feynman electrostatic theorem is useful [10]:

$$F_l = -\nabla_l \left[\sum_{j(\neq i)} \frac{q_i q_j}{r_{ij}} + \int \rho_E(\tau;l)\frac{q_i}{|\tau - r_i|}d^3\tau \right] \tag{14}$$

The quantity q_i is the charge on nucleus i, and $\rho_E(\tau;l)$ is the quantum mechanical electron probability density corresponding to the electronic state l. Note that the right side of Eq. (14) is the force on nucleus i, as it would be calculated from classical electrostatics if $\rho_E(\tau;l)$ was known. Thus, this equation justifies the description of (short-ranged) intra-molecular interactions via empirical force fields. We remark that Eq. (14), at least in principle, allows quantum MD calculations. For fixed positions of the nuclei, $\rho_E(\tau;l)$ may be computed solving Schrödinger's equation numerically. Subsequently, the nuclei are displaced according to Eqs. (1) and (14). However, this procedure is prohibitively slow, and in practice other methods are used.

Empirical potential energy expressions commonly have the form

$$U = U_{valence} + U_{nonbonding} \tag{15}$$

where

$$U_{\text{valence}} = \sum_{\langle ij \rangle} u_{ij}^{\text{bond}} + \sum_{\langle ijk \rangle} u_{ijk}^{\text{valence angle}} + \sum_{\langle ijkl \rangle} u_{ijkl}^{\text{torsion}}$$
$$+ \sum_{\substack{\langle ij \rangle \\ \langle ijk \rangle}} u_{ij,ijk}^{\text{bond} \times \text{valence angle}} + \cdots$$

is the contribution motivated via the Hellmann–Feynman electrostatic theorem. The individual terms contain the potential energy contributions due to small atom groups (here indicated via $\langle \cdots \rangle$) consisting of two to four atoms in most cases as well as couplings between these terms. U_{valence} describes the potential energy change due to small distortions with respect to the local molecular equilibrium geometry (e.g., bond length or valence angle distortions usually are modeled via simple harmonic oscillator potentials). The second term in Eq. (15) is

$$U_{\text{nonbonding}} = \sum_{i<j} \left(u_{ij}^{\text{overlap}} + u_{ij}^{\text{dispersion}} + u_{ij}^{\text{Coulomb}} \right)$$

In this case the first two sums contain contributions describing pairwise additive atom–atom interactions due to repulsion at short distances and dispersion attraction at large distances. The latter follows from a perturbative treatment of the fluctuations in $\rho_E(\tau; l)$. The Lennard–Jones potential is an often used approximation for these contributions. The last term in the above equation describes monopole–monopole interactions between partial charges due to the different electron affinities of the elements. Note that the assumption of purely pairwise additive interactions neglect dynamic polarization of the electronic charge distribution. The latter can be included on a phenomenological level, however, if desired.

Force fields contain numerous parameters. Parameterization procedures employ training sets, for example, amino acids in the case of proteins or relevant monomer–analogous molecules in the case of technical polymers. The parameters are adjusted using thermodynamic, spectroscopic, or structural data, and increasingly quantum chemical calculations (most importantly in the case of the torsion potential in the valence part of a force field) are available for the training set. Probably the most difficult parameters are the partial charges, q_i, usually located on the nuclei. This is because they are affected by a comparatively large number of atoms in their vicinity. One may start with the vacuum values for the q_i determined for the training molecules via an ESP procedure (ElectroStatic Potential fitting) based on fast semiempirical methods like AM1 (Austin Model 1). Polarization effects due to the solvent are included roughly in a mean field sense via a scaling factor multiplying the charges. The latter may be determined by comparison with suitable experimental data (e.g., the density of a solution consisting of the solvent of interest and a monomer–analogous solute as a function of temperature and solute concentration). The same scaling factor is subsequently applied to the charges on the oligomer, which can also be determined in vacuum using AM1 (possibly including conformational averaging, which, if it is done correctly, of course requires information not available at this point). A more systematic, and perhaps in the long run a more promising approach, is the fluctuating charge method [11] (and references therein). Here the q_i are dynamic variables, akin to the positions, r_i.

The above potential energy expressions are currently used to describe systems containing up to a few thousand atoms. One should realize that a single-point calculation of U is an $O(N)$ operation if only short-range interactions (e.g., LJ) are included. Inclusion of long-range interactions increases the computational effort to $O(N \ln N)$. Here N is the number of atoms or interaction sites. MD allows to describe the dynamics of such a system for roughly 10 nsec, where the integration time step is around 1 fsec.

The art of designing phenomenological molecular potential functions is described in Ref. 12, which we recommend despite its age. A general introductory text on molecular modeling is described in Ref. 13 (see also Refs. 14 and 15).

C. Coarse-Grained Systems

To decrease the spatial and temporal limitation imposed by detailed phenomenological potential energy expressions one may attempt to "coarse grain" the system. A basic simplification is the united atom approach. This means that CH-, CH_2-, and CH_3-groups are replaced by a single united atom. Clearly, as long as crystalline packing is not of interest, this approximation will be a good one. So why not create united atoms formed by larger groups of atoms? Obviously, this is a more difficult approximation likely to affect the entropic as well as the enthalpic part of the free energy in ways, which alter the system behavior of interest. Thus far, there is no systematic procedure to follow, and the approximations used in this field are rather ad hoc.

Instead of coarse graining the system uniformly, it sometimes is appropriate to consider a high level of detail in one part and a low level of detail in another part. For instance, the solvent may be taken into account via a potential of mean force U_{eff}. A commonly used MD method along this line is Langevin dynamics. Langevin dynamics simulates molecules in contact with a heat bath or solvent without considering the explicit structure of the solvent, that is, the expensive integration of the solvent trajectories is omitted — as well as solvent specific effects, of course. Here the equations of motion are given by

$$\ddot{r}_i = m_i^{-1} \left(-\nabla_i U_{eff} + Z_i \right) - \eta_i \dot{r}_i \tag{16}$$

U_{eff} is a potential of mean force of the solvent molecules including the explicit intra- and inter-molecular interactions of the solute. Z_i is a stochastic force simulating collisions of solute site i with the solvent molecules. It is assumed that Z_i is uncorrelated with the positions and the velocity of the sites i. Moreover Z_i has a Gaussian distribution centered on zero with variance $\langle Z_i(t) \cdot Z_j(t') \rangle = 6m_i\zeta_i k_B T_B \delta(t - t')\delta_{ij}$, where T_B is the temperature of the solvent, and ζ is a friction parameter. In practice the components α of Z_i are extracted from a Gaussian distribution with $\langle Z_{i,\alpha}(t) \cdot Z_{j,\alpha}(t') \rangle = 2m_i\zeta_i k_B T_B/\Delta t$, where Δt is the integration time step. We may solve Eq. (16) numerically by replacing F_i in Eq. (6) by $-\nabla_i U_{eff} + Z_i(t)$. It is important to note that Langevin dynamics in its above form is useful for sampling the conformation of large molecules, but it can be shown not to yield the correct dynamics. A detailed analysis of different algorithms in the context of Langevin dynamics may be found in Ref. 16 (see also Ref. 17).

Thus far, there is no textbook dealing exclusively with coarse-graining techniques. A good starting point, however, is the chapter on bridging length and timescales in Ref. 9. Another noteworthy reference is 18, which reviews simulation of self-assembling systems with an emphasis on linking molecular dynamics to mesoscopic and macroscopic length and timescales.

D. Lattice Simulations

Instead of just coarse graining the molecules of interest one may chose to coarse grain space itself, for example, the physical system may be mapped onto a lattice. The lattice-gas model probably is the most well-known representative of this approach. Reference 19 is a recent example where a lattice gas with two- and three-body couplings is used as a model for amphiphilic aggregation. Here, occupied lattice sites are amphiphile molecules whereas empty sites represent water molecules. Metropolis MC simulations are carried out based on moves which attempt to exchange a randomly chosen amphiphile with one nearest-neighbor water. In this fashion the authors determine the critical micelle concentration (cf. later), the micellation temperature and other quantities. An advantage of highly simplified lattice models like this one is, that many studies (e.g., in the context of critical

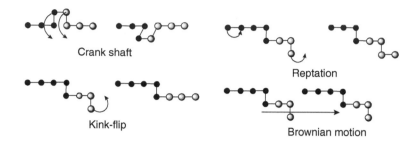

Figure 2   Schematic depiction of MC moves suitable for oligomers on a 3D cubic lattice.

phenomena), using also other techniques, have compiled much information which can be used to interpret the simulation results.

If the real experimental monomers possess significant conformational freedom, like in the case of a linear oligomer, they usually are modeled as a path connecting the nodes of a regular lattice. The total energy of such a system may be represented by

$$\beta E = \sum_{\langle ss' \rangle} n_{ss'} \epsilon_{ss'} \tag{17}$$

where $n_{ss'}$ is the number of ss' contacts, and $\epsilon_{ss'}$ is the contact energy in units of $k_B T$. The sum is the over all distinct contacts. Usually only nearest-neighbor bonds form a contact. Figure 2 depicts a set of possible MC moves which may be used for oligomers [20]. The efficiency of the algorithm can be controlled through the relative weight of these moves during the simulation. However, care should be taken that a particular algorithm fulfills the detailed balance condition.

The above MC moves allow the formation of micellar structures from stable monomers. In the case of simpler reversible aggregating systems like linear living or equilibrium polymers[1] one may integrate scission–fusion moves directly into the algorithm. We cannot discuss the different algorithms at this point and refer the reader to the literature (e.g., Ref. 21 is a good starting point).

III.   REVERSIBLY ASSEMBLING MOLECULAR SYSTEMS

A.   Low Concentration Limit — Basic Concepts [22–25]

Here we want to familiarize the reader with some of the basic theoretical concepts used to study reversibly assembling systems at low concentrations. This is important for a better understanding of the computer simulation results discussed in the next section.

Consider a solution containing different types, t, of monomer molecules M_t ($t = A, B, \ldots$). The monomer chemical potential is given by

$$\mu_t = \bar{\mu}_t + k_B T \ln[x_t \gamma_t] \tag{18}$$

where x_t denotes the mole fraction t-monomer, and γ_t denotes the respective activity coefficient. The quantity $\bar{\mu}_t$ "lumps together" the remaining dependencies of the chemical potential.

Starting point for a quantitative description of the distribution of material in the system may be the assumption of independent reactions, $s_A M_A + s_B M_B + \cdots \rightleftharpoons A_{s_A, s_B, \ldots}$, for every type of aggregate (closed association model [23]) or equivalently the stepwise association,

---

[1]If monomers can be added at the ends of polymeric units only one refers to the polymers as living polymers; if polymeric chains can undergo scission and fusion anywhere along their contour they are called equilibrium polymers.

$M_t + A_{s_A,s_B, \dots, s_t - 1, \dots} \rightleftharpoons A_{s_A,s_B, \dots, s_t, \dots}$ [25]. The independent reaction model is equivalent to $s_A \mu_A + s_B \mu_B + \cdots = \mu_{s_A,s_B,\dots}$, where

$$\mu_{s_A,s_B,\dots} = \bar{\mu}_{s_A,s_B,\dots} + k_B T \ln \left[\frac{1}{s_\zeta} x_{\zeta;s_A,s_B,\dots} \gamma_{s_A,s_B,\dots} \right] \qquad (19)$$

is the chemical potential of an s_A, s_B, ... aggregate in solution, and $x_{\zeta;s_A,s_B,\dots} = s_\zeta x_{s_A,s_B,\dots}$ is the mole fraction ζ-monomer contained in the aggregate. Note that the activity coefficient is again of the above form. We thus obtain

$$x_{\zeta;s_A,s_B,\dots} \gamma_{s_A,\dots} = s_\zeta \prod_t (x_t \gamma_t)^{s_t} \exp[s_\zeta \Phi_{\zeta;s_A,s_B,\dots}] \qquad (20)$$

with

$$\Phi_{\zeta;s_A,s_B,\dots} = \frac{1}{k_B T} \left[\sum_t \left(\frac{s_t}{s_\zeta} \bar{\mu}_t \right) - \frac{1}{s_\zeta} \bar{\mu}_{s_A,\dots} \right] \qquad (21)$$

which completely characterizes the material distribution and the structure of the system at equilibrium. Note that the $x_{t;s_A,s_B,\dots,s_t}$ are coupled both via the monomer activity coefficients and the mass conservation condition

$$x_{\text{solute}} = \sum_t x_t + \sum_t \sum_{s_A,s_B,\dots} x_{t;s_A,s_B,\dots} \qquad (22)$$

Provided that expressions for the $\bar{\mu}$ and the activity coefficients do exist, which generally is not the case due to the complicated dependence of these quantities on the thermodynamic variables and the systems composition, one may attempt the (iterative numerical) solution of the coupled set of Eqs. (20) to (22).

It is instructive to consider solutions containing one type of monomer only. At low concentrations the monomer and aggregate activity coefficients are (close to) unity, and Eqs. (20) to (22) simplify to

$$x_s = s\beta^s \qquad (\beta = xe^{\Phi_s}) \qquad (23)$$

$$\Phi_s = \frac{1}{k_B T} \left(\bar{\mu} - \frac{1}{s} \bar{\mu}_s \right) \qquad (24)$$

$$x_{\text{solute}} = x + \sum_{s=m}^{\infty} x_s \qquad (25)$$

Here x is the mole fraction of free monomers, x_s is the monomer mole fraction in s-aggregates, and m is a minimum aggregation number, which depends on the monomer "architecture." The latter governs the s-dependence of Φ_s.

It is at this point that models for $\bar{\mu}_s$ must be constructed describing the growth and the aggregate form at low concentrations. A simple example is the one-parameter expression, $\bar{\mu}_s = s\bar{\mu}_{\text{bulk}} + k_B T \delta s^{(d-1)/d}$.[2] The first term assigns an s-independent free enthalpy, $\bar{\mu}_{\text{bulk}}$, to every monomer in the "bulk" of an aggregate. The second term is a surface term expressing the surface to volume ratio in terms of the aggregation number, where $\delta > 0$[3] is a constant. Here

---

[2]Note that some references (e.g., Ref. 22) discuss $\bar{\mu}_s/s$ instead of $\bar{\mu}_s$, which accounts for the different exponent of s.
[3]Disfavoring the surface compared to the bulk.

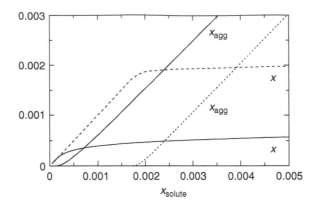

Figure 3 Mole fraction free monomer, x, and aggregated monomer mole fraction, $x_{agg} = \sum_{s=m}^{\infty} x_s$, versus solute mole fraction x_{solute}. Solid lines: $m = 5$ and $\Phi_* = 6$; dashed lines: $m = 50$ and $\Phi_* = 6$. Note that the sudden break in the increase of x, which corresponds to x_{CAC}, can be easily identified for the large m-value, whereas for $m = 5$ this is much less the case.

$d = 1$ corresponds to linear aggregates (e.g., supramolecular polymers), where the end monomers possess a different free enthalpy compared to the "bulk" monomers. Similarly, $d = 2$ corresponds to disklike aggregates, where monomers along the rim feel a different environment. Finally, $d = 3$ corresponds to spherelike aggregates. Inserting this expression for $\bar{\mu}_s$ into Eq. (24) yields $\beta = x \exp[\Phi_* - \delta s^{-1/d}] \approx x \exp[\Phi_*]$,[4] where $\Phi_* = (\bar{\mu} - \bar{\mu}_{bulk})/k_B T$. In combination with Eqs. (23) and (25) we obtain $x_{solute} \approx x + \beta^m (m - (m-1)\beta)/(1-\beta)^2$ if $\beta < 1$. Thus, if m is large (in micellar systems m ranges from 50 to 100) we have $x_{solute} \approx x$, because $\beta^m \ll 1$. However, for β very close to unity there must be an abrupt change, because otherwise the right side of the equation for x_{solute} diverges. Increasing the monomer concentration beyond this point leads to a spontaneous formation of aggregates while x remains essentially constant (cf. Figure 3). The mole fraction of free monomers at which this happens, $x_{CAC} \approx \exp[-\Phi_*]$, is called the critical aggregate (or micelle) concentration. Simple as it is, this model for reversible assembly captures experimental observations in surfactant systems like the dependence of x_{CAC} on hydrocarbon chain lengths ($\ln x_{CAC} \approx -\Phi_* \propto -n_{CH_2}$, where n_{CH_2} is the number of methylene units) or on ionic strength (e.g., figure 4.12 in Ref. 23).

What can be said about the sizes and the shapes of the aggregates above the CAC? For $d > 1$, still referring to the above model, one can show that as soon as the aggregates form they will grow without bound akin to a phase transition [22,25]. Generally, however, this is not the case, because the model does not include details of the monomer architecture which greatly affect the aggregate shape. Simple surfactants, for instance, may form spherical micelles, but they are not described by the $d = 3$ model, because their diameter is essentially fixed by the tail length.[5] Only by forming vesicles, which is not compatible with the above simple $d = 3$ model, they can expand into larger spherical aggregates. In the literature, complex models for $\bar{\mu}_s$ are discussed, which are tailored according to the specific system [22–25]. However, it is beyond the scope of this article to discuss details. Instead, we return to the above model for $d = 1$, that is, reversibly assembled linear polymers.

An s-polymer is stabilized by its $(s-1)$ monomer–monomer contacts each contributing a free enthalpy Φ_0 (in units of $k_B T$), that is, Φ_s is given by $\Phi_s = \Phi_0(s-1)/s$ (here: $\Phi_0 = \delta = (\bar{\mu} - \bar{\mu}_{bulk})/k_B T$ and $m = 2$). The particle size distribution in terms of the aggregation number is $x_s = sx^s \exp[\Phi_0(s-1)]$. A straightforward calculation yields the maximum of the distribution

---

[4]Not necessarily a good approximation, but it simplifies the discussion.
[5]The shape of surfactant aggregates largely depends on the critical packing parameter, that is, the quantity $b/(Ra_h)$, where b is the volume of the surfactant molecule, R is its tail length, and a_h is the head group area. Certain values of $b/(Ra_h)$ produce spherical aggregates ($< \frac{1}{3}$), cylinders (between $\frac{1}{3}$ and $\frac{1}{2}$), bilayers (≈ 1), continuous cubic structures (≥ 1), etc.; see, for example, figure 3.27 in Ref. 24.

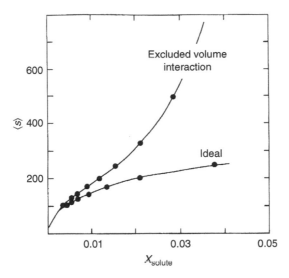

Figure 4 Mean aggregation number, $\langle s \rangle$, versus solute concentration, x_{solute}, for polydisperse rodlike aggregates. Ideal means that no intermicellar interactions are included. The excluded volume interaction is calculated based on the assumption that the aggregates are spherocylinders using the so-called "y-expansion." (Adapted from W.M. Gelbart, A. Ben-Shaul, W.E. McMullen, and A. Masters. *J. Phys. Chem.* 88:861–866, 1984. With permission.)

at $s_{\text{M}} = (\ln[\langle s \rangle + 1]/[\langle s \rangle - 1])^{-1}$, the average aggregation number, $\langle s \rangle = \sum_{s=1}^{\infty} sx_s / \sum_{s=1}^{\infty} x_s = (1 + 4x_{\text{solute}} \exp[\Phi_0])^{1/2}$ (cf. the experimental work in Ref. 26; see also figure 18 in Ref. 27), and the polydispersity, $\sigma = (\langle s^2 \rangle - \langle s \rangle^2)^{1/2} = (2x_{\text{solute}} \exp[\Phi_0])^{1/2}$.[6] Note that for large $\langle s \rangle$ we have $s_{\text{M}} \sim (x_{\text{solute}} \exp[\Phi_0])^{1/2}$ and $\sigma \sim \sqrt{2} s_{\text{M}}$. It is worth emphasizing the strong dependence of $\langle s \rangle$ on the contact free enthalpy, Φ_0. Note that in rodlike aggregates with large diameter (e.g., many of the biological rodlike aggregates) growth is significantly enhanced due to the increase of the effective monomer coordination number.

Figure 4 compares the ideal behavior, that is, $\langle s \rangle \sim x_{\text{solute}}^{1/2}$, to the result when excluded volume interactions are included. We can understand the shape of the excluded volume curve in terms of our simple model for linear aggregates. Using $\ln \gamma_s \approx (a_1 s + a_2) x_{\text{solute}}$ from Ref. 28 (second virial approximation neglecting polydispersity) and $x_s = s(x\gamma_1 \exp[\Phi_s])^s \gamma_s^{-1}$ (cf. Eq. [20]) yields $\langle s \rangle \sim (x_{\text{solute}} \exp[\Phi_0 + a_2 x_{\text{solute}}])^{1/2}$ and $\sigma \sim \langle s \rangle$ for large $\langle s \rangle$,[7] that is, we obtain the same excluded volume induced growth behavior as shown in Figure 4. Note that excluded volume affects the size through a_2, which means that the aggregate ends drive the growth.

Again we stress that our present discussion is based on assuming "reasonable" approximations for Φ_s, because it is prohibitively complicated to calculate $\bar{\mu} - \bar{\mu}_s/s$ exactly for any real molecular system. Another simple model of a reversibly assembling linear aggregate is that of monomers at positions r_i subject to the constraints $\langle l_1^2 \rangle = \langle l_{s-1}^2 \rangle = l^2$, $\langle \sum_{i=2}^{s-2} l_i^2 \rangle = (s-3)l^2$, and $\langle \sum_{i=1}^{s-2} l_i \cdot l_{i+1} \rangle = (s-2)l^2 t$, where $l_i = r_i - r_{i-1}$, and l is a constant [29,30]. The quantity t controls the stiffness, that is, $t = 0$ corresponds to wormlike flexibility, whereas $t = 1$ creates a stiff chain. Unfortunately, the nonlocal form of the constraints reduces the transparency of the t-dependence. Nevertheless, the resulting aggregate size distribution is $x_s \sim sx^s \exp[(E_{\text{sc}}/k_{\text{B}}T)(s-1)](1-t^2)^{3(s-2)/2}$ for $s \geq 2$, where E_{sc} is a bond scission energy analogous to the above quantity Φ_0. Thus, this model introduces an additional

---

[6] In the polymer field, polydispersity often is characterized via the polydispersity coefficient, $U = \langle s \rangle_{\text{w}}/\langle s \rangle_{\text{n}} - 1$, where $\langle s \rangle_{\text{w}}$ is the aggregate weight average, and $\langle s \rangle_{\text{n}}$ is the aggregate number average, that is, $U = 0$ in the monodisperse limit. Here $\langle s \rangle_{\text{w}} = \langle s \rangle$ and $\langle s \rangle_{\text{n}} = \sum_{s=1}^{\infty} x_s / \sum_{s=1}^{\infty} x_s/s$ so that $\langle s \rangle = 2\langle s \rangle_{\text{n}} - 1$ and $U = (\langle s \rangle - 1)/(\langle s \rangle + 1)$.

[7] Here we use $\sum_{s=1}^{\infty} s^{\lambda+1} \beta^s \approx \int_0^{\infty} s^{\lambda+1} \exp[s \ln \beta] \, ds = \Gamma[\lambda + 2](\ln \beta)^{\lambda+2}$ with $\beta \approx 1 - z$ and $z \ll 1$ assuming on average large aggregation numbers.

intra-aggregate interaction which results in the t-dependent factor even in the isotropic phase. For $t = 0$, that is, wormlike aggregates, this factor is unity and the above ideal linear aggregate model is obtained. For $t > 0$, that is, with increasing stiffness, a strong suppression of the larger aggregates result, that is, $\langle s \rangle \sim (1 - t^2)^{3/2}(x_{\text{solute}} \exp[E_{\text{sc}}/k_B T])^{1/2}$ and $\sigma \sim \langle s \rangle$ for large $\langle s \rangle$. It is worth noting that the nonlocal form of the above constraints is motivated by the desire to obtain a simple analytic partition function — at the expense of unphysical behavior in the limit $t \to 0$. For instance, for $s = 3$ it is easy to see that $t = 1$ implies $l_1 = l_2$, which means complete internal rigidity of the trimer. The question is, therefore, how much of the pronounced coupling between aggregate flexibility and aggregate size predicted by this model can be observed in real experimental systems.

Another model of the aggregate interior takes an opposite viewpoint. It assumes shape persistent aggregates whose (single aggregate) partition function is approximated by the product $Q_s = q_t(s)q_r(s)q_v(s)$, that is, the aggregate is a "solid" possessing translational (t), rotational (r), and vibrational (v) degrees of freedom. Each term contributes an s-dependence, and thus couples aggregate size and shape. Consider, for example, the rotational partition function $q_r(s) \propto (I_A I_B I_C)^{1/2}$, where the I are the three main moments of inertia of the aggregate [31]. For cylinders we obtain $q_r(s) \propto s^{7/2}$, that is, $I_A \sim I_B \propto ss^2$ and $I_C \propto s$, whereas for spheres we have $q_r(s) \propto s^{5/2}$. Thus $\bar{\mu}_s$ now contains a contribution $\propto -\ln q_r(s)$. In general terms $\sim -\lambda \ln s$ in $\bar{\mu}_s$ yield a mean aggregate size $\langle s \rangle \sim (x_{\text{solute}} \exp[\Phi_0])^{1/(2+\lambda)}$ (in the ideal solution) and $\sigma \approx \langle s \rangle / \sqrt{2 + \lambda}$ reducing the $\frac{1}{2}$ power in the above $\sqrt{x_{\text{solute}}}$ growth behavior.[8] We do not want to expand this point, but rather refer the reader to the more extensive discussion in Ref. 25.

The reason for this comparatively elaborate discussion of the mean aggregate size is that it has been studied not only in analytical work (e.g., in a recent mean field calculation based on an extended Flory–Huggins lattice model [32]) but also in numerous computer simulations [21,33,34]. A noteworthy aspect is the continuous growth of the average aggregation number with increasing solute concentration as predicted by all calculations and simulations (to the best of the authors' knowledge). In contrast to this a carefully studied experimental example exists, for which a pronounced maximum of the average aggregation number as function of solute concentration at constant temperature is observed (cf. figure 3.15 in Ref. 35; the micelles are discotic in this case). Both the height of the maximum as well as $\langle s \rangle$ at fixed x_{solute} depend on T. Both decrease upon heating. In Section III.B we mention a computer simulation study, where a similar behavior is found in a model for rodlike micelles.

B. Self-Assembly and Liquid-Crystalline Ordering

The same mechanisms leading to liquid-crystalline phases in inert lyotropic systems are also responsible for liquid crystallinity in reversibly assembling systems. Here, the new and interesting aspect is the coupling between the size and shape distribution of the aggregates and the formation of ordered phases. Earlier extensive studies were carried out by Gelbart and coworkers, using continuum models along the lines of the Onsager approach [36–38] (for a more complete listing see Ref. 25), and by Herzfeld and coworkers, using lattice models [39–41] (for a more complete listing see Refs. 42 and 43). A critical discussion of the various approaches can be found in Ref. 44.

In the following we briefly discuss the continuum models of the Onsager type. The development of continuum models describing structural phase behavior in the context of reversible assembly, largely parallels the Onsager approach to ordinary lyotropics and its extensions. In the simplest case we may write Eq. (20) as

$$x_s(\Omega) = \exp\left[(\Phi_s + \ln[x\gamma_1])s + \ln s - 2\rho_{\text{total}} \sum_{s'} \int_{\Omega'} \frac{d\Omega'}{4\pi} \frac{x_{s'}(\Omega')}{s'} B_{s,s'}(\Omega, \Omega') + \cdots \right] \quad (26)$$

---

[8] Again evaluating the summations via the above integral approximation.

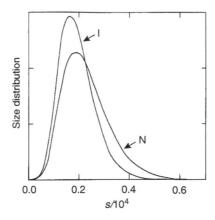

Figure 5 Aggregation number distribution for rodlike aggregates at isotropic–nematic coexistence. (Taken from W.W. McMullen, W.M. Gelbart, and A. Ben-Shaul. *J. Chem. Phys.* 82:5616–5623, 1985. With permission.)

where $x_s(\Omega)$ is the mole fraction monomer in s-aggregates with orientation Ω relative to an external coordinate system,[9] and ρ_{total} is the particle number density including solvent molecules. The quantity $x\gamma_1$ is the activity of the free monomers, and $B_{s,s'}(\Omega, \Omega')$ is the second virial coefficient for interacting s- and s'-aggregates. This extension of Eq. (23) can be solved numerically. Figure 5 shows a plot of x_s comparing the "typical" size distributions on the isotropic and the nematic side of the coexistence region for a system forming spherocylindrical micelles [36]. The calculation is based on Eq. (26) assuming, however, a detailed model for the micelle interior. Note that the general behavior, that is, the shift of s_M to higher s-values and the corresponding broadening of the distribution, is in accord with our above simple model calculations (neglecting the effects of anisotropy of course).

Calculations based on the full size distribution, $x_s(\Omega)$, are tedious. A number of authors, therefore, replace $x_s(\Omega)$ by $x_L(\Omega)$, where L is a mean aggregate length. Using this monodispersity assumption in conjunction with Onsager's trial function approach, Odijk [45] has investigated the effect of micellar flexibility on the isotropic-to-nematic phase transition in solutions of linear aggregates. Φ_L is modeled in terms of the simple $d = 1$ model discussed above. In addition Odijk replaces the $\ln s$-term in Eq. (26) by $6 \ln s$, where the additional $5 \ln s$ are due to translational ($\frac{3}{2}$) and rotational ($\frac{7}{2}$) degrees of freedom of the rodlike L-polymers (cf. the above discussion of the intra-micellar partition function Q_s). Finally, the orientational confinement of the persistent flexible polymers is included via the method of Semenov and Khokhlov [46] (and references therein; see also Ref. 47 for a more recent discussion). The main result of this investigation is the strong inhibition of growth of flexible rodlike micelles in the anisotropic phase. This calculation was extended by Hentschke [48] taking the possibility of hexagonal ordering[10] at high concentrations into account. The main result of this calculation is the possible existence of an isotropic–nematic–hexagonal columnar triple point in the P-x_{solute} plane, the reduction of the persistence length, P,[11] suppresses the nematic phase, which intervenes between the isotropic and the columnar phase, as shown in Figure 6. Note that compared with the (metastable) nematic phase under identical conditions the aggregate ends are more tightly packed in the columnar phase, whereas the average side by side separation is increased. The resulting reduction of undulatory confinement favors the columnar phase. On a qualitative level this prediction is in accord with experimental observations (cf. Ref. 50). A triple point of this type is

[9]We consider one type of monomer only. In addition, we have cylindrical aggregates in mind, and thus $(4\pi)^{-1} \int_\Omega d\Omega = 1$. In general Ω includes all three Euler angles.

[10]A discussion of micellar growth in hexagonal phases can be found in Ref. 49.

[11]P is defined via $\langle n(0) \cdot n(\tau) \rangle = \exp[-\tau/P]$, where n denotes unit vectors tangential to the contour of the aggregate separated by a distance τ along its contour.

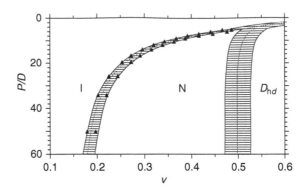

Figure 6 Theoretical phase diagram of a system of monodisperse, self-assembling linear aggregates. Ref. 51.
I: isotropic phase; N: nematic phase; D_{hd}: hexagonal columnar phase. Here P is the persistence
length, D is the aggregate diameter, and v is the solute volume fraction. Shaded areas indicate phase
coexistence. The dotted lines mark the crossover of the free energies describing separately the ordering
in the various phases. The triangles are experimental coexistence volume fractions, which were fitted
by adjusting certain model parameters. (Taken from R. Hentschke, P.J.B. Edwards, N. Boden, and
R. Bushby. *Macromol. Symp.* 81:361–367, 1994. With permission.)

found, for instance, in the phase diagram of a stacking triphenylene derivative, which was studied in
detail by Boden and coworkers [51]. One problem of course is that the temperature does not appear
explicitly in the model. It enters indirectly via the (unknown) T-dependencies of P and Φ_0.[12] The
model of Ref. 48 was refined by van der Schoot and coworkers [53,54] taking polydispersity into
account. Qualitatively, however, the results remain the same.

The functional dependence of aggregate size on alignment and flexibility, within the above simple
polymeric aggregate model, can be understood roughly as follows. Rewriting Eq. (26) yields

$$x_s = \int_\Omega \frac{d\Omega}{4\pi} f_s'(\Omega) \exp[-\ln f_s'(\Omega) + [\cdots]] \tag{27}$$

where $f_s'(\Omega)$ is the orientation distribution of the (rodlike) aggregates $((4\pi)^{-1} \int_\Omega d\Omega f_s'(\Omega) = 1)$,
and $[\cdots]$ denotes the term in corresponding brackets in Eq. (26). Using the cumulant expansion,
$\langle e^A \rangle = e^{\langle A \rangle}(1 + \cdots)$, we may write

$$x_s \approx \exp\left[-\sigma_s(f_s') + \int_\Omega \frac{d\Omega}{4\pi} f_s'(\Omega)[\cdots]\right] \tag{28}$$

To include flexibility we replace $\sigma_s(f_s')$ by

$$\sigma_s(f_s') = \begin{cases} \int_\Omega \frac{d\Omega}{4\pi} f_s'(\Omega) \ln f_s'(\Omega) + \mathcal{O}\left(\frac{L_s}{P}\right) = \ln\alpha - 1 + \cdots & \frac{L_s}{P} \ll 1 \\[2ex] \frac{L_s}{P} \frac{1}{8} \int_\Omega \frac{d\Omega}{4\pi} \frac{\partial f_s'(\Omega)}{\partial\theta} \frac{\partial \ln f_s'(\Omega)}{\partial\theta} + \mathcal{O}(1) = \frac{L_s}{4P}\alpha + \cdots & \frac{L_s}{P} \gg 1 \end{cases}$$

This expansion in terms of the aggregate contour length, L_s, divided by its persistence
length, P, of which we reproduce the leading order terms only, is due to Semenov and
Khokhlov [46] (see also Ref. 55 for an insightful discussion; a derivation of the leading terms

---

[12]For the above triphenylene derivative the T-dependence of P and Φ_0 was studied in a computer simulation [52], which
showed a strong decrease of both quantities with increasing temperature (280 K $\leq T \leq$ 300 K). For an experimental study
of $P(T)$, see Ref. 50.

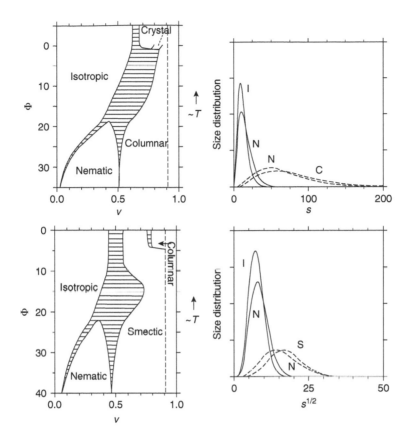

Figure 7 Left: Φ–v phase diagrams of self-assembling rodlike (top) and disklike (bottom) systems (Adopted from M.P. Taylor and J. Herzfeld. *Phy. Rev. A* 43:1892–1905. With permission). Shaded areas indicate phase coexistence. The vertical dashed lines mark the respective close packing limits. Right: Corresponding orientationally averaged aggregation number distributions for $\Phi = 25$ (top) and $\Phi = 30$ (bottom) at the indicated transition boundaries (I: isotropic; N: nematic; C: columnar; S: smectic).

on an elementary level can also be found in Ref. 56). In addition we have replaced f'_s by the trial function $\alpha \exp[-\alpha\theta^2/2]$ assuming large α (strong alignment).[13] Correspondingly $2(4\pi)^{-1}\rho_{\text{total}} \sum_{s'}(x_{s'}/s') \int_{\Omega,\Omega'} d\Omega \, d\Omega' f'_s(\Omega) f'_{s'}(\Omega') B_{s,s'}(\Omega,\Omega') \approx (b_1 s\alpha^{-1/2} + b_2)x_{\text{solute}},$[14] where b_1 and b_2 are positive constants. Analogous to the above isotropic case we thus obtain $\langle s \rangle \sim (x_{\text{solute}} \exp[\Phi_0 + b_2 x_{\text{solute}} + \ln\alpha - 1])^{1/2}$ in the limit of slightly flexible aggregates and $\langle s \rangle \sim (x_{\text{solute}} \exp[\Phi_0 + b_2 x_{\text{solute}} + \ln[\alpha/4]])^{1/2}$ in the limit of wormlike aggregates. (Note that the $\ln[\alpha/4]$ term is due to the $\mathcal{O}(1)$ term in the above expansion.) Because α increases upon increasing alignment, we find that alignment progressively increases the aggregate size, where the effect of flexibility enters indirectly through α. Note that the minimization of the free energy with respect to α, that is, $d(\sigma_s + \text{const } Lx_{\text{solute}}\alpha^{-1/2})/d\alpha = 0$, yields $\alpha \sim (Lx_{\text{solute}})^2$ for stiff aggregates, where L is the mean aggregate size, and $\alpha \sim (Px_{\text{solute}})^{2/3}$ for wormlike aggregates.

We remark that a similar isotropic–nematic–columnar triple point as mentioned above is also obtained in the Φ–v plane of the phase diagram for hard, rigid, polydisperse, self-assembling rods [57]. The same reference also studies the phase behavior of disklike aggregates. Both phase diagrams are shown in Figure 7. As indicated, Φ is expected to decrease with increasing temperature.

---

[13] We also assume that all s have the same orientation distribution. Note that with this trial function $(4\pi)^{-1} \int d\Omega f'_s(\Omega) \approx 2\alpha \int_0^\infty d\theta \, \theta \exp[-\alpha\theta^2/2]$!

[14] Here $(4\pi)^{-2} \int_{\Omega,\Omega'} d\Omega \, d\Omega' |\sin(\Omega,\Omega')| \approx (\pi/\alpha)^{1/2}$ is the leading contribution due to the angular dependence in $B_{s,s'}$ for cylinders and large α [55].

Note the qualitative agreement of the size distributions at the isotropic-to-nematic transition in the rodlike system with the result of Figure 5.

Here, we have considered effective one-component systems only. Lyotropic ternary mixtures containing self-assembling proteins forming rodlike aggregates, nonaggregating protein, and solvent have been considered by Herzfeld and coworkers [58] (and references therein). The technical approach is the same as discussed above. The additional computational effort, however, is considerable. Another example of a binary mixture containing polymer and surfactant which form complexes exhibiting lyotropic behavior is discussed in Ref. 59 in terms of scaling arguments.

IV. COMPUTER SIMULATIONS OF REVERSIBLE ASSEMBLY AND STRUCTURE FORMATION IN THE BULK

Because supramolecular structuring based on reversible assembly is rather common, it is not surprising that a literature survey reveals contributions from researchers interested in quite different systems. There are those whose primary focus is on reversibly assembling systems in biology (such as, tubulin, actin, sickle cell hemoglobin, . . .) or on technical surfactants. Then there is the polymer community, that is, those who are interested in the phase behavior of living/equilibrium polymers or block copolymer systems. And then there are those who study low-molecular weight fluids with strong anisotropic interactions (e.g., ferrofluids). Here we cannot represent all of these equally, because we are limited to a few examples only.[15]

A. MD Simulation

A typical example of this approach is Ref. 52, where the authors study a single rodlike aggregate immersed in water via straightforward MD (cf. Figure 8). The monomer is a triphenylene derivative, and the aggregates are polydisperse rods formed by stacking monomers. The authors study the aggregate-water interface, that is, the effective diameter of the rods, the persistence length of the aggregates, and the monomer–monomer contact free enthalpy, which is computed via thermodynamic integration. The combined information may be used in an analytic calculation aimed at computing the relative stability of liquid-crystalline phases (isotropic, nematic, and hexagonal columnar) formed by the aggregates (cf. p. **119**).[16]

Another example is Ref. 63 focussing on a biological model system. The authors study the stability of single n-dodecyl phosphate micelles and bilayer segments in aqueous solution depending on temperature and pH via atomistic MD. Typical number of atoms are around 2×10^4. The length of these simulations are between 3 and 8 nsec.

Even though the simulation timescale accessible with atomistic MD is short, some authors succeed in modeling micelle formation starting from an initially random distribution of monomers [64]. Provided the ability to frequently repeat this computer experiment, that is, formation of an isolated micelle (or cluster), with varying number of monomers in the simulation cell, it is possible to observe the dependence of micelle shape on monomer concentration. Usually such simulations are carried out for coarse-grained molecules [65]. But even then equilibration is difficult. The collapse of randomly distributed monomers into a condensed structure may be fast; however, the transformation of this structure into an equilibrium structure may be very slow. In addition, finite size effects may strongly influence the results, because often the characteristic dimensions of the observed structures usually are comparable to the size of the simulation volume.

In the following two examples reversible assembly of coarse-grained polymeric systems is considered. In Ref. 66 the authors use straightforward Langevin dynamics (cf. p. **113**) to study the

---

[15] A still recent review discussing computer simulation of surfactant solutions is Ref. 60. A short review on the simulation of block copolymers including micelle formation is Ref. 61. A general review article on the thermodynamics of dipolar fluids, which includes a discussion of computer simulation results is Ref. 62.

[16] The experimental phase diagram was determined by Boden and coworkers (cf. figure 3.9 in Ref. 35).

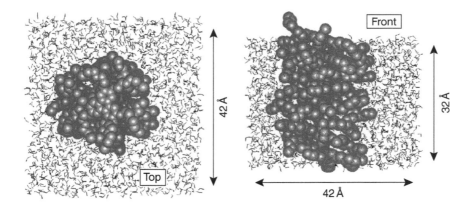

Figure 8 Snapshot of the simulation box at $T = 300$ K containing a stack of eight monomers (2,3,6,7,10,11-hexa-(1,4,7-trioxaoctyl)-triphenylene) in van der Waals representation surrounded by water molecules in stick representation. The left panel shows a view along the micelle's axis, whereas the right panel shows a side view. Note that due to the application of periodic boundary conditions we model a segment of eight monomers within a virtually infinite micelle.

micellation of different types of π-shaped copolymers. Results are obtained for the dependence of the CMC and the radial structure of the micelles on the molecular architecture of the monomers. In Ref. 67 the author performs mesoscopic simulation of polymer surfactant aggregation. Here a modified form of Langevin dynamics is used. The focus, however, is on bridging the gap between atomistic and mesoscopic simulation. The idea is to perform simulations of molecular fragments retaining atomistic detail to derive suitable Flory–Huggins χ-parameters, which are then used in a coarse-grained model. In this fashion a rough phase diagram may be obtained.

The next example, which we discuss in some detail, is a MD study of the self-assembly and liquid-crystal phase behavior of reversibly aggregating linear model polymers. In this model, a detailed description can be found in Ref. 68, the particles, that is, the monomers, interact via a Lennard–Jones-like potential with anisotropic attraction:

$$u_{ij} = 4\epsilon_{ij} \left[\left(\frac{\sigma_{ij}}{r_{ij}} \right)^{12} - \mu \left(\frac{(\mathbf{n}_i \cdot \mathbf{r}_{ij})(\mathbf{n}_j \cdot \mathbf{r}_{ij})}{r_{ij}^2} \right)^{\nu} \left(\frac{\sigma_{ij}}{r_{ij}} \right)^{6} \right] \qquad (29)$$

where ϵ_{ij} and σ_{ij} are the usual Lennard–Jones parameters. $\mathbf{n}_i$ is a unit vector assigning an orientation to monomer i. Note that the factor $(\cdots)^{\nu}$ is unity if both $\mathbf{n}_i$ and $\mathbf{n}_j$ are parallel to $\mathbf{r}_{ij}$, the vector connecting the monomers (or sites) i and j. If either $\mathbf{n}_i$ or $\mathbf{n}_j$ or both are perpendicular to $\mathbf{r}_{ij}$ then $(\cdots)^{\nu}$ is zero. If either $\mathbf{n}_i$ or $\mathbf{n}_j$ is antiparallel to $\mathbf{r}_{ij}$ then $(\cdots)^{\nu}$ depends on ν, that is, $(\cdots)^{\nu} = -1$ for ν odd and $(\cdots)^{\nu} = 1$ for ν even. Note that the magnitude of ν controls the angular width of the attraction, whereas μ controls the strength of the attraction. Both parameters thus control the stiffness, that is, the persistence length, of the polymer aggregates. Figure 9 shows simulation snapshots of the system at three different densities (here $*$ indicates Lennard–Jones units). Below a certain monomer density the system is isotropic, whereas at higher densities the aggregates are aligned, and the average aggregation number increases. Further increase of the density drives the system into a hexagonal columnar phase. Figure 10 illustrates the dependence of the mean aggregate size on monomer density and on the parameter ν. Also shown is the partial T^* versus ρ^* phase diagram of this system in Figure 11.[17]

Recently this model was extended to include explicit solvent particles interacting via a simple Lennard–Jones potential with the associating monomers. In this system, where we now employ

---

[17]Note that there is an obvious similarity of this system with dipolar liquids, which may also exhibit chain formation and ordering behavior (general review: see Ref. 62).

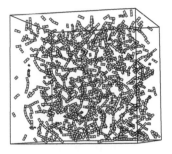
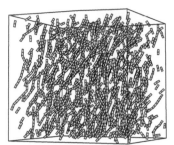
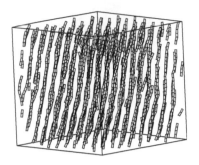

Figure 9   Snapshots of an equilibrium polymer system at different densities. Left: $\rho^* = 0.34$ (isotropic ordering). Middle: $\rho^* = 0.47$ (nematic ordering). Right: $\rho^* = 0.67$ (hexagonal ordering). The potential parameters are $\sigma = 1, \epsilon = 1, \mu = 3.3$, and $\nu = 2$.

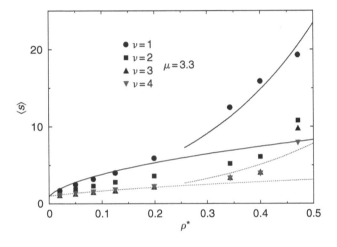

Figure 10   Mean aggregation number, $\langle s \rangle$, versus monomer density, ρ^*, for different ν. For $\nu = 1$ (solid lines) and for $\nu = 4$ (dashed lines) the $\langle s \rangle$ values are fitted via $\langle s \rangle = (1 + 4x_{\text{solute}} \exp[\delta])^{1/2} = (1 + 4K_1\rho^*)^{1/2}$ (ideal behavior), and above $\rho^* = 0.25$ via $\langle s \rangle = (1 + K_1\rho^* \exp[K_2\rho^*])^{1/2}$ (cf. Figure 4).

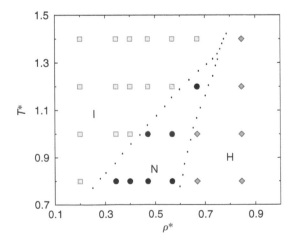

Figure 11   Phase diagram T^* versus ρ^* for the model system $\mu = 3.3$ and $\nu = 2$. Squares: Isotropic ordering (I); circles: nematic ordering (N); diamonds: hexagonal columnar ordering (H). For clarity we separate the different regions by dotted lines.

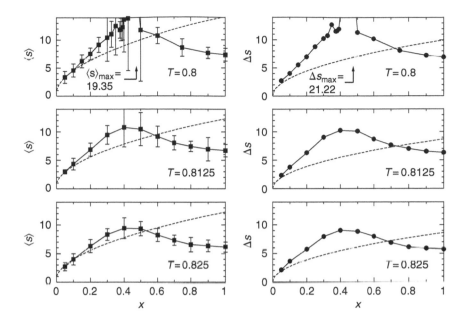

Figure 12 Left: Mean aggregate length (aggregation number), $\langle s \rangle$, versus solute mole fraction, x, at the indicated temperatures. Solid lines connect the simulation results, whereas dashed lines correspond to the simple approximation valid for low concentration as used in Figure 10. Right: Δs ($\equiv \sigma$) versus x corresponding to the panels on the left. Solid lines connect the simulation results. The dashed lines again correspond to the simple low concentration approximation explained in Section III.A.

constant pressure MD, the mean aggregate size as function of solute mole fraction may exhibit a maximum (cf. above). Figure 12 shows examples at three different temperatures. An additional mean field calculation based on an extended Flory–Huggins model, which includes three different nearest-neighbor interactions $\epsilon_{ss'}$ (s: solvent, free monomer, bound monomer) in addition to intra-aggregate monomer–monomer contacts, suggests that a favorable interaction between free monomers leads to the observed maximum. As a consequence the phase diagram in Figure 11 becomes more complex [69].

There are numerous simulation studies, both MD and MC, of what we call $H_l T_m / W_n$, systems (with $l, m = 1, 2, \ldots$; $n = 0, 1$) in the literature. Here, H stands for "head" and T for "tail"; W is a solvent (usually water). A $H_4 T_4$ monomer, for instance, consists of eight linearly connected interaction sites. All H sites are identical, and likewise all T sites are identical. Usually the H–W interactions are favorable, whereas the T–W interactions are less favorable. A $H_l T_m$ monomer may correspond to a simple technical surfactant. However, as l and m increase the monomers become simple diblock copolymers. Figure 13 shows a simulation snapshot showing the formation of a lamellar structure obtained at high solute concentration [70]. In general, it is difficult to equilibrate such systems in MD simulations, and lattice MC methods, as discussed in Section IV.B, are better suited.

B. MC Simulation

Even though MC is not limited to coarse-grained models or lattices we do not discuss atomistic MC, because the latter does not offer significant advantages over MD in the present context. In addition we confine ourselves to lattice MC results.[18]

---

[18]Note, however, two recent review papers, one on off-lattice MC methods for coarse-grained models of polymeric materials [71] and one on MC applied to block copolymer systems [61], in which off-lattice simulations pertaining to self-assembly are discussed.

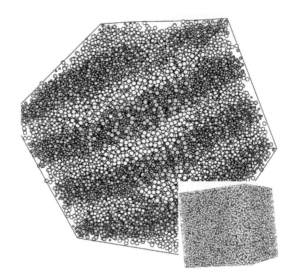

Figure 13   MD simulation of a surfactant system using a coarse-grained model. A surfactant molecule consists of four head (light gray) and four tail sites (white); a water molecule is represented by one site (dark gray). The inset on the lower right shows the random distribution of the molecules at the beginning of the simulation.

Our first example is Ref. 72. The authors compare self-assembly in off-lattice and lattice H_1T_4 and H_4T_1 systems. The off-lattice simulations are MD simulations using constraint bonds (springs) between the interaction sites (beads). The authors perform simulations with explicit solvent as well as without (using Brownian dynamics). The results are compared to MC simulations, using the configurational bias method to move the molecules, on an incompressible lattice. The lattice size is chosen to yield the same surfactant density and number of molecules as in the off-lattice simulations. From their simulations the authors conclude that compressibility and solvent, excluded volume effects, strongly influence the self-assembly behavior of the above model system — except in the weak aggregation regime. Thus, some care must be exercised interpreting the results obtained on lattices in relation to real systems.

A study of H_lT_m/W_1 systems is documented in a series of papers by Larson (see Ref. 73 and references therein). The author carries out a systematic study of the phase behavior for $4 \leq l, m \leq 16$ using a MC lattice model with each head or tail site occupying one site on a cubic lattice. Likewise the solvent (water, oil) is modeled as a single site. At high concentrations he finds a sequence of liquid-crystalline phases illustrated in Figure 14, like the hexagonal cylindrical, the bicontinuous gyroid, and the lamellar phase. The general dependence of the calculated phase diagrams on amphiphile volume fraction, temperature, and amphiphile architecture (head-to-tail ratio) are found to be in good accord with corresponding experimental surfactant systems. Of course, effects tied to the details of the molecular architecture are not captured by the lattice model.

More recently, dilute H_lT_m systems, with chain lengths ranging from 5 to 22, and with various head and tail proportions, were studied by Kenward and Whitmore [20]. The authors are interested in studying the size and shape distributions of the aggregates, and the structure of the aggregates' cores and surfaces. Note that these simulations as well as related numerical work (e.g., [74]) are discussed elsewhere in this book. There is significant interest in self-assembly and surfactant phase behavior in supercritical solvents. Reference 75 discusses a lattice MC study based on an extension of Larson's model. The authors perform canonical MC simulations at a supercritical temperature. Again the number of surfactant head and tail segments as well as the solvent (carbon dioxide) density and solute concentration is varied. Various properties such as the critical micelle concentration, the aggregate size distribution, and the size and shape of the micelles are evaluated and pseudophase

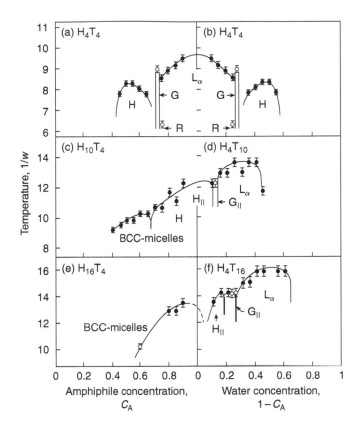

Figure 14   Phase diagrams of (a) H_4T_4, (c) $H_{10}T_4$, and (e) $H_{16}T_4$, and the complementary surfactants (b) H_4T_4, (d) H_4T_{10}, and (f) H_4T_{16} in "water." H and L_α are hexagonal and lamellar, G and R are cubic gyroid and rhombohedral-like mesh intermediate phases. BCC-micelles is a body-centered cubic packing of spherical micelles. The index$_{||}$ indicates inverse symmetry. (Taken from R.G. Larson. Monte Carlo simulation of the phase behaviour of surfactant solutions. *J. Phys. II France* 6:1441, 1996. With permission.)

diagrams are constructed. Obvious extensions of the H_lT_m systems are triblock model surfactants or triblock copolymers. The phase behavior and micellization of several model lattice diblock and triblock surfactants (HTH as well as THT) have been investigated by grand canonical MC in Refs. 76 and 77. Finally, we mention an example of lattice MC, where the authors study phase separation and liquid-crystal self-assembly in ternary surfactant-inorganic-solvent systems [78]. Monomer chain lengths in all examples considered, thus far, are rather short. Longer chains ($N = 500$) are used in Ref. 79 to simulate the thermodynamics of reversibly associating polymer solutions on a cubic lattice. The polymers associate via "stickers" evenly distributed along their contour length, and the solvent is modeled as empty lattice sites. For this system the authors investigate the sol–gel transition as function of temperature.

V.   Conclusion

Our intention was to illustrate the current level of molecular modeling applied to structure formation in reversibly assembling bulk systems. Clearly, analytical models can describe simple systems with likewise simple aggregate architectures, and thus may guide our understanding of the more complex systems, that is, the simple models do provide important insight regarding the relation between molecular interaction and aggregate phase behavior as it couples to size and shape of the aggregates. Nevertheless, quantitative prediction of properties and function of complex multicomponent supramolecular assemblies in biological systems or in technical applications is not possible based on analytic theories alone. Using a number of examples we have illustrated the current state of computer modeling as it applies to this field. The feeling is that we are rapidly approaching size and timescales, not only due to advances in computer technology but also due to new ideas for novel algorithms [80], which will allow to understand the mesoscopic and macroscopic behavior of supramolecular assemblies based on their molecular interactions even for complicated systems.

REFERENCES

1. H. Ringsdorf, B. Schlarb, and J. Venzmer. *Angew. Chem. Int. Ed. Engl.* 27:113–158, 1988.
2. J.-M. Lehn. *Supramolecular Chemistry*, New York: VCH, 1995.
3. A. Ciferri. *Prog. Polym. Sci.* 20:1081–1120, 1995.
4. A. Ciferri. *TRIP* 5:142–146, 1997.
5. A. Ciferri. *Liq. Cryst.* 26:489–494, 1999.
6. M.P. Allen and D.J. Tildesley. *Computer Simulation of Liquids*, Oxford: Clarendon Press, 1990.
7. D. Frenkel and B. Smit. *Understanding Molecular Simulation*, New York: Academic Press, 1996.
8. D.P. Landau and K. Binder. *A Guide to Monte Carlo Simulations in Statistical Physics*, Cambridge: Cambridge University Press, 2000.
9. M.J. Kotelyanskii and D.N. Theodorou. *Simulation Methods for Polymers*, New York: Marcel Dekker, 2004.
10. R.P. Feynman. *Phys. Rev.* 56:340–343, 1939.
11. S.W. Rick and B.J. Berne. *J. Am. Chem. Soc.* 118:672–679, 1996.
12. U. Burkert and N.L. Allinger. *Molecular Mechanics*, ACS Monograph 177, Washington, DC: American Chemical Society, 1982.
13. A.R. Leach. *Molecular Modeling*, Harlow: Addison-Wesley, 1996.
14. K.B. Lipkowitz and D.B. Boyd, Eds., *Reviews in Computational Chemistry*, Weinheim: VCH, 1990.
15. P. von Ragué Schleyer, Ed., *Encyclopedia of Computational Chemistry*, New York: Wiley, 1988.
16. R.W. Pastor, B.R. Brooks, and A. Szabo. *Mol. Phys.* 65:1409–1419, 1988.
17. P.H. Hünenberger and W.F. vanGunsteren. In: W.F. vanGunsteren, P.K. Weiner, and A.J. Wilkinson, Eds., *Computer Simulation of Biomolecular Systems — Theoretical and Experimental Applications*, Vol. 3. Dordrecht: Kluwer Academic Publishers, 1997.
18. R. Rajagopalan. *Curr. Opin. Colloid Interface Sci.* 6:357–365, 2001.
19. M. Girardi and W. Figueiredo. *Physica A* 324:621–633, 2003.
20. M. Kenward and M.D. Whitmore. *J. Chem. Phys.* 116:3455–3470, 2002.
21. Y. Rouault. *Eur. Phys. J. B* 6:75–81, 1998.
22. J.N. Israelachvili. *Intermolecular and Surface Forces*, New York: Academic Press, 1992.
23. D.F. Evans and H. Wennerström. *The Colloidal Domain*, New York: VCH, 1994.
24. B. Jönsson, B. Lindman, K. Holmberg, and B. Kronberg. *Surfactants and Polymers in Aqueous Solution*, New York: John Wiley & Sons, 1998.
25. A. Ben-Shaul and W.M. Gelbart. In: W.M. Gelbart, A. Ben-Shaul, and D. Roux, Eds., *Micelles, Membranes, Microemulsions, and Monolayers*, New York: Springer-Verlag, 1994.
26. J.M. Biltz and M.R. Fisch. *Langmuir* 11:3595–3597, 1995.
27. A. Ciferri. *J. Macromol. SCI-Polym. Rev. C* 43:271, 2003.
28. W.M. Gelbart, A. Ben-Shaul, W.E. McMullen, and A. Masters. *J. Phys. Chem.* 88:861–866, 1984.
29. R.G. Winkler, P. Reineker, and L. Harnau. *J. Chem. Phys.* 101:8119–8129, 1994.
30. W. Carl and Y. Rouault. *Macromol. Theory Simul.* 7:497–500, 1998.
31. T.L. Hill. *An Introduction to Statistical Thermodynamics*, New York: Dover, 1960.
32. J.T. Kindt and W.M. Gelbart. *J. Chem. Phys.* 114:1432–1439, 2001
33. A. Milchev and D.P. Landau. *Phys. Rev. E* 52:6431–6441, 1995
34. M. Kröger and R. Makhloufi. *Phys. Rev. E* 53:2531–2536, 1996
35. N. Boden. In: W.M. Gelbart, A. Ben-Shaul, and D. Roux, Eds., *Micelles, Membranes, Microemulsions, and Monolayers*, New York: Springer-Verlag, 1994.
36. W.M. McMullen, W.M. Gelbart, and A. Ben-Shaul. *J. Chem. Phys.* 82:5616–5623, 1985.
37. W.M. McMullen, W.M. Gelbart, and A. Ben-Shaul. *J. Phys. Chem.* 88:6649–6654, 1984.

38. W.M. Gelbart, W.M. McMullen, and A. Ben-Shaul. *J. Phys.* 46:1137–1144, 1985.
39. J. Herzfeld. *J. Chem. Phys.* 76:4185–4190, 1982.
40. J. Herzfeld and R.W. Briehl. *Macromolecules* 14:1209–1214, 1981.
41. J. Herzfeld and M.P. Taylor. *J. Chem. Phys.* 88:2780–2787, 1988.
42. M.P. Taylor and J. Herzfeld. *J. Phys. Condens. Matter* 5:2651–2678, 1993.
43. J. Herzfeld. *Acc. Chem. Res.* 29:31–37, 1996.
44. T. Odijk. *Curr. Opin. Colloid Interface Sci.* 1:337–340, 1996.
45. T. Odijk. *J. Phys.* 48:125–129, 1987.
46. A.N. Semenov and A.R. Khokhlov. *Sov. Phys. Usp.* 31:627–659, 1989.
47. B. Mulder. *Macromol. Symp.* 81:329–331, 1994.
48. R. Hentschke. *Liq. Cryst.* 10:691–702, 1991.
49. P. Mariani and L.Q. Amaral. *Phys. Rev. E* 50:1678–1681, 1994.
50. J.R. Mishic, R.J. Nash, and M.R. Fisch. *Langmuir* 6:915–919, 1990.
51. R. Hentschke, P.J.B. Edwards, N. Boden, and R. Bushby. *Macromol. Symp.* 81:361–367, 1994.
52. T. Bast and R. Hentschke. *J. Phys. Chem.* 100:12162–12171, 1996.
53. P. van der Schoot and M.E. Cates. *Europhys. Lett.* 25:515–520, 1994.
54. P. van der Schoot *J. Chem. Phys.* 104:1130–1139, 1996.
55. T. Odijk. *Macromolecules* 19:2313–2329, 1986.
56. R. Hentschke. *Statistische Mechanik*, Weinheim: Wiley-VCH, 2004.
57. M.P. Taylor and J. Herzfeld. *Phys. Rev. A* 43:1892–1905, 1991.
58. D.T. Kulp and J. Herzfeld. *Biophys. Chem.* 57:93–102, 1995.
59. G.H. Fredrickson. *Macromolecules* 26:2825–2831, 1993.
60. J.C. Shelly and M.Y. Shelly. *Curr. Opin. Colloid Interface Sci.* 5:101–110, 2000.
61. K. Binder and M. Müller. *Curr. Opin. Colloid Interface Sci.* 5:315–323, 2000.
62. P.I.C. Teixeira, J.M. Tavares, and M.M. Telo da Gama. *J. Phys. Condens. Matter* 12:R411–R434, 2000.
63. L.D. Schuler, P. Walde, P.L. Luisi, and W.F. van Gunsteren. *Eur. Biophys. J.* 30:330–343, 2001.
64. S.J. Marrink, D.T. Tieleman, and A.E. Mark. *J. Phys. Chem. B* 104:12165–12173, 2000.
65. P.K. Maiti, Y. Lansac, M.A. Glaser, and N.A. Clark. *Langmuir* 18:1908–1918, 2002.
66. K.H. Kim, S.H. Kim, J. Huh, and W.H. Jo. *J. Chem. Phys.* 119:5705–5710, 2003.
67. R.D. Groot. *Langmuir* 16:7493–7502, 2000.
68. B. Fodi and R. Hentschke. *J. Chem. Phys.* 112:6917–6924, 2000.
69. P.B. Lenz and R. Hentschke. to be submitted; P.B. Lenz and R. Hentschke. *J. Chem. Phys.* 121:10809–10813, 2004.
70. B. Fodi and R. Hentschke. *Langmuir* 16:1626–1633, 2000.
71. K. Binder and A. Milchev. *J. Comp. Aid. Design* 9:33–74, 2003.
72. D. Bedrov, G.D. Smith, K.F. Freed, and J. Dudowicz. *J. Chem. Phys.* 116:4765–4768, 2002.
73. R.G. Larson. *J. Phys. II Fr.* 6:1441–1463, 1996.
74. M.P. Pepin and M.D. Whitmore. *Macromolecules* 33:8644–8653, 2000.
75. M. Lisal, C.K. Hall, K.E. Gubbins, and A.Z. Panagiotopoulos. *J. Chem. Phys.* 116:1171–1184, 2002.
76. A.Z. Panagiotopoulos, M.A. Floriano, and S.K. Kumar. *Langmuir* 18:2940–2948, 2002.
77. S.H. Kim and W.H. Jo. *J. Chem. Phys.* 117:8565–8572, 2002.
78. F.R. Siperstein and K.E. Gubbins. *Langmuir* 19:2049–2057, 2003.
79. S.K. Kumar and A.Z. Panagiotopoulos. *Phys. Rev. Lett.* 82:5060–5063, 1999.
80. G.H. Fredrickson, V. Ganesan, and F. Drolet. *Macromolecules* 35:16–39, 2002.

Chapter 5

Supramolecular Low-Molecular Weight Complexes and Supramolecular Side-Chain Polymers

Takashi Kato

Contents

I. Introduction

The concept of supramolecular chemistry and self-assembly has become one of the central topics in current chemistry [1–7]. A wide variety of supramolecular systems composed of multicomponents have been prepared in crystalline, liquid-crystalline, and solution states. The introduction of this concept to the design of soft matters such as polymers, liquid crystals, colloids, and amphiphiles has attracted a great deal of attention because dynamically functional materials can be obtained by this approach [1–33]. A new class of functional materials, supramolecular liquid crystals [7–23], and supramolecular polymers [8–16,18,20–32] have been studied, since supramolecular complexation has been shown to be useful for building the molecular structures of functional materials in neat states. Kato and Fréchet [17] and Lehn and coworkers [19] reported that liquid-crystalline low-molecular weight complexes with well-defined structures can be built through the formation of hydrogen bonds between different molecules. In these materials, molecular self-assembly processes lead to the preparation of dynamically functional molecular complexes. New properties are induced by the complexation of several molecular species. For example, generation or enhancement of liquid-crystalline behavior is observed for the mesogenic complexes [7–10]. This concept initiated the field of supramolecular polymers [8–16,18,20–32].

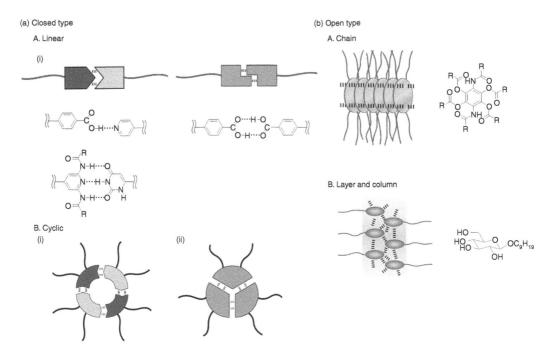

Figure 1 (a) Hydrogen-bonded complexes with closed structures: (A) linear hydrogen-bonded complexes between (i) different and (ii) identical molecules; (B) cyclic hydrogen-bonded association of (i) different and (ii) identical molecules. (b) Open type hydrogen-bonded structures: (A) chain and (B) layer and column.

Supramolecular polymers were first reported by Kato and Fréchet [18] and Lehn and coworkers [20]. These materials were developed based on the concept that functional molecular structures can be built by the complexation of molecular components by specific molecular interactions [17,19]. Such approaches have been applied to build a variety of supramolecular polymeric materials.

This chapter first focuses on the closed type of supramolecular low-molecular weight complexes forming linear and cyclic structures (Figure 1(a)). Most of these supramolecular materials are built through hydrogen bonding. Hydrogen bonding is useful for this approach because of its moderate stability and directionality [33]. There are two types of complexes [11]: (i) consisting of different (complementary) molecules; and (ii) consisting of identical (self-complementary) molecules (Figure 1(a), (i) and (ii)). Only the closed type of assemblies are discussed in this chapter. Hydrogen-bonded materials of the open type (Figure 1(b), A and B) of assembly, chain [34–36], and layer and column [37,38], which are successive hydrogen-bonded assemblies are not mentioned.

Furthermore, the structures and properties of supramolecular side-chain polymers are described. Supramolecular polymers can be divided into two classes: (a) side-chain and (b) main-chain supramolecular polymers (Figure 2). Supramolecular main-chain polymers will be discussed in Chapter 6.

In the initial stage of the research into supramolecular materials, liquid-crystalline properties were extensively studied. These materials are one of the representative supramolecular materials, which exhibit unique properties by combination of anisotropic molecular order and supramolecular structures. Moreover, the induction of the liquid crystallinity is a useful indicator of the effect of supramolecular complexation. On the other hand, nonliquid-crystalline supramolecular materials have been developed in crystal and solid states. Therefore, for clarity, liquid-crystalline materials and some representative nonliquid-crystalline supramolecular materials in the neat states are described in separate sections.

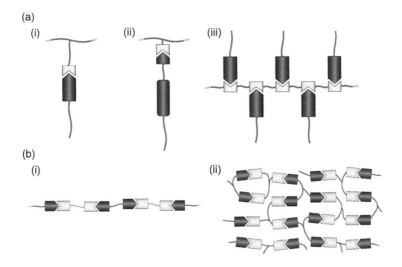

Figure 2 Schematic illustration of two classes of supramolecular polymers: (a) side-chain supramolecular polymers; (b) main-chain supramolecular polymers.

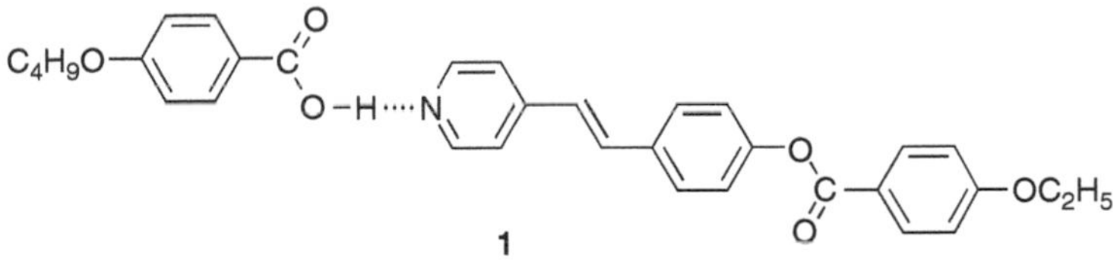

Figure 3 Early examples of supramolecular mesogenic complexes **1** and **2** formed through complementary hydrogen bonding.

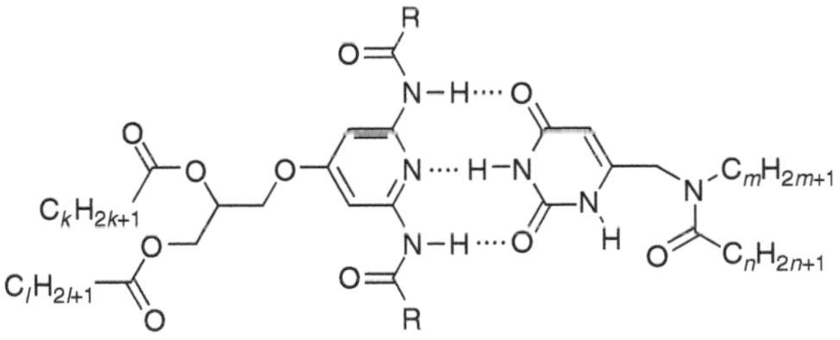

II. SUPRAMOLECULAR LOW-MOLECULAR WEIGHT COMPLEXES

A. Liquid-Crystalline Complexes

The first examples of supramolecular mesogenic complexes are shown in Figure 3. The single hydrogen bond formed between carboxylic acid and pyridine moieties is used to build dissimilar mesogen **1** with well-defined structures [17]. Complex **1** shows nematic and smectic phases and its liquid-crystalline state is stable over 200°C, which is higher than that of each component. Hydrogen-bonded mesogenic complex **2** obtained through the formation of triple hydrogen bonds between uracil and 2,6-diaminopyridines exhibits columnar phases [19], where two complexes form one disk side by side due to two lipophilic chains for each component.

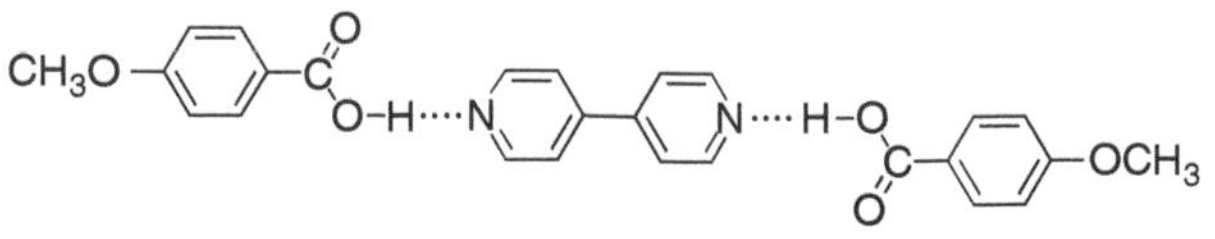

Figure 4   Rodlike mesogenic complexes **3–5** prepared by using carboxylic acid/pyridine hydrogen bonding.

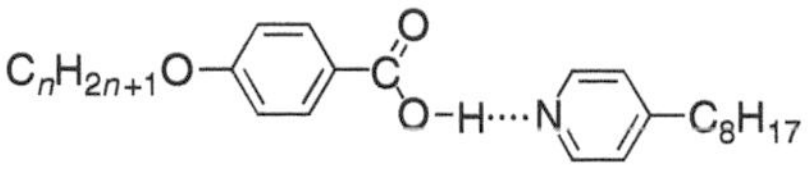

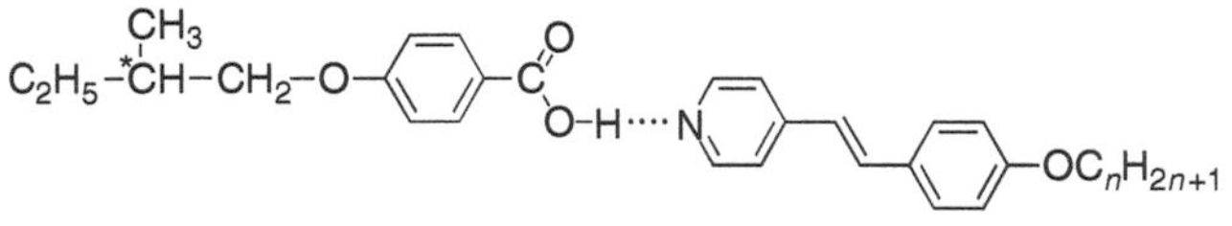

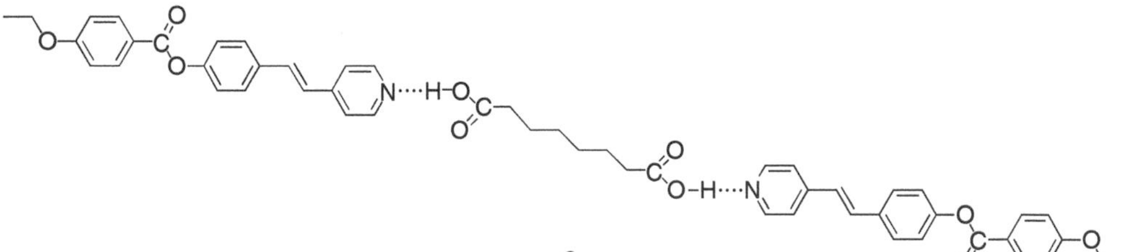

Figure 5   Dimeric hydrogen-bonded complex **6** with a flexible aliphatic spacer.

A variety of rodlike mesogenic structures were prepared by using the carboxylic acid/pyridine hydrogen bond (Figure 4) [39–57]. A liquid-crystalline phase is observed for complex **3**, which is prepared by 2:1 (molar ratio) complexation of 4-methoxybenzoic acid and 4,4′-bipyridine [39,40]. It shows a nematic phase over 160°C. In this case, the complexation of nonmesomorphic components leads to the induction of liquid-crystalline phases. The molecular order stabilizes the interaction [40,44]. When the hydrogen-bonded mesogens become shorter, the complexes show lower liquid-crystalline temperature ranges [41–48]. Hydrogen-bonded complex **4** ($n = 6$–10) exhibits nematic phases at room temperature [41,42]. Chiral smectic phases, which are ferroelectric, are induced by incorporating chiral molecular components [45,46]. Chiral complex **5** ($n = 8$) exhibits a chiral smectic C phase, while the chiral single component does not show ferroelectric liquid crystal behavior. Dimeric hydrogen-bonded complexes having two rod units connected by flexible spacers were prepared [49]. An odd–even effect is observed for transition temperatures of dimeric complex **6**, comprising of aliphatic dicarboxylic acid (Figure 5). Several dimeric structures [49–52], which is considered to be a kind of precursor of main-chain supramolecular complex were prepared by this complexation approaches. Supramolecular hydrogen-bonded complexes can behave as the single component of liquid crystals. Single crystal structure studies of pyridine/carboxylic acid complexes also support the formation of stable linear complexes [53]. One advantage of the noncovalently bonded supramolecular complexes is that they show dynamic functional properties such as responsiveness to external stimuli due to the dissociation–association behavior or the change of interaction patterns [7,31,32]. Theoretical approaches had been studied to these supramolecular liquid crystals [58–61].

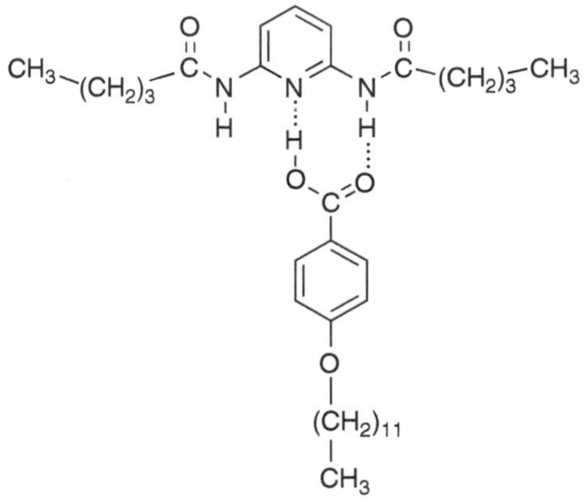

Figure 6   Double hydrogen-bonded supramolecular complex **7**.

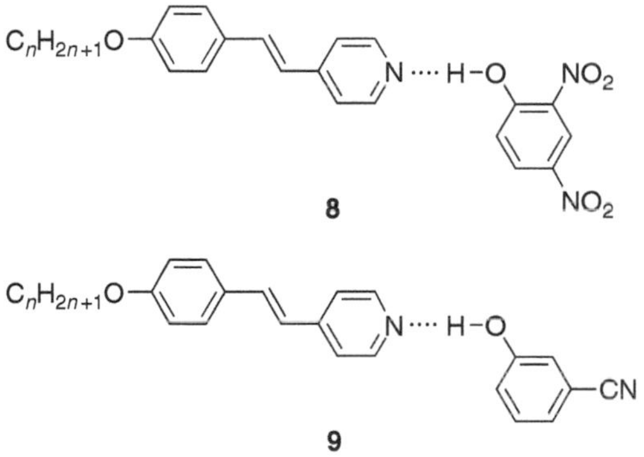

Figure 7   Supramolecular mesogenic complexes **8** and **9** using hydrogen bonding between phenol and pyridine.

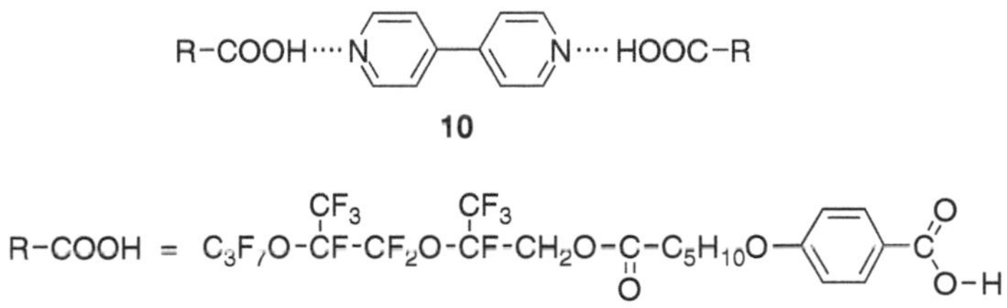

Figure 8   Hydrogen-bonded complex **10** exhibiting a cubic phase.

Some of hydrogen-bonded mesogenic structures that are not simple linear rod shape have been shown to be liquid-crystalline. Double hydrogen bonds between carboxylic acid and aminopyridine moieties are employed to prepare H-bonded complex **7** (Figure 6) [62], which exhibits a monotropic crystal B phase. The single hydrogen bond between phenol and pyridine can also be used for the formation of supramolecular mesogens although the pyridine–phenol hydrogen bond does not give linear complex structures [63,64]. Phenols containing electron-withdrawing groups which enhance the acidity of the hydroxyl moiety, can form complexes **8** and **9** exhibiting stable LC phases (Figure 7). A cubic phase is induced for the hydrogen-bonded complex of **10** prepared from fluoroalkyl substituted benzoic acid and 4, 4′-bipyridine (Figure 8) [65].

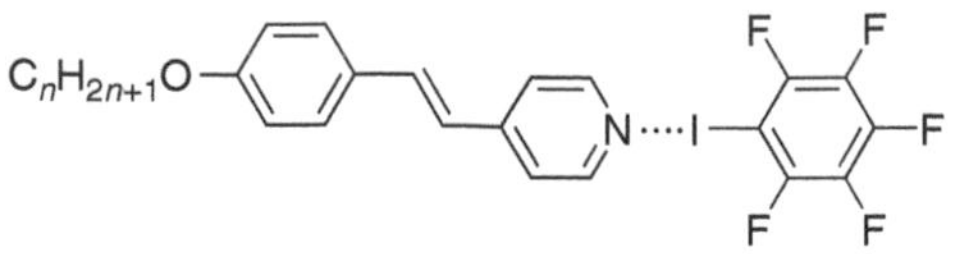

Figure 9 Supramolecular mesogenic complex **11** using halogen/pyridine interactions.

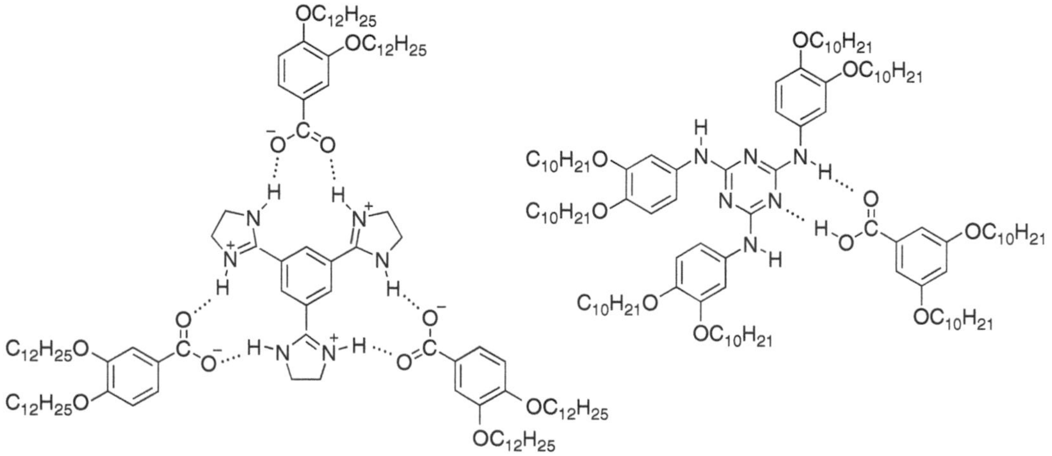

Figure 10 Supramolecular complexes **12** and **13** formed by disklike association of different molecules.

Not only hydrogen bonding but also halogen/pyridine interactions can yield a mesogenic complex [66]. A rodlike supramolecular structure has been built by "halogen bonding." Complex **11** ($n = 6$) exhibits monotropic nematic and smectic phases (Figure 9).

Supramolecular disklike complexes can also be designed and prepared (Figure 10). The 3:1 complexation of benzoic acid and tris (imidazoline) base gives complex **12**, which exhibits a columnar phase [67]. A melamine core is used for the preparation of supramolecular disklike liquid crystals [68]. In this case, the 1:1 supramolecular complex of **13** is obtained through the formation of double hydrogen bonds.

Since the early 20th century, hydrogen-bonded rodlike complexes of identical molecules, 4-substituted benzoic acid dimers, have long been known to be liquid crystals [69–71]. These classical dimers are not categorized within supramolecular complexes. Recently, association of more elaborate identical molecules are often referred to as supramolecular self-assembly because new elegant structures are formed and new functions are expected for these aggregates [72,73]. Disklike liquid-crystalline complexes exhibiting columnar phases are obtained by self-complementary dimeric association of half-disk structures of pyridone molecule **14** (Figure 11) [74]. Trimeric hydrogen-bonded complexation is observed for columnar liquid crystal **15** [73]. In this structure tautomerization of phthalhydrazide is a key for self-assembly (Figure 11).

Biomolecules provide a variety of hydrogen-bonding components even for synthetic approaches. Recently, hydrogen-bonded assemblies of folate [75] and guanosine [76] have attracted much attention. Gottarelli and coworkers [77,78] reported that folic acid **16** self-assemble into tetramer, which form columnar structures in aqueous potassium salt solution (Figure 12). In this case, potassium cation promotes the formation of lyotropic columnar liquid-crystalline phase (Figure 12).

Thermotropic liquid-crystalline folic acid derivatives have been designed and prepared by the introduction of lipophilic moieties onto the glutamic moieties [79–82]. Folic acid derivatives exhibit

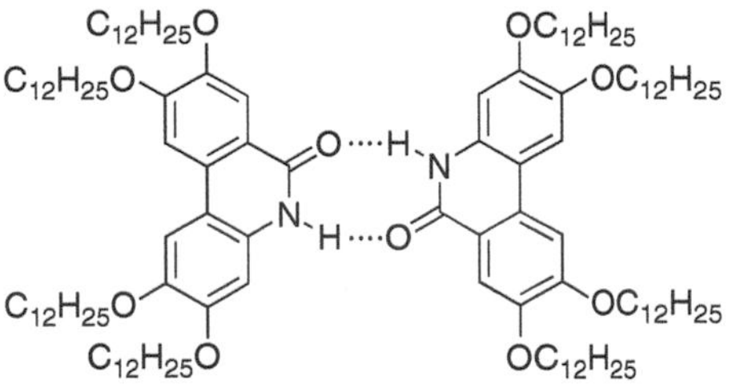

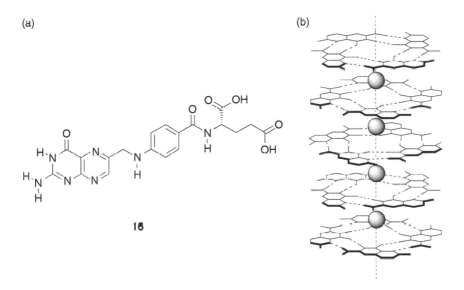

14

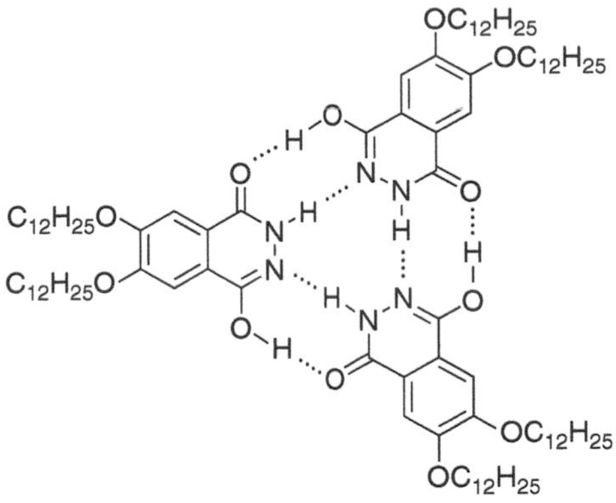

15

Figure 11   Disklike liquid-crystalline complexes **14** and **15** consisting of identical molecules.

(a)

(b)

16

Figure 12   (a) Structure of folic acid **16**. (b) Self-assembled columnar structure of **16** formed by stacking of tetramers of the pterin rings in the presence of cations.

liquid-crystalline phases over wide temperature ranges: smectic phases for **17** ($n = 6, 11$) and columnar phases for **17** ($n = 14, 18$) (Figure 13). The pterin ring can form two self-assembled patterns: cyclic disklike aggregation (Figure 14(a)) and linear ribbonlike aggregation (Figure 14(b)). The smectic phases are formed due to the aggregation of the ribbon structure, while the columnar phase is formed by the stacking of the disklike aggregates. These two hydrogen-bonded patterns are selected by the steric effects of the outer bulky lipophilic groups. Compound **18** ($n = 6, 11, 18$) having an oligo (glutamic acid) moiety bears eight lipophilic chains (Figure 13). These molecules self-assemble only into disklike aggregates due to the bulkiness of the lipophilic part (Figure 14(a)), leading to the induction of hexagonal columnar and micellar Pm3n cubic liquid-crystalline phases. It is considered that the segmentation of the columns results in the transition to the micellar cubic phase (Figure 15). Supramolecular chirality is induced for the compounds in hexagonal columnar and micellar cubic phases in the presence of alkali metal salts (Figure 15).

The cation-templated aggregation has been studied for guanosine (G) derivatives. Compound **19** (Figure 16(a)) shows lyotropic cholesteric and hexagonal columnar phases in hexane in the presence

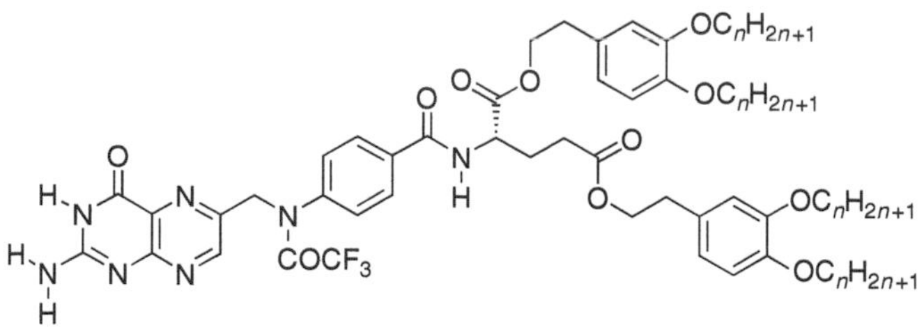

17

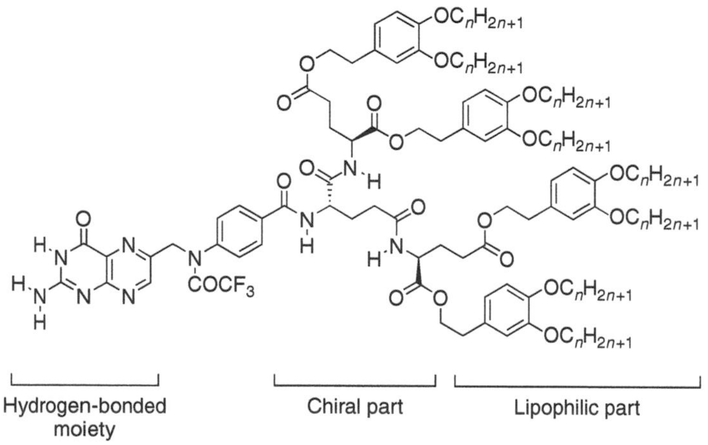

| Hydrogen-bonded moiety | Chiral part | Lipophilic part |

18

Figure 13 Structures of thermotropic liquid-crystalline folic acid derivatives **17** and **18**.

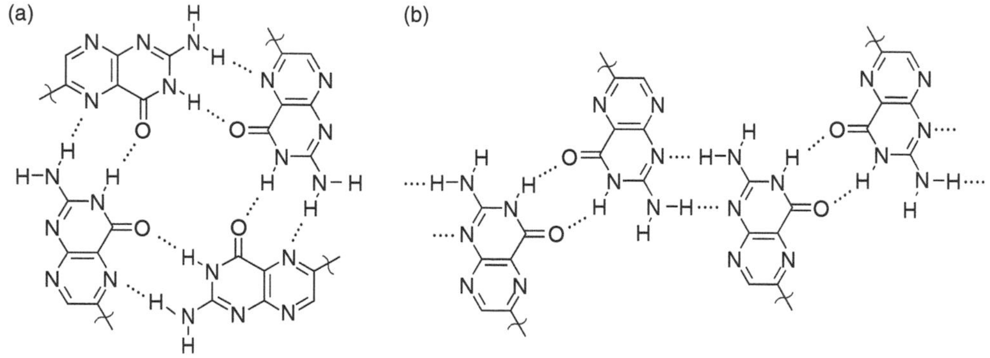

Figure 14 Hydrogen-bonded structures of the pterin rings of folic acid: (a) cyclic disklike aggregation, and (b) linear ribbonlike aggregation.

of potassium picrate [83,84]. The disklike structure shown in Figure 16(b) leads to the induction of such mesomorphic behavior.

B. Nonliquid-Crystalline Complexes

Some representative examples of the formation of closed complex structures in the solid state are described in this section. Carboxylic acid dimerization is a representative example of self-complimentary complexation. Stronger dimerization has been shown for ureidopyrimidone dimer **20**

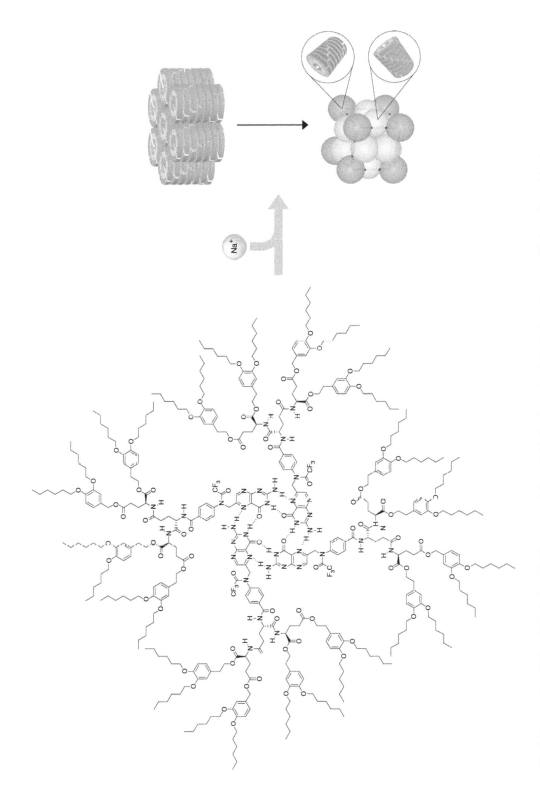

Figure 15 Self-assembly of folic acid derivative **18**. The segmentation of chiral columns leads to the transition to a chiral micellar cubic phase.

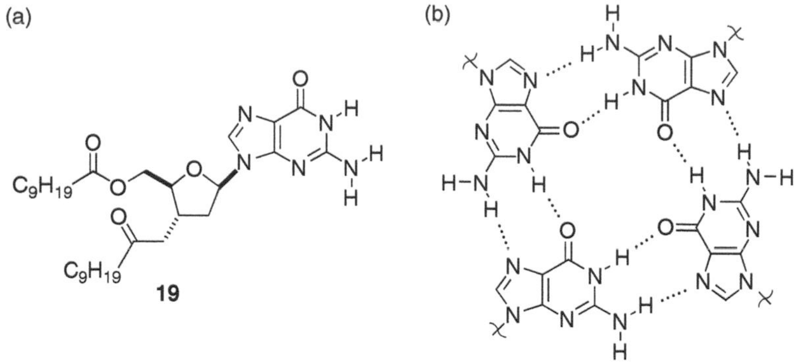

(a)   (b)

Figure 16 (a) Structure of lipophilic guanosine derivative **19**. (b) Cyclic hydrogen-bonded quartet of the guanine residues.

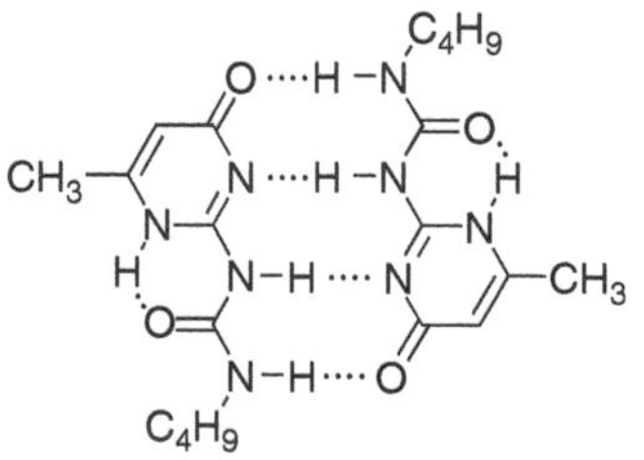

Figure 17 Quadruple hydrogen-bonded dimer **20** of ureidopyrimidones.

obtained through the formation of quadruple hydrogen bonding (Figure 17) [85]. This interaction has been used for the preparation of main-chain supramolecular polymers [86,87].

Benzoic acid dimerization was also applied for the building of closed complex structures in crystalline states. For example, 5-substituted isophthalic acid derivatives form well-defined structures of cyclic hexameric aggregate **21** (Figure 18) [88]. Supramolecular cyclic hydrogen-bonded complex **22** is formed from cyanuric acid and melamine (Figure 19) [89]. A cyclic structure is formed by 1:1 complexation.

Crystal structures were determined for the guanosine assembly by x-ray. The results demonstrate that for **23** with K^+ picrate, K^+ is sandwiched between disks (Figure 20) [90].

III. SUPRAMOLECULAR SIDE-CHAIN POLYMERS

A. Liquid-Crystalline Polymeric Complexes

Supramolecular polymers are classified into two classes: side chain (Figure 2(a)) and main chain (Figure 2(b)). There are three types of side-chain polymers (Figure 2(a)): (i) polymers bearing supramolecular mesogen in the side chain attached via flexible spacers; (ii) "grafted" polymers complexed with end-functionalized "amphiphilic" molecules incorporating mesogens; and (iii) polymers complexing with rigid molecules on their backbones. Main-chain polymers are divided into two types: (i) linear main-chain polymers and (ii) networks.

The initial approach in 1989 to yield supramolecular polymers involved complexation of a stilbazole molecule to a polyacrylate having 4-oxybenzoic acid moieties in the side-chain (Figure 21) [18]. The nematic phase of supramolecular side-chain polymer **24** was observed to be

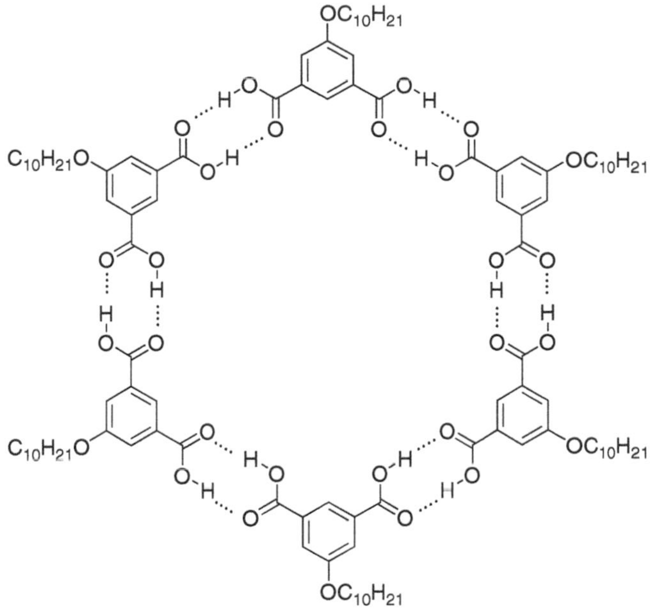

Figure 18 Cyclic hexamer **21** of Isophthalic acid derivatives.

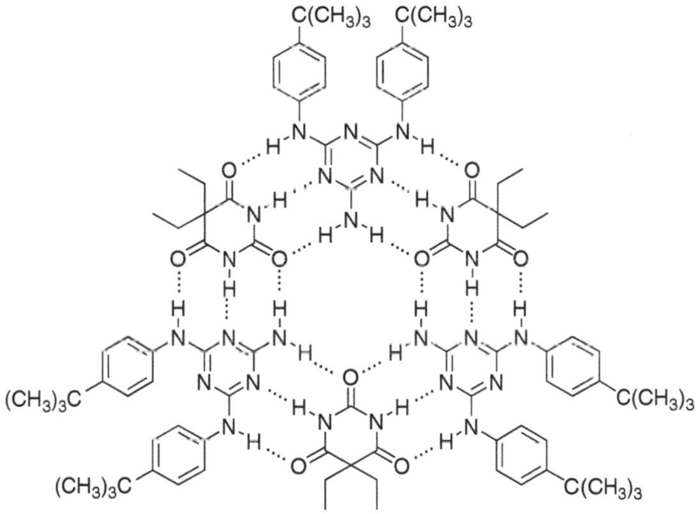

Figure 19 Cyclic hydrogen-bonded complex **22**.

over 250°C for the 1:1 complexes. A variety of structures based on polyacrylates and polysiloxanes have since been prepared (Figure 2(a) [i], Figure 21) [22,43,91–96]. A smectic phase is induced for **25** by the complexation of nonmesomorphic components. A ferroelectric chiral smectic C phase is observed for complex **26** consisting of functionalized polysiloxanes and chiral stilbazoles [92]. A double hydrogen-bonded mesogen is formed in the side-chain of the functionalized polyacrylate and polymethacrylate. A liquid-crystalline columnar phase is seen for complex **27** (R = $-CH_3$) (Figure 22) [12].

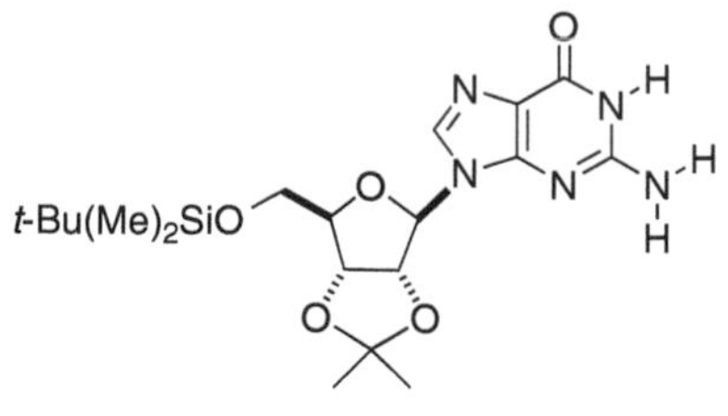

23

Figure 20 Structure of lipophilic guanosine derivative **23**.

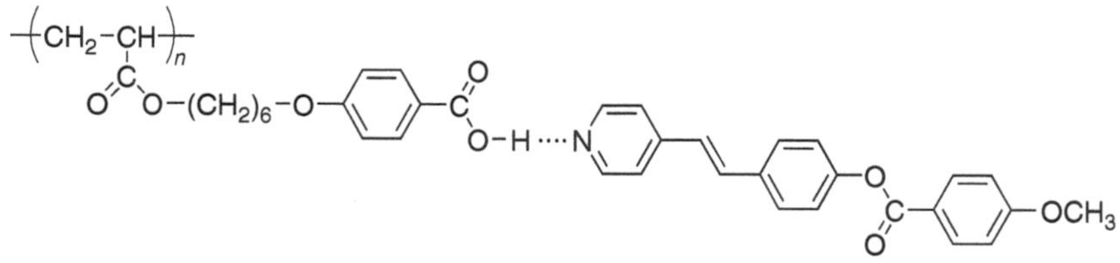

24

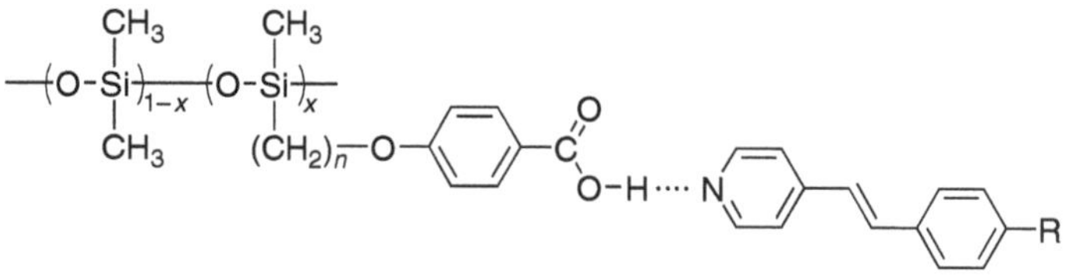

25 R = —OCH$_2$CH$_3$

26 R = —O—CH$_2$—$\overset{\text{OCH}_3}{\text{CH}}$—CH$_3$

Figure 21 Supramolecular side-chain liquid-crystalline polymers **24–26**.

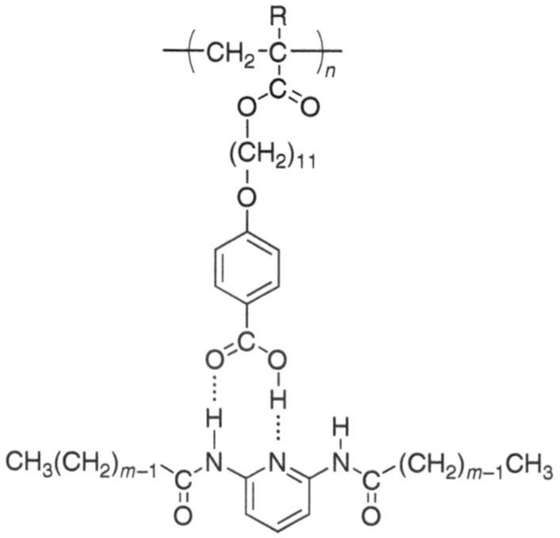

R = –H, –CH$_3$

27

Figure 22 Supramolecular side-chain polymer **27** using double hydrogen bonding.

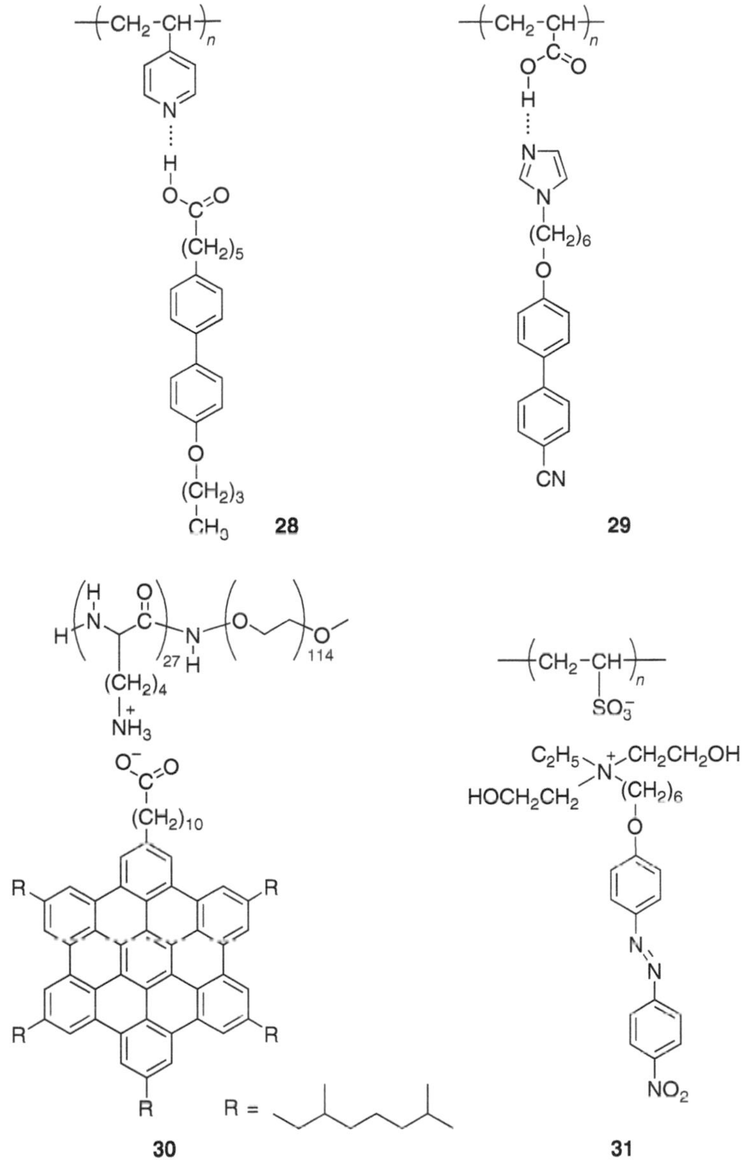

Figure 23 Supramolecular side-chain liquid-crystalline polymers: hydrogen-bonded complexes **28** and **29**; ionic complexes **30** and **31**.

End-functionalized mesogenic molecules are used to form the second type of supramolecular side-chain polymers (Figure 2(a) [ii], Figure 23) [97–108]. Polymeric complex **28** has been prepared based on poly(4-vinylpyridine) [97]. The hydrogen-bonding formation between imidazole and carboxylic acid moieties yields supramolecular side-chain polymers of **29** [103], which exhibit smectic A phases. The interactions of carboxylic acid/dialkylamine [104–106], phenol/amine [107], and hydroxyl/pyridine [108] were used for the preparation of the side-chain mesogenic complexes.

Ionic interactions can be used for the preparation of supramolecular side-chain polymers [109–111]. The interactions between $-SO_3^-$ and $-NH_3^+$ and those between $-COO^-$ and $-NH_3^+$ give liquid-crystalline polymeric complexes. The complex of poly(ethylene oxide)-*block*-poly(L-lysine) and a carboxyl-functionalized hexa-peri-haxabenzocornene (**30**) shows columnar structures (Figure 23) [109]. Polymeric complex **31** forms a stable smectic phase [111]. This complex shows

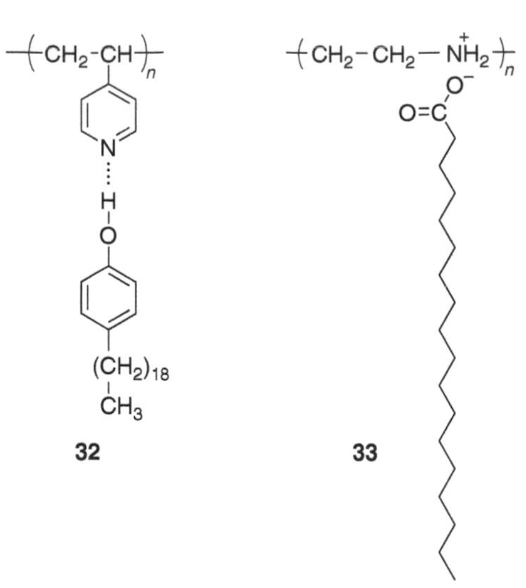

Figure 24 Supramolecular liquid-crystalline polymers **32** and **33** without aromatic mesogens.

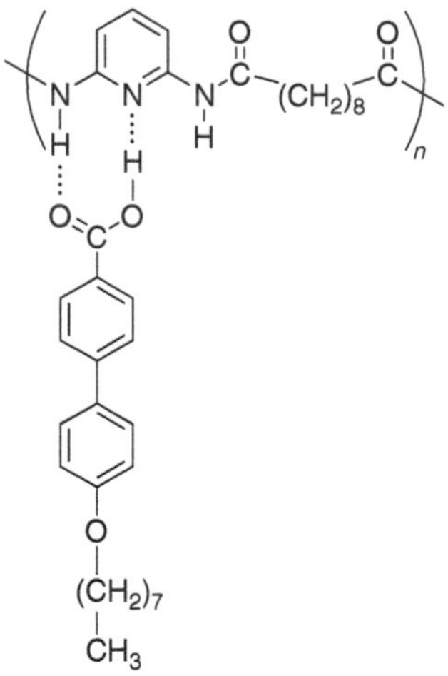

Figure 25 Supramolecular liquid-crystalline polyamide **34**.

homeotropic alignment on glass substrate due to the introduction between ionic moieties and the glass surfaces.

Supramolecular polymeric complexes without aromatic mesogens can exhibit liquid-crystalline phases due to microphase segregation [112–117]. Complex **32** consisting of poly(4-vinylpyridine) complexed with alkylphenol shows a layered mesophase (Figure 24) [112]. Ionic supramolecular complex **33** based on poly(ethyleneimine) and alkanoic acids also forms layered mesophases (Figure 24) [115].

Polymer backbones can be functionalized for supramolecular complexation. The introduction of a 2, 6-diaminopyridine moiety into polyamide structures enables the polymer to recognize hydrogen-bonded molecules on their backbones (Figure 25) [23,118–120]. Supramolecular polyamide **34** formed by complexation of the polyamide and the biphenyl carboxylic acid exhibits a stabilized smectic A phase up to 350°C [23]. Cooperation of hydrogen bonds of supramolecular complexation

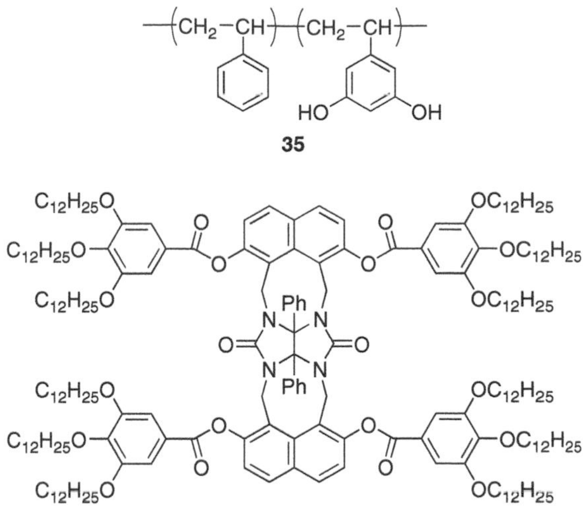

Figure 26 Hydroxyl-functionalized polystyrene **35** and compound **36** forming inclusion polymeric complexes.

between the aminopyridine and the benzoic acid, amide hydrogen-bond chain, and aromatic stacking may effectively stabilize the ordered states. Alkoxybenzoic acids substituted at the 3-position are also complexed with the polyamide to yield stable supramolecular complexes [118–120]. Supramolecular inclusion complexes have been prepared from hydroxyl-functionalized polystyrene **35** and molecule **36** (Figure 26) [121]. Liquid-crystalline properties are observed for the complexes. Side-chain polymers based on polyurethanes have been obtained [122]. These supramolecular complexes can be referred to as a type of side-chain supramolecular polymers (Figure 2(a) [iii]).

Physical cross-linking is easily incorporated into the structures of supramolecular side-chain polymers by using bi- or multifunctional molecules [9,31,93]. The complexation of the functionalized polyacrylates, a stilbazole, and bipyridine molecules form supramolecular side-chain cross-linked polymer **37** (Figure 27) [31,93]. Bis-imidazolyl compounds are also used for such cross-linking [105]. Reversible association–dissociation of the hydrogen bonding leads to the reversible thermal order–disorder transitions, which is an important feature of supramolecular materials. Liquid-crystalline main-chain supramolecular networks have been prepared from multifunctional components [32,123–126], which will be described in Section IV.

B. Nonliquid-Crystalline Polymeric Complexes

Hydrogen-bonding moieties such as diaminotriazine and diaminopyridine units have been incorporated into the side chain of polymers for versatile approaches to polymer functionalization (Figure 28) [127–130]. For example, ring opening metathesis polymerization has been used for the synthesis of hydrogen-bonded polymer **38** because of its compatibility with functional groups and the possibility of structural control for block copolymers [127]. Random copolymer **39** possessing metal complexes and hydrogen-bonding units has also been prepared [128]. Supramolecular polymer **40** has been formed by using halogen/pyridine interactions (Figure 29) [131].

IV. FUNCTIONALIZATION OF COMPLEXES

Functional supramolecular polymeric materials are designed and prepared by incorporation of functional species such as chiral compounds, dyes, and conductive molecules. For these

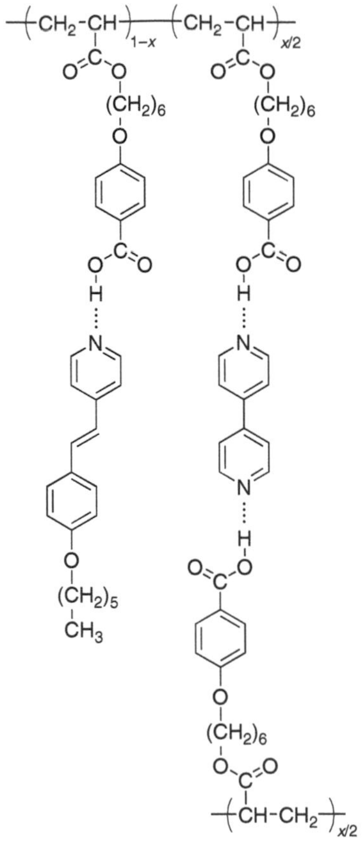

Figure 27 Cross-linked supramolecular mesogenic polymer **37**.

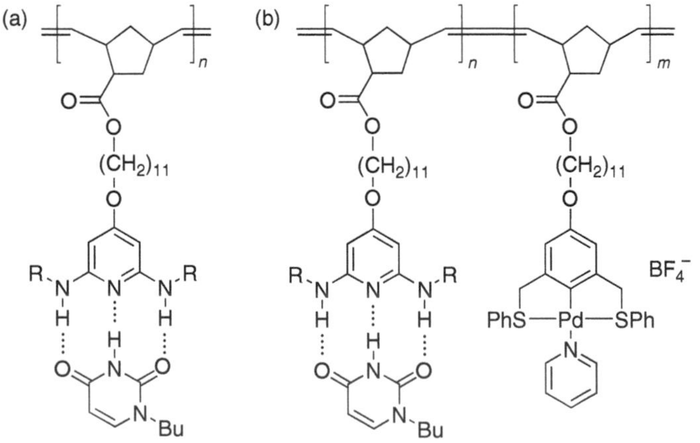

Figure 28 (a) Hydrogen-bonded polymeric complex **38** and (b) supramolecular random copolymer **39**.

materials more dynamic functions are expected from their structures. Ferroelectric chiral smectic C phases are induced by the hydrogen-bond complexation of chiral nonmesomorphic benzoic acid and achiral stilbazoles [45]. It should be noted that stable ferroelectric switching in electric fields is observed even for the hydrogen-bonded mesogens [45,46]. Ferroelectricity is also

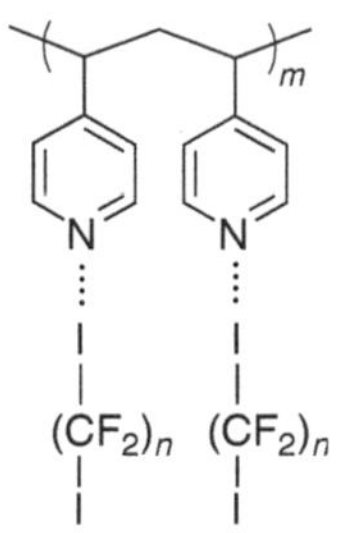

40

Figure 29 Comb-shaped supramolecular polymer **40** using halogen/pyridine interactions.

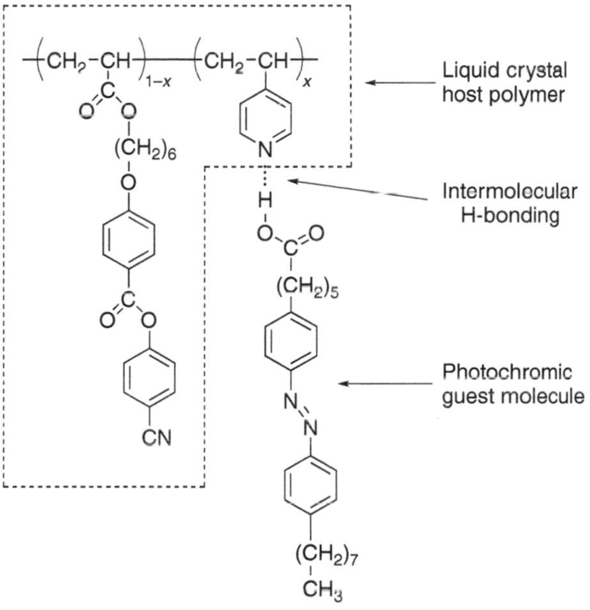

41

Figure 30 Supramolecular functional guest–host polymer **41**.

seen for carboxyl-functionalized polysiloxanes with nonmesomorphic chiral stilbazoles [92]. The incorporation of photochromic dyes such as an azobenzene moiety into liquid-crystalline polymers through the noncovalent interactions gives new type of host–guest liquid-crystalline polymer systems (Figure 30) [102]. Complex **41** is a photoresponsive nematic material that exhibits isothermal nematic–isotropic transitions by *cis–trans* photochemical isomerization of the azobenzene guest. In this case, simple mixing of interacting species enhances their transition properties.

Control of macroscopic orientation of the microphase-segregated structures leads to efficient anisotropic conduction of proton and ions [132–139]. Proton conductive supramolecular materials have been prepared by using microphase segregation of supramolecular block polymers **42** consisting of poly(styrene)-*block*-(4-vinylpyridine), toluene sulfonic acid, and 3-pentadecyl phenol (Figure 31) [137]. Ionic conductive side-chain polymers have been obtained by complexation of oligo(ethylene oxide)sulfonic acid with poly(styrene)-*block*-(4-vinylpyridine) [139].

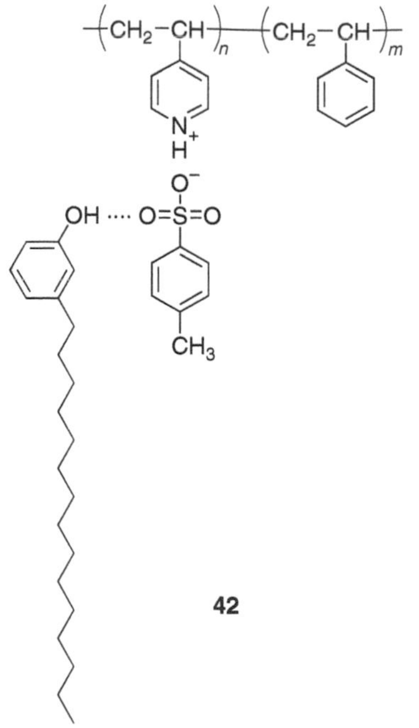

Figure 31 Supramolecular polymeric complex **42** prepared from a polystyrene-*block*-poly(4-vinylpyridine) copolymer, pentadecylphenol, and toluenesulfonic acid.

V. Conclusion

The concept for the design of supramolecular liquid crystals and supramolecular polymers has opened new fields in materials and polymer science, which are ever expanding. New stable and dynamic structures are generated by self-organization of these materials. Related functional polymeric materials such as dendrimers [140–143], block copolymers [144], polymer blends [145,146], rotaxanes [147–149], anisotropic gels [150–152], metallo-supramolecular polymers [153,154], nanoobjects [155,156] as well as supramolecular polymers are also obtained by self-assembly of multicomponents through noncovalent interactions.

References

1. Special section on supramolecular chemistry and self-assembly. *Science* 295:2395–2421, 2002.
2. M. Fujita, Ed. *Structure & Bonding*, Vol. 96. Berlin: Springer, 2000.
3. J.-M. Lehn. *Supramolecular Chemistry*. Weinheim: VCH, 1995.
4. G.M. Whitesides, J.P. Mathias, and C.T. Seto. *Science* 254:1312–1319, 1991.
5. D.S. Lawrence, T. Jiang, and M. Levett. *Chem. Rev.* 95:2229–2260, 1995.
6. P. Timmerman, D.N. Reinhoudt, and L.J. Prins. *Angew. Chem. Int. Ed. Engl.* 40:2383–2426, 2001.
7. T. Kato. *Science* 295:2414–2418, 2002.
8. T. Kato. Hydrogen-bonded systems, in *Handbook of Liquid Crystals*, Vol. 2B. D. Demus, J.W. Goodby, G.W. Gray, H.-W. Spiess, and V. Vill, Eds., Weinheim: Wiley-VCH, pp. 969–979, 1998.
9. T. Kato and J.M.J. Fréchet. *Macromol. Symp.* 98:311–326, 1995.
10. J.-M. Lehn. *Makromol. Chem. Macromol. Symp.* 69:1–17, 1993.
11. T. Kato. *Struct. Bonding (Berlin)* 96:95–146, 2000.
12. T. Kato, N. Mizoshita, and K. Kanie. *Macromol. Rapid Commun.* 22:797–814, 2001.
13. T. Kato. *Supramol. Sci.* 3:53–59, 1996.

14. C.M. Paleos and D. Tsiourvas. *Angew. Chem. Int. Ed. Engl.* 34:1696–1711, 1995.
15. C.T. Imrie. *Trends. Polym. Sci.* 3:22–29, 1995.
16. N. Zimmerman, J.S. Moore, and S.C. Zimmerman. *Chem. Ind. (London)* 604–610, 1998.
17. T. Kato and J.M.J. Fréchet. *J. Am. Chem. Soc.* 111:8533–8534, 1989.
18. T. Kato and J.M.J. Fréchet. *Macromolecules* 22:3818–3819, 1989 [Errata: T. Kato and J.M.J. Fréchet. *Macromolecules* 23:360, 1990.
19. M.-J. Brienne, J. Gabard, J.-M. Lehn, and I. Stibor. *J. Chem. Soc. Chem. Commun.* 1868–1870, 1989.
20. C. Fouquey, J.-M. Lehn, and A.-M. Levelut. *Adv. Mater.* 2:254–257, 1990.
21. C. Alexander, C.P. Jariwala, C.M. Lee, and A.C. Griffin. *Macromol. Symp.* 77:283–294, 1994.
22. U. Kumar, T. Kato, and J.M.J. Fréchet. *J. Am. Chem. Soc.* 114:6630–6639, 1992.
23. T. Kato, Y. Kubota, T. Uryu, and S. Ujiie. *Angew. Chem. Int. Ed. Engl.* 36:1617–1618, 1997.
24. S.I. Stupp, V. LeBonheur, K. Walker, L.S. Li, K.E. Huggins, M. Keser, and A. Amstutz. *Science* 276:384–389, 1997.
25. M. Muthukumar, C.K. Ober, and E.L. Thomas. *Science* 277:1225–1232, 1997.
26. A. Ciferri, Ed. *Supramolecular Polymers.* New York: Marcel Dekker, 2000.
27. L. Brunsveld, B.J.B. Folmer, and E.W. Meijer. *MRS Bull.* 25:49–53, 2000.
28. L. Brunsveld, B.J.B. Folmer, E.W. Meijer, and R.P. Sijbesma. *Chem. Rev.* 101:4071–4097, 2001.
29. P. Bladon and A.C. Griffin. *Macromolecules* 26:6604–6610, 1993.
30. C.-M. Lee and A.C. Griffin. *Macromol. Symp.* 117:281–290, 1997.
31. T. Kato, H. Kihara, U. Kumar, T. Uryu, and J.M.J. Fréchet. *Angew. Chem. Int. Ed. Engl.* 33:1644–1645, 1994.
32. H. Kihara, T. Kato, T. Uryu, and J.M.J. Fréchet. *Chem. Mater.* 8:961–968, 1996.
33. G.A. Jeffrey. *An Introduction to Hydrogen Bonding.* Oxford: Oxford University Press, 1997.
34. Y. Matsunaga and M. Terada. *Mol. Cryst. Liq. Cryst.* 141:321–326, 1986.
35. Y. Kobayashi and Y. Matsunaga. *Bull. Chem. Soc. Jpn.* 60:3515–3518, 1987.
36. J. Malthéte, A.M. Levelut, and L. Liébert. *Adv. Mater.* 4:37–41, 1992.
37. G.A. Jeffrey. *Mol. Cryst. Liq. Cryst.* 110:221–237, 1984.
38. J.W. Goodby. *Mol. Cryst. Liq. Cryst.* 110:205–219, 1984.
39. T. Kato, P.G. Wilson, A. Fujishima, and J.M.J. Fréchet. *Chem. Lett.* 2003–2006, 1990.
40. T. Kato, J.M.J. Fréchet, P.G. Wilson, T. Saito, T. Uryu, A. Fujishima, C. Jin, and F. Kaneuchi. *Chem. Mater.* 5:1094–1100, 1993.
41. M. Fukumasa, T. Kato, T. Uryu, and J.M.J. Fréchet. *Chem. Lett.* 65–68, 1993.
42. T. Kato, M. Fukumasa, and J.M.J. Fréchet. *Chem. Mater.* 7:368–372, 1995.
43. T. Kato, H. Kihara, T. Uryu, A. Fujishima, and J.M.J. Fréchet. *Macromolecules* 25:6836–6841, 1992.
44. T. Kato, T. Uryu, F. Kaneuchi, C. Jin, and J.M.J. Fréchet. *Liq. Cryst.* 14:1311–1317, 1993.
45. T. Kato, H. Kihara, T. Uryu, S. Ujiie, K. Iimura, J.M.J. Fréchet, and U. Kumar. *Ferroelectrics* 148:161–167, 1993.
46. H. Kihara, T. Kato, T. Uryu, S. Ujiie, U. Kumar, J.M.J. Fréchet, D.W. Bruce, and D.J. Price. *Liq. Cryst.* 21:25–30, 1996.
47. S. Machida, T. Urano, K. Sano, and T. Kato. *Langmuir* 13:576–580, 1997.
48. M. Fukumasa, K. Takeuchi, and T. Kato. *Liq. Cryst.* 24:325–327, 1998.
49. T. Kato, A. Fujishima, and J.M.J. Fréchet. *Chem. Lett.* 919–922, 1990.
50. J.-W. Lee, J.-I. Jin, M.F. Achard, and F. Hardouin. *Liq. Cryst.* 30:1193–1199, 2003.
51. M.J. Wallage and C.T. Imrie. *J. Mater. Chem.* 7:1163–1167, 1997.
52. A. Takahashi, V.A. Mallia, and N. Tamaoki. *J. Mater. Chem.* 13:1582–1587, 2003.
53. D.J. Price, H. Adams, and D.W. Bruce. *Mol. Cryst. Liq. Cryst.* 289:127–140, 1996.
54. K. Willis, J.E. Luckhurst, D.J. Price, J.M.J. Fréchet, H. Kihara, T. Kato, G. Ungar, and D.W. Bruce. *Liq. Cryst.* 21:585–587, 1996.
55. H. Bernhardt, W. Weissflog, and H. Kresse. *Angew. Chem. Int. Ed. Engl.* 35:874–876, 1996.
56. R. Deschenaux, F. Monnet, E. Serrano, F. Turpin, and A.M. Levelut. *Helv. Chim. Acta.* 81:2072–2077, 1998.
57. Z. Sideratou, D. Tsiourvas, C.M. Paleos, and A. Skoulios. *Liq. Cryst.* 22:51–60, 1997.
58. M. Shoji and F. Tanaka. *Macromolecules* 35:7460–7472, 2002.
59. F. Tanaka. *Polym. J.* 34:479–509, 2002.
60. B.A. Veytsman. *Liq. Cryst.* 18:595–600, 1995.
61. R.P. Sear and G. Jackson. *Phys. Rev. Lett.* 74:4261–4264, 1995.

62. T. Kato, Y. Kubota, M. Nakano, and T. Uryu. *Chem. Lett.* 1127–1128, 1995.
63. K. Willis, D.J. Price, H. Adams, G. Ungar, and D.W. Bruce. *J. Mater. Chem.* 5:2195–2199, 1995.
64. D.J. Price, K. Willis, T. Richardson, G. Ungar, and D.W. Bruce. *J. Mater. Chem.* 7:883–891, 1997.
65. E. Nishikawa, J. Yamamoto, and H. Yokoyama. *Chem. Lett.* 454–455, 2001.
66. H.L. Nguyen, P.N. Horton, M.B. Hursthouse, A.C. Legon, and D.W. Bruce. *J. Am. Chem. Soc.* 126:16–17, 2004.
67. A. Kraft, A. Reichert, and R. Kleppinger. *Chem. Commun.* 1015–1016, 2000.
68. D. Goldmann, R. Dietel, D. Janietz, C. Schmidt, and J.H. Wendorff. *Liq. Cryst.* 24:407–411, 1998.
69. A.C. de Kock. *Z. Phys. Chem.* 48:129, 1904.
70. D. Vorlaender. *Ber. Dtsch. Chem. Ges.* 41:2033, 1908.
71. G.W. Gray and B. Jones. *J. Chem. Soc.* 236–244, 1955.
72. S.C. Zimmerman, F. Zeng, D.E.C. Reichert, and S.V. Kolotuchin. *Science* 271:1095–1098, 1996.
73. M. Suarez, J.-M. Lehn, S.C. Zimmerman, A. Skoulios, and B. Heinrich. *J. Am. Chem. Soc.* 120:9526–9532, 1998.
74. R. Kleppinger, C.P. Lillya, and C. Yang. *Angew. Chem. Int. Ed. Engl.* 34:1637–1638, 1995.
75. G. Gottarelli, G.P. Spada, and A. Garbesi. Templating, self-assembly and self-organization, in *Comprehensive Supramolecular Chemistry*, Vol. 9. J.P. Sauvage and M.W. Hosseini, Eds., Oxford: Pergamon, pp. 483–506, 1996.
76. J.T. Davis. *Angew. Chem. Int. Ed.* 43:668–698, 2004.
77. S. Bonazzi, M.M. de Morais, G. Gottarelli, P. Mariani, and G.P. Spada. *Angew. Chem. Int. Ed. Engl.* 32:248–250, 1993.
78. F. Ciuchi, G. Di Nicola, H. Franz, G. Gottarelli, P. Mariani, M.G.P. Bossi, and G.P. Spada. *J. Am. Chem. Soc.* 116:7064–7071, 1994.
79. K. Kanie, T. Yasuda, S. Ujiie, and T. Kato. *Chem. Commun.* 1899–1900, 2000.
80. K. Kanie, T. Yasuda, M. Nishii, S. Ujiie, and T. Kato. *Chem. Lett.* 480–481, 2001.
81. K. Kanie, M. Nishii, T. Yasuda, T. Taki, S. Ujiie, and T. Kato. *J. Mater. Chem.* 11:2875–2886, 2001.
82. T. Kato, T. Matsuoka, M. Nishii, Y. Kamikawa, K. Kanie, T. Nishimura, E. Yashima, and S. Ujiie. *Angew. Chem. Int. Ed.* 43:1969–1972, 2004.
83. E. Mezzina, P. Mariani, R. Itri, S. Masiero, S. Pieraccini, G.P. Spada, F. Spinozzi, J.T. Davis, and G. Gottarellli. *Chem. Eur. J.* 7:388–395, 2001.
84. S. Pieraccini, G. Gottarelli, P. Mariani, S. Masiero, L. Saturni, and G.P. Spada. *Chirality* 13:7–12, 2001.
85. F.H. Beijer, R.P. Sijbesma, H. Kooijman, A.L. Spek, and E.W. Meijer. *J. Am. Chem. Soc.* 120:6761–6769, 1998.
86. R.P. Sijbesma, F.H. Beijer, L. Brunsveld, B.J.B. Folmer, J.H.K.K. Hirschberg, R.F.M. Lange, J.K.L. Lowe, and E.W. Meijer. *Science* 278:1601–1604, 1997.
87. J.H.K. K. Hirschberg, A. Ramzi, R.P. Sijbesma, and E.W. Meijer. *Macromolecules* 36:1429–1432, 2003.
88. J. Yang, J.-L. Marendaz, S.J. Geib, and A.D. Hamilton. *Tetrahedron Lett.* 35:3665–3668, 1994.
89. J.A. Zerkowski, C.T. Seto, and G.M. Whitesides. *J. Am. Chem. Soc.* 114:5473–5475, 1992.
90. S.L. Forman, J.C. Fettinger, S. Pieraccini, G. Gottarelli, and J.T. Davis. *J. Am. Chem. Soc.* 122:4060–4067, 2000.
91. T. Kato and J.M.J. Fréchet. *The Polymeric Materials Encyclopedia, Synthesis, Properties and Applications.* J.C. Salamone, Ed., Boca Raton: CRC Press, p. 815, 1996.
92. U. Kumar, J.M.J. Fréchet, T. Kato, S. Ujiie, and K. Iimura. *Angew. Chem. Int. Ed. Engl.* 31:1531–1533, 1992.
93. T. Kato, H. Kihara, S. Ujiie, T. Uryu, and J.M.J. Fréchet. *Macromolecules* 29:8734–8739, 1996.
94. T. Kato, M. Nakano, T. Moteki, T. Uryu, and S. Ujiie. *Macromolecules* 28:8875–8876, 1995.
95. K. Araki, T. Kato, U. Kumar, and J.M.J. Fréchet. *Macromol. Rapid Commun.* 16:733–739, 1995.
96. E. Barmatov, S. Grande, A. Filippov, M. Barmatova, F. Kremer, and V. Shibaev. *Macromol. Chem. Phys.* 201:2603–2609, 2000.
97. C.G. Bazuin, F.A. Brandys, T.M. Eve, and M. Plante. *Macromol. Symp.* 84:183–196, 1994.
98. C.G. Bazuin and F.A. Brandys. *Chem. Mater.* 4:970–972, 1992.
99. F.A. Brandys and C.G. Bazuin. *Chem. Mater.* 8:83–92, 1996.
100. D. Stewart and C.T. Imrie. *J. Mater. Chem.* 5:223–228, 1995.
101. D. Stewart and C.T. Imrie. *Macromolecules* 30:877–884, 1997.
102. T. Kato, N. Hirota, A. Fujishima, and J.M.J. Fréchet. *J. Polym. Sci. Polym. Chem. Ed.* 34:57–62, 1996.
103. T. Kawakami and T. Kato. *Macromolecules* 31:4475–4479, 1998.

104. R.V. Tal'roze, S.A. Kuptsov, T.I. Sycheva, V.S. Bezborodov, and N.A. Platé. *Macromolecules* 28:8689–8691, 1995.

105. R.V. Tal'roze, S.A. Kuptsov, T.L. Lebedeva, G.A. Shandryuk, and N.D. Stepina. *Macromol. Symp.* 117:219–228, 1997.

106. C.G. Bazuin and A. Tork. *Macromolecules* 28:8877–8880, 1995.

107. X. Wu, G. Zhang, and H. Zhang. *Macroml. Chem. Phys.* 199:2101–2105, 1998.

108. S. Malik, P.K. Dhal, and R.A. Mashelkar. *Macromolecules* 28:2159–2164, 1995.

109. A.F. Thünemann, S. Kubowicz, C. Burger, M.D. Watson, N. Tchebotareva, and K. Müllen. *J. Am. Chem. Soc.* 125:352–356, 2003.

110. S. Ujiie and K. Iimura. *Chem. Lett.* 411–414, 1991.

111. S. Ujiie and K. Iimura. *Macromolecules* 25:3174–3178, 1992.

112. J. Ruokolainen, J. Tanner, O. Ikkala, G. ten Brinke, and E.L. Thomas. *Macromolecules* 31:3532–3536, 1998.

113. M.C. Luyten, G.O.R.A. van Ekenstein, J. Wildeman, G. ten Brinke, J. Ruokolainen, O. Ikkala, M. Torkkeli, and R. Serimaa. *Macromolecules* 31:9160–9165, 1998.

114. C.M. Paleos. *Mol. Cryst. Liq. Cryst.* 243:159–183, 1994.

115. S. Ujiie, S. Takagi, and M. Sato. *High Perform. Polym.* 10:139–146, 1998.

116. M. Antonietti and J. Conrad. *Angew. Chem. Int. Ed. Engl.* 33:1869–1879, 1994.

117. C.K. Ober and G. Wegner. *Adv. Mater.* 9:17–31, 1997.

118. O. Ihata, H. Yokota, K. Kanie, S. Ujiie, and T. Kato. *Liq. Cryst.* 27:69–74, 2000.

119. T. Kato, O. Ihata, S. Ujiie, M. Tokita, and J. Watanabe. *Macromolecules* 31:3551–3555, 1998.

120. O. Ihata and T. Kato. *Sen'i Gakkaishi* 55:274–278, 1999.

121. J.L.M. van Nunen, B.F.B. Folmer, and R.J.M. Nolte. *J. Am. Chem. Soc.* 119:283–291, 1997.

122. G. Ambrozic and M. Zigon. *Macromol. Rapid Commun.* 21:53–56, 2000.

123. C.B. St Pourcain and A.C. Griffin. *Macromolecules* 28:4116–4121, 1995.

124. H. Kihara, T. Kato, T. Uryu, and J.M.J. Fréchet. *Liq. Cryst.* 24:413–418, 1998.

125. H. Kihara, T. Kato, and T. Uryu. *Trans. Mater. Res. Soc. Jpn.* 20:327–330, 1996.

126. P.V. Shibaev, S.L. Jensen, P. Andersen, K. Schaumburg, and V. Plaksin. *Macromol. Rapid Commun.* 22:493–497, 2001.

127. L.P. Stubbs and M. Weck. *Chem. Eur. J.* 9:992–999, 2003.

128. J.M. Pollino, L.P. Stubbs, and M. Weck. *J. Am. Chem. Soc.* 126:563–567, 2004.

129. J.M. Pollino, L.P. Stubbs, and M. Weck. *Macromolecules* 36:2230–2234, 2003.

130. F. Ilhan, M. Gray, and V.M. Rotello. *Macromolecules* 34:2597–2601, 2001.

131. R. Bertani, P. Metrangolo, A. Moiana, E. Perez, T. Pilati, G. Resnati, I. Rico-Lattes, and A. Sassi. *Adv. Mater.* 14:1197 1201, 2002.

132. T. Ohtake, M. Ogasawara, K. Ito-Akita, N. Nishina, S. Ujiie, H. Ohno, and T. Kato. *Chem. Mater.* 12:782–789, 2000.

133. M. Yoshio, T. Mukai, K. Kanie, M. Yoshizawa, H. Ohno, and T. Kato. *Adv. Mater.* 14:351–354, 2002.

134. K. Kishimoto, M. Yoshio, T. Mukai, M. Yoshizawa, H. Ohno, and T. Kato. *J. Am. Chem. Soc.* 125:3196–3197, 2003.

135. M. Yoshio, T. Mukai, H. Ohno, and T. Kato. *J. Am. Chem. Soc.* 126:994–995, 2004.

136. O. Ikkala and G. ten Brinke. *Mater. Res. Soc. Symp. Proc.* 775:213–223, 2003.

137. R. Mäki-Ontto, K. de Moel, E. Polushkin, G.A. van Ekenstein, G. ten Brinke, and O. Ikkala. *Adv. Mater.* 14:357–361, 2002.

138. J. Ruokolainen, R. Mäkinen, M. Torkkeli, T. Mäkelä, R. Serimaa, G. ten Brinke, and O. Ikkala. *Science* 280:557–560, 1998.

139. H. Kosonen, S. Valkama, J. Hartikainen, H. Eerikainen, M. Torkkeli, K. Jokela, R. Serimaa, F. Sundholm, G. ten Brinke, and O. Ikkala. *Macromolecules* 35:10149–10154, 2002.

140. D.A. Tomalia and J.M.J. Fréchet. *J. Polym. Sci. Part A: Polym. Chem.* 40:2719–2728, 2002.

141. J.M.J. Fréchet. *Proc. Natl. Acad. Sci. USA* 99:4782–4787, 2002.

142. V. Percec, C.-H. Ahn, G. Ungar, D.J.P. Yeardley, M. Möller, and S.S. Sheiko. *Nature* 391:161–164, 1998.

143. A.W. Bosman, R. Vestberg, A. Heumann, J.M.J. Fréchet, and C.J. Hawker. *J. Am. Chem. Soc.* 125:715–728, 2003.

144. V. Abetz. Assemblies, in *Complex Block Copolymer Systems in Supramolecular Polymers.* A. Ciferri, Ed., New York: Marcel Dekker, pp. 215–262, 2000.

145. A. Sato, T. Kato, and T. Uryu. *J. Polym. Sci. Part A: Polym. Chem.* 34:503–505, 1996.

146. R.F.M. Lange and E.W. Meijer. *Macromolecules* 28:782–783, 1995.

147. A. Harada, J. Li, and M. Kamachi. *Nature* 356:325–327, 1992.

148. H.W. Gibson, N. Yamaguchi, and J.W. Jones. *J. Am. Chem. Soc.* 125:3522–3533, 2003.

149. H.-R. Tseng, S.A. Vignon, P.C. Celestre, J. Perkins, J.O. Jeppesen, A. Di Fabio, R. Ballardini, M.T. Gandolfi, M. Venturi, V. Balzani, and J.F. Stoddart. *Chem. Eur. J.* 10:155–172, 2004.

150. T. Kato, T. Kutsuna, K. Hanabusa, and M. Ukon. *Adv. Mater.* 10:606–608, 1998.

151. N. Mizoshita, K. Hanabusa, and T. Kato. *Adv. Mater.* 11:392–394, 1999.

152. M. Moriyama, N. Mizoshita, T. Yokota, K. Kishimoto, and T. Kato. *Adv. Mater.* 15:1335–1338, 2003.

153. J.-F. Gohy, Bas G.G. Lohmeijer, and U.S. Schubert. *Chem. Eur. J.* 9:3472–3479, 2003.

154. H. Kihara, T. Kato, and T. Uryu. *Macromol. Rapid Commun.* 18:281–286, 1997.

155. T. Shimizu. *Polym. J.* 35:1–22, 2003.

156. E.R. Zubarev, M.U. Pralle, E.D. Sone, and S.I. Stupp. *J. Am. Chem. Soc.* 123:4105–4106, 2001.

Chapter 6

Hydrogen-Bonded Supramolecular Polymers: Linear and Network Polymers and Self-Assembling Discotic Polymers

Perry S. Corbin and Stephen C. Zimmerman

CONTENTS

I. INTRODUCTION

Of the innumerable chemical advances in the 20th century, one could argue that the development of polymers has had the most dramatic impact upon everyday living. Spurred by an overwhelming accumulation of fundamental knowledge in chemistry and physics, polymer science has blossomed into a field that has shaped, in part, both industry and academia. As part of the evolution of

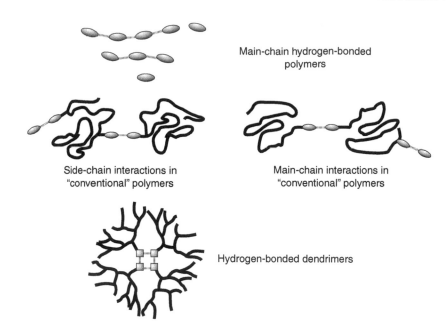

Figure 1 A schematic representation of different classes of hydrogen-bonded, supramolecular polymers.

this field, numerous strategies have been developed for synthesizing polymers. These polymers, viewed simplistically, consist of covalently linked repeat units derived from single molecule precursors (monomers). However, recently there have been reports of "supramolecular polymers" [1] constructed from noncovalently associated monomers. Certainly, these studies challenge Carothers' original notion that structural repeat units "are not capable of independent existence" [2]. Taken more broadly, supramolecular polymers encompass specific, noncovalent interactions between repeat units and side chains of classical, covalent polymers, as well as interactions of these polymers with small molecules or metal ions.

One of the most formidable challenges in developing supramolecular polymers is the design of "sticky" monomers, or polymer substituents, that associate strongly in a predictable manner. Because of their strength and directionality, hydrogen bonds and transition metal coordination are ideal "glues" for assembling such polymers. Although a variety of elegant hydrogen-bonded supermolecules [3] and metal-scaffolded assemblies [4] have been reported, ranging from solution aggregates to engineered crystals, the study of supramolecular polymers is still in its infancy. Progress in this area is reported herein. For related reviews of supramolecular polymers, see Ref 5.

One should be aware that studies of the type described are multidisciplinary, representing an opportune merging of organic, inorganic, and materials chemistry [5a]; and, thus, a report of this nature could be written from any of several different perspectives. We have taken the relatively broad view of supramolecular polymers alluded to above. However, this account will primarily focus on linear hydrogen-bonded polymers and networks (Figure 1) that have been rationally designed to form in liquid-crystalline, isotropic solution, and bulk phases and will be preceded by a general description of hydrogen-bonding subunits for supramolecular polymer construction. Cyclic, hydrogen-bonded liquid crystals and dendrimers, as well as the metal-directed assembly of supramolecular polymers in isotropic solution are also briefly introduced.[1] Although hydrogen-bonded tapes, ribbons, helices, and sheets are relatively common structural motifs found in crystals, and such solid-state aggregates may be classified as supramolecular polymers, an in-depth description of these structures is not given herein. Readers are referred to pertinent reviews for a discussion of this topic [6].

---

[1] Portions of this chapter were published previously in the first edition of *Supramolecular Polymers*.

II.   Hydrogen-Bonding Subunits for Supramolecular Polymer Construction

A.   General Considerations

One cannot initiate an account on hydrogen bonding without first recognizing the significance of this noncovalent, primarily electrostatic, force in biological systems [7]. From a molecular standpoint, biology offers several excellent examples of supramolecular, hydrogen-bonded polymers, including double- and triple-helical DNA, as well as protein β-sheets [8]. As architects of artificial assemblies, supramolecular chemists will always be challenged to contemplate and create systems that are as intricate and complex, yet as simple and efficient in function, as those found in nature.

When considering the assembly of nonnatural, supramolecular, hydrogen-bonded polymers, two questions immediately arise: (1) Why use hydrogen bonds? and (2) What types of building blocks should be used? As mentioned above, hydrogen bonds are relatively strong and directional, but hydrogen bonding is also reversible. Therefore, from a practical standpoint, hydrogen bonds are ideal for supramolecular polymer construction because they facilitate the assembly of polymers with substantial lengths and with well-defined, thermodynamically-controlled, primary structures. Moreover, because of the reversible nature of hydrogen bonding, polymers can be produced with lengths and, thus, properties that are highly dependent upon concentration, temperature, pH, and additives. As a result, supramolecular, hydrogen-bonded polymers represent an important, new class of functional materials.

Addressing the second question, a variety of building blocks can be envisioned to "synthesize" hydrogen-bonded polymers. Singly hydrogen-bonded complexes have been remarkably effective in affording liquid-crystalline polymers (*vide infra*). Typically, the subunits of these complexes have the advantage of being easy to synthesize. Moreover, individual hydrogen bonds, although weak, may sum in multivalent contacts to provide high stability. Nonetheless, robust, multiply hydrogen-bonded complexes are also attractive targets. It is especially critical to use such building blocks when designing solution aggregates because unfavorable entropic forces must be overcome in dilute solution. Furthermore, multiply hydrogen-bonded, liquid-crystalline polymers may have more stable mesophases than their singly hydrogen-bonded counterparts. Polymers may also be created from multiply hydrogen-bonding subunits that are stable at relatively high temperatures and under significant strain, but yet have reversible, thermally-dependent bulk properties. Solution studies of some recently reported hydrogen-bonding building blocks are discussed below.

B.   Models for Predicting the Stability of Multiply Hydrogen-Bonded Complexes

Before describing specific building blocks, it is important to make note of several issues that influence complex stability. Although it is difficult to predict the strength (i.e., association constant, K_{assoc}) of a multiply hydrogen-bonded complex *a priori*, theoretical and experimental investigations have aided in the dissection of components that affect binding strength in solution. Primary hydrogen-bond strength, that is, the energy of a single hydrogen bond between a hydrogen-bond acceptor (A) and donor (D), is determined by a number of factors, including the acidity and basicity of A and D, and geometry [3e,9,10].

Thus, complexes with the same number of hydrogen bonds have been found to have very different stabilities. In many cases, the differences depend upon the arrangement of the donors and acceptors within the complex. In short, complex strength does not depend solely upon the number of hydrogen bonds in the complex. This observation, along with computational studies, led Jorgensen and Pranata to propose that secondary electrostatic interactions are critical in determining complex stability [11,12]. Their hypothesis is illustrated in Figure 2, with hydrogen-bond donors represented as positive charges and hydrogen-bond acceptors as negative charges. The model specifically suggests that in addition to attractive primary electrostatic interactions there are important secondary interactions (attractive or repulsive) between neighboring hydrogen-bonding sites. This model explained

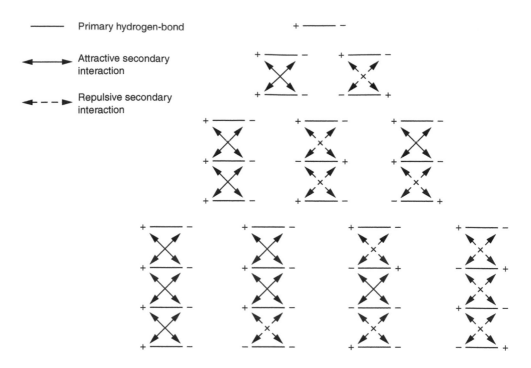

Figure 2 A schematic representation of Jorgensen's secondary hydrogen-bonding hypothesis for complexes containing one to four primary hydrogen bonds (adapted from W.L. Jorgensen and J. Pranata *J. Am. Chem. Soc.* 112:2008–2010, 1990 and J. Pranata, S.G. Wierschke, and W.L. Jorgensen. *J. Am. Chem. Soc.* 114:4010–4011, 1991.)

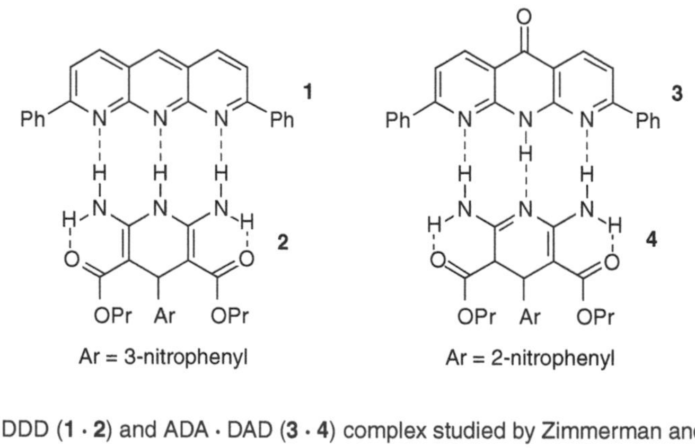

Figure 3 An AAA · DDD (**1** · **2**) and ADA · DAD (**3** · **4**) complex studied by Zimmerman and Murray [9,13].

the observation that K_{assoc} for a guanine:cytosine (AAD · DDA) base pair in CDCl$_3$ is over 100 times larger than that for a uracil:2,6-diaminopyridine (ADA · DAD) complex. It further predicted the general stability series, AAA · DDD > AAD · DDA > ADA · DAD, for triply hydrogen-bonded complexes with equivalent primary interactions.

Experimental studies on a series of triply hydrogen-bonded complexes by Zimmerman and Murray provided empirical support for the secondary hydrogen-bonding hypothesis [9,13]. For example, AAA · DDD complex **1** · **2**, has a K_{assoc} in chloroform that is more than 10^4 times larger than the K_{assoc} for **3** · **4**, an ADA · DAD complex (Figure 3). The primary hydrogen bonds in **1** · **2** and **3** · **4** are expected to be similar in nature. Therefore, the difference in stability may arise because of the varied A/D arrangement and, thus, differences in secondary hydrogen-bonding interactions.

Subsequently, Sartorius and Schneider have proposed that the free energy of association (ΔG_{assoc}) for a hydrogen-bonded complex in chloroform can be described by two increments, one representing primary interactions and another representing secondary interactions [14]. Based upon this linear free energy relationship, the energetic contribution of a single primary hydrogen bond is estimated to be 7.9 kJ/mol (1.9 kcal/mol), and the contribution from a secondary interaction, either attractive or repulsive, is estimated to be approximately 2.9 kJ/mol (0.7 kcal/mol). In an earlier analysis, the energy of a hydrogen bond was estimated to be 5 ± 1 kJ/mol using only a single increment [15].

Recent theoretical investigations have questioned the validity of the secondary hydrogen-bond hypothesis and the calculation of complexation energies via empirical increments [16]. These models continue to be debated [17]. Moreover, based upon studies of flexible tetrapeptide analogs, Gardner and Gellman have demonstrated that secondary interactions do not necessarily determine the optimal arrangement of hydrogen-bonding sites in flexible systems [18].[2] Thus, one cannot expect all hydrogen-bonded complexes to have predictable complexation energies. Experimental studies have, however, led to a database of stability constants and synthetic procedures for the supramolecular polymer chemist.

C. Examples of Multiply Hydrogen-Bonding Building Blocks

Recently, there have been several reports of building blocks that are capable of forming exceptionally robust complexes [3e,21]. Examples include $1 \cdot 2$ and the complexes shown in Figure 4. A description of singly hydrogen-bonded complexes is reserved for the discussion on hydrogen-bonded, liquid-crystalline polymers. Being cued by nature, the majority of subunits, including those in Figure 4, are heterocycles. Two structurally related self-complementary hydrogen-bonded dimers of interest are ureidopyrimidine dimer $5 \cdot 5$ reported by Meijer, Sijbesma and coworkers [22–24] and the dimers formed from ureidopyridopyrimidine 6 and its tautomers, which was studied by Zimmerman and coworkers [25,26]. Dimer $5 \cdot 5$ associated strongly in chloroform with a dimerization constant (K_{dim}) $> 10^6 \, M^{-1}$ and was also formed in the solid state as evidenced in the crystal structure of 5 [23]. More accurate dimerization constants, $K_{dim} = 2 \times 10^7 \, M^{-1}$ in chloroform and $K_{dim} \sim 10^8 \, M^{-1}$ in toluene, were determined using a fluorescently-tagged derivative of 5 [24]. Meijer and coworkers have demonstrated that ditopic monomers containing 5 are useful for constructing hydrogen-bonded polymers with intriguing properties (*vide infra*).

Corbin and Zimmerman have pointed out that prototropy (i.e., tautomerism involving proton shifts) can be detrimental to hydrogen-bonded complex formation [25]. Therefore, heterocycle 6 was designed to contain a self-complementary hydrogen-bonding array irrespective of its protomeric form. The dimer formed from the N-3(H) protomer, $6 \cdot 6$, is shown in Figure 4. An N-1(H) homodimer and N-1(H)/N-3(H) heterodimer were also formed. Interestingly, all three of these dimers maintained a similar spatial arrangement of an alkyl substituent, and association was strong, with $K_{dim} \geq 10^7 \, M^{-1}$ estimated from ^{1}H NMR dilution studies in chloroform. Fluorescence studies analogous to those of Sijbesma and coworkers [24], provided a lower limit for the dimerization constant, $K_{dim} \geq 5 \times 10^8 \, M^{-1}$, of a pyrene-substituted derivative of 6 in dried chloroform [26]. Dimerization of the N-3(H) protomeric form was observed in the x-ray crystal structure of a trityl-substituted derivative of 6.

In addition to modules 5 and 6, compound 7 contains a self-complementary ADAD hydrogen-bonding array [27]. Although, dimerization of 7 is weaker ($K_{dim} = 2 \times 10^4 \, M^{-1}$) than that of 5 and 6 — as theoretically expected based upon secondary hydrogen-bonding interactions — ditopic monomers containing 7 form helical supramolecular polymers via a combination of hydrogen-bonding and solvophobic effects (*vide infra*).

---

[2]For a study of the effect of "freely rotating bonds" upon host–guest complexation see Ref. 19; for a study of the effect of these bonds upon hydrogen-bond complexation see Ref. 20.

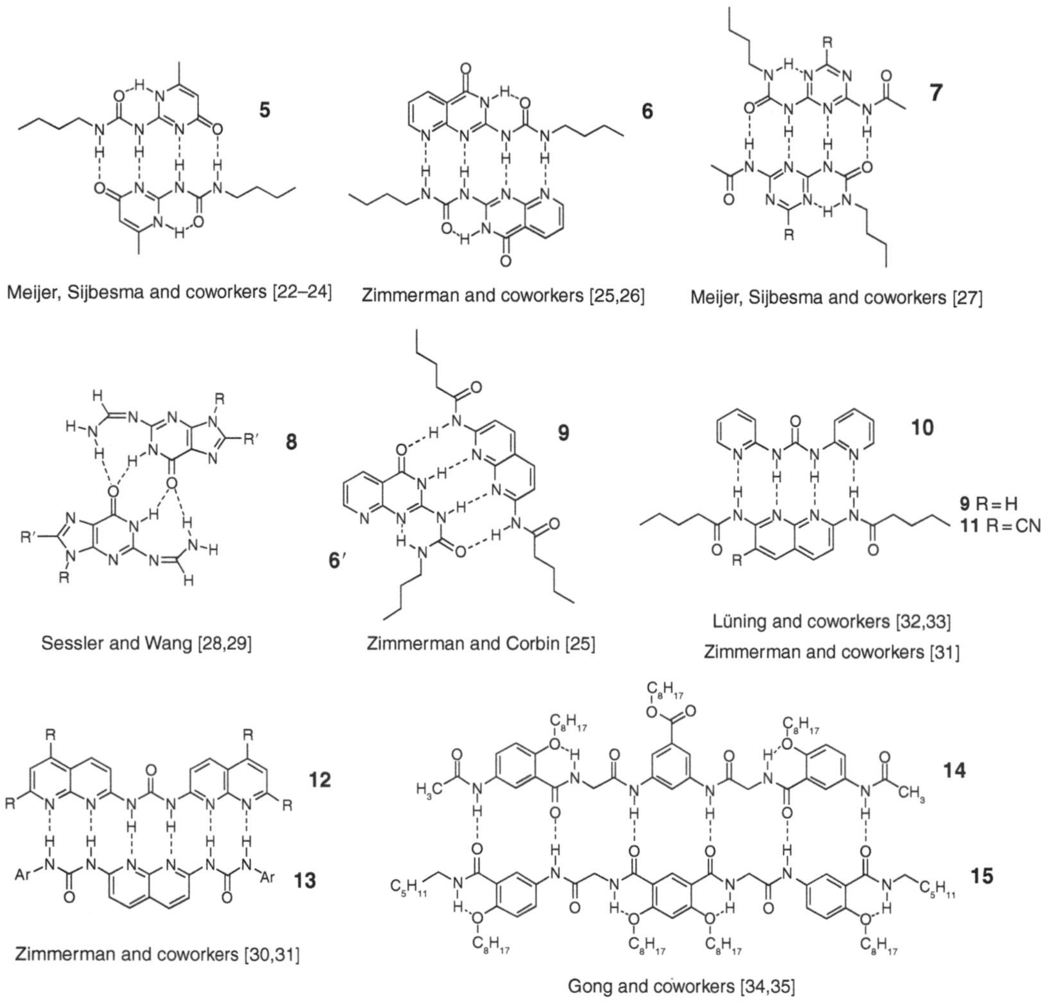

Figure 4 Recently reported examples of multiply hydrogen-bonded complexes.

Another example of a self-complementary hydrogen-bonding subunit is modified guanosine **8** reported by Sessler and Wang [28,29]. Upon linking two units of **8** with a rigid spacer, a robust "tetrameric" array forms in a number of solvents.

Although the dimers described thus far form by self-complementary hydrogen bonding, heterodimeric complexes may also be effective for supramolecular polymer construction. Such complexes include **1 · 2, 6' · 9, 10 · 11(9), 12 · 13**, and **14 · 15**. Complex **6' · 9** (K_{assoc} = 3500 M^{-1}, 5% DMSO-d_6/CDCl$_3$) is formed by a conformational switch in ureidopyridopyrimidine **6**.[3] Likewise, unfolding and association of AADDAA module **12** and DDAADD module **13** provides heterodimer **12 · 13**, which is held together by six hydrogen bonds [30,31].

Nonheterocyclic modules **14** and **15** associate via amide donor–acceptor interactions reminiscent of those responsible, in part, for the ordered secondary structures of oligopeptides and proteins [34,35]. Association of **14** and **15** is exceptionally strong in chloroform, K_{assoc} = 1.3 × 10^9 M^{-1}, as indicated by isothermal microcalorimetry. Gong and coworkers have also assembled self-complementary hydrogen-bonded dimers from compounds analogous to **14** and **15** [38].

---

[3]Li and Chen have recently described a heterodimer formed from modules analogous to **5** and **9** [36,37]. The authors have also studied ADDA · DAAD complexes formed from diamidonaphthyridine and hydrazide-based modules [36].

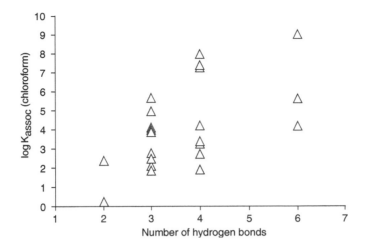

Figure 5 A graphical representation of the variation of K_{assoc} (chloroform) with the number of hydrogen bonds for some recently reported complexes. K_{assoc} values include those for aggregates in Figure 4 and additional complexes described in Refs. 3e, 31, 39, and 40.

The hydrogen-bonding modules described herein are just a few examples of the types of building blocks that have been or have the potential to be useful in the construction of supramolecular polymers. A plot relating the log K_{assoc} for a series of complexes to the corresponding number of hydrogen bonds within the complexes is shown in Figure 5.

Most of the studies described in this section are solution studies. Thus, when using such units to design supramolecular aggregates in different states, for example, liquid-crystalline and melt phases, one must consider whether solution studies provide adequate precedent for the formation of hydrogen-bonded complexes in these phases. Indeed, if similar hydrogen-bonded complexes are formed both in the solid state and in solution, it seems reasonable to conclude that comparable complexes may be formed, for example, in liquid-crystalline mesophases. However, one should be aware of this "phase-problem."

III. HYDROGEN-BONDED, SUPRAMOLECULAR LIQUID-CRYSTALLINE POLYMERS

A. General Considerations

The general premise of using hydrogen bonds to construct low-molecular weight [41] mesogenic supermolecules (e.g., dimers and trimers) has been demonstrated in a number of accounts and was initially evoked to explain the mesomorphic behavior of carboxylic acids,[4] for example, 4-alkoxybenzoic acids [43]. In this case, as well as other examples described below, intermolecular hydrogen bonding between two subunits is believed to afford a rigid core which, along with a flexible side chain(s), promotes liquid crystallinity.

Analogously, Fréchet and coworkers [44–46] and Lehn and coworkers [47] have demonstrated the ability to construct heterodimeric mesogens from independent, complementary hydrogen-bonding units (Figure 6). An example of Fréchet's and Kato's work is the mesogenic unit arising from the singly hydrogen-bonded complex of 4-butoxybenzoic acid (**16**) and stilbazole (**17**) [44].[5] A 1:1 mixture of these compounds exhibited a nematic mesophase over a significantly broader temperature range than did **16** and **17** alone and had a smectic phase that was not observed for the individual

---

[4]For additional reviews of hydrogen-bonded liquid crystals (including polymers) see Ref. 42.
[5]For an early review of Fréchet's work on hydrogen-bonded supramolecular liquid crystals see Ref. 45 and references therein.

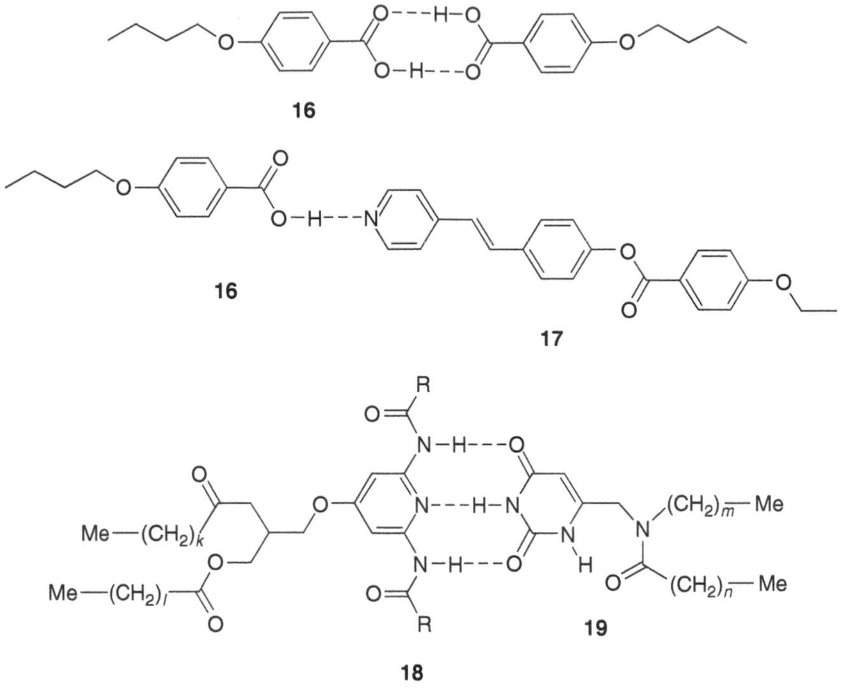

16

16

17

Me—(CH₂)ₖ

Me—(CH₂)ₗ

18

R

(CH₂)ₘ—Me

(CH₂)ₙ—Me

19

Figure 6   Typical examples of low-molecular weight, hydrogen-bonded, liquid-crystalline complexes [43–47].

molecular components. Hydrogen-bonded complex formation was verified by infrared spectroscopy (IR) [46]. Furthermore, Lehn's triply hydrogen-bonded complex **18 · 19** exhibited a "metastable" mesophase depending upon the length of the side chains [47].[6] This particular liquid-crystalline complex is attractive because it is formed from nonmesomorphic subunits.

The examples above clearly demonstrated the viability of using hydrogen bonds to construct novel mesogens and raised the question as to whether hydrogen-bonding units could be used to construct mesomorphic, hydrogen-bonded polymers. Indeed, much of the work in the area of supramolecular, hydrogen-bonded polymers, to date, has been devoted to the study of thermotropic liquid crystals. These polymers can be generically classified as either main-chain or side-chain polymers [5d]. A third class consisting of networked, hyperbranched, and dendritic polymers may also be considered [5a]. Examples of these structures are dispersed within the discussions of main-chain and side-chain, liquid-crystalline systems.

B. Main-Chain Polymers

One could envision assembling supramolecular polymers from ditopic units containing hydrogen-bonding complements A and B similar to those described in the preceding sections. The ditopic subunits could, conceivably, be constructed with two units of either A or B (i.e., A–A and B–B) or with a single unit containing both A and B (i.e., A–B). Hydrogen bonding between A and B would then serve as the glue, or perhaps more appropriately the molecular "velcro" (suggesting a greater ease in reversibility) that holds the polymer together (Figure 7). Ditopic units containing a self-complementary hydrogen-bonding component C could be used in a similar manner (Figure 7). Polymers similar to those in Figure 7 are what we and others refer to as main-chain polymers [5d,49].

---

[6]For studies of a related complex in the solid state and solution see Ref. 48 and references therein.

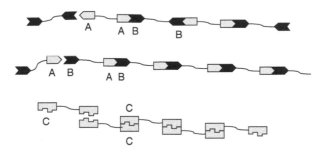

Figure 7 A schematic representation of main-chain, hydrogen-bonded polymers constructed from ditopic building blocks.

As an example, Griffin and coworkers have described liquid-crystalline, main-chain supra-molecular polymers constructed from hydrogen-bonded dipyridyl and diacid units (Figure 8) [41,49]. Complex 8(a), for instance, showed a nematic, as well as a monotropic smectic phase. Maximum transition temperatures were obtained with a 1:1 mixture of the two components, which the authors suggest is an indication of linear chain formation and reflects a step-growth polymerization process [41]. Further evidence (IR, small-angle x-ray scattering, etc.) supporting the formation of liquid-crystalline, hydrogen-bonded polymers from dipyridyl and diacid units has been recently reported [50]. However, it is difficult to estimate the average degree of polymerization (DP) for the 1:1 liquid-crystalline complexes.

As an extension of the dimeric liquid-crystalline complex **18 · 19** described above, Lehn and colleagues have characterized polymeric systems assembled from ditopic units containing these two components [1,51]. Multiple hydrogen-bonding interactions between the diamidopyridine and uracil building blocks leads to polymeric structures (Figure 8(b)) that have thermotropic (hexagonal columnar) mesophases over broad temperature ranges from below room temperature to greater than 220°C. The isotropization temperature for the complex is significantly higher than the isotropic to metastable mesophase transition temperature, 73°C, observed for **18·19** and appears to support supra-molecular polymer formation. This is in analogy with classical, liquid-crystalline polymers, which display increasing isotropization temperatures with increasing molecular weight [52].[7] Moreover, x-ray diffraction studies led to the proposal of a triple-helical superstructure for the polymer in which both ditopic units contain the linker derived from L(+) tartaric acid [51]. Likewise, Lehn has reported the self-assembly of rigid rods from a related ditopic pyridine and uracil unit with rigid, aromatic spacers (Figure 8(c)) [53]. A lyotropic mesophase is formed upon cooling a hot solution of the two components (1:1 mixture, 50 mM in 1,1,2,2,-tetrachloroethane) and can be attributed to hydrogen-bonded polymer formation.

The polymers represented in Figure 7 are based upon ditopic hydrogen-bonding subunits. However, multisite (>2 sites) building blocks might also be used. A schematic representation of an aggregate resulting from interaction of hydrogen-bonding complementary tritopic and ditopic components is shown in Figure 9. If the linker of one or both molecular components is flexible then cross-linking should arise and networked polymers would be produced. On the other hand, if rigid linkers are used then cross-linking should be inhibited and a hyperbranched structure should form. Along these lines, Fréchet has reported the formation of liquid-crystalline network polymers from triacids and stilbazole derivatives (Figure 10). For example, mixtures of **20** and **21** were found to have a smectic A phase [54]. The fluidity of the proposed liquid-crystalline networks has been ascribed to the dynamic nature of hydrogen bonding, that is, the breaking and reforming of hydrogen bonds [54].

---

[7]For a description of the theory of main-chain liquid-crystalline polymer formation see Chapters 2 and 3.

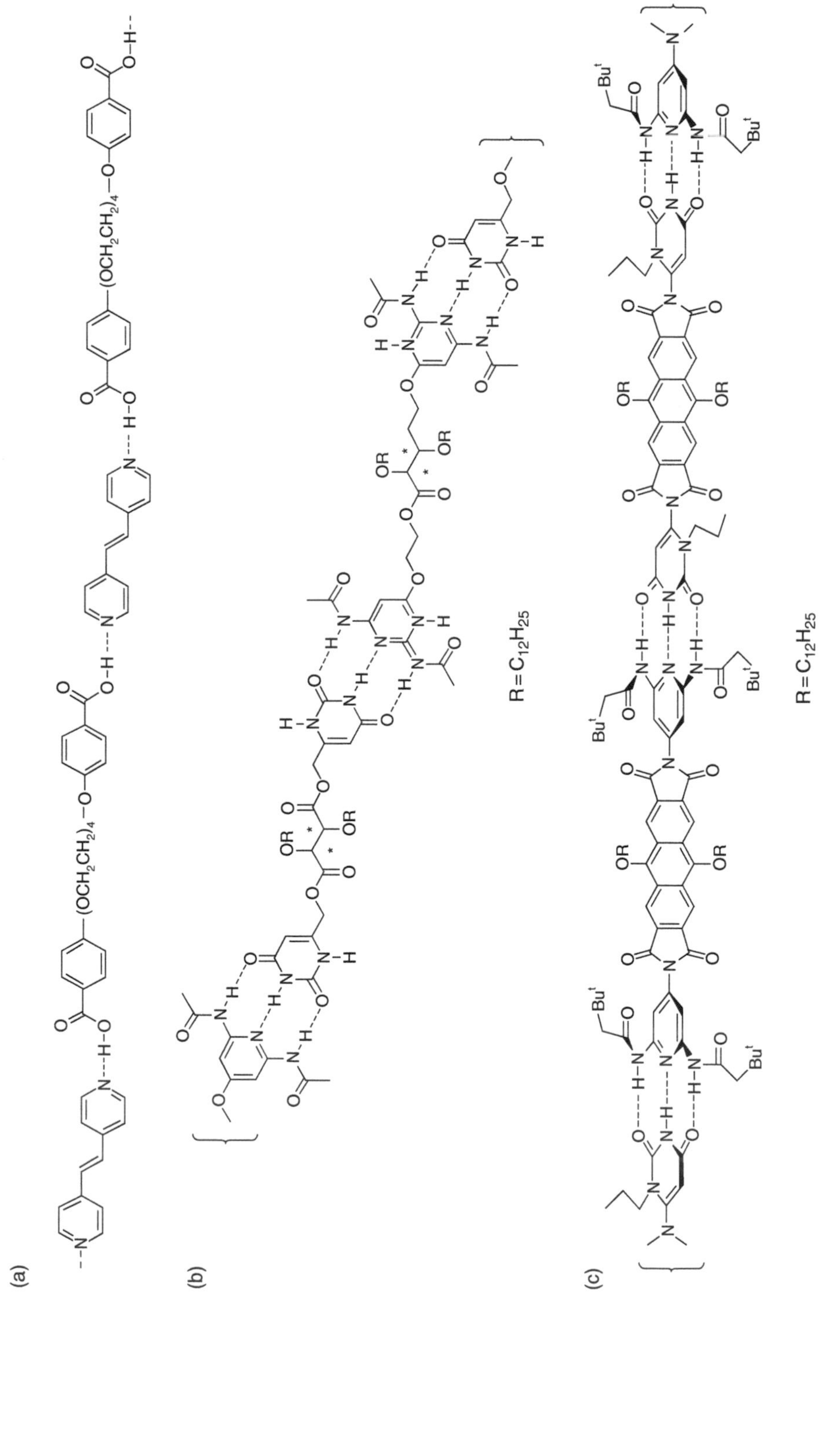

Figure 8 Representative examples of main-chain, hydrogen-bonded, liquid-crystalline polymers [41,49–53].

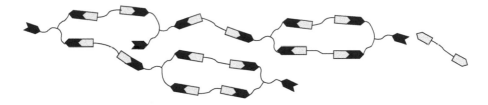

Figure 9 A schematic representation of cross-linked polymers formed from a ditopic and tritopic hydrogen-bonding unit (adapted from H. Kihara, T. Kato, T. Uryu, and J.M.J. Frechet *Chem. Mater.* 8:961–968, 1996.).

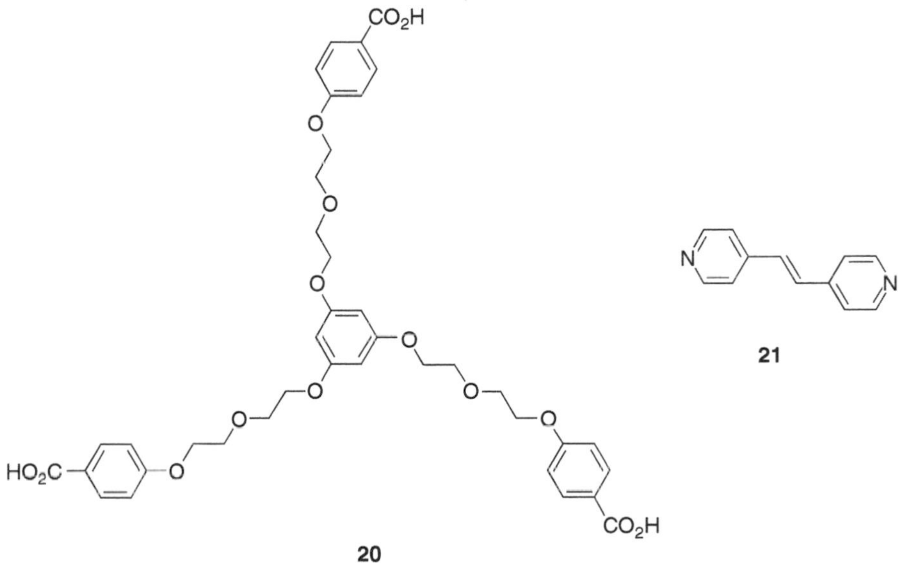

20

21

Figure 10 Subunits used by Fréchet and coworkers to construct liquid-crystalline, network polymers [54].

In related work, Griffin and coworkers have reported hydrogen-bonded polymers constructed from **22**, **23**, and **24** (Figure 11) [55]. In these studies, a mixture of **22** and **24** was found to have temperature-dependent rheological properties consistent with the formation of reversible hydrogen-bonded networks. Furthermore, complex **23 · 24** exhibited a mesophase, which has been attributed to a ladder-type structure in which **23** is in a rodlike conformation [55]. Interestingly, the mesogenicity of the complex is lost upon extensive heating above the isotropization temperature. This is believed to arise because of a conformational switch of **23** from the rodlike form to a "pseudotetrahedral" conformation — thus, leading to the formation of extended hydrogen-bonded networks.

C. Side-Chain Polymers

The second generic class of hydrogen-bonded, liquid-crystalline polymers is side-chain polymers. The general design of these structures entails incorporation of hydrogen-bonding sites into the side chains of covalent polymers, and is based upon the desire to enhance the "mesomorphicity" of small molecule mesogens and polymers by hydrogen-bonding interactions between the two [56]. A schematic representation of this process is shown in Figure 12. The concept of liquid-crystalline side-chain polymers, and thus most of the work in this area, is a product of Fréchet and Kato.

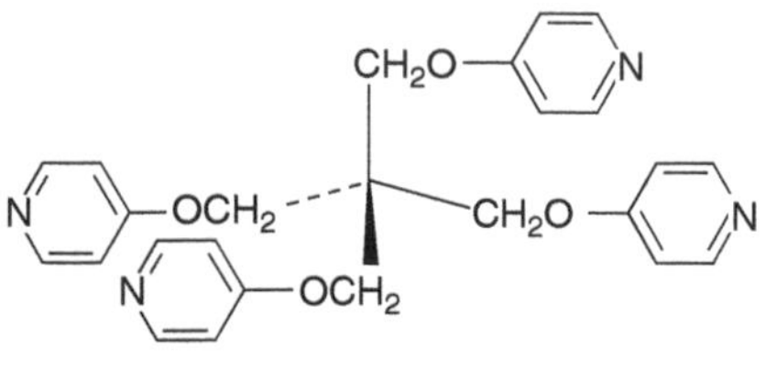

22

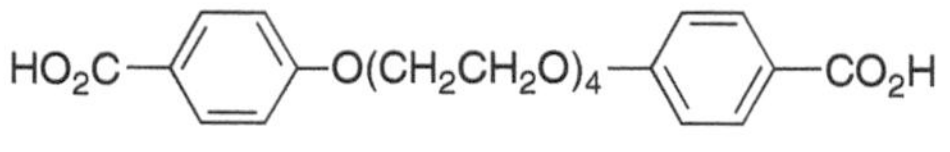

24 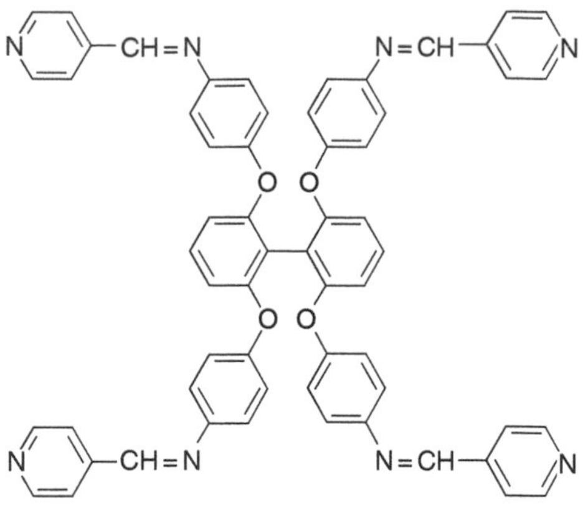   **23**

Figure 11 Subunits used by Griffin and coworkers to construct supramolecular networks [55].

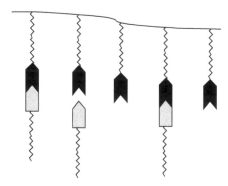

Figure 12 A schematic representation of side-chain/small molecule hydrogen-bonding interactions in liquid-crystalline polymers.

Subsequently, an excellent summary of these types of aggregates has been reported elsewhere [45][8]; therefore, an example will suffice to demonstrate this concept.

The first hydrogen-bonded liquid-crystalline, side-chain polymer reported by Fréchet used a polyacrylate covalent polymer with pendant, hexyloxybenzoic acid side chains (**25**), and stilbazole **26** (Figure 13) [56]. As a result, a 1:1 mixture (with respect to the side chain) formed a nematic phase over a broad temperature range from 140°C to 252°C, exceeding that of the individual mesomorphic components. This strategy has been extended to polysiloxane polymers, as well as to systems with varying small molecule mesogens [45]. Additionally, when using a ditopic hydrogen-bonding unit, network polymers, represented schematically in Figure 14, have been proposed to form [57].

Related to side-chain polymers, liquid crystallinity may also be afforded by the interaction of small molecules with hydrogen-bonding groups that are incorporated into the backbone of a covalent polymer. Liquid crystals of this type, represented schematically in Figure 15(a), have been referred to as "combined" polymers [42c]. An example is the liquid-crystalline polymer arising from hydrogen bonding between the pyridyl amide groups of polyamide **27** and the carboxylic acid group of **28** (Figure 15(b)) [42c,58].

---

[8]For further discussion of liquid-crystalline side-chain polymers see Chapters 5 and 6.

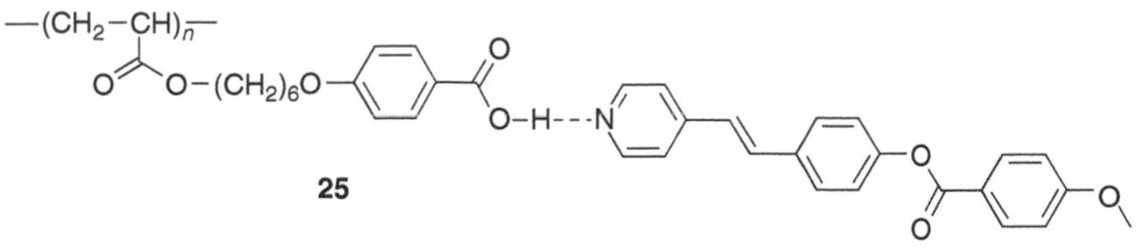

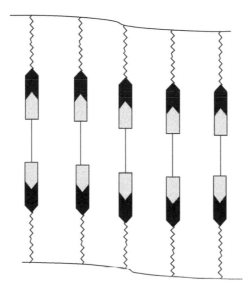

Figure 13 A representative example of a side-chain liquid-crystalline polymer [45,56].

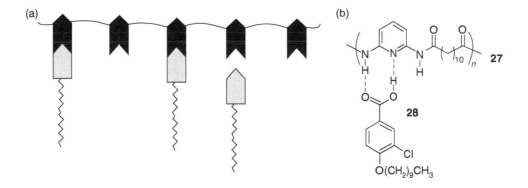

Figure 14 A schematic representation of cross-linking in liquid-crystalline side-chain polymers (adapted from T. Kato, H. Kihara, U. Kumar, T. Uryu, and J.M.J Frechet. *Angew. Chem. Int. Ed.* 33:1644–1645, 1994. With permission).

Figure 15 A schematic representation of main-chain/small molecule hydrogen-bonding interactions in liquid-crystalline polymers; (b) a representative example of a "combined" liquid-crystalline polymer [42c,58].

D. Liquid Crystals from Cyclic, Hydrogen-Bonded Aggregates

In addition to the liquid-crystalline polymers described in previous sections, liquid crystals may also form by the self-organization of discrete, cyclic, hydrogen-bonded aggregates. Along these

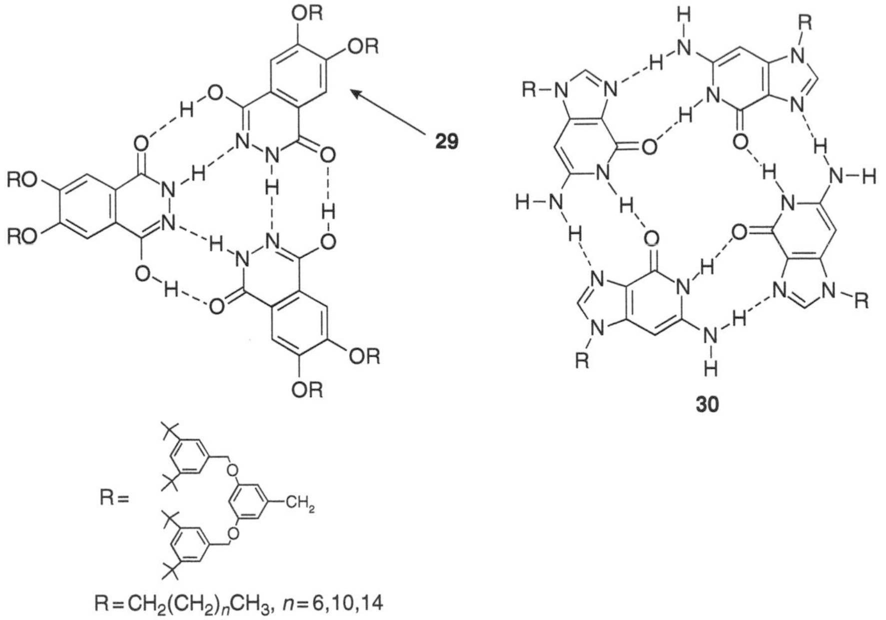

Figure 16 Examples of liquid-crystalline, disk-shaped aggregates [62,63].

lines, there have been several excellent examples of cyclic, hydrogen-bonded aggregates in solution [3b,59–61] and in mesophases. Stacking of the hydrogen-bonded disks in these systems affords supramolecular polymers. Two examples of hydrogen-bonded disks that form mesophases are described below.

The first example to be discussed, reported by Lehn, Zimmerman, and coworkers is based upon a phthalhydrazide subunit [62]. In these studies, lactim–lactam protomer **29** (Figure 16), containing Fréchet-type dendrons, trimerized in solution, as evidenced by ^{1}H NMR and size-exclusion chromatography (SEC). Furthermore, columnar, discotic mesophases were observed for alkyl-substituted versions of **29**. Depending upon the length of the peripheral side chains on the heterocycle, hexagonal, and/or rectangular columnar mesophases were observed. The formation of these thermotropic, discotic liquid crystals can be attributed to the formation of hydrogen-bonded trimers and subsequent stacking of these disks.

Similarly, Gotarelli, Spada, Garbesi, Mariani, and coworkers have described lyotropic mesophases formed from stacked, guanosine tetramers (**30**) in water (Figure 16) [63]. This strategy has been extended to a series of oligomers containing two to five guanosines [63–66]. Lyotropic phases have also been reported to form from an oligomer, d($G_pG_pA_pG_pG$), in which the central guanosine is replaced by an adenosine [67].

IV. HYDROGEN-BONDED, SUPRAMOLECULAR POLYMERS IN ISOTROPIC SOLUTION

A. Main-Chain Polymers

The same design principles used to describe hydrogen-bonded, liquid-crystalline polymers, may also be used to classify hydrogen-bonded polymers in isotropic solution. The first examples described herein are main-chain polymers constructed from ditopic monomers (Figure 7).

As pointed out by Griffin and coworkers, the production of main-chain, hydrogen-bonded polymers can be considered a step-growth process [49], with the number-average DP defined by the

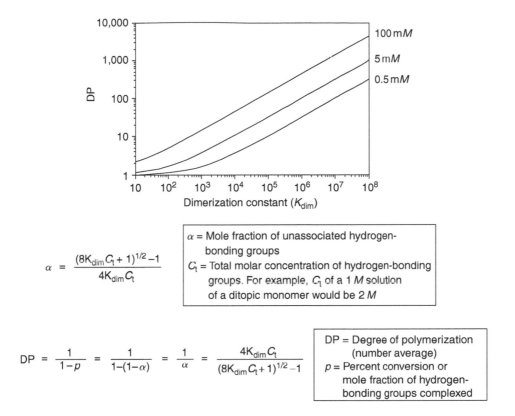

$$\alpha = \frac{(8K_{dim}C_t + 1)^{1/2} - 1}{4K_{dim}C_t}$$

| α = Mole fraction of unassociated hydrogen-bonding groups |
| C_t = Total molar concentration of hydrogen-bonding groups. For example, C_t of a 1 M solution of a ditopic monomer would be 2 M |

$$DP = \frac{1}{1-p} = \frac{1}{1-(1-\alpha)} = \frac{1}{\alpha} = \frac{4K_{dim}C_t}{(8K_{dim}C_t + 1)^{1/2} - 1}$$

| DP = Degree of polymerization (number average) |
| p = Percent conversion or mole fraction of hydrogen-bonding groups complexed |

Figure 17 Equations and plot relating the number-average DP to the dimerization constant for a self-complementary ditopic monomer at varying concentrations (see also Refs. 5e and 5f). Concentrations are shown at the right of the plot. Equations were derived as in Ref. 68.

Carothers' equation (Figure 17). If the stepwise polymerization of ditopic units is noncooperative (i.e., the association constant for each step is equal) and the association constant for the corresponding monotopic subunits is known and assumed to be similar to that of the hydrogen-bonding groups within the ditopic monomer, then the percent conversion (complexation) and average DP can be estimated. A theoretical plot relating the DP of a self-complementary ditopic monomer (i.e., C–C in Figure 7) at varying concentrations to the K_{dim} of the hydrogen-bonding groups within the monomer is shown in Figure 17. To obtain a high DP in "dilute" solution, the association constant of the monomers should be large.

One example of hydrogen-bonded polymers with significant lengths in isotropic solution is the supramolecular polymers of Meijer and coworkers [22]. Along these lines, polymers were assembled from the ditopic unit **31** (Figure 18), which contains the quadruply hydrogen-bonding subunit **5** (Figure 4).[9] As mentioned previously, ureidopyrimidine **5** associates very strongly in chloroform. Likewise, the viscosities of solutions of **31** in chloroform were high and exhibited a concentration dependence that could be attributed to changes in polymer length with varying monomer concentration. The viscosity of a solution of **31** was also lowered upon addition of **5**, as the result of **5** acting as a "capping" agent to decrease the average DP.[10] These observations were consistent with a reversible formation of hydrogen-bonded polymers in chloroform. The average DP for **31** was

[9]Ureidopyrimidine polymers are introduced herein. Readers are referred to Chapter 14 (Sijbesma et al.) for a more thorough discussion of this important class of supramolecular polymers.

[10]For a discussion of photogenerated end-capping groups see Ref. 69.

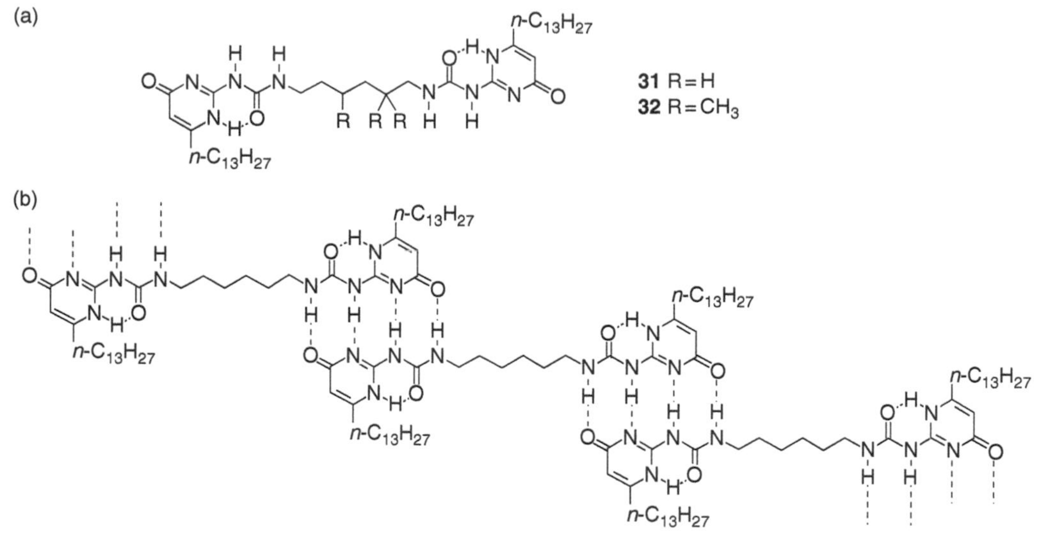

(a)

31 R = H
32 R = CH₃

(b)

$(31)_n$

Figure 18 (a) Ditopic ureidopyrimidine monomers; (b) the supramolecular polymer formed from **31** [22].

especially notable — estimated to be 700 at a concentration of 40 mM — and is consistent with the large dimerization constant observed for **5**.

In addition to supramolecular polymers formed from **31**, ditopic monomer **32** (Figure 18), incorporating a less flexible trimethylhexyl linker, exists in an equilibrium between cyclic and linear aggregates in solution as evidenced by viscosity studies and diffusion-ordered ^{1}H NMR spectroscopy [70,5g].[11] Interestingly, the specific viscosity of a 145 mM solution of **32** in chloroform, became larger with increasing temperature (range = 255 to 325 K), suggesting that an entropy driven ring opening of less viscous cyclic structures occurred to produce linear polymers.

Meijer, Sijbesma, and coworkers have also reported studies of ditopic monomers containing hydrogen-bonding subunit **7** ([71,72], Chapter 14). Along these lines, polymeric aggregates of **33** formed in dodecane via ureidotriazine dimerization (Figure 19). Further solvophobic organization (stacking) of the disklike hydrogen-bonded dimers produced columnar structures (represented schematically in Figure 19(b)), as evidenced by small-angle neutron scattering (SANS) analysis. Circular dichroism studies revealed a helical bias in columns formed from ditopic ureidotriazine monomers containing chiral solubizing groups (e.g., **34**). Moreover, monomers incorporating ethylene oxide side chains formed helical polymers in water [72]. In this case, the hydrophobic environment afforded by discotic aggregation, facilitated hydrogen bonding in the highly polar solvent.

In addition to the preceding main-chain polymers, Rebek and coworkers have described the assembly of ditopic calixarene units into polymeric capsules, "polycaps," in chloroform, as a result of homo- and heterodimerization of urea and sulfonyl-urea groups on the upper rim of the calixarenes [73–75]. Each calixarene dimer along the polymer chain encapsulates solvent or other small molecule guests (e.g., 4-fluorodibenzene and o-dichlorobenzene). A schematic representation of a calixarene dimer and homopolymer is shown in Figure 20.[12]

Polycap formation was supported by ^{1}H NMR spectroscopic analysis [73,74]. Moreover, a solution of ditopic monomer in o-dichlorobenzene exhibited a concentration-dependent viscosity

---

[11] Ring-chain equilibria are briefly introduced herein. Readers are referred to Ref. 5g for a more thorough discussion of this phenomenon. Ditopic monomers that are preorganized to form cyclic aggregates are described in Section IV.D of this chapter.
[12] For a report of the formation of a calixarene dimer through multiple hydrogen-bonding interactions as in **5** see Ref. 76.

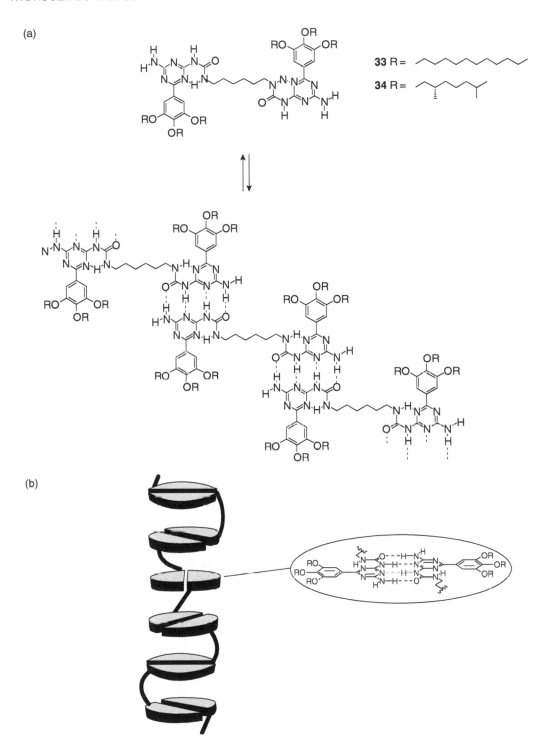

Figure 19 (a) Supramolecular polymer formation from ditopic ureidotriazine monomers **33** and **34**; (b) a schematic representation of the columnar aggregates formed from **33** and **34**. Only one of two possible helices is shown (adapted from J.H.K. Ky Hirschberg, L. Brunsveld, A. Ramzi, J.A.J.M. Vekemans, R.P. Sijbesma, and E.W. Meijer. Nature 407:1167–1170, 2000 and L. Brunsveld, J.A.J.M. Vekemans, J.H.K.K. Hirschberg, R.P. Sijbesma, and E.W. Meijer. *Proc. Natl. Acad. Sci. USA.* 99:4977–4982, 2002.).

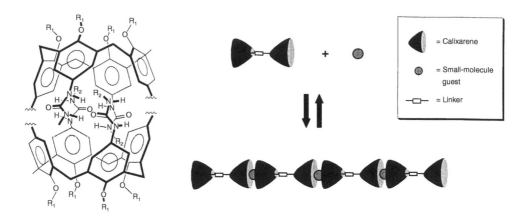

Figure 20   Calixarene dimer (left) and homopolymer (right) (adapted from R.K. Castellano, D.M. Rudkerich, and J. Rebek *Jr. Proc. Natl. Acad. Sci. USA.* 94:7132–7137, 1997 and R.K. Castellano and J. Rebek, Jr. *J. Am. Chem. Soc.* 120:3657–3663, 1998. With permission).

consistent with the formation of linear supramolecular polymers [75]. When a small amount of protic solvent, methanol, was added to a 2.8% (w/w) solution of homopolymer in *o*-dichlorobenzene, hydrogen bonding was disrupted, and solution viscosity decreased considerably. However after evaporation of methanol at 50°C, high molecular weight polymer chains reformed and high viscosity was reestablished.

Mechanical analysis of sulfonyl-urea containing homopolymer solutions indicated that the polycaps could withstand significant forces without completely disrupting the hydrogen-bonded assembly [75]. When a polycap solution was subjected to a large steady shear rate (500 sec^{-1}), the viscosity decreased somewhat; however, high visocities were reestablished upon subsequent application of lower shear rates. Fibers with relatively high tensile strengths were drawn from chloroform solutions of the polycaps. Moreover, cross-linked networks were formed from mixtures of a ditopic monomer and a tetravalent calixarene cross-linker.

Although the aforementioned hydrogen-bonded polymers were prepared from self-complementary ditopic monomers, supramolecular polymers have also been prepared in isotropic solution by the interaction of hydrogen-bonding complements A and B within ditopic monomers A–A and B–B (Figure 7). Specifically, Lehn and coworkers have reported the synthesis and study of ditopic monomers **35** and **36** [77], which, respectively, incorporate complementary cyanuric acid (ADA–ADA) and diamidopyridine (DAD–DAD) subunits (Figure 21). Hydrogen bonding between **35** and **36** produced linear polymers, as verified by ^{1}H NMR spectroscopic studies in tetrachlorethylene, solution viscosity studies, and electron microscopy. Fibers were observed in a micrograph of a sample prepared from a 1:1 mixture of **35** and **36** in tetrachloroethylene. The lengths observed were consistent with the formation of hydrogen-bonded polymers with high degrees of polymerization.

A further example of the supramolecular polymerization of heterocomplementary ditopic monomers are the polymers of Craig and coworkers ([78], Chapter 9), which were assembled by base pairing within oligonucleotides (e.g., **37** and **38**, Figure 22). Compounds **37** and **38** are monomers of the general type A–B and C–D, in which A is a hydrogen-bonding complement of C and B, in turn, is a hydrogen-bonding complement of D. Melting curves obtained for mixtures of oligonucleotides supported duplex formation. Moreover, the viscosities of mixtures of **37** and **38** were greater than that of the component monomers, suggesting that hydrogen-bonded polymers were formed. Supramolecular polymers have also been prepared from monomers of the type A–B, in which A and B are complementary oligonucleotide sequences.

The ability to vary the linker and, thus, the orientation of hydrogen-bonding groups in ditopic building blocks similar to those described is very powerful. For instance, appropriate linkers may lead

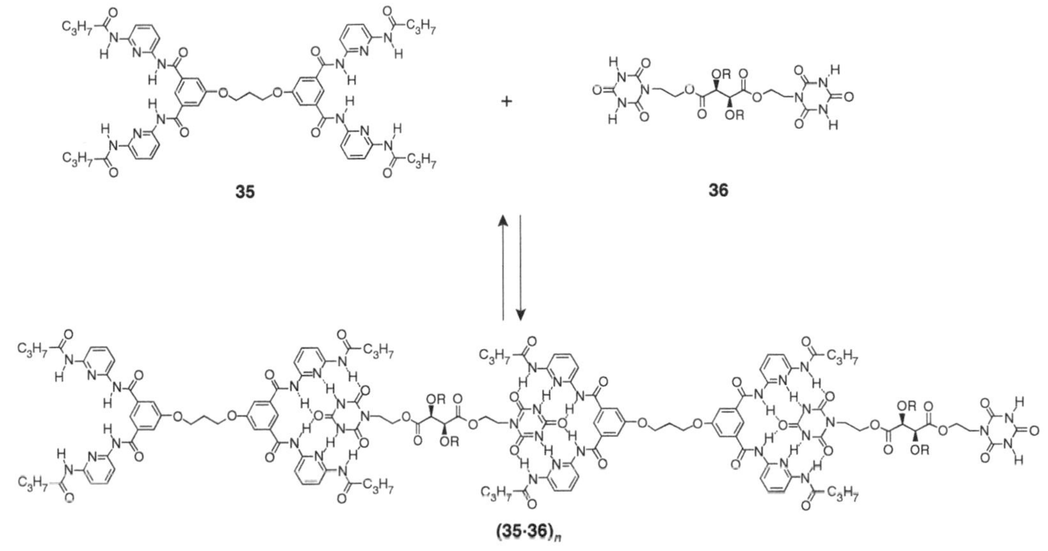

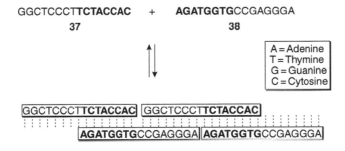

Figure 21 Supramolecular polymerization of heterocomplementary monomers **35** and **36** [77].

GGCTCCCTT**CTACCAC**   +   **AGATGGTG**CCGAGGGA
        37   **38**

A = Adenine
T = Thymine
G = Guanine
C = Cytosine

GGCTCCCTT**CTACCAC** GGCTCCCTT**CTACCAC**
AGATGGTGCCGAGGGA **AGATGGTG**CCGAGGGA

Figure 22 Supramolecular polymerization of oligonucleotides **37** and **38** [78].

to polymers with well-defined, potentially predictable, secondary structures. The aggregate formed from *bis*-2-amidopyridine unit **39** and pimelic acid (**40**) is such an example [79]. With regard to this aggregate, Hamilton and coworkers have reported a helical superstructure in the solid state that is formed by hydrogen bonding between pimelic acid and **40** in a "syn–syn" orientation (Figure 23).[13] ^{1}H NMR chemical shifts (5% THF-d_8/CD$_2$Cl$_2$), as well as variable temperature ^{1}H NMR and NOE investigations, are consistent with the formation of a helical aggregate in solution. However, these studies are not conclusive, and the DP has not been estimated. The cyclic-peptide nanotubes of Ghadiri, which form in the solid state and in lipid bilayers, also have intriguing secondary structures and can be regarded as linear supramolecular polymers [80,81].

B. Hydrogen-Bonding Telechelic Polymers

At this point, it is worth mentioning that the possibility exists to assemble main-chain, hydrogen-bonded polymers from covalent polymers with sticky ends, that is, hydrogen-bonding telechelic polymers. For example, Lenz and coworkers have proposed that the liquid-crystalline behavior of

---

[13]For a further discussion of helical supramolecular polymers see Chapters 2 and 3.

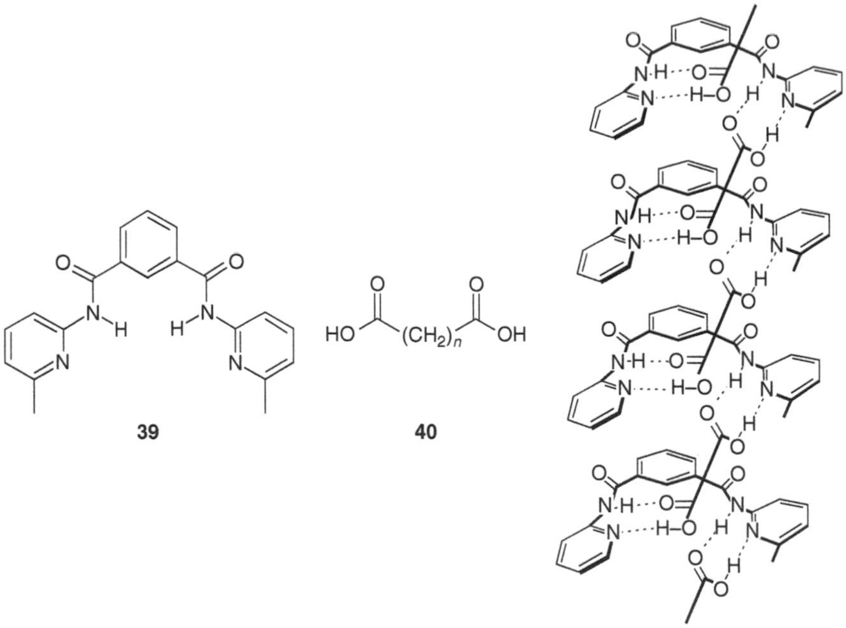

Figure 23 Schematic representation of a helical aggregate formed from *bis*-amidopyridine **39** and pimelic acid (**40**) [79].

Figure 24 A schematic representation of hydrogen-bonding interactions between the end groups of telechelic polymers (adapted from B.J.B. Folmer, R.P. Sijbesma, R.M. Versteegen, J.A.J. van der Rijt, and E.W. Meijer. *Adv. Mater.* 12:874–878, 2000.).

polymeric glycols terminated with diacids might be explained by dimerization of the carboxylic-acid termini [82]. Lillya and coworkers have also demonstrated that dimerization of the carboxyl termini of modified poly(tetramethyloxides) has interesting effects upon the telechelic polymer's bulk properties [83]. Recently, there have been reports of telechelic polymers containing heterocyclic end groups capable of forming robust, hydrogen-bonded complexes. Solution and bulk studies of such polymers are described herein. Certainly, these investigations point to the power and potential of using multiply hydrogen-bonding subunits to construct supramolecular polymers with useful material properties.

As delineated in the previous section, ditopic monomer **31** consists of two ureidopyrimidine subunits connected by a short, hexyl linker. Meijer and coworkers have also studied telechelic oligomers/polymers containing ureidopyrimidine **5** as end groups ([22,84], Chapter 14). Solution viscosity studies and bulk rheological measurements with **41** indicated that a high molecular weight supramolecular polymer formed from the oligomeric building block (Figure 24 and Figure 25). Likewise, polysiloxane **42** exhibited viscoelastic bulk properties that differed from a non-hydrogen-bonding polysiloxane of similar molecular weight, as a result of association of the end groups. In this case, the viscoelastic properties of the telechelic polymers were attributed to the entangle-ment of high molecular weight hydrogen-bonded polymer chains as opposed to physical cross-links

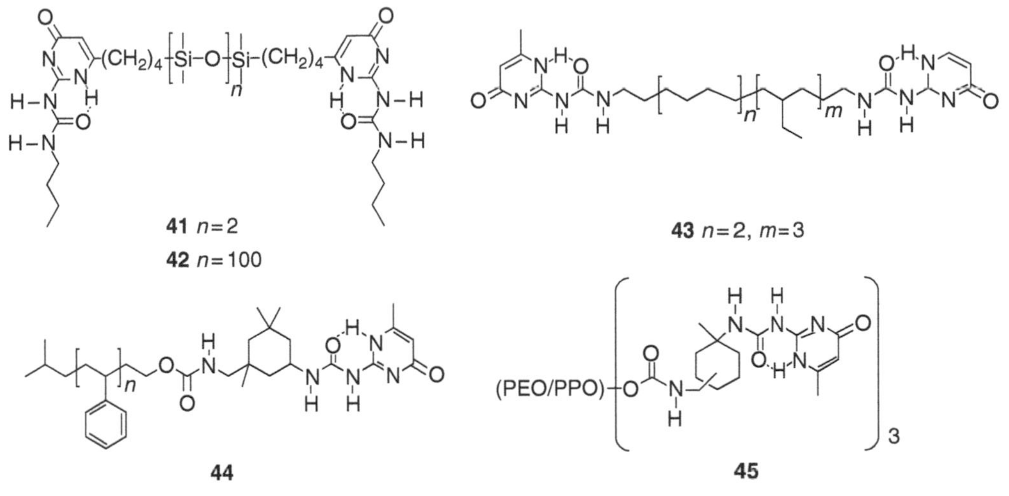

Figure 25 Examples of telechelic, hydrogen-bonded polymers containing ureidopyrimidine 5 [22,84–87]. *Note*: PEO/PPO in structure **45** is a trifunctionalized star block copolymer.

afforded by aggregation of the hydrogen-bonding groups into microcrystalline domains. Additional examples of telechelic polymers with terminal ureidopyrimidine moieties include the bifunctional polyethylene/isobutylene copolymer **43** and monofunctional polystyrene **44** [85,86].

Networks have also been assembled from trifunctionalized copolymer **45** (Figure 25). Solution viscosity studies, including capping studies, indicated that a reversible network formed in chloroform [87]. Likewise, the polymer had interesting bulk mechanical properties — for example, a higher plateau modulus in dynamic mechanical analysis than a covalent network polymer — as a result of the formation of reversible, hydrogen-bonded cross-links, and, thus a thermodynamically stable network.

Examples of telechelic polymers incorporating heterocyclic end groups other than **5** are the nucleotide-terminated polystyrenes (**46** and **47**, Figure 26) of Long and coworkers [88]. Reversible hydrogen bonding between **46** and **47** — which, in turn, contain complementary adenine and thymine end groups — was verified by variable temperature ^{1}H NMR spectroscopic analysis of blends of the two polymers in toluene.

C. Side-Chain Polymers

Hydrogen-bonding interactions between the side groups of covalent polymers (*vide supra*) has been shown to lead to network polymers with interesting properties (Figure 9). For example, polybutadienes modified with 4-(3,5-dioxo-1,4,4-triazlidine-4-yl) benzoic acid (**48**) form such structures. The mechanical and thermal properties of these polymers differ from unmodified polybutadienes as the result of hydrogen-bonded cross-links (Figure 27) and further organization of the "association polymers," through dipole–dipole interactions, into ordered domains [89]. Cross-linked polymers have also been reported to form from polybutadienes derivatized with a related phenyl-substituted urazole; see Ref. 90 and references therein.

Covalent polymers bearing ureidopyrimidine **5** as pendant side groups have been reported. Along these lines, Coates and Long have, respectively, reported studies of polyolefin copolymers **49** and polyacrylate copolymers **50** [91,92].[14] Solution viscosity studies, including capping studies, suggested that random copolymer **49** (Figure 28) forms reversible hydrogen-bonded networks in

[14]A further example of hydrogen-bonding side-chain polymers are the diaminotriazine and diaminopyridine-functionalized polymers of Weck and coworkers, which are prepared by ring-opening metathesis polymerization [93].

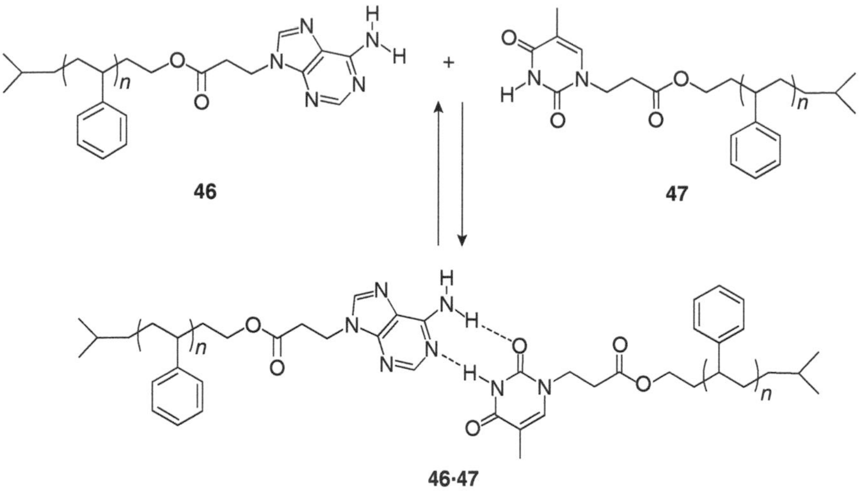

Figure 26 Supramolecular polymerization of oligonucleotide-terminated polystyrenes **46** and **47** [88].

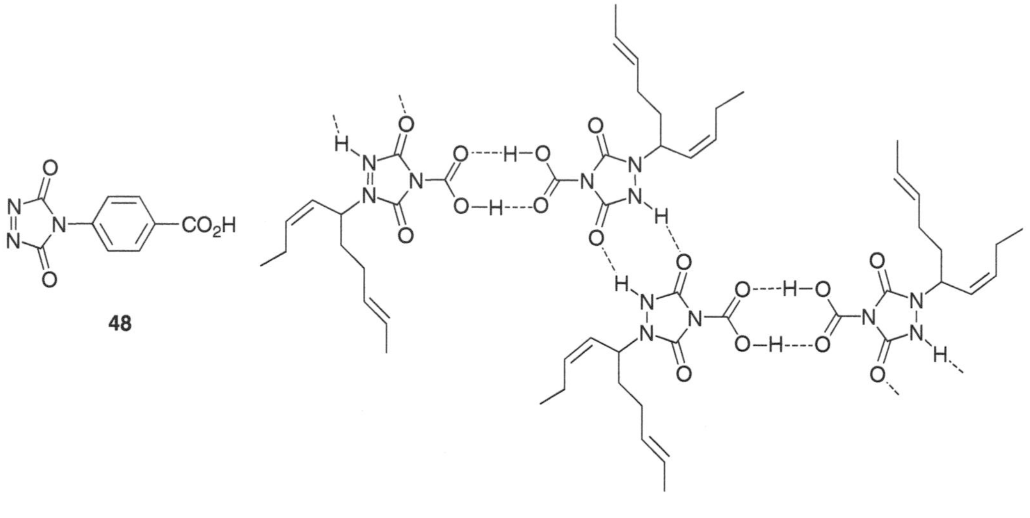

Figure 27 When polybutadiene is modified with **48** by an ene reaction, network polymers arise that rely on hydrogen-bond cross-links (structure to the right) and dipole–dipole interactions to form ordered domains [89].

toluene [91]. Stress–strain profiles obtained from mechanical analysis of the bulk polymer, revealed that the polymer had elastomeric properties consistent with the formation of hydrogen-bonding cross-links. Likewise, ^{1}H NMR spectroscopic analysis and solution viscosity studies indicated that hydrogen bonding occurs between the chains of copolymer **50** in toluene and chloroform [92]. A dissociation temperature of 80°C was estimated for the bulk polymer in melt viscosity studies. Interestingly, copolymer **50** exhibited significantly better adhesive properties than a related unadorned homopolymer.

It should be mentioned that the interaction of hydrogen-bonding side groups has also been used in the preparation of polymer blends and is especially effective in improving the miscibility of polymer components within a blend.[15] Numerous accounts of hydrogen-bonded polymer blends have been

---

[15]For a general description of polymer blends see Ref. 94.

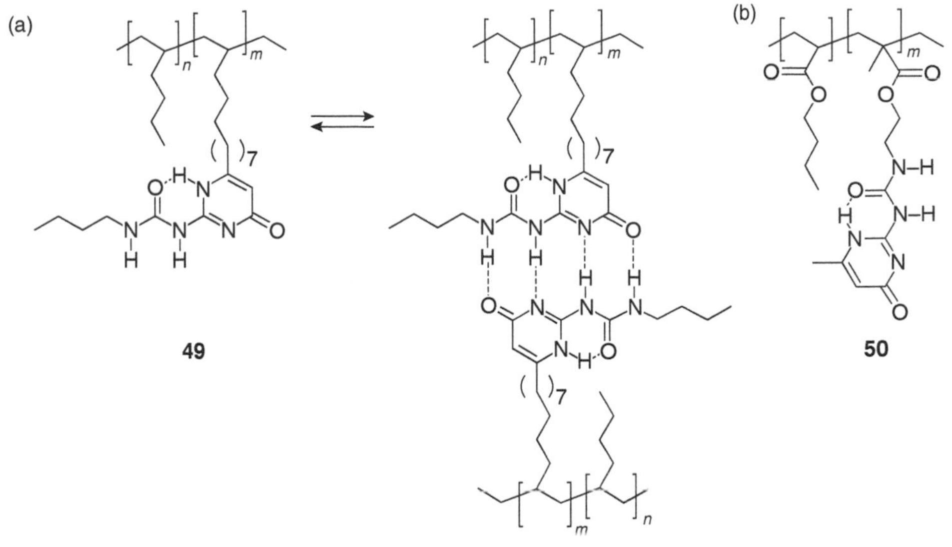

Figure 28 Examples of hydrogen-bonding side-chain polymers [91,92].

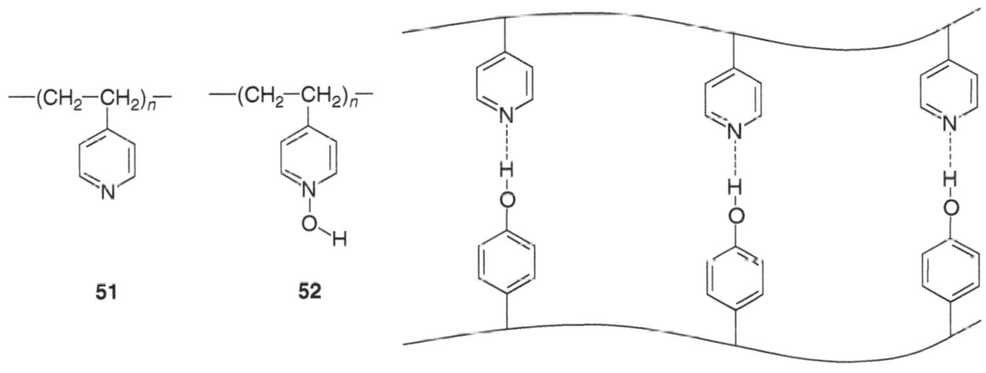

Figure 29 A schematic representation of hydrogen-bonding interactions between polymers **51** and **52** in a polymer blend [95].

reported, and a complete description is beyond the scope of this review. However, a few examples are mentioned to illustrate the concept.

Meftahi and Fréchet have described polymer blends prepared from poly(4-vinylpyridine) (**51**) and poly(4-hydroxystyrene) (**52**) [95]. In these studies polymer **51** was immiscible with various weight percentages of polystyrene, a polymer lacking hydrogen-bond acceptor groups, as evidenced by the observation of two, discrete glass-transition temperatures. To the contrary, a 1:1 mixture of **51** and **52** precipitated from methanol — a solvent in which the individual components are soluble — is miscible and displays a single glass-transition temperature higher than that of the individual polymers [95]. Hydrogen bonding between the two polymers, as represented schematically in Figure 29, is believed to give rise to blend compatibility. In addition to this example, there are many other blends prepared from phenolic-type polymers and polymers containing a variety of hydrogen-bond acceptors [96–98]. Meijer and coworkers have also briefly described blends prepared from maleimide and cyanuric acid polymers that rely on multiple hydrogen-bonding interactions between repeat units [99].

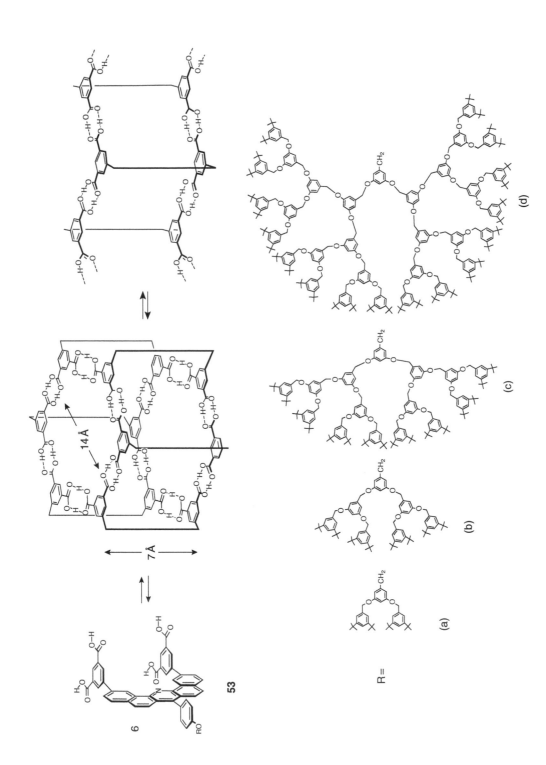

Figure 30 Self-assembled dendrimers studied by Zimmerman and coworkers [104–106].

D. Hydrogen-Bonded Dendrimers

The synthesis and study of dendrimers has received significant attention recently [100]. As part of the growth of this field, there has been sharpened interest in the self-assembly of dendrimers and their use in molecular recognition ([101–103], Chapter 11). Accordingly, hydrogen bonding has proven to be very effective for constructing self-assembled dendrimers. The aggregates arising from the assembly of *bis*-isophthalic acid units **53(a)–(d)** are such examples [104–106] (Figure 30). The *bis*-isophthalic-acid unit **53** is particularly intriguing because subunits **53(b)–(d)**, containing second- through fourth-generation Fréchet-type dendrons [104,106], formed cyclic hexamers in nonpolar solvents as a result of carboxylic-acid dimerization. On the other hand, the *bis*-isophthalic acid unit **53(a)** containing a less bulky, first-generation dendron is believed to form polymeric aggregates. Based upon these observations and modeling, it has been suggested that peripheral dendron interactions, specifically those that would arise in the linear aggregates of the second- through fourth-generation isophthalic acid subunits, aid in the formation of hexameric species. Discrete aggregates such as those described rival the size (MW of the fourth-generation hexamer = 34.6 kDa), shape, and, potentially, function of small proteins [104]. For example, one could envision catalysis in the core of a dendrimer similar to the action of an enzyme, as well as reversible guest complexation reminiscent of holoprotein complex formation.

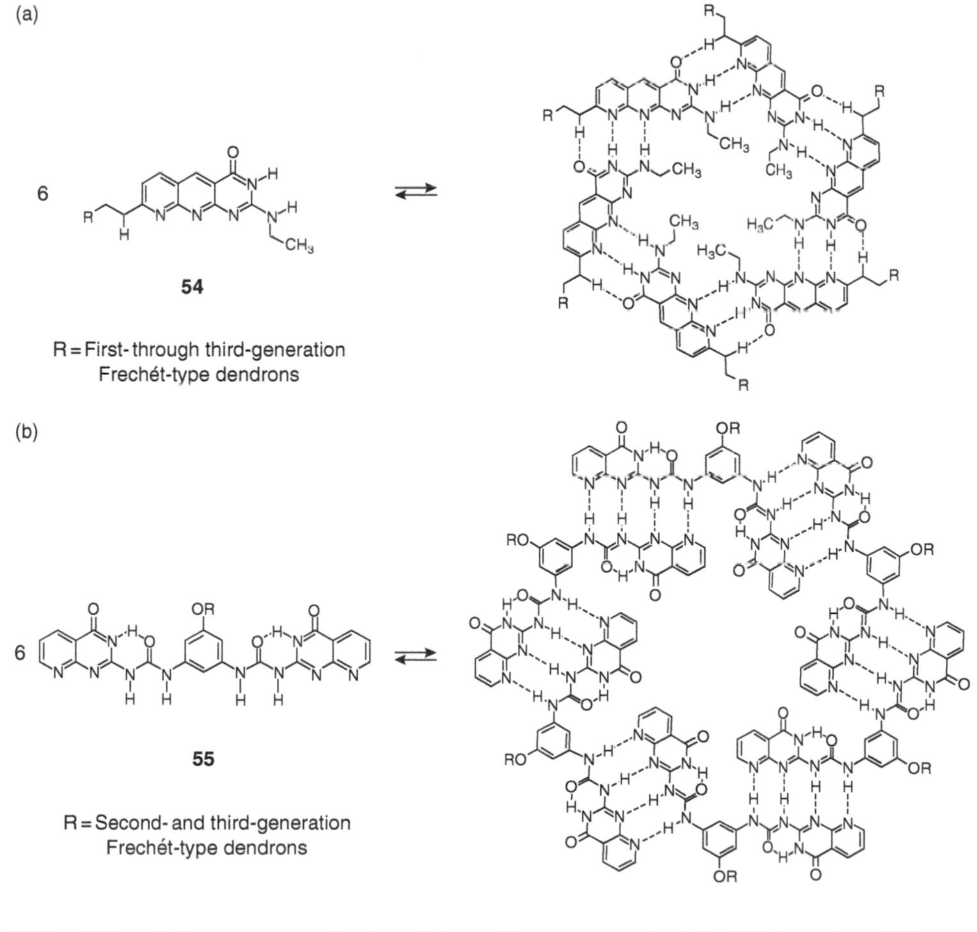

Figure 31 (a) A cyclic hexamer formed by the self-assembly of preorganized dendritic monomers (**54**) containing complementary DDA and AAD hydrogen-bonding arrays [107,108]; (b) a cyclic hexamer formed by self-assembly of a ditopic monomer (**55**) containing ureidopyridopyrimidine 6 [26].

Robust self-assembled dendrimers have also been constructed from heterocyclic subunits [26,107–109]. For example, compound **54** containing complementary DDA and AAD hydrogen-bonding arrays is preorganized to form a cyclic hexamer in solution (Figure 31(a)). A building block related to those described in preceding sections is ditopic monomer **55** (Figure 31(b)), which has two quadruply hydrogen-bonding subunits **6** linked via a rigid aromatic spacer [26]. Monomers bearing second- and third-generation Fréchet-type dendrons formed cyclic hexamers (Figure 30(b)) in toluene and chloroform. A larger cyclic aggregate is suspected to form from monomer **55** containing a first-generation dendron.

V. FROM DISPERSIONS TO MONOLAYERS

In addition to the solution aggregates described in the previous section, stable dispersions of supramolecular, hydrogen-bonded polymers have also been reported. One example is the aggregate arising from diimide **56** and melamine **57** (Figure 32(a)) [110].[16] Kimizuka and coworkers have reported

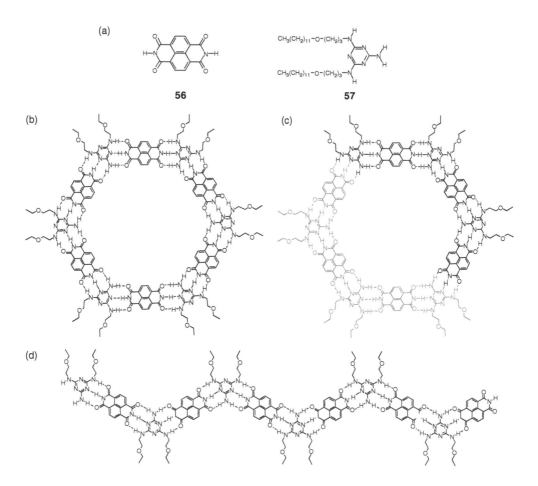

Figure 32   (a) Components of the 1:1 complex; (b) possible cyclic dodecamer; (c) possible helical aggregate; (d) possible linear tape. Peripheral alkyl chains are truncated in (b)–(d) for clarity [110].

[16]For a slightly varied example see Ref. 111, and for an example of a gel constructed from related building blocks see Ref. 112.

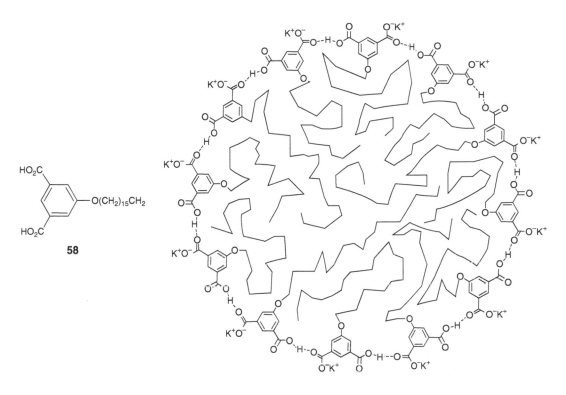

Figure 33 Schematic representation of a proposed cyclic aggregate that self-organizes into a fibrous material [113,114].

that a dispersion of a 1:1 complex of **56** and **57** forms in cyclohexane. Interestingly, bundles of long strands were observed in both transmission- and scanning-electron micrographs. The exact nature of the supramolecular aggregate that leads to these strands has not been determined. However, the authors have suggested three possible supramolecular structures (Figure 32): (1) stacks of dodecameric disks; (2) helical tapes, or (3) extended, linear tapes. In a somewhat related report, Menger and coworkers have reported the formation of fibers from the monopotassium salt of **58** when "precipitating" the salt from water [113,114]. In this case, hydrogen-bonding and hydrophobic interactions are proposed to lead to disks (Figure 33), which form stacks that, subsequently, produce the fibrous material.

Kimizuka and coworkers have extended their own work, to the "assembly" of bilayer membranes. For example, multiple hydrogen-bonding interactions between **57** and cyanuric acid derivative **59** containing a polar head group leads to hydrogen-bonded tapes that organize into bilayers [115,116].[17] The proposed supramolecular structure of the bilayer, represented schematically in Figure 34, was verified by IR and x-ray diffraction studies of films cast from an aqueous dispersion of the bilayer, as well as by electron microscopy of the dispersions.

In related studies, Kunitake and coworkers have described the preparation of monolayers consisting of hydrogen-bonded tapes formed from a substituted melamine and barbituric acid (Figure 35) [120]. Hydrogen bonding in the resulting Langmuir–Blodgett film was verified by IR and x-ray photoelectron spectroscopy. This strategy has recently been extended to the preparation of monolayers from tritopic melamine units, with hopes of producing network structures within the monolayer [121].

---

[17] For related studies see Refs. 117–119.

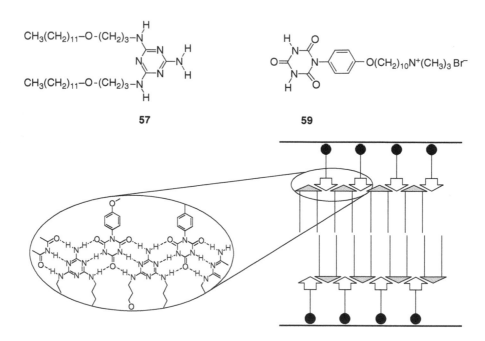

Figure 34   Schematic representation of the formation of a bilayer membrane (adapted from N. Kimizuka, T. Kawasaki, and T. Kunitake. *J. Am. Chem. Soc.* 115:4387–4388, 1993 and N. Kimizuka, T. Kawasaki, K. Hirata, and T. Kunitake. *J. Am. Chem. Soc.* 120:4094–4104, 1998. With permission).

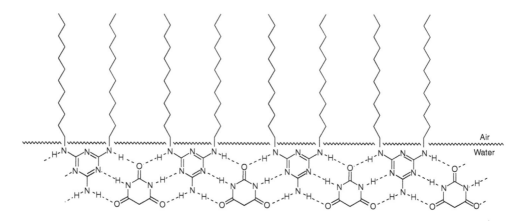

Figure 35   Schematic representation of monolayer formation from a substituted melamine and barbituric acid [120].

VI.   REVERSIBLE COORDINATION POLYMERS

Although the chief focus of the present review is hydrogen-bonded supramolecular polymers, the use of metal coordination in the design and construction of supramolecular polymers should be briefly mentioned. Analogous to the preparation of hydrogen-bonded polymers, reversible linear coordination polymers may be assembled by interaction of ditopic ligand receptors and metals of varying coordination numbers (represented schematically in Figure 36). Networks, in turn, may be prepared from tritopic monomers. Such polymers have the potential to function as responsive materials that are sensitive to changes in pH, concentration, and temperature, as well as to electrochemical and photochemical stimuli.

Figure 36 A schematic representation of linear coordination polymers formed from ditopic monomer units and metal ions.

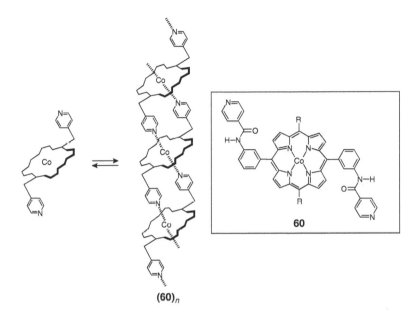

Figure 37 A schematic representation of supramolecular polymer formation via porphyrin/cobalt coordination (adapted from U. Michelsen and C.A. Hunter. *Angew. Chem. Int. Ed.* 39:764–767, 2000.).

The stability constant for the metal/ligand (monomer) complexation should be large to afford high molecular weight polymers in isotropic solution. Moreover, the complexation should be readily reversible (kinetically labile) to impart dynamic properties to the polymers. Two recent examples of soluble, reversible coordination polymers are the self-assembled porphyrins of Michelsen and Hunter [122] and the high molecular weight copper(II) coordination polymers of Leize, Lehn, and coworkers [123].

In the former case, polymers were assembled by complexation of a pyridyl-substituted porphyrin with the hexacoordinate metal ion, cobalt(II). The resulting supramolecular polymers, $(60)_n$ (Figure 37), were characterized by UV/vis spectroscopy, NMR diffusion studies, and SEC. Polymers with concentration-dependent molecular weights up to 136 kDa (DP $\sim$ 100) formed, as indicated by SEC data. The dynamic nature of polymer formation was implied in the concentration dependence of the molecular weights as well as in capping studies with a porphyrin chain stopper [122].

The aforementioned copper(II) polymers were prepared by mixing equimolar amounts of ditopic ligand **61** (Figure 38) and copper(II) triflate in acetonitrile [123]. Supramolecular polymers, with molecular weights up to approximately 33 kDa (DP $\sim$ 47) were detected by electrospray mass spectrometry. Dissociation of the polymers occurred in solutions containing small amounts of water, and the polymers did not form below ligand/copper concentrations of 10^{-4} M. Bipyridyl moieties on successive monomer units are believed to chelate copper(II) ions — thus producing the observed linear polymers (represented schematically in Figure 38(b)).

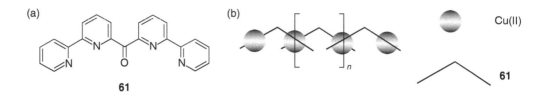

Figure 38 (a) Monomer used in preparing copper(II) coordination polymers; (b) schematic representation of the supramolecular polymer formed by complexation of copper(II) with **61** (adapted from H. Nierengarten, J. Rojo, E. Leize, J.-M. Lehn, and A. van Dorsselaer. *Eur. J. Inorg. Chem.* 573–579, 2002.).

VII. CONCLUSION

The study of supramolecular polymers is, clearly, a burgeoning area of research. A set of building blocks is continuing to be compiled, which will facilitate expansion of this field and allow the construction of even more new and interesting polymers. One can assume that further development will lead to new assemblies and materials with applications in a number of different areas. Studies of this nature will also continue to answer important, fundamental questions concerning how and why molecules interact. Whether the development of these "supramolecular polymers" will have as great an impact as the development of conventional polymers remains to be seen. However, the field of supramolecular polymers, unquestionably, has a promising future.

REFERENCES

1. C. Fouquey, J.-M. Lehn, and A.-M. Levelut. *Adv. Mater.* 2:254–257, 1990.
2. W.H. Carothers. *J. Am. Chem. Soc.* 51:2548–2559, 1929.
3. (a) D.S. Lawrence, T. Jiang, and M. Levett. *Chem. Rev.* 95:2229–2260, 1995. (b) G.M. Whitesides, E.E. Simanek, J.P. Mathias, C.T. Sato, D.N. Chin, M. Mammen, and D.M. Gordon. *Acc. Chem. Res.* 28:37–44, 1995. (c) M.M. Conn and J. Rebek Jr. *Chem. Rev.* 97:1647–1668, 1997. (d) M.J. Krische and J.-M. Lehn. *Struct. Bonding* 96:3–29, 2000. (e) S.C. Zimmerman and P.S. Corbin. *Struct. Bonding* 96:63–94, 2000. (f) L.J. Prins, D.N. Reinhoudt, and P. Timmerman. *Angew. Chem. Int. Ed.* 40:2382–2426, 2001. (g) E.A. Archer, H. Gong, and M.J. Krishce. *Tetrahedron* 57:1139–1159, 2001.
4. (a) S. Leininger, B. Olenyuk, and P.J. Stang. *Chem. Rev.* 100:853–907, 2000. (b) M. Fujita, *Struct. Bonding* 96:177–201, 2000. (c) S.R. Seidel and P.J. Stang. *Acc. Chem. Res.* 35:972–983, 2002.
5. (a) J.-M. Lehn. *Makromol. Chem. Macromol. Symp.* 69:1–17, 1993. (b) J.-M. Lehn. *Pure Appl. Chem.* 66:1961–1966, 1994. (c) V. Percec, J. Heck, G. Johannson, D. Tomazos, M. Kawasumi, and G. Ungar. *J. Macromol. Sci., Pure Appl. Chem.* A 31:1031–1070, 1994. (d) N. Zimmerman, J.S. Moore, and S.C. Zimmerman. *Chem. Industry* 15:604–610, 1998. (e) L. Brunsveld, B.J.B. Folmer, E.W. Meijer, and R.P. Sijbesma. *Chem. Rev.* 101:4071–4097, 2001. (f) A. Ciferri. *Macromol. Rapid Commun.* 23:511–529, 2002. (g) A.T. ten Cate and R.P. Sijbesma. *Macromol. Rapid Commun.* 23:1094–1112, 2002. (h) U.S. Schubert and C. Eschbaumer. *Angew. Chem. Int. Ed.* 41:2892–2926, 2002.
6. J.C. Macdonald and G.M. Whitesides. *Chem. Rev.* 94:2383–2420, 1994.
7. G.A. Jeffrey. *An Introduction to Hydrogen Bonding.* New York: Oxford University Press, 1997, pp. 184–212.
8. A. Aggeli, M. Bell, N. Boden, J.N. Keen, P.F. Knowles, T.C.B. McLeish, M. Pitkeathly, and S.E. Radford. *Nature* 386:259–262, 1997.
9. S.C. Zimmerman and T.J. Murray. *Philos. Trans. R. Soc. Lond. A* 345:49–56, 1993.
10. C.S. Wilcox, E. Kim, D. Romano, L.H. Kuo, A.L. Burt, and D.P. Curran. *Tetrahedron* 51:621–634, 1995.
11. W.L. Jorgensen and J. Pranata. *J. Am. Chem. Soc.* 112:2008–2010, 1990.
12. J. Pranata, S.G. Wierschke, and W.L. Jorgensen. *J. Am. Chem. Soc.* 113:2810–2819, 1991.
13. T.J. Murray and S.C. Zimmerman. *J. Am. Chem. Soc.* 114:4010–4011, 1992.
14. J. Sartorius and H.-J. Schneider. *Chem. Eur. J.* 2:1446–1452, 1996.
15. H.-J. Schneider, R.K. Juneja, and S. Simova. *Chem. Ber.* 122:1211–1213, 1989.

16. (a) P.L.A. Popelier and L. Joubert. *J. Am. Chem. Soc.* 124:8725–8729, 2002. (b) O. Lukin and J. Leszczynski. *J. Phys. Chem. A* 106:6775–6782, 2002.
17. H.-J. Schneider. *J. Phys. Chem. A* 107:9250, 2003.
18. R.B. Gardner and S.H. Gellman. *J. Am. Chem. Soc.* 117:10411–10412, 1995.
19. S.C. Zimmerman, M. Mrksich, and M. Baloga. *J. Am. Chem. Soc.* 111:8528–8530, 1989.
20. F. Eblinger and H.-J. Schneider. *Angew. Chem. Int. Ed.* 37:826–829, 1998.
21. R.P. Sijbesma and E.W. Meijer. *Chem. Commun.* (1): 5–16, 2003.
22. R.P. Sijbesma, F.H. Beijer, L. Brunsveld, B.J.B. Folmer, J.H.K.K. Hirschberg, R.F.M. Lange, J.K.L. Lowe and E.W. Meijer. *Science* 278:1601–1604, 1997.
23. F.H. Beijer, R.P. Sijbesma, H. Kooijman, A.L. Spek, and E.W. Meijer. *J. Am. Chem. Soc.* 120:6761–6769, 1998.
24. S.H.M. Söntjens, R.P. Sijbesma, M.H.P. van Genderen, and E.W. Meijer. *J. Am. Chem. Soc.* 122:7487–7493, 2000.
25. P.S. Corbin and S.C. Zimmerman. *J. Am. Chem. Soc.* 120:9710–9711, 1998.
26. P.S. Corbin, L.J. Lawless, Z. Li, Y. Ma, M.J. Witmer, and S.C. Zimmerman. *Proc. Natl. Acad. Sci. USA* 99:5099–5104, 2002.
27. F.H. Beijer, H. Kooijman, A.L. Spek, R.P. Sijbesma, and E.W. Meijer. *Angew. Chem. Int. Ed.* 37:75–78, 1998.
28. J.L. Sessler and R. Wang. *Angew. Chem. Int. Ed.* 37:1726–1729, 1998.
29. J.L. Sessler and R. Wang. *J. Org. Chem.* 63:4079–4091, 1998.
30. P.S. Corbin and S.C. Zimmerman. *J. Am. Chem. Soc.* 122:3779–3780, 2000.
31. P.S. Corbin, S.C. Zimmerman, P.A. Thiessen, N.A. Hawrluk, and T.J. Murray. *J. Am. Chem. Soc.* 123:10475–10488, 2001.
32. U. Lüning and C. Kühl. *Tetrahedron Lett.* 39:5735–5738, 1998.
33. U. Lüning, C. Kühl, and A. Uphoff. *Eur. J. Org. Chem.* (23):4063–4070, 2002.
34. B. Gong. *Synlett.* 5:582–589, 2001.
35. H. Zeng, R.S. Miller, R.A. Flowers II, and B. Gong. *J. Am. Chem. Soc.* 122:2635–2644, 2000.
36. X. Zhao, X.Z. Wang, X.K. Jiang, Y.Q. Chen, Z.T. Li, and G.J. Chen. *J. Am. Chem. Soc.* 125:1518–15139, 2003.
37. X.Z. Wang, X.Q. Li, X.B. Shao, X. Zhao, P. Deng, X.K. Jiang, Z.T. Li, and Y.Q. Chen. *Chem. Eur. J.* 9:2904–2913, 2003.
38. B. Gong, Y. Yan, H. Zeng, E. Skrzypczak-Jankunn, Y.W. Kim, J. Zhu, and H. Ickes. *J. Am. Chem. Soc.* 121:5607–5608, 1999.
39. S. Brammer, U. Lüning, and C. Kühl. *Eur. J. Org. Chem.* (23):4054–4062, 2002.
40. S.-K. Chang and A.D. Hamilton. *J. Am. Chem. Soc.* 110:1318–1319, 1988.
41. P. Bladon and A.C. Griffin. *Macromolecules* 26:6604–6610, 1993.
42. (a) C.M. Paleos and D. Tsiourvas. *Angew Chem. Int. Ed.* 34:1696–1711, 1995. (b) T. Kato. *Struct. Bonding* 96:95–146, 2000. (c) T. Kato, N. Mizoshita, and K. Kanie. *Macromol. Rapid. Commun.* 22:797–814, 2001. (d) C.M. Paleos and D. Tsiourvas. *Curr. Opin. Coll. Interface Sci.* 6:257–267, 2001.
43. E. Bradfield and B. Jones. *J. Chem. Soc.* 2660–2661, 1929.
44. T. Kato and J.M.J. Fréchet. *J. Am. Chem. Soc.* 111:8533–8534, 1989.
45. T. Kato and J.M.J. Fréchet. *Macromol. Symp.* 98:311–326, 1995.
46. T. Kato, T. Uryu, F. Kaneuchi, C. Jin, and J.M.J. Fréchet. *Liq. Cryst.* 14:1311–1317, 1993.
47. M.-J. Brienne, J. Gabard, J.-M. Lehn, and I. Stibor. *J. Chem. Soc. Chem. Commun.* (24):1868–1870, 1989.
48. A.D. Hamilton and D. Van Engen. *J. Am. Chem. Soc.* 109:5035–5036, 1987.
49. C. Alexander, C.P. Jariwala, C.M. Lee, and A.C. Griffin. *Macromol. Symp.* 77:283–294, 1994.
50. C. He, A.M. Donald, A.C. Griffin, T. Waigh, and A.H. Windle. *J. Polym. Sci. Part B: Polym. Phys.* 36:1617–1624, 1998.
51. T. Gulik-Krzywicki, C. Fouquey, and J.-M. Lehn. *Proc. Natl. Acad. Sci. USA* 90:163–167, 1993.
52. A. Sirigu. *Liquid Crystallinity in Polymers.* A. Ciferri, Ed., New York: VCH Publishers, 1991, pp. 261–313.
53. M. Kotera, J.-M. Lehn, and J.-P. Vigneron. *J. Chem. Soc. Chem. Commun.* (2):197–199, 1994.
54. H. Kihara, T. Kato, T. Uryu, and J.M.J. Fréchet. *Chem. Mater.* 8:961–968, 1996.
55. C.B. St. Pourcain and A.C. Griffin. *Macromolecules* 28:4116–4121, 1995.

56. (a) T. Kato and J.M.J. Fréchet. *Macromolecules* 22:3818–3819, 1989. (b) T. Kato and J.M.J. Fréchet. *Macromolecules* 23:360, 1990.

57. T. Kato, H. Kihara, U. Kumar, T. Uryu, and J.M.J. Fréchet. *Angew. Chem. Int. Ed.* 33:1644–1645, 1994.

58. T. Kato, O. Ihata, S. Ujiie, M. Tokita, and J. Watanabe. *Macromolecules* 31:3551–3555, 1998.

59. A. Marsh, M. Silvestri, and J.-M. Lehn. *Chem. Commun.* (13):1527–1528, 1996.

60. P. Mathias, E.E. Simanek, and G.M. Whitesides. *J. Am. Chem. Soc.* 116:4326–4340, 1994.

61. S.C. Zimmerman and B.F. Duerr. *J. Org. Chem.* 57:2215–2217, 1992.

62. M. Suarez, J.-M. Lehn, S.C. Zimmerman, A. Skoulios, and B. Heinrich. *J. Am. Chem. Soc.* 120:9526–9532, 1998.

63. G. Gottarelli, G.P. Spada, and A. Garbesi. *Comprehensive Supramolecular Chemistry*, J.-P. Sawage and A.W. Hosseini, Eds., Vol 9. New York: Pergamon Press, 1996, pp. 483–506.

64. P. Mariani, C. Mazabard, A. Garbesi, and G.P. Spada. *J. Am. Chem. Soc.* 111:6369–6373, 1989.

65. S. Bonazzi, M. Capobianco, M.M. de Morais, A. Garbesi, G. Gottarelli, P. Mariani, M.G.P. Bossi, G.P. Spada, and L. Tondelli. *J. Am. Chem. Soc.* 113:5809–5816, 1991.

66. G. Gottarelli, G. Proni, G.P. Spada, S. Bonazzi, A. Garbesi, F. Ciuchi, and P. Mariani. *Biopolymers* 42:561–574, 1997.

67. G. Gottarelli, G. Proni, and G.P. Spada. *Liq. Cryst.* 22:563–566, 1997.

68. R.B. Martin. *Chem. Rev.* 96:3043–3064, 1996.

69. B.J.B. Folmer, E. Cavini, R.P. Sijbesma, and E.W. Meijer. *Chem. Commun.* (17):1847–1848, 1998.

70. B.J.B. Folmer, R.P. Sijbesma, and E.W. Meijer. *J. Am. Chem. Soc.* 123:2093–2094, 2001.

71. J.H.K.K. Hirschberg, L. Brunsveld, A. Ramzi, J.A.J.M. Vekemans, R.P. Sijbesma, and E.W. Meijer. *Nature* 407:1167–1170, 2000.

72. L. Brunsveld, J.A.J.M. Vekemans, J.H.K.K. Hirschberg, R.P. Sijbesma, and E.W. Meijer. *Proc. Natl. Acad. Sci. USA* 99:4977–4982, 2002.

73. R.K. Castellano, D.M. Rudkevich, and J. Rebek Jr. *Proc. Natl. Acad. Sci. USA* 94:7132–7137, 1997.

74. R.K. Castellano and J. Rebek Jr. *J. Am. Chem. Soc.* 120:3657–3663, 1998.

75. R.K. Castellano, R. Clark, S.L. Craig, C. Nuckolls, and J. Rebek Jr. *Proc. Natl. Acad. Sci. USA* 97:12418–12421, 2000.

76. J.J. González, P. Prodos, and J. de Mendoza. *Angew. Chem. Int. Ed.* 38:525–528, 1999.

77. V. Berl, M. Schmutz, M.J. Krische, R.K. Khoury, and J.-M. Lehn. *Chem. Eur. J.* 8:1227–1243, 2002.

78. E.A. Folgeman, W.C. Yount, J. Xu, and S.L. Craig. *Angew. Chem. Int. Ed.* 41:4026–4028, 2002.

79. S.J. Geib, C. Vicent, E. Fan, and A.D. Hamilton. *Angew. Chem. Int. Ed.* 32:119–121, 1993.

80. M.R. Ghadiri, J.R. Granja, R.A. Milligan, D.E. McRee, and N. Khazanovich. *Nature* 366:324–327, 1993.

81. H.S. Kim, J.D. Hartgerink, and M.R. Ghadiri. *J. Am. Chem. Soc.* 120:4417–4424, 1998.

82. H. Hoshino, J.-I. Jin, and R.W. Lenz. *J. Appl. Polym. Sci.* 29:547–554, 1984.

83. C.P. Lillya, R.J. Baker, S. Hütte, H.H. Winter, Y.-G. Lin, J. Shi, L.C. Dickinson, and J.C.W. Chien. *Macromolecules* 25:2076–2080, 1992.

84. J.H.K.K. Hirschberg, F.H. Beijer, H.A. van Aert, P.C.M.M. Magusin, R.P. Sijbesma, and E.W. Meijer. *Macromolecules* 32:2696–2705, 1999.

85. B.J.B. Folmer, R.P. Sijbesma, R.M. Versteegen, J.A.J. van der Rijt, and E.W. Meijer. *Adv. Mater.* 12:874–878, 2000.

86. K. Yamauchi, J.R. Lizotte, D.M. Hercules, M.J. Vergne, and T.E. Long. *J. Am. Chem. Soc.* 124:8599–8604, 2002.

87. R.F.M. Lange, M. Van Gurp, and E.W. Meijer. *J. Polym. Sci. Part A: Polym. Chem.* 37:3657–3670, 1999.

88. K. Yamauchi, J.R. Lizotte, and T.E. Long. *Macromolecules* 35:8745–8750, 2002.

89. C. Hilger and R. Stadler. *Macromolecules* 25:6670–6680, 1992.

90. R. Stadler and L. de Lucca Freitas. *Colloid Polym. Sci.* 264:773–778, 1986.

91. L.R. Rieth, R.F. Eaton, and G.W. Coates. *Angew. Chem. Int. Ed.* 40:2153–2156, 2001.

92. K. Yamauchi, J.R. Lizotte, and T.E. Long. *Macromolecules* 36:1083–1088, 2003.

93. L.P. Stubbs and M. Weck. *Chem. Eur. J.* 9:992–999, 2003.

94. J.A. Manson and L.H. Sperling. *Polymer Blends and Composites*. New York: Plenum Press, 1976.

95. M.V. de Meftahi and J.M.J. Fréchet. *Polymer* 29:477–482, 1988.

96. C.J.T. Landry, D.J. Massa, D.M. Teegarden, M.R. Landry, P.M. Henrichs, R.H. Colby, and T.E. Long. *Macromolecules* 26:6299–6307, 1993.

97. C.J.T. Landry, D.J. Massa, D.M. Teegarden, M.R. Landry, P.M. Henrichs, R.H. Colby, and T.E. Long. *J. Appl. Polym. Sci.* 54:991–1011, 1994.

98. D.J. Massa, K.A. Shriner, S.R. Turner, and B.I. Voit. *Macromolecules* 28:3214–3220, 1995.

99. R.F.M. Lange and E.W. Meijer. *Macromolecules* 28:782–783, 1995.

100. G.R. Newkome, C.N. Moorefield, and F. Vögtle. *Dendritic Molecules — Concepts, Syntheses, Perspectives*. New York: VCH Publishers Inc., 1996.

101. F. Zeng and S.C. Zimmerman. *Chem. Rev.* 97:1681–1712, 1997.

102. S.C. Zimmerman. *Curr. Opin. Colloid Interface Sci.* 2:89–99, 1997.

103. S.C. Zimmerman and L.J. Lawless. *Top. Curr. Chem.* 217:95–120, 2001.

104. S.C. Zimmerman, F. Zeng, D.E.C. Reichert, and S.V. Kolotuchin. *Science* 271:1095–1098, 1996.

105. P. Thiyagarajan, F. Zeng, C.Y. Ku, and S.C. Zimmerman. *J. Mater. Chem.* 7:1221–1226, 1997.

106. F. Zeng, S.C. Zimmerman, S.V. Kolotuchin, D.E.C. Reichert, and Y. Ma. *Tetrahedron* 58:825–843, 2002.

107. S.V. Kolotuchin and S.C. Zimmerman. *J. Am. Chem. Soc.* 120:9092–9093, 1998.

108. Y. Ma, S.V. Kolotuchin, and S.C. Zimmerman. *J. Am. Chem. Soc.* 124:13757–13769, 2002.

109. Y. Wang, F. Zeng, and S.C. Zimmerman. *Tetrahedron Lett.* 38:5459–5462, 1997.

110. N. Kimizuka, T. Kawasaki, K. Hirata, and T. Kunitake. *J. Am. Chem. Soc.* 117:6360–6361, 1995.

111. N. Kimizuka, S. Fujikawa, H. Kuwahara, T. Kunitake, A. Marsh, and J.-M. Lehn. *J. Chem. Soc. Chem. Commun.* (20):2103–2104, 1995.

112. K. Hanabusa, T. Miki, Y. Taguchi, T. Koyama, and H. Shirai. *J. Chem. Soc. Chem. Commun.* (18):1382–1384, 1993.

113. F.M. Menger and S.J. Lee. *J. Am. Chem. Soc.* 116:5987–5988, 1994.

114. F.M. Menger, S.S. Lee, and X. Tao. *Adv. Mater.* 7:669–671, 1995.

115. N. Kimizuka, T. Kawasaki, and T. Kunitake. *J. Am. Chem. Soc.* 115:4387–4388, 1993.

116. N. Kimizuka, T. Kawasaki, K. Hirata, and T. Kunitake. *J. Am. Chem. Soc.* 120:4094–4104, 1998.

117. N. Kimizuka, T. Kawasaki, and T. Kunitake. *Chem. Lett.* (1):33–36, 1994.

118. N. Kimizuka, T. Kawasaki, and T. Kunitake. *Chem. Lett.* (8):1399–1402, 1994.

119. T. Kawasaki, M. Tokuhiro, N. Kimizuka, and T. Kunitake. *J. Am. Chem. Soc.* 123:6792–6800, 2001.

120. H. Koyano, Y. Kanami, K. Ariga, T. Kunitake, Y. Oishi, O. Kawano, M. Kuramori, and K. Suehiro. *Chem. Commun.* (15):1769–1770, 1996.

121. H. Koyano, P. Bissel, K. Yoshihara, K. Ariga, and T. Kunitake. *Langmuir* 13:5426–5432, 1997.

122. U. Michelsen and C.A. Hunter. *Angew. Chem. Int. Ed.* 39:764–767, 2000.

123. H. Nierengarten, J. Rojo, E. Leize, J.-M. Lehn, and A. Van Dorsselaer. *Eur. J. Inorg. Chem.* (3):573–579, 2002.

Chapter 7

Dendrimeric Supramolecular and Supra*macro*molecular Assemblies

Donald A. Tomalia

Contents

I. INTRODUCTION

If all scientific knowledge were lost in a cataclysm, what single statement would preserve the most information for the next generations of creatures? How could we best pass on our understanding of the world? [I might propose:] ... All things are made of atoms — little particles that move around in perpetual motion, attracting each other when they are a little distance apart, but repelling upon being squeezed into one another .... In that one sentence, you will see, there is an enormous amount of information about the world, if just a little imagination and thinking are applied. Richard P. Feynman [1]

In this simple quotation, Feynman has perhaps described nature's ultimate example of a minimalist self-assembly. This is most certainly not a molecular level self-assembly; nonetheless atoms serve to remind us that self-organization of fundamental subatomic entities occurred to give us the most basic building blocks of the universe [2]. These self-assembling events were consummated some 10 to 13 billion years ago and marked a unique moment in time from which *first order* was forever derived from chaos. This was the genesis of the long, unrelenting evolutionary journey to more complex forms of natural matter.

The earliest events involved the assembly of subatomic particles into roughly spherical entities reminiscent of core-shell type architecture. First, lighter elements were formed followed by nuclear synthesis leading to the heavier elements. These discrete, quantitized core-shell assemblies of electrons and nuclei were so precise, dependable, and indestructible in chemical reactions that they have functioned as the fundamental building blocks of the universe. Within these elements, nature successfully organized nuclei and electrons to control atomic space "at the subpicoscopic level (i.e., $<10^{-12}$ m) as a function of: *size* (atomic number), *shape* (bonding directionality), *surface stickiness* (valency), and *flexibility* (polarizability). These variables may, thus, be considered *critical atomic design parameters*" — CADPs. This new order set the primordial stage for all evolutionary patterns that followed. These patterns seemed to follow the simple principle; "*order begets order from chaos*" [3–6].

The next phase in this evolutionary sequence involved the natural combination of these reactive elements to produce a bewildering array of simple molecular combinations derived from these core-shell atomic spheroids (i.e., NH_3, CH_4, urea, etc.) followed by the formation of more complex, but yet small molecules that included α-amino acids, nucleic acids, sugars, hydrocarbons, etc. Combinations and permutations of specific CADPs at the atomic level articulated molecular level architectures and incipient properties. One path led to *abiotic* molecular evolution (inorganic chemistry); whereas, the other initiated the *biotic* molecular evolution (organic chemistry) and ultimately life as we recognize it today.

The biotic molecular evolution was defined by the respective CADPs of the combined (atoms) required to produce this new molecular level order. It is now known that within this hierarchical level, *new sizes, shapes, surface chemistries* (functional groups/nonbonding interactions), *flexibilities* (conformations), and *topologies* (architectures) arise. These parameters may be visualized by the various shapes, valencies, and polarizabilities associated with the element carbon in its well-known sp, sp^2, or sp^3 hybridized states. We define these unique features as "*critical small molecule design parameters*" — CSMDPs. Molecular entities in this domain are generally <1000 atomic mass units, thus they occupy space of up to approximately 10 Å (1 nm) in diameter, when normalized as spheroids. They may be thought of as subnanoscale in dimension.

A. Suprachemical Categories and Dimensions

The rich patterns of electronegative and electropositive domains found in these small atom and molecule combinations allowed nature to devise new rules and strategies for advancement to the next higher levels of ordered complexity by nonbonding interactions (suprachemistry) (e.g., *supramolecular: higher in organization or more complex than a molecule; often composed of many molecules*) [7]. These strategies may be roughly categorized into several major types, namely:

Category I — Supraatomic (exo-). Those assemblies involving small clusters of metal atoms or elements with subnano and nanoscale dimensions (e.g., quantum dots, etc.).

Category II — Supramolecular (endo-). Those assemblies leading to small- and medium-sized supramolecular structure. These examples include primary convergent-type binding compounds (i.e., spheroidal guest–host structures, macrocyclic, carcerands, etc.).

Category III — Supramolecular (exo-). Those assemblies involving amphiphilic monomers that lead to medium–large supramolecular structures. These assemblies tend to function as transport entities, barriers, membranes, and container-type structures (i.e., micelles, liposomes, lipid bilayers, etc.).

Category IV — Supramacromolecular (exo-). Those assemblies leading to precise, three-dimensional (3D), structure-controlled, noncovalently bound macromolecules. These supra*macro*molecular structures are derived from more complex, but precisely controlled macromolecular structures capable of information storage, expression, amplification, and used as functional or structural building blocks (e.g., protein folding, DNA–histone complexes, DNA expression, etc.). Figure 1 illustrates these suprachemical categories as a function of dimensions.

The objective of this account is to examine abiotic examples and parameters related to larger supramolecular and supra*macro*molecular dendritic structures analogous to those found in Categories II, III, and IV.

B. Progress in the Science of Abiotic Synthesis

Whereas, nature has been evolving the complexity of matter over the past 10 to 13 billion years, mankind formally began its journey directed at the "*science of abiotic synthesis*" only approximately 200 years ago. Beginning with Lavoisier's "atom hypothesis" [8], Dalton's "molecular hypothesis" [9] followed by the initiation of "organic chemistry" with Wöhler's work [10], the progress of man-made synthetic evolution appears to be in its infancy compared with nature's evolution [11].

Based on the various hybridization states of carbon and other elements in the periodic table, small molecule synthesis has led to at least four major architectural patterns. The major architectural classes may be visualized as described in Figure 2 and includes (I) Linear, (II) Bridged, (III) Branched, and (IV) Dendritic (cascade) architectures.

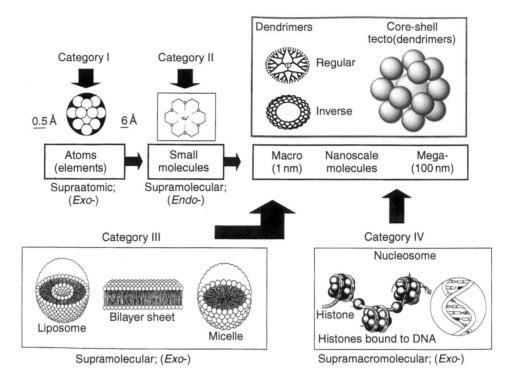

Figure 1 Suprachemistry categories/dimensions.

C. Progress in the Science of Abiotic Supramolecular Chemistry

> It seemed clear to me now, that the sodium (potassium) ion had fallen into the hole at the center of the molecule. C.J. Pedersen (Aldrichchim. Acta, 4, 1 [1971]).

This simple, but very bold statement made by C.J. Pedersen in the early 1970s literally ushered in the era of supramolecular chemistry based on further elaborations of these basic small molecule architectural classes. J.M. Lehn developed terms to describe two major categories of supramolecular receptors [12]. They are broadly defined as *endo-* and *exo-*receptors. The former present interactive sites that converge on a central locus leading to complexations that are cyclic or severely bent molecules containing, clefts, or cavities. These neutral organic ligands are generally referred to as convergent binding molecules. They include small molecule endo-receptors such as crown ethers, cryptands, podands, and spherands. This work has been reviewed extensively by many workers [13–19]. Exo-complexation typically involves noncyclic molecules containing interactive sites that communicate outwardly. Considerably fewer examples of this type of complexation have been reported.

Recently, however, exciting new reports are appearing, which describe the rapid self-assembly of highly ordered linear as well as two- and three-dimensional molecular structure [20]. These strategies usually involve exo-recognition between small components to produce tapes [21], squares [22–24], rosettes [25], and other interesting topologies [26–32].

The only topological type in this small molecule area that has not been exploited extensively as a neutral ligand in supramolecular chemistry was the Class IV dendritic (cascade)-type architecture. Vögtle introduced this fourth major small molecule architecture in the late 1970s in a single isolated communication [33]. These low molecular weight dendritic (cascade) molecules served as

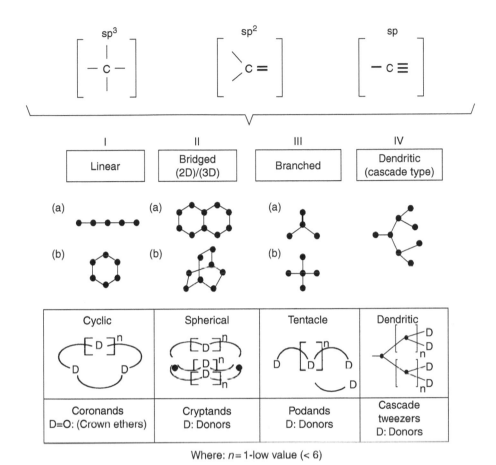

Figure 2 Major small molecule architectures defined as Classes I to IV with comparisons to known donor-type ligands.

precursors to macromolecules, which are now recognized as *dendrons, dendrimers,* and *hyperbranched polymers* [34–36]. Interest in the supramolecular aspects of dendritic systems has occurred only recently. As described earlier [37], they offer the potential for either endo- or exo-complexation and will be the focus of this account.

It was only during the past several decades that substantial progress has been made in the use of amphiphilic reagents (monomers) in the construction of nonbonding macroassemblies. These supramolecular constructs are of the Category II-type and usually lead to entities that are medium to large in dimension. Amphiphilic reagents used for these constructions are generally derived from Class I (linear) or Class III (branched) small-molecule architectures. They include a variety of surfactants, phospholipids, and amphiphilic oligomers as described in Figure 2, Section I.A. This area has received considerable attention and has been extensively reviewed [38–40].

Many of the reactive linear and branched small molecule reagents were derived during the rich and prolific era of organic chemistry in the 19th century [11]. These developments provided many of the necessary reagents (monomers) upon which H. Staudinger based his work. With these building blocks, Staudinger initiated the synthetic macromolecular evolution some 70 years ago, as he introduced his "macromolecular hypothesis" [41]. This evolution has led to three major macromolecular architectures; namely *linear, cross-linked (bridged),* and *branched* types. These architectural classes are recognized as traditional polymers [42]. In all cases, these structures/architectures are produced by largely random, uncontrolled polymerization processes. These processes produce polydispersed (i.e., M_w/M_n > 2–10) products of many different molecular weights. In general, these *are not* structure-controlled macromolecular architectures such as one observes in biological systems. However, considerable progress has occurred recently in the areas of living-anionic [43], cationic [44], and radical polymerizations [45,46].

D. Structure-Controlled Macromolecules by Abiotic Synthesis

Structure-controlled abiotic synthesis of macromolecules that mimicked biological polymers were first reported by Merrifield nearly 30 years ago. Abiotic synthesis of poly(peptides) by use of solid-phase synthesis was reported as early as 1963 [47]. This synthesis strategy was soon extended to the structure-controlled synthesis of poly(amides), poly(nucleotides), and poly(saccharides) [48]. Simply stated, the growing chain in all cases is covalently anchored with a cleavable linker to an insoluble substrate. Monomers are sequenced by means of protect/deprotection methods using *linear genealogical synthesis schemes* [49]. As early as 1979, we discovered simple synthetic strategies that have allowed us to produce structure-controlled macromolecules in ordinary laboratory glassware [50–53]. These strategies do not require the need for biological components or immobilized substrate reactions. Utilizing traditional organic reagents and monomers such as, ethylenediamine (EDA) and alkyl acrylates we are now able to routinely synthesize commercial quantities (kilograms) of controlled macromolecular structures with polydispersities of 1.0005 to 1.10. These new structures are referred to as *dendrons* or *dendrimers* [36,54].

Although the mimicry is less elegant and more minimalistic than that found in nature, these synthetic strategies appear to mimic the four pervasive patterns devised by nature for the structure control of natural macromolecules, for example, DNA, RNA, and proteins (Figure 3).

1. Primary atomic (CADPs) or molecular (CMDPs) information is defined and stored in the initiator core or seed. This information that includes its size, shape, multiplicity (N_c), and chemistry (valency) is presented in an exo-fashion and communicated to a "template polymerization region," namely, the reactive terminal groups.
2. Appropriate "feedstock monomers" such as acrylates, acrylonitrile, and other organic reagents are defined and adapted to various "template polymerization" schemes. This introduces geometrical amplification at the termini. These amplification values are defined by the multiplicity (N_b) of the branch junctures. Protect/deprotection procedures allow control of complementary chemical reactivities in the "template polymerization region." The chemistry is designed to assure high yield chemical bond formation at each iterative growth stage (generation = G) in an effort to avoid defects or digressions from geometrically ideal branch amplification.

"Regio-selective control of the template polymerization" is obtained by transfer of genetic (i.e., CADP or CMDPs) information through the hierarchy of chemical bond connectivity involved in the dendrimer construction. This includes the following: [Initiator core] (*DNA mimic*) — [*transcription*] → [interior branch cells] (*RNA mimic*) — [*translation*] → [surface branch cells] — (*ribosome mimic*) → (terminal surface groups [Z] — (*template polymerization region*). In this minimalistic, abiotic core-shell construction, the architectural mimicry of a biological cell is readily apparent by the following comparison: [initiator core] ≅ [biotic cell nucleus], [interior branch cells] ≡ [biotic cell cytoplasm], and [surface branch cells + terminal groups] = [biotic cell membrane].

"Self replication" of the primary genetic information (i.e., the initiator core) may occur according to a geometrically driven (2^G) amplification process as described in Figure 3. This specific "self-replicating" amplification process produces "interior branch cells," "surface branch cells," and "surface functional groups" according to the geometrically progressive values illustrated. These familiar doubling values reflect the well-known (2^G) amplification values associated with *biological cell* division (mitosis) or DNA *amplification* (polymerase chain reactions, PCR). For this

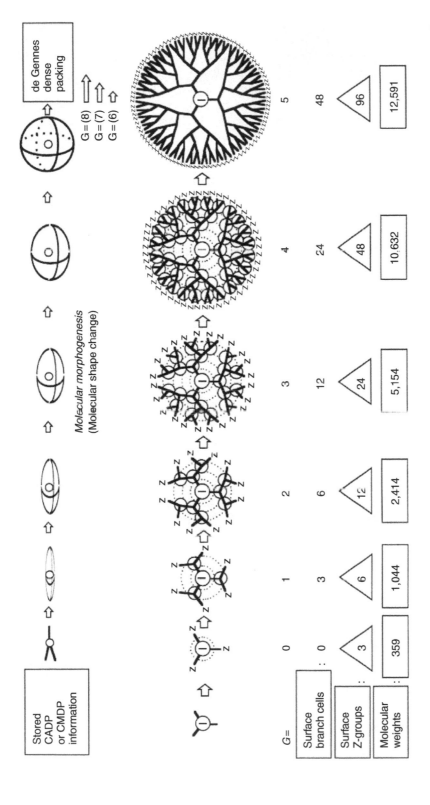

Figure 3 A dendrimer series with core multiplicity ($N_C = 3$) and branch-cell multiplicity ($N_C = 2$) illustrating molecular shape changes, surface branch cells, surface Z-groups, and molecular weights as a function of generation.

poly(amidoamine) PAMAM dendrimer family (Figure 3), initiated from an ammonia core ($N_c = 3$) with a branch cell multiplicity ($N_b = 2$), the expected mass values of 359, 1,044, 2,414, 5,154, 10,632, and 21,591 are obtained for generations 0 to 5, respectively. These values are verified routinely by electrospray or matrix-assisted, laser desorption mass spectroscopy (MALDI-TOF) methods. Polydispersity values (M_w/M_n) are obtained ranging from 1.000002 to 1.005 for this series. Over 100-different dendrimer families possessing compositionally different branch cells (i.e., carbon, nitrogen, silicon, sulfur, phosphorous [55], metals, etc.) and multiplicity values of $N_c = 1$–100 and $N_b = 2$–5 have been synthesized and characterized. Of course, there may be some errors or defects (mutations) in these *divergent dendrimer constructions*, just as there are well-known genetic defects and mutations in all *biotic genealogically directed processes.* This simply adds to the richness of the comparison between these *abiotic cells/organisms* and *biotic cells/organisms.*

Biological cells may be thought of as core-shell-like microscale reactors designed to manufacture both structural and functional building blocks (proteins). Similarly, dendrimers may be thought of as nanoscale, core-shell models with certain analogies to biological cells. In each case, core-shell characteristics and growth patterns are determined by a central library of information, which flows from the respective cores to the shell-like surfaces [49]. The major differences between the two systems are (1) scaling (Figure 9) and (2) the mode of information transfer.

II. THE DENDRITIC STATE

Dendritic architecture has been widely recognized as a fourth major class of macromolecular architecture [51–53]. The signature for such a distinction is the unique repertoire of new properties manifested by this class of polymers. New properties and applications for this polymer class have been reviewed elsewhere [34,36,54]. Within the realm of the macromolecular structure, dendritic architecture may be viewed as an intermediary architectural state, which resides between linear (thermoplastic) structures and cross-linked (thermoset) systems [56,57]. As such, the dendritic state may be visualized as advancement from a lower order to a higher level of structural complexity [58–60]. Furthermore, recent developments in the synthetic control of macromolecular structure now suggest these transitions may involve various levels of structural control. In fact, these transformations may occur via (i) *statistical*, (ii) *Semicontrolled*, or (iii) *Controlled* pathways. It is widely recognized that dendrons/dendrimers constitute a significant subclass of "dendritic polymers" and represent a unique combination of very high structural complexity, together with extraordinary structural control. The focus of this section will be confined to the dendritic supramolecular and supra*macro*molecular aspects of assembling those entities indicated in the boxed area of Figure 1. Moreover, a cursory examination of the supramolecular properties of these dendritic structures will be made as they relate to nanotechnology and other issues of theoretical interest.

The assembly of reactive monomers [50], branch cells [14,36,54], or dendrons [61,62] around atomic or molecular cores to produce dendrimers according to divergent/convergent dendritic branching principles is well-demonstrated [63]. Such systematic filling of space around cores with branch cells as a function of generational growth stages (branch cell shells) to give discrete, quantized bundles of mass have been shown to be mathematically predictable [49,64]. Predicted theoretical molecular weights have been confirmed by mass spectroscopy [65–67] and other analytical methods [68,69]. In all cases their growth and amplification is driven by the following general mathematical expressions (Figure 4).

A. A Comparison of Traditional Polymer Science with Dendritic Macromolecular Chemistry

It is appropriate to compare covalent bond formation in traditional polymer chemistry with that in dendritic macromolecular chemistry. This allows one to fully appreciate the implications and

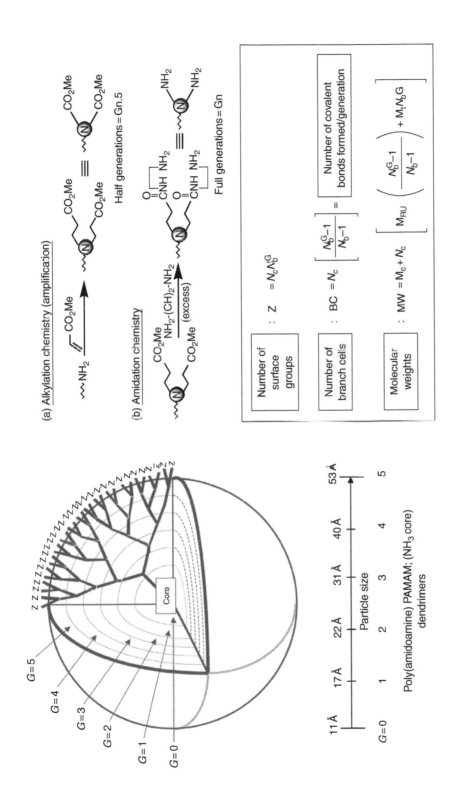

Figure 4 Core-shell dendrimer architecture with mathematics defining number of surface-groups (Z), number of branch cells (BC), theoretical molecular weights (MW), dimensions (Å) as a function of generation (G), branch-cell multiplicity (N_b), core multiplicity (N_c), and core molecular weight (M_c), (a) alkylation chemistry, (b) amidation chemistry.

differences between the two areas in the context of supramolecular polymerization. Covalent synthesis in traditional polymer science has evolved around the use of reactive modules (AB-type monomers) that can be engaged in multiple covalent bond formation to produce single molecules. Such multiple bond formation is driven either by chain reaction or by poly(condensation) schemes. Staudinger first introduced this paradigm in the 1930s by demonstrating that reactive monomers could be used to produce a statistical distribution of one-dimensional molecules with very high molecular weights (i.e., $>10^6$ Da) [41,70]. These covalent synthesis protocols underpin the science of traditional polymerizations. As many as 10,000 or more covalent bonds can be formed in a single chain reaction of monomers. Although megamolecules with nanoscale dimensions may be attained, relatively little control can be exercised to precisely manage critical molecular design parameters, such as: sizes, atom positions, covalent connectivity (i.e., other than linear topologies), or molecular shapes [44–46].

These polymerizations usually involve AB-type monomers based on substituted ethylenes, strained small ring compounds, or AB-type monomers which may undergo polycondensation reactions. The chain reactions may be initiated by free radical, anionic, or cationic initiators. Multiple covalent bonds are formed per chain sequence wherein, the average lengths are determined by monomer to initiator ratios. Generally, polydispersed structures, which are statistically controlled, are obtained. If one views these polymerizations as extraordinarily long sequences of individual reaction steps, the average number of covalent bonds formed per chain may be visualized in Figure 5. All three classical polymer architectures, namely, Class I — *linear*, Class II — *Cross-linked* (bridged), and Class III — *Branched* topologies can be prepared by these methods (Figure 11), keeping in mind that simple introduction of covalent bridging bonds between polymer chains (Class I-type) are required to produce Class II cross-linked (thermoset) type systems [36,51,56,71] (see Figure 5).

In the case of dendron/dendrimer syntheses, one may view those processes leading to those structures as simply sequentially staged (generations), quantized polymerization events. Of course, these events involve the polymerization of AB_2 monomer units around a core to produce arrays of covalently bonded branch cells that may amplify up to the shell saturation limit as a function of generation.

Mathematically, the number of covalent bonds formed per generation (reaction step) in a dendron/dendrimer synthesis varies as a power function of the reaction steps (Figure 6). This analysis shows that covalent bond amplification occurs in all dendritic strategies. This feature clearly differentiates dendritic processes from covalent bond synthesis found in traditional organic chemistry or polymer chemistry [49]. Polymerization of AB_2 or AB_x monomers leading to hyperbranched systems also adhere approximately to these mathematics, however, in a more statistical fashion.

It is interesting to note that this same mathematical analysis may be used to predict the amplification of DNA by PCR methods or the proliferation of biological cells by mitosis as a function of generation. This comparison will be described later.

B. Dendrimer Synthesis Strategies

Beginning in 1979 [35,50,51,72–76], two major strategies have evolved for dendrimer synthesis. The first was the *divergent method* wherein, growth of a dendron (molecular tree) originates from a core site (root) (Figure 7). During the 1980s, virtually all dendritic polymers were produced by construction from the root of the molecular tree. This approach involved assembling monomeric modules in a radial, branch upon branch motif according to certain dendritic rules and principles [53]. This divergent approach [51,77] is currently the preferred commercial route used by worldwide producers; such as, Dendritic NanoTechnologies, Inc. (U.S.A.), DSM (Netherlands), and Perstorp (Sweden).

A second method, which was pioneered by Fréchet et al. in 1989, is referred to as the convergent growth process [78] and proceeds in the opposite direction inward from what will become the

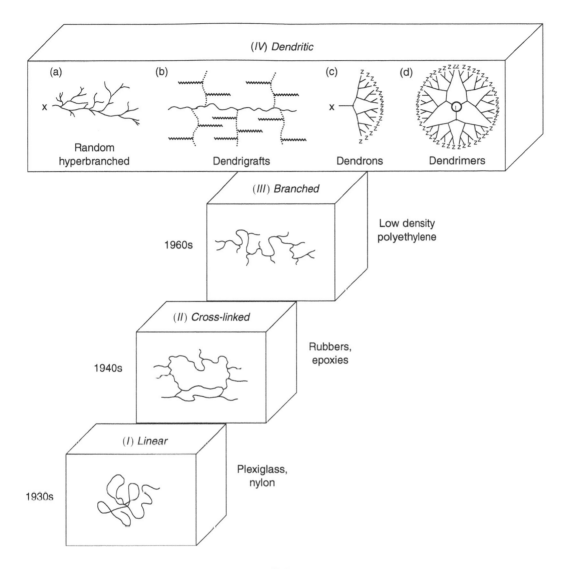

Figure 5 Four major macromolecular architectures: (I) linear, (II) cross-linked, (III) branched, (IV) dendritic.

dendrimer molecular surface to a focal point (root). In this matter, the latter results in formation of a single dendron, in order to obtain a multidendron structure one additional step is required (i.e., an anchoring reaction). In this reaction, several dendrons are connected via a covalently coupling reaction with a multifunctional core to yield the desired dendrimer product. Using these two major synthetic strategies, over 100 compositionally distinct dendrimer families have been synthesized and are reviewed elsewhere [14,35–37].

C. A Comparison of Divergent Abiotic Synthesis with the Biotic Strategy

Dendrimers may be thought of as nanoscale information processing devices. It is now well recognized that they possess the facility for transcribing and translating their core information into a wide variety of dimensions, shapes, and surfaces by organizing abiotic monomeric/branch-cell reagents into the dendritic architecture [49]. Furthermore, these transcription and translational events are accompanied by mathematically defined dendritic amplifications. The syntheses may be thought of as sequence-staged, quantized polymerizations orchestrated to the amplification patterns dictated by

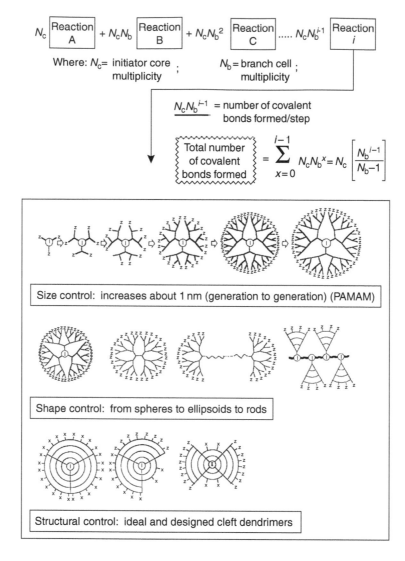

Figure 6 Dendritic macromolecular chemistry.

the core (N_c) and branch-cell (N_b) multiplicities. Presently, there appears to be very little limitation to the linear or branched polymerization units that may be used for these constructions. The limitations are determined primarily by the ability to control the regio-specific reactivity of these construction units. Over 100 different types of polymerizable units have been used to make at least that many different compositional dendrimer families. These reactive monomeric reagents are processed by *terminal group directed transcription* into precisely defined peripheral assemblies. Such assemblies of polymerizable units organized in the terminal group region constitute well-defined shells (generations) [49,72]. These self-assembled shells are frozen genealogically by covalent fixing (translation) of these organized arrays.

Phenomenologically speaking, these molecular level events are reminiscent of those that occur during biotic protein synthesis. Within biotic cells, the objective is to provide a micron-sized compartment (reactor) for the synthesis and amplification of a requisite population of *structural* and *functional* proteins — the building blocks of life. These steps involve the dynamic transfer of information through space within the cell by specific carrier molecules (i.e., RNA, etc.). This journey begins with a supramolecular transfer of information in the nucleus (core) and terminates with

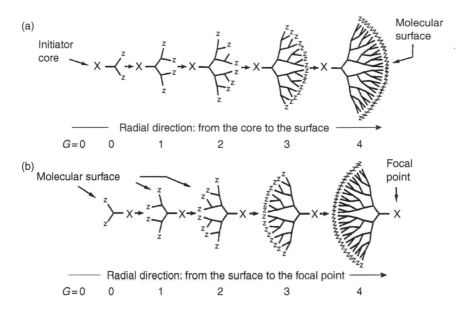

Figure 7 Two principal synthetic methods for building up hyperbranched dendritic macromolecules: (a) divergent and (b) convergent methods.

a supramolecular codon transfer of this information at the ribosome sites. These supramolecular events are covalently fixed as *linear, genealogically directed sequences* catalyzed by polymerases to produce the primary structure of proteins. Similarly, within the nanoscale environ of a dendrimer the information journey begins at the core of the dendrimer. Information such as *shape, size, multiplicity*, and *directionality* manifested at the dendrimer nucleus is transferred by means of presumed supramolecular events, followed by covalent bond formation at the terminal groups of the dendrimer construct. These informational parameters are processed by molecular level events that are dauntingly analogous to those that occur in biological cells. It should be apparent that the phenomena of transcription, translation, and amplification of this information has occurred unimolecularly within the space occupied by the dendrimer. Documentation of this information transfer is ultimately frozen and amplified at each generational level ($G = 1$ to 5) in the form of dendritic covalent connectivity.

Although these comparisons of information transfer phenomena have been made between vastly different dimensional scales, the analogies appear to be remarkably consistent. Biotic cells manufacture precise protein building blocks for either structural or functional purposes. Similarly one may view the parallel products of dendrimeric growth as both structural (i.e., interior branch cells) and functional (i.e., surface branch cells) to the dendrimer. It should be noted that the final dendrimer product possesses dimensions (i.e., poly(amidoamine) (PAMAMs)) that scale very closely to many important life supporting protein building blocks, as illustrated in Figure 53. In fact, recent attention has been directed toward the mimicry of both globular proteins [79–82] and protein assemblies [83]. Articulation of the subtle shape and regio-specific chemistry control found in proteins has not yet been attained; however, crude mimicry with certain dendrimer constructs has been reported [79], [83], [84].

D. Supramolecular Aspects of the Classical Divergent Dendrimer Synthesis

1. The "All or Nothing" Observation

The divergent synthesis process has been the objective of much speculation and curiosity. Questions often asked are the following: Does the controlled generational growth of a dendrimer have any supramolecular characterizations? Is it an example of exo-molecular recognition and self-assembly, followed by covalent bond formation? The answer at this time is *very likely*. However, the evidence at this time is indirect and not unequivocal. Undoubtedly, the amphiphilic nature, the complementary shapes of the termini and the reagents as well as the processing conditions will determine the degree of supramolecular character one might expect in these transformations.

First, the molecular recognition character at the terminal groups is largely determined by the complementary reactivities (communication) between the reagents and these terminal sites. The question remains — is there any evidence for enhanced or catalytical reactivity at these termini, which might suggest preorganization followed by covalent bond formation? The strongest evidence in support of this contention is the so-called "*all or nothing reactivity*" of dendrimer surfaces, which has been observed in our laboratory, as well as Meijer's laboratory (Eindhoven University) with certain amphiphilic reagents. This observation is made routinely when allowing amphiphilic fatty acid chlorides to react with primary amine-terminated dendrimers in the presence of an acid acceptor.

When using stoichiometric amounts of a fatty acid chloride in the presence of triethylamine with either amine-terminated poly(amidoamine) PAMAM or poly(trimethyleneamine) POPAM-type dendrimers one observes only *fully unreacted* and *fully reacted dendrimer* products as illustrated in Figure 8. This observation may be interpreted as a manifestation of regio-specific, self-assembly, followed by covalent amide bond formation, since the reactions are performed under kinetically favored mild conditions (i.e., 25 to 30°C). Alternatively, one might invoke some unidentified catalytic neighboring group effect. This catalytic effect may favor very rapid complete modification of those dendrimer surfaces, which are initially substituted. However, very recent work by Froehling [85] supports the self-assembly hypothesis. When these same reactions were performed with fatty acids under thermodynamically driven (i.e., 130 to 150°C) azetroping conditions, a statistical distribution of substituents was observed in all cases. This is in complete contrast to the "all or nothing" reaction products, which are invariably obtained under more mild, kinetically driven conditions.

2. Sterically Induced Stoichiometry (SIS)

It was interesting to note that when these authors attempted to amidate higher generation poly(propyleneimine) (PPI) dendrimers under these same conditions, they observed regio-specific, positional preference for *single substitution* at the available amine dyad termini. This observation is in complete agreement with the so-called "*sterically induced stoichiometry*" (SIS) hypothesis we have proposed previously [36,37,49,86]. This phenomenon is uniquely related to the size of a proposed surface modification reagent and the tethered congestion conditions that a dendrimer manifests at its surface. This becomes significant especially at higher generations. It was first predicted in 1983 [87], and occurs as a manifestation of de Gennes dense packing. As indicated in this experimental example, these congestion phenomena can affect dendrimer surface substituent patterns and reaction rates. We shall describe later how this congestion phenomenon literally determines the shell saturation levels for the *supramolecular polymerization of dendrimers* into core-shell tecto(dendrimers).

3. Dendrimer Structure Ideality — A Signature for Self-Assembly and for "de Gennes Dense Packing"

Further evidence to support the role of self-assembly in divergent dendrimer synthesis can be found in the construction of poly(amidoamine) PAMAM dendrimers. First, ideality and fidelity of structure is usually a signature of a self-assembly event. This is a universal observation throughout biological systems. In all cases, nearly ideal PAMAM structures are obtained *only* under mild conditions favoring such organizational events as a prerequisite to bond formation [88]. Attempts to impose more severe reaction conditions (i.e., reaction temperature >40°C) *dramatically reduces* the ideality of dendrimer structures even in the early generations. Under mild, kinetically favorable reaction conditions however, nearly ideal dendrimer structures are obtained up to the onset

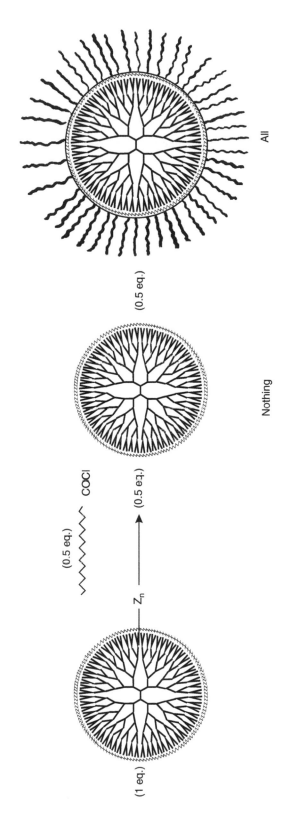

Figure 8 Reaction of substoichiometric amounts of acid chloride reagents with dendrimers to produce the "all or nothing" phenomena.

of "de Gennes dense packing." Observation of the ideal structure is consistent with amplification predictions based on core- and branch-cell multiplicities as well as observed mass defects by mass spectral analysis.

In order to understand the parameters that cause sterically induced defects in dendrimer synthesis better, it is appropriate to review the contributing events. First, normal divergent growth occurs precisely as an ideal molecular structure consistent with molecular weights that predictably obey geometrical mathematics, as described earlier. It should be noted that the experimentally determined radius of a dendrimer molecule increases in a linear fashion as a function of generation; whereas, the terminal groups amplify according to geometric progression laws. Therefore, ideal dendritic growth cannot extend indefinitely. Such a relationship produces a critical congestion state at some generational level. This creates a significant dilemma at a reacting dendrimer surface as a result of inadequate space to accommodate all of the mathematically required new monomer or branch-cell units. This congested generational growth stage is referred to as the *de Gennes dense packed state* [87]. At this stage, the dendrimer surface has become so crowded with terminal functional groups that it is sterically prohibited from reacting completely to give ideal branching and dendrimer growth. The "de Gennes dense packed state" is the point in dendritic growth; wherein, the average free volume available to the reactive surface groups decreases below the molecular volume required for the transition state of the desired reaction to extend ideal branching growth to the next generation. Nevertheless, the onset of the de Gennes dense packed state in divergent synthesis does not preclude further dendritic growth beyond this point, however, it does mean that there will be notable mass defect digressions from ideality.

(a) *Dramatic changes in dendrimer container properties coincidental with the de Gennes dense packing*

We have previously described the dramatic influence that core multiplicity (N_c) and branch-cell multiplicity (N_b) manifested on the onset of the "de Gennes dense packing" [35,37,87]. As shown in Figure 9(a) and (b), it can be seen that for ammonia and EDA core PAMAM dendrimers, respectively, nearly ideal molecular weight masses are observed by mass spectroscopy up to generations 5 and 4, respectively. It should be noted that a systematic pattern of mass defects is observed at a critical generation, in each case, due to the SIS induced by the "de Gennes dense packing" phenomenon [35,37,87]. For example, in the case of ammonia core ($N_c = 3$; $N_b = 2$) dendrimer mass defects are not observed until $G = 5$ (Figure 9(a)). At that generation, a gradual digression from theoretical masses occurs for $G = 5$–8, followed by a substantial break (i.e., $\Delta = 23\%$) between $G = 8$ and 9. This *discontinuity in shell saturation is interpreted as the signature for de Gennes dense packing.* It should be noted that shell saturation values continue to decline monotonically beyond this breakpoint down to a value of 35.7% of theoretical saturation at $G = 12$. A similar mass defect trend is noted for the EDA core, PAMAM series ($N_c = 4$; $N_b = 2$), however, the shell saturation inflection point occurs *at least one generation earlier* (i.e., $G = 4$–7, (Figure 9(b))) for the higher multiplicity, EDA core. This suggests that the onset of "de Gennes dense packing" may be occurring between $G = 7$ and 8. These latter data are completely consistent with metal ion probe experiments [89]. It has been shown that *the interior* of hydroxyl terminated (EDA core) PAMAM dendrimers $G = 1$–6 is completely accessible to Cu^{++} hydrate. However, attempts to drive $Cu^{++} \cdot 6H_2O$ (metal ion hydrate) into the interior of $G = 7$–10 did not occur even under forcing conditions. Subsequent treatment of these respective Cu^{++} as a function of $G = 1$–10 (chelated) solutions with hydrogen sulfide manifested three different behaviors as a function of generation. Copious precipitates were obtained from $Cu^{++}/G = 1$–3. In contrast, completely soluble solutions were obtained for $Cu^{++}/G = 4$–6. Finally, a totally different precipitate was obtained for $Cu^{++}/G = 7$–10. TEM analysis of the latter two sets of metal/dendrimer combinations confirmed that $G = 4$–6 dendrimers had incarcerated copper sulfide and were functioning as host container molecules to the copper sulfide guest aggregation.

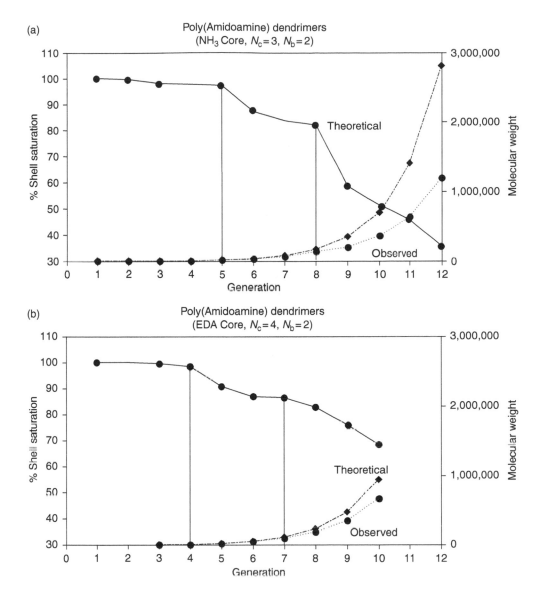

Figure 9 Shell saturation level (%), theoretical molecular weights, and observed molecular weights (mass spectroscopy) as a function of generation for (a) NH_3 core ($N_c = 3$, $N_b = 2$) and (b) EDA core ($N_c = 4$, $N_b = 2$) poly(amidoamine) PAMAM dendrimers.

Similar analysis of the last set indicated $G = 7$–10 were functioning as surface scaffolding with virtually no metal incorporation into the interior [90,91]. This clearly defines a congestion periodicity pattern, as illustrated in Figure 10, and reveals a unique pattern of dendrimer/metal ion relationships as a consequence of generational surface congestion. Obviously, these metal ion probe experiments are consistent with and support the mass spectral "de Gennes dense packing" signature hypothesis [36].

In summary, it is readily apparent that these abiotic, amplified genealogically directed polymerizations of AB_2 monomers have many things in common with biological systems. For example, certain biological processes, such as PCR [92] or cell mitosis [8–10] may be thought of as analogous examples of amplification, but at much larger dimensional scales than are found in dendrimers. Most importantly, branch-cell amplification and DNA amplification may be viewed as special

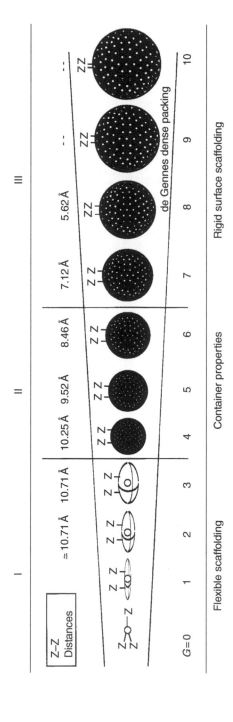

Figure 10 Periodic congestion patterns for a PAMAM EDA core ($N_c = 4$, $N_b = 2$) dendrimer series ($G = 0$ to 10) defining (I) interior scaffolding, (II) container-like properties, and (III) surface scaffolding properties when combined with an aqueous Cu^{+2} probe.

examples of *supramolecular-directed polymerization leading in each case to very highly controlled macromolecular structures.*

E. Strategies for the Supramolecular Assembly of Components to Produce Dendrimers

One of the major themes in biological systems is the noncovalent assembly of large structures from smaller components. Self-assembly is usually rapid, requires minimized energy for synthesis, and guarantees reproducible construction of complex products with high fidelity [8–10]. The presumed supramolecular construction of dendrimers from its constitutional components provides prime examples of abiotic supramolecular polymerizations to give well-defined structures with notable complexity. More recent developments in this area may be broadly categorized into those involving the supramolecular assembly of *"dendron"* components and those involving the assembly of subdendron units or *"branch cells."*

1. Supramolecular Assembly of Dendrimers Based on Focal Point Functionalized Dendrons

This construction strategy evolved from the pioneering work of Zimmerman et al. 1996 [93]. Their approach involved the hydrogen bond directed aggregation of Fréchet-type dendrons, which were terminated with two isophthalic diacid moieties at their focal point. In appropriate solvents, these assemblies were stable enough to be characterized by size exclusion chromatography (SEC), laser light scattering (LLS), and vapor pressure osmometry (VPO). Dendrons as high as $G = 4$ could be self-assembled by this process (Figure 11). A variety of other single-site, focal point functionalized dendrons have been reviewed [62] and reported recently [62,94,95].

An extension of this strategy involves a very clever assembly of two dendritic hemispheres bearing crown ethers at their focal points as reported by Dykes and Smith [96]. They invoked traditional supramolecular bridging of the dendron focal points with ditopic ammonium cation functionalized templates. These constructs could be reversibly assembled and disassembled exhibiting gel phase type properties.

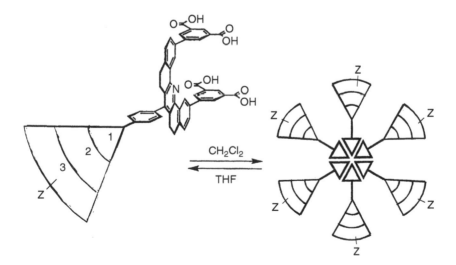

Figure 11 The self-assembly of a hexameric aggregate of wedge-like dendrons that are functionalized at their focal points by two isophthalic acid moieties. (From O.A. Matthews. *Prog. Polym. Sci.* 23:1–56, 1998. With permission from Elsevier Science.)

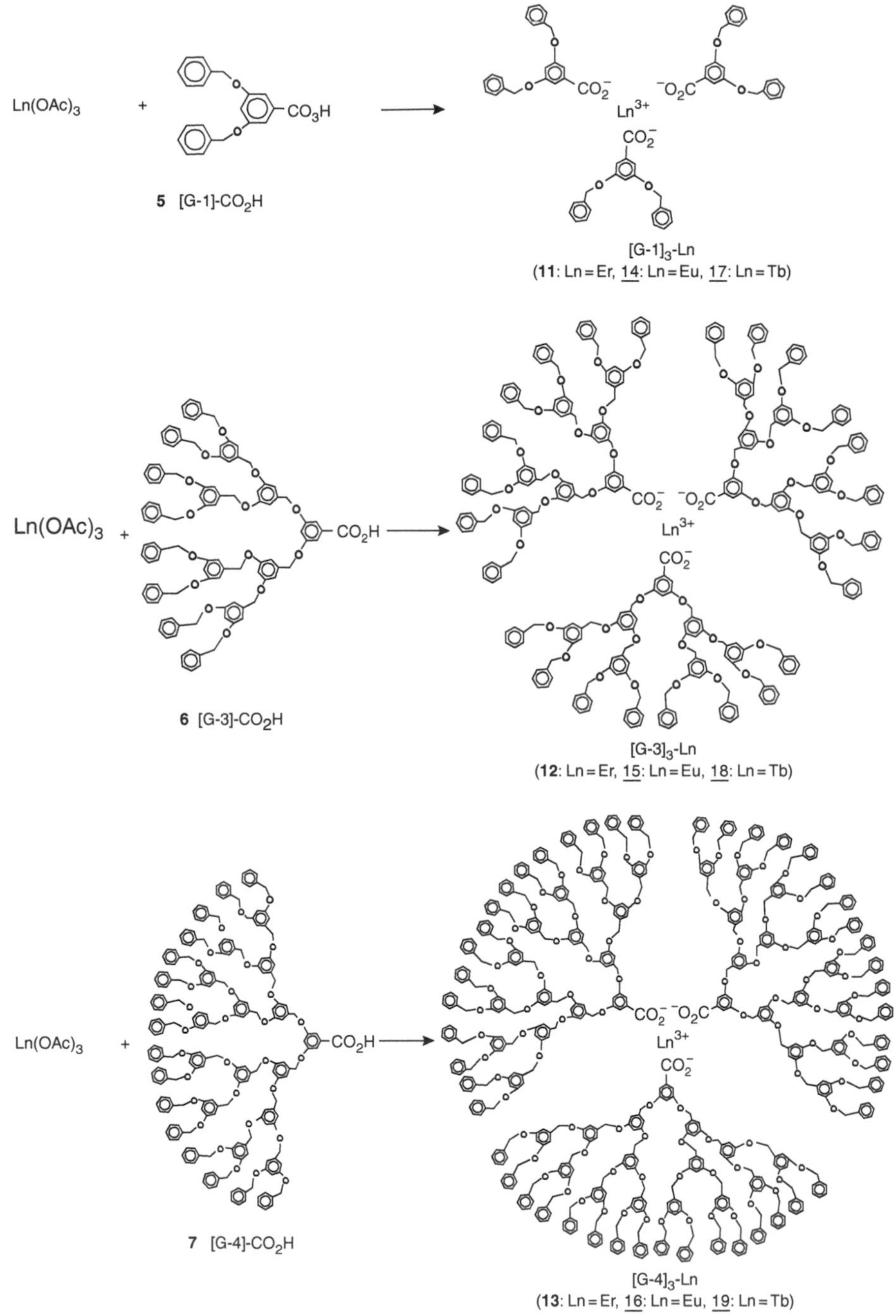

Figure 12 Self-assembly of focal point functionalized dendrons around metal cations according to Fréchet et al. (J.M.J. Fréchet et al. *Chem. Mater.* 10:287, 1998. Copyright 1998 American Chemical Society.)

Fréchet and coworkers [314] have recently described a similar self-assembly of benzyl ether dendrons, possessing carboxylic acid substituents at their focal points by metal–ligand coordination around a core of trivalent lanthanide metals (e.g., Er^{+3}, Eu^{+3}, or Tb^{+3}). These self-assembled dendrimers were isolated by using ligand exchange reactions to produce structures derived from metal–ligand ionic interactions as shown in Figure 12. As a consequence, these self-assembled dendrons served as a dendritic shell, which shielded the lanthanide atoms from one another. It was significant to note that these dendritic structures produced a substantial decrease in the rate of self quenching (an energy transfer process) between metal atoms. This effect has been referred to as a dendrimer based "*site isolation*" phenomenon [97]. The importance of this effect was clearly demonstrated as one progressed from lower to higher generations; wherein, the dendrimers exhibited vastly enhanced luminescence activity over the lower generation dendrimers.

Recent work by Tomalia et al. [79] describes the syntheses of a new class of poly(amidoamine) (PAMAM) dendrimers possessing a disulfide function at its core. Traditional reduction chemistry produces various single site, sulfhydryl functionalized dendrons (i.e., $G_{en} = 0$–6) possessing amino, hydroxyl, acetamido, or dansyl surface groups. These focal point, thiol functionalized dendrons were found to readily self-assemble around a variety of metallic nanoparticles (i.e., especially gold) to produce dendronized nano-clusters and quantum dots possessing designable chemical surfaces. Analogous efforts by Peng et al. [98], Fox et al. [99], and Astruc et al. [100] have demonstrated the versatility of this self-assembly strategy.

More recently, single site, functionalized, single-stranded (ssDNA) *dendri*-poly(amidoamine) (PAMAM) di-dendrons have been synthesized by covalently conjugating complementary 32 base pair oligonucleotides to single-site, thiol functionalized *dendri*-PAMAM di-dendrons possessing neutral or anionic surface groups as shown in Figure 13 [101]. Combining complementary (ss-DNA) functionalized PAMAM di-dendrons at appropriate assembly temperatures produced Crick–Watson base paired (dsDNA) cores, surrounded by four PAMAM dendrons. These novel core-shell nano-structures represent a new class of precise monodisperse, *linear-dendritic architectural copolymers*. Using comparative gel electrophoresis, it was demonstrated that these self-assembled (di-dendron) dendrimers could be hemispherically differentiated as a function of surface chemistry as well as generational size. This new supramacromolecular approach offers a very facile and versatile strategy for the combinatorial design of size, shape, and surface substituents for both homogeneous and

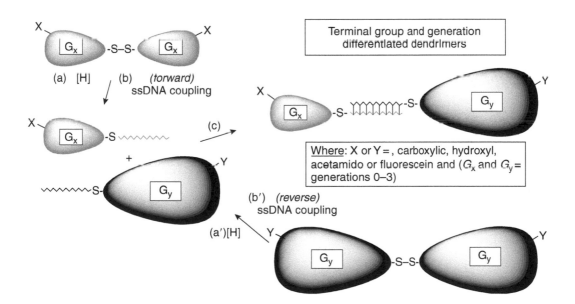

Figure 13 Scheme describing (a)/(a′) reduction of disulfide, thiol-functionalized *dendri*-(PAMAM) di-dendrons (b)/(b′) bioconjugation of (ssDNA) (forward/reverse) to produce respective, single-site (ssDNA)-(PAMAM) di-dendrons, and (c) self-assembly of respective (ssDNA)-(PAMAM) di-dendron conjugates by Crick–Watson base pairing to produce (dsDNA) core-(tetra-dendron) PAMAM dendrimers with differential sizes (i.e., generations: G_x or G_y) and surface groups (X or Y). (From C.R. DeMattei. NanoLetters 4:771–777, 2004. Copyright 2004, American Chemical Society.)

differentiated dendritic nanostructures [101]. Supramolecular, self-assembly around functional cores to produce a variety of new and unique dendrimers has been reviewed recently by Diedrich and Smith [102,103] and Fréchet [104].

2. Supramolecular Assembly of Dendrimers Based on the Assembly of Branch Cells

Numerous examples have been reported describing the use of hydrogen-bonding between complementary groups to induce self-assembly [61,105]. Another approach is to use metal-induced coordination chemistry as the driving force in the assembly process. Metal-based coordination-driven methodology allows the rapid, facile formation of discrete structures with well-defined shapes and sizes. Metal–ligand dative bonds are stronger than hydrogen bonds and have more directionality than other weak interactions such as π–π stacking; electrostatic, hydrophobic–hydrophilic interactions; and even hydrogen-bonding. One metal–ligand interaction can therefore replace several hydrogen bonds in the construction of supramolecular species. Perhaps the earliest work reported in this area, was that described by Balzani et al. [106–112]. Referred to by Balzani as the "complexes as ligands and complexes as metals" strategy, the synthesis involved the divergent assembly of dendrimer structures as described in Figure 14.

In a similar fashion, Puddephatt and coworkers [113–115] described the use of platinum coordination chemistry involving a convergent strategy. By this method, they reported the synthesis of a $G = 4$ dendrimer containing up to 28 coordination centers.

Perhaps the only true, reversible coordination, self-assembly for the synthesis of dendritic systems is that described in a series of papers by Reinhoudt and coworkers. They first reported the reversible assembly of hyperbranched spheres from palladium substituted compounds by replacing a labile, coordinating nitrile ligand with a kinetically stable moiety located on an AB_2 monomer [116,117]. Subsequently, they used a combination of hydrogen-bonding and noncovalent coordination chemistry to construct unique rosette-type structure [118]. More recently, this group reported related chemistry that allows either a divergent or a convergent approach [118] to dendrimer synthesis [119,120]. This strategy involves first the synthesis of three basic building units as described in Figure 15. The controlled divergent assembly requires first the core unit containing three Pd–Cl pincer complexes, which are activated by removing chloride ion with $AgBF_4$. Subsequent addition of three equivalents of nitrile containing building block

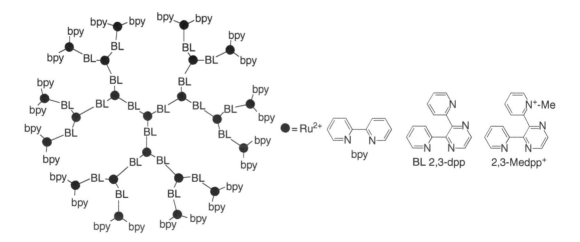

Figure 14 Dendritic ruthenium complex reported by Balzani and coworkers. (Courtesy *Chem. Rev.* 97:1706, 1997. Copyright 1997 American Chemical Society.)

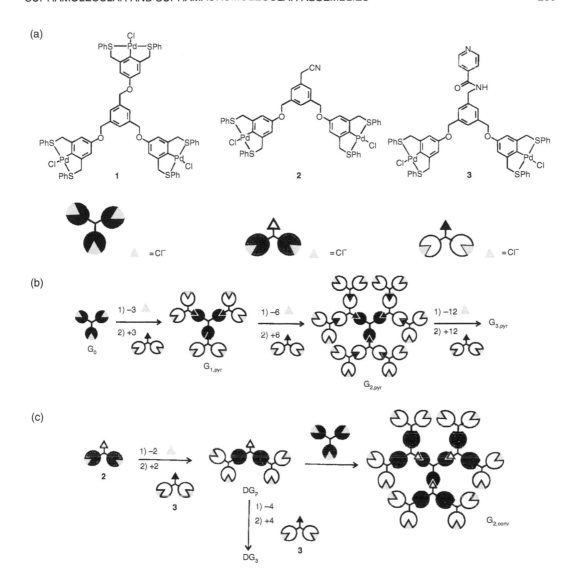

Figure 15 Self-assembly of Pd contained metallodendrimers using (a) building blocks 1 to 3 (b) divergent method, and (c) convergent method according to Reinhoudt et al. (Courtesy *J. Am. Chem. Soc.* 120:6242–6243, 1998. Copyright 1998 American Chemical Society.)

2 yields a first generation, metallo(dendrimer) possessing nitrile ligands coordinated to the palladium centers. Using this concept, they were able to synthesize metallo(dendrimers) up to generation 5 [116]. They also described [120] the synthesis of building unit 3 with pyridine instead of nitrile ligand and were able to use this intermediate for the preparation of more stable metallodendrimers, due to the stronger coordination of pyridine ligands to the palladium centers.

The convergent route (Figure 15(c)) begins with the synthesis of dendrons using building block 2 as a carrier unit and building block 3 as a building unit. The three dendrons were coupled to a tri-functional core 1 to form the dendrimer structure. Figure 15(c) illustrates a dendron (DG_2 and DG_3) constructed *via* controlled convergent assembly. Activation of 2 with $AgBF_4$ and subsequent addition of two equivalents of 3 gave DG_2 after coordination of the pyridine ligands. Reiterating these steps led to higher generation metallo(dendrons). Evidence for the formation of all of these metallo(dendrimers), was obtained by IR, H^1 NMR, ESMS, and MALDI-TOF mass spectrometry.

III. SUPRAMOLECULAR AND SUPRA*MACRO*MOLECULAR CHEMISTRY OF DENDRIMERS

In the very first papers published [72–76] it was apparent from electron microscopy data presented that dendritic architecture would offer rich possibilities in the area of supramolecular chemistry. First, it was virtually impossible to observe individual dendrimer modules due to their great propensity to form clusters or aggregates even when sampled from very dilute solutions. Similar observations were made by Newkome, as he published his early work on the related arborols [121]. In contrast, samples from more concentrated solutions displayed breathtaking dendritic arrays of microcrystallites upon drying [72,75]. Recent work by Amis et al. [122] using cryo-TEM and other methods has shown that dendrimers exhibit a great propensity to self-organize even in solution.

In the same early publications [72–76], the reported observation that copper sulfate could be solubilized in organic solvents with ester terminated PAMAM dendrimers to produce deep blue transparent chloroform solutions, offered very early evidence for the unique *endo*-receptor (i.e., unimolecular inverse micellar) properties of these macromolecules. These properties are discussed in the following section.

Some of the earliest reports describing supramolecular properties of PAMAM dendrimer interiors and surfaces were published by Turro et al. and have been reviewed extensively [123,124]. Utilizing both photophysical and photochemical probes many important supramolecular properties such as dendrimeric encapsulation, cooperative/noncooperative surface aggregations, and dendrimeric molecular morphogenesis were characterized and rationalized.

A. The Dualistic Role of Dendrimers as Either *Endo-* or *Exo-*Receptors

The field of dendritic supramolecular chemistry is young [104,125]. As recently as three years ago fewer than a half dozen papers could be found on the subject [126]. Since that time, this field has expanded dramatically. As early as 1990, we commented extensively on the dualistic property of dendrimers [37]. At that time, it was noted that dendrimers could function as unimolecular *endo*-receptor-type ligands manifesting noncovalent chemistry reminiscent of traditional regular or inverse micelles or liposomes. Furthermore, it was noted that dendrimers also exhibited a very high propensity to cluster or complex in an *exo*-fashion with a wide variety of biological polymers (i.e., DNA, proteins) or metals. In the following sections, we would like to describe nonexhaustive samplings of recent work illustrating the supramolecular and supra*macro*molecular behavior of dendrimers that connects them to all three categories of supra-chemical types observed in biological systems (Figure 1).

B. Dendrimers as Unimolecular Nanoscale Cells or Container Molecules

1. *Evolution of Abiotic Container Molecules — From Carcerands/Carceplexes to Dendrimers/Dendriplexes*

The abiotic evolution of "container-type" molecules may be traced from the original synthesis of cubane [127], pentaprismane [128], and dodecahedrane [129]. These platonic solids are all closed surface compounds; however, their interiors are too small to host organic or inorganic compounds. It was not until (1983) that Cram et al. [130] reported a container-like hydrocarbon referred to as carcerand that was large enough to host simple organic compounds inorganic ions or gases [130].

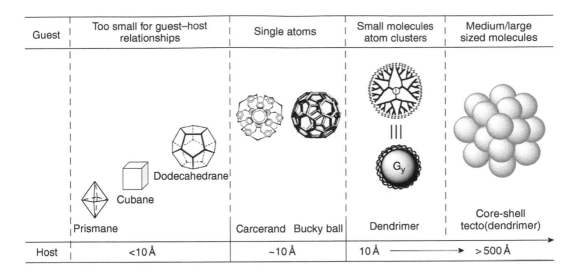

Guest	Too small for guest–host relationships	Single atoms	Small molecules atom clusters	Medium/large sized molecules
	Prismane Cubane Dodecahedrane	Carcerand Bucky ball	Dendrimer	Core-shell tecto(dendrimer)
Host	<10 Å	~10 Å	10 Å ⟶ >500 Å	

Figure 16 Evolution of abiotic container molecules as a function of dimensions and complexity.

This discovery has led to a vast array of small container-type molecules, which have been extensively reviewed [131]. Quite remarkably, this was approximately the same time that ester terminated PAMAM dendrimers were noted to dissolve and incarcerate copper salts to produce "blue chloroform" solutions due to their unimolecular, inverse micelle properties. This observation was publicly reported in 1984–1985 [72], during the same year that Smalley and Kroto et al. described the first synthesis of buckminsterfullerene [132]. Of course, it is well-known that "bucky balls" will host a variety of metals. Thus, it is apparent that the emergence of *"container molecules"* has followed the systematic enhancement of the organic host structure dimensions as illustrated in Figure 16.

(a) Incarceration of zero-valent metals and metal compounds

Very recently, a novel and versatile method has been reported for the construction of stable, zero-valent metal quantum dots, using dendrimers as well-defined nano-template/containers. The concept involves the use of dendrimers as hosts to preorganize small molecules or metal ions, followed by a simple *in situ* reaction, which immobilizes and incarcerates these nanodomains (Figure 17) [89]. The size, shape, size distribution, and surface functionality of the dendritic nanocomposites are determined and controlled by the architecture. Dendrimer-based nanocomposites display unique physical and chemical properties as a consequence of the atomic/molecular level interactions between the guest and host components within the dendrimer as well as the dendrimer interaction with various solvents and media.

Preparation of stable, zero-valent metallic copper solutions were demonstrated in either water or methanol [89]. After complexation within various surface modified poly(amidoamine) (PAMAM) dendrimers, copper(II) ions were reduced to zero valence metallic copper thus providing a bronze, transparent dendrimer–metal nanocomposite soluble in water. Solubility of the metal domains is determined by the surface properties of the dendrimer host molecules, however, their solutions still display characteristic optical properties associated with metal domains. Both aqueous and methanolic solutions of copper clusters were stable for several months in the absence of oxygen. Similar work and results were also reported by Crooks et al. [133].

More recent work describing SAN characterization of copper sulfide PAMAM dendrimer nanocomposites [134] and extensions to silver [135] and gold-based dendrimer nanocomposites and

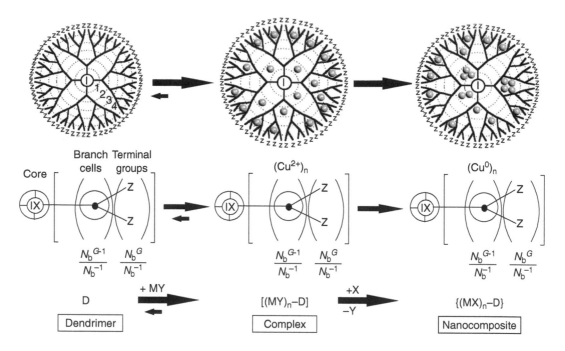

Figure 17 Construction of dendrimer nanocomposites by reactive encapsulation. Y (Cu^{+2}) denotes ligands after complex formation. X (Cu^0) represents zero-valent metal incarcerated within the dendrimer interior after reduction. (Courtesy *J. Am. Chem. Soc.* 120:7355, 1998. Copyright American Chemical Society.)

their optical properties [136] have been reported [315]. This latter work confirms and demonstrates the periodic container properties described in Figure 10.

(b) Incarceration of organic compounds — unimolecular encapsulation — the dendritic box

As early as 1989, Tomalia et al. [137] demonstrated with carbon-13, spin-lattice relaxation (T_1) techniques that small molecules such as acetylsalicyclic acid or 2,4-dichlorophenoxyacetic acid could be encapsulated within the interior of carboxymethyl ester terminated PAMAM dendrimers exhibiting guest:host stoichiometries of ~4:1 by weight or ~3:1 on a molar basis. These encapsulations appear to be driven by simple ion pairing of the acidic guest molecules with the tertiary amine sites located within the PAMAM dendrimer host. More recently, extensive work by Twyman et al. [138,139] and others [140] has shown that dendrimers may function as very versatile host molecules for drug delivery systems. Figure 18 illustrates the incorporation of benzoic acid within the interior of a hydroxyl terminated PAMAM dendrimer.

In another elegant study, Meijer and coworkers [141,142] skillfully enhanced an earlier concept for producing artificial cells by modifying dendrimer surfaces to induce "unimolecular encapsulation" behavior [37]. They referred to this new construction as the "dendrimer box." Surface-modifying $G = 5$, poly(propylene imine) (PPI) dendrimers [142] with 1-phenylalanine or other amino acids [143] induced dendrimer encapsulation by forming dense, hydrogen-bonded surface shells with almost solid-state character. Small guest molecules were captured in such dendrimer interiors and were unable to escape even after extensive dialysis [141]. The maximum number of entrapped guest molecules per dendrimer box was directly related to the shape and the size of the guests, as well as to the number, shape, and size of the available internal dendrimer cavities. For example, 4 large guest molecules (e.g., Bengal Rose, Rhodamide B, or New Coccine) and 8 to 10 small guest

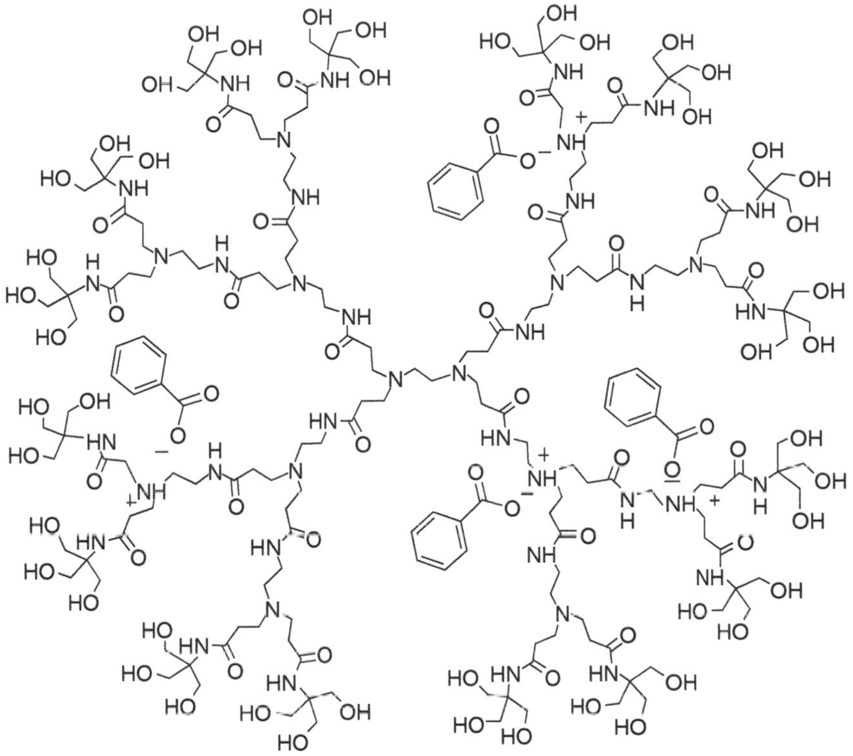

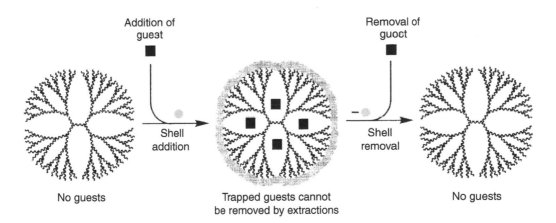

Figure 18 Incorporation of benzoic acid within the branched interior of an OH-terminated PAMAM dendrimer. (With permission of Taylor and Francis Ltd.)

Figure 19 Principle of the dendritic box. (From O.A. Matthews. *Prog. Polym. Sci.* 23:1–56, 1998. With permission from Elsevier Science.)

molecules (e.g., *p*-nitrobenzoic acid, nitrophenol, etc.) could be simultaneously encapsulated within these PPI dendrimers containing 4 large and 12 smaller cavities (Figure 19). Quite remarkably, this dendrimer box could also be opened to release either all or only some of the entrapped guest molecules [141]. For example, partial hydrolysis of the hydrogen-bonded shell liberated only small guest molecules, whereas total hydrolysis (with 12*N* HCl; 2 h at reflux) released all sizes of entrapped molecules.

2. Mimicry of Classical Regular Micelles

Based on qualitative evidence, Newkome et al. first hypothesized the analogy between dendrimers and regular micelles in 1985 [121]. Simultaneously, Tomalia et al. reported the direct observation of sodium carboxylated PAMAM dendrimers by electron microscopy in 1985 [72] and 1986 [144], which experimentally supported the fact that dendrimers clearly possessed topologies reminiscent of regular classical micelles. At that time, it was also noted from electron micrographs that a very high population of individual dendrimers possessed hollowness presumably due to the peripheral stacking association of terminal head groups. This was experimentally confirmed [145] by noting the importance of *branch-cell symmetry* as a requisite for interior void space. Such solvent filled interior hollowness is observed in essentially all symmetrically branched dendrimers, but does not appear to exist in asymmetrically branched dendrimers such as those described by Denkewalter [146–149]. Furthermore, experimentally determined hydrodynamic dimensions [145], shape confirmations, and comparisons of dendrimer termini (surface areas) as a function of generation with traditional micelle head groups added further support to this hypothesis in 1985–1987 [144,145,150]. This unique dualist property of micelle topology and interior void space normally associated with liposomes was noted by Tomalia in 1989 [151]. Subsequent NMR studies and computer-assisted simulations by Goddard et al. in 1989 [137], molecular inclusion work by Newkome [152] in 1991, as well as extensive photochemical probe experiments by Turro et al. [123,153–155], have now unequivocally demonstrated the fact that symmetrically branched dendrimers may be viewed as unimolecular micelles (nanoscale container molecules).

Based on a study of dendrimer–surfactant complexation probed by the pyrene fluorescence, lower generations behaved as ordinary electrolytes and higher generations acted as a novel type of polyelectrolytes. For the later generations, the addition of dodecyltrimethylammonium bromide (DTAB) first occurred in a noncooperative manner *via* electrostatic binding, but at higher concentration, cooperative binding caused aggregation of surfactants to form micellar structures on the dendrimer surface. This cooperative binding was believed to result from alkyl chain association induced by the closely packed charged groups on the surface of higher generation dendrimers. Overall, the break in the dendrimer behavior sensed by pyrene coincided with the predicted change in the morphology of the dendrimer structures by simulation [37,137].

Depending on the nature of their surface groups and interiors these dendrimers will manifest behavior reminiscent of either traditional regular or inverse micelles, however, with unique differences and advantages.

3. Mimicry of Classical Inverse Micelles

The first examples and observed dendritic inverse micelle properties were noted in the initial paper on poly(amidoamine) dendrimers published in 1984–1985 [35,50,72–76]. At that time, it was observed that methylene chloride or chloroform solutions of the methyl/alkyl ester modified dendrimers readily extracted copper ion (Cu^{+2}) from water into the organic phase. Beautiful "blue chloroform" solutions were obtained that were completely transparent and did not scatter light. It was assumed that the copper ions had been chelated into the interior and were being compatibilized by the more hydrophobic sheathing of the dendrimer surface groups. Variations of this work were both patented [156] and ultimately published [157]; wherein, PAMAM dendrimers were hydrophobically modified with alkyl epoxides and used to extract metal ions into toluene, styrene monomer, or a variety of other hydrophobic solvents.

Meanwhile, other examples of unimolecular dendritic, inverse micelles have been reported by Meijer et al. [158] and DeSimone et al. [159]. In the latter case, surface perfluorinated *dendri*-PPI dendrimers were used, which have a high affinity for liquid CO_2. Thus, the hydrophilic dendrimer core provides a favorable environment for hydrophilic guests such as ionic methyl orange. Thus while

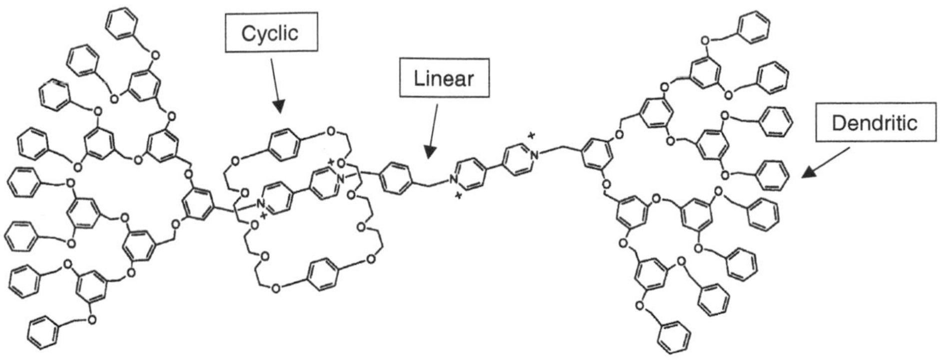

Figure 20 A rotaxane with dendritic stoppers. An example of a dendritic-linear-cyclic architectural copolymer. (Courtesy *Chem. Rev.* 97:1701, 1997. Copyright 1997 American Chemical Society.)

the $-CO_2$-philic dendrimer shell allows the micelle to dissolve in CO_2, the ionic dye can be effectively transferred from a water layer into liquid CO_2.

C. Dendrimers as Nanoscale Amphiphiles

1. *Dendritic Architectural and Compositional Copolymers*

Perhaps one of the most elegant and complex examples of an *architectural copolymer* [160] that was constructed by both self-assembly and covalent synthesis is that reported by Stoddart et al. [161,162]. Stoddart and coworkers prepared a series of rotaxanes with dendritic stoppers using a "threading approach" (Figure 20). Within this single rotaxane structure possessing dendritic stoppers one finds that all three architectural types are represented. Although no amphiphilic properties were reported for this system, it has been shown that both compositional, as well as architectural dendritic copolymers do manifest interesting amphiphilic properties leading to self-assembly processes.

It is very well know that the general area of traditional amphiphilic structure–property rela-tionships can be broadly divided into two major types, namely, (a) *small molecule surfactants* and (b) *amphiphilic block copolymers*. Each of these classes has been the subject of extensive studies [38,163–166]. A detailed examination of the influence that increasing head/tail sizes and shapes have on the nature of aggregation has been limited to traditional structures. Such structures have included only low molecular weight surfactants, possessing compact polar head groups/tails, or classical linear, block copolymers.

With the advent of dendritic synthesis, the precise construction of nanoscale sizes and shapes have allowed the synthesis of new larger, dimensioned amphiphilic structures. Perhaps the first well-documented example of dendrimers exhibiting *exo*-receptor, supramolecular behavior was that reported by Friberg and Tomalia in 1988 [167]. It is well known that one may evolve a rich variety of lyotropic mesophases by merely coordinating the relative sizes of the amphiphilic components of a system (i.e., hydrophobic head groups versus hydrophobic tail). This was accomplished by combining a $G = 2$, poly(ethyleneimine) (PEI) dendrimer (hydrophilic head group) with octanoic acid (hydro-phobic tail) to give a lamellar liquid-crystalline assembly. In effect, a nanoscale amphiphile was constructed *in situ* by the acid–amine reaction between octanoic acid and $G = 2$ *dendri*-PEI–$(NH)_{12}$ to produce an amphiphilic salt.

Additional examples of dendritic-type surfactants were reported in the early patent literature as hydrophobic core functionalized PAMAM dendrons [35,168]. For example, when the core was a 12 carbon chain, the first two generation PAMAM dendrons exhibited hydrocarbon solubility (TMF).

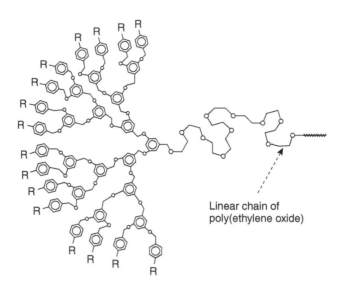

Figure 21 Hybrid dendritic linear polymer obtained by attaching poly(ethylene oxide) to the focal point of structure G-4.

However, beyond generation 2, the amphiphilic dendrons were water-soluble. This demonstrates the effect one observes as the hydrophilic head group is amplified through the HLB crossover point. The result is a dendron possessing a dominant hydrophilic head group, which undoubtedly engulfs the tethered hydrocarbon core in aqueous medium. Many of these products were useful as amphiphilic reagents to produce novel dendrimer-type micelles [168]. Among the many interesting properties exhibited by these micelles was the ability to sequester hardness ions (i.e., Mg^{+2}, Ca^{+2}, etc.) thus manifesting self-building surfactant properties within unimolecular structures.

Later extensions of this general concept were reported by Chapman et al. [169] and Fréchet [54]; wherein, they reversed the amphiphilic components to produce a functionalized dendron possessing a hydrophobic head and a hydrophilic tail. The Chapman dendritic amphiphiles were derived from alkylene oxide tails and BOC terminated, Denkewalter-type dendritic heads. The Fréchet amphiphiles were obtained by attaching poly(ethylene oxide) to the focal point of a hydrophobic poly(ether) dendron as illustrated in Figure 21 [170,171].

Nearly simultaneously, Fréchet et al. pioneered [54] the synthesis and development of a new type of amphiphilic dendrimers; wherein, the unimolecular architecture was differentiated. In these instances, Fréchet demonstrated that his poly(ether) dendrimers derived *via* convergent synthesis could be either homogeneously terminated or differentiated into hydrophobic and hydrophilic hemispheres, as illustrated in Figure 22(a) and (b). These dendrimers could be oriented at an interface, as illustrated, or under the influence of an external stimulus.

Indeed amphiphilic dendrimers form monolayers at the air–water interface and such dendrimers [172] are useful in forming interfacial liquid membranes or in stabilizing aqueous–organic emulsions. A dendrimer constructed from two segregated types of chain ends (i.e., half electron dominating and half electron withdrawing) can be oriented in an electrical field and have been shown to exhibit very large dipole moments [173].

One of the most comprehensive examinations of dendritic nanoscale amphiphiles involved structures that mimicked traditional surfactants. This work was reported in a thesis by van Hest et al. [126] [166]. In this study, extensive hydrophobic tails were precisely constructed by the living anionic polymerization of styrene, followed by termination with a primary amine function. Dendronization of these amine functionalized poly(styrenes) produced focal point functionalized, hydrophobically substituted PPI dendrons, as illustrated later (Figure 23). More specifically, the living polystyrene anion is terminated with ethyleneoxide, thus providing a hydroxyl terminated poly(styrene) with

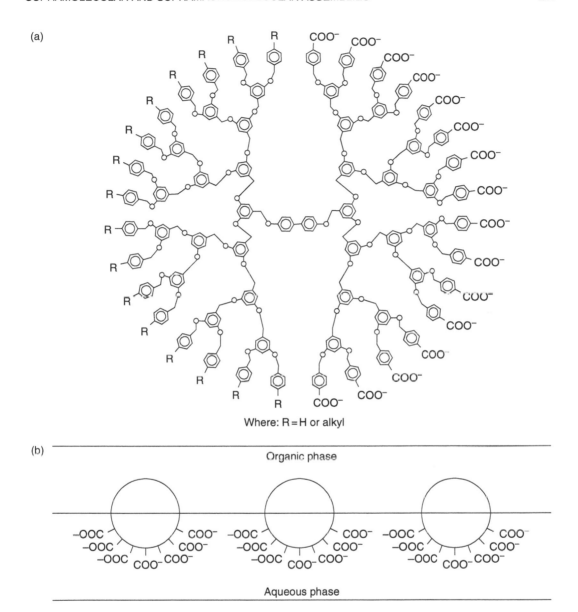

Figure 22 (a) Amphiphilic dendrimer obtained by hemispherical functionalization of half the terminal groups with carboxylic anions; (b) liquid membrane of amphiphilic dendrimer at the interface between water and an immiscible organic solvent.

molecular weights between 3000 and 5000 and a polydispersity <1.05. In a phase-transfer reaction the addition of acrylonitrile to the alcohol, followed by a (RaneyCo–H$_2$) hydrogenation yields the primary amine-terminated poly(styrene). Subsequent dendrimer construction is performed by the sequential Michael addition of acrylonitrile and heterogeneously catalyzed hydrogenation to form a variety of "*super amphiphiles*" with different sized nanoscale headgroups.

These new "architectural copolymers" derived from both *linear* and *dendritic* architectures exhibited very profound properties as nanoscale amphiphiles. In general, as one increases the head group size from PS–*dendri*–PPI–(NH$_2$)$_4$ to PS–*dendri*–PPI–(NH$_2$)$_{32}$, the aggregation topologies change from planar bilayers to vesicles to rod-like micelles to finally spherical micelles. Since only the head group size and not the chemical composition of the amphiphilic structure is being changed, this

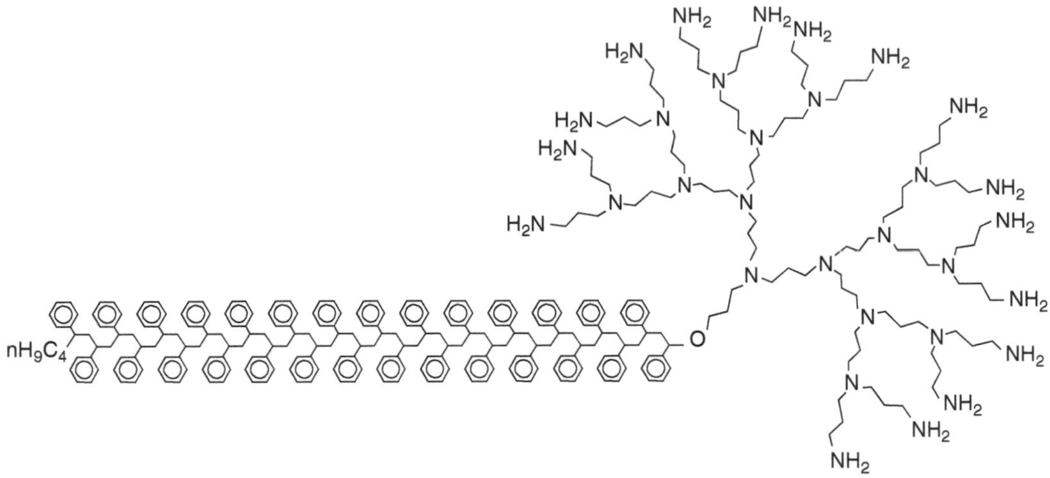

Figure 23 Linear-poly(styrene), focal point functionalized poly(propyleneimine) (PPI) dendrons; $G = 4.0$.

study offers excellent proof for the validity of Israelachvili's theory [174,175] of shape-dependent aggregation behavior.

Changing the head group functionality from primary amine terminated to either carboxylic acid or quaternary amine terminated produced amphiphilic behavior, which was very rational in the context of traditional theory. A major difference in all cases was the substantially larger dimensions of the aggregates compared to those obtained from traditional small molecule amphiphiles. In fact, one might visualize these dendritic amphiphiles as amplified forms of their traditional analogs. Aggregates and assemblies derived from these linear, dendritic architectural copolymers are roughly five times the size of those obtained from traditional surfactant molecules.

One of the earliest examples demonstrating the self-assembly of dendrimers based on inherent amphiphilic characters was that reported by Newkome et al. [176]. It was found that dendri-poly(amidoalcohols); $G = 1$–2 attached to various alkyne and hydrocarbon cores (Figure 24), produced thermally reversible gels in aqueous solutions. It was postulated that formation of these rod-shaped aggregations was driven by a combination of hydrogen-bonding of the hydroxyl terminal groups and the hydrophobic bonding of the core substituents in an orthogonal fashion. Various helical morphologies were proposed to account for the extraordinarily large diameters (i.e., $\cong 600$ Å) that were observed by electron microscopy [152]. It is quite likely that multiples of these rod aggregates may have self-assembled much as has been observed recently for covalently fixed, linear-dendritic architectural polymer rods reported recently by Tomalia et al. [160].

D. Dendrimers as Nanoscale Scaffolding

1. Dendritic Rods and Cylinders

(a) Congestion-induced morphogenesis (shape control)

Cylindrical, rod-shaped dendrimer assemblies were first synthesized by Tomalia et al. as early as 1987 [35,37,177]. These structures represent some of the first examples of hybridized dendritic architecture. Since they possess a *linear* polymeric core and *dendritic* arms, they are called "*architectural copolymers*" or "*dendronized, linear polymers.*" This work was recently reported in detail [160]. The method involved the divergent dendronization of *linear* poly(ethyleneimine) (PEI) cores

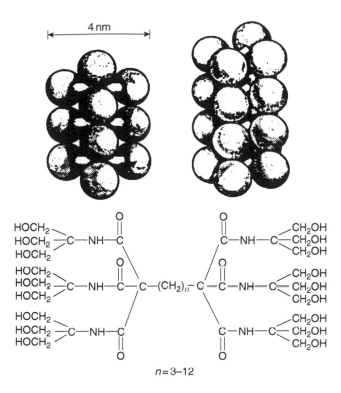

Figure 24 Molecular rods derived from micellarization of dendri-poly(amidoalcohols).

(Scheme 1). These cores, which had DPs of $\cong$100–500, were dendronized by iterating from the active hydrogens on the backbone of the *linear* poly(ethyleneimine) (PEI) with standard PAMAM chemistry. The first step involved the reaction of methyl acrylate followed by an excess of EDA to give first generation monodendrons along the PEI backbone. Reiteration of these steps led to the development of higher generations ($G = 1$–4). These final products were characterized by FTIR, C^{13} NMR, SEC, HPLC, MALDI-TOF MS, and transmission electron microscopy. Electron microscopy of sodium carboxylated forms of $G = 1$–3 revealed nondescript random coil topologies. However, as the $G = 3$ structure was advanced to the next generation ($G = 4$), a remarkable congestion-induced shape change occurred to produce rigid, rod-like cylinders as determined by electron microscopy. The rod diameters varied between 25 and 32 Å with lengths ranging between 500 and 3000 Å. Furthermore, additional supra*macro*molecular assembly appeared to be occurring to give parallel clusters of the rods (Figure 25(a) and (b)).

As early as 1996, Schluter et al. [178] reported their first efforts to synthesize *linear-dendritic* architectural copolymers [179]. In general, their approaches involved two general strategies. (i) The first method involved the synthesis of a linear core, followed by the coupling of preformed dendrons to produce the architectural copolymers; (ii) the second method involved the synthesis of dendrons possessing polymerizable functionality at their focal points. These dendritic macromonomers were then polymerized to give a polymeric main chain with pendant dendrons. In each case, *linear-dendritic architectural copolymers* would be obtained. They would be expected to exhibit rod-like or cylindrical properties if the backbone core possessed a high degree of polymerization and the dendritic component was highly congested.

With the first method, Schluter [180,181] utilized the Suzuki reaction to produce a poly(*p*-phenylene)-type backbone possessing reactive *ortho*-hydroxy-methyl substituents. These substituents were subsequently used to couple a variety of Fréchet-type monodendrons along the backbone as described in Figure 26. In general, the lower generations (i.e., $G = 1$) coupled with

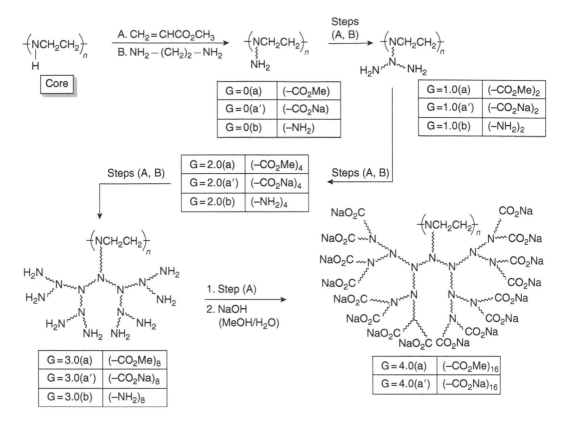

Scheme 1 Dendronization scheme for conversion of *linear* poly(ethyleneimine) cores to *dendri*-poly(amidoamine) hybrids. (Courtesy of *J. Am. Chem. Soc.* 1998, 120, 2679. Copyright 1998 American Chemical Society.)

linking efficiencies as high as 95%. However, coupling efficiencies for $G = 2$ or 3 decreased dramatically (i.e., 50 to 60%) unless appropriate 2-hydroxyethyl spacers were introduced to relieve steric problems [180].

The second approach involved the covalent attachment of Fréchet-type monodendrons (i.e., $G = 1–3$) to appropriate polymerizable reagents such as styrenes [178,181] or methacrylates [179] (Figure 27). Suitable radical catalysts were used to polymerize these dendritic macromonomers into structures that were observed to be cylindrical, rod-like architectures [179,181,182]. The area of dendronized, linear polymers has been extensively reviewed by Schluter et al. [183].

Percec and coworkers [184] utilized a similar strategy for the conversion of perfluorinated alkylene functionalized 3,4,5-trihydroxy benzoic acid-type dendrons into methyl methacrylate functionalized dendritic macromonomers. Characterization of the resulting *linear-dendritic* architectural copolymers involved DSC, x-ray diffraction, and thermal optical polarized microscopy. It was concluded that the self-assembly of the pendant dendritic mesogens forced the linear backbone into a tilted, helical ribbon-type structure. The self-assembly behavior was largely controlled by the multiplicity, composition, and molecular weights of the pendant dendritic mesogens.

Using a variation of this dendritic macromonomer method, Percec et al. [185,186] synthesized a variety of dendritic 7-oxanorbornene macromonomers and polymerized them with ROMP catalysts [186] as shown in Figure 28, as well as with free radical or anionic catalysts.

Yields were dependent upon the route used. Products were characterized by DSC, ^{13}C-NMR, and x-ray scattering. It was proposed that the resulting structures were mesogen assembled supramolecular columns possessing single chain relicity.

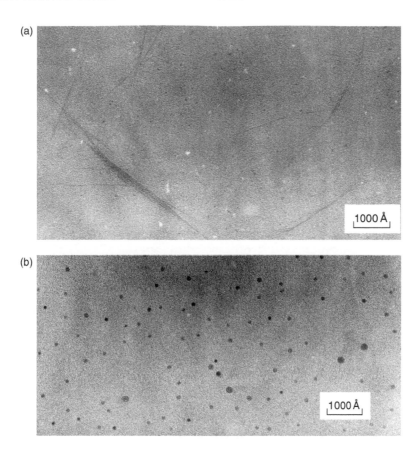

Figure 25 (a) Electron micrograph (TEM) of *linear* poly(ethyleneimine) (PEI) cores; *dendri*-poly(amidoamine) PAMAM; $G = 4$(a); $Z = (-CO_2Na)_{16}$; $N_C = 300$ to 500 (*Note*: self-organization of dendrimer rods into parallel arrays). (b) Electron micrograph (TEM) of ammonia (NH_3) core; *dendri*-poly(amidoamine) PAMAM; $G = 4$(a); $Z = (-CO_2Na)_{48}$; $N_C = 3$ (*Note*: self-organization of dendrimer spheroids into clusters). (Courtesy *J. Am. Chem. Soc.* 120:2679, 1998. Copyright 1998 American Chemical Society.)

(b) Dendrons as mesogens — dendromesogens

In addition to main chain and side chain liquid-crystalline polymers based on linear polymeric architecture [187], it was apparent as early as 1992 that new displays of mesogens presented on alternate architectures were possible. At that time, Ringsdorf et al. [188] and Percec et al. [189–192] pioneered extensive work in the area of dendritic mesogens. In general, this activity involved the examination of liquid-crystalline properties for both hyperbranched and dendron-type architecture. For example, Percec et al. [193,194] synthesized low generation perfluorinated dendrons possessing a 15-crown-5-ether and their focal point. These unique structures were found to self-assemble into cylinders and ultimately into hexagonal columnar supramolecular arrays as illustrated in Figure 29.

Lattermann et al. [195] first described the attachment of mesogens to the terminal groups of poly(propyleneimine) dendrimers as a function of generation (i.e., $G = 0$–4). These new structures were referred to as "*dendromesogens*." The authors noted that for the lower generations (i.e., $G = 0$–3), hexagonal columnar mesophases (Col_n) were observed. As the mesogen-induced congestion maximized at $G = 4$, mesomorphism disappeared and it was hypothesized that the dendromesogen transformed into a globular structure. More recent work by Serrano et al. [196–199] has extended these observations to the poly(amidoamine) dendrimer series as well. These observations appear to parallel those of Tomalia [71,160] and Percec [200,201].

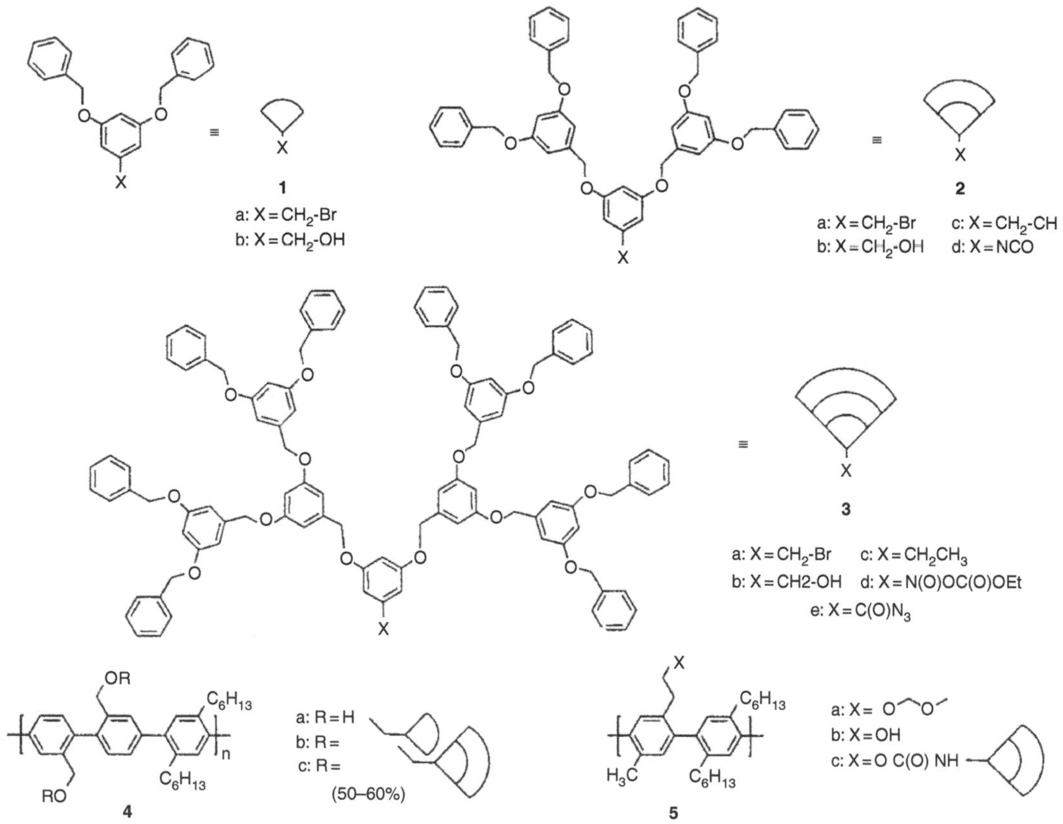

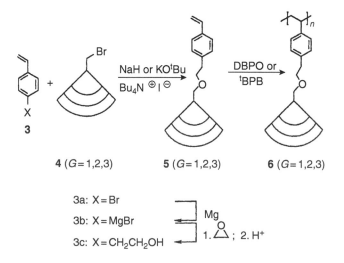

1

a: X = CH₂-Br
b: X = CH₂-OH

2

a: X = CH₂-Br   c: X = CH₂-CH
b: X = CH₂-OH   d: X = NCO

3

a: X = CH₂-Br   c: X = CH₂CH₃
b: X = CH2-OH   d: X = N(O)OC(O)OEt
e: X = C(O)N₃

a: R = H
b: R =
c: R =
(50–60%)

4

a: X = O⌒O⌒
b: X = OH
c: X = O C(O) NH

5

Figure 26 Fréchet-type monodendrons (generations 1, 2, and 3) possessing various reactive groups at the focal point. Rod-shaped, dendritic-linear architectural copolymers obtained by coupling respective dendrons to linear poly(*para*-phenylene) (PPP) backbone. (Courtesy *J. Am. Chem. Soc.* 119:3297, 1997. Copyright 1997 American Chemical Society.)

NaH or KO^tBu
Bu₄N ⊕I⊖

DBPO or
^tBPB

3   **4** (*G* = 1,2,3)   **5** (*G* = 1,2,3)   **6** (*G* = 1,2,3)

3a: X = Br
3b: X = MgBr
3c: X = CH₂CH₂OH

Mg
1. △ ; 2. H⁺

Figure 27 The synthetic sequence to macromonomers 5 (*G* = 1, 2, 3) carrying dendritic fragments of the first (*G* = 1), second (*G* = 2), and third (*G* = 3) generation and corresponding polymers 6 (*G* = 1, 2, 3). (Courtesy *Macromol. Rapid Commun.* 17:519, 1996. With permission of Wiley-VCH Verlag GmbH Germany.)

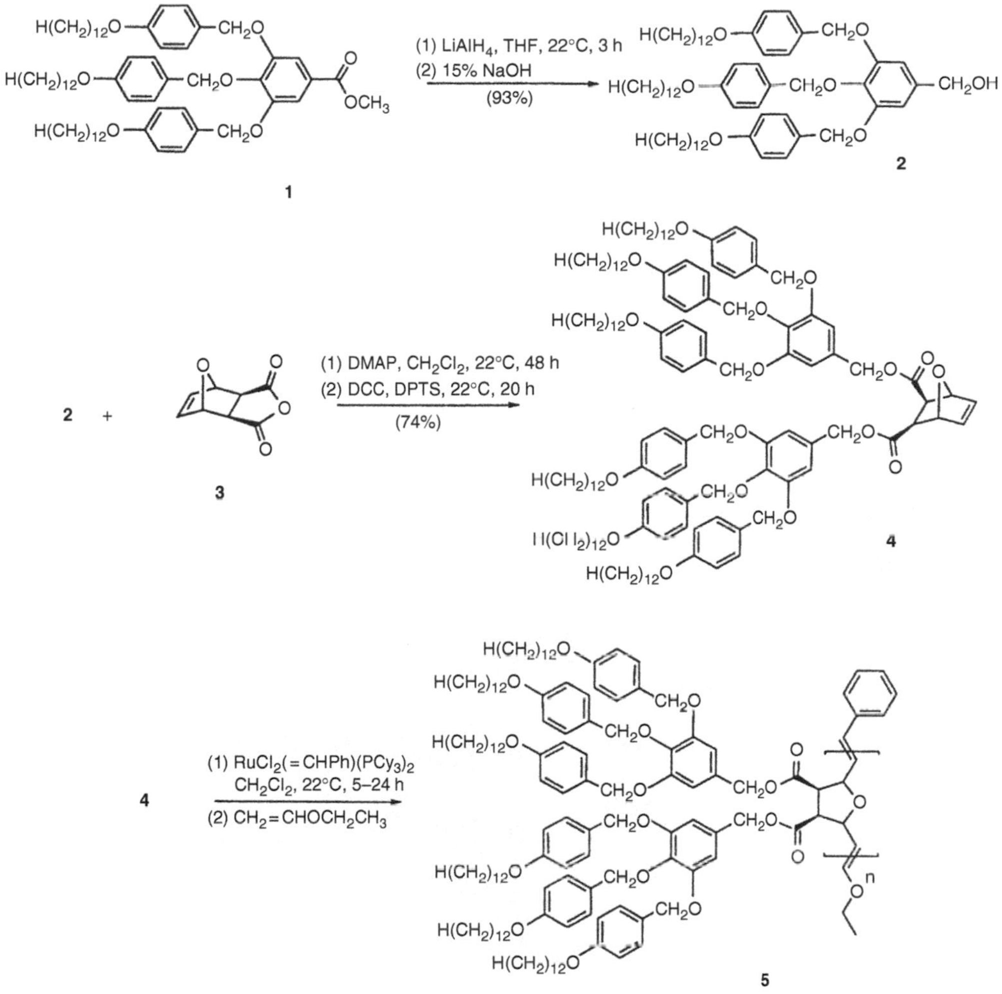

Figure 28 Synthesis of 7-oxanorbornene macromonomer derived from two ($G = 1$) dendrons followed by ROMP polymerization into a linear dendritic hybrid architecture. (Courtesy of *Macromolecules*, 39(19):5786, 1997. Copyright 1997 American Chemical Society.)

Dendrimer-based thermotropic liquid crystal (LC) dendrimers are of special interest and have been widely studied since they are able to self-assemble into large organized assemblies. They offer a rich array of supramolecular properties since their architecture combines two opposite tendencies; namely, structural anisometric units, which will preferentially favor anisotropic order (enthalpic gain) and flexibility in the dendritic architecture from which the branches radiate isotropically, leading to pseudospherical morphology (entropic gain). Thus many unique supramolecular properties are derived from such mesomorphic features (i.e., phase type, transition temperatures/thermodynamic stability), which are highly dependent on the enthalpy/entropy balance (i.e., core conformation), the degree of chemical incompatibility, nanoscale sizes of the different dendrimer generations (i.e., the nano-segregated domains), as well as the characteristics of the mesogenic unit itself and its location on the dendrimeric scaffolding. For these reasons the field of LC dendrimers has properly expanded and literature has been reviewed in the following dendritic compositional areas: hyperbranched polymers with mesogenic end groups [202], poly(siloxane) LC dendrimers [203], poly(carbosilane) LC dendrimers [204–217], poly(amidoamine) LC dendrimers, [196,198,218,219], and poly(propyleneimine) LC dendrimers [195,197,220–224].

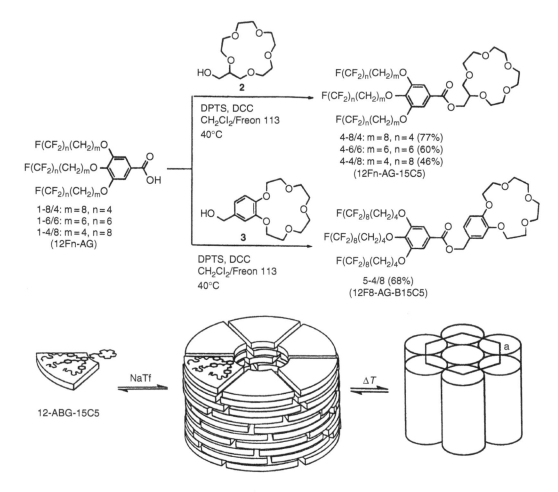

Figure 29   (a) Synthesis of monodendritic building unit with 15-crown-5 ether in the focal point; (b) self-assembly of cylindrical building blocks into a hexagonal columnar supramolecular architecture. (Courtesy *J. Am. Chem. Soc.* 118(41):9858, 1996. Copyright 1996 American Chemical Society.)

(c)   Quasi-equivalence of dendritic surfaces (coats)

An emerging direction in polymer chemistry is biomimicry of biological structures by using appropriate polymeric architectures, building blocks, and functionality. Quasi-equivalent building blocks or subunits are defined as chemically identical subunits, which may control their shape by switching between different conformational states during the process of self-assembly [225,226]. Classic biological examples include the flat-tapered and conical protein subunits that are required for sheathing the nucleic acid component of a tobacco-mosaic virus. By comparing very similar protein subunits it is apparent they will assume conical-type conformation to provide an appropriate sheathing for the protection of genetic material in an icosahedral virus, respectively [227]. In the former case, flat-tapered proteins self-assemble in the presence of a nucleic acid to generate rod-like viruses, which have a helical symmetry. The cylindrical shape of this assembly induces a helical conformation to the nucleic acid (e.g., a classic rod-like virus: tobacco mosaic virus [TMV]). On the other hand, icosahedral viruses are approximated by a spherical shape and are constructed from cone-shaped proteins. In the case of icosahedral viruses the nucleic acid adopts a random-coil conformation.

Using totally abiotic dendron subunits, Percec et al. [200,228] built structures that adapted the shape of either a rod-like virus with helical symmetry or an icosahedral virus with cone-shaped symmetry. Figure 30 outlines each of these viruses and their respective abiotic mimics.

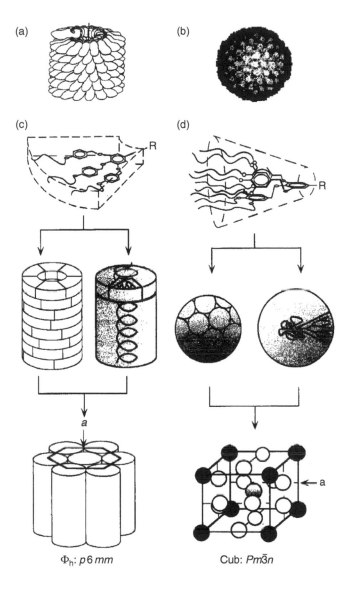

Figure 30 Quasi-equivalency of natural and synthetic supramolecular systems with cylindrical and spherical shapes: (a) TMV, (b) icosahedral virus; synthetic analog of (a) and (b) have been self-assembled from (c) tapered, and (d) conical monodendrons. (Courtesy of *Nature* 1998, 391, 8, 161. Copyright 1998 Macmillan Magazines Limited.)

In several seminal papers, Percec et al. [194,200] reported the demonstration of abiotic quasi-equivalency by controlling the degree of polymerization of various polymerizable dendrons described earlier. For example, twelve (12) second-generation conical monodendrons (DP = 12) produced a spherical dendrimer which results from self-organization into a cubic P$m3n$ three-dimensional (3D) liquid-crystalline (LC) lattice. The attachment of methacrylate (12G2-AG-MA) or styrene (12G2-AG-S) monomer moieties to the dendron followed by radical initiation yields spherical (DP < 20) or cylindrical (DP > 20) polymers depending on the degree of polymerization (DP). The spherical polymers self-organize into the same cubic (P$m3n$) 3D lattice while the cylindrical ones self-assemble into a p6mm columnar hexagonal (Figure 31) (2D LC lattice). Furthermore, Percec found that the radical polymerization of the styrene or methylacrylic functionalized dendrons exhibited a very dramatic self-acceleration in polymerization kinetics. These kinetics are presumably enhanced

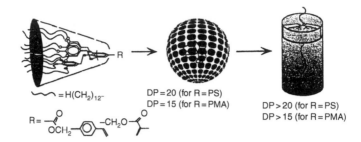

Figure 31   Scheme describing conversion of dendritic macromonomers to either spheroidal or cylindrical architectures depending on the degree of polymerization (DP). (Courtesy of *J. Am. Chem. Soc.* 1997, 199, 12978. Copyright 1997 American Chemical Society.)

by the unique self-assembly events that accompany these polymerizations [229]. As illustrated in Figure 31, the principles for the design of such macromolecular structures are outlined [71,201].

2.   Supramolecular Chemistry at the Dendrimer Surface

Kim et al. [230] developed very interesting pseudrotaxane type methodology that demonstrated the ability to supramolecularly modify the periphery of poly(propyleneimine) (PPI) dendrimers bearing diaminobutane substituents. These periphery substituents then served as "guest sites" that quantitatively self-assembled cucurbituril host molecules around the diaminobutane groups as illustrated in Figure 32.

3.   Hyper-Valency/Hyper-Cooperativity

Biological systems have long exploited the advantages of multivalent recognition events in the development of *exo*-supramolecular and supra*macro*molecular structures. Most notable is the power of multivalent, hyper-cooperative binding associated with biological cell adhesion processes. It is now well recognized that this biological strategy is very versatile and ubiquitous. In the case of carbohydrate recognition at cell surfaces the preferred recognition mode is to involve many polyvalent soft recognition events which are geometrically optimized to provide hyper-cooperativity as opposed to single, isolated events with harder recognition parameters. The power of this concept is illustrated in Figure 33, below and eloquently described by Kiessling and Pohl [231] and further elaborated upon more recently by Whitesides et al. [232].

Simply stated, once a ligand has attached itself to a cell at one site, it suffers a smaller entropy loss by binding at neighboring sites. Mimicry of these biological adhesion parameters can be exquisitely modeled and tested with dendrimeric systems. Although a substantial number of linear, poly(valence) polymeric architectures (e.g., poly[acrylamides], etc.) have been tested with some success, [233–236] efforts toward the use of dendrimer technology to create multivalent ligands for these purposes are in their infancy.

(a)   Glyco(dendrimers)

While searching for inhibitors of influenza virus hemagglutinin, Roy et al. [237–241] pioneered the synthesis and use of carbohydrate-substituted dendrimers, using solid-phase methodology to synthesize sialic acid decorated dendrimers. These dendrimers, containing 2, 4, 8, or 16 sialic acid residues, all inhibited the agglutination of erythrocytes by influenza virus (which is caused by hemagglutinin-mediated cross-linking of the erythrocytes) in the micromolar concentration range (Figure 34).

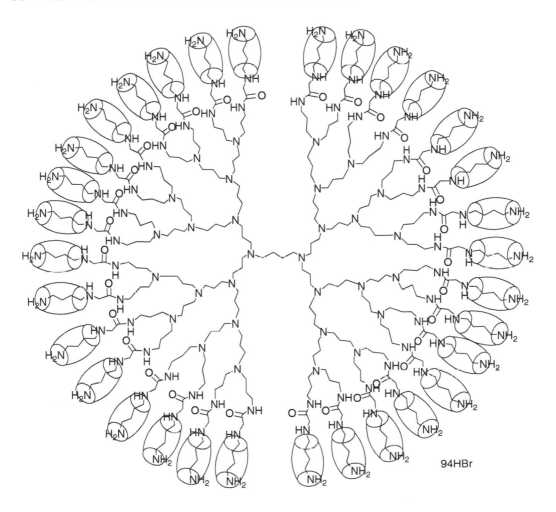

94HBr

Figure 32 DAB dendrimer, functionalized at the periphery with pseudorotaxane formed from cucurbituril. (Courtesy of *Angew. Chem. Int. Ed.*) 40:746–749, 2001.

It has been shown that specific subclasses [242] of dendritic architecture offer distinct advantages as scaffolding for presenting C-sialoside groups. For example, use of dendrigraft architecture [243] as scaffolding for the presentation of sialic acid groups was found to be more than 1000 times more effective than the corresponding linear architecture for the inhibition of influenza A viruses (see Figure 35).

In recent work [244–246], a series of poly(amidoamine) (PAMAM) and poly(lysine) dendrimers were developed and tested as antiviral agents. This dendrimers series contained aryl-sulfonic acid salts on the outer surface and exhibited antiviral/antimicrobial activity against a broad range of viruses and other microorganisms. It has been postulated that these dendrimers mimic several of the biological activities of heparin (G. Holan, personal communication) [247] but do not possess some of the attributes such as anticoagulant activity, which limit the usefulness of heparin as an anti-infective agent (C. Virgona, personal communication) [248]. Witvrouw et al. [244,249], utilized "time of addition" studies to determine the mode of action of these dendrimers and found that specific PAMAM dendrimers inhibit both viral attachment to cells, as well as the action of HIV viral reverse transcriptase and virally encoded integrase, which occur intracellularly during the virus replicative cycle. It is interesting to note that some dendrimers however, showed inhibition of viral attachment only.

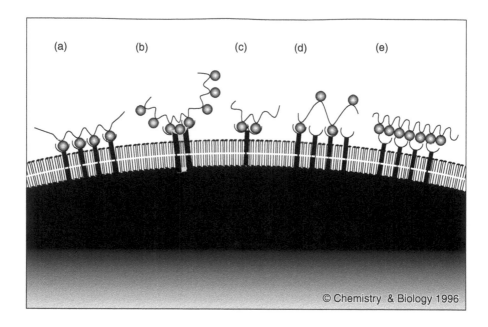

© Chemistry & Biology 1996

Figure 33 Specific recognition in multivalent interactions. Cells can use several strategies to bind to a multivalent ligand: (a) forming a cluster of many monovalent receptors on a small area of cell surface; (b) using oligomeric receptors; or (c) using receptors with more than one saccharide-binding site. In all such systems, multivalent saccharide ligands bind more tightly to the cell than their monovalent counterparts. For a divalent ligand, the free energy of binding to a multivalent receptor array will be greater than the sum of the contributions of each individual site. This primarily results from the fact that once the ligand has attached itself to a cell by one site, it is closer to the second site and will suffer a smaller entropy loss by binding to it. Multivalent ligands with incompatible relative orientations (d) or spacing (e) of the saccharide units in the multivalent array will not bind tightly. (Courtesy of *Chem. Biol.* 1996, 3(2), 72. Copyright Current Biology Ltd. London.)

In addition to inhibition of HIV viral attachment, the dendrimers also inhibit adsorption of certain enveloped viruses such as herpes simplex virus 2 (HSV) [250] and respiratory syncytial virus (RSV) [251,252] in cell culture at very low concentrations (0.1 to 1 μg/ml). In recent work, PAMAM and poly(lysine) dendrimers were also shown to inhibit human cytomegalovirus (HCMV), Ebola Virus, Hepatitis B virus (HBV), Influenza A and B, Epstein Barr virus (EbV), Adeno- and Rhino-viruses [245]. Several of these surface functionalized dendrimers are presently entering clinical studies as dendrimer-based antiviral nanodrugs.

E.   Dendrimers as Nanoscale Tectons (Modules)

1.   Two-Dimensional Dendritic Assemblies

The self-assembly of dendrimers in two-dimensions has been studied at both the air–water interface (i.e., Langmuir–Blodgett [LB] films) as well as at the air–bulk solid interface (i.e., self-assembled monolayers [i.e., SAM films]).

(a)   At the air–water interface (LBs)

Some of the earliest work was reported by Fréchet et al. in 1993 [253]. It involved the examination of LB films derived from the spreading of amphiphilic hydroxyl functionalized poly(arylether) dendrons as a function of generation level (i.e., $G = 1$–5) (Figure 36).

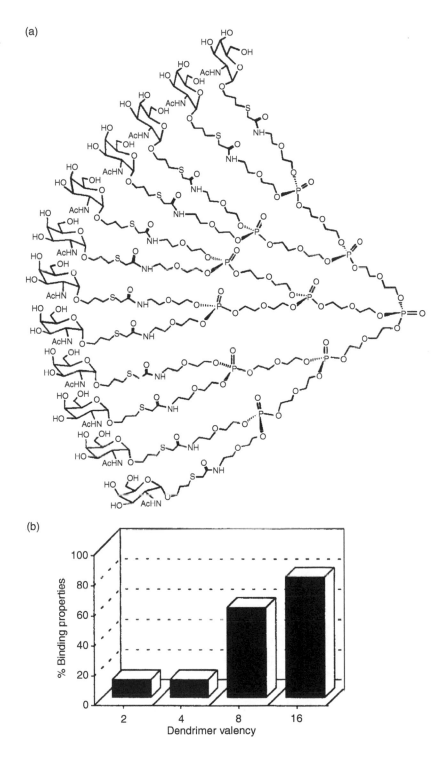

Figure 34   (a) Structure of phosphotriester GalNAC glycodendrimer. (b) Enzyme-linked lectin assays (ELLA) of L-lysine-based sialic acid dendrimers used as coating antigens in microtiter plates and detected with horseradish peroxidase labeled wheat germ agglutinin (HRPO-WGA). Binding activity as a function of dendrimer valency. (Courtesy of *Polymer News*, 1996, 21(7), 230. Copyright Gordon and Breach Publishers, Switzerland.)

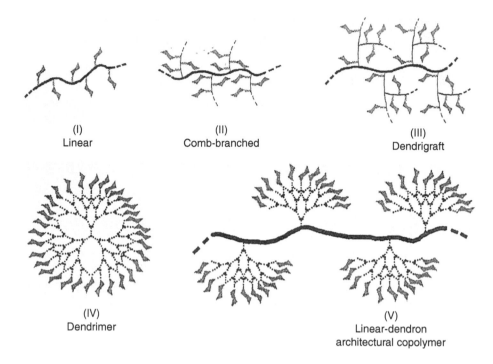

(I)	(II)	(III)
Linear	Comb-branched	Dendrigraft

(IV)
Dendrimer

(V)
Linear-dendron
architectural copolymer

Figure 35 Model representations of sialic-acid-conjugated polymeric inhibitor subunits used to inhibit viral infection. Structures are not drawn to scale. (Courtesy of *Bioconjugate Chem.* 1999, (10)2, 273. Copyright 1999 American Chemical Society.)

There was a strong dependence of the isotherm on molecular weight (generation level). The lower generations (i.e., $G = 1$–4) exhibited an increase in surface pressure through the liquid expanded phase (LE) followed by a peaked collapse transition, indicating a nucleation and growth event leading to a liquid condensed phase (LC). A sharp transition in behavior occurred in progressing from $G = 4$ to 5. Advancement from $G = 4$ to 5, caused envelopment and isolation of the hydrophilic dendrimer focal point group from the water phase, thus proceeding directly to the condensed solid phase. In summary for generations 1–4, the dendrimers behave as classical surfactant molecules on a Langmuir trough. The isotherms of $G = 5$–6, however, manifest nonsurfactant behavior, once again reflecting the surface congestion-induced properties known for higher generation dendrimers. Compression of the fourth generation poly(ether) dendrimer results in the formation of a stable bilayer. In this bilayer, the dendrimers are compressed laterally in respect to the surface normal, producing an ellipsoid shape which is twice as high as it is broad (Figure 37(a)). Neutron scattering studies on analogs with perdeuterated end groups indicate that the terminal benzyl groups are located at the top of the lower layer [254]. More stable monolayers were formed when oligo(ethylene glycol) tails were used as core functionality [255,256].

Very interesting amphiphilic behavior at the air–water interface was observed by van Hest [126] for hydrophobe (focal point) modified poly(propyleneimine) (PPI) dendrons (i.e., poly(styrene)–*dendri*–PPI $(NH_2)_n$; where: $n = 2, 4, 8)(6)$. These dendritic amphiphiles were essentially a reverse version of the Fréchet examples (see above). Only PS–*dendri*–PPI–$(NH_2)_n$ with $n = 8$ and 16 (i.e., $G = 3$ and 4) exhibited normal pressure-area isotherms. The lower generations (i.e., $G = 1$ and 2) all displayed isotherm curves indicating they transitioned directly to solid state behavior.

Poly(propyleneimine) dendrimers functionalized with hydrophobic alkyl chains (palmitoyl chains or alkyloxyazobenzene chains) assembled into stable monolayers at the air–water interface [157,257]. In the assemblies, the dendrimers adopt a cylindrical, amphoteric shape, in which the ellipsoidal

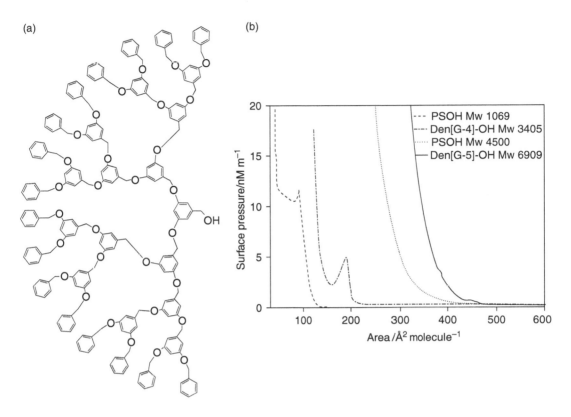

Figure 36 (a) Fourth generation [G4] dendron, based on 3,5-dihydroxy benzyl alcohol core. (b) Isotherms of [$G = 4$] and [$G = 5$] dendrons showing differences due to compression rate and pause time before compression. (Courtesy of *J. Phys. Chem.* 1993, 97, 294. Copyright 1993 American Chemical Society.)

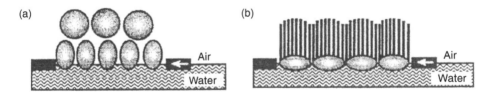

Figure 37 (a) Compressed dendrimer Langmuir bilayer; (b) dendrimer Langmuir monolayer.

dendritic moiety acts as a polar headgroup and the alkyl chains arrange in a parallel fashion to form an apolar tail (Figure 37(b)). This representation is based on the observation that the molecular area of a dendritic molecule increases linearly with the number of end groups in this molecule.

Amphiphilic PAMAM dendrimers comparable in design to those reported for the PPI dendrimers, have been studied at the air–water interface by Tomalia et al. [157]. The PAMAMs with aliphatic core groups of varying lengths (6, 8, 10, and 12 carbon atoms) also display the linear behavior between the molecular area at the compressed state and the number of end groups per molecule. Tomalia et al. explain their findings using a model in which the lower generations are asymmetric like the PPI dendrimers, while the higher generations act as hydrophobic spheroids floating on the air–water interface. Since no indication for the latter behavior is found, it is proposed here that the amphiphilic PAMAM dendrimers of high generations when disposed on air–water interfaces, are also highly distorted with all aliphatic end groups pointing upwards (Figure 37(b)).

Most notable was the fact that metal-loaded (Cu^{+2}) gold dendrimers can be readily organized into two-dimensional layers [90]. Langmuir isotherms obtained for both unloaded and metal-loaded dendrimers in this series differed substantially from those observed by Fréchet. It was reassuring to note that radii measurements obtained from limiting area Langmuir–Blodgett film studies compared very favorably with radii determined by SEC measurements.

(b) At the air–bulk phase interface (SAMs)

Very early observations [258,259] indicated the amine-terminated PAMAM dendrimers exhibited tenacious adhesion to a variety of substrates. Early indications were that they formed SAMs on glass, silicon, or metal surfaces (i.e., gold, etc.). In a pioneering effort by Mansfield [260], it was predicted that dendrimers would exhibit various deformation modes on surfaces depending on generation and adsorption strength. This Monte Carlo simulation considered the adsorption of dendrimers on a surface at different interaction strengths. The calculations show a flattening of the dendrimer shape with increasing adsorption strengths. As reflected in the "phase" diagram (Figure 38), the mode of adsorption of the dendrimers is dependent on adsorption strength and on the generation number (higher generation dendrimers have more interaction sites per molecule, and therefore, these dendrimers have a better chance to be adsorbed).

A wide variety of dendrimer adhesion modes have been defined experimentally that are beginning to fulfil the Mansfield predictions. They vary from self-assembled monolayers to multilayer assemblies as described in Figure 39(a) to (e). Perhaps, one of the first published works to clearly demonstrate the ability to construct mono/multi-layers was reported by Regen et al. [261]. They fabricated multilayers by repetitive activation with K_2PtCl_4 on a silicon wafer surface possessing primary amine groups, followed by deposition of PAMAM dendrimer, illustrated in Figure 39(d). Reiteration of this sequence produced film thickness that were shown by elipsometry to increase linearly as a function of the number of cycles performed. After 12 to 16 cycles, multilayers with a thickness close to 80 nm were obtained by using $G = 8$ or 6 PAMAM dendrimers, respectively. The elemental composition was confirmed by photoelectron spectroscopy (XPS), which both demonstrates the incorporation of PAMAM dendrimers as well as the necessity of the Pt^{+2} as a requisite component in the growth cycle.

An interesting type of dendrimer deformation has been reported by Crooks et al. [133,262,263]. Monolayers of PAMAM dendrimers adsorbed on a gold surface flatten due to multiple Au–amine interactions, but subsequent submission of alkanethiols to the surface results in a mixed monolayer in

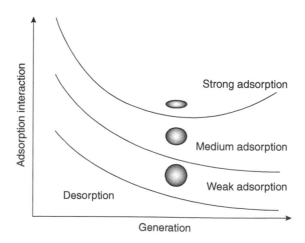

Figure 38   A "phase" diagram that shows how the shape of dendrimers in absorbed monolayers depends on the strength of the adsorption interaction and the dendrimer generation. The data are based on calculations by Mansfield. (From M.L. Mansfield et al. *J. Chem. Phys.* 105, 3245–3249, 1996.)

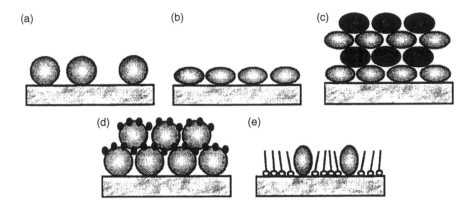

Figure 39 Schematic representation of the different modes of adsorption of dendrimers on surfaces: (a) adsorbed noninteracting dendrimers; (b) adsorbed dendrimers with surface-interacting end groups; (c) interacting multilayers dendrimer films; (d) multilayer dendrimer films with ionic shielding; (e) mixed monolayers.

which the PAMAMs acquire a prolate configuration due to the shear exerted by the thiols (Figure 39(e)). The shear originates from the stronger thiol–Au interaction as compared with the amine–Au interaction. If the adsorption time of the dendrimer monolayer is rather short (45 sec instead of 20 h), exposure to hexadecanethiol results in piling up of the dendrimers to vacate the surface in favor of the thiols [133,262,263]. Eventually, this leads to complete desorption of the dendrimers from the surface.

Tsukruk et al. [264–266] reported on the alternating electrostatic layer-by-layer deposition of PAMAM dendrimer up to generation $G = 10$. This concept is illustrated in Figure 39(c). First an amine-terminated dendrimer is adsorbed onto a SiO_2 surface followed by a carboxylic acid terminated dendrimer which deposits on top of a full generation. The cycle is repeated and multilayers are formed. They measured the layer thickness and found it to be below the theoretical one, which also supports the elastic deformation of dendrimer to obtain the most favorable energy balance. Molecular dendrimer dimensions follow scaling laws as a function of molecular weight of dendrimer with exponent of 0.27 (for spheres it is $\frac{1}{3}$). Cast and spin-coated films exhibit large sensitivity toward condition during their formation. Thermal annealing can change dendrimer shape from oblate toward spherical. Change in surface characteristics (post or prefunctional group modification) can largely influence the quality and properties of SAMs. Phase diagrams based on molecular modeling defined the correlation between generation number, interaction strength, and shape (i.e., at strong interactions PAMAM $G = 4$ on Au dendrimer will be oblate on a surface, but at strong repulsion ($C_{16}SH$, PAMAM; $G = 8$ on Au) dendrimer will be prolate. With weak interactions ($G = 0$, $G = 2$) dendrimer can retain sphere-like shapes.

Perhaps one of the most remarkable breakthroughs concerning dendrimeric SAMs is that reported recently by Fréchet et al. [267]. They have demonstrated that monolayers of dendritic polymers can be prepared by covalent attachment to a silicon wafer surface (Figure 40(a)). These ultra-thin polymer films can serve as effective resists for high-resolution lithography using the scanning probe microscope. These dendrimer films may be patterned using the SPL to create features with dimensions below 60 nm. Although very thin, the dendrimer films are resistant to an aqueous HF etch, allowing the production of a positive tone image as the patterned oxide relief features are selectively removed.

The patterned oxide relief features can be selectively removed under aqueous hydrofluoric acid (50:1, 60 sec) etching conditions resulting in a pattern transfer of raised oxide relief features into positive tone images. Figure 40(b) is a two-dimensional AFM image of the same patterns (from Figure 40(c)) after etching the wafer for 60 sec in 1 M HF. The dark regions represent depressions ca. 2 nm deep into the silicon. Under these conditions, the dendrimer monolayer clearly resists the etch as evidenced from the lack of any line broadening or pitting in the unpatterned regions.

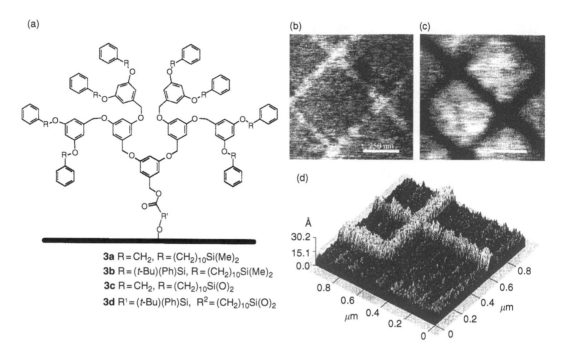

(a)

3a R = CH$_2$, R = (CH$_2$)$_{10}$Si(Me)$_2$
3b R = (t-Bu)(Ph)Si, R = (CH$_2$)$_{10}$Si(Me)$_2$
3c R = CH$_2$, R = (CH$_2$)$_{10}$Si(O)$_2$
3d R^1 = (t-Bu)(Ph)Si, R^2 = (CH$_2$)$_{10}$Si(O)$_2$

Figure 40   Covalent attachment of dendritic monolayers to a silicon wafer surface. (Courtesy of *Cur. Opin. Colloid Interface Sci.*, T. Emrick, J.M.J. Fréchet, "Self-Assembly of Dendritic Structures," 4, 1999. With permission from Elsevier Science.)

2. Three-Dimensional Dendritic Assemblies

(a) Statistical structures

Earlier work by Tomalia et al. [60] has shown that dendrimers may be used as reactive (modules) building blocks to construct statistical three-dimensional covalent networks and gels [268]. In contrast to traditional cross-linked polymer networks [66], one is able to observe within cross-linked dendrimer networks unique conserved order in the form of characteristic topological features that have been observed in electron micrographs of the dendritic system. Examples of *linear* (*catenanes*), *bridged, radially bridged* (*clefts*), and *macrocyclic bridged-type* topologies were observed as shown in Figure 41.

Several of these topological types (i.e., linear, cyclic, clusters) are remarkably reminiscent of those that are obtained by noncovalent self-assembly of proteins [8]. This offers a glimpse of the possible covalent topologies; however, virtually no examples of dendrimer self-assembly by non-covalent, electrostatic type interactions have been reported until recently. Such abiotic, noncovalent PAMAM ($G = 9$) dendrimer self-assemblies have now been observed by atomic force microscopy (AFM) on freshly cleaved mica surfaces [268] (Figure 42).

Other examples have been reported by Aida et al. [269], wherein they reported an electro-statically, directed assembly of porphyrin core dendrimers to produce large infinite aggregates as illustrated below (Figure 43). Energy transfer constants were obtained and used to determine distances between acceptors and donors as demonstrated that communication occurred between those domains.

Self-assembled, three-dimensional dendritic network structures possessing critical π-type surface groups have been reported to exhibit unique electrical conducting properties. Miller/Tomalia and coworkers [270] observed unusually high conductivities (i.e., 18 sec/cm at 90% humidity) for

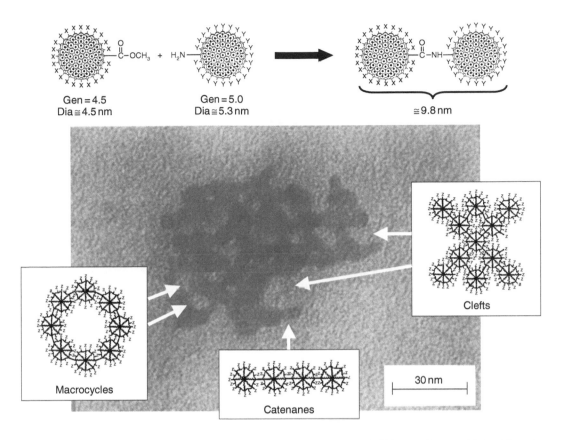

Figure 41 TEM of osmium tetraoxide stained clusters presented on a grid. This sample was obtained from the high molecular weight fraction of methanol soluble decantate.

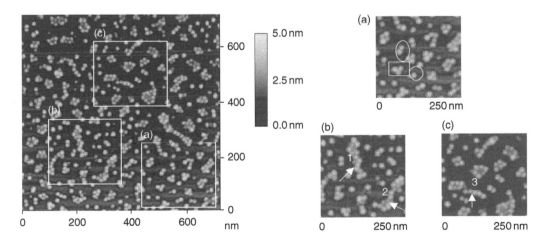

Figure 42 Tapping mode AFM images of G9 PAMAM dendrimer molecules on mica surface (left) and magnified images (right).

$G = 3$ PAMAM dendrimers surface modified with cationically substituted naphthalene diimides (Figure 44(a) and (b)). In all cases, the conductivity was electronic and isotropic. Near infrared spectra showed the formation of extensive π-stacking, which presumably favored electron hopping via a three-dimensional network.

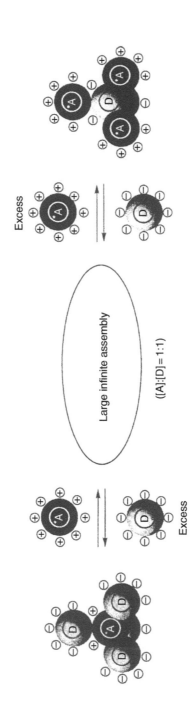

Figure 43 Schematic representation of the electrostatic interaction between negatively and positively charged dendrimer electrolytes. D represents a donor and A an acceptor in energy transfer. (Courtesy of *Angew. Chem. Int. Ed.* 1998, 37(11) 1533. Copyright Wiley-VCH Verlag GmbH, Weinheim, Germany.)

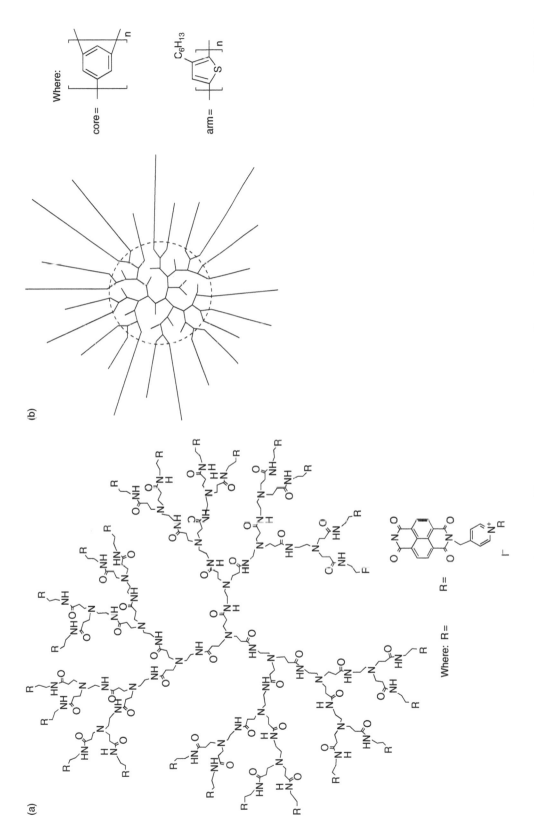

Figure 44 Examples of electrically conducting (a) dendrimeric and (b) hyperbranched star polymers. (Courtesy of *J. Am. Chem. Soc.* 1997, 119, 1006 and (b) 11106. Copyright 1997 American Chemical Society.)

Similarly, Wang and coworkers [271] reported that hyperbranched dendritic structures possessing poly(3-alkylthiophene) arms self-assemble into thin films with morphological features, as well as electrical and optical properties that reveal a surprising degree of structural order. Typical conductivities varied between 42 and 65 sec/cm.

Controlled interparticle spacing of gold nanoparticles has recently been demonstrated by utilizing PAMAM dendrimer ($G = 0$ to 6) mediated self-assembly to provide a "bricks and mortar" approach for constructing designable dendrimer-based nanocomposites [272].

(b) Structure-controlled core-shell tecto(dendrimers)

Core-shell architecture is a very recognizable concept in the lexicon of science. Beginning with the first observations by Galileo concerning the heliocentricity of the solar system [273] to the planetary model first proposed by Rutherford [1] and expounded upon by Bohr [274], such architecture has been broadly used to describe the influence that a central focal point component may exercise on its surrounding satellite components. Such has been the case recently at the sub-nanoscale level. Rebek et al. have [275] described the influence that a guest-molecule may have on self-assembling components that are directed by hydrogen-bonding preferences and the filling of space. It was shown that hydrogen-bonding preferences combined with spacial information such as molecular curvature can be used to self-assemble a single core-shell structure, as shown below (Figure 45) at the subnanoscopic level.

At the nanoscale dimensional level it was shown by Hirsch et al. [276] that [60]-fullerenes could be used as a core tecton to construct a core-shell molecule with T_b-symmetrical C_{60} core and an extraordinarily high branching multiplicity of 12.

As described earlier, the assembly of reactive monomer [50] branch cells [14,54] around atomic or molecular cores to produce dendrimers according to divergent/convergent dendritic branching principles is well-demonstrated [63]. The systematic filling of molecular space around dendrimer cores with monomer units or branch cells as a function of generational growth stages (dendrimer shells) to give discrete, quantized bundles of mass has proven to be mathematically predictable [64,277]. This generational mass relationship has been demonstrated experimentally by mass spectroscopy [65,66] and other analytical methods [68,69,122]. These synthetic strategies have allowed the systematic control of molecular structure as a function of size [122], shape [160], and surface/interior functionality [37]. Such synthetic strategies have allowed construction of dendrimeric structures with dimensions that extend well into the lower nanoscale regions (i.e., 1 to 30 nm).

Since the very first reports on dendrimers in 1984 [72–76], we have proposed the use of these entities as fundamental building blocks for the construction of higher complexity structures on

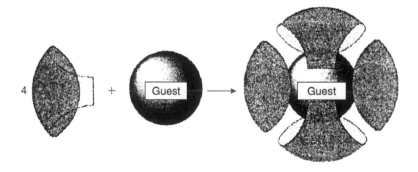

Figure 45   Self-assembly of a core-shell structure involving hydrogen-bonding and complementary shape preference according to Rebek et al. (From T. Martin et al. *Science* 281, 1842–1845, 1998. With permission.)

(a)

(b)

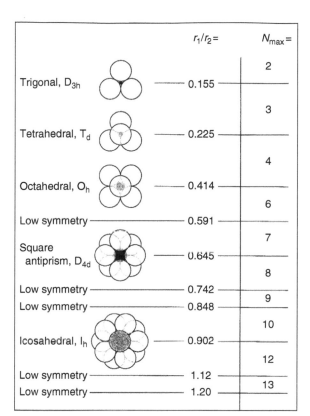

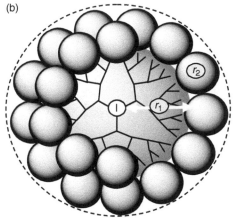

Symmetry	$r_1/r_2 =$	$N_{max} =$
		2
Trigonal, D_{3h}	0.155	
		3
Tetrahedral, T_d	0.225	
		4
Octahedral, O_h	0.414	
		6
Low symmetry	0.591	
		7
Square antiprism, D_{4d}	0.645	
		8
Low symmetry	0.742	
Low symmetry	0.848	9
		10
Icosahedral, I_h	0.902	
		12
Low symmetry	1.12	
Low symmetry	1.20	13

r_1 = radius of core dendrimer
r_2 = radius of shell dendrimer

N_{max} = total theoretical number of shell-like spheroids with radius r_2 that can be ideally parked around core spheroid with radius r_1

(c)

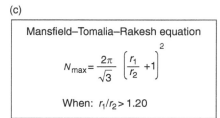

Mansfield–Tomalia–Rakesh equation

$$N_{max} = \frac{2\pi}{\sqrt{3}}\left[\frac{r_1}{r_2}+1\right]^2$$

When: $r_1/r_2 > 1.20$

Figure 46 (a) Symmetrical properties for core-shell structures where $r_1/r_2 < 1.20$. (b) SIS based on respective radii (A) and (B) of dendrimers. (c) Mansfield–Tomalia–Rakesh equation for calculation of maximum shell filling when $r_1/r_2 > 1.20$.

numerous occasions [35,64,277,278]. Early electron microscopy studies [60] and other analytical methods [68,69] indicated that the supramacromolecular assembly leading to formation of dimers, trimers, and other multimers of dendrimers occurred almost routinely, however, these were largely uncontrolled events.

Recent studies have shown that poly(amidoamine) PAMAM dendrimers are indeed very well defined, systematically sized spheroids [122] as a function of generation level. Furthermore, evidence has been obtained by small angle neutron scattering studies to show that these PAMAM dendrimers behave as originally described by de Gennes [87]. The terminal groups remain largely at the periphery and are *exo*-presented with virtually no backfolding [279–281].

Anticipating the use of these nanoscale modules in a variety of construction operations we examined the random parking of spheres upon spheres [282]. From this study, it was rather surprising and pleasing to find that at low values of radii ratios (i.e., <1.2), absolutely beautiful symmetrical properties appeared as illustrated in Figure 46. However, at higher radii ratios, the mathematics resolved into the following Mansfield–Tomalia–Rakesh general expression. From this relationship, it is possible to calculate the number of spheroidal dendrimers one could possibly place in a shell around a core dendrimer as a function of their respective radii (Figure 47).

Inspired by these derived values for shell filling around a "central dendrimer core," we devised several synthetic approaches to test this hypothesis. The first method involved the direct covalent reaction of a dendrimer core with an excess of dendrimer shell reagent; referred to as the (i) Direct

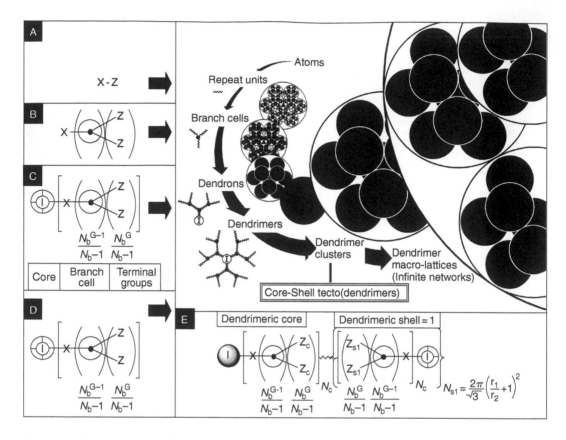

Figure 47 Hierarchy of empirical construction components: (A) monomers, (B) branch cells, (C) dendrons, and (D) dendrimers leading to (E) core-shell tecto(dendrimers). (With permission from John Wiley & Sons Ltd.)

Covalent Method. The second method involved self-assembly by electrostatic neutralization of the dendrimer core with excess shell reagent to give the (ii) Self Assembly with Sequential Covalent Bond Formation Method. These strategies are described in Section IV. In each case, relatively monodispersed products are obtained. We call these new dendritic architectures; *core-shell (tecto)dendrimers* [52,83,283,284].

IV. SUPRA*MACRO*MOLECULAR POLYMERIZATION OF DENDRIMERS INTO PRECISE CORE-SHELL TECTO(DENDRIMERS)

A. Direct Covalent Method

This route produces partially filled shell structures and involves the reaction of a nucleophilic dendrimer core reagent with an excess of electrophilic dendrimer shell reagent as illustrated in Figure 48 [83,285].

Various poly(amidoamine) PAMAM dendrimer core reagents (i.e., either amine or ester functionalized) were allowed to react with an excess of appropriate PAMAM dendrimer shell reagents. The reactions were performed at 40°C in methanol and monitored by FTIR, ^{13}C NMR, size exclusion chromatography (SEC), and gel electrophoresis. Conversions in step (A) were followed by the formation of shorter retention time, higher molecular weight products using SEC. Additional evidence was gained by observing the loss of migratory band associated with the dendrimer core reagent

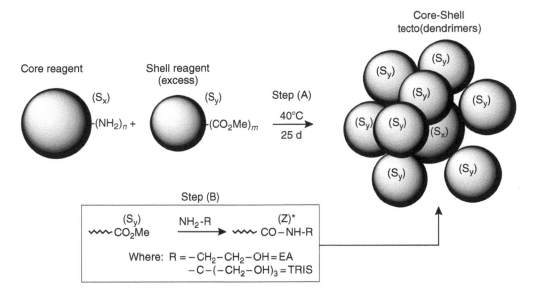

Figure 48 Synthetic scheme for core-shell tecto(dendrimers), where Z = surface functionality after reduction with R-type reagents.

Table 1 Analytical Evidence for Core-Shell Tecto (Dendrimers)

X;[(Y)(Z*)]$_n$	G4[(G3);(EA)]$_n$	G5(G3);(TRIS)$_n$	G6(G4); (TRIS)$_n$	G7(G5); (TRIS)$_n$
Theoretical Shell Sat. Levels (n)	9	15	15	15
Observed Shell Sat. Levels ($n*$)	4	8–10	6–8	6
Percent Theoretical Shell Sat. Levels	44%	53–66%	40–53%	40%
MALDI-TOF-MS (MW):	56,496	120,026	227,606	288,970
PAGE (MW)	58,000	116,000	233,000	467,000
AFM: Observed Dimensions:	25 × 0.38 nm (D,H)	33 × 0.53 nm (D,H)	38 × 0.63 nm (D,H)	43 × 1.1 nm (D,H)
CALC. (MW):	56,000	136,000	214,000	479,000

present in the initial reaction mixture, accompanied by the formation of a higher molecular weight product, which displayed a much shorter migratory band position on the electrophoretic gel. In fact, the molecular weights of the resulting core-shell tecto(dendrimer) could be estimated by comparing the migratory time of the core-shell product ([Polyacrylamide] gel electrophoresis [PAGE] results, Table 1) with the migration distances of the PAMAM dendrimers (e.g., $G = 2$–10) used for their construction [68,69].

It was important to perform capping reactions on the surface of the resulting ester terminated core-shell products in order to pacify the highly reactive surfaces against further reaction. Preferred capping reagents were either 2-amino-ethanol or *tris*-hydroxymethyl aminomethane.

The capping reaction, step (B), was monitored by following the disappearance of an ester band at 1734 cm^{-1}, using FTIR. Isolation and characterization of these products proved that they were indeed relatively mono(dispersed) spheroids as illustrated by AFM. It was very important to perform the AFM analysis at very high dilution to avoid undesirable core-shell molecular clustering. In spite of these efforts, a small amount of clustering is still observed.

A distinct core-shell dimensional enhancement was observed as a function of the sum of the core-shell generation values used in the construction of the series (e.g., $G4/G3$, $<G5/G3$, $<G6/G4$, and $<G7/G5$). This is in sharp contrast to nondescript polydispersed

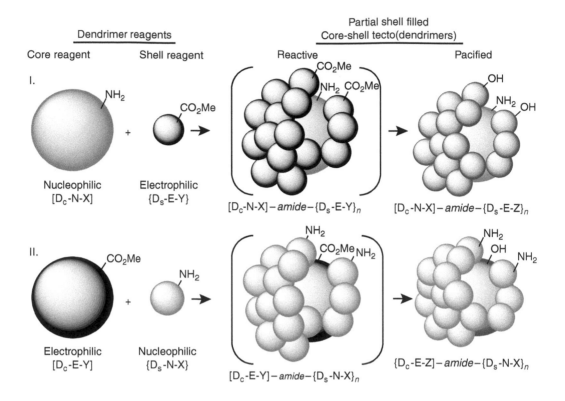

Figure 49 Two routes to partial shell filled tecto(dendrimers). Route I involves amidation of a limited amount of nucleophilic core dendrimer with an excess of electrophilic shell dendrimer reagent to produce *reactive* (PS:CST)–[D_c-N-X]–*amide*–{D_s-E-Y}$_n$. These products may be pacified by reacting with 2-aminoethanol (EA) or *tris*(hydroxymethyl) aminomethane (*tris*) to produce shell *pacified*–(PS:CST)–[D_c-N-X]–*amide*–{D_s-E-Z}$_n$. Route II involves amidation of a limited amount of electrophilic core dendrimer with an excess of nucleophilic shell dendrimer reagent to produce *reactive*-(PS:CST)–[D_c-E-Y]–*amide*–{D_s-N-X}$_n$. These products may be converted to pacified forms by reacting with 2-aminoethanol (EA) or an excess of EDA to produce core *pacified*-(PS:CST)–[D_c-E-Z]–*amide*–{D_s-N-X}$_n$.

dendrimer cluster/gel formation observed for 1:1 reaction ratios described in our earlier work [60].

Molecular weights for the final products were determined by MALDI-TOF-MS or PAGE. They were corroborated by calculated values from AFM dimension data (Table 1).

Calculations based on these experimentally determined molecular weights allowed the estimation of shell filling levels for respective core-shell structures within this series. A comparison with mathematically predicted saturated shell structures reported earlier [282], indicates these core-shell structures are only partially filled (i.e., 40 to 66% of fully saturated values, see Table 1).

Functional group differentiated nanoscale clefts and cusps produced on the surfaces of these partially filled tecto(dendrimer) core-shell structures have exhibited extraordinary "autoreactive" properties unless pacified as described recently [83,289] (Figure 49).

B. Self-Assembly with Sequential Covalent Bond Formation Method

The chemistry used in this approach involved the combination of an amine-terminated core dendrimer with an excess of carboxylic acid terminated shell reagent dendrimer (Figure 50) [283,285]. These two charge differentiated species were allowed to equilibrate and self-assemble into the electrostatically driven core-shell tecto(dendrimer) architecture followed by covalent fixing of these charge neutralized dendrimer contacts with carbodiimide reagents. Reactions were readily monitored by SEC, gel

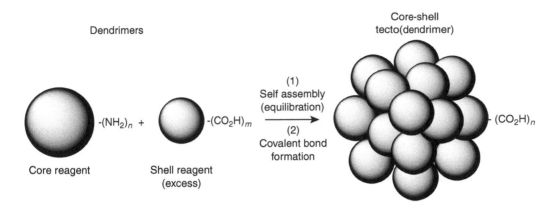

Figure 50 Self-assembly of core and shell dendrimers by charge neutralization (1) followed by covalent bond formation (2).

Table 2 Comparison of % Shell Filling as a Function of Core and Shell Dendrimers, Using the Supramolecularly Assisted Method Followed by Covalent Bond Formation

Core Reagent	Shell Reagent	Observed No. of Parked Dendrimers by Mass Spec.	Ideal No. of Parked Dendrimers[a]	Shell Filling (%)
G5	G3)-COOH	10	12	83
G6	G3)-COOH	13	15	87
G7	G3)-COOH	15	19	79
G7	G5)-COOH	9	12	75

[a] Calculated using Mansfield–Tomalia–Rakesh equation for the maximum parking problem.

$$N_{max} = \frac{2\pi}{\sqrt{3}} \left(\frac{r_1}{r_2} + 1 \right)^2$$

Source: From M.L. Mansfield et al. *J. Chem. Phys.* 105:3245–3249, 1996. With permission.

electrophoresis, AFM, and MALDI-TOF mass spectroscopy. As might be expected, preliminary data shows that the self-assembly method provides for more efficient parking of the dendrimer shell reagents around the core to yield very high saturation levels as shown in Table 2. Our present experimentation indicates that this method should allow the assembly of additional shells in a very systematic fashion to produce precise nanostructures that transcend the entire nanoscale region (1 to 100 nm) [36,286]. In contrast to the unpacified partial shell filled core-shell tecto(dendrimers), these structures do not exhibit "autoreactive" behavior.

V. OVERVIEW — PRESENT APPLICATIONS — DENDRITIC NANODEVICES

A. Overview of the Dendritic State

In summary, rheological (fluidity) investigations in our laboratories indicate that dendrimers behave like soft spherical bodies surrounded by relatively hard surface shells (i.e., like "core-shell" type entities) [287,288]. According to this emerging picture, the interiors of dendrimers may be deformable or rigid depending on the character of the monomers used in construction. The interiors may contain cavities, voids, and well-defined space, capable of accommodating many small guest molecules such as solvents, dyes, oligomers (intermediates between monomers and polymers). Guest molecules may enter or exit the dendrimer depending on their size and shape.

On the other hand, dendrimer surfaces appear to be impenetrable to other dendrimer molecules or natural and synthetic macromolecules, especially at higher generations. Dendrimers can thus be envisioned as "unimolecular containers" for "encapsulations" [37] or so-called "dendrimer boxes" [142]. Potential uses include molecular delivery agents, transport vehicles, unimolecular micelles, molecule ball bearings, interphase catalysts, flow regulators in fluids, and highly monodispersed "dwarf latexes" for coatings, etc. If applications in these areas can be realized, the future of dendrimers and other polymers is very bright indeed.

Another fascinating area of dendrimer applications is based on their high surface functionality. No other class of synthetic or natural compounds contains so many reactive terminal groups per molecule as do the dendrimers. This provides two major directions for possible exploitation. First, dendrimers can be modified in various ways with reagents of small molecular weight. It is thus possible to produce dendrimers with so-called *exo*-modified or differentiated surfaces. For example, attachment of catalytic or biological receptor sites suggests many possible applications. Furthermore, the dendrimer interiors may be modified in many yet specific ways. Interior differentiated dendrimers with different combinations of radial layers or segments may be prepared by using different dendrons or parts of dendrons in their construction. Incarceration of zero-valent metals or their salts (i.e., iron, copper, silver, palladium, platinum, or cadmium sulfide, etc.) suggest many uses as catalysts, magnetic dendrimers, and quantum dots.

Equally significant, is the possibility of using the dendrimer surface reactivity to open a new branch of synthetic chemistry, namely, *"nanoscale chemistry"* [289]. This would involve using dendrimers as building blocks for the preparation of even larger compounds with nanoscale and microscopic dimensions [64,277,290,291]. Such mega-molecules could result from reactions between dendrimers or with appropriate biological macromolecules yielding covalently bonded nanoscopic structures with hybridized architectures (i.e., linear, branched, cross-linked, or dendritic-types) [160]. Of course, this section describes the numerous possibilities based upon various molecular recognition and noncovalent bonding processes, which include both supra and supra*macro*molecular interactions.

Perhaps most exciting is the emerging role that dendritic architecture is playing in the role of commodity polymers. Recent reports by Guan et al. [292] have shown that ethylene monomers polymerize to dendrigraft poly(ethylene) at low pressures. This occurs when using late transition metal or Brookhart-type catalyst (Figure 51). Furthermore, these authors also stated that small amounts of dendrigraft poly(ethylene) architecture may be expected from analogous early transition metal–metallocene catalysts.

B.   Dendritic Nanodevices

Abiotic–biotic hybrids composed of dendritic polymers and natural biopolymers have already found application as nanodevices. In many instances, these dendritic nanodevices are used in abiotic–biotic molecular recognition events involving suprachemistry [104,293,294]. One prime example is the conjugation of poly(amidoamine) dendrimers to IgG antibodies for use in diagnostic immunoassay [295–297]. In that work, the architecturally precise dendrimers act as a replacement for a secondary antibody by spacing the primary antibody away from the solid phase. The replacement of a secondary binding antibody dramatically reduces lot rejections in manufacture and minimizes nonspecific interactions between analytes and the immobilized antibodies. In this nanodevice, the dendrimer acts both as an antibody replacement in contact with the solid phase as well as a macromolecular spacer to hold the antibody away from the solid phase.

Dendrimers have also found application as carriers of genetic materials into cells [84,298–300]. In this usage, the dendrimers can be thought of as a histone mimetic. With appropriately charged surface groups the polycationic dendrimers forms a supra*macro*molecular complex with the polyanionic nucleic acid biopolymer. This dendrimeric nanodevice transports the genetic material across

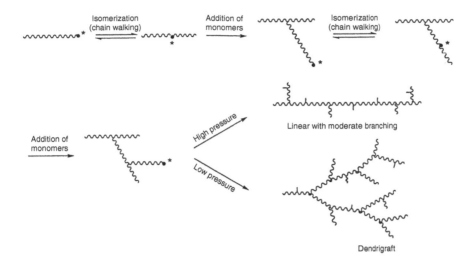

Figure 51 (a) Condensation polymerization of AB$_2$ monomers to give a hyperbranched polymer. Condensation of each monomer increases the active site from one to two. (b) "Self-condensing" polymerization of vinyl monomers to give a hyperbranched polymer. Addition of new monomer increases the active site from one to two. (c) Proposed new approach to make hyperbranched polymer. The active site isomerizes to the internal backbone, and addition of monomers leads to branching.

cell membranes; wherein, the DNA is eventually incorporated into the cellular expression machinery to generate protein.

Dendrimer-based nanodevices have also found application in magnetic resonance imaging [301–304]. Dendrimer-chelate conjugates have been prepared that are extraordinarily robust complexes of gadolinium ions. The size of the dendrimer–gadolinium complexes allows sufficiently long residence times in the bloodstream for appropriate imaging studies. Furthermore, the nuclear relaxation parameters are dramatically enhanced when compared with conventional MRI contrast systems. Higher complexity conjugates have been prepared with target directors to introduce biotic molecular recognition functionality into these conjugates [301].

Dendritic polymers may also have application in the delivery of anticancer agents. Dendrimers have been shown to selectively deliver a high payload of traditional chemotherapeutic agents, such as cisplatin, to a tumor [305]. Blood vessels in tumors have a higher permeability and poorer lymphatic drainage than vessels in normal tissue. This enhanced permeability and retention effects (EPR effect) allows selective delivery of a drug to a tumor, and has been demonstrated with nondendritic polymer–antitumor agent conjugates [306,307].

Dendrimer–saccharide conjugates are unique nanodevices in their ability to tightly bind lectins and other proteins with specific saccharide recognition capabilities [242]. The multivalency of dendrimer surfaces allows for a cooperative binding effect. Where single saccharide–receptor interactions are relatively weak, the multiplicity of poly(valent) interactions introduced by conjugating a large number of sugars onto a single dendrimer particle yields a cooperative unimolecular binding event that is thermodynamically much more favorable.

Dendritic polymers offer the potential to serve as amplifiers in a variety of applications. Researchers in boron neutron capture therapy have investigated the potential of dendrimers to deliver a large payload of boron to tumor sites by conjugating a large number of boron clusters to the surface of a dendrimer [308]. When irradiated with a neutron beam, the high cross-section of boron captures sufficient energy to produce a secondary radiation of sufficient energy to damage cells in the immediate vicinity of the boron. Again, conjugates with higher specificity can be prepared by introduction of target receptors [21].

Although many of the above examples are in the early stages of development, they have clearly demonstrated concepts that illustrate the current potential for dendrimers in a variety of nanodevices,

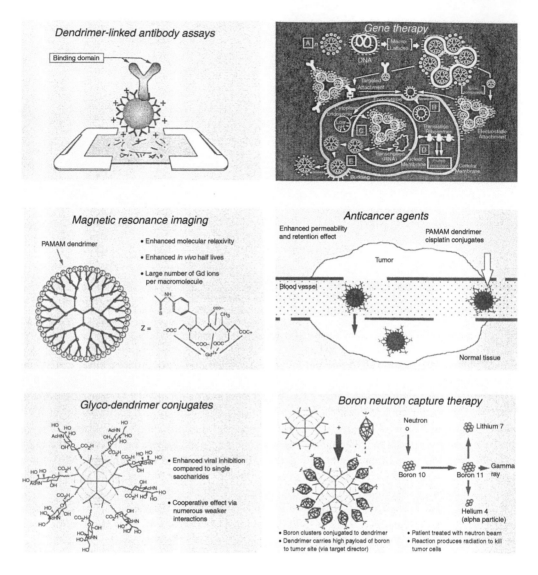

Figure 52   Some current nanobiological devices utilizing dendritic polymers.

many of them supramolecular or supra*macro*molecular in nature. Figure 52 is a pictorial summary of some of these nanobiological devices. Further information on the individual concepts may be found in the literature [34,309,310].

VI.   THE FUTURE/CONCLUSIONS

Nature is a very strange affair, and the strangeness already encountered by our friends the physicists are banalities compared to the queer things being glimpsed in biology, and the much queerer things that lie ahead.

   As these turn up. . . they will inevitably change the way the world looks. And when this happens, the view of life itself will also shift; old ideas will be set aside; the *look of a tree will be a different look*; the connectedness of all the parts of nature will become a reality for everyone, not just the mystics, to think about; painters will begin to paint differently; music will change from what it is to something new and unguessed at; poets will write stranger poems; and the culture will begin a new cycle of change.
Lewis Thomas, 1985

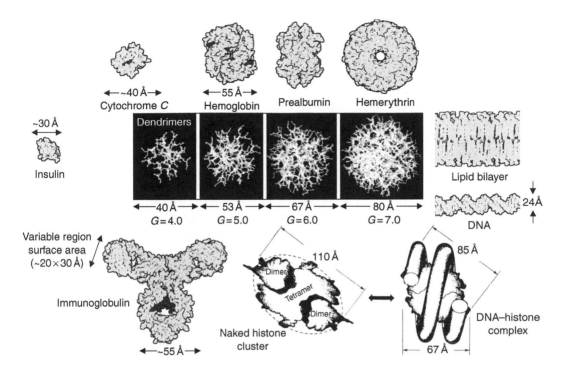

Figure 53 Scale comparison of tridendron (NH_3 core) poly(amidoamine) dendrimers; $G = 4$ to 7, sizes and shapes with various proteins. (Courtesy of *J. Mater. Chem.* 1997, 7(7), 1199. Copyright Royal Society of Chemistry.)

This quotation conveys the daunting prophesies that may be expected from the convergence of supramolecular and supra*macro*molecular chemistry with nanoscale building blocks such as dendrimers or other nanoscale objects in the biological world.

It was both remarkable and surprising to find that many of these abiotic, structure-controlled, macromolecules (dendrimers, dendrigrafts, etc.) possess topologies, function, and dimensions that scale very closely to a wide variety of important biological polymers and assemblies. Figure 53 compares poly(amidoamine) PAMAM dendrimers as a function of generation with important biological structures that are both conserved and essential for life in the plant, as well as the animal kingdoms. History has shown that the introduction of traditional synthetic polymer architectures (i.e., (I) linear, (II) cross-linked, (III) branched) by Staudinger, Carothers, Flory, and others [42,311] provided the basis for the replacement of many natural polymers (i.e., silk, rubber, cotton, etc.) accompanied by many new properties and advantages associated with each new macromolecular architecture.

This fourth new dendritic architecture is already bearing fruit with new properties never before observed for traditional polymeric architectures. These new observations are referred to as "dendritic effects" [312]. With the present understanding of the covalent, supramolecular, and supra*macro*molecular chemistry of both traditional and dendrimeric systems, it is now possible to visualize new architectural possibilities. This area will undoubtedly provide the enabling science for the replacement of many biological nanostructures such as histones, antibodies, hemoglobin, etc. In fact, recent developments are driving this new and emerging area of science, which is now being referred to as *nanoscale chemistry, biologic nanotechnology*, or *nanobiology*. Briefly defined, it is the science of understanding covalent, structure-controlled synthesis, characterization, nanoscale combining rules, and the supramolecular/supra*macro*molecular dynamics of molecules and assemblies that manifest complexity and dimensions greater than traditional chemistry [20,313]. This new chemical science will undoubtedly evolve the necessary molecular level understanding to create requisite bridges to biology. These new connections to biology may lead one to expect many new paradigm shifts in the treatment and remediation of acute/chronic diseases, genetic defects, and enhancements of longevity (aging) leading to improvements in the human condition, as well as the biological environment in general.

Recent successes in the development of dendritic nanodevices reveal progress in this endeavor. It will not be surprising to see the evolution of synthetic *immunosystems, gene expression protocol based on artificial histones, or the treatment of cancer with dendritic nanodevices* in the foreseeable future.

As proposed earlier, dendritic biomimicry especially as it relates to dimensional size scaling, macromolecular structure control, shape/topological matching, genealogical sequencing, exponential amplification polyvalency (hypercooperativity), site isolation, unimolecular encapsulation, and self-assembly are now possible with dendritic structures and architecture. In many cases, this mimicry has already been demonstrated in preliminary form (Figure 52). Recent demonstration of these analogies, not only has fulfilled some unexpected prophecies but also more importantly has set the stage (or the table) for an unimaginable banquet of new discoveries and properties. Hopefully, these endeavors will provide solutions to problems that presently plague and hamper the enhancement of the human condition.

Furthermore, we predict that the *dendritic state* [312] will undoubtedly provide the *"quintessential scientific bridge"* by which both abiotic molecular level scientists will be able to communicate and collaborate on such critical issues as disease control, increased agricultural production and finally longevity with enhanced quality of life. It is with these thoughts and premises that we should be very excited about the future of dendritic macromolecular technology and the role its supra-properties will provide as we enter the next millennium.

ACKNOWLEDGMENTS

The authors would like to thank the U.S. Army Research Laboratory (ARL/Central Michigan University Dendritic Polymer Center of Excellence), especially Dr. S. McKnight and Dr. B. Fink for financial support of this research (DADD19-03-2-0012) and Linda S. Nixon for manuscript and graphics preparation.

REFERENCES

1. B. Pullman. *The Atom in the History of Human Thought*. Oxford University Press: New York, 1998.
2. S.E. Mason. *Chemical Evolution*. Clarendon Press, Oxford University Press: Oxford, 1991.
3. M. Eigen, W. Gardiner, P. Schuster, and R. Winkler-Oswatitsch. *Evolution Now*. Freeman: New York, 1982.
4. M. Eigen. *Naturwissenschaften* 10:465, 1971.
5. H. Kuhn and J. Waser. *Angew. Chem. Int. Ed.* 20:500–520, 1981.
6. I. Prigogine. *Phys. Today* 25(12):38–44, 1972.
7. *Webster's Third New International Dictionary*. G & C Merriam Co.: Springfield, MA, 1981.
8. B. Alberts, D. Bray, J. Lewis, M. Raff, K. Roberts, and J.D. Watson. *Molecular Biology of the Cell*. Garland Publishing Inc.: New York, 1994.
9. J. Darnell, D. Baltimore, and H. Lodish. *Molecular Cell Biology*. Scientific American Books, Inc.: New York, 1986.
10. G.F. Joyce. *Sci. Am.* 267(6):90, 1992.
11. H.W. Salzburg. *From Caveman to Chemist*. American Chemical Society: Washington, DC, 1991.
12. J.M. Lehn. *Angew. Chem. Int. Ed.* 27:89, 1988.
13. G.W. Gokel. *Crown Ethers and Cryptands*. The Royal Society of Chemistry: Cambridge, 1991.
14. G.R. Newkome, C.N. Moorfield, and F. Vögtle. *Dendritic Molecules*. VCH: Weinheim, 1996.
15. D. Philip and J.F. Stoddart. *Synletters* 7:445, 1991.

16. P.R. Ashton, N.S. Isaacs, F.H. Kohnke, J.P. Mathias, and J.F. Stoddart. *Angew. Chem. Int. Ed.* 28:1258, 1989.
17. P.R. Ashton, N.S. Isaacs, F.H. Kohnke, G.S. D'Alcontres, and J.F. Stoddart. *Angew. Chem. Int. Ed.* 28:1261, 1989.
18. F.H. Kohnke, J.P. Mathias, and J.F. Stoddart. Proceedings of the International Symposium on Chemical and Biochemical Problems in Molecular Recognition, Exeter, 17–21 April 1989, Ed. S.M. Roberts. Special Publication No. 78, Royal Society of Chemistry, Cambridge, 1989.
19. F.H. Kohnke, J.P. Mathias, and J.F. Stoddart. *Angew. Chem. Int. Ed.* 28:1103, 1989.
20. M. Freemantle. *Chem. Eng. News* (April 19) 16:51–58, 1999.
21. C.M. Drain, F. Nifiatis, A. Vasenko, and J.D. Batteas. *Angew. Chem. Int. Ed.* 37:2344, 1998.
22. C.M. Drain and J.M.J. Lehn. *Chem. Commun.* 19:2313, 1994.
23. R.W. Wagner, J. Seth, S.I. Yang, D. Kim, D.F. Bocian, D. Holten, and J.S. Lindsey. *J. Org. Chem.* 63:5402, 1998.
24. R.V. Slone and J.T. Hupp. *Inorg. Chem.* 36:5422, 1997.
25. C.M. Drain, K.S. Russell, and J.M. Lehn. *Chem. Commun.* 3:337, 1996.
26. E. Alessio, M. Macchi, S. Heath, and L.G. Marzilli. *J. Chem. Soc. Chem. Commun.* 12:1411, 1996.
27. E. Alessio, M. Macchi, and S. Heath. *J. Chem. Commun.* 36:5614, 1997.
28. Y. Kobuke and H. Miyaji. *Bull. Chem. Soc. Jpn.* 69:3563, 1996.
29. R.T. Stibrany, J. Vasudevan, S. Knapp, J.A. Potenza, T. Emge, and H.J. Schugar. *J. Am. Chem. Soc.* 118:3980, 1996.
30. H. Tamiaki, T. Miyatake, R. Tanikaga, A.R. Holzwarth, and K. Schaffner. *Angew. Chem. Int. Ed.* 35:772, 1996.
31. D.B. Amabilino, C.O. Dietrich-Bucheker, and J.-P. Sauvage. *J. Am. Chem. Soc.* 188:3285, 1996.
32. S. Anderson, H.L. Anderson, and J.K.M. Sanders. *Acc. Chem. Res.* 26:469, 1993.
33. E. Buhleier, W. Wehner, and F. Vögtle. *Synthesis* 405:155–158, 1978.
34. D.A. Tomalia and R. Esfand. *Chem. Ind.* 11:416–420, 1997.
35. D.A. Tomalia and H.D. Durst. Topics in current chemistry, Vol. 165, in *Supramolecular Chemistry I — Directed Synthesis and Molecular Recognition*. E.W. Weber, Ed. Springer Verlag: Berlin, Heidelberg, pp. 193–313, 1993.
36. J.M.J. Fréchet and D.A. Tomalia. *Dendrimers and Other Dendritic Polymers*. John Wiley & Sons, Ltd.: West Sussex, 2001.
37. D.A. Tomalia, A.M. Naylor, and W.A. Goddard III. *Angew. Chem. Int. Ed. Engl.* 29:138–175, 1990.
38. J. Fendler. *Membrane Mimetic Chemistry: Characterization and Applications of Micelles, Microemulsions, Monolayers, Bilayers, Vesicles, Host Guest Systems and Polyions*. Wiley Interscience: Chichester, 1982.
39. C. Tschierske. *J. Mater. Chem.* 8(7):1485–1508, 1998.
40. C. Tanford. *The Hydrophobic Effect: Formation of Micelles and Biological Membranes*. John Wiley & Sons: New York, 1973.
41. H. Staudinger, Organic Chemistry of Macromolecules, New York, Wiley, 1985.
42. H. Morawetz. *Polymers. The Origin and Growth of a Science*. John Wiley & Sons: New York, 1985.
43. H.L. Hsieh and R.P. Quirk. *Anionic Polymerization: Principles and Practical Applications*. Marcel Dekker: New York, 1996.
44. K. Matyjaszewski. *Cationic Polymerizations: Mechanisms, Synthesis and Applications*. Marcel Dekker: New York, 1996.
45. K. Matyjaszewski. Controlled radical polymerization. In *ACS Symposium Series 685*, Vol. 685. Washington DC, 1997.
46. K. Hatada, T. Kitayama, and O. Vogl. *Macromolecular Design of Polymeric Materials*. New York, Dekker 1997.
47. R.B. Merrifield. *J. Am. Chem. Soc.* 85:2149–2154, 1963.
48. R.B. Merrifield. *Science* 232:341–347, 1986.
49. M.K. Lothian-Tomalia, D.M. Hedstrand, and D.A. Tomalia. *Tetrahedron* 53:15495–15513, 1997.
50. D.A. Tomalia. *Sci. Am.* 272:42–46, 1995.
51. D.A. Tomalia and J.M.J. Fréchet. *J. Polym. Sci.: Part A: Polym. Chem.* 40:2719–2728, 2002.
52. D.A. Tomalia, R. Esfand, L.T. Piehler, D.R. Swanson, and S. Uppuluri. *High Perform. Polym.* 13:S1–S10, 2001.
53. D.A. Tomalia. *Macromol. Symp.* 101:243–255, 1996.

54. J.M.J. Fréchet. *Science* 263:1710–1715, 1994.

55. C.-O. Turrin, V. Maraval, J. Leclaire, E. Dantras, C. Lacabanne, and A.-M. Caminade. *Tetrahedron* 59:3965–3973, 2003.

56. K. Dusek and M. Duskova-Smrckova. Formation, structure and properties and the crosslinked state relative to precursor architecture, in *Dendrimers and Dendritic Polymers*. John Wiley & Sons Ltd.: West Sussex, J.M.J. Frechet and D.A. Tomalia Eds. pp. 111–145, 2001.

57. K. Dusek. *TRIP* 5(8):268–274, 1997.

58. D.A. Tomalia and J.M.J. Fréchet. Introduction to the dendritic state, in *Dendrimers and Other Dendritic Polymers*. J.M.J. Frechet and D.A. Tomalia, Eds. John Wiley & Sons: West Sussex, pp. 3–44, 2001.

59. S. Uppuluri, D.R. Swanson, H.M. Brothers II, L.T. Piehler, J. Li, D.J. Meier, G.L. Hagnauer, and D.A. Tomalia. *Polym. Mater. Sci. Eng. (ACS)* 80:55–56, 1999.

60. D.A. Tomalia, D.M. Hedstrand, and L.R. Wilson. Dendritic polymers, in *Encyclopedia of Polymer Science Engineering*, 2nd edn. New York, John Wiley & Sons, 1990.

61. F. Zeng and S.C. Zimmerman. *Chem. Rev.* 97:1681–1712, 1997.

62. S.C. Zimmerman and L.J. Lawless. Supramolecular chemistry of dendrimers, in *Topics in Current Chemistry*, Vol. 217. F. Voegl, C.A. Scharrey, eds. Springer Verlag: Berlin, pp. 95–120, 2001.

63. O.A. Matthews, A.N. Shipway, and J.F. Stoddart. *Prog. Polym. Sci.* 23:1–56, 1998.

64. D.A. Tomalia. *Adv. Mater.* 6:529–539, 1994.

65. G.J. Kallos, D.A. Tomalia, D.M. Hedstrand, S. Lewis, and J. Zhou. *Rapid Commun. Mass Spectrom.* 5:383–386, 1991.

66. P.R. Dvornic and D.A. Tomalia. *Macromol. Symp.* 98:403–428, 1995.

67. J. Peterson, V. Allikmaa, J. Subbi, T. Pehk, and M. Lopp. *Eur. Polym. J.* 39:33–42, 2003.

68. C. Zhang and D.A. Tomalia. Gel electrophoresis characterization of dendritic polymers, in *Dendrimers and Other Dendritic Polymers*. J.M.J. Fréchet and D.A. Tomalia, Eds. John Wiley & Sons: West Sussex, pp. 239–252, 2001.

69. H.M. Brothers II, L.T. Piehler, and D.A. Tomalia. *J. Chromatogr. A* 814:233–246, 1998.

70. H. Staudinger. *From Organic Chemistry to Macromolecules*. Wiley-Interscience: New York, 1970.

71. D.A. Tomalia. *Nat. Mater.* 2:711–712, 2003.

72. D.A. Tomalia, H. Baker, J. Dewald, M. Hall, G. Kallos, S. Martin, J. Roeck, J. Ryder, and P. Smith. *Polym. J. (Tokyo)* 17:117–132, 1985.

73. D.A. Tomalia. *Sixth Biennial Carl S Marvel Symposium — Advances in Synthetic Polymer Chemistry.* Tucson, AZ, 1985.

74. D.A. Tomalia. *Akron Polymer Lecture Series*. Akron, OH, April 1984.

75. D.A. Tomalia, J.R. Dewald, M.J. Hall, S.J. Martin, and P.B. Smith. *Preprints of the 1st SPSJ International Polymer Conference.* Society of Polymer Science: Kyoto, Japan, August, p. 65, 1984.

76. D.A. Tomalia. *ACS Great Lakes/Central Regional Meeting*. Kalamazoo, MI, May 1984.

77. R. Esfand and D.A. Tomalia. Laboratory synthesis of poly(amidoamine) (PAMAM) dendrimers, in *Dendrimers and Other Dendritic Polymers*. J.M.J. Fréchet and D.A. Tomalia, Eds. John Wiley & Sons: West Sussex, pp. 587–604, 2001.

78. J.M.J. Fréchet, H. Ihre, and M. Davey. Preparation of Frechet-type polyether dendrons and aliphatic polyester dendrimers by convergent growth: an experimental primer, in *Dendrimers and Other Dendritic Polymers*. J.M.J. Fréchet and D.A. Tomalia, Eds. John Wiley & Sons: West Sussex, pp. 569–586, 2001.

79. D.A. Tomalia, B. Huang, D.R. Swanson, H.M. Brothers II, and J.W. Klimash. *Tetrahedron* 59:3799–3813, 2003.

80. M.R. Rauckhorst, P.J. Wilson, S.A. Hatcher, C.M. Hadad, and J.R. Parquette. *Tetrahedron* 59:3917–3923, 2003.

81. B. Huang and J.R. Parquette. *Org. Lett.* 2:239–242, 2000.

82. B. Huang and J.R. Parquette. *J. Am. Chem. Soc.* 123:2689–2690, 2001.

83. D.A. Tomalia, H.M. Brothers II, L.T. Piehler, H.D. Durst, and D.R. Swanson. *Proc. Natl. Acad. Sci. USA* 99(8):5081–5087, 2002.

84. W. Chen, N.J. Turro, and D.A. Tomalia. *Langmuir* 16:15–19, 2000.

85. P.E. Froehling and H.A.J. Linssen. *Macromol. Chem. Phys.* 199:1691–1695, 1998.

86. A.B. Padias, H.K. Hall Jr., and D.A. Tomalia. *J. Org. Chem.* 52:5305, 1987.

87. P.G. de Gennes and H.J. Hervet. *J. Physique-Lett. (Paris)* 44:351–360, 1983.

88. A.D. Meltzer, D.A. Tirrell, A.A. Jones, P.T. Inglefield, and D.M. Hedstrand. *Macromolecules* 25:4541, 1992.

89. L. Balogh and D.A. Tomalia. *J. Am. Chem. Soc.* 120:7355, 1998.
90. J.-A. He, R. Valluzzi, K. Yang, T. Dolukhanyan, C. Sung, J. Kumar, S.K. Tripathy, L. Samuelson, L. Balogh, and D.A. Tomalia. *Chem. Mater.* 11:3268–3274, 1999.
91. L. Balogh, R. Valluzzi, K.S. Laverdure, S.P. Gido, G.L. Hagnauer, and D.A. Tomalia. *J. Nanoparticle Res.* 1:353–368, 1999.
92. K.B. Mullis and F.A. Faloona. *Meth. Enzymol.* 155:335, 1987.
93. S.C. Zimmerman, F. Zeng, E.C. Reichert, and S.V. Kolotuchin. *Science* 271:1095–1098, 1996.
94. F. Zeng, S.C. Zimmerman, S.V. Kolotuchin, E.C. Reichert, and Y. Ma. *Tetrahedron* 58:825–843, 2002.
95. Y. Ma, S.V. Kolotuchin, and S.C. Zimmerman. *J. Am. Chem. Soc.* 124:13757–13769, 2002.
96. G.M. Dykes and D.K. Smith. *Tetrahedron* 59:3999–4009, 2003.
97. S. Hecht and J.M.J. Fréchet. *Angew. Chem. Int. Ed.* 40(1):74–91, 2001.
98. W. Guo, J. Li, A. Wang, and X. Peng. *J. Am. Chem. Soc.* 125:3901–3909, 2003.
99. K.R. Gopidas, J.K. Whitesell, and M.A. Fox. *J. Am. Chem. Soc.* 125:6491–6502, 2003.
100. M.-C. Daniel, J. Ruiz, S. Nlate, J.-C. Blais, and D. Astruc. *J. Am. Chem. Soc.* 125:2617–2628, 2003.
101. C.R. DeMattei, B. Huang, and D.A. Tomalia. *NanoLetters* 4:771–777, 2004.
102. F. Diederich and B. Felber. *Proc. Natl. Acad. Sci.* 99:4778–4781, 2002.
103. D.K. Smith and F. Diederich. *Topics Curr. Chem.* 210:183–227, 2000.
104. J.M.J. Fréchet. *Proc. Natl. Acad. Sci.* 99:4782–4787, 2002.
105. M. Conn and J. Rebek Jr. *Chem. Rev.* 97:1647–1668, 1997.
106. V. Balzani, G. Denti, S. Serroni, S. Campagna, V. Ricevuto, and A. Juris. *Proc. Indian Acad. Sci. Chem. Sci.* 105:421–434, 1993.
107. V. Balzani, S. Campagna, G. Denti, A. Juris, S. Serroni, and M. Venturi. *Coord. Chem. Rev.* 132:1–13, 1994.
108. V. Balzani, S. Campagna, G. Denti, A. Juris, S. Serroni, and M. Venturi. *Sol. Energy Mater. Sol. Cells* 38:159–173, 1995.
109. S. Campagna, G. Denti, S. Serroni, A. Juris, M. Venturi, V. Ricevuto, and V. Balzani. *Chem. Eur. J.* 1:211–221, 1995.
110. A. Juris, M. Venturi, L. Pontoni, I.R. Resino, V. Balzani, S. Serroni, S. Campagna, and G. Denti. *Can. J. Chem.* 16:1875–1882, 1995.
111. S. Serroni, G. Denti, S. Campagna, A. Juris, M. Ciano, and V. Balzani. *Angew. Chem. Int. Ed.* 31:1493–1495, 1992.
112. S. Serroni, S. Campagna, A. Juris, M. Venturi, and V. Balzani. *Gazz. Chim. Ital.* 124:423–427, 1994.
113. S. Achar and R.J. Puddephatt. *J. Chem. Soc., Chem. Commun.* (16):1895–1896, 1994.
114. S. Achar and R.J. Puddephatt. *Angew. Chem. Int. Ed.* 33:847–849, 1994.
115. S. Achar, J.J. Vittal, and R.J. Puddephatt. *Organometallics* 15:43–50, 1996.
116. W.T.S. Huck, F.C.J.M. VanVeggel, and D.N. Reinhoudt. *Angew. Chem. Int. Ed.* 35:1213–1215, 1996.
117. W.T.S. Huck, F.C.J.M. Van Veggel, B.L. Kropman, D.H.A. Blank, E.G. Keim, M.M.A. Smithers, and D.N. Reinhoudt. *J. Am. Chem. Soc.* 117:8293–8294, 1995.
118. W.T.S. Huck, R. Hulst, P. Timmerman, F.C.J.M. Van Veggel, and D.N. Reinhoudt. *Angew. Chem. Int. Ed.* 36:1006–1008, 1997.
119. W.T.S. Huck, F.C.J.M. Van Veggel, and D.N. Reinhoudt. *J. Mater. Chem.* 7:1213–1219, 1997.
120. W.T.S. Huck, L.J. Prins, R.H. Fokkens, N.M.M. Nibbering, F.C.J.M. Van Veggel, and D.N. Reinhoudt. *J. Am. Chem. Soc.* 120:6240–6246, 1998.
121. G.R. Newkome, Z.-Q. Yao, G.R. Baker, and V.K. Gupta. *J. Org. Chem.* 50:2003–2004, 1985.
122. J.L. Jackson, H.D. Chanzy, F.P. Booy, B.J. Drake, D.A. Tomalia, B.J. Bauer, and E.J. Amis. *Macromolecules* 31:6259–6265, 1998.
123. N.J. Turro, J.K. Barton, and D.A. Tomalia. *Acc. Chem. Res.* 24(11):332–340, 1991.
124. N.J. Turro, W. Chen, and M.F. Ottaviani. Characterization of dendrimer structures by spectroscopic techniques, in *Dendrimers and Other Dendritic Polymers*. J.M.J. Fréchet and D.A. Tomalia, Eds. John Wiley & Sons: West Sussex, pp. 309–330, 2001.
125. J.M. Lehn. *Proc. Natl. Acad. Sci.* 99:4763–4768, 2002.
126. J.C.M. van Hest. Ph.D. New Molecular Architetures Based on Dendrimers dissertation, Eindhoven University, The Netherlands, 1996.
127. P.E. Eaton and T.W. Cole. *J. Am. Chem. Soc.* 86:962–964, 1964.
128. P.E. Eaton, Y.S. Or, S.J. Branka, and B.K.R. Shanker. *Tetrahedron* 42:1621, 1986.
129. L.A. Paquette, R.J. Ternansky, D.W. Balogh, and G. Krentgen. *J. Am. Chem. Soc.* 105:5446, 1983.
130. D.J. Cram. *Science* 219:1177, 1983.

131. D.J. Cram and J.M. Cram. *Container Molecules and their Guests.* The Royal Society of Chemistry: Cambridge, England, 1994.
132. R.F. Curl and R.E. Smalley. *Science* 242:1017–1022, 1988.
133. M. Zhao, L. Sun, and R.M. Crooks. *J. Am. Chem. Soc.* 120:4877–4878, 1998.
134. N.C. Beck Tan, L. Balogh, S.F. Trevino, D.A. Tomalia, and J.S. Lin. *Polymer* 40:2537–2545, 1999.
135. L. Balogh, D.R. Swanson, D.A. Tomalia, G.L. Hagnauer, and A.T. McManus. *NanoLetters* 1(1):18–21, 2001.
136. R.G. Ispasoiu, L. Balogh, O.P. Varnavski, D.A. Tomalia, and T. Goodson III. *J. Am. Chem. Soc.* 122:11005–11006, 2000.
137. A.M. Naylor, W.A. Goddard III, G.E. Keifer, and D.A. Tomalia. *J. Am. Chem. Soc.* 111:2339–2341, 1989.
138. P.J. Gittins and L.J. Twyman. *Supramol. Chem.* 15:5–23, 2003.
139. A.E. Beezer, A.S.H. King, I.K. Martin, J.C. Mitchell, L.J. Twyman, and C.F. Wain. *Tetrahedron* 59:3873–3880, 2003.
140. C. Kojima, K. Kono, K. Maruyama, and T. Takagishi. *Bioconjugate Chem.* 11:910–917, 2000.
141. J.F.G.A. Jansen, E.W. Meijer, and E.M.M. de Brabander-van den Berg. *J. Am. Chem. Soc.* 117:4417–4418, 1995.
142. J.F.G.A. Jansen, E.M.M. de Brabander-van den Berg, and E.W. Meijer. *Science* 266:1226–1229, 1994.
143. E.M.M. de brabander-van den Berg, A. Nijenhuis, M. Mure, J. Keulen, R. Reintjens, F. Vandenbooren, B. Bosman, R. DeRaat, T. Frijns, and S. Wal. *Macromol. Symp.* 77:51–62, 1994.
144. D.A. Tomalia, H. Baker, J. Dewald, M. Hall, G. Kallos, S. Martin, J. Roeck, J. Ryder, and P. Smith. *Macromolecules* 19:2466–2468, 1986.
145. D.A. Tomalia, M. Hall, and D.M. Hedstrand. *J. Am. Chem. Soc.* 109:1601–1603, 1987.
146. R.G. Denkewalter, J.F. Kole, and W.J. Lukasavage. U.S. Pat. 4, 410688, 1983.
147. R.G. Denkewalter, J.F. Kole, and W.J. Lukasavage. *Chem. Abst.* 100, 103907, 1984.
148. S.M. Aharoni, C.R. Crosby III, and E.K. Walsh. *Macromolecules* 15:1093–1098, 1982.
149. S.M. Aharoni and N.S. Murthym. *Polym. Commun.* 24:132, 1983.
150. D.A. Tomalia, V. Berry, M. Hall, and D.M. Hedstrand. *Macromolecules* 20:1164–1167, 1987.
151. D.A. Tomalia, D.M. Hedstrand, L.R. Wilson, and D.M. Downing. Starburst dendrimers: size, shape and surface control of macromolecules, in *Frontiers of Macromolecular Science, 32nd IUPAC Proceedings.* T. Saegusa, T. Higashimura, and A. Abe, Eds. Blackwell Scientific Publications, Oxford, 1989.
152. G.R. Newkome, C.N. Moorfield, G.R. Baker, A.L. Johnson, and R.K. Behera. *Angew. Chem. Int. Ed.* 30(9):1176–1180, 1991.
153. G. Moreno-Bondi, G. Orellana, N.J. Turro, and D.A. Tomalia. *Macromolecules* 23:910–912, 1990.
154. K.R. Gopidas, A.R. Leheny, G. Caminati, N.J. Turro, and D.A. Tomalia. *J. Am. Chem. Soc.* 113:7335–7342, 1991.
155. D.M. Watkins, Y. Sayed-Sweeet, J.W. Klimash, N.J. Turro, and D.A. Tomalia. *Langmuir* 13:3136–3141, 1997.
156. D.M. Hedstrand, B.J. Helmer, and D.A. Tomalia. U.S. Pat. 5,560, 929, 1996.
157. Y. Sayed-Sweet, D.M. Hedstrand, R. Spindler, and D.A. Tomalia. *J. Mater. Chem.* 7(7):1199–1205, 1997.
158. S. Stevelmans, J.C.M. Van Hest, J.F.G.A. Jansen, D.A.F.J. Van Boxtel, E.M.M. de Brabander-van den Berg, and E.W. Meijer. *J. Am. Chem. Soc.* 118:7398–7399, 1996.
159. A.I. Cooper, J.D. Londono, G. Wignall, J.B. McClain, E.T. Samulski, J.S. Lin, A. Dobrynin, M. Rubinstein, A.L.C. Burke, J.M.J. Fréchet, and J.M. DeSimone. *Nature* 389:368–371, 1997.
160. R. Yin, Y. Zhu, and D.A. Tomalia. *J. Am. Chem. Soc.* 120:2678–2679, 1998.
161. D.B. Amabilino, P.R. Ashton, M. Belohradsky, F.M. Raymo, and J.F. Stoddart. *J. Chem. Soc., Chem. Commun.* 7:751–753, 1995.
162. D.B. Amabilino, P.R. Ashton, V. Balzani, C.L. Brown, A. Credi, and J.M. J. Fréchet. *J. Am. Chem. Soc.* 118:12012–12020, 1996.
163. F.M. Menger and C.A. Littau. *J. Am. Chem. Soc.* 115:10083–10090, 1993.
164. E.H. Lu, E. Reynders. *Anionic Surfactants, Physical Chemistry of Surfactant Action. Surfactant Series,* Vol. 11. Marcel Dekker: New York, 1981.
165. S. Buckingham, C. Garvey, and G. Warr. *J. Phys. Chem.* 97:10236–10244, 1993.
166. J. Israelachvili, D. Mitchell, and B. Ninham. *J. Chem. Soc., Faraday Trans. II* 72:1525, 1976.
167. S.E. Friberg, M. Podzimek, D.A. Tomalia, and D.M. Hedstrand. *Mol. Cryst. Liq. Cryst.* 164:157–165, 1988.
168. H. Smith and D.A. Tomalia. U.S. Patent 5,331,100, 1994.

169. T. Chapman, G. Hillyer, E. Mahan, and K. Schaffer. *J. Am. Chem. Soc.* 116:11195–11196, 1994.

170. I. Gitsov, K.L. Wooley, and J.M.J. Fréchet. *Angew. Chem. Int. Ed.* 31(9):1200–1202, 1992.

171. I. Gitsov, K.L. Wooley, C.J. Hawker, P.T. Ivanova, and J.M.J. Fréchet. *Macromolecules* 26:5621–5627, 1993.

172. C.J. Hawker, K.L. Wooley, and J.M.J. Fréchet. *J. Chem. Soc., Perkin Trans.* I:1287–1297, 1993.

173. K.L. Wooley, C.J. Hawker, and J.M.J. Fréchet. *J. Am. Chem. Soc.* 115:11496–11505, 1993.

174. J.N. Israelachvili, D. Mitchell, and B. Ninham. *Biophys. Acta* 470:150, 1977.

175. J.N. Israelachvili, S. Marcelja, and R. Horn. *Rev. Biophys.* 13:121, 1980.

176. G.R. Newkome, G.R. Baker, S. Arai, M.J. Saunders, P.S. Russo, K.J. Theriot, C.N. Moorefield, L.E. Rogers, J.E. Miller, T.R. Lieux, M.E. Murray, B. Philips, and L. Pascal. *J. Am. Chem. Soc.* 112:8458–8465, 1990.

177. D.A. Tomalia and P.M. Kirchoff. U.S. Patent 4,694,064, 1987.

178. I. Neubert, E. Amoulong-Kirstein, and A.-D. Schluter. *Macromol. Rapid Commun.* 17:517–527, 1996.

179. I. Neubert, R. Klopsch, W. Claussen, and A.-D. Schluter. *Acta Polym.* 47:455–459, 1996.

180. B. Karakaya, W. Claussen, K. Gessler, W. Saenger, and A.-D. Schluter. *J. Am. Chem. Soc.* 199:3296–3301, 1997.

181. W. Stocker, B. Karakaya, L.B. Schurmann, P.J. Rabe, and A.-D. Schluter. *J. Am. Chem. Soc.* 120:7691–7695, 1998.

182. J.M.J. Fréchet and I. Gitsov. *Macromol. Symp.* 98:441–465, 1995.

183. A.-D. Schluter and P.J. Rabe. *Angew. Chem. Int. Ed.* 39:865–883, 2000.

184. G. Johansson, V. Percec, G. Ungar, and J.P. Zhou. *Macromolecules* 29:646–660, 1996.

185. V. Percec, D. Schlueter, J.C. Ronda, G. Johansson, G. Ungar, and J.P. Zhou. *Macromolecules* 29:1464–1472, 1996.

186. V. Percec and D. Schlueter. *Macromolecules* 30:5783–5790, 1997.

187. D. Demus. *Mol. Cryst. Liq. Cryst.* 364:25–91, 2001.

188. S. Bauer, H. Fischer, and H. Ringsdorf. *Angew. Chem. Int. Ed.* 32:1589, 1993.

189. V. Percec, P. Chu, G. Johansson, D. Schlueter, J.C. Ronda, and G. Ungar. *Polym. Preprint* 37:68, 1996.

190. V. Percec and M. Kawasumi. *Macromolecules* 25:3843, 1992.

191. V. Percec, P. Chu, and M. Kawasumi. *Macromolecules* 27:4441, 1994.

192. V. Percec, P. Chu, G. Ungar, and J. Zhou. *J. Am. Chem. Soc.* 117:11441, 1995.

193. V. Percec, G. Johansson, G. Ungar, and J.P. Zhou. *J. Am. Chem. Soc.* 118:9855–9866, 1996.

194. S.D. Hudson, H.-T. Jung, V. Percec, W.-D. Cho, G. Johansson, G. Ungar, and V.S.K. Balagurusamy. *Science* 278:449–452, 1997.

195. J.H. Cameron, A. Facher, G. Latterman, and S. Diele. *Adv. Mater.* 9:398–403, 1997.

196. J. Barbera, M. Marcos, and J.L. Serrano. *Chem. Eur. J.* 5:1834–1840, 1999.

197. J. Barbera, M. Marcos, A. Omenat, J.L. Serrano, J.I. Martinez, and P.J. Alonso. *Liq. Cryst.* 27:255–262, 2000.

198. M. Marcos, R. Gimenez, J.L. Serrano, B. Donnio, B. Heinrich, and D. Guillon. *Chem. Eur. J.* 7:1006–1013, 2001.

199. J.-M. Rueff, J. Barbera, B. Donnio, D. Guillon, M. Marcos, and J.L. Serrano. *Macromolecules* 36:8368–8375, 2003.

200. V. Percec, C.-H. Ahn, G. Unger, D.J.P. Yeardly, and M. Moller. *Nature* 391:161–164, 1998.

201. X. Zeng, G. Ungar, Y. Liu, V. Percec, A.E. Dulcey, and J.K. Hobbs. *Nature* 428:157–160, 2004.

202. A. Sunder, M.F. Quincy, R. Mulhaupt, and H. Frey. *Angew. Chem. Int. Ed.* 38:2928–2930, 1999.

203. S.A. Ponomarenko, E.A. Rebrov, N.I. Boiko, N.G. Vasilenko, A.M. Muzafarov, Y.S. Freidzon, and V.P. Shibaev. *Polym. Sci. Ser. A* 36:896–901, 1994.

204. S.A. Ponomarenko, E.A. Rebrov, Y. Bobronsky, N.I. Boiko, A.M. Muzafarov, and V.P. Shibaev. *Liq. Cryst.* 21:1–12, 1996.

205. K. Lorenz, D. Holter, R. Mulhaupt, and H. Frey. *Adv. Mater.* 8:414–416, 1996.

206. M.C. Coen, K. Lorenz, J. Kressler, H. Frey, and R. Mulhaupt. *Macromolecules* 29:8069–8076, 1996.

207. K. Lorenz, H. Frey, B. Stuhn, and R. Mulhaupt. *Macromolecules* 30:6860–6868, 1997.

208. B. Stark, B. Stuhn, H. Frey, C. Lach, K. Lorenz, and B. Frick. *Macromolecules* 31:5415–5423, 1998.

209. S.A. Ponomarenko, E.A. Rebrov, N.I. Boiko, A.M. Muzafarov, and V.P. Shibaev. *Polym. Sci. Ser. A* 40:763–774, 1998.

210. E.I. Ryumtsev, N.P. Evlampieva, A.V. Lezov, S.A. Ponomarenko, N.I. Boiko, and V.P. Shibaev. *Liq. Cryst.* 25:475–476, 1998.

211. B. Trahasch, B. Stuhn, H. Frey, and K. Lorenz. *Macromolecules* 32:1962–1966, 1999.

212. R.M. Richardson, S.A. Ponomarenko, N.I. Boiko, and V.P. Shibaev. *Liq. Cryst.* 26:101–108, 1999.

213. D. Terunuma, T. Kato, R. Nishio, Y. Aoki, M. Nohira, K. Matsuoka, and H. Kuzuhara. *Bull. Chem. Soc. Jpn.* 72:2129–2134, 1999.

214. R.M. Richardson, I.J. Whitehouse, S.A. Ponomarkeno, N.I. Boiko, and V.P. Shibaev. *Mol. Cryst. Liq. Cryst.* 330:167–174, 1999.

215. S.A. Ponomarkenko, N.I. Boiko, E.A. Rebrov, A.M. Muzafarov, I.J. Whitehouse, R.M. Richardson, and V.P. Shibaev. *Mol. Cryst. Liq. Cryst.* 332:43–50, 1999.

216. S.A. Ponomarenko, N.I. Boiko, V.P. Shibaev, R.M. Richardson, I.J. Whitehouse, E.A. Rebrov, and A.M. Muzafarov. *Macromolecules* 33:5549–5558, 2000.

217. S.A. Ponomarenko, N.I. Boiko, V.P. Shibaev, and S.N. Magonov. *Langmuir* 16:5487–5493, 2000.

218. N.I. Boiko, X. Zhu, R. Vinokur, E.A. Rebrov, A.M. Muzafarov, and V.P. Shibaev. *Ferroelectrics* 243:59–66, 2000.

219. K. Suzuki, O. Haba, R. Nagahata, K. Yonetake, and M. Ueda. *High Perform Polym.* 10:231–240, 1998.

220. U. Stebani and G. Lattermann. *Adv. Mater.* 7:578–581, 1995.

221. M. Seitz, T. Plesnivy, K. Schimossek, M. Edelmann, H. Ringsdorf, H. Fischer, H. Uyama, and S. Kobayashi. *Macromolecules* 29:6560–6574, 1996.

222. M.W.P.L. Baars, S.H.M. Sontjens, H.M. Fischer, H.W.I. Peerlings, and E.W. Meijer. *Chem. Eur. J.* 4:2456–2466, 1998.

223. K. Yonetake, K. Suzuki, T. Morishita, R. Nagahata, and M. Ueda. *High Perform Polym.* 10:373–382, 1998.

224. K. Yonetake, T. Masuko, T. Morishita, K. Suzuki, M. Ueda, and R. Nagahata. *Macromolecules* 32:6578–6586, 1999.

225. D.L.D. Gasper. *Biophys. J.* 32:103–138, 1980.

226. J.D. Watson. *Molecular Biology of the Gene.* W. A. Benjamin, Inc., Menlo Park, Ca, 1976.

227. A.J. Levine. *Viruses.* W. H. Freeman: New York, 1992.

228. V. Percec, C.-H. Ahn, W.-D. Cho, G. Johansson, and D. Schlueter. *Macromol. Symp.* 118:33–43, 1997.

229. V. Percec, C.-H. Ahn, and B. Barboiu. *J. Am. Chem. Soc.* 119:12978–12979, 1997.

230. J.W. Lee, Y.H. Ko, S.-H. Park, K. Yamaguchi, and K. Kim. *Angew. Chem. Int. Ed.* 40:746–749, 2001.

231. L. Kiessling and N.L. Pohl. *Chem. Biol.* 3:71–77, 1996.

232. M. Mammen, S.-K. Choi, and G.M. Whitesides. *Angew. Chem. Int. Ed.* 37:2754–2794, 1998.

233. S.K. Choi, M. Mammen, and M. Whitesides. *Chem. Biol.* 3:97–104, 1996.

234. K.H. Mortell, M. Gingras, and L.L. Kiessling. *J. Am. Chem. Soc.* 116:12053–12054, 1994.

235. C. Fraser and R.H. Grubbs. *Macromolecules* 28:7248–7255, 1995.

236. K.H. Mortell, R.V. Weathermen, and L.L. Kiessling. *J. Am. Chem. Soc.* 118:2297–2298, 1996.

237. R. Roy and J.M. Kim. *Tetrahedron* 59:3881–3893, 2003.

238. R. Roy, D. Zanini, S.J. Meunier, and A. Romanowska. *J. Chem. Soc. Commun.* 24:1869–1872, 1993.

239. R. Roy, W.K.C. Park, Q. Wu, and S.N. Wany. *Tetrahedron* 36:4377–4380, 1995.

240. R. Roy. *Curr. Opin. Struct. Biol.* 6:692–702, 1996.

241. R. Roy and J.M. Kim. *Angew. Chem. Int. Ed.* 38:369–372, 1999.

242. J.D. Reuter, A. Myc, M.M. Hayes, Z. Gan, R. Roy, D. Qin, R. Yin, L.T. Piehler, R. Esfand, D.A. Tomalia, and J.R. Baker Jr. *Bioconjugate Chem.* 10:271–278, 1999.

243. D.A. Tomalia, D.M. Hedstrand, and M.S. Ferrito. *Macromolecules* 24:1435–1438, 1991.

244. M. Witvrouw, V. Fikkert, W. Pluymers, B. Matthews, K.W. Mardel, D. Schols, J. Raff, Z. Debyser, E. DeClercq, G. Holan, and C. Pannecouque. *Mol. Pharmacol.* 58:1100–1108, 2000.

245. G. Holan, B. Matthews, B. Korba, E. DeClerq, M. Witvrouw, E. Kern, R. Sidwell, D. Barnard, and J. Huffman. *Antiviral Res.* 46:A55, 2000.

246. M. Witvrouw, C. Pannecouque, B. Matthews, D. Schols, G. Andrei, R. Snoek, J. Neyts, J. Desmyter, E. De Clercq, and G. Holan. *Antiviral Res.* 34:A88, 1997.

247. G. Holan. personal communication.

248. C. Virgona. personal communication.

249. M. Witvrouw, C. Pannecouque, B. Matthews, D. Schols, G. Andrei, R. Snoeck, J. Neyts, P. Leyssen, J. Desmyter, J. Raff, E. DeClerq, and G. Holan. *Antiviral Res.* 41:A41, 1999.

250. N. Bourne, L.R. Stanberry, E.R. Kern, G. Holan, B. Matthews, and D.I. Bernstein. *Antimicrob. Agents Chemother.* 44:2471–2474, 2000.

251. D.L. Barnard, R.W. Sidwell, T.L. Gage, K.M. Okleberry, B. Matthews, and G. Holan. *International Conference on Antiviral Research (ICAR).* Atlanta, GA, U.S.A, 1997.

252. D.L. Barnard, J.E. Matheson, A. Morrison, R.W. Sidwell, J.H. Huffman, B. Matthews, and G. Holan. *Proceedings of the Interscience Conference on Antimicrobial Agents and Chemotherapy (ICAAC)* San Diego, Ca, 1998.

253. P.M. Saville, J.W. White, C.J. Hawker, K.L. Wooley, and J.M.J. Fréchet. *J. Phys. Chem.* 97:293–294, 1993.

254. P.M. Saville, P. Reynolds, J.W. White, C.J. Hawker, J.M.J. Fréchet, K.L. Wooley, J. Pemford, and J.R.P. Webster. *J. Phys. Chem.* 99:8283–8289, 1995.

255. J.P. Kampf, C.W. Frank, E.E. Malmstrom, and C.J. Hawker. *Langmuir* 15:227–233, 1999.

256. J.P. Kampf, C.W. Frank, E.E. Malmstrom, and C.J. Hawker. *Science* 282:1730–1733, 1999.

257. A.P.H.J. Schenning, C. Ellisen-Roman, J.-W. Weener, M.W.P.L. Baars, S.J. van der Gast, and E.J. Meijer. *J. Am. Chem. Soc.* 120:8199–8208, 1998.

258. A.K. Naj. *The Wall Street Journal.* New York, p. B1, vol. LXXVII No. 92, Feb. 26, 1996.

259. D.A. Tomalia and G.R. Killat. U.S. Patent 4,871,779, 1989.

260. M.L. Mansfield. *Polymer* 37:3835–3841, 1996.

261. S. Watanabe and S.L. Regen. *J. Am. Chem. Soc.* 116:8855–8856, 1994.

262. M. Zhao, H. Tokuhisa, and R.M. Crooks. *Angew. Chem. Int. Ed.* 36:2596–2598, 1997.

263. A. Hierlemann, J.K. Campbell, L.A. Baker, R.M. Crooks, and A.J. Ricco. *J. Am. Chem. Soc.* 120:5323–5324, 1998.

264. V.V. Tsukruk, F. Rinderspacher, and V.N. Bliznyuk. *Langmuir* 13:2171–2176, 1997.

265. V. Tsukruk. *Adv. Mater.* 10:253–257, 1998.

266. V.N. Bliznyuk, F. Rinderspacher, and V.V. Tsukruk. *Polymer* 39:5249–5252, 1998.

267. T. Emrick and J.M.J. Fréchet. *Curr. Opin. Coll. Interface Sci.* 4:15–23, 1999.

268. D.A. Tomalia, S. Uppuluri, D.R. Swanson, and J. Li. *Pure Appl. Chem.* 72:2343–2358, 2000.

269. N. Tomioka, D. Takasu, T. Takahashi, and T. Aida. *Angew. Chem. Int. Ed.* 37:1531–1534, 1998.

270. L.L. Miller, R.G. Duan, D.C. Tully, and D.A. Tomalia. *J. Am. Chem. Soc.* 119:1005–1010, 1997.

271. F. Wang, R.D. Rauh, and T.L. Rose. *J. Am. Chem. Soc.* 119:11106–11107, 1997.

272. B.L. Frankamp, A.K. Boal, and V.M. Rotello. *J. Am. Chem. Soc.* 124:15146–15147, 2002.

273. G. Galileo. *Dialogue Concerning the Two Chief World Systems: Ptolemaic and Copernican.* University of California Press: Berkeley, CA, 1967.

274. N. Bohr. *Nobel Lecture, The Structure of the Atom.* 1922. http:/www.nobel/.sc/physics/laureates/1922/bohrlecture.html

275. T. Martin, U. Obst, and J. Rebek Jr. *Science* 281.1842–1845, 1998.

276. X. Camps, H. Schonberger, and A. Hirsch. *Chem. Eur. J.* 3(4):561–567, 1997.

277. D.A. Tomalia. *Aldrichim. Acta* 26(4):91–101, 1993.

278. D.A. Tomalia and P.R. Dvornic. Dendritic polymers, divergent synthesis (starburst polyamidoamine dendrimers), in *Polymeric Materials Encyclopedia.* J.C. Salamone, Ed., Vol. 3(D–E). CRC Press: Boca Raton, FL, pp. 1814–1840, 1996.

279. B.J. Bauer, A. Topp, T.J. Prosa, E.J. Amis, R. Yin, D. Qin, and D.A. Tomalia. *Polym. Mater. Sci. Eng. (ACS)* 77:87–88, 1997.

280. A. Topp, B.J. Bauer, J.W. Klimash, R. Spindler, and D.A. Tomalia. *Macromolecules* 32:7226–7231, 1999.

281. A. Topp, B.J. Bauer, D.A. Tomalia, and E.J. Amis. *Macromolecules* 32:7232–7237, 1999.

282. M.L. Mansfield, L. Rakesh, and D.A. Tomalia. *J. Chem. Phys.* 105:3245–3249, 1996.

283. S. Uppuluri, L.T. Piehler, J. Li, D.R. Swanson, G.L. Hagnauer, and D.A. Tomalia. *Adv. Mater.* 12(11):796–800, 2000.

284. D.A. Tomalia and R. Esfand. Laboratory synthesis of poly(amidoamine) (PAMAM) dendrimers, in *Dendrimers and Other Dendritic Polymers.* J.M.J. Frechet and D.A. Tomalia, Eds. John Wiley & Sons, Ltd.: West Sussex, pp. 587–602, 2001.

285. D.A. Tomalia and D.R. Swanson. Laboratory synthesis and characterization of megamers: core-Shell Tecto(dendrimers), in *Dendrimers and Other Dendritic Polymers.* J.M.J. Fréchet and D.A. Tomalia, Eds. John Wiley & Sons: West Sussex, pp. 617–629, 2001.

286. D.A. Tomalia, K. Mardel, S.A. Henderson, G. Holan, and R. Esfand. Dendrimers — An enabling synthetic science to controlled organic nanostructures, in *Handbook of Nanoscience, Engineering and Technology.* W.A. Goddard III, D.W. Brenner, S.E. Lyshevski, and G.J. Iafrate, Eds. CRC Press: Boca Raton, FL, pp. 1–34, 2003.

287. S. Uppuluri, S.E. Keinath, D.A. Tomalia, and P.R. Dvornic. *Macromolecules* 31:4498–4510, 1998.

288. P.R. Dvornic and S. Uppuluri. Rheology and solution properties of dendrimers, in *Dendrimers and Other Dendritic Polymers*. J.M.J. Fréchet and D.A. Tomalia, Eds. John Wiley & Sons: West Sussex, pp. 331–358, 2001.

289. D.A. Tomalia. *Aldrichim. Acta* 37(2):39–57, 2004.

290. D.A. Tomalia. Nanoscopic modules for the construction of higher ordered complexity, in *Modular Chemistry, Proceedings of the NATO Advanced Research Workshop on Modular Chemistry*. J. Michl, Ed. Kluwer Publishers: The Netherlands, pp. 183–191, 1997.

291. P.R. Dvornic and D.A. Tomalia. *Macromol. Symp.* 88:123–148, 1994.

292. Z. Guan, P.M. Cotts, E.F. McCord, and S.J. McLain. *Science* 283:2059–2062, 1999.

293. S.C. Lee et al. Antibodies to PAMAM dendrimers: reagents for immune detection, patterning and assembly of dendrimers, in *Dendrimers and Other Dendritic Polymers*. J.M.J. Fréchet and D.A. Tomalia, Eds. John Wiley & Sons: West Sussex, pp. 559–565, 2001.

294. S.C. Lee et al. Dendrimers in nanobiological devices, in *Dendrimers and Other Dendritic Polymers*, J.M.J. Fréchet and D.A. Tomalia, Eds. John Wiley & Sons: West Sussex, pp. 547–554, 2001.

295. P. Singh. *Bioconjugate Chem.* 9:54–63, 1998.

296. P. Singh. Dendrimer-based biological reagents: preparation and applications in diagnostics, in *Dendrimers and Dendritic Polymers*. J.M.J. Fréchet and D.A. Tomalia, Eds. John Wiley & Sons Ltd.: West Sussex, pp. 463–484, 2001.

297. K.K. Ong, A.L. Jenkins, R. Cheng, D.A. Tomalia, H.D. Durst, J.L. Jensen, P.A. Emanuel, C.R. Swim, and R. Yin. *Anal. Chim. Acta* 444:143–148, 2001.

298. A. Bielinska, J.F. Kukowska-Latallo, J. Johnson, D.A. Tomalia, and J. Baker Jr. *Nucl. Acid Res.* 24:2176–2182, 1996.

299. J.F. Kukowska-Latallo, A.U. Bielinska, J. Johnson, R. Spindler, D.A. Tomalia, and J.R. Baker Jr. *Proc. Natl. Acad. Sci. USA* 93:4897–4902, 1996.

300. J.D. Eichman, A.U. Bielinska, J.F. Kukowska-Latallo, B.W. Donovan, and J.R. Baker Jr. Bioapplications of PAMAM dendrimers, in *Dendrimers and Other Dendritic Polymers*. J.M.J. Fréchet and D.A. Tomalia, Eds. John Wiley & Sons Ltd.: West Sussex, pp. 441–461, 2001.

301. E.C. Wiener, M.W. Brechbiel, H.M. Brothers II, R.L. Magin, O.A. Gansow, D.A. Tomalia, and P.C. Lauterbur. *Magn. Reson. Med.* 31:1–8, 1994.

302. A.T. Yordanov, A. Lodder, E. Woller, M. Cloninger, N. Patronas, D. Milenic, and M.W. Brechbiel. *NanoLetters* 2:595–599, 2002.

303. H. Kobayashi, S. Kawamoto, S.-K. Jo, M.W. Brechbiel, and R.A. Star. *Bioconjugate Chem.* 14:388–394, 2003.

304. H. Kobayashi, S. Kawamoto, N. Sato, P.L. Choyke, M.V. Knopp, R.A. Star, T.A. Waldmann, Y. Tagaya, and M.W. Brechbiel. *Magn. Reson. Med.* 50:758–766, 2003.

305. D.S. Wilbur, P.M. Pathare, D.K. Hamlin, K.R. Buhler, and R.L. Vesella. *Bioconjugate Chem.* 9:813–825, 1998.

306. R. Esfand and D.A. Tomalia. *Drug Discov. Today* 6(8):427–436, 2001.

307. J.R. Baker Jr., L. Quintana, L. Piehler, M. Banazak-Holl, D. Tomalia, and E. Raczka. *Biomed. Microdev.* 3:61–69, 2001.

308. A.H. Soloway, W. Tjarks, B.A. Barnum, F.-G. Rong, R.F. Barth, I.M. Codogni, and J.G. Wilson. *Chem. Rev.* 98:1515–1562, 1998.

309. C. Bieniarz. Dendrimers: applications to pharmaceutical and medicinal chemistry, in *Encyclopedia of Pharmaceutical Technology*, Vol. 18. Marcel Dekker: New York, pp. 55–89, 1998.

310. D.A. Tomalia and H.M. Brothers II. Regiospecific conjugation to dendritic polymers to produce nano-devices, in *Biological Molecules in Nanotechnology.*, S. Lee and L.M. Savage, Eds., Vol. 1927. IBC Library Series: Southborough, MA, pp. 107–120, 1998.

311. M.E. Hermes. *Enough for One Lifetime: Wallace Carothers, Inventor of Nylon.* American Chemical Society and Chemical Heritage Foundation, 1996. U.S.A.

312. D.A. Tomalia. *Mater. Today* (December) 72, 2003. (No. vol.)

313. N. Zimmerman, J.S. Moore, and S.C. Zimmerman. *Chem. Indust.* (August 3) 604–610, 1998.

314. M. Kawa and J.M.J. Frechet, *Chem. Mater.* 10:286–296, 1998.

315. J. Zheng, C. Zhang, P.M. Dickson, *Phys. Rev. Lett.* 93(7):677402–1, 2004.

Chapter 8

Polymers with Interwined Superstructures and Interlocked Structures*

J. Benjamin Beck and Stuart J. Rowan

Contents

I. Mechanically Interlocked Macromolecules

Macromolecules incorporating repeat units connected by covalent bonds are widespread in nature [1]. Synthetic procedures for the construction of their artificial counterparts are well established [2]. Furthermore, the properties of these unnatural macromolecules are now rather well understood [2] and, indeed, polymeric materials have found [2] applications in numerous branches of science and technology. In recent years, synthetic chemists have learnt [3] how to introduce mechanical bonds (Figure 1) into small molecules. Mechanically interlocked "rings," as well as "wheels" mechanically trapped onto "axles," can be constructed efficiently to afford molecular compounds, named catenanes and rotaxanes, respectively. (The term *catenane* derives from the Latin word *catena* meaning chain. The term *rotaxane* derives from the Latin words *rota* and *axis* meaning wheel and axle, respectively.) Metal coordination [4,5], donor/acceptor interactions [6,7], hydrogen bonds [8,9], and/or hydrophobic interactions [10] among appropriate components have all been

---

*This chapter is an update and revision of Chapter 8 which appeared in the first edition of this book and was written by J.F. Stoddart and F.M. Raymo.

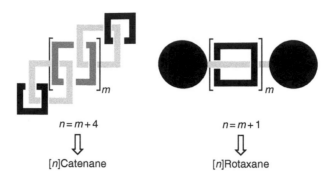

$n = m + 4$ $n = m + 1$

⇓   ⇓

[*n*]Catenane   [*n*]Rotaxane

Figure 1   Schematic representation of an [*n*]catenane and an [*n*]rotaxane.

employed to template [11] the formation of these exotic molecules. Making the transition from simple catenanes and rotaxanes to their macromolecular counterparts — namely, polycatenanes and polyrotaxanes, respectively — offers the possibility of generating a range of novel polymeric materials. Indeed, the fundamental difference between "conventional" macromolecules and these "unconventional" polymers — that is, the presence of mechanical bonds — is expected to confer unusual properties upon them, which could have profound technological implications. Mechanically interlocked macromolecules can be regarded as the molecular-sized counterparts of macroscopic objects, such as abacuses, bearings, chains, and joints, which could become the components of some of the smallest possible devices sometime in the future.

II. POLYCATENANES

Despite some preliminary and highly speculative claims [12], the synthesis of a (linear) [*n*]catenane (Figure 1), incorporating more than five mechanically interlocked rings is still to be achieved. A compound called Olympiadane [13] — that is, a [5]catenane — has been isolated and fully characterized following a two-step procedure (Figure 2) that relies upon the mutual recognition which exists between the components in contiguous rings. Reaction of **1** and **2** in the presence of the macrocyclic polyether **3** as a template leads to the [3]catenane **4** which was isolated in a yield of 10%, after counterion exchange. A small amount of a [2]catenane incorporating cyclobis(paraquat-4,4'-phenylene) and only one macrocyclic polyether component was obtained. Each of the two macrocyclic polyether rings present in **4** is large enough to accommodate, within its cavity, at least one other bipyridinium unit belonging to a cyclophane. Thus, the reaction of **5** and **6**, in the presence of the [3]catenane **4** as a template affords the [5]catenane **7**, with its five mechanically interlocked rings related to each other in a linear fashion, in a yield of 18%. In addition, a [4]catenane, incorporating two different tetracationic cyclophanes and two identical macrocyclic polyether components, was also obtained in a yield of 51%. When the same reaction was carried out under ultrahigh pressure (12 kbar), the yield of the [5]catenane **7** rose to 30%. Furthermore, a [6]catenane and a [7]catenane were also obtained in yields of 28 and 26%, respectively. These higher *branched* catenanes incorporate one and two, respectively, additional cyclobis(paraquat-*p*-phenylene) cyclophane components. The [5]catenane **7** was characterized by liquid secondary ionization mass spectrometry (LSIMS) that revealed signals resulting from the loss of hexafluorophosphate counterions as well as from the loss of the component macrocycles. The mechanical interlocking of the five ring components in this [5]catenane was demonstrated unequivocally by x-ray crystallography. Additionally, the x-ray structural analysis of **7** revealed that the [$\pi \cdots \pi$] stacking interactions between the π-donors and π-acceptors are accompanied by [C–H $\cdots$ O] hydrogen bonds between selected polyether oxygen atoms and certain α-bipyridinium hydrogen atoms, as well as by [C–H $\cdots \pi$] interactions between

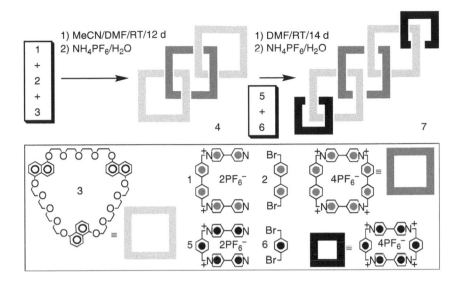

Figure 2   The two-step template-directed synthesis of the [5]catenane **7**.

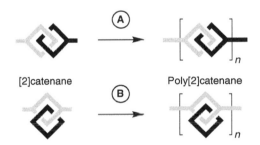

Figure 3   Synthetic strategies for the preparation of poly[2]catenanes.

some of the 1,5-dioxynaphthalene hydrogen atoms and the phenylene rings in the tetracationic cyclophane components.

Poly[2]catenanes (Figure 3) are macromolecules composed of [2]catenane repeating units linked by covalent bonds. The syntheses of poly[2]catenanes have been realized according to the procedures **A** or **B**, depending on whether the starting [2]catenane bears one reactive group on each of the two interlocked rings or two reactive groups on the same ring. [2]Catenanes, incorporating one reactive group attached to each of the two interlocked rings, have been obtained on account of the templating which result from the hydrogen-bonding interactions between the amide functions present in the two rings. Indeed, upon reaction [14] of the bisamine **8** with the bisacid chloride **9**, the [2]catenane **10**, which incorporates an aryl bromine substituent on each of its two ring components, self-assembles (Figure 4) in a yield of 9%. Coupling of this bisfunctionalized [2]catenane in turn with the bisfunctionalized reagent **11** and **12** affords the poly[2]catenanes **13** and **14**, in yields of 84 and 99%, respectively. Gel permeation chromatographic (GPC) analysis of **13** and **14** revealed number-average molecular weights (M_n) of 3.0×10^3 and 3.3×10^3, respectively, with the respective weight-average molecular weights M_w being 3.6×10^3 and 5.0×10^3. Differential scanning calorimetry (DSC) revealed a glass transition at ca. 245°C, even though these polymers are only of moderate molecular sizes.

A similar approach was employed [15] in the construction (Figure 5) of the poly[2]catenanes **19a** and **19b**. The bisamine **15** was reacted with the bisacid chloride **16** in the presence of Et$_3$N to afford the corresponding [2]catenane. Methylation of the amide groups, followed by the hydrolysis

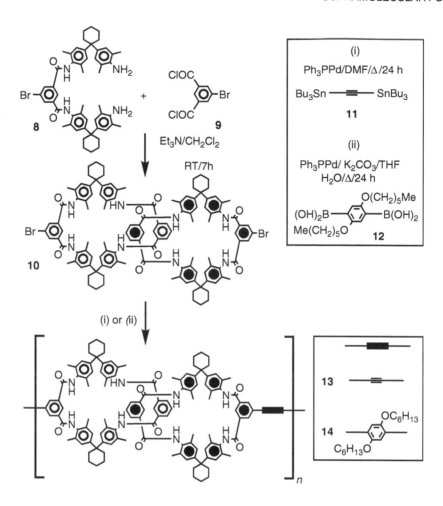

Figure 4 The hydrogen-bonding assisted template-directed synthesis of the [2]catenane **10** and its conversion into the poly[2]catenanes **13** and **14**.

of the benzyloxy groups, afforded the two isomeric [2]catenanes **17a** and **17b** in a yield of 26% and 30%, respectively. It should be noted that these catenanes are conformationally frozen with respect to the rotational mobility of the rings which in turn allowed these two isomeric interlocked species to be separated. **17a** has a conformation in which the hydroxyl substitutent on one of the rings is pointing in to the catenane structure while the other hydroxyl group is pointing out of the catenane structure (IN-OUT). **17b** on the other hand has both hydroxyl groups pointing out of the catenane structure (OUT-OUT). Copolyesterification of the terephthalic acid derivative **18** with either of these compounds under mild conditions afforded the poly[2]catenanes **19a** and **19b**. The formation of ester bonds was confirmed by infrared spectroscopy (IR) which reveals a band characteristic for the [C=O] group of the ester functions at $1754\,cm^{-1}$. The molecular weight distribution was investigated by GPC using a universal calibration. These investigations revealed M_n and M_w values of 25.5×10^3 and 60.9×10^3 for **19a** and 41.5×10^3 and 96.8×10^3 for **19b**, DSC analysis of the phase transition behavior revealed high glass transition temperatures, $T_g = 277°C$ and $207°C$, for **19a** and **19b**, respectively along with a very high thermal stability ($380°C$). Estimations of the Kuhn segment length of these polymers from GPC and viscometry data suggest that 19a ($l_k = 27$Å) forms a more compact structure than 19b ($l_k = 44$Å) which is presumably a result of the difference in geometric shapes of the two rigid catenane repeat units.

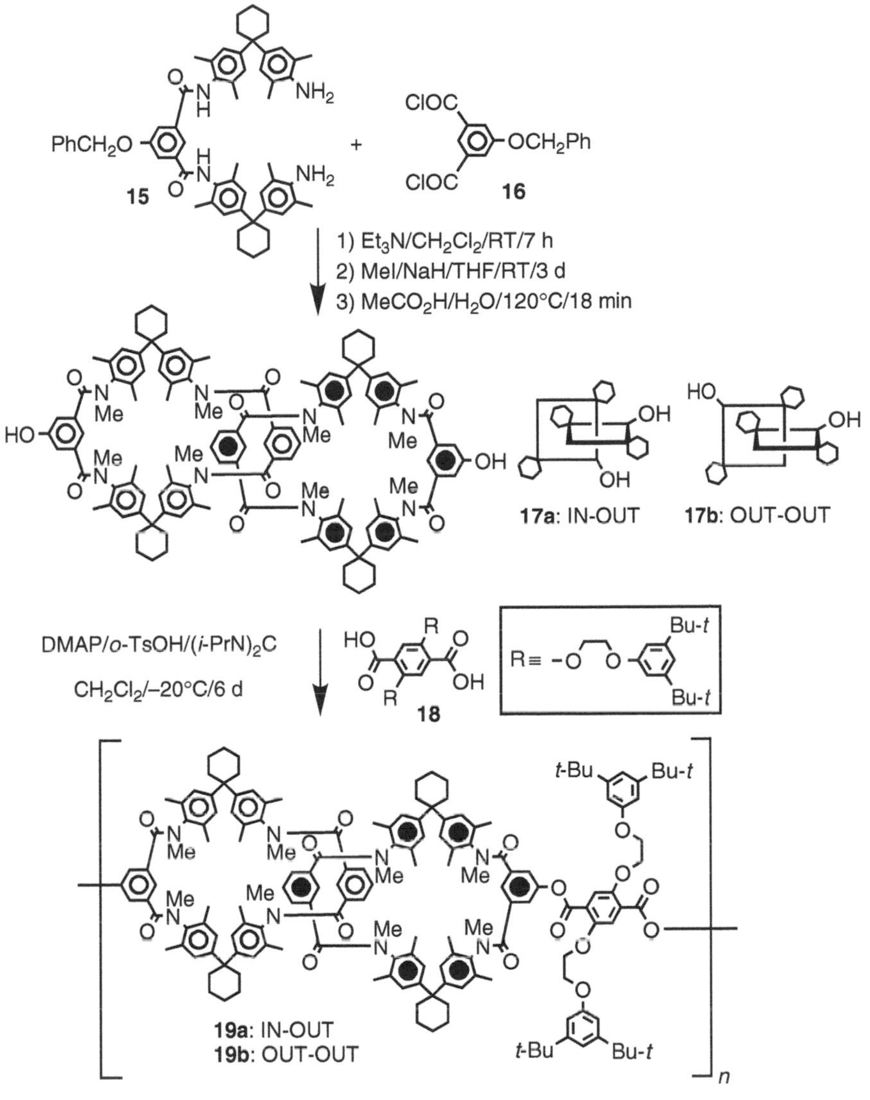

Figure 5 The hydrogen-bonding assisted template-directed synthesis of the [2]catenane **17** and its conversion into the poly[2]catenane **19**.

More recently, poly[2]catenanes derived from the engineering plastic, bisphenol A polycarbonate (PC), were prepared [16] by copolymerizing similar hydrogen-bond directed templated [2]catenanes with oligomers of the bisphenol A PC. The synthetic route taken to these systems involved the solid-state copolymerization of PC oligomers with an appropriate bisphenol catenane. The octa-N-methylbisphenol [2]catenane **20** was prepared using similar chemistry to those described before and the PC oligomers ($M_n = 1300, PDI = 1.7$) were obtained by melt polycondensation of bisphenol A with a slight excess of diphenyl carbonate. This results in PC oligomers that have a high abundance of phenyl chain ends, thus enabling them to be reacted with the catenane **20**, which bears phenol functionalities, and result in the formation of polymer. Catenane **20** (10 to 30% [w/w]) was solution blended with the PC oligomers and, after crystallization, was copolymerized via a solid-state reaction (6×10^{-2} mbar, 190 to 215°C, 28 h) to yield the desired PC/[2]catenane copolymer in which about 2 to 6% of the repeat units are the [2]catenane (Figure 6). The copolymers were characterized by

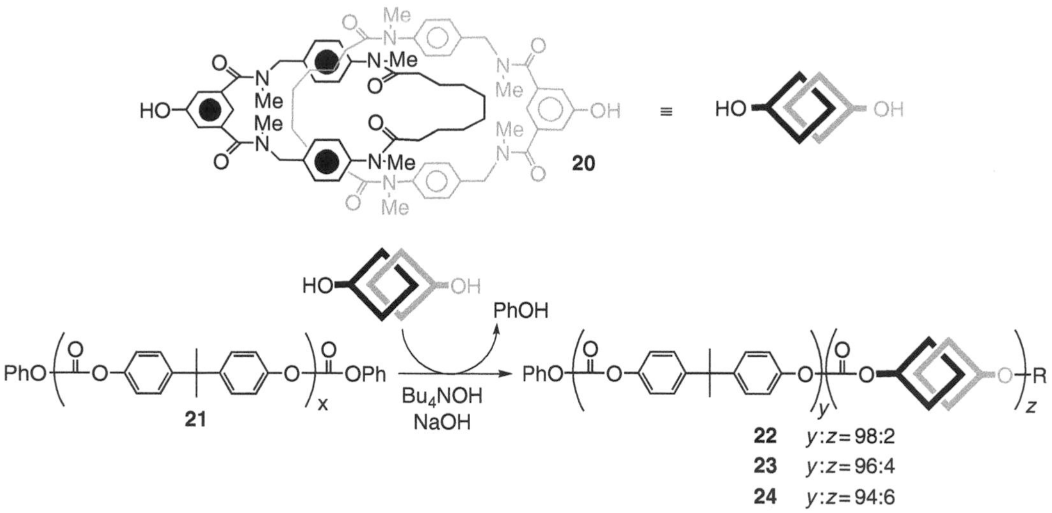

Figure 6   The copolymerization of polycarbonate oligomers with the [2]catenane **20** to form the poly[2]catenanes **22** to **24**.

GPC and by using a PC-based universal calibration, absolute molecular weights were calculated. For example, the PC containing 10% (w/w) catenane was $M_n = 15.9 \times 10^3, \text{PDI} = 2.5$.

Solid-state properties of the catenane copolymer were determined using DSC and dynamic mechanical analysis (DMA) and were compared with pure PC. DSC revealed a glass transition temperature of ca. 150°C for all the catenane copolymers **22** to **24**, which is essentially the same as the pure PC sample of the same molecular weight. This insensitivity of the T_g to the presence of the catenane repeat unit is possibly on account of the considerable flexibility/mobility of these catenanes. DMA of the copolymer containing 20% (w/w) catenane showed three transitions at -100, -6, and $+80$°C. The first and third transitions are observed in PC films while the -6°C was linked to the catenane ring/chain movements.

Poly[2]catenanes in which the catenane component was prepared by utilizing metal-ligend interactions have also been synthesized and studied. Threading of the phenanthroline-containing macrocyclic polyether **25** onto the acyclic phenanthroline-containing diphenol **26** occurs [17] in the presence of Cu$^+$ ions (Figure 7). Reaction of the resulting complex **27** with the diiodide **28**, followed by reduction and demetallation, gave the [2]catenane **29**, carrying one hydroxymethyl group on each of its two ring components. Copolyesterification of this rotationally mobile [2]catenane under mild conditions afforded (Figure 8) the poly[2]catenane **30** with M_n and M_w values of 5.5×10^4 and 180×10^4, respectively. The metal-containing form of **29**, in which the catenane units are now rotationally frozen was also copolymerized under identical conditions to yield a poly[2]catenate having values of 60.0×10^4 and 420×10^4 for M_n and M_w, respectively. Although the ^{1}H-NMR and ^{13}C-NMR spectra revealed resonances characteristic of the [2]catenane subunit as well as of the spacer unit, the signals were significantly broader than those observed in the spectra of the monomeric components. Thermogravimetric analysis of the metalated and demetalated forms of this poly[2]catenane revealed that the thermal stability is ca. 90°C higher in the absence of the metal ions. However, as may be expected the rigid metalated polycatenane has a higher T_g than the more flexible **29** (T_g 80°C vs 75°C). The bisfunctionalized [2]catenate **34** has been prepared [18] from the phenanthroline-containing compounds **31** and **32** (Figure 9). These species form the complex **33** when mixed with Cu(MeCN)$_4$BF$_4$ in solution. Reaction of **33** with the appropriate diiodide gave the [2]catenate **34** which was (i) demetalated, (ii) deprotected, (iii) remetalated, (iv) co-polymerized, and finally (v) demetalated to afford (Figure 10) the poly[2]catenane **35**. The formation of amide bonds was confirmed by IR spectroscopy, which indicated the presence of a band for the [C=O]

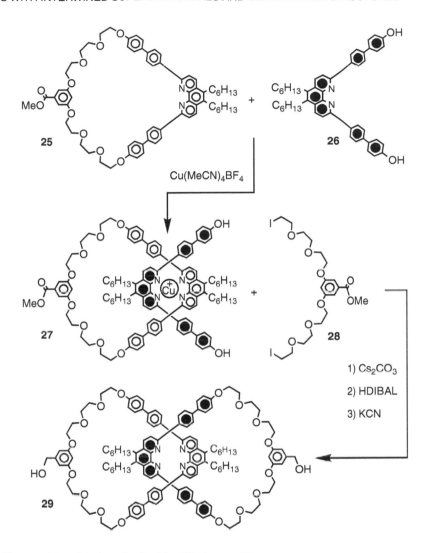

Figure 7   The metal-templated synthesis of the [2]catenane **29**.

groups at 1650 cm^{-1}. GPC analysis of **35** in DMF revealed M_n and M_w values of 8.1×10^5 and 33.2×10^5, respectively, for this polymer with numerous mechanical linkages.

Bisfunctionalized [2]catenanes have also been prepared by employing template directed syntheses that involve the interaction of π-donors and π-acceptors. Reaction (Figure 11) of the dibromide **36** with the dicationic salt **37** in the presence of either **38** or **39** as the macrocyclic polyether component afforded [19] the [2]catenanes **40** and **41**, respectively, after counterion exchange. The aromatic hydroxymethyl group located within the tetracationic cyclophane portion of the [2]catenane **40** was converted [20] into a chloromethyl group by treatment of **40** with 10 M HCl$_{aq}$.. After counterion exchange, the chloromethyl group was transformed into a bromomethyl group since it is reactive enough to enable *in situ* polyesterification in the presence of a base. Comparison of the ^{1}H-NMR spectrum of the resulting poly[2]catenane **42** with that of the monomeric [2]catenane **40** highlighted some significant differences. In particular, the methylene protons of the hydroxymethyl group of the [2]catenane give rise to a singlet. By contrast, the methylene protons of the ester groups of the poly[2]catenane appear as a multiplet. GPC analysis of the chloride salt of poly[2]catenane **42** revealed it to have an M_n value of 35.0×10^3. The [2]catenane **41** has a hydroxymethyl group attached to each of its two ring components. Its copolymerization with the bis-isocyanate **43** afforded [19] a

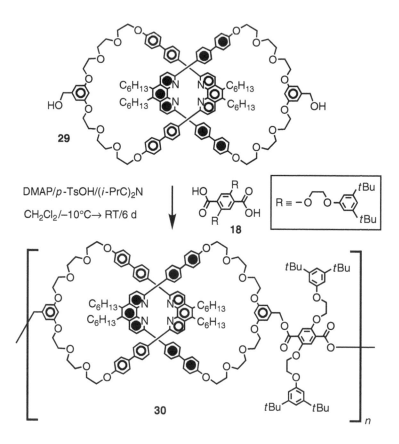

Figure 8 The synthesis of the poly[2]catenane **30**.

poly[2]catenane **44** that can incorporate up to three different "bridging motifs." The diphenylmethane bridge can link together (i) two identical π-electron-rich or (ii) two identical π-electron-deficient macrocyclic components of two adjacent [2]catenane repeating units, or (iii) a π-electron-rich to a π-electron-deficient macrocyclic component. Presumably, the poly[2]catenane **44** is a mixture of constitutionally isomeric macromolecules. The ^{1}H- and ^{13}C-NMR spectra of this poly[2]catenane indicate the presence of signals characteristic of the [2]catenane subunit as well as of the bridging spacer unit. In addition, the IR spectrum of **44** showed bands at 3375 and 1734 cm^{-1} characteristic of the [N–H] and [C=O] groups, respectively, in the urethane linkages. GPC analysis of the chloride salt of the poly[2]catenane **44** revealed an M_n value of 26.5×10^3, suggesting that it incorporates 17 repeating units on average. The bisfunctionalized [2]catenane **47** incorporates two hydroxymethyl groups attached to the hydroquinone ring in its macrocyclic polyether salt component. This [2]catenane was synthesized [20] by reacting (Figure 12) the dicationic salt **37** with the dibromide **46** in the presence of the crown ether derivative **45**, followed by hydrolysis of the two acetyloxy groups, and counterion exchange. Copolymerization of the bisfunctionalized [2]catenane **47** with the bis-isocyanate **43** afforded the poly[2]catenane **48**. The IR spectrum of this poly[2]catenane confirmed the formation of urethane linkages, displaying two bands at 3375 and 1734 cm^{-1} for the [N–H] and [C=O] groups, respectively. GPC analysis of the chloride salt of **48** revealed an M_n value of 27.0×10^3, corresponding to a degree of polymerzation (DP) of 20.

An alternative approach to mechanically interlocked macromolecules involves the copolymerization of bisfunctionalized bis[2]catenanes [21], rather than of simple [2]catenanes carrying two functional groups. The bis[2]catenane **50**, which incorporates two hydroxymethyl groups, was obtained [20] by reacting (Figure 13) the dibromide **36** with the dicationic salt **37** in the presence of

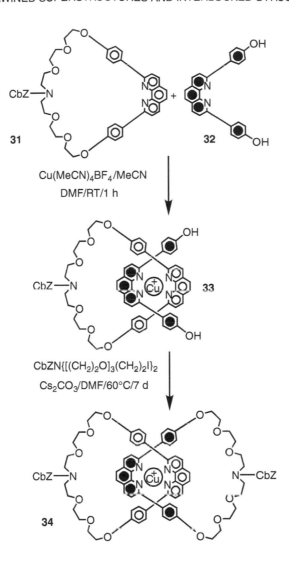

Figure 9 The metal-templated synthesis of the [2]catenane **34**.

bis(macrocyclic polyether) **49**. Copolymerization of the bis[2]catenane **50** and the bis-isocyanate **43** afforded the poly(bis[2]catenane) **51**. Similarly, the bis[2]catenane **53** has been constructed [20] by reacting (Figure 14) the dibromide **36** with the dicationic salt **37** in the presence of bis(macrocyclic polyether) **52**. Copolymerization of the bis[2]catenane **53** with the bis-isocyanate **43** afforded the poly(bis[2]catenane) **54**. GPC analyses of the chloride salts of the poly(bis[2]catenane)s **51** and **54** revealed an M_w value of 45.0×10^3 for both polymers, indicating that they incorporate 15 repeating units on average. Their IR spectroscopic analysis demonstrated the formation of urethane linkages by showing the characteristic bands at 3375 and 1734 cm^{-1} in their spectra. The bis[2]catenane **56** has also been prepared [22] using a template-directed synthetic strategy (Figure 15) relying upon interactions between π-donors and π-acceptors. Reaction of the dibromide **46** with the dicationic salt **55** in the presence of the bis(macrocyclic polyether) **49** afforded the bis[2]catenane **56**. This compound incorporates 2,2'-bipyridine units that are able, in principle, to bind transition metals. Indeed, on mixing the bis[2]catenane **56** with CF$_3$SO$_3$Ag, the poly(bis[2]catenane) **57** self-assembled spontaneously. Comparison of the ^{1}H-NMR spectra of the bis[2]catenane with that of the poly(bis[2]catenane) indicated the presence of significant differences which are particularly evident

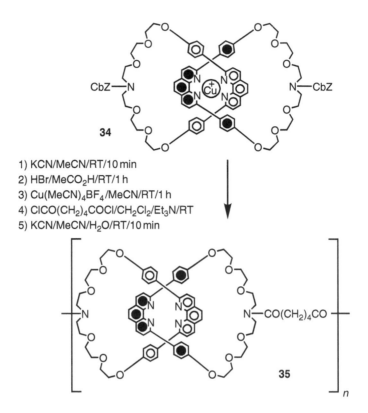

1) KCN/MeCN/RT/10 min
2) HBr/MeCO$_2$H/RT/1 h
3) Cu(MeCN)$_4$BF$_4$/MeCN/RT/1 h
4) ClCO(CH$_2$)$_4$COCl/CH$_2$Cl$_2$/Et$_3$N/RT
5) KCN/MeCN/H$_2$O/RT/10 min

Figure 10   The synthesis of the poly[2]catenane **35**.

for the resonances associated with the 2,2′-bipyridine protons. This particular poly(bis[2]catenane) is held together by a combination of covalent, mechanical, and coordinative bonds.

Electrically conducting poly[2]catenanes of type B (Figure 3) that consist of a conducting polymer backbone (thiophene-phenylene-thiophene) with catenanes formed from similar π-donor and π-acceptor units, have also been prepared [23]. An electropolymerizable [2]catenane **59** that comprises the electron-poor tetracationic cyclophane and a macrocyclic polyether **58** with two reactive (3,4-ethylenedioxy)thiophene groups on either side was synthesized (Figure 16). Subsequent electropolymerization of the [2]catenane **59** affords the conducting poly[2]catenane **61**. The macrocyclic monomer **58** was also polymerized and the electrochemical behavior of each system was monitored. The [2]catenane proved to be slightly more difficult to oxidize than the macrocycle **58** which was expected on account of the repulsive nature of the radical-cation stacked above the tetracationic bipyridinium cyclophane. The conductivity profile of the poly[2]catenane displays a maximum of 0.2 S cm^{-1} at a potential of 0.12 V versus Fc/Fc$^+$ that decays quickly thereafter. This is in contrast to polymer **60** which exhibits a maximum of 11 S cm^{-1} at 0.2 V versus Fc/Fc$^+$ that does not decay at higher voltages. These results suggest that the catenane units act as barriers to conduction along the conjugated backbone with each repeat unit acting as a discrete entity, in a similar manner to a redox conductor.

Another type of polymeric catenane architecture that has recently been reported is that of a polymeric [2]catenane in which the two macrocycles that constitute the catenane are themselves polymers [24]. A polystyrene (PS)-poly(2-vinylpyridine) (P2VP) catenane copolymer with a molecular weight around 10,000, $M_w/M_n = 1.3$ has been isolated and characterized by SEC, ^{1}H-NMR, and fluorescence spectroscopy. The synthesis of this interesting material was achieved by the end-to-end coupling of a linear P2VP dianion lithium salt with 1,4-bis(bromomethyl

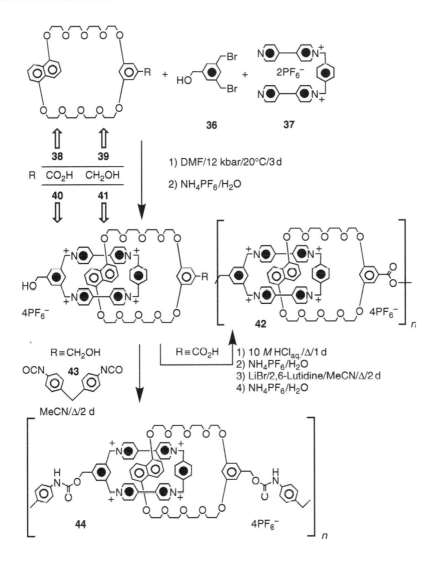

Figure 11 The donor/acceptor template-directed synthesis of the [2]catenanes **40** and **41** and their conversion into the poly[2]catenanes **42** and **44**, respectively.

benzene) in the presence of a PS macrocycle (molecular weight = 4,500). The catenated block copolymer is isolated by multiple precipitation/extraction procedures to give the polymeric catenane PS-cat-P2VP in an overall yield of 0.5%. This low yield is on account of the lack of an enthalpic driving force for the threading of the linear P2VP through the PS macrocycle.

III. MAIN-CHAIN POLYROTAXANES

Main-chain polyrotaxanes can be synthesized (Figure 17) from pseudopolyrotaxanes by attaching covalently bulky groups either at the termini or along the backbone of the acyclic component. Pseudopolyrotaxanes can be prepared following route **C** or **D** either by polymerizing n preformed pseudorotaxanes or by threading n rings onto a preformed acyclic polymer, respectively. The first examples of pseudopolyrotaxanes were prepared [25,26] following route **C** by solution or interfacial copolymerization of diamines and diacid dichlorides in the presence of β-cyclodextrin (β-CD).

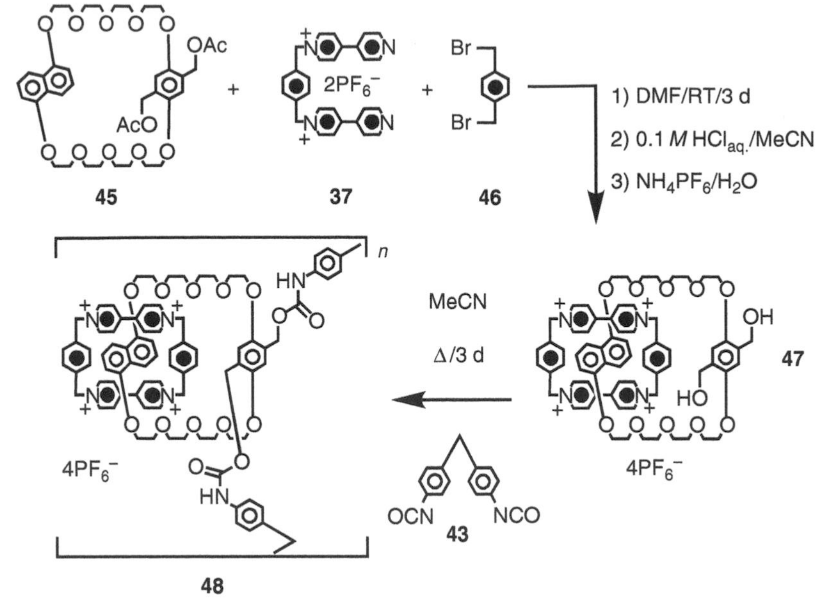

Figure 12 The donor/acceptor template-directed synthesis of the [2]catenane **47** and its conversion into the poly[2]catenane **48**.

A similar approach was employed [27] (Figure 18) to incorporate α-CD (**64**) into the polyrotaxane **65** that was obtained by reacting the diaminobenzidine **62** with the diol **63** in the presence of α-CD and $RuCl_2(PPh_3)_3$. Along its acyclic backbone, this polyrotaxane incorporates benzimidazole units that are too bulky to thread through the cavitiy of α-CD. As a result, each ring component is trapped mechanically between two benzimidazole units and encircles the aliphatic spacer that bridges the two bulky units. GPC analysis of the polyrotaxane **65** afforded M_n and M_w values of 4.0×10^3 and 7.2×10^3, respectively, while ^{1}H-NMR spectroscopic studies showed a ring/repeating unit ratio of 0.163. Interestingly, the rigidity imposed on the polymeric backbone by the encircling α-CD rings imposes a 20°C increase in the T_g of the polyrotaxane relative to the T_g of the polymer in its free form.

Pseudopolyrotaxanes incorporating crown ethers have also been prepared following route C. For example, the reaction of ethylene oxide with the potassium salt of tetraethylene glycol [28], the free-radical polymerization of acrylonitrile or styrene [29], the polycondensation of diols and diacid dichlorides [30], and transesterification polymerizations [31] have all been carried out in the presence of a preformed crown ether, often used as the solvent, to afford pseudopolyrotaxanes or polyrotaxanes. However, it was found [32] that the threading efficiency increases significantly when bulky groups are introduced into the monomers. Thus, reaction (Figure 19) of the tetraarylmethane-based diol **66** with the tetraarylmethane-based diacid chlorides **67** in the presence of the crown ether **68** affords [30d] the polyrotaxane **69** that was isolated pure following consecutive reprecipitations using THF/H₂O mixtures. In this polyrotaxane, each ring component is trapped mechanically between two tetraarylmethane groups encircling some of the aliphatic spacers that bridge pairs of bulky units. The ^{1}H-NMR spectra of the polyrotaxane **69** indicated the presence of signals for the cyclic and acyclic components and, from their relative intensities, a ring/repeating unit ratio of 0.172 was determined. GPC analysis of the polyrotaxane **69** revealed it to have an M_n and M_w values of 25.2×10^3 and 87.3×10^3, respectively. Furthermore, the GPC traces did not contain the characteristic peak associated with the crown ether **68** in its free form, confirming that the acyclic and cyclic components of the polyrotaxane **69** are mechanically interlocked. A further increase in the threading efficiency was observed [32] when poly(urethane)s were employed as the acyclic

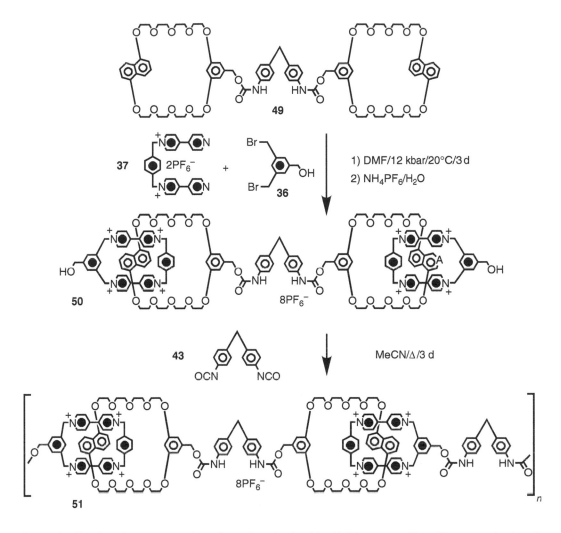

Figure 13   The donor/acceptor template-directed synthesis of the bis[2]catenane **50** and its conversion into the polybis[2]catenane **51**.

components. Indeed, [N–H $\cdots$ O] hydrogen bonds [33] between the urethane hydrogen atoms of the polymer and the polyether oxygen atoms of the ring component are believed to assist the threading process as well as inhibit the dethreading process. Thus, when diols are reacted with bis-isocyanates, pseudopolyrotaxanes incorporating up to 63% by mass of the ring component are obtained [32]. The polyrotaxane **71** was prepared [34] by reacting (Figure 20) the tetraarylmethane-based diol **66** with the bis-isocyanate **70** in the presence of 30-crown-10 (**68**). The absence of free 30-crown-10 was confirmed by GPC analysis that also revealed M_n and M_w values of 23.4×10^3 and 37.8×10^3, respectively. ^{1}H-NMR spectroscopic studies on this polyrotaxane indicated that it has "solvent switchable" microstructures. In CDCl$_3$, [N–H$\cdots$O] hydrogen bonds between the urethane hydrogen atoms and the polyether oxygen atoms encourage the crown ether component to encircle the urethane units, as suggested by the upfield shift for the polyether protons that suffer shielding effects exerted by the diphenylmethane unit. However, the rings move toward the tetraarylmethane groups in (CD$_3$)$_2$SO to allow for solvation of the urethane linkages by this highly polar solvating solvent.

Pseudopolyrotaxanes incorporating cycloalkanes [35], CDs [36,37], cyclourethanes [38], cyclophanes [39], and crown ethers [29] have all been prepared following route **D**. Hydrophobic interactions assist [40] the threading of α-cyclodextrin (α-CD) (**64**) onto polyamine **72** in aqueous

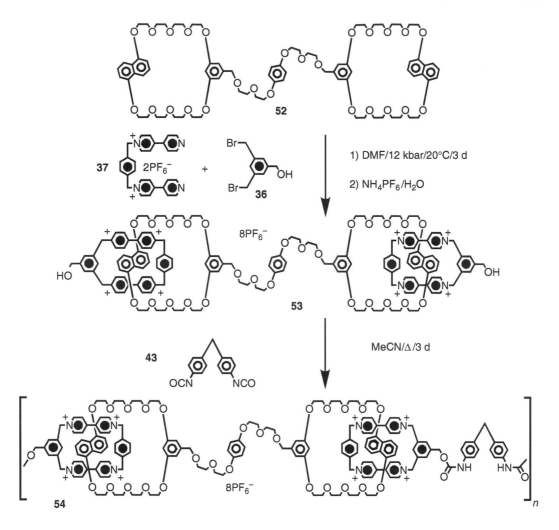

Figure 14 The donor/acceptor template-directed synthesis of the bis[2]catenane **53** and its conversion into the polybis[2]catenane **54**.

solution to yield (Figure 21) pseudopolyrotaxanes. On threading, the specific viscosity (η_{sp}) increases gradually with time and reaches a constant value after 2 h when an equilibrium between complexed and uncomplexed species is reached. The opposite process was investigated by removing the macrocyclic component from the equilibrium by dialysis. The decreasing concentration of α-CD was monitored by optical rotation measurements. Complete dethreading was achieved after 15 h. Upon covalent attachment of nicotinoyl groups to some of the amine groups of this pseudopoly-rotaxane, the threaded rings are trapped mechanically onto the polymeric backbone, affording the polyrotaxane **73**. From the relative intensities of the resonances in the ^{1}H-NMR spectrum of **73**, it was established that this polyrotaxane incorporates 10 mol% of threaded rings. A similar approach was employed in the supramolecularly assisted synthesis of a polyrotaxane incorporating two different types of macrocycles. When γ-CD (**74**) and β-CD (**75**) are combined (Figure 22) with the stilbene-containing polymer **76** and the stilbene-containing "monomer" **77** in H$_2$O, the pseudopolyrotaxane **78** self-assembles [42] spontaneously. On irradiation of an aqueous solution of **78** with ultraviolet (UV) light at 312 nm for 30 h, tetraphenylenecyclobutane blocking groups are formed, yielding the polyrotaxane **79**. The formation of "isolated" phenylene rings was confirmed by the increased absorption at 250 nm. In addition, the signals for the [CH=CH] protons of the stilbene units disappear

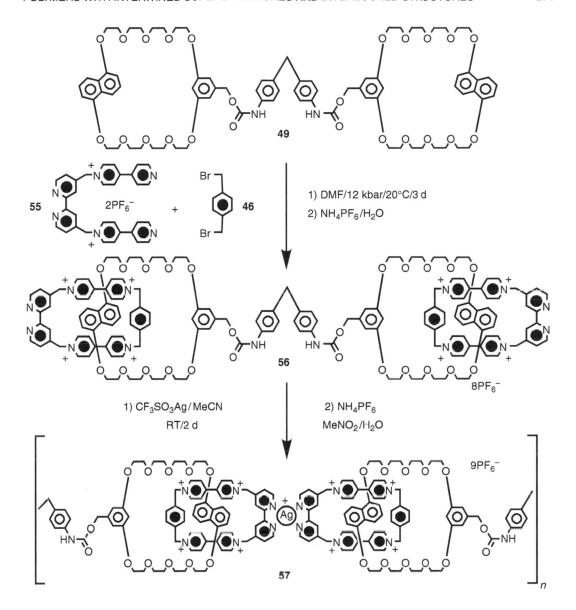

Figure 15 The donor/acceptor template-directed synthesis of the bis[2]catenane **56** and its conversion into the polybis[2]catenane **57**.

in the ^{1}H-NMR spectrum which also demonstrates that this polyrotaxane incorporates 20 mol% of threaded rings. GPC analysis did not show any free or easily removable CD ring, confirming that **79** is a polyrotaxane.

The effect of the ring size on the rate of complexation/decomplexation of pseudopoly-rotaxanes as they adsorb on to a surface has been investigated using the polymer back-bone poly(*N*-dimethyldecamethyleneammonium) (PDDA) with α- and β-CDs [43]. Three pseudopolyrotaxanes were prepared: PDDA with α-CD, PDDA with β-CD, and finally PDDA with β-CD followed by α-CD; and then were adsorbed onto mica surfaces (Figure 23). Adsorption of the PDDA onto the mica was monitored by thermogravimetric analysis, and the mass of PDDA adsorbed was determined by weight loss of the mica between 250 and 450°C. Polyelectrolytes, such as PDDA, adsorb rapidly on layered silicates [44] and within 3 h of exposure, complete adsorption

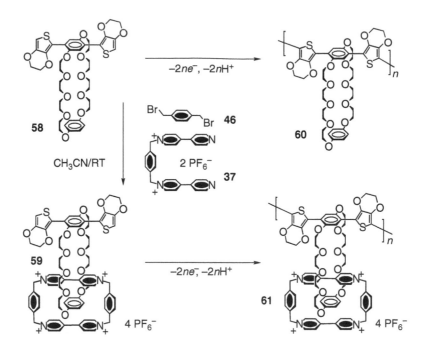

Figure 16   The donor/acceptor template-directed synthesis of the [2]catenane **59** and its electropolymerization to form the side-chain poly[2]catenane **61**.

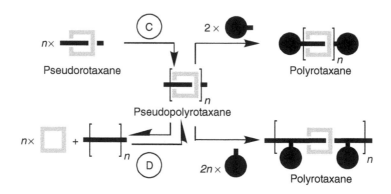

Figure 17   Strategies for the preparation of a pseudopolyrotaxane and its conversion into polyrotaxanes.

was reached. For the pseudopolyrotaxane samples a combination of TGA and optical rotation of the supernatant solution was used to estimate the percentage of CD threaded on the adsorbed PDDA. For the β-CD sample, while it is estimated that the fraction of CDs per repeat unit of PDDA (f) is 60% in solution, the adsorbed material contained less than 4% β-CD consistent with most of the β-CD slipping off the polymer chain before it adsorbs (Figure 23(b)). In solution, the time to reach equilibrium in these systems is much slower for α-CD than with β-CD and these slower binding kinetics means that the amount of α-CD present in the adsorbed polymers is much larger. The formation of the pseudopolyrotaxanes prepared from α-CD takes weeks or months to reach equilibrium in homogenous solutions. After 20 d equilibration in solution the fraction (f) of α-CD present on polymer chain in solution was ca. 10% and roughly the same value of f was found in the adsorbed material suggesting that the removal of the α-CD is slower than the absorption kinetics. The absorption of the pseudorotaxane prepared from both β-CD and α-CD also resulted in the same fraction

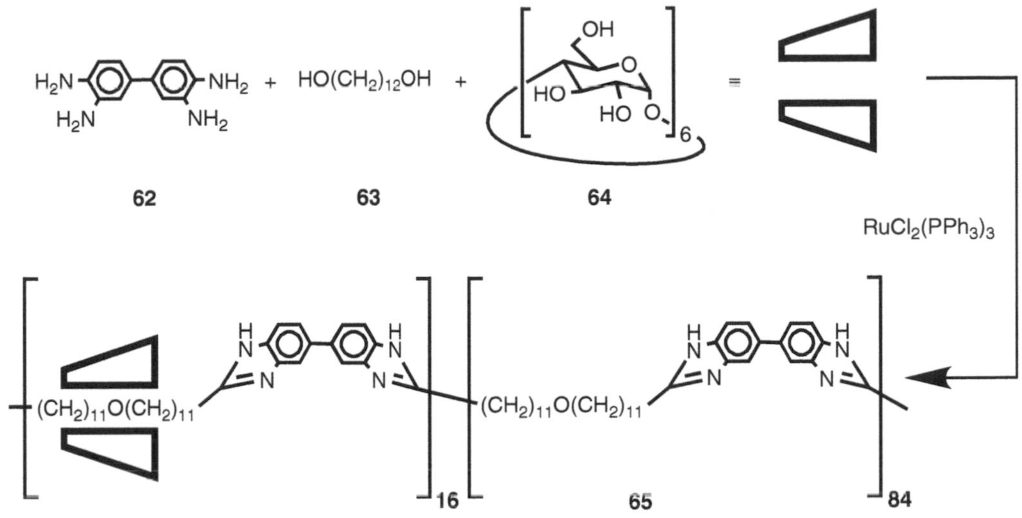

Figure 18   The synthesis of the polyrotaxane **65**.

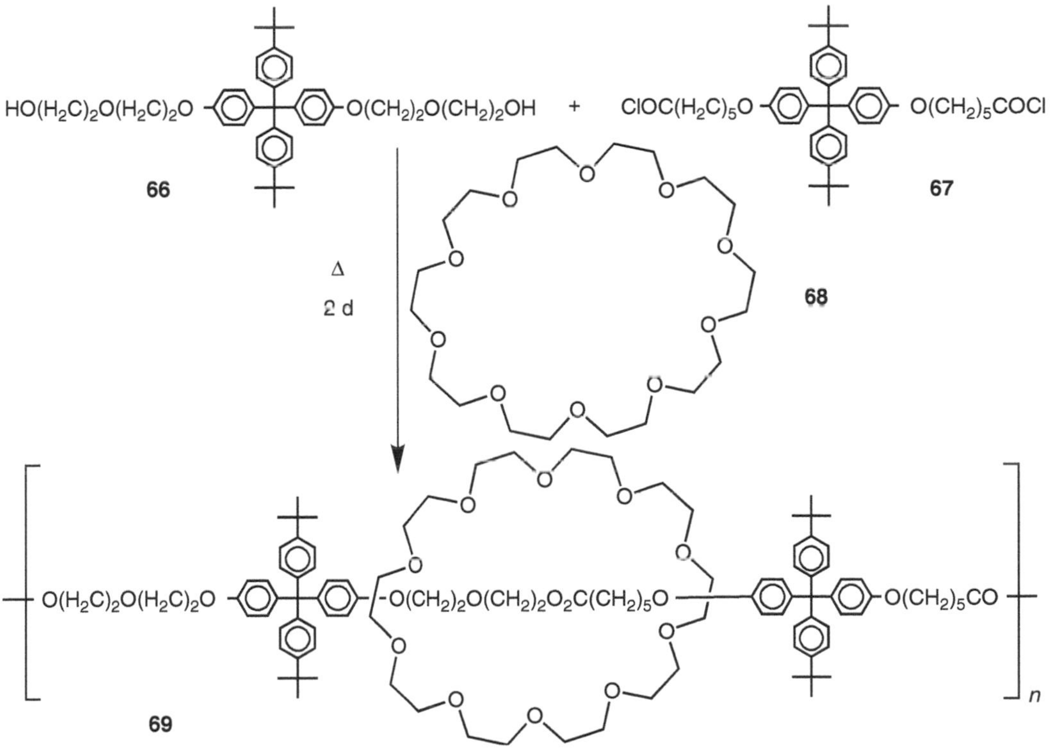

Figure 19   The synthesis of the polyrotaxane **69**.

of CD being present on the surface as was in solution (ca. 55% after 20 d). This interesting result suggests that slower decomplexation kinetics of the smaller α-CDs result in the β-CDs being trapped on the polymer chain preventing them from dethreading. Thus in this case the α-CDs effectively act as a kind of supramolecular stopper.

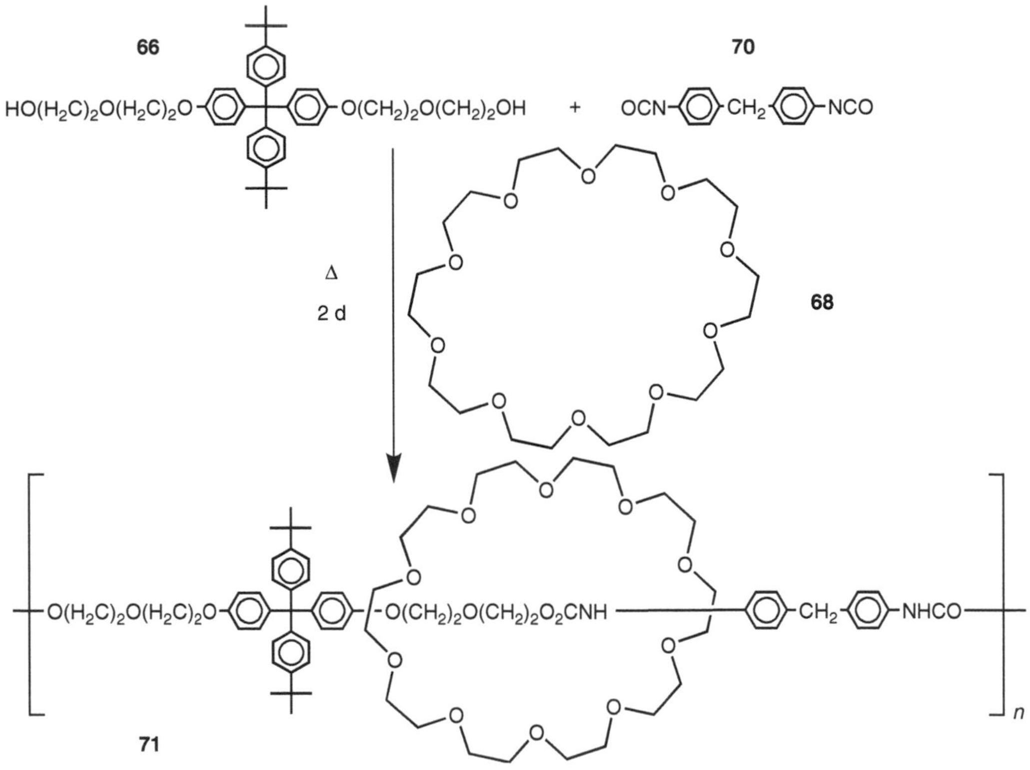

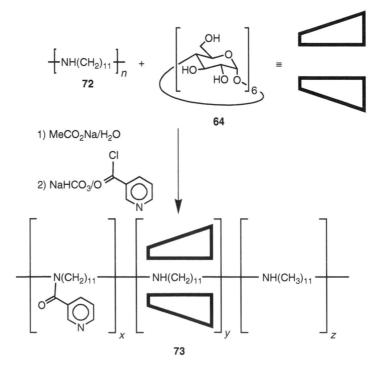

Figure 20 The synthesis of the polyrotaxane **71**.

Figure 21 The synthesis of the polyrotaxane **73**.

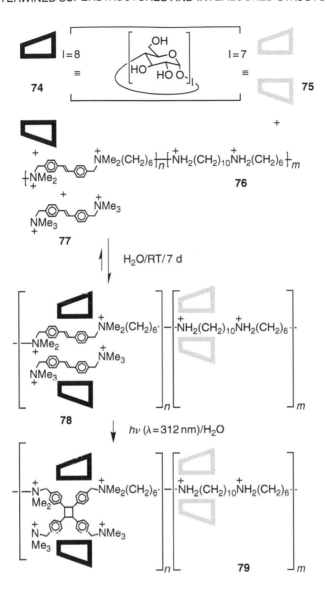

Figure 22 The synthesis of the polyrotaxane **79**.

Threading of α-CD (**64**) onto the poly(ethylene glycol) bisamine **80**, followed by reaction (Figure 24) with 2,4-dinitrofluorobenzene, affords [45] the polyrotaxane **81**. One- and two-dimensional NMR spectroscopic analysis showed signals for the ring, thread, and stopper components and confirmed that the poly(ethylene glycol) chain is inserted through the cavities of the rings. In fact, the presence of hydrogen bonding between the CDs is believed to play an important role in the formation of these and related polyrotaxanes [46]. Indeed the intermolecular hydrogen bonding between the CDs has a cooperative effect on inclusion behavior which helps to result in efficient formation of the polyrotaxane. A combination of ^{1}H-NMR and UV spectroscopic and optical rotation data revealed that this polyrotaxane incorporates approximately one ring for every two bismethylene units, suggesting that the α-CD rings are "close-packed" from end to end along the polymer chain. On average, 15–20 α-CDs are incorporated into this polyrotaxane. However, a polyrotaxane incorporating as many as 37 α-CDs was obtained by the fractionation of the product. Treatment of **81** with epichlorohydrin resulted [47] in the covalent bridging of the threaded α-CD

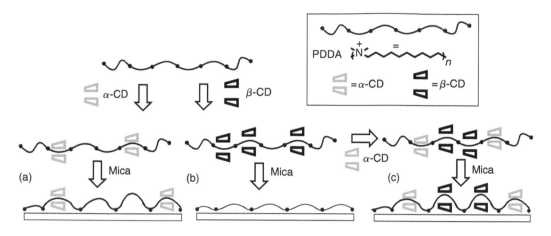

Figure 23 Schematic representation of the formation and adsorption behavior of polymeric inclusion compounds on Li-mica: (a) PDDA/α-CD; (b) PDDA/β-CD; (c) PDDA/β-CD/α-CD.

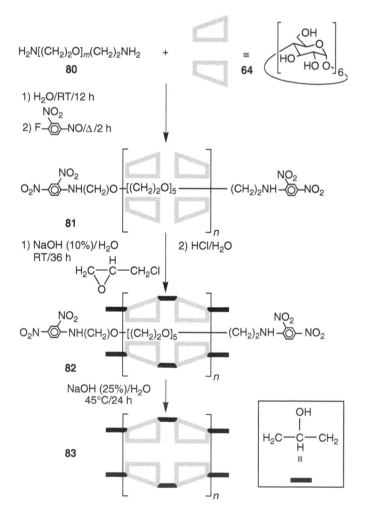

Figure 24 The template-directed synthesis of the CD-based nanotube **83**.

rings. Finally, cleavage of the terminal stoppers of **82** released the intact "molecular tube" **83**, composed of covalently bridged α-CD rings. GPC analysis of this novel material revealed that its average molecular weight is less than 20.0×10^3.

The controlled movement, under ambient conditions, of the α-CDs in polyrotaxane **81**, which was on a MoS_2 substrate, has been demonstrated by using an STM tip to mechanically push the α-CDs along the polyethylene backbone [48]. The nature of the polyrotaxane architecture meant that upon reversing the direction of the STM tip the α-CD retraced its path back to its original position. The STM also allowed for quantitative analysis [49] of intramolecular conformation of the α-CDs along the poly(ethylene oxide) backbone. On account of the CDs having two different faces, one with primary hydroxyl groups and one with secondary hydroxyl groups, there are three possible ways for the α-CDs to interact, tail-to-tail, head-to-head, and head-to-tail. While macroscopic techniques (e.g., NMR) estimated a 100% head–head/tail–tail configuration of the α-CDs in the polyrotaxane, STM found that about 20% of the α-CDs had the head-to-tail placement. Comparison of these results with theoretical models allowed an estimation of relative strength of the primary–primary and the secondary–primary hydrogen bonding to be about 2:1.

β-Cyclodextrin (β-CD) and γ-CD have also been shown to form pseudorotaxanes with low molecular weight inorganic polymer chains such as poly(dimethylsiloxane) (PDMS) [50] and poly(dimethylsilane) [51]. The two CDs showed two different chain-length selectivities; β-CD preferred low molecular weights (<1000 PDMS) while the γ-CD showed some increasing levels of complexation as molecular weight increased (1000–3000 PDMS). Similar behavior was observed for poly(dimethylsilane)s ranging in degree of polymerization (DP) from 1 to 16 (M_n 88–958). NMR data suggests that three $-SiMe_2-$ are included within each CD cavity and powder x-ray data of the crystalline material is consistent with the formation of CD channels in the solid state. Poly(dimethylsilane)s show significant delocalization of σ-electrons along its backbone and as a consequence show strong electronic absorption and emission bands that are somewhat sensitive to the conformation of the backbones. The solution fluorescence spectra of the poly(dimethylsilane) backbone shows a red shift of about 10 nm upon addition of γ-CD, suggesting that the CD increases the σ conjugation of the polymer backbone, consistent with the CD forcing the backbone into an all *trans*-conformation (Figure 25).

Cyclodextrin-based rotaxanes of conducting polymers takes advantage of the unique topological architecture to allow the formation of "insulated" molecular wires [52,53]. Here the CDs act as an insulating sheath around the conducting polymer preventing intermolecular interactions occurring through the π-systems. However, these rotaxanes still allow charge transport along the chain and the polymers can retain their semiconducting properties. A series of polyrotaxanes have been prepared that have conjugated light-emitting polymers as their backbones (Figure 26) and their subsequent optical and morphological properties studied [54]. Polyrotaxane **84** was synthesized by aqueous Suzuki coupling of the paraphenylene monomer in the presence of β-CD using the hydrophobic effect to drive threading. 1-Iodonapthalene-3,6-disulfonate disodium salt was used as both a chain

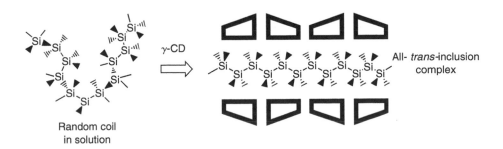

Figure 25 The formation of an all *trans*-poly(dimethylsilane) inclusion comlplex with γ-CD.

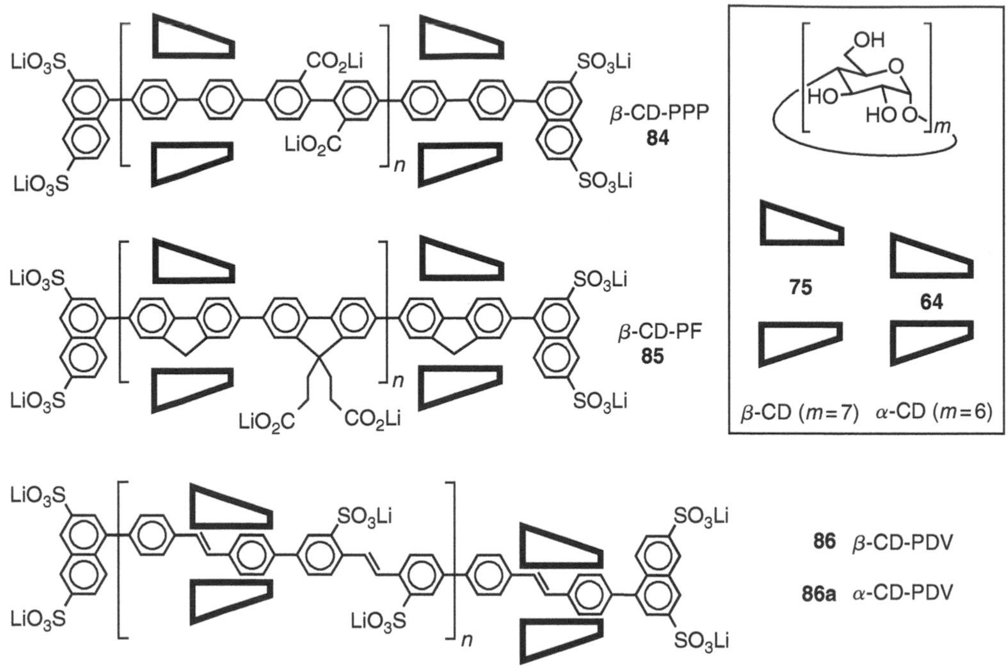

Figure 26 Chemical structures of cyclodextrin polyrotaxanes **84** to **86** threaded with the conjugated polymers poly(paraphenylene) (PPP), polyfluorene (PF), and poly(diphenylene vinylene) (PDV).

terminator to control the DP and as a bulky stopper moiety to prevent dethreading. Polyrotaxanes **85** and **86** which contain either a polyfluorene, PF, or a poly(4,4′-diphenylenevinylene), PDV, backbone, respectively, and β-CD were prepared using similar conditions. In addition, the PDV core of **86** is also slender enough to allow the synthesis of the α-CD-based rotaxane **86a**. The average number of macrocycles per repeat unit for **84** to **86** was found by NMR to be between 0.9 and 1.6, which is slightly less than full coverage of the backbone (ca. 2 CDs per repeat unit). The photoluminescence spectra of thin films of all three β-CD polyrotaxanes **84** to **86** are blue-shifted with respect to their "naked" nonthreaded polymers, with differences ranging from 10 nm for PPP-derivative **84** up to 30 nm for PDV derivative **86**. This is consistent with either a reduction in the main-chain intermolecular interactions or an increase in the twist angle between adjacent aromatic units. Higher photoluminescent efficiencies are also observed for the rotaxanes, for example, polyrotaxane **86** showed a 17% efficiency as compared to 4.3% for the naked PDV. Time-resolved fluorescence spectroscopy showed agreement with the efficiency data with an average decay time of 0.4 nsec for polyrotaxane **86** (at 520 nm) versus 0.1 nsec for the uninsulated cores.

To determine the effect of the supramolecular architecture on the electroluminescence of these materials, the polymers were spin coated onto ITO anodes and fabricated into a standard polymer LED with a Ca cathode. Successful LED operation was achieved and the polyrotaxanes showed improved emission efficiency over the noninsulated polymers. The threaded CDs did not prevent charge transport in the conjugated material because they did not completely prevent all interaction between the cores. Charge-hopping could occur across the exposed regions of the individual chains. However, on account of the rotaxane architecture, the long axes of the chains cannot be aligned. AFM studies on the polyrotaxanes and analogs showed that the insulated cores while densely packed are distinguishable from each other, whereas the uninsulated cores π–π stack to form self-assembled domains. This confirmed that the CDs did indeed reduce the tendency for the conjugated polymers to aggregate.

Cyclodextrin-based polyrotaxanes have also been shown to have useful properties for biological and medical applications. In such systems the CDs can be functionalized, for example, with a drug [55], to enhance or influence the appropriate biological response and/or the stopper groups can be biodegradable, for example, enzyme degradable, to control the release of the CD from the polymer backbone [56]. Maltose functionalized α-CDs threaded on to a poly(ethylene glycol) polymer results in polyrotaxanes that show enhanced binding to the maltose-binding protein Concanavalin A (Con A) [57]. As well as the multivalent nature of the polyrotaxane, which has been shown in other polymeric architectures to enhance binding to Con A [58], it was suggested that the unique mobility of the CD along the polymer backbone further enhances this interaction by reducing any spatial mismatches that may occur between the maltose and Con A binding sites. Polyrotaxanes consisting of hydroxypropylated α-CDs threaded on to a poly(ethylene glycol) polymer and end-capped with L-phenylalanine result in materials that accelerate physicochemical interactions with blood cells and inhibit thrombin-induced platelet activation [59]. Similar polyrotaxane systems based on sulfonated α-CDs, PEG-b-PPG-b-PEG triblock copolymer backbone and end-capped with L-phenylalanine have been shown to act as blood-compatible, synthetic anti-coagulants [60].

The formation of pseudopolyrotaxanes incorporating cyclobis(paraquat-p-phenylene) (**87**) and π-electron-rich polyethers has been achieved [39] through a reliance on a combination of [C–H$\cdots$O] hydrogen bonds, [$\pi\cdots\pi$] stacking, and [C–H$\cdots\pi$] interactions. The bipyridinium-based tetracationic cyclophane **87** binds [6] dioxyarene-based guests, producing complexes with pseudorotaxane geometries. When **87** is mixed with one of the dioxyarene-based polyethers **88–92** in solution, a purple color develops [61] as a result of the charge-transfer interactions between the π-electron-rich and -deficient recognition sites present in the thread and ring components of the pseudopolyrotaxanes **93** to **97** (Figure 27). Detailed ^{1}H-NMR spectroscopic investigation revealed that the amount of

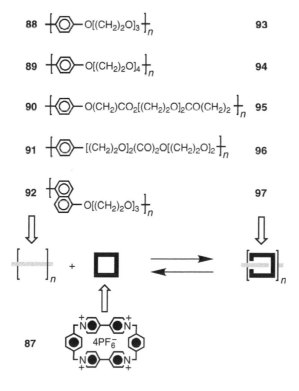

Figure 27   The self-assembly of pseudopolyrotaxanes incorporating the bipyridinium-based tetracationic cyclophane **87**.

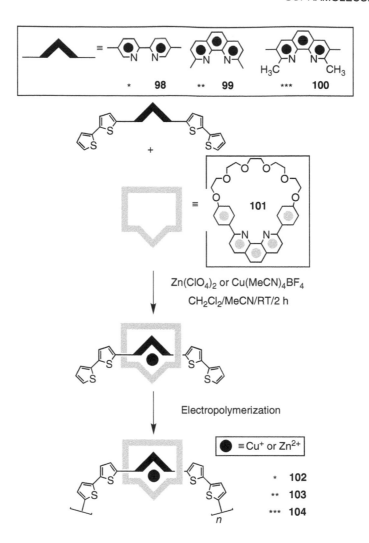

Figure 28   The synthesis of the metal-containing pseudopolyrotaxanes **102**, **103**, and **104**.

threaded rings is affected significantly (i) by the nature of the spacer unit separating the dioxyarene recognition sites and (ii) by the temperature.

Threading of the phenanthroline-based macrocycle **101** onto either the bipyridine-based or the 2,9-disubstituted 1,10-phenanthroline-based bis-thiophene derivatives **98** and **99**, respectively, occurs [62] (Figure 28) spontaneously in the presence of metal ions (e.g., Cu^+) which can be coordinated tetrahedrally. Electropolymerization of the resulting metal-containing pseudorotaxanes afforded the metal-containing pseudopolyrotaxanes **102** and **103**. The decomplexation/complexation of Zn^{2+} ions can be achieved reversibly in the case of **102**. By contrast, the reinsertion of Cu^+ ions into **103** is only possible when the demetalation is performed in the presence of Li^+ ions which prevent the collapse of the polymer by coordinating to the phenanthroline units. The redox and conducting properties of **102** are affected dramatically by the coordinated metal ions that produce charge localization and participate in conduction by means of redox processes. Placement of the thiophene moieties on the 3 and 8 positions of the 1,10-phenanthroline ligand (**100**) results in the conjugated polyrotaxane **104** after electropolymerization [63]. This change in the substitution of the phenanthroline ligand results in a more linear polymer backbone with better π-conjugation. The enhanced electronic delocalization of these systems is confirmed by their electrochemical response being shifted to cathodic potentials (0.80 and 1.04 V) relative to **103** (0.96 and 1.28 V), which has the U-shaped ligand **99**. In addition, the presence of the methyl groups on the 2,9 positions of the ligand allows the decomplexation and partial reinsertion of Cu^+ ions without Li^+ ions being used as a transient scaffold, although more efficient recomplexation is obtained if Li^+ ions are present. The demetallated coordination sites can also be remetallated with another metal ion such as Zn^{2+} or Co^{2+}. *In situ* conductivity and ESR measurements were performed on all these metallopolyrotaxanes. These studies show that the electrochemical responses of the components of the polyrotaxanes are well separated indicating charge localization on both the macrocycle and the tetrathienylene units. In addition, the conductivity of the polyrotaxane architectures (σ ca. 2×10^{-5} S/cm) is greatly reduced compared to the free linear polymer (σ ca. 5×10^{-3} S/cm) suggesting that the coordinating macrocyclic component acts as a barrier decreasing charge carrier mobility.

A complex three-stranded conducting ladder polymer was made possible by the electropolymerization of a metallorotaxane monomer that contains two independently polymerizable groups [64]. The metallorotaxane monomer **107** has thiophene moieties attached to an electron-poor bipyridine unit as the thread component **106** and electron-rich ethyelenedioxythiophene units, which are easily oxidatively polymerized at low potentials, on the macrocycle **105**. Thus controlled oxidation of the monomer causes electropolymerization of the macrocycle unit while leaving the thread unreacted yielding the side-chain pseudopolyrotaxane **108**. Applying a higher potential to **108** results in the polymerization of the threads thus yielding a three-stranded ladder polymer **109** in which all three chains are comprised of polythiophene derivatives (Figure 29). Conductivity and cyclic voltammetry

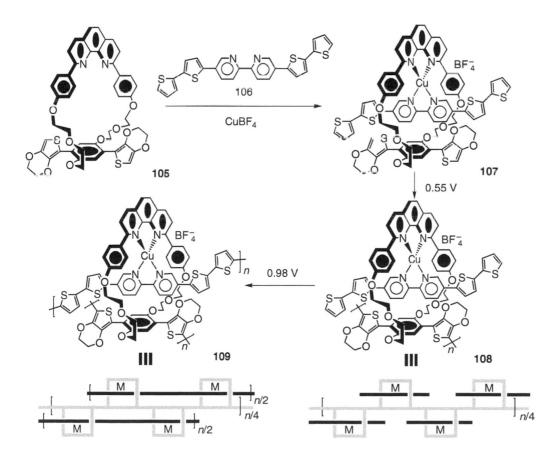

Figure 29   The metal-templated synthesis of the pseudorotaxane **107** and its two-step conversion into the ladder pseudopolyrotaxane **109**.

of the electropolymerized films suggest that the central polymer chain (the one derived from the ethylenedioxythiophene units) is not only the main contributor to the conductivity of the ladder polymers but is also partially isolated when the outer two polymer strands are in their undoped (insulating) state. In addition, the Cu centers are also thought to play a role in the conduction of this system by aiding interchain conduction. This is especially important at low potentials when the outer chains are insulating.

A different strategy to the synthesis of main-chain polyrotaxanes has recently been reported. Here the macrocyclic component for the polyrotaxane acts as a catalyst for the polymerization reaction. Cucurbituril is known to catalyze the 1,3-dipolar cycloadditions within its cavity. Therefore, this macrocycle was added to a reaction mixture containing diazide and dialkyne monomer units, which both have stopper units as the core of the monomer. As a consequence, the reaction between the alkyne and azide groups that occur within the macrocyclic cavity will also necessarily result in the formation of the rotaxane. This reaction is accelerated by a factor of 10^5 when bound within the cucurbituril cavity and therefore each new repeat unit will also contain exactly one macrocycle. GPC of the resulting polyrotaxane found that $M_n = 5.1 \times 10^3$, $M_w = 9 \times 10^3$ with ^{1}H-NMR estimating M_n to be 13×10^3.

IV. Side-Chain Polyrotaxanes

Side-chain polyrotaxanes incorporating CDs have been prepared [65] by reacting (Figure 30, route **E**) a preformed semirotaxane with a polymer having reactive groups on its side chains. Upon mixing the methylated β-CD **110** with an acyclic compound bearing a trityl group at one end, the semirotaxane **111** precipitates [65a] out and can be separated from its components by filtration. This semirotaxane was characterized by fast atom bombardment mass spectrometry (FABMS) that revealed peaks corresponding to the "molecular" ion of **111**. Reaction of **111** with the preformed polymer **112** affords (Figure 31) the side-chain polyrotaxane **113**. The incorporation of β-CD rings into the polymer **113** was demonstrated by IR and ^{1}H-NMR spectroscopy: no free β-CD was detected by chromatography. The presence of interlocked β-CD rings in the polyrotaxane **113** affects significantly its solubility and viscosity. For example, this polyrotaxane is soluble in Et$_2$O, whereas the parent polymer is not. Also, the viscosity of the polyrotaxane is lower than that of the parent polymer and shows a very different temperature dependence.

A similar strategy was employed [66] to synthesize (Figure 32) side-chain polyrotaxanes by connecting two semirotaxane subunits to the same side chain. The semirotaxane **111** was reacted with the preformed polymer **114**, bearing two reactive groups per side chain, to afford the polyrotaxane **115**. The ^{1}H-NMR spectrum of the polyrotaxane **115** revealed signals broader than those observed for the parent polymer. These observations indicate that each CD ring moves rapidly on the ^{1}H-NMR timescale along the aliphatic chain at ambient temperature. However, the rings are located preferentially next to the terminal bulky groups at $-10°$C. DSC measurements showed an increase of the glass transition temperature of ca. 19°C for the polyrotaxane relative to the parent polymer. Another

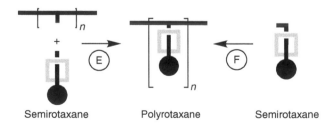

Semirotaxane   Polyrotaxane   Semirotaxane

Figure 30   Synthetic strategies for the construction of side-chain polyrotaxanes.

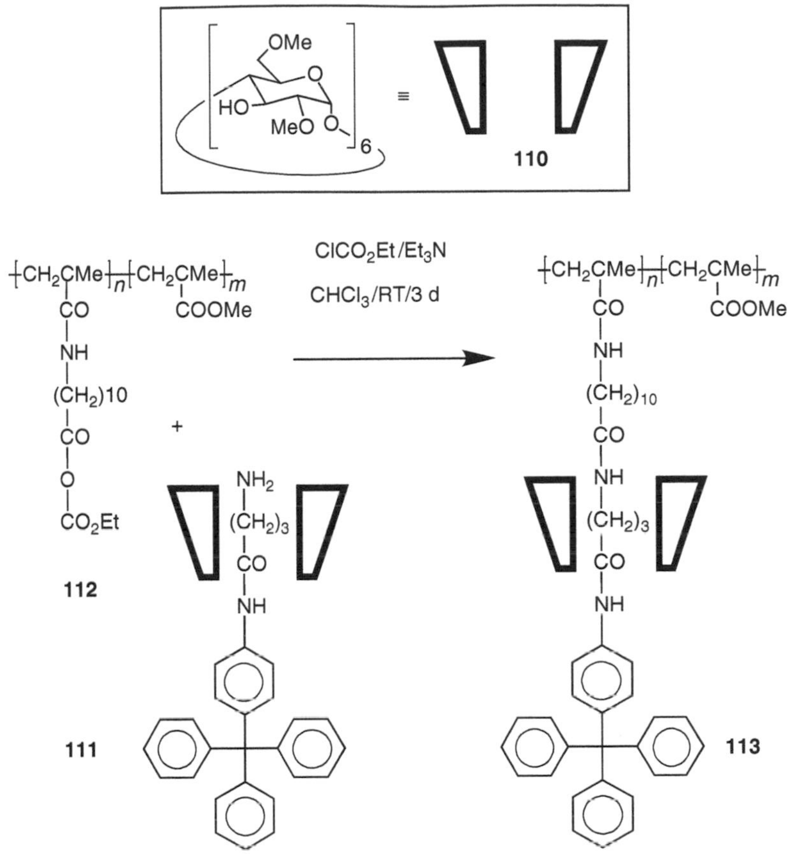

Figure 31 The synthesis of the polyrotaxane **113**.

example of a side-chain polyrotaxane is illustrated in Figure 33. In this instance, the semirotaxane **117**, incorporating the β-CD derivative **116**, was reacted [67] with the poly(benzimidazole) **118** in the presence of sodium hydride. The IR spectra of the resulting polyrotaxane **119** displayed the same bands which are observed for the parent polymer with the addition of a signal at 1040 cm^{-1} for the [C–O] groups of the CD rings. The amount of threaded macrocycles was determined by ^{1}H-NMR spectroscopy which indicated that 57% of the side chains of the polyrotaxane **119** are encircled by β-CD rings. GPC analysis of this material revealed M_n and M_w values of 25.4×10^3 and 42.1×10^3, respectively.

A different strategy that allows access to side-chain polyrotaxane systems is the polymerization of a semirotaxane (or rotaxane) (Figure 30, route **F**). In this case an inert stopper is placed at one end of the monomer while a polymerizable end group is placed at the other end [68]. An example of such a synthesis is shown in Figure 34. In this case hydrogen bonding is the driving force for the formation of the semirotaxane **120** with dibenzo[24]crown8 as the macrocyclic component and a secondary ammonium salt bearing an acrylate functionality at one end as the threadlike monomer unit. Free-radical polymerization of the semirotaxane **120** followed by acylation of the ammonium centers yields **121**, a copolymer of the rotaxane and free "thread" repeat units [69]. The ammonium centers were acetylated to "switch-off" the interaction between the crown and side chain. The percentage of side chains that had the rotaxane architecture was found to vary with solvent and temperature with a maximum value of ca. 50%.

An alternative approach to side-chain (pseudo)polyrotaxanes involves the threading of acyclic monomeric components through the cavities of large rings appended to a polymeric backbone.

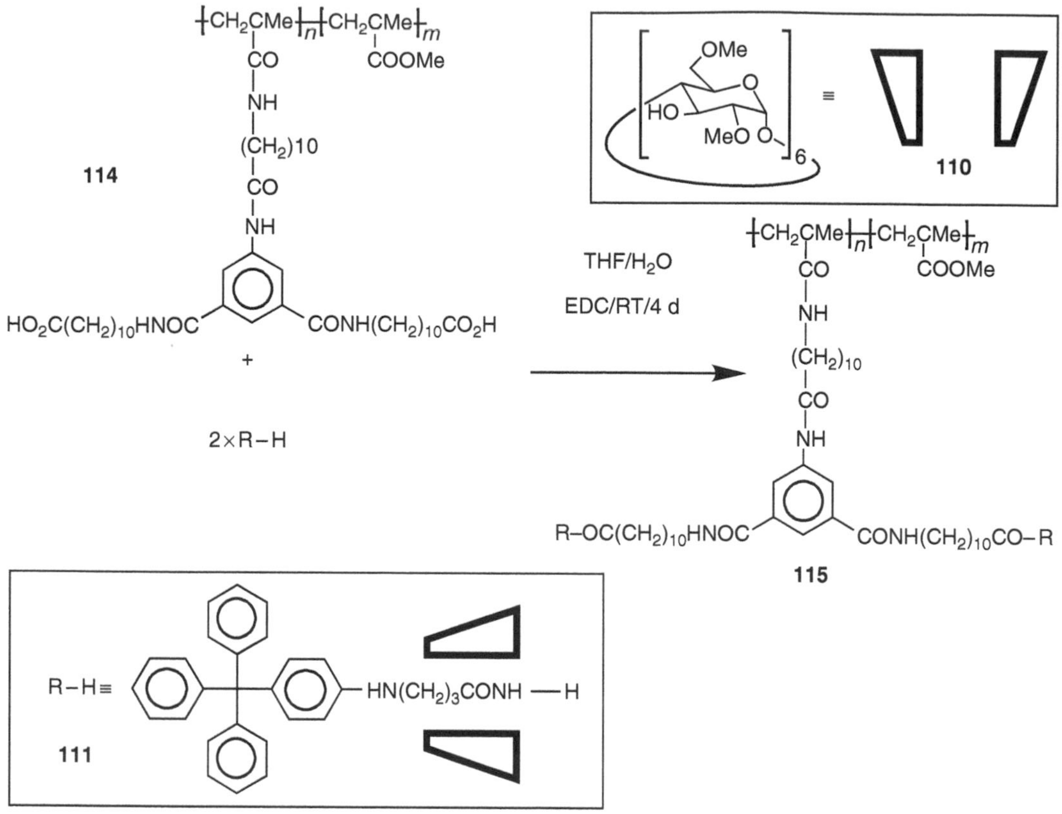

Figure 32 The synthesis of the polyrotaxane **115**.

This strategy was employed [70] to generate (Figure 35) side-chain pseudopolyrotaxanes. The polymers **122** to **124** were obtained by polymerizing preformed macrocyclic monomers. As a result of [C–H···O] hydrogen bonds and [π···π] stacking interactions, insertion of the bipyridinium-based guests **125** and **126** inside the cavities of the appended rings of **122** to **124** affords side-chain pseudopolyrotaxanes. Interestingly, electrostatic perturbation upon binding induces a significant decrease of carrier mobility and, as a result, conductivity in **122**. Similarly, a reduction of the fluorescence intensity of **123** and **124** is observed on binding. It is believed that the excitations diffuse along the polymer backbone and are quenched by the bipyridinium-based guests inserted through the appended macrocycles. Thus, in all instances, pseudopolyrotaxane formation can be easily detected by monitoring the change of the properties associated with the polymer backbone.

V. MECHANICALLY INTERWOVEN POLYMER CHAINS

A mechanically interwoven polyrotaxane differs from other polyrotaxane architectures in that the backbone of the polymer comprises both mechanical and covalent bonds. Figure 36 shows some different possible architectures that have both these features present along their polymer backbone. There are a variety of possible synthetic routes to such systems. An obvious strategy is the self-assembly of the pseudopolyrotaxane system and then, if desired, attachment of the stopper to yield the polyrotaxane. Following this strategy essentially means that the DP of the polymer will be influenced by the strength of the interaction between the ring and the threadlike components. As a result these pseudorotaxane architectures, which are formed through a supramolecular

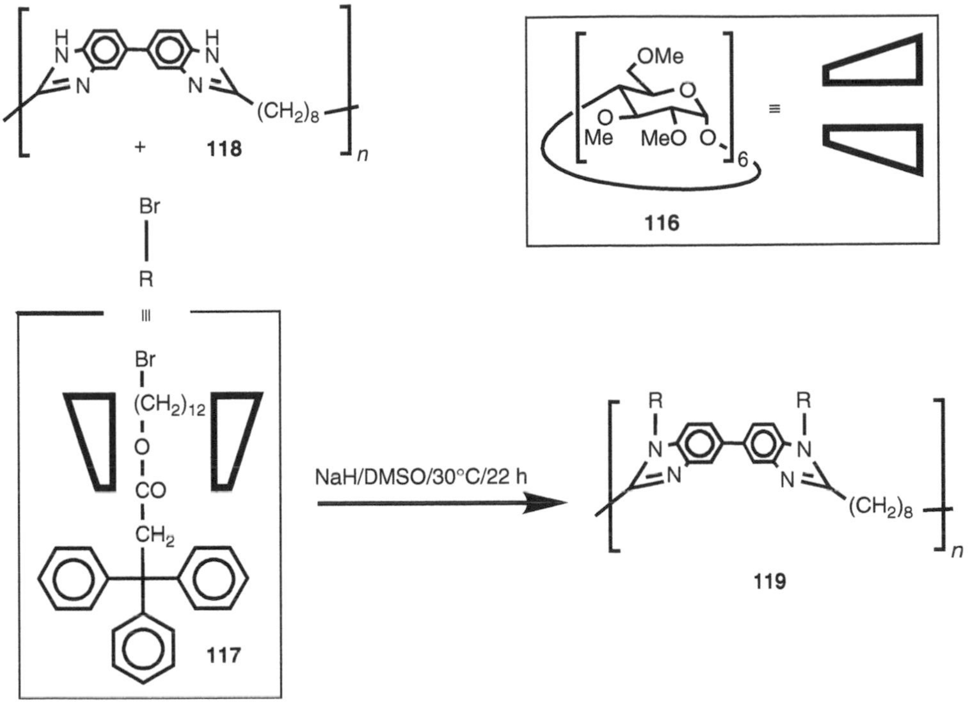

Figure 33   The synthesis of the polyrotaxane **119**.

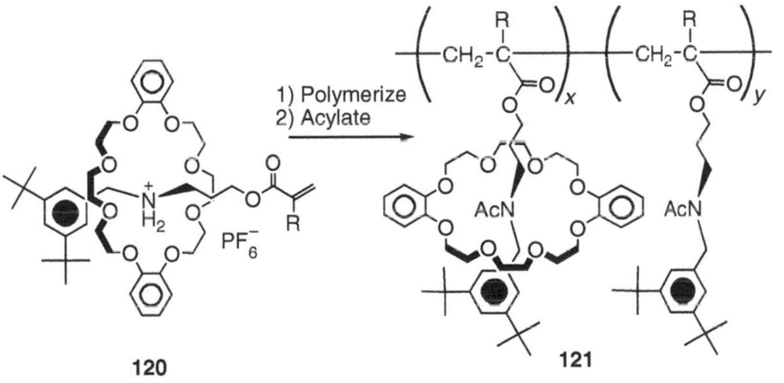

Figure 34   The synthesis of the polyrotaxane **121**.

polymerization process, are extremely sensitive to external environmental factors. Namely, any factor (e.g., temperature, etc.) which changes the degree of interaction between the two components can significantly alter the DP of the aggregate. The conceptually simplest form of the mechanically interwoven (pseudo)polyrotaxanes is the so-called "daisy-chain" (pseudo)polyrotaxanes, where a single heteroditopic monomeric unit, which combines both the ringlike and rodlike components, can self-associate into polymeric architectures. A number of attempts have been made to realize such structures using this methodology. Early examples have shown that careful monomer design is required in order to inhibit the entropically favored formation of macrocyclic structures [71]. Figure 37 shows some structures that have been shown to form macrocyclic products in solution extensively. However, it should be noted that with such dynamic systems it can be expected that an increase in the monomer concentration will favor the formation of polymeric systems. Monomer **127**,

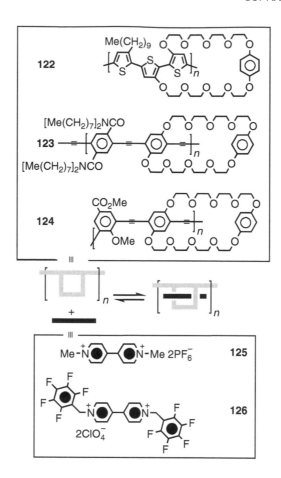

Figure 35 The self-assembly of pseudopolyrotaxanes incorporating π-electron-rich and -deficient components.

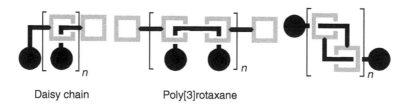

Daisy chain   Poly[3]rotaxane

Figure 36 Schematic representation of some possible mechanically interwoven polyrotaxane constructs.

which uses the paraquat/π-electron-rich crown ether binding motif, shows a dramatic increase in solution viscosity at ca. 2 M in acetone, which along with the ability to obtain fibers from these solutions suggest that **127** does form polymeric aggregates at high concentrations [72]. Another system which has been investigated with the goal of obtaining daisy-chain type polymeric structures utilizes α- or β-CDs functionalized with an aromatic ring on one of the primary hydroxyl groups. A series of six-substituted CDs (with either cinnamoyl or hydrocinnamoyl groups attached) have been investigated [73]. Both the size of the CD (α or β) and the rigidity of the spacer group determine the nature of the aggregation. With α-CD no small cyclic or linear oligomers are observed. With the larger β-CD intramolecular complexes result if the spacer is flexible. However, **129** which has the more rigid cinnamoyl substituent, is insoluble in water unless a competing guest, which

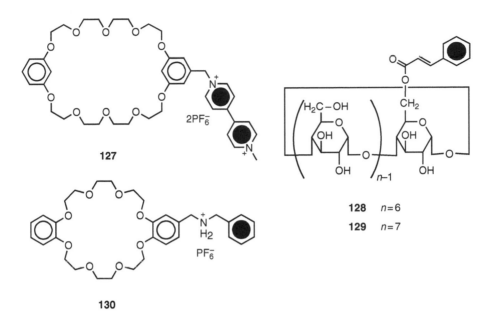

127

128   n=6
129   n=7

130

Figure 37   Heteroditopic monomeric units that can self-associate into "Daisy-Chain" architectures.

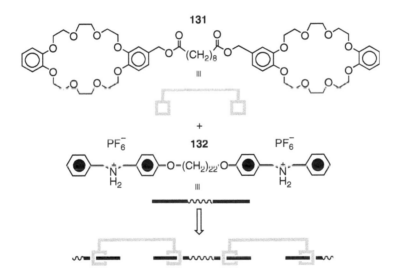

Figure 38   Formation of supramolecular polymers from the homoditopic units **131** and **132**.

would effectively act as a chain terminator, is added. Wide angle x-ray diffraction data of **129** are consistent with this compound forming supramolecular polymers in the solid state with the CDs forming a layered structure.

An alternative architecture (Figure 38) is obtained if two complementary homoditopic monomers are mixed together. Supramolecular polymers of this type, which uses the dibenzo[24]crown8 and dibenzylammonium molecular recognition motif, have been investigated [74]. For example, the two component system comprised of an equimolar amount of the bis(crown ether) homoditopic host **131** and bisammonium salt homoditopic guest **132** [75] in a concentrated ($>0.5\ M$) acetone/chloroform (1/1) solution results in reversible chain extension thus yielding the linear supramolecular polymers. ^{1}H-NMR data (end-group analysis) suggests that at 2.0 M the M_{n} of the aggregate is $\sim$18,000, which

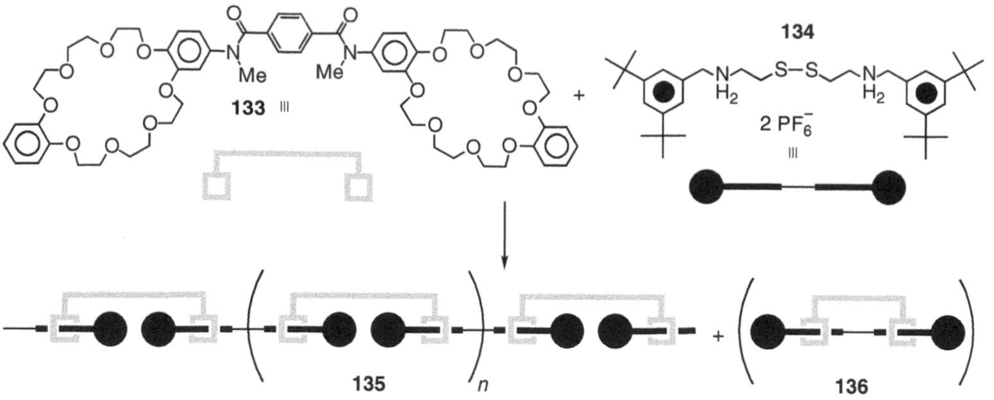

Figure 39   Dynamic-covalent synthesis of the poly[3]rotaxane **135**.

corresponds to a DP of ~9. Furthermore, solution viscosity experiments of the supramolecular assemblies showed a behavior characteristic of the formation of large linear aggregates, and the preparation of films and fibers from concentrated solutions further displays the polymeric nature of these materials. It is important for the formation of polymeric aggregates in these systems that the distance between the two binding sites in the complementary homoditopic monomers is such that the formation of entropically favored 1:1 cyclic species is inhibited. These supramolecular polymers are of course pseudopolyrotaxanes. Attempts have also been reported to prepare rotaxane versions of these systems, the so-called poly[3]rotaxanes, using the same binding motif [76]. The synthetic route used to access such systems utilizes reversible (dynamic) covalent chemistry [77]. A bis(crown ether) homoditopic monomer **133** and dumbbell-like bisammonium homoditopic monomer **134**, which is terminated with bulky stoppers, were used to construct the desired interlocked species (Figure 39). The bisammonium monomer possesses a disulfide link within the molecule that can reversibly exchange with a thiol moiety. The set up of such an equilibrium results temporarily in the removal of the bulky stopper groups thus allowing the crown ether units to "slip" onto the backbone and bind to the ammonium functionality. Equimolar amounts of **133** and **134** were reacted with a catalytic amount of benzenethiol in a mixture of d-chloroform/d-acetonitrile. When reacted at low concentration (0.1 M) the resulting products are cyclic oligomers with the main product being the [1 + 1]cyclic[2]rotaxane **136** which was isolated as a white solid by preparative GPC. Increasing the monomer concentration to 1.0 M resulted in a bimodal molecular weight distribution that was observed by GPC. The low molecular weight fraction corresponded to the cyclic oligomers. The higher molecular weight material (**135**) was isolated in 64% yield by GPC and end-group analysis of this fraction by ^{1}H-NMR estimated a DP of 29 corresponding to an M_n value of 28×10^3.

VI.   INTERWOVEN DENDRIMERS

Conceptually there are a variety of ways that an interlocked structure can be incorporated in to a dendritic polymer. For example, the interlocked structure could be at the dendrimer core, the dendrimer surface, as part of the branching units [78], or within the dendritic arms (Figure 40).

[4]Pseudorotaxane dendrimers have been self-assembled [79] from a trisammonium homotritopic core **137** and a series of monotopic crown ethers functionalized with different generations of benzyl ether dendrons to yield dendritic species that have the interlocked architecture at the dendrimer core (Figure 41). Combining the core **137** with dibenzo[24]crown8 derivatives with first, second, or third generation benzyl ether dendrons attached to the ring (**138b** to **d**) in acetone resulted in the formation of [2]-, [3]-, and [4]pseudorotaxane dendrimeric species as observed by ^{1}H-NMR. Determination of

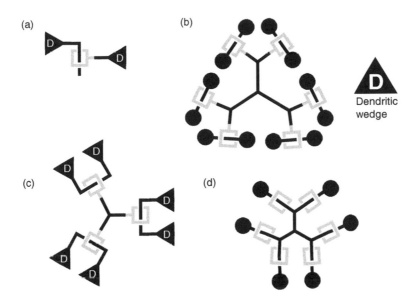

Figure 40 Schematic representation of some possible dendrimers containing interlocked structures: (a) at the core, (b) at the dendrimer surface, (c) as the branching units, and (d) within the dendritic arms.

the three binding constants for the formation of the these pseudorotaxane architectures, for example, for **139d** $K_1 = 250$; $K_2 = 140$; $K_3 = 210\,M^{-1}$, demonstrated that they exceeded the statistical ratio of 3:1:0.33 expected for independent binding which suggests that positive cooperativity is driving the self-assembly of the [4]pseudorotaxane dendrimers. In addition, it was shown that the extent of cooperativity increases with the size of the dendron which was attributed to the ability of the dendron to screen the ionic core from the nonpolar solvent. Furthermore, the dendritic [4]pseudorotaxanes are formed almost exclusively (>90%) in chloroform-d. A structurally similar dendritic species has been prepared using a similar binding motif and the so-called "slippage" approach [80]. This thermodynamic approach to rotaxanes relies on the fact that while under certain conditions, for example, room temperature and nonpolar solvents, the stopper groups (in this case cyclohexyl groups) are sufficiently large enough to prevent dethreading of the macrocyclic component (here dibenzo[24]crown8). However, under different conditions, for example, higher temperatures (40°C) in this system or polar solvents (DMSO), the stoppers are small enough to allow an equilibrium to be established between the "free" components and the rotaxane. A dibenzo[24]crown8 with two benzyl ether dendrons attached to its aromatic rings was assembled at 40°C in CH_2Cl_2 with a benzyl ether dendron, which has a cyclohexylmethyl ammonium at its focal point. After 90 d the dendritic [2]rotaxane was isolated by column chromatography in ca. 20% yield. This dendritic species could be converted back to its "free" components by dissolving in d^6-DMSO. The dissociation was slow showing a half life, $t_{1/2}$, of 17.7 h.

Pseudorotaxane-functionalized dendrimers, where the pseudorotaxane architecture is on the surface of the dendrimer, have also been investigated [81]. The primary amines of first and third generation PPI dendrimers were converted into isocyanates and then coupled to 5-hydroxymethyl-1,3-phenylene-1,3-phenylene[32]crown10, which is known to bind to paraquat derivatives, to yield "crowned" PPI dendrimers **140** and **141** (93 and 83% yields, respectively) shown in Figure 42. The binding of the paraquat diol **142** to the multicrowned dendrimers was not only reduced compared with a model crown system but also the sequential binding of the guests to these complementary polytopic dendrimers was found to be anticooperative. Interestingly, when the tertiary amines of the dendrimers are protonated with trifluoroacetic acid, the binding of **142** was found to be independent of the number of bound sites and the average association constant was found to be similar to the model crown system. This difference in binding ability of the crowned dendrimer upon protonation

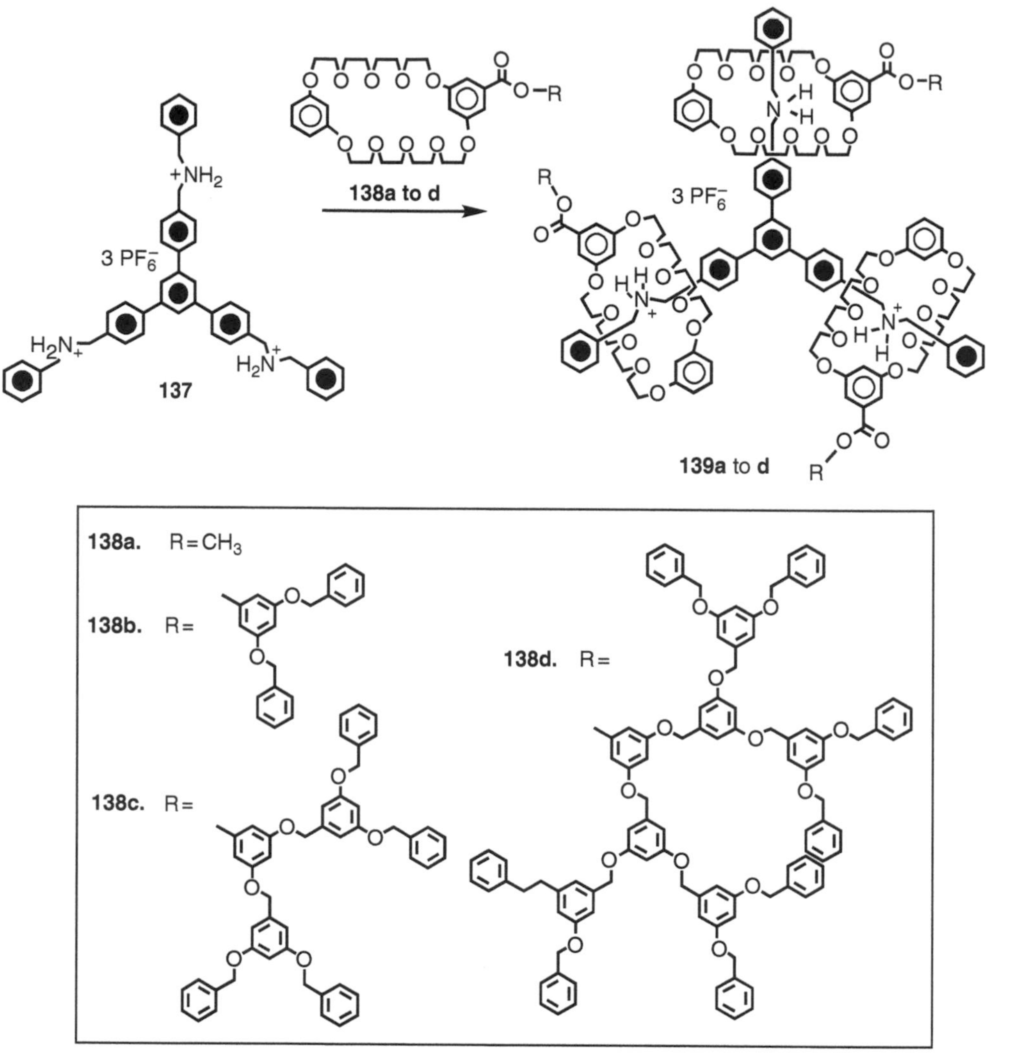

Figure 41   The donor/acceptor template-directed formation of first to third generation dendrimers **139b** to **d**.

was explained by a rigidification of the protonated dendrimer that results in the crown ether binding sites being less crowded and more accessible.

VII.   POLYROTAXANES VIA POST-ASSEMBLY MODIFICATION

One conceptually easy way to access polyrotaxane structures is to prepare a rotaxane monomer which has reactive groups on (or as) its stoppers and/or on its macrocyclic component and then subsequently carry out a conventional polymerization reaction. This potentially opens up access to a wide range of interlocked polymeric architectures from simple main-chain polyrotaxanes to molecular necklaces, the daisy-chain systems and even dendritic interlocked species.

There are a few reports outlining the construction of larger interlocked structures by synthesizing known rotaxane systems with functional groups attached to their stopper and ring components, which can then be converted into the desired material via a subsequent conventional polymerization reaction [82]. An elegant example of such a strategy is the use of rotaxane monomers bearing stoppers that contain blocked isocyanate functionalities that has allowed access to main-chain, side-chain,

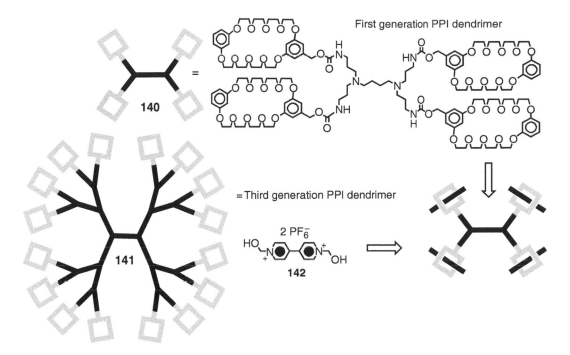

Figure 42   The formation of "crowned" dendrimers containing interwoven structures at the surface.

and crosslinked network polyrotaxanes [83]. The different caprolactam blocked isocyanate rotaxane building blocks were prepared using amide hydrogen bond templating around a series fumaramide derived dumbbell components. Three different rotaxane monomers were prepared (Figure 43) that have either one blocked isocyanate group on each stopper, **143**, two blocked isocyanate groups on one stopper, **144**, or two blocked isocyanate groups on each stopper, **145**. These monomers were polymerized in the bulk with the appropriate amounts of a bisamine polyether comonomer, Jeffamine ($M_n = 400$), at 175°C to yield a main-chain, side-chain, and crosslinked network polyrotaxanes, respectively. The linear polymers were measured by GPC and were found to possess on average 96 rotaxane units for the side-chain ($M_w = 165 \times 10^3$) and 46 for the main-chain polyrotaxanes ($M_w = 86 \times 10^3$). The crosslinked polymer was insoluble, however, ^{1}H-NMR performed on a CDCl$_3$ swollen sample of the material confirmed that the rotaxane blocks remained intact during the reaction. Dynamic mechanical testing (DMTA) measurements on a film of this polyrotaxane gave a plateau modulus of 10^6 Pa and its elongation at break was found to be $>100\%$ confirming its rubbery nature.

An example of rotaxanes in which the functional (or reactive) group is the stopper are the triphenyl phosphonium-stopped rotaxanes [84]. These "surrogate"-stopped rotaxanes can be easily reacted with a variety of aldehydes, via standard Wittig chemistry, resulting in a direct stopper exchange process to yield larger rotaxane aggregates without any dethreading of the macrocyclic component [85]. Using this methodology a variety of daisy-chain systems [86], linear main-chain oligorotaxanes, molecular necklaces [87], and dendritic rotaxanes in which one of the branching points is a mechanical bond [88] have all been prepared (Figure 44).

VIII.   PROPERTIES

Since the synthetic strategies for the preparation of poly[2]catenanes have been developed [14–20,22] only very recently, their properties are just beginning to be investigated. By contrast, the properties of pseudopolyrotaxanes and polyrotaxanes have been studied extensively and are understood in some

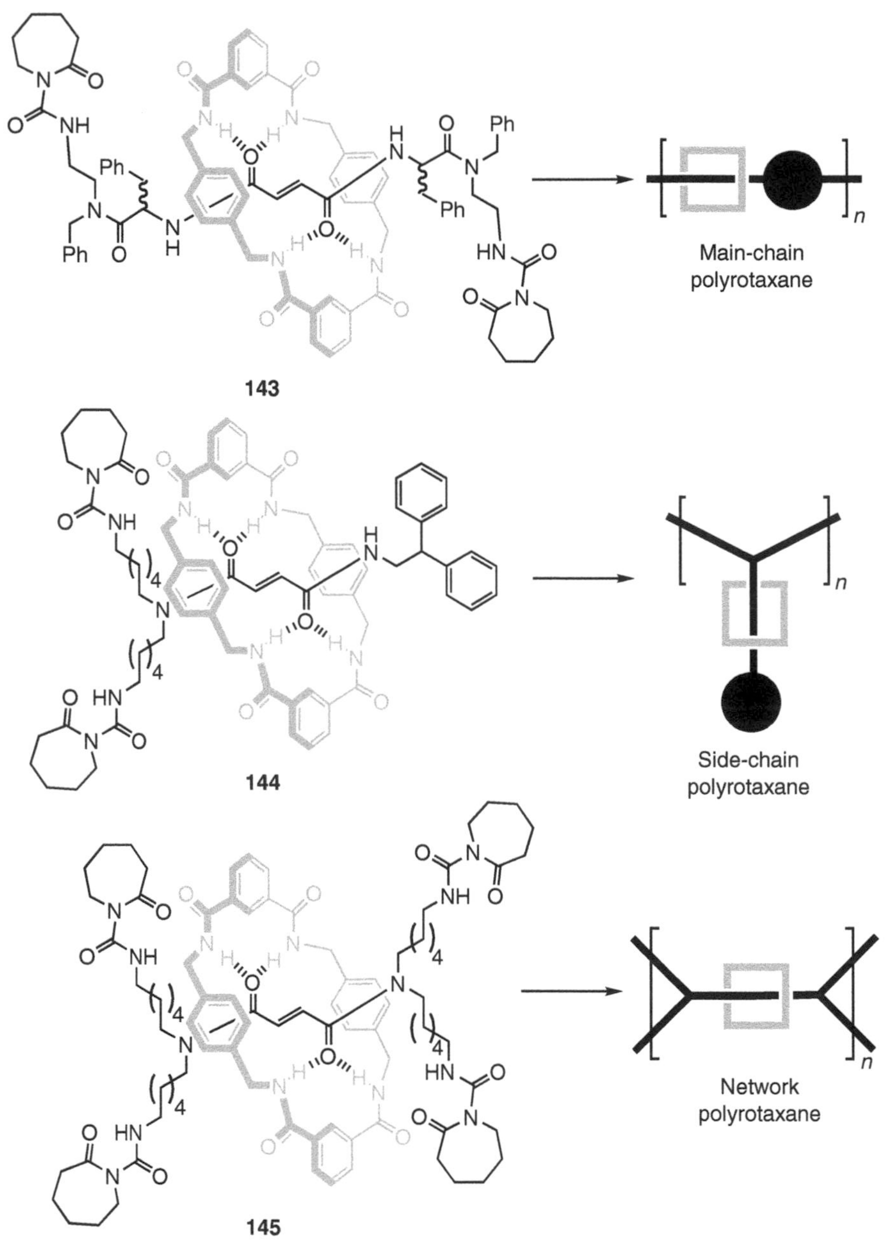

143

144

145

Figure 43 Polymerization of caprolactam-blocked rotaxanes **143** to **145** to form main-chain, side-chain, and network polyrotaxanes.

considerable detail already. Indeed, it has been established that solubilities, phase transition behavior, hydrodynamic volumes, and the intrinsic viscosities of pseudopolyrotaxanes and polyrotaxanes are significantly different from those of their separate components. Solubilities can either be depressed or enhanched as a result of interlocking. For example, poly(ethylene glycol) and α-CD are soluble [37] in H_2O whereas, the pseudopolyrotaxane formed on threading poly(ethylene glycol) through many α-CD rings is not soluble in water. On the other hand, poly(acrylonitrile) is insoluble [40c] in MeOH but it becomes soluble when encircled by 60-crown-20. In most instances, two distinct melting points, slightly lower than those of the separate acyclic and cyclic components, are associated with

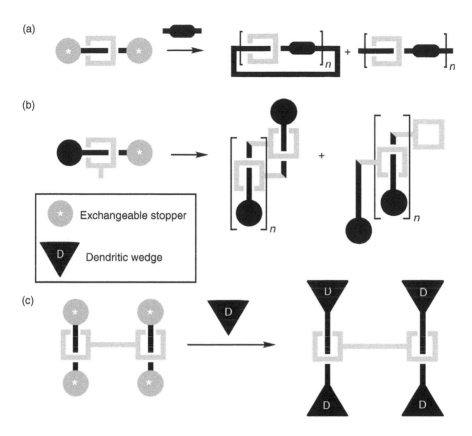

Figure 44 Schematic representation of the interlocked structures (a) molecular necklaces, (b) daisy-chains, and (c) interlocked dendrimers which have been prepared using the phosphonium stopper exchange methodology.

pseudopolyrotaxanes. This behavior is observed normally for those pseudopolyrotaxanes where the ring components are sufficiently mobile to aggregate, nucleate, and crystallize. Significant changes in the T_g values are observed when phase mixing of the mechanically interlocked components occurs. In polyurethane-based pseudopolyrotaxanes, the T_g corresponds [32c] to a weighted average of those of the separate components, while the T_gs of side-chain polyrotaxanes tend [66a] to be often higher than those of the parent components. Extensive GPC analyses has revealed [89] that hydrodynamic volumes and, as a result, intrinsic viscosities increase on mechanical interlocking. Indeed, the change of intrinsic viscosity on threading has been exploited [41a] to monitor the formation of CD-containing pseudopolyrotaxanes.

IX. CONCLUSIONS

Polycatenanes and polyrotaxanes are macromolecules held together by a combination of covalent and mechanical bonds, in some instances aided and abetted by noncovalent bonding interactions. This 'unconventional' combination of bonds imposes upon these polymeric systems unique topologies and unusual properties, making them particularly attractive synthetic targets. Not surprisingly, a number of synthetic strategies for the construction of mechanically-interlocked macromolecules have been developed. Most of these synthetic approaches rely upon the assistance of noncovalent bonding interactions and borrow recognition motifs widely employed for the template-directed syntheses of monomeric catenanes and rotaxanes. Thus, poly[2]catenanes, as well as main- and side-chain polyrotaxanes, are now more than just simple chemical curiosities: they can be prepared easily and efficiently even in large quantities! Investigation of the current generation of systems has revealed that their physical and mechanical behavior can be drastically affected by the nature and frequency of the interlocking structural elements. The continuing study of mechanically-bonded polymers might well result in the development of materials which exhibit a new matrix of polymer properties not found in more conventional architectures and therefore, find some important applications in the near future. For example, the development of interlocked structures that can be "switched" between different states has received a lot of attention in the field of molecular electronics, devices and machines.[90,91,92,93,94,95,96] Specially designed rotaxanes have been shown to reversibly "shuttle" under certain conditions, and redox-controllable catenanes will "flip" like a switch at different potentials. The incorporation of such switchable catenanes and rotaxanes, sensitive to a variety of stimuli e.g., temperature, pH, light, electrical fields, into a polymeric architecture opens up new routes to stimuli-responsive materials. For example, such polymeric systems could conceivably result in materials which can expand/contract in response to an external stimulus, not unlike a muscle, or be utilized as coatings for smart surfaces whose properties (e.g. wetting, adhesion, etc) can be switched on demand. In fact, research in the field has begun to move past the stage of pure synthesis, into a new regime of applied science and engineering, in which the researcher now looks to other disciplines to evolve these novel architectures into nano- and macro-scale devices.

ACKNOWLEDGMENTS

The authors would like to thank the NSF (CAREER: CHE-0133164), the U.S. Army Research Office, and the NIH (NIBIB) for financial support.

REFERENCES

1. MacGregor, E.A. and Greenwood, C.T. *Polymers in Nature*, Wiley, New York, **1980**.
2. Salamone, J.C., Ed. *The Polymeric Materials Encyclopedia*, CRC Press, Boca Raton, FL, **1996**.
3. For books and reviews on mechanically interlocked molecules and macromolecules, see: (a) Schill, G. *Catenanes, Rotaxanes and Knots*, Academic Press, New York, **1971**. (b) Walba, D.M. *Tetrahedron* **1985**, *41*, 3161–3212. (c) Dietrich-Buchecker, C.O. and Sauvage, J.-P. *Chem. Rev.* **1987**, *87*, 795–810. (d) Dietrich-Buchecker, C.O. and Sauvage, J.-P. *Bioorg. Chem. Front.* **1991**, *2*, 195–248. (e) Chambron, J.-C., Dietrich-Buchecker, C.O., and Sauvage, J.-P. *Top. Curr. Chem.* **1993**, *165*, 131–162. (f) Gibson, H.W. and Marand, H. *Adv. Mater.* **1993**, *5*, 11–21. (g) Gibson, H.W., Bheda, M.C., and Engen, P.T. *Prog. Polym. Sci.* **1994**, *19*, 843–945. (h) Amabilino, D.B., Parsons, I.W., and Stoddart, J.F. *Trends Polym. Sci.* **1994**, *2*, 146–152. (i) Amabilino, D.B. and Stoddart, J.F. *Chem. Rev.* **1995**, *95*, 2725–2828. (j) Gibson, H.W. *Large Ring Molecules*, Ed. Semlyen, J.A., Wiley, New York, **1996**, 191–202. (k) Belohradsky, M., Raymo, F.M., and Stoddart, J.F. *Collect. Czech. Chem. Commun.* **1996**, *61*, 1–43. (l) Raymo, F.M. and Stoddart, J.F. *Trends Polym. Sci.* **1996**, *4*, 208–211. (m) Belohradsky, M., Raymo, F.M., and Stoddart, J.F. *Collect. Czech. Chem. Commun.* **1997**, *62*, 527–557. (n) Jäger, R. and Vögtle, F. *Angew. Chem. Int. Ed. Engl.* **1997**, *36*, 930–944. (o) Sauvage, J.-P. and Dietrich-Buchecker, C.O., Eds. *Catenanes, Rotaxanes and Knots*, VCH-Wiley, Weinheim, **1999**.
4. (a) Sauvage, J.-P. *Acc. Chem. Res.* **1990**, *23*, 319–327. (b) Chambron, J.-C., Dietrich-Buchecker, C.O., Hemmert, C., Khemiss, A.K., Mitchell, D., Sauvage, J.-P., and Weiss, J. *Pure Appl. Chem.* **1990**, *62*, 1027–1034. (c) Chambron, J.-C., Chardon-Noblat, S., Harriman, A., Heitz, V., and Sauvage, J.-P. *Pure Appl. Chem.* **1993**, *65*, 2343–2349. (d) Chambron, J.-C., Dietrich-Buchecker, C.O., Nierengarten, J.-F., and Sauvage, J.-P. *Pure Appl. Chem.* **1994**, *66*, 1543–1550. (e) Chambron, J.-C., Dietrich-Buchecker, C.O., Heitz, V., Nierengarten, J.-F., Sauvage, J.-P., Pascard, C., and Guilhem, J. *Pure Appl. Chem.* **1995**, *67*, 233–240. (f) Chambron, J.-C., Dietrich-Buchecker, C.O., and Sauvage, J.-P. *Comprehensive Supramolecular Chemistry*, Vol. 9, Eds. Hosseini, M.W. and Sauvage, J.-P., Pergamon, Oxford, **1996**, pp. 43–83.

5. (a) Bickelhaupt, F. *J. Organomet. Chem.* **1994**, *475*, 1–14. (b) Fujita, M. and Ogura, K. *Coord. Chem. Rev.* **1996**, *148*, 249–264. (c) Fujita, M. *Comprehensive Supramolecular Chemistry*, Vol. 9, Eds. Hosseini, M.W. and Sauvage, J.-P., Pergamon, Oxford, **1996**, pp. 253–282. (d) Jeon, Y.M., Whang, D., Kim, J., and Kim, K. *Chem. Lett.* **1996**, *25*, 503–504. (e) Whang, D., Jeon, Y.M., Heo, J., and Kim, K. *J. Am. Chem. Soc.* **1996**, *118*, 11333–11334. (f) Whang, D., Heo, J., Kim, C.A., and Kim, K. *Chem. Commun.* **1997**, 2361–2362. (g) Whang, D. and Kim, K. *J. Am. Chem. Soc.* **1997**, *119*, 451–452. (h) Whang, D., Park, K.M., Heo, J., Ashton, P.R., and Kim, K. *J. Am. Chem. Soc.* **1998**, *120*, 4899–4900. (i) Roh, S.G., Park, K.M., Park, G.J., Sakamoto, S., Yamaguchi, K., and Kim, K. *Angew. Chem. Int. Ed.* **1999**, *38*, 638–641.

6. (a) Amabilino, D.B. and Stoddart, J.F. *Pure Appl. Chem.* **1993**, *65*, 2351–2359. (b) Pasini, D., Raymo, F.M., and Stoddart, J.F. *Gazz. Chim. Ital.* **1995**, *125*, 431–435. (c) Langford, S.J. and Stoddart, J.F. *Pure Appl. Chem.* **1996**, *68*, 1255–1260. (d) Amabilino, D.B., Raymo, F.M., and Stoddart, J.F. *Comprehensive Supramolecular Chemistry*, Vol. 9, Eds. Hosseini, M.W. and Sauvage, J.-P., Pergamon, Oxford, **1996**, pp. 85–130. (e) Raymo, F.M. and Stoddart, J.F. *Pure Appl. Chem.* **1997**, *69*, 1987–1997. (f) Gillard, R.E., Raymo, F.M., and Stoddart, J.F. *Chem. Eur. J.* **1997**, *3*, 1933–1940. (g) Raymo, F.M. and Stoddart, J.F. *Chemtracts* **1998**, *11*, 491–511.

7. (a) Hamilton, D.G., Sanders, J.K.M., Davies, J.E., Clegg, W., and Teat, S.J. *Chem. Commun.* **1997**, 897–898. (b) Try, A.C., Harding, M.M., Hamilton, D.G., and Sanders, J.K.M. *Chem. Commun.* **1998**, 723–724. (c) Hamilton, D.G., Davies, J.E., Prodi, L., and Sanders, J.K.M. *Chem. Eur. J.* **1998**, *4*, 608–620. (d) Hamilton, D.G., Feeder, N., Prodi, L., Teat, S.J., Clegg, W., and Sanders, J.K.M. *J. Am. Chem. Soc.* **1998**, *120*, 1096–1097.

8. (a) Hunter, C.A. and Purvis, D.H. *Angew. Chem. Int. Ed. Engl.* **1992**, *31*, 792–795. (b) Hunter, C.A. *J. Am. Chem. Soc.* **1992**, *114*, 5303–5311. (c) Hunter, C.A. *Chem. Soc. Rev.* **1994**, *23*, 101–109. (d) Carver, F.J., Hunter, C.A., and Shannon, R.J. *J. Chem. Soc. Chem. Commun.* **1994**, 1277–1280. (e) Adams, H., Carver, F.J., and Hunter, C.A. *J. Chem. Soc. Chem. Commun.* **1995**, 809–810. (f) Brodesser, G., Güther, R., Hoss, R., Meier, S., Ottens-Hildebrandt, S., Schmitz, J., and Vögtle, F. *Pure Appl. Chem.* **1993**, *65*, 2325–2328. (g) Vögtle, F., Dünnwald, T., and Schmidt, T. *Acc. Chem. Res.* **1996**, *29*, 451–460. (h) Vögtle, F., Jäger, R., Händel, M., and Ottens-Hildebrandt, S. *Pure Appl. Chem.* **1996**, *68*, 225–232. (i) Johnston, A.G., Leigh, D.A., Pritchard, R.J., and Deegan, M.D. *Angew. Chem. Int. Ed. Engl.* **1995**, *34*, 1209–1212. (j) Johnston, A.G., Leigh, D.A., Nezhat, L., Smart, J.P., and Deegan, M.D. *Angew. Chem. Int. Ed. Engl.* **1995**, *34*, 1212–1216. (k) Leigh, D.A., Moody, K., Smart, J.P., Watson, K.J., and Slawin, A.M.Z. *Angew. Chem. Int. Ed. Engl.* **1996**, *35*, 306–310. (l) Johnston, A.G., Leigh, D.A., Murphy, A., Smart, J.P., and Deegan, M.D. *J. Am. Chem. Soc.* **1996**, *118*, 10662–10663. (m) Leigh, D.A., Murphy, A., Smart, J.P., and Slawin, A.M.Z. *Angew. Chem. Int. Ed. Engl.* **1997**, *36*, 728–732. (n) Lane, A.S., Leigh, D.A., and Murphy, A. *J. Am. Chem. Soc.* **1997**, *119*, 11092–11093. (o) Leigh, D.A., Murphy, A., Smart, J.P., Deleuze, M.S., and Zerbetto, F. *J. Am. Chem. Soc.* **1998**, *120*, 6458–6467.

9. (a) Kolchinski, A.G., Busch, D.H., and Alcock, N.W. *J. Chem. Soc. Chem. Commun.* **1995**, 1289–1291. (b) Kolchinski, A.G., Alcock, N.W., Roesner, R.A., and Busch, D.H. *Chem. Commun.* **1998**, 1437–1438. (c) Glink, P.T., Schiavo, C., Stoddart, J.F., and Williams, D.J. *Chem. Commun.* **1996**, 1483–1490. (d) Glink, P.T. and Stoddart, J.F. *Pure Appl. Chem.* **1998**, *70*, 419–424. (e) Fyfe, M.C.T. and Stoddart, J.F. *Coord. Chem. Rev.* **1999**, *183*, 139–155. (f) Fyfe, M.C.T. and Stoddart, J.F. *Adv. Supramol. Chem.* **1999**, *5*, 1–53.

10. (a) Stoddart, J.F. *Angew. Chem. Int. Ed. Engl.* **1992**, *31*, 846–848. (b) Ogino, H. *New J. Chem.* **1993**, *17*, 683–688. (c) Wenz, G., Wolf, F., Wagner, M., and Kubik, S. *New J. Chem.* **1993**, *17*, 729–738. (d) Isnin, R. and Kaifer, A.E. *Pure Appl. Chem.* **1993**, *65*, 495–498. (e) Harada, A. *Polym. News* **1993**, *18*, 358–363. (f) Harada, A., Li, J., and Kamachi, M. *Proc. Jpn. Acad.* **1993**, *69*, 39–44. (g) Wenz, G. *Angew. Chem. Int. Ed. Engl.* **1994**, *33*, 802–822. (h) Harada, A. *Coord. Chem. Rev.* **1996**, *148*, 115–133. (i) Harada, A. *Large Ring Molecules*, Ed. Semlyen, J.A., Wiley, New York, **1996**, pp. 406–432. (j) Harada, A. *Supramol. Sci.* **1996**, *3*, 19–23. (k) Harada, A. *Adv. Polym.* **1997**, *133*, 142–191. (l) Harada, A. *Carbohydrate Polym.* **1997**, *34*, 183–188. (m) Harada, A. *Acta Polym.* **1998**, *49*, 3–17. (n) Nepogodiev, S.A. and Stoddart, J.F. *Chem. Rev.* **1998**, *98*, 1959–1976.

11. For accounts and reviews on template-directed syntheses, see: (a) Busch, D.H. and Stephenson, N.A. *Coord. Chem. Rev.* **1990**, *100*, 119–154. (b) Lindsey, J.S. *New J. Chem.* **1991**, *15*, 153–180. (c) Whitesides, G.M., Mathias, J.P., and Seto, C.T. *Science* **1991**, *254*, 1312–1319. (d) Philp, D. and Stoddart, J.F. *Synlett* **1991**, 445–458. (e) Busch, D.H. *J. Inclusion Phenom.* **1992**, *12*, 389–395. (f) Anderson, S., Anderson, H.L., and Sanders, J.K.M. *Acc. Chem. Res.* **1993**, *26*, 469–475. (g) Cacciapaglia, R. and Mandolini, L. *Chem.*

Soc. Rev. **1993**, *22*, 221–231. (h) Hoss, R. and Vögtle, F. *Angew. Chem. Int. Ed. Engl.* **1994**, *33*, 375–384. (i) Schneider, J.P. and Kelly, J.W. *Chem. Rev.* **1995**, *95*, 2169–2187. (j) Philp, D. and Stoddart, J.F. *Angew. Chem. Int. Ed. Engl.* **1996**, *35*, 1155–1196. (k) Raymo, F.M. and Stoddart, J.F. *Pure Appl. Chem.* **1996**, *68*, 313–322. (l) Fyfe, M.C.T. and Stoddart, J.F. *Acc. Chem. Res.* **1997**, *30*, 393–401.

12. (a) Karagounis, G. and Pandi-Agathokli, I. *Prakt. Akad. Athenon* **1970**, *45*, 118–126. (b) Karagounis, G., Pandi-Agathokli, J., and Kondaraki, E. *Chim. Cronika* **1972**, *1*, 130–147. (c) Karagounis, G., Pandi-Agathokli, I., Petassis, E., and Alexakis, A. *Folia Bioch. Biol. Graeca* **1973**, *10*, 31–41. (d) Karagounis, G., Kontakari, E., and Petassis, E. *Prakt. Akad. Athenon* **1973**, *49*, 118–126. (e) Karagounis, G., Pandi-Agathokli, I., Kontakari, E., and Nikolelis, D. *Prakt. Akad. Athenon* **1974**, *49*, 501–513. (f) Karagounis, G., Pandi-Agathokli, I., Kontakari, E., and Nikoleis, D. *IUPAC Colloid Surf. Sci. Int. Conf. Selected Papers A* **1975**, *1*, 671–678. (g) Karagounis, G. and Pandazi, M. *Proc. 5th Int. Conf. Raman Spectroscopy, Freiburg im Breisgau, Freiburg* **1976**, pp. 72–73.

13. (a) Amabilino, D.B., Ashton, P.R., Reder, A.S., Spencer, N., and Stoddart, J.F. *Angew. Chem. Int. Ed. Engl.* **1994**, *33*, 1286–1290. (b) Amabilino, D.B., Ashton, P.R., Boyd, S.E., Lee, J.Y., Menzer, S., Stoddart, J.F., and Williams, D.J. *Angew. Chem. Int. Ed. Engl.* **1997**, *36*, 2070–2072. (c) Amabilino, D.B., Ashton, P.R., Boyd, S.E., Lee, J.Y., Menzer, S., Stoddart, J.F., and Williams, D.J. *J. Am. Chem. Soc.* **1998**, *120*, 4295–4307.

14. Geerts, Y., Muscat, D., and Müllen, K. *Macromol. Chem. Phys.* **1995**, *196*, 3425–3435.

15. (a) Muscat, D., Witte, A., Köhler, W., Müllen, K., and Geerts, Y. *Macromol. Rapid Commun.* **1997**, *18*, 233–241. (b) Muscat, D., Köhler, W., Rädes, H.J., Martin, K., Mullirs, S., Müller, B., Müller, K., Geerts, Y. Macromolecules 1999, *32*, 1737–1745.

16. Fustin, C.-A., Bailly, C., Clarkson, G., De Groote, P., Galow, T., Leigh, D., Robertson, D., Slawin, A., and Wong, J. *J. Am. Chem. Soc.* **2003**, *125*, 2200–2207.

17. Weidmann, J.L., Kern, J.M., Sauvage, J.P., Geerts, Y., Muscat, D., and Müllen, K. *Chem. Commun.* **1996**, 1243–1244.

18. Shimada, S., Ishiwara, K., and Tamaoki, N. *Acta Chem. Scand.* **1998**, *52*, 374–376.

19. Menzer, S., White, A.J.P., Williams, D.J., Belohradsky, M., Hamers, C., Raymo, F.M., Shipway, A.N., and Stoddart, J.F. *Macromolecules* **1998**, *31*, 295–307.

20. Hamers, C., Raymo, F.M., and Stoddart, J.F. *Eur. J. Org. Chem.* **1998**, 2109–2117.

21. For examples of bis[2]catenanes, see: (a) Ashton, P.R., Reder, A.S., Spencer, N., and Stoddart, J.F. *J. Am. Chem. Soc.* **1993**, *115*, 5286–5287. (b) Ashton, P.R., Preece, J.A., Stoddart, J.F., and Tolley, M.S. *Synlett* **1994**, 789–792. (c) Amabilino, D.B., Ashton, P.R., Preece, J.A., Stoddart, J.F., and Tolley, M.S. *Am. Chem. Soc. Div. Polym. Chem. Polym. Prepr.* **1995**, *36*, 587–588. (d) Ashton, P.R., Huff, J., Parsons, I.W., Preece, J.A., Stoddart, J.F., Williams, D.J., White, A.J.P., and Tolley, M.S. *Chem. Eur. J.* **1996**, *2*, 123–136. (e) Huff, J., Preece, J.A., and Stoddart, J.F. *Macromol. Symp.* **1996**, *102*, 1–8. (f) Ashton, P.R., Horn, T., Menzer, S., Preece, J.A., Spencer, N., Stoddart, J.F., and Williams, D.J. *Synthesis* **1997**, 480–488.

22. Hamers, C., Kocian, O., Raymo, F.M., and Stoddart, J.F. *Adv. Mater.* **1998**, *10*, 1366–1369.

23. Simone, D.L. and Swager, T.M. *J. Am. Chem. Soc.* **2000**, *122*, 9300–9301.

24. Gan, Y., Dong, D., and Hogen-Esch, T.E. *Macromolecules* **2002**, *35*, 6799–6803.

25. Ogata, N., Sanui, K., and Wada, J. *J. Polym. Sci. Polym. Lett. Ed.* **1976**, *14*, 459–462.

26. For other early examples of pseudopolyrotaxanes prepared following route *C*, see: (a) Maciejewski, M. and Smets, G. *Pr. Nauk. Inst. Technol. Organicz. Tworz* **1975**, *16*, 57–69. (b) Maciejewski, M. and Panasiewicz, M. *J. Macromol. Sci. Chem.* **1978**, *A12*, 701–718. (c) Maciejewski, M. *J. Macromol. Sci. Chem.* **1979**, *A13*, 77–85. (c) Maciejewski, M., Gwizdowski, A., Peczak, P., and Pietrzak, A. *J. Macromol. Sci. Chem.* **1979**, *A13*, 87–109.

27. Yamaguchi, I., Osakada, K., and Yamamoto, T. *J. Am. Chem. Soc.* **1996**, *118*, 1811–1812.

28. (a) Agam, G., Graiver, D., and Zilkha, A. *J. Am. Chem. Soc.* **1976**, *98*, 5206–5214. (b) Agam, G. and Zilkha, A. *J. Am. Chem. Soc.* **1976**, *98*, 5214–5216.

29. (a) Engen, P.T., Lecavalier, P.R., and Gibson, H.W. *Am. Chem. Soc. Div. Polym. Chem. Polym. Prepr.* **1990**, *31*, 703–704. (b) Gibson, H.W., Engen, P.T., Lee, S.-H., Liu, S., Marand, H., and Bheda, M.C. *Am. Chem. Soc. Div. Polym. Chem. Polym. Prepr.* **1993**, *34*, 64–65. (c) Gibson, H.W. and Engen, P.T. *New J. Chem.* **1993**, *17*, 723–727. (d) Lee, S.H., Engen, P.T., and Gibson, H.W. *Macromolecules* **1997**, *30*, 337–343.

30. (a) Gong, C. and Gibson, H.W. *Macromolecules* **1996**, *29*, 7029–7033. (b) Gong, C. and Gibson, H.W. *Macromol. Chem. Phys.* **1997**, *198*, 2321–2332. (c) Gibson, H.W., Liu, S., Gong, C., Ji, Q., and Joseph, E. *Macromolecules* **1997**, *30*, 3712–3727. (d) Gong, C., Ji, Q., Glass, T.E., and Gibson, H.W.

Macromolecules **1997**, *30*, 4807–4813. (e) Gong, C. and Gibson, H.W. *Macromolecules* **1997**, *30*, 8524–8525.

31. (a) Wu, C., Bheda, M.C., Lim, C., Shen, Y.X., Sze, J., and Gibson, H.W. *Polym. Commun.* **1991**, *32*, 204–207. (b) Gibson, H.W., Liu, S., Lecavalier, P., Wu, C., and Shen, Y.X. *J. Am. Chem. Soc.* **1995**, *117*, 852–874.

32. (a) Shen, Y.X., Lim, C., and Gibson, H.W. *Am. Chem. Soc. Div. Polym. Chem.* **1991**, *32*, 166–167. (b) Shen, X.Y. and Gibson, H.W. *Macromolecules* **1992**, *25*, 2058–2059. (c) Shen, X.Y., Xie, D., and Gibson, H.W. *J. Am. Chem. Soc.* **1994**, *116*, 537–548.

33. Marand, E., Hu, Q., Gibson, H.W., and Veystman, B. *Macromolecules* **1996**, *29*, 2555–2562.

34. (a) Gong, C. and Gibson, H.W. *Angew. Chem. Int. Ed. Engl.* **1997**, *36*, 2331–2333. (b) Gong, C., Glass, T.E., and Gibson, H.W. *Macromolecules* **1998**, *31*, 308–313.

35. Harrison, I.T. *J. Chem. Soc. Chem. Commun.* **1977**, 384–385.

36. (a) Wenz, G. and Keller, B. *Macromol. Symp.* **1994**, *87*, 11–16. (b) Steinbrunn, M.B. and Wenz, G. *Angew. Chem. Int. Ed. Engl.* **1996**, *35*, 2139–2141. (c) Meier, L.P., Heule, M., Caseri, W.R., Shelden, R.A., Suter, U.W., Wenz, G., and Keller, B. *Macromolecules* **1996**, *29*, 718–723. (d) Kräuter, I., Herrmann, W., and Wenz, G. *J. Incl. Phenom.* **1996**, *25*, 93–96. (e) Weickenmeier, M. and Wenz, G. *Macromol. Rapid Commun.* **1997**, *18*, 1109–1115. (f) Herrmann, W., Keller, B., and Wenz, G. *Macromolecules* **1997**, *30*, 4966–4972. (g) Wenz, G., Steinbrunn, M.B., and Landfester, K. *Tetrahedron* **1997**, *53*, 15575–15592.

37. (a) Harada, A. and Kamachi, M. *J. Chem. Soc. Chem. Commun.* **1990**, 1322–1323. (b) Harada, A. and Kamachi, M. *Macromolecules* **1990**, *23*, 2821–2823. (c) Harada, A., Li, J., and Kamachi, M. *Chem. Lett.* **1993**, 237–240. (d) Harada, A., Li, J., Suzuki, S., and Kamachi, M. *Macromolecules* **1993**, *26*, 5267–5268. (e) Harada, A., Li, J., and Kamachi, M. *Macromolecules* **1993**, *26*, 5698–5703. (f) Li, J., Harada, A., and Kamachi, M. *Bull. Chem. Soc. Jpn.* **1994**, *67*, 2808–2818. (g) Harada, A., Li, J., and Kamachi, M. *Macromolecules* **1994**, *27*, 4538–4543. (h) Harada, A., Li, J., and Kamachi, M. *Nature* **1994**, *370*, 126–128. (i) Li, J., Harada, A., and Kamachi, M. *Polym. J.* **1994**, *26*, 1019–1026. (j) Harada, A., Okada, M., Li, J., and Kamachi, M. *Macromolecules* **1995**, *28*, 8406–8411. (k) Harada, A., Suzuki, S., Okada, M., and Kamachi, M. *Macromolecules* **1996**, *29*, 5611–5614. (l) Pozuelo, J., Mendicuti, F., and Mattice, W.L. *Macromolecules* **1997**, *30*, 3685–3690. (m) Harada, A., Li, J., Kamachi, M., Kitagawa, Y., and Katsube, Y. *Carbohydr. Res.* **1998**, *305*, 127–129.

38. (a) Lipatova, T.E., Kosyanchuk, L.F., Gomza, Y.P., Shilov, V.V., and Lipatov, Y.S. *Dokl. Akad. Nauk. SSSR, Engl. Trans.* **1982**, *263*, 140–143. (b) Lipatova, T.E., Kosyanchuk, L.F., and Shilov, V.V. *J. Macrol. Sci. Chem.* **1985**, *A22*, 361–372. (c) Lipatova, T.E., Kosyanchuk, L.F., Shilov, V.V., and Gomza, Y.P. *Polym. Sci. USSR* **1985**, *27*, 622–629.

39. (a) Sun, X., Amabilino, D.B., Parsons, I.W., and Stoddart, J.F. *Am. Chem. Soc. Div. Polym. Chem. Polym. Prepr.* **1993**, *34*, 104–105. (b) Sun, X., Amabilino, D.B., Ashton, P.R., Parsons, I.W., Stoddart, J.F., and Tolley, M.S. *Macromol. Symp.* **1994**, *77*, 191–207. (c) Owen, G.J. and Hodge, P. *Chem. Commun.* **1998**, 11–12.

40. (a) Wenz, G. and Keller, B. *Angew. Chem. Int. Ed. Engl.* **1992**, *31*, 197–199. (b) Wenz, G. and Keller, B. *Am. Chem. Soc. Div. Polym. Chem. Polym. Prepr.* **1993**, *34*, 62–63.

41. Herrmann, W., Schneider, M., and Wenz, G. *Angew. Chem. Int. Ed. Engl.* **1997**, *36*, 2511–2514.

42. Kelch, S., Caseri, W., Shelden, R., Suter, U., Wenz, G., and Keller, B. *Langmuir* **2000**, *16*, 5311–5316.

43. Theng, B.K.G. *Formation and Properties of Clay–Polymer Complexes*, Elsevier, Amsterdam, **1979**.

44. (a) Harada, A., Li, J., and Kamachi, M. *Nature* **1992**, *356*, 325–327. (b) Harada, A., Li, J., Nakamiysu, T., and Kamachi, M. *J. Org. Chem.* **1993**, *58*, 7524–7528. (c) Harada, A., Li, J., and Kamachi, M. *J. Am. Chem. Soc.* **1994**, *116*, 3192–3196.

45. (a) Okumura, Y., Ito, K., and Hayakawa, R. *Polym. Adv. Technol.* **2000**, *11*, 815–819. (b) Kamitori, S., Matsuzaka, O., Kondo, S., Muraoka, S., Okuyama, K., Noguchi, K., Okada, M., and Harada, A. *Macromolecules* **2000**, *33*, 1500–1502.

46. (a) Harada, A., Li, J., and Kamachi, M. *Nature* **1993**, *364*, 516–518. (b) Harada, A. *Am. Chem. Soc. Div. Polym. Chem. Polym. Prepr.* **1995**, *36*, 570–571.

47. Shigekawa, H., Miyake, K., Sumaoka, J., Komiyama, M., and Harada, A. *J. Am. Chem. Soc.* **2000**, *122*, 5411–5412.

48. Miyake, K., Yasuda, S., Harada, A., Sumaoka, J., Komiyama, M., and Shigekawa, H. *J. Am. Chem. Soc.* **2003**, *125*, 5080–5085.

49. Okumura, H., Kawaguchi, Y., and Harada, A. *Macromolecules* **2001**, *34*, 6338–6343.

50. Okumura, H., Kawaguchi, Y., and Harada, A. *Macromolecules* **2003**, *36*, 6422–6429.

51. Yoshida, K., Shimomura, T., Ito, K., and Hayakawa, R. *Langmuir* **1999**, *15*, 910–913.

52. Nepal, D., Samal, S., and Geckeler, K.E. *Macromolecules* **2003**, *36*, 3800–3802.

53. Cacialli, F., Wilson, J.S., Michels, J.J., Daniel, C., Silva, C., Friend, R.H., Severin, N., Samori, P., Rabe, J.P., O'Connell, M.J., Taylor, P.N., and Anderson, H.L. *Nat. Mater.* **2002**, *1*, 160–164.

54. Ooya, T. and Yui, N. *J. Control. Release* **1999**, *58*, 251–269.

55. (a) Ooya, T., Arizono, K., and Yui, N. *Polym. Adv. Technol.* **2000**, *11*, 642–665. (b) Ooya, T., Eguchi, M., and Yui, N. *Biomacromolecules* **2001**, *2*, 200–203.

56. Ooya, T., Eguchi, M., and Yui, N. *J. Am. Chem. Soc.* **2003**, *125*, 13016–13017.

57. For example, see: Gestwicki, J.E., Cairo, C.W., Strong, L.E., Oetjen, K.A., and Kiessling, L.L. *J. Am. Chem. Soc.* **2002**, *124*, 14922–14933.

58. Yui, N., Ooya, T., and Kumeno, T. *Bioconjugate Chem.* **1998**, *9*, 118–125.

59. Park, H.D., Lee, W.K., Ooya, T., Park, K.D., Kim, Y.H., and Yui, N. *J. Biomed. Mater. Res.* **2002**, *60*, 186–190.

60. (a) Mason, P.E., Parsons, I.W., and Tolley, M.S. *Angew. Chem. Int. Ed. Engl.* **1996**, *35*, 2238–2241. (b) Mason, P.E., Parsons, I.W., and Tolley, M.S. *Polymer* **1998**, *39*, 3981–3991.

61. (a) Zhu, S.S., Carroll, P.J., and Swager, T.M. *J. Am. Chem. Soc.* **1996**, *118*, 8713–8714. (b) Zhu, S.S. and Swager, T.M. *J. Am. Chem. Soc.* **1997**, *119*, 12568–12577. (c) Vidal, P.L., Billon, M., Divisia-Blohorn, B., Bidan, G., Kern, J.M., and Sauvage, J.P. *Chem. Commun.* **1998**, 629–630.

62. (a) Sauvage, J.-P., Kern, J.-M., Bidan, G., Divisia-Blohorn, B., and Vidal, P.-L. *New J. Chem.* **2002**, *26*, 1287–1290. (b) Divisia-Blohorn, B., Genoud, F., Borel, C., Bidan, G., Sauvage, J.-P., and Kern, J.-M. *J. Phys. Chem. B* **2003**, *107*, 5126–5132.

63. Buey, J. and Swager, T.M. *Angew. Chem. Int. Ed.* **2000**, *39*, 608–612.

64. (a) Born, M. and Ritter, H. *Makromol. Chem. Rapid Commun.* **1991**, *12*, 471–476. (b) Born, M., Koch, T., and Ritter, H. *Acta Polym.* **1994**, *45*, 68–73. (c) Ritter, H. *Macromol. Symp.* **1994**, *77*, 73–78. (d) Born, M., Koch, T., and Ritter, H. *Macromol. Chem. Phys.* **1995**, *196*, 1761–1767. (e) Born, M. and Ritter, H. *Macromol. Rapid. Commun.* **1996**, *17*, 197–202. (f) Noll, O. and Ritter, H. *Macromol. Rapid Commun.* **1997**, *18*, 53–58. (g) Noll, O. and Ritter, H. *Macromol. Chem. Phys.* **1998**, *199*, 791–794.

65. (a) Born, M. and Ritter, H. *Angew. Chem. Int. Ed. Engl.* **1995**, *34*, 309–311. (b) Born, M. and Ritter, H. *Adv. Mater.* **1996**, *8*, 149–151.

66. Yamaguchi, I., Osakada, K., and Yamamoto, T. *Macromolecules* **1997**, *30*, 4288–4294.

67. (a) Noll, O. and Ritter, H. *Macromol. Chem. Phys.* **1998**, *199*, 791–794. (b) Jeromin, J. and Ritter, H. *Macromolecules* **1999**, *32*, 5236–5239.

68. Takata, T., Kawasaki, H., Kihara, N., and Furusho, Y. *Macromolecules* **2001**, *34*, 5449–5456.

69. (a) Marsella, M.J., Carrol, P.J., and Swager, T.M. *J. Am. Chem. Soc.* **1994**, *116*, 9347–9348. (b) Swager, T.M., Marsella, M.J., Newland, R.J., and Zhou, Q. *Am. Chem. Soc. Div. Polym. Chem. Polym. Prepr.* **1995**, *36*, 546–547. (c) Zhou, Q., Ezer, M.R., and Swager, T.M. *Am. Chem. Soc. Div. Polym. Chem. Polym. Prepr.* **1995**, *36*, 607–608. (d) Zhou, Q. and Swager, T.M. *J. Am. Chem. Soc.* **1995**, *117*, 7017–7018. (e) Marsella, M.J., Carroll, P.J., and Swager, T.M. *J. Am. Chem. Soc.* **1995**, *117*, 9832–9841. (f) Zhou, Q. and Swager, T.M. *J. Am. Chem. Soc.* **1995**, *117*, 12593–12602.

70. (a) Cantrill, S.J., Youn, G.J., Stoddart, J.F., and Williams, D.J. *J. Org. Chem.* **2001**, *66*, 6857–6872. (b) Ashton, P.R., Baxter, I., Cantrill, S.J., Fyfe, M.C.T., Glink, P.T., Stoddart, J.F., White, A.J.P., and Williams, D.J. *Angew. Chem. Int. Ed.* **1998**, *37*, 1294–1297. (c) Ashton, P.R., Parsons, I.W., Raymo, F.M., Stoddart, J.F., White, A.J.P., Williams, D.J., and Wolf, R. *Angew. Chem. Int. Ed.* **1998**, *37*, 1913–1916.

71. Yamaguchi, N., Nagvekar, D.S., and Gibson, H.W. *Angew. Chem. Int. Ed.* **1998**, *37*, 2361–2364.

72. (a) Harada, A., Kawaguchi, Y., and Hoshino, T. *J. Incl. Phenom. Macrocyd. Chem.* **2001**, *41*, 115–121. (b) Harada, A., Miyauchi, M., and Hoshino, T. *J. Polym. Sci. A* **2003**, *41*, 3519–3523.

73. Yamaguchi, N. and Gibson, H.W. *Angew. Chem. Int. Ed.* **1999**, *38*, 143–147.

74. Gibson, H.W., Yamaguchi, N., and Jones, J.W. *J. Am. Chem. Soc.* **2003**, *125*, 3522–3533.

75. Oku, T., Furusho, Y., and Takata, T. *J. Polym. Sci. A* **2002**, *41*, 119–123.

76. Rowan, S.J., Cantrill, S.J., Cousins, G.R.L., Sanders, J.K.M., and Stoddart, J.F. *Angew. Chem. Int. Ed.* **2002**, *41*, 899–952.

77. Alvarez-Parrilla, E., Cabrer, P.R., Al-Soufi, W., Meijide, F., Núñez, E.R., and Tato, J.V. *Angew. Chem. Int. Ed.* **2000**, *39*, 2856–2858.

78. Gibson, H.W., Yamaguchi, N., Hamilton, L., and Jones, J.W. *J. Am. Chem. Soc.* **2002**, *124*, 4653–4665.

79. Elizarov, A.M., Chang, T., Chiu, S.-H., and Stoddart, J.F. *Org. Lett.* **2002**, *4*, 3565–3568.

80. Jones, W., Bryant, W.S., Bosman, A.W., Janssen, R.A.J., Meijer, E.W., and Gibson, H.W. *J. Org. Chem.* **2003**, *68*, 2385–2389.

81. For some recent examples, see: (a) Osswald, F., Vogel, E., Safarowsky, O., Schwanke, F., and Vögtle, F. *Adv. Synth. Catal.* **2001**, *343*, 303–309. (b) Belaissaoui, A., Shimada, S., Ohishi, A., and Tamaoki, N. *Tetrahedron Lett.* **2003**, *44*, 2307–2310. (c) Werts, M.P.L., van den Boogaard, M., Tsivgoulis, G.M., and Hadziioannou, G. *Macromolecules* **2003**, *36*, 7004–7013.

82. Kidd, T.J., Loontjens, T.J.A., Leigh, D.A., and Wong, J.K.Y. *Angew. Chem. Int. Ed.* **2003**, *42*, 3379–3383.

83. Rowan, S.J., Cantrill, S.J., and Stoddart, J.F. *Org. Lett.* **1999**, *1*, 129–132.

84. (a) Rowan, S.J. and Stoddart, J.F. *J. Am. Chem. Soc.* **2000**, *122*, 164–165. (b) Rowan, S.J. and Stoddart, J.F. *Polym. Adv. Technol.* **2002**, *13*, 777–787. (c) Chiu, S.-H., Rowan, S.J., Cantrill, S.J., Stoddart, J.F., White, A.J.P., and Williams, D.J. *Chem. Eur. J.* **2002**, *8*, 5170–5183.

85. (a) Rowan, S.J., Cantrill, S.J., and Stoddart, J.F. *Org. Lett.* **2000**, *2*, 759–762. (b) Chiu, S.-H., Rowan, S.J., Cantrill, S.J., Stoddart, J.F., White, A.J.P., and Williams, D.J. *Chem. Commun.* **2002**, 2948–2949. (c) Chiu, S.-H., Rowan, S.J., Cantrill, S.J., Stoddart, J.F., White, A.J.P., and Williams, D.J. *Chem. Eur. J.* **2002**, *8*, 5170–5183.

86. Chiu, S.-H., Rowan, S.J., Cantrill, S.J., Ridvan, L., Ashton, P.R., Garrell, R.L., and Stoddart, J.F. *Tetrahedron* **2002**, *58*, 807–814.

87. Elizarov, A.M., Chiu, S.-H., Glink, P.T., and Stoddart, J.F. *Org. Lett.* **2002**, *4*, 679–682.

88. Gibson, H.W., Liu, S., Shen, Y.X., Bheda, M., Lee, S.H., and Wang, F. *Molecular Engineering for Advanced Materials*, Eds. Becher, J. and Schaumburg, K., Kluwer Academic Publishers, Dordrecht, **1995**, pp. 41–58.

89. Ballardini, R., Balzani, V., Credi, A., Gandolfi, M.T., Venturi, M. *Acc. Chem. Res. 34*, 445–455, 2001.

90. Balzani, V., Credi, A., Venturi, M. *Pure Appl. Chem. 75*, 541–547, 2003.

91. Harada, A. *Acc. Chem. Res. 34*, 456–464, 2001.

92. Balzani, V., Credi, A., Raymo, F.M., Stoddart, J.F. *Angew. Chem. Int.* Ed. *39*, 3348–3391, 2000.

93. Pease, A.R., Jeppesen, J.O., Stoddart, J.F., Luo, Y., Collier, C.P., Heath J.R. *Acc. Chem. Res. 34*, 433–444, 2001.

94. Dietrich-Buchecker, C., Jimenez-Molero, M.C., Sartor, V., Sauvage J-P. *Pure Appl. Chem. 75*, 1383–1393, 2003.

95. Collin, J-P., Dietrich-Buchecker, C., Gavina, P., Jimenez-Molero, M.C., Sauvage, J-P. *Acc. Chem. Res. 34*, 477–487, 2001.

Chapter 9

Theory of Block Copolymers

Mark D. Whitmore

CONTENTS

I. INTRODUCTION

A *block copolymer* is a macromolecule that is composed of distinct sections of different chemical species that are chemically bonded together to form a single molecule. The simplest kind is a diblock copolymer, with two blocks, such as poly(styrene)–*b*–poly(isoprene), or PS–*b*–PI. A general label for a diblock copolymer is A–*b*–B, or simply AB. There are also triblock copolymers that can contain either two or three different species, for example, ABA or ABC, linear multiblocks, for example, ABAB. . ., and more complicated architectures, such as starblocks or graft copolymers. These are illustrated schematically in Figure 1.

The phase behavior of block copolymers has been the object of considerable interest. In most cases, the effective interactions between monomers of different species are repulsive. In blends of different homopolymers, this normally causes the components to phase separate, into domains that are typically on the order of microns in size. This can be called *macro*phase separation. However, copolymers cannot macrophase separate because the different components are bonded together in each molecule. Instead, they can *micro*phase separate, forming domains whose dimensions are comparable with the sizes of the constituent blocks. In both cases, the phase separation is driven by the accompanying decrease in interaction energy, but opposed by the resultant decrease in entropy. The equilibrium phase behavior and morphology are governed by a balance between these factors.

The transition from a disordered phase (D) to a microphase is often referred to as the *microphase separation transition*, or MST. A variety of domain structures occurs, which can form either ordered or disordered arrays. In most cases, the material within each domain is amorphous, but it can sometimes be crystalline. The simplest structure consists of alternating layers, or lamellae. This is shown schematically for both amorphous and semicrystallizable copolymers in Figure 2. The phase behavior is relatively simple for diblock copolymers, but can be far richer for triblocks.

Copolymers are often blended, with solvents, homopolymers, other copolymers, or combinations of them. Depending on the interactions with the different copolymer blocks, the solvent can be

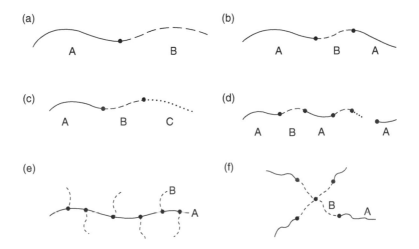

Figure 1   Schematic diagram of common copolymer architectures: (a) AB diblock copolymers; (b) linear ABA triblocks; (c) linear ABC triblocks; (d) linear ABAB . . . multiblocks, which can terminate with either an A or a B block; (e) graft copolymers, illustrated with B side chains grafted to an A backbone; (f) multiarm star copolymers, illustrated as a four-arm star with the B blocks joined at the center.

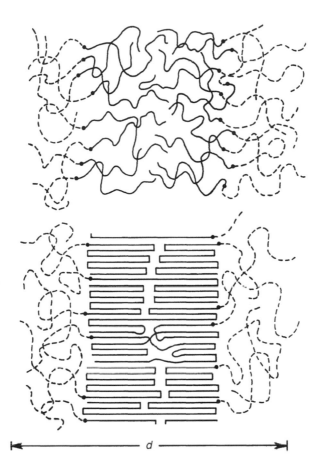

Figure 2 Schematic diagram of layered structures of diblock copolymers. The upper and lower panels illustrate amorphous and semicrystallizable copolymers, respectively. Both panels correspond to strong segregation, with each block segregated into its respective subdomain, and the copolymer joints localized within narrow interphase regions between them. As indicated near the center of the lower panel, the crystallizable block is generally semicrystalline, and the chain folding can occur both in each amorphous/crystalline interphase and in the interiors of the crystalline layers.

classified as selective or nonselective. In all these blends, both microphase and macrophase separation can occur. In systems with only low copolymer concentrations, micelles can form. In other cases, the phases are similar to those of the neat copolymers, with the solvent or homopolymer solubilized within the microdomains.

The formation of microphases is also often referred to as self-assembly. The domains can be thought of as supramacromolecules; the self-assembly is driven by intermolecular forces, and the polymers in each domain are not chemically linked to each other. In the ordered microphases, the order persists over large length scales compared with each domain size.

In this book, Chapter 10 by Abetz presents the experimental knowledge of block copolymers and blends, including a summary of the synthesis of the molecules, potential applications, and the observed microphases. The primary purpose of this chapter is to provide an overview of the theories of these systems and the insight that they provide. The focus is on the equilibrium phases, the domain sizes, and their internal structure. The most successful approach to date is the numerical self-consistent field theory (NSCFT), and the chapter concentrates on it. Weak and strong segregation theories, their range of applicability, and their relationship to the NSCFT, are described. Most of the theoretical results in the literature are for two-species copolymers, for example, AB diblocks or stars, and the chapter concentrates on them. NSCFT also provides a basis for understanding the dominant physical factors that control these systems, and it can be used to develop simple predictive relationships, some of which are also discussed. Going beyond mean field theories, the theoretical treatments of fluctuations are noted, and some Monte Carlo simulations discussed.

II. INTRODUCTION TO THE MICROPHASE DIAGRAMS OF AMORPHOUS BLOCK COPOLYMERS

In order to introduce the terminology and phenomena, it is useful to summarize the microphase behavior of the simplest system, which is the amorphous, conformationally symmetric, diblock AB copolymers. Diblocks are characterized by their total degree of polymerization, N_C, the block volume fractions, f_A and $f_B = 1 - f_A$, the statistical segment lengths and pure component densities of each block, and the Flory A–B interaction parameter, χ. This parameter normally decreases with increasing temperature, T, and increases with decreasing T. Conformational symmetry and asymmetry are discussed in Section V. At this point it suffices to note that, for a conformationally symmetric copolymer, if the monomers in each block are defined to have equal volume, then the statistical segment lengths are also equal.

The phase behavior of neat copolymers is often summarized in diagrams such as Figure 3, which shows the various equilibrium phases as a function of the relevant quantities as discussed throughout

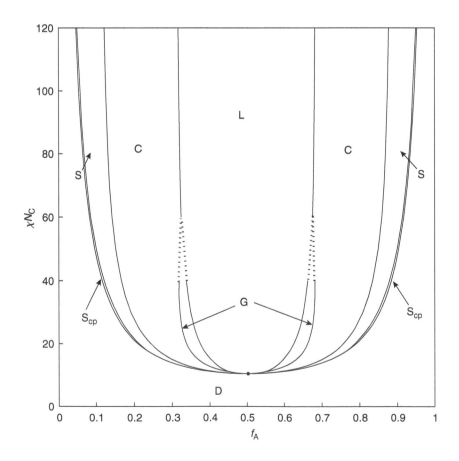

Figure 3 Microphase diagram for conformationally symmetric, diblock copolymers calculated using NSCFT [1]. The MST separates the disordered phase, (D), from the microphases. The microphases are lamellae (L), gyroid (G), cylinders on an hexagonal lattice (C), spheres on a bcc lattice (S), and close-packed spheres (S_{cp}). The dotted lines indicate extrapolated boundaries.

this chapter. These diagrams are usually called phase diagrams, although some authors prefer the term phase map because, at a minimum, all points corresponding to different volume fractions f_A actually refer to different molecules.

Figure 3 was calculated using NSCFT by Matsen and Bates [2]. It shows the equilibrium phases for diblocks with total degree of polymerization N_C and Flory interaction parameter χ. An important, general feature of block copolymers is that the corresponding diagrams for many two-species copolymers are very similar in structure to this one, with only small shifts in the phase boundaries that depend on molecular architecture and species [2–14]. As described in Chapter 10, the phase diagrams of triblocks can be more complex.

Even for diblocks, there are a number of microphases that can form. The simplest one is the alternating layer structure of Figure 2, in which each layer is relatively rich in one of the components, for example, A–B–A–B–$\cdots$. We will denote this phase simply as L. Another consists of parallel cylinders, each rich in the minority component, arranged on a hexagonal lattice. We denote these by C, or by C_A and C_B where the subscript labels the primary component of the interiors of the cylinders. A third one consists of A or B rich spheres, which can be arranged on a body-centered cubic (bcc-S) or a close-packed (cp-S) lattice, or can be disordered. The L, C, and bcc-S phases have traditionally been labeled as the "classical" phases. The best understood "exotic" structure is the bicontinuous gyroid phase, G, in which the minority component forms two interweaving, threefold coordinated lattices [15,16]. The perforated lamellar, or PL phase, has also been observed [12,13,17,18]. It is lamellar, but with the minority-component layers perforated by "holes" that are filled by the majority component. However, for neat diblocks it is metastable [19,20]. The bicontinuous double-diamond structure has also been reported in the literature [21,22], but it is now believed that those systems were actually in G phases [23]. As seen in Figure 3, NSCFT theory predicts regions of stability for the L, G, C, bcc-S, and cp-S microphases.

An important prediction of mean field theory, which is implicit in Figure 3, is that the phase diagrams for conformationally symmetric diblocks depend on only two parameters, f_A and the product χN_C. The diagrams are symmetric about $f_A = 0.5$, as well. For small χN_C, the system is disordered, (D). At sufficiently large χN_C, the system microphase separates. For a perfectly symmetric copolymer, $f_A = 0.5$, the MST is predicted to be a second-order transition from the disordered state to the L phase, and occur at [24]

$$\chi N_C = 10.5 \tag{1}$$

At $f_A \neq 0.5$, the transition is first order. The point defined by $f_A = 0.5$ and $\chi N_C = 10.5$ is a mean field critical point. Systems near this point are weakly segregated, with broad interphases between domains and mixing of each component within each one.

For compositionally asymmetric molecules, that is, $f_A \neq 0.5$, the MST moves to higher χN_C, and the order–order phase boundaries develop within the phase separated region. In strong segregation, the behavior is relatively simple. There is no G phase, and the phase boundaries are nearly vertical on this diagram, meaning they depend on f_A but only very weakly on χN_C. The lamellae persist over $f_A \simeq 0.3$ to 0.7. The S phases are formed by highly asymmetric copolymers, that is, from the MST up to $f_A \simeq 0.1$ on one side of the phase diagram, and from $f_A \simeq 0.9$ to the MST on the other. Each S region consists of a very narrow cp-S strip adjacent to the MST, and a larger bcc-S region, although there is some numerical evidence that the cp-S phase disappears at very large χN_C. The C phase forms between the S and L phases.

The behavior changes as χN_C decreases. At $\chi N_C \simeq 60$, the G appears between the L and C phases, extending over a composition range of a few percent on each side of the diagram, but no other ordered phase becomes stable. As χN_C decreases further, all phase boundaries curve toward $f_A = 0.5$. The G phase terminates at triple points at $\chi N_C = 11.14$ and $f_A = 0.452$ and 0.548, where it coexists with the L and C phases. Similarly, the cp-S phase terminates at triple points at $\chi N_C = 17.67$ and $f_A = 0.235$ and 0.765 where it coexists with the D and bcc-S phases. However, the L, C, and S phases all extend to the critical point where the D/S, S/C, and C/L boundaries all merge.

The structure of the MST can be understood from simple thermodynamics. In the disordered phase, each monomer interacts with both like and unlike monomers. For fixed composition, that is, f_A, the number of repulsive contacts per molecule varies linearly with N_C, and hence the repulsive energy per molecule, inducing the MST, scales as χN_C (actually, as $k_B T \chi N_C$ since χ is in units of $k_B T$). The MST is accompanied by a decrease in entropy, but the associated change in free energy per molecule is essentially independent of molecular weight. This entropic contribution increases with temperature, and the net result is that the disordered phase is stable for sufficiently high temperature and low molecular weight, but the system microphase separates for decreased temperature or increased molecular weight. Because the number of A–B contacts per molecule in the disordered phase also varies as the product $f_A f_B$, the MST moves to larger χN_C on either side of $f_A = f_B = 0.5$. Understanding the details of the phase diagram within the microphase separated region, that is, the order–order boundaries, is one of the prime objectives of block copolymer theories.

III. NSCFT of Amorphous Block Copolymers and Blends

A. General Theory for Gaussian-Like Chains

Numerical SCFT has been used to study the equilibrium microphase diagrams and domain structures of neat copolymers, and copolymers blended with other copolymers, homopolymers, and solvent. Historically, it was developed first for strongly [25–31] and weakly [24,32–34] segregated systems with a focus on diblocks and the classical structures, then for all degrees of segregation [35], and extended to the more complex phases [2,3,6,36–38]. Approximately in parallel with the work on neat copolymers, the theory has been extended and applied to various kinds of blends [39–52] and different architectures [4,5,7,8,53–59]. The roots of all this work can be traced to the the early SCF treatment of polymers by Edwards and Dolan [60,61]. A number of reviews currently exist in the literature [1,62,63]. Here, we present a very general version of the theory, and then relate it to theories of weak and strong segregation.

We consider an arbitrary mixture of copolymers, C, homopolymers, H, and solvent molecules, S. We denote the number of molecules of type κ by $\tilde{N}_\kappa$. There can be one or more kinds of each of these. Different solvents correspond to different chemical species. If, for example, there are two solvent species, then $S = 1$ and 2, each with corresponding numbers $\tilde{N}_S$. Different kinds of polymers could correspond to different species, or the same species with different molecular weight, relative block sizes, or tacticity. Polydispersity can, therefore, be included in this way, but we will focus here on monodisperse polymers.

Each solvent species is characterized by its pure component number density, ρ_{0S}. Each amorphous polymer or polymer block is modeled as the continuous limit of a Gaussian chain with degree of polymerization N_κ, statistical segment length b_κ, and pure component density $\rho_{0\kappa}$. The total degree of polymerization of a copolymer C is the sum of the block degrees of polymerization, for example, $N_C = N_{CA} + N_{CB}$ for the simplest case of an AB diblock. Each configuration of a chain or block is represented by a space curve $r(\tau)$, with τ ranging from zero to N_κ. It is assumed that the volume of the system is constant, that is, independent of the degree of mixing, which is equivalent to assuming that the system is incompressible.

The theory begins with a very general expression for the partition function. A given configuration of the entire system is specified by sets of solvent molecule coordinates $\{r_{Si}\}$ with $i = 1, \ldots, \tilde{N}_S$ for each S, homopolymer space curves $\{r_{Hj}(\tau)\}$ for $j = 1, \ldots, \tilde{N}_H$ for each H, and copolymer space curves for each block of each copolymer. In the first instance, we consider only diblock copolymers, so we need $\{r_{CAk}(\tau)\}$ and $\{r_{CBk}(\tau)\}$ for $k = 1, \ldots, \tilde{N}_C$ for each C. The partition function can be written very generally as

$$
\mathcal{Z} = \left(\prod_\kappa^C \frac{z_\kappa^{\tilde{N}_\kappa}}{\tilde{N}_\kappa!} \right) \int \left(\prod_S \prod_{i=1}^{\tilde{N}_S} \mathrm{d}r_{Si} \right) \left(\prod_H \prod_{j=1}^{\tilde{N}_H} P_H[r_{Hj}(\cdot)] \, \mathrm{d}[r_{Hj}(\cdot)] \right)
$$

$$
\times \left(\prod_C \prod_{k=1}^{\tilde{N}_C} P_{CA}[r_{CAk}(\cdot)] P_{CB}[r_{CBk}(\cdot)] \, \mathrm{d}[r_{CAk}(\cdot)] \, \mathrm{d}[r_{CBk}(\cdot)] \right.
$$

$$
\left. \times \delta[r_{CAk}(N_{CA}) - r_{CBk}(N_{CB})] \right) \delta\left[1 - \sum_\kappa \phi_\kappa(r) \right] e^{-\beta \hat{V}} \tag{2}
$$

The "C" over the first product sign indicates that each copolymer is to be treated as a single molecule for that product. Each factor z_κ is the kinetic energy contribution of one molecule of type κ. The underlying Gaussian nature of the chains is reflected in the Wiener weights

$$
P_\kappa[r(\cdot)] \propto \exp\left\{ -\frac{3}{2b_\kappa^2} \int_0^{N_\kappa} \dot{r}(\tau)^2 \, \mathrm{d}\tau \right\} \tag{3}
$$

The integrations in Eq. (2) are over all solvent molecule positions and all chain shapes and positions, subject to two constraints. The first one reflects the fact that the two blocks of each copolymer are bonded together and so must always occupy the same point in space. This is included via the first delta function appearing in Eq. (2). The second constraint is imposed by the second delta function. It is the incompressibility condition, requiring the local volume fractions to always sum to unity everywhere. The local volume fraction of each component at a point r is

$$
\phi_\kappa(r) = \frac{\langle \hat{\rho}_\kappa(r) \rangle}{\rho_{0\kappa}} \tag{4}
$$

where the $\langle \hat{\rho}_\kappa(r) \rangle$ is the ensemble average number density of component κ at r. These averages are calculated from the corresponding number densities for each configuration, which are given by

$$
\hat{\rho}_S(r) = \hat{\rho}_S(r; \{r_{Si}\})
$$

$$
= \sum_{i=1}^{\tilde{N}_S} \delta[r - r_{Si}] \tag{5}
$$

for the solvent, and

$$
\hat{\rho}_P(r) = \hat{\rho}_P(r; \{r_{Pj}(\cdot)\})
$$

$$
= \sum_{j=1}^{\tilde{N}_P} \int_0^{N_P} \delta[r - r_{Pj}(\tau)] \, \mathrm{d}\tau \tag{6}
$$

for $P = H$ for the homopolymers, or $P = CA$ or CB for each block of the copolymers. Each block is treated as a separate component in this expression, and in the corresponding summation in Eq. (2). This notation emphasizes the fact that quantities with carets over them depend on the system configuration, and hence the molecular coordinates.

The final contribution is the Boltzmann factor due to the interactions. Assuming the interactions between any pair of components can be represented by two-body interactions then, for a given configuration, this energy can be expressed as

$$
\begin{aligned}
\beta \hat{V} &= \beta \hat{V}(\{r_{Si}\}, \{r_{Pj}(\cdot)\}) \\
&= \frac{1}{2} \sum_{\kappa\kappa'} \int \hat{\rho}_\kappa(r) W_{\kappa\kappa'}(r - r') \hat{\rho}_{\kappa'}(r') \, dr \, dr' \\
&\equiv \hat{W}
\end{aligned}
\tag{7}
$$

with $\beta = 1/k_B T$.

These results are easily extended to other copolymer architectures, by modifying the factors attributable to the copolymers in Eq. (2). For example, for an ABC copolymer, shown in Figure 1(c), we need to insert an additional factor

$$
P_{CC}[r_{CCk}(\cdot)] \, d[r_{CCk}(\cdot)]\delta[r_{CBk}(0) - r_{CCk}(N_{CC})]
\tag{8}
$$

into Eq. (2). The additional δ function ensures that the $\tau = 0$ end of the B block is attached to the $\tau = N_{CC}$ end of the C block. As a second example, we consider a star copolymer, each with Σ arms, and each arm an AB diblock, all joined at the ends of the B blocks, as illustrated in Figure 1(f). In this case, we need a Wiener weight for each of the 2Σ blocks, a δ function joining the A and B blocks in each arm, and one more δ function connecting the other ends of each of the B blocks. Thus, we make the replacement

$$
P_{CA}[r_{CAk}(\cdot)]P_{CB}[r_{CBk}(\cdot)] \, d[r_{CAk}(\cdot)] \, d[r_{CBk}(\cdot)]\delta[r_{CAk}(N_{CA}) - r_{CBk}(N_{CB})]
$$

$$
\rightarrow \left(\prod_{\mu=1}^{\Sigma} P_{CA}[r_{CAk\mu}(\cdot)]P_{CB}[r_{CABk\mu}(\cdot)] \, d[r_{CAk\mu}(\cdot)] \, d[r_{CBk\mu}(\cdot)]\delta[r_{CAk\mu}(N_{CA}) - r_{CBk\mu}(N_{CB})] \right)
$$

$$
\times \left(\prod_{\mu=2}^{\Sigma} \delta[r_{CBk\mu}(0) - r_{CBk1}(0)] \right)
\tag{9}
$$

in Eq. (2). In the notation here, CA and CB mean A and B blocks of copolymer of type C, k labels the molecule, and μ labels each arm in molecule k. Generalizations of the partition function to other architectures can be made with similar changes.

It is convenient at this stage to introduce a number of changes of variable. For each independent function $\hat{\rho}_\kappa(r)$, we can introduce a Dirac δ function, and write the interaction Boltzmann factor as

$$
e^{-\hat{W}} = \int \left(\prod_\kappa d\rho_\kappa(\cdot)\delta[\rho_\kappa(\cdot) - \hat{\rho}_\kappa(\cdot)] \right) e^{-W(\{\rho_\kappa(r)\})}
\tag{10}
$$

where $W(\{\rho_\kappa(r)\})$ is defined as in Eq. (7), but now for continuous functions $\{\rho_\kappa(r)\}$. We can then use a Fourier representation for each of the δ functions in Eq. (10), which introduces a second new set of field variables $\{\omega_\kappa(r)\}$. Finally, we make a similar transformation of the δ function in Eq. (2) that imposes the incompressibility condition, and this introduces a final additional field, $\eta(r)$.

With these changes, the partition function, Eq. (2), can be transformed into

$$
\mathcal{Z} = \left[\prod_\kappa^C \frac{z_\kappa^{\tilde{N}_\kappa}}{\tilde{N}_\kappa!} \right] \int d\eta(\cdot) \left[\prod_\kappa d\rho_\kappa(\cdot) \, d\omega_\kappa(\cdot) \right] \left[\prod_\kappa^C Q_\kappa^{\tilde{N}_\kappa} \right]
$$

$$
\times \exp\left\{ \int dr \left[\eta(r)\left(1 - \sum_\kappa \frac{\rho_\kappa(r)}{\rho_{0\kappa}}\right) + \sum_\kappa \omega_\kappa(r)\rho_\kappa(r) \right] - W[\{\rho_\kappa(r)\}] \right\}
\tag{11}
$$

where

$$Q_S = \int d\mathbf{r}\, e^{-\omega_S(\mathbf{r})} \tag{12}$$

$$Q_H = \int d\mathbf{r}(\tau) \exp\left\{ -\int_0^{N_H} \left[\frac{3}{2b_H^2}\dot{r}(\tau)^2 + \omega_H(r(\tau)) \right] d\tau \right\} \tag{13}$$

and

$$Q_C = \int d\mathbf{r}_A(\tau)\, d\mathbf{r}_B(\tau)\delta[\mathbf{r}_A(N_{CA}) - \mathbf{r}_B(N_{CB})] \exp\left\{ -\int_0^{N_{CA}} \left[\frac{3}{2b_{CA}^2}\dot{r}_A(\tau)^2 + \omega_{CA}(r_A(\tau)) \right] d\tau \right\}$$

$$\times \exp\left\{ -\int_0^{N_{CB}} \left[\frac{3}{2b_{CB}^2}\dot{r}_B(\tau)^2 + \omega_{CB}(r_B(\tau)) \right] d\tau \right\} \tag{14}$$

for the diblock copolymer. This last result is easily generalized to other architectures. In each generalization, the revised Eq. (14) needs to include functional integrations over all positions and shapes of all sections of every molecule, δ functions to ensure appropriate block connections and, for every possible configuration, an integral of the appropriate Wiener weight and ω_κ along every block.

For nonpolar systems, the long-range interactions are normally modelled by contact interactions. Finite range corrections are also included sometimes, but they have negligible effect on the phase diagrams [35,41] and we do not include them here. The result, as will become clear below, is that effective interactions in incompressible systems can be modeled by simple Flory parameters, $\chi_{\kappa\kappa'}$. The situation is more complicated in compressible systems.

Equations (2) to (14) comprise a very general theory of Gaussian-chain polymers in incompressible systems. Up to this point, no mean field approximation has been made. The equations cover ordered and disordered phases, and weak or strong segregation. They assume two-body interactions, often, but not necessarily, represented by Flory parameters. Even though the system is incompressible, there can still be fluctuations in the individual volume fractions. There is no assumption about equal statistical segment lengths, and hence no restriction to conformational symmetry. However, it is assumed that the chains are long enough to be modeled by Gaussians.

The problem, of course, is that the integrations required for these expressions cannot be evaluated exactly. To do so, one would have to consider all possible fields and all possible configurations and, for each possibility, perform contour integrals for them all. Approximations must be made.

B. Numerical Self-Consistent Field Theory

It is convenient to first express the partition function as

$$\mathcal{Z} = \mathcal{N} \int d\eta(\cdot) \left[\prod_\kappa d\rho_\kappa(\cdot)\, d\omega_\kappa(\cdot) \right] \exp\{-\mathcal{G}[\{\rho_\kappa(\mathbf{r})\}, \{\omega_\kappa(\mathbf{r})\}, \eta(\mathbf{r})]\} \tag{15}$$

with

$$\mathcal{G}[\{\rho_\kappa(\mathbf{r})\}, \{\omega_\kappa(\mathbf{r})\}, \eta(\mathbf{r})] = \mathcal{F}[\{\rho_\kappa(\mathbf{r})\}, \{\omega_\kappa(\mathbf{r})\}, \eta(\mathbf{r})] + \int \eta(\mathbf{r}) \left[1 - \sum_\kappa \frac{\rho_\kappa(\mathbf{r})}{\rho_{0\kappa}} \right] d\mathbf{r} \tag{16}$$

and

$$\mathcal{F}[\{\rho_\kappa(\boldsymbol{r})\}, \{\omega_\kappa(\boldsymbol{r})\}, \eta(\boldsymbol{r})] = W[\{\rho_\kappa(\boldsymbol{r})\}] - \int d\boldsymbol{r} \sum_\kappa \omega_\kappa(\boldsymbol{r})\rho_\kappa(\boldsymbol{r}) + \sum_\kappa^C \tilde{N}_\kappa \left[\ln\left(\frac{\tilde{N}_\kappa}{\mathcal{Q}_\kappa z_\kappa}\right) - 1 \right] \tag{17}$$

Stirling's approximation is the only approximation in this.

The assumption underpinning mean field theory is that there is one (perhaps inhomogeneous) set of fields that dominates the integrations of Eq. (15), as well as similar integrations for expressions such as the equilibrium density distributions, and that all other contributions can be ignored. The computational task is then to find and calculate these fields which, initially, we denote $\{\rho_\kappa^0(\boldsymbol{r})\}$, $\{\omega_\kappa^0(\boldsymbol{r})\}$, and $\eta^0(\boldsymbol{r})$. The equilibrium structural and thermodynamic properties can then be calculated from them. In this approximation, the quantities of interest reduce to their extremal values. For example, each density distribution reduces to

$$\langle \hat{\rho}_\kappa(\boldsymbol{r}) \rangle \rightarrow \rho_\kappa^0(\boldsymbol{r}) \tag{18}$$

It is worth noting that, in this NSCFT, chains are not restricted to their most probable configuration. As will be seen below, all possible chain configurations are included, weighted by appropriate probabilities.

The extremal values of the fields are found by minimizing $\mathcal{G}$. Applying

$$\frac{\delta}{\delta\eta(\boldsymbol{r})}\mathcal{G} = 0 \tag{19}$$

to Eqs. (15) and (16) recovers the incompressibility condition, and then $\mathcal{G}^0 = \mathcal{F}^0$, and we have

$$\mathcal{F}^0 = W[\{\rho_\kappa^0(\boldsymbol{r})\}] - \int d\boldsymbol{r} \sum_\kappa \omega_\kappa^0(\boldsymbol{r})\rho_\kappa^0(\boldsymbol{r}) + \sum_\kappa^C \tilde{N}_\kappa \left[\ln\left(\frac{\tilde{N}_\kappa}{\mathcal{Q}_\kappa^0 z_\kappa}\right) - 1 \right] \tag{20}$$

From this point onwards, we drop the superscripts "0," but all quantities refer to these mean field values.

The other minimization conditions give prescriptions for calculating the density distributions and fields. For a solvent, $\phi_S(\boldsymbol{r})$ and $\omega_S(\boldsymbol{r})$ are related by

$$\phi_S(\boldsymbol{r}) = \bar{\phi}_S e^{-\omega_S(\boldsymbol{r})} \tag{21}$$

where $\bar{\phi}_S$ is the overall volume fraction of solvent S. We also need $\mathcal{Q}_S$, which is given by Eq. (12).

For a homopolymer and each block of a copolymer, we need to solve a modified diffusion equation for propagators $Q_p(\boldsymbol{r}, \tau|\boldsymbol{r}')$,

$$\left[-\frac{b_p^2}{6}\nabla^2 + \omega_p(\boldsymbol{r}) \right] Q_p(\boldsymbol{r}, \tau|\boldsymbol{r}') = -\frac{\partial}{\partial\tau}Q_p(\boldsymbol{r}, \tau|\boldsymbol{r}') \tag{22}$$

subject to the initial condition

$$Q_p(\boldsymbol{r}, 0|\boldsymbol{r}') = \delta(\boldsymbol{r} - \boldsymbol{r}') \tag{23}$$

The local volume fractions and factors are calculated from convolutions of these propagators. For homopolymer,

$$\phi_{\mathrm{H}}(r) = \frac{\tilde{N}_{\mathrm{H}}}{\mathcal{Q}_{\mathrm{H}}} \int_0^{N_{\mathrm{H}}} d\tau \int dr_1 dr_2 Q_{\mathrm{H}}(r_1, N_{\mathrm{H}} - \tau | r) Q_{\mathrm{H}}(r, \tau | r_2) \tag{24}$$

with

$$\mathcal{Q}_{\mathrm{H}} = \int dr_1 dr_2 Q_{\mathrm{H}}(r_1, N_{\mathrm{H}} | r_2) \tag{25}$$

For the local volume fraction of A at r due to the CA blocks of a copolymer, the corresponding expressions are

$$\phi_{\mathrm{CA}}(r) = \frac{\tilde{N}_{\mathrm{C}}}{\mathcal{Q}_{\mathrm{C}}} \int_0^{N_{\mathrm{CA}}} d\tau \int dr_1 dr_2 Q_{\mathrm{CA}}(r_1, N_{\mathrm{CA}} - \tau | r) \tilde{Q}_{\mathrm{CA}}(r, \tau | r_2) \tag{26}$$

with

$$\tilde{Q}_{\mathrm{CA}}(r, \tau | r_2) = \int dr_3 Q_{\mathrm{CA}}(r, \tau | r_2) Q_{\mathrm{CB}}(r_2, N_{\mathrm{CB}} | r_3) \tag{27}$$

and

$$\mathcal{Q}_{\mathrm{C}} = \int dr_1 dr_2 dr_3 Q_{\mathrm{CA}}(r_1, N_{\mathrm{CA}} | r_2) Q_{\mathrm{CB}}(r_2, N_{\mathrm{CB}} | r_3) \tag{28}$$

The expressions for $\phi_{\mathrm{CB}}(r)$ are trivially obtained from these by exchanging the A and B labels.

Many of these expressions are often written in terms of functions $q_{\mathrm{p}}(r, \tau)$ and $\tilde{q}_{\mathrm{p}}(r, \tau)$, which are just spatial integrals of the propagators, for example,

$$q_{\mathrm{p}}(r, \tau) = \int dr_1 Q_{\mathrm{p}}(r, \tau | r_1) \tag{29}$$

They also satisfy the diffusion equation, but the initial conditions, which follow trivially from integrating Eqs. (23) and (27), are

$$q_{\mathrm{p}}(r, 0) = 1 \tag{30}$$

and

$$\tilde{q}_{\mathrm{CA}}(r, 0) = q_{\mathrm{CB}}(r, N_{\mathrm{CB}}) \tag{31}$$

Equations (20) to (31) give the density distributions and free energies for systems with solvent, homopolymer, and diblock copolymer, as long as all the interaction potentials and the potential fields $\omega_{\mathrm{p}}(r)$ are known. Expressions for other copolymer architectures are similar, and in the literature.

The remaining ingredient of the theory is the set of potentials, $\omega_{\mathrm{p}}(r)$, which depend on the system, and are functions of the interactions and the Lagrange multiplier field, $\eta(r)$. As noted earlier, it is common in the literature to assume that the interactions can be represented by Flory parameters. If there is a solvent present, then it follows from Eq. (21) that

$$\omega_{\mathrm{S}}(r) = \omega_{\mathrm{S}}(r_0) + \ln\left(\frac{\phi_{\mathrm{S}}(r_0)}{\phi_{\mathrm{S}}(r)}\right) \tag{32}$$

Like all effective potentials, $\omega_S(r)$ is arbitrary to within an additive constant; this is reflected here by the $\omega_S(r_0)$ which is an arbitrary value at an arbitrary reference point r_0.

The expressions for the effective fields acting on the polymers depend on what components are present. If there is a solvent present, then $\eta(r)$ can be expressed in terms of the $\phi_S(r)$ and eliminated from the expressions for the potentials. As a specific example, if there are solvent plus two types of monomers, A and B, then the potential acting on a monomer of type A at r can be expressed as

$$\omega_A(r) = \omega_A(r_0) + \frac{\rho_{0R}}{\rho_{0A}} \left\{ \ln\left(\frac{\phi_S(r_0)}{\phi_S(r)}\right) + 2\chi_{AS}[\phi_S(r) - \phi_S(r_0)] + 2\chi_{AB}[\phi_B(r) - \phi_B(r_0)] \right\}$$

(33)

The $\omega_B(r)$ is obtained from $\omega_A(r)$ from the usual exchange A $\leftrightarrow$ B. For these fields, it does not matter if the polymers are homopolymers or copolymers. In this case, the $\phi_B(r)$ is the total volume fraction of B monomer at r, which may be due to both B homopolymer and B blocks of copolymers.

Flory χ parameters are defined using a reference density, and their values depend on the value chosen. This density is denoted by ρ_{0R} in Eq. (33). In polymer–solvent systems, the solvent density is usually used for this, that is, $\rho_{0R} = \rho_{0S}$. Less well appreciated is the fact that all Flory parameters have this dependence. For example, the χ_{AB} in this expression depends on this choice. As a result, the value of χ_{AB} would differ in different solvents, even for the *same* pair of polymers. In the absence of solvent, one must still specify the density used in the definition of the interaction parameter. For these reasons, ρ_{0R} has been maintained here to sustain generality.

Equation (33) is easily generalized to other blends. For example, if there are only solvent and one polymer, say polymer A, then $\phi_B(r) = 0$ everywhere and the last interaction term disappears. Alternatively, if a third polymer species is present, say a homopolymer C, a C block of a copolymer, or both of these, then Eq. (33) is augmented by an additional interaction term, that is, $2\chi_{AC}[\phi_C(r) - \phi_C(r_0)]$.

If there is no solvent, then there is no expression corresponding to Eq. (32) and the $\eta(r)$ cannot be eliminated from the effective fields. In the case of A and B species present, for example, neat AB copolymers, the $\omega_A(r)$ is

$$\omega_A(r) = \omega_A(r_0) + \frac{\rho_{0R}}{\rho_{0A}} \left\{ 2\chi_{AB}[\phi_B(r) - \phi_B(r_0)] + \frac{\eta(r)}{\rho_{0R}} \right\}$$

(34)

The equations above are not quite complete; they must be augmented by the incompressibility condition. For example, when solvent and A and B monomers are present, this is

$$\phi_S(r) + \phi_A(r) + \phi_B(r) = 1$$

(35)

everywhere. In the absence of solvent, the combined requirements of self-consistency and incompressibility suffice to determine the fields $\eta(r)$.

The expressions presented above comprise a complete set for the NSCFT of polymers and polymer–solvent blends, applicable to both homopolymers and diblock copolymers. The generalizations to other architectures are generally straightforward. In all cases, one must solve a diffusion equation for polymer propagators, from which the polymer volume fractions are calculated. When solvent is present, its distribution is then calculated from the incompressibility condition. However, in order to solve the diffusion equation, the potential fields $\omega_p(r)$ are needed, and they depend on the volume fraction profiles. As a result, the problem must be solved self-consistently.

Once a self-consistent solution has been found, it can be inserted into Eq. (20). If, as in this chapter, we are interested in ordered microphases, it is convenient to express all free energies relative to the free energy of the disordered phase, that is,

$$F = F_{\mathrm{D}} + \Delta F \tag{36}$$

It is straightforward to calculate all necessary quantities for the disordered phase, and then calculate F_{D} and subtract it from Eq. (20). It is convenient to express this difference as a free energy per unit volume, and in dimensionless units. The result can be written

$$\frac{\Delta F}{\rho_{0\mathrm{R}} k_{\mathrm{B}} T V} = \frac{1}{V} \int_V d\mathbf{r} \left\{ \frac{1}{2} \sum_{\kappa\kappa'} \chi_{\kappa\kappa'} [\phi_\kappa(\mathbf{r}) \phi_{\kappa'}(\mathbf{r}) - \bar{\phi}_\kappa \bar{\phi}_{\kappa'}] - \sum_\kappa \frac{\rho_{0\kappa}}{\rho_{0\mathrm{R}}} \omega_\kappa(\mathbf{r}) \phi_\kappa(\mathbf{r}) \right\}$$

$$- \sum_\kappa^C \frac{\rho_{0\kappa}}{\rho_{0\mathrm{R}}} \frac{\bar{\phi}_\kappa}{N_\kappa} \ln \left(\frac{\mathcal{Q}_\kappa}{V} \right) \tag{37}$$

The only new quantity here is the effective monomer density for the copolymer, $\rho_{0\mathrm{C}}$. It is calculated via a weighted average of the effective volumes of the monomers, which are proportional to the inverse densities. For example, for a diblock copolymer, it is given by

$$\frac{1}{\rho_{0\mathrm{C}}} = \frac{1}{N_{\mathrm{C}}} \left[\frac{N_{\mathrm{CA}}}{\rho_{0\mathrm{A}}} + \frac{N_{\mathrm{CB}}}{\rho_{0\mathrm{B}}} \right] \tag{38}$$

This reduces trivially to $\rho_{0\mathrm{C}} = \rho_{0\mathrm{A}} = \rho_{0\mathrm{B}}$ in the case of $\rho_{0\mathrm{A}} = \rho_{0\mathrm{B}}$.

The spatial integrations in Eq. (37) are over the macroscopic volume of the system, V. These integrals, as well as $\mathcal{Q}_\kappa$, are of order V. In practice, we can restrict the integrations to a single unit cell and replace V by the unit cell volume, hence rendering the integrals (and $\mathcal{Q}_\kappa$) finite.

C. Finding the Equilibrium Morphology

The NSCFT formalism comprises as a set of self-consistent equations for the equilibrium density distribution and effective field for each component. For a given system, there can be multiple solutions. The original, and still most-used, procedure is to find solutions for each possible microphase and, for each structure, a range of lattice parameters (domain sizes). The equilibrium morphology is then the phase and lattice parameter that have the lowest free energy. Metastable phases can also be identified.

More recently, a "real space" numerical method has been developed that does not need any information about potential equilibrium morphologies [64–66]. This has great potential for the future, but is still at an early stage of development.

D. Unit Cells and Unit Cell Approximations

Each microphase has its own symmetry and unit cell, and a full solution to the NSCF problem must be consistent with them. For example, the unit cell for the cylindrical phase is hexagonal, that for bcc is the appropriate Wigner–Seitz cell, and so on. The early NSCFT calculations approximated the unit cells for the latter two phases by cylinders and spheres, respectively. Matsen and Schick eliminated this unit cell approximation (UCA), by expanding each function of position in a series of orthonormal basis functions, each of which has the full symmetry of the phase under consideration. This provides a more accurate description of the C and S phases and, most importantly, permits the treatment of the more complex, nonclassical phases.

The NSCFT phase diagram calculated using the UCA is shown in Figure 4. The effects of the UCA can be seen by comparing this figure with Figure 3. The G phase is, of course, absent, but

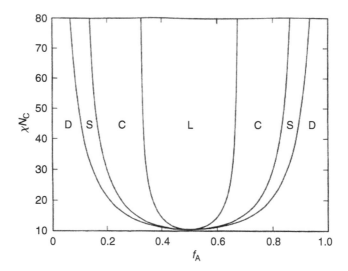

Figure 4 Microphase diagram for conformationally symmetric, diblock copolymers calculated using NSCFT, but with the UCA [35].

the remaining changes are small. The region near the critical point is unchanged, with all phase boundaries curving and meeting there. The MST and the remaining order–order boundaries are qualitatively the same, with very small quantitative shifts [67]. For example, eliminating the UCA shifts the S-C boundary toward slightly greater asymmetry, by about $\delta f_A \lesssim 0.01$. If one compares the L and C free energies directly, without considering the G phase, one finds that the UCA shifts the L/C boundary in the same direction, but by an even smaller amount, $\delta f_A \simeq 0.005$.

IV. WEAK AND STRONG SEGREGATION THEORIES

Beginning in 1975, Helfand and coworkers developed NSCFT for strongly segregated copolymers [25–31]. In 1980, Leibler published a weak segregation theory of conformationally symmetric, incompressible, neat diblock copolymers, applicable to both the disordered phase and weakly segregated ordered phases near the MST [24]. At about the same time, Erukhimovich also developed a weak segregation theory [33] and, a short time later, Hong and Noolandi developed a perturbative solution to the SCF equations presented in Section III.B, comprising a weak segregation theory generalized to blends [34]. The Leibler, Helfand, and more recent strong segregation theories have been widely used, and it is instructive to compare them with the full NSCFT. For these purposes, we use the NSCFT results for the classical phases only, since those were the ones treated by the limiting theories.

A. Weak Segregation Theories

Leibler introduced an order parameter

$$\psi(r) = \left\langle \frac{\hat{\rho}(r)}{\rho_{0A}} \right\rangle - f_A \tag{39}$$

which is equivalent to the variation in the local volume fraction $\phi_A(r)$ about its spatial average, $\delta\phi_A(r) = \phi_A(r) - \bar{\phi}_A$. Either $\delta\phi_A(r)$ or $\delta\phi_B(r)$ could be used, since $\delta\phi_B(r) = -\delta\phi_A(r)$ due to incompressibility.

He then analyzed the MST by developing an expansion of ΔF in powers of $\psi(r)$, using a random phase approximation. In this approximation, the effective potentials acting on the monomers take into account the monomer–monomer interactions, but the response of the system to external potentials is calculated assuming that the response functions are the same as those of ideal chains.

It is convenient to work with Fourier transforms. For neat copolymers, the leading nonzero term in the free energy expansion, which is second order, can be written as

$$\frac{\Delta F_2}{\rho_0 k_B TV} = \frac{1}{2V} \int \frac{dq}{(2\pi)^3} S^{-1}(q)|\delta\phi_A(q)|^2 \tag{40}$$

where $\delta\phi_A(q)$ is the Fourier transform of $\delta\phi_A(r)$, and $S(q)$ is the structure factor. In the disordered phase, $S(q)$ describes fluctuations in the density of each component about its average, and is given by

$$S(q) = \frac{N_C}{F(x,f) - 2\chi N_C} \tag{41}$$

where $f = f_A$ or f_B, and

$$x = q^2 R_g^2 = N_C b^2/6, \tag{42}$$

$$F(x,f) = \frac{g(1,x)}{g(f,x)g(1-f,x) - (1/4)[g(1,x) - g(f,x) - g(1-f,x)]^2} \tag{43}$$

and $g(f,x)$ are the correlation functions of ideal noninteracting chains,

$$g(f,x) = \frac{2}{x^2}(e^{-fx} - 1 + fx) \tag{44}$$

The structure factor is proportional to the scattering intensity in small angle x-ray scattering or small angle neutron scattering measurements. It has a maximum occurring at $x^* = 3.873$, which corresponds to a wavenumber

$$q^* = \sqrt{3.873}/R_g \tag{45}$$

This maximum is small at small χN_C, but increases with increasing χN_C. For perfectly symmetric polymers, $f = 0.5$, $S(q)$ diverges at $\chi N_C = 10.5$, signaling a second order, order–disorder phase transition at that point. This transition is to the lamellar phase.

This behavior implies that the Fourier series expansions for the order parameter and free energy for each possible structure can be approximated by including terms of just one wavenumber, $|q| = q^*$, but summed over different directions determined by the symmetry of the phase. The simplest phase is the lamellar one, for which this reduces to

$$\psi(r) = \Psi \cos(q^* x) \tag{46}$$

The use of this single wavenumber greatly simplifies the exploration of the MST and competing microphases. The general approach is to substitute Eq. (46) or its generalization to other structures into the free energy expression, using all terms in ΔF up to fourth order, minimize each resulting expression with respect to q^* and the amplitudes, and then compare the results for the different structures. The one with the lowest free energy is the predicted equilibrium structure. This is valid only if the interactions are weak enough that the amplitude of each density variation is small, and the profiles are simply cosine-like. These limitations will be explored below.

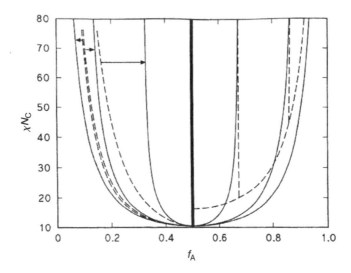

Figure 5 Comparison of the NSCFT phase diagram (solid lines) with limiting theories, for conformationally symmetric, diblock copolymers [35]. For $f_A < 0.5$, the dashed curves are calculated using the weak
segregation theory of Leibler [24], with the arrows showing the corresponding phase boundaries
in two approaches. For $f_A > 0.5$, the dashed curves are calculated using the narrow interphase
approximation [31].

The phase diagram predicted by this approach, along with the NSCFT result calculated using the
UCA, are shown in left-hand panel of Figure 5. The agreement is essentially perfect very near the
critical point at $f_A = 0.5$. However, although the overall topology is the same, there are significant
differences away from this point. The L/C and C/S boundaries are shifted too far toward high
asymmetry, and the S phase occupies only a very narrow strip on either side of the phase diagram.
At large χN_C, the predicted L region is very broad, for example, spanning from $f_A < 0.2$ to
$f_A > 0.8$, when $\chi N_C > 50$. The width of the C phase is reasonable, but it is generally located too far
toward each side of the diagram. The domain of agreement between the NSCFT and Leibler theory
is limited to about $0.45 \lesssim f_A \lesssim 0.55$, and $\chi N_C \lesssim 15$. Beyond this region, large differences are not
unexpected, since they are the result of applying a weak segregation theory to large χN_C.

However, there are two important features that emerge from this comparison that are often
overlooked. First, most of the MST is also shifted, and the shift can be large. For example, at
$f_A = 0.1$ (or 0.9), NSCFT predicts the MST at $\chi N_C \simeq 50$, whereas this approximate theory predicts
it to be at $\chi N_C \simeq 80$, a shift of $\delta \chi N_C \simeq 30$. The second important difference is the narrowness of
the S region. Both these features indicate that these systems are not weakly segregated, even near
the MST, except for small χN_C. The differences develop at $\chi N_C \simeq 15$, which suggests that this
is the upper limit on the range of "weak segregation."

One of the reasons for the differences just described is the single-wavenumber approximation,
Eq. (46). One way in which weak segregation theory has been refined is by incorporating higher-
order wavenumbers in the calculation of the density profiles and free energy, but within a fourth-order
approximation to the free energy [68–71]. This has been referred to as the "multiwavenumber" or
"multiharmonic" approximation. It turns out that the additional wavenumbers become significant
beyond about $\chi N_C \simeq 15$, which is another indication that this marks, qualitatively, the limit of weak
segregation.

B. Strong Segregation Theories

The other extreme is the strong segregation limit (SSL), for which Helfand and coworkers developed
an NSCFT over a period of years beginning in 1974, culminating with the publication of a phase

diagram for PS–PBD in 1982 [31]. In their numerical work, they explicitly included the requirement that the copolymer joints be confined to a narrow AB interphase, that is, the "narrow interphase approximation" [25,27].

Since our primary purpose here is exploring the limits of SSL theory, we show the results obtained by applying their approach to conformationally symmetric polymers, in the right-hand panel of Figure 5. The results are qualitatively the same for their PS–PBD diagram. It is clear from the figure that the phase boundaries agree very closely with the full NSCFT results for large χN_C, although some small differences remain for the MST even up to $\chi N_C = 80$. However, there are qualitative differences at small and intermediate χN_C. Instead of all phase boundaries curving and meeting at a critical point, the order–order phase boundaries remain almost vertical, and intersect the MST. There are direct ∂/L transitions ranging from about $f_A = 0.35$ to 0.65, ∂/C transitions over $f_A \simeq 0.15$ to 0.35 and 0.65 to 0.85, and ∂/S transitions for $f_A \lesssim 0.15$ and $f_A \gtrsim 0.85$. The MST for symmetric polymers occurs at about $\chi N_C \simeq 16$.

Work on the SSL has continued since the pioneering work of Helfand et al. Semenov [72] and Likhtman and Semenov [73] developed an analytic theory for the strong segregation regime using the UCA. They predicted that the S/C and L/C phase boundaries for conformationally symmetric diblocks are at

$$f_{S/C} = 0.1172 + C_1(\chi N_C)^{-1/3} \tag{47}$$

and

$$f_{L/C} = 0.2911 + C_2(\chi N_C)^{-1/3} \tag{48}$$

where C_1 and C_2 are constants. In the limit of infinite segregation, these boundaries are thus $f_{S/C} = 0.1172$ and $f_{L/C} = 0.2911$. Matsen [74] has done very careful NSCFT calculations, and determined the L/C and S/C boundaries up to $\chi N_C = 1400$, with an estimated numerical uncertainty in the phase boundaries of better than 10^{-6}. He then plotted the phase boundaries against $(\chi N_C)^{-1/3}$, as suggested by Eqs. (47) and (48). Plotted in this way, the L/C boundary retained considerable curvature, at least up to $\chi N_C \simeq 600$. Beyond that, it became reasonably straight. A linear extrapolation of the last four points gave a limit just above $f_{L/C} \to 0.300$. Given the estimated uncertainties in the extrapolation, the results are not inconsistent with the limit of 0.2911 predicted by Likhtman and Semenov, although the persistence of the curvature up to such large χN_C was noted. For the S/C boundary, Matsen's results were reasonably linear when plotted against $(\chi N_C)^{-1/3}$. However, they are difficult to reconcile with the SSL theory. They decreased monotonically with increasing χN_C over the full range of the calculations, reaching a minimum of $f_{S/C} = 0.1166$ at $\chi N_C = 1400$. This is below the SSL prediction of 0.1172. They extrapolate to about $f_{S/C} = 0.108$, which is even further below the SSL prediction. In order for the NSCFT results to agree with the SSL value, the monotonic decrease would have to reverse, and then $f_{S/C}$ would have to *increase* beyond $\chi N_C \geq 1400$. Neither the discrepancy between the extrapolated value and the SST limit, nor the possibility of a reversal in this monotonic decrease, is currently understood. Although these small discrepancies would be very difficult to detect experimentally, they are of interest because they suggest that some piece of physics may be missing from the current understanding of the SSL.

V. CONFORMATIONAL SYMMETRY AND ASYMMETRY

Many theoretical treatments, and all results presented here to this point, apply to the idealized conformationally symmetric case of equal pure component densities and statistical segment lengths for each block. In fact, a diblock copolymer is characterized by seven independent quantities, the pure component densities, degrees of polymerization and statistical segment lengths of each block, and the interaction parameter. The monomers can be defined so as to have equal pure component densities, but that still leaves six independent characteristics. We turn now to the incorporation of these additional features.

With equal pure component densities, $\rho_{0A} = \rho_{0B}$, the relative volume fraction of each block is the same as the relative degree of polymerization, that is,

$$f_A = \frac{N_{CA}}{N_C} \tag{49}$$

If the statistical segment lengths are equal, $b_A = b_B$, then the unperturbed radii of gyration of the blocks satisfy

$$\frac{R_{A,g}^2}{R_{B,g}^2} = \frac{N_{CA}}{N_{CB}} \tag{50}$$

Equation (49) or (50) can always be satisfied by choice of monomer definition. If *both* hold, then

$$\frac{f_A}{f_B} = \frac{R_{A,g}^2}{R_{B,g}^2} \tag{51}$$

and the copolymer is said to be conformationally symmetric. This is an assumption of many theories, but it occurs rarely, if ever, in nature.

The mismatch can be characterized by a conformational asymmetry parameter defined as [35]

$$\epsilon = \frac{f_A/f_B}{R_{A,g}^2/R_{B,g}^2} \tag{52}$$

which is unity for conformationally symmetric polymers. An equivalent definition is

$$\epsilon = \frac{\rho_{0B} b_B^2}{\rho_{0A} b_A^2} \tag{53}$$

which, if the monomers are chosen to have equal size, reduces to $\epsilon = (b_B/b_A)^2$. An alternative conformational asymmetry parameter sometime used [6,36] is simply b_B/b_A which, with monomer sizes chosen equal, is equivalent to $\sqrt{\epsilon}$.

This parameter represents the degree of mismatch between the volume fractions of each block and their corresponding unperturbed radii of gyration. In defining ϵ, either block can be defined as the A block, so values of ϵ and ϵ^{-1} are physically equivalent under the exchange A $\leftrightarrow$ B. If $\epsilon > 1$, then the B block is the more extended one, that is, $b_B > b_A$. In practice, the range of ϵ is from about $\frac{1}{3}$ to 3.

As we have seen, mean field theory predicts that the phase behavior of conformationaly symmetric diblocks can be described in terms of just two quantities, χN_C and f_A. A general result of the NSCFT is that, as long as each microphase can be described by a single independent lattice parameter, then the entire microphase diagram of a system of diblock copolymers depends on three parameters, which are χN_{eff}, the compositional asymmetry f_A, and the conformational asymmetry ϵ. N_{eff} is an effective degree of polymerization, defined by

$$N_{eff} = \rho_{0R} \left[\frac{N_{CA}}{\rho_{0A}} + \frac{N_{CB}}{\rho_{0B}} \right] \tag{54}$$

where ρ_{0R} is the reference density used in defining χ. If the two reference densities are chosen to be equal, and in turn equal to ρ_{0R}, then $N_{\text{eff}} = N_C$.

Although all copolymer phase diagrams are qualitatively similar, they vary quantitatively from copolymer to copolymer, and they are generally not perfectly symmetric about $f_A = 0.5$. The small differences in phase diagrams for different molecules of the same architectures, for example, diblock PS–PI versus diblock PS–PDB, can thus be characterized in terms of different conformational asymmetries of the polymers.

VI. MICROPHASES AND MICRODOMAINS OF NEAT COPOLYMERS

A. Microphase Diagrams

The NSCFT microphase diagram of conformationally symmetric, diblock copolymers has been presented in Figure 3 and discussed in Section II. In this section, we present a number of other diagrams for different systems, exploring the effects of conformational asymmetry and architectures. We focus on two-species copolymers, which have been the subject of most calculations.

Figure 6 shows NSCFT phase diagrams for conformationally asymmetric diblock copolymers [6]. In the notation used here, they correspond to conformational asymmetry parameters of $\epsilon = 1$ (top panel), 0.44, and 0.25, with the A block having the larger effective statistical segment length, that is, if $\rho_{0A} = \rho_{0B}$, then $b_A/b_B = 1$, 1.5, and 2 in the three panels. The upper panel is the same as the lower half of Figure 3. The lower panel is at, or possibly just beyond, the limit of experimental accessibility. The conformational asymmetry does not alter the topology of the phase diagram, but there are changes in some of the phase boundaries. The MST is barely affected, and the order–disorder transition is still always a first-order transition to spheres, except that the critical point persists, at which the transition is still a second-order one to lamellae. The minimum in the MST remains nearly unchanged, but the critical point is shifted to the right of it and slightly upwards, to just beyond $f_A = 0.6$ in the most asymmetric case (lower panel). The G and cp-S phases still terminate at triple points, and the S/C, C/L, and D/S boundaries still merge at the critical point. However, all the order–order phase boundaries shift to higher f_A, and the diagram loses its symmetry. The regions for each phase on the left side ($f_A < 0.5$) are all broadened, and those on the right are narrowed. In the most extreme case, the C/G and G/L boundaries on the left side of the diagram actually change slope for large enough χN_C, moving to larger f_A with increasing segregation strength.

The shift in the order–order boundaries is easily explained by the conformational asymmetry. The argument is most transparent for layers. In these diagrams, $b_B < b_A$, which implies that the A block is naturally more extended than the B block. If the diblocks have equal volumes, $f_A = f_B$, then, in the lamellar morphology, the A and B subdomains must have the same thickness. However, in this case of $b_B < b_A$, this means that the B block is stretched beyond its unperturbed R_g more than the A block. A corollary to this is that equal stretching will occur for copolymers with $f_A > f_B$. This means that the tendency for these polymers to be driven toward a B-centered morphology is reduced by this conformational asymmetry. Hence, the L/C phase boundary moves toward greater f_A. Similar arguments apply to the other order–order phase boundaries, which all shift in the same direction.

The next level of complexity in architecture is the triblock. The simplest of these is the ABA triblock, with the two end blocks having equal degrees of polymerization. Three NSCFT phase diagrams for this case are shown in Figure 7 [8]. The overall structure is, again, unchanged, with the same stable phases, triple points, and critical point. For the conformationally symmetric case (middle panel), the critical point is very near $f_A = 0.5$ as it is for diblocks, but the phase diagram is not symmetric about this line. There are also small but noticeable shifts in the MST, with the minimum moving to modestly smaller χN_C. The effects of conformational asymmetry are qualitatively the same as before: if $b_A < b_B$ (top panel) the order–order boundaries move to the left, and if $b_A > b_B$

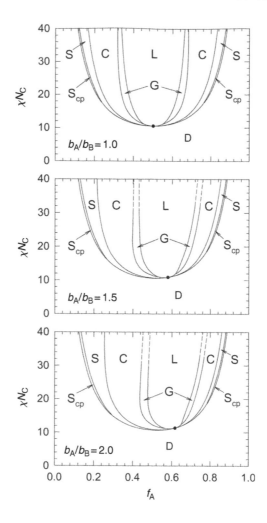

Figure 6 NSCFT phase diagrams for conformationally symmetric (top panel) and asymmetric (lower two panels) diblock copolymers [6].

(bottom panel) they move to the right. Matsen has also examined ABA triblocks with unequal A-block lengths; details are given in Ref. 7.

Moving up in terms of complexity, we next consider multiblock ABABA ... copolymers. Matsen and Schick published NSCFT phase diagrams for conformationally symmetric and asymmetric copolymers for this case, in the limit of an infinite number of blocks [5]. Once again, there was no change to the overall structure of the phase diagrams. For conformationally symmetric polymers, it was symmetric about $f_A = 0.5$, and very similar to the diblock diagram. For the conformationally asymmetric copolymers, the differences were qualitatively the same as for diblocks and triblocks: small shifts in the order–order boundaries but no change in the topology. The most interesting difference is that the MST is moved upward. This shift was analyzed in some detail in Ref. 55 for varying numbers of blocks. For the $f_A = 0.5$, conformationally symmetric case, they calculated the locations of the MST for two families of polymers. In the first family, the degree of polymerization of all blocks was equal; for this family, the A and B volume fractions are unequal if there are odd numbers of blocks, that is, $f_A > f_B$ for ABA, ABABA, ABABABA, etc. In the second family, all A blocks had the same degree of polymerization, as did all B blocks, but they were chosen so that $f_A = f_B$. These two families are identical for polymers with even numbers of blocks, for example, AB, ABAB, ABABAB, etc. The location of the MST varied with the number of

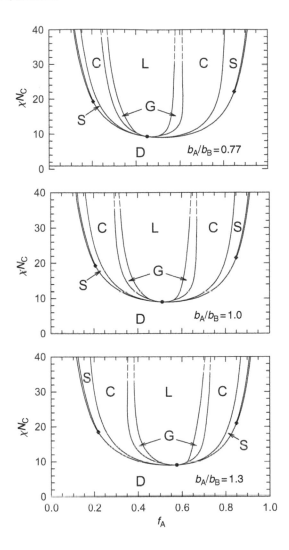

Figure 7 NSCFT phase diagrams for ABA triblock copolymers [8].

blocks in each family. These variations were conveniently expressed in terms of χN_{ave}, where N_{ave} is the average of the block sizes. For diblocks, the MST is at $\chi N_{ave} = 10.5/2 = 5.25$. For many blocks, it increases to $\chi N_{ave} = 7.55$, which is consistent with Ref. 5. For the family of polymers with equal A and B volume fractions, the increase was monotonic. For the other family, it was nonmonotonic, reflecting the alternation between equal and unequal volume fractions for even and odd numbers of blocks. Of course, the results for the two families converged in the limit of many blocks.

Matsen and Schick also did AB starblock copolymers, with three, five, and nine armed-stars [4]. Once again, there was no change in the topology of the phase diagram, but the order–order boundaries were shifted. As in all the other cases, the conformational asymmetry simply shifted the order–order phase boundaries to the left if $b_A < b_B$ or to the right if $b_A > b_B$.

Figure 8 provides the most comprehensive and direct comparison of this theory with experiment [63]. The upper panel is the NSCFT diagram for conformationally symmetric diblocks, and the lower panel is the diagram constructed for PS–PI from experiment. In this construction, the dots represent the observations, and lines are added as visual aids. There are numerous aspects of agreement between the theory and experiment. The stable ordered phases are the L, G, C, and S, and no others.

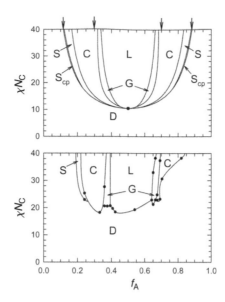

Figure 8 Comparison of theoretical (upper panel) and experimental PS–PI (lower panel) phase diagrams [63]. In the upper panel, the lines are the result of NSCFT for conformationally symmetric polymers, the arrows indicate the L/C and S/C phase boundaries predicted by strong segregation theory, and the dot is the mean field critical point. In the lower panel, the dots are the experimental observations for PS–PI, and the lines serve only as visual aids.

For $\chi N_C \gtrsim 30$, the locations of all the phase boundaries (in terms of f_A) agree well. The general shape of the MST agrees, but there is qualitative disagreement near the MST, particularly for $0.2 \lesssim f_A \lesssim 0.8$. In the experiments, the MST occurs at larger χN_C, by up to nearly a factor of two in some cases. Furthermore, there are direct transitions from the disordered phase to the C and G phases, and the D/L transition occurs over a finite window of f_A from about 0.4 to beyond 0.6. These differences may be attributable to fluctuation effects that are not included in mean field theories, including the NSCFT, as discussed below.

B. Domain Sizes and Density Distributions

A number of calculations have revealed some interesting correlations between the domain size scaling, shapes of the density profiles, and the microphase diagram. We turn now to a brief discussion of these relationships.

We begin by summarizing the two limits. As discussed in Section IV.A, the weak segregation theory describes each density profile as a sinusoidal variation about its mean value, or as a linear combination involving wave vectors of only one magnitude, as summarized in Eqs. (45) and (46). Since the components are only weakly segregated, each subdomain contains both species, and the "interfaces" are broad. The overall domain sizes are, to within a numerical factor,

$$d \sim N_C^{1/2} b \qquad (55)$$

This specific form is for conformationally symmetric polymers. Note that it is independent of the interaction parameter, χ.

In the strong segregation theories, each species is assumed to segregate into the two subdomains, which are separated by narrow interfaces. Expressed in terms of local volume fractions, each density profile is essentially unity in one subdomain, and falls smoothly but rapidly to zero in the other.

In this limit, the domain sizes scale as [72,75]

$$d \sim \chi^{1/6} N_C^{2/3} \tag{56}$$

and the interfacial widths as

$$d_I \sim \frac{2b}{(6\chi)^{1/2}} \left[1 + \frac{c}{(\chi N_C)^{1/3}} \right] \tag{57}$$

where c is a constant.

Between the MST and strong segregation, there are an evolution in the density profiles and an increase in the domain sizes, which are continuous except at the order–order transitions. There is an accompanying increase in the stretching of the polymers, which can become highly stretched in strong segregation, and a narrowing of the interface. The NSCFT predicts the details of these changes, without making a prior assumptions about the shapes, for example, the single-wavenumber or narrow interphase approximations. It can also help to delineate the range of applicability of the limiting theories.

Figure 9 shows the layer thickness calculated for symmetric copolymers in the lamellar phase [2], along with the SSL prediction. In strong segregation, the NSCFT results agree with the SSL scaling, but only beyond $\chi N_C > 100$. Toward weak segregation, the NSCFT results deviate downward from the SSL curve, becoming steeper. This means that the dependence of d on χ and N_C is strongest in weak segregation.

An interesting prediction of NSCFT is that, if one chooses to describe the domain size in terms of power laws, then it obeys the functional form

$$d \propto (\chi N_C)^p N_C^{1/2} \tag{58}$$

everywhere [35]. Equations (55) and (56) for weak and strong segregation are both consistent with this form, with $p = 0$ in the first case and $p = \frac{1}{6}$ in the second.

The numerical results also obey this relationship, with the value of the power varying slowly, from $p = \frac{1}{6}$ in strong segregation and *increasing* in weak segregation. Vavasour and Whitmore found

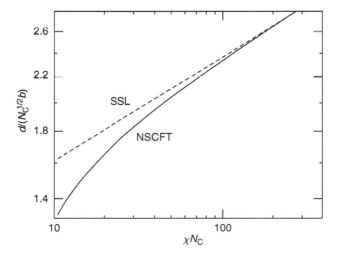

Figure 9 Domain size, d, in units of $N_C^{1/2}b$, for a conformationally symmetric copolymer with $f_A = 0.5$ in the lamellar phase. The solid curve is from the NSCFT [2], and the dashed curve the SSL theory.

that it reached $p = \frac{1}{2}$ at the critical point [35]. They also examined cylinders and spheres, using $f_A = 0.3$ and 0.1, respectively. The results for cylinders showed an increase in the effective power as for lamellae, reaching $p \simeq 0.4$ just prior to the transition to spheres. For spheres, they were prevented from studying systems very close to the MST by the accuracy they could attain in their calculations. In the accessible range, the scaling was the same as in SSL.

There is a complementary approach to studying the effects of asymmetry on the domain scaling, which is to choose a fixed χN_C and vary f_A. This approach revealed new behavior very near the MST [6]. As the asymmetry increases, the domain size initially decreases but, near the MST, this decrease reverses and the domain size increases. The explanation for this is that the short minority blocks are only weakly anchored in their spherical domains, and can escape into the majority domain. In this realm, the spheres continue to shrink, but the overall domains increase in size.

Density profiles for lamellae have been presented in Refs. 2, 5, and 35, for cylinders and spheres in Ref. 35, and for all stable phases in Ref. 37. The results show some interesting common behavior as functions of χN_C. Except for highly asymmetric copolymers, the density profiles are quasi-sinusoidal only if $\chi N_C \lesssim 15$. At higher χN_C, the maximum volume fractions reach unity, and the profiles begin to flatten out, beginning their evolution to the SSL-type profiles. For highly asymmetric copolymers, this occurs at higher χN_C, for example, at about $\chi N_C \gtrsim 20$ for $f_A \lesssim 0.2$.

As noted earlier, it is in approximately the range of $\chi N_C \lesssim 15$ that the NSCFT phase diagram agrees well with the one predicted by the weak segregation theory. Thus, there is a good correlation between the agreement of the phase diagrams and the validity of the single wavenumber, low amplitude description of the density profiles. Both are limited to a rather small part of the phase diagram.

C. Dominant Physics

Much of the interesting behavior can be understood on the basis of some simple principles, which can be illustrated by considering conformationally symmetric AB diblocks with statistical segment length b. In strong segregation, there are three main contributions to the free energy density of a microphase separated state relative to the homogeneous reference state,

$$\Delta f \simeq \Delta f_{\text{int}} + f_{\text{I}} + f_{\text{el}} \tag{59}$$

The first contribution is the change in interaction energy resulting from the reduction in A–B contacts. In the limit of strong segregation, it is

$$\Delta f_{\text{int}} \simeq -\chi f_A f_B \tag{60}$$

This is the driving force inducing microphase separation, but its value is independent of the resulting structure or domain size. The next term is the free energy associated with the interfacial tension at each interphase. For the lamellar structure, it can be approximated as

$$f_{\text{I}} \simeq \gamma \frac{2}{d} \tag{61}$$

where γ is the interfacial tension which can be approximated, to leading order by [76]

$$\gamma \simeq \left(\frac{\chi}{6}\right)^{1/2} b \tag{62}$$

This contribution decreases with increasing d, so drives the system toward large domain sizes. However, large domain sizes cause stretching of the polymer blocks. The third term is the elastic free energy due to this stretching. Again for the lamellar structure, it can be approximated by

$$f_{el} \simeq \frac{1}{2N_C}\left[\alpha^2 + \frac{2}{\alpha} - 3\right] \tag{63}$$

where

$$\alpha = \frac{d}{N_C^{1/2}b} \tag{64}$$

and is a measure of the stretching of the polymer. Using Eqs. (60)–(64) in Eq. (59), and minimizing Eq. (59) with respect to d, yields

$$\Delta f \simeq \chi\left[-f_A f_B + \left(\frac{3}{2\chi N_C}\right)^{2/3}\right] \tag{65}$$

and

$$d \propto \chi^{1/6}N_C^{2/3}b \tag{66}$$

Equation (65) implies that the MST should be governed by two factors, χN_C and f_A (or f_B), and occur approximately at

$$\chi N_C \simeq \frac{3}{2}\left(\frac{1}{f_A(1-f_A)}\right)^{3/2} \tag{67}$$

This also implies that, for a given χN_C, the MST would occur at

$$f_A = 0.5 \pm \frac{1}{2}\left[1 - \left(\frac{12}{\chi N_C}\right)^{2/3}\right]^{1/2} \tag{68}$$

These simple equations are remarkably accurate, especially in strong segregation. At all $\chi N_C \gtrsim 20$, Eq. (68) predicts the MST at values of f_A that agree with the full NSCF result to within 3%. The scaling exponent in Eq. (66) agrees with the numerical results for strong segregation, $p = \frac{1}{6}$. Even in weak segregation, this picture is in near agreement, predicting the MST at $\chi N_C = 12$, instead of 10.5, for $f_A = 0.5$. This discrepancy, of course, reflects the use of SSL approximations in this picture. The overall level of accuracy of this simple picture supports the picture of a balance of the three physical factors identified above as controlling the order–disorder transition and domain sizes, except for $\chi N_C \lesssim 15$.

The order–order boundaries can be understood in terms of more subtle principles [63]. If the blocks are of unequal length, then, in the lamellar phase, the longer blocks are stretched more than the shorter ones, and it becomes favorable for the interface to curve toward the minority domain. This is the mechanism that causes order–order transitions and, of itself, would suggest the phase sequence L → PL → G → OBDD → C → S as f_A changes from 0.5 to zero or unity. Not all of these are stable, and that is because of a further effect known as packing frustration, which is due to differential stretching of different chains in the system. The cylindrical phase serves to illustrate. The unit cell for this phase is hexagonal. Suppose $f_A < f_B$ so the cylinder cores are made of A. If the cores are perfect cylinders, then some of the B chains must stretch farther than others in order to fill up the six "corners" of the hexagon. This can be relieved by the cores' deforming from perfect cylinders, but at the cost of some increase in the interfacial area and energy. The actual shape is determined by a balance between the two, and the mismatch causing it is known as packing frustration. It can be quantified using the mean-squared variation in the curvature of the interface, but the primary point is that high frustration causes high stretching free energy, so low frustration systems are more stable. The OBDD and PL structures have higher frustration than the G phase, and so are not stable.

In summary, the unfavorable AB contacts drive the phase separation but do not control the morphology. The domain sizes are determined primarily by a balance between the interfacial and stretching free energies. In order to balance the stretching of the A and B blocks, the interfaces become curved, and this mechanism causes the order–order transitions. Finally, the shapes are adjusted in detail by the need to minimize the interfacial energy balanced against the need for uniform domains to minimize stretching free energy. Packing frustration, which is a measure of the incompatibility of these last two effects, explains the absence of the PL and OBDD phases.

VII. FLUCTUATIONS AND SIMULATIONS

Although NSCFT is remarkably successful in predicting the stable phases and much of the phase diagram, we have seen that it is not in accord with experiment in the vicinity of the MST for nearly symmetric copolymers, $0.4 \lesssim f_A \lesssim 0.6$. As Leibler pointed out very early [24], this region is near a critical point, where one can anticipate that fluctuations may be important. Two general means of exploring these issues are by incorporating fluctuations into theories, and by numerical simulations.

A. Theory of Fluctuations

Fredrickson and Helfand developed a weak segregation theory with fluctuations [77], starting by mapping the Leibler free energy functional onto one treated earlier by Brazovskii [78]. They used the UCA and treated the three classical phases. This has since been extended to the G phase [79,80] and to triblocks [81], and modified to include a multiharmonic approximation for the density profiles, although not for the G phase [82].

Qualitatively, there are a number of effects, which all depend on N_C, but diminish with increasing N_C and vanish in the limit $N_C \rightarrow \infty$. Hence, for finite N_C, the phase diagram depends on N_C in addition to the two original factors, χN_C and f_A.

The first effect is to shift the MST upward to larger χN_C. For perfectly symmetric copolymers, it is predicted at

$$\chi N_C = 10.5 + \frac{41}{N_C^{1/3}} \tag{69}$$

The second effect is to alter the microphase boundaries near the MST. For finite N_C, some finite D/L, D/C, and D/G windows open near $f_A = 0.5$. This is qualitatively in accord with experiment, but these windows are very small: at $N_C = 10^6$ the D/L window extends over $0.47 < f_A < 0.50$, the D/C window over $0.39 < f_A < 0.47$, and the D/S over $f_A < 0.39$ (with corresponding windows on the other side of the diagram). These D/L and D/C windows are much narrower than what is observed, there is no D/G window, and the S strip along the MST is very narrow. At $N_C = 5 \times 10^3$, the D/L window broadens somewhat, to about $0.435 < f_A < 0.5$, and there is a very narrow D/G window over about $0.425 < f_A < 0.435$. However, at this point the S phase completely disappears from the phase diagram.

The third change is that, even in the disordered phase, the chains are stretched, with the stretching increasing as the MST is approached. This prediction is consistent with small angle x-ray [83] and neutron [84] scattering experiments, in which the maximum in the scattering intensity shifts to lower wavenumber (q^*) with decreasing temperature, at temperatures above the MST.

The upward shift in the MST can be significant; for example, Eq. (69) predicts a change of $\delta\chi N_C \simeq 3$ for $N_C = 5000$, or more for smaller N_C. This is in qualitative agreement with what is observed, and would explain at least part of the differences apparent in Figure 8. However, it also means that most of the interesting region is beyond where the weak segregation theory is expected to apply. All this suggests that fluctuations are important to the topology of the phase diagram near the MST, but it is unrealistic to expect quantitative accuracy for a theory grounded in a fourth-order free energy functional.

Another approach has been to incorporate fluctuations into the full NSCFT, and there has been work in this direction by Shi and coworkers [85–87]. They have investigated the nature of the fluctuations in the ordered phases and the stability of the phases, but have not yet calculated the phase diagram.

B. Simulations

Simulations provide another approach that eliminates the mean field approximations [88–100]. They are, however, limited by the computational resources available. For an accurate simulation of a real system, one would need to model many molecules of realistic length, and use a system size much greater than any of the microdomain length scales. Dense systems are particularly difficult to handle, since any Monte Carlo move of a single monomer is precluded by the presence of other monomers.

There is a related, very large challenge facing simulations of copolymers. Typically, one performs the simulation by first annealing at very high temperature, and then slowly lowering the temperature, that is, increasing the interaction parameter. When a microphase forms and the microdomains become progressively more strongly segregated, it becomes very difficult for molecules to migrate from one domain to another. However, the equilibrium properties vary smoothly with temperature which, in general, requires interdomain migration. Thus, investigating the temperature dependence of equilibrium properties of microphase separated systems is very challenging.

It is beyond the scope of this chapter to review all the simulations in the literature, but recent work by Banaszak et al. [98,99] and Vassiliev and Matsen [100] is illustrative. Banaszak et al. did Monte Carlo simulations of ABA triblocks. Their systems contained 900 chains densely packed on a face-centered cubic (fcc) lattice. The chains were moved by cooperative rearrangements, called the cooperative motion algorithm: since all sites are occupied, a segment can be moved only if other segments are moved simultaneously. Their chains all had 30 monomers, and they explored three different compositions, each with equal end block lengths. The first had 3 units of A on each end and 24 units of B in the middle; this was labeled 3-24-3. The others were 7-16-7 and 10-10-10. The overall A-volume fractions for these are $f_A = 0.2$, 0.46, and 0.67.

For all three systems, they performed simulations in which the systems were first equilibrated in the high temperature, athermal limit. A jump was then made to finite temperature at which the system was still disordered, and then gradually cooled down. At each temperature, they performed 2×10^6 Monte Carlo steps. For the 7-16-7 system, they also studied other thermal treatments: instead of slow cooling, they quenched to each required temperature from the athermal state, to a total of 39 different temperatures. As well, starting from the lowest temperature probed, they slowly reheated the system. In all cases, they monitored the polymer mean-squared end-to-end distances R_{rms}, the specific heat C_V, and the energy per lattice site. Systems were always cooled into a microphase separated state.

In the athermal state, the chains closely followed ideal Gaussian statistics. As T was lowered, both R_{rms} and C_V slowly increased. At a certain point, C_V rose to a peak and then decreased monotonically. In contrast with this, R_{rms} continued to increase past this point, but then leveled off. The peak in C_V can be taken as signaling the MST. For the 3-24-3 polymers, the microphase that formed at the MST was weakly segregated spheres, but it then transformed to cylinders at a lower temperature.

For the 7-16-7 polymers, the microphase was lamellar. Both these sets of results are consistent with the NSCFT predictions for these compositions. The 10-10-10 chains formed the bicontinous double-diamond structure. This contrasts with the NSCFT prediction that the gyroid phase is stable, but it is modest disagreement given that the NSCFT predicts that the double-diamond structure is metastable, and the energy difference between the two is small.

The increase in R_{rms} as the MST is approached from above is consistent with the fluctuation theory just discussed. Its leveling off at low temperatures is probably a result of the cooling procedure. Banaszak et al. suggested that, once a microphase forms at some orientation, it becomes locked in, and this prevents the period from varying as much as it would if equilibrium were attained at each temperature.

They explored this conjecture in the second paper, where they applied the quenching and reheating treatment to the 7-16-7 polymers. At the temperatures above the MST, the results were the same as in the first processes, but there were differences below the MST. The energy per site was lower, and the rms end-to-end distance was greater, than in the slow cooling simulation, but the specific heat was the same, down to temperatures below which those changes developed. However, well below the MST, the specific heat developed a second peak, signally a second transition. Both the ordered phases were lamellar, but the nature of the second phase needs more investigation.

In the microphase separated regime, the R_{rms} results in the quenching calculations were fitted to a power law dependence on temperature, $R_{rms} \propto T^{\mu}$. The result was $\mu \simeq -0.21$. This is in very good agreement with the NSCFT prediction of $\mu \simeq -0.20$ except very near the MST where NSCFT theory predicts $\mu < -0.20$. It also compares with the SCF prediction of $\mu = -\frac{1}{6} \simeq -0.17$ for very strongly segregated systems.

Vassiliev and Matsen [100] modeled symmetric polymers with $N_C = 20$, 30, and 40, and made a very detailed comparison with NSCFT. In order to make as direct a comparison as possible, they used the lattice-based implementation of the NSCFT of Scheutjens and Fleer [101], and used the same fcc lattice for both the NSCFT and MC calculations. They did not use cooperative rearrangements of the molecules. Instead, they used copolymer volume fraction of $\phi_C = 0.8$, again in both sets of calculations. Like Banaszak et al., they found a peak in C_V at the MST, and that, in the disordered phase, the radius of gyration of the molecules increased as the MST was approached. They examined this last result very closely, and concluded that, rather than undergoing uniform stretching, the increase in R_g was due to the centers of mass of the A and B blocks spreading apart without each changing size. The non-mean field behavior occurred at all temperatures above the MST, not just near it. They also found a large change in the location of the MST; it occurred in their MC simulations at a value of χN_C, which was 2.7 times larger than in the their NSCFT calculations, and the differences appeared to increase with N_C, not decrease as one would normally expect.

VIII.   POLYMER–POLYMER BLENDS

Copolymers can be mixed with other copolymers, homopolymers, or solvents. Broadly speaking, there are three general problems related to the phase behavior of these blends: the microphase behavior, the interplay between microphase and macrophase separation, and micelle formation at low copolymer concentration. The mixtures sometimes form one phase, which can be either ordered or disordered, and sometimes they separate into two macrophases. In the latter case, each of the macrophases can be ordered or disordered. For example, there could be coexisting phases of spheres and cylinders. The phase behavior of a two component system can be summarized by a temperature–composition phase diagram [102,103], that of a three component system by a series of ternary phase diagrams, and so on. Space permits touching only briefly on a small sample of these possibilities in this chapter. Copolymers can also be used as surfactant in homopolymer–homopolymer blends, but that topic is beyond the scope of this chapter.

A. Copolymer–Copolymer Blends

Matsen applied NSCFT to binary mixtures of perfectly symmetric copolymers, some long with N_l units, and some short with N_s units [46]. He found that, if $N_s \gtrsim N_l/5$, then they were completely compatible, forming one phase that was either disordered or lamellar. For more disparate block lengths, that is, $N_s \lesssim N_l/5$, then the blend can phase separate into two lamellar microphases, or one lamellar and a disordered phase. In the first case, the short-period lamellae are almost pure short copolymers, but the long-period lamellae can contain a substantial amount of short polymers. In the second case, the disordered phase is primarily short polymers, and the lamellar phase is a mixture of short and long polymers, with the short ones located preferentially at the AB interfaces.

With blends of asymmetric copolymers, there is a wealth of possible microphase and macrophase behavior, which has been explored to some degree. Matsen and Bates [47] treated polymers of the same N_C but different fractions f_A in each component, which they labeled f_1 and f_2. For polymers that are closely matched in composition, they found that the phase behavior is very similar to that of a single component characterized by χN_C and the average composition $\bar{f}_A = \bar{\phi}_1 f_1 + \bar{\phi}_2 f_2$, where $\bar{\phi}_1$ and $\bar{\phi}_2 = 1 - \bar{\phi}_1$ are the overall volume fractions of the two components. For highly dissimilar polymers, two-phase regions develop. Koneripalli et al. [48] examined such blends both experimentally and theoretically. Among their findings is that blending diblocks with $f_A < 0.5$ with ones with $f_A > 0.5$ can result in a single lamellar phase, even if the individual components would form nonlamellar morphologies.

An interesting noncentrosymmetric (NCS) phase can develop in blends. This is a lamellar phase that lacks a center of symmetry. As described in Chapter 10 by Abetz, it has been observed in triblock–triblock and diblock–triblock blends [104], and in ABCA tetrablocks [105]. Wickham and Shi examined the case of ABC triblocks blended with ac diblocks using NSCFT [106]. Here, the "A" and "a" refer to the same species, but the upper and lower cases distinguish between the triblocks and diblocks, with similar meaning for "C" and "c." They found a narrow NCS region in various phase diagrams, with a layer structure of alternating triblocks and diblocks, ... ABCcaABCcaABCca ..., occurring at a composition of about 40% diblocks.

B. Copolymer–Homopolymer Blends

The phase diagrams of copolymer–homopolymer blends can be complex. Matsen has examined binary blends of AB copolymers with A homopolymers, that is, AB/A, and presented phase diagrams for various values of f_A for the copolymer, examining the overall phase behavior, including the stability of different microphases and how they are affected by the homopolymer [44,45]. A notable conclusion of this work is that the addition of homopolymer can stabilize the OBDD and PL phases. This is attributed to the relief of packing frustration, discussed above in Section VI.C, which results from the nonuniform distribution of the homopolymer, with preferential localization at the "corners" of the unit cells.

There has also been some theoretical work on AB/A/B ternary blends. Among it is a prediction by Naughton and Matsen of a nonperiodic lamellar phase in ternary AB/A/B blends, consisting of a random sequence of thin and thick layers [50].

Conceptually, microphase formation is an inherently block copolymer phenomenon. Interestingly, however, copolymers in the disordered phase can sometimes be made to microphase separate by the addition of homopolymer [107–109]. This phenomenon is called induced microphase separation, and it was predicted by the weak segregation theory [107] of blends prior to its observation [108]. It occurs if the copolymers are near the MST, and if the added homopolymer has sufficiently high molecular weight. Relatively low molecular weight homopolymer has the opposite effect, tending to destabilize the microphase. When induced microphase separation occurs, the added homopolymer is solubilized within the domains.

High and low molecular weight homopolymers also have opposite effects on the domain size. Experimentally [110–114], the solubilization of very low molecular weight homopolymer causes either a decrease or a very slight increase in the domain size. The addition of high molecular weight homopolymer causes an increase in domain size, up to the solubilization limit of the homopolymer.

Since induced microphase formation occurs at the MST, it can be amenable to analysis using weak segregation theory. A detailed analysis was carried out for binary AB/A and ternary AB/A/B blends [109,115]. Two threshold effects associated with added homopolymer were predicted, one associated with induced microphase separation and one with its effects on the layer thickness. For AB/A binary blends with symmetric copolymers, ($f_A = 0.5$ and $\epsilon = 1$), the threshold for homopolymer-induced microphase formation was predicted to be

$$\frac{N_H}{N_C} \simeq \frac{1}{4} \tag{70}$$

where N_H and N_C are the degrees of polymerization of the hompolymer and copolymer, respectively. Added homopolymer with $N_H \lesssim N_C/4$ tends to destabilize a microphase; otherwise, it tends to stabilize, or induce, the microphase. The theory also predicts homopolymer-induced microphase separation in ternary A–b–B/A/B blends, with the same threshold as for binary blends, Eq. (70), independent of the relative volume fractions of A and B homopolymer.

The predicted threshold associated with the layer thickness for these blends near the MST is

$$\frac{N_H}{N_C} \simeq \frac{1}{5} \tag{71}$$

Added homopolymer with N_H less than this threshold causes the layer thickness to decrease, whereas it increases with greater N_H. This prediction was subsequently generalized to stronger segregation and finite homopolymer concentrations, becoming [49]

$$\frac{N_H}{N_C} \simeq \frac{1}{\chi N_C}(1.39\bar{\phi}_C + 0.68) \tag{72}$$

where $\bar{\phi}_C$ and $\bar{\phi}_H = 1 - \bar{\phi}_C$ are the overall volume fractions of copolymer and homopolymer already present when additional homopolymer is added. Equation (72) reduces to Eq. (71) when homopolymer is added to neat copolymer very near the MST, that is, $\bar{\phi}_C = 1$ and $\chi N_C \simeq 10.5$. However, there is a strong dependence on the degree of segregation; for example, the threshold decreases by a factor of about five if χN_C changes from near the MST to $\chi N_C \simeq 50$. It also decreases as additional homopolymer is added.

These results are in qualitative agreement with the experiments, but the degree of quantitative agreement is difficult to determine. Qualitatively, experiment and theory both indicate that there is such a threshold effect and that it occurs at values of N_H/N_C similar to those of Eqs. (71) and (72). Equation (72) even predicts the kind of nonmonotonic dependence observed by Winey, where the addition of certain homopolymers initially caused a decrease in layer thickness, and then an increase. To understand this, consider a simple example of neat copolymers with $\chi N_C = 20$, to which are added homopolymers with $N_H/N_C = 0.09$. This is just below the threshold of 0.10 predicted by Eq. (72) for these copolymers, so adding these homopolymers will initially cause a modest reduction in layer thickness. However, the added homopolymer reduces the threshold, which in this case reaches 0.09 at $\bar{\phi}_C \simeq 0.8$. This means that adding homopolymer up to $\bar{\phi}_H \simeq 0.2$ will cause a decrease in thickness, but beyond that will begin to cause an increase. This is in agreement with the observed behavior when PS was added to nearly symmetric PS–PI. Very low molecular weight PS induced an initial decrease and then an increase in thickness, whereas higher molecular weight PS caused only an increase.

The quantitative comparisons are more problematic. One needs to know the degree of conformational asymmetry and, most importantly, the χ parameter. As has been noted elsewhere, determining χ can be very difficult, particularly if it is extracted from experiments in which fluctuations may be important. For example, if χ is determined by fitting scattering data, then the value obtained from the fit can depend, to quite a large degree, on the theory. If the Leibler mean field theory is used, then fluctuation effects will reduce the fitted value of χ but, presumably, this reduced value should not be used in stronger segregation where fluctuation effects are not significant. If the Fredrickson–Helfand theory is used, then the fitted value will be larger, but affected by the approximations in that theory.

In Ref. 49, the use of literature values of χ produced good quantitative agreement between experiment and theory for the effects of homopolymers on the domain size for some [112], but not all [111], experiments. Therefore, for the case of relatively poor agreement, a new value of χ was tried. This new value was obtained by fitting the NSCFT-calculated layer thickness to the observed layer thickness for the *neat* copolymer. When this χ was used, the theoretical description of the blends agreed quantitatively with experiment. This χ was larger than the literature values, and corresponded to $\chi N_C \simeq 50$ for the copolymer. This is a relatively strong segregation, where fluctuations are not important. Thus, one value of χ gave a consistent picture of both the neat copolymer and blends, as long as they were both well away from the MST.

The similarity of the two thresholds described above, and details of the density profiles that are calculated in NSCFT, provide an overall picture of the effects of these additives. Adding low molecular weight homopolymer tends to destabilize the microphase, reduce the domain thickness, and move the system in the direction of weaker segregation with broader interphases. The homopolymer tends to be distributed throughout the compatible subdomain. Higher molecular weight homopolymer tends to stabilize the microphase, cause the domain size to increase, and drive the system toward stronger segregation and narrower interphases. Of course, the actual extent to which this happens is limited by the solubilization limit of the homopolymer.

IX. COPOLYMER–SOLVENT BLENDS

Conceptually, it is useful to consider two regimes of polymer–solvent blends. In the first one, the copolymers comprise a small fraction of the system, typically 5% volume fraction or less. In these systems, which are considered in Section X, the polymers often form small isolated domains called micelles. In the second regime, considered in this section, the polymers are a major, even majority, component. Experimentally, mixing copolymer with solvent provides access to a far greater range of segregation strength than can be achieved by neat copolymer, particularly if the so-called dilution approximation holds.

It is straightforward to apply the full NSCFT formalism of Section III to any copolymer–solvent blend, as long as one knows the various polymer and solvent characteristics, and χ parameters. What is less straightforward is understanding the limits of the theory: since it is a mean field theory, it loses its validity in the case of a dilute solution of polymer in good solvent. These limits are not yet quantitatively understood.

A. Nonselective Solvent

A diblock copolymer–solvent blend has three interaction parameters, χ_{AB}, χ_{AS}, and χ_{BS}. The difference between χ_{AS} and χ_{BS} is a measure of the selectivity of the solvent for each of the copolymer blocks. We consider first the idealized case of a perfectly nonselective solvent, with $\chi_{AS} = \chi_{BS}$.

Vavasour and Whitmore used NSCFT to calculate five microphase diagrams [41], exploring N_C ranging from 100 to 1000, χ_{AB} from 0.05 to 0.4, and $\bar{\phi}_C$ from 0.8 all the way down to 0.1, which is probably beyond the limits of mean field theory. They neglected the possibility of two-phase coexistence. The first interesting result was that these phase diagrams were almost identical to each other, and to that of neat copolymers, as long as the polymer–polymer interaction parameter was replaced by an effective parameter,

$$\chi_{eff} = \bar{\phi}_C \chi_{AB} \tag{73}$$

so that the phase diagram for these blends is virtually identical to Figure 4, as long as the vertical axis is replaced by $\chi_{eff} N_C$.

This numerical result was a confirmation of the dilution approximation, proposed in 1971 by Helfand and Tagami [76]. In this approximation, the solvent density is assumed to be uniform throughout the system, and its only role is to screen the A–B interactions. The uniformity of the solvent density has been explicitly probed in NSCFT [50,62,116]. The calculated solvent density profiles were, indeed, nearly uniform, except for a very small excess in the A–B interphase region, even in strong segregation where the A and B polymer densities vary sharply through the interphase. In the absence of the small inhomogeneities in the solvent density, the full NSCFT reduces exactly to the dilution approximation. The domain size scaling also maps directly onto that for neat polymers, becoming

$$d \propto (\chi_{eff} N_C)^p N_C^{1/2}$$
$$\propto (\bar{\phi}_C \chi_{AB})^p N_C^{p+1/2} \tag{74}$$

so that d varies with polymer concentration in the same way that it does with χ_{AB}, and the dependence is strongest in weak segregation. This means that the addition of nonselective solvent causes the equilibrium domain size to *decrease*. Equation (74), which is strictly true for the dilution approximation, is largely although not exactly consistent with the NSCFT results.

A final point to note is that these results apply as long mean field theory holds, and the solvent is perfectly nonselective, $\chi_{AS} = \chi_{BS}$, irrespective of their actual values. The solvent could be Θ, good or even athermal.

Experimentally, Hashimoto et al. [117] and Lodge et al. [118] have studied PS–b–PI in dioctyl phthalate (DOP) and toluene, which are neutral, good solvents for PS and PI. For systems in equilibrium, the layer thickness [117] scaled as

$$d \propto \bar{\phi}_C^{1/3} \left(\frac{1}{T}\right)^{1/3} N_C^{2/3} \tag{75}$$

These exponents were determined without distinguishing between strong and weak segregation, and the concentration dependence approximately followed a $\frac{1}{3}$ power all the way to the MST; in fact, the results were consistent with a slight increase in the power near the MST. All this is consistent with the NSCFT predictions, as long as the interaction parameters are independent of composition and inversely proportional to temperature.

The dilution approximation has also been successful in predicting the order–order transitions [119–121]. However, the predictions for the MST itself do not agree with experiment. The NSCFT calculations, with or without the dilution approximation, give the simple result that the MST should obey the functional form

$$(\bar{\phi}_C \chi_{AB} N_C)_{MST} = F(f_A) \tag{76}$$

where $F(f_A)$ is the MST for pure copolymer. This implies that the MST varies with concentration according to

$$(\chi_{AB}N_C)_{MST} \propto \frac{1}{\bar{\phi}_C} \qquad (77)$$

In good solvent such as this, chain swelling should be included for the semidilute regime. In this case, Olvera de la Cruz [122] and Fredrickson and Leibler [123] predicted the form

$$(\bar{\phi}_C^{1.61}\chi_{AB}N_C)_{MST} = F(f_A) \qquad (78)$$

in the semidilute regime. The order–order boundaries in very strong segregation are predicted to be unaffected by chain swelling [124].

Equation (78) implies

$$(\chi_{AB}N_C)_{MST} \propto \frac{1}{\bar{\phi}_C^{1.61}} \qquad (79)$$

or, equivalently

$$(\bar{\phi}_C)_{MST} \propto \frac{1}{(\chi_{AB}N_C)^{0.62}} \qquad (80)$$

Lodge et al. [118] examined six PS–PI copolymers in the neutral solvent DOP, and found the MST obeyed $(\bar{\phi}_C)_{MST} \propto N_C^{-0.60}$, in virtually perfect agreement with Eqs. (79) and (80). Subsequently, Hanley and Lodge [120] found that the MST scaled as $(\chi_{AB})_{MST} \propto \bar{\phi}_C^{-1.4}$, in better agreement with Eq. (79) than with the dilution approximation. Perhaps the most interesting result, however, is that these scaling rules were obeyed all the way up to the melt, not just in the semidilute regime. This is not yet fully understood.

B. Selective Solvent

One would rarely, if ever, expect a solvent to be perfectly nonselective. In reality, binary copolymer–solvent blends offer many degrees of freedom: the N_C and f_A of the copolymers, the three interaction parameters and their temperature dependences, the size of the solvent molecules, and the overall composition. Qualitatively, there are a number of possible solvent selectivities. The solvent could be almost neutral; it could be selective but good for both blocks; it could be good for one block but a Θ solvent or nonsolvent for the other one. There is, in fact, a continuum of degrees of selectivity, and a very large range of possibilities to explore.

There have been a number of experimental and NSCFT studies of copolymers in solvents of varying degrees of selectivity [50,118,120,121,125–134]. A detailed review of all the results of those papers is beyond the scope of this chapter, but some general observations can be made.

A selective solvent will be solubilized preferentially into one subdomain. The difference between the solvent densities between subdomains increases as the selectivity increases, but the density within each one is nearly constant. The solvent density difference between subdomains can easily be greater than the solvent excess at the interface found in neutral solvents.

Conceptually, one can think of two qualitative effects resulting from adding solvent. One, explored in detail above, is to dilute the AB interactions. The other, which increases in importance with increasing selectivity, is to alter the relative volume fractions of each subdomain. Hanley et al. [121] found it useful to describe this in terms of effective volume fractions of A and B, say f_A' and f_B'. They then interpreted the observed phase diagrams in terms of an effective χ parameter and the effective volume fractions. The addition of neutral solvent to the system affects only χ_{eff} and is represented by a vertical trajectory on such a diagram. Adding a selective solvent affects both χ_{eff} and f'_A, and is represented by a diagonal trajectory.

In general, the MST is shifted to smaller values of $\chi_{AB}N_C$ as the solvent selectivity is increased. Put another way, the selective solvent has a weaker tendency to destabilize the microphase than does the nonselective one, and the difference can be very significant [127]. It can also strongly affect the ordered phases. For example, inverse cylinders can form, that is, A-centered cores formed of polymers with $f_A > f_B$, if the solvent enhances the effective B-block fraction f'_B. The sequence of phases can be complex, for example, just varying the concentration can induce the sequence $S_A \rightarrow C_A \rightarrow L \rightarrow C_B \rightarrow S_B$. This sequence would not occur with neutral solvent.

The domain size behavior can also vary. As discussed above, adding a nonselective solvent induces a reduction in the layer thickness. Hanley et al. [121] found that the rate of reduction weakens with solvent selectivity; in fact, when the solvent is sufficiently selective, the thickness increases rather than decreases, reaching $d \propto \bar{\phi}_c^{-0.2}$ at $\chi_{BS} = 1.0$. This behavior is captured by the NSCFT calculations. It is also reminiscent of the effects of adding A homopolymer to AB copolymers. If the A homopolymer tends to be segregated in the one subdomain and away from the A–B interface, then it causes an increase in layer thickness.

X. BLOCK COPOLYMER MICELLES

The other regime of interest noted in Section IX is the addition of low concentrations of block copolymers, typically a few percent or less, to highly selective solvent or homopolymer. In this case, the copolymers can form micelles, which are usually but not always approximately spherical. Each core consists primarily of the incompatible copolymer blocks, while the compatible blocks form a corona extending into the host. The micellar sizes are on the same scale as the molecular sizes, typically tens of nanometers. For example, the core of a spherical micelle is comparable to the radius of gyration of the corresponding block, although somewhat stretched as discussed below. Interesting questions in these systems include the conditions under which micelles form, that is, the critical micelle concentration (cmc), the average size and shape of the micelles, the distribution of sizes and shapes, and the micelle number densities.

Although NSCFT has been used to study these systems [128–131], this section summarizes some different approaches, with the goal of providing some complementary perspectives on the theory and results, and illustrating the power of some simple models.

A. Simple Theories of Spherical Micelles

We first consider spherical micelles, using a very simple equilibrium theory that is in the same spirit as the model discussed in Section VI.C. The model presented here has its genesis in mean field models that are best suited to copolymer micelles in a homopolymer matrix [135–138], but it has also, perhaps surprisingly, provided good descriptions of copolymer micelles in selective solvent [139]. Halperin has introduced a "star" model of micelles based on a scaling picture of the corona block [140]. In principle, it should provide a better theory for micelles in good solvent, but in the high N limit. However there is now considerable evidence that mean field theories often provide at least as good a picture of end-tethered polymers as scaling theory does, even in good solvent [141,142]. At any rate, the model presented here, although far from perfect, suffices for the purposes of illustrating the dominant physical effects and results.

As in Eq. (59), we consider the free energy of the microphase separated system, in this case the micellar state, relative to the disordered phase of copolymer dissolved in the host. We can express the important contributions as

$$\Delta F \simeq \Delta F_{\text{int}} + F_{\text{I}} + F_{\text{el}} + F_{\text{loc}} \tag{81}$$

The first three terms are similar to those in Eq. (59). The first is the reduction in interaction energy when the incompatible copolymer blocks avoid contacts with the host by forming the micelle cores. The second term is the interfacial energy, in this case arising primarily from the core–corona interface. The third is the stretching term, which differs for the core and corona blocks but has the same functional form for each one. The final, and new, term is due to the reduction in entropy that occurs when the copolymers are localized to micelles.

As in the copolymer model of the MST, ΔF_{int} is the driving force for forming micelles, but it does not control the micelle sizes. These are, again, controlled primarily by the balance between the interfacial tension and stretching, but this time the stretching of the core block. As a result, for a simplified model system consisting of AB copolymer in an A-homopolymer or A-block compatible solvent, and assuming all micelles are of the same size, the equilibrium core radius is predicted to scale approximately as

$$l_{\text{core}} \simeq \chi^{1/6} N_{\text{CB}}^{2/3} b \tag{82}$$

The χ parameter here is the most important one acting on the B monomers. In the selective solvent, it is $\chi = \chi_{\text{BS}}$. For AB copolymers in a homopolymer-A matrix, it would be $\chi = \chi_{\text{AB}}$.

From Eq. (82) and simple volume considerations, it follows that the number of micelles per unit volume, n_{M}, scales as

$$n_{\text{M}} \propto \frac{1}{\sqrt{\chi}} \frac{\bar{\phi}_{\text{C}}}{N_{\text{CB}}} \tag{83}$$

Equation (82) indicates that the size scaling behavior is similar to that for microphase separated copolymers. The main difference is that micelle core radius scales with the degree of polymerization of the corresponding copolymer *block*, N_{CB}.

However, the conditions under which micelles form are rather different from the conditions under which copolymers microphase separate, for example, Eq. (1). These conditions are described by the cmc, which can be defined as the minimum copolymer volume fraction needed to form micelles. To a first approximation, this model predicts

$$\phi_{\text{C}}^{\text{cmc}} \propto \frac{(\chi N_{\text{CB}})^{2/3}}{f_{\text{B}}} e^{\{-\chi N_{\text{CB}} + \cdots\}} \tag{84}$$

This indicates that the onset of micellization is dominated by an exponential dependence on χN_{CB}. Since χ_{AB} varies approximately inversely with temperature, this model predicts, approximately, that the cmc should vary exponentially with $1/T$ and with N_{CB}. This is quite different from Eq. (1).

Micelles with larger cores can be made using copolymers with longer B blocks. Another potential route to larger micelles is by swelling them with homopolymers that are compatible with the cores. Swollen micelles also provide a potential mechanism for dispersing homopolymers throughout a host in very small domains. In this case, there are two dominant questions: first, to what extent can the homopolymers be solubilized within the micelle cores before separating into a separate macrophase and, second, what is their effect on the sizes and number densities of the micelles?

These questions have also been addressed within a simple model similar to that presented above. In this model, the free energy of the micelles contains the same physics as in the one for the unswollen micelles, for example, Eq. (81), plus an additional contribution accounting for the localization of a fraction of the added homopolymer solubilized within the micelle cores, f_M. This free energy is then minimized with respect to f_M, as well as the micelle sizes and number density. For low homopolymer volume fractions, the model predicts $f_M = 1$. As the homopolymer content increases, it switches abruptly to $f_M = 0$. This means that either all or none of the homopolymer is solubilized within the cores.

This behavior of f_M means that there is a threshold effect: there is a distinct solubilization limit for the homopolymer. The maximum homopolymer volume fraction corresponding to this limit, ϕ_{HB}^{max}, is directly proportional to the total volume fraction of core-forming material of the copolymers, $\bar{\phi}_{CB}$, that is, $\phi_{HB}^{max} \propto \bar{\phi}_{CB}$. The proportionality constant needs to be calculated numerically, and is discussed in Ref. 143. However, it behaves *approximately* as

$$\phi_{HB}^{max} \propto (\chi N_{CB})^{-1/2} \left(\frac{N_{HB}}{N_{CB}} \right)^{-3/2} \bar{\phi}_{CB} \tag{85}$$

This expression captures the dominant behavior of the solubilization limit. First, it is a rapidly decreasing function of the ratio of the degrees of polymerization of the two components of the core, the homopolymer and copolymer. Second, it is a decreasing function of the product χN_{CB}.

Examples of the solubilization limits calculated numerically for the case of PEO swelling PEO–PS micelles in a PS matrix are shown in Figure 10. In most cases, $\phi_{HB}^{max} < \bar{\phi}_{CB}$; ϕ_{HB}^{max} reaches a value comparable with $\bar{\phi}_{CB}$ only if the degree of polymerization of the homopolymer is much smaller than that of the copolymer block, for example, $N_{HB} \lesssim N_{CB}/10$. When the two have comparable degrees of polymerization, $N_{HB} \simeq N_{CB}$, the solubilization limit is very small, with $\phi_{HB}^{max} \ll \bar{\phi}_{CB}$.

In principle, the solubilized homopolymer could induce changes to both the equilibrium sizes and numbers of micelles. However, the theory predicts very simple changes to the system. To a very good approximation, the number of micelles is unaltered, but each micelle is swollen. It follows

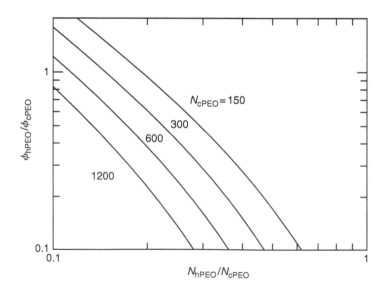

Figure 10 Solubilization limits of homopolymer calculated using a simple mean field model [143]. These calculations are specifically for PS–PEO copolymers forming micelles in a PS matrix, swollen by PEO homopolymer, but results are similar for other systems.

from this that the core radius of the micelle becomes

$$l_{core} = \left(1 + \frac{\bar{\phi}_{HB}}{\bar{\phi}_{CB}}\right)^{1/3} l_{core}^0 \tag{86}$$

where l_{core}^0 is the radius of core of the unswollen micelles, for example, Eq. (82).

It should be emphasized that all the predictions of this section are based on very simple physical models, and a number of results involve further approximations used to obtain analytic results. These results do, however, capture the dominant physical effects; more quantitative results are available in the literature [135–139,143].

B. Monte Carlo Simulations of Copolymer Micelles

As in the simulations discussed in Section VII.B, simulations of these systems avoid the mean field approximations, as well as the additional assumptions underlying the simple model just described. A number of simulations have been performed, with illustrative results presented here [144–154]. Because of the considerations discussed before, the systems are limited in overall size, chain lengths, and temperature range.

As in the simulations of more dense copolymer systems, one typically begins by annealing at very high temperature, and then slowly lowers the temperature, that is, increases the interaction parameter. At first, the copolymers are individually dissolved in the solvent. As the temperature is reduced, very small aggregates begin to form, and then micelles do as well. As the temperature continues to decrease, the equilibrium number of micelles decreases, and each one contains progressively more polymers. These changes require the migration of molecules from micelles to other micelles. However, the energy barrier opposing this motion can be very large, scaling like $\sim \chi N_{CB}$. This can make the extraction of a chain a very slow process. Furthermore, even after a molecule has overcome this barrier, this is only its first step in its migration to a different micelle. From here it will diffuse until it encounters a micelle, in which it will most likely stay for another long period of time. However, the micelle it is most likely to encounter is the nearest one, which is the one it just left. If it is recaptured by that one, then no significant change has occurred. Together, these factors mean that simulation times of real systems, especially in the strongly segregated regimes of most micellar systems, would be extraordinarily long. This is, of course, reflected in the behavior of real systems which can become trapped in metastable states for long periods of time.

In part to address this problem, the simulations discussed here include a careful monitoring of a suite of autocorrelation functions and times at each step of the calculation. This plays two important roles. The first is very practical: it provides guidance on how long the simulations must be carried out at each stage in order to ensure equilibrium is reached. Second, it provides insight into the behavior of the system, especially as micelles form. The correlation times are those associated with relaxation of the end-to-end vectors of the chains, the chain extraction time that measures the time for chain extraction from micelles, the chain exchange time that is associated with chains escaping one micelle and migrating to a different one, the chain diffusion time that monitors its motion through the system, and weighted chain extraction and exchange times. These are similar to the extraction and exchange times, but weighted by the sizes of the aggregates.

In the simulations described below, the molecules are diblocks, with shorter heads and longer tails, immersed in solvent. The Monte Carlo procedure follows the standard Metropolis algorithm on a cubic lattice, using a variety of moves including reptation and local bending of the polymers. They are shown in Figure 11. The A block is compatible with the solvent, and the incompatible B block forms the micelle cores. In other language, the A and B blocks are the heads and tails of the molecules. These micelles, with relatively large cores and thin coronas are sometimes known as "crew-cut" micelles. In principle, there can be three interaction parameters: A–B (head–tail), A–S (head–solvent), and

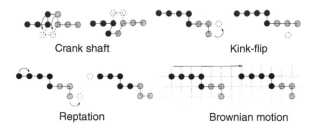

Crank shaft   Kink-flip

Reptation   Brownian motion

Figure 11   Schematic representation of the Monte Carlo moves used in the simulations described in Sections X.B to X.D.

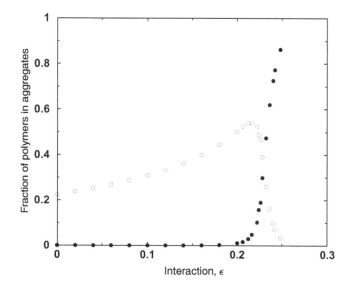

Figure 12   Fraction of polymers in micelles and small aggregates as a function of ϵ, for a system of 1000 molecules with head and tail lengths $N_{CA} = 10$ and $N_{CB} = 50$ at an overall volume fraction of 2%. At all stages of the simulations shown here, $\epsilon_{AB} = \epsilon_{BS} = \epsilon$, and $\epsilon_{AS} = 0$ [152].

B–S (tail–solvent). These interactions are described by three nearest neighbor interaction parameters. In principle, these are all independent but, in the calculations reported here, the A–B and B–S interactions were set equal, say $\epsilon_{AB} = \epsilon_{BS} \equiv \epsilon$, and the A–S interactions were $\epsilon_{AS} = 0$, $-\epsilon$, or -2ϵ. The calculations were done by initially setting $\epsilon = 0$ and letting the systems equilibrate, and then turning on all interactions by slowly increasing the value of ϵ.

C.   Simulations: Size and Shape Distributions

Figures 12 to 14 show some results for one sample case. These calculations are for a system of 1000 polymers with head and tail lengths of $N_{CA} = 10$ and $N_{CB} = 50$, and an overall volume fraction of 2%. Figure 12 shows the fraction of molecules in aggregates of 2 to 10 polymers, and the fraction in aggregates of more than 10 molecules. For reasons that will become apparent, we identify this last category as "micelles" for this system. At $\epsilon = 0$, about 80% of the molecules are individually dissolved in the solvent, about 20% are in small aggregates, and there are no micelles. The 20% in small aggregates is simply due to random, transient contact of the polymers. As ϵ increases, tail–tail contacts become energetically favored, and the fraction of polymers in small aggregates increases modestly. Once a certain value of ϵ is reached, in this case $\epsilon \simeq 0.2$, micelles begin to form. Beyond this point, the fractions of molecules that are individually dissolved or in small aggregates decrease,

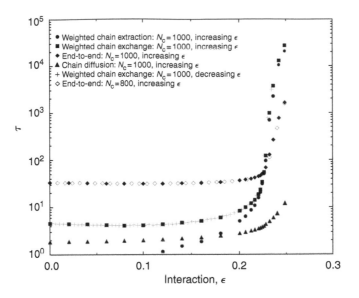

Figure 13 Autocorrelation times for the system shown in Figure 12 [152].

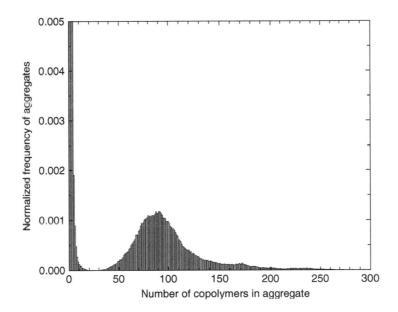

Figure 14 Normalized frequency distribution of aggregates as a function of the number of copolymers in the aggregates [152]. The system is the same one as in Figure 12 and Figure 13, taken at $\epsilon = 0.248$, where about 90% of the copolymers are in micelles.

and the fraction in micelles increases. These changes are rather sharp; for example, the fraction of molecules in micelles increases from essentially zero to about 90% as ϵ increases from 0.2 to 0.25.

The various correlation times for this system, which are shown in Figure 13, reflect this behavior and in particular the formation of the micelles. At small ϵ, all of the calculated times are either constant or slowly varying, and this continues up to the point where micelles begin to form. Then, over the narrow range of ϵ in which 90% of polymers self-assemble into micelles, the maximum correlation time, which we identify as the system relaxation time, increases by about three orders of magnitude. It should be noted that the quantitative details for these times depend on the details of the moves used in the simulations. For example, including nonlocal moves decreases the diffusion time relative to the other times. However, the dominant qualitative feature, that is, a sudden increase in the relaxation time as micelles form, is independent of these details.

Figure 14 shows another aspect of this same system, which is the size distribution of aggregates. This figure shows the relative number of aggregates with aggregation number n, for the same system at the relatively large value of $\epsilon = 0.248$, where about 90% of the molecules are in micelles. There are three parts to this distribution. The first is the small n region, describing the individual molecules and small aggregates. The next is the peak centered at about 80 molecules per aggregates, and the third part is the extended tail reaching out to about 250 molecules per aggregate. This extended tail is the contribution from aggregates that are actually two or more micelles whose cores are in contact, which are counted as a single aggregate in the simulations. (If only the coronas are in contact, they are counted as separate aggregates.) This occurs only if the head group is small relative to the tail, which leads to exposed micelle cores. The deep minimum in the distribution in the neighborhood of $n \simeq 10$ (or sometimes slightly more) is the reason for our using the aggregation number ten for distinguishing between "small" aggregates and micelles.

The width of the distribution can, of course, be characterized by the polydispersity index. There are two obvious choices to calculate it here. One is to use all aggregates with 10 or more molecules. This choice includes the large, multiple-micelle ones identified in the previous paragraph. The second is to use only those that contribute to the single-micelle peak. For the distribution of Figure 14, these two options give polydispersity indices of 1.14 and 1.05, respectively. The latter value compares very favorably with the experimental value of 1.08 that was observed by Zhang et al. for crew-cut micelles [155].

The simulations make no assumption about the shapes of the aggregates. Instead, the shapes can be calculated from the data. This was done by examining the aggregates' principal radii of gyration tensors and, from them, asphericity parameters. This can be done at each temperature, and as a function of aggregation number in each case.

The procedure is as follows. At a given time, we first calculate the radius of gyration tensor for every aggregate. Next, for every aggregate, we diagonalize the matrix to find the three eigenvalues, say R_x^2, R_y^2, and R_z^2, which we then order from smallest to largest. We then find the average value of each ordered eigenvalue, where the average is over all aggregates of a given aggregation number, n, and then take the square roots of these averages. The results are the mean-squared principal radii of gyration of aggregates of n molecules. In what follows, they are denoted $\langle R_i^2 \rangle^{1/2}$ with $i = 1, 2$, and 3. The asphericity parameter is a single number characterizing the shape. It is calculated from the $\langle R_i^2 \rangle$ through

$$\alpha_s = \frac{\sum_{i>j=1}^{3}(\langle R_i^2 \rangle - \langle R_j^2 \rangle)^2}{2\left(\sum_{i=1}^{3}\langle R_i^2 \rangle\right)^2} \tag{87}$$

If the aggregates are all perfect spheres, then all principal radii of gyration are equal, and $\alpha_s = 0$. For a sphere extended in one direction, for example, an ellipsoid, the two smaller ones are equal, and $\alpha_s > 0$; in the limit of infinite cylinder, $\alpha_s = 1$. If the sphere is flattened, then the third one is reduced but, again, $\alpha_s > 0$; in the limit of a large, flat disk, $\alpha_s = \frac{1}{4}$. Conventionally, an aggregate is considered approximately spherical if $\alpha_s \lesssim 0.1$.

Figure 15 and Figure 16 illustrate the results. Since these calculations are computationally demanding, they are restricted to relatively short polymers. Figure 15 shows the size distributions, radii of gyration, and asphericity parameter for a particular system, as described in the figure caption. For the most part, the three radii of gyration are all about equal, and $\alpha_s < 0.1$. This is not the case

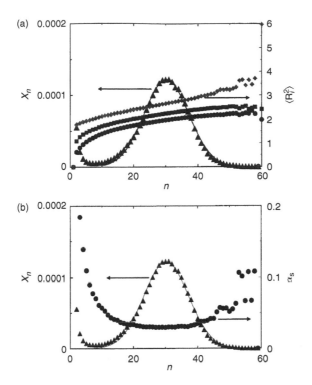

Figure 15 Size and shape distributions for a system of 1000 molecules with $N_C = 14$, $N_{CA} = 4$, and $N_{CB} = 10$, at a volume fraction of $\bar{\phi}_C = 0.025$, with interaction parameters $\epsilon_{AB} = \epsilon_{BS} = 0.7$ and $\epsilon_{AS} = -0.7$ [154]. (a) Monte Carlo (symbols) and fitted size distributions (solid curve) refer to the left-hand axis, and the three principal radii of gyration refer to the right-hand axis and are in units of the bond length. (b) Size distributions (left-hand axis) and asphericity parameter, α_s (right-hand axis).

at very small n, where the small aggregates have irregular shapes. It may not be the case at large n, although the numerical noise is large here because the number of very large aggregates is small.

The case illustrated in Figure 16 presents an interesting contrast. The figure shows three size distributions, for three different values of ϵ, and the principal radii of gyration and asphericity parameter for the largest ϵ. At each ϵ, the size distribution shows a well-defined peak, and it moves to larger n with increasing ϵ. This shift is expected from mean field theory. The three principal radii of gyration that are shown are about equal, and α_s is small, for aggregates up to about the peak micelle size. Larger micelles, however, show a change in shape. The two smaller principal radii of gyration begin to plateau, while the larger one increases more quickly. This implies that, on average, the larger micelles are becoming more elongated without getting any broader, that is, are becoming more cylindrical. There is some evidence of a second change at the largest aggregation numbers. The largest $\langle R_i^2 \rangle^{1/2}$ appears to stop growing, while the intermediate one begins to grow again. This would signal the beginning of a broadening of the cylinders in one direction. However, the data are quite noisy in this range of aggregation number, so these are only tentative interpretations. All this behavior is also reflected in the asphericity parameter. Except at small n (again), it is small up to the point where the growth begins in one direction. It then increases fairly quickly to $\alpha_s \simeq 0.4$, and finally appears to decrease slightly when the growth in the second direction begins.

The size distributions can also be interpreted in terms of coexisting spheres and cylinders. The procedure is based on work of Nelson et al. [156], and is described in Ref. 154. The size distribution is fitted to a linear combination of two functions, one for spheres and one for cylinders. In those cases where there is no extended tail, such as Figure 15, there is essentially no contribution from cylinders. When there is a tail, such as the case just described, then there is a contribution from

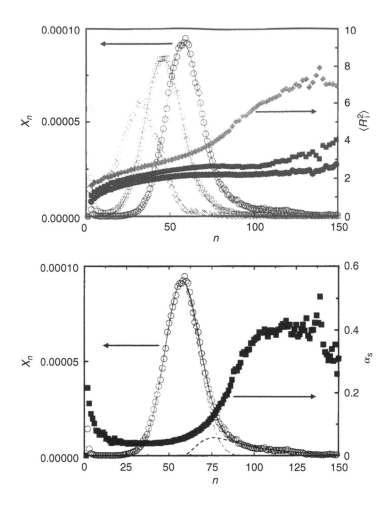

Figure 16 Size and shape distributions for a system of 2000 molecules with $N_C = 10$, $N_{CA} = 2$, and $N_{CB} = 8$, at a volume fraction of $\bar{\phi}_C = 0.027$ [154]. In all cases, $\epsilon_{AB} = \epsilon_{BS} = \epsilon$, and $\epsilon_{AS} = -2\epsilon$. Upper panel, left axis: Monte Carlo distributions (symbols) and fitted size distributions (solid curves) at three values of $\epsilon = 0.50$ (left-hand peak), 0.55 (middle peak), and 0.60 (right-hand peak). Upper panel, right axis: the three principal radii of gyration calculated for $\epsilon = 0.60$, in units of the bond length. Lower panel: size distributions (left axis) and asphericity parameter, α_S (right axis) at $\epsilon = 0.60$. These distributions can be decomposed into two parts, attributable to spheres throughout and cylinders beyond $n = 57$, which are shown by the two dashed curves in the lower panel.

cylinders. The separate contributions from spheres and cylinders for this case are included in the lower panel of Figure 16.

D. Simulations of Swollen Micelles

The swelling of micelles by an added homopolymer has also been studied, to a modest degree, by simulations, and one set of results is shown in Figure 17. It summarizes the outcomes of 11 different sets of extensive calculations done for different copolymers, homopolymers, and volume fractions. In these simulations, autocorrelation times for both the copolymers and the homopolymers were calculated and monitored. In some cases, the homopolymer was solubilized within the cores, and in others it formed a separate, large aggregate containing nearly all the homopolymers. This phase behavior for these 11 systems is summarized in Figure 17. It is consistent with the predictions of

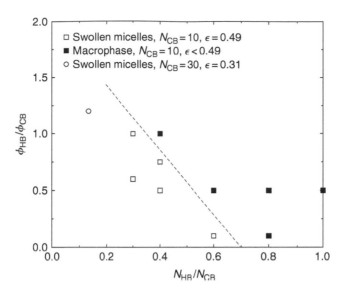

Figure 17 Solubilization limits of homopolymer calculated in the Monte Carlo simulations [153]. The figure is for the chain lengths shown on the figure. These chains are much shorter than those used in the mean field model, shown in Figure 10, but the results are similar [143]. These calculations are specifically for PS–PEO copolymers forming micelles in a PS matrix, swollen by PEO homopolymer, but results are similar for other systems.

the simple mean field theory described above and the experiments: the maximum volume fraction of homopolymer that can be solubilized is proportional to $\bar{\phi}_{CB}$, and is a decreasing function of the fraction N_{HB}/N_{CB}. The quantitative results depend on the interaction parameters, but they are similar. The solubilization limit varies from $\phi_{HB}^{max}/\bar{\phi}_{CB} \simeq 1$ at $N_{HB}/N_{CB} \simeq 0.1$ to $\phi_{HB}^{max}/\bar{\phi}_{CB} < 0.1$ when $N_{HB}/N_{HB} \gtrsim 0.5$.

XI. SEMICRYSTALLIZABLE DIBLOCK COPOLYMERS

A semicrystallizable diblock copolymer has one crystallizable block and one amorphous block. They can form lamellar structures, in which one sublayer consists of folded, semicrystalline blocks, and the other one is amorphous. The structure is illustrated schematically in the lower panel of Figure 2. An example is PS–PEO, in which the PEO block is the semicrystalline one.

As in amorphous polymer systems, it is difficult to achieve equilibrium in these systems. When it is achieved then, as for amorphous copolymers, the equilibrium layer thickness is governed by a balance of thermodynamic driving forces. However, the underlying physics has some differences from the amorphous case. Nonetheless, these structures can still be understood on the basis of a modified NSCFT [62,109].

Reference 109 contains the full NSCFT for both neat copolymers and copolymer–solvent mixtures, but we restrict attention here to the neat copolymer case. Each layer of thickness d contains amorphous and semicrystalline subdomains of thickness d_A and d_B, respectively. All the A–B joints are assumed to be localized within a narrow interphase of thickness a and, except for this interphase, all A and B monomers are assumed to be in their respective subdomains. This assumption is consistent with the strong segregation that occurs in these systems. The amorphous A blocks are treated in an NSCFT. The B blocks are modeled as semicrystalline, folded chains, characterized by the heat of fusion per unit volume, ΔH_f, the degree of crystallinity of the B block, τ_C, and the energy of each fold, E_{fold}.

As before, we focus on the reduced free energy density, relative to a hypothetical uniform melt. It can be expressed as

$$\frac{\Delta F}{\rho_{0A} k_B T V} = f_{int} + f_{loc} + f_{cr} + f_{am} \tag{88}$$

where we have chosen ρ_{0A} for the reference density, ρ_{0R}. The first term is the reduction in the interaction energy due to the formation of the layers. In this case, since the A and B components are completely segregated except at the interfaces, these interactions contribute only to the interfacial tension. Even here, their contribution is small, and it was simply modeled using a χ parameter.

The next term is due to the change in entropy associated with the localization of the joints. It is given by

$$f_{loc} = \frac{\rho_{0C}}{\rho_{0A}} \frac{1}{N_C} \ln \left(\frac{d}{2a} \right) \tag{89}$$

This can be interpreted as a result of the reduction in volume available to the joints; there are two interphase regions of thickness a per domain of thickness d, so the available volume is reduced by the factor $(d/2a)$.

The third term is the crystallization energy of the B block. In this chain folding model, it can be expressed as

$$f_{cr} = \left(\frac{\rho_{0B}}{\rho_{0A}} \right) \frac{f_B}{k_B T} \left[-\frac{\tau_C \Delta H_f}{\rho_{0B}} + \frac{n_{fold}}{N_{CB}} E_{fold} \right] \tag{90}$$

where n_f is the number of folds per molecule. The first and second terms inside the square brackets are equivalent to the average crystallization energy per monomer, and the average fold energy per monomer, respectively. This equation, and others in this section, neglect changes in the density of the B block upon crystallization. The more general expression that includes this change can be found in Ref. 109.

The final term in Eq. (90) is the part that can be associated solely with the amorphous region. It can be calculated via an NSCF calculation, but this time only for one of the blocks, incorporating new initial and boundary conditions that reflect the anchoring of one end (at the joint) to the interface and the complete segregation of the A and B blocks. Details are given in Ref. 109. The result can be expressed as

$$f_{am} = \frac{1}{N_C} g_\gamma(\alpha) \tag{91}$$

where $g_\gamma(\alpha)$ is the free energy associated with the amorphous region *per molecule*. It is a function of two variables.

$$\alpha = \left(\frac{3}{N_{CA}} \right)^{1/2} \frac{d_A}{b_A} \tag{92}$$

and

$$\gamma = \left(\frac{1}{N_{CA}} \right)^{1/2} \frac{a}{b_A} \tag{93}$$

The first of these is a measure of the thickness of the amorphous region relative to unperturbed A blocks. It is the kind of stretching parameter that often appears. The second parameter is a measure

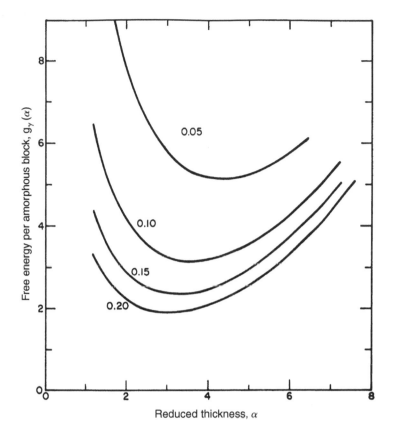

Figure 18 Free energy per molecule due to the amorphous block, as a function of the reduced amorphous region, α. Each curve is labeled by the reduced interfacial thickness, γ. α and γ are defined in Eqs. (92) and (93).

of the thickness of the interphase region relative to unstretched A blocks. Formally, its appearance in the theory can be traced to the size of the interface relative to the A domain, which scales as

$$\frac{a}{d_A} \propto \frac{\gamma}{\alpha} \tag{94}$$

This function g can be used for the amorphous block of any semicrystallizable copolymer. It is shown in Figure 18. For a given value of γ, g is large both for small α, which corresponds to compressed chains, and for large α, which corresponds to stretched chains. In between, there is a minimum that occurs when the root mean-squared thickness of the amorphous block is comparable with its unperturbed end-to-end distance. For a given degree of stretching, g increases with decreasing interface width, and this can be characterized in terms of the single parameter, γ.

The variations with α are the widely understood result of polymer stretching or compression. The dependence on γ is less-widely recognized. It can be understood as a result of the conformational constraints imposed on the molecules by a narrower interphase. The more narrow the interphase, relative to the block size, the greater the constraints and hence the greater the free energy.

The calculated values of g could all be fitted to

$$g \simeq \frac{A}{\sqrt{\gamma}}\alpha^{\mu} + \frac{B}{\alpha\gamma} + C \tag{95}$$

where the constant C is small. The resulting values of μ vary from about $\mu \simeq 2.34$ for $\gamma = 0.2$ to $\mu \simeq 2.11$ for $\gamma = 0.05$. A simple extrapolation to $\gamma \rightarrow 0$, which is equivalent to $N_{CA} \rightarrow \infty$ for a fixed interphase width, gives $\mu \simeq 2.0$, in agreement with limiting theories [72].

The equilibrium layer thickness, d, is obtained by minimizing the free energy, Eq. (88). All of the contributions to ΔF depend on d, but the dependence of both f_{loc} and f_{int} is very weak. Hence, the equilibrium thickness is determined primarily by a balance between f_{cr} and f_{am}.

Changes in f_{cr} with d could occur through two mechanisms. The first would be changes in the degree of crystallinity, τ_C. This would be straightforward to include, but experiments have not indicated any systematic variation of τ_C with d, so we neglect this here. The second mechanism is the chain folding. Folding occurs at the amorphous–crystalline interphases, and could occur in the interiors of the crystalline sublayers. However, these interior folds result in higher free energy, and so should not occur in equilibrium. In either case, the number of times the crystalline block traverses the subdomain scales as

$$n \propto \frac{N_{CB}}{d_B} \tag{96}$$

The number of folds per molecule is $n_{fold} = n - 1$.

Returning to Eqs. (90) and (91) for f_{cr} and f_{am} and assuming τ_C is constant, the dominant contributions to the free energy density that depend on d can be written as

$$f(d) \simeq \frac{1}{N_C} \left[\left(\frac{\rho_{0C}}{\rho_{0A}} \right) \left(\frac{n_{fold} E_{fold}}{k_B T} \right) + g_\gamma(\alpha) \right] \tag{97}$$

This expression illustrates the essential physics of the problem. The number of folds, and hence the free energy of the semicrystalline block, are reduced if the layer thickness increases, but this tendency is opposed by the associated reduction in entropy of the stretched amorphous blocks.

Since the amorphous blocks are stretched at equilibrium, it is reasonable to approximate g by the leading term, $g \propto \alpha^\mu / \sqrt{\gamma}$. Using this approximation in Eq. (97) and then minimizing, yields a scaling relation for the equilibrium domain thickness. In the limit of $\mu \rightarrow 2$, it becomes

$$d_{eq} \propto \left(\frac{E_{fold}}{k_B T} \right)^{1/3} \frac{N_C}{N_{CA}^{5/12}} \tag{98}$$

As well, the equilibrium number of folds scales as

$$n_{fold,eq} \propto \left(\frac{k_B T}{E_{fold}} \right)^{1/3} N_{CA}^{5/12} \tag{99}$$

Prefactors and expressions for $\mu \neq 2$ can be obtained from the expressions in Ref. 109.

These results are quite different from the corresponding predictions for amorphous polymers. They imply that equilibrium number of folds of the *crystalline* block depends on the degree of polymerization of the *amorphous* block, N_{CA}, and is independent of N_{CB}.

The exponent of $\frac{5}{12}$ appearing in these two equations is a consequence of both the α and γ dependences of g. Birshtein and Zhulina [124] developed an analytic theory of semicrystallizable copolymers for large N_{CA}, based on the Semenov approach. Their free energy for the amorphous block varied as $g \propto \alpha^2$ with no dependence on γ, and they obtained the value of $\frac{1}{3}$ rather than $\frac{5}{12}$ for the exponent. The difference can be traced directly to the dependence of g on γ in the numerical theory presented here. If this dependence is neglected, then minimizing Eq. (97) gives an exponent of $\frac{1}{3}$ instead of $\frac{5}{12}$.

Turning to experiment, Douzinas et al. [157] studied eight (ethylene-co-butylene)–ethylethylene copolymers with total and amorphous block degrees of polymerization ranging from about 760 to 2900, and 110 to 1600, respectively. They found that the equilibrium thickness scaled as N_C/N_{CA}^{ν}, with a best fit value for the power of $\nu = 0.42 \pm 0.02$, agreeing the value of $\frac{5}{12} \simeq 0.417$ to within experimental accuracy. Rangarajan et al. performed a similar study of seven ethylene-(ethylene-*alt*-propylene) copolymers [158]. Their best fit gave a slightly stronger inverse dependence, $d \propto N_C/N_{CA}^{0.45}$. They speculated that the small difference might be due to the presence of ethyl branches in the E block. These results support the predicted effects of γ. Earlier work on copolymers with solvent qualitatively also supports the theory and physical picture described here, but this interpretation is obscured by nonequilibrium effects [159–166].

XII. CONCLUDING COMMENTS

Our understanding of the equilibrium microphase behavior of block copolymers, and certain aspects of block copolymer blends, is well advanced. The NSCFT provides an excellent description of the equilibrium phases and domains of block copolymers of various architectures and species, but with one important exception: the region near the order–disorder transition for nearly symmetric molecules. It appears that understanding this region will require a theory that includes both the high order effects incorporated within the NSCFT, and fluctuation effects. The theory can also provide detailed descriptions of the amorphous regions of semicrystallizable copolymers.

An important contribution of the theory is a physical understanding of the most important mechanisms governing this behavior, as described in Section VI.C. These ideas can be used to develop simple and approximate, but powerful and predictive, physical models, such as those described in Section X.

This chapter has also touched on approaches that go beyond mean field theories such as NSCFT, in Section VII. There is much more work to be done in this area, but it has the potential to provide insights into the effects of fluctuations in polymer systems in general, and the behavior of block copolymers near the MST in particular.

ACKNOWLEDGMENTS

The author is particulary indebted to J.D. Vavasour and M. Pépin for research contributions, and to M. Matsen for numerous helpful discussions, and contributions to Figures 3 and 6 to 9.

REFERENCES

1. M.W. Matsen and M. Schick. *Curr. Opin. Colloid Interface Sci.* 1: 329–336, 1996.
2. M.W. Matsen and F.S. Bates. *Macromolecules* 29: 1091–1098, 1996.
3. M.W. Matsen and M. Schick. *Phys. Rev. Lett.* 16: 2660–2663, 1994.
4. M.W. Matsen and M. Schick. *Macromolecules* 27: 6761–6767, 1994.
5. M.W. Matsen and M. Schick. *Macromolecules* 27: 7157–7163, 1994.
6. M.W. Matsen and F.S. Bates. *J. Polym. Sci. B: Polym. Phys.* 35: 945–952, 1997.
7. M.W. Matsen. *J. Chem. Phys.* 113: 5539–5544, 2000.
8. M.W. Matsen and R.B. Thompson. *J. Chem. Phys.* 111: 7139–7146, 1999.
9. F.S. Bates. *Science* 251: 898–905, 1991.
10. F.S. Bates, M.F. Schulz, A.K. Khandpur, S. Förster, J.H. Rosedal, K. Almdal, and K Mortensen. *Faraday Discuss.* 98: 7–18, 1994.
11. F.S. Bates and G.H. Fredrickson. *Phys. Today* 52: 32–38, 1999.

12. S. Förster, A.K. Khandpur, J. Zhao, F.S. Bates, I.W. Hamley, A.J. Ryan, and W. Bras. *Macromolecules* 27: 6922–6935, 1994.

13. A.K. Khandpur, S. Förster, I.W. Hamley, A.J. Ryan, W. Bras, K. Almdal, and K. Mortensen. *Macromolecules* 28: 8796–8806, 1995.

14. S.-M. Mai, W. Mingvanish, S.C. Turner, C. Chaibundit, J.P.A. Fairclough, F. Heatley, M.W. Matsen, A.J. Ryan, and C. Booth. *Macromolecules* 33: 5124–5130, 2000.

15. D.A. Hajduk, P.E. Harper, S.M. Gruner, C.C. Honeker, G. Kim, E.L. Thomas, and L.J. Fetters. *Macromolecules* 27: 4063–4075, 1994.

16. M.F. Schulz, F.S. Bates, K. Almdal, and K. Mortensen. *Phys. Rev. Lett.* 73: 86–89, 1994.

17. I.W. Hamley, K.A. Koppi, J.H. Rosedale, F.S. Bates, K. Almdal, and K. Mortensen. *Macromolecules* 26: 5959–5970, 1993.

18. I.W. Hamley, M.D. Gehlsen, A.K. Khandpur, K.A. Koppi, J.H. Rosedale, M.F. Schulz, F.S. Bates, K. Almdal, and K. Mortensen. *J. Phys. II Fr.* 4: 2161–2186, 1994.

19. D.A. Hajduk, H. Takenouchi, M.A. Hillmyer, F.S. Bates, M.E. Vigild, and K. Almdal. *Macromolecules* 30: 3788–3795, 1997.

20. M.E. Vigild, K. Almdal, K. Mortensen, I.W. Hamley, J.P.A. Fairclough, and A.J. Ryan. *Macromolecules* 31: 5702–5716, 1998.

21. S.L. Aggarwal. *Polymer* 17: 938–956, 1972.

22. E.L. Thomas, D.B. Alward, D.J. Kinning, D.C. Martin, D.L. Handlin, and L.J. Fetters. *Macromolecules* 19: 2197–2202, 1986.

23. D.A. Hajduk, P.E. Harper, S.M. Gruner, C.C. Honeker, E.L. Thomas, and L.J. Fetters. *Macromolecules* 28: 2570–2573, 1995.

24. L. Leibler. *Macromolecules* 13: 1602–1617, 1980.

25. E. Helfand. In *Recent Advances in Polymer Blends, Grafts, and Blocks*, L.H. Sperling, Ed., Plenum: New York, pp. 141–155, 1974.

26. E. Helfand. *Macromolecules* 8: 552–556, 1975.

27. E. Helfand and Z.R. Wasserman. *Macromolecules* 9: 879–888, 1976.

28. E. Helfand and Z.R. Wasserman. *Polym. Eng. Sci.* 17: 582–586, 1977.

29. E. Helfand and Z.R. Wasserman. *Macromolecules* 11: 960–966, 1978.

30. E. Helfand and Z.R. Wasserman. *Macromolecules* 13: 994–998, 1980.

31. E. Helfand and Z.R. Wasserman. In *Developments in Block Copolymers*, I. Goodman, Ed., Elsevier: New York, Vol. 1, pp. 99–125, 1982.

32. L. Leibler and H. Benoit. *Polymer* 22: 195–201, 1981.

33. I.Ya. Erukhimovich. *Polymer Sci. USSR* 24: 2223–2232, 1982.

34. K.M. Hong and J. Noolandi. *Macromolecules* 16: 1083–1093, 1983.

35. J.D. Vavasour and M.D. Whitmore. *Macromolecules* 25: 5477–5486, 1992.

36. M.W. Matsen and M. Schick. *Macromolecules* 27: 4014–4015, 1994.

37. M.W. Matsen and F.S. Bates. *J. Chem. Phys.* 106: 2436–2438, 1997.

38. S.-M. Mai, J.P.A. Fairclough, N.H. Terrill, S.C. Turner, I.W. Hamley, M.W. Matsen, A.J. Ryan, and C. Booth. *Macromolecules* 31: 8110–8116, 1998.

39. L. Leibler. *Makromol. Chem. Rapid Commun.* 2: 393–400, 1981.

40. L. Leibler. *Macromolecules* 15: 1283–1290, 1982.

41. J.D. Vavasour and M.D. Whitmore. *Macromolecules* 25: 2041–2045, 1992.

42. A.-C. Shi and J. Noolandi. *Macromolecules* 27: 2936–2944, 1994.

43. A.-C. Shi and J. Noolandi. *Macromolecules* 28: 3103–3109, 1995.

44. M.W. Matsen. *Phys. Rev. Lett.* 74: 4225–4228, 1995.

45. M.W. Matsen. *Macromolecules* 28: 5765–5773, 1995.

46. M.W. Matsen. *J. Chem. Phys.* 103: 3268–3271, 1995.

47. M.W. Matsen and F.S. Bates. *Macromolecules* 28: 7298–7300, 1995.

48. N. Koneripalli, R. Levicky, F.S. Bates, M.W. Matsen, S.K. Satija, J. Ankner, and H. Kaiser. *Macromolecules* 31: 3498–3508, 1998.

49. J.D. Vavasour and M.D. Whitmore. *Macromolecules* 34: 3471–3483, 2001.

50. J.R. Naughton and M.W. Matsen. *Macromolecules* 35: 5688–5696, 2002.

51. J.R. Naughton and M.W. Matsen. *Macromolecules* 35: 8926–8928, 2002.

52. M.W. Matsen. *Macromolecules* 36: 9647–9657, 2003.

53. M. Olvera de la Cruz and I.C. Sanchez. *Macromolecules* 19: 2501–2508, 1986.

54. H. Benoit and G. Hadziioannou. *Macromolecules* 21: 1449–1464, 1988.

55. T.A. Kavassalis and M.D. Whitmore. *Macromolecules* 42: 5340–5345, 1991.

56. M.W. Matsen and M. Schick. *Macromolecules* 27: 187–192, 1994.

57. M.W. Matsen. *J. Chem. Phys.* 102: 3884–3887, 1995.

58. M.W. Matsen. *J. Chem. Phys.* 108: 785–796, 1998.

59. T.A. Shefelbine, M.E. Vigild, M.W. Matsen, D.A. Hajduk, M.A. Hillmyer, E.L. Cussler, and F.S. Bates. *J. Am. Chem. Soc.* 121: 8457–8465, 1999.

60. S.F. Edwards. *Proc. Phys. Soc.* 85: 613–624, 1965.

61. A.K. Dolan and S.F. Edwards. *Proc. R. Soc. Lond. A* 343: 427–442, 1975.

62. M.D. Whitmore and J.D. Vavasour. *Acta Polym.* 46: 341–360, 1995.

63. M.W. Matsen. *J. Phys. Condens. Matter* 14: R21–R47, 2002.

64. N. Marits and J.G.E.M. Fraaije. *J. Chem. Phys.* 107: 5879–5889, 1997.

65. F. Drolet and G.H. Fredrickson. *Phys. Rev. Lett.* 83: 4317–4320, 1999.

66. G.H. Fredrickson, V. Ganesan, and F. Drolet. *Macromolecules* 35: 16–39, 2002.

67. M.W. Matsen and M.D. Whitmore. *J. Chem. Phys.* 105: 9698–9701, 1996.

68. A.M. Mayes and M. Olvera de la Cruz. *Macromolecules* 24: 3975–3976, 1991.

69. J. Melenkevitz and M. Muthukumar. *Macromolecules* 24: 4199–4205, 1991.

70. M. Olvera de la Cruz, A.M. Mayes, and B.W. Swift. *Macromolecules* 25: 944–948, 1992.

71. M. Banaszak and M.D. Whitmore. *Macromolecules* 25: 2757–2770, 1992.

72. A.N. Semenov. *Sov. Phys. JETP* (English translation) 61: 733–742, 1985.

73. A.E. Likhtman and A.N. Semenov. *Europhys. Lett.* 51: 307–313, 2000.

74. M.W. Matsen. *J. Chem. Phys.* 114: 10528–10530, 2001.

75. A.N. Semenov. *Macromolecules* 26: 6617–6621, 1993.

76. E. Helfand and Y. Tagami. *J. Chem. Phys.* 56: 3592–3601, 1972.

77. G.H. Fredrickson and E. Helfand. *J. Chem. Phys.* 87: 697–705, 1987.

78. S.A. Brazovskii. *Sov. Phys. JETP* 41: 85–89, 1975.

79. V.E. Podneks and I.W. Hamley. *JETP Lett.* 64: 617–624, 1996.

80. I.W. Hamley and V.E. Podneks. *Macromolecules* 30: 3701–3703, 1997.

81. A.M. Mayes and M. Olvera de la Cruz. *J. Chem. Phys.* 95: 4670–4677, 1991.

82. M. Olvera de la Cruz. *Phys. Rev. Lett.* 67: 85–88, 1991.

83. J.N. Owens, I.S. Gancarz, J.T. Koberstein, and T.P. Russell. *Macromolecules* 22: 3380–3387, 1989.

84. K. Almdal, J.H. Rosedale, F.S. Bates, G.D. Wignall, and G.H. Fredrickson. *Phys. Rev. Lett.* 65: 1112–1115, 1990.

85. A.-C. Shi, J. Noolandi, and R.C. Desai. *Macromolecules* 29: 6487–6504, 1996.

86. M. Laradji, A.-C. Shi, R.C. Desai, and J. Noolandi. *Phys. Rev. Lett.* 78: 2577–2580, 1997.

87. A.-C. Shi. *J. Phys. Condens. Matter* 11: 10183–10197, 1999.

88. H. Fried and K. Binder. *J. Chem. Phys.* 94: 8349–8366, 1991.

89. H. Fried and K. Binder. *Europhys. Lett.* 16: 237–242, 1991.

90. K. Binder and H. Fried. *Macromolecules* 26: 6878–6883, 1993.

91. A. Weyersberg and T.A. Vilgis. *Phys. Rev. E* 48: 377–390, 1993.

92. R.G. Larson. *Macromolecules* 27: 4198–4203, 1994.

93. U. Micka and K. Binder. *Macromol. Theory Simul.* 4: 419–447, 1995.

94. T. Dotera and H. Hatano. *J. Chem. Phys.* 105: 8413–8427, 1996.

95. A. Hoffman, J.-U. Sommer, and A. Blumen. *J. Chem. Phys.* 106: 6709–6721, 1997.

96. A. Hoffman, J.-U. Sommer, and A. Blumen. *J. Chem. Phys.* 107: 7559–7570, 1997.

97. T. Pakula, K. Karatasos, S.H. Anastasiadis, and G. Fytas. *Macromolecules* 30: 8463–8472, 1997.

98. M. Banaszak, S. Woloszczuk, T. Pakula, and S. Jurga. *Phys. Rev. E* 66, 031804–1 to 031804–7, 2002.

99. M. Banaszak, S. Woloszczuk, and S. Jurga. *J. Chem. Phys.* 119: 11451–11457, 2003.

100. O.N. Vassiliev and M.W. Matsen. *J. Chem. Phys.* 118: 7700–7713, 2003.

101. J.M.H.M. Scheutjens and G.J. Fleer. *J. Phys. Chem.* 83: 1619–1635, 1979.

102. W.-C. Zin and R.-J. Roe. *Macromolecules* 17: 183–188, 1984.

103. R.-J. Roe and W.-C. Zin. *Macromolecules* 17: 189–194, 1984.

104. T. Goldacker, V. Abetz, R. Stadler, I.Y. Erukhimovich, and L. Leibler. *Nature* 398: 137–139, 1999.

105. A. Takano, K. Soga, J. Suzuki, and Y. Matsushita. *Macromolecules* 36: 9288–9291, 2003.

106. R.A. Wickham and A.-C. Shi. *Macromolecules* 34: 6487–6494, 2001.

107. J. Noolandi and K.M. Hong. *Macromolecules* 16: 1083–1093, 1983.

108. R.E. Cohen and J.M. Torradas. *Macromolecules* 17: 1101–1102, 1984.
109. M.D. Whitmore and J. Noolandi. *Macromolecules* 21: 1482–1496, 1988.
110. X. Quan, I. Gancarz, J.T. Koberstein, and G.D. Wignall. *Macromolecules* 20: 1431–1434, 1987.
111. T. Hashimoto, T. Tanaka, and H. Hasegawa. *Macromolecules* 23: 4378–4386, 1990.
112. K.I. Winey, E.L. Thomas, and L.J. Fetters. *Macromolecules* 24: 6182–6188, 1991.
113. H. Tanaka, H. Hasegawa, and T. Hashimoto. *Macromolecules* 24: 240–251, 1991.
114. H. Tanaka and T. Hashimoto. *Macromolecules* 24: 5713–5720, 1991.
115. M. Banaszak and M.D. Whitmore. *Macromolecules* 25: 249–260, 1992.
116. M.D. Whitmore and J. Noolandi. *J. Chem. Phys.* 93: 2946–2955, 1990.
117. T. Hashimoto, M. Shibayama, and H. Kawai. *Macromolecules* 16: 1093–1101, 1983.
118. T.P. Lodge, C. Pan, X. Jin, Z. Liu, J. Zhao, W.W. Maurer, and F.S. Bates. *J. Polym. Sci. Polym. Phys. Ed.* 33: 2289–2293, 1995.
119. S. Sakurai, T. Hashimoto, and L.J. Fetters. *Macromolecules* 29: 740–747, 1996.
120. K.J. Hanley and T.P. Lodge. *J. Polym. Sci. Polym. Phys. Ed.* 36: 3101–3113, 1998.
121. K.J. Hanley, T.P. Lodge, and C.-I. Huang. *Macromolecules* 33: 5918–5931, 2000.
122. M. Olvera de la Cruz. *J. Chem. Phys.* 90: 1995–2002, 1989.
123. G.H. Fredrickson and L. Leibler. *Macromolecules* 22: 1238–1250, 1989.
124. T.M. Birshtein and E.B. Zhulina. *Polymer* 31: 1312–1320, 1990.
125. T.P. Lodge, M.W. Hamersky, K.J. Hanley, and C.-I. Huang. *Macromolecules* 30: 6139–6149, 1997.
126. C.-I. Huang and T.P. Lodge. *Macromolecules* 31: 3556–3565, 1998.
127. M. Banaszak and M.D. Whitmore. *Macromolecules* 25: 3406–3412, 1992.
128. P. Linse. *J. Chem. Phys.* 97: 13896–13902, 1993.
129. P. Linse. *Macromolecules* 26: 4437–4449, 1993.
130. P. Linse. *Macromolecules* 27: 6404–6417, 1994.
131. M. Svensson and P. Linse. *Macromolecules* 31: 1427–1429, 1998.
132. M. Svensson, P. Alexandridis, and P. Linse. *Macromolecules* 32: 637–645, 1999.
133. M. Svensson, P. Alexandridis, and P. Linse. *Macromolecules* 32: 5435–5443, 1999.
134. N.P. Susharina, P. Linse, and A.R. Khokhlov. *Macromolecules* 33: 8488–8496.
135. L. Leibler, H. Orland, and J.C. Wheeler. *J. Chem. Phys.* 79: 3550–3557, 1983.
136. J. Noolandi and K.M. Hong. *Macromolecules* 15: 482–492, 1982.
137. J. Noolandi and K.M. Hong. *Macromolecules* 16: 1443–1448, 1983.
138. M.D. Whitmore and J. Noolandi. *Macromolecules* 18: 657–665, 1985.
139. T. Bluhm and M.D. Whitmore. *Can. J. Chem.* 63: 249–252, 1985.
140. A. Halperin. *Macromolecules* 20: 2943–2946, 1987.
141. M.S. Kent, L.T. Lee, B.J. Factor, F. Rondelez, and G.S. Smith. *J. Chem. Phys.* 103: 2320–2342, 1995.
142. M.S. Kent. *Macromol. Rapid Commun.* 21: 243–270, 2000.
143. M.D. Whitmore and T.W. Smith. *Macromolecules* 27: 4673–4683, 1994.
144. C.M. Wijmans and P. Linse. *Langmuir* 11: 3748–3756, 1995.
145. T. Haliloglu and W.L. Mattice. *Polym. Prepr.* 34: 460–461, 1993.
146. K. Rodrigues and W.L. Mattice. *Polym. Bull.* 25: 239–243, 1991.
147. K. Rodrigues and W.L. Mattice. *J. Chem. Phys.* 95: 5341–5347, 1991.
148. K. Rodrigues and W.L. Mattice. *J. Langmuir* 8: 456–459, 1992.
149. Y. Wang, W.L. Mattice, and D.H. Napper. *Langmuir* 9: 66–70, 1993.
150. Y. Zhan and W.L. Mattice. *Macromolecules* 27: 677–682, 1994.
151. Y. Zhan and W.L. Mattice. *Macromolecules* 27: 683–688, 1994.
152. M. Pépin and M.D. Whitmore. *Macromolecules* 33: 8644–8653, 2000.
153. M. Pépin and M.D. Whitmore. *Macromolecules* 33: 8654–8662, 2000.
154. M. Kenward and M.D. Whitmore. *J. Chem. Phys.* 116: 3455–3470, 2002.
155. L. Zhang, R.J. Barlow, and A. Eisenberg. *Macromolecules* 28: 6055–6066, 1995.
156. P.H. Nelson, G.C. Rutledge, and T.A. Hatton. *J. Chem. Phys.* 107: 10777–10781, 1997.
157. K.C. Douzinas, R.E. Cohen, and A.F. Halasa. *Macromolecules* 24: 4457–4459, 1991.
158. P. Rangarajan, R.A. Register, and L.J. Fetters. *Macromolecules* 26: 4640–4645, 1993.
159. M. Gervais and B. Gallot. *Makromol. Chem.* 171: 157–178, 1973.
160. M. Gervais and B. Gallot. *Makromol. Chem.* 174: 193–214, 1973.
161. M. Gervais and B. Gallot. *Makromol. Chem.* 178: 1577–1593, 1978.
162. M. Gervais and B. Gallot. *Makromol. Chem.* 178: 2071–2078, 1978.
163. J.-J. Herman, R. Jérome, P. Teyssié, M. Gervais, and B. Gallot. *Makromol. Chem.* 179: 1111–1115, 1978.
164. M. Gervais and B. Gallot. *Makromol. Chem.* 180: 2041–2044, 1979.
165. M. Gervais and B. Gallot. *Polymer* 22: 1129–1133, 1981.
166. M. Gervais, B. Gallot, R. Jérome, and P. Teyssié. *Makromol. Chem.* 182: 989–995, 1981.

Chapter 10

Assemblies in Complex Block Copolymer Systems

Volker Abetz

Contents

I. Introduction

Block copolymers are macromolecules, composed of two or more polymer blocks of chemically different monomers, which are linked together by chemical bonds. The resulting chain topologies

In memory of Prof. Dr. Reimund Stadler

Table 1   Examples of Binary and Ternary Block Copolymers with Different Block Distributions

Binary block copolymers

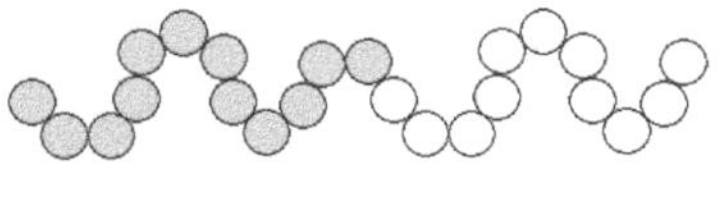	AB diblock copolymer	ϕ_A, χ_{AB}
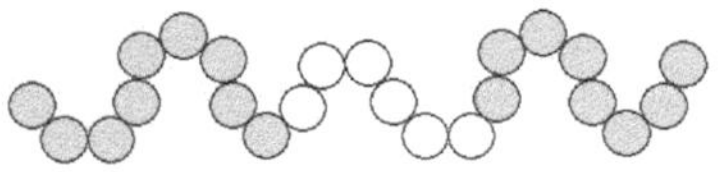	ABA triblock copolymer	ϕ_{A1}, ϕ_{A2}, χ_{AB}
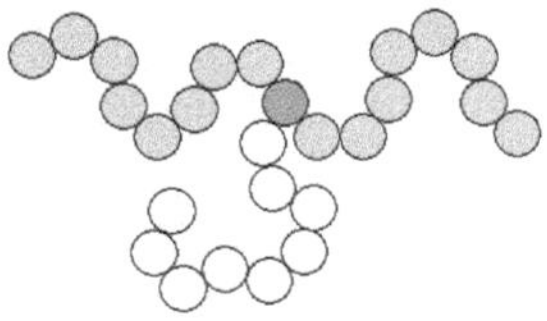	A_2B miktoarm star copolymer	ϕ_{A1}, ϕ_{A2}, χ_{AB}
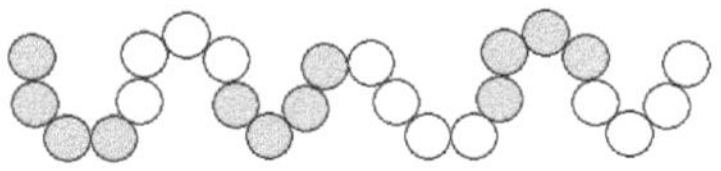	AB multiblock copolymer	ϕ_A, χ_{AB}
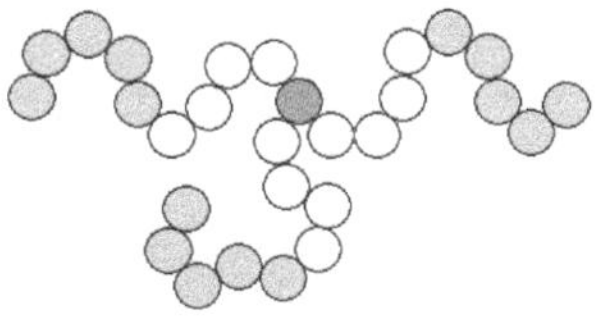	AB 3-arm starblock copolymer	ϕ_A, χ_{AB}

Ternary block copolymers

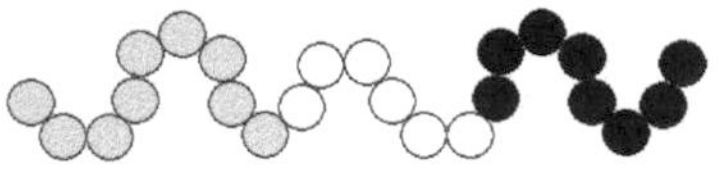	ABC triblock terpolymer	ϕ_A, ϕ_B, χ_{AB}, χ_{BC}, χ_{AC}
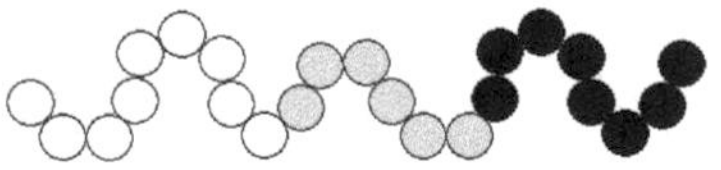	BAC triblock terpolymer	ϕ_A, ϕ_B, χ_{AB}, χ_{BC}, χ_{AC}
	ACB triblock terpolymer	ϕ_A, ϕ_B, χ_{AB}, χ_{BC}, χ_{AC}
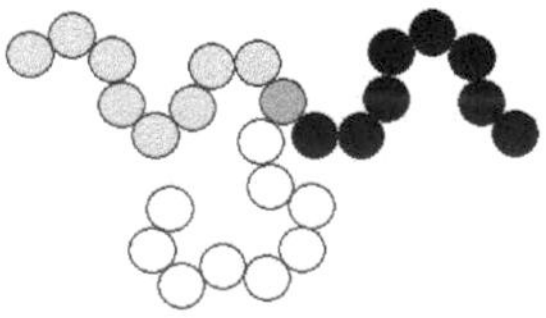	ABC 3-miktoarm star terpolymer	ϕ_A, ϕ_B, χ_{AB}, χ_{BC}, χ_{AC}

Note: Third column represents independent system variables — ϕ: volume fraction; χ: segmental interaction.

can be linear, branched, or cyclic. The total number of monomers (or more correctly, repeating units) in a block copolymer is the degree of polymerization N. Table 1 shows a few examples for block copolymers containing two or three different types of monomers. Systematic studies of these materials became possible through the developments in polymerization techniques, which made possible the synthesis of well-defined block copolymers with a very small polydispersity.

BCC　　　　H　　　　G　　　　L　　　　G　　　　H　　　　BCC

Figure 1 Schemes for different diblock copolymer morphologies. From left to right the volume fraction of one component increases. The morphologies are body-centered cubic spheres (BCC), hexagonally packed cylinders (H), gyroid (G), lamellae (L).

During the last decade block copolymers have attracted increasing interest both from scientific and commercial points of view [1–6]. This is due to their unique morphological behavior, namely the formation of crystal-like order on a length scale in the range between a few nanometers up to several hundreds of nanometers. Figure 1 shows some of the typical morphologies found in amorphous diblock copolymers, where the different blocks self-assemble into different microphases (microdomains). The dispersed microdomains (spheres, cylinders) form the "core" and the surrounding matrix forms a "shell" or "corona." Note that the crystal-like order occurs on a supramolecular length scale in these systems. On a molecular level the blocks may be disordered (liquid-like, amorphous), but they can also be liquid-crystalline or crystalline.

These morphologies arise because the different incompatible blocks have to find a way to reduce repulsive interactions without reducing the conformational entropy of the blocks too much. The conformational entropy has its maximum in the disordered state. In order to minimize the free energy such a system has to find a compromise between reduction of repulsive enthalpic interactions between chemically different segments, and the reduction of conformational entropy of the blocks. The enthalpic interactions between different segments are often described by the Flory–Huggins–Staverman interaction parameter χ [7–9] (a positive value of χ indicates repulsion, while a negative value indicates attraction between the segments, see also Chapters 2, 9, and 11). The conformational entropy of the block copolymer chain is maximum, when $\chi = 0$, that is, when the contact enthalpies between chemically different segments and chemically similar segments are the same. Since, usually chemically different segments show repulsive interactions, the conformational entropy will be reduced, because the different segments self-assemble into different domains, the so-called microphases. The junction points between blocks are located on a common interface or within an interfacial region of finite thickness. The larger the thickness of the interface, the larger the conformational entropy. Systems with sharp interfaces (where the thickness of the interface is much smaller than the domain size or long period of the morphology) are called strongly segregated systems. The concentration profile perpendicular to the microphase domain boundary can be described by a hyperbolic tangent (or less precisely by a step function). The two different microdomains contain only one type of segments, that is, the microdomains are composed of only one component. If the interfacial width is comparable with the length scale of the long period and the concentration profile is only a weak fluctuation from the mean value, the system is called a weakly segregated system and the concentration profile perpendicular to the interface can be described by a sine function. The two extrema are thus called the strong segregation limit (SSL) and the weak segregation limit (WSL), respectively [1]. Schemes of the concentration profiles perpendicular to an interface between two blocks of a block copolymer in the bulk or diluted state are shown in Figure 2.

Different microphase separated morphologies occur depending on the relative compositions of the different components and the total molecular weight as expressed by the degree of polymerization. In addition, the aggregation state of the blocks (amorphous, liquid-crystalline, or crystalline) or the presence of solvents (formation of micelles, vesicles) largely influences the morphology as well.

Thus microphase separated block copolymers can be considered as a class of supramolecular polymers that form large regular structures via a self-assembling process without being chemically

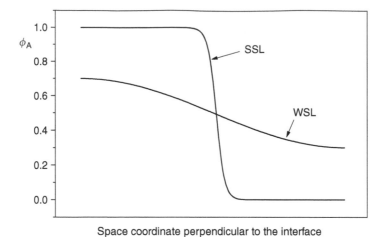

Space coordinate perpendicular to the interface

Figure 2 Scheme of the concentration profile perpendicular to a microdomain boundary. SSL: strong segregation limit, WSL: weak segregation limit.

linked to each other. In this chapter, most exciting morphologies of such supramolecular assemblies will be shown.

Their thermodynamic properties make block copolymers interesting for applications such as thermoplastic elastomers [2,10], surfactants [11], compatibilizers in multiphase polymer blends [12], or topologically controlled hosts for catalysts (e.g., transition metal complexes) [13], colloidal metals [14–19], nonlinear optical moieties [20,21], light emitting devices [22], or photonic materials [23], etc. Liu et al. [24] synthesized nanofibers by cross-linking the cylindrical domains of a diblock copolymer in the bulk and subsequent dissolution of the system leading to a kind of "hairy rods." Ikkala et al. synthesized diblock copolymers showing microphase separation on two different length scales. By using hydrogen-bonding complexes they could attach low molecular weight amphiphiles to a polar block of a microphase separated diblock copolymer, leading to a second microphase separation between the hydrogen-bonding complexes and the hydrophobic parts of the amphiphiles within the domains of the polar block [25,26]. Block copolymers have also attracted interest recently as templates to create mesoscale structures in inorganic materials, which never would self-organize in such a way by themselves. Templin et al. [27] have synthesized ceramic materials with spherical, cylindrical, and disk-like structures. Krämer et al. [28] used block copolymers as templates to create nanoporous silica.

In this chapter, the following topics will be addressed. After a short overview of block copolymer synthesis and their morphological characterization, we will discuss the morphological behavior in the bulk state, followed by the behavior in the swollen and diluted state. Then we will consider blends of block copolymers, block copolymers with crystalline blocks, and finally we address the influence of external fields on block copolymers.

II. SYNTHESIS OF BLOCK COPOLYMERS

There are different ways to synthesize block copolymers. They have been reviewed recently [29] and here we will just give a brief description. Which chemical route can be chosen depends upon the particular monomer. The chemical route also determines the polydispersity of the product. Principally, a block copolymer can be formed by subsequently adding the various types of monomers to the chain, as it happens in some chain reactions. Or it can be formed by linking different polymers with each other, as it happens in step-growth reactions. This can be achieved by using polymers carrying functional groups at their chain ends or somewhere along the chain. In the first case linear block copolymers will be obtained and in the latter case grafted or star-like block copolymers will occur. Often this type of block copolymers is the result of a polycondensation reaction. Polycondensation is important for commercial multiblock copolymers, like polyurethanes, polyureas, and polyesters with hard and soft segments, but leads to polydisperse materials. Some recent work in the field of polycondensation was related to linear and hyperbranched block copolymers [30–33]. Blocking is usually performed by reacting a telechelic prepolymer carrying hydroxyl or amino end groups with other telechelic prepolymers carrying isocyanate, acid or ester end groups. The overall molecular weight thereby depends on the stoichiometry between the two different functional groups as well as upon the extent of the reaction. Removing the side product (water or alcohol in the cases of polyesters and polyamides) moves the chemical equilibrium towards the formation of the block copolymers.

In chain reactions the different types of monomers can be added subsequently to an active chain end. The most important techniques here are sequential living polymerization techniques, such as anionic or cationic polymerization. Certain metallocenes can be used in coordination polymerization of olefins leading to stereo block copolymers, like polypropylene where crystalline and amorphous blocks alternate with each other due to the change of tacticity along the chain [34]. In comparison to living polymerization techniques, free radical and coordination polymerization lead to rather polydisperse materials in terms of the number of blocks and their degree of polymerization.

Living cationic polymerization has been used for the synthesis of block copolymers, which are often based on polyisobutylene [35–37], polysiloxanes [38], or polytetrahydrofurane [39,40]. Cationic polymerization also was used to synthesize block copolymers containing organic and phosphazene blocks [41]. Also a lot of progress has been achieved in the field of controlled radical polymerization [42–44], but the chain lengths are more limited than in living cationic or anionic polymerization. However, living anionic polymerization is still the most important way to synthesize well-defined block copolymers and most basic investigations on the properties of linear and star copolymers are based on systems prepared by living anionic polymerization [45–60]. In recent years, combinations of various polymerization techniques have been used in order to synthesize block copolymers of different monomers, which could not be polymerized by the same technique. Besides combining cationic and radical polymerization [44,61], anionic and radical polymerization [44], also combinations of anionic and cationic polymerization [44,62,63], anionic polymerization with enzymatic polymerization [64], or anionic polymerization and polycondensation [65] have been used.

Since in all the techniques based on living polymerization the different monomers are added sequentially, deactivation of some living chains cannot be suppressed completely when the next monomer is added (due to impurities). These deactivated polymers (homopolymers in case of diblock copolymers or homo and diblock copolymers in case of triblock copolymers) can sometime be separated from the desired block copolymer by fractionation or preparative size exclusion chromatography.

III.   METHODS FOR INVESTIGATION OF MORPHOLOGICAL PROPERTIES AND PHASE TRANSITIONS

It was established at an early stage of the research on block copolymers that the morphology of the sample can be influenced by the solvent, when films are cast from solution [66–68]. While for binary block copolymers a nonselective solvent could be used for film casting, in the case of block copolymers with more than two chemically different species a solvent will always be selective. This results in a different degree of swelling of the different blocks in solution and thus their effective volume fractions are modified at the point where the morphology forms during evaporation of the solvent, that is, when microphase separation occurs. Thus, great care must be taken when discussing thermodynamic stability of the observed morphologies. Possibilities for assessing the thermodynamic stability include annealing of samples or using solvents with different selectivities and comparing the obtained morphologies. While the first strategy might work for block copolymers with relatively low molecular weights and low number of entanglements, it usually fails for samples with large molecular weights.

Different techniques have been used for the investigation of the phase behavior of block copolymers. Small angle x-ray scattering (SAXS) [69] and small angle neutron scattering (SANS) [70] have been widely used for the investigation of the order–disorder transition temperature (ODT) [71–75], and the characterization of the ordered morphology. In addition, SANS was used for the investigation of the single-chain behavior in the disordered [76] and ordered melt [76–78]. In the latter, it was found that the block chains are stretched along the normal of a lamella, while they are not stretched in the disordered state. Neutron reflection has been used for studies of the interfacial widths of block copolymers with lamellar morphology [79]. NMR spin-diffusion experiments were also used for the investigation of the interfacial width, that is, to determine the degree of segregation between different blocks [80]. Dynamic mechanical analysis has been proven to be very sensitive for the ODT [81–84]. Storage (G') and loss (G'') moduli follow the scaling laws of homopolymers as a function of frequency ω in the terminal flow regime in the disordered phase ($G' \sim \omega^2$, $G'' \sim \omega^1$), while different exponents apply for the ordered phases. Balsara et al. [85] used time resolved depolarized light scattering to follow the ordering kinetics of a cylindrical diblock copolymer after a temperature quench from the disordered state and found a nucleation and growth mechanism, which they could explain by a Ginzburg–Landau model. Most important for the characterization of the microphase separated morphologies are transmission electron microscopy (TEM) [69,86], scanning electron microscopy [87,88], and scanning force microscopy [89–91].

IV. BLOCK COPOLYMERS IN THE BULK STATE

A. Linear Block Copolymers

1. Binary Block Copolymers

In the field of amorphous, linear, binary block copolymers a lot of work has been done and extensively reviewed [1,2,5,92–97]. Among the binary, linear, block copolymers the *diblock copolymers* have been studied in great detail. They can be considered as model systems for more complicated block copolymers, like block copolymers with more than two components, or block copolymers with other block distributions. Here we will address experimental work in the field of their phase behavior and mention only some theoretical work in the following. For a detailed theoretical description see Chapter 5.

The phase behavior of diblock copolymers was described theoretically within the WSL by Leibler [98] and Erukhimovich [99]. The latter expanded the description also for ABA triblock and multiblock and multigraft copolymers [100]. Within the SSL, after basic works by Meier [101] and Helfand et al. [102–104], the phase behavior was also described by Semenov who gave an analytical solution for the elastic part of the free energy originated by the reduced conformational entropy of the different blocks [105]. Since strong segregation theories overestimate the degree of stretching of the blocks, their prediction of the scaling behavior of the degree of polymerization N with the long period of the morphology L cannot be a correct one. Khokhlov and coworkers introduced the super strong segregation limit (SSSL), in which the scaling predictions of the SSL theories ($L \propto N^{2/3}$ as compared to $L \propto N^{1/2}$ in the WSL) become correct due to a very large degree of incompatibility ($\chi N \gg 100$) [106]. Matsen studied the phase behavior in the intermediate segregation regime [107] and also gave a self-consistent description of the whole phase behavior from the disordered state through the weakly segregated to the strongly segregated state [107,108] (Figure 3). The influence of dissimilarities between the segmental lengths of the different blocks also has been studied and it was found that it only affects the symmetry of the phase diagram, with no change in its topology [109].

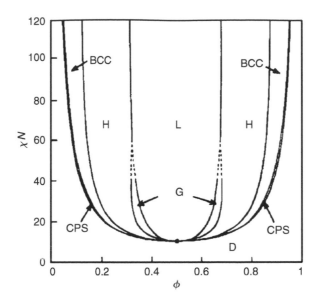

Figure 3 Phase diagram of diblock copolymers with equal segmental lengths and segmental volumes of both block components. χ: Flory–Huggins–Staverman interaction parameter, N: degree of polymerization, ϕ: volume fraction, D: disordered phase, CPS: close packed spheres, BCC: body-centered cubic spheres, H: hexagonally packed cylinders, G: gyroid, L: lamellae. (From M.W. Matsen and M. Schick. *Macromolecules* 29:1091, 1996, Copyright 1996 American Chemical Society.)

While the descriptions mentioned so far only apply for block copolymers with coiled blocks, work has also been done on block copolymers with liquid crystalline or crystalline blocks. If one of the two blocks is liquid crystalline, spherical and cylindrical morphologies are predicted to become more stable with the liquid crystalline block forming the core domains at larger values of χN [110].

Well known are the results on the morphology of polystyrene–block–polyisoprene (S–I) diblock copolymers by means of TEM as a function of composition [1]. Three different morphologies were found: spheres on a cubic lattice, hexagonally arranged cylinders, and lamellae (see Figure 1). In 1972, Aggarwal [111] found a new morphology, which in 1986 was identified as a "cocontinuous morphology" [112], where the minority component forms two interpenetrating networks. This morphology is observed within a relatively small composition range between the lamellar and cylindrical region of the phase diagram. The cocontinuous networks, which were observable both in linear and star block copolymers, were believed to have the symmetry of a diamond and thus the morphology was named "ordered bicontinuous double diamond" (OBDD) structure [112]. However, a few years later investigations by SAXS on star block copolymers showed that rather than two interpenetrating tetrapod-diamond lattices two interpenetrating tripod-lattices with a mirror-symmetry characterize this structure, which was then named gyroid structure [113] (see Figure 1). A lot of theoretical and experimental work has been carried out on this type of cocontinuous morphology [107,114–117]. It could be shown theoretically that there exists a whole class of gyroid morphologies rather than only one type [118,119]. However, these morphologies differ only in details from each other and so far no experimental results on different gyroid morphologies have been reported for binary block copolymers.

Other morphologies like perforated lamellae [120–122] or core-shell cylinders [123], which have been discovered more recently, are considered to be metastable [108]. Khandpur et al. [122] carefully measured the phase diagram of S–I diblock copolymers. The results are in reasonable agreement with the predictions by Matsen, although the experimentally found perforated lamellae is not an equilibrium morphology [108]. Vigild et al. [115] studied the reversibility of microphase transitions, especially the transitions from and to the gyroid phase on a poly(ethylene-propylene)–block–poly(dimethylsiloxane) diblock copolymer. When quenching the system from the cylindrical phase to the gyroid phase, hexagonally perforated lamellae were found as an intermediate state, from which the gyroid phase occurs via epitaxial growth (compare with Figure 3). Thus the formation of the hexagonally perforated lamellae is kinetically favored and then this metastable phase transforms into the gyroid phase. In a poly(styrene–block–(ethylene-alt-propylene)–block–styrene) triblock copolymer the thermally reversible transition between lamellae and perforated lamellae was reported [124].

Lodges' group [125] and Hashimotos' group [126,127] investigated the order–order transition (OOT) between cylindrical and spherical phases of S–I–S triblock copolymers and found them to be reversible.

Although there are uncertainties in the calculations in a certain range of incompatibility χN as expressed by dashed lines in Figure 3, it follows from these calculations that the gyroid morphology should be only stable within a finite range of χN. Further increase of incompatibility leads to a lamellar morphology. A problem with most block copolymers is the limited range of χN, which can be realized by changing the temperature. Thus experimental verification of the theoretical predictions is difficult. To overcome this problem, in a recent work Davidock and coworkers used a polymer analogous reaction, which enabled them to change gradually the incompatibility between polystyrene and (modified) polyisoprene without changing the chain flexibility (statistical segment length) of the latter. Using this approach they were able to obtain data in a large range of the $\chi N/\phi$-diagram [128]. From these experiments it follows that the gyroid phase is stable up to significantly larger values of χN than predicted by the self-consistent field theory (SCFT) of Matsen [107], but the topology of the phase diagram remains unaltered.

2. Ternary Block Copolymers

In comparison with binary block copolymers much less work on ternary block copolymers has been published so far. There are more independent system variables in ternary block copolymers as compared with binary block copolymers. This leads to a richer phase diagram. In addition, the block sequence can be changed, which introduces another tool to influence the morphology [129]. Among the ternary block copolymers the triblock terpolymers are the ones that have been mostly investigated. As mentioned before in the case of diblock copolymers, systematic studies of triblock terpolymers became possible with the development of sequential polymerization techniques with living anionic polymerization being still the most important one. Before discussing some experimental results on the morphology of triblock terpolymers, we give a short overview on theoretical works of these systems.

Riess [60] gave a first description of the phase behavior of linear ternary triblocks. While in microphase separated diblock copolymers only one structural feature can exist like lamella, cylinder, or sphere, ternary block copolymers can simultaneously exhibit different features in the microphase separated state (e.g., spheres within a cylinder or lamella, etc.).

Kane and Spontak [130] developed a self-consistent field theory (SCFT) for lamellar ABC triblock terpolymers based on Semenov's approach for diblock copolymers [105]. They also described the scaling behavior of the periodicity L with the degree of polymerization N, which was found to be similar to diblock copolymers ($L \propto N^{2/3}$). The periodicity of an ABC triblock terpolymer was found to be slightly larger than the periodicity of an AC-diblock copolymer with the same overall degree of polymerization. These theoretical results confirm systematic SAXS and SANS studies by Mogi et al. [131] on lamellar I–S–VP block copolymers. Zheng and Wang [132] published a theoretical description of different morphologies in ABC triblock terpolymers based on a strong segregation approach following earlier work by Ohta and Kawasaki [133], and Nakazawa and Ohta [134]. Lyatskaya and Birshtein [135] described different cylindrical and lamellar morphologies within the SSL, including the possibility of different segmental volumes and persistence lengths of the different blocks. They could show that these segmental properties greatly influence the border line of stability between different morphologies, a result which also had been obtained by Matsen and Schick [109] for diblock copolymers. Phan and Fredrickson [136] studied symmetric ABC triblock terpolymers ($\phi_A = \phi_C$) and confirmed former results of Nakazawa and Ohta [134], according to which the square lattice arrangement of A and C cylinders in a B-matrix should be more stable than a hexagonal arrangement. Also the CsCl-type of packing for A and C spheres in a B matrix was confirmed to be more stable than other types of spherical morphologies. They found that even for ABC-triblock terpolymers the gyroid phase should be more stable than the ordered tricontinuous double diamond (OTDD) morphology, however, both cocontinuous morphologies are unstable with respect to cylindrical or lamellar morphologies in the SSL. Matsen showed the gyroid morphology to be stable for symmetric ABC triblock terpolymers with a B matrix in the intermediate segregation regime [137]. Its stability range is extended towards the cylindrical region, because the tetragonal packing of A and C cylinders leads to a stronger chain frustration (and thus larger energy). Based on his theory, Matsen simulated different projections of TEM images of the gyroid and found strong evidence for the morphology found before by Mogi et al. [138,139] (and identified as an OTDD) to be gyroid. An approach based on SCFT without preliminary choice of the symmetry of the morphology was proposed by Bohbot-Raviv and Wang to describe both linear triblock terpolymers and miktoarm star terpolymers [6,140]. The idea of this approach is to allow the finding of new, unpredicted morphologies. Also Fredrickson et al. [141,142] have developed theoretical methods along this direction.

Mogi et al. [138] and Gido et al. [143] studied triblock terpolymers based on polystyrene (S), polyisoprene (I), and poly(2-vinylpyridine) (VP) with different block sequences. The difference in block sequence resulted in a different morphology for a similar overall composition of the systems. While polyisoprene-block–polystyrene-block–polyvinylpyridine I–S–VP with similar amounts of all three components forms lamellar stacks (Figure 4(a)) [138], polystyrene-block–polyisoprene-block–polyvinylpyridine S–I–VP forms hexagonally packed core-shell cylinders (Figure 4(b)) [143].

This behavior can be understood as a consequence of the competition between the different interfacial tensions between adjacent blocks: while the interfacial tensions between S and I on one side and S and VP on the other side are of approximately similar magnitude, the interfacial tension

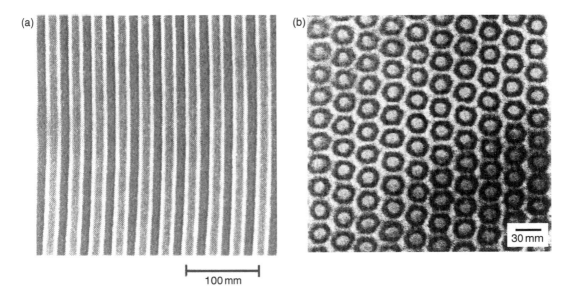

(a)

(b)

30 mm

100 mm

Figure 4 Transmission electron micrographs of (a) polyisoprene-block–polystyrene-block–poly(2-vinylpyridine) I–S–VP. (From Ref. 138, Copyright 1994 American Chemical Society.) (b) Polystyrene-block–polyisoprene-block–poly(2-vinylpyridine) S–I–VP. (From Ref. 143, Copyright 1993 American Chemical Society.) Polyisoprene appears black due to staining with OsO_4.

between I and VP is much larger than between I and S. As a consequence the system favors a smaller interface between I and VP as compared with I and S, which leads to a morphology with different interfacial areas on both ends of the middle block in the case of S–I–VP. In comparison, the system I–S–VP will form lamellae due to the fact that the interfacial areas on both ends of the middle block are of approximately the same size.

Hückstädt et al. [144] investigated the influence of the block sequence in polystyrene-block–polybutadiene-block–poly(2-vinylpyridine) (S–B–VP) and B–S–VP triblock terpolymers. For a given B–S and S–B precursor series of triblock terpolymers with varying amount of VP were studied. These systems behave similar to the S–I–VP and I–S–VP triblock terpolymers discussed before. While B–S–VP formed lamellae for various compositions, S–B–VP formed core-shell cylinders, core-shell double gyroid and lamellae with increasing volume fraction of VP. Bates group later used the same strategy for polystyrene-block–polyisoprene-block–poly(ethylene oxide) (S–I–EO) and I–S–EO triblock terpolymers with similar amounts of S and I ($\phi_S = \phi_I$), which they studied close to the ODT [145,146]. For S–I–EO in the range of $0 < \phi_{EO} < 0.33$ they found the morphological sequence of S–I lamella, perforated lamellae, core-shell cylinder, core-shell double gyroid, perforated lamellae, S–I–EO lamellae [145]. Changing the block sequence to I–S–EO lead to another sequence of morphologies from I–S lamellae via a noncubic triple network structure to the I–S–EO lamellae upon increase of the EO content ($0 < \phi_{EO} < 0.33$) [146].

The influence of the block sequence on the morphology has also been observed for triblock terpolymers based on polystyrene S, polybutadiene B, and poly(methyl methacrylate) M. While an S–B–M triblock terpolymer with similar amounts of the three components forms lamellae, the corresponding B–S–M triblock terpolymer shows a core-shell double gyroid morphology [144,147,148]. The same morphology with the same composition was observed for polystyrene-block–poly(2-vinylpyridine)-block–poly(t-butyl methacrylate) (S–VP–tBMA) [149], where the relative interfacial tensions should be comparable to the ones in B–S–M. The interfacial curvature in a gyroid morphology is lower than in a cylindrical morphology. Thus the dissimilarity of interfacial tensions between middle block and outer blocks is less pronounced in B–S–M or S–VP–tBMA than in S–I–VP.

The morphologies discussed so far differ from the morphologies known for diblock copolymers basically by a layer of the third component between the two other blocks, that is, they can be considered as core-shell analogs of the diblock copolymer morphologies shown in Figure 1. In the following paragraphs, we will present morphologies, which are more complex.

Stadler et al. studied triblock terpolymers based on polystyrene S, polybutadiene B, and poly(methyl methacrylate) M and a number of new morphologies were discovered [6,129,150–155]. For symmetric systems with the block sequence S–B–M, that is, where the end blocks have similar size, lamellar morphologies were found. Varying the volume fraction of the middle block from 0.03 up to approximately 0.3 forms spheres, cylinders, or a lamella between the lamellae of the outer blocks [150] (Figure 5).

Upon further increase of the volume fraction of the middle block it forms the matrix embedding cylindrical [155] or spherical domains of the S and M end blocks. Depending on the molecular

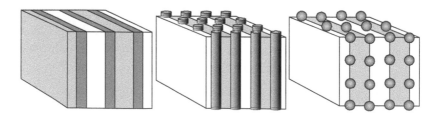

Figure 5   Schemes for different lamellar morphologies of ABC triblock terpolymers. Upon decreasing the volume fraction of the middle block it changes from lamellae via cylinders to spheres.

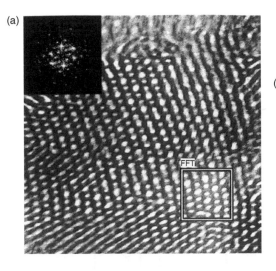

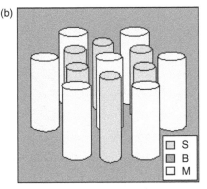

Figure 6 Polystyrene-block–polybutadiene-block–poly(methyl methacrylate) S–B–M with a B matrix embedding hexagonally packed cylindrical domains of S and M. (a) Transmission electron micrograph stained with OsO₄, (b) Scheme of the morphology. (From S. Brinkmann et al. *Macromolecules* 31:6566, 1998, Copyright 1998 American Chemical Society.)

weight, the endblocks can be either located in different domains or form mixed microdomains. This should have consequences on the symmetry of the morphology and, more importantly, on the mechanical properties. While in the case of mixed microdomains the middle block can form either loops (both end blocks of a particular chain are located within the same microdomain), it can only form links (bridges) between two different microdomains, when the end blocks stay incompatible. A comparison between S–B–S and S–B–M triblock copolymers of same molecular weight with lamellar or cylindrical morphology shows in fact a higher modulus of the S–B–M systems [156].

As mentioned before, the casting solvent might influence the morphology of ABC triblock terpolymers [68]. In the case of an S–B–M triblock terpolymer with a B matrix a new hexagonal morphology was found, which has not been predicted before by theory[155]. While a square-like ordering of the S and M cylinders may be expected for an S–B–M triblock terpolymer with similar amounts of S and M, in this case a hexagonally packed array of S cylinders was found, where each S cylinder is surrounded by six M cylinders (Figure 6). This special morphology was obtained from a solution with mixed solvents (30% benzene and 70% cyclohexane). A similar morphology was found for an S–B–A triblock terpolymer (A being poly(methacrylic acid)). In that system the difference in the solubilities of A on one side and B and S on the other side are much more pronounced than in S–B–M and lead to a sequential microphase separation, where first A separates from the solution in tetrahydrofuran, before B and S separate from each other [157].

When changing the relative composition in S–B–M triblock terpolymers in such a way that the volume fraction of one endblock is above ca. 0.6, lamellar morphologies are no longer obtained. In such case this large endblock forms the matrix and the other endblock forms a core sphere or core cylinder. Depending on the relative volume fraction of the middle block with respect to the core forming endblock, different morphologies are found [153], which are schematically shown in Figure 7.

The most spectacular morphology in Figure 7 is the helical morphology [151], which is the only noncentrosymmetric morphology in ABC triblock terpolymers observed so far. However, due to helix reversals, the sample contains about the same number of right- and left-handed helices and thus is not chiral on a large scale. An electron micrograph of that morphology is shown in Figure 8.

Besides changes of the volume fractions discussed so far, chemical modification of one or two blocks is another way to influence morphology. For example, by hydrogenation of the

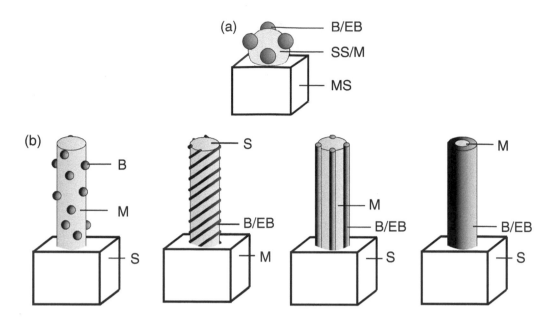

Figure 7 Schemes for (a) the spheres-on-sphere morphology. (From U. Breiner et al. *Polym. Bull.* 40:219, 1998, Copyright 1998 Springer-Verlag GmbH & Co. KG.) (b) From left to right: spheres-on-cylinder, helices-around-cylinder, cylinders-at-cylinder, core-shell-cylinder morphology. (From U. Breiner et al. *Macromol. Chem. Phys.* 198:1051, 1997, Copyright 1997 Wiley-VCH Verlag GmbH, Weinheim.)

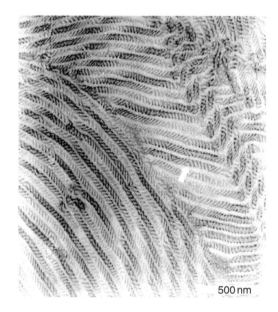

Figure 8 Transmission electron micrograph of a polystyrene-block–polybutadiene-block–poly(methyl methacrylate) S–B–M showing the helical morphology. Cylindrical domains of S are surrounded by black stained helices of B. M forms the matrix. The arrow points to a helix reversal. (From U. Krappe et al. *Macromolecules* 28:4558, 1995, Copyright 1995 American Chemical Society.)

B block in symmetric S–B–M block copolymers, where B forms spheres [150,158] or cylinders at the lamellar interface between S and M, the corresponding polystyrene-block–poly(ethylene-co-butylene)-block–poly(methyl methacrylate) S–EB–M forms a hexagonal morphology, where S cylinders are surrounded by EB rings in an M matrix [150,159] (Figure 9).

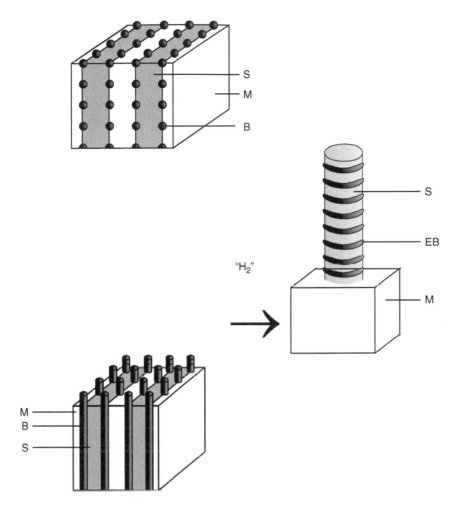

Figure 9 Scheme for the change of the morphological behavior of a polystyrene-block–polybutadiene-block–poly(methyl methacrylate) S–B–M induced by hydrogenation of B to poly(ethylene-co-butylene) EB: from a lamellar morphology with B spheres or B cylinders between S and M lamellae in an S–B–M triblock terpolymer to a hexagonal morphology, where EB rings surround S cylinders in an M matrix after hydrogenation.

This morphological transition is induced by a change of the interfacial tensions between the middle block and the end blocks. While the interfacial tension between S and B is close to the one between B and M, the situation changes strongly for S and EB, and EB and M. This leads to a displacement of the spheres or cylinders at the lamellar interface in the S–B–M block copolymers and induces curvature into the interface between the outer blocks. This scenario is schematically shown for an ABC triblock terpolymer in Figure 10.

A similar observation was made for an S–B–M triblock terpolymer with spheres at the lamellar interface, where the B block was partly modified with different transition metal complexes [13]. In that case a morphological transition to a cocontinuous morphology or a hexagonal morphology was observed. Comparison of these findings with the morphological behavior of diblock copolymers leads to the conclusion that the stability region of different morphological classes (e.g., like the cylindrical morphology) may be extended to block copolymers with almost symmetric compositions by adjusting the relative interactions of a short middle block with respect to the end blocks.

A very interesting example of morphological change induced by hydrogenation of the B block in an S–B–M triblock terpolymer was obtained for symmetric systems with ca. 27% B. While the

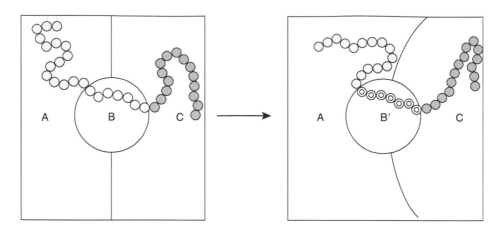

Figure 10 Scheme for the change of curvature of the intermaterial dividing surfaces by changing the relative interactions between the middle and the outer blocks in an ABC triblock terpolymer via chemical modification of B to B' ($\chi_{AB} = \chi_{BC}$, $\chi_{AB'} < \chi_{B'C}$).

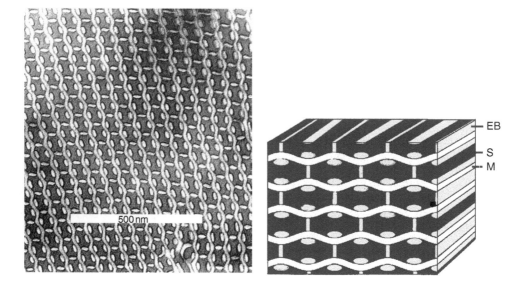

Figure 11 Polystyrene-block–poly(ethylene-co-butylene)-block–poly(methyl methacrylate) S–EB–M showing the "knitting pattern" morphology. (a) Transmission electron micrograph stained with RuO$_4$ (S appears dark, M and EB appear light). (b) Scheme. (From U. Breiner et al. *Macromol. Rapid Commun.* 17:567, 1996, Copyright 1996 Wiley-VCH Verlag GmbH, Weinheim.)

S–B–M triblock terpolymer shows a lamellar morphology, the analogous S–EB–M self-organizes into the so-called "knitting pattern" morphology [160–162] (Figure 11).

This is a nice example of a block copolymer morphology with a highly nonconstant mean curvature of the interfaces between the domains. This morphology is supposed to be located between the lamellar morphology where all blocks are localized in lamellae, and the morphology where the middle block forms a cylinder at the lamellar interface between the outer blocks (compare with Figure 5). A support for this assumption is the dependence of this morphology upon the casting solvent: while the knitting pattern morphology is obtained when casting the polymer film from chloroform, a lamellar morphology is obtained from toluene solution. In other samples with a larger volume fraction of the B block where lamellae were obtained, the choice of the solvent had no

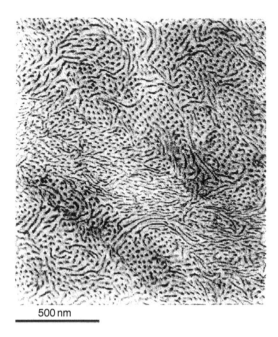

500 nm

Figure 12 Transmission electron micrograph of polybutadiene-block–polystyrene-block–poly(methyl metha-crylate) B–S–M forming the "banana" morphology. Due to staining with OsO$_4$ only the curved B cylinders are visible. (From K. Jung et al. *Macromolecules* 29:1076, 1996, Copyright 1996 American Chemical Society.)

influence on the morphology and thus a larger distance from a morphological transition within the phase diagram can be assumed.

Another interesting feature can be observed in triblock terpolymers with two short, incompatible outer blocks, which tend to form dispersed domains in the matrix formed by the middle block. Such a situation is found in a B–S–M triblock terpolymer when the volume fractions of both B and M are approximately 0.2, and S forms the matrix. Both B and M tend to form cylinders. However, due to the dissimilarity of the interfacial tensions between the end blocks and the S matrix, the interfacial areas between B/S and S/M are not the same. This leads to the formation of cylinders with different diameters and different long periods for both the lattices of B and M cylinders. Since such kind of simultaneous packing into different hexagonal or tetragonal lattices with different long periods is impossible, an irregular microphase separated "banana"-shaped morphology is obtained [152] (Figure 12).

That situation can be regarded as a mesoscale glass, generated by the competing tendencies within this block copolymer to form two cylindrical sublattices with different periodicity.

An interesting question relating AC diblock and ABC triblock terpolymers is the influence of the B block on the microphase separation between A and C. Annighöfer and Gronski [163,164], as well as Hashimoto et al. [165], reported on the morphological properties of ABC triblock terpolymers where B consisted of a random or tapered block of A and C. Kane and Spontak [130] found in their theoretical work that a random A/C middle block can enhance the mixing of the outer blocks due to an increase of the conformational entropy of the middle block. A similar result was obtained for symmetric ABC triblock terpolymers, where B forms either spheres, cylinders, or a lamella between the lamellae of the A and C blocks [166]. Erukhimovich et al. [167] studied the influence of a very short strongly incompatible C block on the ODT of an ABC and ACB block copolymer within the WSL. It was found that in both cases, for certain compositions and certain relative incompatibilities between C and the other two blocks, a stabilization of the disordered phase can occur as compared to the pure AB diblock copolymer.

While only the influence of a very short strongly interacting third block on the microphase separation of a diblock copolymer was theoretically investigated, Neumann et al. studied the influence of curvature between an incompatible C block on the phase behavior of the adjacent AB diblock copolymer. These studies were performed on four different poly(1,4-isoprene)-block–poly(1,2-butadiene)-block–polystyrene (I–B–S) [168] and their hydrogenated analogues, poly(ethylene-alt-propylene)-block–poly(ethyl ethylene)-block–polystyrene (EP–EE–S) [57,84]. In these block copolymers the volume fractions of the two elastomeric components I and B or EP and EE were similar and only the relative amount of polystyrene was changed with respect to the other two blocks. It is well known that blends of I and B are highly compatible [169] and thus the I–B diblock copolymer is disordered. In the I–B–S triblock terpolymers a phase behavior similar to that of a diblock copolymer was found, where polystyrene forms one microphase and the two elastomeric components together form the other microphase. In the EP–EE–S triblock terpolymers the behavior was different. While the EP–EE diblock precursors were found to show an ODT around room temperature (as shown by dynamic mechanical properties), the corresponding EP–EE–S triblock terpolymers showed a more complicated picture. While I–B–S with 16% of S shows a spherical morphology, the corresponding EP–EE–S shows a cylindrical morphology, which is orthorhombic, as evidenced by SAXS. I–B–S with 26% of S shows a hexagonally packed cylindrical morphology, and the corresponding EP–EE–S shows a cocontinuous morphology. These morphological transitions via hydrogenation indicate an increase of the incompatibility between the two elastomeric blocks and also a shift of the ODT of EP–EE toward higher temperatures, when being grafted on an incompatible surface (polystyrene domains).

3. Block Copolymers with More than Three Components

Also block copolymers with more than three components have been studied. In principle, such systems can show more complex structures than the ones presented so far. A first systematic study has been published for tetrablock quaterpolymers, that are linear block copolymers with four chemically different blocks. In this case core-shell shell matrix morphologies analogous to the spheres, cylinders, and double gyroid known for diblock copolymers and lamellae (compare with Figure 1) were found [170,171]. As a nice example, in Figure 13 core-shell shell cylinders are shown.

B. Star Copolymers

Polymers with a star-like topology have attracted interest for many years. The rheological behavior in the melt and in solution of starpolymers differs from the behavior of linear polymers [172]. Polystyrene starpolymers with selectively deuterated core or corona chains were investigated by SANS and it was found that the chains are more stretched within the core (or close to the branching point), while the outer parts of the chains follow the single chain behavior of linear polymers [173]. This result confirmed theoretical predictions by Daoud and Cotton [174] and Birshtein et al. [175]. A similar behavior was found for the chain conformations in star-like block copolymer ionomer micelles, which were studied by SANS, too [176].

There are different types of star copolymers. In one kind of these polymers, several block copolymers are connected at one of their chain ends (starblock copolymers). In other systems different homopolymers are connected at a chain end (miktoarm star copolymers). Some different situations are schematically shown in Table 1.

One of the basic questions is how the topological restriction of the different blocks by a common junction point influences the morphology. While in microphase separated binary miktoarm star copolymers the common junction points between different blocks are located on a common interface, in completely microphase separated ternary miktoarm star copolymers (miktoarm star terpolymers) an interfacial line might be expected rather than an interfacial surface.

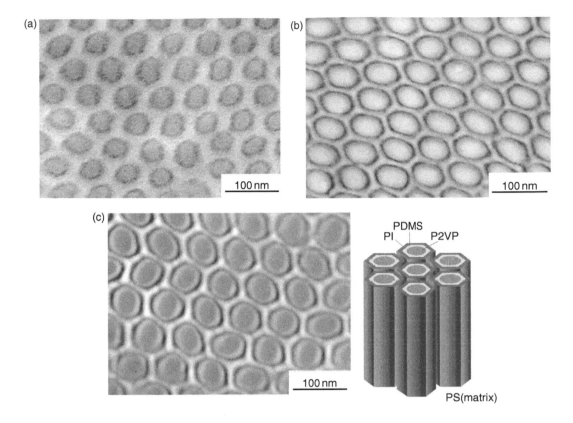

Figure 13 (a) Transmission electron micrograph of polystyrene-block–polyisoprene-block–polydimethylsiloxane-block–poly(2-vinylpyridin) without staining, (b) stained with OsO$_4$, and (c) stained first with OsO$_4$ and subsequently with CH$_3$I, (d) scheme. (From K. Takahashi et al. *Macromolecules* 35:4859, 2002, Copyright 2002 American Chemical Society.)

As in the case of linear block copolymers, we concentrate here on the phase behavior and morphologies of amorphous miktoarm star copolymers.

1. Miktoarm Star Copolymers

Erukhimovich [100] and Olvera de la Cruz and Sanchez [177] developed a theory for the ODT of AB$_2$-miktoarm star copolymers, which predicts that branched block copolymers have a larger disordered phase as compared with linear block copolymers of the same overall degree of polymerization and composition. The same result was obtained for linear (AB)$_n$ multiblock copolymers and A$_n$B$_n$ miktoarm star copolymers [177]. The phase diagrams of monodisperse miktoarm star copolymers were developed in the WSL by Dobrynin and Erukhimovich [178]. It was also found that for star copolymers with diblock copolymer arms the general topological behavior of the phase diagram is similar to that for linear diblock copolymers and reveals the same succession of the transitions disordered state — body-centered cubic phase — hexagonally packed cylindrical phase — lamellar phase [178]. For the miktoarm star copolymers with different homopolymer arms A$_n$B$_m$ this behavior prevails only if the arm numbers m and n are not too large ($m, n < 5$). In the opposite case the hexagonal and body-centered cubic morphologies were found to be less stable than the lamellar one for compositions close to symmetrical. This was explained by a rather specific behavior of the 4-point-correlations involving the monomers of many miktoarm star copolymers. Phase diagrams of star copolymers consisting of four diblock copolymers were also studied [179] for which a restricted

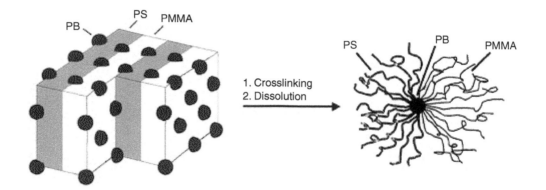

Figure 14 Schematic representation of the Janus micelles' synthesis (left-hand side: sketch of SBM ls-morphology). (From R. Erhardt et al. *Macromolecules* 34:1069, 2001, Copyright 2001 American Chemical Society.)

composition range of a stable gyroid phase was predicted via a WSL theoretical approach taking into account the higher harmonics contribution. The fluctuation corrections showed to affect mainly the body-centered cubic and hexagonal phases.

The first binary star copolymers were synthesized by Fetters group more than two decades ago [180]. The group of Hadjichristidis has done a lot of work in the field of star copolymers with various architectures [50–52,181].

A few studies on miktoarm star copolymers A_nB_m have been carried out in order to compare their morphological behavior with that of linear diblock copolymers having similar composition. Whereas similar morphology was found for $n = m$ [182,183], differences were found for $n \neq m$. The finding was attributed to the bending energy arising from the reduction of chain stretching of the component with more arms [184]. Hadjichristidis and coworkers [185] synthesized compositionally symmetric star copolymers of various types and found that a tricontinuous morphology could be induced by the chain architecture. This is in agreement with the results on compositionally similar AB, AB_2, or AB_3 block copolymers, where the increasing asymmetry of the numbers of chemically different blocks also leads to an increasing curvature of the interface between the different domains [49], thus shifting the system in the direction from lamellar toward spherical or even disordered morphology. Matsushita et al. [186] found a similar morphological behavior for 4- and 12-armed miktoarm star copolymers of polystyrene and polyisoprene as compared with linear diblock copolymers of the same composition. For a polystyrene–polyisobutylene–polystyrene triblock copolymer S–IB–S and a triarm starblock copolymer with IB core and S corona of the same composition the same morphology was also found [187]. A cocontinuous morphology in a star copolymer, where each arm consists of a diblock copolymer, is attributed to be stable in a wider composition range as compared with a linear diblock copolymer [188].

A special case of a miktoarm star copolymer with many arms are so-called Janus Micelles, which are formed by cross-linking the short middle block of a triblock terpolymer in the microphase separated bulk state, in which the center block self-assembles in spherical [189,190] or cylindrical domains [191]. By this procedure the two different outer blocks are oriented to the two opposite hemicoronas around the center block domain and subsequent dissolution leads to amphiphilic particles (Figure 14). While spherical Janus Micelles form superstructures in solution, the cylindrically shaped Janus Micelles seem to have a lower tendency of self-aggregation to higher superstructures.

2. Miktoarm Star Terpolymers

The first miktoarm star terpolymers containing three different arms were presented by Fujimoto et al. [54]. They used a special technique to combine three different blocks (*t*-butyl methacrylate, styrene, and dimethylsiloxane), which were all synthesized by living anionic polymerization. PDMS was synthesized by initiation with a functionalized 1,1-diphenyl ethylene (DPE). In a second step, living polystyrene anions were reacted with the DPE linked to the PDMS, before the t-butyl methacrylate was added to give the 3-miktoarm star terpolymer. Several years later the morphological properties of these star copolymers were published as well. TEM and SAXS support a threefold symmetry where the junction points between the three incompatible blocks are confined on one line [192].

Iatrou and Hadjichristidis synthesized a 3-miktoarm star terpolymer containing polystyrene, polyisoprene, and polybutadiene by terminating these living polyanions in a sequential mode by a trichlorosilane. In these systems, the microphase separation seems mainly to occur between polystyrene and the two elastomers [193]. Stadler's group investigated a 3-miktoarm star terpolymer of styrene, butadiene, and methyl methacrylate, where three glass transition temperatures were found indicating three different microphases [55]. Due to the asymmetric composition a hexagonal morphology was found, where the polybutadiene block forms the cylinders in a PMMA matrix (unpublished). Most likely the PS block forms a shell around the PB cylinders, due to a reduction of interfacial energy.

Hadjichristidis' group worked later on miktoarm star terpolymers of polyisoprene, polystyrene, and poly(methyl methacrylate), where they also investigated an asymmetric system. They found a cylindrical morphology with a nonconstant mean curvature, that is, the cylinders were deformed to rhombohedral structures [194,195]. In symmetric systems they found threefold morphologies [196], similar to the results by the group of Hashimoto [192] (Figure 15(a,b)). It is interesting to note that some of their electron micrographs exhibit features similar to those of a linear polystyrene-block–poly((4-vinylbenzyl)dimethylamine)-block–polyisoprene triblock terpolymer with almost equal amounts of the three components, when cast from benzene [68] (Figure 15(c,d)). Also a 4-miktoarm star quaterpolymer consisting of polystyrene, polyisoprene, polybutadiene, and poly(4-methylstyrene) was reported, but no morphological characterization was given [52].

V. MIXTURES OF BLOCK COPOLYMERS WITH LOW MOLECULAR WEIGHT SOLVENTS

Besides the investigations on pure block copolymers in the bulk state, also work on block copolymers in solution has been published. Rather closely related to the phase and rheological behavior of block copolymers are the corresponding properties of low molecular weight surfactants in solution [197–199]. Most works deal with the phase behavior of amphiphilic block copolymers (micellization) [200–208] and the rheological behavior [209–211]. The reason for the large interest in this direction is that amphiphilic molecules play a fundamental role in biology and find also widespread technological applications because of their unique ability to self-organize at interfaces, which leads to modification of interfacial properties and enhances the compatibility between two phases. Amphiphiles are also the subject of other chapters in this volume.

Figure 16 shows schematically various morphologies observed in micellar solutions of amphiphilic diblock copolymers in a water–oil system. The micelles self-assemble into similar morphologies as they are observed in the bulk state of amorphous diblock copolymers. While in the so-called open micelles single macromolecules can leave the micelle to the solution and vice versa, in the so-called frozen micelles the structure is not dynamic, for example, due to a glass transition temperature of the core above the system's temperature.

Systematic investigations on the phase behavior of polystyrene–*block*–polyisoprene (S–I) diblock copolymers in solution have been carried out by Lodge's group. The swelling of the blocks depends on the selectivity of the solvent. A nonselective solvent swells the different blocks in a similar way, which leads to an effective reduction of the segmental repulsion between incompatible blocks, that is, the ODT is occurring at lower temperatures with increasing dilution [212]. The polymer concentration dependence of $\chi_{ODT}(\chi_{ODT} \sim \phi^{-1.3-1.6})$ does not follow a simple dilution law, indicating a stabilization of the disordered state by fluctuations. OOT, on the other hand, were found to follow

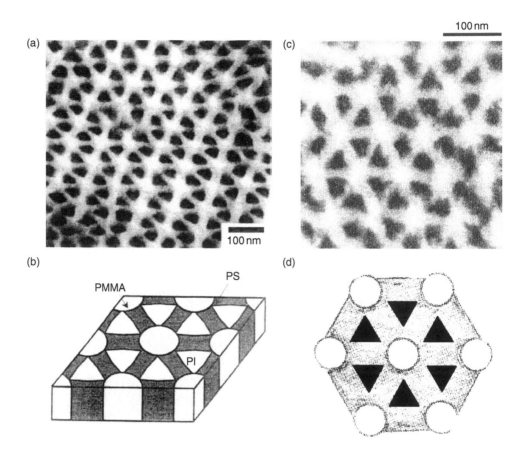

Figure 15   (a) Transmission electron micrograph of a polyisoprene-arm–polystyrene-arm–poly(methyl methacrylate) miktoarm star terpolymer stained with OsO$_4$, (b) scheme of the threefold symmetry. (From
S. Sioula et al. *Macromolecules* 31:8429, 1998, Copyright 1998 American Chemical Society.),
(c) transmission electron micrograph of a polystyrene-block–poly((4-vinylbenzyl)dimethylamine)-
block–polyisoprene, (d) scheme of the threefold symmetry. (From Y. Matsushita et al. *Macromolecules*
16:10, 1983, Copyright 1983 American Chemical Society.)

a simple dilution law ($\chi_{OOT} \sim \phi^{-1}$). Using selective solvents, the degree of swelling differs for
the different blocks and as a result of this the morphology might change [213]. Figure 17 shows
the phase diagrams of an asymmetric S–I diblock copolymer in different solvents, as they were
determined by SAXS. All phase transitions were found to be reversible and showed some hysteresis
on temperature scans.

Below a certain concentration the block copolymers will form micellar aggregates in a selective
solvent and upon further dilution the block copolymer chains will dissolve completely.

The sheet-like double layers in Figure 16(a) form vesicles and it has been speculated if
these vesicles are equilibrium structures, or not. Luo and Eisenberg [207] gave evidence for the
equilibrium nature of vesicles by performing a sequence of dilution experiments of polystyrene-
block–poly(methacrylic acid) in THF/Dioxane and water. The glass transition temperature was
always below the temperature of the experiment, and depending on the amount of water the size of
the micelles changed reversibly, thus indicating a thermodynamic property rather than a kinetic one
(Figure 18). In the same paper, these authors presented images of vesicles at various stages of fusion
or fission, which display some similarity to biological cells (Figure 19).

Also triblock terpolymers have been studied in solution by several groups [214–216]. Gohy et al.
described a system with the sequence of polystyrene, poly(4-vinylpyridine), poly(ethylene oxide),

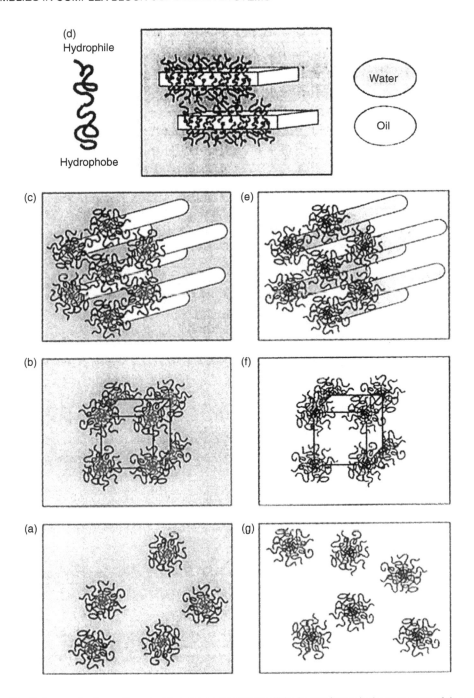

Figure 16   Schemes for the self-assemblies of an amphiphilic diblock copolymer in the presence of the solvents water and oil selective for the two blocks. (a) Micellar solution, (b) micellar cubic lyotropic liquid crystal (LLC), (c) hexagonal LLC, (d) lamellar LLC, (e) reverse hexagonal LLC, (f) reverse cubic LLC, (g) reverse micellar solution. (From P. Alexandridis et al. *Langmuir* 13:23, 1997, Copyright 1997 American Chemical Society.)

in which micelles are formed with a polystyrene core. Bieringer et al. and Liu et al. described systems with a sequence of a polyamine, polystyrene, and poylacid. These latter systems show inversions of micelles or vesicles dependent on the pH-value of the solution (Figure 20). This could be a nice way to carry and release guest molecules from one area to another with different pH-value.

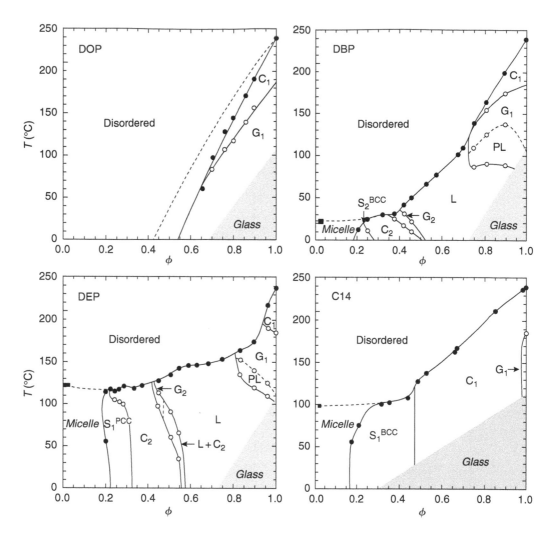

Figure 17   Phase diagram for an S–I diblock copolymer ($M_n = 33$ kg/mol, 31 vol % PS) as a function of temperature (T) and polymer volume fraction (ϕ) for solutions in dioctyl phthalate (DOP), di-n-butyl phthalate (DBP), diethyl phthalate (DEP), and n-tetradecane (C14). Filled and open circles identify ODTs and OOTs, respectively. The dilute solution critical micelle temperature (cmt) is indicated by a filled square. The ordered phases are denoted by: C, hexagonal-packed cylinders; G, gyroid; PL, perforated lamellae; L, lamellae; S, cubic packed spheres. The subscript 1 identifies the phase as "normal" (PS chains reside in the minor domains) or "inverted" (PS chains located in the major domains). The phase boundaries are drawn as a guide to the eye, except for DOP in which the OOT and ODT phase boundaries (solid lines) show the previously determined scaling of the SI interaction parameter ($\chi_{ODT} \sim \phi^{-1.4}$ and $\chi_{ODT} \sim \phi^{-1}$); the dashed line corresponds to the "dilution approximation" ($\chi_{ODT} \sim \phi^{-1}$). (From K.J. Hanley et al. *Macromolecules* 33:5918, 2000, Copyright 2000 American Chemical Society.)

VI.   BLENDS OF BLOCK COPOLYMERS

A.   Block Copolymers as Compatibilizers in Polymer Blends

Most blends of different polymers are incompatible, that is, do not form a homogeneous phase. This is due to the usually repulsive enthalpic interactions between the different species, which can easily balance the entropy of mixing, which is small as compared to low molecular weight materials. Thus

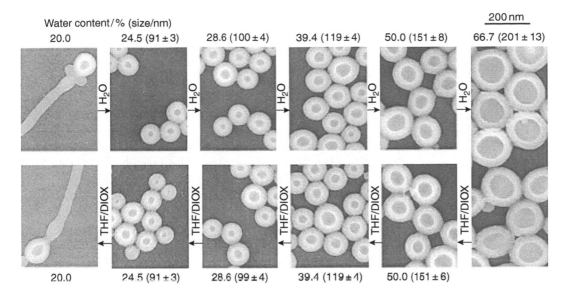

Figure 18 Reversibility of vesicle sizes in response to increasing or decreasing water contents for PS$_{300}$–b–PAA$_{44}$ vesicles in a THF/Dioxane (44.4/55.6) solvent mixture. (From L. Luo and A. Eisenberg. *Langmuir* 17:6804, 2001, Copyright 2001 American Chemical Society.)

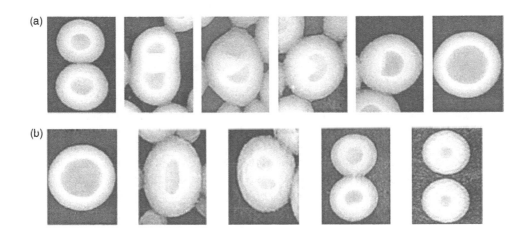

Figure 19 Possible mechanisms of (a) fusion of vesicles and (b) fission of a vesicle. (From L. Luo and A. Eisenberg. *Langmuir* 17:6804, 2001, Copyright 2001 American Chemical Society.)

Figure 20 Schematic of the inversion of vesicles from PAA outside to P4VP outside. The charges are on the chains, not counterions. (From F. Liu and A. Eisenberg. *J. Am. Chem. Soc.* 125:15059, 2003, Copyright 2003 American Chemical Society). (a) Vesicles with PAA outside (before inversion), (b) vesicles with both PAA and P4VP outside (after 2 h), and (c) vesicles with P4VP outside (after 8 h).

macrophase separation occurs with a minimization of the interfacial area between the components, if the system is in thermodynamic equilibrium (see Chapter 2). Reduction of the interfacial tension between the components may reduce the extent of macrophase separation and this can be achieved by addition of compatibilizers such as block copolymers. Adding an AB diblock copolymer to a blend of A and B will not necessarily lead to an increase of the internal surface between A and B, because the diblock copolymer can also self-aggregate into micelles inside the A and B phases. In fact, in these cases there are no enthalpic attracting forces between the block copolymer and the other blend components, thus there is hardly any driving force for the block copolymer to localize itself at the interface between A and B. A way out of this problem is to synthesize *in situ* during processing graft block copolymers at the interface between A and B. A very nice example for this strategy is the formation of a cocontinuous morphology in a blend of polyethylene and polyamide [217].

Stadler and coworkers used S–M diblock copolymers to compatibilize blends of poly(styrene-stat-acrylonitrile) (SAN) and poly(2,6-dimethyl phenylene ether) (PPE). In this case, there are enthalpic attractive forces between S and PPE on one side, and between M and SAN on the other side. It could be shown by TEM-investigations that in fact the domain sizes in these blends are much smaller as compared with the SAN/PPE-blend without S–M diblock copolymer [218]. To improve the mechanical properties of the blend, the S–M diblock copolymer was replaced by S–EB–M triblock terpolymers, where the poly(ethylene-co-butylene) EB middle block avoids fracture at the domain boundaries between PPE and SAN [12].

Balsara et al. [219] studied by SANS the thermodynamic behavior of a blend of AB-diblock with A and B homopolymers in the homogeneous disordered melt. They found the random phase approximation for multicomponent systems [220,221] to work for their systems of polyolefines, where only van-der-Waals interactions are present.

B. Blends of Block Copolymers with a Homopolymer

The work on blends of block copolymers with an emphasis on blends with triblock terpolymers has been reviewed a few years ago [148]. In this section, we concentrate on the question of how to control the morphology of microphase separated block copolymers by adding a homopolymer or another block copolymer. Some of the work on blends of diblock copolymers with a homopolymer being chemically identical to one of the blocks was motivated by the investigation of the stability range of the ordered bicontinuous phase between the cylindrical and lamellar phase. Winey et al. [222] investigated S–I and S–B diblock copolymers, which were blended with corresponding homopolymers. The stability window of the cocontinuous morphology (in their paper still erroneously named OBDD) was found to be comparable to the stability window of the diblock copolymers at a corresponding overall composition. Macrophase separation was shown to occur when the homopolymer was of larger size than the corresponding block in the copolymer. Macrophase separation was also observed when a large amount of the homopolymer corresponding to the shorter block was added. Koberstein et al. blended a short poly(ethylene-co-butylene) with different degrees of deuteration into an S–EB–S triblock copolymer and studied the location of the homopolymer within the midblock domain with SANS and SAXS. They found an enrichment of the homopolymer in the center of the midblock domain, where the triblock most likely has more loop chains, and a uniform distribution within the midblock domain, and where the midblock mostly forms tie-chains between neighbored PS-domains [223]. Hashimoto et al. investigated blends of a lamellar starblock copolymer (S–B)$_4$, with a linear polystyrene having a smaller molecular weight as compared to the S-blocks in the starblock copolymer. They found two new morphologies with mesh and strut topology [224] that are having hyperbolic interfaces and are similar to perforated lamellae and cocontinuous networks, respectively. Hashimoto et al. also investigated blends of S–I with PPE [225]. Using an S–I diblock copolymer forming I-spheres, the addition of PPE (which homogeneously mixes with the S corona) does not change the morphology and no macrophase separation occurs from solution in toluene.

Blends of a lamellar S–I diblock copolymer with PPE, however, show besides microphase separation of the I also a long wavelength fluctuation of the morphology, indicating a macroscopic demixing into regions with variable ratios of the two polymers. Xie et al. studied S–B–S triblock copolymers with S cylinders, which they blended with poly(vinyl methylether) (PVME). The PVME mixed only with the S domains and a change from cylindrical to a cocontinuous morphology was found [226].

Blending of an ABC triblock terpolymer of S–I–VP (showing the honeycomb-like core-shell morphology with a nonconstant mean curvature of the interface between I and S earlier discussed by Gido et al. [143]) with linear polystyrene leads to a morphology with a constant mean curvature, that is, a core-shell cylindrical morphology [227]. Thus the added homopolymer is preferentially located in the corners of the former honeycomb structure and by this allows all the block copolymer chains to be similarly stretched. If the degree of polymerization of the linear polystyrene is too large, however, macrophase separation occurs because the free polystyrene chains cannot swell the S-corona of the triblock terpolymer (dry brush regime). This is due to the competition of mixing entropy (favoring the swelling of the brushes) and additional stretching of the tethered blocks by swelling (reduction of conformational entropy and as a consequence of this, favoring expulsion of homopolymers from the brushes).

C. Blends of Two Block Copolymers

Ways to control morphology of block copolymers have also been investigated via blending of different block copolymers. Hadziioannou and Skoulios investigated various blends of linear S–I, S–I–S, and I–S–I block copolymers. They found mixing at a molecular level and observed a dependence of the morphology on the overall chemical composition, rather on chain architecture [228]. Hashimoto et al. [229] investigated blends of two S–I diblock copolymers with different molecular weights, but mainly symmetric compositions. When the molecular weights differed by more than a factor of 10, macrophase separation into partly mixed phases of long and short chains was observed. The domains with larger chains contained a larger amount of small chains, and vice versa. It was found that the presence of one asymmetric composed block copolymer influences the interface curvature, thus leading to morphologies different from the lamellar one for the mixed system. Mayes et al. studied thin films of blends containing symmetric long and short S–M diblock copolymers by neutron reflectometry. They found the short chains at the lamellar interface, while the segments of the larger chains also fill the center regions of the lamellar domains [79]. In macroscopically demixed lamellar phases, the phase boundaries between the lamellae with larger and smaller long periods were investigated and the possibility of a macrophase separation initiated by microphase separation was discussed [230]. Spontak et al. [231] studied a blend of a symmetric S–I diblock copolymer and a (S–I)₄ multiblock copolymer with the same overall molecular weight by electron tomography and found macroscopic phase separation into two different lamellar phases. Vilesov et al. [232] mixed two antisymmetric cylindrical S–B diblock copolymers with each other and obtained a lamellar phase. Schulz et al. [233] did a similar experiment by blending a cylindrical and a lamellar S–VP diblock copolymer to obtain a gyroid morphology. Spontak et al. [234] used S–I diblock copolymers of various composition but similar molecular weight to tune the morphology by changing the relative amounts of the two block copolymers. Sakurai et al. [235,236] studied a similar system and in addition they found an OOT between the lamellar and the gyroid phase in a block copolymer blend upon increasing temperature.

The mixing behavior of two diblock copolymers with different relative composition has been treated theoretically. Sakurai and Nomura analyzed the phase behavior of diblock mixtures using the random phase approximation. They found the disordered phase to be suppressed in mixed diblock blends and a tendency to macrophase separation for such blends when the composition of the involved blocks becomes very asymmetric. They conclude that there are limitations in the comparison between blends of diblock copolymers with a diblock copolymer having the same overall composition, since the χ-parameter determined for the blend may differ from the one determined for a single block copolymer [237]. Shi and Noolandi [238] developed phase diagrams for different ratios of two diblock copolymers and found regions, where single mixed phases should be stable, besides regions where two different phases should coexist. A similar result was obtained by Matsen and Bates [239], who found a considerably large two phase region. Birshtein et al. [232,240] and Borovinskii and Khokhlov [241] had also studied blends of diblock copolymers in the SSL.

In the last part of this section results on blends of ABC with BC, or ABC with AC block copolymers will be presented. As mentioned before, the way of preparing the blend may have a great influence on the final morphology (choice of casting solvent, casting speed, etc.). When blending ABC with BC block copolymers with equal lengths of all blocks, different situations can occur, which are schematically shown in Figure 21.

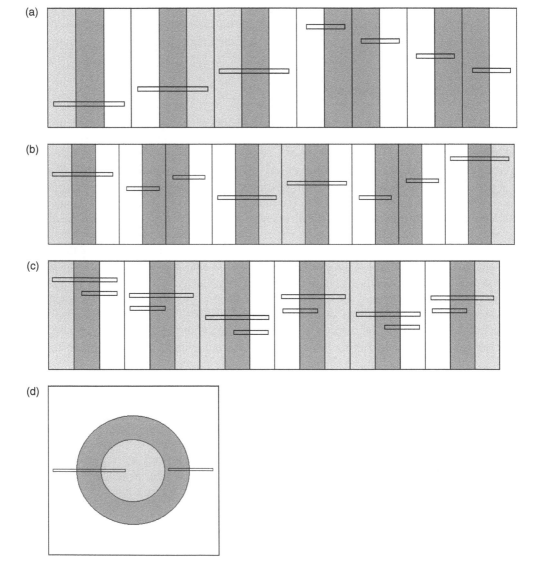

Figure 21   Schemes for blends of lamellar ABC and BC block copolymers (A: gray, B: dark, C: white). (a) Macro-phase separation between BC diblock copolymer and ABC triblock terpolymer, (b) centrosymmetric double layers of BC and ABC, (c) centrosymmetric mixed layers of BC and ABC, (d) nonlamellar superstructure of BC and ABC.

Besides a macrophase separation between the two block copolymers, also blends with the sequence ABC CB BC CBA can occur (a centrosymmetric structure of double layers of both diblock and triblock terpolymers). Another possibility is a kind of random sequence between BC and ABC block copolymers, which will occur when the C blocks of both diblock and triblock terpolymer do not show any preferential mixing with either C block. In this case, an aperiodic super-structure will be obtained. Another possible superstructure is the incorporation of the BC diblock with the same molecular orientation into the ABC structure, which will lead to a real effective increase of the volume fractions of both C and B with respect to A. Thus, a lamellar superstructure may only be expected for small volume fractions of diblock copolymer. For larger amounts of diblock chains a lamellar superstructure will be disfavored with respect to a superstructure with curved intermaterial dividing surfaces, such as cocontinuous, cylindrical, or spherical morphologies. Experiments on blends of lamellar S–B–M with lamellar B–M [242], and also of lamellar S–B–T with lamellar B–T, when all blocks were of about the same length proved the existence of the last case in Figure 22.

Depending on the blend ratio either a core-shell cylindrical morphology (Figure 22(a,c)) or a core-shell cocontinuous morphology (Figure 22(b,d)) were obtained for both blend systems. The driving

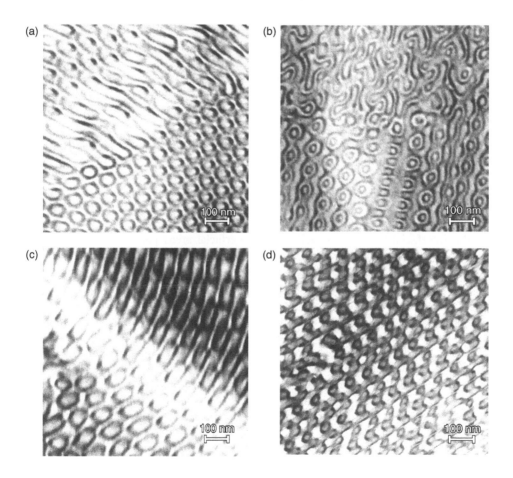

Figure 22 Transmission electron micrographs (stained with OsO$_4$). Blends of polystyrene-block–polybutadiene-block–poly(methyl methacrylate) S–B–M with polybutadiene-block–poly(methyl methacrylate) B–M. (a) 40 mol% S–B–M: 60 mol% B–M, (b) 70 mol% S–B–M: 30 mol% B–M. Blends of polystyrene-block–polybutadiene-block–poly(tert.butyl methacrylate) S–B–T with polybutadiene-block–poly(tert.butyl methacrylate) B–T. (c) 40 mol% S–B–T: 60 mol% B–T, (d) 70 mol% S–B–T: 30 mol% B–T.

force for the formation of this kind of blends is of entropic origin. Due to the same chemical nature of the mixing blocks there is no favored enthalpic interaction in this kind of blends. In recent theoretical work it could be shown that the swelling of the AB domains of an ABC triblock with an AB diblock copolymer can reduce the stretching of the blocks, in addition to the gain of mixing entropy for the system [243].

In contrast to the blends of S–B–M with B–M, blends of S–B–M and S–B with similar block lengths macrophase separate. The reason for this behavior is a stronger incompatibility between B and M as compared to S and B: it was found theoretically that blends of ABC with BC having similar block lengths form common superlattices only if the incompatibility between B and C is larger than between A and B, that is, if the diblock copolymer has the same interfacial tension as the more incompatible interface of the triblock terpolymer [243]. As in the case of homopolymer blends, lowering of molecular weight of one block can enhance the formation of common superlattices of two block copolymers. In the case of S–B–M and S–B the corresponding blends with an S–B diblock copolymer having a shorter B-block form common superlattices [148].

In blends of ABC with AB block copolymers a mixing at the molecular level of both blocks of the diblock copolymer with the corresponding blocks of the triblock terpolymer leads to a centrosymmetric superstructure. However, in the case of a blend of ABC and AC block copolymers a noncentrosymmetric structure is obtained when all diblock and triblock terpolymer molecules prefer to form common domains with the corresponding block of the other species. Before discussing this situation in some detail, other general possibilities for blend formations need to be considered. Besides the trivial case of macrophase separation and the random sequence of ABC and AC block copolymers due to equal energetic situations between the different A blocks and C blocks, two different centrosymmetric double-layered structures are also possible.

Either the A blocks prefer the formation of common microdomains with A blocks of the other species, while the corresponding C blocks tend to phase separate from each other, or vice versa. The different possibilities are shown in Figure 23. The formation of mixed domains of different A blocks or C blocks is due to an entropic gain caused by the reduction of chain stretching for one species. This leads to a depression of the overall free energy as compared to the macrophase separated state (where only similar blocks from the same species form common microdomains). In fact, in ABC triblock terpolymers there are AB and BC interfaces, while in AC diblock copolymers there is an AC interface.

Thus the interfacial tensions as well as the interfacial areas per chain are different, resulting in a different degree of chain stretching. It is known for the bulk state that the free chain ends interdiffuse with each other, causing a certain interpenetration between blocks from opposite layers (these blocks are having one free chain end and are connected with the other end to an interface; they can be thus considered as polymer brushes). Note that this situation differs from corona chains of diblock copolymer micelles in good solvents, which do not interdiffuse with similar chains of other micelles. The interpenetration zone will be symmetric, when the end blocks coming from opposite sides belong to similar block copolymers, and asymmetric when end blocks from AC and ABC block copolymers form common microdomains. Leibler et al. [244] have shown theoretically that a free energy gain results from the formation of this noncentrosymmetric layering.

Figure 24(a) shows a transmission electron micrograph of a blend of S–B–T and S–T block copolymers. The different sequences of colors on the left and right part of that picture indicate opposite polarities. A special proof for the existence of the noncentrosymmetric superlattice is the defect in the middle of the TEM micrograph, also shown in Figure 24(b). The system avoids additional interfaces between S and T domains by breaking up the B lamellae. This superlattice shows for the first time the spontaneous formation of a noncentrosymmetric lamellar morphology with a periodicity length scale in the range of 60 nm [245]. The formation of this blend demonstrates in a nice way that superstructures with new symmetries can be obtained by self-assembling of block copolymers in a blend.

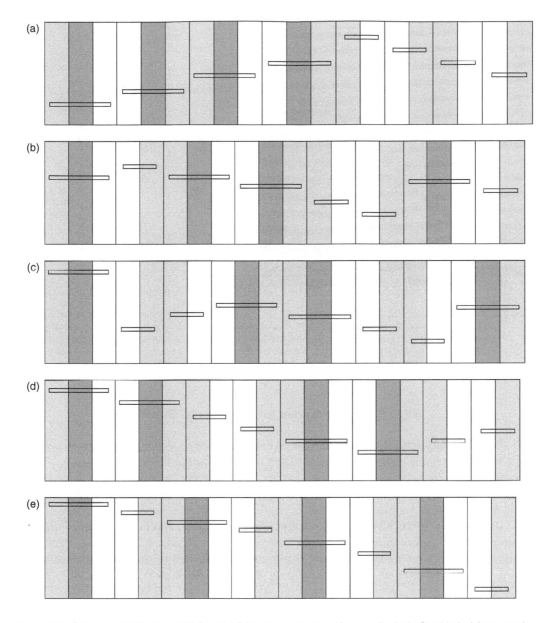

Figure 23 Schemes for blends of ABC and AC block copolymers (A: gray, B: dark, C: white). (a) macrophase separation, (b) random sequence of layers of diblock and triblock terpolymers, (c) and (d) centrosymmetric double layers of diblock and triblock terpolymers, (e) noncentrosymmetric array of alternating layers of diblock and triblock terpolymers.

Also blends of two triblock terpolymers have been studied, in which just the middle block differs in length. While in the case of two S–B–T triblock terpolymers a noncentrosymmetric superlattice was found [148], a blend of two S–B–M triblock terpolymers showed a knitting pattern morphology similar to the one shown in Figure 11 [246].

Stimulated by the finding of noncentrosymmetric superlattices in block copolymer blends also the possibility to generate this morphology by a pure A–B–C–A tetrablock terpolymer was investigated and found [247]. Here the difference between the interfacial tensions between A, B and A, C might be the major controlling parameter for the formation of this morphology.

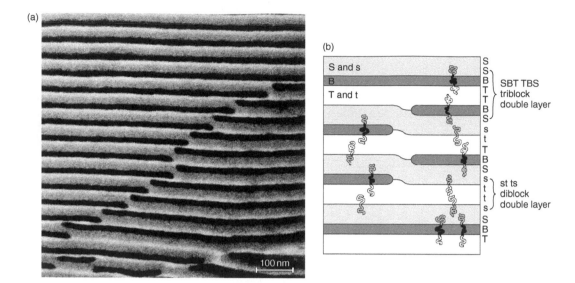

Figure 24   Periodic noncentrosymmetric lamellar superstructure of a blend of polystyrene-block–polybutadiene-block–poly(tert.butyl methacrylate) S–B–T with polystyrene-block–poly(tert.butyl methacrylate) S–T with the composition 50 mol% S–B–T: 50 mol% S–T. (a) Transmission electron micrograph (stained with OsO₄), (b) scheme of the characteristic defect proving the periodic noncentrosymmetry. (From T. Goldacker et al. *Nature* 398:137, 1999, Copyright 1999 Macmillan Magazines Ltd.)

VII. BLOCK COPOLYMERS WITH CRYSTALLINE BLOCKS

So far we discussed systems with amorphous block copolymers only. A second level of self-assembly is the formation of liquid crystals or crystals within microdomains of block copolymers.

For linear *semicrystalline* block copolymers the reader is referred to experimental works carried out on binary [248–254] and ternary [255–261] block copolymers where the kinetics of crystallization and the influence of morphology on crystallization (and vice versa) has been investigated. The crystallization process can disturb an already formed microphase separated structure, inhibit the formation of self-organized microphases on larger length scales, or induce morphological transitions. Crystallization also can be suppressed and larger undercoolings are required in dispersed microdomains (spheres) in comparison to continuous crystallizable domains (interconnected cylinders, lamellae, matrix-forming blocks). Nojima et al. [251] investigated a diblock copolymer of polybutadiene and polycaprolactone, which microphase separated above the crystallization temperature of polycaprolactone. Lowering the temperature leads to a change from spherical polycaprolactone domains into lamellar ones upon crystallization of the latter. Cross-linking of the polybutadiene matrix above the crystallization temperature suppresses a morphological transition from spheres to lamellae and only a deformation of the spheres upon crystallization of polycaprolactone could be observed. Balsamo et al. [258] studied a linear triblock copolymer of polystyrene, polybutadiene, and polycaprolactone (S–B–Cl) with a polystyrene matrix and a short polycaprolactone block (16 wt.%). Here the crystalline polycaprolactone block self-organizes into cylindrical domains, surrounded by a polybutadiene shell. Due to the crystallinity of the polycaprolactone block, the cylinders displays a noncircular cross-section. Figure 25 shows this morphology. Very interesting are the mechanical properties of this material: most likely due to the crystalline nature of the core-block, the whole material behaves ductile and can be extended to large strains (~900%) in a tensile testing experiment at room temperature, although the polystyrene matrix is glassy. It was also found that the annealing time at high temperatures can have an influence on the final morphology in S–B–Cl triblock terpolymers [259].

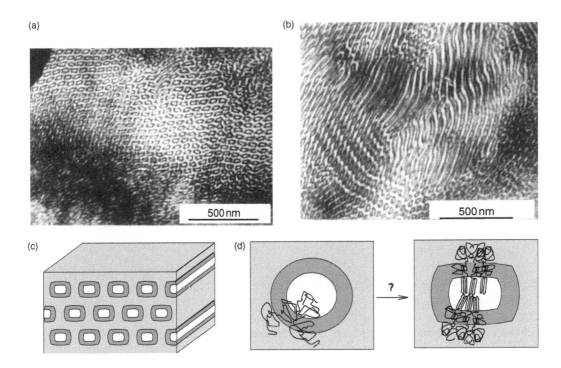

Figure 25 Transmission electron micrographs of polystyrene-block–polybutadiene-block–polycaprolactone (S–B–Cl), stained with OsO$_4$. B appears dark, Cl appears light, and S appears gray: (a) top view onto the cylindrical domains; (b) side view onto the cylindrical domains; (c) scheme of the morphology; (d) scheme of the distortion of round cylinders into cylinders with edges by the crystallization of the Cl-core. (From V. Balsamo et al. *Macromolecules* 32:1226, 1999, Copyright 1999 American Chemical Society.)

Crystallization in star copolymers was recently investigated by Floudas et al. [262]. They studied systems with two crystallizable blocks and could show that depending on the thermal treatment the crystallization of one block can be largely suppressed in favor of the other one.

Theoretical treatments of the problem of crystallization in microdomains have been presented by DiMarzio et al. [263] and Noolandi et al. [264].

The first work on block copolymers with a block capable of forming *liquid-crystalline* phases was published by the group of Gronski [265,266]. Möller's group investigated block copolymers containing mesomorphous polysiloxane blocks in the bulk [267] and the groups of Ober and Thomas [268] investigated the influence of the block composition on the phase transition between liquid-crystalline and isotropic state in a rigid/coil block copolymer. A theoretical description of the phase diagram of diblock copolymers consisting of a liquid-crystalline and an amorphous block was given by Williams and Halperin [269]. The same authors also investigated the mechanical properties of ABA triblock copolymers with glassy A and a B block capable of forming liquid-crystalline phases [270,271]. More rod–coil block copolymers are described in detail in Chapter 12.

VIII. BLOCK COPOLYMERS IN EXTERNAL FIELDS

In this section, we will present some works dealing with the influence of mechanical or electrical fields on block copolymers. Upon the mechanical fields the influence of hydrostatic pressure on the ODT has been investigated, or the influence of a directional force like an extensional or shear force on the orientation of block copolymer domains was investigated.

A. Pressure Dependent Phase Behavior

Earlier we have seen that the incompressibility condition was used for the theoretical description of the temperature-dependent phase behavior. In this section, however, some experimental results will be presented, showing that block copolymers are to some extent compressible, that is, their phase behavior can be manipulated by the application of pressure. So far not many studies on this problem have been carried out. Some authors reported on pressure-induced miscibility of diblock copolymers, both for systems with an upper critical [272–274] and lower critical disordering temperature [275]. For some systems with an UCDT it was shown that a minimum of the ODT occurs upon increasing pressures [276,277]. Stühn explained this effect by the creation of free volume in the mixing process, forming the disordered state at lower pressures. At higher pressures the larger compressibility of the ordered state reduces the mixing tendency and thus leads to the microphase separated state.

B. Orientation by Shear

One basic goal in polymer science is to relate macroscopic properties to the molecular structure or to the morphology. In the case of microphase separated block copolymers it is necessary to use macroscopically aligned systems, that is, single crystals without defects. Morphological defects might otherwise govern the macroscopic properties and prevent insight into the inherent anisotropic properties of the morphology. In this part, we report some important results for structuring of block copolymers by external shear fields. The group of Thomas developed a method called "roll casting," which enables preparation of films with a size of several square centimeters by a slow solvent evaporation while the solution is subjected to a shear field between two rotating cylinders [278]. In a study comparing different methods for macroscopic alignment of block copolymers, roll casting proved to be most effective [279]. However, this method must be optimized for each block copolymer system by using the right solvent, rotation speeds of the cylinders, gap width between the cylinders, etc.

Block copolymers subjected to large shear fields in the bulk state have been of interest since the first studies by Keller et al. [280]. They oriented a cylindrical S–B–S triblock terpolymer of polystyrene (S) and polybutadiene (B) by extrusion. The cylindrical domains of the polystyrene blocks were found to orient parallel to the direction of shear flow. Then several years passed before the influence of shear on the alignment and phase behavior of block copolymers became the subject of further scientific investigations. The experimental and theoretical works on diblock copolymers subjected to shear have been reviewed recently [96,97].

Using large amplitude oscillatory shear (LAOS), Koppi et al. [281] investigated the orientation behavior of lamellar poly(ethylene-alt-propylene)–block–poly(ethyl ethylene) EP–EE diblock copolymers and found at lower frequencies a parallel alignment of the lamellae to the shear field, while at larger frequencies the lamellae and their normals orient perpendicular to the shear field. Winey et al., Riise et al., and Chen et al. [282–284] made similar experiments on lamellar S–I diblock copolymers and found just the opposite behavior. Wiesner investigated the influence of LAOS as a function of shear amplitude and frequency on the orientation behavior of lamellar S–I diblock copolymers of various molecular weights and confirmed that both the above-described situations may occur [285]. At low strain amplitudes parallel alignment seems to be favored all over the frequency regime. However, at larger strains an intermediate frequency regime exists for the perpendicular orientation. Kornfields group investigated the time resolved strain birefringence of S–I diblock copolymers in an oscillatory shear field at various strain amplitudes and frequencies. They studied the morphology after various times of the orientation process by SAXS and TEM [286] and found at a low frequency (1 rad/sec) a perpendicular alignment of the lamellae and a monotonic increase of the birefringence (i.e., mainly form birefringence). At a larger frequency (4 rad/sec) the form birefringence went through a maximum and at a very large frequency (100 rad/sec) it went through a minimum. Under these conditions also transient perpendicular (4 rad/sec) and transversal

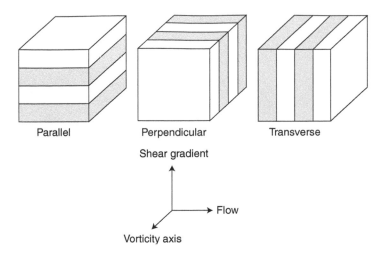

Figure 26 Scheme of parallel, perpendicular, and transverse orientations of a lamellar diblock copolymer in a shear field.

(100 rad/sec) orientations were found on the pathway to a parallel alignment of the lamellae to the shear flow direction. Figure 26 shows the different orientations with respect to the shear field.

It was pointed out in their work that morphological defects play an important role during the orientation process [284]. While S–I diblock copolymers showed different orientations under LAOS for various frequencies, a lamellar S–I–S triblock copolymer showed only perpendicular orientation [283]. The perpendicular orientation was found to be only a transient state for EP–EE–EP triblock copolymers [287]. The perpendicular orientation was also found in blends of a triblock and a pentablock copolymer [288] indicating the important contribution of bridging blocks through adjacent lamellae on the orientation behavior. In the case of triblock with higher viscous outer blocks and more pronounced in pentablock copolymers the sliding of adjacent lamellae, as possible in diblock copolymers, becomes hindered due to the anchoring by bridging blocks. This prevents the system from a parallel orientation to the shear field. The perpendicular orientation was also found in an S–I–M triblock terpolymer system [289]. Moreover, the ODT itself is influenced by shear. In some triblock copolymers, like cylindrical S–I–S [290] or lamellar EP–EE–EP [287], shear stabilizes the disordered state, in other systems like cylindrical S–B–S it stabilizes the ordered phase [291]. Jackson and coworkers also found indications of a martensitic-like transition at very high shear rates for a cylindrical S–B–S triblock copolymer [291,292]. Shear fields were also used to generate liposome-like superstructures in a blend of S–I–M triblock terpolymer with PMMA by Jérôme's group [293]. Thereby the molecular weight of the PMMA homopolymer equals the corresponding molecular weight of the PMMA block in the triblock terpolymer, that is, the system is in the dry brush regime (Figure 27).

C. Orientation by Electric Field

The application of an electric field is another way to macroscopically align microphase separated block copolymers. While in the bulk state this technique is limited by the viscosity of the system, swollen block copolymers with a larger molecular weight (range of 100 kg/mol) can be aligned by an electric field as well [294]. Russell and coworkers used an electric field to align a S–M diblock copolymer with cylindrical morphology in the bulk state. The cylinders thereby align parallel to the electric field [295,296]. Studies on lamellar systems revealed that these systems also align in such a way that the interface is parallel to the electric field [297]. However, the second direction of the lamellar planes is not fixed and therefore a random orientation of lamellar stacks in the plane

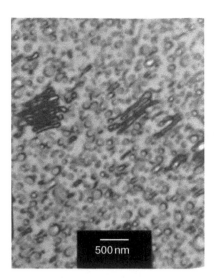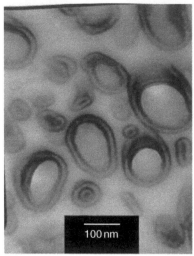

Figure 27   Transmission electron micrographs at two different magnifications for the PMMA/S–I–M (80/20 wt/wt) solvent-cast blend modified by LAOS. (Copyright Ref. 293, Copyright 2004 American Chemical Society)

perpendicular to the electric field is obtained, unless other effects like surface effects of the sample holder influence the orientation. The kinetics of the alignment in solutions of lamellar S–I diblock copolymers could be investigated in some more detail by *in situ* synchrotron SAXS measurements. The alignment occurs quite fast, typical time constants for the alignment range from seconds to minutes [298]. Another interesting feature is the mechanism of alignment: while weakly segregated block copolymers seem to align via a "melting–recrystallization" mechanism, that is, nonparallel oriented domains dissolve and reform with the correct orientation, in strongly segregated systems the whole domains rotate [299]. These findings could be confirmed by computer simulation [298].

IX.   Concluding Remarks

This chapter presented an overview of the fascinating variety of microphase morphologies found in different types of block copolymers. The increase in the number of components was shown to lead to an enormous increase of outstanding morphologies consistently with the increase of independent parameters of the system: composition and segmental interaction. Chain topology and crystallization of single blocks were also shown to affect morphology. Blends of these materials further enrich the population of self-assembled microphases with regular periodic arrays of the component copolymers.

All the above morphologies are the natural result of molecular recognition of two or more incompatible blocks connected by single chemical bonds. These systems self-assemble solely due to rather weak van-der-Waals interactions. Also specific interactions like hydrogen bonding of ionic interactions can be used to generate supramolecular structures of block copolymers [26,300]. Self-assembly of pure block copolymers or mixtures of block copolymers with other components may thus be regarded as a supramolecular polymerization of similar or different macromolecules occurring in a precursor solution, or directly in the bulk.

References

1. F.S. Bates and G.H. Frederickson. *Annu. Rev. Phys. Chem.* 41:525, 1990
2. G. Riess. In *Thermoplastic Elastomers. A Comprehensive Review*, N.R. Ledge, G. Holden, H.E. Schroeder, Eds.; Hanser: Munich, 1987

3. F.S. Bates and G.H. Fredrickson. *Phys. Today* 52:32, 1999
4. M.W. Matsen and M. Schick. *Curr. Opin. Colloid Interface Sci.* 1:329, 1996
5. I.W. Hamley. *The Physics of Block Copolymers*; Oxford University Press: Oxford, 1998
6. V. Abetz. In *Encyclopedia of Polymer Science and Technology*, 3rd edn, J.I. Kroschwitz, Ed., Vol. 1; John Wiley & Sons, Inc.: New York, 2003, p. 482
7. J.P. Flory. *Principles of Polymer Chemistry*; Cornell University Press: Ithaca, 1953
8. M.L. Huggins. *J. Chem. Phys.* 9:440, 1941
9. A.J. Staverman. *Rec. Trav. Chim.* 60:640, 1941
10. R.J. Spontak and N.P. Patel. *Curr. Opin. Colloid Interface Sci.* 5:334, 2000
11. J. Hernández-Barajas and D.J. Hunkeler. *Polymer* 38:449, 1997
12. C. Auschra and R. Stadler. *Macromolecules* 26:6364, 1993
13. L. Bronstein, M. Seregina, P. Valetsky, U. Breiner, V. Abetz, and R. Stadler. *Polym. Bull.* 39:361, 1997
14. Y.N.C. Chan, R.R. Schrock, and R.E. Cohen. *Chem. Mater.* 4:24, 1992
15. M. Antonietti, E. Wenz, L. Bronstein, and M. Seregina. *Adv. Mater.* 7:1000, 1995
16. J.P. Spatz, A. Roescher, and M. Möller. *Adv. Mater.* 8:337, 1996
17. J.P. Spatz, S. Sheiko, and M. Möller. *Macromolecules* 29:3220, 1996
18. L. Bronstein, E. Kraemer, B. Berton, C. Burger, S. Foerster, and M. Antonietti. *Chem. Mater.* 11:1402, 1999
19. S. Förster and M. Antonietti. *Adv. Mater.* 10:195, 1998
20. G.S.W. Craig, R.E. Cohen, R.R. Schrock, C. Dhenaut, I. LeDoux, and J. Zyss. *Macromolecules* 27:1875, 1994
21. G.S.W. Craig, R.E. Cohen, R.R. Schrock, R.J. Silby, G. Puccetti, I. LeDoux, and J. Zyss. *J. Am. Chem. Soc.* 115:860, 1993
22. Y. Heischkel and H.-W. Schmidt. *Macromol. Chem. Phys.* 199:869, 1998
23. M. Maldovan, W.C. Carter, and E.L. Thomas. *Appl. Phys. Lett.* 83:5172, 2003
24. G.-Liu, L. Quiao, and A. Guo. *Macromolecules* 29:5508, 1996
25. J. Ruokolainen, R. Mäkinen, M. Torkkeli, T. Mäkelä, R. Serimaa, G. ten Brinke, and O. Ikkala. *Science* 280:557, 1998
26. J. Ruokolainen, M. Saariaho, O. Ikkala, G. ten Brinke, E.L. Thomas, M. Torkkeli, and R. Serimaa. *Macromolecules* 32:1152, 1999
27. M. Templin, A. Franck, A. Du Chesne, H. Leist, Y. Zhang, R. Ulrich, V. Schädler, and U. Wiesner. *Science* 278:1795, 1997
28. E. Krämer, S. Förster, C. Göltner, and M. Antonietti. *Langmuir* 14:2027, 1998
29. N. Hadjichristidis, S. Pispas, and G. Floudas. *Block Copolymers: Synthetic Strategies, Physical Properties, and Applications*; John Wiley and Sons, Inc.: Hoboken, NJ, 2003
30. H.R. Kricheldorf and T. Stukenbrok. *J. Polym. Sci. Polym. Chem.* 36:31, 1998
31. U. Schulze and H.-W. Schmidt. *Polym. Bull.* 40:159, 1998
32. M.V. Pandya, M. Subramaniam, and M.R. Desai. *Eur. Polym. J.* 33:789, 1997
33. Y. Bourgeois, Y. Charlier, and R. Legras. *Polymer* 37:5503, 1996
34. J.C.W. Chien, Y. Iwamoto, M.D. Rausch, W. Wedler, and H.H. Winter. *Macromolecules* 30:3447, 1997
35. S. Jacob and J.P. Kennedy. *Polym. Bull.* 41:167, 1998
36. R. Faust and D. Li. *Macromolecules* 28:4893, 1995
37. Y.C. Bae and R. Faust. *Macromolecules* 31:2480, 1998
38. N. Mougin, P. Rempp, and Y. Gnanou. *Macromol. Chem.* 194:2553, 1993
39. G.-H. Hsiue, Y.-L. Liu, and Y.-S. Chiu. *J. Polym. Sci. Polym. Chem.* 31:3371, 1993
40. P. Van Caeter, E.J. Goethals, and R. Velichkova. *Polym. Bull.* 39:589, 1997
41. J.M. Nelson, A.P. Primrose, T.J. Hartle, and H.R. Allcock. *Macromolecules* 31:947, 1998
42. D.A. Shipp, J.-L. Wang, and K. Matyjaszewski. *Macromolecules* 31:8005, 1998
43. A. Mühlebach, S.G. Gaynor, and K. Matyjaszewski. *Macromolecules* 31:6046, 1998
44. H. Mori and A.H.E. Müller. *Prog. Polym. Sci.* 28:1403, 2003
45. M. Szwarc. *J. Polym. Sci. Polym. Chem.* 36:IX, 1998
46. J.M. Yu, Y. Yu, and R. Jérôme. *Polymer* 38:3091, 1997
47. G. Hild and J.P. Lamps. *Polymer* 36:4841, 1995
48. R.P. Quirk, T. Yoo, and B. Lee. *JMS Pure Appl. Chem.* A31:911, 1994
49. N. Hadjichristidis, Y. Tselikas, H. Iatrou, V. Efstratiadis, and A. Avgeropoylos. *JMS Pure Appl. Chem.* A33:1447, 1996

50. N. Hadjichristidis. *J. Polym. Sci. A: Polym. Chem.* 37:857, 1999
51. H. Iatrou and N. Hadjichristidis. *Macromolecules* 25:4649, 1992
52. H. Iatrou and N. Hadjichristidis. *Macromolecules* 26:2479, 1993
53. S. Sioula, Y. Tselikas, and N. Hadjichristidis. *Macromolecules* 30:1518, 1997
54. T. Fujimoto, H. Zhang, T. Kazama, Y. Isono, H. Hasegawa, and T. Hashimoto. *Polymer* 33:2208, 1992
55. H. Hückstädt, V. Abetz, and R. Stadler. *Macromol. Rapid Commun.* 17:599, 1996
56. C. Auschra and R. Stadler. *Polym. Bull.* 30:257, 1993
57. C. Neumann, V. Abetz, and R. Stadler. *Polym. Bull.* 36:43, 1996
58. O. Lambert, S. Reutenauer, G. Hurtrez, G. Riess, and P. Dumas. *Polym. Bull.* 40:143, 1998
59. O. Lambert, P. Dumas, G. Hurtrez, and G. Riess. *Macromol. Rapid Commun.* 18:343, 1997
60. G. Riess, M. Schlienger, and S. Marti. *J. Macromol. Sci. Polym. Phys. Ed.* 17:355, 1980
61. E. Yoshida and A. Sugita. *J. Polym. Sci. Polym. Chem.* 36:2059, 1998
62. E. Ruckenstein and H. Zhang. *Macromolecules* 31:2977, 1998
63. J. Feldthusen, B. Iván, and A.H.E. Müller. *Macromolecules* 31:578, 1998
64. K. Loos and R. Stadler. *Macromolecules* 30:7641, 1997
65. H. Schmalz, V. Abetz, R. Lange, and M. Soliman. *Macromolecules* 34:795, 2001
66. H. Funabashi, Y. Miyamoto, Y. Isono, T. Fujimoto, Y. Matsushita, and M. Nagasawa. *Macromolecules* 16:1, 1983
67. Y. Isono, H. Tanisugi, K. Endo, T. Fujimoto, H. Hasegawa, T. Hashimoto, and H. Kawai. *Macromolecules* 16:5, 1983
68. Y. Matsushita, K. Yamada, T. Hattori, T. Fujimoto, Y. Sawada, M. Nasagawa, and C. Matsui. *Macromolecules* 16:10, 1983
69. S. Sakurai. *Curr. Trends Polym. Sci.* 1:119, 1996
70. K. Mortensen. In *Scattering in Polymeric and Colloidal Systems*, W. Brown and K. Mortensen, Eds.; Gordon & Breach Science Publishers: Amsterdam, 2000, pp. 413–456
71. B. Stühn, R. Mutter, and T. Albrecht. *Europhys. Lett.* 18:427, 1992
72. T. Wolff, C. Burger, and W. Ruland. *Macromolecules* 26:1707, 1993
73. K. Almdal, F.S. Bates, and K. Mortensen. *J. Chem. Phys.* 96:9122, 1992
74. B. Holzer, A. Lehmann, B. Stühn, and M. Kowalski. *Polymer* 32:1935, 1991
75. V.T. Bartels, V. Abetz, K. Mortensen, and M. Stamm. *Europhys. Lett.* 27:371, 1994
76. V.T. Bartels, M. Stamm, V. Abetz, and K. Mortensen. *Europhys. Lett.* 31:81, 1995
77. G. Hadziioannou, C. Picot, A. Skoulios, M.-L. Ionesu, A. Mathis, R. Duplessix, Y. Gallot, and J.-P. Lingelser. *Macromolecules* 15:263, 1982
78. H. Hasegawa, T. Hashimoto, H. Kawai, T.P. Lodge, E.J. Amis, C.J. Glinka, and C.C. Han. *Macromolecules* 18:67, 1985
79. A.M. Mayes, T.P. Russell, V.R. Deline, S.K. Satija, and C.F. Majkrzak. *Macromolecules* 27:7447, 1994
80. W.Z. Cai, K. Schmidt-Rohr, N. Egger, B. Gerharz, and H.-W. Spiess. *Polymer* 34:267, 1993
81. M.D. Gehlsen, K. Almdal, and F.S. Bates. *Macromolecules* 25:939, 1992
82. J.H. Rosedale, F.S. Bates, K. Almdal, K. Mortensen, and G.D. Wignall. *Macromolecules* 28:1429, 1995
83. C.D. Han, D.M. Baek, and J.K. Kim. *Macromolecules* 23:561, 1990
84. C. Neumann, D.R. Loveday, V. Abetz, and R. Stadler. *Macromolecules* 31:2493, 1998
85. N.P. Balsara, B.A. Garetz, M.Y. Chang, H.J. Dai, M.C. Newstein, J.L. Goveas, R. Krishnamoorti, and S. Rai. *Macromolecules* 31:5309, 1998
86. D.L. Handlin and E.L. Thomas. *Macromolecules* 16:1514, 1983
87. P.B. Himelfarb and K.B. Labat. *Scanning* 12:148, 1990
88. H. Elbs, C. Drummer, V. Abetz, and G. Krausch. *Macromolecules* 35:5570, 2002
89. W. Stocker, J. Beckman, R. Stadler, and J. Rabe. *Macromolecules* 29:7502, 1996
90. H. Ott, V. Abetz, V. Altstädt, Y. Thomann, and A. Pfau. *J. Microsc.* 205:106, 2002
91. D.A. Chernoff and S. Magonov. *Atomic Force Microscopy. Comprehensive Desk Reference of Polymer Characterization and Analysis*, 2003, pp. 490–531
92. G.E. Molau. In *Colloidal and Morphological Behaviour of Block Copolymers*, G.E. Molau, Ed.; Plenum Press: New York, 1971
93. D.J. Meier, Ed., *Block Copolymers: Science and Technology*; Gordon & Breach Science Publisher: Tokyo, 1983
94. G. Riess, G. Hurtrez, and P. Bahadur. Block Copolymers. In *Encyclopedia of Polymer Science and Engineering*, Vol. 2; John Wiley & Sons: New York, 1985

95. K. Binder. *Adv. Polym. Sci.* 112:182, 1994
96. G.H. Fredrickson and F.S. Bates. *Annu. Rev. Mater. Sci.* 26:501, 1996
97. R.H. Colby. *Curr. Opin. Colloid Polym. Sci.* 1:454, 1996
98. L. Leibler. *Macromolecules* 13:1602, 1980
99. I.Ya. Erukhimovich. *Polym. Sci. U.S.S.R.* 24:2223, 1982
100. I.Ya. Erukhimovich. *Polym. Sci. U.S.S.R.* 24:2232, 1982
101. D.J. Meier. *J. Polym. Sci.* C26:81, 1969
102. E. Helfand and A.M. Sapse. *J. Chem. Phys.* 62:1327, 1975
103. E. Helfand and Z.R. Wasserman. *Macromolecules* 9:879, 1976
104. E. Helfand and Z.R. Wasserman. *Macromolecules* 11:960, 1978
105. A.N. Semenov. *Sov. Phys. JETP* 61:733, 1985
106. I.A. Nyrkova, A.R. Khokhlov, and M. Doi. *Macromolecules* 26:3601, 1993
107. M.W. Matsen and F.S. Bates. *J. Chem. Phys.* 106:2436, 1997
108. M.W. Matsen and F.S. Bates. *Macromolecules* 29:1091, 1996
109. M.W. Matsen and M. Schick. *Macromolecules* 27:4014, 1994
110. D.R.M. Williams and A. Halperin. *Phys. Rev. Lett.* 71:1557, 1993
111. S.L. Aggarwal. *Polymer* 17:938, 1972
112. E.L. Thomas, D.B. Alward, D.J. Kinning, D.C. Martin, D.L. Handlin, and L.J. Fetters. *Macromolecules* 19:2197, 1986
113. D.A. Hajduk, P.E. Harper, S.M. Gruner, C.C. Honeker, E.L. Thomas, and L.J. Fetters. *Macromolecules* 28:2570, 1995
114. H. Hasegawa, T. Hashimoto, and S.T. Hyde. *Polymer* 37:3825, 1996
115. M.E. Vigild, K. Almdal, K. Mortensen, I.W. Hamley, J.P. Fairclough, and A.J. Ryan. *Macromolecules* 31:5702, 1998
116. I.Ya. Erukhimovich. *JETP Lett.* 63:460, 1996
117. S.T. Milner and P.T. Olmsted. *J. Phys. II* 7:249, 1997
118. W. Gozdz and R. Holyst. *Macromol. Theory Simul.* 5:321, 1996
119. K. Grosse-Brauckmann. *J. Colloid Interface Sci.* 187:418, 1997
120. I.W. Hamley, K.A. Koppi, J.H. Rosedale, F.S. Bates, K. Almdal, and K. Mortensen. *Macromolecules* 26:5959, 1993
121. S. Förster, A.K. Khandpur, J. Zhao, F.S. Bates, I.W. Hamley, A.J. Ryan, and W. Bras. *Macromolecules* 27:6922, 1994
122. A.K. Khandpur, S. Förster, F.S. Bates, I.W. Hamley, A.J. Ryan, W. Bras, K. Almdal, and K. Mortensen. *Macromolecules* 28:8796, 1995
123. I.I. David, S.P. Gido, K. Hong, J. Zhou, J.W. Mays, and N.B. Tan. *Macromolecules* 32.3216, 1999
124. S. Mani, R.A. Weiss, M.E. Cantino, L.H. Khairallah, S.F. Hahn, and C.E. Williams. *Eur. Polym. J.* 36:215, 2000
125. C.Y. Ryu and T.P. Lodge. *Macromolecules* 32:7190, 1999
126. K. Kimishima, T. Koga, Y. Kanazawa, and T. Hashimoto. *ACS Symp. Series* 739:514, 2000
127. N. Sota, N. Sakamoto, K. Saijo, and T. Hashimoto. *Macromolecules* 36:4534, 2003
128. D.A. Davidock, M.A. Hillmyer, and T.P. Lodge. *Macromolecules* 37:397, 2004
129. V. Abetz and R. Stadler. *Macromol. Symp.* 113:19, 1997
130. L. Kane and R.J. Spontak. *Macromolecules* 27:663, 1994
131. Y. Mogi, K. Mori, H. Kotsuji, Y. Matsushita, I. Noda, and C.C. Han. *Macromolecules* 26:5169, 1993
132. W. Zheng and Z.-G. Wang. *Macromolecules* 28:7215, 1995
133. T. Ohta and K. Kawasaki. *Macromolecules* 19:2621, 1989; 23:2413, 1990
134. H. Nakazawa and T. Ohta. *Macromolecules* 26:5503, 1993
135. Yu.V. Lyatskaya and T.M. Birshtein. *Polymer* 36:975, 1995
136. S. Phan and G.H. Fredrickson. *Macromolecules* 31:59, 1998
137. M.W. Matsen. *J. Chem. Phys.* 108:785, 1998
138. Y. Mogi, M. Nomura, H. Kotsuji, K. Ohnishi, Y. Matsushita, and I. Noda. *Macromolecules* 27:6755, 1994
139. Y. Mogi, K. Mori, Y. Matsushita, and I. Noda. *Macromolecules* 25:5412, 1992
140. Y. Bohbot-Raviv and Z.-G. Wang. *Phys. Rev. Lett.* 85:3428, 2000
141. F. Drolet and G.H. Fredrickson. *Phys. Rev. Lett.* 83:4317, 1999
142. G.H. Fredrickson, V. Ganesan, and F. Drolet. *Macromolecules* 35:16, 2002
143. S.P. Gido, D.W. Schwark, E.L. Thomas, and M.C. Goncalves. *Macromolecules* 26:2636, 1993

144. H. Hückstädt, A. Göpfert, and V. Abetz. *Polymer* 41:9089, 2000
145. T.S. Bailey, H.D. Pham, and F.S. Bates. *Macromolecules* 34:6994, 2001
146. T.S. Bailey, C.M. Hardy, T.H. Epps, and F.S. Bates. *Macromolecules* 35:7007, 2002
147. K. Jung. Doctoral Thesis, Universität Mainz, 1996
148. V. Abetz and T. Goldacker. *Macromol. Rapid Commun.* 21:16, 2000
149. E. Giebeler. Doctoral Thesis, Universität Mainz, 1996
150. R. Stadler, C. Auschra, J. Beckmann, U. Krappe, I. Voigt-Martin, and L. Leibler. *Macromolecules* 28:3080, 1995
151. U. Krappe, R. Stadler, and I.-G. Voigt-Martin. *Macromolecules* 28:4558, 1995
152. K. Jung, V. Abetz, and R. Stadler. *Macromolecules* 29:1076, 1996
153. U. Breiner, U. Krappe, V. Abetz, and R. Stadler. *Macromol. Chem. Phys.* 198:1051, 1997
154. U. Breiner, U. Krappe, T. Jakob, V. Abetz, and R. Stadler. *Polym. Bull.* 40:219, 1998
155. S. Brinkmann, R. Stadler, and E.L. Thomas. *Macromolecules* 31:6566, 1998
156. S. Brinkmann, V. Abetz, R. Stadler, and E.L. Thomas. *Kautschuk Gummi Kunststoffe* 52:806, 1999
157. V. Abetz, K. Markgraf, and V. Rebizant. *Macromol. Symp.* 177:139, 2002
158. C. Auschra, J. Beckmann, and R. Stadler. *Macromol. Rapid Commun.* 15:67, 1994
159. C. Auschra and R. Stadler. *Macromolecules* 26:2171, 1993
160. U. Breiner, U. Krappe, E.L. Thomas, and R. Stadler. *Macromolecules* 31:135, 1998
161. U. Breiner, U. Krappe, and R. Stadler. *Macromol. Rapid Commun.* 17:567, 1996
162. H. Ott, V. Abetz, and V. Altstädt. *Macromolecules* 34:2121, 2001
163. F. Annighöfer and W. Gronski. *Colloid Polym. Sci.* 261:15, 1983
164. F. Annighöfer and W. Gronski. *Macromol. Chem.* 185:2213, 1984
165. T. Hashimoto, Y. Tsukahara, K. Tachi, and H. Kawai. *Macromolecules* 16:648, 1983
166. V. Abetz, R. Stadler, and L. Leibler. *Polym. Bull.* 37:135, 1996
167. I. Erukhimovich, V. Abetz, and R. Stadler. *Macromolecules* 30:7435, 1997
168. C. Neumann, V. Abetz, and R. Stadler. *Colloid Polym. Sci.* 276:19, 1998
169. C.M. Roland and C.A. Trask. *Macromolecules* 22:256, 1989
170. K. Takahashi, H. Hasegawa, T. Hashimoto, V. Bellas, H. Iatrou, and N. Hadjichristidis. *Macromolecules* 35:4859, 2002
171. H. Hasegawa, K. Takahashi, T. Hashimoto, V. Bellas, H. Iatrou, and N. Hadjichristidis. *Polym. Mater. Sci. Eng.* 88:156, 2003
172. M. Doi and S.F. Edwards. *The Theory of Polymer Dynamics*; Clarendon Press: Oxford, 1989
173. C.W. Lantman, W.J. MacKnight, J.F. Tassin, L. Monnerie, and L.J. Fetters. *Macromolecules* 23:836, 1990
174. M. Daoud and J.P. Cotton. *J. Phys. (Les Ulis Fr.)* 43:539, 1982
175. T.M. Birshtein, E.B. Zhulina, and O.V. Borislov. *Polymer* 27:1078, 1986
176. M. Moffitt, Y. Yu, D. Nguyen, V. Graziano, D.K. Schneider, and A. Eisenberg. *Macromolecules* 31:2190, 1998
177. M. Olvera de la Cruz and I.C. Sanchez. *Macromolecules* 19:2501, 1986
178. A.V. Dobrynin and I.Y. Erukhimovich. *Vysokomol Soyedin* 33A:1100, 1991
179. G. Floudas, S. Pispas, N. Hadjichristidis, T. Pakula, and I. Erukhimovich. *Macromolecules* 29:4142, 1996
180. L.-K. Bi and L.J. Fetters. *Macromolecules* 9:732, 1976
181. Y. Tselikas, H. Iatrou, N. Hadjichristidis, K.S. Liang, K. Mohanty, and D.J. Lohse. *J. Chem. Phys.* 105:2456, 1996
182. C.M. Turner, N.B. Sheller, M.D. Foster, B. Lee, S. Corona-Galvan, R.P. Quirk, B. Annis, and J.-S. Lin. *Macromolecules* 31:4372, 1998
183. F.L. Beyer, S.P. Gido, Y. Poulos, A. Avgeropoylos, and N. Hadjichristidis. *Macromolecules* 30:2373, 1997
184. S. Milner. *Macromolecules* 27:2333, 1994
185. Y. Tselikas, N. Hadjichristidis, R.L. Lescanec, C.C. Honeker, M. Wohlgemuth, and E.L. Thomas. *Macromolecules* 29:3390, 1996
186. Y. Matsushita, T. Takasu, and K. Yagi. *Polymer* 35:2862, 1994
187. R.F. Story, B.J. Chisholm, and Y. Lee. *Polymer* 34:4330, 1993
188. M.W. Matsen and M. Schick. *Macromolecules* 27:6761, 1994
189. R. Erhardt, A. Böker, H. Zettl, H. Kaya, W. Pyckhout-Hintzen, G. Krausch, V. Abetz, and A.H.E. Müller. *Macromolecules* 34:1069, 2001

190. R. Erhardt, M. Zhang, A. Böker, H. Zettl, C. Abetz, P. Frederik, G. Krausch, V. Abetz, and A.H.E. Müller. *J. Am. Chem. Soc.* 125:3260, 2003

191. Y. Liu, V. Abetz, and A.H.E. Müller. *Macromolecules* 36:7894, 2003

192. S. Okamoto, H. Hasegawa, T. Hashimoto, T. Fujimoto, H. Zhang, T. Kazama, A. Takano, and Y. Isono. *Polymer* 38:5275, 1997

193. N. Hadjichristidis and H. Iatrou. *Macromolecules* 26:5812, 1993

194. S. Sioula, Y. Tselikas, and N. Hadjichristidis. *Macromolecules* 30:1518, 1997

195. S. Sioula, N. Hadjichristidis, and E.L. Thomas. *Macromolecules* 31:5272, 1998

196. S. Sioula, N. Hadjichristidis, and E.L. Thomas. *Macromolecules* 31:8429, 1998

197. H. Hoffmann, S. Hofmann, and U. Kästner. *Adv. Chem. Ser.* 248:219, 1996

198. G. Schmidt, S. Müller, P. Lindner, C. Schmidt, and W. Richtering. *J. Phys. Chem. B* 102:507, 1998

199. M.W. Matsen and F.S. Bates. *Macromolecules* 29:7641, 1996

200. K. Mortensen, Y. Talmon, B. Gao, and J. Kops. *Macromolecules* 30:6764, 1997

201. G. Yu and A. Eisenberg. *Macromolecules* 31:5546, 1998

202. P. Alexandridis and B. Lindman, Eds., *Amphiphilic Block Copolymers: Self-Assembly and Applications*; Elsevier Science B.V.: The Netherlands, 1997

203. P. Alexandridis, U. Olsson, and B. Lindman. *Langmuir* 13:23, 1997

204. O. Glatter, G. Scherf, K. Schiller, and W. Brown. *Macromolecules* 27:6046, 1994

205. G. Wanka, H. Hoffmann, and W. Ulbricht. *Macromolecules* 27:4145, 1994

206. Z. Tuzar and P. Kratochvil. *Surface and Colloid Science*, E. Matijevic Ed., Vol. 15; Plenum Press: New York, 1993

207. L. Luo and A. Eisenberg. *Langmuir* 17:6804, 2001

208. S. Förster, V. Abetz, and A.H.E. Müller. *Adv. Polym. Sci.* 166:173, 2004

209. G. Schmidt, W. Richtering, P. Lindner, and P. Alexandridis. *Macromolecules* 31:2293, 1998

210. H. Soenen, H. Berghmans, H.H. Winter, and N. Overbergh. *Polymer* 38:5653, 1997

211. A. Gast. *Langmuir* 12:4060, 1996

212. T.P. Lodge, K.J. Hanley, B. Pudil, and V. Alahapperuma. *Macromolecules* 36:816, 2003

213. K.J. Hanley, T.P. Lodge, and C.-I. Huang. *Macromolecules* 33:5918, 2000

214. J.-F. Gohy, N. Willet, S. Varshney, J.-X. Zhang, and R. Jérôme. *Angew. Chem. Int. Ed.* 40:3214, 2001

215. R. Bieringer, V. Abetz, and A.H.E. Müller. *Eur. Phys. J. E Softmatter (EPJ E)* 5:5, 2001

216. F. Liu and A. Eisenberg. *J. Am. Chem. Soc.* 125:15059, 2003

217. H. Pernot, M. Baumert, F. Court, and L. Leibler. *Nat. Mater.* 1:54, 2002

218. C. Auschra, R. Stadler, and I.G. Voigt-Martin. *Polymer* 34:2081, 1993

219. N.P. Balsara, S.V. Jonnalagadda, C.C. Lin, C.C. Han, and R. Krishnamoorti. *J. Chem. Phys.* 99:10011, 1993

220. A.Z. Akcasu and M. Tombakoglu. *Macromolecules* 23:607, 1990

221. H. Benoît, M. Benmouna, and W.L. Wu. *Macromolecules* 23:1511, 1990

222. K.I. Winey, E.L. Thomas, and L.J. Fetters. *Macromolecules* 25:422, 1992

223. S.-H. Lee, J.T. Koberstein, X. Quan, I. Gancarz, G.D. Wignall, and F.C. Wilson. *Macromolecules* 27:3199, 1994

224. T. Hashimoto, S. Koizumi, H. Hasegawa, T. Izumitani, and S.T. Hyde. *Macromolecules* 25:1433, 1992

225. T. Hashimoto, K. Kimishima, and H. Hasegawa. *Macromolecules* 24:5704, 1991

226. R. Xie, B. Yang, and B. Jiang. *Macromolecules* 26:7097, 1993

227. R.L. Lescanec, F.J. Fetters, and E.L. Thomas. *Macromolecules* 31:1680, 1998

228. G. Hadziioannou and A. Skoulios. *Macromolecules* 15:267, 1982

229. T. Hashimoto, K. Yamasaki, S. Koizumi, and H. Hasegawa. *Macromolecules* 26:2895, 1993

230. T. Hashimoto, S. Koizumi, and H. Hasegawa. *Macromolecules* 27:1562, 1994

231. R.J. Spontak, C.J. Fung, M.B. Braunfeld, J.W. Sedat, D.A. Agard, A. Ashraf, and S.D. Smith. *Macromolecules* 29:2850, 1996

232. A.D. Vilesov, G. Floudas, T. Pakula, E.Yu. Melenevskaya, T.M. Birshtein, and Yu.V. Lyatskaya. *Macromol. Chem. Phys.* 195:2132, 1994

233. M.F. Schulz, F.S. Bates, K. Almdal, and K. Mortensen. *Phys. Rev. Lett.* 73:86, 1994

234. R.J. Spontak, J.C. Fung, M.B. Braunfeld, J.W. Sedat, D.A. Agard, L. Kane, S.D. Smith, M.M. Satkowski, A. Ashraf, D.A. Hajduk, and S.M. Gruner. *Macromolecules* 29:4494, 1996

235. S. Sakurai, H. Irie, H. Umeda, S. Nomura, H.H. Lee, and J.K. Kim. *Macromolecules* 31:336, 1998

236. S. Sakurai, H. Umeda, C. Furukawa, H. Irie, S. Nomura, H.H. Lee, and J.K. Kim. *J. Chem. Phys.* 108:4333, 1998

237. S. Sakurai and S. Nomura. *Polymer* 38:4103, 1997

238. A.-C. Shi and J. Noolandi. *Macromolecules* 28:3103, 1995

239. M.W. Matsen and F.S. Bates. *Macromolecules* 28:7298, 1995

240. T.M. Birshtein, Yu.V. Lyatskaya, and E.B. Zhulina. *Polymer* 33:2750, 1992

241. A.L. Borovinskii and A.R. Khokhlov. *Macromolecules* 31:1180, 1998

242. T. Goldacker and V. Abetz. *Macromolecules* 32:5165, 1999

243. T.M. Birshtein, E.B. Zhulina, A.A. Polotsky, V. Abetz, and R. Stadler. *Macromol. Theory Simul.* 8:151, 1999

244. L. Leibler, C. Gay, and I.Ya. Erukhimovich. *Europhys. Lett.* 46:549, 1999

245. T. Goldacker, V. Abetz, R. Stadler, I.Ya. Erukhimovich, and L. Leibler. *Nature* 398:137, 1999

246. T. Goldacker and V. Abetz. *Macromol. Rapid Commun.* 20:415, 1999

247. A. Takano, K. Soga, J. Suzuki, and Y. Matsushita. *Macromolecules* 36:9288, 2003

248. R. Unger, D. Beyer, and E. Donth. *Polymer* 32:3305, 1991

249. P. Rangarajan, R.A. Register, and L.J. Fetters. *Macromolecules* 26:4640, 1993

250. S. Nojima, K. Kato, S. Yamamoto, and T. Ashida. *Macromolecules* 25:2237, 1992

251. S. Nojima, K. Hashizume, A. Rohadie, and S. Sasaki. *Polymer* 38:2711, 1997

252. D.J. Quiram, R.A. Register, G.R. Marchand, and D.H. Adamson. *Macromolecules* 31:4891, 1998

253. L. Zhu, S.Z.D. Cheng, B.H. Calhoun, Q. Ge, R.P. Quirk, E.L. Thomas, B.S. Hsiao, F.J. Yeh, and B. Lotz. *J. Am. Chem. Soc.* 122:5927, 2000

254. Y.L. Loo, R.A. Register, A.J. Ryan, and G.T. Dee. *Macromolecules* 34:8968, 2001

255. V. Balsamo, A.J. Müller, and R. Stadler. *Macromolecules* 31:7756, 1998

256. V. Balsamo, A.J. Müller, F. von Gyldenfeldt, and R. Stadler. *Macromol. Chem. Phys.* 199:1063, 1998

257. V. Balsamo and R. Stadler. *Macromol. Symp.* 117:153, 1997

258. V. Balsamo, F. von Gyldenfeldt, and R. Stadler. *Macromolecules* 32:1226, 1999

259. V. Balsamo, G. Gil, C. Urbina de Navarro, I.W. Hamley, F. von Gyldenfeldt, V. Abetz, and E. Canizales. *Macromolecules* 36:4515, 2003

260. H. Schmalz, A. Knoll, A.J. Müller, and V. Abetz. *Macromolecules* 35:10004, 2002

261. H. Schmalz, A.J. Müller, and V. Abetz. *Macromol. Chem. Phys.* 204:111, 2003

262. G. Floudas, G. Reiter, O. Lambert, and P. Dumas. *Macromolecules* 31:7279, 1998

263. E.A. DiMarzio, C.M. Guttman, and J.D. Hoffman. *Macromolecules* 13:1194, 1980

264. M.D. Whitmore and J. Noolandi. *Macromolecules* 21:1482, 1988

265. J. Adams and W. Gronski. *Macromol. Chem. Rapid Commun.* 10:553, 1989

266. J. Sänger, W. Gronski, H. Leist, and U. Wiesner. *Macromolecules* 30:7621, 1997

267. A. Molenberg, M. Möller, and T. Pieper. *Macromol. Chem. Phys.* 199:299, 1998

268. G. Mao, J. Wang, S.R. Clingman, C.K. Ober, J.T. Chen, and E.L. Thomas. *Macromolecules* 30:2556, 1997

269. D.R.M. Williams and A. Halperin. *Phys. Rev. Lett.* 71:1557, 1993

270. A. Halperin and D.R.M. Williams. *Macromolecules* 26:6652, 1993

271. A. Halperin and D.R.M. Williams. *Phys. Rev. E Rapid Commun.* 49:R986, 1994

272. V.T. Bartels, M. Stamm, and K. Mortensen. *Polym. Bull.* 36:103, 1996

273. H. Frielinghaus, D. Schwahn, and T. Springer. *Macromolecules* 29:3263, 1996

274. H. Ladynski, W. De Odorico, and M. Stamm. *J. Non-Crystall. Solids* 235/237:491, 1998

275. M. Pollard, T.P. Russell, A.V. Ruzette, A.M. Mayes, and Y. Gallot. *Macromolecules* 31:6493, 1998

276. B. Steinhoff, M. Rüllmann, M. Wenzel, M. Junker, I. Alig, R. Oser, B. Stühn, G. Meier, O. Diat, P. Bösecke, and H.B. Stanley. *Macromolecules* 31:36, 1998

277. D. Schwahn, H. Frielinghaus, K. Mortensen, and K. Almdal. *Phys. Rev. Lett.* 77:3153, 1996

278. R.J. Albalak and E.L. Thomas. *J. Polym. Sci. Polym. Phys.* 31:37, 1993

279. C.C. Honeker and E.L. Thomas. *Chem. Mater.* 8:1702, 1996

280. A. Keller, E. Pedemonte, and F.M. Willmouth. *Nature* 225:538, 1970

281. K.A. Koppi, M. Tirrell, F.S. Bates, K. Almdal, and R.H. Colby. *J. Phys. II* 2:1941, 1992

282. K.I. Winey, S.S. Patel, R.G. Larson, and H. Watanabe. *Macromolecules* 26:2542, 1993

283. B.L. Riise, G.H. Fredrickson, R.G. Larson, and D.S. Pearson. *Macromolecules* 28:7653, 1995

284. Z.-R. Chen, J.A. Kornfield, S.D. Smith, J.T. Grothaus, and M.M. Satkowski. *Science* 277:1248, 1997

285. U. Wiesner. *Macromol. Chem. Phys.* 198:3319, 1997

286. Z.-R. Chen, A.M. Issaian, J.A. Kornfield, S.D. Smith, J.T. Grothaus, and M.M. Satkowski. *Macromolecules* 30:7096, 1997

287. T. Tepe, D.A. Hajduk, M.A. Hillmyer, P.A. Weimann, M. Tirrell, F.S. Bates, K. Almdal, and K. Mortensen. *J. Rheol.* 41:1147, 1997

288. Y. Mori, L.S. Lim, and F.S. Bates. *Macromolecules* 36:9879, 2003

289. S. Stangler and V. Abetz. *Rheologica Acta* 42:569, 2003

290. H.H. Winter, D.B. Scott, W. Gronski, S. Okamoto, and T. Hashimoto. *Macromolecules* 26:7236, 1993

291. A.I. Nakatani, F.A. Morrison, J.F. Douglas, J.W. Mays, C.L. Jackson, M. Muthukumar, and C.C. Han. *J. Chem. Phys.* 104:1589, 1996

292. C.L. Jackson, K.A. Barnes, F.A. Morrison, J.W. Mays, A.I. Nakatani, and C.C. Han. *Macromolecules* 28:713, 1995

293. C. Koulic and R. Jérôme. *Macromolecules* 37:888, 2004

294. J. Le Meur, J. Terrisse, C. Schwab, and P. Goldzene. *J. Phys. Colloq.* 32:C5a–301, 1971

295. T. Thurn-Albrecht, J. Schotter, G.A. Kastle, N. Emley, T. Shibauchi, L. Krusin-Elbaum, K. Guarini, C.T. Black, M.T. Tuominen, and T.P. Russell. *Science* 290:2126, 2000

296. T. Thurn-Albrecht, J. DeRouchey, T.P. Russell, and H.M. Jaeger. *Macromolecules* 33:3250, 2000

297. K. Amundson, E. Helfand, X. Quan, S.D. Hudson, and S.D. Smith. *Macromolecules* 27:6559, 1994

298. A. Böker, H. Elbs, H. Hänsel, A. Knoll, S. Ludwigs, H. Zettl, A.V. Zvelindovsky, G.J.A. Sevink, V. Urban, V. Abetz, A.H.E. Müller, and G. Krausch. *Macromolecules* 36:8078, 2003

299. A. Böker, H. Elbs, H. Hänsel, A. Knoll, S. Ludwigs, H. Zettl, G. Krausch, V. Urban, V. Abetz, and A.H.E. Müller. *Phys. Rev. Lett.* 89:135502, 2002

300. S. Jiang, A. Göpfert, and V. Abetz. *Macromolecules* 36:6171, 2003

Chapter 11

Microstructure and Crystallization of Rigid-Coil Comblike Polymers and Block Copolymers

Katja Loos and Sebastián Muñoz-Guerra

CONTENTS

I. INTRODUCTION

Copolymer systems based on blocks which behave in a coil-like fashion (including di- and triblock copolymers) have been widely studied (Chapter 9). Coil–coil multiblock systems build of incompatible coil segments have been found to exist in a wide range of microphase separated supramolecular structures, such as spheres, cylinders, double diamond (DD), double gyroid (DG), and lamella. Their phase behavior mostly results from the packing constraints imposed by the connectivity of each block

In memory of Prof. Dr. Reimund Stadler.

and by the mutual repulsion of the dissimilar blocks. Phase separation and therefore the resulting stable morphology in diblock systems is greatly influenced by the total degree of polymerization (DP) ($N = N_A + N_B$), the Flory–Huggins χ parameter and the composition expressed by volume fractions $f_A, f_B, \ldots$.

This chapter focuses on block copolymers and comblike polymers in which at least one component is based on a conformationally rigid segment [1–3]. A measurement of the stiffness of a polymer is afforded by the so-called persistence length that gives an estimate of the length scale over which the tangent vectors along the contour of the chains backbone are correlated. Typical values for persistence lengths in synthetic and biological systems can be several orders of magnitude larger than for flexible, coil-like polymers. Rodlike polymers have been found to exhibit lyotropic liquid-crystalline ordered phases such as nematic and/or layered smectic structures with the molecules arranged with their long axes nearly parallel to each other. Supramolecular assemblies of rodlike molecules are also capable of forming liquid-crystalline phases. The main factor governing the geometry of supramolecular structures in the liquid-crystalline phase is the anisotropic aggregation of the molecules (Chapters 2, 3).

By combining these different classes of polymers a novel class of self-assembling materials can be produced since the molecules share certain general characteristics typical of diblock molecules and thermotropic calamitic molecules. The difference in chain rigidity of rodlike and coil-like blocks is expected to greatly affect the details of molecular packing in the condensed phases and thus the nature of thermodynamically stable morphologies in these materials. The thermodynamic stable morphology should be originated as the result of the interdependence of microsegregation and liquid crystallinity. From this point of view it is very fascinating to compare the microstructures originating in solution and in the bulk for such materials.

Practical applications in which the block copolymers are characterized by some degree of structural asymmetry have been suggested. For instance a flexible block may be chosen as it donates a flexural compliance, whereas the more rigid portion offers tensile strength. Besides mechanical properties, the orientation order and electrical conductance of certain rigid blocks could be exploited in optical and electrical devices. Furthermore, one could utilize the special properties of a constituent rod segment like, for instance, taking advantage of the chiral information in helical rods or even using certain hybrid systems (copolymers with polypeptides or polysaccharides) as enhanced biocompatible materials in medical technology.

II. COMBLIKE POLYMERS: HELICAL MAIN CHAIN WITH FLEXIBLE SIDE CHAINS

A. Polypeptides with Covalently Attached Long Alkyl Side Chains

References to polypeptides bearing long alkyl side chains are almost confined to the esters of poly(α,L-glutamic acid) (PγAG-n) and poly(α or β,L-aspartic acid) (PβAA-n or PαAA-n) whose chemical structures are depicted in Figure 1. In these compounds the alkyl side chain is connected to the main chain by a carboxylate group that is directly anchored to the backbone in the case of poly(β-peptide)s and through a methylene or ethylene spacer in poly(α-aspartate)s and poly(α-glutamate)s, respectively. Most of work on comblike polypeptides has been done with polyglutamates with some sporadic incursions in the polyaspartate area. N-acyl substituted poly(L-lysine)s constitute the only system investigated other than polyaspartates and polyglutamates. Although solution studies revealed that comblike poly(N-acyl-L-lysine)s are in the α-helical conformation when dissolved in hydrocarbons [4] these polypeptides show a strong tendency to adopt the β-folded conformation in the solid state [5,6].

Until very recently, the α-helical conformation with 3.6 residues per turn (Figure 2(a)) was considered to be exclusive of poly(α-peptide)s. In these last two decades it has been demonstrated however that poly(β-peptide)s, specifically those derived from aspartic acid, are also well suited for adopting folded secondary structures stabilized by intramolecular hydrogen bonds [7,8]. In these

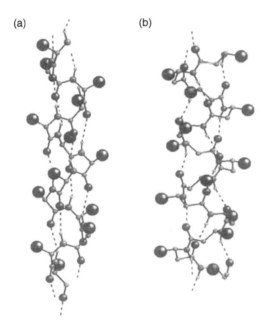

Figure 1 Chemical structures of comblike poly(γ-alkyl-α,L-glutamate) (PγAG-n), poly(β-alkyl-α,L-aspartate)s (PβAA-n), and poly(α-alkyl-β,L-aspartate)s (PαAA-n); n indicates the number of carbons in the poly-methylene side chain. The carbonyl position relative to the nitrogen atom is indicated by Greek letters.

Figure 2 (a) The 18/5 right-handed α-helix typical of poly(α,L-peptide)s. (b) The 13/4 right-handed pseudo-α-helix observed in poly(α-alkyl-β,L-aspartate)s. In both helices the hydrogen bond scheme is set between every third amide group but counted at opposite directions with respect to the directionality of the main chain. The larger spheres represent the side chains and hydrogen bonds are indicated as dotted lines.

systems, hydrogen bonds are set in a similar way as they are in the genuine α-helix although the existence of one additional methylene in the backbone of the repeating unit gives rise to a 13/4 helix with 3.25 residues per turn (Figure 2(b)). The stability of these pseudo-α-helices has been shown to be comparable to that of the α-helix in spite that several polymorphs have been identified in the solid state for different side chain groups. Nevertheless, features relevant to the assembling of supramolecular structures, such as polarity, chain stiffness, and molecular shape, are substantially identical for all these types of helices.

A great number of flexible polymers bearing linear long alkyl side chains has been examined and much evidence has been collected to conclude that the commonplace structure of these systems in the solid state is a biphasic arrangement of alternating layers of main chain and side chains [9]. A detailed x-ray diffraction study on poly(α-olefin)s carried out by Turner-Jones [10] demonstrated that the polymethylene side chain is crystallized only if it contains seven or more carbon atoms and that the crystal structure adopted may be either orthorhombic or monoclinic. Today, it is widely accepted

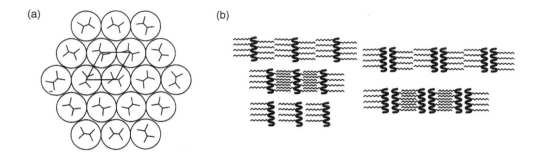

Figure 3   Packing of comblike polymers bearing polymethylene side chains. (a) Scheme of the hexagonal packing of polymethylene chains as viewed down the chain axis. (b) One-layer (left) and two-layer (right) models for the comblike polymer lamellar structure.

that the critical chain length for the onset of crystallization depends on the chemical constitution of the polymer and it turns out to be around 8–10 carbon atoms. In many comblike polymers, such as polyacrylates, polyethers, or poly(*N*-alkyl acrylamide)s [11], the side chain tends to crystallize in a hexagonal lattice with chains lying parallel to each other and the zigzag chain planes oriented at random about the chain axis (Figure 3(a)). Such a structure is similar to that found in *n*-paraffins at temperatures near above the melting point. This structure is described as a roughly hexagonal packing of cylinders with an average distance of 4.6–4.8 Å [12]. The number of methylenes included in the side chain crystallites (which may be estimated by calorimetry using heat fusion data available for *n*-alkanes [13]) is variable and may be only a small fraction of the whole side chain. This is thought to be due to the conformational distortions created when side chains approximate to each other and to the restrictions to mobility imposed by the functional bridging groups.

Two types of arrangements corresponding to one-layer and two-layer models with side chains interdigitated to a more or less extent can be observed in comblike polymers (Figure 3(b)). Side chains may be lying either normal or tilted with respect to the layers although the former is the model most frequently observed. The arrangement adopted is mainly determined by the constitutional nature of the polymer and its stereochemistry, although crystallization conditions may play a decisive role too.

It is well known that the polymethylene chain in the amorphous state tends to retain a considerable degree of ordering. In the axial direction, molten chains are described as an average *trans* sequence of 3 to 4 backbone bonds whereas the radial packing consists of flexible chains arranged in a random manner. In comblike polymers the regular spacing imposed by the main chain sequence contributes additionally to retain some ordering in the molten state, making the fusion process particularly complex. As a result, polymethylene side chains of comblike polymers at temperatures near above melting tend to be arranged in a quasi-hexagonal structure with chains in a more or less extended conformation. Such quasi-ordered structure is thought to act as a very efficient nucleating agent for rapid crystallization usually taking place in these systems upon cooling.

1.   Poly(γ-Alkyl-α,L-Glutamate)s (PγAG-n)

(a)   Isotropic and liquid-crystalline solutions

In spite of the fact that the higher homologs of the PγAG-*n* series are widely soluble in organic solvents, the behavior of these polymers in solution has not received much attention. This is partially explained because the ability of these systems to form crystal liquid phases in the bulk has directed much of the interest to the study of the solid state. Smith and Woody [14] carried out a viscosimetric and DC-ORD study on poly(γ-*n*-dodecyl-α,L-glutamate) (PγAG-12) dissolved in hydrocarbons and in strong breaking H-bond solvents, such as dichloroacetic acid or trifluoroacetic acid (TFA). Like it happens in polypeptides made of natural occurring amino acids, the conformation of PγAG-*n* in solution was shown to be highly dependent upon both solvent and temperature. More recently Poché [15] examined the solution properties of polydisperse poly(γ-stearyl-α,L-glutamate) (PγAG-18) in tetrahydrofuran (THF). Light scattering and capillary viscosimetry measurements were used to establish a Mark–Houwink equation with an exponent $a \approx 1.3$ applicable in the 35–250 K molecular weight domain. Such value is indicative of a stiff chain, although less stiff than poly(γ-benzyl-α,L-glutamate) and other rodlike polymers for which a has been estimated to be much closer to the theoretical value of 1.8. The apparent diameter of mutual exclusion calculated with the Onsager–Zimm–Schulz equation [16] was 37 ± 6 Å, substantially higher than the diameter of 23 Å computed by treating the polymer as a solid cylinder. This implies that the side chain sheath wrapping the helical backbone is perfectly solvent permeable. It was concluded furthermore that interrod distances less than 30 Å are prevented by the dynamics of the system that includes both side chain motion and molecular spinning about the rod axis. Cholesteric phases are described for toluene solutions of PγAG-18 with molecular weight in the 60–200 K range at concentrations above 16% [17]. No significant differences with those described for classical polypeptides such as poly(γ-benzyl-α,L-glutamate) are apparent. Homeotropic structures displaying a colorful background are reported to be occasionally observed. The cholesteric arrangement was found to collapse upon heating into a nematic structure [18].

Iizuka et al. [19] have examined the solution properties of a series of comblike poly(γ-*p*-alkylbenzyl-α,L-glutamate)s with alkyl chains containing from 18 up to 22 carbon atoms. In these systems a *p*-benzylene unit has been inserted between the carboxylate group and the polymethylene chain. When these polymers are dissolved in THF they show the high ellipticity characteristic of α-helix conformation with a helical content similar to that present in poly(γ-benzyl-α,L-glutamate)s of comparable molecular weights. Their concentrated solutions display, under crossed polarizers, equally spaced retardation lines characteristic of cholesteric mesophases. A new interesting observation reported by these authors is that line-width and therefore the pitch of the supramolecular helix is a function of the length of the alkyl side chain although no correlation is apparent. Moreover, the concentration dependence of the cholesteric pitch diminishes as the alkyl side chain becomes longer and vanishes when the number of carbons is larger than 16 (Figure 4). Solvent molecule exclusion from side interchain space is claimed to be the reason for these striking experimental results.

(b) Crystallized structures

The first account on the structural behavior in the solid state of comblike polyglutamates was reported by Thierry and Skoulios in a short letter published in 1978 [20]. They identified three ordered phases for PγAG-12 in the bulk as a function of temperature. The phase occurring at temperatures below 25°C was described as a layered structure with the polypeptide chain in the β-sheet conformation and the alkyl side chain crystallized in a separate phase with a periodicity of 24 Å. The high temperature phase appears in the proximity of 125°C and was thought to be a nematic phase with the α-helix polypeptide packed in a hexagonal array. Nothing was said about the structure adopted by the polymer at intermediate temperatures.

Watanabe et al. [21] carried out a systematic research on the structure in the solid state of PγAG-*n* embracing from pentyl to octadecyl side chains. The study included x-ray diffraction, differential scanning calorimetry (DSC) and dynamic mechanical methods. They concluded that the main chain assumes the α-helix conformation for all the members within the whole range of temperatures examined. Members with side chains containing ten or more carbon atoms were found to behave according to the pattern described by Thierry and Skoulious [20] for the dodecyl derivative. In such cases, side chains are long enough to crystallize separately upon cooling and the crystallized phase induces a layered arrangement with the α-helices aligned in sheets and the paraffin crystallites placed in between (Figure 5).

Information provided by the x-ray diffraction recorded in the low angle region revealed that side chains are oriented near normal to the sheet planes and that they have to interdigitate in more or less

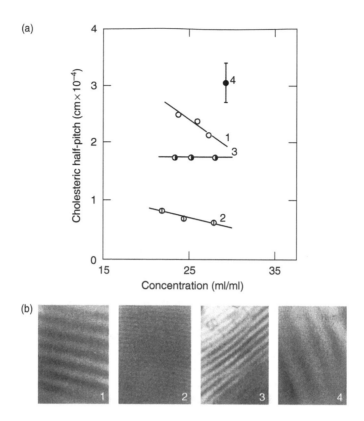

Figure 4 Changes in the cholesteric pitch of poly(γ-p-alkylbenzyl-α,L-glutamate)s concentrated solutions: (a) dependence of the half-pitch on polymer concentration; (b) polarizing optical micrographs (1: octyl, 2: dodecyl, 3: hexadecyl, 4: docosyl). (Adapted from E.I. Iizuka, K. Abe, K. Hanabusa, and H. Shirai. In: R.M. Ottenbrite, Ed. *Current Topics in Polymer Science*, Munich: Carl Hanser, 1987, p. 235. With permission.)

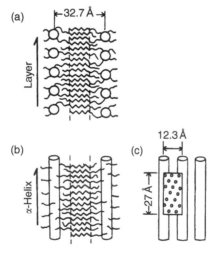

Figure 5 Model for the packing of PγAG-18 at low temperatures: (a) view parallel to the main chain axes; (b) view parallel to the side chain axis and perpendicular to the main chain axis; (c) view perpendicular to both the layer and the chain axis of the α-helices. (Taken from J. Watanabe, H. Ono, I. Uematsu, and A. Abe. *Macromolecules* 18:2141, 1985. With permission.)

extension according to their lengths. Unfortunately no information concerning the packing of the α-helices along the sheets could be obtained due to shear disorder affecting the stacking of the sheets. The melting temperature of the paraffin crystallites increased from −24 to +62°C when the number of carbon atoms in the alkyl side chain increased from 10 to 18. Wide-angle x-ray diffraction data indicated that crystallization of the side chains proceeded like in low molecular weight n-alkanes, so that a triclinic unit cell with the polymethylene chain in a fully extended conformation is adopted. The crystallinity of the system was evaluated by DSC which showed that the enthalpy of fusion increased linearly with the length of the side chain. The number of crystallized methylene units was calculated to vary from a minimum of 0.4 up to a maximum of 7.7 for the decyl and the octadecyl derivatives, respectively. These results are also supported by dynamic mechanical and dielectric measurements [21,22]. The fact that the γ-relaxation commonly attributed to small amplitude motions of alkyl chains was observed irrespective of temperature, whereas the presence of the β-relaxation arising from whole chain movements appeared restricted to temperatures at which the side chains are in the molten state is in agreement with the occurrence of partial crystallization of the side chain.

(c) Liquid crystallinity in the bulk

In the pioneer paper published by Thierry and Skoulios [20], it was advanced that polypeptides with sufficient long polymethylene side chains could display liquid-crystalline phases upon melting of the side chain. They in fact reported a high temperature nematic phase that reversed upon cooling into a lower temperature phase, whose structure was not elucidated. Since then, a good amount of research has been made on comblike polyglutamates, with the focus mainly toward the development of thin films with thermochromic properties. The liquid-crystal structure anticipated for these systems should be cholesteric according to the chiral nature of the main chain α-helix. However other mesomorphs have been described under certain conditions and for certain side chain compositions.

Watanabe et al. [23] dedicated great efforts to understand the high temperature behavior of PγAG-n. However many aspects still remain undisclosed, in particular, those concerning the new phase that is formed when the lamellar structure is heated just above the melting point of crystallites formed by the alkyl side chains. A hexagonal lattice made of a two-strand coiled-coil conformation was put forward for such a phase in PγAG-12 on the basis of both density measurements and low angle x-ray diffraction [24]. Since no similar evidence has been found so far for any other member of the series [21,25], such situation seems to be rather exceptional and needs corroboration. In this regard, one should mention the DSC studies carried out by Daly et al. [18], which show that melting of the lamellar state is strongly influenced by the existence of microheterogeneity in the polymer composition and that the structure of the phase formed after melting is largely conditioned by the thermal history of the sample.

PγAG-18 is likely the comblike polypeptide most thoroughly investigated. The most outstanding feature displayed by PγAG-18, shared also by other PγAG-n with $n \geq 10$, is the formation of liquid-crystalline phases upon side chain melting. However, all structural work published so far on the behavior of this polymer is phenomenological and restricted to phase identification and thermal transition characterization. The properties displayed by such phases are reported to be largely dependent on polymer size [22]. In samples with molecular weights above 100 K, a cholesteric phase is formed immediately after melting to ~50°C, which is then converted into a columnar liquid-crystal phase upon heating at higher temperatures [26]. The latter phase exhibits an extremely high viscosity and can be distinguished by a characteristic fan-shaped texture. X-ray diffraction indicated that such structure is made of a hexagonal two-dimensional lattice of α-helices. The columnar phase is actually viewed as a kind of micellar structure derived from the differences in polarity and shape existing between the side chains and the main chain. This structure had been theoretically predicted to appear as a result of excluded volume effects associated to the lateral packing of hard rods [27,28]. The columnar phase is closely related to the smectic-B phase from which it differs in the translational disorder existing along the direction of the helix axis (Figure 6(a)). The transition from cholesteric to columnar occurs in the vicinity of 200°C for samples with molecular weights above 60 K. For smaller sizes the transition temperature increases to attain values beyond the decomposition temperature when molecular weights fall below 20 K (Figure 6(b)).

In fact the liquid-crystalline behavior of low molecular weight PγAG-18 appears to be much more complex [29]. Up to three types of textures were observed under the polarizing optical microscope for a 20 K sample of this polymer as a function of temperature (Figure 7). The low temperature lamellar

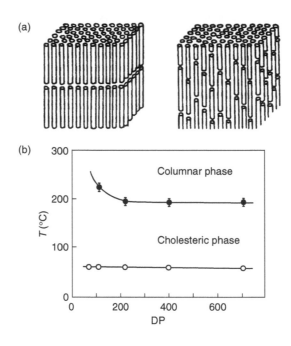

Figure 6 (a) Models for the smectic-B and hexagonal columnar phases of PγAG-18. (b) Variation in the cholesteric-to-columnar transition temperature with the polymerization degree. (Taken from J. Watanabe and Y. Takashina. *Macromolecules* 24:3423, 1991. With permission.)

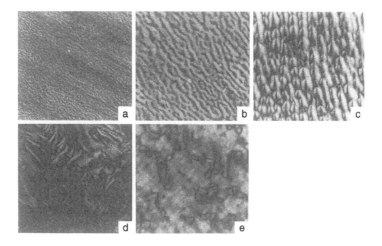

Figure 7 Liquid-crystal textures observed for PγAG-18 under the polarizing microscope as a function of temperature: (a) and (b) wormlike textures found in the 60°C to 110°C range; (c) the pseudofocal conic texture observed at 110°C; (d) the transition phase occurring at 130°C, and (e) the cholesteric fingerprint texture observed at 180°C. (Taken from J. Watanabe and Y. Takashina. *Polym. J.* 24:709, 1992. With permission.)

crystal phase is found to melt at 60°C into a biphasic wormlike texture composed of alternating dark and bright stripes. Above 110°C the polymer exhibits a pseudofocal conic texture, which is attributed to a smectic-A structure, and at 180°C a typical cholesteric phase is revealed by the appearance of the characteristic fingerprint pattern.

Stable liquid-crystallinity phases in the bulk may be also attained by combining short and medium size alkyl groups in the same polypeptide. Like in the case of homopolypeptides, the flexible side chains in the molten state play the role of the solvent in lyotropic compounds. Early observations on these systems were made on poly(γ-methyl-α,D-glutamate-co-γ-hexyl-α,D-glutamate)s [30], which are able to develop mesophases for comonomer compositions ranging from 30 to 70%. Similar results obtained in the study of other copolymers led to the conclusion that differences in length of the comonomer side chains of at least five methylene units are required to induce such behavior. Other comblike random copolymers showing a similar interesting behavior are poly(γ-benzyl-α,D-glutamate-co-γ-dodecyl-α,D-glutamate)s [31,32]. An additional interest afforded by these systems is that introduction of irregularities under control in the side chain wrap makes it difficult or even suppresses side chain crystallization upon solidification, while the lamellar structure is retained [23].

2. Poly(α-Alkyl-β,L-Aspartate)s (PαAA-n)

In the last few years a sustained research has been made on the structure of poly(β-peptide)s derived from L-aspartic acid. This research takes benefit of the expertise accumulated in the investigation of comblike poly(α-peptide)s and has brought in evidence of the close resemblance between the two families of polypeptides in both structure and properties. The research covers more than 20 poly(β,L-aspartic) esters differing either in the length or in the shape of the side chain. All these compounds were prepared by ring-opening polymerization of optically active β-lactams and have molecular weights above 100 K. The monomers were synthesized by a general methodology that makes use of cyclization and ester exchange reactions and starts from natural-occurring L-aspartic acid [33]. The most outstanding feature is that an α-helix-like conformation is adopted by the whole series both in solution and in the solid state [34,35]. As it happens with PγAG-n the degree of ordering achieved in the packing of the helices and the model of assembly adopted by the poly(β-peptide) is determined by the length of the alkyl side chain. This behavior is clearly characterized by x-ray diffraction of both isotropic and uniaxially oriented samples (Figure 8). Consequently PαAA-n may be classified in three structural groups corresponding to alkyl side chains of short ($n \leq 5$), medium ($n = 6$ and 8), and large ($n \geq 12$) sizes. The structural and conformational aspects of these three groups in connection with the supramolecular properties exhibited by each of them will be discussed in the following paragraphs.

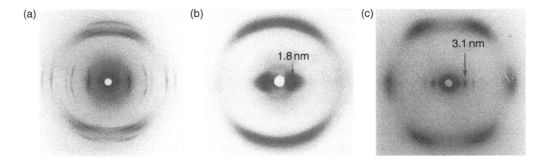

Figure 8 X-ray diffraction patterns of PαAA-n fibers: (a) short side chain (PαAA-3); (b) medium side chain (PαAA-6); and (c) long side chain (PαAA-18).

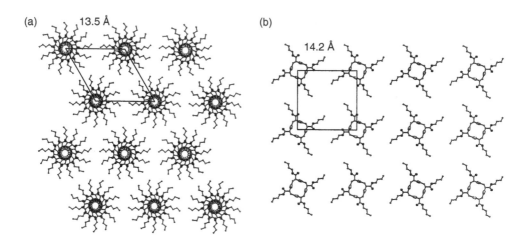

Figure 9 Projection along the chain axis of the crystal lattices of PαAA-4: (a) hexagonal form; (b) tetragonal form. Note that in (a) the helices are up and down whereas in (b) all the chains are pointing to the same direction. (For clarity the lattice scale is exaggerated with respect to the motif.)

(a)   Short side chain systems

Alkyl esters of poly(β,L-aspartic acid) ranging from methyl to pentyl in the solid state tend to be arranged in three-dimensionally ordered arrays with chains in the α-helix-like conformation. Two crystal forms are known to prevail over a few others of minor importance which show only slight differences with respect to the former [36]. The main crystal form is a pseudohexagonal packing of 13/4 helices arranged in antiparallel fashion while the second major form is a tetragonal array of parallel 4/1 helices (Figure 9). In both cases, the structure consists of a tied assembly of packed cylinders with lattice positions and radial orientations precisely determined by the interactions taking place between alkoxycarbonyl side groups.

In analogy with PγAG-n the conformation of PαAA-n in nonpolar solvents is helical and transition to the random coil state is promoted by addition of strong acids. Crystallization of the isobutyl derivative from chloroform dilute solutions yields hexagonal- or square-shaped single crystals 200 to 400 Å thick with helices arranged normal to the surface of the crystals [37]. On the other hand, the slow evaporation from chloroform produces complex microcrystalline aggregates with morphological features suggestive of a supramolecular helical structure (Figure 10(a)). Consistently, a continuous mesophase displaying the fingerprint pattern characteristic of cholesteric liquid crystals is generated upon storage of concentrated dichloromethane solutions (Figure 10(b)) [38].

(b)   Medium size side chain systems

PαAA-6 failed to crystallize in a three-dimensionally ordered structure. X-ray diffraction from solid films of this polymer indicated a rough hexagonal packing of 13/4 helices with an average interchain distance of 17 Å lacking azimuthal order [39]. Computer simulations based on molecular mechanics calculations revealed that helices tend to tilt with respect to the molecular axis when they are forced to approach each other in a hexagonal lattice with a parameter consistent with experimental data (Figure 11(a)). Also, the side chain becomes distorted from the all-*trans* conformation with torsion angles adopting values as low as 120°. Therefore, both calculations and experimental results support the conclusion that n-hexyl side chains are too long to allow the polymer to crystallize in a three-dimensional lattice but too short to crystallize themselves in a separated phase.

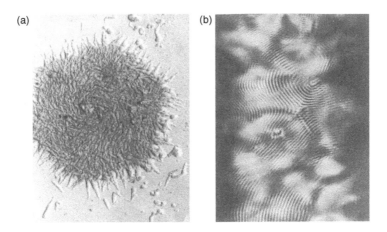

Figure 10 (a) Complex entities of poly(α-isobutyl-β,L-aspartate)s grown upon slow evaporation of a dilute solution in chloroform. (b) Fingerprint pattern displayed by a 23% (v/v) dichloromethane solution. (Taken from J.M. Montserrat, S. Muñoz-Guerra, and J.A. Subirana. *Makromol. Chem. Macromol. Symp* 20/21:319, 1988. With permission.)

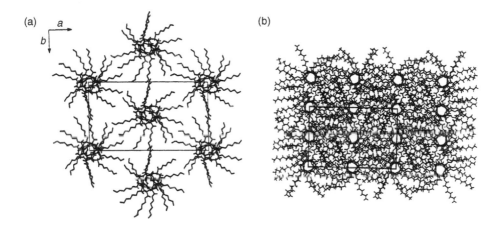

Figure 11 Projection down the *c*-axis of crystallized structures of medium size PαAA-*n* obtained by molecular mechanics calculations: (a) the hypothetically hexagonal structure of PαAA-6; note that the helix is tilted about 30° with respect to the *c*-axis; (b) the crystal lattice model of PαAA-8 including 16 chains; note that overlapping of side chains is preferred along the direction normal to the layers (*a*-axis). (Taken from J.J. Navas, C. Alemán, F. López-Carrasquero, and S. Muñoz-Guerra. *Polymer* 38:3477, 1997. With permission.)

The behavior of PαAA-8 was found to be remarkably different from that of PαAA-6 in spite of the fact that the alkyl side chain is only two methylenes longer [39]. Casting from chloroform solutions produces a partially disordered structure (similar to that adopted by PαAA-6) that could be crystallized by annealing. The crystal structure consists of a layered arrangement of chains with an orthorhombic unit cell in the $P2_12_12$ space group. The intersheet distance is 18 Å and the distance separating the helices along the sheet is 12 Å. Model calculations showed that the side chain trajectories in the interhelical space are almost fully extended and oriented normal to the helix sheet planes whereas those located within the sheets are folded. The degree of interdigitation into the molecular space of neighboring molecules greatly depends on the position of the residue along the 13/4 helix; whereas side chains protruding normal to the sheet are intercalated by about 50% of their lengths, those coming out within the sheet plane show an almost negligible overlapping (Figure 11(b)).

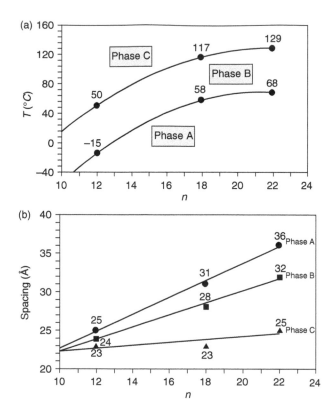

Figure 12   (a) Temperature domains of the three phases characterized in PαAA-*n*. (b) X-ray spacings of PαAA-*n* as a function of the number of carbon atoms contained in the alkyl side chain. (Adapted from F. López-Carrasquero, S. Montserrat, A. Martínez-Ilarduya, and S. Muñoz-Guerra. *Macromolecules* 28:5535, 1995. With permission.)

(c)   Long size side chain systems

The combined DSC and x-ray analysis of PαAA-*n* with $n \geq 12$ revealed the occurrence of two first-order transitions at temperatures T_1 and T_2 separating three structurally distinct phases [40]. Both transition temperatures and small-angle x-ray spacings characteristic of the phases decrease steadily with the number of methylenes in the side chain (Figure 12).

Not only T_1 and T_2 values, but also the enthalpy associated with transitions increase with the length of the alkyl side chain. The low temperature phase existing at $T < T_1$ is described as a layered structure with side chains crystallized in a separated hexagonal lattice. The conformation of the side chain in this phase is all-*trans* and the main chain is in the familiar 13/4 helical conformation. The number of methylene units crystallized ranges from about 2 in PαAA-12 up to 12 in PαAA-22, representing in all cases a small fraction of the side chains. The fact that side chains prefer a hexagonal packing in the paraffinic crystallites over the monoclinic one usually adopted in poly(γ-glutamate)s is related to the restriction to chain mobility derived from the lacking of the flexible ethylene spacer linking the alkoxycarbonyl group to the main chain. This feature however seems to have no significant influence on the molecular assembling of the helices.

T_1 is therefore attributed to the melting of the paraffinic phase composed of alkyl side chains. The slight contraction of the intersheet distance observed for this transition is consistent with a small deviation of the C–C torsion angles from the all-*trans* conformation. Therefore, the phase appearing above T_1 is regarded as a structure with the side chains in a conformation not very different from that present at low temperatures. Solid-state nuclear magnetic resonance (NMR)

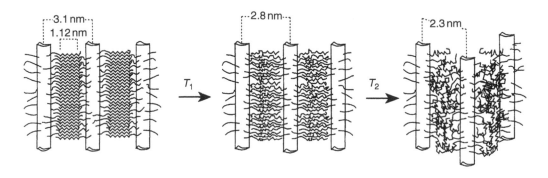

Figure 13 Schematic model illustrating the structural changes that take place in PαAA-*n* for *n* ≥ 12 upon heating. Both transitions are reversible with that occurring at low temperatures displaying a much faster switching. Distances are specified for the particular case of PαAA-18. (After López-Carrasquero et al. [40])

measurements gave strong support to these observations and corroborated that the α-helix conformation is retained by the polymer in this new phase. A detailed study carried out by DSC showed that crystallization of alkyl side chains upon cooling from a temperature above T_1 does not obey the Avrami model [41], but proceeds instead at unexpectedly high initial rates. This fact can be explained by assuming that a very efficient self-nucleation is operating; such mechanism is fully consistent with the arrangement accepted for the molten phase with side chains located very near to their crystallized positions [42]. Further insight into the structure of the phase existing between T_1 and T_2 was provided by polarizing optical microscope observations. Striking beautiful colors changing from red to blue were displayed by uniaxially oriented films of PαAA-12 and PαAA-18 upon heating from T_1 up to T_2 to finally exhibit uniform white birefringence when the latter temperature was surpassed. Such sequence of colors was found to reverse upon cooling. These effects are interpreted as due to selective reflection of circularly polarized light having a wavelength similar to the half-pitch of the supramolecular helical structure that gradually varies with temperatures [43].

The structural changes taking place at T_2 entail a considerable shortening in the periodicity of the lamellar structure. Differences with the low temperature phase ($T < T_1$) of about 11 Å for PαAA-18 and 8 Å for PαAA-16 are measured. Such differences imply that the layered structure is abandoned and helices are rearranged in a new phase characterized by an average interchain distance of 23 to 25 Å. In this high temperature, phase side chains are probably in the random coil conformation allowing the main chain helices to be loosely positioned in the space.

In summary, three structural phases differing in both the arrangement of the main chain helical rods and the conformation of the side chain are taken up by PαALA-*n* with *n* ≥ 12 as a function of temperature (Figure 13). The phase existing between T_1 and T_2 is assumed to be cholesteric with ability to crystallize upon cooling below T_1, and to convert into a nematic phase upon heating above T_2. Since the layered phase existing at low temperatures may be envisioned as a quenched smectic structure with helices immobilized by the crystallized side chains, the structural transitions occurring in these systems appear to follow a smectic ↔ cholesteric ↔ nematic sequence.

B. Ionically Attached Long Alkyl Side Chain Polypeptides

Assemblies bonded by means of noncovalent interactions such as hydrogen bonding or charge transfer, offer a number of advantages that makes these systems particularly interesting as building blocks for generating supramolecular structures [44]. The preparation of such systems does not involve complicated chemical procedures, structure equilibrium may be realized and rearrangement of the structure is feasible by adequate adjustment of the external conditions. Nevertheless, it is

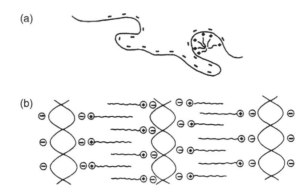

Figure 14 Scheme of polyelectrolyte–surfactant complexes in solution: (a) flexible polyelectrolyte chain; (b) rigid polyelectrolyte chain. (Taken from A. Ciferri. *Macromol. Chem. Phys.* 195:457, 1994. With permission.)

remarkable that these assemblies are able to reproduce the mesophases known to exist in covalently linked comblike polymers. In particular, the superimposition of electrostatic free energy to soft and hard interactions characteristic of uncharged rodlike polymers gives rise to supramolecular arrangements of relevance to self-assembly of complex biological structures [45].

Formation of complexes from two polyelectrolytes of complementary charge has been recently reviewed [46]. Systems investigated so far include natural compounds as polysaccharides as well as other synthetic pairs. Cell-like thin spherical membranes and other shapes displaying higher anisotropy can be also fabricated by selecting the adequate techniques. Nevertheless, the most studied self-assembling polymeric systems based on electrostatic interactions are complexes made of synthetic polyelectrolytes and oppositely charged low molecular weight amphiphilic molecules. The recent discovery of the ability of these systems to dissolve in organic solvents and spontaneously form well-defined supramolecular structures has motivated broad interest in this area. Today it is known that electrostatic complexes may generate stable assemblies with lamellar or cylindrical shape depending on the charge density of the polymer and the chemical nature of both the polymer and the surfactant [47].

Low molecular weight surfactants are known to form spherical micelles at the critical concentration. On the other hand, complexes of different architecture (Figure 14) are formed by coupling surfactant with polyelectrolytes according to the conformation assumed by the polymer [48]. Flexible polymer chains and surfactant molecules in solution are envisioned as discrete clusters of surfactant enclosed by polymer loops. In the case of rigid polymers the basic micelle is assumed to consist of a cylindrical assembly with ionically bounded molecules aligned perpendicular to the axis of the polymer cylinder. These basic units may self-aggregate either longitudinally or side-by-side to yield nematic phases and gels. The assembly can be maintained in solution due to the presence of uncompensated charges at the cylinder skin. The regular distribution of rigid (polymer chain) and flexible (surfactant chain) components reproduces the arrangement found in both high-performance composites and biological complexes where highly oriented polypeptide or polysaccharide chains are embedded in an amorphous matrix of cross-linked proteins.

Complexes of α-helical polypeptides and oppositely charged low molecular weight surfactants constitute a novel class of comblike assemblies, which are able to form lamellar structures in the solid state similar to those observed for covalently linked flexible–rigid polymers. Complexes of poly(γ,L-glutamic acid) [49,50] and poly(L-lysine) [51,52] with surfactants of opposite charge (Figure 15) have been recently investigated by the Tirrell's group. A brief account of the most remarkable features of the structures displayed by these complexes is given in Section II.B.1 and Section II.B.2.

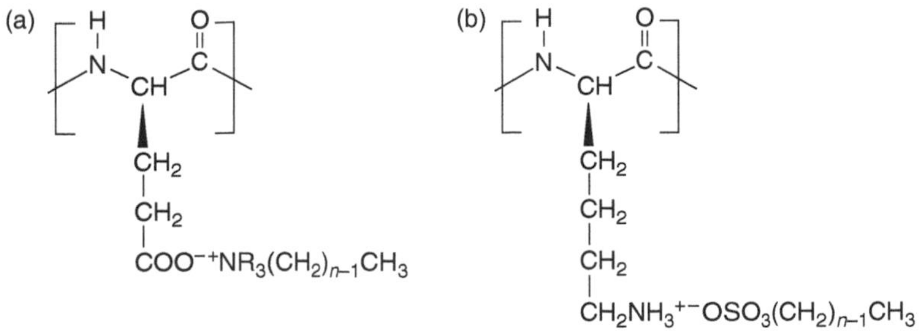

Figure 15 Chemical structures of (a) poly(α,L-glutamic acid)–alkyltrimethylammonium cation complexes and (b) poly(L-lysine)–alkyl sulfate anion complexes.

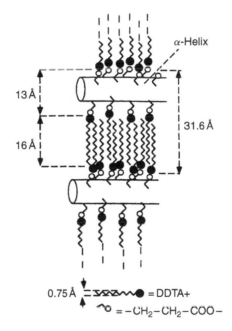

Figure 16 Scheme of the lamellar structure of the poly(γ,L-glutamic acid)–dodecyltrimethylamonium complexes viewed parallel to the layers and perpendicular to the α-helices axes. (Taken from E.A. Ponomarenko, A.J. Waddon, K.N. Bakeev, D.A. Tirrell, and W.J. MacKnight. *Macromolecules* 29:4340, 1996. With permission.)

1. Poly(α,L-Glutamic Acid)–Alkyltrimethylammonium Cation Complexes

Stoichiometric ionic complexes of poly(γ,L-glutamic acid) and alkyltrimethylammonium cations with the long linear alkyl group being dodecyl, hexadecyl, and octadecyl were obtained as solid precipitates when equal amounts of aqueous solutions of sodium poly(γ,L-glutamate) and surfactant were mixed at room temperature at pH 8 [49,50]. The analysis of solid films of these complexes by FTIR and CD indicated that the polypeptide chain is predominantly in the α-helical conformation and low angle x-ray diffraction revealed that all the complexes adopt a lamellar structure made of alternating layers of polypeptide and surfactant (Figure 16). The shorter surfactant alkyl chains containing 12 and 16 carbon atoms are disordered in the complex, whereas the surfactant chain of 18 carbon atoms is crystallized in a hexagonal lattice.

A linear dependence of the long period of the lamellae on the number of carbon atoms in the surfactant chain was found. The slope of the line is about 1.3 Å per methylene indicating that surfactant chains are nearly fully extended, interdigitated and oriented nearly normal to the lamellar surfaces. The paraffin crystallites of the octadecyltrimethylammonium complex melt at 48°C with an associated fusion heat of about 2.1 kcal/mol. No other thermal transition was found to take place in the temperature range of 10 to 170°C. The complexes do not show ordered melt and do not flow upon heating at temperatures below decomposition. They undergo irreversible changes in structure and properties upon storage, a process that is accompanied by generation of a small amount of crystalline surfactant. Such alteration is assumed to be due to unbinding of the alkyltrimethylammonium cations caused by uncontroled hydrolysis.

2. Poly(L-Lysine)–Alkyl Sulfate Complexes

Stoichiometric complexes of poly(L-lysine) cations and dodecyl sulfate anions were investigated by Ponomarenko et al. [51,52]. The complexes are prepared by the same method used in the preparation of poly(γ,L-glutamic acid)–alkyltrimethylammonium complexes. The secondary structure adopted by the polypeptide in these complexes was found to be very susceptible to environmental conditions. In chloroform solution containing 1 to 2% (v/v) TFA the polypeptide chain is in the α-helical conformation whereas a disordered state is present at TFA concentrations above 5%. It was interesting to observe that a noticeable decrease in the ^{1}H spin-lattice relaxation time accompanied the helix-coil transition whereas no changes were detected in T_1 of protons attached to the surfactant counterparts. It was concluded therefore that in both states the surfactant alkyl chains are nearly extended and exposed to the solvent shielding both the polypeptide backbone and the joining ionic groups (Figure 17).

In the solid state the poly(L-lysine)–dodecyl sulfate complex is invariably organized in a lamellar structure consisting of layers of polypeptide chains alternating with double layers of surfactant molecules arranged tail to tail. However, both the secondary structure adopted by the polypeptide and the packing of the polymethylene chains in the paraffinic phase depend upon the history of the sample. In the powder isolated from synthesis, the structure of the complex is predominantly of β-sheet with side chains in a liquid-like state characterized by only short-range order. On the contrary, films cast from a chloroform solution in which the polypeptide is in the disordered state exhibit a α-helix conformation of the main chain with side chains also uncrystallized but with the mobility more restricted than in the former case.

When a mixture of ethyl and octadecyl sulfates are mixed with poly(L-lysine) hydrobromide, the complex that is formed contains exclusively octadecyl sulfate anions. The selective binding of the long surfactant chains is attributed to the hydrophobic driving forces that are determinant for micellization. Conversely, complexes with the composition reflecting that of the mixture were obtained when two long chain surfactants were used. By this means it was feasible to prepare complexes of poly(L-lysine) with either octyl or octadecyl sulfates, and with mixtures of both surfactants for a variety of compositions. The method turns to be unique in providing ionically bonded comblike copolymers with a distribution predominantly in blocks. It seems that the hydrophobic self-association of the surfactants is combined with the ability of the electrostatically attached side chains to move along the main chain. The microstructure of these copolymers is therefore distinguished from covalently linked copoly(γ-alkyl-α,L-glutamate)s in which the attached side chains invariably have a random distribution.

In agreement with other covalently linked systems previously discussed, the electrostatically bounded supramolecular complexes of poly(L-lysine)–surfactant complexes adopt lamellar structures too, although the polypeptide is predominantly in the β-sheet conformation. A striking feature of these mixed systems is that the molecular organization in the supramolecular lamellar structure is governed not only by the surfactant chain length but also by the composition of the complex (Figure 18). Thus the assembly generated by the poly(L-lysine)–octadecyl sulfate complex has side chains crystallized in a hexagonal lattice with the chains interdigitated and oriented perpendicular to the sheet plane. Conversely, in the poly(L-lysine)–octyl sulfate complex, the chains are in the disordered state and arranged tail-to-tail. In complexes with mixed surfactants, the lamellar organization approaches one or the other extreme according to composition. Assemblies containing more than

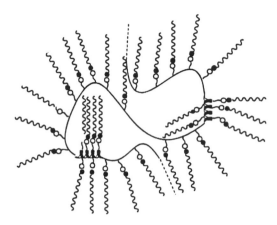

Figure 17 Schematic representation of the structure of the poly(L-lysine)–dodecyl sulfate complex with the polypeptide chain mainly in the coil conformation. (Taken from E.A. Ponomarenko, D.A. Tirrell, and W.J. MacKnight. *Macromolecules* 31:1584, 1998. With permission.)

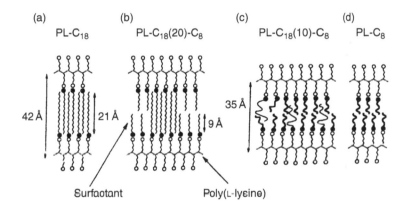

Figure 18 Arrangement of the surfactant molecules within the lamellae of complexes of poly(L-lysine)–alkyl sulfates containing mixed alkyl chains. Figures in brackets indicate the composition of the system. (Taken from E.A. Ponomarenko, D.A. Tirrell, and W.J. MacKnight. *Macromolecules* 31:1584, 1998. With permission.)

20% of octadecyl chains adopt a partially crystalline structure similar to the poly(L-lysine)–octadecyl sulfate complex whereas a disordered arrangement is preferred at lower contents.

III. BLOCK COPOLYMERS: ROD–COIL MAIN CHAIN POLYMERS

Even though most experimental studies on the behavior of rod–coil block copolymers were published only recently, previously a lot of theoretical work had been done for this interesting class of polymers. In one of the earliest theoretical work Semenov and Vasilenko introduced a microscopic model that accounts for the steric interactions among the rods, the stretching of the coils and the unfavorable rod–coil interactions. They used the so-called strong segregation limit (SSL) approach. Theories that deal with block copolymer phase behavior can be divided into two categories: (i) SSL (Flory–Huggins parameter $[\chi N] \gg 10$) and (ii) weak segregation limit (WSL, $\chi N \leq 10$) (Chapters 9, 10). In the SSL theories, well-developed microdomain structures are assumed to occur with relatively sharp interfaces and chain stretching is explicitly accounted for. The WSL theories are premised on lower amplitude sinusoidal composition profiles and unperturbed Gaussian coils (neglected chain stretching) [53]. Assuming that the rods are strictly aligned in the direction normal to a lamellar-like assembly, Semenov and Vasilenko considered a nematic phase and a smectic-A lamellar phase where the rods remain perpendicular to the built sheets. The transition from a monolayer to a bilayer lamellar phase was also analyzed [54]. Semenov extended the work including smectic-C phases, with the possibility for the rods to tilt by an angle θ to the lamellar normal [55]. In a subsequent work multiblock linear macromolecules were considered. The authors constructed a theory for nematic–smectic and smectic-A–smectic-C transitions in the melt [56]. Williams and Frederickson extended these calculations and incorporated nonlamellar phases, where the rods form finite-sized disks ("hockey pucks") surrounded by a corona of coil molecules. These pucks are predicted to pack and form a three-dimensional superstructure [57]. In the WSL spatially uniform planes of diblocks had been predicted by Holyst and Schick [58]. The effect on the lamellar phases was studied by Matsen by varying the stiffness of the rod component [59]. Schick and coworkers introduced anisotropic interactions between the semiflexible blocks and studied the effects of external fields furthermore [60]. They also studied spatially ordered phases in a rod–coil diblock copolymer melt by applying self-consistent field equations in the WSL or with a brushlike approximation in the SSL [61]. Matsen and Barrett extended the Semenov–Vasilenko model by analyzing it with self-consistent field techniques for lamellar morphologies [62].

The phase behavior in solution of rod–coil polymers has been analyzed by Halperin [63,64] and by de Gennes and coworkers [65]. By incorporating enthalpic contributions from the solvent, "tilted" phases were predicted for copolymers in a selective solvent for the coil component by Halperin. Microdomains resembling "plates," "fences," and "needles" were instead considered by de Gennes. Sevick et al. examined the morphology of rod–coil polymers grafted to repulsive surfaces by the end of the flexible segment in poor solvents. The predictions had to be altered in this case because the lack of flexibility of the rod component becomes more significant as the dimensionality is reduced from three (bulk) to two (surface). The micelles forming on a surface were predicted to be turnip and jellyfish-like [66].

In a very recent publication Düchs and Sullivan report on dilute solutions of rod–coil block copolymers by self-consistent field theory considering just the difference in chain rigidity of diblock copolymers. The theory was developed in one as well as in three dimensions and the one-dimensional approach was used to compute the complete phase diagram of rod–coil block copolymers in solution [67].

Li and Gersappe used a two-dimensional self-consistent field lattice model to study the phase diagram of rod–coil block copolymer melts. With this model they observed not only lamellar morphologies but also structures found in experiments, such as the "zigzag" morphology and elliptical cross-sectional cylinders. Nevertheless, free energy analysis suggests that these morphologies are metastable [68].

The phase diagram of rod–coil block copolymer melts was also studied by Reenders and ten Brinke. The authors were able to distinguish up to seven different morphologies depending on molecular architecture, temperature, and relative strength of interaction. For this they derived a Landau expansion of the free energy up to the fourth order in the two-order parameters ψ and S, representing the compositional ordering (driven by Flory–Huggins interaction) and the nematic ordering (driven by Maier–Saupe interactions), respectively. The obtained phase diagram is in good accordance with experimental results [69].

A. Helical-Rod Systems

1. Copolymers with Polypeptides

Polypeptides are classical examples of polymer molecules whose rigid regular conformation is stabilized by strong intramolecular hydrogen bonds. They can be found in a wide range of

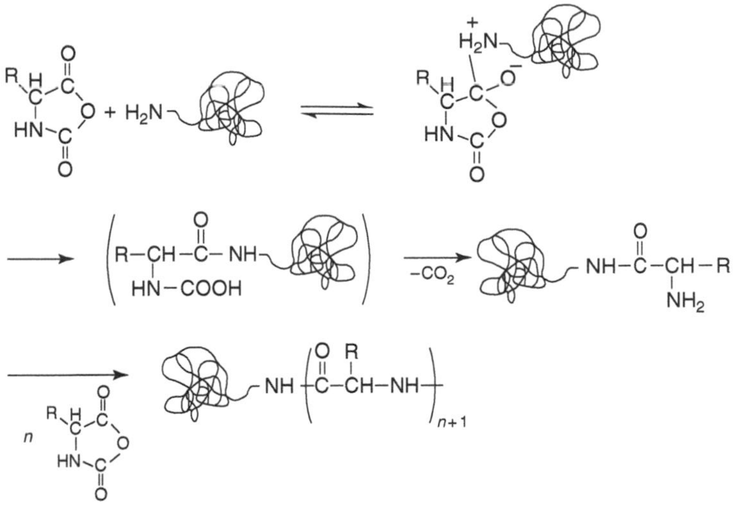

Figure 19 Exemplary synthetic scheme for the polymerization of *N*-carboxyanhydrides from amino-functionalized polymers.

conformations from which the two major ones are a right-handed α-helix and a sheet structure (see Section II.A). The α-helix is a rodlike structure in which the polypeptide main chain is coiled and forms the inner part of the rod; the side chains extend outward in a helical array. The helix is stabilized by hydrogen bonding between the NH and the CO groups of the main chain. The β-pleated sheet conformation differs markedly from the α-helix because it rather behaves like a sheet than a rod. Copolymers that incorporate polypeptide segments are not only attractive because of their rod–coil behavior but also because they are biocompatible and biodegradable, which results from their natural or nature identical component. For a recent review see Ref. 70.

Gallot and coworkers reported on polybutadiene–block–poly(γ-benzyl-L-glutamate) block copolymers that were synthesized by polymerization of the *N*-carboxyanhydrides of the corresponding amino acids from amino-terminated polybutadiene [71] (see Figure 19 for an exemplary synthetic route).

From this polymer the authors also obtained polybutadiene–block–poly(N^5-hydroxypropyl-glutamine) by the reaction with 3-amino-1-propanol. By studying the resulting block copolymers by x-ray diffraction the authors found lamellar structures for a wide range of peptide content (20 to 75%). This result was also confirmed by studying film samples with electron microscopy. For that purpose mesomorphic gels were prepared in 2,3-dichloro-1-propene and then subjected to UV irradiation, ultramicrotomed, and stained by OsO$_4$. The structural parameters obtained by both methods are in good agreement with each other. By analyzing the samples with x-rays in an angle domain corresponding to the periodicity between 3 and 50 Å the authors found that the polypeptide chains are in the α-conformation and packed in a hexagonal array. Furthermore, they are folded into sheets showing a very uniform lamellar thickness even though the polypeptide blocks have a high polydispersity. The authors proposed an interdigitated (where the chains fold after crossing all the thickness of the polypeptide layer) and bilayer arrangement (see Figure 20), whereas the first assembly seems to be favored by energetical considerations.

Polystyrene–block–poly(L-lysine), polybutadiene–block–poly(L-lysine) [72], and block copolymers of polystyrene or polybutadiene with poly(carboxybenzoxy-L-lysine) [73] were also synthesized by Gallot and coworkers and evidence for lamellar assemblies were presented. Polysaccharide–block–polypeptide rod–coil systems were obtained by polymerizing the *N*-carboxyanhydrides of γ-benzyl-L-glutamate from the asparagine α-amino function of carbohydrate fractions of ovomucoid (a glycoprotein extracted from hen egg white). Lamellar assemblies in concentrated solutions and in

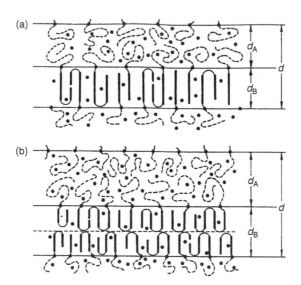

Figure 20 Schematic representation of the lamellar structure of block copolymers based on polypeptides as proposed by Gallot and coworkers; (a) the polypeptide chains fold after crossing all the thickness of the polypeptide layer and (b) the polypeptide chains fold after crossing only half of the thickness of the layer. (Reprinted from B. Perly, A. Douy, and B. Gallot. *Makromol. Chem.* 177:2569, 1976. Copyright [1976] Wiley-VCH. With permission.)

the dry state were again observed [74]. The chain folding of the polypeptides in the lamellar structures was studied for these block copolymers with x-ray diffraction and IR-spectroscopy. It became obvious that the number of folds depends upon the nature of the two blocks (the vinyl polymer and the polypeptide); for instance, in block copolymers with polystyrene blocks the number of folds of the polypeptide chains is greater than with polybutadiene blocks. Also poly-L-lysine seems to be more rigid than, for example, poly-γ-L-glutamate that is represented in the number of folds [75].

Schlaad et al. reported in greater detail on the solid-state phase behavior of linear and bottle-brush shaped polystyrene–block–poly(Z-L-lysine) in a wide range of molecular weight, composition, and architectures [76]. According to earlier reports on rod–coil block copolymers with polypeptides, mainly (undulated) lamellar morphologies were observed (see Figure 21).

Poly(γ-benzyl-L-glutamate)–block–polyisoprene diblock copolymers were reported by Tirrell and coworkers [77]. The synthesis was based on coupling of carboxy-terminated polyisoprene and amino-terminated poly(γ-benzyl-L-glutamate) prepolymers. By dynamic light scattering studies in dichloromethane they found that aggregates were observed in less than 24 h and remained stable over several days. In dimethyl formamide (DMF) two relaxation modes were observed that the authors assigned to translational diffusion of the micelles (slow mode) and to moleculary dispersed diblock molecules (fast mode). Since the radius of the sphere in DMF was found to be too large to be assigned to a simple micelle, the authors proposed a multilayered complex sphere built by a simple sphere and successive diblock layer adsorption. By studying the system in a selective solvent for the rod segment (DMF) the observed micelles increased in size if the coil segment was extended. In contrast to this an extension of the coil segment does not influence the aggregate in dichloromethane.

Recently, two different groups reported independently in greater detail on the vesicular structures found in solutions of polybutadiene–block–poly(γ-sodium-L-glutamate) [78] and polybutadiene–block–poly(γ-L-glutamic acid) [79,80], respectively, shown in Figure 22. In both cases the ring opening polymerization of the N-carboxyanhydride of γ-benzyl-L-glutamate was initiated from amino end functionalized polybutadiene (anionic polymerization). The γ-benzyl-groups were removed from the resulting block copolymers by palladium catalyzed hydrogenation to result in polybutadiene-block-poly(γ-L-glutamic acid) [79] and after further reaction with NaOH

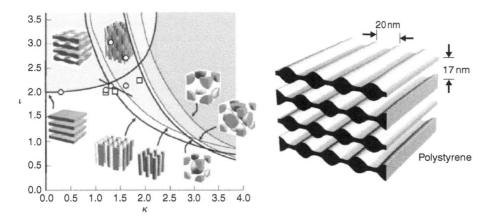

Figure 21 Generalized phase diagram including morphologies found in polystyrene–block–poly(Z-L-lysine), □ linear block copolymers, ○ bottle-brush shaped block copolymers, and schematic presentation of the undulated lamellar morphology. (Reprinted from H. Schlaad, H. Kukula, B. Smarsly, M. Antonietti, and T. Pakula. *Polymer* 43:5321, 2002. Copyright [2002], with permission from Elsevier Science.)

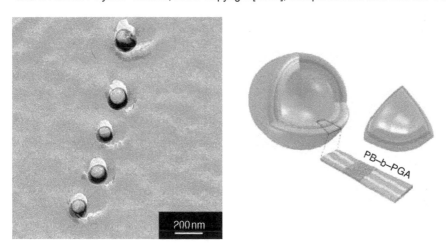

Figure 22 Freeze fracture TEM image of vesicles found in aqueous solutions of polybutadiene–block–poly(γ-sodium-L-glutamate) and schematic model of vesicles found in polybutadiene–block–poly(γ-L-glutamic acid). (Reprinted from H. Kukula, H. Schlaad, M. Antonietti, and S. Förster. *J. Am. Chem. Soc.* 124:1658, 2002. Copyright [2002 and 2003], with permission from American Chemical Society and Springer. F. Chécot, S. Lecommandoux, H.A. Klok, and Y. Gnanou. *Eur. Phys. J. E* 10:25, 2003.)

polybutadiene-block-poly(γ-sodium-L-glutamate) was obtained [78]. In both cases aggregates could be detected in aqueous solution (dynamic and static light scattering, small angle neutron scattering, freeze fracture TEM) that cannot be assigned to micelles but to a vesicle type morphology. (see Figure 22). The secondary structure of the polypeptide corona could be altered from α-helical to random coil by variation of the pH (as detected by CD measurements) but the vesicular nature of the aggregates was not destroyed by this.

Lecommandoux and coworkers were able to manipulate the vesicular structures of polybutadiene–block–poly(γ-L-glutamic acid) in solution with varying pH and ionic strength. By cross-linking the diene units of the block copolymers they obtained shape persistent stimuli responsive nanoparticles [80]. Borsali et al. reported on rod–coil to coil–coil transition of poly(styrene-d_8)–block–poly(γ-benzyl-L-glutamate) in different solvents with small angle neutron scattering [81] and discussed the static and dynamic scattering properties of rod–coil block copolymers via random phase approximation [82].

Hayashi and coworkers [83] reported on A–B–A-type block copolymers consisting of poly(γ-benzyl-L-glutamate) as the A component and polybutadiene as the B component prepared by polymerization of γ-benzyl-L-glutamate N-carboxyanhydrides from *bis*-amino-terminated *trans*-polybutadiene. Circular dichroism spectroscopy in 1,2-dichloroethane and IR spectroscopy in the solid state showed that the peptide chains form an α-helix. In the solid state the polybutadiene chains were in a random coil conformation forming cylindrical domains embedded in the polypeptide matrix phase, as shown by wide-angle x-ray diffraction and electron microscopy. The temperature dependence of the dynamic and loss moduli of the triblocks showed that the dynamic mechanical properties can be explained by the observed microseparated structure. By studying block copolymers with poly(γ-benzyl-D,L-glutamate) the helical content of the D,L-copolymers in a helicogenic solvent and in solid state was estimated on the basis of ORD and IR spectra. The content of the right-handed α-helices and left-handed α-helices and random coil conformations of the polypeptide chains was determined. From wide-angle x-ray diffraction measurements, it was shown that the hexagonal crystalline phase almost disappeared for D,L-copolymer membranes indicating a breakdown of the α-helical conformation in the main chains. The morphological structure of these block copolymer membranes was studied with electron microscopy and the D,L-block copolymers membranes were found to build spherical structures when the volume fractions of the polybutadiene portion were rather high [84].

The authors also reported on A–B–A-type block copolymers composed of poly(γ-benzyl-L-glutamate) as the A component and polyisoprene as the B component. By using wide-angle x-ray diffraction it could be shown that the block copolymers exhibit mesophase behavior in different solvents. In this case polyisoprene chains are in a random coil conformation and form domains embedded in the matrix phase consisting of poly(γ-benzyl-L-glutamate) chains in the α-helix conformation. Model analysis of the complex modulus of the membrane cast from solution suggested the occurrence of spherical and cylindrical domain structures in the membrane [85]. Morphological study of these triblock copolymers using pulsed NMR spectroscopy and electron microscopy revealed that the solid block copolymers are in microphase separated state. The NMR measurement was used to detect the interface of the block copolymer. Spin-lattice relaxation time T_1 of the block copolymer revealed a domain size, which is in good agreement with the result obtained from electron microscopy [86].

Gervais et al. used x-ray photoelectron spectroscopy (XPS) to study the surface properties of polypeptide–block–polybutadiene–block–polypeptide triblock copolymers and found that the lamellar structures (which is as well observed by x-ray diffraction for these triblock systems) are perpendicular to the air–polymer interface. For more hydrophilic polypeptides they found that the surface is richer with polybutadiene whereas for hydrophobic polypeptides the surface resembled the block copolymer composition [87].

Hayashi et al. also analyzed the membrane surfaces of A–B–A triblock copolymers containing poly(ε-N-benzyloxycarbonyl-L-lysine) as the A component and polybutadiene as the B component. The surface characteristics of membranes built from the triblock were investigated by XPS, replication-electron microscopy, and wettability measurements. By investigating the surface with XPS measurements the composition of the outermost surface proved to be quite different from the bulk composition; contact-angle measurements indicated the existence of an interfacial region between the α-helical polypeptide domains and the polybutadiene domains at the surfaces of the triblock membranes [88]. The authors found that the permeability of membranes prepared from their triblock systems was much higher than that of the polypeptide homopolymer and increased in proportion to the interfacial area between the domains. This suggests that residues near the end of the peptide chain and terminal residues of the diene block, which are located in the interfacial region between the domains, are responsible for water permeability, since the amine and carbonyl groups do not form intramolecular hydrogen bonds (see Figure 23 for a schematic drawing of a membrane built from a diblock copolymer) [89]. The authors also took very great care studying the biocompatibility, biodegradability, and the membrane efficiency of their polymers [90–95].

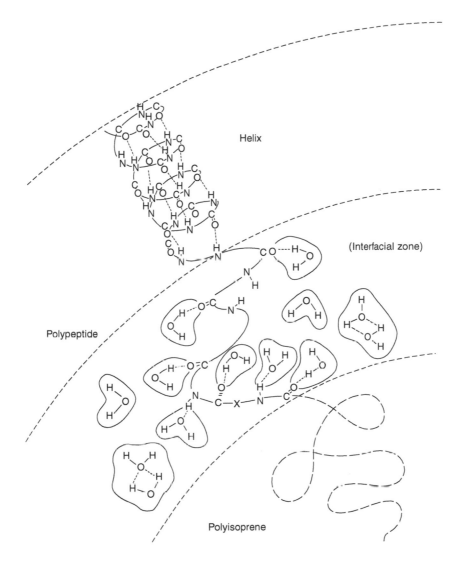

Helix

(Interfacial zone)

Polypeptide

Polyisoprene

Figure 23 Schematic representation of a membrane formed by a polyisoprene–block–polypeptide diblock copolymer. In the interfacial zone (which is thought to be the main structural property that allows hydraulic permeability in this system) hydrogen-bonded water and clusters of water are drawn. (Reprinted from R. Yoda, S. Komatsuzaki, and T. Hayashi. *Eur. Polym. J.* 32:233, 1996. Copyright [1996], with permission from Elsevier Science.)

Anderson and coworkers reported on rod–coil–rod triblock copolymers basing on poly(γ-benzyl-L-glutamate) as the rod segment and poly(butadiene/acrylonitrile) as the coil segment. In films prepared from dioxane (selective for the polypeptide block) they observed a lamellar structure whereas in films cast from chloroform (not preferential for either block) a nearly homogeneous phase mix became obvious. An interesting time dependency of the morphologies was found in films processed from xylene: films from freshly prepared solutions showed a well-defined lamellar morphology, whereas in films cast from solutions that had been stored for a minimum of 2 days a poorer domain formation occurred. The alternation of microphase structure may be due to gelation resulting in aggregation of the polypeptide helices [96].

Recently, Klok and coworkers studied the behavior of poly(γ-benzyl-L-glutamate)–block–poly(ethylene glycol)–block–poly(γ-benzyl-L-glutamate) rod–coil–rod triblock copolymer melts. It became obvious that the self-assembly of the copolymers strongly depends on the peptide volume

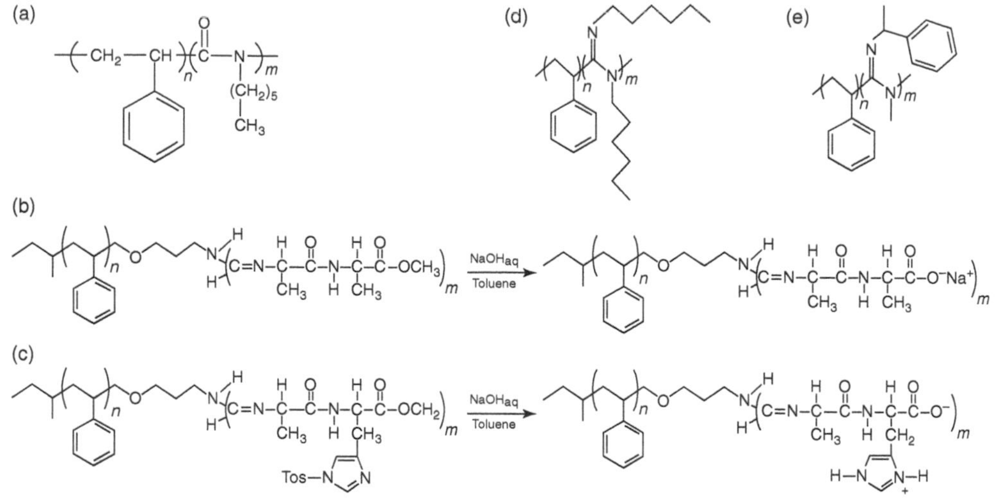

Figure 24 Chemical structures of (a) polyhexylisocyanate; (d) poly(N, N'-di-n-hexylcarbodiimide); (e) poly[(R/S)-
N-(1-phenylethyl)-N'-methylcarbodiimide)]; (b) and (c) shows the synthetic scheme of the removal
of the ester functions of poly(isocyano-L-alanine-L-alanine) and poly(isocyano-L-alanine-L-histidine),
respectively.

fraction ranging from microphase separated assemblies with defined secondary structure in the
peptide blocks to structures showing interfacial mixtures with less coherent peptide secondary
structures [97].

2. Copolymers with Polyisocyanates, Polyisocyanide, and Polycarbodiimides

In contrast to polypeptides the conformational properties and chain rigidity of polyisocyanates,
polyisocyanide, and polycarbodiimide is determined by their primary chemical structure rather
then by intramolecular hydrogen bonding. This results in a greater stability in the conformational
characteristics in solution as compared to polypeptides. Furthermore, it is possible to achieve high
molecular weights in these systems and with certain synthetic methods (anionic synthesis, transition
metal catalysis) even the polydispersity can be reduced, a reason that makes block copolymers based
on these rigid segments very attractive.

Chen et al. [98,99] reported on diblock copolymers containing polystyrene and polyhexylisocyanate, respectively as the coil and the rodlike components (see Figure 24(a)). The copolymers
were synthesized anionically in a solvent mixture of THF and toluene and initiated by n-butyllithium.
In their first studies on this system, the authors used a copolymer with a total molecular weight of
135.000 g/mol and a weight fraction of polystyrene of 0.2; the total polydispersity was 1.36. By
studying concentrated (>15%) solutions of the block copolymer by optical polarized microscopy, it
became obvious that the solution contained liquid-crystalline regions. If the solution was sheared with
a razor blade and allowed to relax, a banded texture was seen with the observed strips perpendicular to
the shearing direction. By studying the samples with transmission electron microscopy (TEM) (thin
films cast from toluene on carbon support; stained by RuO_4) a "zigzag" morphology was observed
with a very high long-range order and a high degree of spatial correlation between adjacent layers
(see Figure 25(a)). The domain spacing (as monitored by TEM analysis) of the polystyrene block
was ~250 Å and of the polyhexylisocyanate domain ~1800 Å. A wide-angle electron diffraction
pattern showed that the polyhexylisocyanate domains are crystalline showing a very high orientation
order. The high number of reflections pointed out a high degree of crystallinity. For wide-angle
x-ray scattering (WAXS) a polymorphism for the crystal data of the polyhexylisocyanate domain
became obvious even though most of the WAXS data suggested hexagonal packing. This result

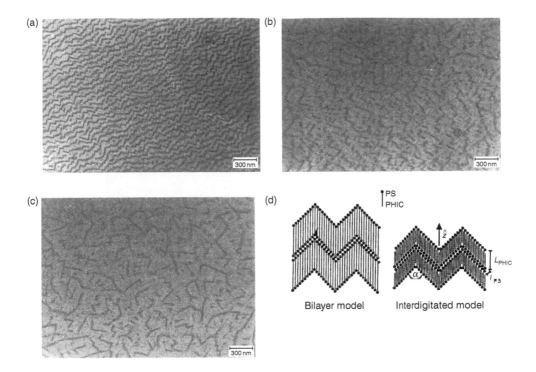

Figure 25 Bright field TEM images of a polyhexylisocyanate–block–polystyrene block copolymer with a molecular weight of 135.000 g/mol and a polystyrene weight fraction of 0.2 (thin films; stained with RuO$_4$); cast from (a) toluene, (b) CHCl$_3$, and (c) CCl$_4$. Whereas (a) shows a zigzag morphology with very high long-range order, this order is lost for the cast from CHCl$_3$ (b) and even more from CCl$_4$ (c). The proposed packing arrangements (for the zigzag morphology) either in an interdigitated or a bilayer fashion are schematically drawn in (d). (Reprinted from J.T. Chen, E.L. Thomas, C.K. Ober, and S.S. Hwang. *Macromolecules* 28:1688, 1995. Copyright [1995] American Chemical Society. With permission.)

was in accordance with the data obtained for pure polyhexylisocyanate. By analyzing the intensity for each layer in the electron diffraction pattern a 8_3 or 8_5 helix with a 1.95 Å translation and a rotation of 135° per monomeric unit along the *c*-axis was suggested. Electron diffraction pattern also showed that the polyhexylisocyanate rods are oriented parallel to the pointing direction of the zigzags. The zigzag morphology is schematically shown in Figure 25(d) either for a case where the rods are stacked in an interdigitated fashion or as a bilayer model. Calculation of the rod domain size for a copolymer with this composition suggests that the interdigitated model offers a better representation of the reported data. The authors also emphasized the tilting in the zigzag morphology in line with theoretical predictions of Halperin by judging the effects of the appropriate interfacial and deformational forces. The results they obtained are in good agreement with the observed morphology. The effect of changing the casting solvent on the morphology is quite dramatic. In the case of CCl$_4$ the authors observed fragmented polystyrene zigzags surrounded by a crystalline polyhexylisocyanate matrix with actually no long-range order. The morphology also revealed micellar-like assemblies having polystyrene as the core and polyhexylisocyanate as the corona (see Figure 25(c)). On casting from CHCl$_3$ short zigzags were observed which showed a little more long-range order than in the CCl$_4$ case. This morphology also includes micellar polystyrene inclusions (see Figure 25(b)).

The authors proposed a different mechanism for the formation of the above morphologies. During toluene casting the solution seems to separate into a solvent-rich phase and a nematic copolymer-rich phase. After further evaporation, a second transition to a smectic phase appears to occur and because toluene is a better solvent for the polystyrene block the coil part can stretch and the zigzag morphology can be build by the tilted assembly of the rods and eventual "locking" of the structure by crystallization of the polyhexylisocyanate. The nonpolar CCl$_4$ is a better solvent for the polyhexylisocyanate and

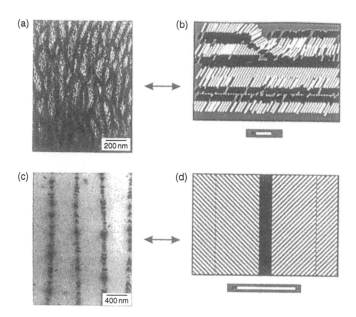

Figure 26 TEM micrograph of polyhexylisocyanate–block–polystyrene with the respective molecular weights (a) 73.000 and 104.000 g/mol; (c) 245.000 and 9.000 g/mol (thin film, cast from toluene and stained with RuO$_4$). The observed wavy lamellar morphology (a) can be explained by the schematic model in (b); the scheme (d) resembles the arrow-head morphology seen in (c). (Adapted from J.T. Chen, E.L. Thomas, C.K. Ober, and G.-P. Mao. *Science* 273:343, 1996.)

by evaporation the swelling of the polyhexylisocyanate mostly hinders the formation of a long-range structure. This also explains the micellar assembly of the polystyrene domains in this morphology. The morphology obtained by casting from CHCl$_3$ is an intermediate between that obtained from toluene and CCl$_4$ because CHCl$_3$ is not preferential to either block.

The authors also reported on the effect of varying composition of the block copolymers. A wavy lamellar structure was observed for a copolymer with a molecular weight of 104.000 g/mol for the polystyrene block and of 73.000 g/mol for the polyhexylisocyanate block (Figure 26(a)). The polyhexylisocyanate domain had an average width of ∼60 nm. By calculating the predicted rod length for the appropriate molecular weight and assuming interdigitated stacking the authors predicted that the rods are tilted by approximately 60° with respect to the lamellar normal. Because of the relative high volume fraction of the polystyrene the system fails to show long-range order. Samples with a polystyrene block of 14.000 g/mol and a polyhexylisocyanate block of 36.000 g/mol form instead the already observed zigzag morphology. For this composition the authors proposed a bilayer arrangement of the rods because: (i) beam damage occurred from the middle of the rod domain (depolymerization from the chain ends), (ii) atomic force microscopy (AFM) studies showed a narrow dip in the middle of the polyhexylisocyanate domain, and (iii) the domain spacing was in good accordance with the predicted rod length in a bilayer arrangement for the appropriate chain length.

For copolymers with even shorter polystyrene blocks another fascinating morphology was observed. In this morphology, the polystyrene domains built arrow-heads whose pointing direction flipped up and down with every other polystyrene-rich layer (see Figure 26(c)). The morphology had an enormous long-range order over tens of micrometers. The electron diffraction pattern indicated that the orientation of the polyhexylisocyanate chain axis in adjacent layers alternated between 45° and −45° with respect to the normal layer. By comparing the predicted size of the polyhexyliso-cyanate domains with the one actually found in TEM, an interdigitated model seems to best resemble a polymer system with a molecular weight of 386.000 g/mol for the polyhexylisocyanate block and of 7.100 g/mol for the polystyrene block. Instead, a bilayer model offered a better representation for

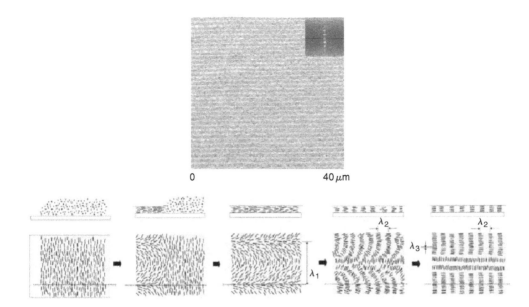

Figure 27 Scanning force microscopy (SFM) phase image of polystyrene–block–poly(3-[triethoxysilyl]propyl isocyanate) film cast on glass and schematic drawing of the evaporation-induced hierarchical morphology. The initial nematic phase reorders into a smectic-layered state and finally crystallizes creating 90° tilt boundaries. (Reprinted from J.-W. Park and E.L. Thomas. *Adv. Mater.* 15:585, 2003. Copyright [2003] Wiley-VCH. With permission.)

a copolymer system with a molecular weight of 245.000 g/mol for the polyhexylisocyanate block and of 9.300 g/mol for the polystyrene block.

For the first time, Park and Thomas observed a Néel wall pattern formation in a block copolymer system, namely polystyrene–block–poly(3-[triethoxysilyl]propyl isocyanate) copolymer (PS–b–PIC; PIC weight fraction 39%; total molecular weight 62 kg/mol), via directional casting of films [100]. The directions were applied by tilting of the casting substrate and/or different exposure to the atmosphere. More sophisticated methods might include magnetic or electric fields. A grating like domain wall pattern with very high long-range order along the direction of solvent evaporation (see Figure 27) became obvious. After detailed studies, the authors propose the mechanism schematically shown in Figure 27 for the assembly of the observed morphologies. As the solvent evaporates the system undergoes changes to a lyotropic nematic state → solid-like film with high viscosity and periodic Néel inversion walls → subsequent microphase separation that stabilize the wall patterns against relaxation resulting in the final morphology.

The authors also utilized these block copolymers to obtain polymer brushes on Si surfaces, as the nanoscale domains present in solution can be covalently attached to the surface by the multiple pending triethoxysilane groups. Patterning of these nanodomains was possible by microcontact printing [101].

Pearce and coworkers studied the possibility to produce stable blends of rod–coil block copolymers consisting of polyhexylisocyanate as the rod segment and of poly(ethylene glycol) as the coil segment (PHIC–b–PEG) that were synthesized via living titanium (IV) catalyzed coordination polymerization [102]. While blends with polyvinylacetate exhibited liquid-crystalline-like domains in the solid state, blends with polymethylmethacrylate showed a fairly good compatibility when the fraction of poly(ethylene glycol) was high in the blend.

Recently, Lee and coworkers reported on polyhexylisocyanate–block–polystyrene–block–polyhexylisocyanate and polyhexylisocyanate–block–polisoprene–block–polyhexylisocyanate triblock copolymers (PHIC–b–PS–b–PHIC and PHIC–b–PI–b–PHIC) that were synthesized via a one pot anionic polymerization by bifunctional initiation with sodium naphthalenide (addition of sodium tetraphenylborate) [103,104].

Charged polystyrene–block–poly(isocyanodipeptides) were reported by Nolte and coworkers [105]. These block copolymers were synthesized by polymerization under catalysis of a Ni-complex with amino-terminated polystyrene using the isocyanides of dipeptides as monomers. Copolymers with different compositions of either isocyano-L-alanine-L-alanine (IAA) and isocyano-L-alanine-L-histidine (IAH) were prepared (see Figure 24(b) and (c)). CD spectroscopy revealed a helical conformation for the isocyanide block (right-handed for poly[isocyano-L-alanine-L-alanine] and left-handed for poly[isocyano-L-alanine-L-histidine]). The removal of the ester functions of the block copolymers resulted in negatively charged helical headgroups for the polystyrene–block–poly(isocyano-L-alanine-L-alanine) and zwitterionic headgroups for polystyrene–block–poly(isocyano-L-alanine-L-histidine) (see Figure 24(b) and (c)). Because of the aggregation behavior of the amphiphilic blocks, recording of CD spectra proved to be very difficult. To study these systems with electron microscopy the authors sonicated a 0.1 weight% dispersion in water (pH 7) for 1 h at 70°C and after this buffered with sodium acetate to pH 5.6. Under these conditions (half of all acid functions protonated for polystyrene–block–poly[isocyano-L-alanine-L-alanine] and zwitterionic headgroups for polystyrene–block–poly[isocyano-L-alanine-L-histidine]), the block copolymers with a DP of 40 for the polystyrene block and of 20 for the poly(isocyanide) block showed rodlike structures with a diameter of 12 nm (see Figure 28(a)). These structures are thought to be micellar aggregates (core of polystyrene ~8 nm and corona of poly[isocyanide] ~ 2 nm) (see Figure 28(b)). The rodlike structures assembled to a zigzag like structure on mica and polyvinyl formaldehyde plastic plates as monitored by TEM and AFM. For longer poly(isocyanide) blocks (DP = 30) no clear morphologies could be seen, by making the poly(isocyanide) blocks shorter (DP ~ 10) the polystyrene–block–poly(isocyano-L-alanine-L-alanine) copolymer showed bilayer type structure in addition to collapsed vesicles, bilayer filaments, and left-handed superhelices (diameter 180 nm and pitch 110 nm) (see Figure 28(c)). Contrary to this a polystyrene–block–poly(isocyano-L-alanine-L-histidine) with a DP of 15 for the poly(isocyanide) blocks forms a right-handed superhelix with smaller dimensions (diameter 28 nm and pitch 19 nm) probably build of coiled rods.

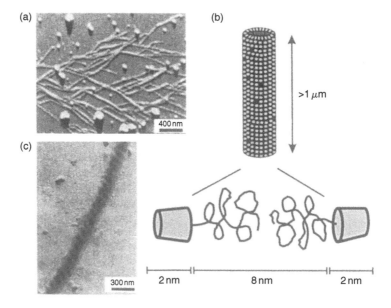

Figure 28   (a) AFM image of polystyrene–block–poly(isocyano-L-alanine-L-alanine) with a DP of 40 for the polystyrene block and of 20 for the poly(isocyano-L-alanine-L-alanine) block. The observed rod-like structures can be explained by a schematic model as seen in (b). For a DP of 40 for the polystyrene block and of 10 for the poly(isocyano-L-alanine-L-alanine) block, a left-handed super-helix can be seen in TEM (c). (Both images in pH 5.6 sodium acetate buffer [0.2 m*M*].) (Adapted from J.J.L.M. Cornelissen, M. Fischer, and R.J.M. Nolte. *Science* 280:1427, 1998.)

Vesicular structures became obvious in rod–coil block copolymers with thiophene modified rod segments (polystyrene–block–poly(3-[isocyano-L-alanyl-amino-ethyl]-thiophene)) [106] (see Figure 29). The vesicles are quite polydisperse with sizes ranging from 2 to 22 μm with wall thicknesses of 30 nm and fused together over time. Via cross-linking of the thiophene units stable vesicles could be produced, which might be suitable as microreactors and the authors already showed very promising results by incorporating an enzyme (Candida Antarctica Lipase-B) into the vesicles. The entrapped enzyme was capable of catalyzing the transesterification of esters that permeated through the walls of the vesicles that were monitored by fluorescence measurements.

Rowan and coworkers attached even bulkier side groups to the isocyanide backbone (porphyrins) and succeeded in producing nanometer long well-defined arrays of porphyrin molecules [107]. The authors did not yet report on rod–coil block copolymers with this rod segment, but these could prove very useful as synthetic antenna systems.

In a recent communication, Sommerdijk and coworkers reported on structures in which the linear part of a rod–coil block copolymer is exchanged to a flexible bulky and apolar dendritic carbosilane segment (poly[isocyano-L-alanine-L-alanine] as the rod segment) [108]. With dynamic light scattering, not only micellar structures but also larger aggregates of ∼90 nm became obvious in chloroform solutions. The authors succeeded in nucleating AgBF$_4$ in these aggregates and obtained silver nanowires with this technique (see Figure 30). Interestingly, the crystalline character of the silver cluster varied with the composition of the hybrid structures (lower polyisocyanide content results in oriented silver [109], whereas nanowires with higher polyisocyanide content showed polycrystallinity).

Polybutadiene–block–polyisocyanide block copolymers synthesized by living transition metal catalysis were reported by Deming and Novak [110]. The catalyst used first polymerized butadiene

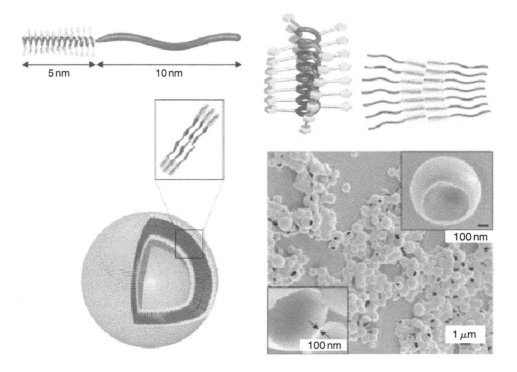

Figure 29 TEM images of vesicles of (polystyrene–block–poly(3-[isocyano-L-alanyl-amino-ethyl]-thiophene)) formed in CHCl$_3$ and schematic drawings on the way the rod–coil polymers form these vesicles. (Reprinted from D.M. Vriezema, J. Hoogboom, K. Velonia, K. Takazawa, P.C.M. Christianen, J.C. Maan, A.E. Rowan, and R.J.M. Nolte. *Angew. Chem. Int. Ed.* 42:772, 2003. Copyright [2003] Wiley-VCH. With permission.)

(a)   (b)

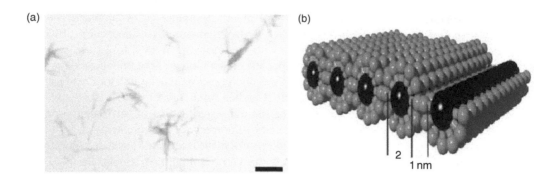

Figure 30   TEM image and schematic drawing of silver nanorods nucleated in rod–coil hybrids build of flexible
bulky and apolar dendritic carbosilane segment as the coil and of poly(isocyano-L-alanine-L-alanine) as
the rod segment. (Reprinted from J.J.L.M. Cornelissen, R. van Heerbeek, P.C.J. Kamer, J.N.H. Reek,
N.A.J.M. Sommerdijk, and R.J.M. Nolte. *Adv. Mater.* 14:489, 2002. Copyright [2002] Wiley-VCH. With
permission.)

and, after addition of methyl benzyl isocyanide, polymerized this second monomer without further
including butadiene segments. A morphology based on spherical rod domains (in the order of 0.2 μm)
within a polybutadiene matrix was observed using scanning electron microscopy (SEM). Block
copolymers with a total molecular weight of 140.000 g/mol and a rod to coil ratio of 30%/70%
were prepared. In a subsequent publication the synthesis of broken rods and also rod–coil–rod triblock copolymers by living bimetallic initiators was reported [111]. Novak and coworkers [109]
also reported on polystyrene–block–poly(N, N'-di-n-hexylcarbodiimide) and polystyrene–block–
poly([R/S]-N-[1-phenylethyl]-N'-methylcarbodiimide) block copolymers (see Figure 24(d) and (e))
which were synthesized by sequential anionic polymerization in benzene with *sec*-butyl lithium as initiator. The polymers were characterized by IR-spectroscopy and the composition of the diblocks was
monitored by thermogravimetric analysis. In these polymers the rod segment is the major component
($\sim$83%). Samples were cast from different selective solvents and the microstructure (monitored by
TEM) was found to be sphere-like in hexane, which is a good solvent for the polycarbodiimide and
a nonsolvent for polystyrene. The morphology was found to be lamellar for benzene (good solvent
for polystyrene and θ-solvent for polycarbodiimide) and cyclohexane (good solvent for polycarbodiimide and θ-solvent for polystyrene). In THF, which is a good solvent for both blocks, no clear
morphology became obvious. The authors confirmed the microstructure produced from benzene
solutions with AFM.

3. Copolymers with Polysaccharides

Polysaccharides are classified on the basis of their main monosaccharide components and the
sequences and linkages between them, as well as the anomeric configuration of linkages, the ring
size (furanose or pyranose), the absolute configuration (D- or L-), and any other substituents present.
Certain structural characteristics, such as chain conformation and intermolecular associations, will
influence the properties of polysaccharides. Certain polysaccharides show a helical conformation in
solution and in bulk with high persistence length, amylose being a prominent example. Amylose
consists of α(1–4)-linked glucose residues in ${}^4C_1$-chair conformation and can form three forms:
double-helical A- and B-amyloses and the single-helical V-amylose. The latter contains a channel-like
central cavity that is able to include molecules, "iodine's blue" being the best-known representative.
V-amylose is the predominant form in solution and consists of a left-handed single helix (it contains
six glucoses per turn with 7.91 to 8.17 Å pitch height). The helical conformation of V-amylose is
very stable as all monomers are in *syn* orientation, the secondary hydroxyl groups are hydrogen
bonded $O(3)_n \cdots O(2)_{n+1}$, and the V-amylose helix is additionally stabilized by hydrogen bonds

$O(6)_n \cdots O(2)_{n+6}$ between the turns. There are a couple of rod–coil systems in which amylose is the rod segment. Most copolymer systems with amylose are based on enzymatic polymerization with potato phosphorylase which was pioneered by Beate Pfannemüller who reported on linear, star, and comblike polymers carrying amylose chains [112–115].

Stadler and coworkers reported rod–coil systems composed of amylose blocks with polysiloxanes [116–118] (comblike structures) and polystyrene [119] (block-like structures). Amylose–block–polystyrene is too brittle to obtain block copolymer films to study the morphological behavior but shows a very interesting solution behavior. Müller and coworkers [120,121] report on up to four different species that can be found in THF solutions and can not be assigned to classical micelles. Scaling laws could be obtained with varying compositions of the block copolymers but the system does not seem to be in an equilibrium state. "Pseudo crew-cut" micelles in water could be produced by a single solvent/elevated pressure approach. These aggregates are more defined and observe scaling laws predicted by theory.

Akiyoshi et al. reported on the behavior of amylose–block–polyethylene oxide in solution. Their results show that the amylose helix is quite expanded in chloroform solutions and/or a complete different microenvironment is involved [122,123]. These results are also supported by CD studies of the interaction of methyl orange with the amylose helices.

Bosker et al. studied the interfacial behavior of polystyrene–block–maltodextrins at the air–water interface via Langmuir–Blodgett technique [124]. Owing to the slow adsorption/desorption of the polysaccharide chains and block copolymer aggregation, the systems show interfacial pressure hysteresis between consecutive compression and expansion cycles of a monolayer.

Haddleton and Ohno reported on maltodextrin modified polymers via copper(I)-mediated living radical polymerization [125]. Comblike structures based on polystyrene with amylose entities were synthesized by Kobayashi et al. [126] and bottle-brush-like structures by Kakuchi and coworkers [127].

B. Nonhelical Rod Systems

1. Copolymers with Rigid Aromatic Rod Segments

As already mentioned, the rigidity, for instance, in polypeptides is caused by the building of helical superstructures due to hydrogen bonding. Another reason for chain rigidity can be found in para-linked aromatic polyamides. The high resonance energy in the amide group leads to quasiconjugation and coplanarity in the structure that decreases the flexibility of the chains. Many of these structures exhibit a high equilibrium rigidity and can form nematic mesophases in concentrated solutions. The reason for the chain rigidity found in structures, such as poly(p-phenyleneethynylene) and poly(p-phenylene), is the symmetry of the bonds along the chain. The linear bonds of the phenylene units in poly(p-phenylene) for instance only tolerate rotation of the repeating units around the polymer axis (e.g., of some nonhelical rigid polymers refer to Figure 31).

The work by Ciferri and coworkers [128–130] focused on the supramolecular organization of aromatic rod–coil copolyamides in diluted (isotropic) and moderately concentrated (lyotropic) phases. Their diblock copolymers were based on a rigid block of poly(p-benzamide) (PBA) having a DP about 100, and different comparable lengths of flexible blocks, such as poly(m-phenylene isophthalamide) (MPD-I), poly(m-benzamide) (MBA), or poly(ethylene glycol) (PEG). The use of end-capped prepolymers and selective extraction techniques assured a strict two-block sequence, the absence of free homopolymers not strongly bound to the copolymers, and a fractionation in terms of the rigid/flexible compositional distribution ratio.

From viscosity, light scattering, and critical concentration data in H_2SO_4 and in N, N-dimethylacetamide (DMAc)/3% LiCl, they concluded that in the latter solvent poly(p-benzamide) is not molecularly dispersed but rather occurs as supramolecular aggregates composed likely of seven polymer chains with a side-by-side shift of one-fourth of the molecular length. Since the

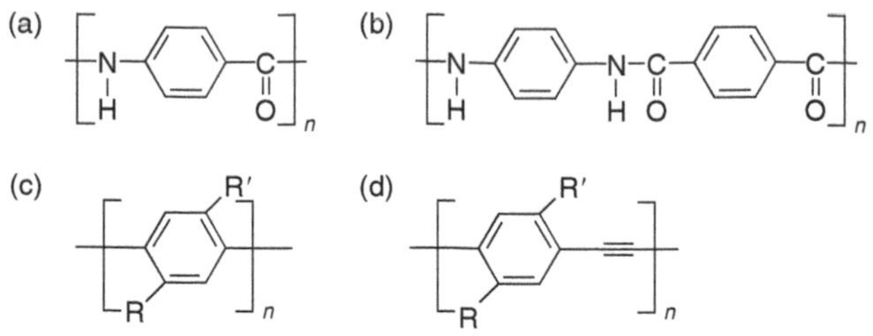

Figure 31   Chemical structures of rigid aromatic polymers: (a) poly(p-benzamide); (b) poly(p-phenyleneterephthalamide); (c) poly(p-phenylene) and (d) poly(p-phenyleneethynylene).

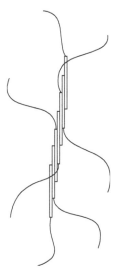

Figure 32   Schematic representation of a nematic aggregation of copolymer assemblies involving seven molecules each with a side-by-side shift of one-fourth of the length and alternating orientations. (Reprinted from P. Cavalleri, A. Ciferri, C. Dell'Erba, M. Novi, and B. Purevsuren. *Macrolmolecules* 30:3513, 1997. Copyright [1997] American Chemical Society. With permission.)

critical concentration at which the mesophase appears for poly(p-benzamide)–poly(m-phenylene isophthalamide) and poly(p-benzamide)–poly(m-benzamide) copolymers in DMAc/3% LiCl and it was not much affected by the flexible blocks, they suggested that the organization within the mesophase could be represented by assemblies based (Figure 32) on the above indicated aggregates of poly(p-benzamide) segments and on suitably oriented flexible blocks.

Solubility data (Figure 33(a)), showing that the DMAc/3% LiCl solvent is a good solvent for both poly(p-benzamide) and poly(m-benzamide) (or poly[m-phenylene isophthalamide]) blocks, were interpreted to mean that in isotropic solutions a dispersion of assemblies such as that schematized in Figure 9 occurs. Quite a different conclusion was reached in the case of the poly(p-benzamide)–poly(ethylene glycol) copolymer for which solubility data (Figure 33(b)) showed a worsening of the solvent quality of DMAc toward poly(ethylene glycol) as the LiCl concentration was increased from 0.5 to 3.0%. A bilayer-like organization of the type envisioned by Halperin [64] was suggested. In the DMAc/0.5% LiCl solvent, the better solvated poly(ethylene glycol) segments screen from the solvent (Figure 33(b)-left) relatively large clusters of poly(p-benzamide) aggregates. However, in the DMAc/3% LiCl solvent (Figure 33(b)-right) the poly(p-benzamide) aggregates are exposed to the solvent, screening the less solvated flexible blocks.

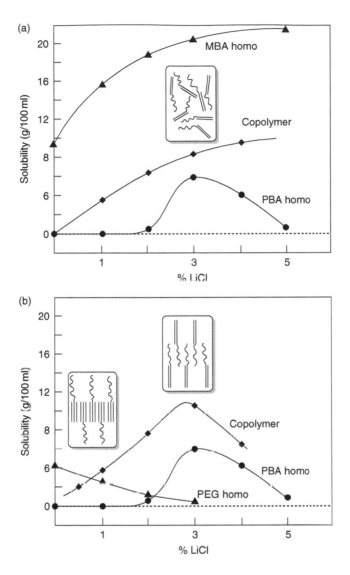

Figure 33 Solubility of copolymers and corresponding homopolymers in DMAc with varying LiCl concentrations. (a) Copolymer of poly(p-benzamide) and poly(m-benzamide); (b) Copolymer of poly(p-benzamide) and poly(ethylene glycol). Possible supramolecular organizations in the isotropic solution are schematized in the insert. (Reprinted from A. Gabellini, M. Novi, A. Ciferri, and C. Dell'Erba. *Acta Polym.* 50:127, 1999. Copyright [1999] Wiley-VCH. With permission.)

The above structures were postulated to undergo further level of organization upon increasing copolymer concentration until the formation of liquid-crystalline and crystalline phase. However, no morphological studies have yet been reported. The copolymers of Ciferri and coworkers are of interest in the areas of mechanical and transport properties, and may represent a novel approach to the preparation of self-assembling high performance materials that avoids the current cumbersome route of engineered composites.

Jenekhe and coworkers investigated the effect of supramolecular structure and morphology on the photophysical properties in polymer systems. For this purpose, they used rod–coil di- and triblock copolymers and polymer blend systems.

The authors investigated the electronic energy transfer from a light absorbing energy donating rod–coil system to a rigid-rod polymer (randomly dispersed in this polymer blend) (see Figure 34(a)).

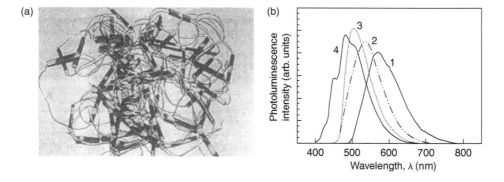

Figure 34 Chemical structures of (a) a blend of a light absorbing energy donating rod–coil system and a rigid-rod homopolymer; (b) poly(1,4-phenylenebenzo*bis*thiazole-co-decamethylenebenzo*bis*thiazole); (c) poly([1,4-phenylenedivinylene]benzo*bis*thiazole-co-decamethylenebenzo*bis*thiazole); and (d) poly(phenylquinoline)–block–polystyrene.

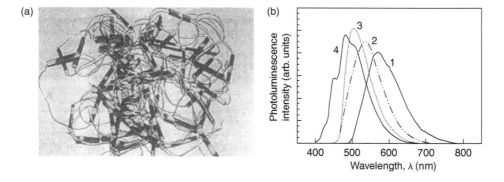

Figure 35 (a) Model of the supramolecular structure and morphology of the rod–coil copolymers by Jenekhe and Osaheni. (b) Shows the photoluminescence spectrum of poly(benzo*bis*thiazole-1,4-phenylene) and copolymers: 1–100% PBZT; 2–40%; 3–20%; 4–5%. (Reprinted from S.A. Jenekhe and J.A. Osaheni. *Chem. Mater.* 6:1906, 1994. Copyright [1994] American Chemical Society. With permission.)

The rod–coil copolymer was varied by the number of methylene groups in the coil segments that resulted in a changed supramolecular structure and directly affected the Förster energy transfer efficiency (an increase for this efficiency for 12 methylene units as compared to 7 methylene units) [131].

For another type of rod–coil copolymers [132] the authors varied the fraction of the rod segment and analyzed the photoluminescence in this system. The random rod–coil system is built from the rigid segment poly(*p*-phenylenebenzo*bis*thiazole) and the flexible poly(benzo*bis*thiazoledecamethylene) (see Figure 34(b)).

By studying the systems with x-ray diffraction and polarized optical microscopy, the authors found that the morphology of their system was amorphous. A proposed picture of the morphology is seen in Figure 35(a). It became obvious that the photoluminescence emission changed with the

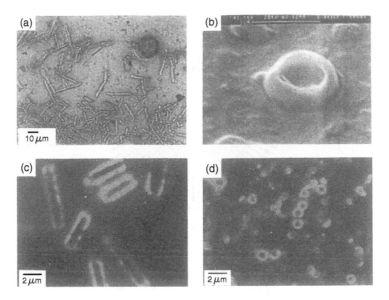

Figure 36 (a) Optical micrograph of cylinders built from poly(phenylquinoline)–block–polystyrene (9:1 TFA:DCM, 25°C); (b) scanning electron micrograph of giant vesicles built from the same block copolymer (1:1–1:4 TFA:DCM, 25°C); (c) and (d) shows the fluorescence photomicrographs of the cylinders and the vesicles respectively. (Adapted from S.A. Jenekhe and X.L. Chen. *Science* 279:1903, 1998.)

composition from yellow (100% rod) to UV (<5% rod). The corresponding photoluminescence spectra are seen in Figure 35(b).

For two series of rod–coil copolymers [133] poly(1,4-phenylenebenzobisthiazole-co-decamethylenebenzobisthiazole) (see Figure 34(b)) and poly([1,4-phenylenedivinylene]benzobis-thiazole-co-decamethylenebenzobisthiazole) (see Figure 34(c)) the authors also investigated the photophysical properties with varying composition. In these cases the photoluminescence quantum yield reached over seven-fold higher values than for a pure conjugated homopolymer. Furthermore, the emission color was tunable in the visible region by varying rod fraction.

A certain class of rod–coil copolymers studied by Jenekhe and Chen (poly[phenylquinoline]–block–polystyrene) [134] (see Figure 34(d)) proved to built fascinating supramolecular structures. For a block copolymer in which poly(phenylquinoline) had a DP of 50 and the DP of polystyrene was 300 the authors observed micellar assemblies in the form of spheres, lamella, cylinders, and vesicles (see Figure 36). The detected form of the supramolecular aggregates was depended on the type of solvent or solvent mixture and the drying rate (the solvent mixtures were either TFA and dichloromethane or TFA and toluene [selective for the rod segment]). Very rapid drying resulted in micellar assemblies with diameters of 0.5–10 μm whereas by slow evaporation of the solvent nonmicellar structures, such as lamella, cylinders, etc., became predominant (with diameters of 5–30 μm for lamellar aggregates; 0.5–1 μm for vesicles and 1–3 μm for the cylinders [length of 5–25 μm]). These huge assemblies can not result from core–shell micelles as observed in other polymer systems because they are ~10 to 15 times larger than these micelles build by a rod–coil copolymer with the same composition. Therefore the authors proposed that these structures are build by hollow cavities. The cover of these cavities are highly ordered as they show crystalline features. By studying the structures with fluorescence microscopy it became obvious that the fluorescinating rod blocks are on the outermost surface of the observed assemblies and photoluminescence studies showed that the photophysical properties strongly depend on the supramolecular structure of the photoactive rod segments (see Figure 36).

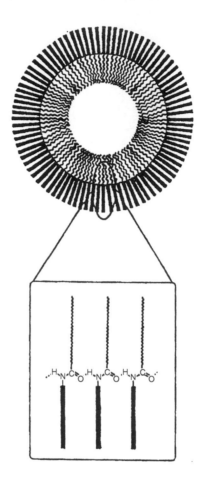

Figure 37   Schematic illustration of the self-assembly of poly(phenylquinoline)–block–polystyrene diblock copolymers into hollow aggregates. (Adapted from S.A. Jenekhe and X.L. Chen. *Science* 279:1903, 1998.)

The stability of these giant self-organized structures is thought to result from strong inter-molecular hydrogen bonding caused by the linking amide unit in the (poly[phenylquinoline]–block–polystyrene) copolymer (see Figure 37). It was also proposed that the stable aggregate structure results from efficient packing of the rod blocks. By photoluminescence excitation measurements, emission spectroscopy, and fluorescence microscopy a J-like aggregation of the polymer blocks became obvious [135].

The authors also found that their rod–coil systems proved to be feasible to encapsulate fullerens into the cavities. Compared to usual solvents for C_{60} and C_{70}, such as dichloromethane or toluene, the solubility was enhanced by up to 1000 times when the molecules were encapsulated into rod–coil micelles. The solubilization of up to 10 billion fullerenes into one micelle increased the aggregation number to 10^9 and the obtained diameters up to 30 μm [136]. Such "micro containers" could also be used for an encapsulation process of other substances and so one can think of applications, such as drug delivery, emulsions, etc.

Francois and coworkers [137,138] reported on the synthesis of polystyrene–block–poly(p-phenylene) block copolymers via the anionic polymerization to polystyrene–block–poly(1,3-cyclohexadiene) and subsequent aromatization. The micellar assemblies and molecular dispersed free chains could be separated by size exclusion chromatography and with this method, the size of the spherical assemblies was found to be 30–50 Å for the poly(p-phenylene) core, and 3–15 nm for the full micelle. By evaporating the solvent from solutions in carbon disulfide in moist air,

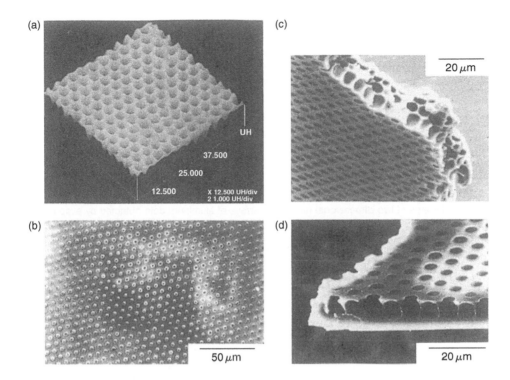

Figure 38 Scanning electron micrograph of polystyrene–block–poly(p-phenylene) (respective molecular weights 30.000 and 7.000 g/mol); (a) side view, (b) plan view, and (c) micrograph of a monolayer of pores in the copolymer film. AFM also manifests the honeycomb morphology (d). (Reprinted from G. Widawski, M. Rawiso, and B. Francois. *Nature* 369:387, 1994. Copyright [1994] Macmillan Magazines Ltd. With permission.)

a hexagonal array of empty cells was observed from optical and electron microscopy. By removing the superficial skin of these cells an internal honeycomb-like structure became evident (see Figure 38)

The size of the cell walls (as studied by AFM) was influenced by the length of the polystyrene block in the copolymer; for molecular masses larger than 50.000 the regularity in the structure gradually disappeared. The authors related the mechanism of formation of this structure to the classical "phase inversion" process for the production of polymeric membranes [139]. In a later work the authors reported on the possibility to monitor this honeycomb structure in different polymeric systems including polystyrene–block–poly(p-phenylene), star branched polystyrenes, polystyrene–block–poly-3-hexylthiophene, and polystyrenes with polar endgroups and polymer blends [140]. Possible applications for such fascinating structures are in the area of polymeric membranes and optical devices.

Hadziioannou and coworkers utilized the honeycomb morphology build by the light emitting rod–coil block copolymer polystyrene–block–poly(2,5-dioctyloxy-p-phenylene vinylene) (PS–b–DOOPPV) to obtain two-dimensional hexagonal arrays of aluminum dots. For this purpose the top layer of the honeycomb morphology was first photo cross-linked and subsequently aluminum deposited on this structure. After the removal of the top layer with adhesive tape the polymer was completely removed with solvent and a highly ordered array of aluminum dots became obvious (see Figure 39) [141].

Francois and coworkers reported on the synthesis of polystyrene–polythiophene block and graft copolymers [142–144]. These materials incorporate the unique properties of polythiophenes with ease of solubilization and processing. The reported block copolymers showed nearly the same spectral characteristics as pure polythiophene and the authors proved that their block

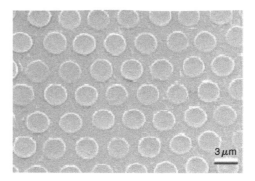

Figure 39 SEM image of two-dimensional hexagonal arrays of aluminum dots obtained by evaporation of aluminum on the honeycomb morphology build by polystyrene–block–poly(2,5-dioctyloxy-*p*-phenylene vinylene). (Reprinted from B. de Boer, U. Stalmach, H. Nijland, and G. Hadziioannou. *Adv. Mater.* 12:1581, 2000. Copyright [2000] Wiley-VCH. With permission.)

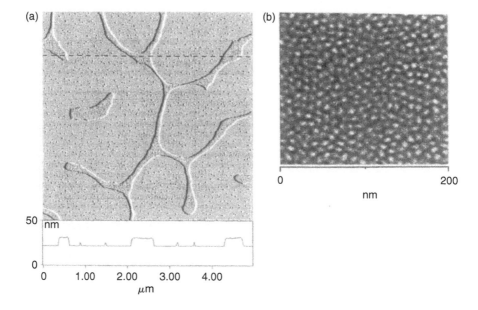

Figure 40 (a) Phase SFM image of a polystyrene–block–oligothiophene–block–polystyrene triblock copolymer cast on mica from a solution in toluene (1 mg/ml). (b) Micelles built by the same triblock system as observed in a closed film by phase SFM (after Fourier filtering). (Reprinted from M.A. Hempenius, B.M.W. Langeveld-Voss, J.A.E.H. van Haare, R.A.J. Janssen, S.S. Sheiko, J.P. Spatz, M. Moeller, and E.W. Meijer. *J. Am. Chem. Soc.* 120:2798, 1998. Copyright [1998] American Chemical Society. With permission.)

copolymers were still soluble after doping. On films of their copolymers subjected to a pyrolysis step they reported different morphologies (spheres and fibrils) depending on volume fraction and film thickness.

Hempenius et al. [145] reported on polystyrene–oligothiophene–polystyrene triblock copolymers. Size exclusion chromatography revealed the occurrence as self-assembled clusters as confirmed by scanning force microscopy (SFM) of a dilute solution evaporated on mica. Incomplete covering of the mica surface allowed the observation of single droplets which can be attributed to single micelles (see Figure 40(a)). The TEM image confirms the occurrence of micelles having a diameter of 12 nm corresponding to an aggregate of about 60 molecules, which is in line with results obtained from GPC (see Figure 40(b)).

(a) 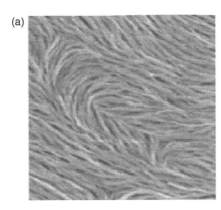 (b)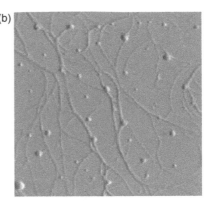

Figure 41 SFM images of nanoribbons found in thin films of poly(p-phenyleneethynylene)–block–poly(dimethylsiloxane). (Reprinted from P. Leclère, A. Calderone, D. Marsitzky, V. Francke, K. Müllen, J.L. Brédas, and R. Lazzaroni. *Synth. Met.* 121:1295, 2001. Copyright [2001] Elsevier. With permission.)

Films of gold stained micelles obtained by chemical oxidation of the cores with $HAuCl_4$ in toluene were monitored by TEM (the micellar cores appear as dark spots) and these doped micelles were still soluble. The electronic properties of the triblocks are in accordance with the ones of associated unsubstituted oligothiophenes.

Müllen and coworkers reported on rod–coil block copolymers with poly(p-phenyleneethynylene) or poly(p-phenylene) as the rod segment synthesized by condensation of mono-functionalized homopolymer blocks [146,147]. In thin films of poly(p-phenyleneethynylene)–block–poly(dimethylsiloxane), nanoribbons were observed (see Figure 41) [148].

Godt and coworkers [149] reported on the synthesis of poly(p-phenyleneethynylene)–block–polyisoprene and the transformation of such rod–coils into corresponding coil–rod–coil triblock copolymers. Luminescent rod–coil or coil–rod–coil di- or triblock copolymers were synthesized by mono- or difunctional atom transfer radical polymerization from oligo(p-phenyleneethynylene) [150].

2. Copolymers with Short Monodisperse Rod Segments

In this section, the work on systems incorporating small organic molecules as the rigid segment is included eventhough these materials can not be strictly considered as true block copolymers. Still these systems possess the major advantage of a monodisperse rod and therefore the resulting polymeric material has a very narrow polydispersity.

Stupp and coworkers reported on rod–coil systems containing a monodisperse rod part basing on an azo dye bound to a rigid monomer [151–153]. For these systems, the authors synthesized polyisoprenes anionically and terminated the living chains with carbon dioxide resulting in carboxylated polyisoprenes that were then coupled to the rigid block (see Figure 42(a) for the end structure).

Copolymers with varying volume fraction of the rod segment were prepared and the resulting morphological changes were monitored by TEM. For the copolymer with the highest rod volume fraction ($f_{rod} = 0.36$) the authors observed (after annealing a thin film at $140°C$ for 12 h) a morphology that seemed to consist of alternating strips of coils and rods (5–6 and 6–7 nm thickness respectively, see Figure 43(a)). Thicker films (cast from more concentrated solutions) showed a terrace-like assembly. By studying these samples by electron tomography, the authors found that the copolymers self-assembled into layered two-dimensional superlattices and ordered three-dimensional morphologies. For a rod volume fraction $f_{rod} = 0.36$, the same morphology as in the thin films was found for odd numbers of steps, but for even numbers of steps no contrast was observable. Slices parallel,

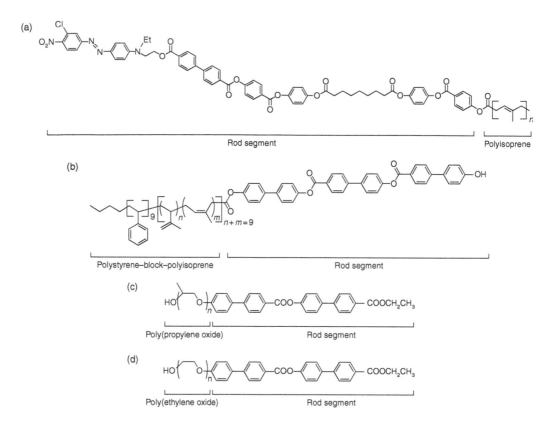

Figure 42   Chemical structures of rod–coil block copolymers with short, monodisperse rod segments. (a) Block copolymer with polyisoprene and a rod segment built of an azo dye and a rigid monomer and (b) miniaturized triblocks from Stupp and coworkers. (c) and (d) Show the chemical structures of block copolymers with either poly(propylene oxide) or poly(ethylene oxide) and a short rod segment as reported by Lee et al.

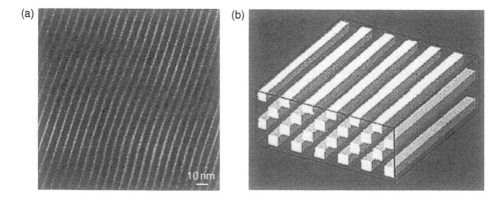

Figure 43   (a) TEM micrograph of the rod–coil block copolymers reported by Stupp et al. with a rod volume fraction of $f_{rod} = 0.36$ (thin film cast from cyclohexane, stained with OsO$_4$, and enhanced by translational Fourier filtering) and (b) a schematic model for the observed morphology of alternating strips. (Reprinted from L.H. Radzilowski and S.I. Stupp. *Macromolecules* 27:7747, 1994 and L.H. Radzilowski, B.O. Carragher, and S.I. Stupp. *Macromolecules* 30: 2110, 1997. Copyright [1994, 1997] American Chemical Society. With permission.)

(a)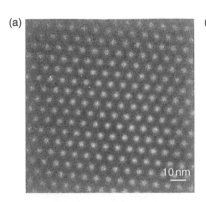

(b)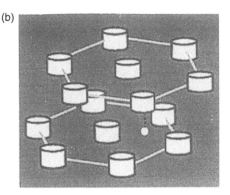

Figure 44 (a) Filtered TEM image of rod–coils with a rod volume fraction of $f_{rod} = 0.25$; (b) scheme for the hexagonal superlattice (build by rod aggregates). (Reprinted from L.H. Radzilowski and S.I. Stupp. *Macromolecules* 27:7747, 1994 and L.H. Radzilowski, B.O. Carragher, and S.I. Stupp. *Macromolecules* 30: 2110, 1997. Copyright [1994, 1997] American Chemical Society. With permission.)

perpendicular, and orthogonal to the film plane showed strip morphology built of discrete channel-like objects in layers parallel to the surface (which are displaced to the strip above and below) with the long axis also parallel to the plane of the film. The loss in contrast for the even numbers of layers was attributed to the effectively uniform electron density across the plane of the sample. For odd numbers the alternating rod domains vary by one and a contrast is observable. The terraces in thick films are build of layers that contain the channel-like objects. The strip morphology is schematically illustrated in Figure 43(b).

The authors correlated this morphology to cylinder phases in coil–coil diblock copolymers. However, the dimension of the strips (6–7 nm) do not correspond to a radial arrangement of cylinders, but rather to interdigitated bilayers. For a rod volume fraction $f_{rod} = 0.25$ the rod domains organize into a superlattice with hexagonal order measuring ∼7 nm in diameter and a domain spacing of 15 nm (see Figure 44(a)). On studying the thick films with electron tomography, the layers showed a continuous network of small domains. Slices parallel, perpendicular, and orthogonal to the film plane showed that the rod domains were discrete objects with roughly the same dimensions in all directions. The aggregates are ordered in layers parallel to the surface. This morphology is schematically described in Figure 44(b). The sketch resembles a closed-packed stacking of hexagonal superlattices.

An intermediate mix of the ribbon-type and the discrete aggregate-like morphology was observed in thin films bearing a rod volume fraction $f_{rod} = 0.3$ (see Figure 45(a)), which was as well found for up to three layers as monitored by electron tomography. However, by looking at more layers a network pattern based on hexagonal fused cells with a discrete rod aggregate as the nucleus was evident (see Figure 45(c)). Cross-sectional slices showed that the hexagonal network is build by discrete rod aggregates having dimensions similar to those found for a rod volume fraction of 0.25. Moreover, the "nucleus" (with higher contrast) is build by columns spawning the film thickness and discrete rod aggregates building the "cell walls" as seen in Figure 45(b). On looking at slices of the reconstructed volume, the authors observed a two-dimensional hexagonal superlattice for each slice that is rotated by 30° with respect to the lattice above and below. The observed morphology was interpreted by a certain rotation of the hexagonal superlattices to each other and a certain distortion in the hexagonal packing of the rod domains. This is schematically shown in Figure 45(d).

The effect of annealing was also studied by the authors. While temperatures below 130°C showed no effect, increasing the temperature above 140°C caused the loss of long-range order. Because the crystal to liquid-crystal transition of the unconnected rod segment is around 140°C it was suggested that only at temperatures between 130 and 140°C the molecules have the needed mobility to form the observed morphologies. For a copolymer with a rod volume fraction $f_{rod} = 0.19$, the sample looses the long-range order at 140°C and annealing at 100°C resulted in a microphase separated structure.

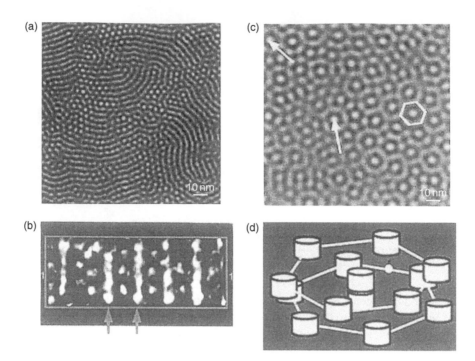

Figure 45   (a) Filtered TEM micrograph of a rod–coil block copolymer with a rod volume fraction of $f_{rod} = 0.3$.
For three and more stacked layers (as monitored by electron tomography) the morphology (c) can
be observed (the arrows point to the "cell walls" and the "cell nuclei" of the observed structure
respectively). A cross sectional slice from the reconstructed volume (b) shows that the "cell nuclei"
are formed by rod aggregates spawning the thickness of the film; the morphology is explained by the
schematic drawing in (d) which shows the stacked and rotated hexagonal superlattices. (Reprinted
from L.H. Radzilowski and S.I. Stupp. *Macromolecules* 27:7747, 1994 and L.H. Radzilowski,
B.O. Carragher, and S.I. Stupp. *Macromolecules* 30: 2110, 1997. Copyright [1994, 1997] American
Chemical Society. With permission.)

This temperature effect on polymers which are more asymmetric in copolymer composition was
interpreted as a hint that the loss of long-range order at higher temperatures is caused by an order–
disorder transition. In the cast state, samples with rod volume fractions $f_{rod} = 0.36$ and $f_{rod} = 0.3$,
respectively, showed ribbons and small aggregates of rod segments. The strips assemble parallelly
in some regions of the sample, but remain isolated in other parts and curve in the later regions. The
thickness of ribbons was found to be 7 nm with length of up to 1 μm. For a nonannealed film of
a copolymer with rod volume fraction $f_{rod} = 0.25$ only small aggregates were observed; for a rod
volume fraction $f_{rod} = 0.19$ nanophase separation was detected.

Stupp and coworkers also reported on miniaturized triblock copolymers containing
oligopolystyrene–block–oligoisoprene and a rigid structure [154–156]. The oligopolystyrene–block–
oligoisoprene copolymer was synthesized via anionic polymerization and terminated by carbon
dioxide. The carboxy-functionalized diblock was afterward coupled to the rod segment (biphenylester block) (see Figure 42(b)). Thin films cast from chloroform (no staining) showed uniform
nanosized aggregates in the TEM. The electron diffraction pattern of the sample indicated crystalline regions in the morphological organization of the rod segments. The wide-angle x-ray pattern
indicated that rod segments are orientated normal to the film plane. The authors emphasized the fact
that their copolymers are composed of a uniform rod part (the biphenylester block) and a chemically
diverse coil part (the oligopolystyrene–block–oligoisoprene diblock $\Rightarrow$ Poisson distribution for the
anionic synthesis, atactic structure of the oligostyrene, 1.4 to 3.4 structure of the oligoisoprene, etc.).
The diverse coil structure is thought to prevent the three-dimensional organization of the structure,
eventhough, the rod segment has a very strong tendency to aggregate via π–π overlap. The authors

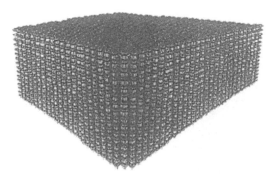

Figure 46 Schematic representation of the proposed structure build of the miniaturized triblocks reported by Stupp et al. The mushroom-shaped nanostructures are assembled to a three dimensional structure in a "cap to stem" fashion. (Reprinted from S.I. Stupp, M. Keser, and G.N. Tew. *Polymer* 39:4505, 1998. Copyright [1998], with permission from Elsevier Science.)

proposed (after intensive molecular modeling calculations) that these mini copolymers build supramolecular units in the form of mushrooms containing 100 triblock units with a molar mass of about 200 kD which assemble in a "cap to stem" fashion. By studying solution cast films in the TEM, a layered (lamellar-like) morphology was detected. This result supported the proposed mushroom-like assembly of the blocks because the dimensions of the bands can be correlated to either the extended rod (dark bands) or to the nonextended oligopolystyrene–block–oligoisoprene diblock (lighter band). The samples were found to be layered terrace-like with a step-size under 100 Å, confirming that the sample is organized in monolayers rather than in bilayers. The structure is suggested to be formed by the mushrooms containing the mini blocks (see Figure 46).

Recently Stupp and coworkers introduced the fascinating class of so-called dendron rod–coils (DRCs) that are assembled of a coil segment, a rod segment, and a dendritic structure (see Figure 47). The molecules start to gel even at very low concentrations suggesting a strong aggregation behavior [157,158]. In highly diluted solutions of the DRCs in dichloromethane the authors observed nanoribbons, that could get as long as 10 μm and had a very uniform thickness of 10 nm (see Figure 47). By studying the crystal structures of model compounds the authors were able to explain the self-assembly of the DRCs by hydrogen bonding and π–π stacking interactions (see schematic drawing in Figure 47)

In studies of the aggregation of the DRC molecules in solution with static and dynamic light scattering, it became obvious that the molecules aggregate in the matter of minutes (in 2-propanol) or hours to days (in ethyl acetate) to concentration dependent nanoribbons with a persistence length of 45 nm [159]. In a very thorough study on the self-assembly of rod–dendron and rod–dendron–rod architectures varying in composition and the dendron generation [160] by optical microscopy, DSC, and x-ray scattering smectic, columnar or cubic phases with symmetries dominated by the anisotropic rod–coil interactions were observed.

Lee et al. also reported on rod–coil systems with a small rod segment. Their copolymers consisted of poly(ethylene oxide) or poly(propylene oxide) and 4-(4′-oxy-4-biphenylcarbonyloxy)-4′-biphenylcarboxylate, synthesized by coupling of the rod segment to monotosylated poly(ethylene oxide) or poly(propylene oxide) (see Figure 42(c) and [d] for the end structures). As alkali metal salts are selectively soluble in poly(ethylene oxide) and poly(propylene oxide), the authors also investigated the phase change of their copolymers by the addition of $LiCF_3SO_3$. These polymers were characterized by DSC, optical polarized microscopy and x-ray scattering. First a rod–coil system with a poly(ethylene oxide) block with a DP of 12 was reported [161,162]. By complexation with 0.05–0.2 mol $LiCF_3SO_3$ per ethylene oxide unit a smectic-A phase was the high temperature mesophase. For 0.2 mol salt per repeating unit an additional cubic phase was observed.

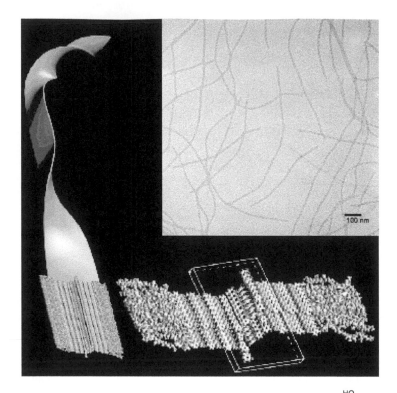

Figure 47   TEM image of twisted nanoribbons formed by DRCs and schematic drawing of the proposed
structure of the nanoribbons. (Reprinted from E.R. Zubarev, M.U. Pralle, E.D. Sone, and S.I. Stupp.
J. Am. Chem. Soc. 123:4105, 2001. Copyright [2001] American Chemical Society. With permission.)

For 0.25 mol salt per ethylene oxide unit, the smectic-A phase disappeared and only a cubic mesophase remained. For 0.3 mol $LiCF_3SO_3$ a cylindrical micellar phase was the high temperature
mesophase. For 0.4–0.7 mol salt per repeating unit, the system only occurred in a cylindrical micellar mesophase. At higher salt concentrations, the samples became amorphous. It was also noticed
that increasing the salt concentration increases the mesomorphic–isotropic transition temperature and
decreases the crystal melting temperature with the result of a higher thermal stability of the smectic
phases. The authors also compared the above block copolymers based on poly(ethylene oxide) with
copolymers based on poly(propylene oxide) having the same DP (DP = 12). The latter copolymer
did not exhibit a smectic A but showed a hexagonal columnar mesophase [163]. The differences
between these similar systems is mostly caused by the steric hindrance of the additional methyl side
group in poly(propylene oxide) and resulting different smectic ordering of the rods due to different
coil packing effects. Lee and coworkers also reported on blends of poly(ethylene oxide) with their
rod–coil block copolymers [164] and on coil–rod–coil triblock copolymers (with poly[propylene
oxide] as the coil segment) and their complexes with $LiCF_3SO_3$ [165].

In a very thorough work on the influence of the length of the coil segment on the phase behavior,
the authors studied block copolymers with poly(propylene oxide) with different DP but the same

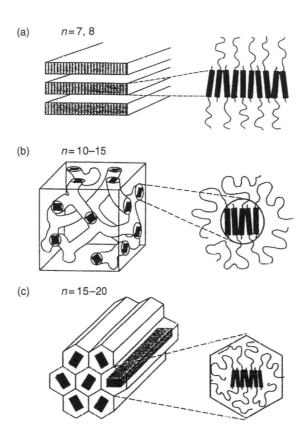

Figure 48 Schematic representation of self-assemblies in the rod–coil diblock copolymer system by Lee et al. (a) The lamellar smectic A; (b) the bicontinuous cubic, and (c) the hexagonal columnar mesophase (*n* represents the DP of the polyisoprene coil segment). (Reprinted from M. Lee, B.-K. Cho, H. Kim, J.Y. Yoon, and W.C. Zin. *J. Am. Chem. Soc.* 120:9168, 1998. Copyright [1998] American Chemical Society. With permission.)

rod segment as above [166,167]. Molecules with seven and eight propyleneoxide units showed, either in DSC traces as in the polarization microscopy, a crystalline phase that melted (coil segment) and recrystallized into a second crystal phase (rod segment) followed by a smectic-C phase. The latter transformed to a smectic-A phase on second heating. For block copolymers with 10 and 12 repeating units no birefringence between crossed polarizers was detected. DSC traces showed a melt transition and an additional phase transition accompanied by a decrease of viscosity suggesting a cubic mesophase. For 15 propylene oxide units a transition to an optical isotropic cubic mesophase occured after crystalline melting followed at higher temperature by a hexagonal columnar meso-phase which underwent isotropization at 46°C. Only a hexagonal columnar phase after crystalline melting was observable for block copolymers with 17 and 20 propylene oxide units. The authors also investigated the microstructure of the copolymers by x-ray diffraction either in the crystalline solid or in the liquid-crystalline phase. For molecules with 7–12 repeating units a lamellar structure with interdigitated rods in the crystalline state became obvious. With 15–17 repeating units, the rods were tilted relative to the layer normal inside the lamellar structure. For 7 and 8 repeating units one can observe smectic-C and smectic-A phases, for 10–15 repeating units the system shows a bicon-tinuous cubic mesophase. For 15–20 repeating units a hexagonal columnar mesophase was observed (see Figure 48).

In a rod–coil block copolymer with poly(propylene oxide) as the coil segment and a slightly elongated rod segment (4′-biphenylcarboxy-4′-biphenyl-4-*n*-propyloxybenzoate) the already mentioned honeycomb morphology (see Figure 49) was found as well [168].

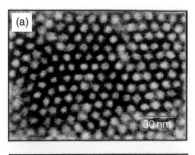

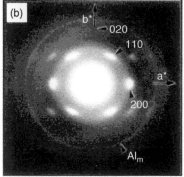

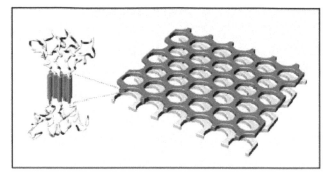

Figure 49 TEM image, ED pattern, and schematic representation of the honeycomb morphology build by a rod–coil block copolymer with poly(propylene oxide) as the coil segment and (4′-biphenylcarboxy-4′-biphenyl-4-n-propyloxybenzoate) as the rod segment. (Reprinted from M. Lee, B.-K. Cho, K.J. Ihn, W.-K. Lee, N.-K. Oh, and W.-C. Zin. *J. Am. Chem. Soc.* 123:4647, 2001. Copyright [2001] American Chemical Society. With permission.)

Lee et al. also studied aggregates formed by a series of linear rod–coil multiblock oligomers with identical rod to coil volume fraction and found that they can regulate the supramolecular structure formed by the systematic variation of the number of repeating units [169]. This work shows very nicely the dependence of the microstructure on molecular composition as the mesophase shows a phase transition from lamellar to columnar phase with a bicontinuous cubic phase with Ia3d symmetry as the intermediate regime. Such a behavior has already been observed within the phase diagram of certain coil–coil block copolymers (Chapters 9, 10).

In addition to the work on copolymers with uniform and monodisperse small rod molecules that is included in this chapter there is a lot of literature available on even smaller rodlike segments. A review was presented by Poser et al. [170].

Recently, a new class of rod–coil block copolymers with nonhelical rod segment was introduced by Zhou and coworkers based on so-called "mesogen jacketed liquid-crystal polymers" (MJLCPs) [171–173]. In the rod segment, mesogenic groups are laterally attached to the polymer backbone through their center of gravity without any or with a very short spacer. The mesogens consist of 2,5-disubstituted vinyl hydroquinones and the polymers are in all reported cases synthesized via living radical polymerization techniques. Initial studies on the solution [174,175] regarding morphological and rheological [176] properties have been reported recently.

ACKNOWLEDGMENTS

This chapter is dedicated to the memory of Prof. Reimund Stadler whose sad and untimely death was a terrible loss for the scientific community. The authors are much indebted to Prof. Alberto Ciferri for his advise and revision of the manuscript. Many thanks to Volker Abetz, Carlos Alemán, Alexander Böker, Thomas Breiner, Jennifer David, Gerd Mannebach, Francisco López-Carrasquero, Antxon Martínez de Ilarduya-Saéz de Asteasu, Salvador León, Gabriele Rösner-Oliver, David Schlitzer, Holger Schmalz, and David Zanuy for their hints, help with the computer, and proof-reading. Sebastián Muñoz-Guerra is grateful to Spanish Ministerio de Cienciay Tecnología for financial support with Grant MAT2003-06955-CO2-1.

References

1. S.I. Stupp. *Curr. Opin. Colloid. Interface Sci.* 3:20, 1998.
2. M. Lee, B.-K. Cho, and W.-C. Zin. *Chem. Rev.* 101:3869, 2001.
3. H.-A. Klok and S. Lecommandoux. *Adv. Mater.* 13:1217, 2001.
4. V.P. Shibaev, M. Palumbo, and E. Peggion. *Biopolymers* 14:73, 1975.
5. V.P. Shibaev, V.V. Chupov, V.M. Laktionov, and N.A. Platé. *Vysokomol. Soedin B* 16:332, 1974.
6. V.V. Chupov, V.P. Shibaev, and N.A. Platé. *Vysokomol. Soedin A* 21:218, 1979.
7. J.M. Fernández-Santín, J. Aymamí, S. Muñoz-Guerra, A. Rodríguez-Galán, and J.A. Subirana. *Nature* 311:53, 1984.
8. S. Muñoz-Guerra, F. López-Carrasquero, J.M. Fernández-Santín, and J.A. Subirana. In: J.C. Salamone, Ed. *Encyclopedia of Polymeric Materials*, Vol. 6, Boca Raton, FL: CRC Press, 1996, p. 4694.
9. N.A. Platé and V.P. Shibaev. *J. Polym. Sci. Macromol. Rev.* 8:117, 1974.
10. A. Turner-Jones. *Makromol. Chem.* 71:1, 1964.
11. N. Morosoff, H. Morawetz, and B. Post. *J. Am. Chem. Soc.* 87:3035, 1965.
12. A.Z. Golik, A.F. Skyshevskii, and I.I. Adamenko. *Zh. Strukt. Khim.* 8:1015, 1967.
13. E.F. Jordan, D.W. Feldeisen, and A.N. Wrigley. *J. Polym. Sci. A-1* 9:1835, 1971.
14. J.C. Smith and R.W. Woody. *Biopolymers* 12:2657, 1973.
15. D.S. Poché. Synthesis and Characterization of Linear and Star-Branched Poly(γ-stearyl-α,L-glutamate). Ph.D. Dissertation, Louisiana State University, 1990.
16. L. Onsager. *Ann. NY. Acad. Sci.* 51:627, 1949.
17. W.H. Daly, I.I. Negulescu, P.S. Russo, and D. Poche. In: P. Shoever and A.C. Balazs, Eds. *Macromolecular Assemblies in Polymer Systems. ACS. Symp. Ser.* 493:292–299, Washington: American Chemical Society, 1992.
18. W.H. Daly, D. Poché, and I. Negulescu. *Prog. Polym. Sci.* 19:79, 1994.
19. E.I. Iizuka, K. Abe, K. Hanabusa, and H. Shirai. In: R.M. Ottenbrite, Ed. *Current Topics in Polymer Science*, Munich: Carl Hanser, 1987, p. 235.
20. A. Thierry and A. Skoulios. *Mol. Cryst. Liq. Cryst.* 41:125, 1978.
21. J. Watanabe, H. Ono, I. Uematsu, and A. Abe. *Macromolecules* 18:2141, 1985.
22. F.J. Romero, J.L. Gómez, and J.M. Barrales-Rienda. *Macromolecules* 27:5004, 1994.
23. J. Watanabe. Thermotropic Liquid Crystals in Polypeptides. *Proceedings of the OUMS'93 on Ordering in Macromolecular Systems*. Osaka, 1993, A. Teramoto, M. Kobayashi, and T. Norisuye, Eds. Springer-Verlag, p. 99.
24. J. Watanabe and H. Ono. *Macromolecules* 19:1079, 1986.
25. F.J. Romero, J.L. Gómez, J. Lloveras-Macia, and S. Muñoz-Guerra. *Polymer* 32:1642, 1991.
26. J. Watanabe and Y. Takashina. *Macromolecules* 24:3423, 1991.
27. D. Frenkel. *Liq. Cryst.* 5:929, 1989.
28. M. Hoshimo, H. Nakano, and H. Kimura. *J. Phys. Soc. Jpn.* 46:1709, 1990.
29. J. Watanabe and Y. Takashina. *Polym. J.* 24:709, 1992.
30. J. Watanabe, Y. Fukuda, R. Gehani, and I. Uetmasu. *Macromolecules* 17:1004, 1982.
31. J. Watanabe, M. Gotoh, and T. Nagase. *Macromolecules* 20:298, 1987.
32. J. Watanabe and T. Nagase. *Polym. J.* 19:781, 1987.
33. F. López-Carrasquero, M. García-Alvarez, and S. Muñoz-Guerra. *Polymer* 35:4502, 1994.
34. M. García-Alvarez, S. León, C. Alemán, J.L. Campos, and S. Muñoz-Guerra. *Macromolecules* 31:124, 1998.
35. A. Martínez de Ilarduya, C. Alemán, M. García-Alvarez, and S. Muñoz-Guerra. *Macromolecules* 32:3257, 1999.

36. J.M. Fernández-Santín, S. Muñoz-Guerra, A. Rodríguez-Galán, J. Aymami, J. Lloveras, and J.A. Subirana. *Macromolecules* 20:62, 1987.
37. S. Muñoz-Guerra, J.M. Fernández-Santín, C. Alegre, and J.A. Subirana. *Macromolecules* 22:1540, 1989.
38. J.M. Montserrat, S. Muñoz-Guerra, and J.A. Subirana. *Makromol. Chem. Macromol. Symp.* 20/21:319, 1988.
39. J.J. Navas, C. Alemán, F. López-Carrasquero, and S. Muñoz-Guerra. *Polymer* 38:3477, 1997.
40. F. López-Carrasquero, S. Montserrat, A. Martínez-Ilarduya, and S. Muñoz-Guerra. *Macromolecules* 28:5535, 1995.
41. M.J. Avrami. *Phys. Chem.* 7:1103, 1939.
42. Y. Calventus, P. Colomer, J. Malêk, S. Montserrat, F. López-Carrasquero, A. Martínez-Ilarduya, and S. Muñoz-Guerra. *Polymer* 40:801, 1999.
43. F. López-Carrasquero. Síntesis, Estructura y Propiedades de Poli(α-alquil-β,L-aspartatos). Ph.D. Dissertation, Universidad Politécnica de Catalunya, 1995.
44. J.M. Lehn. *Angew Chem. Int. Ed. Engl.* 29:1304, 1990.
45. A. Ciferri. *Prog. Polym. Sci.* 20:1081, 1995.
46. E. Tsuchida. *J. Macromol. Sci. Pure Appl. Chem.* A31:1, 1994.
47. M. Antonietti and J. Conrad. *Angew. Chem. Int. Ed. Engl.* 33:1869, 1994.
48. A. Ciferri. *Macromol. Chem. Phys.* 195:457, 1994.
49. E.A. Ponomarenko, A.J. Waddon, K.N. Bakeev, D.A. Tirrell, and W.J. MacKnight. *Macromolecules* 29:4340, 1996.
50. E.A. Ponomarenko, A.J. Waddon, D.A. Tirrell, and W.J. MacKnight. *Langmuir* 12:2169, 1996.
51. E.A. Ponomarenko, D.A. Tirrell, and W.J. MacKnight. *Macromolecules* 29:8751, 1996.
52. E.A. Ponomarenko, D.A. Tirrell, and W.J. MacKnight. *Macromolecules* 31:1584, 1998.
53. F.S. Bates. *Science* 251:898, 1991.
54. A.N. Semenov and S.V. Vasilenko. *Sov. Phys. JETP* 63:70, 1986.
55. A.N. Semenov. *Mol. Cryst. Liq. Cryst.* 209:191, 1991.
56. A.N. Semenov and A.V. Subbotin. *Sov. Phys. JETP* 74:690, 1992.
57. D.R.M. Williams and G.H. Fredrickson. *Macromolecules* 25:3561, 1992.
58. R. Holyst and M.J. Schick. *Chem. Phys.* 96:727, 1992.
59. M.W.J. Matsen. *Chem. Phys.* 104:7758, 1996.
60. R.R. Netz and M. Schick. *Phys. Rev. Lett.* 77:302, 1996.
61. M. Mueller and M. Schick. *Macromolecules* 29:8900, 1996.
62. M.W. Matsen and C.J. Barrett. *Chem. Phys.* 109:4108, 1998.
63. A. Halperin. *Europhys. Lett.* 10:549, 1989.
64. A. Halperin. *Macromolecules* 23:2724, 1990.
65. E. Raphael and P.G. de Gennes. *Makromol. Symp.* 62:1, 1992.
66. E.M. Sevick and D.R.M. Williams. *Science* 129/130:387, 1997.
67. D. Düchs and D.E. Sullivan. *J. Phys. Condens. Matter.* 14:12189, 2002.
68. W. Li and D. Gersappe. *Macromolecules* 34:6783, 2001.
69. M. Reenders and G. ten Brinke. *Macromolecules* 35:3266, 2002.
70. H. Schlaad and M. Antonietti. *Eur. Phys. J. E* 10:17, 2003.
71. B. Perly, A. Douy, and B. Gallot. *Makromol. Chem.* 177:2569, 1976.
72. J.-P. Billot, A. Douy, and B. Gallot. *Makromol. Chem.* 177:1889, 1976.
73. J.-P. Billot, A. Douy, and B. Gallot. *Makromol. Chem.* 178:1641, 1977.
74. A. Douy and B. Gallot. *Makromol. Chem.* 178:1595, 1977.
75. A. Douy and B. Gallot. *Polymer* 23:1039, 1982.
76. H. Schlaad, H. Kukula, B. Smarsly, M. Antonietti, and T. Pakula. *Polymer* 43:5321, 2002.
77. D. Vernino, D. Tirrell, and M. Tirell. *Polymer Mater. Sci. Eng.* 71:496, 1994.
78. H. Kukula, H. Schlaad, M. Antonietti, and S. Förster. *J. Am. Chem. Soc.* 124:1658, 2002.
79. F. Chécot, S. Lecommandoux, Y. Gnanou, and H.A. Klok. *Angew. Chem. Int. Ed.* 114:1395, 2002.
80. F. Chécot, S. Lecommandoux, H.A. Klok, and Y. Gnanou. *Eur. Phys. J. E* 10:25, 2003.
81. J.S. Crespo, S. Lecommandoux, R. Borsali, H.A. Klok, and V. Soldi. *Macromolecules* 36:1253, 2003
82. R. Borsali, S. Lecommandoux, R. Percora, and H. Benoit. *Macromolecules* 34:4229, 2001.
83. A. Nakajima, T. Hayashi, K. Kugo, and K. Shinoda. *Macromolecules* 12:840, 1979.
84. T. Hayashi, G.W. Chen, and A. Nakajima. *Polym. J. (Tokyo)* 16:739, 1984.
85. R. Yoda, S. Komatsuzaki, E. Nakanishi, and T. Hayashi. *Eur. Polym. J.* 31:335, 1995.

86. R. Yoda, M. Shimoda, S. Komatsuzaki, T. Hayashi, and T. Nishi. *Eur. Polym. J.* 33:815, 1997.
87. M. Gervais, A. Douy, B. Gallot, and R. Erre. *Polymer* 29:1779, 1988.
88. K. Kugo, Y. Hata, T. Hayashi, and A. Nakajima. *Polym. J. (Tokyo)* 14:401, 1982.
89. A. Nakajima, K. Kugo, and T. Hayashi. *Polym. J.* 11:995, 1979.
90. Y. Yoshida, K. Makino, T. Ito, Y. Yamakawa, and T. Hayashi. *Eur. Polym. J.* 32:877, 1996.
91. R. Yoda, S. Komatsuzaki, and T. Hayashi. *Eur. Polym. J.* 32:233, 1996.
92. R. Yoda, S. Komatsuzaki, and T. Hayashi. *Biomaterials* 16:1203, 1995.
93. R. Yoda, S. Komatsuzaki, E. Nakanishi, H. Kawaguchi, and T. Hayashi. *Biomaterials* 15: 944, 1994.
94. H. Sato, A. Nakajima, T. Hayashi, G.W. Chen, and Y.J. Noishiki. *Biomed. Mater. Res.* 19:1135, 1985.
95. G.W. Chen, T. Hayashi, and A. Nakajima. *Polym. J. (Tokyo)* 16:805, 1984.
96. S. Barenberg, J.M. Anderson, and P.H. Geil. *Int. J. Biol. Macromol.* 3:82, 1981.
97. G. Floudas, P. Papadopoulos, H.A. Klok, G.W.M. Vandermeulen, and J. Rodriguez-Hernandez. *Macromolecules* 36:3673, 2003.
98. J.T. Chen, E.L. Thomas, C.K. Ober, and S.S. Hwang. *Macromolecules* 28:1688, 1995.
99. J.T. Chen, E.L. Thomas, C.K. Ober, and G.-P. Mao. *Science* 273:343, 1996.
100. J.-W. Park and E.L. Thomas. *Adv. Mater.* 15:585, 2003.
101. J.-W. Park and E.L. Thomas. *J. Am. Chem. Soc.* 124:514, 2002.
102. J. Wu, E.M. Pearce, and T.K. Kwei. *Macromolecules* 34:1828, 2003.
103. J.-H. Ahn and J.-S. Lee. *Macromol. Rapid. Commun.* 24:571, 2003.
104. J.-H. Ahn, Y.-D. Shin, S.-Y. Kim, and J.-S. Lee. *Polymer* 44:3847, 2003.
105. J.J.L.M. Cornelissen, M. Fischer, and R.J.M. Nolte. *Science* 280:1427, 1998.
106. D.M. Vriezema, J. Hoogboom, K. Velonia, K. Takazawa, P.C.M. Christianen, J.C. Maan, A.E. Rowan, and R.J.M. Nolte. *Angew. Chem. Int. Ed.* 42:772, 2003.
107. P.A.J. de Witte, M. Castriciano, J.J.L.M. Cornelissen, L.M. Scolaro, R.J.M. Nolte, and A.E. Rowan. *Chem. Eur. J.* 9:1775, 2003.
108. J.J.L.M. Cornelissen, R. van Heerbeek, P.C.J. Kamer, J.N.H. Reek, N.A.J.M. Sommerdijk, and R.J.M. Nolte. *Adv. Mater.* 14:489, 2002.
109. J.L. David, S.P. Gido, and B.M. Novak. *Polymer Preprints* 39/2:433, 1998.
110. T.J. Deming and B.M. Novak. *Macromolecules* 24:5478, 1991.
111. B.M. Novak and T.J. Deming. *Macromol. Symp.* 77:405, 1994.
112. G. Ziegast and B. Pfannemüller. *Makromol. Chem. Rapid. Commun.* 5:373, 1984.
113. B. Pfannemüller and T. Dengler. *Makromol. Chem.* 189:1965, 1988.
114. W.N. Emmerling and B. Pfannemüller. *Colloid Polymer Sci.* 261:677, 1983.
115. G. Ziegast and B. Pfannemüller. *Carbohydr. Res.* 160:185–204, 1987.
116. V.v. Braunmühl, G. Jonas, and R. Stadler. *Macromolecules* 28:17, 1995.
117. V.v. Braunmühl and R. Stadler. *Macromol. Symp.* 103:141, 1996.
118. K. Loos, G. Jonas, and R. Stadler. *Macromol. Chem. Phys.* 202: 3210, 2001.
119. K. Loos and R. Stadler. *Macromolecules* 30:7641, 1997.
120. K. Loos and A.H.E. Müller. *Biomacromolecules* 3:368, 2002.
121. K. Loos, A. Böker, H. Zettl, M. Zhang, G. Krausch, and A.H.E. Müller. *Macromolecules*, 2005, in press.
122. K. Akiyoshi, N. Maruichi, M. Kohara, and S. Kitamura. *Biomacromolecules* 3:280, 2002.
123. K. Akiyoshi, M. Kohara, K. Ito, S. Kitamura, and J. Sunamoto. *Macromol. Rapid. Commun.* 20:112, 1999.
124. W.T.E. Bosker, K. Ágoston, M.A. Cohen Stuart, W. Norde, J.W. Timmermans, and T.M. Slaghek. *Macromolecules* 36:1982, 2003.
125. D.M. Haddleton and K. Ohno. *Biomacromolecules* 1:152, 2000.
126. K. Kobayashi, S. Kamiya, and N. Enomoto. *Macromolecules* 29:8670, 1996.
127. A. Narumi, K. Kawasaki, H. Kaga, T. Satoh, N. Sugimoto, and T. Kakuchi. *Polym. Bull.* 49:405, 2003.
128. P. Cavalleri, A. Ciferri, C. Dell'Erba, M. Novi, and B. Purevsuren. *Macromolecules.* 30:3513, 1997.
129. P. Cavalleri, A. Ciferri, C. Dell'Erba, A. Gabellini, and M. Novi. *Macromol. Chem. Phys.* 199:2087, 1998.
130. A. Gabellini, M. Novi, A. Ciferri, and C. Dell'Erba. *Acta Polym.* 50:127, 1999.
131. C.-J. Yang and S.A. Jenekhe. *Supramol. Sci.* 1:91, 1994.
132. S.A. Jenekhe and J.A. Osaheni. *Chem. Mater.* 6:1906, 1994.
133. J.A. Osaheni and S.A. Jenekhe. *J. Am. Chem. Soc.* 117:7389, 1995.
134. S.A. Jenekhe and X.L. Chen. *Science* 279:1903, 1998.

135. S.A. Jenekhe and X.L. Chen. *J. Phys. Chem. B* 104:6332, 2000.
136. X.L. Chen and S.A. Jenekhe. *Langmuir* 15:8007, 1999.
137. X.F. Zhong and B. Francois. *Makromol. Chem.* 192:2277, 1991.
138. B. Francois and X.F. Zhong. *Synth. Met.* 41–43:955, 1991.
139. G. Widawski, M. Rawiso, and B. Francois. *Nature* 369:387, 1994.
140. B. Francois, O. Pitois, and J. Francois. *Adv. Mater.* 7:1041, 1995.
141. B. de Boer, U. Stalmach, H. Nijland, and G. Hadziioannou. *Adv. Mater.* 12:1581, 2000.
142. T. Olinga and B. Francois. *Makromol. Chem. Rapid. Commun.* 12:575, 1991.
143. T. Olinga and B. Francois. *J. Chim. Phys. Phys. Chim. Biol.* 89:1079, 1992.
144. B. Francois and T. Olinga. *Synth. Met.* 57:3489, 1993.
145. M.A. Hempenius, B.M.W. Langeveld-Voss, J.A.E.H. van Haare, R.A.J. Janssen, S.S. Sheiko, J.P. Spatz, M. Moeller, and E.W. Meijer. *J. Am. Chem. Soc.* 120:2798, 1998.
146. D. Marsitzky, T. Brand, Y. Geerts, M. Klapper, and K. Müllen. *Macromol. Rapid Commun.* 19:385, 1998.
147. V. Francke, H.J. Raeder, Y. Geerts, and K. Müllen. *Macromol. Rapid. Commun.* 19:275, 1998.
148. P. Leclère, A. Calderone, D. Marsitzky, V. Francke, K. Müllen, J.L. Brédas, and R. Lazzaroni. *Synth. Met.* 121:1295, 2001.
149. H. Kukula, U. Ziener, M. Schoeps, and A. Godt. *Macromolecules* 31:5160, 1998.
150. P.K. Tsolakis, J.K. Kallitsis, and A. Godt. *Macromolecules* 35:5758, 2002.
151. L.H. Radzilowski, J.L. Wu, and S.I. Stupp. *Macromolecules* 26:879, 1993.
152. L.H. Radzilowski and S.I. Stupp. *Macromolecules* 27:7747, 1994.
153. L.H. Radzilowski, B.O. Carragher, and S.I. Stupp. *Macromolecules* 30: 2110, 1997.
154. S.I. Stupp, V. LeBonheur, K. Walker, L.S. Li, K.E. Huggins, M. Keser, and A. Amstutz. *Science* 276:384, 1997.
155. S.I. Stupp, M. Keser, and G.N. Tew. *Polymer* 39:4505, 1998.
156. G.N. Tew, L. Li, and S.I.J. Stupp. *Am. Chem. Soc.* 120:5601, 1998.
157. M. Sayar and S.I. Stupp. *Macromolecules* 34:7135, 2001.
158. E.R. Zubarev, M.U. Pralle, E.D. Sone, and S.I. Stupp. *J. Am. Chem. Soc.* 123:4105, 2001.
159. B.J. de Gans, S. Wiegand, E.R. Zubarev, and S.I. Stupp. *J. Phys. Chem. B* 106:9730, 2002.
160. S. Lecommandoux, H.-A. Klok, M. Sayar, and S.I. Stupp. *J. Polym. Sci. A* 41:3501, 2003.
161. M. Lee, N.-K. Oh, H.-K. Lee, and W.-C. Zin. *Macromolecules* 29:5567, 1996.
162. M. Lee, N.-K. Oh, and M.-G. Choi. *Polym. Bull.* 37:511, 1996.
163. M. Lee, N.-K. Oh, and W.-C. Zin. *Chem. Commun.* 15:1787, 1996.
164. S.H. Ji, W.-C. Zin, N.-K. Oh, and M. Lee. *Polymer* 38:4377, 1997.
165. M. Lee and B.-K. Cho. *Chem. Mater.* 10:1894, 1998.
166. M. Lee, B.-K. Cho, H. Kim, and W.-C. Zin. *Angew. Chem.* 110:661–663, 1998.
167. M. Lee, B.-K. Cho, H. Kim, J.Y. Yoon, and W.C. Zin. *J. Am. Chem. Soc.* 120:9168, 1998.
168. M. Lee, B.-K. Cho, K.J. Ihn, W.-K. Lee, N.-K. Oh, and W.-C. Zin. *J. Am. Chem. Soc.* 123:4647, 2001.
169. M. Lee, B.-K. Cho, N.-K. Oh, and W.-C. Zin. *Macromolecules* 34:1987, 2001.
170. S. Poser, H. Fischer, and M. Arnold. *Prog. Polym. Sci.* 23:1337, 1998.
171. X. Wan, Y. Tu, D. Zhang, and Q. Zhou. *Polym. Int.* 49:243, 2000.
172. H. Zhang, Y. Tu, X. Wan, Q.-F. Zhou, and E.M. Woo. *J. Polym. Res.* 9:11, 2002.
173. Y. Yi, X. Wan, X. Fan, R. Dong, and Q. Zhou. *J. Polym. Sci. A* 41:1799, 2003.
174. Y. Tu, X. Wan, D. Zhang, Q. Zhou, and C. Wu. *J. Am. Chem. Soc.* 122:10201, 2000.
175. Y. Tu, X. Wan, H. Zhang, X. Fan, X. Chen, Q.-F. Zhou, and K. Chau. *Macromolecules* 36:6565, 2003.
176. P. Gopalan, Y. Zhang, X. Li, U. Wiesner, and C.K. Ober. *Macromolecules* 36:3357, 2003.

Part II

Properties and Functions

Chapter 12

DNA Structures and Their Applications in Nanotechnology

Jun Xu, Thom H. LaBean, and Stephen L. Craig

CONTENTS

As one of the four major macromolecules native to living cells, DNA plays a central role in gene storage and transfer. With advances in synthetic and biotechnology routes to its *ex vivo* production, however, DNA has found increasing use in unnatural settings. Among these recent uses are a host of applications in supramolecular polymer systems. It is somewhat fitting that DNA has become a molecule of choice for supramolecular materials and chemistry; the elegance of the DNA duplex design and the fidelity of molecular recognition within the double helix are an oft-cited inspiration for much of supramolecular chemistry. In the context of supramolecular polymerizations, DNA is interesting because it can be used and studied as a closed SP (supramolecular polymer) and also as an open SP of either a linear, a planar, or a three-dimensional variety. In this chapter, the fundamental molecular recognition properties of DNA and a picture of the scope of its utility will be presented. The recent explosion in the use of DNA-templated materials prohibits an exhaustive review of the field, but the overview of recent examples should provide a flavor for the possibilities available to the field.

I.   **DNA** AS A CLOSED **SP**

A.   DNA Structure

The properties and applications of DNA-based SPs are intrinsically tied to the structure and properties of DNA itself. To a chemist, DNA is a linear polymer formed from deoxyribonucleotide monomers. The repeating monomer unit, also called a nucleotide, consists of a $2'$-deoxyribose sugar ring, a phosphate group at the $5'$ position, and one of the four bases — adenine (A), guanine (G), cytosine (C), and thymine (T) — attached at the $1'$ position. The former two bases are purines and the latter two are pyrimidines; these names reflect the biochemical heritage of the molecules. The monomers are linked through phosphodiester bonds between the phosphate of one nucleotide and the $3'$ hydroxyl group of another. As such, DNA has a structure that resembles a random copolymer of the A, G, C, and T nucleotide monomers.

The beauty of DNA, of course, is tied to the fact that the primary structure is far from random. Erwin Chargaff [1] discovered in the late 1940s that in any DNA sample, the adenine and thymine were always present in equimolar amounts, and so too were guanine and cytosine. This relationship (the so-called Chargaff's rule) was true regardless of the absolute concentrations of A/T or G/C, and it was the first evidence of the special complementarity within base pairs. Coupled with the x-ray diffraction studies by Wilkins [2] and Franklin [3], these observations led James Watson and Francis Crick to propose the well-known DNA double helix structure in 1953 [4]. In the double helix, each of two linear single strands of DNA coils around a central axis. This duplex structure is held together by a specific pattern of hydrogen bonds between the bases–adenines are preferentially bound to thymines, and cytosines are preferentially bound to guanines. There are two hydrogen bonds between A and T, and three hydrogen bonds between C and G, as shown in Figure 1(a). The complementarity of the hydrogen bonding patterns is the basis for the selectivity in binding between DNA single strands, as discussed quantitatively below. It should be noted that there is another unusual type of base pairing, called Hoogsteen base pairing (Figure 1(b)). To form Hoogsteen base pairings,

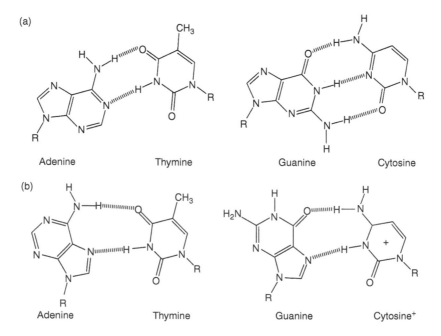

Figure 1   Base pairing of adenine–thymine and guanine–cytosine.   (a) Watson–Crick type;   and (b) Hoogsteen type.

the purines rotate by 180° around the central axis and adopt a *syn* instead of an *anti* conformation relative to the sugar. Another difference is that cytosine should be protonated at N3 and only two H-bonds are formed between G–C$^+$.

The double helix of DNA exists in any of the several conformations, three of which are the dominant forms: B-, A-, and Z-form (Figure 2). The most common conformation is that proposed in the original Watson–Crick model, B-form DNA, and the majority of DNA-mediated supramolecular assemblies are based on this conformer. Two right-handed strands wind around each other in opposite (antiparallel) directions and are held by hydrogen bonds between purines in one strand and corresponding pyrimidines in the other (A–T or C–G). Ideally, the plane of base pairs is perpendicular to the helix axis. The backbone distances between A–T and C–G base pairs are virtually the same (1.1 nm), precluding the possibility of purine — purine (requiring more space) or pyrimidine–pyrimidine pairings (distance is long between two bases). The hydrophilic sugar–phosphate backbone is exposed to the aqueous solution, while the hydrophobic bases stack on each other in the interior. Each base stacks on another base with a distance of 0.34 nm, and rotates 36° with respect to the next, accommodating ten base pairings per turn in the right-handed helix. The distance per turn is therefore 3.4 nm, called the pitch. Since the glycosidic bonds (the bonds link the sugar and the base) are not opposite to each other, two unequal grooves — the major groove and the minor groove — can be created. In certain circumstances such as low humidity, B-form DNA undergoes a conformational change to A-form, which is also right-handed but is flatter and wider than B-form. A-form accommodates 11 base pairs per turn with a pitch of 2.8 nm, and more strikingly, the planes of base pairs are not perpendicular but tilted 20° to the helix axis. Z-form is a left-handed helix and probably occurs with alternating purines and pyrimidines sequences, such as d(CGCGCG), at high salt concentrations. Z-form is longer and thinner than B-form, accommodating 12 base pairs per turn with a pitch of 4.5 nm. It is recognized that the repeating unit in Z-form is dinucleotides, in contrast to mononucleotides in B- and A-forms. These structural features described above are familiar to all biologists and nearly all chemists, but it is important to have them firmly in mind as one considers using DNA as a building block for supramolecular polymerization. Major structural features of three DNA forms are compared and summarized in Table 1.

In addition to its structure, the thermodynamics of DNA base pairing are critical to its use in supramolecular systems. Several factors contribute to the stability of the DNA double helix. It originally was assumed that hydrogen bonding is of central importance, as evidenced by the fact that G–C base pairs are more stable than A–T (three hydrogen bonds compared to two). However, it is now appreciated that hydrogen bonding is far from the entire determinant of the stability of double helix structures [5] because, during denaturation (the process in which double strands dissociate into two individual single strands), the hydrogen bonds between bases are replaced by hydrogen bonds between the bases and water. Therefore, there should be other factors which play a dominant role

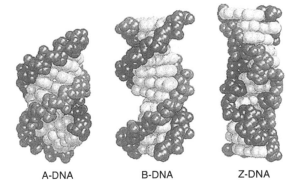

A-DNA   B-DNA   Z-DNA

Figure 2 Cartoons of double helix structures of A-, B-, and Z-DNA.

Table 1   Structural Features of A-, B- and Z-DNA.

	A	B	Z
Helix rotation sense	Right-handed	Right-handed	Left-handed
Base pairs per turn	11	10	12 (6 dimers)
Rise per base pair	2.3 Å	3.4 Å	3.7 Å
rotation per base pair	33°	36°	60°/dimer
Rise per turn (pitch)	25 Å	34 Å	45 Å
Diameter	26 Å	20 Å	18 Å
Base tilt to the helix axis	20°	6°	7°
Major groove	Narrow/deep	Wide/deep	Flat
Minor groove	Wide/shallow	Narrow/deep	Narrow/deep
Glycosyl bond	*Anti*	*Anti*	*Anti* for purimidines; *Syn* for purines

in DNA stability, and these have been established. For B-form DNA, one can see that hydrophobic interactions force the π-rich bases into the interior of the double helix. In this arrangement, there is a concomitant and beneficial separation of the negatively charged phosphate groups of one strand from those of the other. Additive stacking interactions between these bases further strengthen the stability of double helix. In addition, the cations in a buffer solution screen the negative phosphate groups. This association weakens the repulsion between the negatively charged groups and stabilizes the double helix. This shielding effect is more pronounced with bivalent cations (such as Mg^{2+}) than with monovalent cations (such as Na^+). The stability of the DNA double helix is therefore dependent not only on the complementarity of the base pairing, although this certainly plays a significant role, but also on the nature of the solvent and the structure and ionic strength of the aqueous buffer.

The conformations of DNA discussed so far are linear double helices. In living cells, however, many naturally occurring DNAs are circular (Figure 3(a)). For example, plasmids and almost all bacterial chromosomes are circular DNA duplexes. Furthermore, many of these circular DNAs are supercoiled (Figure 3(b)), like a twisted rope. These supercoils are formed when double strands of a circular DNA are underwound or overwound. Hairpins (Figure 3(c)) are also common structures in naturally occurring DNAs. They are formed when a single strand is folded back onto itself to form intrastrand base pairings. Similar structures that occur in double strands are called cruciforms (Figure 3(d)). It should be noted that triple helices are also found, for example in so-called H-DNAs. The H-DNA has one all-purine strand and the other all-pyrimidine strand (Figure 3(e)). Besides the normal Watson–Crick base pairing, an extra Hoogsteen base pairing is involved in the triplex, generally TAT or CGC triplets.

B.   DNA Thermodynamics

The stability of a DNA duplex depends on, besides the chemical and physical structure of the DNA itself, many environmental factors such as solvent, temperature, and salt composition and concentration. For example, when a stable DNA duplex is heated to higher temperatures, the double helix structure will be gradually denatured into two coiled single strands. When cooled to room temperature, the two single strands can reassociate to again form the duplex, a process called DNA renaturation (sometimes also called DNA hybridization or annealing). These processes are reversible, as illustrated in the following equilibrium:

$$SS_1 + SS_2 \overset{K_{eq}}{\rightleftharpoons} DS \tag{1}$$

where SS_1 and SS_2 represent two complementary single strands, DS represents DNA duplex formed from SS_1 and SS_2, and K_{eq} is the equilibrium constant for the duplex formation.

As described in the following sections, supramolecular polymerizations of DNA will reflect the thermodynamics and kinetics of duplex formation. There are generally two major methods

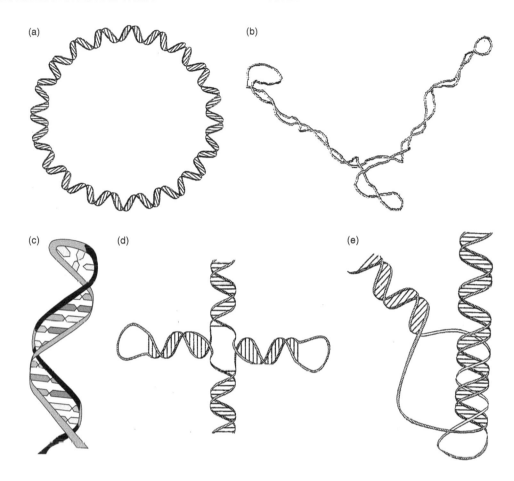

Figure 3 Diagrams of alternative DNA conformations. (a) Circular DNA duplex; (b) supercoiled DNA; (c) hairpin formed from a single strand; (d) cruciform; (e) H-DNA.

widely applied to determine thermodynamic parameters. The first one is thermal melting followed by UV spectroscopy. The transition from DNA duplex to single strands can be monitored by a UV spectrophotometer, because the disruption of base stacking in the duplex leads to an increase in UV absorbance at 260 nm — the hyperchromism effect. As a DNA duplex solution is heated, the UV absorbance versus temperature provides a thermal melting curve. The stability of a DNA duplex is given by the melting temperature T_m, the temperature at which half of the duplexes have been melted to single strands ($[DS] = [SS_1] = [SS_2]$). For short DNA duplexes, oligonucleotides are typically either single strands or double strands, and the concentration of intermediate states during the melting transition is negligible. The unequal molarities on either side of the equilibrium mean that the melting temperature depends on the concentration of DNA, and as described below this concentration dependence provides facile access to more complete thermodynamic data. Under these "two-state" conditions, and assuming $\Delta H°$ and $\Delta S°$ are independent of temperature, K_{eq} and T_m are related to the total concentration of single strands (C_t) by Eqs. (2) and (3):

$$1/T_m = (R/\Delta H°)\ln(C_t) + (\Delta S° - R\ln(4))/\Delta H° \qquad (2)$$

$$K_{eq}(T_m) = 4/C_t \qquad (3)$$

Because T_m depends on C_t, $\Delta H°$, $\Delta S°$, and K_{eq} at various temperatures, it can be extrapolated readily from a plot of $1/T_m$ versus $\ln(C_t)$. It should be noted that, in the case of duplex formation from self-complementary single strands, the above equations need slight modification [6,7].

Although the method mentioned above has been used generally to extrapolate thermodynamic parameters, its accuracy is contingent on the validity of the two-state assumption and temperature independence of $\Delta H°$ and $\Delta S°$. Breslauer [6,8] in particular has pointed out the limitations of these assumptions, and, while useful, UV melting curves should always be interpreted with them in mind. In contrast, calorimetric methods provide a direct, model-independent (and hence more reliable) measure of the enthalpy and entropy changes of the association–dissociation process. Differential scanning calorimetry (DSC) can be used to measure the excess heat capacity (ΔC_p) as a function of temperature. Since $\Delta H° = \int \Delta C_p \, dT$, the area under the baseline-corrected curve corresponds to the enthalpy change during the transition. Similarly, the entropy change can also be determined once $\Delta C_p/T$ as a function of temperature is derived from the same calorimetric curve, since $\Delta S° = \int \Delta C_p/T \, dT$. The free energy ($\Delta G° = \Delta H° - T\Delta S°$) is therefore readily derived from $\Delta H°$ and $\Delta S°$. Calorimetry not only provides accurate thermodynamic data, but in conjunction with UV studies it provides a handle to evaluate the validity of the two-state assumption.

The wealth of experimental thermodynamic data has been used to develop a powerful predictive capacity. As early as 1986, Breslauer [9] illustrated a useful empirical method to predict DNA duplex stability based on nearest-neighbor interactions in the base sequence. $\Delta H°$ and $\Delta S°$ of all 10 possible nearest-neighbor interactions were extrapolated from a small library of 28 DNA duplexes. The overall $\Delta H°$ and $\Delta S°$ of DNA duplex formation could be predicted from the sum of the appropriate empirical values. This rough but robust method has been verified and further refined by more and more available databases of DNA thermodynamics [10,11]. The predictive power of this empirical methodology is a powerful tool for those interested in using DNA in supramolecular (or other) applications for which specific thermodynamics are an important criterion.

Finally, it should be noted that duplex formation is driven by an enormous and favorable enthalpy change that overrides the entropic penalty associated with organizing the single strands. Thus, duplex stability is very temperature dependent, meaning that even duplexes that are very strong at room temperature ($K_{eq} > 10^{20} \, M^{-1}$) melt at easily accessible temperatures (70 to 80°C at μM concentrations). Repetitive annealing therefore provides a facile and rapid route to self-assembled structures that are kinetically inert under ambient conditions.

C. DNA Synthesis and Enzymology: Tools of the Trade

To be a useful tool for nanotechnology, DNA-based monomers must be produced on the appropriate scale with relative ease. Two technological advances have fueled the recent boom in DNA-mediated supramolecular assembly. The polymerase chain reaction (PCR) was invented two decades ago, and it provides very efficient amplification of small quantities of DNA fragments with specified sequences. Comprehensive reviews can be found elsewhere [12,13], but the general principle of PCR is illustrated in Figure 4(a). Two oligonucleotide primers, usually 16–20 bases long, are designed to be complementary to the two ends of the target base sequence to be amplified. The primers are mixed in great excess with the original DNA double strands. Following the heating and cooling to denature and anneal the original double helix, the primers are bound to the ends of the target DNA fragments. Polymerases then extend the base sequences from $3'$ to $5'$ starting right after the last base of each primer. When the cycle of denaturation, annealing, and polymerase extension is repeated, the target DNA fragments are amplified exponentially (2^n after n cycles). In practice, it only takes a few hours to finish 50 cycles for a target DNA fragment of up to several kilobases, and the efficiency is such that a total amplification of 10^{10} can be obtained after 50 cycles. It should be noted that it is unnecessary to add DNA polymerase during each thermocycle (Figure 4(b)) if the enzyme survives at high temperatures. The availability of thermolabile enzymes, such as *Taq* polymerase, thus makes it possible to automate and widely use the PCR technique.

Synthetic advances complement the utility of PCR. Synthetic access to nucleoside phosphoramidites (Figure 5) and their efficient coupling has led to an automated, high yield syntheses. Again, detailed reviews can be found elsewhere [14,15]. Briefly, the building block nucleoside is modified

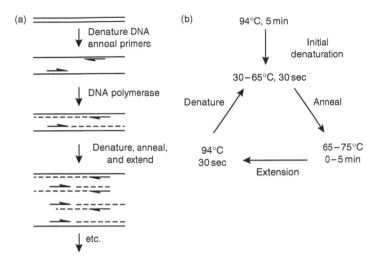

Figure 4 Polymerase chain reaction (PCR). (a) Principle of DNA amplification in PCR; (b) a typical thermocycle in PCR. (Adapted from R.A. Gibbs. *Anal. Chem.* 62:1202–1214, 1990. With permission.)

with a 5′-dimethoxytrityl (DMT) protecting group, a cyanoethyl-protected 3′-phosphite group, and typically a benzoyl or isobutyryl group protecting any primary amine in the base. The synthesis is initiated from a starting monomer immobilized on a solid-support, usually controlled pore glass (CPG) or polystyrene, and involves four major steps during a synthetic cycle: deprotecting, coupling, capping, and oxidizing, as schematically illustrated in Figure 5.

The target sequence is achieved by repeating this cycle $(n-1)$ times when n is the number of bases in the target oligonucleotide, followed by simple cleavage and deprotection procedures. Generally, the efficiency in each cycle is >99%, for example, a total 83% yield for synthesizing a 20-mer, and this method can be applied for syntheses of oligonucleotides of over 100 bases, although the yield might be as low as 30%. The final product includes the full-length desired oligonucleotide as well as a certain population of failure sequences such as $(n-1)$-mer, $(n-2)$-mer, etc. Further purification is therefore required for syntheses of rather long oligonucleotides. Two widely used methods are polyacrylamide gel electrophoresis (PAGE) and high performance liquid chromatography (HPLC). Both methods have their own advantages and may surpass one another depending on modifications and desired usage. For example, PAGE has better separation but is only suitable for small amounts of oligonucleotides (1 μmol scale or less), while HPLC is applicable to larger synthesis scale but it may not remove some failure sequences lacking, for example, only the final base, effectively.

Many modified oligonucleotides are also commercially available. These modifications include, but are not limited to, biotin-, amino-, and thiol-modification, phosphorylation, abasic sites and unnatural spacers, fluorophores, and quenchers. As discussed below, all of these modifications provide important functionality in supramolecular polymerizations. For example, biotin-, amino-, and thiol-modified oligonucleotides can be used as surface attachment points or mechanisms for coupling to proteins or synthetic molecules. 5′-Phosphorylated oligonucleotides are required for use as substrates for DNA ligase, one of many enzymes that play an important role in the wide application of DNA to supramolecular polymerizations. DNA ligase catalyzes the formation of a phosphodiester bond between the 5′-phosphate of one strand and 3′-hydroxyl of another. Thus, it can covalently join two double-stranded DNAs with blunt or cohesive ends, or repair single-stranded nicks in double-stranded DNA. The covalent casting of noncovalently templated structures is therefore possible. On the other hand, endo- and exo-nucleases act like scissors, cutting DNA double strands or single strands from the inside and from the ends respectively, in a direction either from 5′ → 3′ or from 3′ → 5′. Restriction endonucleases cut DNA at specific sequences, and provide an even more precise tool for molecular engineering. A brief summary of some common enzymes and their functions can be found elsewhere

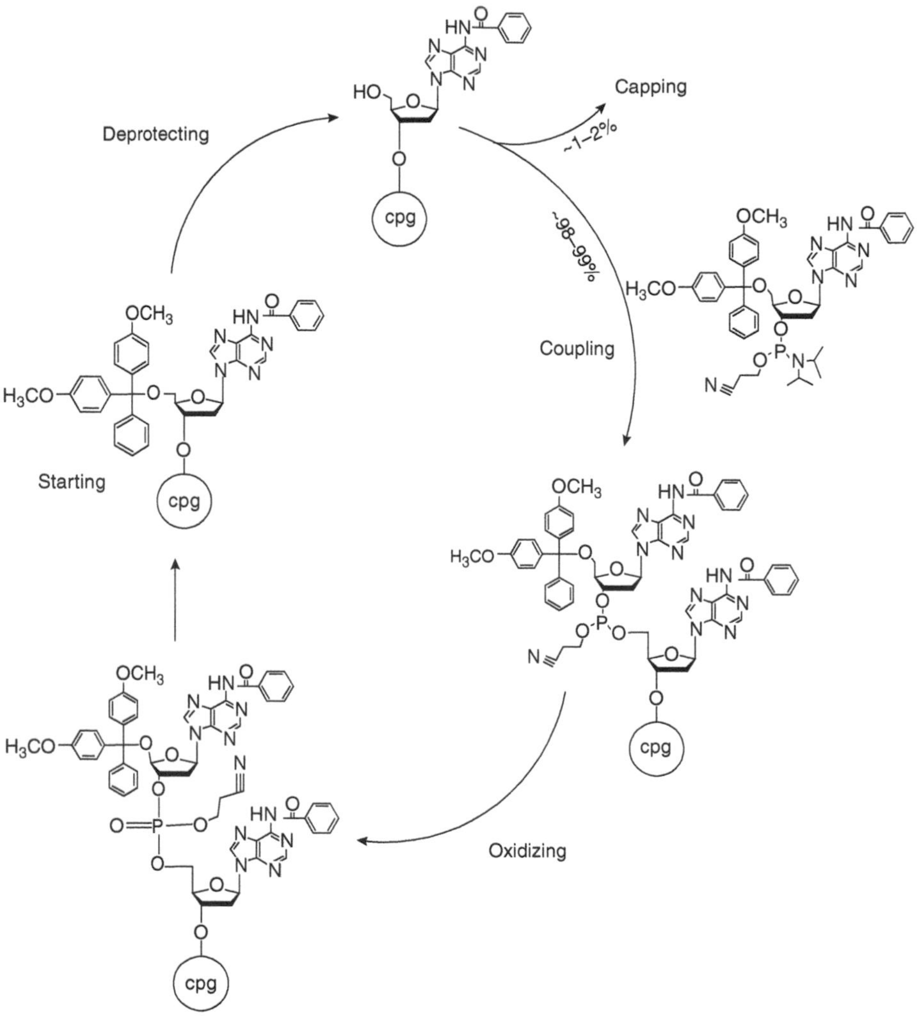

Figure 5   Solid-phase oligonucleotide synthesis cycle based on phosphoramidite chemistry. CPG is the solid
support. (Adapted from http://www.idtdna.com, with permission).

[16]. Another important type of enzyme which is not mentioned there is topoisomerase, an enzyme
which changes the topology but not the covalent structure of circular DNA. Topoisomerase combines
the functions of a nuclease and a ligase, in that it cuts DNA to enable the change of topology and then
seals the nicks. Types I and II topoisomerases act on single- and double-stranded DNA, respectively.
As successfully demonstrated by Seeman [17–19], a combination of ligase, nuclease, and/or topoisomerase makes it realistic to construct very complex objects from rather simple DNA components.

II.   LINEAR, DNA-BASED SUPRAMOLECULAR POLYMERIZATIONS

A.   SPs Based on Modified Oligonucleotides

While the specificity of molecular base pairing, intrinsic stiffness, predictable thermodynamics,
access to programmable sequences, and a versatile enzymatic tool box make DNA a powerful
construct for supramolecular polymerizations, even the individual nucleobases themselves have

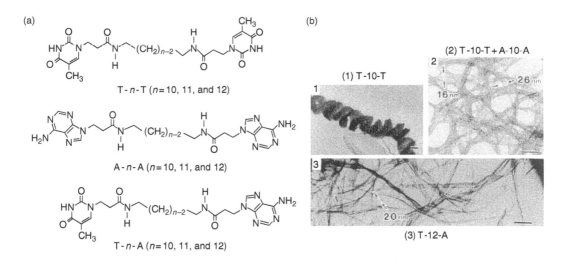

Figure 6 Formation of double helical ropes and nanofibers. (a) Monomer structures; (b) images under energy-filtering transmission electron microscopy (EF-TEM). (1) Double helical ropes formed from T-10-T (bar = 1 μm); (2) nanofibers formed from equimolar T-10-T and A-10-A (bar = 200 nm); and (3) nanofibers formed from T-12-A (bar = 1 μm). (Adapted from T. Shimizu, R. Iwaura, M. Masuda, T. Hanada, and K. Yase. *J. Am. Chem. Soc.* 123:5947–5955, 2001. With permission.)

been used in supramolecular self-assembly. Shimizu [20] has reported the formation of microsized supramolecular fibers directed by internucleobase interactions. Three types of monomeric units were synthesized in which 1-(2-carboxyethyl)thymine (T) or 9-(2-carboxyethyl)adenine (A) were homo-ditopically or heteroditopically attached to both ends of alkyl chains (Figure 6(a)). Supramolecular fibers were formed through the homoassembly of heteroditopic T-12-A (where 12 denotes the number of methylenes spacing the bases) or through heteroassembly of a 1:1 mixture of T-10-T and A-10-A, or unexpectedly through the homoassembly of homoditopic T-10-T (Figure 6(b)). More surprisingly, the unexpected supramolecular fibers from T-10-T were very stable double helical ropes of 1 to 2 μm in width and several hundred microns in length, which was believed to be the first example of micron-sized double helices self-assembled from achiral components. Additional discussions of SP helices can be found in Chapter 16. Using NMR, CD, and UV spectroscopy, this unusual double helix formation was attributed to the formation of stereoisomers through thymine photodimerization under natural light. In a mixture of T-10-T and A-10-A, thymine photodimerization was hindered by favorable A–T base pairing, leading to no helical rope formation. X-ray diffraction further revealed that all homo- or hetero-assemblies adopted stable layered structures due to vertical base stacking.

The variable spacers between nucleobases end groups provide a versatile handle for supra-molecular self-assembly. For example, the length of the spacer can be tuned from none, to short (e.g., oligo[ethylene]), to polymeric, and the flexibility of the spacer can be tuned simultaneously. Work by Rowan [21] has demonstrated the utility of nucleobase pairing to induce the self-assembly of macromonomers into polymeric aggregates. Specifically, a low-molecular weight poly(tetrahedrofuran) ($M_n \sim 1400$ g/mol) was terminated with two thymines or N^6-(4-methoxybenzoyl)adenines at both ends (T2T and $A^{An}2A^{An}$, respectively, Figure 7(a)). N^6-Amine in adenine was protected here to exclude the possibility of forming nonspecific Hoogsteen base pairing. The filmlike materials were formed through hydrogen bonding between $A^{An}2A^{An}$ and T2T, when mixed at 1:1 molar ratio, as evidenced by DSC and FTIR. Surprisingly, fibers formed in the solid state when only $A^{An}2A^{An}$ was present (Figure 7(b)), in contrast to the case of pure T2T in which no fibers were formed. Considering the extremely low homodimerization constant of adenine ($<3 M^{-1} n$ CDCl$_3$), it is unlikely that the supramolecular polymerization is of the type described, for example, in Chapter 6. The authors attributed the unexpected increase in mechanical properties to other possible driving forces such as π-stacking or physical cross-linking. This

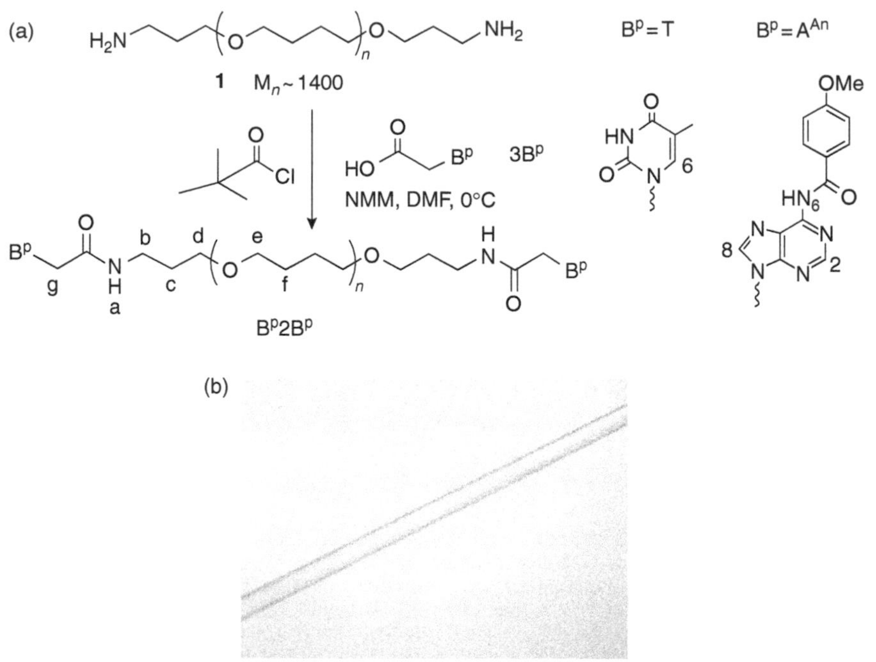

Figure 7 Self-assemblies of nucleobase terminated macromonomers. (a) Synthesis of macromonomers; and (b) optical microscopy (100×) of a fiber formed from A^{An}-2-A^{An}. (Adapted from S.J. Rowan, P. Suwanmala, and S. Sivakova. *J. Polym. Sci. Part A: Polym. Chem.* 41:3589–3596, 2003. With permission.)

conclusion was supported by the addition of up to 20% of dodecyl-A^{An} as a chain terminator to $A^{An}2A^{An}$ system. If the polymerization was due to dimerization of end groups, the degree of polymerization should drop upon addition of the chain-terminator. No appreciable decomposition of the fibers formed by $A^{An}2A^{An}$ was observed, a result that indicated that phase aggregation, instead of chain-terminator-sensitive supramolecular polymerization, may account for the unusual mechanical behavior of $A^{An}2A^{An}$.

The examples mentioned above illustrate the utility of the individual nucleobases. DNA oligonucleotides, however, are far more extensively employed for supramolecular self-assembly. Oligonucleotides have the potential to direct self-assembly of molecules into precise spatial orientations. Linear, one-dimensional assemblies controlled by oligonucleotide complementarity have been used to characterize DNA curvature and ring closure probabilities [22,23]. Early examples of using this concept expressly for the purpose of forming supramolecular polymeric nanostructures were reported by Ohya and Takenaka. Ohya et al. [24] synthesized two homodimeric monomers, called $(T_4)_2Bz$ and $(A_4)_2Bz$ (Figure 8(a)), through coupling of tetrathymidine or tetraadenosine to *p*-hydroxybenzene. These two monomers, when mixed equivalently, led to the formation of high-molecular weight aggregates. Concurrently, Takenaka et al. [25] reported the self-assembly of a two-dimensional net through A–T base bonding of two tetrameric monomers with a ferrous–bipyridine complex core (Figure 8(b)). More recently, McLaughlin [26] demonstrated the precise arrangement of DNA-tethered ruthenium complexes in a linear array. The lengths of the linear arrays can be tuned by the molar ratio of A- and B-complex or by the addition of small amount of chain terminators.

Bunz [27] extended the approach to a more general, modular system. Oligonucleotide modified organics (OMOs) are formed by attaching different oligonucleotides to different functional core groups, for example, cyclobutadiene cobalt complex or a phenyleneethynylene trimer (Figure 9(a)) which might be potentially used in optical or semiconductor materials. Typical OMO designs and

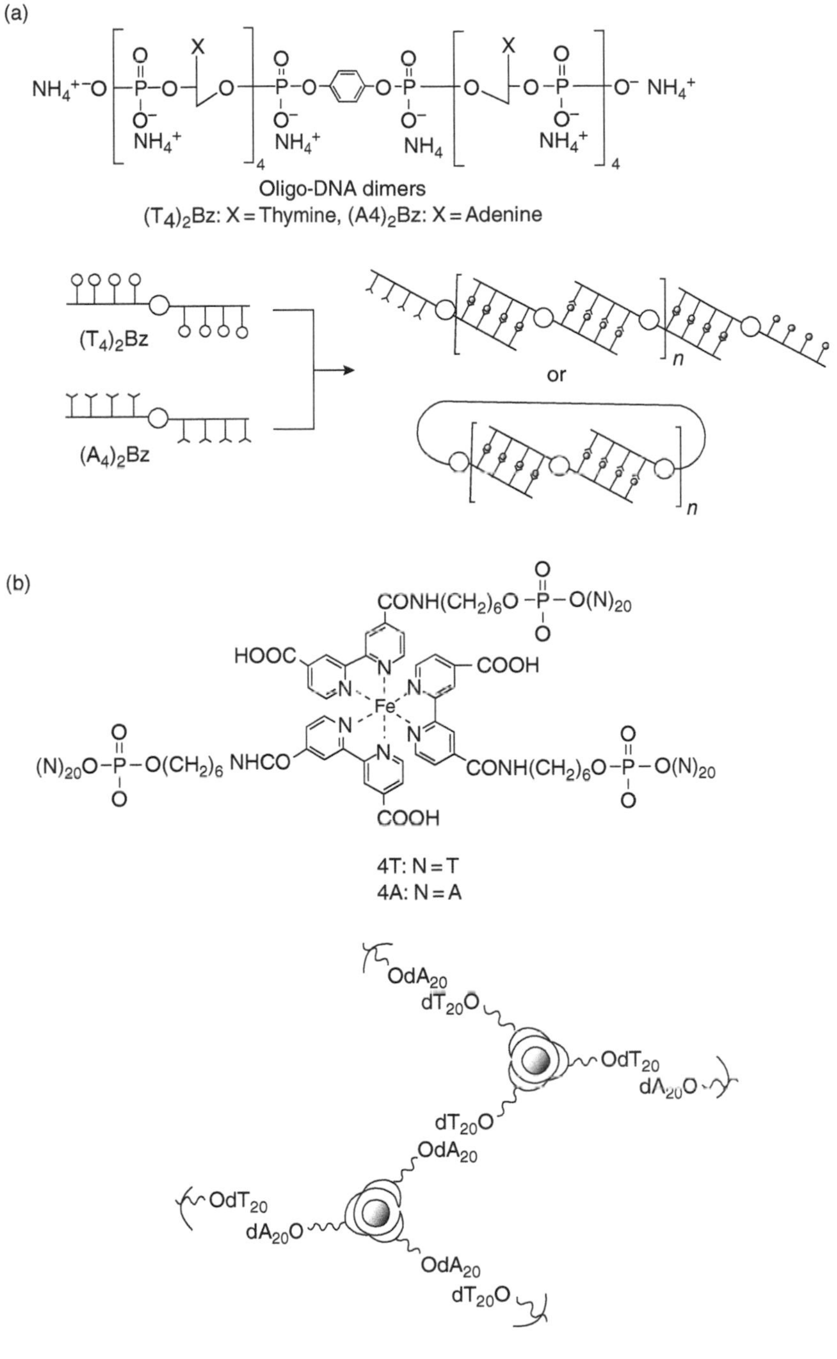

(a)

Oligo-DNA dimers
(T$_4$)$_2$Bz: X = Thymine, (A4)$_2$Bz: X = Adenine

(b)

4T: N = T
4A: N = A

Figure 8 Cartoons of early examples of supramolecular self-assembly showing the formation of (a) linear or cyclic self-assemblies; and (b) a two-dimensional net. (Adapted from Y. Ohya, H. Novo, M. Komatsu, and T. Ouchi. *Chem. Lett.* 25:447–448, 1996 and S. Takenaka, Y. Funatsu, and H. Kondo. *Chem. Lett.* 25:891–892, 1996. With permission.)

their corresponding oligomeric or polymeric assemblies are schematically illustrated in Figure 9. An important aspect of these structures is that they have a supramolecular sequence that is defined by multiple pairwise complementary base sequences. The ability to use DNA to code mutually exclusive pairing interactions within a single motif is a powerful tool for SPs. As revealed in the

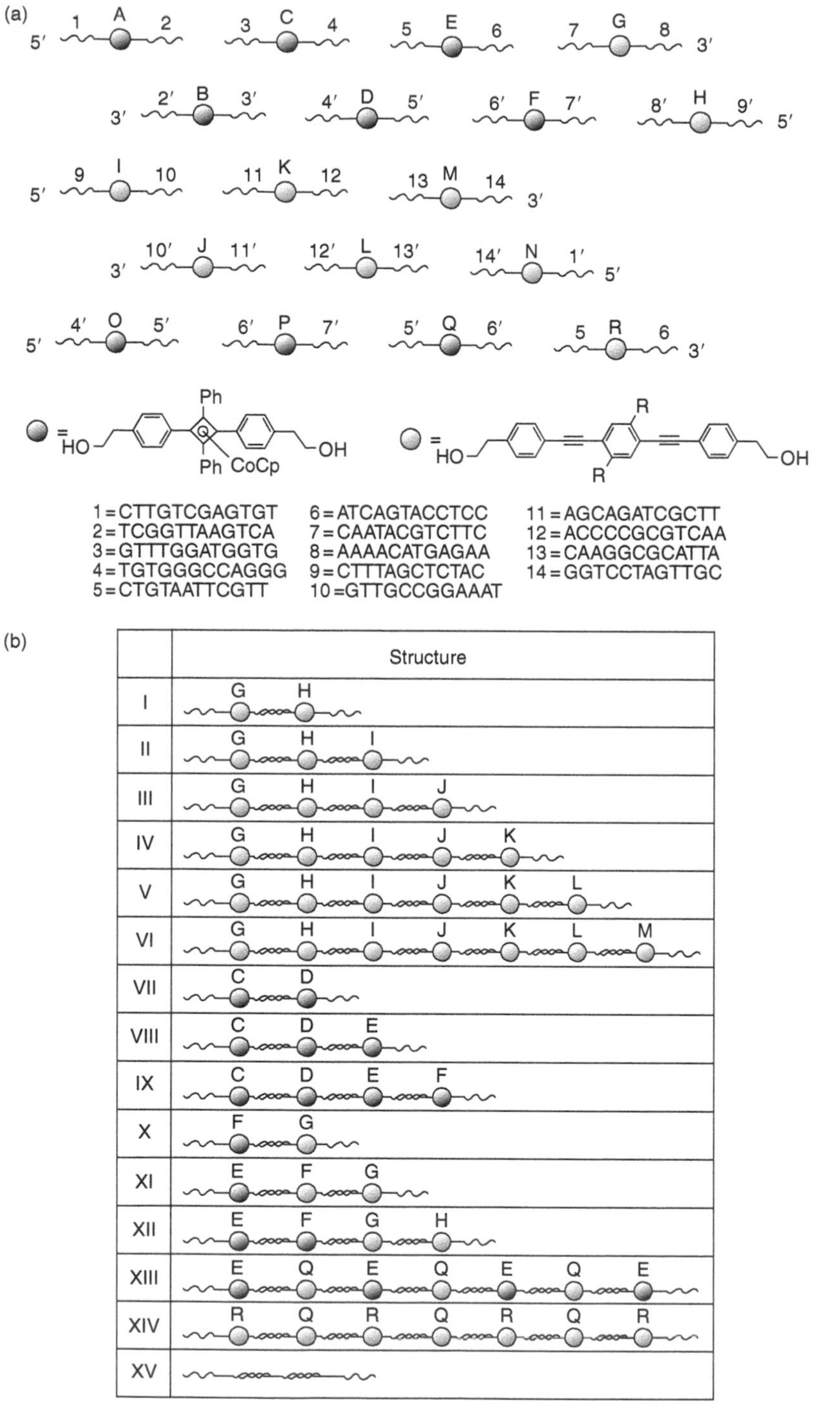

Figure 9 Oligonucleotide modified organics (OMOs). (a) Typical monomer design; and (b) their corresponding oligomeric or polymeric assemblies. (Adapted from S.M. Waybright, C.P. Singleton, K. Wachter, C.J. Murphy, and U.H.F. Bunz. *J. Am. Chem. Soc.* 123:1828–1833, 2001. With permission.)

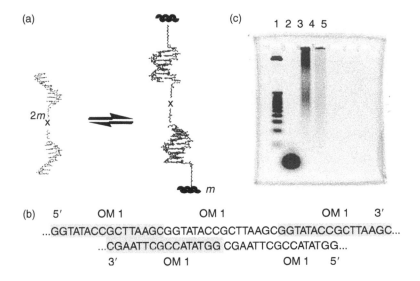

(b)   5'   OM 1   OM 1   OM 1   3'
...GGTATACCGCTTAAGCGGTATACCGCTTAAGCGGTATACCGCTTAAGC...
...CGAATTCGCCATATGG CGAATTCGCCATATGG...
       3'   OM 1   OM 1   5'

Figure 10 DNA-based modular system to study reversible polymerizations. (a) Schematic of DNA-based module. X represents a possible linker, which might be null, or $-(CH_2)_3-$, or $-(CH_2CH_2O)_6-$. $2m$ and m represent the number of monomers (left) and the number of repeating unit (right); (b) reversible polymerization pattern of a typical monomer (OM 1, 5'-GGTATACCGCTTAAGC-3'); (c) denaturing gel electrophoresis of OM 1 samples Lane 1: 25-bp step ladder; 2: OM 1; 3: ligated polymer of OM 1; 4: ligated and exonucleased OM 1; 5: exonucleased OM 1.

sections below, this aspect of DNA-based SPs provides access to a range of otherwise inaccessible structures and structural probes. The OMOs not only demonstrate the utility of DNA hybridization to create well-defined oligomeric or polymeric nanostructures with functional components, but more importantly, shows that the modularity of the methodology is applicable to any organic building blocks bearing hydroxyl end groups. In the words of those authors, the resulting nanostructures are therefore "interest independent."

These linear structures are subtypes of the supramolecular polymers described in Chapter 6, and Fogleman et al. [28] recently reported that they provide a well-behaved and modular model system in that context. As schematically shown in Figure 10(a), a typical oligonucleotide monomer (OM) consists of two independent short oligonucleotides (8-mer in this case) that are covalently linked directly or through an abasic spacer. Each 8-mer is designed in such a way that the base sequence from 5' to 3' is self-complementary to that from 3' to 5'. A head-to-tail association between monomers might therefore lead to the formation of a duplex which is reversible along the main chain (Figure 10(b)). Viscosity and light scattering studies verified the formation of a semiflexible polymer structure, similar to a regular DNA duplex. To take a snapshot of transient polymer structures, Quick DNA ligase was used to covalently integrate reversible polymer chains (gel electrophoresis result in lane 3, Figure 10(c)). Digestion by Lambda exonuclease removed any ligated linear polymers but left behind cyclic structures (black band in lane 4, Figure 10(c)), providing a measure of structural features that is otherwise inaccessible. Moreover, the size and shape of DNA-based assemblies can be quantitatively controlled by the monomer concentration, association strength, and spacer flexibility (e.g., "X" from none to hexa(ethoxyl) in Figure 10(a)) [29]. These extensive, modular studies of DNA-based reversible polymerization not only provide a useful guideline for rational synthesis of materials with desired properties, but also make possible to further study reversible polymeric behaviors in more complex environments, such as at surfaces or between interfaces [30].

Besides construction of linear arrays, DNA oligonucleotides are also powerful tools for more complex architectures. In work that preceded many of the linear systems described above,

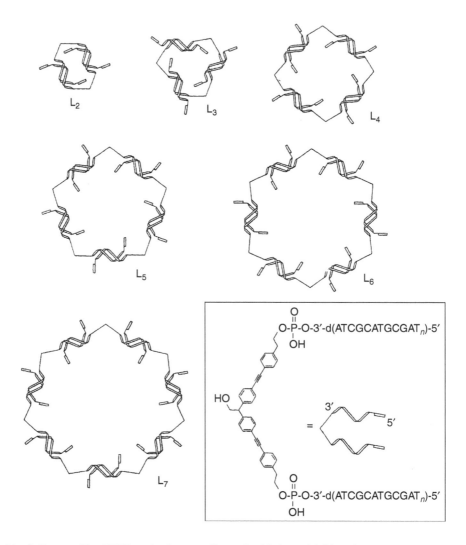

Figure 11 Self-assembly of DNA cycles from an oligonucleotide-based rigid vortices.

Bergstrom [31] designed rigid vertices whose subunits were attached by different DNA oligonucle-otides as connectors to form complex architectures. In principle, up to four different arms can be constructed for a tetrahedral vertex, leading to different types of DNA structures. A simple example of this idea was to create a vertex containing two oligonucleotide arms to form discrete cyclic supramolecular structures. The molecular component of interest (Figure 11) comprised of two p-(2-hydroxyethyl) phenylethynylphenyl spacers, each coupled to self-complementary oligo-nucleotides with 2 or 4 bases as dangling ends. Native PAGE was used to differentiate discrete macrocycles ranging from dimer to beyond heptamer. The presence of 2 or 4 dangling bases did not hinder the formation of macrocycles, indicating a potential strategy of using these sticky ends to form higher order structures.

It is worth noting that the rigidity of the DNA duplexes is a critical element in the successful construction of these two-dimensional arrays. The angles subtended by the rigid phenylethynylphenyl spacers are efficiently transmitted through the duplex DNA linkers, so that a reasonably narrow distribution of pseudocyclic nanostructures is observed. The persistence length of duplex DNA is on the order of 50 nm, and the linkers employed here were ∼5 nm long. At this length, the short duplexes are effectively very rigid rods with high directionality. It is likely that most of the size

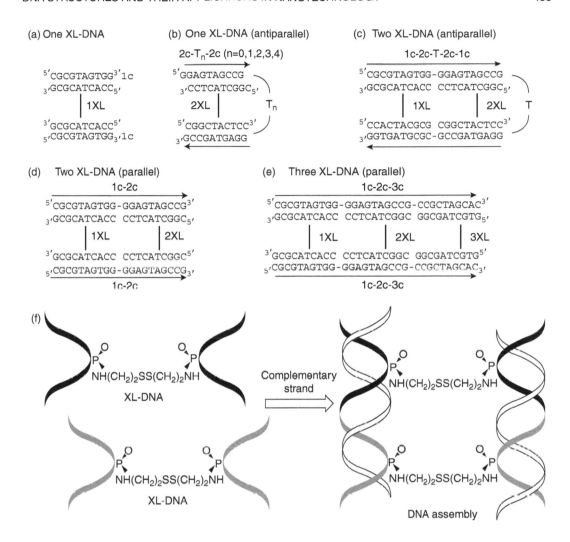

Figure 12 Cross-linked oligonucleotides (XL-DNAs). ([a]–[e]) XL-DNA design and corresponding assemblies with complementary single strands (antiparallel or parallel). Strands 1c, 2c, and 3c are complementary to 1XL, 2XL, and 3XL, respectively. (f) A cartoon to show how complementary strands assemble two XL-DNAs into a double helix. (Adapted from M. Endo and T. Majima. *J. Am. Chem. Soc.* 125:13654–13655, 2003. With permission.)

distribution results from the conformational flexibility of a few bonds right at the DNA-spacer junction.

Unlike Bergstrom's approach to attach oligonucleotides to rigid vertices, Majima [32] recently reported a simple method to increase the rigidity of DNA components to form instead, for example, cross-linked DNA (XL-DNA) molecules. A disulfide diamine cross-linker connected two single strands via coupling to an internal phosphate group in each strand (Figure 12(f)). A- and B-diastereomers were generated during this step and can be separated through reverse phase HPLC. Multiple cross-linked oligonucleotides spontaneous assembled side by side along appropriate DNA templates, except for two identical XL-DNAs, which preferred to form a hairpin instead. For example, the complementary 2c-T_n-2c ($n = 0$–4) formed a hairpin structure with one 2XL in an antiparallel way (Figure 12(b)), and the hairpin was the most stable in the case of 2c-T_1-2c. Two or three XL-DNAs with either parallel or antiparallel assembly patterns were achieved (Figure 12(c)–(e)). Diastereochemical effects were also noted, in that assemblies with B-diastereomers showed higher thermal stabilities than those with A-diastereomers in all cases.

B.   SPs Based on DNA-Polymer Conjugates

The promise of DNA as a tool for supramolecular science is furthered by the conjugation of DNA with other polymers, such as proteins. The Niemeyer group [33–35] has reported nano-structured DNA–streptavidin conjugates (DNA–STV), in which oligomeric DNA–protein networks were formed between biotinylated dsDNA and oligonucleotide-modified or unmodified streptavidin (Figure 13). Although streptavidin is a tetravalent protein with four identical biotin binding sites, in this instance it acts primarily as a bivalent linker to form oligomeric aggregates (aggregates **1** in Figure 13(a)). This result contrasts the expected formation of a three-dimensionally linked network. The low population of tris- and tetra-adducts (10 and 5% respectively) was attributed to the elec-trostatic and steric repulsions of the conjugated DNA fragments. Gel electrophoresis at a variety of molar ratios of DNA and STV revealed that the largest aggregates were formed at an equimolar stoichiometry, while excess DNA or STV resulted in relatively smaller oligomers. The supramolecu-lar aggregation was rather fast (within minutes). Moreover, once the aggregate formed, exchange of dsDNA between aggregates was very slow as long as the molar ratio of dsDNA to STV was equal to or lower than 1. When this ratio was increased to higher than 2, however, a fast exchange of dsDNA was observed, indicating the additive dsDNA interrupted the preformed *bis*-adducts and caused pronounced dissociation.

By manipulating the environmental conditions, the aforementioned linear aggregates can be transformed to supramolecular cycles [36,37]. For example, when aggregates **1** were denatured at 95°C for 2 min followed by rapid cooling, well-defined supramolecular nanocircles were observed (see atomic force microscopy [AFM] image in Figure 13(a)). The amount of nanocircles ranged from 20 to 60% in the mixture, depending on the educt concentration and cooling rate. The length of the dsDNA influenced the cyclization probability; shorter segments of dsDNA tended to form fewer cycles due to their increased intrinsic molecular rigidity.

Fabrication of these supramolecular structures provides a route to technical applications. The noncovalent DNA–STV conjugates **1** and **3** (Figure 13) are useful building blocks for the creation of more complex, functional nanostructures. Additional biotin binding sites provide a handle, by which to selectively position other functional moieties such as semiconductors, fluorophores, or antibodies. The attached oligonucleotide further provides a second handle for such functionalization; a range of multifunctional aggregates might easily be formed from the preformed DNA–STV linear or cyclic aggregates. A specific example is the biotinylated immunoglobulin (IgG) attached to the linear aggregates. It can be coupled directly to the ends of linear aggregates **1** (Figure 13(a)) or through hybridization between pendent oligonucleotides in **3** and the complementary partners in modified IgG (Figure 13(b)). The resulting aggregates **2** and **4** are then suitable reagents for immuno-PCR (IPCR), which is an ultrasensitive detection method for proteins and other antibodies based on the coupling of specific antibodies to a DNA reporter fragment to be amplified by PCR. It has been demonstrated that employing such supramolecular reagents can enhance the detection sensitivity to over 100-fold compared to conventional IPCR protocol [35].

Not only biomacromolecules but also organic polymers are suitable backbones to form DNA–polymer conjugates. This type of conjugate has been increasingly addressed in recent years because the combination of properties from both the polymer and DNA offer functionalities not available in the individual components. Unlike DNA–protein conjugates where both compo-nents are located in the main chain, oligonucleotides in organic polymer conjugates are generally coupled to polymer backbones through postpolymerization modification. Garnier and cowork-ers [38] reported the design of a new biosensor using electroactive polypyrrole functionalized with an oligonucleotide probe (Figure 14). It should be noted that direct electropolymerization of oligonucleotide-modified pyrrole monomers was inaccessible due to strong steric repulsions among oligonucleotides during polymerization. A precursor copolymer bearing a good leaving group *N*-hydroxyphthalimide was therefore preformed, followed by the amide linkage between oligonucleotides and pyrroles. The high electroactivity of the conducting polypyrroles was retained

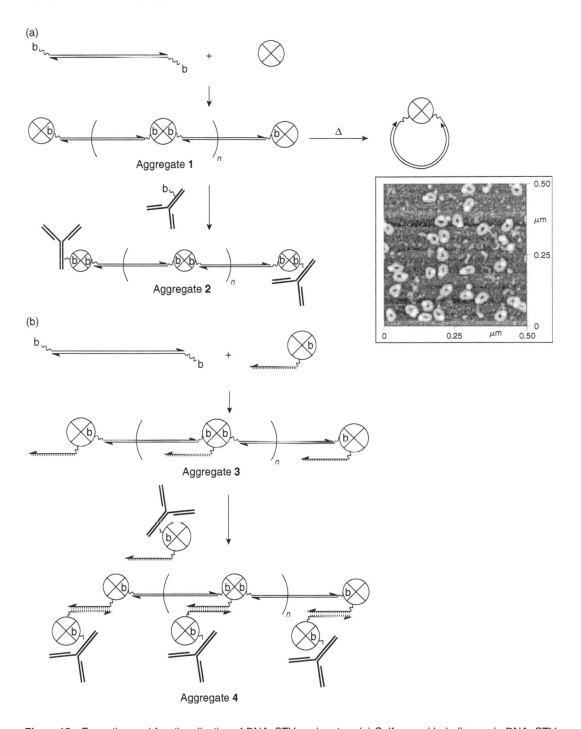

Figure 13 Formation and functionalization of DNA–STV conjugates. (a) Self-assembled oligomeric DNA–STV conjugate **1**. Extra biotin binding sites enable the functionalization of aggregate **1** with biotinylated antibodies. Antibodies are basically coupled to the ends of aggregate **2**. Thermal treatment of aggregate **1** leads to the formation of nanocycles, as visualized by AFM. (b) Formation of new DNA–STV conjugates **3** and **4** by using covalent oligonucleotide-modified straptavidin. Pendent oligonucleotides in **3** provide another handle to functionalize the conjugate in both the ends and the middle. (Adapted from C.M. Niemeyer. *Curr. Opin. Chem. Biol.* 4:609–618, 2000 and C.M. Niemeyer, M. Adler, S. Gao, and L. Chi. *Bioconjug. Chem.* 12:364–371, 2001. With permission.)

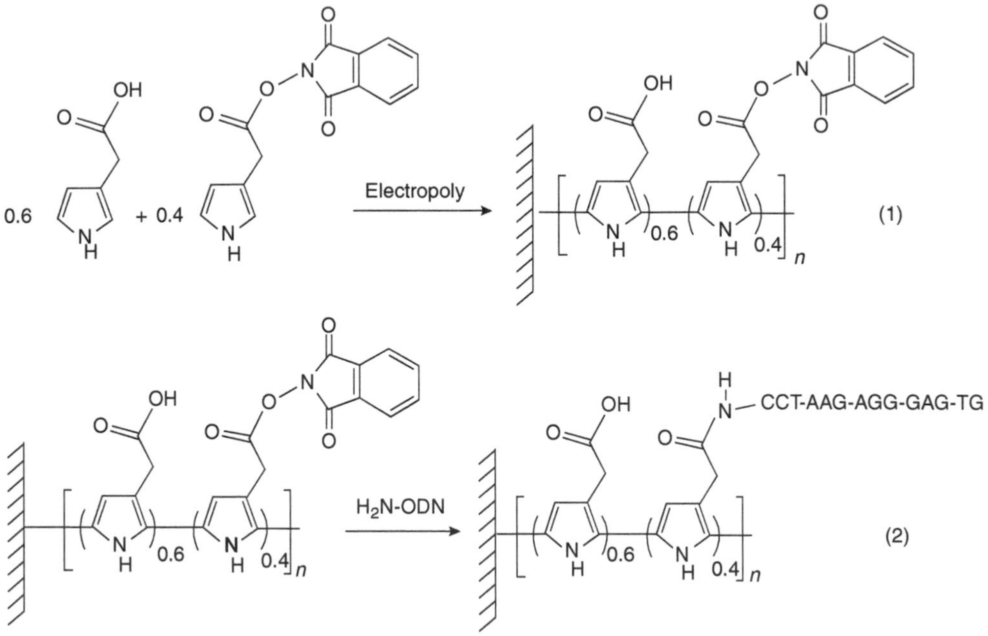

Figure 14 Synthesis of oligonucleotide-modified polypyrrole. (Adapted from H. Korri-Youssoufi, F. Garnier, P. Srivastava, P. Godillot, and A. Yassar. *J. Am. Chem. Soc.* 119:7388–7389, 1997. With permission.)

after oligonucleotide modification, making it possible to selectively detect oligonucleotides based on an electrochemical response, as follows. Specifically, when a complementary oligonucleotide hybridized to the probe dsDNA, the formation of bulky and stiff duplexes in the side chain makes more oxidation energy required for polymer transformation, thus leading to a remarkable change in its voltammetric response, whereas noncomplementary oligonucleotides showed no effect on the signal. It was also recognized that higher sensitivity of oligonucleotide recognition can be achieved by increasing the oligonucleotide complementary length. Potential applications for this type of approach are promising once a practical sensitivity can be reached.

Another example of coupling DNA to organic polymers was illustrated by Nguyen and coworkers [39], who attached oligonucleotides to a well-defined polymer backbone formed from ring-opening metathesis polymerization (ROMP). A practical advantage of this method is that the Grubb's catalyst $Cl_2Ru(PPh_3)_2$=CHPh is commercially available and can initiate a living polymerization unsusceptible to many functional groups, for example, a hydroxyl group in the side chain, which leaves room for DNA coupling. Moreover, the living end of the polymerization is maintained so that DNA-block copolymer conjugates can be easily obtained by the addition of a second monomer, such as nobornenyl-modified ferrocene in Figure 15. This strategy thus paves a facile path to novel macromolecular hybrid materials which can hardly be achieved by regular methods.

Moreover, new amphiphilic PEG–b–nucleotide copolymers were recently reported by Gross and coworkers. [40] These nucleotide-based copolymers self-assembled into polymeric micelles and nanoparticles, whose structures were further stabilized by A–T hydrogen bonding. Such core-shell supramolecular structures may be potentially useful for applications in drug delivery.

III. TWO- AND THREE-DIMENSIONAL DNA SPs

The aforementioned examples illustrate DNA components that are extended in one dimension to form longer linear or cyclic self-assemblies. To enhance the versatility of DNA design and the ability

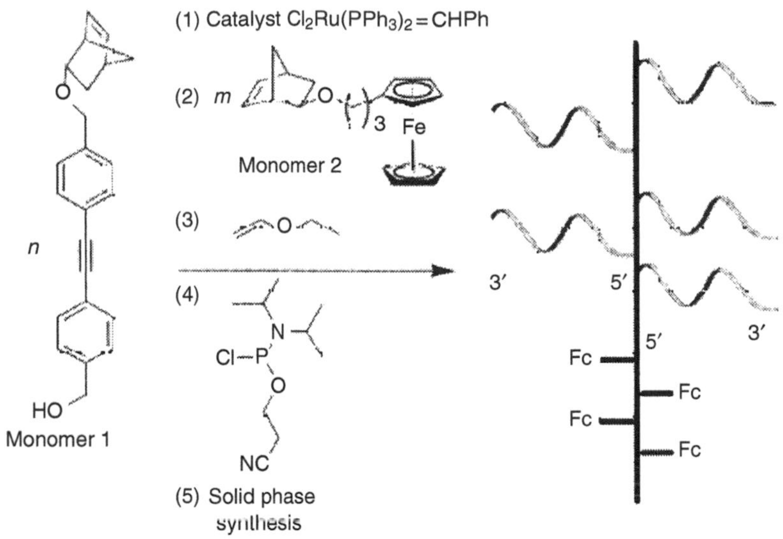

Figure 15 Synthesis of DNA-block copolymer conjugates. ROMP of monomer 1 followed by monomer 2 leads to the formation of a block copolymer, which can be further functionalized by DNA solid phase synthesis. (Adapted from K.J. Watson, S. Park, J. Im, S.T. Nguyen, and C.A. Mirkin. *J. Am. Chem. Soc.* 123:5592–5593, 2001. With permission.)

to fulfill complex functions, two- and three-dimensional supramolecules self-assembled from DNA are essential. Seeman has extensively explored the use of DNA to create complex nanostructures over the past 20 years. From a historical perspective, it is important to note that his work was seminal in its appreciation for the broad organizing power of DNA; many of the "simpler" examples were inspired by his earlier work. The interested reader is referred to several excellent and detailed reviews, which can be summarized only briefly here [41–46]. Branched DNAs were first recognized as powerful building blocks for the creation of complex multi-dimensional nanostructures. This idea was inspired by the naturally occurring Holliday junctions (Figure 16(a)), where four connected double helices flank a branch point. Due to the twofold symmetry of base sequences, the branch point of Holliday junctions may shift around the symmetric center to form isomers, a process called branch migration (Figure 16(a)). To inhibit the branch migration, deliberately designed DNA sequences without twofold symmetry were employed to create immobile Holliday junction analogues (Figure 16(b)) [47]. Furthermore, after extending each double helix arm by a short single-strand overhang (called a sticky end), four individual immobile junctions might self-assemble into a quadrilateral structure (Figure 16(c)). This strategy is applicable to variably armed junctions, for example, from three- to six-armed junction building blocks, and enzymatic ligation can be used to covalently close the final aggregates. Ideally, this assembly can be infinitely extended in two dimensions. This goal, however, cannot be achieved in reality because ligation-closure assays have revealed that the angles between 2 arms of the junction are flexible [48]. These variable angles may result in, for instance, an eight-member cyclic complex (Figure 16(d)). Although this flexibility of angles precludes ideal periodic arrays, branched DNAs are still powerful building blocks for fabrication of multi-dimensional individual nanostructures. This type of construction, however, can only be controlled at a topological level, not a geometrical level. Here, geometrical control means that one can control precisely both the topology and the exact structure (shape).

The first example of fabrication of three-dimensional DNA nanostructures is the synthesis of a DNA cube (Figure 17) [49]. Two squares bearing two turns on the edge are ligated through C–C′ and D–D′ sticky ends to generate a ladderlike intermediate containing three squares, named L (left), F (front), and R (right) respectively. This intermediate, once purified, is ready for cyclization through A–A′ and B–B′ to form a DNA cube. Due to a double ligation in the first step, however, the separation

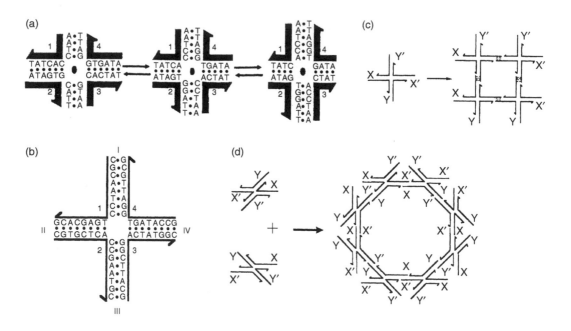

Figure 16 Branched DNAs as building blocks for DNA self-assembly. (a) A Holliday junction and its branch migration; (b) an example of immobile junctions; (c) formation of an ideal two-dimensional lattice from an immobile junction; (d) formation of a cyclic complex from eight immobile junctions due to variable angles between the arms. (Adapted from N.C. Seeman. *Angew. Chem. Intl. Ed.* 37:3220–3238, 1998, and N.C. Seeman. *Biochemistry* 42:7259–7269, 2003. With permission.)

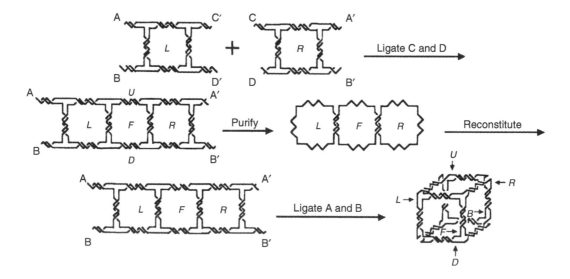

Figure 17 Synthesis of a DNA cube. Ten chemically synthesized strands, two 80-mers and eight 40-mers, are hybridized to form two quadrilaterals in the first step. (Adapted from N.C. Seeman. *Acc. Chem. Res.* 30:357–363, 1997. With permission.)

of the intermediate from byproducts was unsuccessful under nondenaturing condition. A denaturing process was required to produce a triple catenane, which was readily separated from others and reconstituted back to the final precursor of DNA cube.

To achieve a higher control over the synthesis of the intermediates and thus the final target, a solid-support methodology (Figure 18(a)) was developed [17]. The synthesis of a quadrilateral starts with

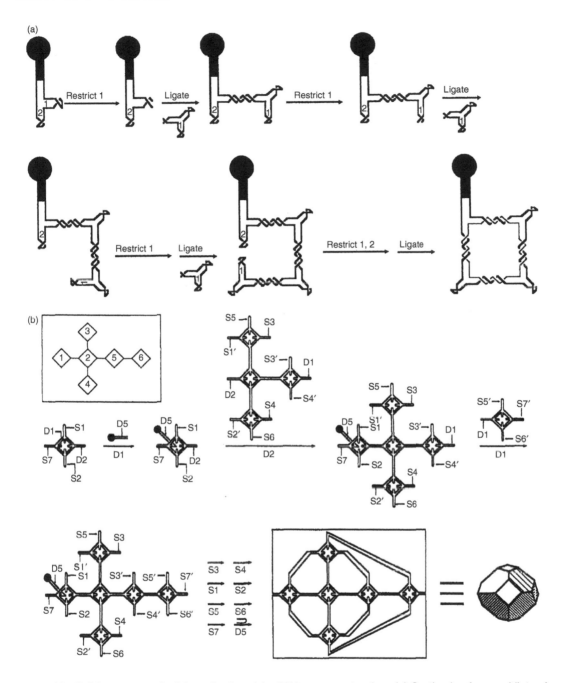

Figure 18 Solid-support methodology developed for DNA nanoconstruction. (a) Synthesis of a quadrilateral; and (b) synthesis of a truncated octahedron on a solid-support. (Adapted from N.C. Seeman. *Angew. Chem. Intl. Ed.* 37:3220–3238, 1998. With permission.)

a close junction bound to a solid support. Each cycle of restriction followed by ligation on position 1 creates one edge at a time. After three cycles, a square can be created if the restriction occurs on both positions 1 and 2 which are then ligated. Advantages of this methodology are quite obvious. The isolation of growing intermediates excludes the possibility of intermolecular ligation. Each edge is formed individually, further enhancing the control over intermediate products. Moreover, the growing intermediates are closed molecules bound to a solid support, so that removal of reactants, catalysts,

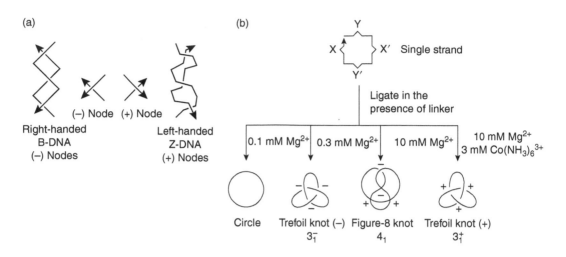

Figure 19 Synthesis of DNA knots. (a) Positive and negative nodes generated from Z- and B-DNAs, respectively; (b) formation of four topological targets by varying solution conditions. (Adapted from N.C. Seeman. *Acc. Chem. Res.* 30:357–363, 1997. With permission.)

and byproducts can be easily achieved by means of physical washing or biological exonuclease digestion.

More complex three-dimensional targets can be created by means of the solid-support methodology, such as truncated octahedrons (Figure 18(b)) [18]. A truncated octahedron contains six squares and eight hexagons, which can be constructed from 4-arm branched junctions. As illustrated in Figure 18(b), the first square was bound to a solid-support, a preformed tetra-square complex is then attached to the first square, followed by adding the sixth square to the penta-square intermediate. Up to this step, six squares are complete but all hexagons need to be created from outer strands. This is achieved by a series of successive restriction/ligation manipulations at positions S1/S1′ to S7/S7′. Cleavage from the solid-support gives the final truncated octahedron.

In comparison to the aforementioned nanostructures fabricated from DNA branches, single-stranded DNA constructions are seemingly straightforward but equally important. This importance is highlighted from their potential in biotechnology, because single strands can be amplified by PCR while branched DNAs cannot, thus giving hope to biological synthesis of complex target molecules. A variety of topological objects have been constructed by single strands, including DNA knots [50–52] and borromean rings [53]. The fundamental unit of knotted molecules is the node. There are two types of nodes, positive and negative nodes (Figure 19(a)). Right-handed B-DNAs generate a negative (−) node during the crossing of each half-turn, whereas left-handed Z-DNAs generate a positive (+) node every half-turn. Commonly, a double helix will adopt a B-conformation. But the B–Z transition can be induced under the appropriate conditions, for example at high salt concentration, with dCdG repeating units, or in the presence of special cations such as $[Co(NH_3)_6]^{3+}$. Using the same single strand might therefore generate several different structures [52], as illustrated in Figure 19(b). The complementary regions X and X′ form a single turn of a double helix and may undergo a B–Z transition, as do Y and Y′. When Mg^{2+} concentration increased from 0.1 to 10 mM, ligation products changed from a simple circle, to a trefoil knot (−) (both domains are B-DNAs), and finally to a figure-8 knot (one B-DNA and one Z-DNA). Addition of $[Co(NH_3)_6]^{3+}$ promoted the B–Z conversion. These species, once formed, can also be interconverted by topoisomerases under suitable conditions [19].

Borromean rings [53,54] are more complex than DNA knots. A remarkable characteristic of borromean rings is that no pairs of rings are linked, which means once any ring is opened, all other rings will break up into their individual constituents. To achieve this topology, positive and negative nodes should be placed around the link. In an example of a three-ring link (Figure 20), the synthesis

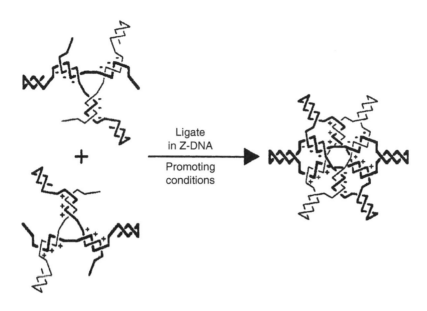

Figure 20 Synthesis of a three-member borromean ring. (Adapted from N.C. Seeman. *Annu. Rev. Biophys. Biomol. Struct.* 27:225–248, 1998. With permission.)

involves a 3-arm B-DNA junction and 3-arm Z-DNA junction with each arm bearing two single strands; one is a near-complete hairpin and the other is a short single strand providing the remainder of another near-complete hairpin. Under conditions favoring Z-form, the final ring link is achieved through the ligation of hairpin nicks.

One ultimate goal of DNA nanoconstruction is the self-assembly of periodic arrays, a goal that is not met by the examples cited above. As opposed, the construction of individual objects, in which there exists necessary specificity to each unique sticky end pair, periodic arrays require a high degree of symmetry. As pointed out by Seeman [55], three prerequisites are essential to this type of construction: predictable intermolecular interactions, predictable local structures, and component rigidity. Utility of the sequence-specific sticky ends may satisfy the first two criteria. However, even if the symmetry is met, they are far less rigid than required. New rigid DNA components are therefore imperative.

Double crossover DNAs (DX-DNA) are the appropriate complexes to satisfy the rigidity requirement. These complexes contain two crossovers that connect two double helices. There are five possible isomers of DX-DNAs (Figure 21(a)): three parallel (DPE, DPOW, and DPON) and two antiparallel (DAE and DAO). Symbols represented here are: D means double crossover, A means antiparallel P means parallel, E and O means even and odd numbers of half-turn, respectively, W and N indicate whether the extra half-turn in DPO corresponds to a major (wide) or minor (narrow) groove. Besides five isomers, another useful DX-DNA complex, called DAE+J (Figure 21(a)), is a derivative of DAE by adding a bulged junction. Due to its higher stability, antiparallel DX-DNAs are more appropriate building blocks than parallel molecules. Moreover, when DAE or DAE+J oligomers are ligated, a long reporter strand results (thick line in the top and bottom of Figure 21(b)). Cyclization events are rarely observed, a fact that testifies to the inherent rigidity of the complexes [56]. Although concatenation complicates the characterization of ligated DAO (thick line in the middle of Figure 21(b)), DAO tiles are still likely to be stiff. Antiparallel DX-DNAs therefore have been widely explored for constructing two- or three-dimensional periodic arrays.

Figure 22 shows how DX and DX+J molecules self-assemble into periodic two-dimensional striped lattices by means of sticky end association [57]. The extra helix in DX+J (B* and D* in Figure 22), symbolized as a filled black circle, provides a good topographic marker for AFM. For example, the size of the repeating tile is around 4 × 16 nm, so the resulting strips were

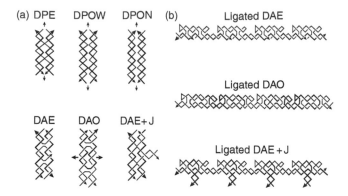

Figure 21 DX-DNA molecules and their ligated products. (a) Five isomers from DPE to DAO and one derivative DAE+J; (b) ligated products of three antiparallel molecules. Report strands result in ligated DAE and DAE+J, shown as thick lines, but ligated DAO contains polycatenanes. (Adapted from N.C. Seeman. *Annu. Rev. Biophys. Biomol. Struct.* 27:225–248, 1998. With permission.)

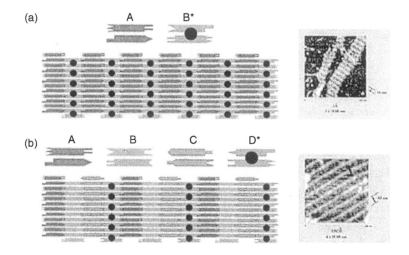

Figure 22 Diagrams of constructing two-dimensional periodic arrays. (a) A conventional DX molecule, A, and a DX+J molecule, B*, self-assemble into two-dimensional lattices. The size of each molecule is 4 × 16 nm. The extra domain in DX-J, marked as a black circle, gives stripes whose distance is 32 nm. (b) As a control, three DX and one DX+J molecules self-assemble into similar lattices, but now the strip distance is 64 nm, as visualized under AFM. (Adapted from N.C. Seeman. *Biochemistry* 42:7259–7269, 2003. With permission.)

separated by a distance of 32 nm from bimolecular self-assembly (Figure 22(a)). As a control experiment, self-assembly of four DX molecules with only one DX+J led to a strip distance ∼64 nm (Figure 22(b)). As usual, enzymatic modifications of the surface of these two-dimensional DNA crystals are available, thus increasing the diversity of patterns [58].

DX-DNA tiles have also been used to assemble two-dimensional lattices displaying aperiodic patterns by a method of directed nucleation on a long scaffold DNA strand [59]. Figure 23 shows a schematic representation and an AFM image of a section of such lattice. The aperiodic surface pattern, due to the placement of DX-DNA tiles containing a stem-loop of DNA projected out of the tile plane (DX+2J), is clearly distinguishable by AFM imaging. The significance of this barcode lattice is that it takes one-dimensional information encoded on the scaffold strand and propagates it into a second dimension demonstrating the ability to organize nanomaterials into structures with higher dimensionality. Similar systems could be used not only for specifically patterning

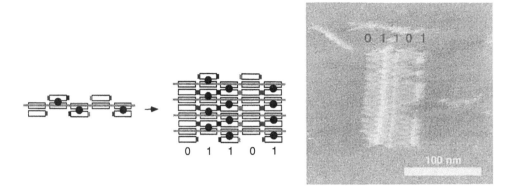

Figure 23 A periodic barcode lattice assembled from DX-DNA tiles by directed nucleation on a DNA scaffold strand. Schematic (left panel) showing one layer (five tiles on a long scaffold strand) forming a lattice of stacked layers (four layers shown). Tiles with black dots represent DX+2J motifs. Right panel shows an AFM image of an example barcode lattice fragment with stripes formed by the extra stem-loops of the DX+2J tiles clearly visible.

other materials (e.g., nanoparticles, proteins, carbon nanotubes, etc.) for use in nanofabrication (see Section VI) but also as a visual readout method for algorithmic computational assemblies (see Section V).

Another way to create periodic arrays is to incorporate DX-DNAs into triangles or deltahedra (polyhedra having all triangular faces). DX-DNAs herein serve as edges of the triangles. For example, it has been shown that triangles bearing two edges of DAE complexes self-assembled into a one-dimensional periodic arrays [60] (Figure 24(a)). Attached biotins enabled the easy removal of unrestricted triangles through strong bindings between biotin and streptavidin beads. Ligation assay and AFM showed no evidence of cyclization. This success provides hope for the construction of two- and three-dimensional periodic arrays of DNA. For example, two-dimensional lattices can be formed from equilateral triangles whose sides consist of DX-DNAs (Figure 24(b)). Also, if starting with an octahedron bearing DX-DNAs in three edges (Figure 24(c)), these DX molecules are not coplanar, and self-assembly of these arms through sticky ends may lead to a large cubic complex face-centered by the octahedron subunits. Such constructions, however, have not yet been achieved.

Triple crossover DNAs (TX-DNA) are also useful building blocks for periodic arrays [61]. In principal, all strategies applied to DX-DNAs are also applicable to TX-DNAs. TX-DNAs are composed of three double helices arranged with their helix axes parallel and coplanar. The helices are connected by strand exchange at four crossover points, two between each adjacent pair. TAO and TAE tiles (Figure 25), which are named following nomenclature conventions given for DX tiles, have been experimentally demonstrated [61–63]. TX tiles display thermal stability comparable to or slightly greater than DX tiles, and they are able to form large lattices containing tens of thousands of unit tiles. Besides periodic arrays TX-DNAs are also useful as building blocks for computational assemblies (as described in Section V).

A novel DNA tile structure known as 4 × 4 (because it contains four 4-arm DNA branched junctions) has recently been reported [64]. The 4 × 4 tile has a square aspect ratio and paired double helical arms pointing in four directions (north, south, east, and west in the tile plane) (Figure 26(a)). It is composed of nine strands, with one central strand participating in all four junctions. Bulged TTTT loops were placed at each of the four corners inside the tile core in order to decrease the probability of stacking interactions between adjacent 4-arm junctions and to allow the arms to point to four different directions. Although individual branched junctions are expected to be fairly flexible, when constrained by three additional junctions within the tile and a further junction in a neighboring tile within lattices, the structure is sufficiently rigid to act as a building block in larger

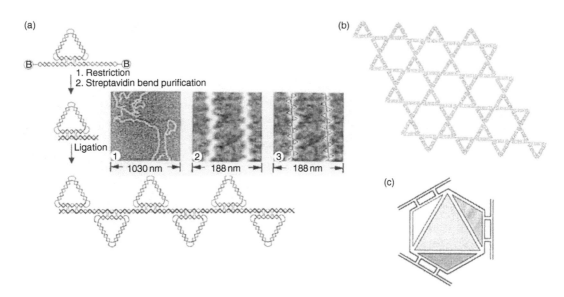

Figure 24 Examples of DX-DNAs incorporated into triangles or deltahedra. (a) Ligation of DNA triangles containing DX-DNAs leads to a linear array with the zigzag character, as visualized by AFM (images 1 and 2). Image 2 is a zoom of the image 1, and is similar to the expected ligation product (see the superposition in image 3). (b) Hypothetical construction of a two-dimensional lattice from a series of equilateral triangles whose sides consist of DX-DNAs, assuming these triangles will retain their angular distributions. (c) An octahedron containing DX-DNAs in three edges. Three noncoplanar edges, once connected, might yield a three-dimensional cubic complex face-centered by the octahedron. (Adapted from X. Li, X. Yang, J. Qi, and N.C. Seeman. *J. Am. Chem. Soc.* 118:6131–6140, 1996 and X. Yang, L.A. Wenzler, J. Qi, X. Li, and N.C. Seeman. *J. Am. Chem. Soc.* 120:9779–9786, 1998. With permission.)

Figure 25 Strand traces through TAE (left) and TAO (right) DNA tiles. Helix axes are drawn horizontal; crossover junctions are shown as paired vertical lines. The TAE contains 6 strands, 3 of which (black) traverse the entire width of the tile. The example TAO, containing 4 strands, is shown with hairpins at both ends of the central helix but could be designed with additional sticky ends instead. Note that minor grooves are designed to pack against major grooves of neighboring double helices.

superstructures. Characterization of the annealed structure by nondenaturing gel electrophoresis and thermal transition analysis has shown the 4 × 4 tile to be stable and well-behaved.

Two versions of the 4 × 4 tile with sticky ends were prepared, assembled into lattice superstructures, and visualized by AFM. Both designs resulted in lattices containing periodic square cavities. Interestingly, one strategy (the original design) produced a high preponderance of uniform width ribbon structures (Figure 26(b)). In this design the distance between adjacent tile centers is an even number of helical half-turns (4 full-turns) so that the same face of each tile points toward the same lattice face. Self-assembly following the original design resulted in long (∼5 μm on average) ribbonlike lattices with uniform width (∼60 nm). The regularity of the periodic cavities is striking, as well as the observation that some of the nanoribbons revealed a single layer flat grid lattice unrolled at the open end of the ribbon. This observation strongly suggests that the ribbon structure result from tubelike structures that flatten when the sample is deposited on mica. The formation of

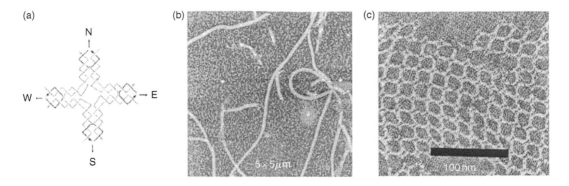

Figure 26 4×4 DNA tile and lattices. (a) Schematic strand trace of a 4×4 tile. (b) AFM image of nanoribbons formed from tubules of 4×4 tile lattice collapsed flat onto mica surface. (c) AFM image of flat lattice nanogrid formed from 4×4 with corrugated tile orientations.

tubelike lattices could be owing to the fact that each component tile is designed to be oriented in the same direction in the lattice plane, therefore any incidental curvature resident in the tiles would accumulate and cause circularization of the lattice. This hypothesis is tested and supported by the corrugated design described below.

The second design strategy aimed to eliminate the lattice curvature and produce larger pieces of flat lattice with square aspect ratios. This strategy, referred to as the corrugated design, caused adjacent tiles to associate with one another such that the same face of each tile is oriented up and down alternately in neighboring tiles, therefore the surface curvature inherent in each tile should be canceled out within the assembly. Figure 26(c) shows an AFM image of the self-assembled lattice with the corrugated design. The designed distance between adjacent tile centers is 4.5 helical turns plus two DNA helix diameters, totaling ∼19.3 nm. The AFM measured distance from the center to the center of adjacent tiles is ∼19 nm, in good agreement with the design. Large lattice pieces, up to several hundred nanometers on each edge, were observed in which the cavities appeared square. A two-dimensional lattice displaying a square aspect ratio should be useful for forming regular pixel grids for information readout from nanoarrays, for example, by encoding information in a pattern of topographic markers.

Another type of DNA-based three-dimensional supramolecular structure is the DNA dendrimer (highly branched macromolecules composed of layers of nucleic acid, see Figure 27(b)–(e)). Compared to conventional hyperbranched polymers, DNA dendrimers may offer extra advantages and applications within a biological context, such as biocompatibility, drug delivery vehicles, and signal amplification [65]. Damha had initially reported both "convergent-growth" [66] and "divergent-growth" [67] syntheses of DNA dendrimers. His approaches, based on standard phosphoramidite chemsitry, are covalent linkages of oligonucleotides step by step, making the synthesis a tough and complicated task. Nilsen and coworkers [68] proposed a theoretical synthesis strategy for assembling nucleic acid dendrimers through intermolecular hybridization. This process involves five different building blocks (A, B′, B″, C′, and C″ in Figure 27(a)). Each building block is hybridized from two oligonucleotides to form a heterodimer possessing a central double-stranded core and four single-stranded arms for binding. These four single-stranded arms are deliberately designed so that a(−) is only complementary to a, b(−) is only complementary to b, etc. Therefore, when starting with the initiator core A, two B′, and two B″ molecules can bind specifically to initiator A to form the first layer of a DNA dendrimer (Figure 27(b)). Successive addition of six C′ and six C″ leads to the formation of the second layer (Figure 27(c)). To create the third layer, another 18 B′ and 18 B″ molecules are required. Clearly, the sequence of multilayers is A–B–C–B–C–··, and there are 12 arms in the first layer, 36 arms in the second, 108 arms in the third, etc. Unfortunately, to the best of our knowledge, this appealing methodology has not yet been reduced to practice.

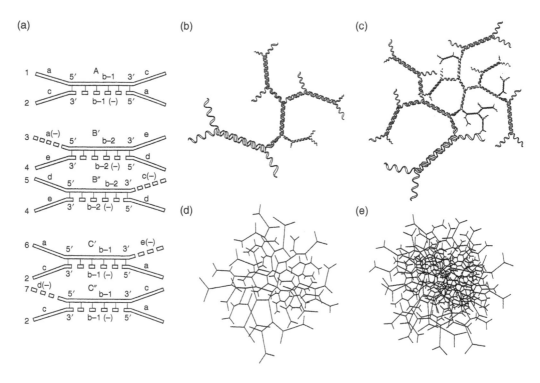

Figure 27 Hypothetic strategy to synthesize DNA dendrimers by spontaneous hybridization. (a) Five potential building blocks; and ([b]–[e]) are one to four generations of DNA dendrimers. (Adapted from T.W. Nilsen, J. Grayzel, and W. Prensky. *J. Theor. Biol.* 187:273–284, 1997. With permission.)

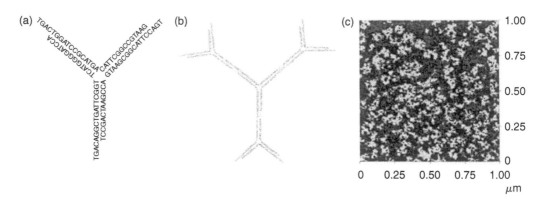

Figure 28 Y-shaped DNAs as potential building blocks for DNA dendrimers. (a) Y-shaped DNA with sticky ends; (b) formation of dendritic structures through hybridization of the core molecule in (a) and other complementary Y-shaped DNAs; and (c) AFM images of dendritic structures. (Adapted from D. Luo. *Mater. Today* 6:38–43, 2003. With permission.)

Recently, the Luo group reported the design of 3-arm branched DNA as the potential building blocks for construction of DNA dendrimers [16]. The so-called Y-shaped DNAs (Figure 28(a)) contain sticky ends which, once specifically designed, can be covalently bonded (via ligation) to specific sites of other Y-shaped DNAs to form a dendritic structure (Figure 28(b)). The process is likely achieved in a controlled way, and self-assembled DNA dendrimer structures can be visualized under AFM (Figure 28(c)).

IV. DNA ACTUATORS

Controlled mechanical motion in supramolecular devices is one of the key goals of nanotechnology. DNA is an excellent candidate in construction of such devices owing to the specificity of base pairing and its robust physicochemical properties. A variety of DNA-based molecular devices displaying rotational and open/close movements have recently been demonstrated. Reversible shifting of equilibrium between two conformational states can be triggered by changes in experimental conditions (ion concentration) or by the addition of a "DNA fuel strand" that provides the driving force for such changes. Several of these DNA actuators will be described in this section.

The first actuated DNA device made use of the controllable switching of a special DNA sequence between right-handed B-form DNA and left-handed Z-form DNA conformations [69]. The proto-Z sequence, $d(CG)_{10}$, was suspended between two rigid DX tile complexes such that when Z-form solution conditions existed (0.25 mM $Co(NH_3)_6Cl_3$, 100 mM $MgCl_2$, 100 mM NaCl) the DX tiles were rotated relative to one another. This induced motion brought a fluorophore and a quencher into close proximity so that operation of the device could be monitored by loss of the fluorescent signal. Return to standard solution conditions (10 mM $MgCl_2$, 100 mM NaCl, and no $Co(NH_3)_6Cl_3$) reestablished B-form DNA conformation and increased the fluorescent signal. The B–Z device was shown to be repeatedly switchable between the two stable states. A more recent example of DNA state switching controlled by changing ionic concentration was demonstrated in networks composed of biotinylated dsDNA bridges linking streptavidin protein molecules [70]. The streptavidin molecules were shown to be, on average, drawn closer together in 40 mM magnesium buffer compared to 4 mM. AFM imaging revealed not only tighter average spacing but also thickening of the DNA bridges which was attributed to supercoiling of adjacent dsDNA bridges.

The first example of a DNA actuator based on intermolecular DNA hybridization was referred to as molecular tweezers because its two states resemble the opened and closed states of macroscopic tweezers [71]. This construction in its open state contained three oligonucleotides, one of which is doubly labeled with a fluorophore and a quencher like so: fluorophore-A-hinge-B-Quencher. Where A and B represent 18-base sequences and hinge represents a flexible 4-base dsDNA region. The other two oligonucleotides have sequences A′–C and D–B′, where A′ represents the Watson–Crick complement of sequence A. In the open state of the tweezers A and B are hybridized to A′ and B′ (respectively) while C and D are single-stranded and dangle flexibly in solution. To close the device another strand (closing fuel) is added; closing fuel has sequence C′–D′–E. The C′ and D′ regions hybridize with C and D, thus closing the device and bringing the fluorophore and quencher together in space, thus decreasing the fluorescence. In the closed state, the E sequence (an 8 base region referred to as a toe hold) dangles unhybridized in solution. To reopen the tweezers, opening fuel E′–D–C is added, which grabs the toe hold E and strips the closing fuel off of the device thus allowing the hinge to relax and the fluorophore and quencher to diffuse apart. Each round of closing and reopening produces a double helical waste product of the fully complemented opening and closing fuel strands. The opening/closing process is fully reversible, and the switching time for the machine is less than 20 sec with second-order rate constants for opening and closing being approximately equal. Interestingly, thermodynamic calculations suggest a closing force of the tweezer of about 15 pN, which is at the upper end of the range of measured forces exerted by some small protein motors.

Two more recent examples of DNA actuation by hybridization made use of the compact state of special sequences able to form clusters of guanine bases known as G-tetrads or G-quadruplex [72,73]. In the intramolecular quadruplex form, a fluorophore and a quencher are held in close proximity thus producing a low fluorescence signal. To open the device and turn on the fluorescence signal, another oligonucleotide is added which contains the full-length complement of the first strand plus a short overhanging single-strand region to act as the toe hold for the closing step. Of course, the closing step involves addition of the full-length complement of the opening fuel strand and a long double helical waste product is produced.

Another strategy for DNA device state switching by hybridization which made use of two topological isomers was recently demonstrated [74]. In this study it was shown that dsDNA fragments can be used to control and fuel a DNA device cycle by inducing the interconversion between two topological motifs — paranemic crossover (PX) DNA and its topoisomer (JX2) DNA. The two crossover motifs can be converted into each other by removal of internal strands with the aid of biotinylated fuel strands, which could subsequently be removed from the solution with streptavidin-coated magnetic beads. The cycling was observed by modifying the PX/JX2 device with half-hexagon motifs as topographical markers. Switching of the DNA devices from the PX to the JX2 state led to conversion of the arrangement of the half-hexagons from a *cis* to a *trans* configuration which was observable by AFM imaging.

Finally, the incorporation of DNA devices into tiling arrays followed by demonstration of controllable state switching of the entire lattice has now been reported [75]. A DNA lattice was constructed using parallelogram tiles (described in Section III) with an extra 3-arm junction built into the helices of the tile sides. The two states were defined by hybridization of two different oligonucleotides across the 3-arm junction sites. The compact state was induced by the use of an oligonucleotide which caused the third arm to loop out from the tile helix while the elongated state was induced by annealing an oligonucleotide which fully complemented the tile strand thus forming a long double helix and eliminating the 3-arm junction. The two states of the lattice were imaged by AFM. The fuel stand and toe hold tactics described above were also used in the actuator lattice implementation. In future versions of the device lattice, if individual devices can be addressed separately, a large number of complex structural states suitable for nanorobotic applications could become available.

The examples of DNA-based nanoactuators reviewed here demonstrate a wide variety of approaches and strategies for incorporation of controllable motions into DNA-based supramolecular complexes. This type of switchable molecular assembly will likely find future use in combination with other nanomaterials such as carbon nanotubes, metallic and semiconducting nanoparticles, and other electrically active components for construction of nanoelectronics for sensors applications, as one possible example.

V. DNA-Based Computation

The first experimental proof of the feasibility of DNA-based computing came from Adleman, when he used DNA to encode and solve a simple instance of a hard combinatorial search problem [76]. He demonstrated the use of artificial DNA to generate all possible solutions to a Hamiltonian path problem (given a set of nodes connected by a set of one-way edges, answer the question of whether or not there exists a path which goes through each node only once). For large graphs, the problem can be very difficult for an electronic computer to solve since there are an astronomical number of possible paths and there is no known algorithm for finding the correct answer. Adleman's approach was to assign a 20 base DNA sequence to each node in an example graph, then to synthesize edge strands containing the complement to the $3'$ half of a starting node fused with the complement to the $5'$ half of the ending node for each valid edge in the graph. The sets of oligonucleotides encoding nodes and edges were annealed and ligated, thereby generating long DNA strands representing all possible paths through the graph. Non-Hamiltonian paths were then discarded from the DNA pool, first by size separation of the path DNA (discard strands greater than or less than the length of a Hamiltonian path, which is equal to the product of the number of nodes times the length of the node sequence), and second by a series of sequence-based separation steps involving DNA probes complementary to each node sequence (discard path sequences if they failed to contain any one of the required nodes). By this experimental protocol, Adleman was able to recover DNA strands encoding the Hamiltonian path through the example graph.

The primary contributions of Adleman's seminal paper were the revolutionary concepts that synthetic DNA could be made to carry information in nonbiological ways and that the inherent massive parallelism of molecular biology operations could be harnessed to solve computationally hard problems. Since, soon after the excitement following Adleman's first experiment, limits have been noted on the size of combinatorial search problems which can be implemented in DNA especially due to the exponential growth of search spaces and the volume constraints on wet computing techniques [77]. In addition to volume constraints, approaches involving biochemical manipulation for massively parallel computation suffer from rather inefficient and tedious hands-on laboratory steps, the total number of which increase at least linearly with problem size. These concerns have been sidestepped by more recent theoretical and experimental advances including the development of various autonomous biochemical systems and by the strategy of computation by algorithmic self-assembly.

Self-assembly, in general, can be defined as the spontaneous formation of ordered structure by components which not only act as structural building blocks but also encode information specifying how the building blocks should be organized into the final construction. In other words, a set of structural units carrying smart address labels which specify neighbor relations are able to self-assemble into a desired structure. In DNA computing by self-assembly, a diverse library of address labels is available via the programmable molecular recognition afforded by Watson–Crick complementary sequence matching. As we shall see, DNA can be used not only as the information carrying "smart glue" but also as the structural material from which the building blocks are formed.

Following Adleman's first experiment, biochemical reactions like anneal, ligate, and separate were mapped to computational primitives, and sets of primitives were evaluated theoretically to determine their inherent computational power. The fundamental insight which spawned DNA computing by self-assembly was made by Winfree when he noted that DNA annealing by itself was theoretically capable of performing universal computation [78,79]. He went further and recognized that certain stable DNA structures being developed by Seeman for nanoengineering and crystallography could serve as physical incarnations of a mathematical model known as Wang tiles which had already been shown to be capable of Turing universal computation. In Wang tiling, unit tiles are labeled with symbols on each edge such that tiles are allowed to associate only if their edge symbols match [80]. Tile sets have been designed which successfully simulate computing devices known as Turing Machines and are therefore capable of universal computation [81]. The recognition that DNA tiles, exemplified by DX and TX complexes, could represent Wang tiles in a physical system, where edge symbols are encoded in the base sequence of sticky ends, led to proofs that DNA tilings are capable of universal computation [82]. Computation by self-assembly of DNA tiles is a significant advance over biochemical manipulation computing schemes because self-assembly involves a single-step procedure in which the computation occurs during the annealing of carefully designed oligonucleotides. Contrast this with Adleman's experiment in which the annealing step generated all possible solutions and where a long series of laboratory steps is required to winnow the set by discarding incorrect answers. Self-assembly (without errors) will theoretically only allow formation of valid solutions during the annealing step, thereby eliminating the laborious phase involving a large number of laboratory steps.

The first published report of a successful computation by DNA self-assembly demonstrated example cumulative-XOR calculations on fixed input strings [62]. XOR (equivalent to the addition operation without the carry-bit) was performed using binary-valued tiles designed to assemble an input layer, which then acted as a foundation upon which output tiles specifically assembled, based on the values encoded on the input tiles. The system made use of TX tiles composed of three double helical domains linked by strand exchange at four crossover points, specifically TAO tiles as shown in Figure 25 [61]. The implementation required eight tile types — two input tiles, four output tiles, and two corner tiles which connected the input layer to the output. The binding slot for the first output tile was formed from a sticky end from the first input tile and a sticky end from a corner tile. Subsequent output binding slots were formed from another end from the next input tile and another from the previous output tile (as shown in Figure 29). In this way, the cumulative XOR was calculated over the encoded input string. The example in Figure 29 shows an input

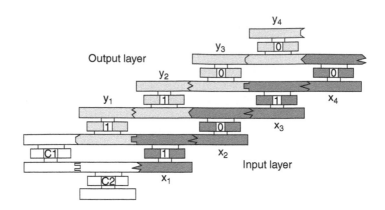

Figure 29 XOR computational assembly constructed from TAO tiles. Input layer (x_1 to x_4) and corner tiles (c_1 and c_2) assemble first, followed by specific assembly of output tiles (y_1 to y_4) into binding slots formed by one input tile and one output tile.

string of 1010 and its cumulative XOR of 1100 on the output string. The described system also demonstrated the use of readout from a reporter strand which was formed by ligation of strands carrying a single bit-value from each tile in the superstructure. Technically, the computation occurred during the self-assembly step, but the readout stage required ligation of the reporter strand segments, PCR amplification, restriction cleavage, and separation of the fragments by polyacrylamide gel electrophoresis. Algorithmic aperiodic self-assembly requires greater fidelity than periodic self-assembly, because correct tiles must compete with partially correct tiles. For example, if the output binding slot displays sticky ends for inputs of 1 and 0, two output tiles would match the first input and two would match the second input but only one output tile type would match both input sticky ends. An error rate of between 2 and 5% was observed on the readout gels, presumably due to a low probability of incorrect output tiles being trapped in the final complex. The tiling system presented is capable of performing parallel computations on multiple, randomly assembled input layers; however the examples implemented were prototype calculations on fixed inputs.

The first parallel molecular computation using DNA tiling self-assembly in which a large number of distinct inputs were simultaneously processed has now been described [83]. The system followed the concept of "string tile" assembly which derives from the observation that by allowing neighboring tiles in an assembly to associate by sticky ends on each side, one could increase the computational complexity of languages generated by linear self-assemblies [84]. Surprisingly, sophisticated calculations can be performed with single-layer linear assemblies when contiguous strings of DNA trace through individual tiles and the entire assembly multiple times [85]. In essence, a "string tile" is the collapse of a multilayer assembly into a simpler superstructure by allowing individual tiles to carry multiple bits of information on multiple segments of the reporter strands, thereby allowing an entire row of a truth table (both input and output bits) to be encoded within each individual tile. "String tile" arithmetic implementations allow input and output strings to assemble simultaneously with each pair-wise operation being directly encoded in the structure of a tile, while control of information flow in multidigit calculations is handled by the tile-to-tile associations in the tile assembly superstructure. The experimentally implemented "string tile" arithmetic used linear self-assembly of DNA DX tiles to perform pair-wise XOR calculations [83]. Experimental execution proceeded as follows: first, the set of purified oligonucleotides for each individual tile type (four tile types — one for each combination of possible inputs) were slowly annealed in separate tubes from 95 to 65°C to ensure the formation of valid tiles before parallel self-assembly of multitile superstructures. Separately annealed computational and corner tiles were then mixed together at 65°C and further annealed to 16°C. Then the reporter strand segments were ligated to one another to produce reporter strands that contain the inputs and outputs of the calculations. Read-out of the calculations was accomplished by first selectively purifying ligated reporter strands corresponding to example (four-bit) computations by extracting the band from a denaturing polyacrylamide gel of the ligation products. The purified reporter strands (containing in this case 24 possible calculations) were then amplified by PCR. Purified PCR products of the proper length were then ligated into a cloning vector. Finally, dideoxy sequencing was performed on mini-prep DNA from five randomly selected clones. The sequencing results obtained from the five randomly selected clones matched the designed sequence words and encoded valid computations.

Theoretical assessments of DNA-based computation by self-assembly indicate computation power far beyond that which has yet been experimentally implemented. Still, it remains to be seen if DNA computers will ever find a computational niche in which they can efficiently compete with electronic computers. One realm in which algorithmic DNA self-assembly appears to provide very great promise is in the nanofabrication of specific aperiodic structures for templating of nanoelectronics devices. Such nanopatterned materials could be used not only for communications and computational devices but also for sensors, biosensors, medical diagnostics, and molecular robotics applications.

VI. DNA-BASED NANOFABRICATION

A major motivating application of self-assembled DNA nanostructures is the need to form complex patterned objects at the molecular scale. The field of computer science is facing a critical challenge in the immediate future, namely the scale limitations of known fabrication techniques for microelectronics. DNA self-assembly offers a potential bottom-up technique to address and overcome this challenge. The microelectronics industry currently uses optical lithography for constructing microelectronics devices on silicon chips. Because of wavelength resolution limits, it is unlikely that optical lithography will scale much below a few nanometers. It has been projected that in approximately 15 years, this top-down method for patterning microelectronics will reach its ultimate limits and rate of progress in miniaturization of microelectronics will either halt or be very much reduced. In addition, the cost of construction and maintenance of modern fabrication facilities is becoming prohibitive. DNA self-assembly is a leading candidate to assist in the further reduction of electronic circuit dimensions.

Programmed self-assembly of DNA objects promises further advances not only in biomolecular computation but also in nanofabrication as a means of creating complex, patterned structures for use as templates or scaffolds for imposing desired structures on other materials. As discussed in Section III, simple, periodic patterns have been successfully implemented and observed on superstructures formed from a variety of different DNA tiles including DX tiles [57], TX tiles [61], triangular tiles [60], parallelogram tiles [86], and 4×4 tiles [64]. Large lattice superstructures formed by such systems have been observed (at least 10 μm by 3 or 4 μm and containing tens of thousands of tiles).

Larger tiles sets with more complicated association rules are currently being developed for the assembly of aperiodic patterns which will be used in the fabrication of patterned objects useful for nanotechnology applications. Two-dimensional tile arrays can be thought of as molecular fabric or tapestry which contain a large number of addressable pixels. Individual tiles can carry one or more pixels depending upon the placement of observable features or binding sites. Overall connectivity can be programmed either with unique sticky ends defined for each tile in the array or by assembly of crossover junctions which specifically stitch together distant segments of a single long scaffold strand as shown in Figure 30.

Implicit in the discussion of DNA self-assemblies as templates for specific patterning of other materials is the need for attachment chemistries capable of immobilizing these materials onto DNA arrays. Materials of interest might include metal nanoparticles, peptides, proteins, other nucleic acids, and carbon nanotubes among others. A variety of strategies and chemistries are being developed including thiols (–SH), activated amino groups, biotin-avidin association, and annealing of prebound,

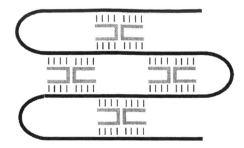

Figure 30 Schematic of a DNA nanostructure composed of a long scaffold strand held together by short oligonucleotides which hybridize to the scaffold and form crossover junctions.

complementary DNA. Oligonucleotides, chemically labeled with a thiol group on either the 5′ or the 3′ end readily bind to gold and have already been used via simple complementary DNA annealing to impart three-dimensional ordering on gold nanospheres [87–89] and gold nanorods [90]. In those studies, gold was labeled with multiple copies of a single DNA sequence, then linear dsDNA was formed between complementary strands attached to adjacent gold particles. More specific chemistries are available including nanogold reagents which make use of 1.4 nm diameter gold clusters, each functionalized with a single chemical moiety for specific reaction with a thiol or a free amino group. These reagents have been used to target the binding of single gold nanoparticles to specific locations on DNA nanoassemblies [63,91]. DNA strands and structures have also been used as selective templates for the construction of very thin (100–15 nm) metallic wires by electroless deposition techniques using a wide variety of different metals [63,64,92–97]. It has recently been reported that metal deposition on a dsDNA template in conjunction with targeted binding of a carbon nanotube can be used to assembly a back-gated transistor [98]. Ongoing studies focus on formation of metal wires laid out in specific patterns on two-dimensional tile lattices. The long-term goal of these metallization studies is the self-assembly of electronic components and circuits using highly parallel techniques at length scales below those available by lithography. Supramolecular constructions self-assembled from DNA strands offer a wide variety of structures and functions and hold great promise for future applications in molecular computation, nanofabrication, and other techniques which can make use of complementary DNA sequence recognition reactions.

REFERENCES

1. S. Zamenhof, G. Brawerman, and E. Chargaff. *Biochim. Biophys. Acta* 9:402, 1952.
2. M.H.F. Wilkins, A.R. Stokes, and H.R. Wilson. *Nature* 171:738–740, 1953.
3. R.E. Franklin and R.G. Gosling. *Nature* 171:740–741, 1953.
4. J.D. Watson and F.H. Crick. *Nature* 171:737–738, 1953.
5. B.A. Schweitzer and E.T. Kool. *J. Am. Chem. Soc.* 117:1863–1872, 1995.
6. K.J. Breslauer. *Meth. Enzymol.* 259:221–242, 1995.
7. L.A. Marky and K.J. Breslauer. *Biopolymers* 26:1601–1620, 1987.
8. G.E. Plum, K.J. Breslauer, and R.W. Roberts. *Compr. Nat. Prod. Chem.* 7:15–53, 1999.
9. K.J. Breslauer, R. Frank, H. Blocker, and L.A. Marky. *Proc. Natl. Acad. Sci. USA* 83:3746–3750, 1986.
10. J. SantaLucia, Jr., H.T. Allawi, and P.A. Seneviratne. *Biochemistry* 35:3555–3562, 1996.
11. N. Sugimoto, S. Nakano, M. Yoneyama, and K. Honda. *Nucleic Acids Res.* 24:4501–4505, 1996.
12. R.A. Gibbs. *Anal. Chem.* 62:1202–1214, 1990.
13. J. Bell. *Immunol. Today* 10:351–355, 1989.
14. M.H. Caruthers. *Acc. Chem. Res.* 24:278–284, 1991.
15. M.J. Gait. *Oligonucleotide Synthesis: A Practical Approach.* IRL Press, Oxford, UK, 1984.
16. D. Luo. *Mater. Today* 6:38–43, 2003.
17. Y. Zhang and N.C. Seeman. *J. Am. Chem. Soc.* 114:2656–2663, 1992.

18. Y. Zhang and N.C. Seeman. *J. Am. Chem. Soc.* 116:1661–1669, 1994.

19. S.M. Du, H. Wang, Y. Tse-Dinh, and N.C. Seeman. *Biochemistry* 34:673–682, 1995.

20. T. Shimizu, R. Iwaura, M. Masuda, T. Hanada, and K. Yase. *J. Am. Chem. Soc.* 123:5947–5955, 2001.

21. S.J. Rowan, P. Suwanmala, and S. Sivakova. *J. Polym. Sci. Part A: Polym. Chem.* 41:3589–3596, 2003.

22. D. Shore, J. Langowski, and R. Baldwin. *Proc. Natl. Acad. Sci. USA* 78:4833–4837, 1981.

23. M. Dlakic and R. Harrington. *J. Biol. Chem.* 270:29945–29952, 1995.

24. Y. Ohya, H. Noro, M. Komatsu, and T. Ouchi. *Chem. Lett.* 25:447–448, 1996.

25. S. Takenaka, Y. Funatsu, and H. Kondo. *Chem. Lett.* 25:891–892, 1996.

26. K.M. Stewart and L.W. McLaughlin. *Chem. Commun.* 2934–2935, 2003.

27. S.M. Waybright, C.P. Singleton, K. Wachter, C.J. Murphy, and U.H.F. Bunz. *J. Am. Chem. Soc.* 123:1828–1833, 2001.

28. E.A. Fogleman, W.C. Yount, J. Xu, and S.L. Craig. *Angew. Chem. Intl. Ed.* 41:4026–4028, 2002.

29. J. Xu, E.A. Fogleman, and S.L. Craig. *Macromolecules* 37:1863–1870, 2004.

30. F.R. Kersey, G. Lee, P. Marszalek, and S.L. Craig. *J. Am. Chem. Soc.* 126:3038–3039, 2004.

31. J. Shi and D.E. Bergstrom. *Angew. Chem. Intl. Ed.* 36:111–113, 1997.

32. M. Endo and T. Majima. *J. Am. Chem. Soc.* 125:13654–13655, 2003.

33. C.M. Niemeyer. *Curr. Opin. Chem. Biol.* 4:609–618, 2000.

34. C.M. Niemeyer, M. Adler, S. Gao, and L. Chi. *Bioconjug. Chem.* 12:364–371, 2001.

35. C.M. Niemeyer, M. Adler, B. Pignataro, S. Lenhert, S. Gao, L. Chi, H. Fuchs, and D. Blohm. *Nucleic Acids Res.* 27:4553–4561, 1999.

36. C.M. Niemeyer, M. Adler, S. Gao, and L. Chi. *Angew. Chem. Intl. Ed.* 39:3056–3059, 2000.

37. C.M. Niemeyer, R. Wacker, and M. Adler. *Angew. Chem. Intl. Ed.* 40:3169–3172, 2001.

38. H. Korri-Youssoufi, F. Garnier, P. Srivastava, P. Godillot, and A. Yassar. *J. Am. Chem. Soc.* 119:7388–7389, 1997.

39. K.J. Watson, S. Park, J. Im, S.T. Nguyen, and C.A. Mirkin. *J. Am. Chem. Soc.* 123:5592–5593, 2001.

40. B. Kalra, A. Kumar, W. Gao, T. Glauser, M. Ranger, J. Hedrick, C.J. Hawker, and R.A. Gross. *Polym. Prepr. (Am. Chem. Soc., Div. Polym Chem.)* 43:720–721, 2002.

41. N.C. Seeman. *Acc. Chem. Res.* 30:357–363, 1997.

42. N.C. Seeman. *Angew. Chem. Intl. Ed.* 37:3220–3238, 1998.

43. N.C. Seeman. *Annu. Rev. Biophys. Biomol. Struct.* 27:225–248, 1998.

44. N.C. Seeman. *Nano Lett.* 1:22–26, 2001.

45. N.C. Seeman. *Nature* 421:427–431, 2003.

46. N.C. Seeman. *Biochemistry* 42:7259–7269, 2003.

47. N.C. Seeman. *J. Theor. Biol.* 99:237–247, 1982.

48. (a) R.I. Ma, N.R. Kallenbach, R.D. Sheardy, M.L. Petrillo, and N.C. Seeman. *Nucleic Acids Res.* 14:9745–9753, 1986; (b) M.L. Petrillo, C.J. Newton, R.P. Cunningham, R.I. Ma, N.R. Kallenbach, and N.C. Seeman. *Biopolymers* 27:1337–1352, 1988.

49. J. Chen and N.C. Seeman. *Nature* 350:631–633, 1991.

50. J.E. Mueller, S.M. Du, and N.C. Seeman. *J. Am. Chem. Soc.* 113:6306–6308, 1991.

51. S.M. Du and N.C. Seeman. *J. Am. Chem. Soc.* 114:9652–9655, 1992.

52. S.M. Du, B.D. Stollar, and N.C. Seeman. *J. Am. Chem. Soc.* 117:1194–1200, 1995.

53. C. Mao, W. Sun, and N.C. Seeman. *Nature* 386:137–138, 1997.

54. C. Liang and K. Mislow. *J. Math. Chem.* 16:27–35, 1994.

55. B. Liu, N.B. Leontis, and N.C. Seeman. *Nanobiology* 3:177–188, 1994.

56. X. Li, X. Yang, J. Qi, and N.C. Seeman. *J. Am. Chem. Soc.* 118:6131–6140, 1996.

57. E. Winfree, F. Liu, L.A. Wenzler, and N.C. Seeman. *Nature* 394:539–544, 1998.

58. F. Liu, R. Sha, and N.C. Seeman. *J. Am. Chem. Soc.* 121:917–922, 1999.

59. H. Yan, T.H. LaBean, L. Feng, and J.H. Reif. *Proc. Natl. Acad. Sci. USA* 100:8103–8108, 2003.

60. X. Yang, L.A. Wenzler, J. Qi, X. Li, and N.C. Seeman. *J. Am. Chem. Soc.* 120:9779–9786, 1998.

61. T.H. LaBean, H. Yan, J. Kopatsch, F. Liu, E. Winfree, J.H. Reif, and N.C. Seeman. *J. Am. Chem. Soc.* 122:1848–1860, 2000.

62. C. Mao, T.H. LaBean, J.H. Reif, and N.C. Seeman. *Nature* 407:493–496, 2000.

63. H. Li, S.H. Park, J.H. Reif, T.H. LaBean, and H. Yan. *J. Am. Chem. Soc.* 126:418–419, 2004.

64. H. Yan, S.-H. Park, G. Finkelstein, J.H. Reif, and T.H. LaBean. *Science* 301:1882–1884, 2003.

65. (a) J. Wang, G. Rivas, J.R. Fernandes, M. Jiang, J.L.L. Paz, R. Waymire, T.W. Nielsen, and R.C. Getts. *Electroanalysis* 10:553–556, 1998. (b) S. Capaldi, R.C. Getts, and S.D. Jayasena. *Nucleic Acids Res.* 28:e21, 2000.

66. R.H.E. Hudson and M.J. Damha. *J. Am. Chem. Soc.* 115:2119–2124, 1993.

67. R.H.E. Hudson, S. Robidoux, and M.J. Damha. *Tetrahedron Lett.* 39:1299–1302, 1998.

68. T.W. Nilsen, J. Grayzel, and W. Prensky. *J. Theor. Biol.* 187:273–284, 1997.

69. C. Mao, W. Sun, Z. Shen, and N.C. Seeman. *Nature* 397:144–146, 1999.

70. C.M. Niemeyer, M. Adler, S. Lenhert, S. Gao, H. Fuchs, and L. Chi. *ChemBioChem* 2:260–265, 2001.

71. B. Yurke, A.J. Turberfield, A.P. Mills, F.C. Simmel, and J.E. Neumann. *Nature* 406:605–608, 2000.

72. J.J. Li and W. Tan. *Nano Lett.* 2:315–318, 2002.

73. P. Alberti and J.-L. Mergny. *Proc. Natl. Acad. Sci. USA* 100:1569–1573, 2003.

74. H. Yan, X. Zhang, Z. Shen, and N.C. Seeman. *Nature* 415:62–65, 2002.

75. L. Feng, S.H. Park, J.H. Reif, and H. Yan. *Angew. Chem. Int. Ed.* 42:4342–4346, 2003.

76. L.M. Adleman. *Science* 266:1021–1023, 1994.

77. J.H. Reif. In *Proceedings of First International Conference on Unconventional Models of Computation*, C.S. Calude, J. Casti, and M.J. Dinneen, Eds. Springer, pp. 72–93, 1998.

78. E. Winfree. In *Proceedings of 1st DIMACS Workshop on DNA Based Computers*, R. Lipton and E.B. Baum, Eds. Princeton, pp. 199–221, American Mathematical Society, 1996.

79. E. Winfree. *J. Biomol. Struc. Dynam. Convers.* 11:263–270, 2000.

80. H. Wang. *Bell Syst. Tech. J.* 40:1–141, 1961.

81. H. Wang. *Fundam. Math.* 82:295–305, 1975.

82. E. Winfree, X. Yang, and N.C. Seeman. In *Proceedings of DNA Based Computers II: DIMACS Workshop*. American Mathematical Society, 1998.

83. H. Yan, L. Feng, T.H. LaBean, and J.H. Reif. *J. Am. Chem. Soc.* 125:14246–14247, 2003.

84. T. Eng in DNA Based Computers III: DIMACS Workshop, June 23–25, 1997. American Mathematical Soc., 1999.

85. E. Winfree, T. Eng, and G. Rozenberg. In *Proceedings of 6th International Workshop on DNA-Based Computers*, A. Condon and G. Rozenberg, Eds. Leiden, Lecture Notes in Computer Science 2054 Springer, pp. 63–88, 2001.

86. C. Mao, W. Sun, and N.C. Seeman. *J. Am. Chem. Soc.* 121:5437–5443, 1999.

87. A.P. Alivisatos, K.P. Johnsson, X. Peng, T.E. Wilson, C.J. Loweth, M.P. Bruchez, Jr., and P.G. Schultz. *Nature (London)* 382:609–611, 1996.

88. C.A. Mirkin, R.L. Letsinger, R.C. Mucic, and J.J. Storhoff. *Nature* 382:607–609, 1996.

89. R.C. Mucic, J.J. Storhoff, C.A. Mirkin, and R.L. Letsinger. *J. Am. Chem. Soc.* 120:12674–12675, 1998.

90. J.K.N. Mbindyo, B.D. Reiss, B.R. Martin, C.D. Keating, M.J. Natan, and T.E. Mallouk. *Adv. Mater.* 13:249–254, 2001.

91. S. Xiao, F. Liu, A.E. Rosen, J.F. Hainfeld, N.C. Seeman, K. Musier-Forsyth, and R.A. Kiehl. *J. Nanoparticle Res.* 4:313–317, 2002.

92. E. Braun, Y. Eichen, U. Sivan, and G. Ben-Yoseph. *Nature* 391:775–778, 1998.

93. K. Keren, M. Krueger, R. Gilad, G. Ben-Yoseph, U. Sivan, and E. Braun. *Science* 297:72, 2002.

94. F. Patolsky, Y. Weizmann, O. Lioubashevski, and I. Willner. *Angew. Chem. Intl. Ed.* 41:2323–2327, 2002.

95. C.F. Monson and A.T. Woolley. *Nano Lett.* 3:359–363, 2003.

96. W.E. Ford, O. Harnack, A. Yasuda, and J.M. Wessels. *Adv. Mater.* 13:1793–1797, 2001.

97. J. Richter, R. Seidel, R. Kirsch, M. Mertig, W. Pompe, J. Plaschke, and H.K. Schackert. *Adv. Mater.* 12:507–510, 2000.

98. K. Keren, R.S. Berman, E. Buchstab, U. Sivan, and E. Braun. *Science* 302:1380–1382, 2003.

Chapter 13

Soluble Amphiphilic Nanostructures and Potential Applications

Nobuo Kimizuka

CONTENTS

I. INTRODUCTION

Nanoscale materials are coming of age. Traditional chemistry has dealt with the synthesis and transformation of individual molecules in the length scale of ~0.1–10 nm. In contrast, synthesis of nano- to mesoscopic (~10–1000 nm) architectures is a challenging issue in molecular nanotechnology. Biological systems provide fascinating examples of such mesoscopic-scale self-assemblies. Figure 1 illustrates the hierarchy of molecular assemblies both in biological and in synthetic systems. Amino acids are polymerized to give peptides and proteins, which are spontaneously folded into

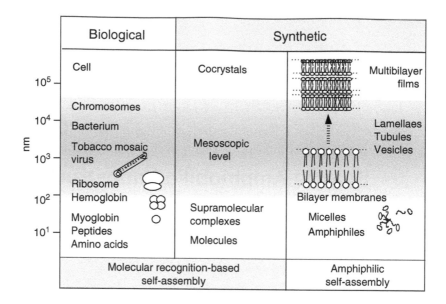

Figure 1 Hierarchy of molecular self-assemblies and the mesoscopic dimension.

nano-sized structures (shown in the left column). These units may further self-assemble into ordered mesoscopic structures, as exemplified by *tobacco mosaic virus*, actin fibers, and microtubles [1]. These mesoscopic supramolecular assemblies are formed by ingeniously employing multiple noncovalent interactions — such as electrostatic interactions, hydrogen bonding, dipole–dipole interactions, and hydrophobic association [1,2]. In this context, weather a nanostructure is comprised of covalent bonding or noncovalent assembly is not an issue of primal importance. Moreover, formation of such supramolecular structures is directed by the aqueous environment. In other words, natural supermolecules possess amphiphilic superstructures which are most stable in water.

The biological self-assemblies inspired the chemists to design molecular recognition-directed supermolecules [3–6]. Multiple hydrogen-bonds between complementary molecular components and metal coordination interactions have been used to design small receptor–guest complexes [3], bulk supramolecular systems such as liquid crystals [7–10], and molecular cocrystals [11–18]. Supramolecular polymers are broadly defined as "polymeric arrays of repeating molecular units which are assembled by reversible and directional noncovalent interactions" [19,20]. This definition refers to their primary structures, and control on their secondary and ternary structures has not been a critical issue.

This chapter describes supramolecular assemblies in mesoscopic dimension and their recent developments. It also compliments earlier reviews [21,22]. The mesoscopic supramolecular assemblies are defined as "hierarchically self-assembled amphiphilic supramolecular structures whose ternary and the higher assembly structures are controlled through solvophilic–solvophobic interactions." Here, pairs of molecules brought by secondary interactions are designed that acquire amphiphilicity upon complexation. They become units of self-assembly and hierarchically grow into mesoscopic-scale supermolecules that are dispersed stably in aqueous or in organic media.

II. SUPRAMOLECULAR MEMBRANES IN WATER

A. Aqueous Bilayer Membranes and Design of Bilayer-Forming Amphiphiles

The studies of soluble, mesoscopic structures have been a subject of colloid chemistry decades before the coining of supramolecular chemistry. It would be instructive, first, to review the molecular

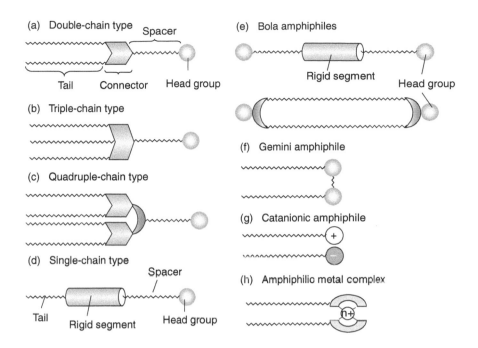

Figure 2 Schematic chemical structures of bilayer-forming amphiphiles.

design developed for bilayer-forming amphiphiles. Since Kunitake's [23,24] pioneering works on synthetic bilayer membranes, a wide variety of mesoscopic-scale structures such as vesicles, tubes, disks, lamellas, and helices are formed from suitably designed amphiphiles (Figure 1, right column). The aggregate morphology and properties of bilayer membranes are controlled by tuning chemical structures of the amphiphiles. This feature draws a sharp line with classical micellar self-assemblies and laid the foundation of today's bottom-up molecular nanotechnology.

To form stable bilayer membranes, it is essential that amphiphilic molecules self-assemble to two-dimensional molecular arrays where the requirement is typically fulfilled by double-chained amphiphiles (Figure 2(a)). Stable bilayers are also formed from triple-chained (Figure 2(b)) [25] or quadruple-chained amphiphiles (Figure 2(c)) [26]. Single-chained amphiphiles also forms bilayers, provided intermolecular interactions are reinforced. For example, introduction of rigid segments such as aromatic chromophores (Figure 2(d)) [27,28] or extremely long alkyl chains [29] promote the formation of bilayers. Single- or double-chained amphiphiles which possess two head groups at both ends form single-layered membranes (Figure 2(e)) [30], which are later popularized as bola-amphiphiles [31]. Amphiphiles with two ionic groups which are connected by a spacer are called gemini surfactants (Figure 2(f)) [32].

The principles obtained for these low-molecular weight amphiphiles are widely applicable. Zhang and coworkers [33], reported the design of surfactant-like N-acetylated peptides with 7 to 8 amino acid residues, which contain aspartic acid at the C terminus (hydrophilic head group) and a lipophilic tail made of hydrophobic amino acids such as valine or leucine. They form nanotubes and vesicles in water. Amphiphilic polymers also form bilayer structures [34]. The diblock copolymer polystyrene–poly(isocyano-L-alanine-L-alanine)$_m$ is among the early examples that form vesicles and chiral nanostructures in water [35]. On the other hand, recent reports on bilayer formation have drastically widened the framework for bilayer-forming amphiphiles. For example, vesicles are formed from porphyrins [36], fullerene-based surfactants [37], dendrimers [38], and even from the wheel-shaped metal oxides [39]. Though careful reinvestigation may be required for some cases, functional groups incorporated in these nonclassical bilayers provide unique opportunity to see bilayer membranes in a new light.

The bilayer-forming amphiphiles discussed so far are single molecules in which solvophilic and solvophobic groups are covalently bound. However, they can also be designed from the combination of molecular components which interact with each other by noncovalent interactions. A typical example of such hybrid amphiphiles is catanionic surfactants (Figure 2(g)) [40]. Mixtures of anionic and cationic surfactants display varied aggregate structures such as globular and rodlike micelles, vesicles, and lamellar phases. These catanionic vesicles are formed spontaneously without external energy input, and they are in contrast to the vesicular systems which require mechanical energy. Amphiphilic metal complexes "metalloamphiphiles" are formed from metal ions and lipophilic ligand molecules (Figure 2(h)) [41]. Metal ions can be incorporated in hydrophilic groups [41,42] or at the hydrophobic interior [43]. In contrast to these hybrid bilayers, when two molecular components are assembled to form bilayer by multiple intermolecular interactions, they can be referred as "supramolecular membranes" [21,22].

B.   Supramolecular Membranes and Related Self-Assemblies in Aqueous Media

Complementary hydrogen bonding is one of the specific interactions involved in chemistry of life. However, hydrogen bonding is not effective in water, because enthalpic gain by the formation of hydrogen-bonds in water is cancelled by the enthalpy to break hydrogen-bonds that have priorly existed between these molecules and water [44]. As water displays high deteriorating action against hydrogen bonding, conventional artificial hydrogen-bond-directed supramolecular assemblies have been designed in aprotic solvents (see Sections I and III). Hydrogen bonding in the aqueous media requires integration of the other noncovalent interactions — such as hydrophobic interactions or aromatic stacking — to compensate the entropic disadvantage. For example, a double helix DNA is an amphiphilic polymer complex, with nucleic bases located on the inside of the right-handed double helix and sugar phosphates on the outside. Complementary base pairs are formed inside the double helix, forming a hydrophobic column of base pairs that are stabilized by stacking [2].

By applying the principle of nucleic acid hybridization, a hydrogen-bond-mediated bilayer membrane can be designed from a pair of complementary molecular subunits. In Figure 3, a hydrophilic subunit **A** possesses a hydrogen-bonding group which is covalently connected to a hydrophilic head group via a spacer unit. A counterpart **B** possesses a complementary hydrogen-bonding group and hydrophobic alkyl chains. If the complementary hydrogen bonding is formed in water, the resultant hydrogen-bond pair **B–A** acquires higher amphiphilicity. The supramolecular amphiphile **B–A** thus generated is expected to hierarchically self-assemble into the four-layered (supramolecular) membranes (Figure 3, right). This is because the equilibrium between the monomeric hydrogen-bond pairs (**B–A**) and supramolecular membranes is expected to shift to thermodynamically more stable tetralayers.

To test this idea, we prepared preformed amphiphilic hydrogen-bond pairs of alkylated melamines (hydrophobic subunits **1**) and quaternary ammonium — derivatized cyanuric acids (hydrophilic subunit **2**) (Figure 4) [45–48]. The pair of melamine and cyanuric acid (or barbituric acid) are employed since they are popularly used to create a variety of supramolecular motifs — such as linear tapes, crinkled tapes, and circular rosette structures (Figure 5) [16,18,49]. The amphiphilic hydrogen-bond pair of **1–2** gave disk-like aggregates with a minimum thickness of $\sim$90 Å (Figure 6), which are comprised of complementary hydrogen-bond-mediated bilayers in water [45]. This is probably the first example of complementary hydrogen-bond-mediated, artificial supramolecular assemblies in water.

The networks of amphiphilic hydrogen-bonds can be spontaneously formed, also by mixing the complementary subunits in water. An azobenzene-containing cyaniuric acid derivative **4** forms globular aggregates when dispersed in water (Figure 7(a)). Absorption λ_{max} of the azobenzene chromophores is located at 357 nm, which is comparable to that in ethanol. Therefore, there are no strong electronic interactions among the azobenzene chromophores in the pure aggregates. On the other hand, melamine-containing chiral subunit **3** alone gave irregular aggregates (Figure 7(b)).

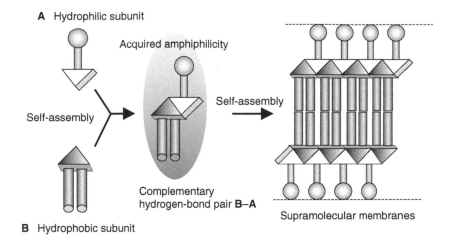

Figure 3 Hierarchical self-assembly of H-bond mediated bilayers in water.

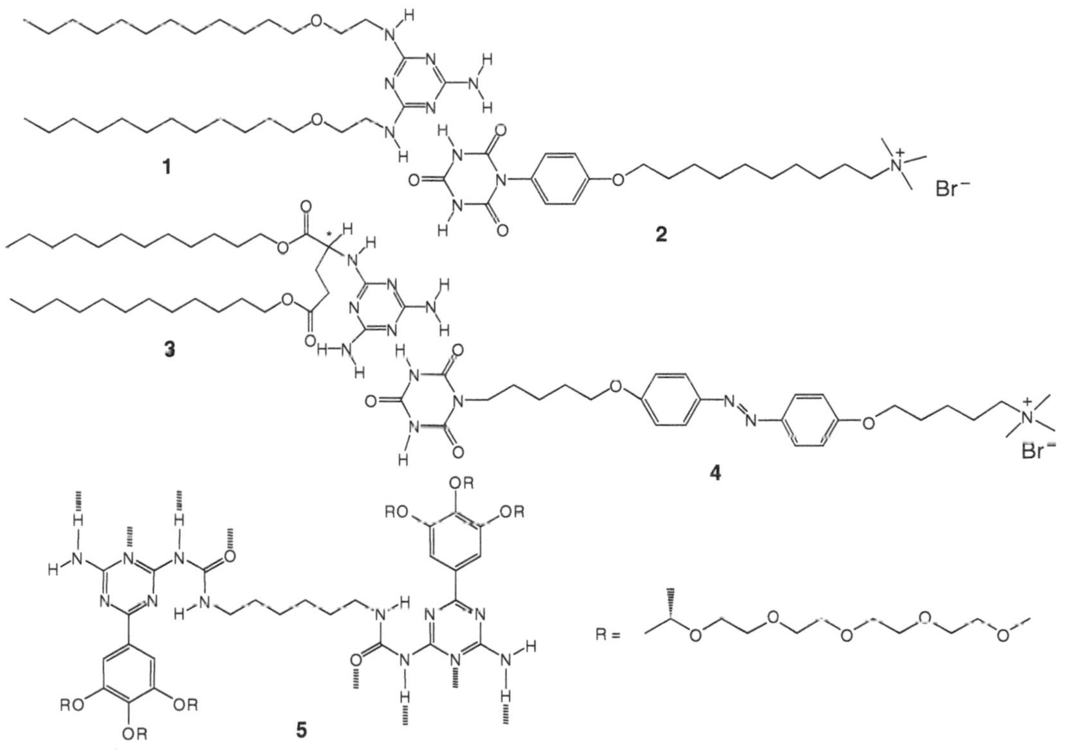

Figure 4 Chemical structures of complementary hydrogen-bond-mediated supermolecules.

Very interestingly, when these subunits were mixed in water, helical superstructures (Figure 7(c), thicknesses: 14–28 nm, widths: 30–50 nm, pitches: 180–430 nm) were immediately formed (association constant; $1.13 \times 10^5 \ M^{-1}$) [50]. These helical structures are typical to the ordered chiral bilayer membranes [51], and are distinct from the aggregate morphology observed for the individual subunits. The observed helical assembly showed blue-shifted absorption of the azobenzene chromophore at 332 nm, with an exciton coupling in circular dichroism. These observations clearly

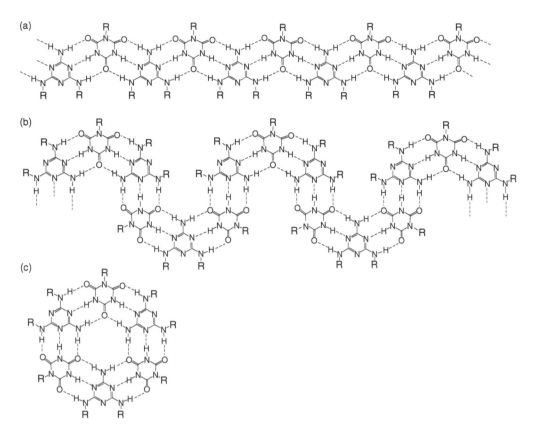

Figure 5 Possible complementary hydrogen-bond networks formed from melamine and cyanuric acid derivatives. (a) Linear tape, (b) crinkled tape, and (c) circular rosette.

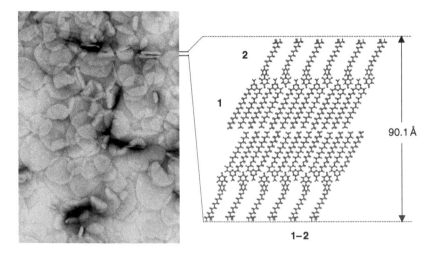

Figure 6 Transmission electron microscopy (TEM) and structure of supramolecular membrane **1–2**.

indicate the presence of excitonic interactions among regularly aligned azobenzene chromophores in the supramolecular membranes **3–4**.

It is noteworthy that the presence of water directs the self-assembly. The hydrophobic melamine subunits are located in the interior, and the ammonium-containing counterparts constitute hydrophilic

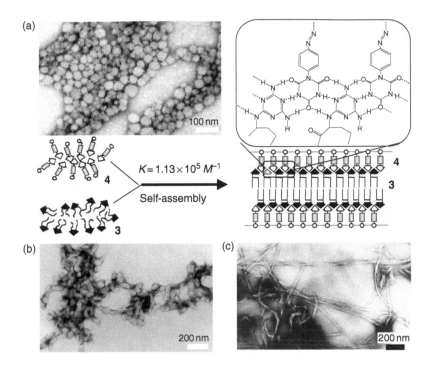

Figure 7 Hierarchical self-assembly of helical superstructures **3–4** in water. Electron micrographs of **4**(a), **3**(b), and **3** and **4**(c).

surface of the assemblies (Figure 7). Among the possible combinations of complementary hydrogen-bond networks such as circular (rosette), crinkled tape, and linear tape (Figure 5) [16,49], only the linear network structure was selected since it provides the most amphiphilic superstructure. This feature is common to the supramolecular hybridization of biological polymers and also to the protein folding. Interestingly, when the azobenzene subunit **4** was photoirradiated and underwent *trans*-to-*cis* photoisomerization, the reconstitution between **4**(*cis*) and **3** did not proceed. Therefore it is important that molecular shape of subunits must be designed not to disturb the ordered molecular orientation. The supramolecular membranes are formed in water when amphiphilic organization of the components satisfies both of the solubility and stabilization of the complementary hydrogen-bonds.

Meijer and coworkers [52] reported formation of hydrogen-bond-mediated polymers in water from bifunctional, self-complementary ureido-*s*-triazines **5**. In water, **5** polymerize into helical columns similarly with the aid of hydrophobic stacking. Shimizu and coworkers [53] reported hydrogen bonding between oligonucleotides and nucleotide-derivatized molecular assembly. Helical nanofibers with a width of 7 to 8 nm were obtained by mixing nucleotide-appended bola amphiphile and oligoadenylic acids. The nanofiber length and their thermal stability are increased when longer oligoadenylic acids are employed, suggesting that the assembly is templated by the oligoadenylic acid [53]. Supramolecular membranes can also be constructed from apoproteins and its cofactor-linked polymers by using a reconstitution technique [54]. A polymeric carboxylic acid was coupled to the ferriprotophophyrin IX via a hydrophilic *bis*(aminoethoxy)ethane spacer, which was reconstituted with the apo-horseradish peroxidase (HRP). Electron microscopy of the dialyzed solution revealed that the enzyme–polymer hybrid formed vesicular aggregates with diameters of 80 to 400 nm. Though the catalytic activity is considerably lowered in these vesicles, the cofactor-reconstitution approach enables the controlled design of protein assemblies. It seems that the progress in site-directed supramolecular modification of proteins hold a key to further development of semiartificial, biomolecular assemblies.

C. Toward the Functional Supramolecular Materials

In Section II.B, amphiphilic pairs of complementary hydrogen bonding subunits are shown to spontaneously self-assemble in water. Giving amphiphilicity to a unit supermolecule is crucial to achieve their medium-dependent hierarchical self-assembly. In this section, this concept is extended to the design of functional supramolecular hydrogels.

1. Hydrogels Formed by Self-Assembly

A wide range of small organic molecules has been found to gel a variety of organic solvents [55]. In contrast, studies on hydrogels formed by self-assembly of small molecules is a growing area of chemistry [56], though formation of fibrous assemblies from amphiphiles has been well documented in literatures [23,31,51]. Figure 8 shows examples of the compounds that are reported to gelatinize water.

Aqueous gel of N-octyl-D-gluconamide **6** is one of the early examples of self-assembling hydrogels [57]. The hydrogel was obtained by cooling the hot aqueous solution ($>70°C$), and was only stable for several hours at room temperature. The dumbbell-shaped arborol **7** forms an interlocking network and gives thermally reversible gels [58]. Similarly, N-alkyl disaccharide amphiphile **12** [59] and other sugar-based amphiphiles [60] displays thermoreversible hydrogel formation. Hydrogels are formed also from charged amphiphiles as exemplified by the fluorinated zwitterionic phosphocholine derivative **8** [61]. The zwitterionic gemini surfactants with hydrocarbon chains also give hydrogels [62]. Combinatorial-style synthesis conducted for more than 40 gemini surfactants showed varied morphology depending on the combinations of two alkyl chain lengths. The gemini surfactant **10** having tartrate counterion also forms hydrogels [63]. It is proposed that the cationic head groups of **10** are held in close contact by the bridging of *bis*anionic

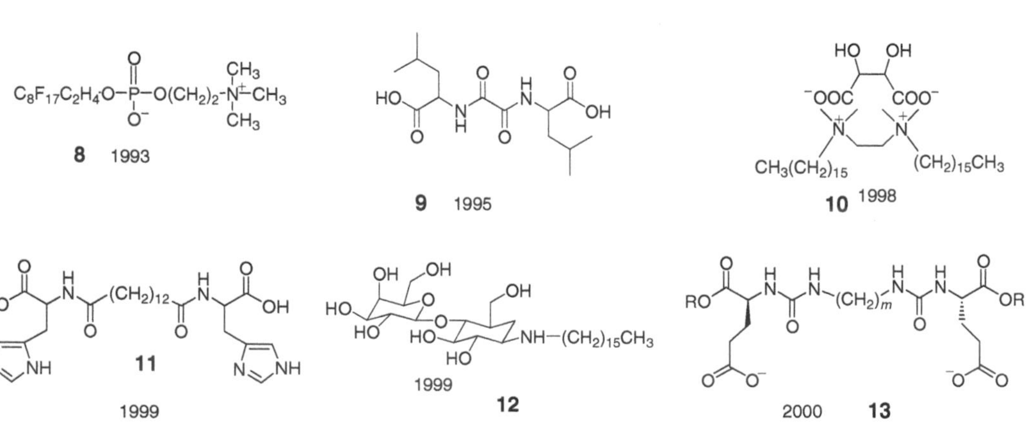

Figure 8 Chemical structure of hydrogel-forming compounds.

tartrate ions which form network of hydrogen-bonds. *Bis*(amino acid) galators **9** [64], **11** [65] and *bis*-urea dicarboxylic acids **13** [66] are nonlipid hydrogelators. Hydrogels are also formed by block copolypeptides [67,68] and *de novo* designed peptides [69]. Though number of reports on self-assembling hydrogels is increasing, they are still focused on the aggregate textures and thermal gel-to-sol transition properties. The development of functional hydrogels based on self-assembly has been largely unexplored.

2. Photofunctional Hydrogels via Combinatorial Approach

In the organization of photosynthetic unit, hundreds of chlorophyll molecules serve as light-harvesting antennae by capturing the sunlight and funneling electronic excitation toward the photosynthetic reaction center (RC). Photosynthetic bacteria evolved a pronounced energetic hierarchy, which funnels electronic excitation from the circular light-harvesting complexes II (LH-IIs) through LH-I to the RC. In the light-harvesting complex II, bacteriochlorophylls (BChls, Figure 9) are noncovalently bound to short peptides which form an octameric aggregate [70]. It is remarkable that LH-II is comprised of the circular self-assembly of a large number of identical transmembrane polypeptide helices that noncovalently hold BChls. BCh is thus a versatile photofunctional unit, but they have neither the capability to self-assemble nor the specifically recognizable functionalities. The peptide subunits determine the intermolecular distance and orientation of chlorophyll molecules which are indispensable for the efficient energy migration.

Inspired by the light-harvesting apparatus of purple bacteria, we designed a light-harvesting supramolecular hydrogel. The principle of peptide-assisted organization of chlorophylls can be simply modeled from a pair of aromatic chromophore and its receptor molecule which forms ordered nanofibrous assemblies upon complexation. Such molecular pairs are self-assembled through multiple intermolecular interactions which may include electrostatic interactions, van der Waals forces, hydrophobic interactions, coordination interactions, and hydrogen-bonds. Here, the ability to gelatinize water by self-assembly is taken as a manifestation of the collective interactions in the supramolecular systems. We employed a combinatorial screening to find the right supramolecular pairs, by recruiting the correct partners from the set of those available [71].

Cationic L-glutamate derivatives **14**, **15** are designed as self-assembling scaffolds which bind anionic fluorophores by the electrostatic and van der Waals interactions (Figure 10) [71]. They are designed by modifying the chemical structure of bilayer-forming amphiphile **16** [72]. The long spacer methylene moieties incorporated in bilayer membranes as in **16** provide hydrophobic clefts for the anisotropic binding of porphyrins [73]. However, equimolar mixtures of conventional

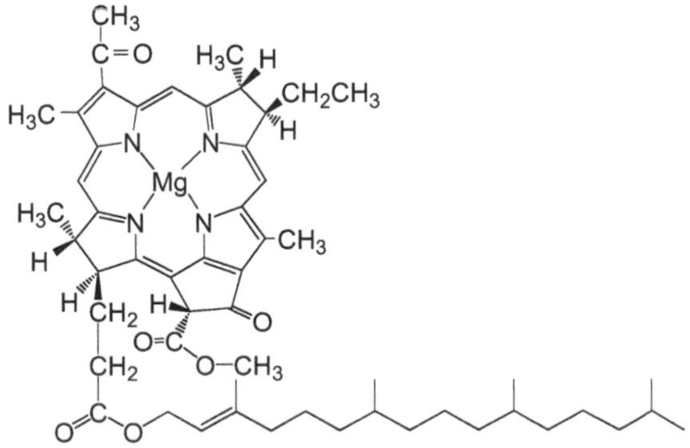

BChl

Figure 9 Chemical structure of bacteriochlorophyll (BChl).

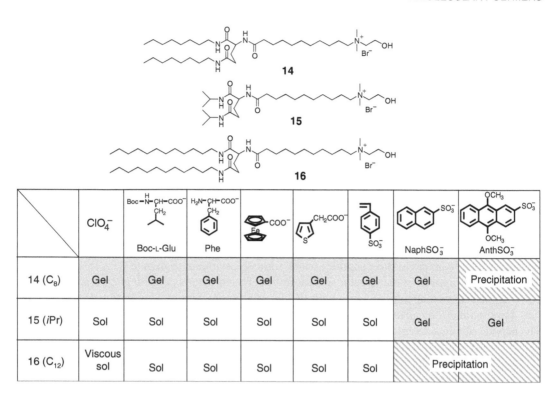

Figure 10 Supramolecular combinations of hydrogel-forming ion pairs.

amphiphiles and oppositely charged molecules tend to precipitate due to the loss of hydration [40a]. To circumvent precipitation, we introduced short octyl and isopropyl chains in the amphiphiles **14** and **15**, respectively. The incorporation of isopropyl groups deviates from the molecular design established for classical bilayer amphiphiles [23,24], since such short tails spoil hydrophobic interactions required for the bilayer formation. Amphiphiles **14**, **15** are soluble in water and transmission electron microscopy (TEM) and differential scanning calorimetry (DSC) showed that **14** forms fibrous bilayer membrane in water. On the other hand, **15** is dispersed in water as monomers or as premicellar aggregates. We anticipated that the lowered molecular interactions can be compensated by the formation of hydrophobic ion pairs, and if moderate supramolecular amphiphilicity is acquired, they may hierarchically self-assemble into hydrogels without precipitation.

The table in Figure 10 shows the result of combinatorial screening for equimolar molecular pairs. Eight kinds of anions, perchlorate, amino acids, and aromatic molecules are employed. Upon mixing aqueous bilayer **16** with aqueous anions, precipitates were formed for 2-naphthalene sulfonate ($NaphSO_3^-$) and sodium 9,10-dimethoxyanthracene 2-sulfonate ($AnthSO_3^-$). Addition of perchlorate ion caused increase in viscosity, but physical gelation was not observed for all the anions employed. On the other hand, upon the addition of $NaphSO_3^-$ to aqueous solution of **14**, transparent hydrogels were formed in a few minutes (concentration, $[\mathbf{14}] = [NaphSO_3^-] = 5$ mM). All the other anions caused formation of hydrogels, except for the most hydrophobic $AnthSO_3^-$ which gave precipitate. Interestingly, hydrogels are even formed from the pairs of **15**/$NaphSO_3^-$ (20 mM) and **15**/$AnthSO_3^-$ (10 mM), though **15** alone does not form ordered self-assemblies. Schematic illustration of the supramolecular hydrogel formation is shown in Figure 11.

Atomic force microscopy (AFM) and TEM of these hydrogels shows the presence of nanofibrous assemblies (minimum width, ~8 nm) and these fibrous structures are comprised of bundles of bilayer membranes. The formation of hydrogels is ascribed to physical cross-linking of these fibrous

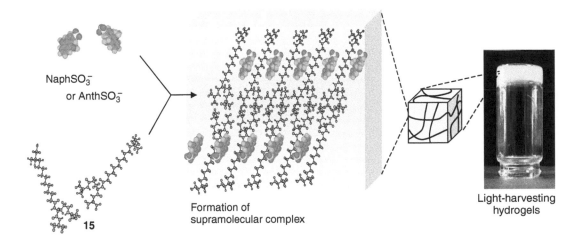

NaphSO$_3^-$
or AnthSO$_3^-$

15

Formation of
supramolecular complex

Light-harvesting
hydrogels

Figure 11   Schematic illustration for the combinatorial self-assembly of light-harvesting hydrogels.

assemblies, and this is promoted by the ion pairing. It effectively reduced electrostatic repulsions that operate between the cationic aggregates, without the transformation into multilamellar precipitates. Interestingly, hydrogel of **14**/NaphSO$_3^-$ showed efficient energy transfer characteristics. Ion-paring-induced water-gelling system was also reported [63], but as far as we are aware, this is the first example of functional hydrogels obtained by self-assembly. The advantage for the supramolecular combinatorial approach compared with the combinatorial covalent synthesis [23,24,62,74] is its simplicity and facile incorporation of the functional molecules.

III.   AMPHIPHILIC HYDROGEN-BONDED NETWORKS IN ORGANIC MEDIA

Soluble supramolecular polymers have been prepared based on hydrogen bonding [19,20], and they are largely classified into main-chain polymers [8,75] and side-chain (liquid-crystalline) polymers [76]. In this section, we focus on the soluble mesoscopic assemblies in organic media, which display directed molecular organization through solvophobic interactions.

As shown in Figure 5, melamine and cyanuric acid can form varied supramolecular structures [16,18,49]. Whitesides and coworkers developed a strategy to selectively synthesize rosette assemblies, by introducing peripheral crowding. This approach introduces bulky substituents on melamine units and minimization of their unfavorable, repulsive steric interactions promotes the exclusive formation of rosettes [16,77]. This steric control-approach allows preorganizing hydrogen-bond donor and acceptor groups in the desired configurations, and it reduces the entropic disadvantages in the assembling process. Unfortunately, however, introduction of such bulky groups inevitably prevents the assembly growth into the mesoscopic dimensions [77]. This situation made the mesoscopic dimension remain unexplored in the conventional supramolecular chemistry.

The development of supramolecular membrane in water [45–48] prompted us to design amphiphilically designed supramolecular assemblies also in organic media [78]. 1,8:4,5-Naphthalene *bis*(dicarboxyimide) **17**, which is hardly soluble in common organic solvents except for dimethyl sulfoxide (DMSO), was complexed with dialkylated melamine **1**. Upon formation of the complementary hydrogen-bond pairs, **17** became soluble in common organic solvents. UV-vis spectrum of the complex **1–17** in methylcyclohexane showed significant broadening of naphthalene absorption, which is ascribed to the stacking of naphthalene chromophores. Electron microscopy revealed the presence of flexible, mesoscopic fibrous structures with the width of ∼10 nm. These observations

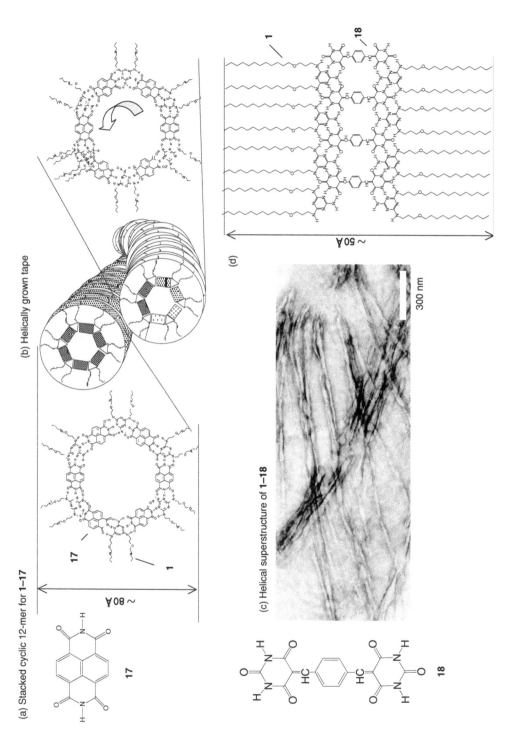

Figure 12 Mesoscopic hydrogen-bond-directed assemblies in organic media. (a) Stacked cyclic 12-mer for **1–17**, (b) helically grown tape for **1–17**, (c) transmission electron micrograph for **1–18**, and (d) a triple-layer membrane formed from the linear hydrogen-bond networks **1–18**.

are compatible with either a stacked cyclic structure (a) or a helically grown structure (b) shown in Figure 12. Formation of these structures has been further supported by AFM observations [79].

The amphiphilic design of supermolecules plays an essential role to direct the packing and folding of complementary subunits, in which a solvophilic moiety (alkyl chains equipped with ether linkages) is oriented toward solvents while a solvophobic moiety (stacked hydrogen-bond networks) is hidden inside the assemblies. The amphiphilic superstructure not only provides solubility in organic media, but also determines the folding structure of hydrogen-bond networks in given media [78]. The concept of mesoscopic supramolecular assemblies has been extended to the combinations of Janus molecules/dialkylated melamines [80] and perylene*bis*imide/dialkylated melamines [81]. Self-aggregates of ureidotriazine derivatives [82] and aqueous rosette nanotubes [83] are similarly designed by considering the supramolecular amphiphilicity. The solvophobicity-driven folding is also highlighted in the single oligomers of phenylene ethynylene reported by Moore and coworkers [84] and those of Lokey and Iverson [85].

Mesoscopic supramolecular assemblies are also obtainable in organic media from amphiphilic linear networks of complementary hydrogen-bonds. An equimolar mixture of *bis*-barbituric acid derivative **18** and **1** gave helical superstructures with a minimum width of ~50 Å (Figure 12(c), note that the components are achiral, but compound **18** displays molecular asymmetry) [86]. The observed thickness corresponds to the three-layered membrane structure schematically depicted in Figure 12(d). Soluble, tapelike supramolecular oligomers can be also prepared, by the covalent preorganization of melamine units [87].

IV. Self-Assembling Nanowires of Organic/Inorganic Superstructures

A variety of supramolecular architectures have also been created by the use of metal coordination. Synthesis of catenanes, knots [88], and double-helical metal complexes (linear or circular helicates) [4] arc few of the prominent examples. Nanostructured supramolecular squares and capsules are also formed by self-assembly (Figure 13) [89].

These supramolecular complexes have been attracting widespread interest. However, they consist of *discrete* metal complexes. Consequently, their applications are largely limited to shape-related functions such as host–guest inclusions. It is only recently that interactions among connecting metal ions are being investigated at low temperatures [90]. On the other hand, control on the magnetism and electronic states in solid metal complexes represent the other important area in inorganic chemistry [91]. It is predicted that the supramolecular complexes in the next generation would integrate these two features and conjugated electronic structures are generated in response to the self-assembly. In this section, the extension of mesoscopic supramolecular systems toward the development of conjugated inorganic molecular wires is introduced.

A. Quasi One-Dimensional Halogen-Bridged Mixed-Valence Complexes

Molecular wires are indispensable elements of future molecular-scale electronic devices, and their fabrication has been attracting much interest. Conventional researches are focused on the synthesis of π-conjugated oligomers and polymers [92]. The application of solid-phase synthesis technique allows strict controlling of the sequence and the molecular length of the π-conjugated oligomers. However, these organic wires intrinsically possess limitations on the kind of elements that can be incorporated. We have recently developed a new strategy to manipulate conjugated metal complexes by "supramolecular packaging" of one-dimensional inorganic complexes $[M^{II}(en)_2][M^{IV}Cl_2(en)_2]$ (M^{II}, M^{IV} = Pt, Pd, and Ni, en: 1,2,-diaminoethane, Figure 14) [93–98].

A family of quasi one-dimensional halogen-bridged M^{II}/M^{IV} mixed-valence complexes $[M^{II}(en)_2][M^{IV}X_2(en)_2]Y_4$ (X = Cl, Br, I; Y: counterions such as ClO_4) has been attracting interest among solid-state physicist and chemists. They display strong electron–lattice interactions and

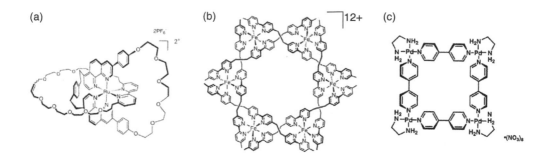

Figure 13   Discrete supramolecular metal complexes. Synthesis of (a) catenanes, knots (adapted from P. Mobian, J.-M. Kern, and J.-P. Sauvage. *J. Am. Chem. Soc.* 125:2016, 2003. With permission). (b) Double-helical metal complexes (adapted from B. Hansenknopf, J.-M. Lehn, N. Boumediene, A.D. Gervais, A.V. Dorsselaer, B. Kneisel, and D. Fenske. *J. Am. Chem. Soc.* 119:10956, 1997. With permission). (c) Nanostructured supra molecular squares (adapted from M. Fujita, J. Yazaki, and K. Ogura. *J. Am. Chem. Soc.* 112:5645, 1990. With permission).

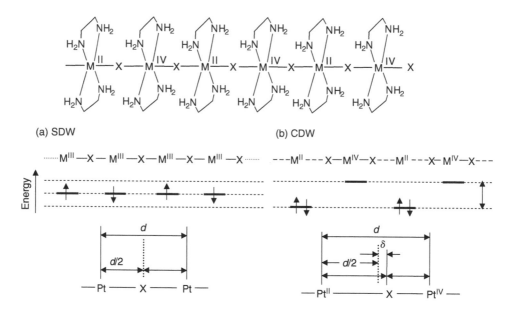

Figure 14   Chemical structure of halogen-bridged mixed-valence complex $[M(en)_2][MX_2(en)_2]^{4+}$ (M = Pt, Pd, Ni; X = Cl, Br, I) and electronic structures. Counter anions are omitted for clarity. (a) Spin density wave state (SDW); (b) charge density wave state (CDW).

unique physicochemical properties such as intense intervalence charge transfer (CT) absorption, semiconductivity, and large third-order nonlinear optical susceptibilities [99]. In these compounds, the one-dimensional electronic structure is comprised mainly of d_{z^2} orbital of the metals and p_z orbital of the bridging halogens, where z-axis is parallel to the chain (Figure 14). The ground state can be expressed as a mono-valence state ([a] spin density wave state, SDW) or a mixed-valence state ([b] charge density wave state, CDW). If the bridging halogen ions X are located at the center of two adjacent metal ions, these metal ions are not distinguishable and adopt M^{III} state. Such a mono-valence (SDW) state has been reported for the Ni compounds [100]. In the case of $[Pt^{II}(en)_2][Pt^{IV}X_2(en)_2]$ compounds, halogen-bridging anions are located with an off-center displacement of δ. This structural change is ascribed to the Peierls transition which is a characteristic of one-dimensional electronic systems.

In spite of these interesting properties, the one-dimensional complexes have not been considered as a candidate for molecular wires, since they exist only in three-dimensional solids. They are not soluble in organic media and when dispersed in water the one-dimensional structure is disrupted and dissociates into the constituent molecular complexes.

B. Self-Assembling Inorganic Molecular Wires by Supramolecular Packaging

We prepared ternary inorganic–organic polyion complexes $[Pt(en)_2][PtCl_2(en)_2](lipids)_4$, as schematically shown in Figure 15. $[Pt(en)_2][PtCl_2(en)_2]$ (**19, 20, 21, 22,** or **23**)$_4$, which displayed yellow (lipid = **19, 22**) or indigo colors (lipid = **20, 21, 23**) depending on the amphiphilic chemical structure is shown in Figure 16 [93–98]. These colors are typical of the intervalence CT ($Pt^{II}/Pt^{IV} \rightarrow Pt^{III}/Pt^{III}$) of halogen-bridged linear complexes. The $[Pt(en)_2][PtCl_2(en)_2](lipids)_4$ complexes can be dispersed in organic media with the maintenance of solid-state colors, indicating that one-dimensional Pt^{II}–Pt^{IV} complexes are maintained in organic media with the help of solvophilic alkyl chains.

Interestingly, in the case of supramolecular complex consisting of dihexadecyl sulfosuccinate **20** and $[Pt(en)_2][PtCl_2(en)_2]^{4+}$, indigo-colored dispersion was obtained at room temperature, whereas the color disappeared upon heating the solution to 60°C. As the CT transition requires the existence of chloro-bridged extended coordination structure, the observed thermochromism indicates disruption of the one-dimensional complex structures at 60°C. The color reappeared reversibly

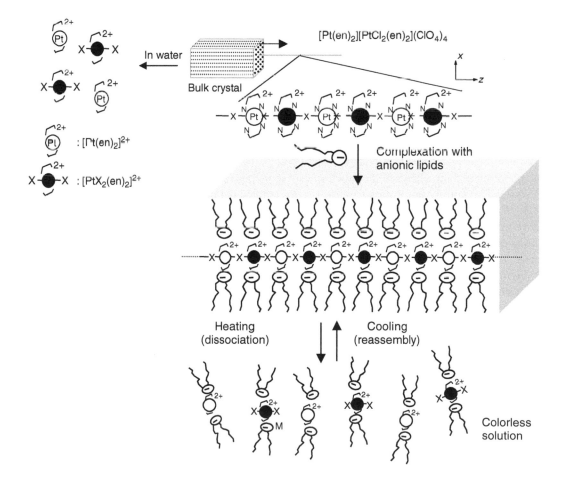

Figure 15 Schematic illustration of self-assembling molecular wires and the supramolecular thermochromism.

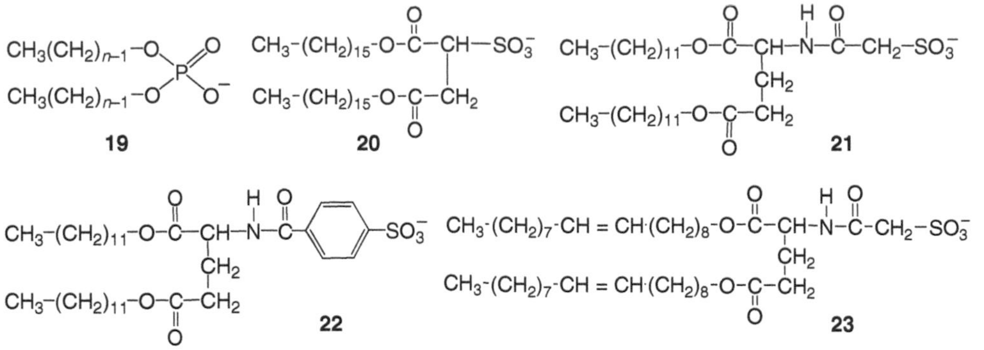

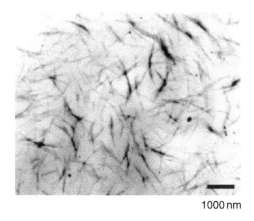

Structure 19: $CH_3(CH_2)_{n-1}-O$, $CH_3(CH_2)_{n-1}-O$ bonded to P with $=O$ and O^-

Structure 20: $CH_3-(CH_2)_{15}-O-\overset{O}{\overset{\|}{C}}-CH-SO_3^-$, $CH_3-(CH_2)_{15}-O-\overset{}{\underset{O}{\overset{\|}{C}}}-CH_2$

Structure 21: $CH_3-(CH_2)_{11}-O-\overset{O}{\overset{\|}{C}}-\overset{}{\underset{CH_2}{CH}}-\overset{H}{N}-\overset{O}{\overset{\|}{C}}-CH_2-SO_3^-$, $CH_3-(CH_2)_{11}-O-\overset{}{\underset{O}{\overset{\|}{C}}}-CH_2$

Structure 22: $CH_3-(CH_2)_{11}-O-\overset{O}{\overset{\|}{C}}-\overset{}{\underset{CH_2}{CH}}-\overset{H}{N}-\overset{O}{\overset{\|}{C}}--SO_3^-$, $CH_3-(CH_2)_{11}-O-\overset{}{\underset{O}{\overset{\|}{C}}}-CH_2$

Structure 23: $CH_3-(CH_2)_7-CH = CH\cdot(CH_2)_8-O-\overset{O}{\overset{\|}{C}}-\overset{}{\underset{CH_2}{CH}}-\overset{H}{N}-\overset{O}{\overset{\|}{C}}-CH_2-SO_3^-$, $CH_3-(CH_2)_7-CH = CH\cdot(CH_2)_8-O-\overset{}{\underset{O}{\overset{\|}{C}}}-CH_2$

Figure 16   Chemical structure of anionic amphiphiles.

1000 nm

Figure 17   TEM of lipid-packaged platinum complex [Pt(en)$_2$][PtCl$_2$(en)$_2$](20)$_4$.

upon cooling and the observed thermochromism indicates that one-dimensional complexes undergo reversible reassembly in solution (Figure 15, supramolecular thermochromism). Figure 17 shows a TEM of [Pt(en)$_2$][PtCl$_2$(en)$_2$](**20**)$_4$ in chloroform, after the heat treatment. Fibrous nanostructures with a minimum width of 18 nm and length of 700 to 1700 nm are abundantly seen [95]. The observed widths of the nanostructures are larger than the bimolecular lengths of **20**, and therefore they must consist of aggregates of amphiphilic supramolecular structure shown in Figure 15.

Figure 18 schematically displays parts of CPK packing model for [Pt(en)$_2$][PtCl$_2$(en)$_2$](**20**)$_4$ in the yz plane (direction of coordination) and in the xy plane, respectively [95]. A minimum distance between aligned anionic head groups of double-chained amphiphiles is estimated as $\sim$5 Å from CPK molecular model, and this value matches well with the Pt^{II}–Cl–Pt^{VI} distance (d) in the one-dimensional complex. Therefore, it is reasonable that the halogen-bridged mixed-valence complex can be well preserved in polyion complexes.

C.   Supramolecular Band Gap Engineering and Solvatochromic Nanowires

Electronic structures of lipid–one-dimensional mixed-valence complexes can be controlled based on the combination of metal ions [94]. On the other hand, the lipids also exert significant effect on the electronic structures. For example, CT absorption peaks observed for [Pt(en)$_2$][PtX$_2$(en)$_2$](**21**)$_4$ in chlorocyclohexane (X = Cl; 2.15 eV, X = Br; 0.87 eV, X = I, <0.56 eV) are peaks significantly red-shifted compared with those observed for crystalline perchlorates [Pt(en)$_2$][PtX$_2$(en)$_2$](ClO$_4$)$_4$

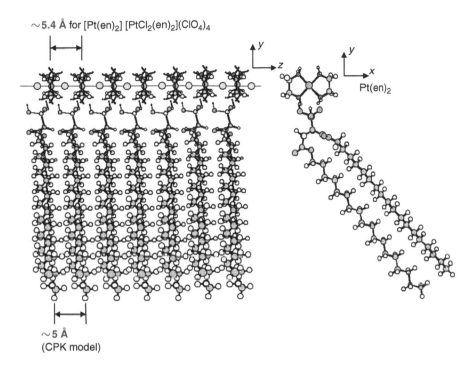

Figure 18 A schematic molecular packing model for [Pt(en)$_2$][PtCl$_2$(en)$_2$](**20**)$_4$.

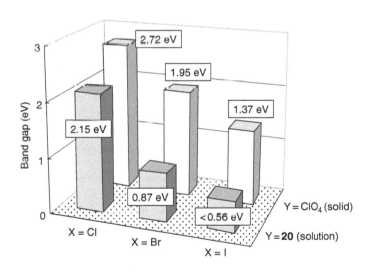

Figure 19 Dependence of HOMO–LUMO band gap on the bridging halogen atoms. Y = ClO$_4$; [Pt(en)$_2$][PtX$_2$(en)$_2$](ClO)$_4$ in the solid-state. Y = **20**;[Pt(en)$_2$][PtX$_2$(en)$_2$](**21**)$_4$ in chlorocyclohexane.

(X = Cl; 2.72 eV, X = Br; 1.95 eV, X = I; 1.37 eV, Figure 19) [96]. The observed red-shift is ascribed to the enhanced delocalization of the excited Pt(III)–Pt(III) states in the coordination chain, which decreases LUMO–HOMO energy gap of the one-dimensional complex. In lipid complexes, it is probable that densely packed sulfonate groups direct the electrostatically bound Pt(en)$_2$ and PtCl$_2$(en)$_2$ complexes to coordinate in higher density. This would cause the increase in overlap between d_{z^2} and p orbitals, depending on the molecular packing and chemical structure of lipids. The shortened interplatinum (Pt^{II}–Cl–Pt^{IV}) distance would promote the excitation delocalization along the conjugated chains (supramolecular band gap engineering) [96]. Another plausible mechanism is enhanced interchain electronic interactions in the lipid complexes. In any case, these phenomena have not been observed in the previous studies on halogen-bridged mixed-valence complexes. The different colors observed for lipid complexes of $[Pt(en)_2][PtCl_2(en)_2]$(**19, 20, 21, 22,** or **23**)$_4$ indicate that the molecular information of organic molecules are converted to electronic structures of the linear conjugated chains. This can be regarded as supramolecular processing of molecular information.

Another unique characteristic of the lipid complexes is solvatochromism. In the case of $[Pt(en)_2][PtCl_2(en)_2]$(**22**)$_4$, intense intervalence CT absorption was observed at 473 nm (ε: 13,846 units $M^{-1}\,cm^{-1}$) in chloroform, whereas it is observed at 513 nm (ε: 15,017 units $M^{-1}\,cm^{-1}$) in dichloromethane [97]. The observed solvatochromism can be ascribed to the altered molecular packing of the lipids which occur as a consequence of the different degree of solvation.

The packaging of low-dimensional inorganic solid enables creation of novel polymer molecules that have not been dealt as molecules. This strategy opens a new dimension in mesoscopic supramolecular assemblies as well as in molecular wire research [98]. The complexation of lipids (or surfactants) and inorganic complexes are recently reported for $[Mo_3Se_3]^-$ chains [101], linear iron(II) complexes [102], wheel-shaped polyoxometalate [103], and also for halogen-bridged platinum complexes [104]. Another class of lipophilic linear complexes are designed by introducing 1-aminoalkanes as ligands in Magnus green salt $[Pt(NH_2R)_4][PtCl_4]$ [105,106]. Field-effect transistors comprising $[Pt(NH_2dmoc)_4][PtCl_4]$ (dmoc, (S)-3,7-dimethylooctyl group) as the active semiconductor layer were produced by dipping a friction-deposited transfer film of poy(tetrafluoroethylene) (PTFE) film into a super-saturated toluene solution of the Pt compound [106].

D. Self-Assembly at Interfaces

One of the important applications of the lipid(surfactant)-inorganic complexes is development of self-assembling molecular or nanoscale electronic devices. The ability to fabricate techno-logically useful architectures on surfaces should be developed on the basis of nonlithographic, self-organization process. The halogen-bridged mixed-valence complexes in organic media allow the formation of thin films, which open applications of these unique nanomaterials.

By spreading chloroform solutions of $[Pt(en)_2][PtCl_2(en)_2]$(**20**)$_4$ at the air–water interface, surface monolayers are obtained [107]. To get stable monolayers which can be deposited on solid surfaces by Langmuir–Blodgett technique, it was necessary to spread the dissociated colorless complexes and let the coordination chains be formed at the air–water interface.

To enhance the lipophilic nature of the lipids, oleyl chains are introduced in the L-glutamate amphiphile **23** [108]. When the complex $[Pt(en)_2][PtCl_2(en)_2]$(**23**)$_4$ is dissolved in chloroform at room temperature, indigo color of the solid is disappeared, due to dissociation of the polymeric complex into component complexes. Upon casting the colorless solution on the quartz plate, purple films were formed immediately after evaporation of the solvent. It is apparent that polymerization of $[Pt(en)_2]$(**23**)$_2$ and $[PtCl_2(en)_2]$(**23**)$_2$ complexes proceeds during the solvent evaporation. Interestingly, when the dilute chloroform solutions are dropped on carbon-coated copper TEM grid under a humid air, regular, double-layered honeycomb architecture were observed in scanning electron microscopy (Figure 20, one side of the hexagons, ~650 to 750 nm; width of the wires, ~100 nm) [108].

The top honeycomb layer is connected to the basal honeycomb layer via perpendicularly oriented pillars (height, ~320 to 370 nm) at the corner of hexagons (Figure 20, right). Self-assembling characteristics of $[Pt(en)_2][PtCl_2(en)_2]$(**23**)$_4$ in the course of rapid solvent evaporation seem to be an essential factor for the formation of such unique stereo-nanostructures.

Formation of two-dimensional honeycomb morphology has been observed in the solvent evap-oration of copolymers [109–111] and fluorinated silver nanoparticles [112]. In these systems, water droplets condensed from moisture on the evaporating solutions act as the template to direct formation

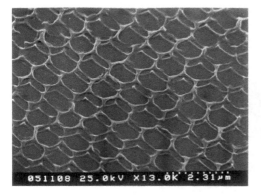 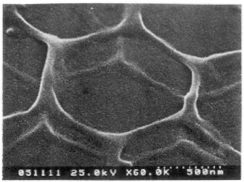

Figure 20   SEM images of stereo-honeycomb nanoarchitectures formed from [Pt(en)$_2$][PtCl$_2$(en)$_2$](**23**)$_4$.

of honeycomb patterns. Similar mechanism would be involved in the present system, where the rigid coordination chains formed during the solvent evaporation stabilize the stereo-honeycomb architectures. A recent report also indicates the importance of rigid polymer structures in the formation of double-layered honeycomb structure [113].

E.   Gel-Like Networks Self-Assembled from Lipophilic Complexes and Unique Thermal Transitions

1,2,4-Triazoles are known as bridging ligands and their linear iron(II) complexes have been rigorously studied, because of the spin crossover characteristics [114–116]. Similar to the case of halogen-bridged mixed-valence complexes, studies on triazole complexes have been limited to the bulk crystalline samples. To develop soluble triazole complexes, we have designed a lipophilic triazole ligand **24** (C$_{12}$OC$_3$Trz) and synthesized their Fe(II) and Co(II) complexes [115,117]. When Fe(**24**)$_3$Cl$_2$ is dissolved in chloroform, yellow gel is formed. It indicates that Fe(II) ion is in high spin state and TEM indicates the presence of fibrous gel-like nanoassemblies as schematically depicted in Figure 21(a) [115]. Similarly, Co(**24**)$_3$Cl$_2$ in chloroform gave a blue gel-like phase at room temperature. The blue color is a characteristic of tetrahedral cobalt(II) ions (T_d, $^4A_2 \rightarrow {}^4T_1(P)$), and networks of fibrous nanoassemblies (width, 5 to 30 nm) are observed in AFM. To our surprise, the blue gel-like phase turns into a pale pink solution *upon cooling*. The pink color indicates the formation of rodlike octahedral (O_h) complexes. The observed sol (lower temperature, O_h complex) — gel (above 25°C, T_d complex) transition is thermally reversible, as schematically shown in Figure 21(b). These features give a clear distinction from the conventional organogels that dissolve upon heating [55]. This is the first example of reversible, heat-set formation of gel-like networks in organic media. The transition from O_h complexes in solution to T_d complexes in the gel-like phase is enthalpically driven [117]. Though the heat-set gelation phenomena have also been observed for polymer hydrogels, they are entropy driven [118]. It is apparent that the amphiphilic one-dimensional coordination systems provide unique self-assembling properties which are not accessible from the conventional inorganic or polymer chemistry.

V.   Amphiphilic Nanostructures Self-Assembled in Ionic Liquids

Room temperature ionic liquids are receiving much attention as environmentally benign solvents. They display negligible vapor pressure, high ionic conductivity, and limited miscibility with water and common organic solvents [119]. The chemistry of ionic liquids is a rapidly expanding field,

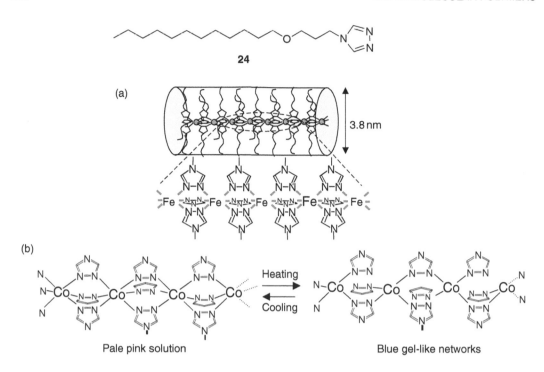

Figure 21 Lipophilic triazole complexes. Schematic illustrations for (a) Fe(II)(**24**)$_3$Cl$_2$ and (b) thermochromic heat-set gelation observed for Co(II)(**24**)$_3$Cl$_2$ in chloroform.

but two major issues have been unexplored. The first was the development of molecularly organized systems or supramolecular assemblies which are stably dispersed in ionic liquids. The second was the development of ionic liquids that molecularly dissolve biomacromolecules such as carbohydrates.

For the purpose of developing ionic liquids which molecularly dissolve biopolymers, we designed ether-containing N, N'-dialkyl imidazolium derivatives **26**, **27** [120]. These compounds are obtained as solid after freeze drying. They eventually become liquid at room temperature in air, probably due to absorption of water from the atmosphere. These ionic liquids were miscible with water. The content of water in these ionic liquids was 2.6 wt% for **26** and 2.5 wt% for **27**, as determined by Karl Fischer's method. To compare the solvent properties of these ionic liquids with conventional ionic liquids, compound **25** was saturated with water (water content, 2.5 wt%) and used as a reference.

Very interestingly, the ether-containing ionic liquids homogeneously dissolve carbohydrates such as β-D-glucose (solubility, 450 mg ml^{-1}), α-cyclodextrin (350 mg ml^{-1}), amylose (30 mg ml^{-1}), and agarose (10 to 20 mg ml^{-1}) by heating [120]. The observed solubility is not ascribed to the water molecules slightly coexisting in these ionic liquids, since amylase can be dissolved, which is a coiled polysaccharide composed of α-1,4-glucosyl units and is only slightly soluble in pure water. A highly glycosylated protein, glucose oxidase (GOD) was also soluble in these ionic liquids (concentration, 1 mg ml^{-1}).

The sugar-philic property observed for these ether-containing ionic liquids **26** and **27** is ascribed to the presence of bromide anions, since the solubility of sugar derivatives was diminished by replacing it with *bis*-trifluoromethane sulfonimide (TFSI) [121]. Bromide ions must be effectively solvating the hydroxyl groups possibly by hydrogen bonding (Figure 22). The dissolution of cellulose in chloride- or bromide-containing ionic liquids by heating was also recently reported [122]. Interestingly, when the heated agarose solution of **26** was cooled to room temperature, physical gelation of the ionic liquid occurred. This can be called as "ionogel," since the commonly used terms of hydrogel or organogel refer to the solvents that are being gelatinized [120]. This is the first example of ionogels

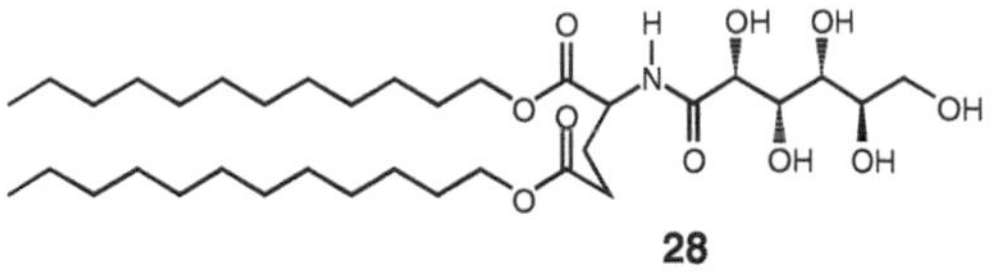

Solvents	28	29
Water	Insoluble	Ionogel formed (10 mM)
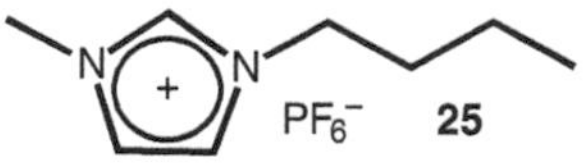 PF₆⁻ 25	Insoluble	Insoluble
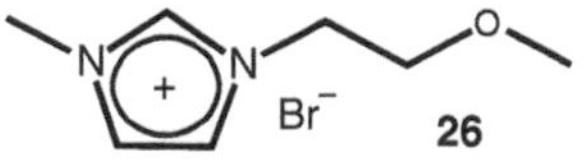 Br⁻ 26	Soluble (10 mM)	Ionogel formed (10 mM)
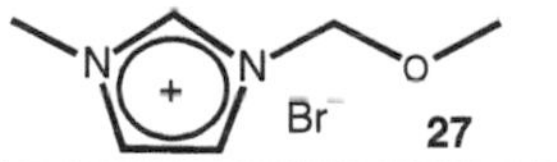 Br⁻ 27	Soluble (10 mM)	Ionogel formed (10 mM)

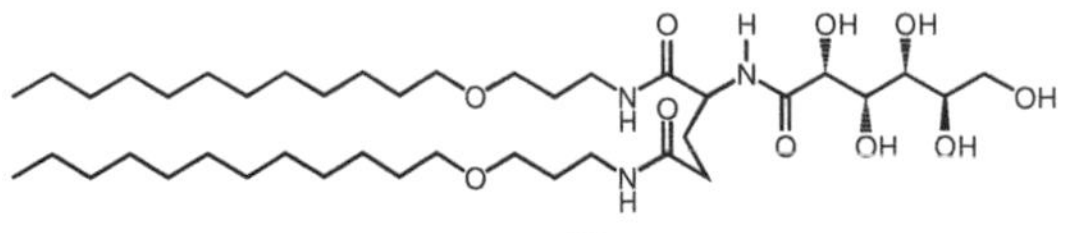

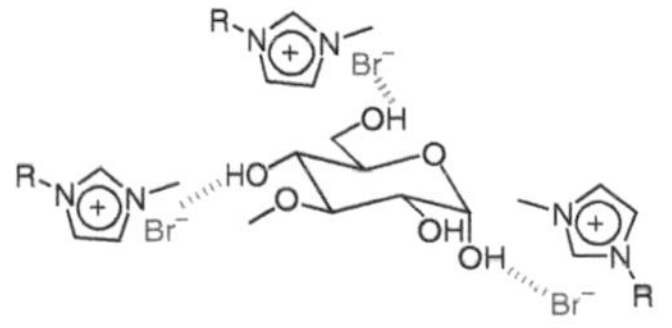

Figure 22 Solubility of synthetic glycolipids in ionic liquids and schematic dissolution mechanism of carbohydrate derivatives in bromide-containing ionic liquids.

formed by biopolymers, though the physical gelation of ionic liquids has been preceded by synthetic polymers [123].

To investigate the formation of ordered molecular assemblies in ionic liquids, solubility of glycolipids **28** and **29** were tested. These glycolipids are not soluble in the conventional ionic liquid **25**. L-Glutamate derivatives **28** ($n = 12, 16$) are insoluble in water even at a concentration of 1 mM, however they can be dispersed in ionic liquids **26** and **27** as microcrystalline aggregates. Very interestingly, physical gelation of these ionic liquids occurred when glycolipid **29**, which contains ether-linked alkyl chains and three amide bonds, was dissolved at concentrations of above 10 mM (Figure 22 and Figure 23). In dark-field optical microscopy, developed fibrous structures whose length extends to several hundred micrometers are abundantly seen (Figure 23(c)). The DSC measurement of **29** in ionic liquid **26** showed a broad endothermic peak at 40°C, at which temperature the ionogel was dissolved. This is the first example of ionogels formed by self-assembly of low-molecular weight compounds [120]. To our surprise, this phase transition was accompanied by a reversible morphological transformation of fibrous aggregate into vesicles (Figure 23(d); diameter 3 to 5 μm).

The formation of bilayer vesicles was also observed for dialkyldimethylammonium bromide **33** in ether-containing ionic liquids **26**, **27** (Figure 24) [124]. Ionic amphiphiles intrinsically possess solvophilic (or ionophilic) groups and electrostatic repulsions between them are highly screened in ionic liquids. In addition, the presence of ether-linkages in ionic liquids increases their net polarity. These factors contribute to enhance the association force of ammonium amphiphiles in ionic liquids and concurrently increase the thermal stability of gel(crystalline) state bilayers [124].

It is established that the molecular assembly in ionic liquids is governed by (i) the balance of ionophilicity and ionophobicity of constituent molecules and (ii) the chemical structure of the ionic liquids. The formation of ordered self-assemblies in ionic liquids [120] has touched off the subsequent

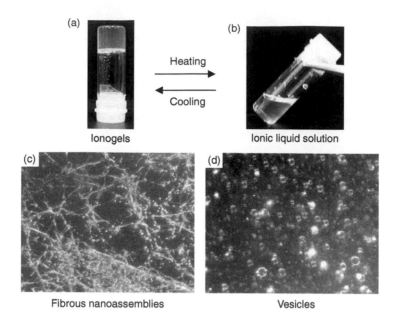

Figure 23 Pictures of **29** in ionic liquid **26**. (a) Ionogel (at room temperature), (b) ionic liquid solution (by heating). Dark-field optical micrographs of (c) ionogels and (d) bilayer vesicles observed for (a) and (b), respectively.

Compounds	Ionic liquids	State of the mixtures	Ref.
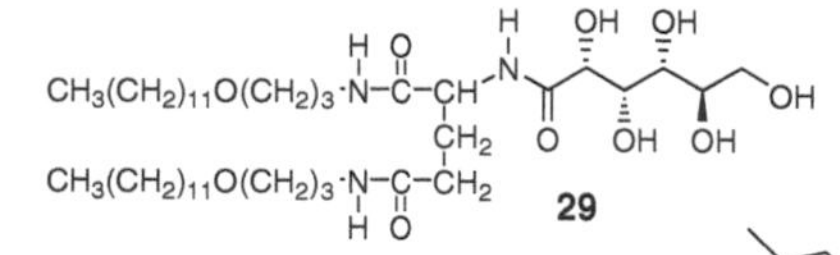 **29**	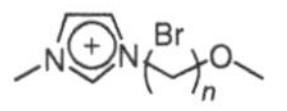	Homogeneous dispersion of bilayer membranes and ionogels (vesicles and fibrous nanoaggregates)	120
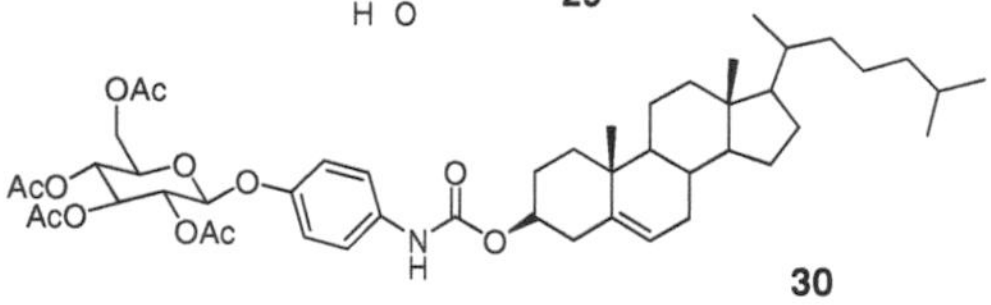 **30**		Ionogels	125
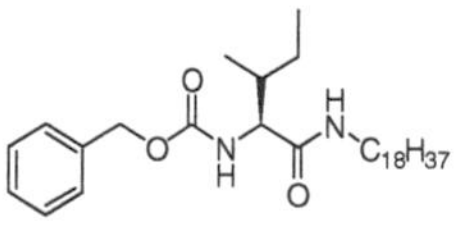 **31**	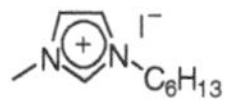	Ionogels	126
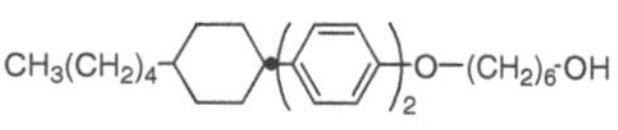 **32**		Phase-separated liquid crystals	128
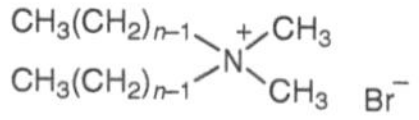 **33**	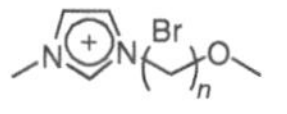	Bilayer membranes (vesicles)	124

Figure 24 Molecular assemblies in ionic liquid.

burst of researches on molecular assembly in ionic liquids (Figure 24). A cholesterol-based organo-gelator **30** was shown to induce gelation of various N,N'-dialkylimidazolium and N-alkylpyridinium salts [125]. An amino acid-derived compound **31** gelatinizes 1-hexyl-3-methylimidazolium iodide, and the ionogel formed was used in the dye-sensitized solar cell [126]. Hydrated samples of 1-decyl-3-methylimidazolium bromide formed liquid-crystalline ionogels [127]. A hydroxyl-terminated mesogenic compound **32** showed limited miscibility with 1-ethyl-3-methylimidazolium tetrafluorobrate and anisotropic ion conduction was observed in the layered liquid-crystalline phase [128].

Formation of molecular assemblies in ionic liquids provide a new dimension in the design of functional supramolecular polymers. It introduces new supramolecular interfaces in ionic liquids, which hold a key to the development of new functions. The introduction of microscopic interfaces also expands the area of ionic liquid research. For example, one-step sol–gel synthesis of hollow TiO_2 microspheres was achieved in ionic liquids by the use of organic microdroplets as template [129]. Now biomolecules, molecular assemblies, organic, and inorganic polymers can be handled in the engineered solvents. A rapid growth of supramolecular chemistry in these particular media is expected.

VI. Conclusions

The integration of molecular-recognition-directed self-assembly and chemistry of bilayer membranes has lead to the development of mesoscopic supramolecular assemblies. The impartment of amphiphilicity to supermolecules drives their hierarchical self-assembly. The solvophilic–solvophobic interactions play a pivotal role in the determination of the supramolecular architecture, and this is a distinct feature from the earlier supramolecular chemistry. The combinatorial supramolecular approach is also effective to develop functional mesoscopic assemblies. In addition, combination of supramolecular polymers and solvent engineering will give a new perspective in the design of mesoscopic materials.

Besides these artificial approaches, biomolecules also provide powerful scaffolds to construct mesoscopic architectures (see Chapters 12 and 17). For example, well-designed nanoarchitectures are synthesized from oligonucleotides [130,131]. We also have recently reported that sticky-end-tagged three way junctions are formed from suitably designed three DNA strands. They further hierarchically self-assemble into mesoscopic particles which possess cagelike architectures [132]. This approach is simple and much easier than those devised for the previous DNA-based supermolecules. Supramolecular assemblies having such biomolecular components will find increased applications.

Finally, the generation of unique electronic structures based on self-assembly is proposed as one of the forthcoming issues in nanochemistry. Studies on lipophilic platinum complexes showed that classical solid-state inorganic complexes can be handled in solutions as self-assembling nanowires. Their electronic structures are tunable with the help of organic molecular assemblies (supramolecular band gap engineering). The heat-set gel-like networks in organic media is another surprise. These unique self-assembling properties are not accessible from conventional polymer or supramolecular chemistry. The growth of electronically conjugated networks is reminiscent of the neural development. We envisage that self-assembling molecular wires would be applied to the design of neuronlike networks, which display growth and self-restoration in response to chemical or electronic stimuli. These self-assembling systems would contribute to the development of chemical learning systems [133]. They may also give an opportunity to embark on *molecular* (or *chemical*) *cybernetics*, which is a new field of molecularly organized chemistry that integrates elements of molecular information, their transmission and translation, amplification, chemical or physical output, and their feedback control.

Acknowledgment

This work was supported by a Grant for 21st Century COE Program from the Ministry of Education, Culture, Sports, Science, and Technology of Japan.

REFERENCES

1. A. Klug. *Angew. Chem. Int. Ed.* 22:565, 1983.
2. B. Alberts, D. Bray, J. Lewis, M. Raff, K. Roberts, and J.D. Watson. *Molecular Biology of the Cell*, 3rd edn, New York: Garland Publishing Inc., 1994.
3. (a) J. Rebek Jr. *Angew. Chem. Int. Ed.* 29:245, 1990. (b) J.L. Awood, J.E.D. Davies, D.M. MacNicol, F. Vögtle, and J.-M. Lehn, Eds. *Comprehensive Supramolecular Chemistry*, Vol. 9, Oxford: Pergamon, 1996.
4. (a) J.-M. Lehn. *Angew. Chem. Int. Ed.* 29:1304, 1990. (b) B. Hansenknopf, J.-M. Lehn, N. Boumediene, A.D. Gervais, A.V. Dorsselaer, B. Kneisel, and D. Fenske. *J. Am. Chem. Soc.* 119:10956, 1997.
5. J.-M. Lehn. *Supramolecular Chemistry—Concepts and Perspectives.* Weinheim: VCH, 1995.
6. G.M. Whitesides, J.P. Mathias, and C.T. Seto. *Science* 254:1312, 1991.
7. P. Mariani, C. Mazabard, A. Garbesi, and G.P. Spada. *J. Am. Chem. Soc.* 111:6369, 1989.
8. (a) C. Fouquey, J.-M. Lehn, and A.-M. Levelut. *Adv. Mater.* 2:254, 1990. (b) J.-M. Lehn. *Macromol. Chem. Makromol. Symp.* 69:1, 1993.
9. M. Kotera, J.-M. Lehn, and J.-P. Vigneron. *J. Chem. Soc. Chem. Commun.* 197, 1994.
10. M. Suárez, J.-M. Lehn, S.C. Zimmerman, A. Skoulios, and B. Heinrich. *J. Am. Chem. Soc.* 120:9526, 1998.
11. D. Voet. *J. Am. Chem. Soc.* 94:8213, 1972.
12. N. Shimizu and S. Nishigaki. *Acta Cryst. B* 38:2309, 1982.
13. M.C. Etter. *Acc. Chem. Res.* 23:120, 1990.
14. J.-M. Lehn, M. Mascal, A. DeCian, and J. Fischer. *J. Chem. Soc. Chem. Commun.* 479: 1990.
15. F.G. Tellado, S.J. Geib, S. Goswami, and A.D. Hamilton. *J. Am. Chem. Soc.* 113:9265, 1991.
16. J.C. MacDonald and G.M. Whitesides. *Chem. Rev.* 94:2383, 1994.
17. K.T. Holman, S.M. Martin, D.P. Parker, and M.D. Ward. *J. Am. Chem. Soc.* 123:4421, 2000.
18. L.J. Prins, D.N. Reinhoudt, and P. Timmerman. *Angew. Chem. Int. Ed.* 40:2383, 2001.
19. A. Chiferri. *Macromol. Rapid. Commun.* 23:511, 2002.
20. L. Brunsveld, B.J.B. Folmer, E.W. Meijer, and R.P. Sijbesma. *Chem. Rev.* 101:4071, 2001.
21. N. Kimizuka. In: N. Yui, Ed. *Supramolecular Design for Biological Applications*, Boca Raton, FL: CRC Press, 2002, p. 373.
22. N. Kimizuka. *Curr. Opin. Chem. Biol.* 7:702, 2003.
23. T. Kunitake. In: J.L. Awood, J.E.D. Davies, D.M. MacNicol, F. Vögtle, and J.-M. Lehn, Eds. *Comprehensive Supramolecular Chemistry*, Vol. 9, Oxford: Pergamon, 1996, p. 351.
24. T. Kunitake. *Angew. Chem. Int. Ed. Engl.* 31:709, 1992.
25. T. Kunitake, N. Kimizuka, N. Higashi, and N. Nakashima. *J. Am. Chem. Soc.* 106:1979, 1984.
26. N. Kimizuka, H. Ohira, M. Tanaka, and T. Kunitake. *Chem. Lett.* 19:29, 1990.
27. T. Kunitake, Y. Okahata, M. Shimomura, S. Yasunami, and T. Kunitake. *J. Am. Chem. Soc.* 103:5401, 1981 and references therein.
28. M. Shimomura and T. Kunitake. *J. Am. Chem. Soc.* 109:5175, 1987 and references therein.
29. F.M. Menger and Y. Yamasaki. *J. Am. Chem. Soc.* 115:3840, 1993.
30. Y. Okahata and T. Kunitake. *J. Am. Chem. Soc.* 101:5231, 1979.
31. J.-H. Fuhrhop. In: J.L. Awood, J.E.D. Davies, D.M. MacNicol, F. Vögtle, and J-M. Lehn, Eds. *Comprehensive Supramolecular Chemistry*, Vol. 9, Oxford: Pergamon, 1996, p. 407.
32. F.M. Menger and J.S. Keiper. *Angew. Chem. Int. Ed.* 39:1907, 2000.
33. (a) S. Vauthey, S. Santoso, H. Gong, N. Watson, and S. Zhang. *Proc. Natl. Acad. Sci.* 99:5355, 2002. (b) S. Zhang, D.M. Marini, W. Hwang, and S. Santoso. *Curr. Opin. Chem. Biol.* 6:865, 2002.
34. (a) D.E. Discher and A. Eisenberg. *Science* 297:967, 2002. (b) M. Antonietti and S. Förster. *Adv. Mater.* 15:1323, 2003.
35. J.J.L.M. Cornelissen, M. Fischer, N.J.M. Sommerdijk, and R.J.M. Nolte. *Science* 280:1427, 1998.
36. (a) J.H. Fuhrhop, U. Bindig, and U. Siggel. *J. Am. Chem. Soc.* 115:11036, 1993. (b) L.M. Scolaro, M. Castriciano, A. Romeo, S. Patanè, E. Cefali, and M. Allegrini. *J. Phys. Chem. B* 106:2453, 2002.
37. (a) H. Murakami, M. Shirakusa, T. Sagara, and N. Nakashima. *Chem. Lett.* 28:815, 1999. (b) A.M. Cassell, C.L. Asplund, and J.M. Tour. *Angew. Chem. Int. Ed.* 38:2403, 1999. (c) M. Brettreich, S. Burghardt, C. Böttcher, T. Bayeri, S. Bayerl, and A. Hirsch. *Angew. Chem. Int. Ed.* 39:1845, 2000. (d) M. Sano, K. Oishi, T. Ishi-i, and S. Shinkai. *Langmuir* 16:3773, 2000. (e) S. Zhou, C. Burger, B. Chu, M. Sawamura,

N. Nagahama, M. Toganoh, U.E. Hackler, H. Isobe, and E. Nakamura. *Science* 291:1944, 2001.
(f) E. Nakamura and H. Isobe. *Acc. Chem. Rev.* 36:807, 2003.

38. K. Tsuda, G.C. Dol, T. Gensch, J. Hofkens, L. Latterini, J.W. Weener, E.W. Meijer, and F.C. De Schryver. *J. Am. Chem. Soc.* 122:3445, 2000.

39. T. Liu, E. Diemann, H. Li, A.W.M. Dress, and A. Müller. *Nature* 426:59, 2003.

40. (a) E.W. Kaler, A.K. Murthy, B. Rodriguez, and J.A. Zasadzinski. *Science* 245:1371, 1989. (b) A. Meister, M. Dubois, L. Belloni, and T. Zemb. *Langmuir* 19:7259, 2003 and references therein.

41. (a) S. Muñoz and G.W. Gokel. *Inorg. Chim. Acta.* 250:59, 1996 and references therein. (b) C. Li, X. Lu, and Y. Liang. *Langmuir* 15:575, 2002.

42. T. Kunitake, Y. Ishikawa, M. Shimomura, and H. Okawa. *J. Am. Chem. Soc.* 108:327, 1986.

43. J. Suh, H. Shim, and S. Shin. *Langmuir* 12:2323, 1996.

44. (a) A.J. Doig and D.H. Williams. *J. Am. Chem. Soc.* 114:338, 1992. (b) M.S. Searle, D.H. Williams, and U. Gerhard. *J. Am. Chem. Soc.* 114:10697, 1992. (c) J.S. Nowick, J.S. Chen, and G. Noronha. *J. Am. Chem. Soc.* 115:7636, 1993.

45. N. Kimizuka, T. Kawasaki, and T. Kunitake. *J. Am. Chem. Soc.* 115:4387, 1993.

46. N. Kimizuka, T. Kawasaki, and T. Kunitake. *Chem. Lett.* 23:33, 1994.

47. N. Kimizuka, T. Kawasaki, and T. Kunitake. *Chem. Lett.* 23:1399, 1994.

48. N. Kimizuka, T. Kawasaki, K. Hirata, and T. Kunitake. *J. Am. Chem. Soc.* 120:4094, 1998.

49. J.A. Zerkowski, J.C. MacDonald, C.T. Seto, D.A. Wierda, and G.M. Whitesides. *J. Am. Chem. Soc.* 116:2382, 1994.

50. T. Kawasaki, M. Tokuhiro, N. Kimizuka, and T. Kunitake. *J. Am. Chem. Soc.* 123:6792, 2001.

51. N. Nakashima, S. Asakuma, and T. Kunitake. *J. Am. Chem. Soc.* 107:509, 1985.

52. L. Brunsveld, J.A.J.M. Vekemans, J.H.K.K. Hirschberg, R.P. Sijbesma, and E.W. Meijer. *Proc. Natl. Acad. Sci.* 99:4977, 2002.

53. R. Iwaura, K. Yoshida, M. Masuda, M.O. Kameyama, M. Yoshida, and T. Shimizu. *Angew. Chem. Int. Ed.* 42:1009, 2003.

54. M.J. Boerakker, J.M. Hannink, P.H.H. Bomans, P.M. Frederik, R.J.M. Nolte, E.M. Meijer, and N.A.J.M. Sommerdijk. *Angew. Chem. Int. Ed.* 41:4239, 2002.

55. P. Terech and R.G. Weiss. *Chem. Rev.* 97:3133, 1997.

56. J.C. Tiller. *Angew. Chem. Int. Ed.* 42:3072, 2003.

57. J.-H. Fuhrhop, S. Svenson, C. Boettcher, E. Rössler, and H.-M. Vieth. *J. Am. Chem. Soc.* 112:4307, 1990.

58. G.R. Newkome, G.R. Baker, S. Arai, M.J. Saunders, P.S. Russo, K.J. Theriot, C.N. Moorefield, L.E. Rogers, J.E. Miller, T.R. Lieux, M.E. Murray, B. Phillips, and L. Pascal. *J. Am. Chem. Soc.* 112:8458, 1990.

59. S. Bhattacharya and S.N.G. Acharya. *Chem. Mater.* 11:3504, 1999.

60. J.H. Jung, G. John, M. Masuda, K. Yoshida, S. Shinkai, and T. Shimizu. *Langmuir* 17:7229, 2001.

61. M.P. Krafft, F. Giulieri, and J.G. Riess. *Angew. Chem. Int. Ed.* 32:741, 1993.

62. F.M. Menger and A.V. Peresypkin. *J. Am. Chem. Soc.* 123:5614, 2001.

63. R. Oda, I. Huc, and S.J. Candau. *Angew. Chem. Int. Ed.* 37:2689, 1988.

64. M. Jokić, J. Makarević, and M. Žinić. *J. Chem. Soc. Chem. Commun.* 1723, 1995.

65. S. Franceschi, Nd. Viguerie, M. Riviere, and A. Lattes. *New J. Chem.* 23:447, 1999.

66. L.A. Estroff and A.D. Hamilton. *Angew. Chem. Int. Ed.* 39:3447, 2000.

67. A.P. Nowak, V. Breedveld, L. Pakstis, B. Ozbas, D.J. Pine, D. Pochan, and T.J. Deming. *Nature* 417:424, 2002.

68. E.R. Wright, R.A. McMillan, A. Cooper, R.P. Apkarian, and V.P. Conticello. *Adv. Func. Mater.* 12:149, 2002.

69. D.J. Pochan, J.P. Schneider, J. Kretsinger, B. Ozbas, K. Rajagopal, and L. Haines. *J. Am. Chem. Soc.* 125:11802, 2003.

70. X. Hu, A. Damjanović, T. Ritz, and K. Schulten. *Proc. Natl. Acad. Sci.* 95:5935, 1998.

71. T. Nakashima and N. Kimizuka. *Adv. Mater.* 14:1113, 2002.

72. N. Kimizuka, M. Shimizu, S. Fujikawa, K. Fujimura, M. Sano, and T. Kunitake. *Chem. Lett.* 27:967, 1998.

73. Y. Ishikawa and T. Kunitake. *J. Am. Chem. Soc.* 113:621, 1991.

74. S. Kiyonaka, K. Sugiyasu, S. Shinkai, and I. Hamachi. *J. Am. Chem. Soc.* 124:10954, 2002.

75. B.J.B. Folmer, E. Cavini, R.P. Sijbesma, and E.W. Meijer. *J. Am. Chem. Soc.* 123:2093, 2001.

76. T. Kato and J.M.J. Fréchet. *Macromol. Symp.* 98:311, 1995.

77. J.P. Mathias, E.E. Simanek, and G.M. Whitesides. *J. Am. Chem. Soc.* 116:4326, 1994.

78. N. Kimizuka, T. Kawasaki, K. Hirata, and T. Kunitake, *J. Am. Chem. Soc.* 117:6360, 1995.

79. M. Kuramori, Y. Oishi, K. Suehiro, N. Kimizuka, T. Kawasaki, and T. Kunitake. *Rep. Prog. Polym. Phys. Jpn.* 39:401, 1996.

80. N. Kimizuka, S. Fujikawa, H. Kuwahara, T. Kunitake, A. March, and J.-M. Lehn. *J. Chem. Soc. Chem. Commun.* 2103, 1995.

81. F. Würthner, C. Thalacker, and A. Sautter. *Adv. Mater.* 11:754, 1999.

82. (a) J.H.K.K. Hirshberg. L. Brunsveld, A. Ramzi, J.A.J.M. Vekemans, R.P. Sijbesma, and E.W. Meijer. *Nature* 407:167, 2000. (b) A.P.H. Schenning, P. Jonkheijm, E. Peeters, and E.W. Meijer. *J. Am. Chem. Soc.* 123:409, 2001.

83. H. Fenniri, P. Mathivanan, K.L. Vidale, D.M. Sherman, K. Hallenga, K.V. Wood, and J.G. Stowell. *J. Am. Chem. Soc.* 123:3854, 2001.

84. (a) J.C. Nelson, J.G. Saven, J.S. Moore, and P.G. Wolynes. *Science* 277:1793, 1997. (b) S. Lahiri, J.L. Thompson, and J.S. Moore, *J. Am. Chem. Soc.* 122:11315, 2000.

85. R.S. Lokey and B. Iverson. *Nature* 375:303, 1995.

86. (a) Y. Takeda, N. Kimizuka, and T. Kunitake. *The 57th Divisional Meeting on Colloid and Interface Chemistry*, CSJ, 3C04b, 1997. (b) Y. Takeda. Master Thesis, Graduate School of Engineering, Kyushu University, 1996.

87. P. Lipkowski, A. Bielejewska, H. Kooijman, A.L. Spek, P. Timmerman, and D.N. Reinhoudt. *Chem. Commun.* 1311, 1999.

88. (a) J.C. Chambron, C.D. Buchecker, and J.-P. Sauvage. In: J.L. Awood, J.E.D. Davies, D.M. MacNicol, F. Vögtle, and J.-M. Lehn, Eds. *Comprehensive Supramolecular Chemistry*, Vol. 9, Oxford: Pergamon, 1996, p. 43. (b) J.-P. Sauvage. *Acc. Chem. Res.* 31:611, 1998. (c) P. Mobian, J.-M. Kern, and J.-P. Sauvage. *J. Am. Chem. Soc.* 125:2016, 2003.

89. (a) M. Fujita, J. Yazaki, and K. Ogura. *J. Am. Chem. Soc.* 112:5645, 1990. (b) M. Fujita. *Acc. Chem. Res.* 32:53, 1999. (c) S. Leininger, B. Olenyuk, and P. Stang. *Chem. Rev.* 100:853, 2000.

90. (a) E. Breuning, M. Ruben, J.-M. Lehn, F. Renz, Y. Garcia, V. Ksenofontov, P. Gütlich, E. Wegelius, and K. Rissanen. *Angew. Chem. Int. Ed.* 39:2504, 2000. (b) M. Ruben, E. Breuning, J.-M. Lehn, V. Ksenofontov, F. Renz, P. Gütlich, and G.B.M. Vaughan. *Chem. Eur. J.* 9:4422, 2003. (c) V.C. Lau, L.A. Berben, and J.R. Long. *J. Am. Chem. Soc.* 124:9042, 2002.

91. D. Gatteschi. *Adv. Mater.* 6:635, 1994.

92. J.M. Tour. *Acc. Chem. Res.* 33:791, 2000.

93. N. Kimizuka, N. Oda, and T. Kunitake, *Chem. Lett.* 27:695, 1998.

94. N. Kimizuka, S.H. Lee, and T. Kunitake. *Angew. Chem. Int. Ed.* 39:389, 2000.

95. N. Kimizuka, N. Oda, and T. Kunitake. *Inorg. Chem.* 12:2684, 2000.

96. S.H. Lee, Y. Hatanaka, and N. Kimizuka. *J. Nanosci.* 5&6:391, 2002.

97. S.H. Lee and N. Kimizuka. *Chem. Lett.* 31:1252, 2002.

98. N. Kimizuka. *Adv. Mater.* 12:1461, 2000.

99. (a) H. Okamoto and M. Yamashita. *Bull. Chem. Soc. Jpn.* 71:2023, 1988. (b) Y. Wada, T. Mitani, M. Yamashita, and T. Koda. *J. Phys. Soc. Jpn.* 54:3143, 1985.

100. K. Toriumi, Y. Wada, T. Mitani, S. Bandow, M. Yamashita, and Y. Fujii. *J. Am. Chem. Soc.* 111:2341, 1989.

101. B. Messer, J.H. Song, M. Huang, Y. Wu, F. Kim, and P. Yang. *Adv. Mater.* 12:1526, 2000.

102. D.G. Kurth, A. Meister, A.F. Thünemann, and G. Förster. *Langmuir* 19:4055, 2003.

103. S. Polarz, B. Smarsly, and M. Antonietti. *Chem. Phys. Chem.* 2:457, 2001.

104. N. Matsushita and A. Taira. *Synth. Metals* 102:1787, 1999.

105. J. Bremi, D. Brovelli, W. Caseri, G. Hähner, P. Smith, and T. Tervoort. *Chem. Mater.* 11:977, 1999.

106. W.R. Caseri, H.D. Chanzy, K. Feldman, M. Fontana, P. Smith, T.A. Tervoort, J.G.P. Goossens, E.W. Meijer, A.P.H.J. Schenning, I.P. Dolbnya, M.G. Debije, M.P. de Haas, J.M. Warman, A.M. van de Craats, R.H. Friend, H. Sirringhaus, and N. Stutzmann. *Adv. Mater.* 15:125, 2003.

107. N. Kimizuka, K. Yamada, and T. Kunitake. *Mol. Cryst. Liq. Cryst.* 342:103, 2000.

108. S.H. Lee and N. Kimizuka. *Proc. Natl. Acad. Sci.* 99:4922, 2002.

109. (a) G. Widawski, M. Rawiso, and B. François. *Nature* 369:387, 1994. (b) B. François, O. Pitois, and J. François. *Adv. Mater.* 7:1041, 1995.

110. (a) O. Karthaus, N. Maruyama, X. Cieren, M. Shimomura, H. Hasegawa, and T. Hashimoto. *Langmuir* 16:6071, 2000. (b) M. Shimomura. In: A. Chiferri, Ed. *Supramolecular Polymers*, New York: Marcel Dekker, 2000, p. 471.

111. L.V. Govor, I.A. Bashmakov, F.N. Kaputski, M. Pientka, and J. Parisi. *Macromol. Chem. Phys.* 201:2721, 2000.
112. T. Yonezawa, S. Onoue, and N. Kimizuka. *Adv. Mater.* 13:140, 2001.
113. H. Yabu, M. Tanaka, K. Ijiro, and M. Shimomura. *Langmuir* 19: 6297, 2003.
114. (a) O. Kahn and C.J. Martinez. *Science* 279:44, 1988. (b) O. Kahn. *Acc. Chem. Res.* 33:647, 2000.
115. N. Kimizuka and T. Shibata. *Polymer Preprints Jpn.* 49:3774, 2000.
116. F. Armand, C. Badoux, P. Bonville, A.R. Teixier, and O. Kahn. *Langmuir* 11:3467, 1995.
117. K. Kuroiwa, T. Shibata, A. Takada, N. Nemoto, and N. Kimizuka. *J. Am. Chem. Soc.* 126:2016, 2004.
118. (a) D.J. Pochan, J.P. Shneider, J. Kretsinger, B. Ozbas, K. Rajagopal, and L. Haines, *J. Am. Chem. Soc.* 125:11802, 2003. (b) E.R. Wright, A. McMillan, A. Cooer, R.P. Apkarian, and V.P. Conticello. *Adv. Funct. Mater.* 12:149, 2002. (c) S. Okabe, S. Sugihara, S. Aoshima, and M. Shibayama. *Macromolecules* 35:8139, 2002.
119. T. Welton. *Chem. Rev.* 99:2071, 1999.
120. N. Kimizuka and T. Nakashima. *Langmuir* 17:6759, 2001.
121. T. Nakashima. Ph.D. Thesis, Graduate School of Engineering, Kyushu University, 2003.
122. R.P. Swatloski, S.K. Spear, J.D. Holbrey, and R.D. Rogers. *J. Am. Chem. Soc.* 124:4974, 2002.
123. R.T. Carlin and J. Fuller. *Chem. Commun.* 1345, 1997.
124. T. Nakashima and N. Kimizuka. *Chem. Lett.* 31:1018, 2002.
125. A. Ikeda, K. Sonoda, M. Ayabe, S. Tamaru, T. Nakashima, N. Kimizuka, and S. Shinkai. *Chem. Lett.* 30:1154, 2001.
126. W. Kubo, T. Kitamura, K. Hanabusa, Y. Wada, and S. Yanagida. *Chem. Commun.* 374, 2002.
127. M.A. Firestone, J.A. Dzielawa, P. Zapol, L.A. Curtiss, S. Seifert, and M.L. Dietz. *Langmuir* 18:7258, 2002.
128. M. Yoshio, T. Mukai, K. Kanie, M. Yoshizawa, H. Ohno, and T. Kato. *Adv. Mater.* 14:351, 2002.
129. T. Nakashima and N. Kimizuka. *J. Am. Chem. Soc.* 125:6386, 2003.
130. N.C. Seeman. *Acc. Chem. Res.* 30:357, 1997.
131. C.M. Niemeyer. *Angew. Chem. Int. Ed.* 40:4128, 2001.
132. K. Matsuura, T. Yamashita, Y. Igami, and N. Kimizuka. *Chem. Commun.* 376, 2003.
133. J.M. Lehn *Science* 295:2400, 2002.

Chapter 14

Supramolecular Properties of Polymers for Plastic Electronics

Davide Comoretto

Contents

I. Introduction

There is currently a wide interest in the academic and industrial community both to reduce the dimensions of the operating optoelectronic devices and to functionalize their operations for specific selected targets. The achievement of these goals would allow several improved facilities such as a faster operation rate, the storage of an increased amount of information, the manipulation of objects at micro- or nanoscale, and the development of new and more specific functionalities. All these items belong to the realm of "Nanotechnology." It is difficult to provide an exhaustive definition of nanotechnology since the opportunity of nanoscale science and engineering is in progress both for the realization of new nanotools and for the growth of new nanostructured materials [1,2]. Examples of these activities are photo- and electron-lithographic techniques applied to bulk semiconductors [3], the development of semiconducting or metallic nanoparticles [4], molecular actuators [5], information storage media [6], molecular machines [7], molecular gates and switches [8], micro- or nano-electro mechanical structures (MEMS or NEMS) [9], molecular electronics [10], self-assembly [11], and photonic crystals [12].

In this chapter, we focus our attention to the topics of organic electronics. In particular, we will describe the opportunity for supramolecular polymer chemistry to improve the functionalities of materials in the fields of polymer electronics (plastic electronics). The development of conjugated polymers having semiconducting characteristics useful for electronics and optoelectronics devices manufacturing are well known, and new products are coming to the market [13]. However, some deficient properties of the present generation of polymers when used in working devices, indicate that new strategies to improve materials are required. Those based on the control of the supramolecular structure seem to be very promising for the improvement of carrier mobility, luminescence efficiency, and device stability.

The aim of this chapter is to introduce the reader into the field of polymer electronics by discussing both fundamental physical properties and the operating principle of simple polymeric devices. For each addressed topic, we will review the research activity reported in literature in order to understand the fundamental science underneath problems and to highlight possible solutions offered by controlling the SP structure. The term "supramolecular polymer" (SP) will be used to indicate either polymers obtained by self-assembling different units through secondary bonds (main-chain SPs, e.g., Class A, Chapter 2) or covalent polymer with supramolecular organization best suitable for the development of polymer electronics (e.g., Class C, Chapter 2). Engineered SPs (Class D, Chapter 2) will also be discussed.

The organization of the chapter is the following. In the Fundamental Physical Properties (Section II), the importance of supramolecular organization of conjugated polymers will be discussed, addressing important items such as polymer chains *orientation* (Section II.A) and *tuning* interchain interactions (II.B), which allow a deeper understanding of the electronics and photophysical properties of the material. In the section Plastic Electronics (Section III), the role of the supramolecular structure on transport properties of metallic and semiconducting polymers (Section III.A), sensors (Section III.B), organic light emitting diodes (OLED, Section III.C), photovoltaic cells (Section III.D), and microdevices (Section III.E) will be reviewed and discussed. In the final section (Section IV), possible perspectives for future work in organic optoelectronics and nanophotonics will be reported.

II. FUNDAMENTAL PHYSICAL PROPERTIES

Conventional plastic materials are electrical insulators. This is due to the presence of saturated σ bonds, which can be excited only for energies above 7 eV [14]. In polyconjugated systems instead, an alternated pattern of single and double bonds is created (Figure 1) thus lowering the excitation energy to $1.5 - 4$ eV, values typical for the semiconducting state. It has been shown that the fully conjugated metallic system is unstable (Peierls dimerization [15]). Then, bond alternation spontaneously takes place stabilizing the system in the semiconducting state. However, semiconducting (conjugated) polymers can be doped to achieve the metallic state by a solid-state redox reaction. Very high conductivities have been obtained for *trans* polyacetylene (10^5 S m^{-1}) [16], but this material was unstable in air due to oxidation processes. The dream of plastics with electrical conductivities comparable to those of copper soon faded, but the preparation of stable conducting polymers with conductivities up to few hundreds of S cm^{-1} is now currently achieved.

Even though polymers are very long chains, π-electrons in conjugated polymers are seldom delocalized along the whole macromolecule but only along discrete segments (conjugation lengths) that are separated each other by chain twisting or defects breaking the electron delocalization. For some aspects, they can be considered similar to molecular solids, whose basic electronic properties

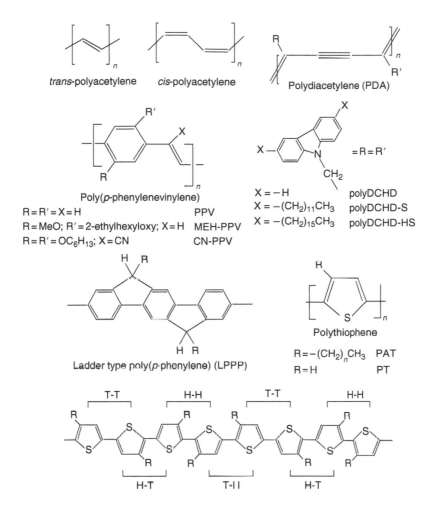

Figure 1 Chemical structure of selected conjugated polymers.

were set in the '60s and '70s. When large delocalization is achieved, conjugated polymers can be considered as natural quantum wires, having π-electron density delocalized along the backbone in a quasi one-dimensional geometry, with lateral confinement of about 0.5 nm. This peculiarity yields a distinctive character to the electronic structure and dynamics of conjugated polymers, that are somewhat in between large organic molecules and low dimensional semiconductors.

Differences between conjugated and nonconjugated polymers are related both to electronic and macromolecular properties. Nonconjugated polymers are constituted by a large number of segments having mutual rotation thus being easily processable from melt or solution. On the contrary, conjugated polymers are stiff and usually not soluble and processable without adding proper side chains or dopants. Moreover, the total chain length and molecular weight distribution, important to confer plastic properties, are strongly dependent on the synthetic method used.

Side groups may play an important role by tuning electronic states of conjugated polymers. The substitution with electron donating or withdrawing groups, for instance methoxy or cyano groups, modifies the ionization potential and the electron affinity of the polymers thus affecting the absorption/emission transition energy and the possibility to inject electrons and holes in the material. All these properties mainly reflect intramolecular characteristics of the polymeric chain. Materials investigated and used as bulk semiconductors usually display random chain assembly. In this state, polymeric chains can be very close to each other and intermolecular interactions and aggregation phenomena may occur, accompanied with different phases or morphologies that modify the electronic and chemical properties of isolated chains. In this context, the control of the supramolecular structure of the polymer plays a major role in tuning the macromolecular functionalities.

Another problem when dealing with macromolecular samples is related to the anisotropy of the physical properties. The synthesis of conjugated polymers provides samples with macromolecules randomly distributed thus masking the intrinsic anisotropy of the intrachain electronic properties. Inside the chain, π-electrons are highly polarizable along the backbone but much less (orders of magnitude) perpendicularly to that. This implies for oriented samples a strong anisotropy of the optical response. Moreover, even a weak interchain interaction could provide some intermolecular polarizability. The identification of these different contributions is very important and requires the investigation of highly oriented conjugated polymers in order to disentangle intrinsic properties with respect to those due to chain misalignment.

The techniques used to orient conjugated polymers may modify their supramolecular structure and then affect their electronic properties. The effect of chain orientation on the study of the optical properties of the systems will be discussed in Section II.A. Interchain interactions usually do not modify too much linear optical properties, but strongly affect photophysical properties of polyconjugated materials. We will discuss two examples in Section II.B. Polydiacetylenes (PDAs) (Section II.B.1), where the choice of side groups having different length allows tuning the nature of the photoexcited states, and regiorandom and regioregular polythiophene (Section II.B.2), where the intramolecular linkage modifies the supramolecular packing of the chains, thus drastically changing their optical and transport properties.

A. Electronic Properties: Chain Orientation

The orientation of conjugated polymers is a very difficult task for which several approaches have been used. The simpler way would be to grow single crystals. Single crystals of conjugated polymers are, however, available only for few selected PDAs due to their peculiar polymerization mechanism (topochemical polymerization, Figure 2). Small conjugated molecules, such as diacetylene monomers, may spontaneously form large area crystals [17]. These diacetylene crystals can then be polymerized by providing enough energy (by heating or by exposure to UV or γ-ray) to overcome the repulsion between the first and fourth carbon atoms of the adjacent monomers (Figure 2). This process is favored when taking place with a minimum movement of the monomer molecules. In this case, the polymer exhibits an extended conjugation (Figure 2) and retains the order of the monomeric structure, thus forming polymeric single crystals. Optical measurements on PDAs single crystals show a very high degree of anisotropy [18]. Recently, this polymerization mechanism has been refined to obtain isolated PDA chains embedded in their monomer single crystal with anisotropy of the $\pi-\pi^*$ transition exceeding 50. These samples allow a detailed study of the single chain intrinsic anisotropy [19].

A different approach to prepare oriented conjugated polymers concerns the use of vacuum evaporation on suitable substrates able to induce epitaxy. The crucial choice of the substrate depends on the molecule to be oriented. For PDAs [20,21] and oligothiophenes [22], very interesting results have been obtained when potassium acid phthalate single crystals are used as a substrate. Interesting results are also obtained when the substrate is covered by a pre-oriented friction-transferred film of Teflon [23]. This technique allows to obtain samples with high absorption dichroism (i.e., high orientation) only for small molecules since the evaporation process cannot be applied to systems having very high molar mass. The peculiar polymerization mechanism of PDAs again makes them the unique conjugated polymer for which this technique may be useful. Recently, vacuum evaporation has been used to orient main-chain SPs, such as hexabenzocoronene discotic liquid crystals (LC) [24]. It was possible not only to obtain orientation of the discotic pillars but also to induce different ordering along the columnar axis, with the aromatic cores either perpendicular or 45° tilted to the substrate surface (see below). The different ordering of the molecular disks strongly modifies the optical properties and the anisotropy of the system [24].

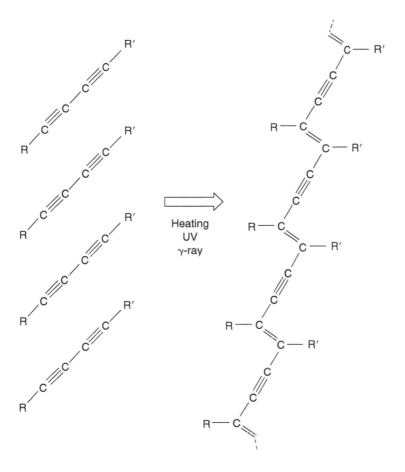

Figure 2 Topochemical polymorization of diacetylenes.

Even though interesting results have been obtained for PDAs, the technique here reported cannot be extrapolated to different conjugated (covalent) polymers. To orient other families of conjugated polymers different methods have been used, for example, Langmuir–Blodgett techniques [25] or shearing in solutions [26] or in liquid-crystalline solutions [27], but with results not suitable for the study of the anisotropical electronic properties of the material. A technique suitable to orient several polymer families is rubbing [28,29]. The degree of backbone orientation achieved with this technique is valuable to obtain polarized emission in light emitting diodes (LEDs) [30], but it is still not enough for detailed fundamental studies on the anisotropical electronic or optical response of the material due to the presence of a large amount of misoriented polymeric chains [31,32]. However, there is room to improve results so far obtained with the rubbing technique. As a matter of fact, some progress has been recently reported by carefully matching the interaction between molecules (diacetylenes and phenylene-ethynilene) and substrates by combining rubbing with the evaporation on preoriented Teflon layers [33].

A technique suitable to provide very good orientation for almost all classes of conjugated polymers is tensile drawing (stretching). When directly applied to insoluble conjugated polymers, some good results have been obtained in the case of Shirakawa's polyacetylene [34]. A better approach is to operate, when possible, on the nonconjugated precursors that are the intermediate species for the synthesis of polyacetylene (Durham–Graz route) [35] or poly(p-phenylenevinylene) (PPV) [36]. In the latter case, very high degree of orientation has been obtained [37]. A variant to this technique can be adopted when dealing with soluble conjugated polymers. In this case, the conjugated polymer is blended with ultra-high molecular weight (UHMW) polyethylene (PE) and subsequently stretched. The orientation of PE matrix induces the orientation of the conjugated polymer guest, thus providing highly oriented samples [38]. The high degree of orientation is evidenced by the large dichroic ratio of the lowest optical absorption and of the photoluminescence emission. The main drawback to this technique is related to light scattering in the blue and UV spectral regions caused by the matrix [39]. However, nice improvements have been recently achieved by covering these films with PE oligomers thus filling the grooves left by the orientation process. This allows refractive index matching, which reduces light scattering and allows the study of the optical anisotropy even in the UV spectral region [40].

Once the covalent polymer chains have been oriented, their supramolecular structure is modified. For instance, in PPV oriented by precursor stretching, it is possible to observe oriented crystallites assembled with hexagonal symmetry, embedded in amorphous regions [40]. Within the crystallites, PPV chains are packed in a herringbone structure [41] while in the amorphous regions random arrangement prevails. These samples provide a unique example of the physical information that can be extracted from optical measurements on highly oriented conjugated polymers. In spite of the impressive number of papers published on PPV, several fundamental issues are still debated. In particular, the role of the electron correlation, Coulomb interactions, interchain interactions, and solid-state effects in general, the impact of chemical substitution, and the role of conjugation chain-ends are not fully understood. The assignment of the optical transitions in the electronic absorption spectrum of this polymer was debated until a few years ago [42,43]. This controversy was due both to limitations of theoretical models adopted to assign the electronic spectra and to the lack of high quality samples, which could allow a deep investigation of the anisotropical electronic properties. Recently, polarized optical spectra in highly stretch-oriented PPV [37] provided a deeper insight into the nature of the PPV optical transitions and have been used as a critical test for the validity of theoretical models [29,44,45]. Figure 3 shows the polarized transmittance and near-normal-incidence reflectance spectra of highly stretch-oriented PPV. In the near-infrared, a progression of well-resolved interference fringes is observed for both kinds of measurements. The presence of the fringes in the spectra is indicative of the high optical quality of oriented films. The different paths of the fringes detected for the two polarizations point out the anisotropy of the refractive index and then the high degree of orientation. Above 2 eV, when the material is no more transparent, several

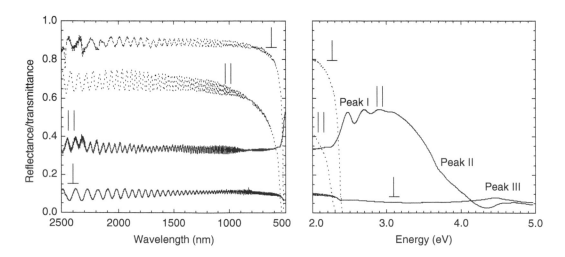

Figure 3   Polarized transmittance (dashed lines) and reflectance (full lines) spectra of highly stretch-oriented PPV free standing films. Elongation ratio 3. Spectra are plotted in nanometers (Left panel) to better display the interference patterns in the near infrared or electron volts (right panel) to show the main electronic transition in the UV-Vis.

electronic transitions are detected in the reflectance spectra. For the parallel component, a strong signal associated with the $\pi-\pi^*$ purely electronic transition of the lowest absorption band is observed at 2.48 eV (peak I), followed by a well-resolved vibronic progression with peaks at 2.70 and 2.95 eV, and by a shoulder at 3.10 eV (for a detailed discussion of the vibronic progression of PPV see Ref. 37). The change in the slope of the spectrum around 3.76 eV indicates the presence of additional transitions (peak II). In the high energy side of the spectrum, a low intensity peak is observed at 4.71 eV accompanied by a shoulder at 4.56 eV (peak III). The lowest optical transition and its vibronic satellites cannot be detected in the perpendicular component of the reflectance spectrum. For this spectrum a transition is revealed at 4.47 eV in the UV spectral region. A fourth additional transition (peak IV), whose polarization is at present not fully characterized, has been also identified at higher energies [37]. From the data reported in Figure 3 and from spectroscopic ellipsometry measurements, the anisotropical complex optical constants (real and imaginary part of the dielectric constant) of oriented PPV have been determined [37,46]. This determination is particularly important from a theoretical point of view because the imaginary part of the dielectric constant is the optical function directly related to the electronic structure of the material [46].

From these data and in particular from the study of the polarization of the optical transition, the assignment of the optical spectra has been performed on the basis quantum chemical calculation extended to include the electronic correlation [29,37]. We summarize here the main results obtained. Peak I (parallel polarized) originates from $\pi-\pi^*$ transitions between delocalized (d) levels. Peak III originates from l $\rightarrow$ d*/d $\rightarrow$ l* (l, localized; d, delocalized) transitions and has a dominant polarization perpendicular to the chain axis while peak IV results from l $\rightarrow$ l* excitations and is polarized parallel to the chain axis. Concerning peak II, quantum chemical calculations suggest that the shoulder observed in the parallel reflectance spectrum at about 3.7 eV (Figure 3) corresponds to an optical transition between delocalized levels (d $\rightarrow$ d*), which is induced by finite-size effects (borders of the conjugated segment) and is polarized along the chain axis, in full agreement with the experimental data. Since the intensity of this peak is expected to decrease when the conjugation chain length is increased [29,37], its weak intensity suggests a reduced contribution of the chain ends and then the presence of relatively long conjugated segments. It is interesting to compare the properties of PPV with those of its alkoxy-substituted derivative (poly(2-methoxy-5-[2'-ethylhexyloxy]-p-phenylene vinylene) [MEH-PPV] see Figure 1) which has a very similar polarized absorption spectrum [39] but a different assignment, in particular regarding peak II. MEH-PPV is characterized by the appearance of new absorption peaks around 3.7 eV again mostly described by l $\rightarrow$ d*/d $\rightarrow$ l* excitations, due to the breaking of the charge conjugation symmetry upon attachment of electroactive substituents to the backbone [29,37]. The main effect of the alkoxy groups is to decrease the separation between the energies of the lowest l $\rightarrow$ d*/d $\rightarrow$ l* and d $\rightarrow$ d* excitations thus enhancing the strength of the interaction between the two types of excitation. This allows for an efficient intensity borrowing from the lowest absorption band that contributes to the intensity of peak II, which is also expected to decrease for growing chain length [29,37]. Quantum–chemical calculations show that the transition dipole moment of these new peaks is governed by their weak d $\rightarrow$ d* character (i.e., the contributions of the l $\rightarrow$ d*/d $\rightarrow$ l* excitations tend to cancel each other), thus yielding a polarization along the chain axis [29,37] in full agreement with the experimental data. It is also interesting to note that when PPV is substituted with alkyl derivatives (i.e., with groups weakly electronically active) its spectrum resembles that of PPV instead of MEH-PPV one [39] thus clearly confirming the role of electrically active substituents to tune the electronic states of the material.

B. Photophysics: Chain Interaction

We briefly describe here the main photophysical concepts used in this chapter for the discussion and interpretation of the photoinduced absorption $(-\Delta T/T)$ spectra, that is, the modification of the absorption spectrum upon photoexcitation (absorption spectrum of the photoexcited state).

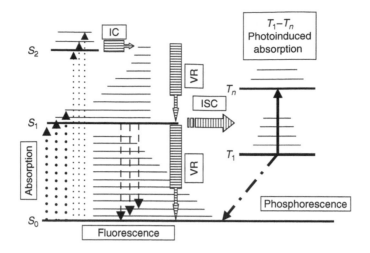

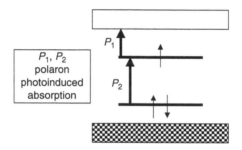

Figure 4   Upper panel: basic photophysical scenario for an isolated molecule. Dotted (dashed) arrows show vibronic absorption (fluorescence); shaded large arrows internal conversion (IC), vibrational relaxation (VR), and intersystem crossing (ISC) processes. Solid black arrow shows photoinduced absorption transitions of triplet excitons taking place in the microsecond time domain. Dot-dashed arrow indicates phosphorescence. Lower panel: schematic representation of negative polaron levels and their spin population within the monoelectronic scheme. Black arrows show photoinduced absorption transitions (P_1 and P_2).

The photophysics of isolated molecules/macromolecules can be summarized as follows (see Figure 4, upper panel) [46–48]. The photoexcitation populates a generic electronic excited state (usually a vibronic state), which has a very short lifetime. Once the lowest singlet state, S_1, is populated, its deactivation occurs radiatively, by photon emission (fluorescence), or radiationless, by internal conversion (IC) vibrational relaxation (VR) or intersystem crossing (ISC) to the triplet manifold. IC is due to nonadiabatic coupling between different isoenergetic vibronic levels and is a purely intramolecular process. VR redistributes excess vibrational energy from optically coupled modes to other molecular modes and finally to the environment. ISC is due to spin-orbit coupling that mixes the spin character of the molecular wavefunction yielding a nonnegligible probability of radiationless energy transfer from the singlet to the triplet manifold. Singlet fission into spin-correlated triplet pair may also take place. It essentially circumvents the spin selection rule and allows generating triplets on the ultrashort timescale. This process is remarkably different with respect to ISC, and triplets this way generated have different characteristics. In particular, singlet and triplet features appear on the same timescale. Such triplets are correlated in pair, and may decay faster then conventional ones, due to mutual "geminate" recombination [46,49]. The populated triplet state is long-lived (μsec to msec) due to the reduced probability of the $T_1 - S_0$ transitions and has its own spectral fingerprint in the photoinduced absorption spectra ($T_1 - T_n$). However, in special cases (see below) radiative de-excitation of triplets (phosphorescence) may occur. Charged states are seldom observed since they do not play a major role in the primary excitation and relaxation processes of isolated molecules.

The photophysics in the condensed phase is much more involved with respect to that of isolated molecules. In fact, there are many open debates on the subject and the nature of the primary excitations cannot be easily assigned. Due to new mechanisms not yet elucidated or identified, singlet, triplet, and charged states are all possible candidates for being observed at short times. Defects and impurities may mask the identification of the molecular states, which can also be substantially modified by intermolecular interactions. Intermolecular deactivation paths may substantially affect IC in the condensed phase. In general, it is not true that the all excited state population reaches S_1. Charge separation is one of the possible channels where energy can go, giving rise to a branching with neutral states. The process seems more favorable from higher lying states. In this case charge-transfer states can be intermediate before complete dissociation. In this case a polaron (radical ion) is formed on the conjugated molecule having its specific photoinduced absorption signature (Figure 4, lower panel).

The presence of molecular aggregates in samples having complex morphology can play an important role in emission, because they can have large radiative rates and behave as energy sinks in the excited material [50]. What seems assessed is that the optimal efficiency for charge photogeneration can be obtained only in mixed systems, donor–acceptor like. Particularly relevant for photovoltaic applications are the composites conjugated polymer–fullerene derivatives, which shows efficient and metastable photo induced electron transfer from conjugated polymer to fullerene acceptor [46]. We notice that the intermolecular interactions are strongly dependent on the polymer supramolecular structure. For this reason, the control of the supramolecular structure provides the way to investigate the fundamental photophysical processes as it will be shown for PDAs and polyalkylthiophenes (PATs).

1. Polydiacetylene

The study of symmetrical PDAs having carbazolyl groups bonded through one methylene spacer to the polymeric backbone (see, for instance, polyDCHD in Figure 1) is very promising since these systems seem to be candidate for fabrication of nonlinear optical devices [50]. However, this form of PDA is nonprocessable, rendering it unsuitable in thin film technology. In order to obtain thin amorphous films prepared via solutions processing, the order achieved in single crystals [18] or epitaxial films [20,21] have to be left behind. To this end, by inserting long aliphatic chains on the 3 and 6 positions of the carbazolyl moiety, soluble substituted polymers have been obtained (polyDCHD-S for dodecyl chains and polyDCHD-HS for hexadecyl chains, Figure 1) [52,53].

The presence of the long alkyl chains on the carbazole rings changes the supramolecular structure of the polymer thus affecting the nature of photoexcitations. In particular it was shown that the presence of long aliphatic chains only slightly affects the electronic absorption and Raman spectra while it dominates the kinetics of photoexcitations. It is indeed the interchain separation among different polydiacetylenic backbones, which is responsible for the photogeneration of charged or neutral states [21,54]. The results of such experiments performed on the blue form of polyDCHD, polyDCHD-S, and polyDCHD-HS are reported in Figure 5(a), upper panel. For these polymers, bands associated with neutral triplet excitons ($T_1 - T_n$ transition) are found at 1.26, 1.45, and 1.45 eV in polyDCHD [55], polyDCHD-S [54], and polyDCHD-HS [53], respectively. In addition, long-lived charged states are photogenerated in both polyDCHD and polyDCHD-S as evidenced by the presence in their spectra of the 0.82 and 0.76 eV peaks and their vibronic replica at 0.96 and 0.88 eV [55]. Note however the different photogeneration yield of the triplet relative to charged species on going from polyDCHD to its alkyl-substituted polymers (Figure 5(a), lower panel). The $-\Delta T/T$ spectra show that the photogeneration of charged states is efficient in polyDCHD, almost inhibited in blue polyDCHD-S, and completely inhibited in polyDCHD-HS. These results suggest the following interpretation: the photogeneration of charged states is strongly reduced in substituted

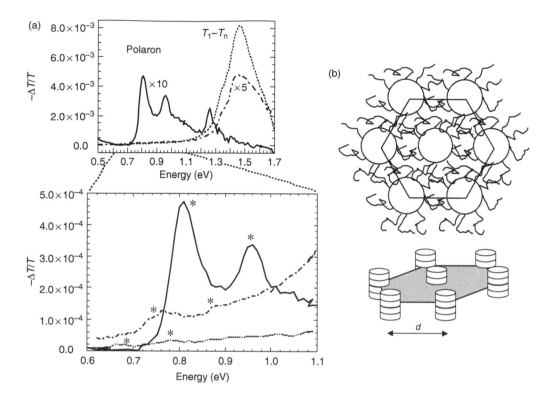

Figure 5 (a) Photoinduced absorption spectra of polyDCHD (full line), polyDCHD-S (dot dashed line), and polyDCHD-HS (dashed line). In the upper panel, the spectral signatures of triplet excitons and charged states are reported. The latter are highlighted (asterisk) in the lower panel thus showing their progressive intensity reduction on going from the polyDCHD to its substituted derivatives having larger interbackbone separations. (b) Schematic representation of the supramolecular hexagonal columnar structure of polyDCHD-S ($d = 32.5$ Å) and polyDCHD-HS ($d = 38.4$ Å). The small cylinders represent the repeating units of the covalent polymer.

polymers relative to polyDCHD because of the larger interchain distance, due to the insertion of the long alkyl chains on the carbazolyl rings, which separates the conjugated backbones. It is well known in fact that formation of charged excitations is favored by the interchain coupling [42]. On the contrary, triplet excitons can be formed along the conjugated backbone (intramolecular excitation) being almost not affected by the intermolecular interaction. This result is also confirmed by x-ray diffraction studies carried out on the red powder of both polyDCHD-S and polyDCHD-HS that indicate the formation of supramolecular hexagonal arrays of long polydiacetylenic columns (Figure 5(b)) whose separation is 32.5 and 38.4 Å, respectively [52]. These large intermolecular distances are indicative of the de-correlation of the PDA backbones that are kept together by the interdigitation of the long alkyl groups.

2. *Polyalkylthiophene*

Another remarkable example of conjugated polymer exhibiting a supramolecular structure deeply affecting the physical and electronic properties of the material is provided by the regiorandom and regioregular polythiophene derivatives.

Polythiophene is insoluble. To increase the conformational entropy of the chain and then promote solubility, it is useful to insert long chains in the 3 or 4 position of the thiophene rings [56]. The presence of substituents modifies the symmetry of the rings (now a head and a tail can be recognized). Three relative orientations are available when thiophene rings are linked in the 2 and 5 positions

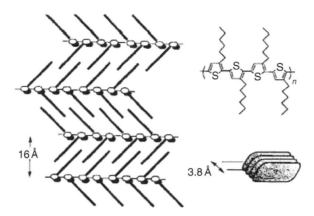

Figure 6 Supramolecular structure of regioregular polyalkylthiophenes. Reprinted with permission from R.D. McCullough, *Adv. Mater.* 10, 93 (1998), Figure 1, copyright Wiley-VCH.

thus playing an important role on the macromolecular and electronic properties of the material. According to the standard synthetic route [57], the rings are randomly joined thus creating a strong steric hindrance when the head-to-head (HH) or tail-to-tail (TT) linkages are achieved (Figure 1). Due to the low planarity of the backbone (thiophene rings have a twist angle of about 40° [58]), the conjugation length for this material is strongly reduced. The enhanced probability of chain bending favors a coil-like organization. Under these conditions, the photoexcited states in the polymer, both neutral or charged, are strongly confined to the conjugated segments that have been created thus having low probability to migrate along the chain. Since singlet excitons cannot migrate on different segments of the macromolecule, they cannot reach defects where their radiative recombination is quenched thus increasing the polymer emission efficiency. However, the performance of electronic devices, in particular transistors, is strongly affected by the mobility of the charge carriers. Since these species are strongly confined, we cannot expect that their mobility will be very high.

Very different is instead the case of regioregular PATs. This linkage can be obtained through polymerization of the dialkyl-bithiophene monomer units [59], that is, a symmetrical monomer, which cannot give rise to HH or TT coupling. The alternative McCulloch [60] or Reike [61] routes rely upon selective bromination of the monomer or upon finely divided zinc associated with selective catalysis to obtain regular polymers having HT or TH enchainment (Figure 1). Even though these routes are not problem free (i.e., metal contamination), they reduce monomer steric hindrance thus increasing the planarity of the conjugated backbone (twist angle 6–9° [58,62]). This fact strongly affects the supramolecular structure of the polymer. The planar conjugated chains have a supramolecular two-dimensional lamellar structure (Figure 6) [60,61] with reduced interchain separation (3.8 Å). The electronic and photophysical properties of the material are correspondingly modified. The mobility of positive carriers increases with respect to typical values found for regiorandom polymers, (from 10^{-5} cm^2 V^{-1} sec^{-1} to 0.1 cm^2 V^{-1} sec^{-1}) [63–65]. The implications for electronic devices will be discussed later in this chapter. As reported in Figure 7, the increased interchain interactions also affect the optical properties. The broad absorption spectrum of regiorandom polymers due to the distribution of conjugation length typical of the coil-like form (Figure 7(a)) is modified for the regioregular lamellar polymer (Figure 7(b)). The absorption spectrum is red shifted, indicating an increase of the average conjugation length. Moreover, the spectrum is better resolved showing its vibronic structure due to the reduced distribution of conjugation length. Similar results are observed for luminescence spectra, with a reduction of its quantum efficiency of more than one order of magnitude for the regioregular polymer [66,68]. The role of the interchain interaction will be further discussed later in connection with the efficiency of LEDs. It is also very interesting to note that the photophysical processes in the same material but with different supramolecular organization are deeply modified. In regiorandom PATs or in regioregular isolated PAT chains, the typical single chain one-dimensional

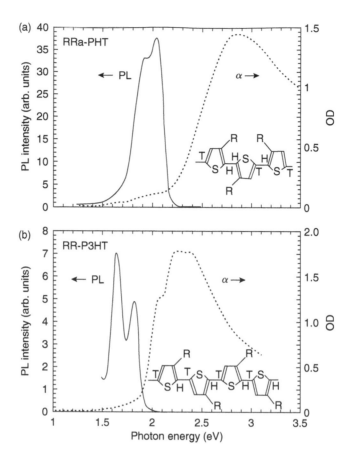

Figure 7 Absorption and photoluminescence spectra of regio-randoam (a) and regio-regular (b) PAT films. Figure reprinted from O.J. Korovyanko, R. Osterbacka, X.M. Jiang, and Z.V. Vardeny, R.A.J. Janssen, *Phys. Rev.* B 64, 235122 (2001). Copyright 2001 by the American Physical Society.

behavior is observed. Charges added to the conjugated chain form spin $\frac{1}{2}$ polarons with associated bond dimerization. Two new localized electronic states appear in the HOMO–LUMO gap of the neutral molecule (Figure 4, lower panel) and new infrared active vibrational modes gain intensity from the Raman-active ones [67]. When neutral excitations are photogenerated in these systems, fluorescence occurs accompanied with ISC from the singlet to the triplet manifold thus creating an excitation having its own photoinduced absorption transition $(T_1 - T_n)$ (Figure 4, upper panel) in addition to those of polarons. In regioregular PAT, due to the increased interchain interactions, both excitons and polarons are delocalized over several conjugated chains thus having a two-dimensional character with reduced electron–phonon coupling with respect to the one-dimensional systems. The spectral signature of the two-dimensional polarons are remarkably different with respect to those observed in the one-dimensional ones for regiorandom polymers [67] and their mobility is strongly increased [68].

III.   PLASTIC ELECTRONICS: THE ROLE OF SUPRAMOLECULAR STRUCTURE

We now discuss the role of supramolecular structure for more specific applications. In particular, we focus our attention to transport properties (both for metallic and semiconducting polymers), sensors, LEDs, transistors, photovoltaic cells, and finally, to the integration into microdevices.

A.   Conducting and Semiconducting Polymers: Transport Properties

The idea to create a one-dimensional metal dates back to the end of the '60s [15,69]. As discussed before, the one-dimensional system spontaneously undergoes a dimerization process, which turns the system from metallic to insulator. Nevertheless, studies on the conductivity of different one-dimensional systems have been reported. Interest in the '70s was attracted by charge transfer salts having donor (e.g., tetrathiafulvalene) and acceptor (e.g., tetracyanoquinodimethane) molecules stacked along a one-dimensional path [70]. This complex can be thought as a main-chain SP with electrons delocalized along the stacking direction. Doping increases the conductivity up to a superconducting state. In the same years, another superconducting system was found to be the polymer $(SN)_x$ [71]. This fact opened new perspectives for the synthesis of carbon-based polymeric conductors. The synthesis of novel forms of polyacetylenes and their doping to metallic state [72] was the boost for the field of conducting/semiconducting polymers. This was the main motivation for awarding the year 2000 Nobel Prize in Chemistry to A.J. Heeger, A. MacDiarmid, and H. Shirakawa [73].

Doping of conjugated polymers is a process different than the corresponding one for inorganic semiconductors. For these latter, substitution of an atom with one having higher (lower) valence adds an extra electron (hole) to the system. In the case of silicon, substitution with phosphorous (boron) produces a negatively (positively) doped semiconductor. In conjugated polymers the doping process is not substitutional but more similar to a redox reaction where a dopant agent becoming closer to the π-electron system provides or withdraws an electron. In this case, the extra electron (hole), self-traps in a novel excitation (soliton, polaron, or bipolaron) depending on the topological nature of the backbone [74], and then can move along the chain that provides a one-dimensional path for its delocalization. The material is now metallic. An extensive description of the metallic properties of conjugated polymers, their anisotropy in oriented samples, and of the metal–insulator transition is beyond the aim of this chapter. These items are discussed in detail in Ref. 75. The transport properties of doped polymers strongly depend on the nature of the dopant molecule, the morphology, and the polymer supramolecular structure. To obtain bulk conductivity charges must not only travel along the backbone but also hop on different chains to continue a macroscopic drift. The interchain hopping is governed by intermolecular interactions that ultimately depend on the supramolecular structure of the polymer. As was discussed before, the synthesis of regioregular PAT [60] provides a reduced interring twist angle inducing a flatter backbone with respect to the regiorandom polymers. The flattening of the backbone can be also obtained by alkoxy substitution that has, in addition to self-electron-donating properties, a reduced van der Waals radius, and consequently a reduced steric hindrance relative to alkyl chains [75]. This aspect has been further developed with the synthesis of poly(3,4-ethylenedioxythiophene) (PEDOT, Figure 8) which gathers conductivity properties (σ up to $300 \, S \, cm^{-1}$ [76]) with reduced optical gap. PEDOT processability is not easy, being the polymer unsoluble. However, the polymerization of ethylenedioxythiophene in aqueous polyelectrolites like polystyrene sulfonic acid (PSS, Figure 8) provides a PEDOT:PSS solution that can be easily processed resulting, after drying, in highly conducting films, that are transparent, mechanically resistant, and insoluble in any common solvent. These properties have been used to produce commercial products for antistatic applications [77], corrosion protection [78], and for the production of hole-injection transparent electrodes for LED devices [79] and batteries [80]. More recently, it was also used as conducting ink in transistors prepared by ink-jet printers [81].

Even though doped conjugated polymers are powerful for several applications, they are still limited to relatively poor mechanical properties. Doped conjugated polymers are therefore blended with two or three tougher polymers in a multiphase system. A matrix of tough insulating polymer confines the doped conjugated one, thus modifying hits supramolecular structure. Conductivity is achieved by percolating the conducting polymer in the insulating hosting matrix. The conductivity of these films depends on the polymer volume fraction at percolation threshold [82]. Even though this volume fraction can be reduced, thus increasing the film conductivity, two drawbacks still remain,

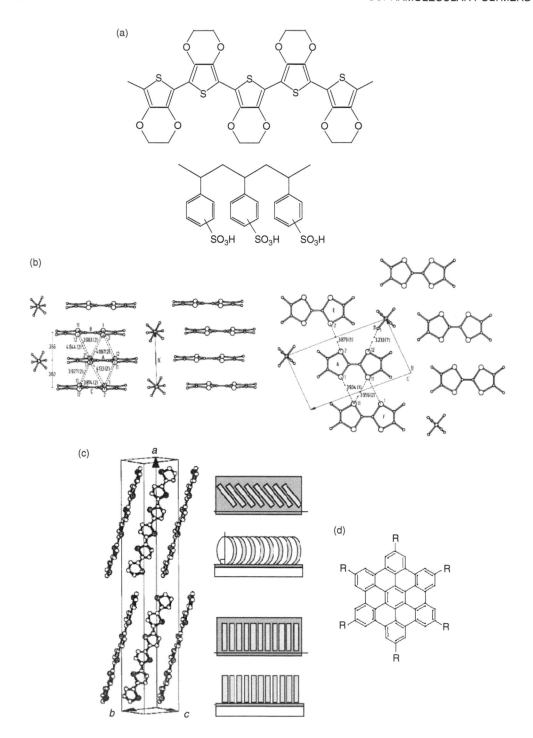

Figure 8   Examples of doped and semiconducting polymers (both CSPs and SPs) with good transport properties. (a) Chemical structure of conductive PEDOT:PPS. (b) Crystal structure of the organic superconductor tetramethyl-tetraselenafulvalenium hexafluorophosphate. (c) Crystalline structure of α-sexithienyl. (d) Hexabenzocoronene oriented discotic liquid crystals. Reprinted with permission from: (b) N. Thorup, G. Rindorf, H. Soling, and K. Bechgaard, *Acta Cryst.* B37, 1236 (1981) Figure 4 and 5, copyright International Union of Crystallography; (c) M. Muccini, E. Lunedei, A. Bree, G. Horowitz, F. Garnier, and C. Taliani, *J. Chem. Phys.* 108, 7327 (1998), Figure 1, copyright 1998, American Institute of Physics; (d) J. Piris, M.G. Debije, N. Stutzmann, A.M. van de Craats, M.D. Watson, K. Mullen, and J.M. Warman, *Adv. Mater.* 15, 1736 (2003), Figure 1, copyright Wiley-VCH.

namely the reduced thickness of the films and its morphological instability for temperatures above the glass transition of the matrix. To solve this problem a three-phase system has been proposed [82]. A high volume fraction nonpercolating polymer matrix is blended with a low volume fraction of a polymer–doped conjugated polymer blend in a double percolating system. It can be shown that the process allows a further reduction of the conducting polymer volume fraction thus providing a remarkable increase of conductivity (three/four orders of magnitude) [82].

In conducting polymers, the extra carriers added upon doping are able to drift under an applied electrical field. In semiconducting polymers, no carriers are available except those thermally excited across the gap. However, negative (positive) carriers can be injected into the material by metallic contacts when the barrier between the metal work function and the LUMO (HOMO) molecular levels is overcome. Then, the injected carriers can move inside the semiconductor if a bias field is applied. Injection of carriers and their transport is a fundamental issue for all electronic devices and transistors in particular. In the following, main transport properties of organic semiconductors (both small molecules and polymers-based) used as active materials in transistors will be reviewed.

An example of main-chain SP is provided by stacked molecular solids. In these systems, π-conjugated molecules are stacked along a direction. π-orbitals of different molecules can then interact thus delocalizing electrons along the stacking direction. There are several examples of these systems, such as charge transfer salts [70] (Figure 8(b)), conjugated oligomers and acenes crystals [83] (Figure 8(c)), and conjugated LCs [84,85] and supramolecular columns [86,87] (Figure 8(d)). Above systems are very different from each other even though they possess intermolecular delocalized π-electrons that increase the intermolecular charge mobility. The details of the relative position of the adjacent molecules strongly affect the electronic properties of the assembly. For the same reason, conjugated LCs provide a strongly anisotropical mobility both in the smectic phase or, more interesting for application to molecular electronics, in the discotic phase. For discotic LCs, for instance those based on hexacoronene, a columnar self-assembled SPs (pillars) is obtained [24,84,85]. Each discotic moiety is sometimes formed by monodendritic building blocks [86]. To further increase the order in the mesophase assembly, it is possible to adopt the epitaxy techniques previously described to orient polymers (deposition on friction-transferred-oriented Teflon layers) [20,23,83] (Figure 8(d)). In this case, the highly oriented discotic assembly shows anisotropical mobility ($\mu_\parallel/\mu_\perp \approx 20$) with maximum value for the component parallel to the stacking direction up to $0.5\,cm^2\,V^{-1}\,sec^{-1}$ [24,88]. From these data, a carrier diffusing length before trapping of $0.5\,\mu m$ is estimated, that is, a value still smaller than the typical channel length in organic field effect transistors (OFETs) where lower values of mobility are measured [89]. An improvement of the order of these SPs may allow to extend by one order of magnitude the carrier diffusing length thus making it larger than typical OTFT channel length. In this case, due to the large carrier mobility of discotic LC, strongly improved OTFT may be envisaged. The exploitation of anisotropical transport properties to improve organic transistor properties has been recently investigated by using Langmuir-Blodgett film deposition. In this case, the anisotropy was induced by polymer orientation achieved during the dipping process [90].

In spite of limitations still occurring, discotic LC supramolecular systems are very promising. The physical origin of the high mobility is related both to the order of the system and to the intermolecular interactions. Since the latter are strongly dependent on the supramolecular structure it is interesting to summarize theoretical predictions for different molecular solids (main-chain SPs) [83,91]. At a microscopic level, the main parameter governing the process is the intermolecular transfer integral (t), which is responsible for removal of single molecule electronic levels degeneracy when packed in a solid. Within monoelectronic band structure calculations (i.e., tight binding or Hückel model), t is responsible for the width (W) of the valence and conduction bands. Since t depends on the intermolecular distance (d) and molecular orientation, it is evident that the supramolecular organization plays a fundamental role in the degree of interaction and then in the mobility properties. Within simple models similar to those adopted to describe inorganic semiconductors, mobility increases with the bandwidth. However, one must remember that in organic semiconductors, in particular in operating devices at room temperature, electron–phonon coupling also plays an important role. This interaction causes both self-trapping of carriers and bandwidth reduction due to scattering with phonons. Charged carriers are then localized on a single molecule and transport is thermally activated by a hopping mechanism. The temperature (T) dependent transfer rate (k_{ET}) from a charged molecule to a neutral one can be described as

$$k_{ET} = \frac{4\pi^2}{h} \frac{1}{\sqrt{4\pi k_B T}} t^2 e^{(-\lambda/4k_B T)} \tag{1}$$

where k_B is the Boltzman constant. The electron-transfer hopping rate is thus determined by t and by the reorganization energy λ, which describes the strength of the electron–phonon coupling, roughly estimated as twice the relaxation energy of the single molecule charge carrier (polaron). Mobility can be evaluated by using the Einstein relation $(\mu = eD/(k_B T)$, where e is the electron charge and D the diffusion coefficient) and the Einstein–Smoluchowski equation $(D = k_{ET} d/2)$ [92]

$$\mu = \frac{e k_{ET} d^2}{2k_B T} = \frac{e\pi^2}{\hbar (k_B T)^{3/2}} d^2 t^2 e^{(\lambda/4k_B T)} \tag{2}$$

thus highlighting the dependence of μ on d, t (depending on d), and λ.

In order to prevent limitations to carrier transport due to self-trapping process, the hopping process must be faster than the self-trapping time, that is, the carrier residency time would be shorter than the time required for carriers to be self-trapped. The residency time τ, is proportional to the inverse of the bandwidth $(\tau \approx \hbar/W \approx 6.58 \times 10^{-16}$ sec W^{-1} [eV]$^{-1}$). The self-trapping time can be roughly estimated as twice the typical phonon frequency of a conjugated system (1500 cm^{-1}, corresponding to about 10^{-14} sec). It turns out that a bandwidth due to intermolecular interaction larger than 0.2 eV needs to allow carrier transport with reduced limitations due to electron–phonon coupling.

To obtain $W \geq 0.2$ eV, a careful engineering of the assembly is required since the intermolecular interactions are governed not only by the intermolecular separation but also by the molecular arrangement. The details of the stacking structure of molecular solids have been investigated for discotic LCs [91] (Figure 8(d)), porphyrin [93], acenes, oligothienylene, and organic solids [83]. For instance, in discotic LCs, the cofacial stacking provides the larger value of t, but rotation of the single units and possible sliding of a unit from the columnar stacking drastically reduce the interchain interactions [91]. The crystal structure of molecular solids must be considered in details to obtain the proper information (Figure 8(c)). Since, to accomplish the calculation for large molecules in extended crystals is very difficult, it is useful to highlight the role of main parameters (intermolecular separation, lateral displacement, chain length, and cluster size) affecting the problem [83]. Without going into details of specific supramolecular structures, some rules of thumb governing the transport properties of molecular assemblies may be understood by modeling clusters of ethylene, the simplest conjugated molecule [83]. The antibonding LUMO level of the isolated molecule has a node in the wavefunction not present in the bonding HOMO level. In a cofacial ethylene dimer, molecules are 4 Å apart and both levels split, the HOMO splitting being about four times than for the LUMO. Qualitatively, for the ideal cofacial case, the lower the number of nodes in the wavefunction for the isolated molecule, the larger is the splitting of the corresponding level in a cluster [83]. In conjugated systems the LUMO wavefunction has at least one node more than the HOMO one, thus explaining why, in general, hole mobility is expected to be larger than the electron mobility. More detailed evaluations of the HOMO and LUMO splittings can be obtained by considering specific molecular dispositions. Theoretical calculations show that between 3 and 4 Å, the typical intermolecular separation in molecular solids [48], strong interactions (and then splittings) are observed. When molecules are displaced to each other along the long or short axis or are rotated along the long axes, thus having a more realistic structure the splitting is in general reduced. However, for peculiar cases, the splitting of the LUMO level may be larger than the HOMO one, thus explaining the reason of the larger electron mobility (with respect to the hole mobility) observed for few selected molecular solids [83]. Finally, we observe that by increasing the number (n) of molecules in a cofacial cluster, the splittings saturate following a simple law ($\cos(\pi/[n+1])$) as expected for the tight binding model in a one-dimensional stack ($W = 4t \cos[\pi/(n+1)]$) indicating that level splitting in the cluster is dominated by the nearest neighbor interaction and that the bandwidth for an infinite system can be estimated as twice the dimer splitting ($2t$).

The scenario above depicted concerns small conjugated molecules in a crystalline state (main-chain SP) for which theoretical modeling is possible. In the case of covalent polymeric systems, due to their larger dimension and intrinsic disorder, the situation is much more complicated. Theoretical suggestions derived for the correspondent oligomers may still be useful for engineering the proper polymeric assembly, but different properties must be considered. A proper tailoring of the intramolecular properties of the conjugated polymer joined with the control of its supramolecular organization is helpful to improve carrier mobility. The trick is to exploit as much as possible both the intramolecular carrier drift along the conjugated backbone and the carrier hopping between different chains. In Section II.B.2 it was shown that regioregular PAT may be the suitable material. In fact, the intrachain regioregularity increases the planarity of the backbone and then the intramolecular transport properties. The planarity also decreases the interchain separation, thus giving rise to the lamellar structure with very short intermolecular separation (3.8 Å), which strongly increases the bulk mobility. Moreover, the electron–phonon coupling in regioregular PAT is reduced with respect to the one-dimensional case (regiorandom PAT) thus reducing the self-trapping that localize carriers. Finally, the possibility to align polymer chains further increases both planarity and intermolecular interactions. In order to explain how these properties works in a real device it is interesting to review the pioneering results obtained on OFET by the Cambridge group [64,66,68,94]. The technological reason for the interest in OFET is related to the development of devices suitable for switching flat-panel or electronic paper displays or for applications to smart cards and electronic identification tags [95]. The use of organic materials for thin film transistors (TFTs) is related also to their low cost, large-area coverage, mechanical flexibility, and low temperature processing [65,96–98]. OFETs with active semiconductor being either small molecules or polymers have been prepared since several years [99]. In spite of the low carrier mobility (μ) values reported for the early devices (about 10^{-4} cm^2 V^{-1} sec^{-1}), a lot of work was done to improve their performances and characteristics and in exploiting the properties of organic materials in the semiconducting and metallic state. FETs with organic electrodes in addition to organic active semiconductors [100] have also been reported, even with complicated circuitry structures [101]. The main drawback for these first generation devices was their low charge mobility. By using small conjugated molecules, such as sexithiophene (Figure 8(c)), which possesses high crystallinity with respect polythiophene derivatives, $\mu \simeq 10^{-2}$ cm^2 V^{-1} sec^{-1} have been reached, still not enough for industrial applications [97]. Nowadays, the use of new generations of conjugated polymers allows to reach μ values comparable with those of amorphous silicon [64,81] and then the interest of several companies in the field is growing [102]. This new developing phase is based on the exploitation of the supramolecular properties of active materials. Anyhow, the use of small molecules with a detailed control of the crystalline properties is also in progress [97,103,104].

As discussed before, the carrier transport in conjugated polymers is dominated by the variable range hopping between disordered chains. This fact limits the charge mobility and the ON/OFF ratio of OFET and therefore prevents the development of an all-polymeric optoelectronics. High quality regioregular poly(hexylthiophene) (P3HT) is suitable for the preparation of OFET showing improved mobility (0.05–0.1 cm^2 V^{-1} sec^{-1}) and ON/OFF ratios in excess of 10^6, much better then previous reported data (10^2–10^4) [64] (see Figure 9). These OFETs can drive organic LEDs (Figure 9) thus demonstrating the possibility to integrate several electronic devices in the same circuit.

Another important issue in polymer electronics is the possibility to prepare integrated circuits with a different approach with respect to conventional semiconductors. Instead of using vacuum deposition and photolithographic techniques, which are very expensive, integrated circuits can be manufactured by cheap methods based on solution processing. In this way, ink-jet printed (IJP) devices with 20 μm

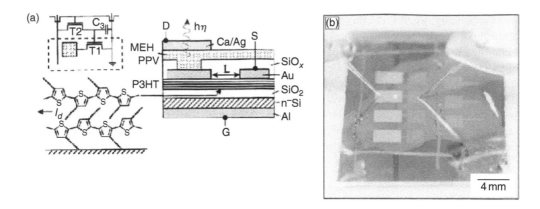

Figure 9 A: Integrated P3HT FET and MEH-PPV LED. B: Photograph of a FET-LED. Reprinted figure with permission from H. Sirringhaus, N. Tessler, and R.H. Friend, *Science* 280, 1741 (1998), Figure 1. Copyright 1998 AAAS.

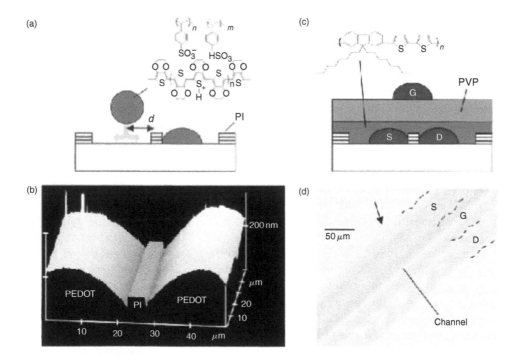

Figure 10 Scheme of prepatterned substrate and PEDOT:PSS drop deposition (A). AFM image of the source and drain electrodes (B). Scheme of IJP organic FET (C). Optical image under crossed polarizers showing the uniaxial aligned domain of F8T2 (D). Reprinted figures with permission from H. Sirringhaus, T. Kawase, R.H. Friend, T. Shimoda, M. Inbasekaran, W. Wu, E.P. Woo, *Science* 290, 2123 (2000), Figure 1. Copyright 2000 AAAS.

spatial resolution have been produced. Even though this resolution is suitable for LED, it is not enough for practical TFTs where channel length shorter than 10 μm are required to reach adequate transport and switching properties in a full polymeric device. The reason for this behavior is the coalescence of polymer ink droplets, which prevents a suitable spatial control of the active material. Sirringhaus and coworkers used hydrophobic and hydrophilic substrates to engineer the device in addition to aligned polymer semiconductors thus obtaining an all-polymeric *IJP* TFT with high mobility and switching ratio [81] (see Figure 10). A 5 μm large polyimide thin film was prepared by photolithography on a glass substrate. Oxygen plasma etching of the exposed glass surface made it hydrophilic, in contrast with the hydrophobic polyimide surface. Water based ink of conducting polymer PEDOT:PSS was then deposited on the glass by a piezoelectric ink-jet printer with coarse alignment of about 5 μm. The hydrophobicity of the polyimide strip prevents the coalescence of PEDOT:PSS drops in the two side of the polyimide thus making spatially defined electrodes (source and drain) with separation of about 5 μm (Figure 10). After drying the PEDOT:PSS ink, a layer of poly(9,9-dioctylfluorene-co-bithiophene) (F8T2) active semiconductor was deposited by spin casting. Then an insulating layer of polyvinylphenol was spin cast and finally a top drop of PEDOT:PSS was deposited as a gate electrode. The immiscibility of different polymer solutions prevented any interpenetration of the different layers thus providing a good definition of the vertical stacking and interfaces. It is interesting to note that the polyimide layer has a dual function. From one side it provides a good lateral definition of the channel in the device (5 μm) and from another, it induces epitaxial orientation of the semiconducting polymer (F8T2). As a matter of fact, when the polyimide layer is mechanically rubbed along the source/drain direction, it induces orientation of the F8T2 chains heated above their liquid-crystalline phase transition temperature (265°C) [81]. In this way, the interchain charge transport is increased and the carrier mobility improved (from 0.01 to 0.02 cm^2 V^{-1} sec^{-1}). It is interesting to notice that the exploitation of supramolecular properties of these polymers (orientation, ink-jet printing, liquid crystallinity) allow a device architecture suitable for integration with other components, even in the vertical direction, through the via-hole connections [81]. Other methods based on the embossing technique have also been used to reduce the channel length of TFT below 1 μm [105].

B. Sensors

Semiconducting and metallic forms of conjugated polymers can be used for sensing applications. We would like to note that the meaning of sensor is very broad and includes devices able to quantify the presence of an analyte, showing an observable response to a solicitation, or having dosimeter functionality [106]. Conjugated polymers have an advantage with respect to small molecule sensors because they exhibit collective response properties, which are sensitive to very small amount of analyte. In general, modification of transport properties, energy migration, light emission, or color changes are used as probing functionalities [107,108]. Variation of conductivity in doped conjugated polymers is used since addition of few charges to the backbone may increase the current several times thus providing a good sensitivity. Potentiometric sensors are based on the detection of chemical potential variations (work function of a conducting polymer). Colorimetric sensors are based on the change of polymer absorption spectra upon exposure to an environment able to change its macromolecular conformation. Also, fluorescence is widely used because it offers several approaches based on intensity variation, energy migration, and lifetime changes joined with a very high sensitivity. Fluorescent sensors based on conjugated polymers are widely used not only as generic biosensors [109] but also for more specific goal, like DNA sequences targeting [110], which seems very promising for industrial applications [111].

Conjugated polymer based fluorescent sensors are unmatched by standard small molecule sensors. It is well known that fluorescence in conjugated polymers is quenched by electron acceptors (see Section III.D for more details). Conjugated polymers substituted with anionic groups in the sidechains are water-soluble and provide a site for cationic quenchers to attach. Photoinduced charge transfer from the polymer to cations efficiently quenches fluorescence. The transfer process exponentially depends on the distance between donor and acceptor [112]. The quenching process is governed by the Stern–Volmer law ($\phi^0/\phi = 1 + K_{SV}C_{quencher}$) where ϕ^0 is the fluorescence intensity without and ϕ with quencher, and $C_{quencher}$ is the quencher concentration [113]. K_{SV} measures the efficiency of the quenching process. Values for K_{SV} of 65 (in inverse concentration units) have been found by using small molecule quenchers [114], but recently conjugated polyelectrolytes showed orders of magnitude larger values (10^7) [109]. Even though the mechanism of this increased efficiency is still not fully understood, it opens new perspectives in the field of biosensors.

Another example of conjugated polymers acting as biosensors is provided by PDAs. PDAs are suitable for sensing application due to their color changes both during the polymerization process and during the blue-to-red phase transition, which is strictly related to the change of their supramolecular properties. Indicator applications of PDAs include monitoring of the time–temperature exposure, temperature, humidity, pressure, radiation exposure, pH, metal ion concentration, and virus or toxic agents [115–117]. Color changes can be mostly divided in two categories: type I, when the change originates from a chemical reaction; and type II when it originates from a physical effect such as matrix strain effects or changes in backbone planarity [115]. The polymerization reaction itself is used for sensing devices since it provides a change in delocalization and a consequent color change from transparent to red/blue. Since the polymerization is triggered by energy absorption (heat, UV light, γ-ray) [118], type I sensors have been able to detect exess exposure to such physical observable. An example of application is LifeLines™ indicator used as smart tags for food labeling [115,116].

As discussed before, PDAs can be prepared in both blue and red forms. In the blue form they have a rodlike conformation with a planar backbone implying an extended π-electrons conjugation. In this form, the HOMO–LUMO transition of PDA is set in the red–green thus conferring them the characteristic blue color. This form allows for an efficient packing of the chain and preserve the ordered three-dimensional structure. When in the red form, even though the rodlike structure seems to be preserved, the PDA backbone is no longer flat and the supramolecular structure is lost. This implies a reduction of the conjugation length and a consequent high energy shift of the absorption

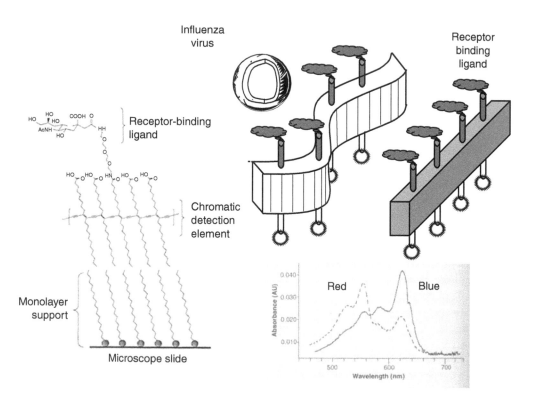

Figure 11 (Left) Scheme of the polymerized bilayer assembly. The blue phase polydiacetylene chromatic detection element is deposited over a support monolayer. The polydiacetylene chain is asymmetrically substituted with urethane side groups partially terminated with receptor-binding ligands. (Right) Absorption spectrum of the blue and red phase PDA with a schematic representation of their chain with (red phase) and without (blue phase) influenza virus attached. Adapted figures with permission from D.H. Charych, J.O. Nagy, W. Spevak, and M.D. Bednarski, *Science* 261, 585 (1993), Figure 2 and 3. Copyright 1993 AAAS.

giving rise to a red coloring. This transition from the blue to the red phase, usually irreversible, can be used to prepare type II sensors, such as defrost indicators [115] or colorimetric detectors for viruses [119], biological toxins [120], and glucose [121]. As an example, the case of influenza virus sensor is reviewed here (Figure 11). The influenza virus (class orthomyoxoviruses) has on the lipid bilayer surface two receptors: hemmagglutinin and neuraminidase. The former is responsible for the binding to the human cells through the sialic acid receptor, while the latter is responsible for the separation from the cells after infection has taken place. Charych and coworkers [119] prepared, over a glass support, a bilayer composed of a self-assembled monolayer (octadecyltrichlorosilane) covered by a Langmuir–Blodget PDA film. This latter film was substituted with a urethane chain, terminated with a binding ligand able to interact with the hemmagglutinin receptors. The PDA backbone in this film is well extended and shows an absorption spectrum with maximum at 620 nm. This is typical for the high ordered form of PDAs in the blue phase where the urethane side groups create a hydrogen bond between the side chains thus conferring to the backbone a high stiffness and an extended electron delocalization (Figure 11). As soon as the film is incubated with the influenza virus, it becomes red. In fact, the hemmagglutinin receptor on the virus binds to the side group ligand, thus tangling the sidechains, breaking the hydrogen bonding, and destroying its ordered structure. This fact strongly reduces the planarity of the backbone guaranteed by the side group [122] and reduces the conjugation length of the polymer thus transforming it into its red form (Figure 11). The presence of the virus can then be easily detected by a color change.

C. Light Emitting Diodes

One of the most impressive optoelectronics application of conjugated polymers and small molecules is provided by OLED technology, which is promising to revolutionize the flat displays market [13,123,124]. Main advantages of this new technology with respect to that based on LCs are wide viewing angle, high brightness, low power consumption, mechanical flexibility [125], and low cost. In order to explain how supramolecular chemistry of conjugated materials (based either on small molecules [126] or polymers [127]) may help in improving their properties, we first discuss how OLEDs work.

In its simplest structure an OLED is an organic semiconductor layer sandwiched between two electrodes having different work functions (Figure 12), supported on a transparent substrate. When the device is forward biased, electron from the cathode and holes from the anode are injected into the device. Opposite carriers drift inside the semiconducting layer, meet each other, and then radiatively recombine thus releasing the excess of energy as photons. The physical processes involved in OLED performances are charge injection, charge transport, charge recombination, light emission, and, eventually, polarized emission [128]. All these processes depend on the film quality and then on the chemical and physical properties of the organic semiconductor, including its supramolecular structure.

Small molecules are usually evaporated in vacuum while polymeric semiconductors are solution processed and deposited, usually, by spin coating. In order to prevent operational failures, grain boundaries and aggregation should be avoided. Amorphous films are usually desired in order to have better performances for devices operating below the polymer glass temperature. As discussed later, oriented systems with polarized emission are requested for specific applications.

In order to inject carriers into the devices, the work function of the different metals should be matched with the HOMO and LUMO levels of the organic semiconductor. The choice of the metals is quite limited. As an anode, transparent metals such as indium tin oxide (ITO) or polyaniline are used. The roughness of ITO may be smoothed by covering with a PEDOT:PSS layer, thus reducing the possibility of accidental shorts. ITO covered by self-assembled monolayers or other molecular layers [129], sometimes improve electrode properties. These metals provide a good matching of the electronic levels for hole injection as well as a good optical transparency for light escaping from the device. It is instead more difficult to find a proper metal as electron injector since a low

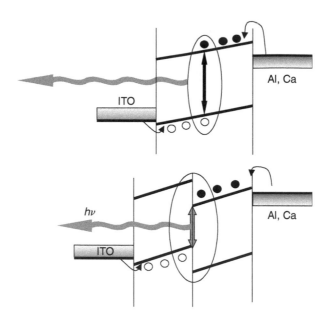

Figure 12 Schematic representation of operating OLED. Full circles: electrons; open circles: holes. Top: monolayer device, bottom: double layer device.

work function is requested. Calcium (work function 2.9 eV) is highly reactive and it is used after covering with another metal. Improvements in OLED performances upon LiF implantation have also been reported [130]. Aluminum is more stable but its higher work function (4.3 eV) reduces the OLED performances. The chemical substitution of conjugated polymers with the proper electron-donating or -withdrawing side groups allows tuning the material ionization potential and electron affinity thus improving the matching with the metals work function as well as changing the color emission [131]. For small molecule OLEDs, the deposition of multilayer structures with molecules having different electronic properties is used to achieve the same results. Both these strategies allow to improve the device efficiency as well as to reduce the operating voltage and then the power consumption.

Once carriers are injected in the device, they must travel along the organic semiconductor layer. Conjugated polymers combine the emission functionality with good holes mobility, while poor transport properties are usually observed for electrons. The unbalanced carrier flux, giving rise to an inefficient recombination close to the cathode, may be equilibrated by using a two layer structure [132] creating a potential barrier where both electrons and holes accumulate before their radiative recombination (Figure 12). In small molecules OLED, the emission and transport functionalities may be fully separated and then proper transport layers can be evaporated. The physical processes governing carriers transport have been previously outlined. We notice however that Supramolecular Chemistry offers further opportunities. As a matter of fact, thermally cross-linked LCs are used to induce order in the transport layer thus improving the carrier mobility and the robustness of room-temperature operating devices [84]. Blending semiconducting polymers with electron deficient molecules may be a solution since opposite carriers may hop on different molecules and then recombine at the inter-molecular interface. However, a limit to this approach is due to the macroscopic phase separation between the two components. Recently, Hadziioannou and coworkers proposed to use copolymers of luminescent polymers and oxidiazole molecules [133]. As a matter of fact, block copolymers are known to give rise to complex nanostructured morphology, which could be used for different applic-ations [134,135]. To properly balance the carriers transport in a diblock copolymer containing an electron transporting moiety and a hole transporting/light emitting block, a very good control of the microphase separation between the two blocks must be achieved. The living radical polymerization method [136] allows a suitable control of the block copolymer structure due to the linear increase of chain length with time, and the possibility of continuing the copolymerization process in different monomer solutions [137].

Next step in the operation of an OLED is charge recombination. This process is responsible for the efficient formation of the emissive centers (singlet excitons) and the exclusion of the parasitic non-radiative decay channels, such as ISC, which leads to the formation of triplet excitons [48,138–140]. The quantum yield of radiative emission upon opposite carriers recombination was supposed to be statistically limited to 25% due to the equivalence of the formation cross-section for singlet (σ_S) and triplet (σ_T) excitons. However, experimental [139] and theoretical [140] investigations showed that $\sigma_S/\sigma_T > 1$ thus explaining observed photoluminescence quantum yields exceeding 25% [141]. Formation of triplet excitons in molecular solids may give rise to phosphorescence [142] seldom observed in conjugated polymers [143]. In polymeric systems where metal atoms are inserted in the main chain, the larger spin-orbit coupling of metals with respect to carbon atoms increases the ISC efficiency and then the phosphorescence emission [144], which in the case of small molecules can be used to produce efficient OLED [142].

The emission efficiency of the material is not only affected by ISC but also by the intermolecular interactions related to the supramolecular structure. Intermolecular interactions quench the isolated molecule luminescence and/or modify its spectral properties [145]. Since conjugated polymers are formed by a distribution of conjugated segments with different length, they should be thought as an ensemble of interacting short molecules rather than an extended one-dimensional semiconductor. For this reason, they are similar to molecular solids and accordingly described [48]. Within this approximation, in the simple case of two cofacial molecules, the intermolecular interactions are responsible for the splitting of degenerate molecular levels, the lower lying being forbidden in the dipole approximation [146]. The vanishing oscillator strength for the lower lying transition is responsible for the decrease of the photoluminescence efficiency in the solid state. Note that when the cofacial arrangement is relaxed, the low energy transition becomes weakly forbidden thus explaining the experimental data for the red shift of photoluminescence upon aggregation [147]. The electron–phonon coupling also contributes to transfer part of the oscillator strength to the lower transition [146]. However, we note that polymers are peculiar in this respect since the splitting and reduced oscillator strength for the low lying transition are theoretically predicted to be nonmonotonically dependent on conjugation length [146]. According to these calculations, two strategies may be pursued to obtain highly luminescent polymers in the solid state: increase separation of the conjugated backbones or synthesize polymers having highly delocalized π-electrons thus mimic an infinite ordered one-dimensional system. From the point of view of isolated molecules properties, rodlike conjugated systems are useful. As a matter of fact, *trans*-stylbene in solution has an emission quantum efficiency ($\sim$5%) significantly higher than *cis*-stylbene (<0.05%). When the *trans*-stylbene flat geometry is blocked by a chemical bridge, its efficiency rise to about 100% [148]. The problem is how to transfer the isolated flat molecule from solution to the bulk inside the device. Ladder-type poly-*p*-phenylene with bulky substituents is a good answer to this problem [149] (see Figure 1). In general, chemical substitution of the conjugated backbone with large side chains, promote the solubility of the polymer and also increases the interchain separation resulting in increased luminescence efficiency, while ladder type structure increases the stiffness of the polymeric backbone thus improving π-electron delocalization.

A different approach has been proposed to improve carrier transport and to increase emission efficiency by exploiting conformational properties of conjugated polymers in different solvents. MEH-PPV in tetrahydrofuran (THF) gives rise to coils that partially suppress interchain interactions. A different situation occurs in chlorobenzene (CB) solutions where a more extended conformation is detected resulting in an efficient aggregates formation [150,151]. Different polymer chain conformations in solutions are likely retained in films spun from them and then exploited to improve LED performances. Since their operation is a complicated balance of transport and emission properties, a multilayer structure with MEH-PPV films spun from different solvents could improve OLED performances. Schwartz and coworkers exploited different conformations of MEH-PPV in CB and THF

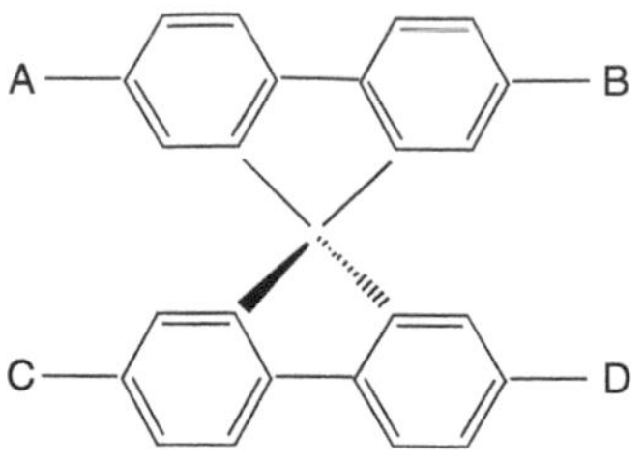

Figure 13 Spyro-linked molecules.

to engineer a new OLED structure. Such devices have three active polymer layers. The central one is cast from THF while those close to the electrodes from CB. Films cast from CB solutions result in a strong polymer aggregation allowing for a good carriers transport near electrodes, while central layer cast from THF traps opposite carriers. Owing to the low interchain interactions in central layer, carriers cannot escape from it, thus being forced to radiatively recombine with great efficiency [152].

Intermolecular interactions may also be suppressed by using spiro-type molecules (Figure 13) [153]. Within this approach two identical small molecules with suitable electronic properties are linked by a common sp^3-carbon atom, which modifies their steric repulsion thus reducing intermolecular interactions of molecules packed in a solid. In this way, polycrystallinity of small molecules films is suppreni thus removing grain boundaries which strongly affect OLED performances. Moreover, the glass transition temperature of low molar mass materials is quite low but can be enhanced by using amorphous spiro-linked emitting molecules. Spiro-molecules can also be functionalized to include in the same system both the emissive and carriers transporting moieties [153]. Although the spiro-approach seems to be particularly suitable for small molecules, it has been recently applied to conjugated polymers with the *spiro-center* inserted in the sidechain thus providing sterically hindered thermal stable PPV derivatives [154].

Another way to prevent intermolecular interactions is provided by threading conjugated polymer within a molecular shield such as cyclodextrin (Figure 14(a)) [155]. These supramolecular wires are separated by the sugar "coating," thus having isolated molecule optical properties. The insulating shield does not modify too much the frontier orbitals of the unthreaded molecule and allows a more uniform protection against interchain interaction with respect to the case of polymers with bulky substitutents. Even though the intermolecular interactions are shielded, they are not fully suppressed to prevent carrier transport. As a matter of fact, OLEDs made by threaded polymers show better performances than those made by the nonthreaded ones [155] (Figure 14(b)). Since sugar-coated conjugated polymers are water soluble, environment friendly processing and biocompatible applications are possible.

The idea to separate conjugated backbones has been further exploited using nanostructured inert matrices (host) having channels where macromolecules (guest) are inserted [156–158]. Conjugated polymers inside the matrix behave as isolated oriented molecules, while those outside the pores have bulk amorphous properties. Nanostructured hosts modify the spontaneous self-organization of macromolecules thus allowing to disentangle intramolecular energy transfer processes (in isolated molecules) with respect to intermolecular ones taking place in macromolecules outside nanopores. Energy transfer processes are very important in natural systems, such as in photosynthetic reaction centers where the light energy harvested in a chromophore is transferred by multiple steps at the proper molecular location. To obtain an efficient process, a control of the electronic functionality and molecular positioning at nanometer scale must be achieved. In devices, the energy transfer process can be exploited by transferring the excitation to a proper emitting center thus allowing to tune color emission. It has also been exploited in metal complexes with conjugated molecules to obtain emission in spectral regions usually not covered by organic materials by using phosphorescent emission of a proper guest molecule or metal ion [142,159–161]. Usually, in

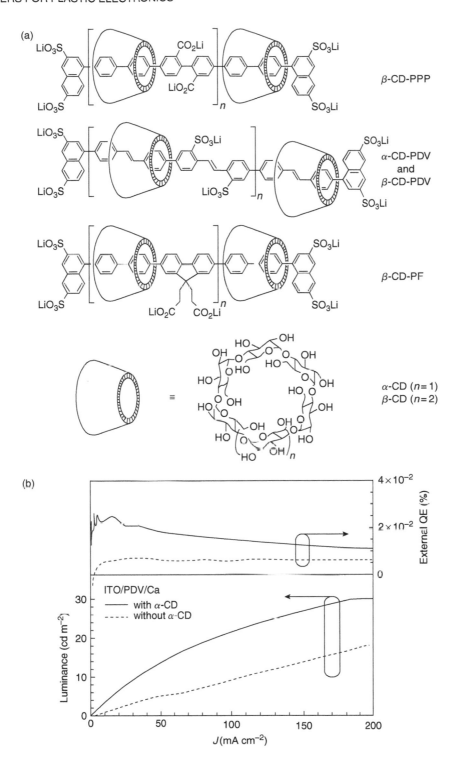

Figure 14 (a) Chemical structures of cyclodextrin-threaded conjugated polyrotaxanes with poly(para-phenylene), poly(4,4-diphenylenevinylene), and polyfluorene cores. (b) External quantum efficiency (top panel) and luminance (lower panel) versus current density characteristics of PDV and CD-PDV-based LEDs. Reprinted figures with permission from F. Cacialli, J.S. Wilson, J.J. Michels, C. Daniel, C. Silva, R.H. Friend, N. Severin, P. Samorì, J.P. Rabe, M.J. O'Connel, P.N. Taylor, and H.L. Anderson, *Nature Materials* 1, 160 (2002), Figure 1a and 3a. Copyright Nature Publishing Group.

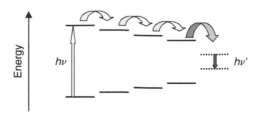

Figure 15   Scheme of the energy transfer process. Excitations move from higher energy conjugated segments (short conjugated segments) to low energy conjugated segments (long conjugated segments). Further transfer to a different molecule where emission may takes place is also shown.

conjugated polymers the driving force for energy transfer is the energy difference among conjugated segments having different length (long conjugated segments have lower energy than shorter ones) [157,162,163]. At the end of the process within the same material, a further step can be eventually associated to a guest molecule (see Figure 15). The interchain energy transfer is governed by the Förster energy transfer mechanism while the intrachain one is much slower since barriers between different conjugated segments must be overcome. Another advantage of the exploitation of the energy transfer process is related to the control of interchain interactions. In fact, when the fast interchain energy transfer is suppressed, the excitation transfer rate to different molecules is very low (low transfer probability) thus reducing the probability for the excitation to find a quenching site and enhancing its emission efficiency. It is interesting to note that the energy transfer in the Förster model, being related to dipole–dipole interactions, is a long-range process whose rate from a donor (D) to an acceptor (A) is given by [158]

$$K_{\text{D-A}} = \frac{1}{\tau_D R^6} \left(\frac{3}{4\pi} \int \left(\frac{c}{\omega n} \right)^4 F_D(\omega) \sigma_A(\omega) \, d\omega \right) \tag{3}$$

where τ_D is the natural lifetime of the donor in the absence of the acceptor, R is the distance between donor and acceptor, n the solvent refractive index, c the speed of the light, $F_D(\omega)$ the normalized fluorescence emission spectrum, and $\sigma_A(\omega)$ the normalized acceptor absorption cross section. This formalism has been then further developed to describe the energy transfer for dipole forbidden transitions like in the case of triplet–triplet energy transfer based on the exchange energy interaction [164]. Note that the different functional dependence on the intermolecular distance existing for the energy transfer and carrier hopping process (Eq. [1]). In fact, typical energy transfer distances in organic materials are tens of angstroms, to be compared with few angstroms for carriers hopping. This implies that a proper engineering of the supramolecular structure of the active system may be useful to improve carrier transport and emission properties of devices.

As outlined before, OLED performances are determined by a variety of physical processes, which can be controlled by organic and supramolecular chemistry through the synthesis of proper small molecules or polymeric systems. Recently, a novel synthetic strategy has been developed based on dendrimers. This topic has been extensively described both in previous and present edition of this book [165] as well as in several other publications [166]. As already noticed, small molecules/polymers with different functionalities useful for OLED can be blended together. However, this approach is very sensitive to the concentration of the guest in the host matrix as well as to phase separation, which can give rise to aggregation followed by luminescence quenching. In order to prevent phase separation, different active molecules/macromolecules should be chemically compatible. For this reason dendritic structures with different electronically active core and the same dendrons may be suitable to prevent both phase separation and luminescence quenching due to aggregates formation [167]. Dendritic molecules suitable for singlet emission have been reported [168]. However, it would be useful to exploit also phosphorescence (75% probability) coming from the triplet excitons which are populated by ISC from the excited singlet states (See Fig. 4).

This has been done by using polymeric systems with heavy atoms (metals) in the backbone [163] thus increasing the ISC rate [169], or by multilayered structures of small molecules and metal complexes [142]. The multiple functionalities needed for the ISC can be inserted in dendrimers having, in comparison to linear systems with the same molecular weight, a lower intrinsic viscosity and a better solubility [170], possibly associated to a control of the glass transition temperature and chemical reactivity [170].

We would like to end this section with an example showing the importance of polarized emission achieved by controlling the SP structure. In LC-based displays, light on/off is switched by a modification of the polarization status of the light caused by optically active cholesteric LCs. Since in this phase, molecules prefer to lie close to each other in a slightly skewed orientation, they induce a helical director configuration rotating through the material. Cholesteric LCs are sandwiched between two crossed polarizers with the molecules at the surfaces aligned such that the director lies parallel to the polarization direction. Without the LC, light entering in the cell would not be transmitted due to crossed polarizers. With the LC, the helical director structure rotates the light polarization direction from the polarizer to the analyzer so that the light may be transmitted. When a voltage is applied across the cell, molecules align with the electric field, thus destroying the helical structure and preventing light transmission [172]. Drawbacks of this structure are poor lateral vision angle and loss of great part of the light filtered by polarizers. The luminance of these devices can be improved using polarized light sources thus reducing light losses. Highly oriented conjugated polymers are suitable for this approach. Recently, polarized emission has been obtained in OLEDs with oriented active material. In such devices, polymer orientation has been achieved by direct rubbing of emitting polymer layer [30] or by epitaxy of small molecules due to oriented PEDOT electrodes [173]. Plastic polarized light sources are then the candidates for a new generation of retro illuminated LC displays possessing a technology intermediate between the present one and the next, fully based on conjugated systems.

D. Photovoltaic Cells

In 1992, the Santa Barbara and Osaka groups reported on the use of conjugated polymers and fullerene for the preparation of organic solar cells [174,175]. Fullerene (C_{60}) is an excellent electron acceptor able to capture several electrons. When mixed with conjugated polymers, photoexcitation of the latter induces an electron transfer to fullerene resulting in a stable positive carrier (polaron or radical cation) in the polymer and an ionized (C_{60}^-) molecule (radical anion) (see Figure 16). The transfer of the electron from the conjugated polymer to the acceptor is ultrafast [176] with a rate almost two order of magnitude larger than competing radiative or nonradiative processes and having efficiency almost one. The back electron transfer to the polymer, although possible, is very slow thus charged excited states survive for milliseconds/seconds [176]. This gives rise to several photophysical processes, which can be exploited in devices, such as luminescence quenching (sensors, see Section III.B), photoconductivity sensitization [177], photovoltaic phenomena [178], and nonlinear processes [179]. Even though steps involved in the electron transfer process are much more complex than those reported in the schematic picture here provided, the critical condition to be fulfilled involves the ionization potential of the excited donor (I_{D*}), the electron affinity of the acceptor (A_A), and the Coulomb attraction of separated radicals including polarization effects (U_C)

$$I_{D*} - A_A - U_C < 0 \tag{4}$$

Even this criterium concerning molecular electronic properties was fulfilled, morphological problems must be addressed for the charge transfer to be effective. Macrophase separation in the donor–acceptor blend affects the process reducing the donor–acceptor wavefunction overlap both due to exceeding intermolecular separation and/or inefficient mutual orientation.

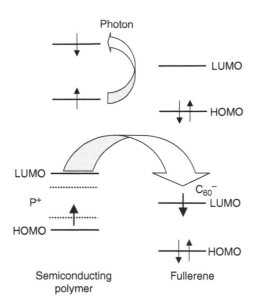

Figure 16   Scheme of photoinduced electron transfer from semiconducting polymers to fullerenes. From the polymer excited states an electron is transferred to fullerene thus creating a positive polaron (radical cation) on the polymer and a radical anion on the fullerene.

The main application of the photoinduced electron transfer process is related to photovoltaic cells, whose market is rapidly expanding [180] and is particularly exciting when materials involved are plastics. The operating principle can be explained with the help of Figure 17. Let us start with one active conjugated polymer layer sandwiched between metals with different work function (for instance, ITO and Al or Ca). As previously described for OLEDs, in forward bias, a rectifying behavior is observed since holes are injected by the high work function metals and electrons from the low work function one. If the injected carrier meet, they can emit light. In reverse bias (Figure 17(a)), injection of carriers from electrodes is strongly inhibited. However, if the reverse biased device is photoexcited thus generating charged carriers (polarons), a current is detected conferring to the device the function of photodetector [181]. Such devices have a dual function working in forward bias like OLEDs and in reverse bias like photodetectors [182]. In the photovoltaic device, no field is applied and the potential due to the different work functions of the electrodes allows to separate the excited states formed upon photon absorption into a pair of oppositely charged carriers. However, since the work function difference is very low, the efficiency of single layer devices is very poor [182]. A drastic improvement was afforded by a bilayer structure (Figure 17(b)), where donor–acceptor materials are superimposed [183]. In this case, enhanced photocurrent for both biasing polarizations is observed due to injections of holes into the polymer and electron into C_{60}. In bilayer cells, in addition to the different electrode work functions, an efficient channel for breaking the excitons into carriers is provided by the charge transfer process that improves their performances. When donor (polymer) and acceptor (fullerene) molecules are very close to each other (i.e., at the interface between the two layers) electrons can jump from the polymer to the fullerene thus increasing the number of charges collected at the electrodes. The main drawback for this configuration is the limited extension of the donor–acceptor interface. The interface can be increased by blending donor and acceptor at a molecular scale. This idea stimulated a specific synthetic work to improve the poor solubility of C_{60} by inserting on the buckyballs suitable substituents, such as in methanofullerenes [184] and fulleropyrrolidine [185]. The new cells, with interpenetrating donor–acceptor network (bulk heterojunction), sketched in Figure 17(c), operate in similar manner to the bilayer one but now with increased efficiency due to the extended donor–acceptor interface [186]. A similar configuration was used for photovoltaic cells based on PPV derivatives with different substituents, conferring them with

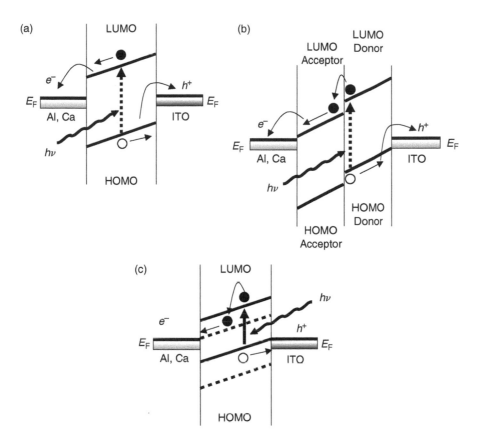

Figure 17   Scheme of working photovoltaic devices. (a): single layer; (b): bilayer; (c): interpenetrating donor/acceptor (bulk heterojunction) photovoltaic cells.

donor–acceptor properties [187]. In this case the advantage is due to the processability of a full plastic device. At present, efficiency of photovoltaic cells with interpenetrating donor–acceptor network is about 3% [188] and industrial effort to bring them to the market is currently pursued [189]. Main drawbacks of these photovoltaic devices are the exploitation of the full solar spectrum and carrier transport. Synthesis of materials with extended absorption spectrum or blending with materials covering a broader solar spectrum section are currently in progress [190]. A big challenge is to improve carrier transport, in particular for low mobility negative carriers. Blends of regioregular PAT, having improved hole mobility (Sections II.B and III.A), and rodlike CdSe nanocrystals, a very good electron mobility material, have been used. In this case, 1.7% power conversion efficiency is obtained under air mass (AM) 1.5 Global solar conditions [191]. Since macrophase separation is a typical problem for blends, supramolecular chemistry strategies may be useful to solve the problem [192]. Improvements in carrier mobility can be reached by using discotic LCs (hexaphenyl-substituted hexabenzocoronene, donor) in combination with perylene derivative (acceptor). In this way, solution processable materials are cast to produce solar cells and photodetectors that exploit the combination of one-dimensional positive carrier path provided by discotic LCs blended with perylene derivatives which provide a crystalline network for negative carrier drift [193]. The blend morphology is remarkably different with respect to that of single components (Figure 18) [193]. Tapping mode atomic force microscopy (AFM) reveals that discotic LC films have a smooth texture with surface roughness below 10 nm (Figure 18(a)). In contrast, pure perylene films show higher roughness with crystallite elongated in the micron-scale (Figure 18(b)). The surface morphology of the films obtained with the materials blended together shows a continuous coverage of the surface and perylene crystallites

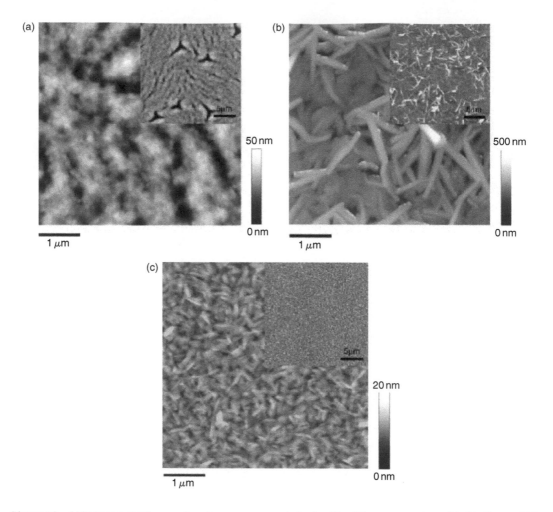

Figure 18 AFM image of (a) a pure hexabenzocoronene derivative film, (b) a pure perylene di-imide film, and (c) a spin-coated film from a blend solution of hexabenzocoronene and perylene derivatives. Reprinted figures with permission from L. Schmidt-Mende, A. Fechtenkotter, K. Mullen, E. Moons, R.H. Friend, and J.D. McKenzie, *Science* 293, 1119 (2001), Figure 2 and 3. Copyright 2001 AAAS.

with reduced dimensions (Figure 18(c)). Further improvements with this approach may be viable for commercial photovoltaic devices [193].

Another supramolecular approach to improve transport properties of devices is based on block copolymers. As noted above, block copolymers give rise to microphase segregation of different blocks [134,135]. Here the idea is either to grab the acceptor inside a cavity formed by a self-assembled rod–coil block copolymer [194], or to synthesize a copolymer with a block possessing the acceptor functionality. Several groups succeeded in this approach [195–199]. The microphase separation achieved in this way strongly enhances the probability of carrier hopping, and may increase the performances of the solar cell. Interesting results have been reported for polymers synthesized by the living radical polymerization technique. Films prepared with these polymers not only show nanophase separation (see Figure 19), but also show a very regular micropatterning on a larger scale [195]. The nanophase separation of these block copolymers does not depend on the presence of the fullerene pendant thus promising to improve the transport properties of carriers in photovoltaic cells [195]. Moreover, the micropatterning obtained in these films when cast in a wet environment, is very regular (Figure 19) and may be useful to improve light collection in solar cells due to reduced reflectance losses caused by the lower refractive index of the honeycoube structure [46,200–202].

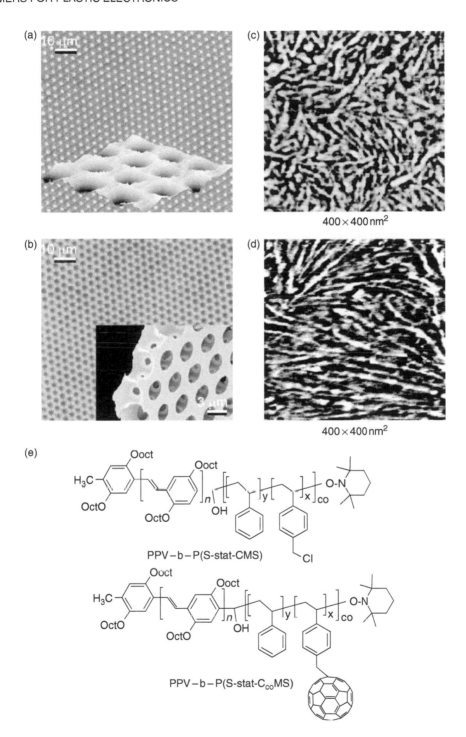

Figure 19   (a) Optical and AFM images and (b) fluorescence and SEM images of the honeycomb structure of PPV-polystyrene copolymer film. AFM images of the morphology of (c) PPV-polystyrene copolymer and (d) PPV-polystyrene-fullerene derivatives. The chemical structure formula for the two copolymers are also reported. Reprinted figures with permission from B. de Boer, U. Stalmach, P.F. van Hutten, C. Melzer, V.V. Krasnikov, and G. Hadziioannou, *Polymer* 42 (2001) 9097. Figure 1, 6, 7. Copyright 2001 Elsevier.

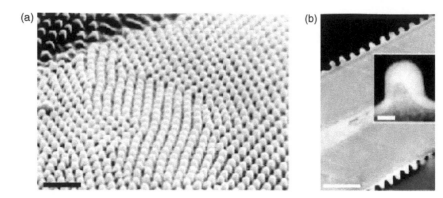

Figure 20   Anti-reflecting ripple arrays on (a) arthropoda ommatidia and (b) transparent wings of a diurnal moth. In the inset, a single ripple is shown. Bars are 1 μm (a); 1 μm (b) and 100 nm (b inset). Reprinted figure with permission from P. Vukusic and J.R. Sambles, *Nature* 424, 852 (2003), Figure 7. Copyright Nature Publishing Group.

Note that similar antireflecting nanostructures are used in the biological world to increase both the photon collection efficiency of arthropodal visual system (Figure 20(a)) and the transparency of diurnal moth wings (Figure 20(b)) [203]. Since porous honeycomb micro- and nanostructures allow tuning the refractive index of the material, they may also be used to increase the escaping light from OLEDs as suggested by recent papers on aerogels [204,205] and microcavities [206]. The possibility to use the above micropatterning for photonic crystals applications has been also reported [207].

E.   Toward Microdevices

As outlined in previous sections, molecular optoelectronics provide the unique opportunity to process active materials from solutions, in particular by using cheap techniques, such as ink-jet printing. The versatility of this approach is of interest not only for conjugated materials but also for ceramic and biological systems [208]. Moreover, it can be used in combination with other chemical approaches to control the nanostructuring of the system with the aim to produce microdevices. These approaches are expected to have an impact on the synthesis of new materials, their self-assembly properties, and the physics of microfluids [209], as well as the work of companies [210] engaged in OLED [81,211,215] and OTFT [64,68,212,213] business.

Optoelectronic devices are usually prepared by vacuum deposition processes followed by patterning with photolithographic techniques. This approach is very fruitful, but is becoming very expensive due to the high demand of material and energy. Alternative approaches need to be considered. The use of active materials in liquid phase associated with direct patterning could be advantageous with respect to traditional technologies. Ink-jet printing technique is used for the realization of microfluidic devices [214]. It requires the synthesis of new inks, the study of droplet formation, and the development of suitable mechanical positioning techniques [215]. Supramolecular chemistry may play a role in the development of active materials for inks and in their droplet positioning control [216]. Since optoelectronic devices must be prepared by subsequent depositions of different materials, it is important to consider the preparation of inks with different functions and solvents. We already noticed that PEDOT:PSS is water soluble and provides good quality hole-injection electrodes [217]. Hydrophobic conjugated polymers can be printed on top of PEDOT:PSS retaining a good quality interface without mixing the two materials. Defined interfaces can also be obtained by using water soluble conjugated polymers since, after drying, PEDOT:PSS becomes insoluble. This is a great advantage since it allows using environment-friendly solvents. We already pointed out the importance of conjugated polyelectrolite (water soluble) for DNA sensors [109,110,112]. The alternative approach based on the preparation of conjugated polymer or molecule nanoparticles should be emphasized.

Several papers have dealt with the preparation of polymer/inorganics [218], polymer/metal [219] nanoparticles, PDAs inside micelles [220], and vesicles carrying both inorganic nanoparticles [221] or fluorescent dyes [222]. The miniemulsion approach seems to be very appealing [223]. Miniemulsions consist of stable droplets (diameter well below 1 μm) obtained by mixing oil, water, surfactant, and hydrophobic compounds. Stable, water soluble conjugated polymer (methyl substituted ladder polymer PPP or polyfluorene) nanospheres can be obtained in spite of the hydrophobicity of the polymer itself [223]. The optical properties of nanoparticles do not depend on particle size and are very similar to those of the same polymer after spin-casting, indicating the lack of any interaction between polymer and surrounding water. OLED made with polymer nanoparticles seem to have slightly better performances than those prepared by traditional methods [223].

Another issue to be addressed is the reduction of the device dimensions. Scaling down the size of optoelectronic components represents a fundamental point that has always been critical for the development of modern technology. We pointed out that the reduction of the channel length in OFET has been realized by lateral nanopatterning by using hydrophobic and hydrophilic substrates to prevent ink drops from spreading [64,94,212]. Additional top-down methods are also used. Nano-imprinting and soft embossing techniques allow to transfer, with very high fidelity, master patterns fabricated by conventional lithography techniques, by exploiting the glass transition of polymers thus allowing the preparation of cavities for organic lasers [224]. Micro-OLEDs have been produced by photoablation of a conjugated polymer without damaging it during the deposition process [225]. Near-field lithography has also been used to produce nanostructured luminescent polymers inhibiting precursor solubility by UV illumination with a SNOM (scanning near-field optical microscopy) having a probe aperture tens of nanometers wide [226]. Large area lithographic masks have been prepared by using nanospheres [227]. Recently, contact lamination, a technique compatible with soft lithography, has been applied to obtain patterned micro-OLED [228]. From the point of view of supramolecular chemistry the bottom-up methods based on molecular recognition appears to be more interesting. Microphase separation in blends of PAT derivatives was observed in active layers of large area OLEDs [229]. When these devices are turned on, different microdomains emitting a single color (depending on the voltage) and composed of a single polymer are observed. Interesting opportunities are also offered by the formation of ordered macroporous films by evaporation of volatile solvent from polystyrene solutions in a moist environment [207]. The evaporation of volatile organic liquids leads to the condensation of water droplets in the micron/submicron size range. Solvent evaporation decreases the temperature at the air–liquid interface thus allowing, below the dew-point, condensation on the surface. Droplets assemble in ordered domains thus imprinting the polymer film (after solvent evaporation). Examples of this patterning are reported in Figure 19 and Figure 21. In particular, in Figure 21(a), an optical microscope image of water bubbles imprinting on a polystyrene film is reported. When the surface of the film is observed by a scanning electron microscope (SEM), two kinds of structures can be observed. A crater left on the film surface by the water sphere after its dissolution or, polystyrene pinnacles, which have been attracted by water bubbles slightly above (few hundreths of nanometers) the surface (Figure 21(b)). These nanostructures can be replicated by soft-lithography with elastomers and then used to repattern conjugated polymers [207] or used like a mask either to create microelectrodes [230] or to pattern the emitting layer of OLEDs [231]. Both these methods are promising for the production of micro-OLEDs.

IV. PERSPECTIVES AND CONCLUSIONS

Examples here reviewed show the opportunity offered by a tight collaboration between supramolecular chemistry and physics to solve problems in emerging technologies such as those concerning OLED, OFET, OPV and sensors. In addition to these items, we suggest additional topics that deserve to be tackled in the near future, namely electrically pumped organic lasers and photonic crystals.

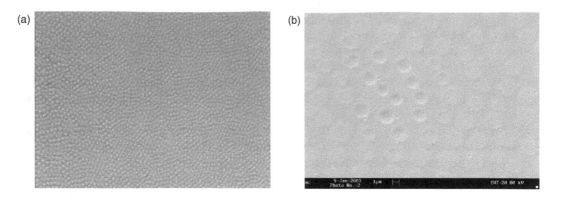

Figure 21   (a) Optical image of the imprint due to moist air (about 5 μm diameter) on polystyrene. (b) SEM image of the film surface.

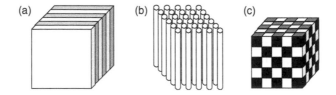

Figure 22   Schematic representation of 1D (a), 2D (b), and 3D (c) photonic crystals.

The high luminescence quantum yield of organic materials stimulated a lot of work on lasing emission, but to date, this goal has been achieved only upon optical pumping. No chance for electrically driven laser is found nowadays [232]. According to Forrest and coworkers, reduction of nonradiative losses and high emission quantum efficiency independent on both current density and electric field are requested to develop an electrically pumped organic laser [232]. Supramolecular chemistry should provide novel ideas to create materials with improved transport properties, enhanced emission efficiency as well as self-essembled sub-micron optical cavities suitable to obtain electrically driven lasing [233].

Another growing interest research field is that of ordered nanostructured materials, called photonic crystals [234]. Photonic crystals are ordered nanostructured materials with periodic variations of the dielectric constant in one, two, or three dimensions (Figure 22). This research field is very broad and complex. Here, we provide only a brief description and few examples of supramolecular architectures that may be competitive with inorganic semiconductors patterned with top-down methods.

The interaction of the electromagnetic radiation with a photonic crystal is described by Maxwell's equations in a periodic dielectric medium for which a Bloch theorem holds, similarly to what happens for electrons in crystals. Accordingly, in these materials photonic bands exist, with allowed and forbidden energy regions. In the energy region of a photonic gap the electromagnetic radiation cannot propagate (and is therefore diffracted), while propagation occurs for energies lying within the photonic band. An accurate design of the photonic crystal structure allows to mould the light signal, especially if the propagation of light is prohibited (within a given frequency range) for all the propagation directions and for any polarization (photonic band gap). In such a structure, the insertion of *ad hoc* defects would provide high bending angle wave guides, in which confinement is not related to the total reflection condition but to the absence of propagation in the surrounding medium. Moreover, the spontaneous emission from an emitting source within the cavity (when the emission frequency falls into the gap) is inhibited. This process can be exploited for the fabrication of thresholdless lasers.

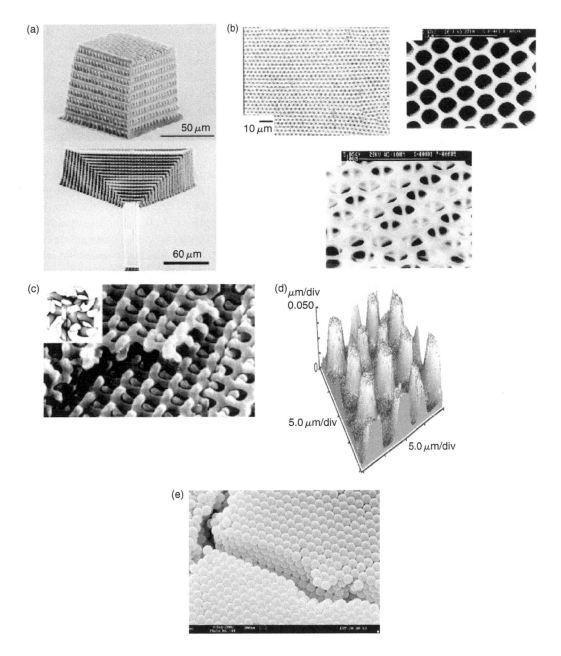

Figure 23 Examples of organic photonic crystals made by engineered SPs. (a) Two-photon absorption poly-merized systems, (b, c) block-copolymers, self-assembly of (d) moist air bubbles or (e) polystyrene nanospheres (opals). Figures (a), (b) and (c) are reprinted with permission respectively from B.H. Cumpston, S.P. Anathavel, S. Barlow, D.L. Dyer, J.E. Ehrlich, L.L. Erskine, A.A. Heikal, S.M. Kuebler, I.Y.S. Lee, D. McCord-Maughon, J.Q. Qin, H. Rockel, M. Rumi, X.L. Wu, S.R. Marder, J.W. Perry, *Nature*, 398, 51 (1999), Figure 3, copyright Nature Publishing Group; S.A. Jenekhe and X.L. Chen, *Science* 279, 1903 (1998), Figure 3, copyright 2001 AAAS; A.M. Urbas, M. Maldovan, P. DeRege, and E.L. Thomas, *Adv. Mater.* 14, 1850 (2002), Figure 2, copyright Wiley-VCH.

The properties of photonic crystals are strongly dependent on their structure and on the dielectric contrast between the different constituent materials. Often, one of these materials is the air in the cavities present in the ordered nanostructured medium. Filling these cavities with a proper material for its dielectric constant and photophysical properties opens intriguing perspectives in the photonic crystal properties. Among active materials of technological interest, conjugated molecules and polymers seem to be the more suitable ones. In fact, though their refractive index (1.5 to 1.8) does not generate a large dielectric contrast, the luminescence yields and nonlinear optical properties of conjugated systems suggest their use in this field. To this end, two different approaches are commonly applied. The first one consists in the growth of a photonic crystal with a large dielectric contrast and subsequent infiltration with nonlinear organic materials. In such a case the effects of the photonic band gap can be modulated by the nonlinearity of the organic material which implies an intensity dependent refractive index. The alternative approach consists in exploiting the emission properties of organics in the photonic crystal to improve its performances and extend its application field through polaritonic effects (interaction between the molecular exciton and the photonic cavity modes), which give rise to a finely tunable emission.

There are two methods that are commonly used to obtain photonic crystals. One is the "top-down" approach which involves lithographic or etching techniques applied to bulk materials (usually inorganic semiconductors and insulators) to obtain the suitable nanopatterning. The other approach, the "bottom-up" is instead based on the preparation of fundamental units which self-assemble into the designed structure. The fabrication of one-dimensional (dielectric multilayers) or two-dimensional (selective etching on a mask) photonic crystals is based on the top-down methods. Instead the creation of three-dimensional photonic crystals is more complex and requires expensive lithographic techniques which are hardly applied to materials different from the conventional ones used in microelectronics (Figure 22). To overcome these problems, alternative techniques have to be used and it is then evident that supramolecular chemistry may play a fundamental role here. At present, techniques relying on supramolecular chemistry used in the preparation of organic photonic crystals are: (i) two-photon absorption induced photopolymerization [235] (Figure 23(a)), (ii) nano- and micro-structures due to block copolymers phase separation [195,236] (Figure 23(b) and (c)), (iii) LCs [87,237], (iv) self-assembly of building blocks such as moist air bubbles in polymers [207] (Figure 23(d)), or (v) nanospheres (opals) [238,239] (Figure 23(e)). The last process seems particularly interesting since based on the spontaneous growth of stable well-defined structures, starting from elements, such as molecules or mesophases interacting through noncovalent bonds.

The topics of photonic crystals and new developments of plastic electronics may be contents suitable for the third edition of this book.

ACKNOWLEDGMENTS

This work is supported by the Italian Ministry of the University and Scientific and Technological research through FIRB 2001 and PRIN 2004 Projects.

REFERENCES

1. M. Ratner and D. Ratner, "*Nanotechnology*," Pearson Education, Delhi (2003).
2. Nanostructure *Science* and Technology by U.S. National Science and Technology Council, www.whitehouse.gov/WH/EOP/OSTP/NSTC (1999).
3. K.A. Valiev, "*The Physics of Submicron Lithography (Microdevices)*," Plenum, New York (1992).
4. S.H. Tolbert and A.P. Alivisatos, *Science* 265, 373 (1994).
5. E.W.H. Jager, O. Inganas, and I. Lundström, *Science* 288, 2335 (2000). R.H. Baughman, C. Cui, A.A. Zakhidov, Z. Iqbal, J.N. Barisci, G.M. Spinks, G.G. Wallace, A. Mazzoldi, D. De Rossi, A.G. Rinzler, O. Jaschinski, S. Roth, and M. Kertsz, *Science* 284, 1340 (1999). http://www.ifm.liu.se/Applphys/ConjPolym/research/micromuscles/micromuscles.html.
6. M. Cavallini, F. Biscarini, S. Leon, F. Zerbetto, G. Bottari, and D.A. Leigh, *Science* 299, 531 (2003), see also references therein.

7. J.-P. Sauvage and C. Dietrich-Buchecker, Eds., *"Molecular Catenanes, Rotaxanes and Knots,"* Wiley-VCH, Weiheim (1999); *Acc. Chem. Res.* 34, 410 (2001) (special issue, J.F. Stoddard. Ed.).

8. C.P. Collier, E.W. Wong, M. Belohradsky, F.M. Raymo, J.F. Stoddard, P.J. Kuekes, R.S. Wolliams, and J.R. Heath, *Science* 285, 391 (1999). C.P. Collier, G. Mattersteig, E.W. Wong, Y. Luo, K. Beverly, J. Sampao, F.M. Raymo, J.F. Stoddard, and J.R. Heath, *Science* 289, 1172 (2000).

9. E.W.H. Jager, O. Inganas, and I. Lundstrom, *Science* 288, 2335 (2000), see also references therein.

10. S.A. Jenekhe, E. Reichmanis, and M.D. Ward, Eds., *"Organic Electronics,"* special issue of *Chemistry of Materials* vol. 16 (2004).

11. See for instance *Science* 295 (2002) (special issue, Supramolecular Chemistry and Self-Assembly).

12. For a review on Photonic Crystals properties see *J. Opt. Soc. Am. B* 10, (1993). (special issue on Photonic crystals)

13. See for instance: http://optics.org/articles/news/8/3/30/1, http://optics.org/articles/ole/6/1/5/1, http://optics.org/articles/ole/8/6/6/1, http://optics.org/articles/ole/8/7/29/1, http://www.spectrum.ieee.org/WEBONLY/publicfeature/sep02/plasticweb.html, http://optics.org/articles/ole/6/7/5/1, http://optics.org/articles/ole/6/3/4/1, http://optics.org/articles/ole/8/4/20/1, http://optics.org/articles/ole/6/3/17/1, http://www.spectrum.ieee.org/publicfeature/aug00/orgs.html, http://optics.org/articles/ole/7/6/9/1.

14. J. Mort and G. Pfister, Eds., *"Electronic Properties of Polymers,"* Wiley, New York (1982).

15. R.E. Peierls, *"Quantum Theory of Solids,"* Oxford University Press, London (1955). Peierls instability is a general property of one-dimensional systems having an unpaired electron in single units. In the undimerized system, a half-filled energy band is created (metallic state). Peierls instability split up the latter band creating a filled valence and an empty conduction band separated by an energy gap (semiconductor), thus allowing the system to gain energy. In fact, the energy gained by the valence band is not compensated by the loss of the conduction band, the latter being empty. The degenerancy of the undimerized system is removed by coupling with vibrational modes having a lower symmetry with respect to the original configuration thus distorting it and removing the degeneracy causing the dimerization.

16. N. Basescu, Z.-Liu, D. Moses, A.J. Heeger, H. Naarman, and N. Theophilou, *Nature* 327, 403 (1987).

17. D. Bloor and R.R. Chance, Eds., *"Polydiacetylenes,"* Nijhoff, Dordrecht (1985).

18. G. Weiser, *Phys. Rev. B* 45, 14076 (1992).

19. S. Spagnoli, J. Berréhar, C. Lapersonne-Meyer, and M. Schott, *J. Chem. Phys.* 100, 6195 (1994). M. Schott, *Synth. Metals* 139, 739 (2003), see also references therein.

20. J. Le Moigne, F. Kajzar, and A. Thierry, *Macromolecules* 24, 2622 (1991). V. Da Costa, J. Le Moigne, L. Ostwald, T.A. Pham, and A. Thierry, *Macromolecules* 31, 1635 (1998).

21. D. Comoretto, I. Moggio, C. Dell'Erba, C. Cuniberti, G.F. Musso, G. Dellepiane, L. Rossi, M.E. Giardini, and A. Borghesi, *Phys. Rev. B* 54, 16357 (1996).

22. A. Borghesi, A. Sassella, R. Tubino, S.Destri, and W. Porzio, *Adv. Mater.* 10, 931 (1998). A. Sassella, A. Borghesi, F. Meinardi, R. Tubino, M. Gurioli, C. Botta, W. Porzio, and G. Barbarella, *Phys. Rev. B* 62, 11170 (2000).

23. J.C. Wittman and P. Smith, *Nature* 352, 414 (1991). M. Fahlman, J. Rasmusson, K. Kaeriyama, D.T. Clark, G. Beamson, and W.R. Salaneck, *Synth. Metals* 66, 123 (1994).

24. J. Piris, M.G. Debije, N. Stutzmann, A.M. van de Craats, M.D. Watson, K. Mullen, and J.M. Warman, *Adv. Mater.* 15, 1736 (2003).

25. R. Pedriali, C. Cuniberti, D. Comoretto, G. Dellepiane, C. Dell'Erba, M. Novi, A. Bolognesi, G. Bajo, and W. Porzio, *Thin Solid Films* 284, 36 (1996). A. Bolognesi, G. Bajo, D. Comoretto, P. Elmino, and S. Luzzati, *Thin Solid Films* 299, 169 (1997).

26. M. Thakur and S. Meyler, *Macromolecules* 18, 2341 (1985). S.J. Gason, D.E. Dunston, T.A. Smith, D.Y.C. Chan, L.R. White, and D. Boger, *J. Phys. Chem. B* 101, 7732 (1997).

27. N.S. Sariciftci, *Synth. Metals* 80, 137 (1997).

28. F. Kajzar, A. Lorin, J. Le Moigne, and J. Szpunar, *Acta Phys. Pol.* 87, 713 (1995).

29. M. Chandross, S. Mazumdar, M. Liess, P.A. Lane, Z.V. Vardeny, M. Hamaguchi, and K. Yoshino. *Phys. Rev. B* 55, 1486 (1997).

30. A. Bolognesi, C. Botta, D. Facchinetti, M. Jandke, K. Kreger, P. Strohriegl, A. Relini, R. Rolandi, and S. Blumstengel, *Adv. Mater.* 13, 1072 (2001). M. Jandke, P. Strohriegl, J. Gmeiner, W. Brüttig, and M. Schwoerer, *Adv. Mater.* 11, 1518 (1999).

31. D. Comoretto, I. Moggio, C. Cuniberti, G.F. Musso, G. Dellepiane, A. Borghesi, F. Kajzar, and A. Lorin, *Phys. Rev. B* 57, 7071 (1998).

32. S. Nagamatsu, W. Takashima, K. Kaneto, Y. Yoshida, N. Tanigaki, K. Yase, and K. Omote, *Macromolecules* 36, 5252 (2003).

33. I. Moggio, J. Le Moigne, E. Arias-Marin, D. Issautier, A. Thierry, D. Comoretto, G. Dellepiane, and C. Cuniberti, *Macromolecules* 34, 7091 (2001).

34. D. Comoretto, G. Dellepiane, G.F. Musso, R. Tubino, R. Dorsinville, A. Walser, and R.R. Alfano, *Phys. Rev. B* 46, 10041 (1992).

35. G. Leising, *Phys. Rev. B* 38 (1988) 10313.

36. T. Ohnishi, T. Noguchi, T. Nakano, M. Hirooka, and I. Murase, *Synth. Metals* 41, 455 (1991). S. Kuroda, T. Noguchi, and T. Ohnishi, *Phys. Rev. Lett.* 72, 286 (1994).

37. D. Comoretto, G. Dellepiane, D. Moses, J. Cornil, D.A. dos Santos, and J.L. Brédas, *Chem. Phys. Lett.* 289, 1 (1998). D. Comoretto, G. Dellepiane, F. Marabelli, J. Cornil, D.A. dos Santos, J.L. Brédas, and D. Moses, *Phys. Rev. B* 62, 10173 (2000). D. Comoretto, G. Dellepiane, F. Marabelli, P. Tognini, A. Stella, J. Cornil, D.A. dos Santos, J.L. Brédas, and D. Moses, *Synth. Metals* 116, 107 (2001). D. Comoretto, F. Marabelli, P. Tognini, A. Stella, J. Cornil, D.A. dos Santos, J.L. Bredas, D. Moses, and G. Dellepiane, *Synth. Metals* 124, 53 (2001). D. Comoretto, F. Marabelli, P. Tognini, A. Stella, and G. Dellepiane, *Synth. Metals* 119, 643 (2001).

38. T.W. Hagler, K. Pakbaz, and A.J. Heeger, *Phys. Rev. B* 49, 10 968 (1994), see also references therein.

39. E.K. Miller, D. Yoshida, C.Y. Yang, and A.J. Heeger, *Phys. Rev. B* 59, 4661 (1999). E.K. Miller, C.Y. Yang, and A.J. Heeger, *Phys. Rev. B* 62, 6889 (2000). E.K. Miller, G.S. Maskel, C.Y. Yang, and A.J. Heeger, *Phys. Rev. B* 60, 8028 (1999).

40. C.Y. Yang, K. Lee, and A.J. Heeger, *J. Mol. Struct.* 521, 315 (2000).

41. D. Chen, M.J. Winokur, M.A. Masse, and F.E. Karasz, *Polymer* 33, 3116 (1992).

42. N.S. Sariciftci, Ed.,"*Primary Photoexcitations in Conjugated Polymers: Molecular Exciton Versus Semiconductor Band Model,*" World Scientific, Singapore (1997).

43. T.A. Skotheim, R.L. Elsembaumer, and J.R. Reynolds, Eds., "*Handbook of Conducting Polymers,*" 2nd ed, Marcel Dekker, New York (1998).

44. N. Kirova, S. Brazovskii, and A.R. Bishop, *Synth. Metals* 100, 29 (1999). A. Köhler, D.A. dos Santos, D. Beljonne, Z. Shuai, J.L. Brédas, A.B. Holmes, A. Kraus, K. Müllen, and R.H. Friend, *Nature* 392, 903 (1998). M. Chandross and S. Mazumdar, *Phys. Rev. B* 55, 1497 (1997). J. Cornil, D.A. dos Santos, D. Beljonne, and J.L. Brédas, *J. Phys. Chem.* 99, 5604 (1995). M. Fahlman, M. Lögdlund, S. Stafström, W.R. Salaneck, R.H. Friend, P.L. Burn, A.B. Holmes, K. Kaeriyama, Y. Sonoda, O. Lhost, F. Meyers, and J.L. Brédas, *Macromolecules* 28, 1959 (1995). Y.N. Gartstein, M.J. Rice, and E.M. Conwell, *Phys. Rev. B* 51, 5546 (1995). M.J. Rice and Y.N. Gartstein, *Phys. Rev. Lett.* 73, 2504 (1994). Y.N. Gartstein, M.J. Rice, and E.M. Conwell, *Phys. Rev. B* 52, 1683 (1995). J. Cornil, D. Beljonne, R.H. Friend, and J.L. Brédas, *Chem. Phys. Lett.* 223, 82 (1994). J. Cornil, D. Beljonne, Z. Suhai, T.W. Hagler, I. Campbell, D.D.C. Bradley, J.L. Brédas, C.W. Spangler, and K. Mullen, *Chem. Phys. Lett.* 247, 425 (1995). J. Cornil, D. Beljonne, C.M. Heller, I.H. Campbell, B.K. Laurich, D.L. Smith, D.D.C. Bradley, K. Müllen, and J.L. Brédas, *Chem. Phys. Lett.* 278, 139 (1997). E. Zojer, Z. Shuai, G. Leising, and J.L. Brédas, *J. Chem. Phys.* 111, 1668 (1999). M. Rohlfing and S.G. Louie, *Phys. Rev. Lett.* 82, 1959 (1999). J.L. Brédas, J. Cornil, and A.J. Heeger, *Adv. Mater.* 8, 447 (1996). D. Beljonne, J. Cornil, R. Silbey, P. Millé, and J.L. Brédas, *J. Chem. Phys.* 112, 4749 (2000). A. Ferretti, A. Ruini, E. Molinari, and M.J. Caldas, *Phys. Rev. Lett.* 90, 086401 (2003). J.-W. Van der Horst, P.A. Bobbert, and M.A.J. Michels, *Phys. Rev. B* 66, 035206 (2002).

45. N. Kirova, S. Brazovskii, and A.R. Bishop, *Synth. Metals* 100, 29 (1999). S. Brazovskii, N. Kirova, A.R. Bishop, V. Klimov, D. McBranch, N.N. Barashkov, and J.P. Ferraris, *Opt. Mater.* 9, 472 (1998). D. Moses, J. Wang, A.J. Heeger, N. Kirova, and S. Brazovski, *Proc. Natl Acad. Sci.* 98, 13496 (2001).

46. D. Comoretto and G. Lanzani, in C. Brabec, V. Dyakonov, J. Parisi, and N.S. Sariciftci, Eds., "*Organic Photovoltaic,*" Springer, Heidelberg (2002).

47. J.B. Birks, "*Photophysics of Aromatic Molecules,*" Wiley, London (1970).

48. M. Pope and C.E. Swemberg, "*Electronic Processes in Organic Crystals and Polymers,*" Oxford University Press, New York (1999).

49. G. Lanzani, G. Cerullo, M. Zavelani-Rossi, S. De Silvestri, D. Comoretto, G.F. Musso, and G. Dellepiane, *Phys. Rev. Lett.* 87, 187402 (2001).

50. P. Mei, M. Murgia, C. Taliani, E. Lunedei, and M. Muccini, *J. Appl. Phys.* 88, 5158 (2000).

51. D. Grando, G.P. Banfi, D. Comoretto, and G. Dellepiane, *Chem. Phys. Lett.* 363, 492 (2002), see also references therein.

52. B. Gallot, A. Cravino, I. Moggio, D. Comoretto, C. Cuniberti, C. Dell'Erba, and G. Dellepiane, *Liq. Cryst.* 26, 1437 (1999). G. Dellepiane, D. Comoretto, and C. Cuniberti, *J. Mol. Struct.* 521, 157 (2000). M. Alloisio, A. Cravino, I. Moggio, D. Comoretto, S. Bernocco, C. Cuniberti, C. Dell'Erba, and G. Dellepiane, *J. Chem. Soc. Perkin Trans.* 2, 146 (2001).

53. D. Comoretto, M. Ottonelli, G.F. Musso, G. Dellepiane, C. Soci, and F. Marabelli, *Phys. Rev. B* 69, 115215 (2004).

54. D. Comoretto, I. Moggio, C. Cuniberti, G. Dellepiane, M.E. Giardini, and A. Borghesi, *Phys. Rev. B* 56, 10264 (1997).

55. G. Dellepiane, C. Cuniberti, D. Comoretto, G.F. Musso, G. Figari, A. Piaggi, and A. Borghesi, *Phys. Rev. B* 48, 7850 (1993).

56. J. Roncali, *Chem. Rev.* 92, 711 (1992); *Chem. Rev.* 97, 369 (1997).

57. R.L. Elsenbaumer, K.Y. Yen, and R. Oboodi, *Synth. Metals* 15, 169 (1986)

58. R.D. McCullough, R.D. Lowe, M. Jayaraman, and D.L. Anderson, *J. Org. Chem.* 58, 904 (1993).

59. M. Zagorska and B. Krischa, *Polymer* 31, 1379 (1990).

60. R.D. McCulloch and R.D. Lowe, *J. Chem. Soc. Chem. Commun.* 70 (1992). T.J. Prosa, M.J. Winokur, and R.D. McCullough, *Macromolecules* 29, 3654 (1996). R.D. McCullough, *Adv. Mater.* 10, 93 (1998).

61. T.A. Chen, X. Wu, and R.D. Rieke, *J. Am. Chem. Soc.* 117, 233 (1995).

62. G. Barbarella, M. Zambianchi, A. Bongini, and L. Antolini, *Adv. Mater.* 6, 561 (1994).

63. Z. Bao, A. Dodabalapur, and A.J. Lovinger, *Appl. Phys. Lett.* 69, 4108 (1996). A. Dodabalapur, Z. Bao, A. Makhija, J.G. Laquindanau, V.R. Raju, Y. Feng, H.E. Katz, and J. Rogers, *Appl. Phys. Lett.* 73, 142 (1998).

64. H. Sirringhaus, N. Tessler, and R.H. Friend, *Science* 280, 1741 (1998). H. Sirringhaus, P.J. Brown, R.H. Friend, M.M. Nielsen, K. Bechgaard, B.M.W. Langeveld-Voss, A.J.H. Spiering, R.A.J. Janssen, E.W. Meijer, P. Horwig, and D.M. de Leeuw, *Nature* 401, 685 (1999).

65. G. Horovitz, *Adv. Mater.* 10, 365 (1998).

66. P.J. Brown, D.S. Thomas, A. Köhler, J.S. Wilson, J.S. Kim, C.M. Ramsdale, H. Sirringhaus, and R.H. Friend, *Phys. Rev. B* 67, 064203 (2003).

67. R. Österbacka, C.P. An, X.M. Jiang, and Z.V. Vardeny, *Science* 287, 839 (2000). O.J. Korovyanko, R. Österbacka, X.M. Jiang, and Z.V. Vardeny, *Phys. Rev. B* 64, 235122 (2001).

68. P.J. Brown, H. Sirringhaus, M. Harrison, M. Shkunov, and R.H. Friend, *Phys. Rev. B* 63, 125204 (2001).

69. H.N. McCoy and W.C. Moore, *J. Am. Chem. Soc.* 33, 273 (1911).

70. K. Bechgaard, C.S. Jacobsen, K. Mortensen, H.J. Pedersen, and N. Thorup, *Solid State Commun.* 33, 1119 (1980). D. Jerome, A. Mazaud, and M. Ribault, *J. Phys. Lett.* 41, L95 (1980). K. Bechgaard, K. Carneiro, F.B. Rasmussen, M. Olsen, G. Rindorf, C.S. Jacobsen, H.J. Pedersen, and J.C. Scott, *J. Am. Chem. Soc.* 103, 2440 (1981). N. Thorup, G. Rindorf, H. Soling, and K. Bechgaard, *Acta Cryst. B* 37, 1236 (1981).

71. R.L. Greene, G.B. Street, and L.J. Suter, *Phys. Rev. Lett.* 34, 577 (1975). V.V. Walatka, M.M. Labes, and J.H. Perlstein, *Phys. Rev. Lett.* 31, 1139 (1973).

72. N. Basescu, Z.-X. Liu, D. Moses, A.J. Heeger, H. Naarman, and N. Theophilou, *Nature* 327, 403 (1987).

73. www.nobel.se/announcement/2000/chemistry.html.

74. W.R. Salaneck, R.H. Friend, and J.L. Brédas, *Phys. Rep.* 319, 231 (1999).

75. R. Kiebooms, R. Menon, and K. Lee, in H.S. Nalwa, Ed., *Handbook of Advanced Electronic and Photonic Materials and Devices, Vol. 8: Conducting Polymers*, Academic Press, Singapore (2001).

76. For a review on PEDOT properties, see L. Groenendaal, F. Jonas, D. Freitag, H. Pielarttzik, and J.R. Reynolds, *Adv. Mater.* 12, 481 (2000).

77. T. Makela, S. Pienimaa, T. Taka, S. Jussila, and H. Isotalo, *Synth. Metals* 85, 1335 (1997). B. Wessling, *Synth. Metals* 85, 1313 (1997). A.E. Wiersma, L.M.A. vd Steeg, and T.J.M. Jongeling, *Synth. Metals* 71, 2269 (1995).

78. T.P. McAndrew, *Trends Polymer Sci.* 5, 7 (1997).

79. Bayer A.G, European Patent 440957 (1991); European patent 553671 (1993); U.S. Patent 686662 (1995); U.S. Patent 5792558 (1996). Agfa Gevaert, European Patent 564911 (1993).

80. T. Nakajima and T. Kawagoe, *Synth. Metals* 28, C629 (1989).

81. H. Sirringhaus, T. Kawase, R.H. Friend, T. Shimoda, M. Inbasekaran, W. Wu, and E.P. Woo, *Science* 290, 2123 (2000).

82. R. Mezzenga, J. Ruokolainen, G.H. Fredrickson, E.J. Kramer, D. Moses, A.J. Heeger, and O. Ikkala, *Science* 299, 1872 (2003).

83. See for instance J.L. Bredas, J.P. Calbert, D.A. da Silva Filho, and J. Cornil, *Proc. Natl Acad. Sci.* 99, 5804 (2002), see also references therein.

84. See for instance, M. O'Neill and S.M. Kelly, *Adv. Mater.* 15, 1135 (2003), see also references therein.

85. See for instance J.-M. Lehn, chap. 1 and A. Ciferri, chap. 2, this volume.

86. V. Percec and D. Schluter, *Macromolecules* 30, 5783 (1997), see also references therein.

87. See for instance W. Zhang and C. Nuckolls, Chap. 16 and J.-M. Lehn, Chap. 1, this volume.

88. A. van de Craats, J.M. Warman, A. Fechtenkotter, J.D. Brand, M.A. Harbison, and K. Mullen, *Adv. Mater.* 11, 1469 (1999).

89. A. van de Craats, N. Stutzmann, O. Bunk, M.M. Nielsen, M.D. Watson, K. Mullen, H.D. Chanzy, H. Sirringhaus, and R.H. Friend, *Adv. Mater.* 15, 495 (2003).

90. D. Natali, M. Sampietro, L. Franco, A. Bolognesi, and C. Botta, *Thin Solid Films* 472, 238 (2005).

91. J. Cornil, V. Lemaur, J.Ph. Calbert, and J.L. Brédas, *Adv. Mater.* 14, 726 (2002).

92. P.W. Atkins, *"Physical Chemistry,"* Oxford University Press, Oxford (1994).

93. Y.-Y. Noh, J.J. Kim, Y. Yoshida, and K. Yase, *Adv. Mater.* 15, 699 (2003).

94. H. Sirringhaus, T. Kawase, R.H. Friend, T. Shimoda, M. Inbasekaran, W. Wu, and E.P. Wo, *Science* 290, 2123 (2000).

95. C. Reese, M. Roberts, M.-M. Ling, and Z. Bao, *Materials Today* September 2004, p. 20.

96. G. Wang, J. Swensen, D. Moses, and A.J. Heeger, *Appl. Phys. Lett.* 93, 6137 (2003).

97. C.D. Dimitrakopoulos and P.R.L. Malenfant, *Adv. Mater.* 14, 99 (2002).

98. C.D. Dimitrakopoulos and D.J. Mascaro, *IBM J. Res. Dev.* 45, 11 (2001).

99. G. Horowitz, X. Peng, D. Fichou, and F. Garnier, *J. Appl. Phys.* 67, 528 (1990). H. Koezuma, A. Tsumura, H. Fuchigami, and K. Kuramoto, *Appl. Phys. Lett.* 62, 1794 (1993). Y. Yang and A.J. Heeger, *Nature* 372, 344 (1994). Z. Bao, A. Dodabalapur, and A.J. Lovinger, *Appl. Phys. Lett.* 69, 4108 (1996). A. Dodabalapur, H.E. Katz, L. Torsi, and R.C. Haddon, *Science* 269, 1560 (1995). Z. Bao, Y. Feng, A. Dodabalapur, V.R. Raju, and A.J. Lovinger, *Chem. Mater.* 9, 1299 (1997).

100. F. Garnier, R. Hajlaoui, A. Yassar, and P. Srivastava, *Science* 265, 1684 (1994).

101. A.R. Brown, A. Pomp, C.M. Hart, and D.M. de Leeuw, *Science* 270, 972 (1995). G. Horowitz, in G. Hadziioannou and P.F. van Hutten, Eds., *"Semiconducting Polymers,"* Wiley-VCH, Weinheim (2000).

102. www.research.philips.com/, www.plasticlogic.com/, http://researchweb.watson.ibm.com/journal/rd/451/dimitrakopoulos.html, http://chester.xerox.com/innovation/focus.html, http://www.siemens.com.

103. A. Dodabalapur, L. Torsi, and H.E. Katz, *Science* 268, 270 (1995). Z. Bao, A.J. Lovinger, and A. Dodabalapur, *Appl. Phys. Lett.* 69, 3066 (1996). R.C. Haddon, *J. Am. Chem. Soc.* 118, 3041 (1996). A. Dodabalapur, H.E. Katz, L. Torsi, and R.C. Haddon, *Appl. Phys. Lett.* 68, 1108 (1996). D.J. Gundlach, Y.Y. Lin, T.N. Jackson, S.F. Nelson, and D.G. Schlom, *IEEE Elect. Dev. Lett.* 18, 87 (1997). F. Garnier, R. Hajlaoui, and M. El Kassmi, *Appl. Phys. Lett.* 73, 1721 (1998). C.D. Dimitrakopoulos, S. Purushothaman, J. Kymissis, A. Callegari, and J.M. Shaw, *Science* 283, 822 (1999).

104. V.C. Sundar, J. Zaumseil, V. Podzorov, E. Menard, R.L. Willett, T. Someya, M.E. Gershenson, and J.A. Rogers, *Science* 303, 1644 (2004).

105. N. Stutzmann, R.H. Friend, and H. Sirringhaus, *Science* 299, 1881 (2003).

106. For a review on Sensor see *Chem. Rev.* 100 (2000) special issue and references therein. T.A. Dickinson, J. White, J.S. Kauer, and D.R. Walt, *Nature* 382, 697 (1996).

107. D.T. McQuade, A.E. Pullen, and T.M. Swager, *Chem. Rev.* 100, 2537 (2000).

108. M.J. Marsella and T.M. Swager, *J. Am. Chem. Soc.* 115, 12214 (1993). R.D. McCullough and S.P. Williams, *J. Am. Chem. Soc.* 115, 11608 (1993).

109. L. Chen, D.W. McBranch, H.-L. Wang, R. Helgeson, F. Wudl, and D.G. Whitten, *Proc. Natl. Acad. Sci.* 96, 12287 (1999). D. Wang, X. Gong, P.S. Heeger, F. Rininsland, G.C. Bazan, and A.J. Heeger, *Proc. Natl. Acad. Sci.* 99, 49 (2002).

110. B.S. Gaylord, A.J. Heeger, and G. Bazan, *J. Am. Chem. Soc.* 125, 896 (2003). S. Wang and G. Bazan, *Adv. Mater.* 15, 1425 (2003). C. Fan, K.W. Plaxco, and A.J. Heeger, *Proc. Natl. Acad. Sci.* 100, 9134 (2003).

111. QTL Biosystems, http://www.qtlbio.com/.

112. P.S. Heeger and A.J. Heeger, *Proc. Natl. Acad. Sci.* 96, 12219 (1999). C. Fan, S. Wang, J.W. Hong, G.C. Bazan, K.W. Plaxco, and A.J. Heeger, *Proc. Natl. Acad. Sci.* 100, 6297 (2003).

113. J.R. Lakowicz, *"Principles of Fluorescence Spectroscopy,"* Kluwer Academic/Plenum, New York (1999).

114. T. Swager, *Acc. Chem. Res.* 31, 201 (1998).

115. R.H. Baughman and R.R. Chance, *Polymer Prep. Am. Chem. Soc.* 27, 67 (1986), see references therein for U.S. patents.

116. D. Birkett and A. Crampton, *Chem. Br.* October, 22 (2003).

117. D.J. Ahn, E.-H. Chae, G.S. Lee, H.-Y. Shim, T.-E. Chang, K.-D. Ahn, and J.-M. Kim, *J. Am. Chem. Soc.* 125, 8976 (2003).

118. D. Bloor, in G. Allen and J.C. Bevington, Eds., *"Comprehensive Polymer Science,"* Vol. 5, Pergamon, Oxford (1989), p. 233.

119. D.H. Charych, J.O. Nagy, W. Spevak, and M.D. Bednarski, *Science* 261, 585 (1993). W. Spevak, J.O. Nagy, D.H. Charych, M.E. Schafer, J.H. Gilbert, and M.D. Bednarski, *J. Am. Chem. Soc.* 115, 1146 (1993).

120. J.J. Pan and D. Charych, *Langmuir* 13, 1365 (1996).

121. Q. Cheng and R.C. Stevens, *Adv. Mater.* 9, 481 (1997).

122. A. Lio, A. Reichert, D.J. Ahn, J.O. Nagy, M. Salmeron, and D. Charych, *Langmuir* 13, 6524 (1997). A. Lio, A. Reichert, J.O. Nagy, M. Salmeron, and D. Charych, *J. Vac. Sci. Technol. B* 14, 1481 (1996).

123. www.research.philips.com/, http://www.covion.com/company/company.html, http://www.cdtltd.co.uk/, http://w4.siemens.de/FuI/en/archiv/zeitschrift/heft2_99/artikel08/, http://www.kodak.com/US/en/corp/display/overview.jhtml, http://www.olight.com/dupont/olight/public/index.cfm, http://chester.xerox,com/innovation/oled.html, http://www-1.ibm.com/technology/ourwork/casestudies/oled.shtml, http://www.lucent.com/news_events/articles/021115.html, http://www.sarnoff,com/products_services/displays/oleds.asp, http://www.universaldisplay.com/, http://www.theclockmag.com/april/articles/oledaboeedit/oledadobeedit.htm.

124. J.K. Borchardt, *Materials Today* September 2004, p. 42.

125. G. Gustafsson, Y. Cao, G.M. Treacy, F. Klavetter, N. Colaneri, and A.J. Heeger, *Nature* 357, 477 (1992).

126. S. Besbes, A. Ltaief, K. Reybier, L. Ponsonnet, N. Jaffrezic, J. Davens, and H. Ben Ouada, *Synth. Metals* 138, 197 (2003).

127. J.R. Burroughes, D.D.C. Bradley, A.R. Brown, N. Marks, K. Mackay, R.H. Friend, P.L. Burn, and A.B. Holmes, *Nature* 347, 539 (1990).

128. I.D. Parker, *J. Appl. Phys.* 75, 1956 (1994).

129. X. Gong, D. Moses, A.J. Heeger, S. Liu, and A.K.-Y. Jen, *Appl. Phys. Lett.* 83, 183 (2003).

130. L.S. Hung, C.W. Tang, and M.G. Mason, *Appl. Phys. Lett.* 70, 152 (1997).

131. J. Cornil, D.A. Dos Santos, D. Beljonne, and J.L. Bredas. *J. Phys. Chem.* 99, 5604 (1995). J.L. Bredas and A.J. Heeger, *Chem. Phys. Lett.* 217, 507 (1994).

132. M. Strukelj, F. Papadimitrakopoulos, T.M. Miller, and I.T. Rothberg, *Science* 267, 1969 (1995).

133. U. Stalmach, B. de Boer, A.D. Post, P.F. van Hutten, and G. Hadziioannou, *Angew. Chem. Int. Ed.* 40, 428 (2001). C. Wang, M. Kilitziraki, L.-O. Pålsson, M.R. Bryce, A.P. Monkman, and I.D.W. Samuel, *Adv. Funct. Mater.* 11, 47 (2001).

134. See for instance V. Abetz, chap. 10, and K. Loos and S. Muñoz-Guerra, chap. 11, this volume.

135. M. Lazzari and M.A. Lopez-Quintela, *Adv. Mater.* 15, 1583 (2003). M.A. Fox, *Acc. Chem. Res.* 32, 201 (1999). G. Krausch and R. Magerle, *Adv. Mater.* 14, 1579 (2002).

136. M. Baumert, H. Frey, M. Hölderle, J. Kressler, F.G. Sernetz, and R. Mülhaupt, *Macromol. Symp.* 121, 53 (1997).

137. B. de Boer, *Ph.D. Thesis*, University of Groningen, The Netherlands (2001).

138. B. Kraabel, D. Moses, and A.J. Heeger, *J. Chem. Phys.* 103, 5182 (1995).

139. M. Wohlgenannt, K. Tandon, S. Mazumdar, S. Ramasesha, and Z.V. Vardeny, *Nature* 409, 494 (2001).

140. Z. Shuai, D. Beljonne, R.J. Silbey, and J.L. Bredas, *Phys. Rev. Lett.* 84, 131 (2000).

141. Y. Cao, I.D. Parker, G. Yu, C. Zhang, and A.J. Heeger, *Nature* 397, 414 (1999). J. Kim, P.K.H. Ho, N.C. Greenham, and R.H. Friend, *J. Appl. Phys.* 88, 1073 (2000).

142. M.A. Baldo, M.E. Thompson, and S.R. Forrest, *Nature* 403, 750 (2000).

143. Y.V. Romanovskii, A. Gerhard, B. Scweitzer, U. Scherf, R.J. Personov, and H. Baessler, *Phys. Rev. Lett.* 99, 1027 (2000).

144. A. Kholer, J.S. Wilson, and R.H. Friend, *Adv. Mater.* 14, 701 (2002). J.S. Wilson, A. Kholer, R.H. Friend, M.K. Al-Suti, M.R.A. Al-Mandhary, M.S. Khan, and P.R. Raithby, *J. Chem. Phys.* 113, 7627 (2000).

145. I.D.W. Samuel, G. Rumbles, and R.H. Friend, in N.S. Sariciftici, Ed., *"Primary Photoexcitations in Conjugated Polymers,"* World Scientific, Singapore (1997), p. 140.

146. J. Cornil, D.A. dos Santos, X. Crispin, R. Silbey, and J.L. Bredas, *J. Am. Chem. Soc.* 120, 1289 (1998). D. Beljonne, J. Cornil, R. Silbey, P. Millé, and J.L. Bredas, *J. Chem. Phys.* 112, 4749 (2000).

147. P. Mei, M. Murgia, C. Taliani, E. Lunedei, and M. Muccini, *J. Appl. Phys.* 88, 5158 (2000).

148. J. Turro, *"Modern Molecular Photochemistry,"* University Science Books California, CF (1991).

149. U. Scherf and K. Mullen, in G. Hadzioannou and P.F. van Hutten, Eds., *"Semiconducting Polymers,"* Wiley-VCH, Weinheim (2000).

150. T.-Q. Nguyen, V. Doan, and B. Schwartz, *J. Chem. Phys.* 110, 4068 (1999).

151. T. Huser, M. Yang, and L.J. Rothberg, *Proc. Natl. Acad. Sci.* 97, 11187 (2000).

152. T.-Q. Nguyen, R.C. Kwong, M.E. Thompson, and B.J. Schwartz, *Appl. Phys. Lett.* 76, 2454 (2000).

153. N. Johansson, D.A. dos Santos, S. Guo, M. Fahlman, J. Salbeck, H. Schenk, H. Arwin, J.L. Bredas, and W.R. Salaneck, *J. Chem. Phys.* 107, 2542 (1997). J. Salbeck, N. Yu, J. Bauer, F. Weissortel, and H. Bestgen, *Synth. Metals* 91, 209 (1997). F. Milota, Ch. Warmuth, A. Tortschanoff, J. Sperling, T. Fuhrmann, J. Salbeck, and H.F. Kauffmann, *Synth. Metals* 121, 1497 (2001). R. Pudzich and J. Salbeck, *Synth. Metals* 138, 21 (2003).

154. D.-C. Shin, Y.-H. Kim, H. You, and S.-K. Kwon, *Macromolecules* 36, 3222 (2003).

155. F. Cacialli, J.S. Wilson, J.J. Michels, C. Daniel, C. Silva, R.H. Friend, N. Severin, P. Samorì, J.P. Rabe, M.J. O'Connel, P.N. Taylor, and H.L. Anderson, *Nat. Mater.* 1, 160 (2002). J.J. Michels, M.J. O'Connel, P.N. Taylor, J.S. Wilson, F. Cacialli, and H.L. Anderson, *Chem. Eur. J.* 9, 6167 (2003).

156. C. Botta, D.R. Ferro, G. Di Silvestro, and R. Tubino in H.S. Nalwa, Ed., *"Supramolecular Photosensitive and Electroactive Materials,"* Academic Press, San Diego (2001), p. 440, see also references therein.

157. T.-Q. Nguyen, J. Wu, S.H. Tolbert, and B.J. Schwartz, *Adv. Mater.* 13, 609 (2001). T.-Q. Nguyen, J. Wu, V. Doan, B.J. Schwartz, and S.H. Tolbert, *Science* 288, 652 (2000).

158. J. Wu, A.F. Gross, and S.H. Tolbert, *J. Phys. Chem. B* 103, 2374 (1999).

159. X. Gong, M.R. Robinson, J.C. Ostrowski, D. Moses, G.C. Bazan, and A.J. Heeger, *Adv. Mater.* 14, 581 (2002). S. Wang and G. Bazan, *Adv. Mater.* 15, 1425 (2003).

160. E.J. List, C. Creely, G. Leising, N. Schulte, A.D. Schluter, U. Scherf, K. Mullen, and W. Graupner, *Chem. Phys. Lett.* 325, 132 (2000).

161. A. Kholer, J.S. Wilson, and R.H. Friend, *Adv. Mater.* 14, 701 (2002), see also references therein.

162. D. Beljonne, G. Pourtois, C. Silva, E. Hennebicq, L.M. Merz, R.H. Friend, G.D. Scholes, S. Setayesh, K. Mullen, and J.L. Bredas, *Proc. Natl. Acad. Sci.* 99, 10982 (2002).

163. L.M. Herz, C. Silva, R.H. Friend, R.T. Philips, S. Setayesh, S. Becker, D. Marsitsky, and K. Mullen, *Phys. Rev. B* 64, 195203 (2001).

164. M. Klessinger and J. Michl, *"Excited States and Photochemistry of Organic Molecules,"* Weinheim, VCH (1995).

165. D.A. Tomalia, chap.7, this volume.

166. D.A. Tomalia, H. Baker, J.R. Dewald, M. Hall, G. Kallos, S. Martin, J. Ryder, and P. Smith, *Polymer J. (Tokyo)* 17, 117 (1985). G.R. Newkome, Z.-Q. Yao, G.R. Baker, and V.K. Gupta, *J. Org. Chem.* 50, 2003 (1985). J.S. Moore, *Acc. Chem. Res.* 30, 402 (1997). C. Devadoss, P. Bharathi, and J.S. Moore, *J. Am. Chem. Soc.* 118, 9635 (1996). V.V. Tsukruk, *Adv. Mater.* 10, 253 (1998). R.G. Duan, L.L. Miller, and D.A. Tomalia, *J. Am. Chem. Soc.* 117, 10783 (1995). H. Tokuhisa, M. Zhao, L.A. Baker, V.T. Phan, D.L. Dermody, M.E. Garcia, R.F. Peez, R.M. Crooks, and T.M. Mayer, *J. Am. Chem. Soc.* 120, 4492 (1998). A.G. Vitukhnovsky, M.I. Sluch, V.G. Ktasovskii, and A.M. Muzafarov, *Synth. Metals* 91, 375 (1997). K. Katsuma and Y. Shirota, *Adv. Mater.* 10, 223 (1998). N. Higashi, T. Koga, and M. Niwa, *Adv. Mater.* 12, 1373 (2000).

167. J.M. Lupton, I.D. Samuel, M.J. Frampton, R. Beavington, and P.L. Burn, *Adv. Funct. Mater.* 11, 287 (2001).

168. X.-T. Tao, Y.-D. Zhang, T. Wada, H. Sasabe, H. Suzuki, T. Watanebe, and S. Miyata, *Adv. Mater.* 10, 226 (1998). P.W. Wang, Y.J. Liu, C. Devadoss, P. Bhaharati, and J.S. Moore, *Adv. Mater.* 8, 237 (1996). M. Halim, J.N.G. Pillow, I.D.W. Samuel, and P.L. Burn, *Adv. Mater.* 11, 371 (1999).

169. D. Beljonne, J. Cornil, R.H. Friend, R.A.J. Janssen, and J.L. Bredas, *J. Am. Chem. Soc.* 118, 6453 (1996).

170. K.L. Wooley, J.M.J. Frechet, and C.J. Hawker, *Polymer* 35, 4489 (1994).

171. C.J. Hawker and F.K. Chu, *Macromolecules* 29, 4370 (1996).

172. http://www.eng.ox.ac.uk/lc/research/introf.html, http://plc.cwru.edu/tutorial/enhanced/files/hindex.html, see also references therein.

173. S.W. Culligan, Y. Geng, S.H. Chen, K.K.K.M. Vaeth, and C.W. Tang, *Adv. Mater.* 15, 1176 (2003).

174. N.S. Sariciftci, L. Smilowitz, A.J. Heeger, and F. Wudl, *Science* 258, 1474 (1992).

175. S. Morita, A.A. Zakhidov, and K. Yoshino, *Solid State Commun.* 82, 249 (1992).

176. N.S. Sariciftci, B. Kraabel, C.H. Lee, K. Pakbaz, A.J. Heeger, and D.J. Sandman, *Phys. Rev. B* 50, 12044 (1994).

177. C.H. Lee, G. Yu, D. Moses, K. Pakbaz, C. Zhang, N.S. Sariciftci, A.J. Heeger, and F. Wudl, *Phys. Rev. B* 48, 15425 (1993).

178. N.S. Sariciftci, D. Braun, C. Zhang, V. Sradanov, A.J. Heeger, G. Stucky, and F. Wudl, *Appl. Phys. Lett.* 62, 585 (1993).

179. M. Cha, N.S. Sariciftci, A.J. Heeger, J.C. Hummelen, and F. Wudl, *Appl. Phys. Lett.* 67, 3850 (1995).

180. J. Nelson, *Mater. Today*, 20 (2002).

181. G. Yu, C. Zhang, and A.J. Heeger, *Appl. Phys. Lett.* 64, 1540 (1994).

182. C. Brabec, in C. Brabec, V. Dyakonov, J. Parisi, and N.S. Sariciftci, Eds., "*Organic Photovoltaic,*" Springer, Heidelberg (2002).

183. N.S. Sariciftci, D. Braun, C. Zhang, V.I. Sradanov, A.J. Heeger, G. Stucky, and F. Wudl, *Appl. Phys. Lett.* 62, 585 (1993).

184. J.C. Hummelen, B.W. Knight, F. Lepec, and F. Wudl, *J. Org. Chem.* 60, 532 (1995). M. Prato, V. Lucchini, M. Maggini, E. Stimpfl, G. Scorrano, M. Eiermann, T. Suzuki, and F. Wudl, *J. Am. Chem. Soc.* 115, 8479 (1993).

185. M. Prato, A. Bianco, M. Maggini, G. Scorrano, C. Toniolo, and F. Wudl, *J. Org. Chem.* 58, 5578 (1993). M. Maggini, G. Scorrano, and M. Prato, *J. Am. Chem. Soc.* 115, 9798 (1993).

186. G. Yu, J. Gao, J.C. Hummelen, F. Wudl, and A.J. Heeger, *Science* 270, 1789 (1995).

187. J.J.M. Halls, C.A. Walsh, N.C. Greenham, E.A. Marseglia, R.H. Friend, S.C. Moratti, and A.B. Holmes, *Nature* 376, 498 (1995).

188. S.E. Shaheen, C.J. Brabec, N.S. Sariciftci, F. Padinger, T. Fromherz, and J.C. Hummelen, *Appl. Phys. Lett.* 78, 841 (2001).

189. http://www.siemens.com, http://www.uni-solar.com/products_markets.html, http://www.bpsolar.com/, http://www.eurosun-solar.de/index.html, http://www.imec.be/, http://www.konarkatech.com/, http://www.pvpower.com/, http://www.powerlight.com/, http://www.shell.com/home/Framework? siteld=shellsolar, http://www.thyssen-solartec.com/uk/index.html, http://www.solarbuzz.com/News/ NewsEUTE11.htm.

190. A. Dhanabalan, J.K.J. van Duren, P.A. van Hal, J.L.J. van Dongen, and R.A.J. Janssen, *Adv. Funct. Mater.* 11, 255 (2001). C. Winder, G. Matt, J.C. Hummelen, R.A.J. Janssen, N.S. Sariciftci, and C.J. Brabec, *Thin Solid Films* 403–404, 373 (2002). C.J. Brabec, C. Winder, N.S. Sariciftci, J.C. Hummelen, A. Dhanabalan, P.A. van Hal, and R.A.J. Janssen, *Adv. Funct. Mater.* 12, 709 (2002). D. Muhlbacher, H. Neugebauer, A. Cravino, N.S. Sariciftci, J.K.J. van Duren, A. Dhanabalan, P.A. van Hal, R.A.J. Janssen, and J.C. Hummelen, *Mol. Cryst. Liq. Cryst.* 385, 205 (2002). C. Winder, D. Muhlbacher, H. Neugebauer, N.S. Sariciftci, C. Brabec, R.A.J. Janssen, and J.K. Hummelen, *Mol. Cryst. Liq. Cryst.* 385, 213 (2002), see also references therein.

191. W.U. Huynh, J.J. Dittmer, and A.P. Alivisatos, *Science* 295, 2425 (2002).

192. See J.-M. Lehn, chap. 1; A. Ciferri, chap. 2; D.A. Tomalia, chap 7, this volume.

193. L. Schmidt-Mende, A. Fechtenkotter, K. Mullen, E. Moons, R.H. Friend, and J.D. MacKenzie, *Science* 293, 1119 (2001).

194. S.A. Jenekhe and X.L. Chen, *Science* 279, 1903 (1998).

195. G. Hadziioannou, *MRS Bull.* 456 (2002). U. Stalmach, B. de Boer, C. Videlot, P.F. van Hutten, and G. Hadziioannou, *J. Am. Chem. Soc.* 122, 5464 (2000). B. de Boer, U. Stalmach, P.F. van Hutten, C. Melzer, V.V. Krasnikov, and G. Hadziioannou, *Polymer* 42, 9097 (2001). J.-F. Eckert, J.-F. Nicoud, J.-F. Nierengarten, S.-G. Liu, L. Echegoyen, F. Barigelletti, N. Armaroli, L. Ouali, V. Krsnikov, and G. Hadzioannou, *J. Am. Chem. Soc.* 122, 7467 (2000).

196. H. Imahori and Y. Sakata, *Adv. Mater.* 9, 537 (1997).

197. E.E. Neuteboom, S.C.J. Meskers, P.A. van Hal, J.K.J. van Duren, E.W. Maijer, R.A.J. Janssen, H. Dupin, G. Pourtois, J. Cornil, R. Lazzaroni, J.-L. Bredas, and D. Beljonne, *J. Am. Chem. Soc.* 125, 8625 (2003).

198. T. Gu and J.F. Nierengarten, *Tetrahedron Lett.* 42, 3175 (2001).

199. D. Chirvase, Z. Chiguvare, M. Knipper, J. Parisi, V. Dyakonov, and J.C. Hummelen, *J. Appl. Phys.* 93, 3376 (2003). E.H.A. Beckers, P.A. van Hal, A.P.H.J. Schenning, A. El-Ghayoury, E. Peeters, M.T. Rispens, J.C. Hummelen, E.W. Meijer, and R.A.J. Janssen, *J. Mater. Chem.* 12, 2054 (2002). E. Peeters, P.A. van Hal, J. Knol, C.J. Brabec, N.S. Sariciftci, J.C. Hummelen, and R.A.J. Janssen, *J. Phys. Chem. B* 104, 10174 (2000). P.A. van Hal, E.H.A. Beckers, E. Peeters, J.J. Apperloo, and R.A.J. Janssen, *Chem. Phys. Lett.* 328, 403 (2000). S.C.J. Meskers, P.A. van Hal, and A.J.H. Spiering, J.C. Hummelen, A.F.G. van der Meer, and R.A.J. Janssen, *Phys. Rev. B* 61, 9917 (2000).

200. R.L. Stolz, O. Inganas, T. Granlund, T. Nyberg, M. Svensson, M.R. Anderson, and J.C. Hummelen, *Adv. Mater.* 12, 189 (2000).

201. See for instance www.ise.fhg.de and references therein.

202. W.M.V. Wan, N.C. Greenham, and R.H. Friend, *J. Appl. Phys.* 87, 2542 (2000). W.M.V. Wan, R.H. Friend, and N.C. Greenham, *Thin Solid Films* 363, 310 (2000). J.S. Kim, P.K.H. Ho, N.C. Greenham, and R.H. Friend, *J. Appl. Phys.* 88, 1073 (2000).

203. P. Vukusic and J.R. Sambles, *Nature* 424, 852 (2003).

204. T. Tsutsui, M. Yahiro, H. Yokogawa, K. Kawano, and M. Yokoyama, *Adv. Mater.* 13, 1149 (2001), see also references therein.

205. K. Meerholz and D.C. Müller, *Adv. Funct. Mater.* 11, 251 (2001).

206. N. Tessler, G.J. Denton, and R.H. Friend, *Nature* 382, 695 (1996). C.E. Finlayson, D.S. Ginger, and N.C. Greenham, *Appl. Phys. Lett.* 77, 2500 (2000). J.M. Lupton, R. Koeppe, J.G. Muller, J. Feldmann, U. Scherf, and U. Lemmer, *Adv. Mater.* 15, 1471 (2003).

207. A. Bolognesi, C. Mercogliano, S. Yunus, M. Civardi, D. Comoretto, and A. Turturro, Langmuir, in press (2005); see therein for an extended reference list.

208. *MRS Bull.* 28 (2003), see also references therein. B.-J. De Gans, P.C. Duineveld, and U.S. Schubert, *Adv. Mater.* 16, 203 (2004).

209. B. Derby and N. Reis, *MRS Bull.* 28, 815 (2003). http://faculty.washington.edu/yagerp/ microfluidicstutorial/basicconcepts/basiccompcets.htm.

210. http://chester.xerox.com/innovation/focus.html, www.plasticlogic.com, www.parc.com/about/pressroom/ news/2003-12-17-flatscreens.html.

211. T.R. Hebner, C.C. Wu, D. Marcy, M.H. Lu, and J.C. Sturm, *Appl. Phys. Lett.* 72, 519 (1998). T.R. Hebner and J.C. Sturm, *Appl. Phys. Lett.* 73, 1775 (1998). J. Bharathan and Y. Yang, *Appl. Phys. Lett.* 72, 2660 (1998).

212. S.E. Burns, P. Cain, J. Mills, J. Wang, and H. Sirringhaus, *MRS Bull.* 28, 829 (2003).

213. Z. Bao, Y. Feng, A. Dodabalapur, V.R. Raju, and A.J. Lovinger, *Chem. Mater.* 9, 1299 (1997). G. Maltezos, R. Nortrup, S. Jeon, J. Zaumseil, and J.A. Rogers, *Appl. Phys. Lett.* 83, 2067 (2003). U. Zschienschang, H. Klauk, M. Halik, G. Schmid, and C. Dehn, *Adv. Mater.* 15, 1147 (2003).

214. D. Pisignano, E. Sariconi, M. Mazzeo, G. Gigli, and R. Cingolani, *Adv. Mater.* 14, 1565 (2002).

215. T. Shimoda, K. Morii, S. Seki, and H. Kiguchi, *MRS Bull.* 28, 821 (2003).

216. A.W. Bosman and R. Sijbesma, chap. 15, this volume.

217. B.L. Groenendal, F. Jonas, D. Freitag, H. Pielartzik, and J.R. Reynolds, *Adv. Mater.* 12, 481 (2000).

218. N. Herron and D.L. Thorn, *Adv. Mater.* 10, 1173 (1998).

219. H.S. Zhou, T. Wada, H. Sasabe, and H. Komiyama, *Synth. Metals* 81, 129 (1996).

220. J.-L. Rehspringer, J. LeMoigne, and J.C. Merle, private communication.

221. H. Parala, H. Winkler, M. Kolbe, A. Wohlfart, R.A. Fischer, R. Schmechel, and H. von Seggern, *Adv. Mater.* 12, 1050 (2000).

222. C.S. Peyratout, H. Mohwald, and L. Dahne, *Adv. Mater.* 15, 1722 (2003).

223. K. Landfester, R. Montenegro, U. Scherf, R. Guntner, U. Asawaapirom, S. Patil, D. Neher, and T. Kietzke, *Adv. Mater.* 14, 651 (2002). T. Piok, S. Gamerith, C. Gadermaier, H. Plank, F.P. Wenzl, S. Patil, R. Montenegro, T. Kietzke, D. Neher, U. Scherf, K. Landfester, and E.J.W. List, *Adv. Mater.* 15, 800 (2003).

224. D. Pisignano, L. Persano, P. Visconti, R. Cingolani, G. Gigli, G. Barbarella, and L. Favaretto, *Appl. Phys. Lett.* 83, 2545 (2003). M. Gaal, C. Gadermeier, H. Plank, E. Moderegger, A. Pongantsch, G. Leising, and E.J.W. List, *Adv. Mater.* 15, 1165 (2003).

225. E.Z. Faraggi, D. Davidov, G. Cohen, S. Noach, M. Golosovsky, Y. Avny, R. Neumann, and A. Lewis, *Synth. Metals* 85, 1187 (1997).

226. R. Riehn, A. Charas, J. Morgado, and F. Cacialli, *Appl. Phys. Lett.* 82, 526 (2003).

227. H.W. Deckman and J.H. Dunsmuir, *Appl. Phys. Lett.* 41, 377 (1982).

228. T.-W. Lee, J. Zaumaseil, Z. Bao, J.W.P. Hsu, and J.A. Rogers, *Proc. Natl. Acad. Sci.* 101, 429 (2004).

229. M. Berggren, O. Inganas, G. Gustafsson, J. Rasmusson, M.R. Andersson, T. Hjeberg, and O. Wennerstrom, *Nature* 372, 444 (1994).

230. B. de Boer, U. Stalmach, H. Nijland, and G. Hadzioannou, *Adv. Mater.* 12, 1581 (2000).

231. A. Bolognesi, C. Botta, and S. Yunus, *Thin Solid Films*, submitted.

232. M.A. Baldo, R.J. Holmes, and S.R. Forrest, *Phys. Rev. B* 66, 035321 (2002), see also references therein.

233. I.D.W. Samuel and G.A. Turnbull, *Materials Today* September 2004, p. 28

234. J.D. Joannopoulos, R.D. Meade, and J.N. Winn, *"Photonic Crystals,"* Princeton University Press, Princeton (1995). E. Yablonovich, *Phys. Rev. Lett.* 58, 2059 (1987). S. John, *Phys. Rev. Lett.* 58, 2486 (1987).

235. B.H. Cumpston, S.P. Anathavel, S. Barlow, D.L. Dyer, J.E. Ehrlich, L.L. Erskine, A.A. Heikal, S.M. Kuebler, I.Y.S. Lee, D. McCord-Maughon, J.Q. Qin, H. Rockel, M. Rumi, X.L. Wu, S.R. Marder, and J.W. Perry, *Nature* 398, 51 (1999).

236. A.M. Urbas, M. Maldovan, P. DeRege, and E.L. Thomas, *Adv. Mater.* 14, 1850 (2002).

237. H. Finkelman, S.T. Kim, A. Munoz, P. Pallfy-Muhoray, and B. Taheri, *Adv. Mater.* 13, 1069 (2001). M. Ozaki, M. Kasano, T. Kitasho, D. Ganzke, W. Haase, and K. Yoshino, *Adv. Mater.* 15, 974 (2003).

238. J.E.G.J. Winhoven and W.L. Vos, *Science* 281, 802 (1998). A.A. Zakhidov, R.H. Baughman, Z. Iqbal, C. Cui, I. Khayrullin, S.O. Dantas, J. Marti, and V.G. Ralchenko, *Science* 282, 897 (1998). P. Jiang, J.F. Bertone, and V.L. Colvin, *Science* 291, 453 (2001).

239. D. Comoretto, D. Cavallo, G. Dellepiane, R. Grassi, F. Marabelli, L.C. Andreani, C.J. Brabec, A. Andreev, and A.A. Zakhidov, *Mater. Res. Soc. Proc.* 708, BB10.19.1 (2002). D. Comoretto, R. Grassi, F. Marabelli, and L.C. Andreani, *Mater. Sci. Eng.* C23, 61 (2003). D. Comoretto, F. Marabelli, C. Soci, M. Galli, F. Pavarini, M. Patrini, and L.C. Andreani, *Synth. Metals* 139, 633 (2003), E. Pavarini, L.C. Andreani, C. Soci, M. Galli, F. Marabelli, and D. Comoretto, *Phys. Rev. B*, submitted.

Chapter 15

Supramolecular Polymers in Action

Anton W. Bosman and Rint P. Sijbesma

CONTENTS

I. INTRODUCTION

As a scientific concept, supramolecular polymers have received increasing attention in recent years. The introduction of concepts from supramolecular chemistry in polymer science have led to the development of polymers in which the main chain is held together by noncovalent interactions such as hydrogen bonds, π-stacking, and coordination interactions. In the present chapter, we will discuss how the scientific developments in a specific class of supramolecular polymers — quadruple hydrogen-bonded supramolecular polymers — is now rapidly leading to applications that take advantage of the unique mechanical properties of these materials.

Many applications of supramolecular polymers are based on the exceptionally strong dependence of their properties on environmental conditions such as temperature and solvent polarity. Because the bonds that keep the unimers of supramolecular polymers together are relatively weak, these parameters affect both the average length of the polymer chain as well as the dynamics of reversible chain breaking/recombination. This results in a viscoelastic response that may be dramatically stronger than in conventional polymers.

Before proceeding to a description of the chemistry that has led to applicable supramolecular polymers, it is useful to discuss the role of directionality in distinguishing supramolecular polymers from other aggregated form of molecular matter, such as colloids and crystals.

In all condensed molecular materials, whether they are liquid, glassy, or (liquid) crystalline, noncovalent interactions with little specificity or directionality are present. Only when highly directional forces dominate the interaction with neighboring unimers, it is meaningful to use the term supramolecular polymer. In ideal linear supramolecular polymers, exactly two of these interactions per unimer lead to the formation of polymeric chains. The presence of linear chains that persist when the material is heated or dissolved is therefore the hallmark of a successful design of strong and directionally interacting functional groups. Once identified, these functional groups may also be used to develop multifunctional unimers capable of forming more complex polymers materials, with dendritic, graft, or network architectures.

In a fundamental research context, the directionality and strength of the interactions are of prime importance. When a single type of interaction dominates the interaction between bifunctional unimers, it is easy to understand and describe the relation between molecular structure and macroscopic properties. While the directionality and strength of the interactions between unimers are also important when developing applications for supramolecular polymers, the use of multifunctional instead of bifunctional building blocks is often an efficient way to achieve the desired properties. However, for applications there are a number of requirements that are sometimes difficult to meet. The most challenging of these are synthetic availability, cost, and stability of appropriate functional groups.

II. QUADRUPLE HYDROGEN BONDS

Among the functionalities that have been used to form supramolecular polymers, multiple hydrogen-bonding units take a prominent position because of the unique combination of specificity, directionality, and strength. However, there is also an intrinsic trade-off between the binding strength and the synthetic accessibility of multiple hydrogen-bonding units. Whereas recent years have seen the development of building blocks associating via five to ten hydrogen bonds, and with association constants in excess of $10^9 \ M^{-1}$, these building blocks are often obtained via lengthy synthesis, and are not available on a large scale. The pioneering work of Lehn et al. has shown that triple hydrogen-bonding units can be used to form supramolecular polymers [1–3]. However, with K_a values below $10^4 \ M^{-1}$ at room temperature, the interactions are not strong enough to give polymers with high degrees of polymerization.

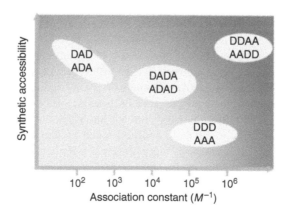

Figure 1   Relation between synthetic accessibility and binding strength of multiple hydrogen-bonding units.

Obviously, the strength of the interaction may be increased by increasing the number of hydrogen-bonding sites from three to four, and a project to find quadruple hydrogen-bonding units that combine high binding strength with synthetic accessibility (Figure 1) was initiated in our group [4]. The principle factors determining association strength between units containing linear arrays of hydrogen-bonding sites, in particular the effect of secondary repulsive and attractive interactions [5] have been discussed in detail by Zimmerman and Corbin in Section II.B of Chapter 6. Several building blocks giving rise to self-complementary DADA arrays bear out the prediction that the dimerization strength of such units with a high number of repulsive secondary interactions is in most cases limited to values $<10^5\ M^{-1}$ [6]. Although the interaction is significantly stronger than for DAD–ADA type triple hydrogen bonds, it is barely sufficient for the formation of linear supramolecular polymers that reach entanglement molecular weights. Moreover, the unit featuring the highest dimerization constant (1) is relatively difficult to synthesize and requires an expensive starting material, diaminopyrimidine. Much higher dimerization constants are expected for dimers of DDAA hydrogen-bonding units, in which the number of repulsive secondary interactions is reduced to 2.

1·1

Eventually, DDAA arrays were realized in ureidopyrimidinones UPy 2 [7]. Initial experiments showed that these molecules form exceedingly stable dimers in CDCl3 solution, which do not show any sign of dissociation in NMR upon dilution to $10^{-4}\ M$, indicating a dimerization constant exceeding $10^5\ M^{-1}$.

2a R = nBu; R′ = Me
2b R = nBu; R′ = Ph

The proposed structure of the dimer was confirmed by the crystal structure of 2a (Figure 2), which shows dimers held together by four hydrogen bonds, between molecules which are preorganized through an intramolecular hydrogen bond from the pyrimidine N–H to the urea carbonyl group. In a similar approach, and almost simultaneous to the development of the UPy unit, Zimmerman

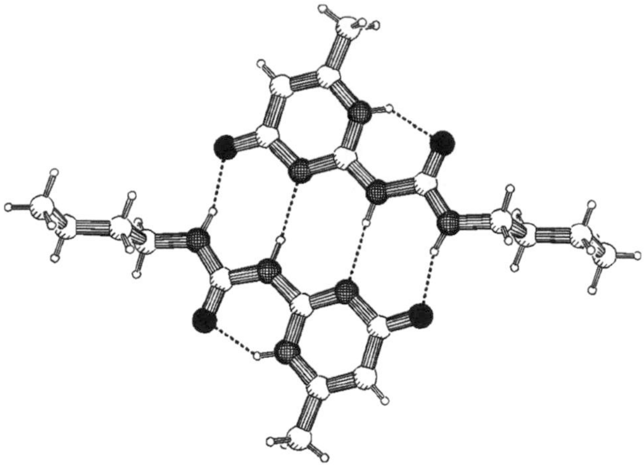

Figure 2   Crystal structure of the dimer of 2-butylureido, 6-methyl pyrimidinone **2a**.

developed the related ureidopyridopyrimidinone unit **3**, which is also capable of dimerizing through a linear array of four hydrogen-bonding sites in a self-complementary fashion [8].

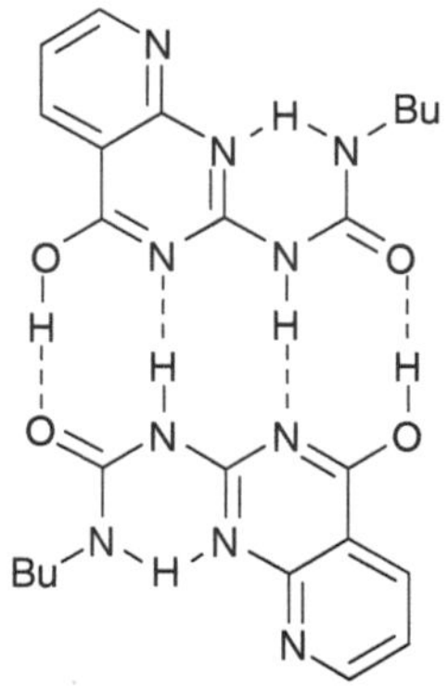

3

Unlike the acylated diaminopyrimidine derivatives with DADA arrays of hydrogen-bonding sites, the simplest UPy derivatives are obtained in a single synthetic step from the cheap starting materials methyl isocytosine and an isocyanate. The combination of strong dimerization and synthetic accessibility made the UPy unit a highly promising candidate for use in supramolecular polymers, and justified further study into the preparation and properties of this building block.

III.   UPy PROPERTIES

Like in many heterocyclic molecules, tautomerism gives rise to different isomers of the UPy molecule, as shown in Scheme 1. In the parent heterocycle, isocytosine, the 3H tautomeric form is the most stable form [9]. In DMSO, a solvent that strongly competes for hydrogen bonding, UPy **2a** is also present in this tautomeric form. However, in chloroform solution, **2a** occurs almost exclusively (>99%) in its 1H tautomeric form, while a very small amount of aromatic pyrimidinol tautomer is observed.

The relative abundance of each tautomer in solution is the combined result of relative energies of the different tautomers in the monomeric form, and their relative dimerization constants. Moreover,

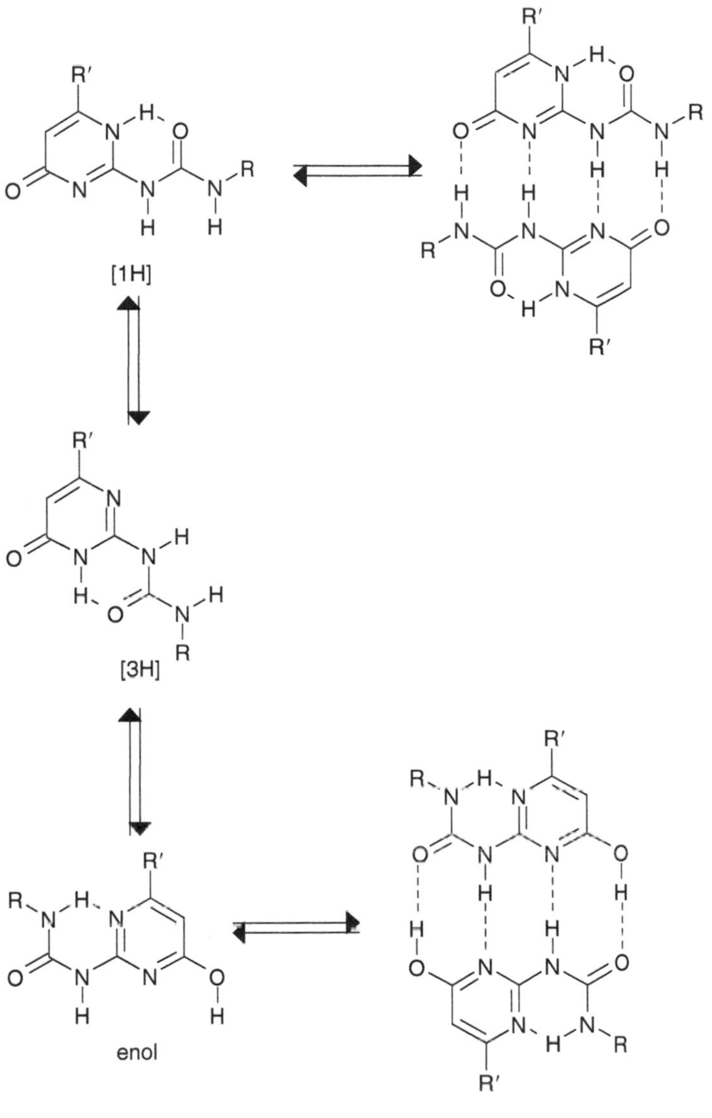

Scheme 1

the equilibrium is dependent upon the nature of the solvent as well as of substituents, with electron withdrawing substituents at the 6-position of the heterocycle favoring the pyrimidinol tautomer. In the 6-phenyl derivative **2b**, the energy of the two dimerizing tautomers is similar, and crystal modifications containing either tautomer were obtained.

The association properties of the UPy unit have been studied quantitatively using Pyrene-labeled derivative **2c** [10]. Concentration dependent excimer fluorescence of this compound was used to determine the degree of association as a function of concentration, and K_{dim} values were derived from these data. The experiments showed that UPy dimerization is very strong indeed; in CDCl$_3$ at 298 K a K_{dim} value of $5 \times 10^7 \, M^{-1}$ was measured, while in toluene this value was ten times higher, $6 \times 10^8 \, M^{-1}$. Surprisingly, the interaction was only moderately weakened by the presence of water, as the K_{dim} value decreased by a factor of 6 ($K_{dim} = 1 \times 10^6$) when the CHCl$_3$ was saturated with water. The lifetime of associated groups is an important parameter for the properties

2c

of supramolecular polymers. For UPy, the exchange rate between homo- and heterodimers in a statistical mixture of **2d** and **2e**, was determined using NMR.

It was established that the pre-exchange lifetime in $CDCl_3$ for the system under study is 170 msec, while in toluene this value is again a factor 10 higher. Comparing the dissociation rates with the thermodynamic K_{dim} values allows the conclusion that the association rate is not diffusion controlled, like it is for other hydrogen bonded complexes. Possibly, tautomerization of the monomer is required before dimerization can occur. Recently, the structure of the dimers in solution was studied using NMR on the ^{15}N-labeled derivative of **2a** [11]. Measurement of the coupling constants across the hydrogen bonds was used to establish that the dimers have a geometry very similar to that in the crystal, thus conclusively confirming the mode of association of the UPy quadruple hydrogen-bonding unit in an environment relevant for application in supramolecular polymers.

In applications of supramolecular polymers, the thermal stability of the UPy functional group is very important. DSC of many UPy supramolecular polymers show that when heated above 180°C for longer periods of time, irreversible changes take place. Like in urea containing polymers, the covalent urea N–C bond in UPy supramolecular polymers is labile and heating leads to scrambling of substituents on the urea. In UPy derivatives, the isocytosine fragment is a relatively good leaving group, and scrambling is expected to become noticeable at lower temperatures than for simple aliphatic or aromatic urea derivatives. Scrambling of *bis*-UPy's results in covalent coupling of linker units, and the formation of highly insoluble molecules such as **4**. It was also shown that above 225°C, the melting point of **2a**, chemical degradation with the formation of gaseous products sets in [12].

IV. SYNTHESIS OF UPY'S

The ureidopyrimidones are easily obtained via the reaction between an isocytosine derivative and aliphatic or aromatic isocyanates, giving access to a wide range of UPy derivatives. The isocytosine

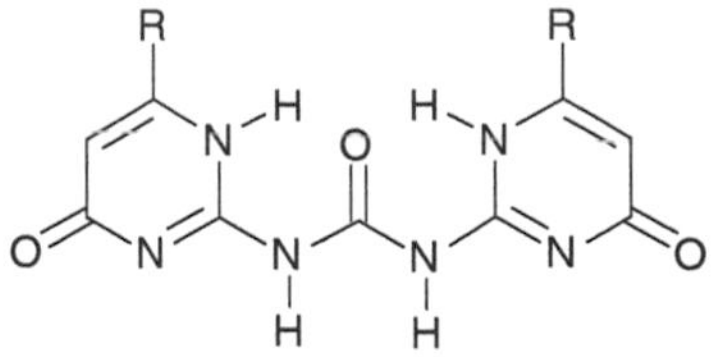

4

building blocks are produced by the gentle condensation of (commercial) β-ketoesters with guanidine-salts, and hence, several isocytosine-derivatives are commercially available in more than gram quantities.

Synthesis of mono UPy's from isocytosine and an isocyanate is also unproblematic, and if the interest in UPy's would be restricted to monofunctional derivatives, there would probably be no need for improved synthetic procedures. However, for the preparation of *bis*-UPy's for linear supramolecular polymers, and for multifunctional UPy derivatives, the addition reaction is not trivial anymore, because molecules with multiple isocyanate groups suffer from hydrolytic instability leading to higher molecular weight species, and therefore are most favorably prepared *in situ* [13].

In the synthesis of a urea group from an isocyanate and an amine, the reactive isocyanate may be part of either of the reaction partners. For the synthesis of monomers containing two or more UPy groups, combination of isocyanato-isocytosine **5** with a multifunctional amine would be a superior synthetic approach, since partial hydrolysis of isocyanato groups before or during reactions can be compensated for by using an excess of the isocyanate, a strategy that is not applicable when multiisocyanates are combined with isocytosine.

However, it was shown some time ago by von Gizicki [14] that isocyanates of aminopyrimidines are unstable, and spontaneously combine to form dimers **6**.

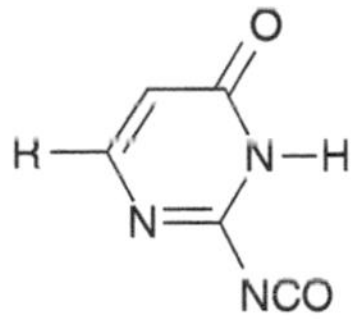

5 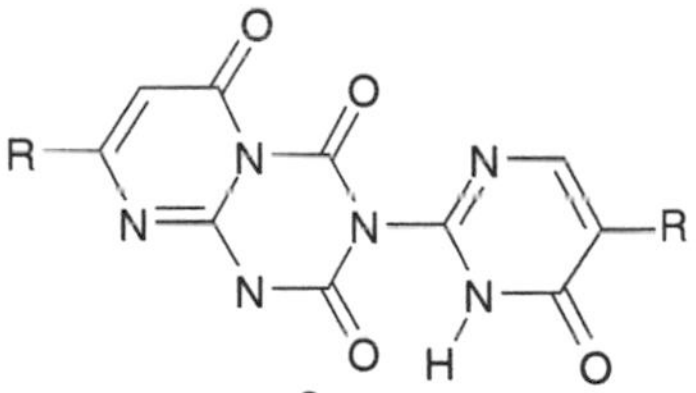   **6**

To circumvent this problem, a reactive UPy synthon was prepared on a large scale by refluxing methylisocytosine in a large excess of diisocyanatohexane, which was used as a solvent [15]. The resulting product (**7**) was isolated by filtration, while the excess of diisocyanatohexane could be recovered by distillation.

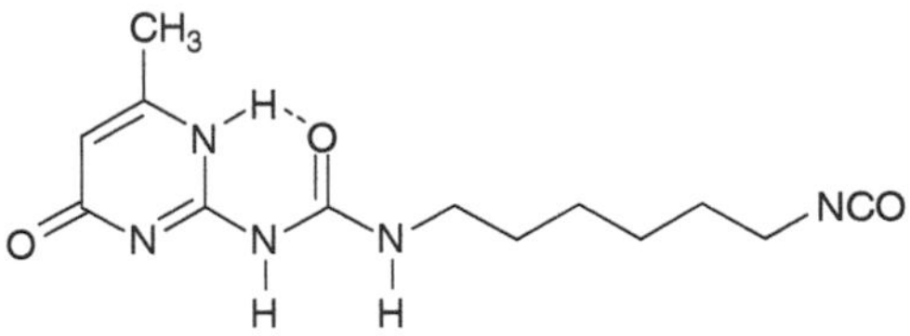

7

Using this procedure, the reactive synthon was obtained in hundreds of grams, and was subsequently used for the functionalization of telechelic polymers. An added advantage of the use of this synthon is that it can be used to functionalize amino as well as hydroxy telechelic supramolecular polymers. A second, very convenient method to circumvent the instability of isocytosine isocyanate was developed recently [16]. This method makes use of the blocked isocyanate equivalent **8**, prepared by reaction of isocytosine with carbonyl diimidazole (CDI), a reagent that has found previous use in the selective formation of asymmetric carbonates. The reactivity differences between the CDI reagent and the blocked isocytosine isocyanate are such that the formation of **8** is complete within 1 h, while the formation of **4** from reaction of the product with isocytosine is not observed. Subsequent reaction with aliphatic and even aromatic amines proceeds with high yields, and is complete within a few hours. The applicability of this reactive isocyanate equivalent in the synthesis of multifunctional UPy's was demonstrated by the preparation of **9** in 84% yield by reaction of triamine **10** with 3.5 equivalents of **8**. The product was readily removed by precipitation in methanol.

Functionalization of UPy derivatives in order to prepare bi- or multifunctional telechelics is not limited to reactions on the ureido-substituent, but may also use a functionality at the 5- or 6-position of the pyrimidine ring (Scheme 2). Olefinic UPy derivatives have been obtained by condensation of an ethylenically unsaturated β-ketoester with guanidine. The resulting UPy **2f** has been used to end-functionalize telechelic poly(dimethylsiloxane) via hydrosilylation with a Karstedt's catalyst [17].

2f n=2
2g n=10

This reaction proceeds in high yields, and has also been used to obtain multifunctional UPy-PDMS from commercially available precursors. Moreover, UPy **2g** has been copolymerized with 1-hexene in a coordination polymerization with Brookhart's catalyst, resulting in UPy-containing polyolefins [18].

V.   UPY-BASED SPs

A.   Low Molecular Weight Linkers

Bifunctional UPy derivatives with low molecular weight have been instrumental in demonstrating the feasibility of linear, reversible supramolecular polymers with high degrees of polymerization.

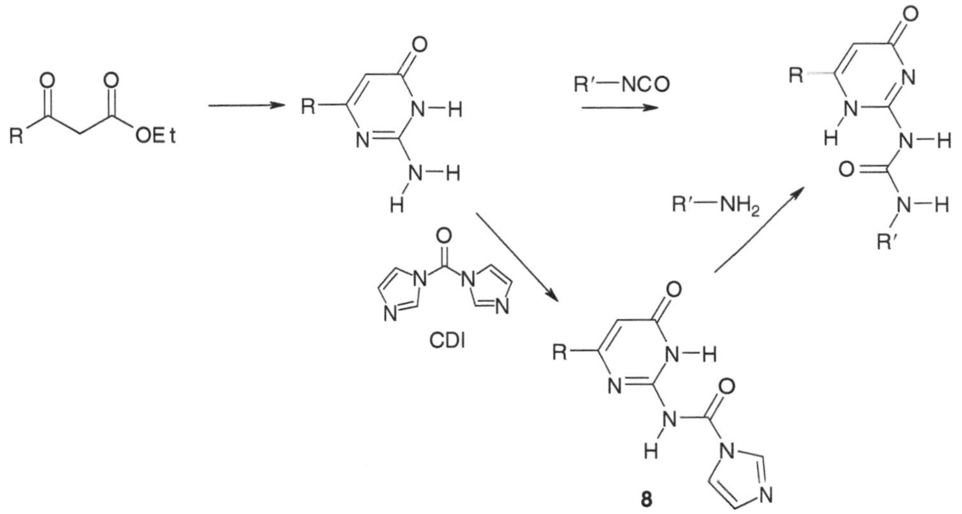

Scheme 2

The monodispersity of these derivatives, and the possibility to purify them with chromatographic methods has been essential for obtaining insight in structure–property relations.

The first UPy-based SPs that were prepared consist of two UPy-groups connected by a C-6 linker unit [19]. This compound forms chloroform solutions with pronounced viscosity. Like in covalent condensation polymers, the degree of polymerization of hydrogen bonded supramolecular polymers is strongly influenced by the presence of monofunctional derivatives, which act as chain stoppers. It was observed that the viscosity of the solutions of **11a** increase with each purification step, an observation that supports the proposed polymerization via hydrogen bonding. Intentional addition of monofunctional stopper molecule **2h** to a 40 mM solution of exhaustively purified **11a** resulted in a noticeable decrease of solution viscosity, even when only 0.1% of this stopper molecule was added (Figure 3).

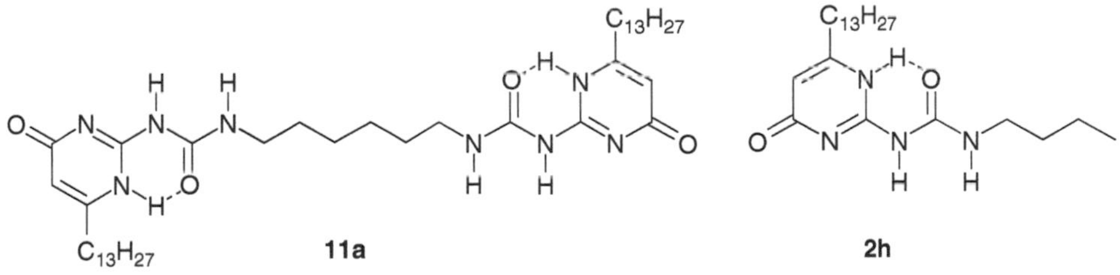

This experiment demonstrates a high selectivity for linear aggregation through hydrogen bonding, because less defined aggregates would not be influenced to the same degree by addition of **2a**. Furthermore, the experiment confirms the reversibility of the supramolecular polymerization, as the reduction of viscosity takes place immediately. Quantitative analysis of the "stopper experiment," using a multistage open association model (see Chapter 1, Section III), and assuming the absence of cyclic species, gave a DP of 550 for a 40 mM solution without added stopper molecules. The effect of impurities and of dissociation of UPy dimers due to a limited K_{dim} cannot be distinguished on the basis of this experiment. However, the value of K_{dim} ($6 \times 10^7 \ M^{-1}$) was known from independent determination on monofunctional model compounds [10], and combination of these data allowed the conclusion that the presence of approximately 0.2% of monofunctional

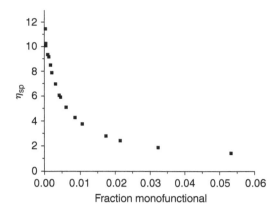

Figure 3   Effect of addition of small amounts of monofunctional UPy derivative **2a** on the specific viscosity of a 40 mM solution of **11a** in chloroform.

impurities in the monomer reduced the degree of polymerization in the chloroform solution from a theoretical value of DP $> 10^4$ to DP $= 550$.

11b R $=$ H; R$'$ $=$ CH$_3$/R $=$ CH$_3$; R$'$ $=$ H

Study of the bulk viscoelastic properties of **11a** are hampered by the crystallinity of the material, even though crystallization is slow. By introducing linkers with a mixed methyl substitution pattern, noncrystallizing supramolecular polymer **11b** was obtained, which was studied using dynamic mechanical thermal analysis (DMTA), rheology, and dielectric relaxation spectroscopy [20].

One of the salient features of the material, with high relevance for applications, is the extremely high activation energy for viscous flow of 105 kJ mol^{-1} (Figure 4), determined with the Andrade–Eyring equation. This strongly results in a temperature dependent melt viscosity, which increases processability of these materials at temperatures just above the melting point or Tg. The high activation energy can be attributed to the contribution of three mechanisms to stress relaxation in sheared melts of supramolecular polymers: One is a mechanism that is shared with covalent polymers — escape from entanglements by reptation [21]. In addition to that, supramolecular polymers have enhanced relaxation at higher temperatures because the chains become shorter. Finally, in a mechanism unique to reversible polymers, the supramolecular chains may lose strain by breaking, followed by recombination of free chain ends without strain [22]. Breaking rates increase with temperature, and contribute to the temperature dependent behavior of supramolecular polymers.

B.   High Molecular Weight Linkers

Supramolecular polymers with short linkers show a range of interesting material properties, but in order to advance the use of UPy-based SPs beyond fundamental science, it should be possible to obtain substantial amounts of material at the lowest possible cost. One way of achieving this goal is to use bifunctional UPy derivatives with long linkers. This will reduce the amount of UPy groups

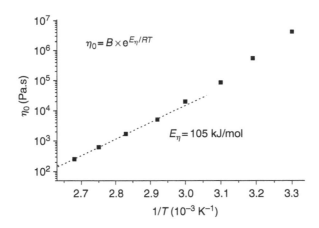

$$\eta_0 = B \times e^{E_\eta/RT}$$

$E_\eta = 105$ kJ/mol

Figure 4 Zero shear viscosity of **11b** versus the reciprocal temperature $(1/T)$; the apparent activation energy is determined according to the Andrade–Eyring equation (dotted line) in which B is a constant and $\bar{E}_\eta$ is the activation energy for viscous flow.

in the material, while maintaining the essential aspects of supramolecular polymers. Following this concept, a series of UPy-based SPs with a variety of oligomeric and polymeric linkers have been prepared [15]. Because the purity of the material is just as important as it is in low molecular weight monomers, complete conversion of telechelic precursors is essential, and to achieve this, reactive UPy synthon **7** was used to functionalize hydroxy telechelic polymers **12–15**, allowing a high degree of conversion.

Reaction of hydroxy telechelic poly(ethylene butylene) (M_n 3500, degree of functionalization: 1.93) with this synthon led to product with <0.2% residual OH groups [23]. The mechanical properties of the product are dramatically different from those of the starting material. Whereas the starting material is a viscous liquid, the UPy-functionalized polymer is a rubber-like material with a Young's modulus of 5 MPa (Figure 5).

Functionalization of more polar hydroxytelechelic polymers with the synthon resulted in comparably spectacular changes in material properties. Functionalized polyether is a material with a broad rubber plateau in DMTA, and a storage modulus of 10 MPa, whereas the starting material is a viscous liquid. The properties of functionalized polyester and functionalized polycarbonate are those of semicrystalline polymers, whereas the starting materials are very brittle.

Viscoelastic properties of covalent polymers can be understood with a model that supposes snakelike reptation of polymer chains through a tube formed by entanglements with other polymer chains. Relaxation of stress imposed on a sample is represented by reptation of a polymer chain out of its tube to a stress-free environment. In this model, the rate of stress relaxation is determined by

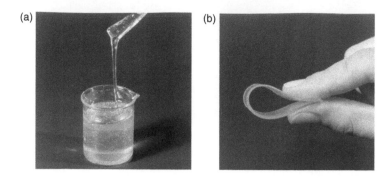

Figure 5   (a) Liquid-like poly(ethylene butylene) precursor, and (b) supramolecular polymer material **12** with elastomeric properties resulting from functionalization with UPy groups. (Reproduced with permission from Folmer, B.J.B. et al., *Adv. Mater.*, 12: 874–878, 2000. Copyright John Wiley & Sons, Inc.)

the dynamics of reptation. In reversible supramolecular polymers, chain breaking plus recombination is an additional modes of stress relaxation, leading to a strongly enhanced temperature dependence of the viscoelastic properties.

Clearly, supramolecular materials based on telechelics combine many of the mechanical properties of conventional macromolecules with the low melt viscosity of low molecular weight organic compounds. The reversibility of supramolecular polymers adds new aspects to many of the principles that are known from condensation polymerizations. For example, a mixture of different supramolecular monomers will yield copolymers, but it is extremely simple to adjust the copolymer composition instantaneously by adding an additional monomer. Moreover, the use of unimers with a functionality of three or more, will give rise to network formation. However, in contrast to condensation networks, the "self-healing" supramolecular network can reassemble to form the thermodynamically most favorable state, thus forming denser networks (Figure 6).

VI.   Applications with Supramolecular Polymers

The strong noncovalent interactions of the quadruple hydrogen-bonding unit combined with the ease of synthesis is probably the main reason that soon after the first publication on this topic was published in 1997 [19], numerous patent applications began to be filed that use these supramolecular architectures in fields ranging from adhesives [24], printing [25–29], cosmetics [30,31], personal care [32], to coatings [33]. The added value that supramolecular polymers are foreseen to give to these daily-life applications are based on (1) their improved processing in the melt or solution while maintaining excellent material properties in the solid state, (2) the lower temperatures needed to obtain tractable materials, (3) the ease of synthesis, (4) the compatibility with existing polymeric systems, and (5) the intrinsic reversibility of supramolecular systems that can be used to make the materials easily removable. In this section, several examples of these applications will be discussed that take advantage of these possibilities originating from the unlocking of the processing properties from the material properties by using supramolecular interactions. Moreover, this unambiguously shows that supramolecular polymers are not restricted to the laboratory anymore.

A.   Inkjet Inks

Yet another application in which the dramatic differences in phase behavior of supramolecular polymers on a relative narrow temperature range can be used, is inkjet printing. In this application, images are created on a substrate (i.e., paper) by the ejection of ink droplets through a small orifice.

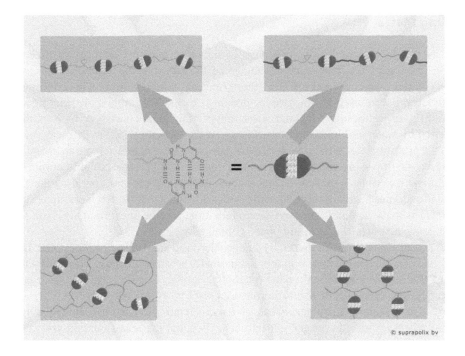

Figure 6 Different architectures possible for supramolecular polymers, top left: linear supramolecular polymer; top right: linear supramolecular block-copolymer; bottom left: supramolecular network via branching; bottom right: supramolecular network via grafting.

Therefore, the ink has to be low viscous in order to be able to eject small droplets. On the other hand, the ink needs to be highly viscous, almost a solid, once the droplet hits the paper. Otherwise, the ink will smear out through capillary action of the paper, resulting in blurry pictures.

Xerox Corporation has filed two patents in which supramolecular polymers are used as binders in ink compositions. One application relates to hot-melt inks, consisting of a colorant and a binder [29]. These inks are solid at temperatures below 50°C, and liquid with a viscosity around 20 cps at 160°C. The binder is a multifunctional low molecular weight compound that has been functionalized with 2 to 5 UPy-groups, resulting in polyether compounds that form supramolecular networks. Mixing these materials at elevated temperatures with other ingredients like UV-stabilizers, antioxidants, and colorants, results in inks that can be used in hot-melt ink printers.

In the other patent application filed by Xerox, aqueous-based inks are formulated with supramolecular polymers [28]. The material properties of the ink (i.e., viscosity) are in this case tuned with heat and polarity of the solvent medium, since those parameters determine the amount of hydrogen bonding between the different UPy's. In this case, the ink consists of aqueous solvent, colorant, and a supramolecular polymeric additive. Because of the high polarity of the solvent, hydrogen bonding will be strongly reduced in the ink solution resulting in a rather low viscosity. In contrast, the viscosity of the ink will rapidly increase during jetting as the solvent will evaporate from the droplet, together with a temperature drop in the ink, resulting in a virtual increase of molecular weight of the polymer additive due to the formation of hydrogen bonding. Consequently, the ink will be prevented from spreading on the paper and a clear image with good permanence characteristics will be produced. The supramolecular compounds used in this application are comparable to the ones for the hot-melt ink, the only difference is the formulation that contains around 10% solids (binder, pigment or dye, and surfactant) in aqueous solvent. Interestingly, Agfa Geveart in Belgium has filed a patent related to ink-jet inks in which the dyes themselves have been modified with UPy-groups, which results in improved lightfastness of the ink [27].

B. Printing Plates

Another way to use supramolecular polymers is shown in a patent application filed by Kodak Polychrome Graphics [26]. They take advantage of the fact that the solubility of supramolecular polymer-coatings increases after heating them (thermal solubilization). This concept is applied in printing plates that are used in lithographic printing processes. Lithographic printing is based on the immiscibility of hydrophobic ink and water; the printing plate consists of hydrophilic surfaces that are wetted by water and hydrophilic surfaces that are wetted by ink. In this way, hydrophobic patterns on the printing plate can be transferred to a substrate (i.e., paper) to create an image. Typically, patterns are created by using a radiation-sensitive coating that becomes soluble upon exposure to radiation, and hence can be removed in a subsequent developing process, revealing the hydrophilic surface that is underneath and that consequently will absorb the ink (positive-working printing plate). If the top layer is made insoluble by exposure to radiation, and the unexposed areas are removed in developing, a negative-working printing plate is obtained.

In these patent applications, a thermally imageable positive-working printing plate is disclosed that makes use of a supramolecular polymer as thermally sensitive coating. The used polymers are obtained by reacting polyfunctional resins (phenolic, acrylic, or polyester) with isocyanate functional UPy. Specific examples of resins are phenol/cresol novolaks and 4-hydroxy-styrene/styrene copolymers. The resulting coating consists of these supramolecular cross-linked polymers and IR-dyes that will transform the laser-light into heat. Because the resins contain several reversible cross-links to reinforce the coating, it is possible to thermally disrupt these cross-links with short ($\sim$100 μsec) IR laser illumination and to solubilize these exposed regions in the following developing step. This procedure eliminates two processing steps normally used in preparing the printing plate (preheating and postdevelopment baking), and superior press-life is obtained when compared with other digitally imaged compositions.

C. Polymerization-Induced Phase Separation

The dynamic flexibility of supramolecular polymers is exploited by Keizer et al. [34]. in polymerization-induced phase separation (PIPS) with hydrogen-bonded supramolecular polymers. In PIPS, a supramolecular polymer is dissolved in a reactive monomer, which is subsequently polymerized to cause phase separation resulting in two polymeric phases with certain morphology. PIPS is currently used to produce multiphase composite materials like high-impact polystyrene, avoiding the use of solvent and consequently resulting in the fast and clean production of polymeric materials.

The rate of phase separation in PIPS is generally limited by the mobility of the dissolved polymer. Supramolecular polymers, however, may dissociate when dissolved in reactive monomer, resulting in strongly enhanced diffusion. Hence, in PIPS of supramolecular polymers macroscopic phase separation can be reached within a very short reaction time (Figure 7).

Keizer et al. [34] report the PIPS of solutions of hydrogen-bonded supramolecular polymers in acrylates, within the short reaction times (0.3 sec) used in UV-curing. The cured films were colorless, transparent, and flexible, and showed macrophase separation in SEM micrographs. Moreover, their mechanical behavior was comparable to high molecular weight polymers. Interestingly, DSM has filed a patent that describes the use of supramolecular polymers in coatings for glass-fibers, a process that needs very fast reaction times indeed. In addition, this strategy could be used for either stratification or patterning via a mold in thin films.

VII. Outlook and the Modular Approach

Ten years ago, the first supramolecular polymers were seen as scientific curiosities. Nowadays, this field of research is generating several technologically important applications. Progress in

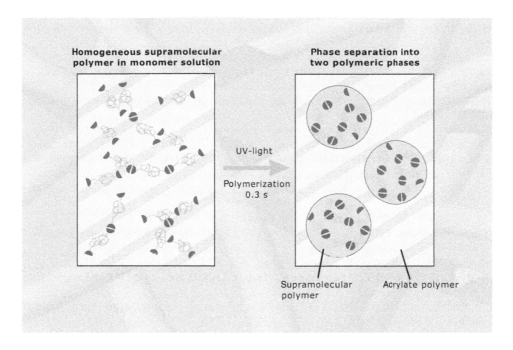

Figure 7 Schematic representation of PIPS using hydrogen-bonded supramolecular polymers; picture on the left depicts situation before polymerization of the acrylate-monomer, picture on the right depicts situation after the polymerization of the acrylate monomer has induced phase separation.

supramolecular chemistry has made it possible to assemble small molecules into polymer arrays, and the created structures possess many of the well-known properties of "traditional" macromolecules. Due to the reversibility in the bonding, these supramolecular polymers are under thermodynamic equilibrium and their properties can be adjusted by external stimuli. Hydrogen-bonded systems have shown to have become of technological relevance and surpass the state of being scientific curiosities only. A large variety of applications are feasible, especially since the chosen approach can also be used for the modification of telechelic oligomers or to modify existing polymers. A highly interesting outlook for supramolecular polymers is the option of the modular approach. By having a box of different and functional UPy-based building blocks, it should be possible to create new functions by just mixing the building blocks in the appropriate amounts. The materials are formed by self-assembly and can be changed by adding another component. Therefore, the possibility to tune the properties by changing the relative ratio of UPy-monomer in, for example, the copolymer feed seems very attractive, while hybrids between blocks of macromolecules and supramolecular polymers are easy to prepare. Therefore, novel thermoplastic elastomers, superglues, hot-melts, and tunable polymeric materials are within reach and currently receive a lot of attention from several industrial research laboratories.

REFERENCES

1. C. Fouquey, J.-M. Lehn, and A.-M. Levelut, *Adv. Mater.* 2: 254–257, 1990.
2. T. Gulik-Krzywicki, C. Fouquey, and J.-M. Lehn, *Proc. Natl. Acad. Sci. USA* 90: 163–167, 1993.
3. M. Kotera, J.-M. Lehn, and J.-P. Vigneron, *J. Chem. Soc. Chem. Commun.* 197–200, 1994.
4. R.P. Sijbesma and E.W. Meijer, *Chem. Commun.* 5–16, 2003.
5. W.L. Jorgensen and J. Pranata, *J. Am. Chem. Soc.* 112: 2008, 1990.
6. F.H. Beijer, H. Kooijman, A.L. Spek, R.P. Sijbesma, and E.W. Meijer, *Angew. Chem. Int. Ed.* 37: 75–78, 1998.

7. F.H. Beijer, R.P. Sijbesma, H. Kooijman, A.L. Spek, and E.W. Meijer, *J. Am. Chem. Soc.* 120: 6761–6769, 1998.

8. P.S. Corbin and S.C. Zimmerman, *J. Am. Chem. Soc.* 120: 9710–9711, 1998.

9. J. Elguero, C. Marzin, A.R. Katritzky, and P. Linda, *The Tautomerism of Heterocycles*. New York: Academic Press, 1976.

10. S.H.M. Söntjens, R.P. Sijbesma, M.H.P. van Genderen, and E.W. Meijer, *J. Am. Chem. Soc.* 122: 7487–7493, 2000.

11. S.H.M. Söntjens, M.H.P.van. Genderen, and R.P. Sijbesma, *J. Org. Chem.* 68:9070–9075, 2003.

12. G. Armstrong and M. Buggy, *Mater. Sci. Eng., C: Biomimetic Supramol. Syst.* C18: 45–49, 2001.

13. R.F.M. Lange, M. van Gurp, and E.W. Meijer, *J. Pol. Sci. A Pol. Chem.* 37: 3657–3670, 1999.

14. U. von Gizycki, *Angew Chem.* 83: 406–408, 1971.

15. B.J.B. Folmer, R.P. Sijbesma, R.M. Versteegen, J.A.J. van der Rijt, and E.W. Meijer, *Adv. Mater.* 12: 874–878, 2000.

16. H.M. Keizer, R.P. Sijbesma, and E.W. Meijer, *EurJOC*: 2553–2555, 2004.

17. J.H.K.K. Hirschberg, F.H. Beijer, H.A. van Aert, P.C.M.M. Magusin, R.P. Sijbesma, and E.W. Meijer, *Macromolecules* 32: 2696–2705, 1999.

18. L.R. Rieth, R.F. Eaton, and G.W. Coates, *Angew. Chem. Int. Ed.* 40: 2153–2156, 2001.

19. R.P. Sijbesma, F.H. Beijer, L. Brunsveld, B.J.B. Folmer, J.H.K.K. Hirschberg, R.F.M. Lange, J.K.L. Lowe, and E.W. Meijer, *Science* 278: 1601–1604, 1997.

20. M. Wübbenhorst, B.J.B. Folmer, J. van Turnhout, R.P. Sijbesma, and E.W. Meijer, *IEEE. Trans. Dielectr. Electr. Insul.* 8: 365–371, 2001.

21. M. Doi and S.F. Edwards, *The Theory of Polymer Dynamics*. Oxford: Clarendon, 1986.

22. M.E. Cates, *Macromolecules* 20: 2289–2296, 1987.

23. H.M. Keizer, R.v. Kessel, R.P. Sijbesma, and E.W. Meijer, *Polymer*, 44:5505–5511, 2003.

24. B. Eling (Huntsman Int), WO 0246260, 2002.

25. Y. Ishizuka, E. Hawakawa, Y. Asawa, and S.P. Pappas (Kodak Polychrome Graphics), WO 02053627, 2002.

26. A. Monk, S. Saraiya, J. Huang, and S.P. Pappas (Kodak Polychrome Graphics), WO 02053626, 2002.

27. J Locculier, L L Vanmaele, E.W. Meijer, P. Fransen, and H. Janssen (Agfa Gevaert), EP1310533, 2003.

28. D.J. Luca, T.W. Smith, and K.M. McGrane (Xerox Corp), U.S. 2003079644, 2003.

29. H.B.S. Goodbrand, D. Popovic, D.A. Foucher, T.W. Smith, and K.M. McGrane (Xerox Corp), U.S. 20030105185, 2003.

30. M.D. Eason, E. Khoshdel, J.H. Cooper, and B.J.L. Royles (Unilever), WO 03032929, 2003.

31. A. Livoreil, J. Mondet, and N. Mougin (L'Oreal), WO 02098377, 2002.

32. D. Kukulj and F. Goldoni (Unilever), WO 02092744 A1, 2002.

33. J.A. Loontjens, J.F.G.A. Jansen, and B.J.M. Plum (DSM), EP 1031589, 2002.

34. H.M. Keizer, R.P. Sijbesma, J.F.G.A. Jansen, G. Pasternack, and E.W. Meijer, *Macromolecules* 36: 5602–5606, 2003.

Chapter 16

Columnar, Helical, and Tubular Supramolecular Polymers

Wei Zhang and Colin Nuckolls

CONTENTS

Self-assembly is a powerful tool to create novel materials with emergent or amplified properties [1–8]. Detailed below are a number of experiments that exemplify an approach where small, information-rich molecules are designed, synthesized, and studied as functional materials. In particular, the emphasis of this chapter will be on assemblies that form *columnar*, *tubular*, or *helical* superstructures, shown schematically in Figure 1. Although nature is a wonderful creator of these types of self-organized structures, it is beyond the scope of this chapter.

I. COLUMNAR MATERIALS

Discotic liquid crystals are a prototypical self-assembled columnar system [9–13]. This class of liquid-crystalline compounds is relatively new, discovered in 1977 by Chandrasekar and coworkers [14,15]. The assembly motif, shown in Figure 2, for this class of compounds has an aromatic core that is surrounded by hydrocarbon chains. These disk-shaped molecules then stack to form columns. These one-dimensional stacks aggregate to form arrayed columns. When the columns have a circular cross-section they typically stack into a hexagonal arrangement as shown in Figure 2(a). Some of the original discotics were hexa-substituted phthalocyanines (**1**), benzenes (**2**), and triphenylenes (**3**), shown in Figure 2(b). The self-assembly of classical discotics will not be presented in depth here because it has been a focus of several comprehensive reviews [9–15].

    The constituent stacks have been likened to molecular-scale wires because the column's interior consists of cofacially stacked π-faces while its exterior is surrounded by an insulating, hydrocarbon wrapper. This arrangement of the aromatic cores has endowed these liquid-crystalline phases with

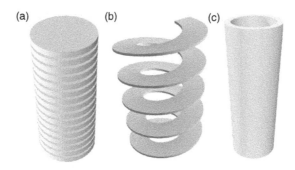

Figure 1 Schematic of supramolecular assemblies that are: (a) columnar, (b) helical, and (c) tubular.

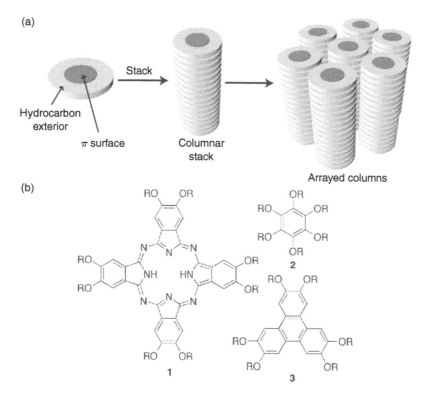

Figure 2 (a) Schematic of stacking of discotic liquid crystals: from disk, to column, to a hexagonal array of columns. (b) Subunits of the classic discotic liquid crystals of hexa-substituted phthalocyanines (**1**), benzenes (**2**), and triphenylenes (**3**).

useful electronic and optic properties [16–26]. Many studies have used time-of-flight electrical mobility measurements and pulsed radiolysis time-resolved microwave conductivity (PR-TRMC) to measure the mobility in discotic samples. Remarkably, the electrical mobility values for some discotics (~1 cm^2/V sec) [19] are approaching the mobilities measured through graphite sheets (3 cm^2/V sec). These values are large enough to be useful in devices if they form the active layer of a thin film transistor. Unfortunately, there are only a limited number of electrical mobility measurements on discotic liquid crystals in thin film transistors [26]. This may be a consequence of a number of factors, either difficulty in injecting charge into the stack, highly resistant grain boundaries, or poor adhesion to the metallic electrodes.

The interplay between the liquid-crystalline properties and the electrical mobility for discotics will likely be a fruitful area of research in future by analogy to calamitic (i.e., rod-shaped) liquid crystals. For calamitic liquid crystals, several research groups have shown that a liquid-crystalline phase — either lyotropic or smectic — has a marked positive effect on their field effect electrical mobility [27–30]. This may be due to the layered structure being able to organize into an arrangement that is ideal for lateral transport in transistor structures, or from the ability to anneal away grain boundaries in a self-healing material. Possibly, the one-dimensional structure of the discotic liquid crystals needs to be nearly perfect to avoid erosion in their properties. To circumvent this problem it is necessary, from a device standpoint to create even smaller devices to match the persistence length of the stacking in liquid-crystalline materials. From a chemistry point of view it is necessary to create discotic liquid crystals that have increased affinity in the stacking direction. These two approaches meet at the nanoscale and drive much of the nanoscience of columnar stacks.

For traditional discotics the affinity between the subunits is low due to the poor electrostatic attraction between the electron-rich π surfaces [31]. This is exemplified by the hexaalkoxyl triphenylenes and their self-assembly. It is not possible for these electron-rich aromatics to form an edge-to-face interaction to give the classic herringbone stack due to the alkyl side chains. These electron-rich aromatics prefer to offset when they stack and often the stacking is dictated by the hydrocarbon exterior that forms a phase of its own from the numerous van der Waals contacts between the alkyl groups. One very elegant method developed by Müllen and coworkers to increase the affinity between discotic subunits is by increasing the surface area of the disks by synthesizing even larger graphite sections [32,33]. The synthesis developed for this class of compounds is general. These graphite sections are substituted on their exterior with solubilizing groups and form highly regular stacks with very promising optical and electrical properties [19].

Instead of increasing the size of the disks, a number of groups have been put in two disks that have complementary structure [34–36]. The recognition between the disks encodes an A–B type stacking shown in Figure 3. Early examples used classic electron acceptors like trinitrofluorenone mixed with the columnar hexaalkoxytriphenylenes as electron donors [34]. The regularity in the stacking of these subunits is less than ideal because of the difference in size and shape of the two subunits. To circumvent this problem, Grubbs and coworkers were successful in forming mixed liquid-crystalline stacks using a perfluorinated triphenylene with a hexaalkoxytriphenylene [35]. Just as with benzene/hexafluorobenzene [37] the mixture forms an interdigitated stack. By using derivatives that incorporate ring opening metathesis monomers, the columnar structure could be locked through polymerization in the mesophase. Another method forming A–B stacks is to use electron-deficient/electron-rich subunits of nearly equal size. Recently, Park and Hamilton have shown that hexa-substituted triphenylenes form A–B columnar liquid-crystalline phases using a threefold symmetric, electron-deficient tri-imide [36]. This interaction is similar to the association of electron-rich and electron-poor naphthyl subunits used by Iverson and coworkers to create foldamers [38].

A number of studies have sought methods of introducing a secondary organization force into the disk to increase the affinity between subunits. One method is to use metal–ligand interactions [39–41]. Shown schematically in Figure 4, these columnar materials are often referred to as metallomesogens.

Complementary subunits

Figure 3 Electrostatic complementarity, A–B stacks.

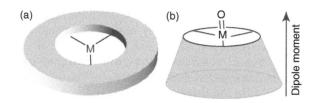

Figure 4 (a) Metallomesogen; (b) polar metallomesogens.

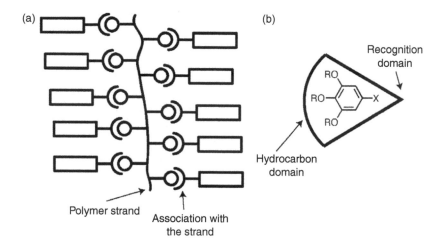

Figure 5 (a) Association with polymer strands. (b) Wedge pieces for association with polymer strands.

This area of research has seen an explosion of interest in recent years and has been well reviewed by others [39–41]. This extra element in the assembly of these mesogens is responsible for a number of interesting polar properties in this class of materials [42]. As shown in Figure 4(b), the oxo-metal compounds provide a dipole moment to many of these mesogens. This allows them to transfer their oxygen atoms between members of the stack to switch the macroscopic polarity [42]. Some have shown ferroelectric behavior [42] that is possibly due to the inability to pack opposing dipoles in a hexagonal lattice [43].

Another strategy for creating well-defined stacks is to use subunits that can recognize a polymer strand. This is shown schematically in Figure 5(a). An example of this strategy showing regular columns is the work of Wegner and coworkers that has used polymeric siloxanes as a central organizational tool to thread through the center of phthalocyanines [44,45]. The aromatic planes that are stacked well within their van der Waals radii and the columns, form a highly regular hexagonal arrangement. Frechet, Kato, Percec, and others have used a postpolymerization assembly of small subunits with a polymer strand to create liquid-crystalline assemblies, similar to the assembly motif of the tobacco mosaic virus [46–48]. The small molecules that recognize the polymer strands have been shown to have a dramatic influence on the mesogeneity and order in these materials [46–48].

A powerful method to organize π stacks and to increase their affinity is to use hydrogen bonds [49–72]. One motif is to use a benzene ring that is substituted with amide functionality shown in Figure 6(a). The crystal structures of benzene and cyclohexane when they are substituted at their 1,3,5-positions with amides have shown a regular columnar structure [53–55]. These crystal structures and similar ones inferred from columnar liquid crystals studied by Matsunaga, Meijer, and others reveal a one-dimensional stacked structure held together by amide hydrogen bonds. Due to a helicity in the hydrogen-bond network, some of these materials show interesting cooperative chirooptic signatures of an all or none, "sergeants and soldiers" motif [64]. Recently, Meijer and

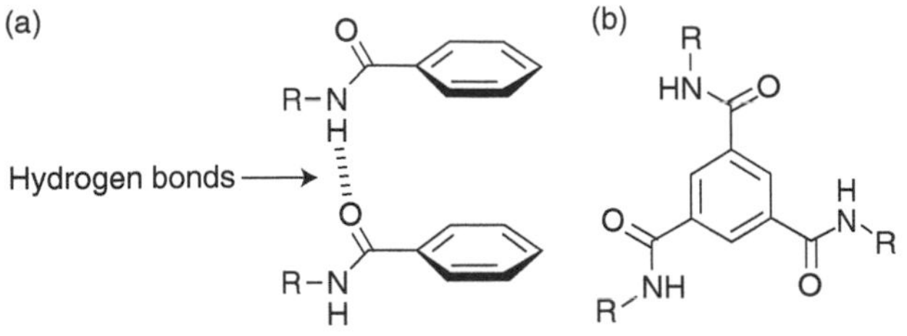

Figure 6 (a) Hydrogen bonds stabilized π stacks; (b) 1,3,5-benzene tris-amides.

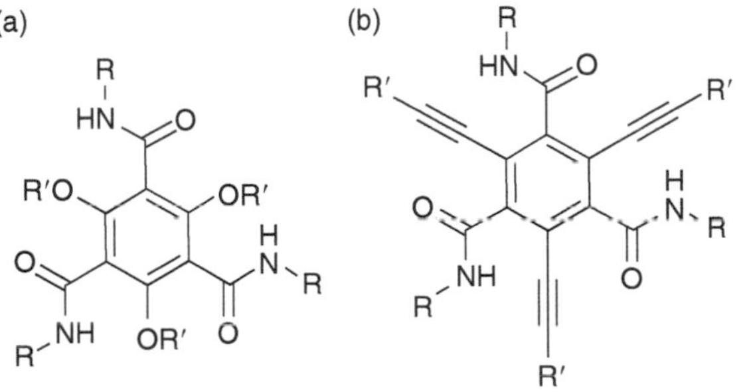

Figure 7 Crowded aromatics with amides flanked by (a) alkoxyls and (b) alkynes.

coworkers have been successful in covalently locking these stacks through a post assembly photo polymerization [65].

For the molecules, in Figure 7, the design principle explored was to determine if the flanking groups, alkoxyl groups in Figure 7(a) and alkynyl substituents in Figure 7(b), would force the amides out of the plane of the central aromatic ring and into a conformation that is *predisposed* to form three intermolecular hydrogen bonds and therefore a stacked structure [66–72]. The size of the flanking substituent determines the angle of twist for the amide out of the central ring plane and consequently modulates the distance between adjacent benzene rings reflecting the relative sizes of the alkoxyl and the alkynyl substituents. Models predict that three helices of hydrogen bonds emerge around the exterior of the columns when assembled. In dilute solution these materials show split CDs and far red-shifted fluorescence indicative of a columnar assembly [68]. At higher concentrations, the material shows highly twisted cholesteric liquid-crystalline phases that reflect light at visible wavelengths [68].

An interesting feature of the models of these subunits is that there is a permanent dipole moment whose direction is perpendicular to the ring plane. Therefore, as the molecules stack the dipoles could sum yielding columns that have a macroscopic dipole moment, similar to the moment that is seen for some of the metallomesogens mentioned above [42] and some conical liquid crystals [43,73]. One indication that these materials are polar comes from bulk samples whose columnar orientation can be directed in an electric field [68,70]. Some of the derivatives in Figure 7(b) show electrooptic switching times of the order of microseconds [70]. These are some of the fastest switching times for columnar discotic liquid crystals [42] including the ones diluted with solvent [87].

The morphology of monolayer films of the mesogens in Figure 7 is a consequence of the association in bulk [69]. Mesogens that do not associate well in bulk yield monolayers that are oriented with their cores parallel to the substrate. Electrostatic force microscopy measurements show that the monolayers are polar with a net negative charge at the surface. This could be due to the surface orienting the dipole moment of each of the subunits or from a charge transfer from the graphite to the subunits. The opposite orientation is observed for the subunits with high affinity in monolayers

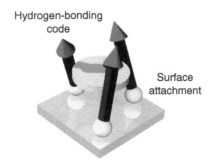

Figure 8   Surface templates for the assembly of π stacks orthogonal to metals surfaces.

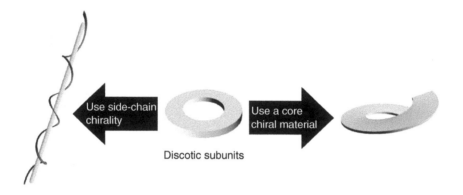

Figure 9   Two ways to produce a chiral stack either with side chains or cores that are chiral.

that are monitored by atomic force microscopy (AFM) and scanning tunneling microscopy (STM) [69,70]. Isolated strands of molecules can be seen in some areas of these films.

When surface-active side chains are put onto the subunits in Figure 7(b) a new type of surface templating agent (shown schematically in Figure 8) is created [72]. When bound to the surface, these molecules form a precise receptor for aromatics of a part size, shape, and hydrogen-bonding code. Through these surface templates, the shortest of columnar stacks, only two high subunits has been produced on gold films and imaged with STM [72]. These stacks may prove useful as a model study for electrical transport through π stacks.

II. HELICAL STACKS

In principle, there are two different motifs of helical stacks commonly employed for columnar materials that are shown in Figure 9. One motif would use the side chains as a means to introduce point chirality that could then be transferred to the cores stacking. This chirality transfer would bias one helical stack versus another. A second motif would start with a structure that has a chiral core and use this as a subunit to stack into a helical column.

Since their discovery as discotic liquid crystals, both triphenylenes and phthalocyanines have been studied as chiral columns [74]. There are no chiroptic signatures such as circular dichroism or optical rotation changes for the triphenylenes that are substituted with chiral side chains. For the phthalocyanines, Nolte and coworkers have shown that when their exterior is substituted with chiral side chains the phthalocyanine core shows an altered CD spectrum in aggregates, in thin films and in concentrated solutions [75]. These helical stacks, wrapped like ropes, can be seen with electron

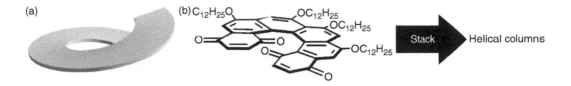

Figure 10 (a) Lock washers. (b) Helicene assembly into a supramolecular helix.

microscopy when they are formed from a side chain chiral phthalocyanine that had appended crown ether functionality [76].

At the interface between side chain chiral and core chiral discotics are ones whose core is conformationally flexible. Some of Swager's metallomesogens twist themselves into a chiral, propeller-like conformation whose handedness is dictated by the side chain's chirality in liquid-crystalline thin films [77]. Moore has shown that the meta-disposed phenylethynylphenylene oligomers cause a helical winding. These oligomers when helically wound show a hexagonal, columnar arrangement and when unwound display a lamellar liquid-crystalline phase [78]. In related work, Lehn and coworkers have used the 2,2'-bipyridal conformation to form giant helices that stack into helical columns [79]. These and related folded structures are the focus of Chapter 2.

To construct helical columns several studies have used a different strategy where the core structure is dissymmetric. Saturated mesogens that are derived from chiral saccharides [80,81] and inositols [82] have a core that is chiral and are substituted with hydrocarbon chains. These materials stack into columnar structures but do not show any macroscopic manifestation of the core chirality. This is likely due to poor chiral information transfer from members of the stack.

One core chiral system that shows dramatic amplification of its chiral structure is the substituted helicenes of Katz and coworkers [83]. In essence, this research cuts the helix into a number of six-helicene subunits that self-assemble (Figure 10). Only when these subunits, which look like lock washers, are prepared in optically pure form the material associates into supramolecular helical columns [84]. The assemblies have been synthesized with different amounts of substitution around the exterior. Depending on the helicenes' substitution, the material exhibits hexagonally ordered soft-crystalline [84] or liquid-crystalline phases [85]. The liquid-crystalline versions of these molecules switch when electric fields are applied to neat and solution-phase samples and have been characterized as a dielectric response [85–87]. Upon association, these materials have enormous changes in their CD intensities and optical rotations [74]. In addition, this supramolecular chirality also significantly enhances the second-order nonlinear optical behavior of these materials in Langmuir–Blodgett films [88].

III. TUBULAR STACKS

To form tubular stacks, there are four main motifs. First, a motif where a number of preformed disks self-associate to form a tube. Second, where wedges come together to form a ring that then further associates to form a tube. Third, where rigid rods are held together lengthwise to form a channel, a barrel-stave motif. Fourth, where the channel is defined by a winding of a polymer chain. The fourth motif is explained in Chapter 2 and beyond the scope of this chapter. In addition, there are countless examples of natural systems that form transmembrane channels, tubes, or disks that will not be presented here.

A number of studies on stacked disks were recently reviewed [89]. Some of the original work focused on stacking preformed disks (Figure 11) that utilize the cation binding potential of crown ethers in concert with the column forming properties of columnar materials [76,90]. Some of these materials have shown utility in ion-transport studies [90]. The pitfall of this approach

Disks

Self-assembly

Tubes

Figure 11   Assembly of cycles into tubes.

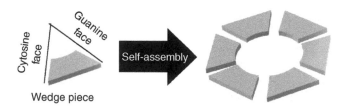

Cytosine face

Guanine face

Self-assembly

Wedge piece

Figure 12   Assembly of wedges with complementary faces into disks.

is the ill-defined and flexible nature of the crown portion and the poor affinity between the subunits.

One of the most well-developed system that exemplifies the stacked-disk motif is the work of Ghadiri and coworkers on cyclic D,L-peptides [89,91–93]. The disks are held together through hydrogen bonds and intermolecular contacts between the peptide side chains. The potential for derivatization and functionalization is great because the disks are made from amino acids. These materials have shown the ability to assemble into lipid bilayers and transport ions across them. In addition these materials have recently been shown to act as self-assembled antibacterial agents with utility as therapeutics [93]. There are other systems that have been synthesized from shape persistent macrocycles to assemble into tubes [91,94].

There is considerable variability in the instruction written into the substructures that make up the second motif for tubes where a wedge assembles into a disk that further assembles into a column (Figure 12). There are materials with no specific interactions between the wedges that have been prepared by Gin [95,96], Kato [97], and Percec [98]. The studies by Percec have demonstrated the ability to control the assembly either into a cubic (spherical) and hexagonal (columnar) morphology based on the constitution of the wedge. Gin has been successful at using the amphiphilic wedges to create inverse micelle phases that form tubular structures. These materials have been useful as analogs for organic zeolites and reaction chambers [95,96]. Recently Kato and coworkers have covalently modified the center of the tube with ionic liquids and shown that they behave as one-dimensional ionic conductors [97]. This strategy was also useful to find electrically conductive tubular stacks [20].

A number of other studies have utilized specific modes of association between the wedges to control the formation of disks. Zimmerman and others have shown that trimesic acid derivatives assemble to form honeycomb arrangements when a filler for the channel is cocrystallized [99]. Early work by Whitesides on the assembly of rosettes from cyanuric acid-melamine [100,101] and their covalent analogs [102] has produced remarkable multicomponent assemblies and one-dimensional superstructures. Building on this research, Reinhoudt and coworkers have investigated a number of new derivatives with remarkable fidelity in the assembly motif and unusual chiral assembly characteristics [103].

In related assembly studies there are several nucleobase constructions that utilize the coordination of natural and unnatural bases around a cationic center [104]. Originally Mascal [105,106], and later Lehn [107], and Zimmerman [108–110], have shown a way to make a single component system that utilizes the assembly motif, shown in Figure 12, is to use molecules that are self-complementary. This

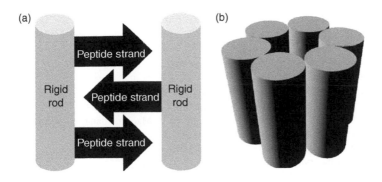

Figure 13 (a) Barrel-stave assembly motif of Matile and coworkers; (b) channel formation.

is shown for the G–C hybrids in Figure 12 where the recognition units are held at the appropriate angle to allow assembly to occur into a hexagonal cycle. This assembly motif forms the basis for tubular materials and has been extensively investigated by Fenniri and coworkers [111–113]. Among other discoveries, they have shown that peptide derived versions assemble due to entropy but not enthalpy. Moreover, Fenniri has shown that they have a helical conformation with interesting chirooptic signatures [111–113].

The final assembly motif that is presented here is the so-called barrel-stave motif. This motif uses a rigid rod that is substituted with recognition domains along its length. These rods then aggregate to form multimers that can have channel architectures (Figure 13(b)). This assembly motif has been reviewed recently. Elegant demonstrations of this concept are the independent studies of Matile [114,115] and Lehn [116]. Lehn and coworkers used metal–ligand interactions to form channel architectures. Matile investigated antiparallel strands of peptides associated on oligophenylenes (Figure 13(a)). These materials due to their interior functionality have been shown to be useful catalysts and transmembrane transport agents.

ACKNOWLEDGMENTS

Colin Nuckolls acknowledges primary financial support from the Chemical Sciences, Geosciences, and Biosciences Division, Office of Basic Energy Sciences, U.S. D.O.E. (#DE-FG02-01ER15264), U.S. National Science Foundation CAREER award (#DMR-02-37860), and the Nanoscale Science and Engineering Initiative of the National Science Foundation under NSF Award Number CHE-0117752 and by the New York State Office of Science, Technology, and Academic Research (NYSTAR). Colin Nuckolls thanks the Beckman Young Investigator Program (2002), the NYSTAR J.D. Watson Investigator Program (2003), The American Chemical Society PRF type G (#39263-G7), and the Dupont Young Investigator Program (2002) for support.

REFERENCES

1. J.-M. Lehn, *Supramolecular Chemistry*, VCH: Weinheim, **1995**.
2. J.-M. Lehn, *Struct. Bond.* **2000**, *96*, 3–29.
3. G.M. Whitesides and B. Grzybowski, *Science* **2002**, *295*, 2418–2421.
4. D.S. Lawrence, T. Jiang, and M. Levett, *Chem. Rev.* **1995**, *95*, 2229–2260.
5. M.M. Conn and J. Rebek, Jr., *Chem. Rev.* **1997**, *97*, 1647–1668.
6. D. Philp and J.F. Stoddart, *Angew. Chem. Int. Ed. Engl.* **1996**, *35*, 1155–1196.
7. M. Muthukumar, C.K. Ober, and E.L. Thomas, *Science* **1997**, *277*, 1225–1232.
8. L.J. Prins, D.N. Reinhoudt, and P. Timmerman, *Angew. Chem. Int. Ed.* **2001**, *40*, 2382–2426.

9. S. Chandrasekhar and G.S. Ranganath, *Rep. Prog. Phys.* **1990**, *53*, 57–84.

10. C. Destrade, P. Foucher, H. Gasparoux, H.T. Nguyen, A.M. Levelut, and J. Malthete, *Mol. Cryst. Liq. Crystl.* **1984**, *106*, 121–46.

11. S. Chandrasekhar, S. Prasad, and J. Krishna, *Contemp. Phys.* **1999**, *40*, 237–245.

12. D. Guillon, *Struct. Bond.* **1999**, *95*, 41–82.

13. S. Chandrasekhar, *Handb. Liq. Crystl.* **1998**, *2B*, 749–780.

14. S. Chandrasekhar, B.K. Sadashiva, and K.A. Suresh, *Pramana* **1977**, *9*, 471–480.

15. S. Chandrasekhar, B.K. Sadashiva, K.A. Suresh, N.V. Madhusudana, S. Kumar, R. Shashidhar, and G. Venkatesh, *J. Phys. Colloq. (Orsay, Fr.)* **1979**, *3*, 120–124.

16. N. Boden and B. Movaghar, *Handb. Liq. Crystl.* **1998**, *2B*, 781–798.

17. J. Simon and C. Sirlin, *Pure Appl. Chem.* **1989**, *61*, 1625–1629.

18. N. Boden, R.J. Bushby, J. Clements, and B. Movaghar, *J. Mater. Chem.* **1999**, *9*, 2081–2086.

19. A.M. Van de Craats, J.M. Warman, A. Fechtenkotter, J.D. Brand, M.A. Harbison, and K. Müllen, *Adv. Mater.* **1999**, *11*, 1469–1472.

20. V. Percec, M. Glodde, T.K. Bera, Y. Miura, I. Shiyanovskaya, K.D. Singer, V.S. Balagurusamy, P.A. Heiney, I. Schnell, A. Rapp, H.-W. Spiess, S.D. Hudson, and H. Duan, *Nature* **2002**, *419*, 384–387.

21. V. Percec, G. Johansson, J. Heck, G. Ungar, and S.V. Batty, *J. Chem. Soc., Perkin Trans. I* **1993**, *13*, 1411–1420.

22. O.E. Sielcken, L.A. van de Kuil, W. Drenth, J. Schoonman, and R.J.M. Nolte, *J. Am. Chem. Soc.* **1990**, *112*, 3086–3093.

23. T. Christ, B. Gluesen, A. Greiner, A. Kettner, R. Sander, V. Stuempflen, V. Tsukruk, and J.H. Wendorff, *Adv. Mater.* **1997**, *9*, 48–52.

24. D. Adam, P. Schuhmacher, J. Simmerer, L. Hänssling, K. Siemensmeyer, K.H. Etzbach, H. Ringsdorf, and D. Haarer, *Nature* **1994**, *371*, 141–143.

25. L. Schmidt-Mende, A. Fechtenkotter, K. Müllen, E. Moons, R.H. Friend, and J.D. MacKenzie, *Science* **2001**, *293*, 1119–1122.

26. A.M. Van de Craats, N. Stutzmann, O. Bunk, M.M. Nielsen, M. Watson, K. Müllen, H.D. Chanzy, H. Sirringhaus, and R.H. Friend, *Adv. Mater.* **2003**, *15*, 495–499.

27. F. Garnier, *Chem. Phys.* **1998**, *227*, 253–262.

28. H.E. Katz and Z. Bao, *J. Phys. Chem. B* **2000**, *104*, 671–678.

29. M. Mushrush, A. Facchetti, M. Lefenfeld, H.E. Katz, and T.J. Marks, *J. Am. Chem. Soc.* **2003**, *125*, 9414–9423.

30. N. Stutzmann, R.H. Friend, and H. Sirringhaus, *Science* **2003**, *299*, 1881–1884.

31. C.A. Hunter and J.K.M. Sanders, *J. Am. Chem. Soc.* **1990**, *112*, 5525–5534.

32. M. Muller, C. Kubel, and K. Müllen, *Chem.—Eur. J.* **1998**, *4*, 2099–2109.

33. J. Wu, M.D. Watson, and K. Müllen, *Angew. Chem. Int. Ed.* **2003**, *42*, 5329–33.

34. H. Bengs, M. Ebert, O. Karthaus, B. Kohne, K. Praefcke, H. Ringsdorf, J.H. Wendorff, and R. Wuestefeld, *Adv. Mater.* **1990**, *2*, 141–144.

35. M. Weck, A.R. Dunn, K. Matsumoto, G.W. Coates, E.B. Lobkovsky, and R.H. Grubbs, *Angew. Chem. Int. Ed.* **1999**, *38*, 2741–2745.

36. L.Y. Park, D.G. Hamilton, E.A. McGehee, and K.A. McMenimen, *J. Am. Chem. Soc.* **2003**, *125*, 10586–10590.

37. J.H. Williams, *Acc. Chem. Res.* **1993**, *26*, 593–598.

38. G.J. Gabriel and B.L. Iverson, *J. Am. Chem. Soc.* **2002**, *124*, 15174–15175.

39. A.G. Serrette, C.K. Lai, and T.M. Swager, *Chem. Mater.* **1994**, *6*, 2252–2268.

40. J.L. Serrano, Ed., *Metallomesogens*, VCH: New York, **1996**.

41. J. Simon and P. Bassoul, *Phthalocyanines: Properties and Applications*, C.C. Leznoff and A.B.P. Lever, Eds., VCH: New York, 1989; Vol. 2, Chapter 6.

42. D. Kilian, D. Knawby, M.A. Athanassopoulou, S.T. Trzaska, T.M. Swager, S. Wrobel, and W. Haase, *Liq. Crystl.* **2000**, *27*, 509–521.

43. H. Zimmermann, R. Poupko, Z. Luz, and J. Billard, *Z. Naturforsch. A: Phys. Chem. Kosmophys.* **1985**, *40A*, 149–160.

44. G. Wegner, *Mol. Cryst. Liq. Crystl.* **1993**, *235*, 1.

45. A. Kaltbeitzel, D. Neher, C. Bubeck, T. Sauer, G. Wegner, and W. Caseri, *Electronic Properties of Conjugated Polymers*, H. Kuzmany, M. Mehring, and S. Roth, Eds., *Springer Ser. Solid State Sci.* **1989**, *91*, 220.

46. T. Kato, H. Kihara, U. Kumar, T. Uryu, and J.M.J. Frechet, *Angew. Chem.* **1994**, *106*, 1728–1730.

47. T. Kato, *Struct. Bond.* (Berlin) **2000**, *96*, 95–146.

48. V. Percec, *Handb. Liq. Crystl. Res.* **1997**, 259–346.

49. Y. Matsunaga, N. Miyajima, Y. Nakayasu, S. Sakai, and M. Yonenaga, *Bull. Chem. Soc. Jpn.* **1988**, *61*, 207–210.

50. J.J. van Gorp, J.A.J.M. Vekemans, and E.W. Meijer, *J. Am. Chem. Soc.* **2002**, *124*, 14759–14769.

51. L. Brunsveld, H. Zhang, M. Glasbeek, J.A.J.M. Vekemans, and E.W. Meijer, *J. Am. Chem. Soc.* **2000**, *122*, 6175–6182 and references cited therein.

52. Y. Yasuda, E. Iishi, H. Inada, and Y. Shirota, *Chem. Lett.* **1996**, *7*, 575–576.

53. M.P. Lightfoot, F.S. Mair, R.G. Pritchard, and J.E. Warren, *Chem. Commun.* **1999**, *19*, 1945–1946.

54. E. Fan, J. Yang, S.J. Geib, T.C. Stoner, M.D. Hopkins, and A.D. Hamilton, *Chem. Commun.* **1995**, *12*, 1251–1252.

55. D. Ranganathan, S. Kurur, R. Gilardi, and I.L. Karle, *Biopolymers* **2000**, *54*, 289–295.

56. C.M. Paleos and D. Tsiourvas, *Angew. Chem. Int. Ed. Engl.* **1995**, *34*, 1696–1711 and references cited therein.

57. M.J. Brienne, J. Gabard, J.-M. Lehn, and I. Stibor, *J. Chem. Soc. Chem. Commun.* **1989**, *24*, 1868–1870.

58. D. Goldmann, R. Dietel, D. Janietz, C. Schmidt, and J.H. Wendorff, *Liq. Crystl.* **1998**, *24*, 407–411.

59. G. Ungar, D. Abramic, V. Percec, and J.A. Heck, *Liq. Crystl.* **1996**, *21*, 73–86.

60. V. Percec, C.-H. Ahn, T.K. Bera, G. Ungar, and D.J.P. Yeardley, *Chem.—Eur. J.* **1999**, *5*, 1070–1083.

61. J. Malthete, A.M. Levelut, and L. Liebert, *Adv. Mater.* **1992**, *4*, 37–41.

62. D. Pucci, M. Veber, and J. Malthete, *Liq. Crystl.* **1996**, *21*, 153–155.

63. R.I. Gearba, M. Lehmann, J. Levin, D.A. Ivanov, M.H.J. Koch, J. Barbera, M.G. Debije, J. Piris, and Y.H. Geerts, *Adv. Mater.* **2003**, *15*, 1614–1618.

64. A.R.A. Palmans, J.A.J.M. Vekemans, E.E. Havinga, and E.W. Meijer, *Angew. Chem. Int. Ed. Engl.* **1997**, *36*, 2648–2651.

65. M. Masuda, P. Jonkheijm, R.P. Sijbesma, and E.W. Meijer, *J. Am. Chem. Soc.* **2003**, *125*, 15935–15940.

66. M.L. Bushey, T.-Q. Nguyen, W. Zhang, D. Horoszewski, and C. Nuckolls, *Angew. Chem. Int. Ed.*, **2004**, *43*, 5446–5453.

67. M.L. Bushey, A. Hwang, P.W. Stephens, and C. Nuckolls, *J. Am. Chem. Soc.* **2001**, *123*, 8157–8158.

68. M.L. Bushey, A. Hwang, P.W. Stephens, and C. Nuckolls, *Angew. Chem. Int. Ed.* **2002**, *41*, 2828–2831.

69. T.-Q. Nguyen, M.L. Bushey, L. Brus, and C. Nuckolls, *J. Am. Chem. Soc.* **2002**, *124*, 15051–15054.

70. M.L. Bushey, T.-Q. Nguyen, and C. Nuckolls, *J. Am. Chem. Soc.* **2003**, *125*, 8264–8269.

71. T.-Q. Nguyen, R. Martel, P. Avouris, M.L. Bushey, C. Nuckolls, and L. Brus, *Chem. Soc.* **2004**, *126*, 5234–5242.

72. G.S. Tulevski, M.L. Bushey, J.L. Kosky, S.J. Toshihiro-Ruter, and C. Nuckolls, *Angew. Chem. Int. Ed.*, **2004**, *43*, 1836–1839.

73. J. Malthete and A. Collet, *J. Am. Chem. Soc.* **1987**, *109*, 7544–7545.

74. L. Brunsveld, E.W. Meijer, A.E. Rowan, and R.J.M. Nolte, *Topics Stereochem.* **2003**, *24*, 373–423.

75. A.E. Rowan and R.J.M. Nolte, *Angew. Chem. Int. Ed.* **1998**, *37*, 63.

76. H. Engelkamp, S. Middelbeek, and R.J.M. Nolte, *Science* **1999**, *284*, 785–788.

77. S.T. Trzaska, H.-F. Hsu, and T.M. Swager, *J. Am. Chem. Soc.* **1999**, *121*, 4518–4519.

78. D.J. Hill, M.J. Mio, R.B. Prince, T.S. Hughes, and J.S. Moore, *Chem. Rev.* **2001**, *101*, 3893–4012.

79. L.A. Cuccia, J.-M. Lehn, J.-C. Homo, and M. Schmutz, *Angew. Chem.* **2000**, *112*, 239.

80. G.A. Jeffrey and L.M. Wingert, *Liq. Crystl.* **1992**, *12*, 179.

81. R. Mukkamala, C.L. Burns, Jr., R.M. Catchings, III, and R.G. Weiss, *J. Am. Chem. Soc.* **1996**, *118*, 9498 and references therein.

82. B. Kohne and K. Praefcke, *Angew. Chem. Int. Ed. Engl.* **1984**, *23*, 82.

83. T.J. Katz, *Angew. Chem. Int. Ed.* **2000**, *39*, 1921–1923.

84. C. Nuckolls, T.J. Katz, G. Katz, P.J. Collings, and L. Castellanos, *J. Am. Chem. Soc.* **1999**, *121*, 79–88.

85. L. Vyklicky, S.H. Eichhorn, and T.J. Katz, *Chem. Mater.* **2003**, *15*, 3594–3601.

86. C. Nuckolls and T.J. Katz, *J. Am. Chem. Soc.* **1998**, *120*, 9541–9544.

87. C. Nuckolls, R. Shao, W.-G. Jang, N.A. Clark, D.M. Walba, and T.J. Katz, *Chem. Mater.* **2002**, *14*, 773–776.

88. T. Verbiest, S. Van Elshocht, M. Karuanen, L. Heliemans, J. Snauwaert, C. Nuckolls, T.J. Katz, and A. Persoons, *Science* **1998**, *282*, 913–915.

89. D.T. Bong, T.D. Clark, J.R. Granja, and M.R. Ghadiri, *Angew. Chem. Int. Ed. Engl.* **2001**, *40*, 988–1011.

90. J.-M. Lehn, J. Malthete, and A.M. Levelut, *Chem. Commun.* **1985**, *24*, 1794–6.
91. M.R. Ghadiri, J.R. Granja, R.A. Milligan, D.E. McRee, and N. Khazanovich, *Nature* **1993**, *366*, 324–327.
92. M.R. Ghadiri, J.R. Granja, and L.K. Buehler, *Nature* **1994**, *369*, 301–304.
93. S. Fernandez-Lopez, H.S. Kim, E.C. Choi, M. Delgado, J.R. Granja, A. Khasanov, K. Kraehenbuehl, G. Long, D.A. Weinberger, K.M. Wilcoxen, and M.R. Ghadiri, *Nature* **2001**, *412*, 452–455.
94. J.S. Moore, *Acc. Chem. Res.* **1997**, *30*, 402–413.
95. S.A. Miller, J.H. Ding, and D.L. Gin, *Curr. Opin. Coll. Int. Sci.* **1999**, *4*, 338–347.
96. D.L. Gin, P. Weiqiang, B.A. Pindzola, and W.-J. Zhou, *Acc. Chem. Res.* **2001**, *34*, 973–980.
97. M. Yoshio, T. Mukai, H. Ohno, and T. Kato, *J. Am. Chem. Soc.* **2004**, *126*, 994–995.
98. S.D. Hudson, H.-T. Jung, V. Percec, W.-D. Cho, G. Johansson, G. Ungar, and V.S.K. Balagurusamy, *Science* **1997**, *278*, 449–452.
99. S.V. Kolotuchin, E.E. Fenlon, S.R. Wilson, C.J. Loweth, and S.C. Zimmerman, *Angew. Chem. Int. Ed. Engl.* **1995**, *34*, 2654–7.
100. D.N. Chin, J.A. Zerkowski, J.C. Macdonald, and G.M. Whitesides, *Mol. Solid State* **1999**, *2*, 185–253.
101. J.A. Zerkowski, C.T. Seto, and G.M. Whitesides, *J. Am. Chem. Soc.* **1992**, *114*, 5473–5475.
102. J.P. Mathias, C.T. Seto, E.E. Simanek, and G.M. Whitesides, *J. Am. Chem. Soc.* **1994**, *116*, 1725–1736.
103. L.J. Prins, D.N. Reinhoudt, and P. Timmerman, *Angew. Chem. Int. Ed.* **2001**, *40*, 2382–2426.
104. C. Roberts, J.C. Chaput, and C. Switzer, *Chem. Biol.* **1997**, *4*, 899–908.
105. M. Mascal, N.M. Hext, R. Warmuth, M.H. Moore, and J.P. Turkenburg, *Angew. Chem. Int. Ed. Engl.* **1996**, *35*, 2204.
106. M. Mascal, N.M. Hext, R. Warmuth, J.R. Arnall-Culliford, M.H. Moore, and J.P. Turkenburg, *J. Org. Chem.* **1999**, *64*, 8479–8484.
107. A. Marsh, M. Silvestri, and J.-M. Lehn, *Chem. Commun.* **1996**, *13*, 1527–1528.
108. S.C. Zimmerman, C. Steven, and S. Perry Corbin, *Struct. Bond. (Berlin)* **2000**, *96*, 63–94.
109. S.K. Kolotuchin and S.C. Zimmerman, *J. Am. Chem. Soc.* **1998**, *120*, 9092.
110. M. Suarez, J.-M. Lehn, S.C. Zimmerman, A. Skoulios, and B. Heinrich, *J. Am. Chem. Soc.* **1998**, *120*, 9526–9532.
111. H. Fenniri, P. Mathivanan, K.L. Vidale, D.M. Sherman, K. Hallenga, K.V. Wood, and J.G. Stowell, *J. Am. Chem. Soc.* **2001**, *123*, 3854–3855.
112. H. Fenniri, B.-L. Deng, A.E. Ribbe, K. Hallenga, J. Jacob, and P. Thiyagarajan, *Proc. Natl. Acad. Sci. (U.S.A)* **2002**, *99*, 6487–6492.
113. H. Fenniri, B.-L. Deng, and A.E. Ribbe, *J. Am. Chem. Soc.* **2002**, *124*, 11064–11072.
114. N. Sakai and S. Matile, *Chem. Commun.* **2003**, *20*, 2514–2523.
115. S. Matile, *Chem. Soc. Rev.* **2001**, *30*, 158–167.
116. P.N.W. Baxter, J.-M. Lehn, G. Baum, and D. Fenske, *Chem.-Eur. J.* **1999**, *5*, 102.

Chapter 17

Crystalline Bacterial Cell Surface Layers (S-Layers): A Versatile Self-Assembly System

Uwe B. Sleytr, Margit Sára, Dietmar Pum, and Bernhard Schuster

CONTENTS

I. INTRODUCTION

Self-assembly of molecules into monomolecular arrays is a new and rapidly growing transdisciplinary scientific and engineering field. Particularly the immobilization of biomolecules in an ordered fashion on solid substrates and their controlled confinement in definite areas of nanometer dimensions are key requirements for many applications in molecular nanotechnology and nanobiotechnology including the development of bioanalytical sensors, molecular electronics, biocompatible surfaces, and signal processing between cells and integrated circuits.

In the last decade several techniques and strategies emerged for creating two-dimensional arrays of proteins on surfaces and for patterning such structures down to the submicrometer scale [1–3].

In this review we describe the basic principles and application potential of crystalline bacterial cell surface layers (S-layers), a self-assembly system optimized through time, selection, and evolution as unique nanostructure [3–6]. S-layers are fascinating model systems for studying the dynamic process of assembly of a biological supramolecular structure [7–10]. The broad application potential of S-layers is based on the specific intrinsic features of the monomolecular arrays composed of identical protein or glycoprotein subunits. Since S-layers are periodic structures they exhibit identical physicochemical properties on each molecular unit down to the subnanometer scale and possess pores identical in size and morphology. Most important, functional groups are aligned on the surface and within the pore areas of S-layer lattices in well-defined position and orientation. Many applications of S-layers depend on the capability of isolated subunits to recrystallize into monomolecular lattices in suspension or on suitable surfaces or interfaces [3,7,8].

Finally, S-layers represent a unique structural basis for generating more complex supramolecular assemblies, involving the essential "building blocks" for life, such as proteins, lipids, carbohydrates, nucleic acids [3,6], and synthesized molecules.

II. LOCATION AND ULTRASTRUCTURE OF S-LAYERS

Among the most commonly observed surface structures on prokaryotic organisms are monomolecular crystalline arrays of proteinaceous subunits termed S-layers [9,11]. S-layers have now been identified in hundreds of different species of every taxonomical group of walled bacteria and are an almost universal feature of archaeal cell envelopes (for compilation see Refs. 8,10–12).

The location and ultrastructure of S-layers have been investigated by different electron microscopical procedures including freeze-etching, ultrathin-sectioning, freeze-drying in combination with metal-shadowing and negative-staining (for review see Refs. 9,13–15).

Although there exists considerable variation in the molecular architecture and supramolecular complexity of prokaryotic envelopes, it is reasonable to classify cell wall profiles containing S-layers into three main groups (Figure 1). In Gram-negative archaea the S-layer represents the only cell wall component and can be so closely associated with the plasma membrane that it is actually integrated into the lipid layer [16]. On both Gram-positive bacteria and archaea S-layers assemble on the surface of the rigid wall matrix (e.g., peptidoglycan or pseudomurein). In the more complex Gram-negative eubacterial cell envelopes S-layers are linked to the surface of the outer membrane.

Unless abundant glycocalyces are present, freeze-etching is the most suitable technique for identifying S-layers on cell surfaces (Figure 2) [9]. High resolution studies on the mass distribution of the lattices are generally performed on negatively stained S-layer fragments or unstained, thin frozen foils [17–19]. More recently, S-layer lattices also have been studied by scanning probe microscopy. Both scanning tunneling microscopy and atomic force microscopy have been applied [3,20–22]. The topographical images obtained strongly resemble the three-dimensional reconstructions of S-layer lattices derived from tilt series of electron microscopical images. Particularly atomic force microscopy in liquid on unfixed S-layer proteins recrystallized on a solid support revealed structural resolution down to the 1 nm range.

S-layer lattices show oblique (p1, p2), square (p4), or hexagonal (p3, p6) symmetry (Figure 3) [8]. Hexagonal symmetry of S-layers is predominant among the archaea [16]. Depending on the lattice type, the morphological units constituting the regular lattices consist of one, two, three, four, and six monomers. The morphological units may have center-to-center spacings from ∼3 to 30 nm. The monomolecular lattices are generally 5 to 25 nm thick. A feature seen with many S-layers is a rather smooth outer and a more corrugated inner surface. In the S-layers of archaea pillar-like domains on the inner corrugated surface are frequently observed which are associated or even integrated in the cytoplasma membrane [16].

Due to their crystalline nature S-layers exhibit uniform pore morphologies. Individual lattices can display more than one pore size. Pore sizes between 2 and 8 nm have been estimated [4,23]. The porosity of the protein meshwork may range approximately from 30 to 70%.

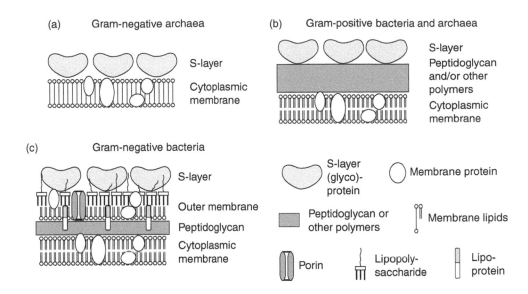

Figure 1   Schematic illustration of the supramolecular architecture of the three major classes of prokaryotic cell envelopes containing crystalline bacterial cell surface layers (S-layers). (a) Cell envelope structure of Gram-negative archaea with S-layers as the only cell wall component external to the cytoplasmic membrane. (b) Cell envelope as observed in Gram-positive archaea and bacteria. In bacteria the rigid wall component is primarily composed of peptidoglycan. In archaea other wall polymers (e.g., pseudomurein or methanochondroitin) are found. (c) Cell envelope profile of Gram-negative bacteria, composed of a thin peptidoglycan layer and an outer membrane. If present the S-layer is closely associated with the lipopolysaccharide of the outer membrane. (Modified after U.B. Sleytr, P. Messner, D. Pum, and M. Sára. *Crystalline Bacterial Cell Surface Proteins*. Austin, TX: R.G. Landes/Academic Press, 1996. With permission.)

III.   Isolation, Chemical Characterization, and Molecular Biology

The subunits of most S-layers interact with each other and with the supporting cell envelope layers through noncovalent forces. In Gram-positive bacteria, a complete disintegration of the protein or glycoprotein lattices into the constituent subunits can be achieved by treatment of intact cells or cell wall fragments with high concentrations of hydrogen-bond breaking agents (e.g., urea or guanidinium hydrochloride) [7,12]. S-layers from Gram-negative bacteria frequently disrupt upon application of metal-chelating agents (e.g., EDTA, EGTA), cation substitution (e.g., Na^+ to replace Ca^{2+}), detergents [8,17,24,25], or by changing the pH value (e.g., pH < 4.0). From conditions required for extraction and disintegration of S-layer lattices it can be derived that bonds holding the subunits together are stronger than those between the crystalline array and the supporting cell envelope layers. There are some S-layers that are very resistant to extraction and disintegration suggesting that the subunits are held together by covalent bonds [26–28]. During removal of the disrupting agents, for example, by dialysis, isolated S-layer subunits frequently reassemble into lattices identical to those observed on intact cells (self-assembly in suspension) [7].

Amino acid analysis of S-layer proteins of organisms from all phylogenetic branches has revealed a rather similar overall composition [4,12,17]. Sequencing of genes encoding the S-layer proteins support data from isoelectric focusing, with a few exceptions (e.g., *Lactobacillus* and *Methanothermus*), where S-layers are composed of an acidic protein or glycoprotein species with an isoelectric point between pH 3 and 6. Accordingly, S-layer proteins have a high amount of glutamic and aspartic acid that together resemble about 15 mol%. The lysine content of S-layer proteins is in the range of 10 mol%. Thus, approximately one-quarter of the amino acids are charged, indicating that ionic bonds play an important role in intersubunit bonding and/or in attaching the S-layer subunits to the underlying cell envelope layer. S-layer proteins have no or only a low content of sulfur-containing amino acids and a high proportion of (40–60 mol%) of hydrophobic amino acids.

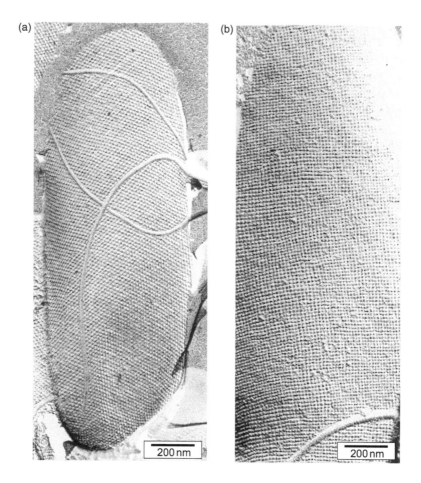

Figure 2 Electron micrographs of freeze-etched preparations of whole cells from (a) *Thermoanaerobacter thermohydrosulfuricus* L111-69 (hexagonal S-layer lattice) and (b) *Desulfotomaculum nigrificans* NCIB 8706 (square S-layer lattice).

Hydrophilic and hydrophobic amino acids do not form extended clusters but instead, hydrophobic and hydrophilic segments alternate with a more hydrophilic region at the very N terminal end [29]. Information regarding the secondary structure of S-layer proteins is either derived from the amino acid sequence or from circular dichroism measurements. In most S-layer proteins, 40% of the amino acids are organized as β-sheet and 10–20% occur as α-helix. Aperiodic foldings and β-turn content may vary between 5 and 45%. Post-translational modifications of S-layer proteins include removal of the signal peptide [29,30], phosphorylation [31], glycosilation [32,33], or processing to cleave the signal peptide and to release two mature S-layer proteins from a single precursor [34]. With the S-layer protein from the extremely halophilic aracheon *Halobacterium halobium* it has been demonstrated for the first time that prokaryotes are capable of producing glycoproteins [35].

Structural analysis of the carbohydrate chains, cloning and sequencing of the S-layer gene, and biosynthesis of this S-layer glycoprotein have been described in detail [36,37]. Chemical analysis and data from nuclear magnetic resonance studies have revealed that the carbohydrate chains from bacterial S-layer glycoproteins consist of up to 50 repeating units. Among the monosaccharide constituents, sugars have been detected which are typical of O antigens of lipopolysaccharides such as quinovosamine, D-rhamnose, *N*-acetylfucosamine, and heptoses [32,33]. Various carbohydrate — amino acid linkage types such as asparagine — rhamnose, tyrosine–galactose, threonine–galactose, or serine–galactose, have been identified [32,38,39].

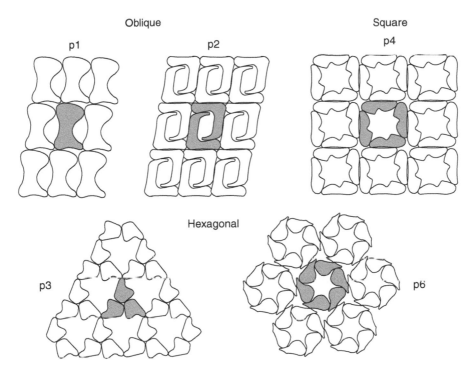

Figure 3 Schematic drawing of different S-layer lattice types. The regular arrays exhibit either oblique (p1, p2), square (p4), or hexagonal (p3, p6) lattice symmetry. The morphological units are composed of one, two, three, four, or six identical subunits. (Modified after U.B. Sleytr, P. Messner, D. Pum, and M. Sára. *Crystalline Bacterial Cell Surface Proteins*. Austin, TX: R.G. Landes/Academic Press, 1996. With permission.)

Although it is quite evident that common structural principles must exist in S-layer proteins (e.g., the ability to form intersubunit bonds and to self-assemble, the formation of hydrophilic pores with low unspecific adsorption, and the interaction with the underlying cell envelope layer), sequencing of S-layer genes from organisms of all phylogenetic branches has led to the conclusion that identities are rare [29,30]. Further sequencing of S-layer genes from strains belonging to the same species, such as *Campylobacter fetus* [40], *Lactobacillus acidophilus* [30], or *Geobacillus stearothermophilus* [41–43], has shown that evolutionary relationship plays an important role for the sequence identity of functionally homologous domains. Actually, for some species the N terminal part of the S-layer proteins represents the conserved structural element being responsible for anchoring the S-layer subunits to the underlying cell envelope layer. For example, the N terminal part of the S-layer proteins from the Gram-negative bacteria *Caulobacter crescentus*, *C. fetus*, and *Aeromonas salmonicida* recognizes specific lipopolysaccharides in the outer membrane as binding site [40,44,45], while the C terminal part comprises surface-located amino acids. The S-layer proteins SbsA and SbsC from two *G. stearothermophilus* wild-type strains [43,46] and SbsD from a temperature-derived variant strain [47] are bound via their N terminal region to an identical type of secondary cell wall polymer (SCWP), which is covalently linked to the peptidoglycan backbone [43,46–49]. The polymer chains consist of tetrasaccharide repeating units that contain glucose, *N*-acetyl glucosamine, and 2,3-dideoxy-2,3-diacetamido-D-mannuronic acid in the molar ratio of 1 to 2 [50]. The N terminal parts of SbsA and SbsC show an identity of 85% and more than 70% of the N terminal 240 amino acids are organized as short α-helices. The sequence identity for the larger part of these S-layer proteins including those domains that are involved in the self-assembly and located on the outer S-layer surface is only 25%. According to the low sequence identity observed for the middle and C terminal parts, SbsA and SbsC assemble into S-layer lattices with either hexagonal or oblique symmetry [42,43,46,48,51–53]. In case of *Corynebacterium glutamicum*, the hydrophobic C terminal part was found to anchor the S-layer subunits to the rigid cell wall layer, which possess a high content of hydrophobic mycolic acids [54]. The S-layer proteins from archaea lacking a rigid cell wall layer integrate with their hydrophobic C terminal part into the cytoplasmic membrane.

By sequence comparison, S-layer homologous (SLH)-domains [55,56] showing an overall identity of 25 to 30% have been identified on the N terminal part of S-layer proteins from various Gram-positive bacteria. Several studies indicate that SLH-domains are responsible for anchoring the S-layer proteins to the rigid cell wall layer by specifically recognizing a distinct type of SCWP [49,57–65]. Despite the structural diversity which can be expected to occur for SCWPs, from different organisms, the common feature for recognizing SLH-domain is most likely the presence of pyruvic acid residues, which have been shown to play a crucial role in the binding process [62,64]. Mesnage et al. [64] have proposed that the binding mechanism between SLH-domains and pyruvylated SCWPs is widespread among procaryotes and has been conserved during evolution. However, the presence of conserved N terminal regions in the S-layer proteins SbsA and SbsC from two *G. stearothermophilus* wild-type strains [43,46,48,49] and in the S-layer protein SbsD from a temperature-derived variant strain [47], which recognizes an identical type of SCWP as binding site in the rigid cell wall layer, strongly indicates that further binding mechanisms could have been conserved during evolution [47,49,65].

Since at a generation time of 20 min about 500 S-layer subunits must be produced per second to keep the bacterial cell surface completely covered with an S-layer lattice, S-layer gene expression must be very efficient and regulatory circuits are necessary to ensure that S-layer protein synthesis is coordinated with cell growth. Actually, most S-layer protein mRNAs have a long leader sequence and an exceptionally high stability [66,67]. In addition, two or even more promoters have been identified for S-layer gene expression that are turned on at different growth stages [68–70]. With the exception of S-layer proteins from *Campylobacter* and *Caulobacter*, all others are produced with a signal peptide suggesting the classical route of secretion [29,30].

Important for understanding S-layer gene regulation is the observation that single bacterial strains can express different (silent) genes. In pathogens such as *C. fetus* S-layer variation can be considered as antigenic variation, which is induced in response to the lytic activity of the immune system and leads to modified cell surface properties. Eight to nine S-layer gene cassettes have been identified in *C. fetus* wild-type strains [40] that are tightly clustered on the chromosome. Several studies have confirmed that only a single promoter exists and that antigenic variation is due to recombination of partial coding sequences. Thereby, the N terminal part of the S-layer protein remains conserved while the C terminal part is exchanged. In the case of *G. stearothermophilus* PV72, S-layer variation can be induced by changing the growth conditions, for example, oxygen supply, during continuous cultivation [71]. Although the exact mechanism of S-layer variation is still unknown, it has been demonstrated that the S-layer gene *sbsB* encoding the S-layer protein SbsB from the variant is located on a megaplasmid in the wild-type strain and is integrated into the *sbsA* expression locus on the chromosome during the switch. The *sbsA* gene encoding the S-layer protein of the wild-type strain is most likely disrupted [72].

IV. DYNAMIC PROCESS OF ASSEMBLY OF S-LAYERS DURING CELL GROWTH

A. Incorporation of New Subunits into Closed Surface Crystals

Numerous *in vitro* and *in vivo* studies have been performed to elucidate the dynamic process of assembly of coherent S-layers during cell growth.

Electron micrographs of freeze-etched rod-shaped bacteria generally reveal a characteristic orientation of the lattice with respect to the longitudinal axis of the cylindrical part of the cell

(Figure 2). This defined alignment of lattices supports the notion that S-layers are "dynamic closed surface crystals" with the intrinsic tendency to assume continuously a structure in a state of low free energy during cell growth [7,73].

Information about the development of coherent S-layer lattices on growing cell surfaces also came from reconstitution experiments with isolated S-layers on cell surfaces from which they had been removed (homologous reattachment) or on those of other organisms (heterologous reattachment) [74,75]. These experiments clearly demonstrated that the formation of the regular patterns entirely resides in the subunits themselves and is not affected by the matrix of the supporting cell envelope layer. Lattices reconstituted on cell envelopes that had maintained their cylindrical shape frequently revealed the orientation with respect to the longitudinal axis of the cell as observed by freeze-etching of intact cells. These *in vitro* experiments confirmed that the curvature of the cylindrical part of the cell induces an orientation of the lattice with least strain between the constituent subunits. On the other hand the spherical curvature on cell poles and septation sites or on the whole surface of coccoidal cells allows a random orientation of the lattices [73,76].

On studying the mechanisms that govern S-layer assembly and recrystallization, it became evident that the only requirement for maintaining the highly ordered monomolecular arrays with no gaps on a growing cell surface is a continuous synthesis of a surplus of subunits and their translocation to sites of lattice growth. Although only limited data are available, S-layers revealed a highly anisotropic charge distribution [23]. For example, in Bacillaceae the inner surface is negatively charged, whereas the outer face is charge neutral. Such dipole characteristics of protomeric units may contribute to the proper orientation during local insertion in the course of lattice growth [77,78].

In most organisms the rate of synthesis of S-layer protein appears to be strictly controlled since only small amounts are detectable in the growth medium. On the other hand, studies on a variety of Bacillaceae have demonstrated that a pool of S-layer subunits, at least sufficient for generating one complete S-layer on the cell surface, may be present in the peptidoglycan containing wall matrix (Figure 1(b)) [79].

Labeling experiments with fluorescent antibodies and colloidal gold/antibody marker methods showed that different patterns of S-layer lattice extensions exist for Gram-positive and Gram-negative bacteria. In Gram-positive bacteria (Figure 1(b)) lattice growth occurs primarily by insertion of multiple bands of S-layers on the cylindrical part of the cell and at new cell poles. In Gram-negative bacteria (Figure 1(c)), however, S-layer lattices grow by insertion of new subunits at diffuse sites over the main cell body [7].

Very few data are available about the accurate incorporation sites of constituent subunits on closed surface crystals. As discussed by Harris and coworkers [80–82] dislocations can serve as sites for incorporation of new subunits in crystalline arrays that grow by "intussusception" (Figure 4). Further on, as a geometrical necessity, closed surface crystals must contain local wedge disclinations (Figure 4), which themselves can act as sources of edge dislocations [80]. Consequently, the rate of growth of a closed surface crystal by the mechanism of nonconservative climb of dislocations will depend on the number of dislocations present and the rate of incorporation of new subunits at these sites. High resolution electron micrographs of freeze-etched preparations reveal the presence of both dislocations and disclinations (Figure 5) on S-layers of intact cells. Nevertheless, the final proof that dislocations are sites of intussusceptive growth of S-layer lattices will require labeling procedures at a resolution where individual S-layer subunits can be detected.

Whereas in Gram-positive bacteria and archaea (Figure 1(b)) the dynamic process of assembly and recrystallization of S-layers occurs on the surface of a rigid cell shape determining a supramolecular envelope structure, Gram-negative archaea (Figure 1(a)) possess S-layers as an exclusive wall component. Analysis of cell morphology and lattice fault distribution in Gram-negative archaea provided strong evidence that S-layers can define cell shape and are involved in cell fission. *Thermoproteus tenax*, an extremely thermophilic archeon, has a cylindrical shape with constant diameter but is variable in length and hemispherical cell poles (Figure 6) [83]. Whereas no dislocations could be observed on the hexagonal array covering the cylindrical part six wedge disclinations could be

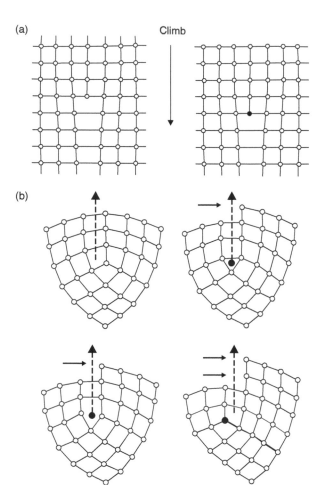

Figure 4 Schematic drawing of dislocations and disclinations as observed on S-layers. (a) Intussusceptive growth of a two-dimensional crystal with square lattice symmetry by nonconservative climb of a dislocation. By adding a new subunit (solid dot) the dislocation climbs one lattice spacing to a new position. (b) Movement of a local wedge disclination in a square lattice. A wedge disclination may be constructed by cutting into the crystal and rotating one face of the cut into the other (positive wedge disclination), or alternatively by inserting a wedge into the cut instead of removing it (negative wedge disclination). When moving, the disclination is shifted diagonally across a distorted square, and in the course of this process generates two edge dislocations (arrows). (Modified from U.B. Sleytr and P. Messner. In: H. Plattner, Ed., *Electron Microscopy of Subcellular Dynamics*. Boca Raton, FL: CRC Press, 1989, pp. 13–31. With permission.)

visualized on each hemispherical cap. This number resembles the minimum of lattice faults required for covering the rounded surface as known from viruses with icosahedral symmetry [84]. Thus it appears feasible that elongation of the cylindrical part of the cell only requires insertion of S-layer subunits at these distinct lattice faults [80]. More detailed studies on the involvement of an S-layer in cell morphology and division has been reported for *Methanocorpusculum sinense* [85]. Cells of this organism reveal, like in many other Gram-negative archeons, a highly lobed cell structure with a hexagonally ordered S-layer. In freeze-etched preparations of intact cells numerous positive and negative 60° wedge disclinations could be detected (Figure 7) which form pentagons and heptagons in the hexagonal array. Since complementary pairs of pentagons and heptagons are the termination points of edge dislocations they can be expected to function both as sites for incorporation of new morphological units into the lattice (Figure 8) in the formation process of the lobed structure, and as initiation points for the cell division process. The latter was shown to be determined by the ratio

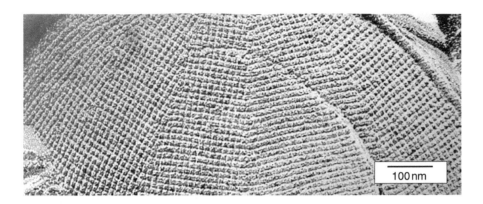

Figure 5 Electron micrograph of a freeze-etched preparation showing several edge dislocations in the S-layer with square lattice symmetry at the cell pole of *Aneurinibacillus thermoaerophilus* DSM 10155. The edge dislocations become visible as line imperfections in the regular array (courtesy P. Messner).

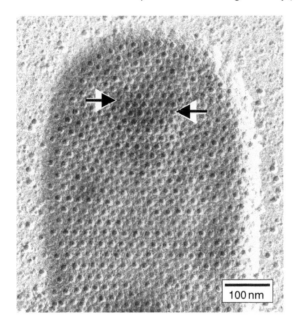

Figure 6 Freeze-dried and shadowed preparation of envelopes of *T. tenax* labeled with polycationic ferritin (PCF). The marker molecule binds to the hexagonal S-layer in a regular fashion. Because one PCF molecule is bound per hexameric unit cell of the S-layer lattice, lattice faults in the cell envelope preparations become clearly visible. Based on theoretical considerations six wedge disclinations have to be present at each cell pole. Two of the six wedge disclination are marked. (Modified after P. Messner, D. Pum, M. Sára, K.O. Stetter, and U.B. Sleytr. *J. Bacteriol.* 166:1046–1054, 1986. With permission.)

between the increase of protoplast volume and the increase in actual S-layer surface area during cell growth [85].

V. SELF-ASSEMBLY OF ISOLATED S-LAYER SUBUNITS

A. Self-Assembly in Suspension

Isolated S-layer subunits from a great variety of bacteria show the ability to self-assemble into lattices identical to those observed on intact cells. Depending on the intrinsic properties of the S-layer protein

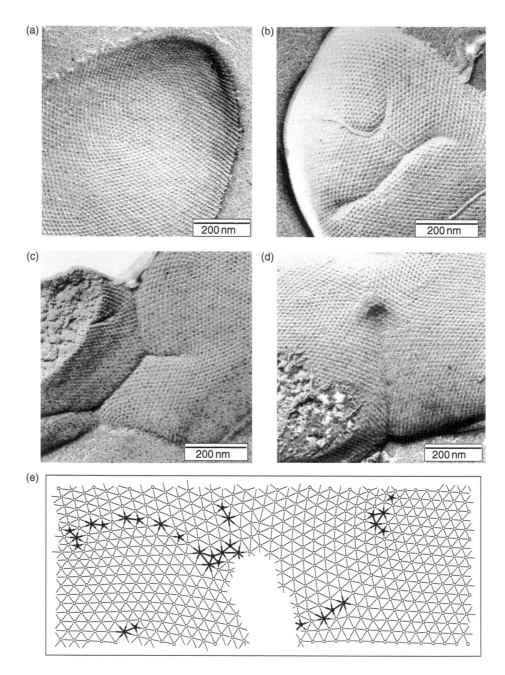

Figure 7   Freeze-etched preparation of *Methancorpusculum sinense* (a) to (d). The hexagonally ordered S-layer shows several lattice faults (a). Wedge disclinations and edge dislocations are seen as point imperfections in the crystalline array. Consecutive stages in the invagination of the cell wall and cell septation are shown in (b) to (d). Initially shallow invaginations are formed (b) which become longer and deeper as new S-layer material is incorporated (c) and (d). The division of deeper invaginations shows that they can also fuse or branch (c). A far advanced stage in the cell fission process is shown in (d). Neighborhood graph of the central region of panel (d) is shown in (e). The alignment of lattice faults (pentagons and heptagons; marked in bold with solid dots) in line with the septation direction indicates the route of the progressing cell septation. (Modified after D. Pum, P. Messner, and U.B. Sleytr. *J. Bacteriol.* 173:6865–6873, 1991. With permission.)

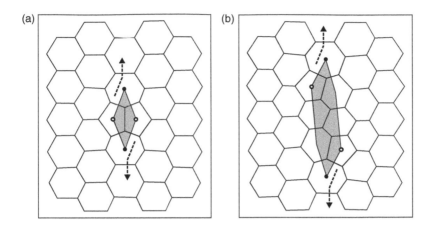

Figure 8 Schematic drawing of the incorporation of a single morphological unit (shaded) in a perfect hexagonal lattice creates a double pair of five and sevenfold wedge disclinations (a). The two pairs move away from each other by gliding or climbing. One possibility is shown in (b) where the incorporation of new morphological units (shaded) along the arrows pushes the two pairs apart and resulting in an invagination which becomes longer and deeper. (Modified after D. Pum, P. Messner, and U.B. Sleytr. *J. Bacteriol.* 173:6865–6873, 1991. With permission.)

and the reassembly conditions (e.g., pH value, ionic strength, ion composition) isolated S-layer subunits may recrystallize into flat sheets, open-ended cylinders, or closed vesicles (Figure 9 and Figure 10) [7,76,86]. Several studies have confirmed that S-layers are entropy-driven self-assembly systems in which all the information for crystallization resides within the individual subunits [74]. Detailed studies on the kinetics of the *in vitro* self-assembly in suspension, the shape of self-assembly products, the charge distribution, and the topographical properties of the outer and inner surface have been carried out with wild-type and recombinant S-layers from Bacillaceae.

Studies on the kinetics of the *in vitro* self-assembly of the S-layer subunits from *G. stearothermophilus* NRS 1536/3c have included light scattering and cross-linking experiments. Thereby, the existence of a rapid initial phase and a slow consecutive process of higher than second order has been observed [87]. The rapid initial phase leads to the formation of oligomeric precursors composed of 12 to 16 subunits ($M_r > 10^6$) that fuse and recrystallize during the second stage. The sheet-like self-assembly products achieve a size of 1 μm and clearly exhibit the square lattice symmetry observed for S-layers on intact cells [87].

Labeling of self-assembly products formed by S-layer proteins of Bacillaceae with polycationic ferritin (PCF) has revealed that they are highly anisotropic structures [88–90]. While the inner surface is frequently capable of binding PCF as a positively charged topographical marker, an equimolar amount of amino and carboxylic acid groups has been found on the outer S-layer surface. After blocking the amino groups on the outer S-layer surface from *Bacillus sphaericus* CCM 2120 with glutaraldehyde, the negative charge density has been determined with 1.6/nm^2 [91,92]. Cross-linking experiments have further demonstrated that amino and carboxylic acid groups from adjacent S-layer subunits are involved in direct electrostatic interactions. Those chemical modification reactions which change the native charge distribution in the S-layer lattice lead to the loss of its structural integrity or to the complete disintegration [88–90]. Most S-layer proteins from *G. stearothermophilus* wild-type strains assemble into mono- or double-layer sheets which is independent of the presence of mono- or bivalent cations and can even be achieved after the addition of metal-chelating agents. In double-layer self-assembly products, the individual monolayers are bound to each other with the outer charge neutral surface [88]. However, an exception has been found for the S-layer glycoprotein from *G. stearothermophilus* NRS 2004/3a, which assembles into an oblique lattice type (Figure 11). In case of this glycosilated S-layer protein, the formation of double-layer self-assembly products is

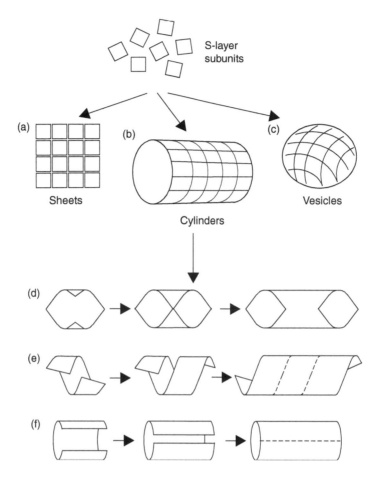

Figure 9   Diagram illustrating different self-assembly routes of S-layer subunits leading to the formation of (a) flat
sheets, (b and d to f) cylinders, and (c) spheres.

strongly dependent on the presence of bivalent cations during the dialysis procedure [93]. Under these
conditions, double-layer sheets or cylinders are formed in which the individual S-layers face each
other with the net negatively charged inner surface (Figure 11). Five different possibilities regarding
the orientation of the individual S-layers with respect to each other have been identified [93]. In the
absence of bivalent cations, this S-layer glycoprotein assembles into two types of monolayer cylinders
in which the charge neutral outer surface is exposed to the ambient environment (Figure 11). For
the S-layer proteins from *B. sphaericus* strains, the self-assembly process is strongly dependent on
the presence of bivalent cations [59,60,94]. In the absence of bivalent cations, most of the S-layer
proteins stay in the water soluble state and can be exploited for recrystallization on solid supports or
lipid films [94].

The production of recombinant S-layer proteins has opened the possibility to identify those
domains that are required for the self-assembly process. Studies on the self-assembly properties have
been performed with N or C terminal truncations of the S-layer proteins SbsC [48], SbsB [62,95–97],
and SbpA [61]. The recombinant full-length S-layer protein rSbsC ($rSbsC_{31-1099}$) self-assembles
into flat double-layer sheets, as well as into open-ended mono- and double-layer cylinders [48]. All
self-assembly products exhibit the oblique lattice structure, which is also found on self-assembly
products formed by the S-layer protein of the wild-type strain. In contrast to full-length rSbsC,
$rSbsC_{258-1099}$ in which the positively charged N terminal part has been deleted, self-assembles only
into open-ended monolayer cylinders, which in negatively stained preparations exhibit a delicate

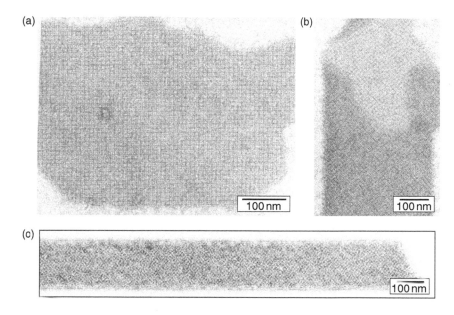

Figure 10 Electron micrographs of negatively stained preparations of S-layer self-assembly products representing double-layers. (a) Flat sheet; (b, c) open-ended cylinders; (a, b) square S-layer lattice; (c) oblique S-layer lattice.

oblique lattice structure. The C terminal truncations $rSbsC_{31-920}$ and $rSbsC_{31-930}$ form self-assembly products with properties comparable to those observed for the full-length rSbsC. Self-assembly products obtained with the C terminal truncation $rSbsC_{31-880}$ or $rSbsC_{31-900}$ do not exhibit a regular lattice structure. This observation is in agreement with the findings that further C terminal truncation leading to $rSbsC_{31-860}$ and $rSbsC_{31-844}$ is linked to the loss in the ability to self-assemble and these rSbsC forms remain in the water soluble state [48]. Considering the water solubility of $rSbsC_{31-860}$ and the formation of first self-assembly products by $rSbsC_{31-880}$, an elongation of only 20 amino acids is sufficient to establish the contact sites necessary for subunit–subunit interactions. Based on the results obtained with $rSbsC_{31-880}$ and $rSbsC_{31-920}$, only 40 additional amino acids in the C terminal part are required to completely restore the oblique lattice structure [48]. The C terminally truncated form $rSbsC_{31-844}$ has been shown to form three-dimensional crystals as required for x-ray analysis [98].

In the case of the S-layer protein SbpA, the deletion of 200 C terminal amino acids is without any effect on its self-assembly properties [60]. For the S-layer protein SbsB it has been demonstrated that the N terminal SLH-domain is not required for the self-assembly process [62,96,97]. Accordingly, the oblique lattice structure formed by $rSbsB_{208-920}$ missing the SLH-domain is identical to that formed by the full-length $rSbsB_{32-920}$ and even the lattice constants are the same [97]. On the other hand, deletion of only 10 amino acids at the C terminal end leads to a completely water soluble rSbsB form [95,96]. The addition of the specific secondary cell wall polymer to the S-layer protein denatured with chaotropic agents has been found to inhibit the *in vitro* self-assembly by keeping the S-layer protein in the water soluble state [99,100]. Interestingly, the soluble S-layer protein is capable of recrystallizing into a monomolecular protein lattice on poly-L-lysine coated electron microscopy (EM) grids to which the subunits bind with their outer charge neutral surface. Several data have indicated that the polymer chains prevent the formation of self-assembly products by acting as spacer between the individual S-layer subunits [99]. According to the findings with the isolated recombinant S-layer proteins, full-length rSbsC and rSbsB form self-assembly products in the cytoplasm of the heterologous expression host [42,43,48,53,95,101], whereas rSbpA, whose self-assembly is strongly dependent on the presence of calcium ions accumulates in inclusion bodies [60].

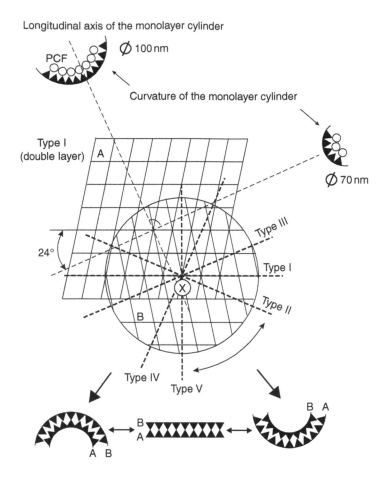

Figure 11   Schematic representation of the formation of mono- and double-layer assembly products as described
with S-layer subunits isolated from *G. stearothermophilus* NRS 2004/3a. This S-layer shows oblique
lattice symmetry with center-to-center spacings of the morphological units of 9.4 and 11.6 nm and a
base angle of 78°. The oblique lattice symmetry allows us to unambiguously determine the orientation
of the constituent monolayer sheets in double-layer self-assembly products. On the oblique monolayer
sheet A the axes of the two types of small (70 and 100 nm diameter) monolayer cylinders are formed as
indicated. One of the axes includes an angle of 24° to the short base vector of the oblique S-layer lattice.
The second axis is perpendicular to the first. Both monolayer cylinders have an identical direction of
curvature. Owing to differences in the charge distribution on both the S-layers, PCF is only bound to the
inner surface of both types of monolayer cylinders. Five types of double-layer self-assembly products
with back-to-back orientation of the inner surface of the constituent monolayers have been found. The
superimposition of sheets A and B in the double-layer assembly products of type I is demonstrated
and the angular displacement of sheet B with respect to A around point X for the assembly products of
type II to V is indicated. (Modified after P. Messner, D. Pum, and U.B. Sleytr. *J. Ultrastruct. Mol. Struct.
Res.* 97:73–88, 1986. With permission.)

B.   Self-Assembly on Lipid Films and at the Air–Water Interface

Reassembly of isolated S-layer proteins at the air–water interface and on lipid films have proven to
be a very useful way to generate coherent S-layer lattices in large-scale (Figure 12).

The anisotropy in the physicochemical surface properties of the S-layer lattices allows us to
control the orientation of the monolayer against different surfaces and interfaces. Since in S-layers
used for recrystallization studies, the outer surface is more hydrophobic than the inner one, the
protein lattices are generally oriented with their outer face against the air–water interface [102,103].
In the case of S-layer protein recrystallized on Langmuir lipid films, the orientation depends on the

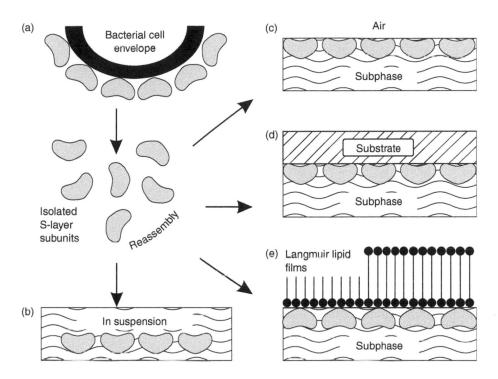

Figure 12 Schematic illustration of recrystallization of isolated S-layer subunits (a) into crystalline arrays. Formation of self-assembly products in suspension (b). The self-assembly process can occur in defined orientation, at the air–water interface (c), on solid supports (d), and on Langmuir lipid films (e). (Modified after U.B. Sleytr, P. Messner, D. Pum, and M. Sára. *Angew. Chem. Int. Ed.* 38:1034–1054, 1999. With permission.)

nature of the lipid headgroup, the phase state of the surface monolayer, as well as the ionic content, and the pH of the subphase. S-layer lattices formed at different interfaces have been studied by electron microscopy and x-ray reflectivity measurements [102–107]. Coherent S-layer monolayers can be obtained on lipid films composed of lipids with zwitterionic headgroups in the presence of calcium ions if the lipid chains possessed a high degree of order, that is, if the lipid films are in the liquid condensed phase. Under these conditions, the S-layer subunits attach to the lipid with their net negatively charged inner surface (Figure 13). In contrast, the S-layer protein recrystallize poorly under most lipids with negatively charged headgroups and under lipids with unsaturated chains. At monolayers composed of cationic lipids, however, the S-layer proteins recrystallize with its outer surface pointing toward the lipid headgroups [104–107].

The orientation of the S-layer lattice with respect to the plane of the interface is routinely determined by high resolution electron microscopical studies of negatively stained S-layer lattices which have been lifted from the air–water interface onto a carbon coated electron microscope grid by horizontal deposition (Langmuir Schaefer technique) [102,103]. The determination of the orientation of S-layers with oblique lattice symmetry is particularly easy since the handedness of the base vector pairs can directly be compared to that observed on the bacterial cell. In case of S-layers with higher lattice symmetry (square or hexagonal) image processing is required to obtain unambiguous results.

Crystal growth of S-layer protein lattices at different surfaces and interfaces has been studied by high resolution electron microscopy and scanning force microscopy (Figure 14) [21,103]. Generally crystal growth is initiated simultaneously at many randomly distributed nucleation points and proceeds in plane until the crystalline domains meet leading to a closed, coherent mosaic of crystalline areas with mean diameters of one to several tens of micrometers. This crystal growth process is commonly observed at liquid–air interfaces, lipid films, and at solid supports [108].

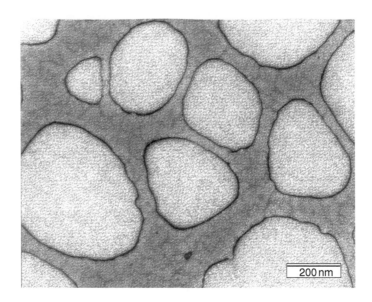

Figure 13   Electron micrograph of an S-layer-supported lipid membrane spanning the holes of a perforated carbon film (on an electron microscope grid). Apertures up to 15 μm in diameter may be covered by composite S-layer/lipid films.

The importance of the ionic strength of the subphase has been demonstrated in the recrystallization of the S-layer protein of *B. sphaericus* CCM 2177, which assembles into a lattice with square symmetry [109]. Depending on the calcium concentration, a broad range of crystal morphologies ranging from tenuous, fractal-like structures to large monocrystalline patches have been found. Although all these structures look like fractals obtained by diffusion limited aggregation, they are not aggregates of randomly oriented protein subunits. Image processing reveals that all morphological units followed the lattice orientation of the crystal lattice.

The recrystallization of S-layer proteins at phosphoethanolamine monolayers on aqueous subphases has been also studied on a mesoscopic scale by dual label fluorescence microscopy and Fourier transform infrared spectroscopy (FTIR) [110]. It has been shown that the phase state of the lipid exerts a marked influence on the protein crystallization. When the surface monolayer is in the phase separated state between fluid and crystalline phase, the S-layer protein is preferentially adsorbed at the boundary line between the two coexisting phases and crystallization proceeded underneath the crystalline phase. Crystal growth is much slower under the fluid lipid and the entire interface is overgrown only after prolonged protein incubation. In turn, as indicated by characteristic frequency shifts of the methylene stretch vibrations on the lipids, protein crystallization affects the order of the alkane chains and drives the fluid lipid into a state of higher order. However, the protein does not interpenetrate the lipid monolayer as confirmed by x-ray reflectivity studies [105–107].

As previously demonstrated, S-layers have a stabilizing effect on the associated lipid layer [103,104,108]. The proportion of lipid molecules in the monolayer that is bound to repetitive domains in the S-layer lattice modulates the lateral diffusion of the free lipid molecules and consequently the fluidity of the whole membrane. Subsequent lipid layers can be deposited on such "semifluid membranes" by standard Langmuir–Blodgett techniques or by fusion of lipid vesicles [108]. The stabilizing effect of S-layers on lipid films has already been demonstrated by covering apertures several micrometers in size on holey carbon grids (Figure 13). These composite structures strongly resemble those archaeal envelope structures that are exclusively composed of an S-layer and a closely associated plasma membrane (Figure 1(a)). Since many of these organisms dwell under extreme environmental conditions (e.g., low pH, high temperatures, high salt concentrations) their S-layers must have a strong stabilizing effect on lipid membranes. Apparently the main reason for this is the

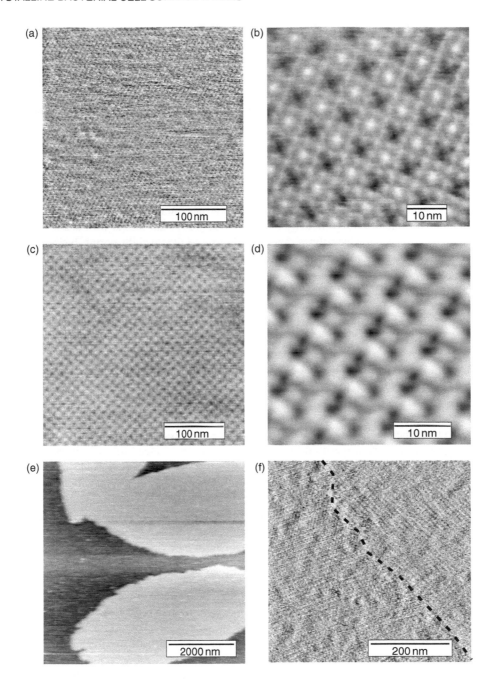

Figure 14   High resolution scanning force microscopical images of S-layers with oblique (p1) (a) and square (p4) lattice symmetry (c) on silicon surfaces. The corresponding computer image reconstructions obtained by cross-correlation averaging are shown in (b) and (d), respectively. Crystal growth is initiated at randomly distributed nucleation points from which crystalline domains grow (e) until the front edges meet and a closed monolayer is formed (f). (Modified after D. Pum and U.B. Sleytr. *Supramol. Sci.* 2:193–197, 1995. With permission.)

reduction of horizontal vibrations which are considered to be the main cause for disintegration of unsupported lipid membranes.

The enzymatic interplay of porcine pancreatic phospholipase A_2 (PLA$_2$) with a zwitterionic lipid monolayer in the absence and presence of a recrystallized S-layer lattice has been monitored by the decay of the surface pressure Π [111]. The same duration of the lag period (i.e., the time period until the enzymatic reaction started) has been observed with S-layer-supported and the corresponding plain lipid monolayers. Thus, the recrystallized S-layer does not induce defects in the lipid monolayer from which the PLA_2 can benefit and the isoporous S-layer lattice represents no barrier for the PLA_2. The observed data strongly suggest that the prevalent proportion of the lipid molecules remains in a morphological and physicochemical state and thus, constitute a good substrate for the PLA_2 [111].

Folded [112] and painted [113] bilayer lipid membranes (BLMs) are generally generated on septa made of Teflon with orifices of 50 to 200 μm and 0.8 to 3 mm in diameter, respectively. The effect of an attached S-layer lattice on the conductance, capacitance, and boundary potential of the lipid membrane is found to be negligible. Thus, the freestanding BLMs are not forced by the attached S-layer lattice to considerable structural rearrangements. In addition, S-layer-supported BLMs have been found to be significantly more viscous [114] than BLMs with absorbed proteins, such as actin [115], or polyelectrolytes, such as high molecular weight poly-L-lysine [116]. The slow opening velocity of pores that have been induced by electroporation and thus, the calculated high viscosity might reflect a high number of contact sites per area (e.g., repetitive domains of the associated S-layer lattice) as it has been suggested for polymers with different density of hydrophobic anchors [117]. The dynamic surface roughness of bilayer membranes upon crystallization of S-layer protein has been investigated too [118]. Crystallization of S-layer protein at both sides of the BLM causes a considerable reduction of the membrane tension whereas the membrane bending energy increases by three orders of magnitude compared to data reported for erythrocytes or vesicular membranes, and is in the same order as a shell composed of a 5 nm thick polyethylene layer [119].

The area expansion upon the application of a hydrostatic pressure has been determined for S-layer-supported lipid membranes and compared with membranes without an S-layer support [120]. Unsupported lipid membranes independent from which side pressurized and S-layer-supported lipid membranes pressurized from the lipid-faced side reveal a pronounced increase in area expansion as measured by the membrane capacitance. By contrast, S-layer-supported lipid membranes pressurized from the protein-faced side reveal only a minute increase in membrane area. The enhanced stability against bulging when a hydrostatic pressure is applied might be an interesting issue for the investigation of mechano-sensitive ion channels [121–124]. For the first time, it might be possible to distinguish between a curvature-induced mechanical activation or a water flow-induced activation [125] and thus, elucidation of the gating mechanism of mechano-sensitive ion channels should be possible.

The electrophysical features of S-layer-supported lipid membranes have been studied both by voltage clamp and black lipid membrane techniques. In the first set of experiments S-layer protein isolated from *Bacillus coagulans* E38-66 has been recrystallized on monolayers of glycerol-dialkyl-nonitol tetraether lipid (GDNT). Voltage clamp examinations have been applied for determining both the barrier function of the lipid film before and after recrystallization of the S-layer protein and the effect of incorporation of the potassium selective ion channel, valinomycin [126]. Upon recrystallization of the S-layer protein, a decrease in conductance of the GDNT-monolayer is observed. Furthermore, it is found that the valinomycin-mediated increase in conductance is less pronounced for the S-layer-supported monolayer than for the plain GDNT-monolayer confirming differences in the accessibility and/or fluidity of the lipid membranes. In contrast to plain GDNT-monolayers, S-layer-supported GDNT-monolayers with high valinomycin-mediated conductance persist over much longer periods of time, indicating enhanced stability of the composite structure.

The effects of a supporting S-layer from *B. coagulans* E38-66 on lipid bilayers has also been investigated by comparative voltage clamp studies on plain and S-layer-supported 1,2-diphytanoyl-*sn*-glycero-3-phosphatidylcholine (DPhPC) bilayers [127]. Upon S-layer recrystallization no significant changes but a slight decrease in conductance can be observed. Thus, both the GDNT-monolayer and DPhPC-bilayer studies indicate that the recrystallized S-layer protein does not penetrate or rupture the lipid film. The effect of a supporting S-layer on the incorporation and self-assembly of a pore-forming protein has also been studied with DPhPC-bilayers formed over a thin Teflon aperture and the staphylococcal exotoxin α-hemolysin [127]. When added to the lipid-exposed side, the assembly of the heptameric α-hemolysin pore is slow compared to unsupported membranes. According to the semifluid membrane model [103] this phenomenon is explained by the altered fluidity of the lipid bilayer. On the other hand, no pore assembly can be detected upon adding α-hemolysin monomers to the S-layer-faced side of the composite membrane. This is due to the intrinsic molecular sieving properties of the S-layer lattice preventing passage of α-hemolysin monomers through the S-layer pores to the lipid bilayer. In comparison to plain lipid bilayers, the S-layer-supported lipid membrane have a decreased tendency to rupture in the presence of α-hemolysin. In a further study, the feasibility to reconstitute α-hemolysin into S-layer-supported lipid membranes at single molecule resolution has been investigated [128]. Indeed, at various applied voltages, current steps can be recorded, each corresponding to the formation of one new lytic α-hemolysin pore.

C. Self-Assembly on Solid Supports

Reassembly of isolated S-layer proteins into larger crystalline arrays can be also induced on solid supports [108,109,129,130]. In particular, the recrystallization of S-layer proteins on technologically relevant substrates such as silicon wafers (Figure 14), carbon-, platinum-, gold-, or silver-electrodes and on synthetic polymers revealed a broad application potential for the crystalline arrays in micro- and nanotechnology [108]. The formation of coherent crystalline arrays depend strongly on the S-layer species, the environmental conditions of the subphase (e.g., temperature, pH, ion compo-sition, and ionic strength), and, in particular, on the surface properties of the substrate. In general, highly hydrophobic surfaces are better suited for the formation of large-scale closed S-protein mono-layers than less hydrophobic or hydrophilic supports [129]. Silanization procedures with different compounds, such as octadecyltrichlorosilane (OTS) or hexamethyldisilazane (HMDS), can be used to produce hydropobic surfaces on silicon or glass [89,90]. Due to practical reasons, most investi-gations have been carried out on silicon wafers with a native or plasma induced oxide layer, or photoresist-coated silicon wafers [129,131]. The recrystallization process follows the same rules as those already described for the formation of coherent large-scale monolayers at liquid surface interfaces. Scanning force microscopy is the only tool which allows to image S-layer protein mono-layers on solid supports at molecular resolution (Figure 14) [21,131]. In particular, scanning force microscopy in contact mode with loading forces in the range of 100–200 pN leads to an image resolution in the subnanometer range (0.5–1.0 nm). All these high resolution investigations are only possible in a fluid cell. But most important, scanning force microscopy allows the investigation of the recrystallization in real time too [130].

Recrystallization of the recombinant S-layer proteins rSbsB and rSbpA on solid supports can be significantly improved when the supports are precoated with the specific SCWP [132]. For coating gold chips with SCWP, the reducing end of the polymer chains is exploited for chemical modifi-cation reactions [62]. Thereby, the latent aldehyde group of the polymer chains is converted into an amino group by reaction with carbohydrazide and subsequently into a thiol group by coupling 2-iminothiolane [62,132].

S-layers which have been recrystallized on solid surfaces are the basis for a variety of applications in nanomanufacturing, and in the development of miniaturized biosensors [108]. For this purpose, supports have to be selected which fulfill specific requirements, such as stiffness or flatness. Further, the availability of functional groups is becoming important if a specific orientation of the recrystal-lized S-layer lattice or cross-linking of the protein array is required. Silicon and gallium arsenide have proven to be the most potent materials for all those applications where S-layer technology will be combined with micromachining, nanoelectronics, or nanooptics [3].

D. Lipid Membranes on Porous and Solid Supports

Phospholipid bilayers or tetraether lipid films incorporating functional molecules (e.g., ion channels, carriers, pore-forming proteins, proton pumps) represent key elements in the development of biomimetic sensor devices. Unfortunately, plain lipid membranes are highly susceptible to damage during manual handling procedures and prolonged storage, and are thus usually not considered for practical devices. Therefore, much effort has been concentrated on the development of stabilized or supported lipid membranes in order to increase their long-term stability [133]. One suggestion is to place soft polymer cushions between the substrate and the functionalized lipid membrane in order to maintain the thermodynamic and structural properties of the system [133]. As an alternative to this approach, S-layers can be used as supports and stabilizing structures for phospholipid bilayers and tetraetherlipid monolayer films (Figure 15 and Figure 16) [10,134–138].

One promising strategy is to use an S-layer ultrafiltration membrane (SUM) with the S-layer as stabilizing and biomimetic layer between the BLM and a porous support. SUMs are isoporous structures with very sharp molecular exclusion limits and were manufactured by depositing S-layer-carrying cell wall fragments on commercial polyamide microfiltration membranes (MFMs) with an average pore size of approximately 0.4 μm [91,92,139]. In general terms, lipid membranes generated on a porous support combine the advantage of possessing an essentially unlimited ionic reservoir on each side of the BLM and of easy manual handling.

The SUM-supported bilayers are tight structures with breakdown voltages well above 500 mV during their whole lifetime of about 8 h [140]. BLMs on a plain polyamide MFM (lifetime about 3 h) rupture at a magnitude of about 210 mV. Specific capacitance measurements and reconstitution experiments have shown that the lipid membrane on the SUM consists of two layers as the pore-forming protein α-hemolysin can be assembled into lytic pores. For the first time, opening and closing behavior of even single α-hemolysin pores has been measured with membranes generated on a porous support. The unitary conductance for single reconstituted α-hemolysin pores is found to be broadly similar when incorporated in folded BLMs and in SUM-supported lipid membranes [140]. The present results indicate that the S-layer lattices of the SUM represent a water-containing and biomimetic layer for the closely attached lipid bilayer and provide also a natural environment for protein domains protruding from the membrane.

Recently, a new and reproducible method to generate stable lipid membranes on SUMs has been described [141]. The membrane-spanning tetraether lipid MPL (main tetraether phospholipid of *Thermoplasma acidophilum*), and also mixtures of MPL with DPhPC at molar ratios of MPL:DPhPC = 1:1 and 5:1, and pure DPhPC are spread at the air–water interface. The monomolecular films are subsequently transferred by one raising step (MPL and mixtures) or by a first lowering and subsequent raising step of the electrolyte (DPhPC) onto the SUM. SUM-supported MPL membranes show a lifetime of about 8 h but an additional monomolecular S-layer lattice recrystallized on the lipid-faced side increases the lifetime of the composite membrane significantly to about 21 h [141].

Measurements on single gramicidin pores have been performed with SUM-supported membranes composed of DPhPC, MPL, and with mixtures of DPhPC and MPL [141]. The conductance of single gramicidin pores reconstituted in folded and SUM-supported membranes is higher bathed in K^+-containing electrolyte than in solutions containing Na^+-cations. The unitary pore conductance, however, decreases with increasing amount of MPL in the membrane, particularly at the folded membranes. This observation might be explained by the structure of the MPL monolayer as MPL molecules are known to form a tilted phase, which in turn may affect the alignment of the two gramicidin subunits. Thus, a mutual offset of the gramicidin subunits due to the tilted MPL molecules might result in a lower single pore conductance. Furthermore, gramicidin pores reveal a higher conductance reconstituted in folded than in SUM-supported membranes. The reduced conductance of gramicidin pores might be caused by a further increase in viscosity induced by frictional forces from the attached porous S-layer lattice [141]. This result is in accordance with a previous study that showed

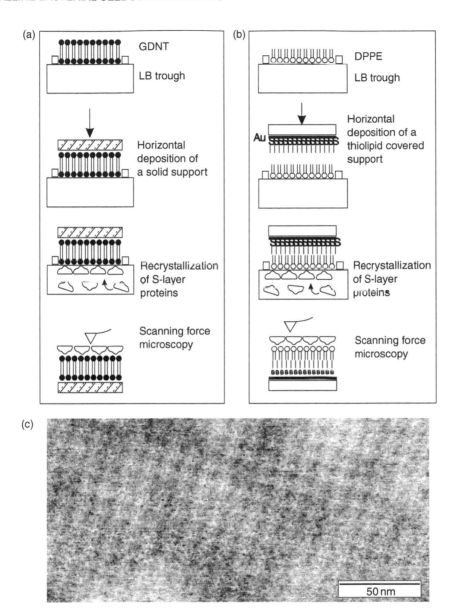

Figure 15 Schematic illustration of the preparation steps required for generating an S-layer stabilized (a) tetraether lipid film and (b) lipid bilayer, on a solid support. Top-to-bottom: After compressing a GDNT-monolayer (a), or a 1,2-dipalmitoyl-*sn*-glycero-3-phosphatidylethanolamine (DPPE) monolayer (b) to a surface pressure of 25 to 30 mN/m, a solid support (silicon wafer: plain or for attachment of thiolipids coated with gold) is placed horizontally onto the monolayer and left in this position until the S-layer protein that was injected afterwards in the subphase has recrystallized as closed monolayer. The whole assembly is subsequently removed, rinsed in deionized water, and transferred into a scanning force microscope. (c) Scanning force micrograph of the surface topography of the outer S-layer face of an S-layer with square lattice symmetry. (Modified after B. Wetzer, D. Pum, and U.B. Sleytr. *J. Struct. Biol.* 119:123–128, 1997. With permission.)

a significantly increased surface viscosity of the phospholipid bilayer upon recrystallization of S-layer proteins [118].

S-layer-stabilized solid-supported lipid membranes have also been fabricated as follows (Figure 15) [21]. After compressing a phospholipid monolayer on a Langmuir trough into the

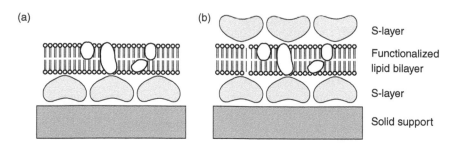

Figure 16 Schematic drawing illustrating S-layer stabilized solid supported lipid membranes. (a) The S-layer functions as a water carrying layer into which membrane integrated molecules may protrude. (b) An S-layer assembled on top of the bilayer that may serve as a protective and stabilizing cover.

liquid-crystalline phase, a thiolipid-coated solid support (glass cover slip or, alternatively, a silicon wafer) is placed horizontally onto the monolayer and left in this position until the S-layer protein, which is subsequently injected into the subphase, has assembled into a closed crystalline monolayer. The whole assembly is then removed from the liquid–air interface (Langmuir–Schaefer technique), and rinsed with deionized water. For demonstrating that recrystallization process is complete the support can subsequently be transferred to a scanning force microscope (Figure 15). The crystallization process of the S-layer proteins follows the same kinetics as described for the reassembly on solid supports.

Silicon substrates have been covered by a closed S-layer lattice and subsequently bilayers have been deposited by the Langmuir–Blodgett technique [142–144]. The lateral diffusion of fluorescently labeled lipid molecules in both layers has been investigated by fluorescence recovery after photobleaching studies [145]. In comparison with hybrid lipid bilayers (lipid monolayer on alkylsilanes) and lipid bilayers on dextran, the mobility of lipids is highest in S-layer-supported bilayers. Furthermore, an S-layer cover causes an enhanced mobility of the labeled molecules in the adjacent lipid layer. In addition, the S-layer cover can also prevent the formation of cracks and other inhomogenities in the bilayer [21,145].

Lipid membranes composed of MPL or DPhPC have also been generated on gold electrodes that have been previously covered by S-layer lattices. These membranes are almost as tight as plain folded BLMs as demonstrated by impedance spectroscopy [146]. Furthermore, membrane-active peptides, such as alamethicin, gramicidin, or valinomycin, can be reconstituted into the lipid membrane resting on the S-layer-covered gold electrode and not only the specific function, but also the interaction with blocking agents, such as amiloride and derivates of amiloride, can be demonstrated [146].

E. Self-Assembly on Liposomes

Artificial lipid vesicles, termed liposomes, are colloid particles in which phospholipid bilayers or tetraether monolayers encapsulate an aqueous medium. Because of their physicochemical properties, liposomes are widely used as model systems for biological membranes and as delivery systems for biologically active molecules. In general, water-soluble molecules are encapsulated within the aqueous compartment whereas water insoluble substances may be intercalated into the liposomal membrane [147].

Isolated monomeric and oligomeric S-layer protein from *B. coagulans* E38-66 [148–150], *B. sphaericus* (Figure 17), and the SbsB from *G. stearothermophilus* PV72/p2 [151,152] can be recrystallized into the respective lattice type on positively charged liposomes composed of 1,2-dipalmitoyl-*sn*-glycero-3-phosphatidylcholine (DPPC), cholesterol, and hexadecylamine. In the case of SbsB, zeta-potential measurements have indicated that the S-layer subunits bind with their outer charge neutral surface to the positively charged liposomes [151], leaving the inner surface with

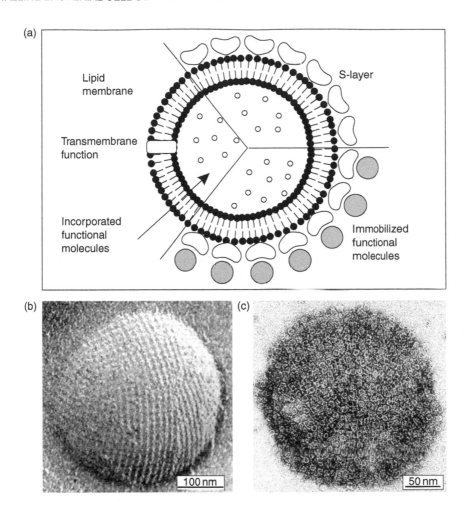

Figure 17 (a) Schematic drawing of a liposome coated with an S-layer lattice that is used as a matrix for immobilizing functional macromolecules. (b) Freeze-etched preparation of a liposome completely covered with a square S-layer lattice (courtesy S. Küpcü). (c) Negative-staining of a liposome coated with an oblique S-layer lattice that was subsequently exploited for the covalent attachment of ferritin.

the positively charged SLH-domain exposed to the ambient environment [95]. This is identical to the orientation of the S-layer lattice obtained by recrystallization of SbsB on poly-L-lysine coated EM grids. On the contrary, SbsB binds with its inner net negatively charged surface to Langmuir films composed of DPPC [100]. Cross-linking of the S-layer lattice on liposomes has revealed that most of the hexadecylamine incorporated into the bilayer can react with the S-layer subunits. This supports the hypothesis of the semifluid membrane model that at least parts of the membrane lipids are fixed to discrete positions of the S-layer subunits [3,103]. Microcalorimetric [149] and sound velocity studies [150] on S-layer coated liposomes have supported the proposed semifluid membrane model in demonstrating that S-layers increase intermolecular order in lipid membranes.

The presence of S-layer lattices significantly enhanced the stability of the liposomes against mechanical stress such as shear forces or ultrasonication and against thermal challenges [151]. The S-layer lattices have been further exploited as a template for chemical modifications and as matrix for the immobilization of macromolecules, such as ferritin (Figure 17) [148]. After biotinylation of histidine or tyrosine residues [153] which do not disturb the structural integrity of the oblique

S-layer lattice formed by SbsB, streptavidin can be immobilized as dense monomolecular layer which is capable of binding biotinylated antibodies in high packing density [152].

VI. PATTERNING OF S-LAYERS RECRYSTALLIZED ON SOLID SUPPORTS

Most applications of S-layers in particularly nanotechnology involve procedures generating S-protein monolayers on solid supports and subsequent patterning of these layers [108]. For example, if biologically functional molecules have to be bound to S-layers recrystallized on electrodes, a procedure must be available that allows in a first step controlled removal of the S-layer in certain areas. Patterning of S-layers by exposure to deep ultraviolet (DUV) radiation has proven to be a powerful technique for imposing structures in recrystallized S-layer lattices on solid supports (Figure 18) [129,131]. In this approach, S-layer protein recrystallized on a silicon wafer is brought in direct contact with a microlithographic mask and exposed to the DUV of a pulsed argon fluoride (ArF) excimer laser (wavelength of emitted light = 193 nm). The S-layer is removed specifically from the silicon surface in the exposed regions but retains its structure and functionality in the unexposed areas. The crucial step in this procedure is the drying of the protein layer. While excess water causing interference fringes in the course of DUV exposure has to be eliminated, enough water has to be retained for the structural integrity of the S-layer protein. Drying the S-layer in a stream of dry nitrogen at room temperature has proven to be a feasible method. The patterning process was performed in several shots of 100–200 mJ/cm^2 each (pulse frequency 1 Hz, pulse duration ~8 nsec). Subsequently, the remaining unexposed S-layer areas can be used either to bind enhancing ligands or to enable electroless metallization in order to form a layer, which allows a final patterning process of the silicon by reactive ion etching. Since S-layer are only 5–10 nm thick and consequently much thinner than conventional resists, a considerable improvement in edge resolution in the fabrication of submicron structures can be expected. As an alternative to the application in the microelectronic sciences, the unexposed S-layer areas may also be used for selectively binding intact cells (e.g., neurons), lipid layers, or biologically active molecules as required for the development of biosensors [154,155].

Finally, it is interesting to note that under exposure to krypton fluoride (KrF) excimer laser radiation supplied in several shots of ~350 mJ/cm^2 the S-layer is not ablated but carbonized, in the exposed areas [109,156]. This result is already used for high resolution patterning of polymeric resists (Figure 18). S-layers that have been formed on top of a spin-coated polymeric resist (on a silicon wafer) have been first patterned by ArF radiation and subsequently served as a mask for a blank exposure of the resist by irradiation with KrF-pulses. This two-step process was possible because S-layers are less sensitive to KrF radiation than polymeric resists. The thinness of the S-layer mask causes very steep side walls in the developed polymeric resist.

Preliminary experiments with electron beam writing or ion beam projection lithography have demonstrated that the S-layer may also be patterned by these techniques in the sub 100 nm range (unpublished results). Most recently, S-layers have also been exploited as matrices for electrochemical deposition and electron beam deposition for generating calibration standards in the 10 nm range for scanning probe microscopy [157].

In addition to the described lithographic procedures, an attractive approach for gaining spatial control over the reassembly of S-layer proteins on surfaces is micromolding in capillaries (MIMIC) [158]. In this technique a mold made of poly(dimethylsiloxane) (PDMS) is brought into conformal contact with a silicon wafer. The mold consists of a pattern of slots forming channels with the support. S-layer protein solution dropped at the openings of the channels is sucked in by capillary forces. Since the PDMS mold forms a tight sealing with the support, the recrystallization of the protein is limited by the overall layout of the mold. Lines of reassembled S-layer proteins with widths down to the submicrometer range can be generated. Fluorescence microscopy has been used to image the filling of the channels with fluorescently labeled S-layer protein in real time. Scanning force microscopy has been used to demonstrate the successful recrystallization of S-layer proteins and the

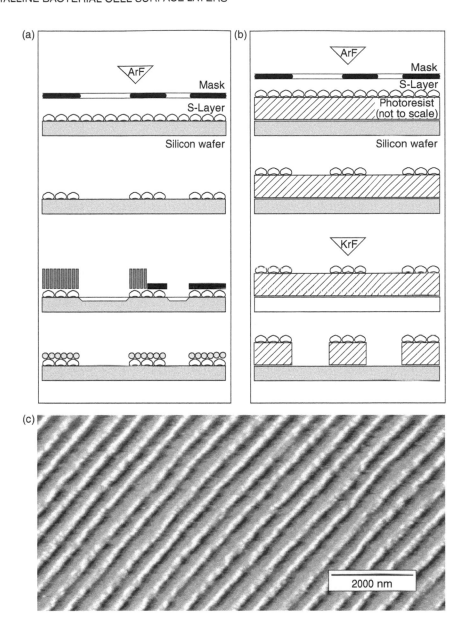

Figure 18 Schematic drawing of patterning S-layers by exposure to DUV radiation. (a) A pattern is transferred onto the S-layer by exposure to ArF excimer laser radiation (wavelength = 193 nm) through a micro-lithographic mask. The S-layer is specifically removed from the silicon surface in the exposed regions but retains its crystalline and functional integrity in the unexposed areas. Unexposed S-layer areas can be used either to bind enhancing ligands or to enable electroless metallization. In both cases a layer is formed which allows a patterning process by reactive ion etching. Alternatively, unexposed S-layers may also be used for selectively binding biologically active molecules that would be necessary for the fabrication of miniaturized biosensors or biocompatible surfaces. (b) In the two-layer resist approach, the S-layer that was formed on top of a spin-coated polymeric resist (on a silicon wafer) is first patterned by ArF radiation and subsequently serves as a mask for a blank exposure of the resist by irradiation with KrF radiation. Due to the thinness of the S-layer very steep sidewalls are obtained in the developed resist. (c) Scanning force microscopical image of a patterned S-layer on a silicon wafer. The ultimate resolution is determined by the wavelength of the excimer laser radiation. (Modified after D. Pum, G. Stangl, C. Sponer, W. Fallmann, and U.B. Sleytr. *Colloids Surf. B* 8:157–162, 1996. With permission.)

good "straightness" of the edges of the pattern. Determination of the step height between protein and support confirms that monolayers have been formed.

VII. S-LAYERS AS MATRICES FOR THE IMMOBILIZATION OF FUNCTIONAL MOLECULES AND FUNCTIONAL S-LAYER-FUSION PROTEINS

The high density and defined position and orientation of surface-located functional groups on S-layer lattices has been exploited for the immobilization of different (macro)molecules [6,8]. For introducing covalent bonds between the S-layer subunits and for enhancing the stability properties, S-layer lattices are cross-linked with homobifunctional amino group specific cross-linkers of different bridge length. For immobilization of foreign (macro)molecules, vicinal hydroxyl groups from the carbohydrate chains of S-layer glycoproteins are either cleaved with periodate or they are activated with cyanogen bromide [159]. Carboxylic acid groups from the S-layer protein are activated with carbodiimide which regarding the modification rate or the binding density turned out to be the best method [159]. Most enzymes such as invertase, glucose oxidase, β-glucosidase, or naringinase form a monolayer of densely arranged molecules on the outer surface of the S-layer lattice (Figure 19) [159,160]. For example, the binding capacity of the square S-layer lattice from *B. sphaericus* for IgG has been determined with 375 ng/cm^2. As derived from the saturation capacity of a planar surface and the molecular dimension of IgG, this corresponds to a monolayer of randomly oriented antibody molecules [161,162]. S-layer microparticles with immobilized protein A are used as escort particles for affinity cross flow filtration for isolating IgG from serum or hybridoma culture supernatants [163,164].

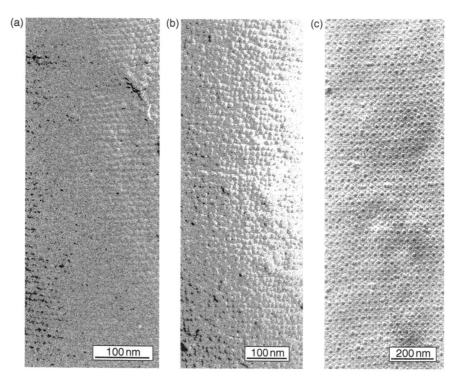

Figure 19 Freeze-etched preparations showing (a, b) ferritin molecules covalently bound or (c) immobilized by electrostatic interactions to S-layer lattices with different lattice symmetry. The immobilized ferritin with a molecular size of 12 nm reflects the periodicity of the underlying S-layer lattice type over a wide range (a, c — hexagonal; b — square). The center-to-center spacings of the morphological units in the S-layer lattices are 14 nm (a), (b) and 30 nm (c).

By depositing S-layers on microporous supports and cross-linking the S-layer lattice with glutaraldehyde, mechanically stable composite structures have become available which after immobilization of various monoclonal antibodies are used as reaction zone for S-layer-based dipsticks [161,165]. Different types of S-layer-based dipsticks have been developed which work according to the principle of solid-phase immunoassays [165]. Such S-layer-based dipsticks are exploited to determine the concentration of tissue type plasminogen activator (t-PA) in whole blood or plasma, IgE as a marker for type I allergies in serum, interleukins in serum or blood to differentiate between septic and traumatic shock, or prion proteins [166]. In general, the advantages of S-layers as immobilization matrix and especially as reaction zone for solid-phase immunoassays in comparison to amorphous polymers can be summarized as follows: (i) S-layers have well-defined surface properties and immobilization of functional macromolecules or the catching antibody can only occur on the outermost surface of the crystal lattice preventing diffusion controlled reactions; (ii) since the functional macromolecules are covalently linked to the S-layer protein, no leakage occurs; (iii) S-layers generally show a low unspecific adsorption and S-layer-based dipsticks can therefore be incubated in whole blood, plasma, or serum without the necessity of blocking steps.

As an alternative approach to functionalize S-layer lattices, S-layer fusion proteins have been produced by means of recombinant technologies [60,95,132,167,168]. In order to obtain S-layer fusion proteins that can bind to SCWPs, that are capable of self-assembling into the respective lattice type and leave the functional sequence exposed on the outer S-layer surface, C terminal fusion proteins have been constructed. Thereby, the N terminal part is exploited for binding to the SCWP which is isolated from the cell wall fragments and linked to solid supports [132,168]. Amino acid positions located on the outer S-layer surface have been identified by fusing Strept-Tag I to C terminally truncated forms of the S-layer protein SbpA and by investigating the binding capacity for streptavidin in the monomeric, self-assembled or recrystallized S-layer protein [60]. By this approach, amino acid position 1068 has been identified to be located on the outer S-layer surface and has been exploited as fusion site. Thereby, the fused functional sequences that are either the major birch pollen allergen [60], heavy chain camel antibodies recognizing lysozyme or prostate-specific antigen [132], or the Fc-binding ZZ-domains of protein A [168] remain exposed to the external environment and therefore accessible for further binding partners. Depending on the properties of the fused functional sequence, the S-layer fusion proteins have been exploited for the development of an antiallergic vaccine [167,169], for generation of sensing layers for label-free detection [132], or for preparing microbeads for isolation of IgG from human plasma from patients suffering from autoimmune disease [168]. For vaccine development, self-assembly products formed by the S-layer fusion protein carrying the sequence of the major birch pollen allergen have been used [169], whereas for all other applications the S-layer fusion protein has been recrystallized on gold chips [132] or microbeads [168] precoated with SCWP.

The S-layer protein SbsB has been chosen as a fusion partner for streptavidin [95]. In total, six different rSbsB–streptavidin fusion proteins have been constructed and produced by heterologous expression in *Escherichia coli*. After isolation from the host cells, the fusion protein is mixed with free streptavidin in the molar ratio of 1:3. By applying a special refolding procedure, functional heterotetramers can be obtained. In the case of the N terminal fusion proteins, the heterotetramers are capable of recrystallizing into the oblique lattice structure on liposomes or on silicon, whereas the C terminal fusion protein recrystallizes only on supports functionalized with SCWP. Monolayers consisting of the rSbsB–streptavidin fusion proteins are capable of binding biotinylated macromolecules in regular manner [95].

VIII. S-LAYERS AS TEMPLATES IN THE FORMATION OF REGULARLY ARRANGED NANOPARTICLES

Currently there is great interest in fabricating nanostructures for the development of a new generation of electronic and optic devices. In particular, the formation of metal clusters for nanoelectronic digital

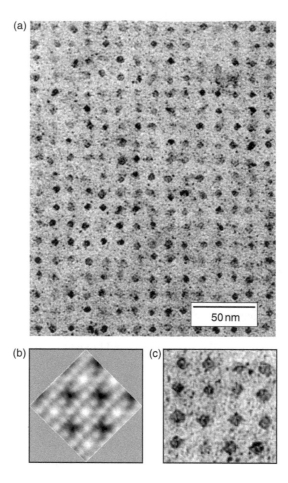

Figure 20   (a) Electron micrograph of a gold superlattice with square lattice symmetry and a lattice constant of 12.8 nm consisting of monodisperse gold nanoparticles with mean diameters of 4 to 5 nm. The point pattern resembles the lattice parameters of the underlying S-layer. (b) Scanning force microscopic image of the native S-layer. (c) Magnified subregion of (a).

circuits require a well-defined size and arrangement of the particles. As an alternative to approaches in which colloidal crystallization has been used to make close-packed nanoparticle arrays [170] the use of S-layers as organic templates allows the synthesis of a wide range of inorganic nanocrystal superlattices [171–178]. Recently, it has been demonstrated that S-layer proteins recrystallize on solid-supports or S-layer self-assembly products which were deposited on such substrates may be used to induce the formation of CdS particles [171] or gold nanoparticles (Figure 20) [172]. CdS inorganic superlattices with either oblique or square lattice symmetries of approximately 10 nm repeat distance have been fabricated by exposing self-assembled S-layer lattices to cadmium ion solutions followed by slow reaction with hydrogen sulfide. Precipitation of the inorganic phase has been confined to the pores of the S-layers with the result that CdS superlattices with prescribed symmetries have been prepared. In a similar procedure a square superlattice of uniform 4 to 5 nm sized gold particles with 13.1 nm repeat distance has been fabricated by exposing a square S-layer lattice in which thiol groups have been introduced before, to a tetrachloroauric (III) acid solution. Reduction of the gold (Au[III]) is either performed by exposing the metallized S-layer to an electron beam in a transmission electron microscope or by slow reaction with H_2S. This technique is already used in the formation of CdS nanoparticles (see above) and further on in the precipitation of Pd- (salt: $PdCl_2$), Ni- ($NiSO_4$), Pt- ($KPtCl_6$), Pb- ($Pb[NO_3]_2$), and Fe- ($KFe[CN]_6$) nanoparticle arrays. Transmission electron microscopical studies show that the gold nanoparticles are formed in the pore region during electron irradiation of an initially grainy gold coating covering the whole S-layer lattice. The shape of the gold particles resembles the morphology of the pore region of the square S-layer lattice. By electron diffraction and energy dispersive x-ray analysis the crystallites have been identified as gold (Au[0]). Electron diffraction patterns reveal that the gold nanoparticles are crystalline but in the long-range order they are not crystallographically aligned.

An alternative and much more flexible way to form highly ordered nanoparticles arrays on S-layers is based on the binding of preformed nanoparticles [179,180]. In a similar approach to the work about the binding of molecules and antibodies onto S-layers in dense packing 4 to 5 nm sized gold and CdS nanoparticles have been bound on the S-layer protein SpbA of *B. sphaericus* CCM2177. In this work electrostatic bonds are used for immobilizing the nanoparticles onto the S-layer template. The formed superlattice of bound nanoparticles reassembles the lattice parameters of the S-layer lattice (square lattice symmetry and a lattice spacing of 13.1 nm). While the gold nanoparticles are citrate stabilized and thus showed a negative surface charge, the CdS particles are amino functionalized yielding a positive surface charge. Since the particular S-layer used exhibits an excess of free amino groups on the inner side the gold nanoparticles are bound on this face. In contrast, the amino functionalized CdS are bound to the free carboxyl groups on the outer face of the S-layer.

A major breakthrough in binding nanoparticles onto S-layers into ordered arrays has been achieved by using the functionality of rSbsB–streptavidin fusion protein lattices as binding matrices [95]. Streptavidin and biotin show the highest affinity interaction between a protein and a ligand known in nature ($K_D = 10^{-14}$ M). In a recent study, C terminal rSbsB–streptavidin fusion proteins have been recrystallized on SCWP containing cell walls and particularly on SCWP coated solid supports. Investigations by transmission electron microscopy demonstrate that biotinylated ferritin binds in dense packing according to the oblique (p1) lattice symmetry of the rSbsB–streptavidin fusion protein [95].

It should be stressed that with S-layers as molecular templates the formation of superlattices with a wide range of interparticle spacings as well as with oblique, square, or hexagonal lattice symmetry becomes possible. This is particularly important for the development of nanometric electronic or optical devices since isolated S-layer subunits have shown the inherent capability to recrystallize on a great variety of solid supports including structured semiconductors [129,131].

IX. CONCLUSIONS

Two-dimensional crystalline arrays of protein or glycoprotein subunits are now recognized as one of the most common cell surface structures in archaea and bacteria. The intrinsic assembly and recrystallization properties of S-layer subunits enable the maintenance of a closed lattice during cell growth and division. From a structural and morphogenetic point of view S-layers represent the simplest biological (protein) membranes developed during evolution. Therefore, it has been suggested that S-layer-like dynamic membranes could have fulfilled barrier and supporting functions as required by self-reproducing systems (progenotes) during the early periods of biological evolution [75,76].

The wealth of information obtained on the general principles of crystalline bacterial cell surface layers, particularly on their structure, assembly, surface, and molecular sieving properties have revealed a broad application potential. Above all, the repetitive physicochemical properties down to the subnanometer-scale make S-layer lattices unique self-assembly structures for functionalization of surfaces and interfaces down to the ultimate resolution limit. S-layers that have been recrystallized on solid substrates can be used as immobilization matrices for a great variety of functional molecules or as templates for the fabrication of ordered and precisely located nanometer-scale particles as required for the production of biosensors, diagnostics, molecular electronics, and nonlinear optics [2,3,6].

Biomimetic approaches copying the supramolecular cell envelope structure in those archaea which have S-layers as the only, and occasionally quite rigid, wall component should lead to new technologies exploiting functional lipid membranes at meso- and macroscopic scale. In particular the technology of S-layer-supported lipid membranes in the long terms may be exploited for generating defined junctions between living cells (e.g., neurons) and integrated circuits for signal processing. Although numerous applications for S-layers have already been demonstrated many further areas may yet emerge. In the near future genetic modifications of S-layer proteins and targeted chemical modifications will significantly influence the development of basic and applied S-layer research [2].

REFERENCES

1. A.S. Blawas and W.M. Reichert. *Biomaterials* 19:595–609, 1998.
2. U.B. Sleytr, P. Messner, D. Pum, and M. Sára. *Angew. Chem. Int. Ed.* 38:1034–1054, 1999.
3. D. Pum and U.B. Sleytr. *Trends Biotechnol.* 17:8–12, 1999.
4. M. Sára and U.B. Sleytr. *Prog. Biophys. Mol. Biol.* 65:83–111, 1996.
5. U.B. Sleytr. *FEMS Microbiol. Rev.* 20:5–12, 1997.
6. U.B. Sleytr and M. Sára. *Trends Biotechnol.* 15:20–26, 1997.
7. U.B. Sleytr and P. Messner. In: H. Plattner, Ed., *Electron Microscopy of Subcellular Dynamics.* Boca Raton, FL: CRC Press, 1989, pp. 13–31.
8. U.B. Sleytr, P. Messner, D. Pum, and M. Sára. *Crystalline Bacterial Cell Surface Proteins.* Austin, TX: R.G. Landes/Academic Press, 1996.
9. U.B. Sleytr. *Int. Rev. Cytol.* 53:1–26, 1978.
10. U.B. Sleytr, M. Sára, D. Pum, and B. Schuster. In: M. Rosoff, Ed., *Nano-Surface Chemistry.* New York: Marcel Dekker, Inc, 2001, pp. 333–389.
11. U.B. Sleytr, P. Messner, D. Pum, and M. Sára. *Crystalline Bacterial Cell Surface Layers.* Berlin: Springer-Verlag, 1988.
12. P. Messner and U.B. Sleytr. *Adv. Microb. Physiol.* 33:213–275, 1992.
13. U.B. Sleytr and A.M. Glauert. In: J.R. Harris, Ed., *Electron Microscopy of Proteins*, Vol. 3. London: Academic Press, 1982, pp. 41–76.
14. S. Hovmöller, A. Sjögren, and D.N. Wang. *Prog. Biophys. Mol. Biol.* 51:131–163, 1988.
15. U.B. Sleytr, P. Messner, and D. Pum. In: F. Mayer, Ed., *Methods in Microbiology*, Vol. 20. London: Academic Press, 1988, pp. 29–60.
16. W. Baumeister and G. Lembke. *J. Bioenerg. Biomembr.* 24:567–575, 1992.
17. T.J. Beveridge. *Curr. Opin. Struct. Biol.* 4:204–212, 1994.
18. W. Baumeister, I. Wildhaber, and B.M. Phipps. *Can. J. Microbiol.* 35:215–227, 1989.
19. S. Hovmöller. In: T.J. Beveridge and S.F. Koval, Eds., *Advances in Bacterial Paracrystalline Surface Layers.* New York: Plenum Press, 1993, pp. 13–21.
20. D.J. Müller, W. Baumeister, and A. Engel. *J. Bacteriol.* 178:3025–3030, 1996.
21. B. Wetzer, D. Pum, and U.B. Sleytr. *J. Struct. Biol.* 119:123–128, 1997.
22. S. Scheuring, H. Stahlberg, M. Chami, C. Houssin, J.L. Rigaud, and A. Engel. *Mol. Microbiol.* 44:675–684, 2002.
23. M. Sára and U.B. Sleytr. *J. Bacteriol.* 169:4092–4098, 1987.
24. D.P. Bayley and S.F. Koval. *Can. J. Microbiol.* 40:237–242, 1993.
25. S.F. Koval. *Can. J. Microbiol.* 34:407–414, 1988.
26. T.J. Beveridge, M. Stewart, R.J. Doyle, and G.D. Sprott. *J. Bacteriol.* 162:728–736, 1985.
27. H. König. *Can. J. Microbiol.* 34:395–406, 1988.
28. S. Schultze-Lam and T.J. Beveridge. *Appl. Environ. Microbiol.* 60:447–453, 1994.
29. B. Kuen and W. Lubitz. In: U.B. Sleytr, P. Messner, D. Pum, and M. Sára, Eds., *Crystalline Bacterial Cell Surface Proteins.* Austin: Landes Company, 1996, pp. 77–111.
30. H. Boot and P. Pouwels. *Mol. Microbiol.* 21:1117–1123, 1996.
31. S.R. Thomas and T.J. Trust. *J. Mol. Biol.* 245:568–581, 1995.
32. P. Messner. *Glycoconj. J.* 14:3–11, 1997.
33. P. Messner and C. Schäffer. In: R.J. Doyle, Ed., *Glycomicrobiology.* New York: Kluwer Academic/Plenum Press, 2000, pp. 93–125.
34. E. Calabi, S. Ward, B. Wren, T. Paxton, M. Panico, H. Morris, A. Dell, G. Dougan, and N. Fairweather. *Mol. Microbiol.* 40:1187–1199, 2001.

35. M.F. Mescher and J.L. Strominger. *J. Biol. Chem.* 252:2005–2014, 1976.
36. J. Lechner and M. Sumper. *J. Biol. Chem.* 262:9724–9729, 1987.
37. M. Sumper. In: T.J. Beveridge and S.F. Koval, Eds., *Advances in Paracrystalline Bacterial Surface Layers.* New York and London: Plenum Press, 1993, pp. 109–118.
38. C. Schäffer, T. Wugeditsch, C. Neuninger, and P. Messner. *Microb. Drug. Resist.* 2:17–23, 1996.
39. C. Schäffer, T. Wugeditsch, H.P. Kählig, A. Scheberl, S. Zayni, and P. Messner. *J. Biol. Chem.* 277(8):6230–6239, 2002.
40. J. Dworkin and M.J. Blaser. *Mol. Microbiol.* 26:433–440, 1997.
41. B. Kuen, U.B. Sleytr, and W. Lubitz. *Gene* 145:115–120, 1994.
42. B. Kuen, M. Sára, and W. Lubitz. *Mol. Microbiol.* 19:495–503, 1995.
43. M. Jarosch, E. Egelseer, D. Mattanovich, U.B. Sleytr, and M. Sára. *Microbiology* 146:273–281, 2000.
44. P. Doig, L. Emödy, and T.J. Trust. *J. Biol. Chem.* 267:43–51, 1992.
45. S.G. Walker, D.N. Karunaratne, N. Ravenscroft, and J. Smit. *J. Bacteriol.* 176:6312–6323, 1994.
46. E. Egelseer, K. Leitner, M. Jarosch, C. Hotzy, S. Zayni, U.B. Sleytr, and M. Sára. *J. Bacteriol.* 180:1488–1495, 1998.
47. E.M. Egelseer, T. Danhorn, M. Pleschberger, C. Hotzy, U.B. Sleytr, and M. Sára. *Arch. Microbiol.* 177:70–80, 2001.
48. M. Jarosch, E.M. Egelseer, C. Huber, D. Moll, D. Mattanovich, U.B. Sleytr, and M. Sára. *Microbiology* 147:1353–1363, 2001.
49. M. Sára and U.B. Sleytr. *J. Bacteriol.* 182:859–868, 2000.
50. C. Schäffer, H.P. Kählig, R. Christian, G. Schulz, S. Zayni, and P. Messner. *Microbiology* 145:1575–1583, 1999.
51. E. Egelseer, I. Schocher, U.B. Sleytr, and M. Sára. *J. Bacteriol.* 178:5602–5609, 1996.
52. C. Schuster, T. Pink, H.F. Mayer, W.A. Hampel, and M. Sára. *J. Biotechnol.* 54:15–28, 1997.
53. B. Kuen, A. Koch, E. Asenbauer, M. Sára, and W. Lubitz. *J. Bacteriol.* 179:1664–1970, 1997.
54. M. Chami, N. Bayan, J.L. Peyret, T. Guli-Krzywicki, G. Leblon, and E. Shechter. *Mol. Microbiol.* 23:483–492, 1997.
55. A. Lupas, H. Engelhardt, J. Peters, U. Santarius, S. Volker, and W. Baumeister. *J. Bacteriol.* 176:1224–1233, 1994.
56. H. Engelhardt and J. Peters. *J. Struct. Biol.* 124:276–302, 1998.
57. E. Brechtel and H. Bahl. *J. Bacteriol.* 181:5017–5023, 1999.
58. S. Chauvaux, M. Matuschek, and P. Beguin. *J. Bacteriol.* 181:2455–2458, 1999.
59. N. Ilk, P. Kosma, M. Puchberger, E.M. Egelseer, H.F. Mayer, U.B. Sleytr, and M. Sára. *J. Bacteriol.* 181:7643–7646, 1999.
60. N. Ilk, C. Völlenkle, E.M. Egelseer, A. Breitwieser, U.B. Sleytr, and M. Sára. *Appl. Environ. Microbiol.* 68:3251–3260, 2002.
61. M. Lemaire, I. Miras, P. Gounon, and P. Beguin. *Microbiology* 144:211–217, 1998.
62. C. Mader, C. Huber, D. Moll, U.B. Sleytr, and M. Sára. *J. Bacteriol.* 186:1758–1768, 2004.
63. S. Mesnage, E. Tosi-Couture, and A. Fouet. *Mol. Microbiol.* 31:927–936, 1999.
64. S. Mesnage, T. Fontaine, T. Mignot, M. Delpierre, M. Mock, and A. Fouet. *EMBO J.* 19:4473–4484, 2000.
65. M. Sára. *Trends Microbiol.* 9:47–49, 2001.
66. S. Chu, S. Chavaignac, J. Feutrier, B. Phipps, M. Kostrzynska, and W.W. Kay. *J. Biol. Chem.* 266:15258–15265, 1991.
67. J.A. Fisher, J. Smit, and N. Agabian. *J. Bacteriol.* 170:4706–4713, 1988.
68. T. Adachi, H. Yamagata, N. Tsukagoshi, and S. Udaka. *J. Bacteriol.* 171:1010–1016, 1989.
69. S. Ebisu, A. Tsuboi, H. Takagi, Y. Naruse, H. Yamagata, N. Tsukagoshi, and S. Udaka. *J. Bacteriol.* 172:614–620, 1990.
70. M. Kahala, K. Savijoki, and A. Palva. *J. Bacteriol.* 179:284–286, 1992.
71. M. Sára, B. Kuen, H.F. Mayer, F. Mandl, K.C. Schuster, and U.B. Sleytr. *J. Bacteriol.* 178:2108–2117, 1996.
72. H. Scholz, E. Riedmann, A. Witte, W. Lubitz, and B. Kuen. *J. Bacteriol.* 183:1672–1679, 2001.
73. U.B. Sleytr and A.M. Glauert. *J. Ultrastruct. Res.* 50:103–116, 1975.
74. U.B. Sleytr. *Nature* 257:400–402, 1975.
75. U.B. Sleytr. In: O. Kiermayer, Ed., *Cell Biology Monographs,* Vol. 8. Wien, New York: Springer-Verlag, 1981, pp. 1–26.

76. U.B. Sleytr and R. Plohberger. In: W. Baumeister and W. Vogell, Eds., *Microscopy at Molecular Dimensions*. Berlin: Springer-Verlag, 1980, pp. 36–47.

77. U.B. Sleytr and P. Messner. *Annu. Rev. Microbiol.* 37:311–339, 1983.

78. U.B. Sleytr and P. Messner. *J. Bacteriol.* 170:2891–2897, 1988.

79. A. Breitwieser, U.B. Sleytr, and K. Gruber. *J. Bacteriol.* 174:8008–8015, 1992.

80. W.F. Harris and L.E. Scriven. *Nature* 228:827–828, 1970.

81. F.R.N. Nabarro and W.F. Harris. *Nature* 232:423–425, 1971.

82. W.F. Harris. *Sci. Am.* 237:130–145, 1977.

83. P. Messner, D. Pum, M. Sára, K.O. Stetter, and U.B. Sleytr. *J. Bacteriol.* 166:1046–1054, 1986.

84. D.L.D. Caspar and A. Klug. *Cold Spring Harbor Symp. Quant. Biol.* 27:1–24, 1962.

85. D. Pum, P. Messner, and U.B. Sleytr. *J. Bacteriol.* 173:6865–6873, 1991.

86. U.B. Sleytr, P. Messner, and D. Pum. *Meth. Microbiol.* 20:29–60, 1988.

87. R. Jaenicke, R. Welsch, M. Sára, and U.B. Sleytr. *Biol. Chem. Hoppe-Seyler* 366:663–670, 1985.

88. M. Sára and U.B. Sleytr. *J. Bacteriol.* 169:2804–2809, 1987.

89. M. Sára, I. Kalsner, and U.B. Sleytr. *Arch. Microbiol.* 149:527–533, 1988.

90. M. Sára, D. Pum, and U.B. Sleytr. *J. Bacteriol.* 174:3487–3493, 1992.

91. S. Weigert and M. Sára. *J. Memb. Sci.* 106:147–159, 1995.

92. S. Weigert and M. Sára. *J. Memb. Sci.* 121:185–196, 1996.

93. P. Messner, D. Pum, and U.B. Sleytr. *J. Ultrastruct. Mol. Struct. Res.* 97:73–88, 1986.

94. U.B. Sleytr, D. Pum, and M. Sára. *Adv. Biophys.* 34:71–79, 1997.

95. D. Moll, C. Huber, B. Schlegel, D. Pum, U.B. Sleytr, and M. Sára. *Proc. Natl. Acad. Sci. U.S.A.* 99:14646–14651, 2002.

96. S. Howorka, M. Sára, Y. Wang, B. Kuen, U.B. Sleytr, W. Lubitz, and H. Bayley. *J. Biol. Chem.* 275(48):37876–37886, 2000.

97. D. Rünzler, C. Huber, D. Moll, G. Köhler, and M. Sára. *J. Biol. Chem.* 279:5207–5215, 2004.

98. T. Pavkov, M. Oberer, E.M. Egelseer, M. Sára, U.B. Sleytr, and W. Keller. *Acta Cryst.* D(59):1466–1468, 2003.

99. M. Sára, C. Dekitsch, H.F. Mayer, E. Egelseer, and U.B. Sleytr. *J. Bacteriol.* 180:4146–4153, 1998.

100. W. Ries, C. Hotzy, I. Schocher, and U.B. Sleytr. *J. Bacteriol.* 179:3892–3898, 1997.

101. S. Howorka, M. Sára, W. Lubitz, and B. Kuen. *FEMS Microbiol. Lett.* 172:187–196, 1999.

102. D. Pum, M. Weinhandl, C. Hödl, and U.B. Sleytr. *J. Bacteriol.* 175:2762–2766, 1993.

103. D. Pum and U.B. Sleytr. *Thin Solid Films* 244:882–886, 1994.

104. B. Wetzer, A. Pfandler, E. Györvary, D. Pum, M. Lösche, and U.B. Sleytr. *Langmuir* 14:6899–6906, 1998.

105. M. Weygand, B. Wetzer, D. Pum, U.B. Sleytr, N. Cuvillier, K. Kjaer, P.B. Howes, and M. Lösche. *Biophys. J.* 76:458–468, 1999.

106. M. Weygand, M. Schalke, P.B. Howes, K. Kjaer, J. Friedmann, B. Wetzer, D. Pum, U.B. Sleytr, and M. Lösche. *J. Mater. Chem.* 10:141–148, 2000.

107. M. Weygand, K. Kjaer, P.B. Howes, B. Wetzer, D. Pum, U.B. Sleytr, and M. Lösche. *J. Chem. Phys. B* 106:5793–5799, 2002.

108. D. Pum and U.B. Sleytr. In: U.B. Sleytr, P. Messner, D. Pum, and M. Sára, Eds., *Crystalline Bacterial Cell Surface Proteins*. Austin, TX: R.G. Landes/Academic Press, 1996, pp. 175–209.

109. D. Pum and U.B. Sleytr. *Colloids Surf. A* 102:99–104, 1995.

110. A. Diederich, C. Hödl, D. Pum, U.B. Sleytr, and M. Lösche. *Colloids Surf. B: Biointerfaces* 6:335–346, 1996.

111. B. Schuster, P.C. Gufler, D. Pum, and U.B. Sleytr. *Langmuir* 19:3393–3397, 2003.

112. M. Montal and P. Mueller. *Proc. Natl. Acad. Sci. U.S.A.* 69:3561–3566, 1972.

113. R. Fettiplace, LG.M. Gordon, S.B. Hladky, J. Requena, H. Zingsheim, and D.A. Haydon. In: E.D. Korn, Ed., *Methods of Membrane Biology*, Vol. 4. New York: Plenum Press, 1975, pp. 1–75.

114. B. Schuster, A. Diederich, G. Bähr, U.B. Sleytr, and M. Winterhalter. *Eur. Biophys. J.* 28:583–590, 1999.

115. M. Lindemann, M. Steinmetz, and M. Winterhalter. *Prog. Colloid. Polym. Sci.* 105:209–213, 1997.

116. A. Diederich, G. Bähr, and M. Winterhalter. *Langmuir* 14:4597–4605, 1998.

117. A. Diederich, M. Strobl, W. Maier, and M. Winterhalter. *J. Phys. Chem.* 103:1402–1407, 1999.

118. R. Hirn, B. Schuster, U.B. Sleytr, and T.M. Bayerl. *Biophys. J.* 77:2066–2074, 1999.

119. E. Sackmann. In: R. Lipowsky and E. Sackmann, Eds., *Structure and Dynamics of Membranes*. Amsterdam: Elsevier Science, 1995, pp. 213–304.

120. B. Schuster and U.B. Sleytr. *Biochim. Biophys. Acta* 1563:29–34, 2002.

121. M.S. Awayda, I.I. Ismailov, B.K. Berdiev, and D.J. Benos. *Am. J. Physiol.* 268:C1450–C1459, 1995.

122. A. Ghazi, C. Berrier, B. Ajouz, and M. Besnard. *Biochimie* 80:357–362, 1998.

123. S. Sukharev, M. Betanzos, and C.S. Chiang. *Nature* 409:720–724, 2001.

124. A. Kloda and B. Martinac. *Eur. Biophys. J.* 31:14–25, 2002.

125. I.I. Ismailov, V.G. Shlyonsky, and D.J. Benos. *Proc. Natl. Acad. Sci.* 94:7651–7654, 1997.

126. B. Schuster, D. Pum, and U.B. Sleytr. *Biochim. Biophys. Acta* 1369:51–60, 1998.

127. B. Schuster, D. Pum, O. Braha, H. Bayley, and U.B. Sleytr. *Biochim. Biophys. Acta* 1370:280–288, 1998.

128. B. Schuster and U.B. Sleytr. *Bioelectrochemistry* 55:5–7, 2002.

129. D. Pum and U.B. Sleytr. *Supramol. Sci.* 2:193–197, 1995.

130. E.S. Györvary, O. Stein, D. Pum, and U.B. Sleytr. *J. Microsc.* 212:300–306, 2003.

131. D. Pum, G. Stangl, C. Sponer, W. Fallmann, and U.B. Sleytr. *Colloids Surf. B* 8:157–162, 1996.

132. M. Pleschberger, A. Neubauer, E.M. Egelseer, S. Weigert, B. Lindner, U.B. Sleytr, S. Muyldermans, and M. Sára. *Bioconj. Chem.* 14:440–448, 2003.

133. E. Sackmann. *Science* 271:43–48, 1996.

134. B. Schuster and U.B. Sleytr. *Rev. Mol. Biotechnol.* 74:233–254, 2000.

135. U.B. Sleytr, M. Sára, D. Pum, and B. Schuster. *Prog. Surf. Sci.* 68:231–278, 2001.

136. U.B. Sleytr, D. Pum, and B. Schuster. *IEEE Eng. Med. Biol.* 22:140–150, 2003.

137. B. Schuster, E.S. Györvary, D. Pum, and U.B. Sleytr. In: T. Vo-Dinh, Ed., *Protein Nanotechnology: Protocols, Instrumentation and Applications.* Totowa, NJ: Humana Press, 2005, pp. 101–124.

138. B. Schuster, P.C. Gufler, D. Pum, and U.B. Sleytr. *IEEE Trans. Nanobiosci.* 2003, 3:16–21, 2004.

139. M. Sára and U.B. Sleytr. *J. Membr. Sci.* 33:27–49, 1987.

140. B. Schuster, D. Pum, M. Sára, O. Braha, H. Bayley, and U.B. Sleytr. *Langmuir* 17:499–503, 2001.

141. B. Schuster, S. Weigert, D. Pum, M. Sára, and U.B. Sleytr. *Langmuir* 19:2392–2397, 2003.

142. K.J. Blodgett. *J. Am. Chem. Soc.* 57:1007–1022, 1935.

143. I. Langmuir and V.J. Schaefer. *J. Am. Chem. Soc.* 59:1406–1417, 1937.

144. J.A. Zasadzinski, R. Viswanathan, L. Madson, J. Garnacs, and K.D. Schwartz. *Science* 263:1726–1733, 1994.

145. E.S. Györvary, B. Wetzer, U.B. Sleytr, A. Sinner, A. Offenhäuser, and W. Knoll. *Langmuir* 15:1337–1347, 1999.

146. P.C. Gufler, D. Pum, U.B. Sleytr, and B. Schuster. *Biochim. Biophys. Acta* 1661:154–165, 2004.

147. D.D. Lasic. *Trends Biotechnol.* 16:307–321, 1998.

148. S. Küpcü, M. Sára, and U.B. Sleytr. *Biochim. Biophys. Acta* 1235:263–269, 1995.

149. S. Küpcü, K. Lohner, C. Mader, and U.B. Sleytr. *Mol. Membr. Biol.* 15:151–175, 1998.

150. T. Hianik, S. Küpcü, U.B. Sleytr, P. Rybar, R. Krivanek, and U. Kaatze. *Colloid Surf. A* 147:331–337, 1999.

151. C. Mader, S. Küpcü, M. Sára, and U.B. Sleytr. *Biochim. Biophys. Acta* 1418:106–116, 1999.

152. C. Mader, S. Küpcü, M. Sára, and U.B. Sleytr. *Biochim. Biophys. Acta* 1463:142–150, 2000.

153. H.J. Lin and J.F. Kirsch. *Meth. Enzymol.* 62D:287–289, 1979.

154. P. Fromherz, A. Offenhäuser, T. Vetter, and J. Weis. *Science* 252:1290–1293, 1991.

155. P. Fromherz and H. Schaden. *Eur. J. Neurosci.* 6:1500–1504, 1994.

156. D. Pum, G. Stangl, C. Sponer, K. Riedling, P. Hudek, W. Fallmann, and U.B. Sleytr. *Microelectron Eng.* 35:297–300, 1997.

157. A. Neubauer, W. Kautek, S. Dieluweit, D. Pum, M. Sahre, C. Traher, and U.B. Sleytr. *PTB Berichte* F-30:188–190, 1997.

158. E.S. Györvary, A. O'Riordan, A. Quinn, G. Redmond, D. Pum, and U.B. Sleytr. *Nano Lett.* 3:315–319, 2003.

159. S. Küpcü, C. Mader, and M. Sára. *Biotechnol. Appl. Biochem.* 21:275–286, 1995.

160. M. Sára and U.B. Sleytr. *Appl. Microbiol. Biotechnol.* 30:184–189, 1989.

161. A. Breitwieser, S. Küpcü, S. Howorka, S. Weigert, C. Langer, K. Hoffmann-Sommergruber, O. Scheiner, U.B. Sleytr, and M. Sára. *BioTechniques* 21:918–925, 1996.

162. S. Küpcü, M. Sára, and U.B. Sleytr. *J. Immunol. Meth.* 196:73–84, 1996.

163. C. Weiner, M. Sára, and U.B. Sleytr. *Biotechnol. Bioeng.* 43:321–330, 1994.

164. C. Weiner, M. Sára, G. Dasgupta, and U.B. Sleytr. *Biotechnol. Bioeng.* 44:55–65, 1994.

165. A. Breitwieser, C. Mader, I. Schocher, U.B. Sleytr, and M. Sára. *Allergy* 53:786–793, 1998.

166. D. Völkel, K. Zimmermann, A. Breitwieser, S. Pable, M. Glatzel, F. Scheiflinger, H.P. Schwarz, M. Sára, U.B. Sleytr, and F. Dorner. *Transfusion* 43:1677–1682, 2003.

167. A. Breitwieser, E.M. Egelseer, D. Moll, N. Ilk, C. Hotzy, B. Bohle, C. Ebner, U.B. Sleytr, and M. Sára. *Protein Eng.* 15:243–259, 2002.

168. C. Völlenkle, S. Weigert, N. Ilk, E.M. Egelseer, V. Weber, F. Loth, D. Falkenhagen, U.B. Sleytr, and M. Sára. *Appl. Environ. Microbiol.* 70:1514–1521, 2004.

169. B. Bohle, A. Breitwieser, B. Zwölfer, B. Jahn-Schmid, M. Sára, U.B. Sleytr, and C. Ebner. *J. Immunol.* 172:6642–6648, 2004.

170. K. Nagayama. *Nanobiology* 1:25–37, 1992.

171. W. Shenton, D. Pum, U.B. Sleytr, and S. Mann. *Nature* 389:585–587, 1997.

172. S. Dieluweit, D. Pum, and U.B. Sleytr. *Supramol. Sci.* 5:15–19, 1998.

173. K. Douglas, N.A. Clark, and K.J. Rothschild. *Appl. Phys. Lett.* 48:676–678, 1986.

174. K. Douglas, G. Devaud, and N.A. Clark. *Science* 257:642–644, 1992.

175. T.A. Winningham, H.P. Gillis, D.A. Choutov, K.P. Martin, J.T. Moore, and K. Douglas. *Surf. Sci.* 406:221–228, 1998.

176. M. Panhorst, H. Brückl, B. Kiefer, G. Reiss, U. Santarius, and R. Guckenberger. *J. Vac. Sci. Technol. B* 19:722–724, 2001.

177. M. Mertig, R. Kirsch, W. Pompe, and H. Engelhardt. *Eur. Phys. J.* 9:45–48, 1999.

178. W. Pompe, M. Mertig, R. Kirsch, R. Wahl, L.C. Ciachi, J. Richter, R. Seidel, and H. Vinzelberg. *Z. Metallkd.* 90:1085–1091, 1999.

179. S.R. Hall, W. Shenton, H. Engelhardt, and S. Mann. *Chem. Phys. Chem.* 3:184–186, 2001.

180. E.S. Györvary, A. Schroedter, D.V. Talapin, H. Weller, D. Pum, and U.B. Sleytr. *J. Nanosci. Nanotechnol.* 4:115–120, 2004.

Chapter 18

Self-Assembled Monolayers (SAMs) and Synthesis of Planar Micro- and Nanostructures

Lin Yan, Wilhelm T.S. Huck, and George M. Whitesides

CONTENTS

I. INTRODUCTION: SAMs AS TWO-DIMENSIONAL POLYMERS

A polymer, by conventional definition, is a macromolecule made up of multiple equivalents of one or more monomers linked together by *covalent bonds* (e.g., carbon–carbon, amide, ester, or ether bonds) [1]. These conventional polymers come in many configurations: for example, linear homopolymers, linear copolymers, block copolymers, cross-linked polymers, dendritic polymers,

Table 1   Comparison Between Conventional Polymers and SAMs, Considered as Two-Dimensional Polymers

		Conventional Polymers	SAMs
Nature of bonding	Short range	Covalent bonding between adjacent monomers	Covalent bonding between head-groups and the substrate; van der Waals, H-bonding, ionic interactions between adjacent monomers
	Long range	van der Waals, H-bonding, ionic interactions between monomeric units proximate in space	No
Structural types		Homopolymers Copolymers (alternating, block, and random) Linear, branched, cross-linked, dendrimeric, etc.	Homogeneous SAMs Mixed SAMs Patterned SAMs
Conformational class		Extended Collapsed Random coiled	Crystalline Disordered Liquid-like

and others. The most common architecture for polymers is based on linear chains that may have other attached chains (branched, grafted, or cross-linked); that is, they are one-dimensional molecules. A few examples have been claimed as two-dimensional sheet polymers.[1]

A supramolecular polymer is a structure in which monomers are organized through *noncovalent interactions* (e.g., hydrogen bonds, electrostatic interactions, and van der Waals interactions) [4]. These less familiar types of polymers also exist in many forms. For example, molecular crystals are large collections of molecules arranged in a three-dimensional periodical lattice through noncovalent intermolecular interactions. Lipid bilayers are two-dimensional structures that exist in water, in which hydrocarbon tails aggregate to form a hydrophobic sheet in the form of a spherical shell, and polar or charged hydrophilic head-groups are exposed to water.

Self-assembled monolayers (SAMs) are highly ordered molecular assemblies that form spontaneously by chemisorption of functionalized molecules on surfaces, and organize themselves laterally, most commonly by van der Waals interactions between monomers [5]. We consider SAMs to be a type of two-dimensional polymer: they are, in a sense, a uniform supramolecular assembly of short hydrocarbon chains covalently grafted onto a macromolecular entity, that is, the surface. In SAMs, individual monomers (usually linear alkyl chains functionalized at one end or both) are not directly linked by covalent bonds to each other, but rather to a common substrate — a metal or a metal oxide surface. SAMs exist in a number of different types: homogeneous SAMs on planar and curved substrates, SAMs on metallic liquids, SAMs on nanoparticles, mixed SAMs, and two-dimensionally patterned SAMs. Table 1 compares some characteristics of SAMs and conventional polymers based on bonding and structural type.

Figure 1 describes the formation of these two types of polymers schematically. A conventional linear polymer is formed by polymerization that links monomers through chemical reactions (e.g., free radical, ionic, and coordination addition, condensation, and ring opening reactions). Polymer chains are often conformationally disordered: in dilute solution, polymer chains are often coiled; in concentrated solution or in bulk, they are entangled. For SAMs, "polymerization" is a spontaneous process involving adsorption that connects monomers to a substrate, and self-organization that orders the system laterally through noncovalent intermolecular interactions. The strong chemical

---

[1] See Refs. 2 and 3 and references therein.

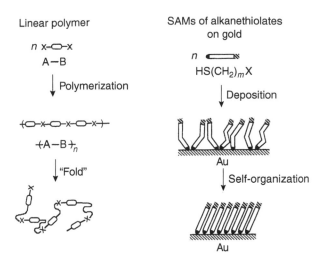

Linear polymer

SAMs of alkanethiolates
on gold

Figure 1   Schematic representation of the formation of conventional polymers and of SAMs.

interaction between the head-group and the substrate renders the "shape" of a SAM two-dimensional, and its "size" the surface area of the substrate. The overall structure of SAMs is determined by the interaction of the head-group and the substrate, the lateral interaction between the neighboring monomers, and the structure of the constituent monomers. SAMs supported on metals or metallic oxides are not soluble, and thus provide no information about the behavior of two-dimensional soluble polymers. They are, however, excellent models for the surface chemistry of insoluble polymers, and among the motivations for the study of SAMs is to understand the physical–organic chemistry of polymer surfaces [6–8], and to develop methods that can be used to control interfacial properties of polymers at the molecular level [9].

This chapter considers SAMs as two-dimensional polymers, and describes the synthesis and structures of SAMs comprising one thiol, and mixed and patterned SAMs comprising more than one thiol (mainly on gold and silver). It reviews some recent studies of chemical transformations of terminal functional groups of SAMs after their assembly, and discusses two potentially useful chemical methods developed in our group for synthesis of mixed SAMs and patterned SAMs, and several of their applications.

II. SYSTEMS OF SAMs

The monomeric units of conventional polymers can be connected by different kinds of chemical bonds; correspondingly SAMs can have various chemical interactions between the head-group and the supporting substrate. A number of different types of SAMs have been explored; several of these systems have been reviewed [5,10–13]. Table 2 categorizes SAMs into groups based on the bonding between the head-group and the surface. The most widely studied systems have been SAMs formed by chemisorption of alkanethiols on gold [14], silver [15], or copper [16]. SAMs of phosphonates have been widely used to synthesize multilayer structures with application in nonlinear optic devices [17–19] and heterogeneous catalysis [20,21]. SAMs of siloxanes on glass and metallic oxides have been studied by Sagiv [22,23] and others [24–27], and is a widely used technologically in surface treatment [28]. SAMs covalently attached to these substrates provide a rugged system for various applications, but certain of these systems — especially those based on $-SiCl_3$ or $-Si(OEt)_3$ head-groups — can be difficult to synthesize, and the reactivity of these head-groups may be incompatible with other functional groups.

Table 2 Different Types of SAMs ($CH_3(CH_2)_nX$)

Head-Group (X)	Precursor	SAMs of	Substrate	Bonding
RS	RSH or (RS)$_2$	Alkanethiolates	Au[14], Ag[15], Cu[16], Pd[29], Fe, Fe$_2$O$_3^{30}$, Hg[31], GaAs[32,33], InP[34]	$RS^- \cdot M_n^{+n}$ or (RS)$_2$; M^n
$RSi{<}^{O}_{O}{-}O$	RSiCl$_3$, RSi(OCH$_3$)$_3$, or RSi(OEt)$_3$	Alkylsiloxanes	SiO$_2$, glass, mica[23,25], Al$_2$O$_3^{35}$, Ga$_2^{37}$O$_3^{32}$, Au[36,37]	Polymeric siloxane
RA$^-$	RCO$_2$H	Acid-functionalized alkanes	Al$_2$O$_3^{35,38-42}$, In$_2$O$_3$/SnO$_2^{43}$, SiO$_2^{41}$, AgO, CuO[40]	Acid base
	RCONHOH		Au, Al$_2$O$_3$, ZrO$_2$, Fe$_2$O$_3$, TiO$_2$, AgO, CuO[44]	
	RSO$_2$H		Au[45]	
	RPO$_3$H$_2$		SiO$_2$, ZrO$_2$, Al$_2$O$_3$, TiO$_2$, mica[46–48]	
RB	R$_2$S, R$_3$P	Base-functionalized alkanes	Au[49–52]	Coordination
	RNC		Au[53], Pt[50]	
R	RCH$_3$	Alkyl groups	Si, graphite[54–56]	Covalent Si–C

SAMs of alkanethiolates that present a wide range of functional groups on thin polycrystalline films of gold and silver are easy to prepare, and have been broadly applied in various fundamental and technological studies. They are well-ordered, and the best characterized systems of organic monolayers presently known. In this chapter, we focus on them.

III. SYNTHESIS OF SAMs ON GOLD AND SILVER

The preparation of SAMs of alkanethiolates on gold and silver is straightforward. The metal substrates are prepared by evaporation of a thin layer of titanium or chromium ($\sim$1–5 nm; this layer of Ti or Cr promotes the adhesion of gold or silver to the supporting substrate) onto silica wafers, glass slides, or other flat surfaces, followed by deposition of gold or silver ($\sim$10–200 nm; in general, $\geq$40 nm is required to achieve a complete coverage of the substrate) [57]. SAMs of alkanethiolates (e.g., X(CH$_2$)$_n$SH, X is a terminal functional group) on gold and silver can be easily generated by immersing the metal substrate in 1–10 mM solutions of alkanethiols at room temperature; ethanol is commonly used as the solvent (Eq. [1]); SAMs can also be generated using vapor phase deposition [58] or electrodeposition [59] of alkanethiols.

$$X(CH_2)_nSH + Au(0)_m \rightarrow X(CH_2)_nS^-Au(I) \cdot Au(0)_{m-1} + \frac{1}{2H_2} \qquad (1)$$

Although formation of SAMs on gold is usually expressed as Eq. (1), the mechanistic details of this reaction is not yet completely understood. It is generally believed that the thiol group forms a thiolate ($RS^-Au(I)$) in its interaction with gold [13]. Some studies using grazing-angle x-ray diffraction have, however, been interpreted to suggest that the interaction of sulfur and gold involves a disulfide ($R_2S_2Au(0)$) [60]. Using molecular dynamics (MD), Gerdy and Goodard have calculated a hypothetical crystal structure for decyl disulfide on Au(111) surface, and found that the resulting structure was energetically stable and the x-ray diffraction pattern derived from such structure was indistinguishable from that observed experimentally [61]. Most of the theoretical studies have, however, been based on the assumption of interactions of thiolates and gold [62]. A conclusive description about the interaction between the sulfur and gold awaits additional experimental and theoretical studies [63]. The fate of the hydrogen atom of the thiol group has also not been resolved. Although there remains a number of uncertainties concerning the structure of the interface between SAMs and gold, most of the interest in SAMs has focused on the structure of the polymethylene $(CH_2)_n$ groups, and on the interaction of the tail-groups with the solution; it is thus immaterial, to some extent, what the binding is between gold and surface.

The kinetics of formation of SAMs on gold has been studied using a range of methods: ellipsometry [64], contact angle [64], quartz crystal microbalance (QCM) [65–69], surface acoustic wave (SAW) [70], surface plasmon resonance (SPR) [71], optical second harmonic generation (SHG) [72], polarized infrared external reflectance spectroscopy (PIERS) [73], scattering Raman spectroscopy [74], and electrochemistry [75]. These studies provide a macroscopic picture of the processes that form SAMs: the growth rate is proportional to the number of unoccupied sites on gold, and can be described as a first-order Langmuir adsorption. Recent atomic force microscopy (AFM) [76] and scanning tunneling microscopy (STM) [77–79] studies depict a three-stage microscopic process:

1. Lattice gas phase — alkanethiols are confined on the surface and diffuse rapidly.
2. Low-density solid phase — molecular axes are aligned with the surface plane and the close pairing of thiol groups is maintained.
3. High-density pseudocrystalline solid phase — alkanethiols are closely packed and their axes are aligned with the surface normal.

Synthesis of SAMs is remarkably convenient; it requires only ambient conditions, and the substrate can be polycrystalline or even electroless gold and silver films [80,81]. Formation of SAMs of a single alkanethiol on gold is known to complete in a few minutes, and may occur in seconds during microcontact printing (μCP), a process in micropatterning [82,83]. SAMs on gold are one of the systems of SAMs most widely used.

IV. Structures of SAMs on Gold and Silver

The molecular structures of SAMs have been studied extensively using various instrumental techniques: PIERS [84–87], x-ray photoelectron spectroscopy (XPS) [64,88], AFM [89–91], grazing-angle x-ray diffraction [92], molecule beam diffraction [93,94], high-energy electron scattering [95], low-energy electron diffraction [57], electrochemistry [96,97], ellipsometry [85,98], and contact angle [87,98–100]. The packing of alkanethiolates on gold is influenced by the spacing of coordination sites and the interaction between the adjacent alkyl chains. Electron diffraction and low-energy helium beam diffraction studies suggest that the sulfur atoms are localized in threefold hallow sites of the Au(111) surface and form a commensurate triangular $\sqrt{3} \times \sqrt{3}R30°$ overlayer lattice (Figure 2) [57,95,101]. In a SAM of n-alkanethiolates, the average cross-sectional area occupied by each thiolate is 21.4 Å^2; this value is larger than that of an alkane chain (18.4 Å^2). The alkyl chains adopt a largely *trans* conformation (for $n \geq 10$) and tilt ~30° with respect to the surface normal in order to maximize van der Waals interactions between adjacent polymethylene chains (the enthalpy of lateral interaction per CH_2 group is ~1.5 kcal/mol). Although the distance between nearest silver atoms (2.89 Å) on the Ag(111) surface is similar to that of gold (2.88 Å), in SAMs

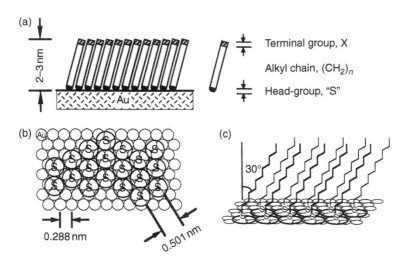

Figure 2 (a) A schematic representation of SAMs of *n*-alkanethiolates on gold. (b) The $\sqrt{3} \times \sqrt{3}$R30° lattice of sulfur atoms on Au(111). (c) Alkyl chains adopt all *trans* conformation and tilt ~30° from the normal of the surface.

of alkanethiolates on silver, the sulfur atoms arrange themselves in a $\sqrt{7} \times \sqrt{7}$R10.9° lattice and the alkyl chains are nearly perpendicular to the surface (tilt ~10° from the normal of the substrate) [102]. Using *ab initio* calculations, Ulman and coworkers suggest that the combination of the lateral discrimination of chemisorption potentials and unfavorable charge–charge interactions between both the thiolates and the underlying Au atoms in SAMs prevent the alkyl chains from packing as densely as those on silver [103]. SAMs formed on freshly prepared silver substrates generally have a lower population of gauche conformations than SAMs on gold, but silver is readily oxidized by oxygen in air and a thick film of silver oxide does not support a well-ordered SAM of alkanethiolates. In general, SAMs of alkanethiolates on freshly prepared silver should be considered more highly ordered than analogous structures on gold.

Although experimental studies have sketched a structural picture of SAMs on gold and silver, the details of this picture remain incompletely defined; among the remaining uncertainties are the exact position of sulfur atoms on gold, the bond angle of the metal-S–C group, the nature of the interaction of the chains with one another, and the nature of lateral movement. Theoretical studies can, in principle, contribute to understanding these issues, although SAMs represent very complex systems for computation. Most of the theoretical work has been carried out on SAMs on gold [62]. Ulman and coworkers have used a very simple model to simulate the thickness of the film, and the molecular orientation and packing of alkyl chains on gold [104]. To a first approximation, they first optimized the geometry of isolated molecules, and then constructed small hexagonal assemblies of these rigid molecules. Considering only van der Waals and electrostatic intermolecular interactions, they examined the interaction energy of a molecule with its neighbor as a function of tilt angles, and found that the calculated thickness and the tilt angles were the same as those established from experimental studies. Klein and coworkers have used MD to investigate the structure and dynamics of alkanethiols on gold [105–108]. They first pinned all the alkyl chains perpendicularly onto a well-defined triangular lattice with the nearest neighbor distance of 4.97 Å and then allowed the system to relax and to evolve into energetically minimal structures. Using both, the united-atom model and the more realistic all-atom model, they found that most of the alkyl chains adopted *trans* conformations, and that the system had fewer gauche conformations when the metal-S–C bonds are colinear than 90°. They also found that the average conformation was temperature dependent; SAMs were less ordered at high temperature, this observation agreed with the results derived from molecular beam studies. Siepman and

McDonald have used Monte Carlo (MC) methods to study the properties of SAMs on gold [109]. Grunze and coworkers have used stochastic global search to explore the configurational space of a SAM of octadecanethiol, $CH_3(CH_2)_{17}SH$, on gold [110]. They used four different force fields and found that several distinct monolayer structures could exist with energy difference less than 1 kcal/mol. Using *ab initio* calculation, Ulman and coworkers found that thiolates prefer the threefold hollow sites over the on-top sites, but Bishop and coworkers used MD and found that the energy difference within the surface corrugation potential is too small to pin sulfur atoms at any particular site [111].

The order of the terminal group and the top part of SAMs is determined not only by the sulfur atoms bound directly to the gold and the intermolecular interaction between the alkyl chains, but also by the size and geometry of the terminal group. STM studies show that SAMs of alkanethiolates on gold are heterogeneous and structurally complex. The alkyl chains form a "superlattice" at the surface of the monolayer of sulfur atoms, that is, a lattice with symmetry and dimensions different from that of the underlying hexagonal lattice formed by sulfur atoms. When alkanethiolates are terminated with end-groups other than the methyl group, the structure of the resulting SAM becomes less predictable. Nelles and coworkers have shown that the superlattice is dependent on the shape of the terminal groups; thiols having terminal groups with relatively spherical cross-sections form hexagonal lattices; thiols with more asymmetric cross-sections form centered rectangular lattices [112]. Sprik and coworkers have used STM and MD to study SAMs terminated with hydrophilic groups such as hydroxyl and amine groups, and found that hydrogen bond induced reconstruction of the top layer and coadsorption of solvents [113].

For SAMs of thiols having tail-groups more complicated than *n*-alkyl chains, the structures of SAMs depend on the size and geometry of these groups. Tao and others have shown that thiols derivatized with aromatic groups form well-ordered SAMs having a packing order different from *n*-alkanethiolates: the sulfur atoms form a $\sqrt{3} \times \sqrt{3}R30°$ lattice on gold, but the aromatic groups adopt a herringbone packing and are perpendicular to the surface [114–118]. SAMs of fluorinated alkanethiolates also pack differently [119,120]. The fluorinated alkyl groups have a van der Waals diameter of 5.6 Å (i.e., larger than that of the 5.0 Å diameter of normal alkyl groups). They form a 2×2 lattice and tilt ~16° from the normal of surface.

Although SAMs are self-assembling systems and tend to reject defects, the presence of defects and pinholes is always observed [121–123]. A variety of factors influence formation and distribution of defects in a SAM, including the molecular structure of the surface, the length of the alkyl chain, and the conditions used to prepare SAMs. Grunze and coworkers have recently described a procedure in which a SAM of alkanethiolates on gold is treated with mercury vapor, then exposed again to a solution of alkanethiol. This procedure seems to heal defects in the SAM by increasing the density of alkyl chains and causing them to reorient to a tilt angle from the normal that resembles that characteristic of copper and silver [124].

V. FUNCTIONAL AND MIXED SAMs ON GOLD

SAMs presenting a variety of functional groups have been applied in a broad range of fundamental studies; representative areas include biocompatibility [125], wetting [126], adhesion [126,127], corrosion [14,128,129], and micro- and nanofabrication [130]. The strong chemo-specific interaction between the thiol group and gold (~24 to 40 kcal/mol) [12] gives SAMs high stability under mild conditions (room temperature) and allows SAMs to display a wide range of organic functionalities in high density (~2×10^{14} molecules/cm^2) on the surface. van der Waals interactions between adjacent polymethylene chains force alkanethiolates to pack at densities approaching those of crystalline poly(ethylene); these lateral interactions make SAMs impermeable to molecules in solution [131–134], and give them the electrical insulating properties similar to that of poly(ethylene) [135]. These characteristics allow the chemical and physical properties of the terminal functional

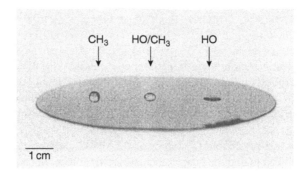

Figure 3 On this 4-in. gold-coated silicon wafer, three different kinds of SAMs were formed. The left part was covered with a SAM of $HS(CH_2)_{15}CH_3$ (a hydrophobic surface), the right part with a SAM of $HS(CH_2)_{11}OH$ (a hydrophilic surface), and the middle area with a mixed SAM of $HS(CH_2)_{15}CH_3$ and $HS(CH_2)_{11}OH$ (formed from a solution containing the two in a ~1:1 ratio). A droplet of water was placed on each of these regions. The shape of the droplet on the mixed SAM was intermediate between that on the hydrophobic and hydrophilic surfaces. This simple experiment shows that the terminal groups on the SAM can control interfacial properties (here, wetting) and also demonstrates the capability of mixed SAMs to tailor the surface properties by controlling the ratio of mixed functionalities on the surface.

groups that are exposed on the surface largely to determine the interfacial properties of a SAM. For example, oligo(ethylene glycol) and oligo(propylene sulfoxide) groups presented on SAMs of undecanethiolates on gold shelter the hydrophobic underlying polymethylene chains effectively from contacting proteins in solution and thus provide a surface that resists nonspecific adsorption of proteins [136–138]. These studies demonstrate that functional SAMs provide well-defined model systems to study surface phenomena.

SAMs that present a mixture of different functional groups — "mixed" SAMs — provide desirable flexibility in design and synthesis of functional SAMs that span a wider range of chemical and physical properties than do pure SAMs [21,139]. It becomes straightforward to tune continuously the interfacial properties of a SAM simply by varying the ratio of compositions of two thiols in the mixed SAM. Synthesis of mixed SAMs on gold is convenient; the gold-coated substrate is immersed in a solution containing two different thiols at a certain ratio, usually for 8 h; the resulting SAM contains a mixture of these two alkyl groups. Figure 3 shows the influence of a terminal group of SAMs on the shape of a droplet of water that contacts the surface. The wetting of water on the surface of a SAM having a ~1:1 mixture of $HS(CH_2)_{15}CH_3$ and $HS(CH_2)_{11}OH$ is intermediate between that of a SAM having $HS(CH_2)_{15}CH_3$ and that having a SAM of $HS(CH_2)_{11}OH$. The ratio of two components in the SAM is often different from that in solution [139]. Although the two organic groups of a mixed SAM are often well mixed [140], thiols that have different properties can phase separate into domains having only one type of thiols; whether this separation is kinetic or thermodynamic is not known [139,141]. AFM and STM studies show that the size of these phase-separated domains is ~50 nm [142–144]. The microscopic heterogeneity does not influence many of the macroscopic chemical and physical properties of the SAMs.

VI. CHEMICAL REACTIONS ON SAMs AFTER THEIR ASSEMBLY

Chemical transformation of terminal functional groups of SAMs *after* their assembly has recently attracted attention. There are several reasons to study chemical reactions on SAMs. They provide (i) controllable well-defined model systems with which to understand the influence of a surface on chemical reactions of functional groups; (ii) alternative methods to functionalize SAMs, to construct multilayers, and to attach molecules and biomolecules to surfaces; (iii) a possible basis for strategies for synthesis of combinatorial libraries of small molecules on a chip; and (iv) synthetic model systems that can be extended to chemical functionalization of polymer surfaces.

A variety of terminal functional groups and their chemical transformations on SAMs have been examined; for example, (i) olefins — oxidation [23,24,131,132], hydroboration, and halogenation [23,24]; (ii) amines — silylation [145,146], coupling with carboxylic acids [22,146], and condensation with aldehydes [22,147]; (iii) hydroxyl groups — reactions with anhydrides [148,149], isocyanates [150], epichlorohydrin [151], and chlorosilanes [152]; (iv) carboxylic acids — formation of acyl chlorides [153], mixed anhydrides [154], and activated esters [148,155]; (v) carboxylic esters — reduction and hydrolysis [156]; (vi) thiols and sulfides — oxidation to generate disulfides [157–159] and sulfoxides [160]; and (vii) aldehydes — condensation with active amines [161]. Nucleophilic displacement on SAMs has also been investigated mainly on SAMs of alkylsiloxanes on Si/SiO$_2$ [146,162–164]. These studies have shown that many organic reactions that work well in solution are difficult to apply to transformations at the surface, because the surface is a sterically hindered environment, and backside reactions (e.g., the S$_N$2 reaction) and reactions with large transition states (e.g., esterification, saponification, Diels–Alder reaction, and others) often proceed slowly. At present, only few synthetic methods are used, but those are capable of introducing a wide range of functional groups onto the surface.

The difficulty of developing useful chemical transformations on the surface is substantially compounded by lack of efficient techniques to identify products and to establish their yields after each reaction. PIERS and XPS are two particularly useful methods to characterize chemical transformations on the surface. PIERS provides direct evidence of transformation of an infrared-active functional group; XPS furnishes evidence for the presence or absence of an element characteristic of the functional group being introduced or eliminated, and is often used to estimate qualitatively the yield of chemical transformation. Other methods, such as ellipsometry, contact angle, secondary ion mass spectroscopy (SIMS) [165], and AFM [166] also provide complementary and valuable support for characterization of chemical reactions on SAMs. Establishing unambiguously the products and yield of a chemical reaction on the surface often requires a combination of information from several techniques.

We have developed two convenient chemical methods that may have general utility for rapid synthesis of functional SAMs. The first procedure has three steps (Figure 4) [167]: preparation of a well-ordered homogeneous SAM of 16-mercaptohexadecanoic acid on gold; conversion of the terminal carboxylic acid groups into interchain carboxylic anhydrides by reaction with trifluoroacetic anhydride; and reaction of the interchain carboxylic anhydride with an alkylamine to give a mixed SAM presenting carboxylic acids and N-alkyl amides on its surface. Figure 5 summarizes the characterization of the product of transformation of terminal carboxylic acids to interchain carboxylic anhydrides, and of anhydrides to a ~1:1 mixture of acids and amides. SAMs of 16-mercaptohexadecanoic acid show C=O stretches at 1744 and 1720 cm^{-1} that are characteristics of terminal carboxylic acid groups presented on a SAM [87]. Treatment of the SAM with trifluoroacetic anhydride gives a SAM having C=O stretches at 1826 and 1752 cm^{-1}; these frequencies are characteristics of a carboxylic anhydride group [168]. XPS studies of the resulting SAM show no fluorine. The combination of these results indicates that the carboxylic acid group is converted into an interchain carboxylic anhydride group by trifluoroacetic anhydride, rather than to a mixed trifluoroacetic carboxylic anhydride. After reaction of the interchain carboxylic anhydride with n-undecylamine (taken as a representative n-alkylamine), the C=O stretches of the anhydride disappear completely, and two new absorption bands appear at 1742 and 1563 cm^{-1}; these bands are assigned as the C=O stretch of a carboxylic acid and an amide II band, respectively. XPS also indicates the presence of nitrogen in the resulting sample. The combination of these results suggests the coupling of amines to the SAM. PIERS further shows that there are two low-frequency shoulder peaks at 2932 and 2859 cm^{-1} in the resulting sample, which are the methylene stretches of the n-undecyl chain; these peak positions suggest that the alkyl chains of the original SAM remain in a *trans* conformation, but that the new alkyl chains are disordered and contain more gauche

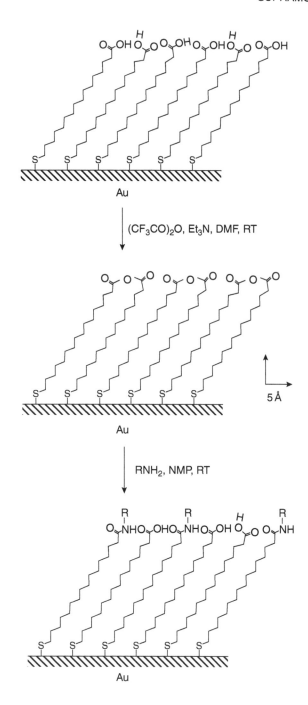

Figure 4 Schematic representation of formation of the interchain carboxylic anhydride and reaction of the anhydride with an alkylamine. The alkyl chains of the original SAM of the carboxylic acid in these SAMs are in *trans* conformation; the methylene groups near the functional groups may adopt gauche conformation and be less ordered. Carboxylic acids in the SAMs of the carboxylic acid, and in the mixed SAMs of amides and carboxylic acids, are hydrogen bonded to neighboring polar groups. The interchain carboxylic anhydrides orient largely parallel to the surface normal. (From L. Yan, C. Marzolin, A. Terfort, and G.M. Whitesides. *Langmuir* 13:6704–6712, 1997. With permission.)

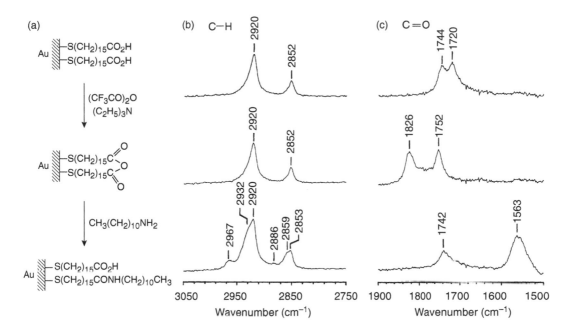

Figure 5 Comparison of PIERS spectra of the SAMs of the carboxylic acid, the interchain carboxylic anhydride, and a mixture of carboxylic acids and *n*-undecylamides on gold: (a) schematic representation of formation of the interchain carboxylic anhydride and the reaction of the anhydride and *n*-undecylamine; (b) the PIERS spectra of these SAMs in the C–H stretching region; (c) the PIERS spectra of these SAMs in the C=O stretching region. (From L. Yan, C. Marzolin, A. Terfort, and G.M. Whitesides. *Langmuir* 13:6704–6712, 1997. With permission.)

conformations [167]. These transformations on the surface occur rapidly, and close to quantitative yield. This new chemical method using interchain carboxylic anhydride as a reactive intermediate has allowed for rapid introduction of many functionalities into SAMs, for example, *n*-alkyl groups [167], perfluorinated *n*-alkyl groups [169], peptides [170], charged groups (sulfonate and guanidine groups) [169], and polymers containing amine groups (e.g., poly[ethylene imine]) [171]; all that is required is a molecule containing the functionality and also an active amine group. This method provides access to SAMs that can be inconvenient or impossible to prepare using the older methods.

Figure 6 compares the second method — the common intermediate method — with the older but more commonly used method. This method also has three steps [172]: formation of a mixed SAM of alkanethiolates on gold derived from the tri(ethylene glycol) ((EG)$_3$OH) terminated thiol (HS(CH$_2$)$_{11}$(OCH$_2$–CH$_2$)$_3$OH; **1**) and the hexa(ethylene glycol)-carboxylic acid ((EG)$_6$CO$_2$H) terminated thiol (HS(CH$_2$)$_{11}$(OCH$_2$CH$_2$)$_6$OCH$_2$CO$_2$H; **2**); generation of activated *N*-hydroxylsuccinimidyl (NHS) esters of thiol **2**; and reaction with proteins, peptides, or small molecules containing active amine groups. These reactions are characterized using PIERS and ellipsometry. Figure 7 shows the PIERS spectra of SAMs of **1**, **2**, of an authentic thiol (HS(CH$_2$)$_{11}$(OCH$_2$CH$_2$)$_6$OCH$_2$CONH(CH$_2$)$_6$NHCOC$_6$H$_4$SO$_2$NH$_2$; **3**), of a mixed SAM comprising **1** and **2**, and of the products of the subsequent reactions. Upon treatment of the mixed SAM with NHS and 1-ethyl-3-(3-dimethylaminopropyl) carbodiimide (EDC), the appearance of bands diagnostic for the NHS ester at 1789 cm^{-1} (symmetric stretch of imide C=O groups) and 1821 cm^{-1} (C=O stretch of the activated ester carbonyl group) indicates the formation of active NHS ester groups on the surface. The complete disappearance of the band at 1631 cm^{-1}, which is assigned as the C=O stretch of a carboxylic acid, suggests near quantitative conversion of carboxylic acid groups to NHS esters. After treatment with a benzenesulfonamide-containing amine (NH$_2$(CH$_2$)$_6$NHCOC$_6$H$_4$SO$_2$NH$_2$; **4**), PIERS shows the appearance of bands at 1550 and 1660 cm^{-1} (characteristic of N–H bending modes) and a weak band at 1733 cm^{-1} (C=O stretch of carboxylic

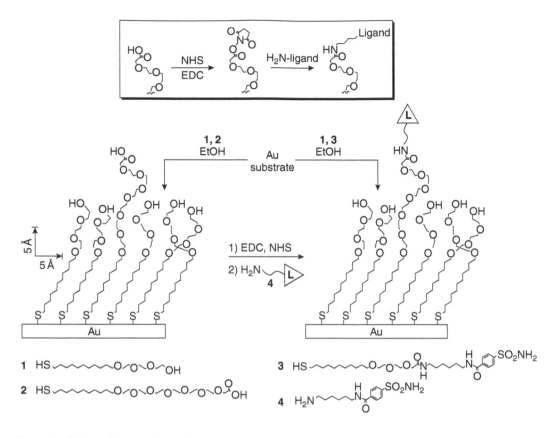

Figure 6 Schematic comparison of the common intermediate method and the older method involving the synthesis of ligand-terminated alkanethiols for preparation of SAMs presenting ligands. In the common intermediate method, a SAM bearing carboxylic acid groups is formed by immersing gold substrates in a mixture of alkanethiols **1** and **2**. This mixed SAM, after activation with NHS and EDC, presents an active NHS ester group on the surface that serves as a common intermediate for the attachment of different ligands by amide bond formation. The upper panel illustrates the chemical transformations involving carboxylic acid groups: (i) activation of carboxylic acid groups with NHS and EDC to generate active NHS esters, and (ii) displacement of the NHS group with an amino group on the ligand (a benzenesulfonamide-containing amine **4** as a representative ligand) or ε-amino groups of lysine residues of proteins to form an amide bond. The polymethylene chains of the alkanethiols in the SAMs are drawn in all-*trans* conformation; this conformation has been observed in SAMs of long chain alkanethiols on gold. The oligo(ethylene glycol) groups are depicted with little or no ordering; the detailed conformation in these SAMs has not been firmly established. (From J. Lahiri, L. Isaacs, J. Tien, and G.M. Whitesides. *Anal. Chem.* 71:777–790, 1999. With permission.)

acid groups). Comparison by PIERS with an authentic SAM of **3** established that benzenesulfonamide ligands were covalently attached to the mixed SAM through amide bonds. Bovine carbonic anhydrase II (CA) can recognize and reversibly bind to the immobilized benzenesulfonamide ligand on the surface [172]. We have also used this procedure to attach proteins to SAMs [172].

VII. PATTERNING OF SAMs ON GOLD IN THE PLANE OF THE MONOLAYER

Patterning SAMs in the plane of the monolayer is useful in determining the two-dimensional distribution of chemical and physical properties on a surface. There are several methods available for generation of patterned SAMs. μCP is a convenient technique that "stamps" a pattern of SAM directly on a surface (Figure 8) [82,83]. In μCP, an elastomeric poly(dimethylsiloxane) (PDMS) stamp — fabricated by casting and curing PDMS against masters that present patterned

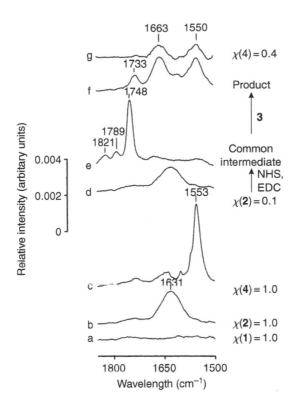

Figure 7   PIERS data for: (i) a homogeneous SAM of **1**; (ii) a homogeneous SAM of **2**; (iii) a homogeneous SAM of **3**; (iv) a mixed SAM comprising **1** and **2** with $\chi(2) = 0.10$; (v) a mixed SAM comprising **1** and **2** with $\chi(2) = 0.10$ after activation with NHS and EDC in H_2O; (vi) a mixed SAM comprising **1** and **2** with $\chi(2) = 0.10$ after treatment with NHS and EDC followed by reaction with **4**; and (vii) a mixed SAM comprising **1** and **3** with $\chi(3) = 0.40$. (From J. Lahiri, L. Isaacs, J. Tien, and G.M. Whitesides. *Anal. Chem.* 71:777–790, 1999. With permission.)

photoresist on silicon wafers — is wetted or inked with an alkanethiol and brought into contact with the gold-coated substrate; SAMs form on the areas that contacted the stamp. SAMs presenting different functional groups can be subsequently formed on the uncontacted areas, either by μCP or by immersion in a solution of another thiol. The edge resolution of the patterns resulting from μCP is ~50 nm [173]. AFM studies show that the structural order characterizing alkanethiolates in SAMs formed by μCP is the same as that deriving from immersion of the gold substrate in a solution of thiol [174]. μCP offers a convenient, low cost, flexible, and nonphotolithographic method to pattern SAMs on large areas [175] and curved substrates [176], and also to pattern SAMs of other systems: alkanethiolates on coinage metals [134], alkylsiloxanes on Si/SiO$_2$ and glass [177,178], and colloids [179,180] and proteins on various substrates [181,182].

Combination of μCP and chemical reaction on a reactive SAM, for example, a SAM presenting interchain carboxylic anhydride groups, can simplify and extend μCP [169]. Figure 9 shows an example: a PDMS stamp with protruding features (squares having ~10 μm on a side) prints *n*-hexadecylamine on the reactive SAM to give a SAM comprising a ~1:1 mixture of *N*-alkyl amides and carboxylic acids. The remaining anhydride groups in the uncontacted regions are allowed to react with another amine, $CF_3(CF_2)_6CH_2NH_2$ to give a patterned SAM having regions presenting *N*-hexadecyl amides and fluorinated *N*-alkyl amides. SEM images (Figure 10) indicate that the edge resolution of these squares is at submicron scale (≤ 100 nm). The high contrast and uniformity in the SEM and SIMS images (Figure 10) suggest that μCP delivered a well-defined pattern of *n*-hexadecylamine to the reactive SAM on both gold and silver. The key chemical reaction — the reaction of amine and surface anhydride — proceeds rapidly and close to quantitative yield under

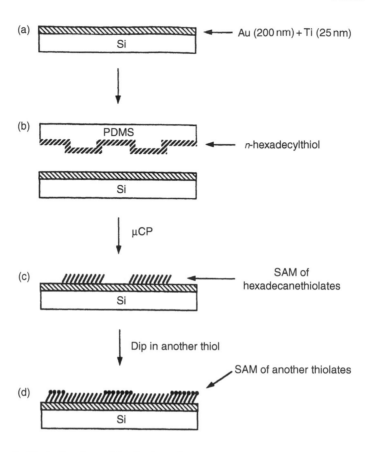

Figure 8 Schematic illustration for μCP of *n*-hexadecanethiol on gold followed by dipping into a solution containing another thiol.

ambient experimental conditions, with good edge definition. This method provides a straightforward route to patterned SAMs that present a variety of functional groups.

VIII. APPLICATIONS

SAMs of alkanethiolates on gold provide excellent model systems for studies on interfacial phenomena (e.g., wetting, adhesion, lubrication, corrosion, nucleation, protein adsorption, cell attachment, and sensing). These subjects have been reviewed previously [125,183–185]. Here we focus on applications that involve using chemical synthesis of functional SAMs after their assembly.

A. Patterning Thin Films of Polymer

Patterned thin films of polymers have many applications, for example, in preventing etching [177], in molecular electronics [186–188], in optical devices [189,190], in biological [191] and chemical sensors [154,192], and in tissue engineering [193]. Thin films of polymers that have reactive functional groups present a surface that can be further modified by chemical reactions [194,195]. There are several methods available for attaching polymers to SAMs: electrostatic adsorption of polyelectrolytes to an oppositely charged surface [196,197], chemisorption of polymers containing reactive groups to a surface [198,199], and covalent attachment of polymers to reactive SAMs [151,154,200,201]. There are presently only a few methods available for patterning thin films of

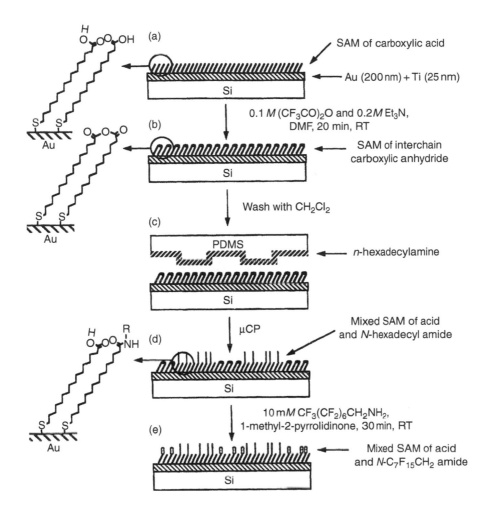

Figure 9 Schematic outline of the procedure for patterning a SAM that presents two different *N*-alkyl amides using μCP and a chemical reaction. The diagram represents the composition of the SAM but not the conformation of the groups in it. (From L. Yan, X.-M. Zhao, and G.M. Whitesides. *J. Am. Chem. Soc.* 120:6179–6180, 1998. With permission.)

polymers on SAMs; these include procedures based on photolithography [200,201], templating the deposition of polymers using patterned SAMs [197,202,203], and templating phase-separation in diblock copolymers [204,205]. Patterned thin films of polymers attached covalently to the surface are more stable than ones only physically adsorbed. Photochemical pattern transfer offers only limited control over the surface chemistry, the properties, and the structure of the modified surfaces.

Combination of μCP and chemical modification of the reactive SAM presenting interchain carboxylic anhydride groups provides a convenient method for patterning thin films of amine-containing polymers having submicron-scale edge resolution on the surface [171]. Figure 11 describes this approach. A reactive SAM presenting interchain carboxylic anhydride groups is prepared using trifluoroacetic anhydride (Figure 4). A PDMS stamp with protruding squares (~10 μm on a side) on its surface is oxidized for ~10 sec with an oxygen plasma. The oxidized PDMS stamp is immediately inked with a 1 wt% solution of poly(ethylene imine) (PEI) in 2-propanol and placed in contact with the substrate. The anhydride groups in the regions that contacted the PDMS stamp react with the amine groups of PEI. Removal of the stamp and hydrolysis of the remaining anhydride groups with aqueous base (pH = 10, 5 min) give a surface patterned with PEI. All these procedures are carried out under ambient conditions; the entire process — from the readily available SAMs

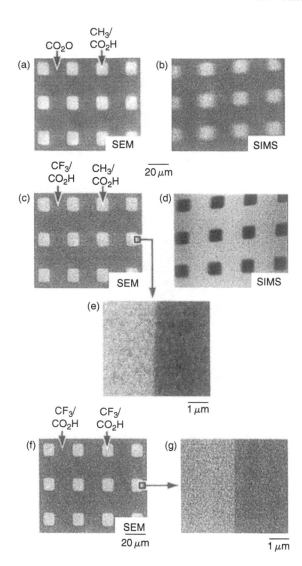

Figure 10   Characterization using SEM and SIMS of the patterned SAMs on Au (a)–(e) and on Ag (f) and (g) generated by μCP of *n*-hexadecylamine on the reactive SAM followed by reaction with $CF_3(CF_2)_6CH_2NH_2$. The light areas in the SEM images were the regions contacted by the stamp. The light squares in SIMS image (b) contained nitrogen while the dark regions did not; the light regions in SIMS image (d) had fluorine while the dark squares did not. The patterns in the SIMS images were distorted because the sample holder was slightly tilted during acquisition of these images. (From L. Yan, X.-M. Zhao, and G.M. Whitesides. *J. Am. Chem. Soc.* 120:6179–6180, 1998. With permission.)

of 16-mercaptohexadecanoic acid to the final patterned PEI films — can be completed in less than one hour.

The AFM images acquired in contact mode show that μCP delivered a well-defined pattern of PEI to the reactive SAM (Figure 12). The resulting thin films of PEI are nearly continuous, but their surfaces are not smooth at the nanometer scale (Figure 12(b)). The roughness of these films is controlled in part by the surface topology of the polycrystalline gold substrate, and probably also by the presence of gel or dust particles in the PEI. Line analysis indicates that the average thickness of the patterned thin films is ~3 nm. Figure 12(c) suggests that the edge resolution of these squares is at the submicron scale (<500 nm); this value is larger than that obtained in μCP of *n*-hexadecylamine on the reactive SAM (<100 nm) and of alkanethiolates on gold (<50 nm). Because

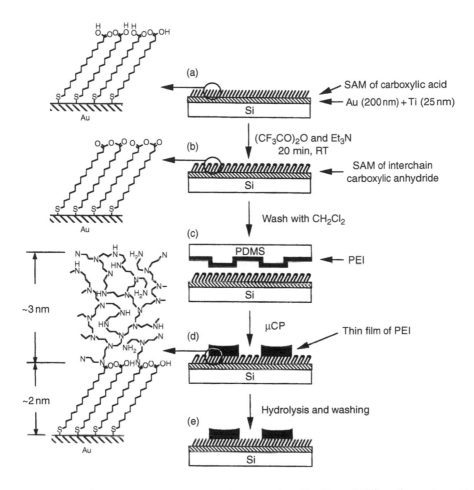

Figure 11 Schematic description of the procedure for patterning thin films of PEI on the surface of a SAM using μCP and a chemical reaction. The scheme suggests the composition of the SAM, but not the conformation of the groups in it; it also makes no attempt to represent either the conformation of the polymer or the distribution of functional groups on the polymer backbone.

PEI is a hydrophilic polymer, it is essential to make the hydrophobic PDMS stamp hydrophilic using oxygen plasma prior to inking in order to form continuous, patterned thin films of PEI on the surface [206]. PIERS studies further show that PEI is covalently linked to the SAM by amide bonds and that the PEI films are thus more stable under both acidic and basic conditions than are polymers physically adsorbed on SAMs of carboxylic acids.

The covalently attached PEI films make a large number of reactive amine groups available for further chemical modification of the surface. These amine groups can react with other functional groups (e.g., acyl chlorides and carboxylic anhydrides) to introduce different organic functionalities into the surface, and to attach polymers that have such organic functional groups. We have shown that the amine groups of the attached PEI film can react with perfluorooctanoyl chloride, palmitoyl chloride, palmitic anhydride, and poly(styrene-*alt*-maleic anhydride) [171].

B. Facile Preparation of SAMs that Present Mercaptan, Charged, and Polar Groups

Chemical reaction provides a straightforward method for the preparation of SAMs that present a variety of functional groups — especially polar, charged, or structurally complex groups, such

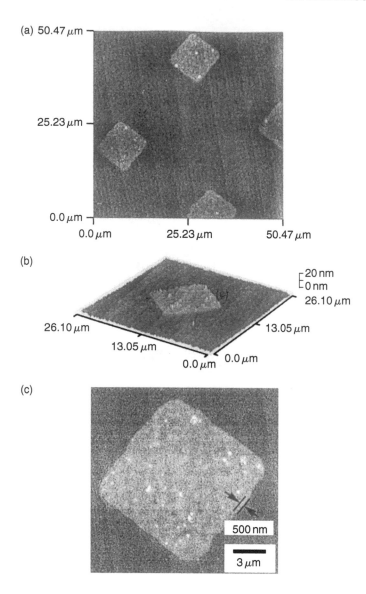

Figure 12 Contact mode AFM characterized the patterned thin films of PEI generated by μCP of PEI on the reactive SAM; these images show a sample patterned by μCP, followed by hydrolysis of the remaining, unreacted interchain carboxylic anhydrides with aqueous base. The light squares in the AFM images (a)–(c) were thin films of PEI on the regions contacted by the PDMS stamp. The AFM image (c) shows that the PEI film was separated by a well-defined boundary (with roughness <500 nm) from the regions presenting carboxylic acid groups. The lines in the images were artifacts generated by the instrument. (From L. Yan, W.T.S. Huck, X.-M. Zhao, and G.M. Whitesides. *Langmuir* 15:1208–1214, 1999. With permission.)

as peptides, polymers, and oligosaccharides — that are difficult to prepare using deposition of thiols terminated at these groups [169].

Figure 13 shows a patterned SAM presenting methyl and thiol groups, and the subsequent assembly of Au nanoparticles in the regions presenting thiol groups. The patterned SAM is generated by μCP of *n*-hexadecylamine on the reactive SAM followed by reaction with cysteamine (HSCH$_2$CH$_2$NH$_2$). Patterned SAMs presenting thiol groups are difficult to prepare by conventional μCP. The resulting patterned substrate is then immersed in an aqueous suspension of Au nanoparticles stabilized with citrate anions. SEM images show that the Au nanoparticles assemble predominately

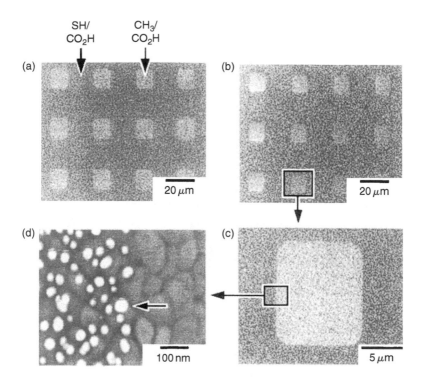

Figure 13 SEM images of (a) a patterned SAM presenting methyl groups (white squares) and thiol groups; (b)–(d) patterned deposition of Au nanoparticles (white dots pointed by an arrow in (d) in the regions that present thiol groups. The background texture in (d) is the "islands" that are formed on evaporating gold under the conditions we used. The mean diameter of the Au nanoparticles was ~20 nm. (From L. Yan, X.-M. Zhao, and G.M. Whitesides. *J. Am. Chem. Soc.* 120:6179–6180, 1998. With permission.)

in the thiol-presenting regions, and the width of the border separating the region having adsorbed nanoparticles from that having none is <100 nm.

We have prepared SAMs presenting sulfonates and guanidines by allowing the reactive SAM to react with 3-amino-1-propanesulfonic acid ($H_2N(CH_2)_3SO_3H$) and agmatine sulfate ($H_2N(CH_2)_4NHC(=NH)NH_2 \cdot H_2SO_4$). For the SAM presenting sulfonates, XPS shows a signal at 168.5 eV; we assign this peak to S(2p) of a sulfonate, and the advancing and receding contact angles of water are both less than 10° (the lowest value we can measure). For the SAM presenting guanidines, XPS shows an N(1s) signal at 400.4 eV, the advancing contact angle of water is 47°, and the receding contact angle 30°.

C. Wetting

Previous extensive studies of carboxylic acid functionalized poly(ethylene) films (PE–CO_2H) [207–209], of SAMs terminated in ionizable acids and bases [100], of mixed SAMs of carboxylic acid- and methyl-terminated alkanethiolates [210], and of SAMs of dialkyl sulfides on gold [211] have established the utility of contact angle titration in characterizing the wetting properties of interfaces. We have determined the advancing contact angle of water θ_a as a function of pH for several mixed monolayers obtained by allowing the interchain anhydride to react with homologous n-alkylamines (n-$C_nH_{2n+1}NH_2$, n = 0, 1, 4, 6, 11, and 18) (Figure 14). In these mixed SAMs, the polar carboxylic acid groups are buried beneath hydrocarbon layers of different thickness.

The values of θ_a for the SAMs derivatized with long n-alkylamines (n = 11 and 18) do not change with pH. These alkyl groups form thick hydrophobic films (~10 and 15 Å, respectively)

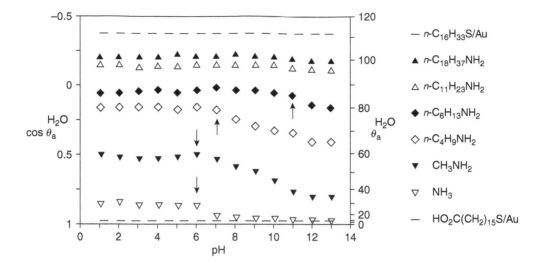

Figure 14 Dependence of the advancing contact angle (θ_a) of buffered aqueous solutions of different values of pH on the SAMs comprising a mixture of carboxylic acids and amides, generated by reactions of the interchain anhydride and alkylamines (n-$C_nH_{2n+1}NH_2$, n = 0, 1, 4, 6, 11, and 18). The curves are labeled by the respective alkylamines on the right side of the plot. Two dashed lines on, the top and the bottom of the plot are reference data for θ_a of $CH_3(CH_2)_{15}S$/Au and $HO_2C(CH_2)_{15}S$/Au, respectively, as indicated. Arrows indicate the onset points of ionization of carboxylic acid groups.

that prevent water from contacting the buried carboxylic acid groups. The values of θ_a change with pH for mixed SAMs derivatized with short n-alkylamines (n = 1, 4, and 6). There are significant features of these data. First, the titration curves do not reach a plateau at high pH; the same behavior is observed for the mixed SAMs of carboxylic acid- and methyl-terminated alkanethiolates on gold prepared from mixtures of $HS(CH_2)_{10}COOH$ and $HS(CH_2)_{10}CH_3$ [211]. By contrast, the carboxylic acid functionalized material obtained by oxidizing poly(ethylene) ($PE–CO_2H$) does achieve plateau values at high pH [207]. Second, the onset points of ionization are approximately pH 11 for the n-hexylamine-modified SAM and approximately pH 7 for the n-butyl amine-modified SAM. The values of θ_a also change with pH for mixed SAMs comprising carboxylic acids and N-methyl amide and primary amides. These titration curves are very similar to that of the mixed SAM of N-butylamides and carboxylic acids. Both the onset points of ionization are approximately at pH 6. Such shifts from the values expected on the basis of titration in aqueous solution are similar to those observed with mixed SAMs terminated with carboxylic acid and methyl groups [211].

Contact angle titrations of these n-alkylamine-modified SAMs suggest that these systems are more like mixed SAMs of methyl- and carboxylic acid-terminated alkanethiolates on gold than they are like $PE–CO_2H$.

D. Patterning Ligands on Reactive SAMs

Patterning ligands on surfaces has several applications: for example, in biosensors using patterned proteins and cells, and in diagnostic tools using patterned DNA fragments, antibodies, and antigens. The techniques used for patterning ligands are often based on photolithographical procedures: these methods may be incompatible with many types of ligands.

Combination of μCP and chemical reaction on a reactive SAM presenting a mixture of active NHS ester groups and oligo(ethylene glycol) groups provides a convenient, inexpensive, and versatile method for patterning ligands on surfaces that can be recognized and bound specifically by biomacromolecules [212]. Figure 15 describes a procedure that has several steps: (i) formation of mixed SAMs presenting thiol **1** and **2**; (ii) activation by immersion in a solution of EDC (0.1 M)

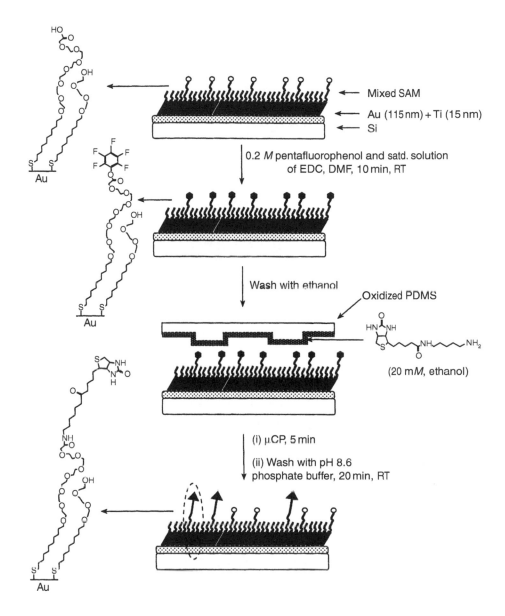

Figure 15 Schematic representation of the procedure used for patterning biotin ligands onto SAMs consisting of activated carboxylic esters. (From J. Lahiri, E. Ostuni, and G.M. Whitesides. *Langmuir* 15:2055–2060, 1999. With permission.)

and pentafluorophenol (0.2 M) for 10 min; and (iii) printing a biotin-containing amine to the reactive SAMs by bringing a freshly oxidized PDMS stamp, inked with the ligand, in contact with the substrate for 5 min. The formation of patterned SAMs presenting biotin ligands was imaged by fluorescence microscopy of substrates that were incubated in a solution of fluorescently labeled antibiotin antibody (Figure 16(a)). The patterns were also detected using a different approach, in which the substrate was incubated sequentially in solutions of streptavidin, biotin-conjugate protein G, fluorescently labeled goat antirabbit IgG, and imaged by fluorescence microscopy (Figure 16(b)). The smallest features resolved in images obtained by these methods were squares with a 5 μm side. The high contrast between the fluorescent regions and nonfluorescent ones indicates that the PDMS stamp delivered biotin ligands to the SAM and that they were bound specifically

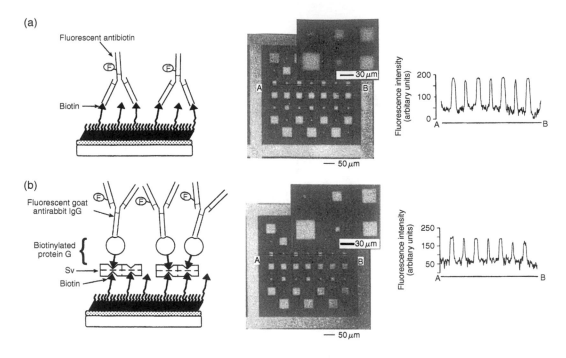

Figure 16   Fluorescence microscopy images of patterned SAMs having biotin groups ($\chi \sim 0.02$) and schematic representations of the surfaces during fluorescence detection. The intensity of fluorescence in regions having biotin groups and not having ones across the line (AB) was analyzed using the NIH image software. (a) Images of fluorescently labeled antibiotin bound to patterned SAMs presenting biotin groups. (b) Images of fluorescently labeled goat antirabbit IgG bound to SAMs presenting biotin groups that were sequentially incubated in solutions of streptavidin and biotin-conjugated protein G. (From J. Lahiri, E. Ostuni, and G.M. Whitesides. *Langmuir* 15:2055–2060, 1999. With permission.)

by its binding proteins. The coupling yields were estimated using SPR to be ~75 to 90% of that obtained by immersion. It was also found that oxidation of the PDMS stamp prior to inking was critical for good coupling yields.

IX.   DEVELOPMENTS FROM 2000 TO 2003

SAMs have been used widely as model surfaces in the controlled modification of surfaces. Recently, well-defined monolayers of perfluorinated trichlorosilane derivatives have been formed on previously notoriously difficult surfaces to functionalize, such as indium tin oxide (ITO), by using supercritical CO_2 as a solvent. The CO_2 weakly adsorb to the surface, thereby stripping the water layer away from the surface. This displacement allows much faster reactions of organosilanes onto these surfaces (a full SAM is formed within hours at 31°C, instead of refluxing in toluene for 7 days) [213,214]. Earlier, we reviewed the unique, two-dimensional (2D) crystalline structure of monolayers on gold and Si/SiO$_2$, as well as the synthetic challenges encountered when translating "standard" organic chemistry to reactions on SAMs. A general lesson learned was that the sterically hindered and relatively inflexible environment considerably slows transformation reactions of terminal functional groups [215,216]. Furthermore, interactions of neighboring groups with the reactive center as well as altered nucleophilicity of reactive groups via interactions with the substrate [217], lead to different reaction rates in SAMs. Several methodologies, such as the interchain anhydride route, have been developed to provide versatile platforms for further modification. In an excellent contribution by Schönherr and coworkers, the difference in reaction kinetics between SAMs, thin polymer films, and solution chemistry is described in detail [218]. The reaction of interest is the hydrolysis of NHS esters groups, which can be described as pseudo-first-order in all cases. The rate constant for the SAM is five times smaller than that for the reaction in thin polymer films, but differs three orders of magnitude compared with similar reactions in solution. These results again demonstrate that, the reactions on surfaces are affected by confinement effects. Especially the difference in activation entropies (-176 J/(mol K) and -59 J/(mol K) for SAMs and thin polymer films, respectively, and calculated according to transition state theory) reveal the steric crowding in the near surface region and the tightness of the transition state.

In this update, we will primarily focus on a number of recent developments where monolayers still provide a guiding platform, but play a larger role in influencing chemistry just above the surface. A striking illustration of the power of SAMs is illustrated by the work of Arias et al. [219], which shows that the external quantum efficiency of photovoltaic devices prepared from blends of semiconducting polymers can be changed by more than an order of magnitude, simply by changing the monolayer from a hydrophilic to a hydrophobic head-group. This change in surface energy leads to a change in the phase separation of the polymer blend from a random morphology to a bilayer structure. In the same context, we will discus the recent introduction of switchable surfaces, where surface chemistry can be changed by varying the functional terminal groups in SAMs via external triggers. From these adaptive surfaces, it is a small step toward surface-initiated polymer brushes, where not the terminal functionality of the SAM, but a tethered polymer chain introduces "smart" surface properties.

A. Switchable Surfaces

In a number of applications (e.g., cell-binding studies, microfluidics, or self-cleaning surfaces) it would be ideal if the surface energy (contact angle) and/or surface chemistry (cell binding, protein adsorption, pH or salt sensitivity, corrosion resistance, etc.) can be modulated *in situ* and relatively fast [220]. Functional monolayers could play a similar role and due to the introduction of chemistry, a broader range of switches in a wider variety of systems. In an early work, Willner and coworkers exploited the photoisomerization of dinitrophenyl spiropyrans to tune the binding of antibodies, specifically raised to recognize spiropyran moieties, to surfaces [221]. Illumination of the spiropyran with 370 nm UV light resulted in the isomerization of the double bond and ring opening of the pyran moiety, thereby switching off antibody binding. Subsequent exposure to 500 nm light reversed this isomerization back to the starting dinitrophenyl spiropyran. Interesting work carried out by Beebe and coworkers [222,223] incorporated 2-nitrobenzyl groups to create photolabile monolayers [224], which were subsequently used to fabricate pressure sensitive switches within microfluidic networks. Upon UV exposure (Figure 17) the *o*-nitro group, in the perfluoro-1-octyl 4-(11-trichlorosilyl-1-oxoundecyloxymethyl) 3-nitrobenzoate SAM (F-SAM), is excited. Intramolecular H-abstraction occurs, and the free acid is formed with the nitroso byproduct. The carboxylic acid groups generated switch the surface to hydrophilic. Surprisingly, the conversion does not go to completion and is rather slow (irradiation times of hours). The final contact angles are still around $60°$, which for supposedly –COOH terminated surfaces indicates incomplete removal of the perfluoro tails is likely.

These SAMs were used to modify the sides of microfluidic channels. It was shown that surfaces with increased advancing water contact angles required higher-pressure fluid streams to overcome the energy barrier to wetting. This implied that surface wettability of the F-SAM could be readily controlled by irradiation time, which can be used to fabricate a pressure sensitive microfluidic gate in which channels are closed to flow until a pressure of fluid is applied. A far more dynamic surface was fabricated by Lahann and coworkers [225]. SAMs carrying large (2-chlorophenyl)diphenyl ester moieties were prepared on gold. These SAMs are not very dense because the large head-groups prevent efficient packing of the alkyl chains. After hydrolysis, the thiols molecules terminated in carboxylate end-groups and these molecules were sufficiently flexible due to the increased special freedom, to bend toward the surface up a change in surface charging.

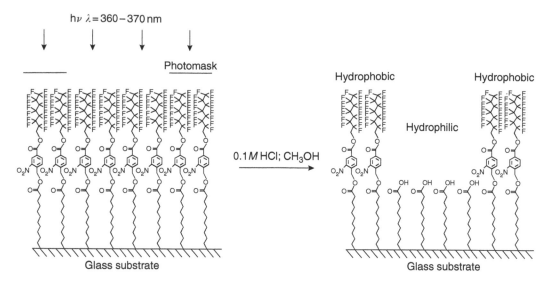

Figure 17 Photodeprotection of F-SAM upon UV irradiation. (Adapted from B. Zhao, J.S. Moore, and D.J. Beebe. *Anal. Chem.* 74:4259, 2002. With permission.)

The Mrksich group recently demonstrated that electrochemistry is a powerful tool not only in characterizing SAMs, but also at the same time chemically modifying the head-groups. SAMs were prepared from alkanethiols terminating in hydroquinone groups, which undergo reversible two-electron electrochemical oxidation to the quinone moiety. This electrochemical switching can be easily carried out in a cyclic voltammetry setup, which enables the time resolved, kinetic study of this process. The quinone undergoes, in contrast to the hydroquinone, a Diels–Alder reaction with cyclopentadiene [226]. The environment for this interfacial Diels–Alder reaction can be "tuned" by making mixed monolayers, where the other thiols terminate in either –OH or –CH3 groups, thereby essentially creating a hydrophilic or hydrophobic environment [227]. The Diels–Alder reaction not only provides new insight into the interfacial reaction and adsorption kinetics, but also facilitates the introduction of a wide range of functional groups via the reaction with functionalized cyclopentadiene derivatives [228]. In this particular example, a RGD oligopeptide was coupled to the SAM to stimulate the attachment of cells such as 3T3 fibroblasts (Figure 18). Very elegantly, the Diels–Alder reaction was combined with patterned surfaces and cell mobility studies. Cells are allowed to bind in areas with RGD-terminated SAMs, with an inert background of an ethyleneglycol-terminated monolayer preventing adhesion outside the pattern. This inert SAMs, however, does contain small amounts of the hydroquinone thiols and after electrochemical activation of the hydroquinone moieties a RGD-cyclopentadiene derivative present in the cell culture medium, coupled to the background, thereby essentially erasing the patterning and allowing the cells to spread over the surface.

A similar strategy was employed to develop SAMs that released groups rather than immobilized ligands. A catechol orthoformate group was used as the electroactive linker that tethered the ligand to the SAM [229]. Upon electrochemical oxidation, the catechol orthoformate was converted to the corresponding orthoquinone plus hydrolyzed formate product. The reaction was complete within a single cyclic voltammetric scan, demonstrating the efficiency and speed of electrochemical reactions on monolayers.

One of the most important recent developments in characterizing SAMs, is the use of MALDI-TOF mass spectrometry to directly "image" molecules chemisorbed to gold surfaces [230]. In a series of articles, Mrksich and coworkers demonstrated convincingly that the technique is also extremely useful in analyzing reactions on monolayers [231]. No special equipment is necessary and a gold-coated substrate can simply be placed in a standard MALDI mass spectrometer. Any

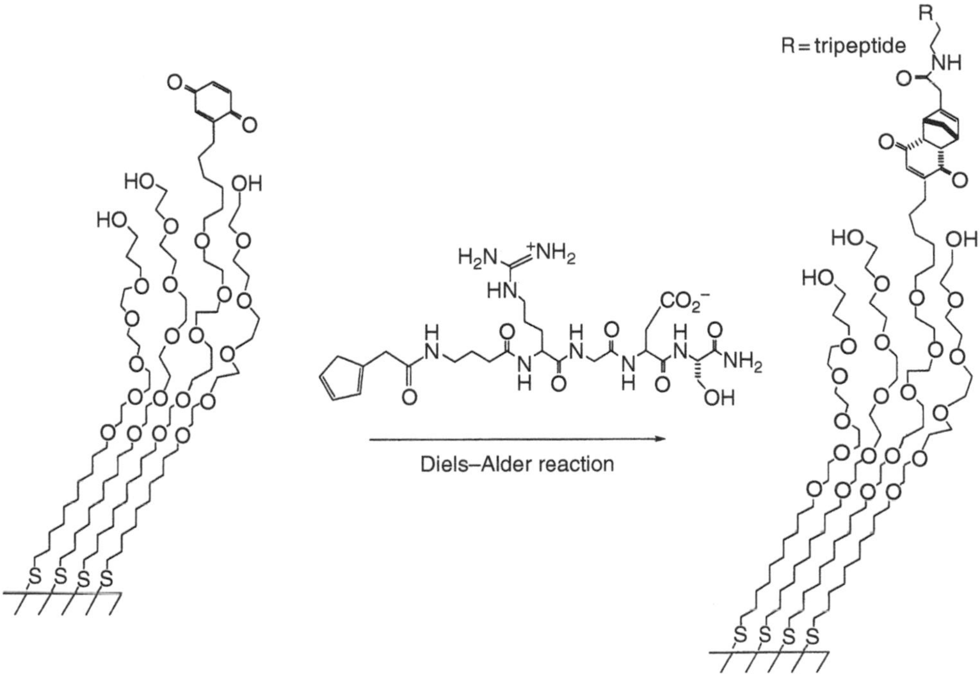

Figure 18 Attachment of ligands to SAMs via Diels–Alder reaction.

chemical reaction on the SAM is easily monitored by analyzing the rations of mixed disulfides in the detector. Because the technique measures mass-to-charge ratio, there is no need for additional labeling. Exchange reactions between SAMs of one alkanethiolate and different alkanethiols in solutions can also be studied [232].

X. POLYMER BRUSHES

We have discussed a great number of organic reactions on planar surfaces with very high densities of functional groups. However, these functional groups are not always available for further functionalization, which leads to incomplete reactions, especially when attempting to link large molecules (e.g., DNA strands) to these surface groups. One possible solution is to construct a "pseudo three-dimensional" surface where the number of reactive available sites is increased. This can be achieved by the modification of surfaces with thin polymer films, which is widely used to tailor surface properties such as wettability, biocompatibility, corrosion resistance, and friction [233,234]. Such thin polymer films can be applied by depositing or spraying a polymeric coating from solution. Alternatively, polymers with reactive end-groups can be grafted onto surfaces, resulting in so-called polymer brushes. The advantage of polymer brushes over other surface modification methods (e.g., self-assembled monolayers) is their mechanical and chemical robustness, coupled with a high degree of synthetic flexibility toward the introduction of a variety of functional groups. There is also an increasing interest of using functional or diblock copolymer brushes for "smart" surfaces, which can change a physical property (hydrophilicity, biocompatibility) upon an external trigger, such as heat (in the case of lower critical solution temperature [LCST] materials), pH, or salt concentration (polyelectrolytes). Commonly, brushes are prepared by grafting polymers to surfaces, either via chemical bond formation between reactive groups on the surface and reactive end-groups, or by physisorption of block copolymers with "sticky" segments. The *grafting to* method is experimentally simple, but has some limitations. It is very difficult to achieve high grafting densities because of steric crowding of reactive surface sites by already adsorbed polymers. Furthermore, film thickness is limited by the molecular weights of the polymer in solution (films in the 100 nm thickness range are inaccessible). Relying on noncovalent adsorption of polymers to surfaces makes the adsorption a reversible process and such brushes are not stable under conditions where high shear forces are involved, for example. The introduction of functional groups might be hampered by the requirements for the physisorption to the surface (electrostatic or hydrophobic interactions), or the formation of a covalent bond via reactive groups on polymer and surface. "Surface-initiated polymerizations" (also called *grafting from*) from initiators bound to surfaces are a powerful alternative to control the functionality, density, and thickness of polymer brushes with almost molecular precision. Recently such systems have been reviewed extensively and here we present some of the highlights as far as they have an impact on supramolecular chemistry [235,236]. First, the substrate of choice (planar or particle) is modified with initiator-bearing self-assembled monolayers. These monolayers can be formed on almost any surface, as long as the anchor functionality is chosen right (e.g., thiols on gold, silanes on glass, Si/SiO_2, and plasma oxidized polymers). The initiator surfaces are then exposed to solutions containing catalyst and monomer (plus solvent if necessary). Ideally, the polymerization is not only surface initiated but also surface confined, that is, no polymerization in solution. In order to achieve maximum control over brush density, polydispersity, and composition, plus at the same time allowing the formation of block copolymers on the surface, a controlled polymerization is highly desirable [237].

There are a number of ways in which polymers can be grown in a controlled manner from surfaces including the use of "living" cationic [238], anionic [239], nitroxide mediated [240–242], ring opening [243], and atom transfer radical polymerization (ATRP) [244–246]. The disadvantages of some of these methods is that even though controlled brush growth can be obtained, they often require high temperatures, addition of sacrificial initiator and long reaction times resulting in polymerization of the monomer in solution [247]. More recently, an alternative method of surface-initiated polymerization was developed in aqueous media offering controlled growth of brushes with a predetermined length and the possibility of block copolymers [248]. The density of these brushes can be controlled by "diluting" the initiator monolayer by mixing with nonreactive CH_3-terminated thiols [249].

A. Smart Surfaces

Polymers can be used to modulate surface characteristics without the need for chemical reactions. Instead, surface properties can be altered by relying on changes in the conformation of the polymer backbone, which can be achieved by changing the solvent conditions for the polymer brush (bad, theta, or good solvent). Such changes can be achieved by changing the solvent or via changes in temperature, salt concentration, or pH. Such response should be much faster than chemical transformations, giving rise to opportunities of fast switching in microfluidic devices. Jones and coworkers have recently reported the growth of poly-N-isopropylacrylamide (PNIPAM) brushes from patterned SAMs [250]. Below the so-called LCST the brushes were hydrophilic and fully hydrated, but by increasing the temperature to 30°C (above the LCST), the brushes collapsed and became hydrophobic, due to their insolubility in water. By lowering the temperature below the LCST, the brushes expanded again into their hydrophilic state. This reversible switching could be used in self-cleaning, antifouling surfaces, where bacteria adsorbed to the hydrophilic polymers detached upon raising the temperature. Böhringer has fabricated devices consisting of microheaters [251] beneath a PNIPAM surface and demonstrated selective protein binding is possible using these heater arrays. Rapid response characteristics to temperature have shown protein adsorption and desorption can take place in less than a second [252]. Another important class of polymer brushes is polyelectrolyte brushes, which carry (fixed) charges along the chain. In pure water, such brushes stretch out to (almost) their maximum length because of repulsion between the monomers. After the addition of salt (or a change in pH leading to neutralization of charges), the charge–charge repulsion is significantly reduced and the elastic energy in the polymer backbone leads to the formation of a more relaxed random coil conformation. Translated to surface properties, this leads to surfaces that expand or collapse depending on salt concentration of pH. This change in thickness could be coupled to movement of inorganic objects, hence converting chemical energy into mechanical energy.

XI. CONCLUSIONS

SAMs can be considered as a form of insoluble, two-dimensional, and grafted polymers; that is, the substrate surface is the backbone, and the attached alkanethiolate (RS^-) groups are the side chains. The backbone already exists before the SAM is formed: the preparation of SAMs is more like postpolymerization modification than it is like polymerization. The packing of the alkyl chains is such that the alkyl chains are highly ordered and crystallize. In SAMs of n-alkanethiolates on gold, the alkyl chains adopt an all-*trans* conformation and tilt ~30° from the normal of the substrate to maximize intermolecular interactions between adjacent chains. There is no comparable phenomenon to this tight packing of side chains in conventional polymers. The high density of side chains leads to anomalous reactivities of functional groups incorporated in the side chains. SAMs provide a sterically highly congested environment. Even terminal functional groups are sterically hindered; their chemical transformations on the surface are often slower than they would be in solution. Because terminal groups are embedded in a sea of hydrophobic polymethylene chains, they are difficult to ionize. On the other hand, the high density of functional groups can result in favorable neighboring and chelating interactions.

The formation of SAMs on gold and silver is simple and convenient. The strong chemoselective interactions between the head-groups and the substrate allow SAMs to present a wide range of functional groups on the surface. Mixed and patterned SAMs deliver flexible control over the lateral distribution of chemical functionalities. Both the interchain carboxylic anhydride and the activated carboxylic esters strategy minimize the amount of organic synthesis required for the preparation of functionalized alkanethiols to prepare SAMs that present a wide range of chemical functionalities. They are straightforward methods and could find many applications.

XII. OUTLOOK

The use of SAMs allows the formation of precisely defined model surfaces and to study a vast number of reactions on such surfaces. This improved understanding of the chemical nature of surfaces has been important in the design of biological microarrays. The attachment of DNA in a reproducible and quantitative way to glass supports is of crucial importance in the production of so-called gene chips, but equally important and far more difficult is the coupling of proteins to surfaces. Also in that area, considerable progress has been made, ensuring both inert surfaces to avoid unfolding of the protein at the interface, while at the same time developing the covalent attachment chemistry which needs to be a high yielding fast reaction. One of the most important breakthroughs in recent years has been the introduction of MALDI-TOF for the characterization of complex molecules on surfaces and the determination of the yield of reactions on SAMs. This technique will undoubtedly find a wide range of applications and allows a far easier monitoring of reaction at interfaces than grazing-angle FT-IR, ellipsometry, or XPS, to name a few standard techniques. These new techniques, coupled with an ever expanding range of synthetic methodologies for coupling complex molecules to surfaces and strategies to switch surface chemistry via external triggers, will lead to further applications of SAMs in sensors, polymer electronics, microfluidics, and array technology. In many of these applications, the next big leap in improved functioning of devices will come from control over the interfacial properties at the nanoscale level and over a period time. SAMs will become crucial parts in such devices, as they provide a controlled interface between biology or (polymer) chemistry and lithographically patterned silicon.

Acknowledgments

This work was supported by DARPA and the ONR. Lin Yan thanks Ned Bowden for technical assistance in obtaining Figure 3.

References

1. P. Flory. *Principles of Polymer Chemistry*. Ithaca, NY: Cornell University Press, 1953.
2. J.-M. Lehn. *Supramolecular Chemistry, Concepts and Perspectives*. Weinheim: VCH, 1995.
3. S.I. Stupp, S. Son, H.C. Lin, and L.S. Li. *Science* 259:59–63, 1993.
4. H. Rehage and M. Veyssié. *Angew. Chem. Int. Ed. Engl.* 29:499–448, 1990.
5. A. Ulman. *Introduction to Thin Organic Films: From Langmuir–Blodgett to Self-Assembly*. Boston, MA: Academic, 1991.
6. C.D. Bain and G.M. Whitesides. *Angew. Chem. Int. Ed. Engl.* 28:506–516, 1989.
7. G.M. Whitesides. *Chimia* 44:310–311, 1990.
8. G.M. Whitesides and C.B. Gorman, Eds. *Self-Assembled Monolayers: Models for Organic Surface Chemistry. Handbook of Surface Imaging and Visualization*. Boca Raton, FL: CRC, 1995, pp. 713–733.
9. G.S. Ferguson, M.K. Chaudhury, H.A. Biebuyck, and G.M. Whitesides. *Macromolecules* 26:5870–5875, 1993.
10. L.H. Dubois and R.G. Nuzzo. *Annu. Rev. Phys. Chem.* 43:437–463, 1992.
11. S. Yitzchaik and T. Marks. *Acc. Chem. Res.* 29:197–202, 1996.
12. A.R. Bishop and R.G. Nuzzo. *Curr. Opin. Coll. Interf. Sci.* 1:127–136, 1996.
13. A. Ulman. *Chem. Rev.* 96:1533–1554, 1996.
14. R.G. Nuzzo, B.R. Zegarski, and L.H. Dubois. *J. Am. Chem. Soc.* 109:733–740, 1987.
15. P.E. Laibinis, M.A. Fox, J.P. Folkers, and G.M. Whitesides. *Langmuir* 7:3167–3173, 1991.
16. P.E. Laibinis and G.M. Whitesides. *J. Am. Chem. Soc.* 114:9022–9027, 1992.
17. H. Lee, L.J. Kepley, H.-G. Hong, and T.E. Mallouk. *J. Am. Chem. Soc.* 110:618–620, 1988.
18. D. Li, M.A. Ratner, T.J. Marks, C. Zhang, J. Yang, and G.K. Wong. *J. Am. Chem. Soc.* 112:7389–7390, 1990.
19. M.E. Thompson. *Chem. Mater.* 6:1168–1175, 1994.
20. S.M.C. Neiva, J.A.V. Santos, J.C. Moreira, Y. Gushikem, H. Vargas, and D.W. Franco. *Langmuir* 9:2982–2985, 1993.
21. M.G.L. Petrucci and A.K. Kakkar. *J. Chem. Soc. Chem. Commun.* 1577–1578, 1995.
22. J. Sagiv. *Isr. J. Chem.* 18:339–345, 1979.
23. J. Sagiv. *J. Am. Chem. Soc.* 102:92–98, 1980.
24. I. Haller. *J. Am. Chem. Soc.* 100:8050–8055, 1978.
25. S.R. Wasserman, G.M. Whitesides, I.M. Tidswell, B.M. Ocko, P.S. Pershan, and J.D. Axe. *J. Am. Chem. Soc.* 111:5852–5861, 1989.
26. S.R. Wasserman, Y.-T. Tao, and G.M. Whitesides. *Langmuir* 5:1074–1087, 1989.
27. T. Nakagawa, K. Ogawa, and T. Kurumizawa. *Langmuir* 10:525–529, 1994.
28. B. Buszewski, R.M. Gadzata-Kopciuch, M. Markuszewski, and R. Kaliszan. *Anal. Chem.* 69:3277–3284, 1997.
29. T.R. Lee, P.E. Laibinis, J.P. Folkers, and G.M. Whitesides. *Pure Appl. Chem.* 63:821–828, 1991.
30. G. Kataby, T. Prozorov, Y. Koltypin, H. Cohen, C.N. Sukenik, A. Ulman, and A. Gedanken. *Langmuir* 13:6151–6158, 1997.
31. K.J. Stevenson, M. Mitchell, and H.S. White. *J. Phys. Chem. B* 102:1235–1240, 1998.
32. C.W. Sheen, J.-X. Shi, J. Martensson, A.N. Parikh, and D.L. Allara. *J. Am. Chem. Soc.* 114:1514–1515, 1992.

33. J.F. Dorsten, J.E. Maslar, and P.W. Bohn. *Appl. Phys. Lett.* 66:1755–1757, 1995.

34. Y. Gu, Z. Lin, R.A. Butera, V.S. Smentkowski, and D.H. Waldeck. *Langmuir* 11:1849–1851, 1995.

35. D.L. Allara and R.G. Nuzzo. *Langmuir* 1:45–52, 1985.

36. H.O. Finklea, L.R. Robinson, A. Blackburn, and B. Richter. *Langmuir* 2:239–244, 1986.

37. D.L. Allara, A.N. Parikh, and F. Rondelez. *Langmuir* 11:2357–2360. 1995.

38. D.L. Allara and R.G. Nuzzo. *Langmuir.* 1:52–66, 1985.

39. P.E. Laibinis, J.J. Hickman, M.S. Wrighton, and G.M. Whitesides. *Science* 245:845–847, 1989.

40. Y.-T. Tao, G.D. Hietpas, and D.L. Allara. *J. Am. Chem. Soc.* 118:6724–6735, 1996.

41. S.H. Chen and C.W. Frank. *Langmuir* 5:978–987, 1989.

42. Y.G. Aronoff, B. Chen, G. Lu, C. Seto, J. Schwartz, and S.L. Bernasek. *J. Am. Chem. Soc.* 119:259–262, 1997.

43. T.J. Gardner, C.D. Frisbie, and M.S. Wrighton. *J. Am. Chem. Soc.* 117:6927–6933, 1995.

44. J.P. Folkers, C.B. Gorman, P.E. Laibinis, S. Buchholz, G.M. Whitesides, and R.G. Nuzzo. *Langmuir* 11:813–824, 1995.

45. J.E. Chadwick, D.C. Myles, and R.L. Garrell. *J. Am. Chem. Soc.* 115:10364–10365, 1993.

46. W. Gao, L. Dickinson, C. Grozinger, F.G. Morin, and L. Reven. *Langmuir* 12:6429–6435, 1996.

47. J.T. Woodward, A. Ulman, and D.K. Schwartz. *Langmuir* 12:3626–3629, 1996.

48. L. Bertilsson, K. Potje-Kamloth, and H.-D. Liess. *J. Phys. Chem. B* 102:1260–1269, 1998.

49. C.D. Bain, H.A. Biebuyck, and G.M. Whitesides. *Langmuir* 5:723–727, 1989.

50. J.J. Hickman, P.E. Laibinis, D.I. Auerbach, C. Zou, T.J. Gardner, G.M. Whitesides, and M.S. Wrighton. *Langmuir* 8:357–359, 1992.

51. K. Uvdal, I. Persson, and B. Liedberg. *Langmuir* 11:1252–1256, 1995.

52. H.A. Biebuyck, C.D. Bain, and G.M. Whitesides. *Langmuir* 10:1825–1831, 1994.

53. A.C. Ontko and R.J. Angelici. *Langmuir* 14:3071–3078, 1998.

54. M.R. Linford and C.E.D. Chidsey. *J. Am. Chem. Soc.* 115:12631–12632, 1993.

55. A. Bansal, X. Li, I. Lauermann, and N.S. Lewis. *J. Am. Chem. Soc.* 118:7225–7226, 1996.

56. J.M. Buriak and M.J. Allen. *J. Am. Chem. Soc.* 120:1339–1340, 1998.

57. P. DiMillaf, J.P. Folkers, H.A. Biebuyck, R. Harter, G. Lopez, and G.M. Whitesides. *J. Am. Chem. Soc.* 116:2225–2226, 1994.

58. L.H. Dubois, R. Zegarski, and R.G. Nuzzo. *J. Chem. Phys.* 98:678–688, 1993.

59. D.E. Weisshaar, B.D. Lamp, and M.D. Porter. *J. Am. Chem. Soc.* 114:5860–5862, 1992.

60. P. Fenter, A. Eberhardt, and P. Eisenberger. *Science* 266:1216–1218, 1994.

61. J.J. Gerdy and W.A. Goodard. *J. Am. Chem. Soc.* 118:3233–3236, 1996.

62. J.I. Siepmann and I.R. McDonald. *Thin Films* 24:205–226, 1998.

63. T. Sawaguchi, F. Mizutani, and I. Taniguchi. *Langmuir* 14:3565–3569, 1998.

64. C.D. Bain, E.B. Troughton, Y.-T. Tao, J. Evall, G.M. Whitesides, and R.G. Nuzzo. *J. Am. Chem. Soc.* 111:321–335, 1989.

65. K. Shimazu, I. Yag, Y. Sato, and K. Uosaki. *Langmuir* 8:1385–1387, 1992.

66. D.S. Karpovich and G.J. Blanchard. *Langmuir* 10:3315, 1994.

67. T.W. Schneider and D.A. Buttry. *J. Am. Chem. Soc.* 115:12391, 1993.

68. Y.-T. Kim, R.I. McCarley, and A.J. Bard. *Langmuir* 9:1941, 1993.

69. W. Pan, C.J. Durning, and N.J. Turro. *Langmuir* 12:4469–4473, 1996.

70. R.C. Thomas, L. Sun, R.M. Crooks, and A.J. Ricco. *Langmuir* 7:620, 1991.

71. T.T. Ehler, N. Malmberg, and L.J. Noe. *J. Phys. Chem. B.* 101:1268, 1997.

72. M. Buck, F. Eisert, J. Fischer, M. Grunze, and F. Traeger. *Appl. Phys. A* A53:552–556, 1991.

73. R.H. Terrill, T.A. Tanzer, and P.W. Bohn. *Langmuir* 14:845–854, 1998.

74. M.A. Bryant and J.E. Pemberton. *J. Am. Chem. Soc.* 113:8284–8293, 1991.

75. C.A. Widrig, C. Chung, and M.D. Porter. *J. Electroanal. Chem.* 310:335–359, 1991.

76. K. Hu and A.J. Bard. *Langmuir* 14:4790–4794, 1998.

77. G.E. Poirier and E.D. Pylant. *Science* 272:1145–1148, 1996.

78. R. Yamada and K. Uosaki. *Langmuir* 14:855–861, 1998.

79. G.E. Poirier. *Langmuir* 15:1167–1175, 1999.

80. Y. Xia, N. Venkateswaran, D. Qin, J. Tien, and G.M. Whitesides. *Langmuir* 14:363–371, 1998.

81. Z. Hou, N.L. Abbott, and P. Stroeve. *Langmuir* 14:3287–3297, 1998.

82. A. Kumar, N.A. Abbott, E. Kim, H.A. Biebuyck, and G.M. Whitesides. *Acc. Chem. Res.* 28:219–226, 1995.

83. J. Tien, Y. Xia, and G.M. Whitesides. *Thin Films* 24:227–253, 1998.

84. R.G. Nuzzo, F.A. Fusco, and D.L. Allara. *J. Am. Chem. Soc.* 109:2358–2368, 1987.

85. M.D. Porter, T.B. Bright, D.L. Allara, and C.E.D. Chidsey. *J. Am. Chem. Soc.* 109:3559–3568, 1987.

86. A.N. Parikh and D.L. Allara. *J. Chem. Phys.* 96:927–945, 1992.

87. R.G. Nuzzo, L.H. Dubois, and D.L. Allara. *J. Am. Chem. Soc.* 112:558–569, 1990.

88. C. Zubraegel, C. Deuper, F. Schneider, M. Neumann, M. Grunze, A. Schertel, and C. Woell. *Chem. Phys. Lett.* 238:308–312, 1995.

89. C.A. Widrig, C.A. Alves, and M.D. Porter. *J. Am. Chem. Soc.* 113:2805–2810, 1991.

90. C.A. Alves, E.L. Smith, and M.D. Porter. *J. Am. Chem. Soc.* 114:1222–1227, 1992.

91. G.E. Poirier and M.J. Tarlov. *Langmuir* 10:2853–2856, 1994.

92. P. Fenter and P. Eisenberger. *Phys. Rev. Lett.* 70:2447, 1993.

93. C.E.D. Chidsey, G.-Y. Liu, P. Rowntree, and G. Scoles. *J. Chem. Phys.* 91:4421–4423, 1989.

94. N. Camillone III, C.E.D. Chidsey, G.-Y. Liu, T.M. Putvinski, and G. Scoles. *J. Chem. Phys.* 94:8493–8502, 1991.

95. L. Strong and G.M. Whitesides. *Langmuir* 4:546–558, 1988.

96. J.J. Hickman, D. Ofer, C. Zou, M.S. Wrighton, P.E. Laibinis, and G.M. Whitesides. *J. Am. Chem. Soc.* 113:1128–1132, 1991.

97. D.M. Collard and M.A. Fox. *Langmuir* 7:1192–1197, 1991.

98. C.D. Bain, E.B. Troughton, Y.-T. Tao, J. Evall, G.M. Whitesides, and R.G. Nuzzo. *J. Am. Chem. Soc.* 111:321–335, 1989.

99. L.H. Dubois, B.R. Zegarski, and R.G. Nuzzo. *J. Am. Chem. Soc.* 112:570–579, 1990.

100. T.R. Lee, R.I. Carey, H.A. Biebuyck, and G.M. Whitesides. *Langmuir* 10:741–749, 1994.

101. C.E.D. Chidsey and D.N. Loiacono. *Langmuir* 6:709, 1990.

102. P.E. Laibinis, G.M. Whitesides, D.L. Allara, Y.-T. Tao, A.N. Parikh, and R.G. Nuzzo. *J. Am. Chem. Soc.* 113:7152–7167, 1991.

103. H. Sellers, A. Ulman, Y. Shnidman, and J.E. Eilers. *J. Am. Chem. Soc.* 115:9389–9401, 1993.

104. A. Ulman, J.E. Eilers, and N. Tillman. *Langmuir* 5:1147–1152, 1989.

105. J. Hautman and M.L. Klein. *J. Chem. Phys.* 91:4994–5001, 1989.

106. J. Hautman and M.L. Klein. *J. Chem. Phys.* 93:7483–7492, 1990.

107. J. Hautman, J.P. Bareman, W. Mar, and M.L. Klein. *J. Chem. Soc. Faraday Trans.* 87:2031–2037, 1991.

108. W. Mar and M.L. Klein. *Langmuir* 10:188–196, 1994.

109. J.I. Siepman and I.R. McDonald. *Mol. Phys.* 79:457–473, 1993.

110. A.J. Pertsin and M. Grunze. *Langmuir* 10:3668–3674, 1994.

111. K.M. Beardmore, J.D. Kress, N. Gronbech-Jensen, and A.R. Bishop. *Chem. Phys. Lett.* 286:40–45, 1998.

112. G. Nelles, H. Schonherr, M. Jaschke, H. Wolf, M. Schaub, J. Kuther, W. Tremel, E. Bamberg, H. Ringsdorf, and H.-J. Butt. *Langmuir* 14:808–815, 1998.

113. M. Sprik, E. Delamarche, B. Michel, U. Rothlisberger, M.L. Klein, H. Wolf, and H. Ringsdorf. *Langmuir* 10:4116–4130, 1994.

114. Y.-T. Tao, M.-T. Lee, and S.-C. Chang. *J. Am. Chem. Soc.* 115:9547–9555, 1993.

115. E. Sabatani, J. Cohen-Boulakia, M. Bruening, and I. Rubinstein. *Langmuir* 9:2974–2981, 1993.

116. J.M. Tour, L.I. Jones, D.L. Pearson, J.J.S. Lamba, T.P. Burgin, G.M. Whitesides, D.L. Allara, A.N. Parikh, and S.V. Atre. *J. Am. Chem. Soc.* 117:9529–9534, 1995.

117. A.-A. Dhirani, R.W. Zehner, R.P. Hsung, P. Guyot-Sionnest, and L.R. Sita. *J. Am. Chem. Soc.* 118:3319–3320, 1996.

118. T.-W. Li, I. Chao, and Y.-T. Tao. *J. Phys. Chem. B* 102:2935–2946, 1998.

119. C.A. Alves and M.D. Porter. *Langmuir* 9:3507–3512, 1993.

120. G.-Y. Liu, P. Fenter, C.E.D. Chidsey, D.F. Ogletree, P. Eisenberger, and M. Salmeron. *J. Chem. Phys.* 101:4301–4306, 1994.

121. K. Edinger, A. Golzhauser, K. Demota, C. Woll, and M. Grunze. *Langmuir* 9:4–8, 1993.

122. C. Schonenberger, J.A.M. Sondag-Huethorst, J. Jorritsma, and L.G.J. Fokkink. *Langmuir* 10:611–614, 1994.

123. P.G. van Patten, J.D. Noll, and M.L. Myrick. *J. Phys. Chem. B* 101:7874–7875, 1997.

124. J. Thome, M. Himmelhaus, M. Zharnikov, and M. Grunze. *Langmuir* 14:7435–7449, 1998.

125. M. Mrksich and G.M. Whitesides. *Ann. Rev. Bio. Phys. Biomol. Struct.* 25:55–78, 1996.

126. J.P. Folkers, P.E. Laibinis, and G.M. Whitesides. *J. Adhesion Sci. Tech.* 6:1397–1410, 1992.

127. M.K. Chaudhury and G.M. Whitesides. *Science* 255:1230–1232, 1992.

128. G.K. Jennings, J.C. Munro, T.-H. Yong, and P.E. Laibinis. *Langmuir* 14:6130–6139, 1998.
129. F.P. Zamborini and R.M. Crooks. *Langmuir* 14:3279–3286, 1998.
130. Y. Xia and G.M. Whitesides. *Angew. Chem. Int. Ed. Engl.* 37:550–575, 1998.
131. R. Maoz and J. Sagiv. *Langmuir* 3:1034–1044, 1987.
132. R. Maoz and J. Sagiv. *Langmuir* 3:1045–1051, 1987.
133. N. Abbott, A. Kumar, and G.M. Whitesides. *Chem. Mater.* 6:596–602, 1994.
134. Y. Xia, X.-M. Zhao, E. Kim, and G.M. Whitesides. *Chem. Mater.* 7:2332–2337, 1995.
135. M.A. Rampi, O.J.A. Schueller, and G.M. Whitesides. *Appl. Phys. Lett.* 72:1781–1783, 1998.
136. K.L. Prime and G.M. Whitesides. *Science* 252:1164–1167, 1991.
137. K.L. Prime and G.M. Whitesides. *J. Am. Chem. Soc.* 115:10714–10721, 1993.
138. L. Deng, M. Mrksich, and G.M. Whitesides. *J. Am. Chem. Soc.* 118:5136–5137, 1996.
139. C.D. Bain and G.M. Whitesides. *J. Am. Chem. Soc.* 110:6560–6561, 1988.
140. L. Bertilsson and B. Liedberg. *Langmuir* 9:141–149, 1993.
141. J.P. Folkers, P.E. Laibinis, G.M. Whitesides, and J. Deutch. *J. Phys. Chem.* 98:563–571, 1994.
142. Y. Li, J. Huang, R.T.J. McIver, and J.C. Hemminger. *J. Am. Chem. Soc.* 114:2428–2432, 1992.
143. Y. Sato, R. Yamada, F. Mizutani, and K. Uosaki. *Chem. Lett.* 26:987–988, 1997.
144. W.A. Hayes, H. Kim, X. Yue, S.S. Perry, and C. Shannon. *Langmuir* 13:2511–2518, 1997.
145. D.G. Kurth and T. Bein. *Angew. Chem. Int. Ed. Engl.* 31:336–338, 1992.
146. D.G. Kurth and T. Bein. *Langmuir* 9:2965–2973, 1993.
147. J.H. Moon, J.W. Shin, S.Y. Kim, and J.W. Park. *Langmuir* 12:4621–4624, 1996.
148. D.A. Hurt and G.J. Leggett. *Langmuir* 13:2740–2748, 1997.
149. S. Pan, D.G. Castner, and B.D. Ratner. *Langmuir* 14:3545–3550, 1998.
150. H.-J. Himmel, K. Weiss, B. Jager, O. Dannenberger, M. Grunze, and C. Woll. *Langmuir* 13:4943–4947, 1997.
151. S. Lofas and B. Johnsson. *J. Chem. Soc. Chem. Commun.* 1526–1528, 1990.
152. A. Ulman and N. Tillman. *Langmuir* 5:1418–1420, 1989.
153. R.V. Ducvel and R.M. Corn. *Anal. Chem.* 64:337–342, 1992.
154. M. Wells and R.M. Crooks. *J. Am. Chem. Soc.* 118:3988–3989, 1996.
155. G.J. Leggett, C.J. Roberts, P.M. Williams, M.C. Davies, D.E. Jackson, and S.J.B. Tendler. *Langmuir* 9:2356–2362, 1993.
156. J. Wang, J.R. Kenseth, V.W. Hones, J.-B. Green, M.T. McDermott, and M.D. Porter. *J. Am. Chem. Soc.* 119:12796–12799, 1997.
157. Y.W. Lee, J. Reed-Mundell, C.N. Sukenik, and J.E. Zull. *Langmuir* 9:3009–3014, 1993.
158. S.M. Amador, J.M. Pachence, R. Fischetti, J.P.J. McCauley, A.B.I. Smith, and J.K. Blasie. *Langmuir* 9:812–817, 1993.
159. P. Kohli, K.K. Taylor, J.J. Harris, and G.J. Blanchard. *J. Am. Chem. Soc.* 120:11962–11968, 1998.
160. N. Tillman, A. Ulman, and J.F. Elman. *Langmuir* 5:1020–1026, 1989.
161. R.C.J. Horton, T.M. Heine, and D.C. Myles. *J. Am. Chem. Soc.* 119:12980–12981, 1997.
162. N. Balachander and C.N. Sukenik. *Langmuir* 6:1621–1627, 1990.
163. T.S. Koloski, C.S. Dulcey, Q.J. Haralson, and J.M. Calvert. *Langmuir* 10:3122–3133, 1994.
164. G.E. Fryxell, P.C. Rieke, L.L. Wood, M.H. Engelhard, R.E. Williford, G.L. Graff, A.A. Campbell, R.J. Wiacek, L. Lee, and A. Halverson. *Langmuir* 12:5064–5075, 1996.
165. A. Benninghoven, B. Hagenhoff, and E. Niehuis. *Anal. Chem.* 65:630A–640A, 1993.
166. S. Akari, D. Horn, H. Keller, and W. Schrepp. *Adv. Mater.* 7:549–551, 1995.
167. L. Yan, C. Marzolin, A. Terfort, and G.M. Whitesides. *Langmuir* 13:6704–6712, 1997.
168. R.G. Cooks. *Chem. Ind.* 142, 1955.
169. L. Yan, X.-M. Zhao, and G.M. Whitesides. *J. Am. Chem. Soc.* 120:6179–6180, 1998.
170. J. Rao, L. Yan, B. Xu, and G.M. Whitesides. *J. Am. Chem. Soc.* 121:2629–2630, 1999.
171. L. Yan, W.T.S. Huck, X.-M. Zhao, and G.M. Whitesides. *Langmuir* 15:1208–1214, 1999.
172. J. Lahiri, L. Isaacs, J. Tien, and G.M. Whitesides. *Anal. Chem.* 71:777–790, 1999.
173. H.A. Biebuyck and G.M. Whitesides. *Langmuir* 10:4581–4587, 1994.
174. N.B. Larsen, H. Biebuyck, E. Delamarche, and B. Michel. *J. Am. Chem. Soc.* 119:3017–3026, 1997.
175. Y. Xia, D. Qin, and G.M. Whitesides. *Adv. Mater.* 8:1015–1017, 1996.
176. R. Jackman, J. Wilbur, and G.M. Whitesides. *Science* 269:664–666, 1995.
177. Y. Xia, M. Mrksich, E. Kim, and G.M. Whitesides. *J. Am. Chem. Soc.* 117:9576–9577, 1995.
178. N.L. Jeon, K. Finnie, K. Branshaw, and R.G. Nuzzo. *Langmuir* 13:3382–3391, 1997.

179. P.C. Hidber, W. Helbig, E. Kim, and G.M. Whitesides. *Langmuir* 12:1375–1380, 1996.
180. P.C. Hidber, P.F. Nealey, W. Helbig, and G.M. Whitesides. *Langmuir* 12:5209–5215, 1996.
181. A. Bernard, E. Delamarche, H. Schmid, B. Michel, H.R. Bosshard, and H. Biebuyck. *Langmuir* 14:2225–2229, 1998.
182. C.D. James, R.C. Davis, L. Kam, H.G. Craighead, M. Isaacson, J.N. Turner, and W. Shain. *Langmuir* 14:741–744, 1998.
183. G.M. Whiteside and P.E. Laibinis. *Langmuir* 6:87–96, 1990.
184. K.R. Stewart, G.M. Whitesides, H.P. Godfried, and I.F. Silvera. *Rev. Sci. Instrum.* 57:1381–1383, 1986.
185. M. Mrksich and G.M. Whitesides, Ed. *Using Self-Assembled Monalayers that Present Oligo(ethylene glycol) Groups to Control the Interactions of Proteins with Surfaces. Poly(ethylene glycol) Chemistry and Biological Applications*, Vol. 680. Washington, DC: American Chemical Society, 1997, pp. 361–373.
186. M. Nishizawa, M. Shibuya, T. Sawaguchi, T. Matsue, and I. Uchida. *J. Phys. Chem.* 95:9042–9044, 1991.
187. P.L. Burn, A. Kraft, D.R. Baigent, D.D.C. Bradley, A.R. Brown, R.H. Friend, R.W. Gymer, A.B. Holmes, and R.W. Jackson. *J. Am. Chem. Soc.* 115:10117–10124, 1993.
188. L. Dai, H.J. Griesser, X. Hong, A.W.H. Mau, T.H. Spurling, Y. Yang, and J.W. White. *Macromolecules* 29:282–287, 1996.
189. P.L. Burn, A.B. Holmes, A. Kraft, D.D.C. Bradley, A.R. Brown, R.H. Friend, and R.W. Gymer. *Nature* 356:47–49, 1992.
190. B.G. Healey, S.E. Foran, and D.R. Walt. *Science* 269:1078–1080, 1995.
191. W. Knoll, M. Matsuzawa, A. Offenhausser, and J. Ruhe. *Isr. J. Chem.* 36:357–369, 1996.
192. Y. Liu, M. Zhao, D.E. Bergbreiter, and R.M. Crooks. *J. Am. Chem. Soc.* 119:8720–8721, 1997.
193. R. Langer and J.P. Vacanti. *Science* 260:920–926, 1993.
194. S.A. Sukhishvili and S. Granick. *Langmuir* 13:4935–4938, 1997.
195. Y. Liu, M.L. Bruening, D.E. Bergbreiter, and R.M. Crooks. *Angew. Chem. Int. Ed. Engl.* 36:2114–2116, 1997.
196. G. Decher. *Science* 277:1232–1237, 1997.
197. P.T. Hammond and G.M. Whitesides. *Macromolecules* 28:7569–7571, 1995.
198. J.M. Stouffer and T.J. McCarthy. *Macromolecules* 21:1204–1208, 1988.
199. T.J. Lenk, V.M. Hallmark, J.F. Rabolt, L. Haussling, and H. Ringsdorf. *Macromolecules* 26:1230–1237, 1993.
200. L. Rozsnyai and M.S. Wrighton. *J. Am. Chem. Soc.* 116:5993–5994, 1994.
201. G. Mao, D.G. Castner, and D.W. Grainger. *Chem. Mater.* 9:1741–1750, 1997.
202. S.L. Clark, M.F. Montague, and P.T. Hammond. *Macromolecules* 30:7237–7244, 1997.
203. E. Kim, G.M. Whitesides, L.K. Lee, S.P. Smith, and M. Prentiss. *Adv. Mater.* 8:139–142, 1996.
204. A. Karim, J.F. Douglas, B.P. Lee, S.C. Glotzer, J.A. Rogers, R.J. Jackman, E.J. Aims, and G.M. Whitesides. *Phys. Rev. E: Stat. Phys. Plasmas, Fluids, Relat. Interdiscip. Top.* 57:R6273–R6276, 1998.
205. M. Boltau, S. Walheim, M. Jurgen, G. Krausch, and U. Steiner. *Nature* 391:877–879, 1998.
206. S. Zhang, L. Yan, M. Altman, M. Lässie, M. Nugent, F. Frankel, D.A. Lauffenburger, G.M. Whitesides, and A. Rich. *Biomaterials* 20:1213–1220, 1999.
207. S.R. Holmes-Farley, R.H. Reamey, T.J. McCarthy, J. Deutch, and G.M. Whitesides. *Langmuir* 1:725–740, 1985.
208. S.R. Holmes-Farley and G.M. Whitesides. *Langmuir* 2:62–76, 1986.
209. S.R. Holmes-Farley, C. Bain, and G.M. Whitesides. *Langmuir* 4:921–937, 1988.
210. C.D. Bain and G.M. Whitesides. *J. Am.Chem. Soc.* 111:7164–7175, 1989.
211. C.D. Bain and G.M. Whitesides. *Langmuir* 5:1370–1378, 1989.
212. J. Lahiri, E. Ostuni, and G.M. Whitesides. *Langmuir* 15:2055–2060, 1999.
213. C. Cao, A.Y. Fadeev, and T.J. McCarthy. *Langmuir* 17:757–761, 2001.
214. C.K. Luscombe, H.W. Li, W.T.S. Huck, and A.B. Holmes. *Langmuir* 19:5273–5278, 2003.
215. V. Chechik, R.M. Crooks, and C.J.M. Stirling. *Adv. Mater.* 12:1161, 2000.
216. T.P. Sullivan and W.T.S. Huck. *Eur. J. Org. Chem.* 17–29, 2003.
217. B. Vaidya, J.H. Chen, M.D. Porter, and R.J. Angelici. *Langmuir* 17:6569, 2001.
218. H. Schönherr, C. Feng, and A. Shovsky. *Langmuir* 19:10843–10851, 2003.
219. A.C. Arias, N. Corcoran, M. Banach, R.H. Friend, J.D. MacKenzie, and W.T.S. Huck. *Appl. Phys. Lett.* 10:1695, 2002.
220. N. Nath and A. Chilkoti. *Adv. Mater.* 14:1243–1247, 2002.

221. I. Willner, R. Blonder, and A. Dagan. *J. Am. Chem. Soc.* 116:9365, 1994.

222. B. Zhao, N.O.L. Viernes, J.S. Moore, and D.J. Beebe. *J. Am. Chem. Soc.* 124:5284, 2002.

223. B. Zhao, J.S. Moore, and D.J. Beebe. *Langmuir* 19:1873, 2003.

224. B. Zhao, J.S. Moore, and D.J. Beebe. *Anal. Chem.* 74:4259, 2002.

225. J. Lahann, S. Mitragotri, T.N. Tran, H. Kaido, J. Sundaram, I.S. Choi, S. Hoffer, G. Somorjai, and R. Langer. *Science* 299:371, 2003.

226. M.N. Yousaf and M. Mrksich. *J. Am. Chem. Soc.* 121:4286–4287, 1999.

227. M.N. Yousaf, E.W.L. Chan, and M. Mrksich. *Angew. Chem. Int. Ed.* 39:1943–1946, 2000.

228. M.N. Yousaf, B.T. Houseman, and M. Mrksich. *Angew. Chem. Int. Ed.* 40:1093–1096, 2001.

229. C.D. Hodneland and M. Mrksich. *Langmuir* 13:6001–6003, 1997.

230. C. Henke, C. Steinem, A. Janshoff, G. Steffan, H. Luftman, M. Sieber, and H. Galla. *Anal. Chem.* 68:3158–3165, 1996.

231. J. Su and M. Mrksich. *Angew. Chem. Int. Ed.* 41:4715–4718, 2002.

232. J. Su and M. Mrksich. *Langmuir* 19:4867–4870, 2003.

233. O. Prucker and J. Ruhe. *Langmuir* 14:6893, 1998.

234. V.V. Tsukruk. *Adv. Mater.* 13:95, 2001.

235. S. Edmondson, V.L. Osborne, and W.T.S. Huck. *Chem. Soc. Rev.* 33:14–23, 2004.

236. J. Ruhe and N. Knoll. *J. Macromol. Sci. Polymer Rev.* 42:91–138, 2002.

237. K. Matyjaszewski, P.J. Miller, N. Shukla, B. Immaraporn, A. Gelman, B.B. Luokala, T.M. Siclovan, G. Kickelbick, T. Vallant, and H. Hoffman. *Macromolecules* 32:8716–8724, 1999.

238. R. Jordon and A. Ulman. *J. Am. Chem. Soc.* 120:243–247, 1998.

239. R. Jordon, A. Ulman, J.F. Kang, M.H. Rafailovich, and J. Sokolov. *J. Am. Chem. Soc.* 121:1016–1022, 1999.

240. M. Husemann, E.E. Malmstrom, M. McNamara, M. Mate, D. Mecerreyes, D.G. Benoit, J.L. Hedrick, P. Mansky, E. Huang, T.P. Russell, and C.J. Hawker. *Macromolecules* 32:1424–1431, 1999.

241. M. Husemann, D.G. Benoit, J. Frommer, M. Mate, W.D. Hinsberg, and C.J. Hawker. *J. Am. Chem. Soc.* 122:1844–1845, 2000.

242. C. Harrison, R.M. Chaikin, D.A. Huse, R.A. Register, D.H. Adamson, A. Daniel, E. Huang, P. Mansky, T.P. Russell, C.J. Hawker, D.A. Egolf, I.V. Melnikov, and E. Bodenschatz. *Macromolecules* 33:857–865, 2000.

243. M. Weck, J.J. Jackiw, R.R. Rossi, P.S. Weiss, and R.H. Grubbs. *J. Am. Chem. Soc.* 121:4088–4089, 1999.

244. J.B. Kim, M.L. Bruening, and G.L. Baker. *J. Am. Chem. Soc.* 122:7616–7617, 2000.

245. R.R. Shah, D. Mecerreyes, M. Husemann, I. Rees, N.L. Abbott, C.J. Hawker, and J.L. Hedrick. *Macromolecules* 33:597–605, 2000.

246. B. Zhao and W.J. Brittain. *Macromolecules* 33:8813–8820, 2000.

247. X. Hunang and M.J. Wirth. *Macromolecules* 32:1694–1696, 1999.

248. D.M. Jones and W.T.S. Huck. *Adv. Mater.* 13:1256–1259, 2001.

249. D.M. Jones, A.A. Brown, and W.T.S. Huck. *Langmuir* 18:1265–1269, 2002.

250. D.M. Jones, J.R. Smith, W.T.S. Huck, and C. Alexander. *Adv. Mater.* 14:1130–1134, 2002.

251. K.F. Böhringer. *J. Micromech. Microeng.* 13:S1–S10, 2003.

252. D.L. Huber, R.P. Manginell, M.A. Samara, B.I. Kim, and B.C. Bunker. *Science* 301:352–354, 2003.

Chapter 19

Layered Polyelectrolyte Assemblies

X. Arys, A.M. Jonas, A. Laschewsky, R. Legras, and F. Mallwitz

CONTENTS

I. INTRODUCTION

In recent years, much progress has been achieved in the preparation and characterization of organic and hybrid organic–inorganic ultrathin multilayers. The growing interest in these systems, both from the fundamental as well as from the applied side, is partly due to the unusual physical properties of nanostructured materials, and to the potential applications resulting from these properties, in particular in the field of integrated molecular optics and electronics. Size quantization, for instance, occurs when electron–hole pairs are confined in domains whose dimensions are comparable to the wavelength of the de Broglie electron, and the mean free path of excitons [1]. This in turn enables the control of the emission color of light emitting diodes (LEDs) [2]. Another reason for the current growing interest in organic multilayer films is the similarity between the methods for obtaining such molecular assemblies and the principles governing the self-organization of organic molecules in natural systems [3,4]. Being able to control the spatial arrangement of biological molecules on a nanometer scale opens the way to numerous applications, among which molecular recognition [5] and multistep catalysis [6] can be cited. At the frontier between biology and physics, molecular bioelectronics is also attracting interest [7]. An example of development in this field is the fabrication of systems that mimic photosynthesis, the goal being to design artificial systems for the efficient conversion of solar energy into chemical or electrical energy [8–10].

The oldest technique for the fabrication of multilayer films has been invented in the 1920s by Langmuir and Blodgett [11–13]. A renaissance of this technique occurred in the 1960s under the impetus of Kuhn and coworkers [14]. Significant progress in the understanding of monolayers assemblies has been gained since the late 1970s [15], in part due to the development of new characterization techniques. In order to build-up Langmuir–Blodgett (LB) films, amphiphilic molecules are spread on water and compressed until they form a solid-like two-dimensional phase. A film of these molecules is then transferred onto a solid surface by dipping a substrate through the air–water interface. Multilayers are realized by repetitive dipping. LB films are highly ordered and have controlled, uniform thickness. However, the requirements for substrates are stringent: they must be smooth, homogeneous [16], and have a regular shape. Furthermore, expensive equipment is required, and the coating of areas larger than 10 cm^2 is demanding. Even more, LB-multilayers are metastable, and as a result have a limited stability against solvents or thermal treatments, although strategies have been designed to overcome these problems [17–19]. Finally, defects formed in a given layer are difficult to cover up by subsequent layers. Although possible, multilayer heterostructures, that is, composed of layers of different amphiphiles, or noncentrosymmetric multilayers are not easily realized. Noncentrosymmetric assemblies are necessary for second-order nonlinear optical (NLO), piezo- and pyro-electric effects.

Another approach to assemble layered structures was presented in 1980s by Netzer and Sagiv [20]. Their strategy is based on a two-step sequence involving the chemisorption of a monolayer followed by chemical activation of the exposed surface, in order to provide polar adsorption sites for the anchoring of the next monolayer. The reagents used had a trichlorosilane head group for the chemisorption step and a reactive functionalized end group for the creation of a hydroxylated surface after the chemical activation step. Examples of such activation reactions are the conversion of double bonds through hydroboration–oxidation [21], LiAlH$_4$ reduction of a surface ester group [22], or photolysis [23]. These covalently linked multilayers are highly ordered, and substrates of any size or shape can be coated by this technique. They are stable against solvents or thermal treatments. Noncentrosymmetric assemblies are easily obtained [24]. Nevertheless, the high specificity of the chemisorption step imposes major requirements on the chemical nature and homogeneity of the substrate; this may limit the practical potential of this technique. The optimum build-up requires a nearly quantitative surface activation step; otherwise defects will appear and grow with layer number. This prerequisite often limits the number of layers to small values. Notwithstanding, it has been reported that, in some instance, thick films with up to 500 layers could be built [22]. Finally, trichlorosilane derivatives are hard to handle, due to their moisture sensitivity. Replacing trichlorosilane derivatives by methoxy- or ethoxy-silanes, allows minimizing this problem, at the expense of a much slower reaction rate [22].

Concomitant to the development of the trichlorosilane technique, a method for the chemisorption of monolayers of disulfides on gold was developed by Nuzzo and Allara [25]. It has been rapidly generalized to the adsorption of monolayers of alkanethiols on noble metal surfaces [22,26]. Using ω-mercaptoalkanoic acid, and converting the acid surface to a copper salt on which another thiol monolayer can adsorb, Ulman and coworkers showed that this chemical self-assembling technique is also useful for multilayers build-up [27,28]. Alternatively α, ω-dithiols and colloids, like CdS nanoparticles, may be used [29].

In 1988, Mallouk and coworkers showed that multilayer films could be prepared simply by sequential complexation of Zr^{4+} and α, ω-bisphosphonic acid [30]. The technique has been extended to the complexation of organophosphonates with a variety of other metals [31,32], and even to completely different metal–ligand systems [33–38]. The multilayers grown by this technique are well ordered, robust, and relatively easy to prepare. Substrates of any size or shape can be covered, and even noncentrosymmetric multilayers may be obtained [39]. Based on very specific interactions, the technique is, however, restricted to a narrow class of chemical compounds.

In 1966, Iler presented a technique for building films of controlled, uniform thickness by the alternate adsorption of positively and negatively charged colloidal particles [40]. Although a few other singular attempts have been reported afterwards [41–43], the development of the electrostatic self-assembly (ESA) technique is mainly due to the work of Decher and coworkers. In their first article, Decher and Hong used anionic and cationic bipolar amphiphiles containing rigid biphenyl cores [44]. The adsorption of such molecules leads to a surface charge reversal, and multilayer assemblies of both compounds can thus be obtained by alternatively dipping a substrate in solutions of the anionic and cationic amphiphiles. In the same year, they published an article extending the method to polyelectrolytes [45]. The crucial factor for successful deposition is the surface charge reversal upon adsorption, which can usually be obtained by a proper choice of deposition parameters. In the meanwhile, many charge-bearing molecules or particles have been deposited by this technique, and numerous examples of successful deposition can be found in the literature. Just to name a few, films have been prepared by ESA from proteins [46], DNA [47], dyes [48], inorganic platelets [49], latex particles [50], dendrimers [51], and even viruses [52]. The thickness of the coatings can be adjusted by changing processing parameters, like ionic strength, pH, or nature of the solvent. Substrates of any size or shape are suited, if they have a charged surface. As methods exist to bring charge to virtually any type of organic or inorganic surface, there is practically no limitation to the nature of the substrate. The versatility of the technique has triggered an abundant and rapidly increasing literature, scattered in a plethora of journals, so that it is opportune to update the existing reviews of the field [53–59].

This success of ESA is probably also linked to the ease with which stable films with precision in the range of a few Ångstrøms can be built by this method, without expensive equipment. Accordingly, this technique can be referred to as a "molecular beaker epitaxy" [60]. There seems to be no limitation to the maximum number of layers that can be deposited, and films with up to 1000 layers have been realized [61]. This is partly due to the self-healing properties of the multilayer build-up, when polyelectrolytes are used. Further, the process is relatively fast, can be easily automated [62,63] (our group has such an automated system at its disposal. It has been purchased from Riegler & Kirstein GmbH [Wiesbaden, Germany]), and is *a priori* environment friendly, since water is usually used as solvent. Finally, aperiodic multilayers are easily prepared [64]. Among the drawbacks of the technique, the structure of the films is in general fuzzy [53], meaning that there is normally an important interpenetration between neighboring layers. The magnitude of the interpenetration is of the order of the layer thickness [65]. Thus, the term "layer pair" should preferably be used instead of "bilayer" to denote a positive–negative pair of molecules, since a structural subunit composed of well-defined bilayers often does not exist [66]. Although some exceptions have been mentioned, films made by ESA are usually centrosymmetric. This limits the scope of potential applications. A possible way to generate noncentrosymmetric coatings and to obtain films made up of only one type of polyelectrolyte is to combine the ESA technique with a chemical activation step [67].

Subsequently, several authors have shown that other interactions besides the electrostatic interactions between oppositely charged molecules could be used to prepare multilayer assemblies (Figure 1). For example, Rubner et al. have demonstrated the successful growth of films made of polyaniline (PAn) alternating with nonionic water-soluble polymers [68]. Four different nonionic water-soluble polymers were used: poly(vinyl pyrrolidone) (PVP), poly(vinyl alcohol) (PVA), poly(acrylamide) (PAAm), and poly(ethylene oxide) (PEO). The authors presumed that the multilayer build-up is due to hydrogen bonding, supporting this hypothesis by infrared (IR) spectroscopy showing the formation of such bonds in the case of the PAn/PVP system. Films have also been prepared with PVP and poly(acrylic acid) (PAA) [69,70], for which again hydrogen bonding interactions were identified by IR spectroscopy. Interestingly, the strong interaction between both polymers led to the formation of a precipitate upon mixing of both solutions. On the other hand, combinations involving polymers with hydrogen bonding capabilities, such as PVP/PVA, PAAm/PVP, PAAm/PVA, and PAAm/PEO were attempted for multilayer build-up, however, without success [71]. Multilayer build-up based on hydrogen bonding interactions was also combined with ESA to produce multilayers [72].

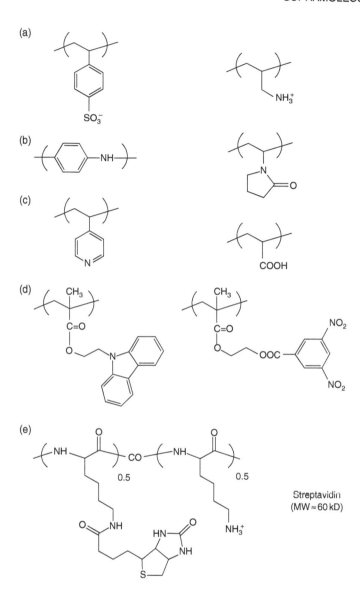

Figure 1   Examples of polymer pairs suited for layer-by-layer assembly. (a) By electrostatic interactions: PSS/PAH; (b) by H-bonding: polyaniline/PVP (from W.B. Stockton and M.F. Rubner. *Mater. Res. Soc. Proc.* 369:587, 1994. With permission); (c) by H-bonding: poly(4-vinyl pyridine)/PAA (from L. Wang, Y. Fu, Z. Wang, Y. Fan, and X. Zhang. *Langmuir* 15:1360, 1999. With permission); (d) by electron-donor–electron-acceptor interaction: poly(2-(9-carbazolyl)ethyl methacrylate)/poly(2-(3,5-dinitrobenzoyl)oxyethyl methacrylate) (from Y. Shimazaki, M. Mitsuishi, S. Ito, and M. Yamamoto. *Langmuir* 14:2768, 1998. With permission); (e) by specific interaction: biotinylated poly(lysine)/streptavidin. (From T. Cassier, K. Lowack, and G. Decher. *Supramol. Sci.* 5:309, 1998. With permission.)

Another approach to multilayer assemblies of polymers was demonstrated by Shimazaki et al. [73,74], who used the sequential adsorption of polymers having, respectively, electron donating and electron accepting pendant groups.

Finally, we cite as examples of multilayers based on interactions other than electrostatic ones, protein multicomponent films, based on specific interactions between the biological molecules or between these biological molecules and polyelectrolytes. For example, immunoglobulin G can be assembled with the anionic poly(styrene sulfonate) (PSS) at pH values above and below its isoelectric point, thus implying that this process is not electrostatically driven [75]. Also, layer-by-layer deposition of avidin and biotin-labeled poly(amine)s leads to the formation of multilayer assemblies through avidin–biotin complexation [46,76–78], even in the case of electrostatic repulsion arising from the net positive charges of avidin (isoelectric point at pH 9.0 to 10) and poly(amine)s [76,77]. The assembly of glucose oxidase multilayers through the use of bispecific antibody interlayer [79], and multilayer build-up of concanavalin A with glycogen [80] are based on the occurrence of specific interactions, too. By contrast, attempts to build multilayers of poly(uridylic acid) and poly(adenylic acid) were unsuccessful, probably because the electrostatic repulsion between these polyanions dominates the binding energy arising from pairing the nuclear bases [81].

Interestingly, ESA bears much formal similarities with the successive ionic layer adsorption and reaction (SILAR) method originally developed in 1985 by Nicolau et al. [82,83]. This technique allows to obtain thin inorganic films of controlled thickness based on successive adsorption from aqueous solutions of small ions on selected substrates.

II. A CASE STUDY: PAH/PSS MULTILAYERS

As an introductory example, we will summarize the literature on the build-up of multilayers made of poly(allylamine hydrochloride) (PAH, Figure 2) and PSS (Figure 2). This polyelectrolyte pair is among the first systems employed for ESA, and has been up to now the most thoroughly studied system [53,65,75,84–116]. The role of this pair of polyelectrolytes as model system is somewhat unfortunate because PAH is a weak polyelectrolyte. Indeed, the dependence of its charge on the experimental conditions, such as pH, renders the comparison and interpretation of the results obtained by different authors difficult. Another complication may result from the hydrogen bonding abilities of the primary amine groups of PAH. Fortunately, no such complications are expected with the strong polyelectrolyte PSS.

The principle of the layer-by-layer growth is illustrated in Figure 2. A substrate is alternately dipped in dilute solutions of PSS and PAH. If the surface charge of the substrate is positive, the cycle is initiated by dipping the substrate in the PSS solution. Conversely, if the surface charge of the substrate is negative, the first dipping occurs in the PAH solution. Note that two polyelectrolyte layers, a polyanionic, and a polycationic one, are deposited per dipping cycle. The samples are rinsed after each immersion in a polyelectrolyte solution. Without the rinsing steps, an adhering layer of solution would be left on the surface of the substrate, leading to coprecipitation upon immersion in the following polyelectrolyte solution, and eventually to the incorporation of precipitated particles in the films [44]. Drying is sometimes applied after each rinsing step.

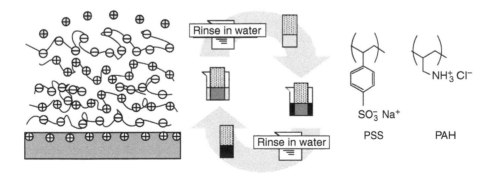

Figure 2 Left: Schematic molecular representation of polyelectrolyte multilayers. Center: Layer-by-layer ESA: films are alternately dipped in a polycation and a polyanion solution. A rinsing step follows every adsorption step. Right: Chemical structure of standard polyelectrolytes PAH and PSS.

As presented in the introduction, the nature of the substrate used for the above-described electrostatic layer-by-layer deposition is not crucial, as long as it is charged. For example, PAH/PSS multilayers have been prepared on plasma-treated glass [84,85], as cleaved mica [86], silicon functionalized by 3-aminopropyldimethoxysilane [87], surface-oxidized poly(4-methyl-1-pentene) [88], chemically modified poly(chlorotrifluoroethylene) (PCTFE) [89,90], plasma-modified poly(tetrafluoroethylene) (PTFE) [91,92], chemically modified as well as untreated poly(ethylene terephtalate) (PET) [93], gold surfaces modified with mercaptopropionic acid [75], and many others. PAH/PSS multilayers have also been grown on chemically modified polystyrene [94] and melamine formaldehyde latex particles [95,96]. Multilayer assembly on latex particles was either accomplished by adsorption from solutions of high polyelectrolyte concentration with intermediate centrifugation cycles in water as a washing procedure, or by adsorption without centrifugation, but at polyelectrolyte concentrations just beyond the onset of ζ-potential saturation (see below).

The deposition process has been followed by UV spectroscopy [87] and quartz crystal microbalance (QCM) [75,97] indicating, for carefully selected process parameters, a linear growth of the amount of material deposited with dipping cycle. Ellipsometry [98], x-ray reflectometry (XRR) [87], and surface plasmon resonance [75] showed a concomitant increase in film thickness, with 1 to 10 nm deposited per dipping cycle, depending on the selected set of processing parameters. The growth of PAH/PSS multilayers on colloidal particles could be followed by dynamic light scattering, single particle light scattering, and fluorescence intensity measurements [94,95]. There seems to be no limit to the maximum number of PAH/PSS layer pairs that can be deposited, and assemblies with up to 200 layers were reported [86].

Although multilayers can be grown on many different supports, some substrates cause reduced growth steps during the first dipping cycles [75,86]. Similarly, the roughness of the first deposited polyelectrolyte layers is close to that of the substrate, but further dipping cycles result in a smoother film surface, at least for carefully chosen processing parameters [86]. In the case of PAH/PSS multilayers, there is no consensus in the literature whether the substrate influences the layer-by-layer deposition process of PAH/PSS multilayers beyond the onset of regular film growth as a function of the number of dipping cycles (see also Section III).

X-ray photoelectron spectroscopy (XPS) has been used to measure the nitrogen–sulfur ratio of the PAH/PSS multilayers, as a function of the number of deposited layers [88,92,93]. This ratio shows a pronounced odd–even trend with the number of deposited layers, being higher when PAH is the outermost layer, and lower when PSS is the outermost layer. This trend is maintained even for high numbers of layers, and reveals an effective stratification of the multilayers. Contact angle measurements [88,92,93] also show a pronounced odd–even trend in the measured angles.

If the films prepared by ESA had a well-defined multilayered structure, reflectometry techniques, like XRR and neutron reflectometry (NR), should allow the determination of the periodically varying refractive index profile along the surface normal [117], due to the well-defined supramolecular structure. Nevertheless, Bragg peaks are observed neither by XRR, nor by NR (when all PSS layers are deuterated) [99,100]. In the latter case at least, the reason for the absence of Bragg peaks cannot be ascribed to the lack of contrast between hydrogenated PAH and deuterated PSS layers, since neutron scattering lengths for hydrogen and deuterium are drastically different [118]. However, when deuterated PSS layers are separated by one or more nondeuterated PSS layers, Bragg reflections are seen by NR [53,65,99,100]. From these measurements, the interfacial width (gaussian width) between neighboring polymer layers was determined to be $\sim$1.9 and $\sim$3 nm, respectively, for films prepared from 2 [100] and 3 M NaCl [65]. This important interfacial width is responsible for the absence of Bragg peaks in XRR measurements and in NR measurements when all layers are deuterated (see Figure 3). The interfacial width must be considered as originating from both the interfacial roughness (or interfacial waviness) and the polymer interdiffusion. Assuming that the interfacial roughness between two neighboring layers is equal to the roughness of the film–air interface, the interdiffusion of two successive polyelectrolyte layers was estimated to be $\sim$1.2 nm for films deposited from 2 M NaCl. Off-specular reflectivity measurements would enable the disclosure of the relative contributions

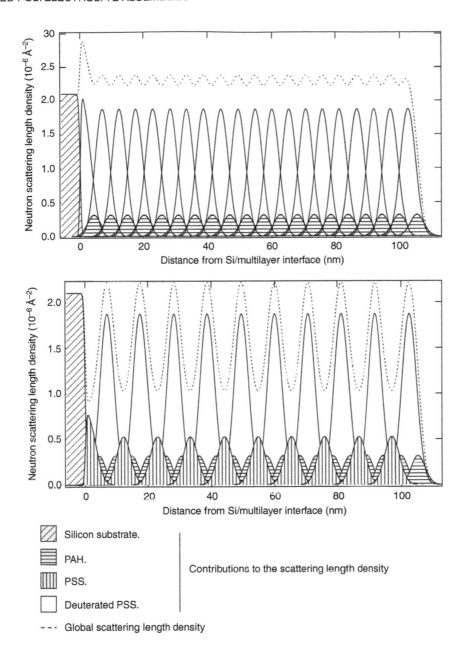

Silicon substrate.

PAH.

PSS.

Deuterated PSS.

Contributions to the scattering length density

--- Global scattering length density

Figure 3 Top: neutron scattering length density profile for a Si/{PSS-deuterated/PAH}$_{20}$ multilayer. Bottom: neutron scattering length density profile for a Si/{PSS-deuterated/PAH/PSS/PAH}$_{10}$ multilayer. No Bragg peaks are observed by NR for the film with all PSS layers deuterated, because oscillations in the scattering length density are too weak. The scattering density profile corresponding to the film with every second PSS layer deuterated shows marked oscillations; accordingly, Bragg peaks are experimentally observed. (Data adapted from J. Schmitt, T. Grünewald, G. Decher, P.S. Pershan, K. Kjaer, and M. Lösche. *Macromolecules* 26:7058, 1993. With permission.)

of interfacial waviness and polymer interdiffusion to the interfacial width [119]. Contact angle measurements [88,92,93] support the hypothesis of strong interpenetration of neighboring layers. Indeed, these data indicate that wettability is controlled by the outermost layer when the layers are sufficiently thick and by at least the two final layers when the layers are thinner, suggesting strong interdiffusion of the layers.

The thickness of the PAH and PSS layers have been determined by NR to be respectively $\sim$2 and $\sim$3.5 nm for films deposited from 2 M NaCl and to be respectively $\sim$1.8 and $\sim$5.8 nm for films prepared from 3 M NaCl [65,100]. Determining a stoichiometry from these results is not straightforward, but it seems that the PAH:PSS ratio is $\geq$1. The stoichiometry was determined from XPS measurements to give an ammonium to sulfate ion ratio ranging from $\sim$1/1 to $\sim$2/1, apparently depending on the nature of the substrate [88,92,93]. We stress that these ratios relate the number of PAH to PSS repeat units, and not directly the number of cationic to anionic groups, which depends on the degree of protonation of PAH.

If the complexation of the anionic and cationic groups is not stoichiometric, some extrinsic charge compensation should occur, and the presence of counterions in the multilayers is thus expected. Unfortunately, NR measurements are not able to determine whether inorganic salts are included in the deposited films [100]. Most of the time their presence could also not be unambiguously determined by XPS [88,92,93], due to problems of sensibility and possible removal of HCl during the measurement, which is performed under ultra-high vacuum. However, 0.8% molar fraction of Na has been reported by an author [101]. This number does not exactly correspond to the molar fraction of Na inside the multilayers, since XPS measurements probe both the inner layers, and the outer layer (this latter layer must contain counterions). Furthermore, XPS is more sensitive to the outer layer than to the inner layers. Measurements of the electrical properties of the multilayers also suggest the presence of mobile ions in the films [101]. Ions can also be trapped in the multilayers, especially in the case of deposition from solutions of high ionic strength. The presence of $MnCl_2$ has indeed been detected by XPS when this salt was added to the PSS solution used for the multilayer build-up [93].

It has been calculated from a detailed analysis of NR data that much water is present in these films [100]: about six water molecules are bound per PSS monomer and one water molecule per PAH monomer, so that water molecules occupy 40% of the film volume. Dehydration of the film leads to shrinking, but a significant part of the volume previously occupied by water seems to remain unfilled by the polyelectrolyte chains.

Of key importance for the electrostatic layer-by-layer deposition process is the occurrence of reproducible surface charge reversal upon polyelectrolyte adsorption, whatever the number of dipping cycles. For PSS concentrations larger than approximately 5 mg/l, it has been shown by surface force measurements (SFA) that the charge of a dioctadecyldimethylammonium covered mica surface was reversed upon PSS adsorption [102]. The thickness of the monolayer was determined to be in the range of 1–3 nm, depending on the molar mass. The adsorption of PAH on pure mica from 0.01 monomol/l solutions containing 1 mM NaCl has also been studied [103]. Adsorbed layers were $\sim$0.7 nm thick in air. In water, mobile chains extended as far as 2.5–4 nm from the surface. Surface charge reversal occurred, as could be demonstrated by measuring the interaction between the polycation monolayer and a bare mica surface. A surface potential of $\sim$34 mV has been estimated by fitting the experimental data. Similarly, surface charge reversal upon PSS adsorption on the PAH-covered mica surface could be evidenced by measuring the interaction between the polycation monolayer and a substrate covered with a polycation–polyanion bilayer. The surface potential was estimated to range from -77 to -98 mV, depending on the NaCl concentration. Approaching two bilayer-covered, PSS-terminated, substrates at sufficiently small distances reveals an attractive segment–segment interaction, most likely resulting from ion pairs formation between anionic and cationic monomer units. This suggests an intermingling of the layers, with polycation and polyanion chains dangling in water. In agreement with the preceding results, a higher charge density when PSS is the outer layer than when it is PAH has been suggested by more pronounced shifts in the apparent pKa value for the titration of dyes inserted in PAH/PSS multilayers [104].

PAH/PSS multilayers deposited on latex particles offer the possibility of electrophoretic mobility measurements. The electrophoretic mobilities were measured as a function of pH and ionic strength (electrophoretic fingerprinting) [105], showing that, for pH values below approximately 8, the surface appears to be positively charged upon PAH adsorption. At higher pH, the negative charge of the PAH-covered polystyrene latex results from deprotonation of PAH, allowing approximately one-third of the original negative surface charge of the latex to be seen. A very pronounced negative surface charge is observed following PSS adsorption for (nearly) all the pH range investigated ($3 \leq pH \leq 10$). At high pH, an apparent increase of the PSS charge of approximately one-third of the value seen at lower pH is observed, due to deprotonation of the PAH at high pH, and subsequent freeing of some of the PSS charged groups. The influence of the next PAH layer is fully consistent with the previous results. The thickness of the top hairy layer was determined to be of the order of 1 nm, and counterion adsorption to the charged groups of the top layer could be observed. pK values for PAH were estimated to be in the range between 9 and 10, and to lay between 3.5 and 4.5 for PSS (which is higher than usually accepted). pK values for PAH are in agreement with values reported by other authors [120,121]. The multilayer film growth was followed by electrophoresis, for films grown from PSS and PAH solutions containing 0.5 M NaCl [94–96]. The measured mobilities were converted into a ζ-potential using the Smoluchowski relation. The ζ-potential showed a pronounced odd–even trend with the number of deposited layers. Irrespective of the number of dipping cycles, ζ-potential values for latex particles having PAH or PSS as the outermost layer were respectively $\sim$40 and $\sim$−40 mV.

Finally, studies of the Förster resonance energy transfer between 6-carboxyfluorescein (6-CF) and rhodamine B-labeled melamine formaldehyde particles covered by PAH/PSS multilayers showed that 6-CF interacts with PAH when the outer polyelectrolyte layer is PAH [96]: upon increasing the 6-CF concentration, an increase of the fluorescence intensity is observed only above a critical concentration, assumed to correspond to the saturation of the charged groups of PAH by the anionic 6-CF. This phenomenon has been used to titrate the number of amino sites of PAH not interacting with PSS in the polyelectrolyte multilayer film; it has been estimated as 1.6/nm^2.

Considering the values of ζ-potential presented by different authors, a wide scatter is observed in the values reported, even in the values reported for PAH and PSS only. This is partly due to the fact that the multilayers were prepared under different conditions, and in part due to difficulties linked to the techniques used for determining ζ-potentials. Nevertheless, charge reversal upon adsorption has been clearly demonstrated.

Obviously, many parameters can be adjusted in the layer-by-layer deposition process: dipping time, polyelectrolyte concentration, pH, ionic strength, nature of the solvent, molar mass of the polyelectrolyte, presence or absence of a drying step.... In the following, we will review the influence of some of these parameters on the growth of PAH/PSS multilayers.

A proper choice of the dipping time presupposes a study of the adsorption kinetics. *In situ* measurements of the adsorption of PAH (from a 0.01 M solution with 0.5 M MnCl$_2$ added) on a PSS-coated optical waveguide showed that the adsorption is practically completed after 3 min [106]. It was shown that the kinetics of adsorption are not transport limited. Similar results hold for subsequent polyelectrolyte adsorption. Ellipsometric measurements of the thickness of adsorbed PSS and PAH layers during film assembly and *in situ* QCM measurements confirm that the adsorption proceeds mainly during the first few minutes [86,97]. *Ex situ* measurements by XPS of the kinetics of adsorption of PAH or PSS on PAH/PSS multilayers deposited on polymer surfaces [88,92,93] also indicated that the adsorption time is shorter than 10 min. Nevertheless, the adsorption of the first PAH layer on nonchemically treated PET and of the first PSS layer on chemically modified PTFE took respectively 30 and 20 min. The adsorption of a PSS layer on a silanized silicon wafer has been followed *ex situ* by atomic force microscopy, by interrupting the adsorption after given periods of time [107,108]. The formation of isolated islands is observed in the initial stage of PSS deposition, especially on defect sites (scratches, microparticles, . . .). For deposition times from 1 to 5 min, a sharp increase of surface coverage by random islands is observed. After 10 min, the surface exhibits a very homogeneous morphology with a roughness as low as 2 nm. These results indicate that the adsorption of PSS is a two-stage process: macromolecular chains are anchored to the surface by some segments during the short initial stage, and then relax to dense packing during the long second stage. In agreement with the above-mentioned *in situ* measurements, this implies that diffusion-limited adsorption mechanisms

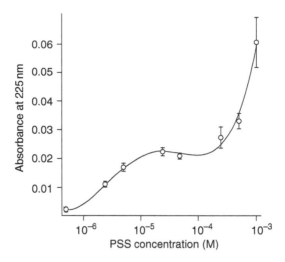

Figure 4   Dependance of the absorbance at 225 nm of ESA PAH/PSS multilayers on the concentration of the
PSS solutions. The PAH concentration is 5×10^{-4} M. (Data taken from K. Shinbo, K. Suzuki, K. Kato,
F. Kaneko, and S. Kobayashi. *Thin Solid Films* 327/329:209, 1998. With permission.)

are not adequate to explain the experimental results. Adsorption of a second PAH layer on top of the
PSS monolayer follows similar tendencies, although a complete PAH film is formed faster [107,108].
Other authors have reported that PAH/PSS multilayers could only be assembled if the duration of
each adsorption cycle is not shorter than 10 min [86], and that films assembled with dipping times
close to this limit showed pronounced roughness. Accordingly, dipping times of 10 min or more are
used in the literature.

Films grown from PSS (0.5 M MnCl$_2$) and PAH (2 M NaCl) with PSS concentrations of 0.1, 0.02,
and 0.004 mol/l showed the same growth step [86], suggesting a weak influence of polyelectrolyte
concentration on the deposition process. The effect of PSS concentration has also been studied
by monitoring the UV-Vis absorbance of PAH/PSS multilayers with 8 polyelectrolyte layers built
from 5×10^{-4} M PAH solution and PSS solutions of various concentrations [101]. Changing the
PSS concentration from 4×10^{-6} to 1×10^{-3} M, increases the adsorbed amount only by a factor
of 3 (Figure 4). Nevertheless, as mentioned before, a lower limit to the PSS concentration must
exist, since surface charge overcompensation has to occur for successful multilayer build-up [102].
The UV absorption is very weak for the multilayers assembled from 5×10^{-7} M PSS solutions
(no salt added), and XRR showed that films could not be grown from 0.001 M PSS (0.5 M MnCl$_2$)
solutions [86].

Changing the ionic strength offers also the possibility to fine-tune the film thickness of the
PAH/PSS multilayers (Figure 5). The addition of electrolytes enables us to adjust the average thickness of each oppositely charged layer pair, with a precision claimed to be as good as 0.05 nm [109].
The available range goes from 1.09 nm (no salt added) to 10.4 nm (3 M NaCl) [99]. The structure of
the PAH/PSS multilayers [110], and in particular the morphology of the film surface [110] and the
interpenetration of the polyelectrolyte layers [109], are also affected by the ionic strength. Screening
of electrostatic interactions is responsible for the observed increase of the film thickness with increasing salt concentration, due to a more coiled conformation of the polyelectrolytes. Interestingly, this
fine-tuning of the layer thickness can be used for the construction of complex film architectures, by
stacking layers of different thickness of PAH and PSS [86]. This is possible because the swelling
of the polyelectrolyte multilayers when immersed in solutions of different ionic strengths is only in
the order of 10 to 15% [111], whereas the influence of the ionic strength on the thickness increment
covers a much wider range.

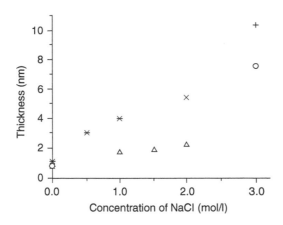

Figure 5 Evolution of the thickness of a layer pair of PAH and PSS in polyelectrolyte multilayers, made from solutions with increasing amounts of NaCl added. (△): NaCl added to the PSS solutions only. (+), (×), (∗), (o): NaCl added to both PSS and PAH solutions. (Data taken from G. Decher and J. Schmitt. *Prog. Colloid. Polym. Sci.* 89:160, 1992 [△]; J. Schmitt, T. Grünewald, G. Docher, P.S. Pershan, K. Kjaer, and M. Lösche. *Macromolecules* 26:7058, 1993 [×]; M. Lösche, J. Schmitt, G. Decher, W.G. Bouwman, and K. Kjaer. *Macromolecules* 31:8893, 1998 (o); D. Korneev, Y. Lvov, G. Decher, J. Schmitt, and S. Yaradalkin. *Physica B* 213/214:954, 1995 (+); R. von Klitzing and H. Möhwald. *Langmuir* 11:3554, 1995 R. von Klitzing and H. Möhwald. *Macromolecules* 29:6901, 1996 (∗). With permission). Thicknesses are determined by XRR. Concentration of the polyelectrolyte solutions: 0.003 *M* or 0.01 *M* (repeat unit). Solutions contain 0, 0.003, or 0.01 *M* HCl.

Deposition with various water–glycerol mixtures showed a two-fold increase in thickness while increasing the glycerol fraction from 20 to 50% [86]. Clearly, changing the solvent offers new processing potentials, although this has not been much explored so far.

The effect of the molar mass of PSS is small: Increasing its degree of polymerization from 900 to 5000, changed the overall thickness by less than 6% [100].

As mentioned above, no Bragg peak is found when PAH/PSS multilayers are characterized by XRR. Nevertheless, if a drying step is inserted every 2, 3, or 4 dipping cycles, Bragg peaks corresponding to these periods were reported [112]. The peaks sharpen when the number of adsorption/drying cycles is increased. The introduction of a drying step after every dipping cycle does not generate Bragg peaks. No effect on the growth step per bilayer is detected. These results have been rationalized by suggesting that the drying step induces denser packing of the outer polyelectrolyte layer, reducing the mutual penetration of chain loops and ends between neighboring layers. Annealing at 90°C did not suppress the supramolecular structure, but a prolonged exposure to water (4 days) did [112].

In another experiment [86], a sample with PSS as outer layer has been dried, and then dipped again for some time in the PSS solution. This drying/adsorption cycle has been repeated three more times. As thickness measurements showed, only upon a total 120 min adsorption did the process attain its saturation. Altogether, the PSS thickness reached 10.4 nm, although for undried samples adsorption did saturate after 10 min to a thickness of 4 nm. Surprisingly, this stepwise treatment had also no influence on the subsequent growth step.

Finally, we would like to stress the stability of the PAH/PSS multilayers, which do not redissolve in pure water (which is very important for the rinsing step of the deposition process) or in solutions of high ionic strength [111]. The thickness of films immersed 10 days in pure water or submitted to water vapor for 2 months present unchanged thickness [115]. Furthermore, films heated for more than 1 week to 200°C did not show any noticeable deterioration, but released some water, which leads to some shrinking of the films. After dehydration of the films over P_2O_5, recovery of water is almost complete (95% of the water molecules return into the film structure) [100]. However, PAH/PSS multilayers delaminate at pH 10, presumably due to deprotonation of the amine groups [116].

III.   Polyelectrolyte Adsorption and Multilayer Build-Up

That polyelectrolyte multilayer build-up is based on electrostatic interactions is intuitively a very reasonable assumption. Taking into account the weak steric requirements of ionic bonds, this assumption explains why so many polycation–polyanion pairs can lead to successful multilayer build-up, with only some rare unsuccessful cases reported (see application range). It also explains why attempts to film deposition based on the alternate adsorption of polyelectrolytes of identical charge were doomed to failure [122,123]. Another implication is that a given protein can be used for multilayer formation either as a polycation or a polyanion by adsorbing it at a pH under and above its isoelectric point respectively. Hemoglobin, for example, is assembled as a positively charged entity at pH 4.5 (with PSS), and as a negatively charged entity at pH 9.2 (with poly[ethylenimine] [PEI]) (Figure 6(a)) [123]. Electrostatic interactions not only control the deposition process, but also determine the stability of polyelectrolyte multilayers: inverting the charge of one of the polyelectrolytes by changing the pH in the film [124], or suppressing the charge of one of the constituent polyelectrolytes by thermochemistry [125] or by changing the pH [116,126] leads to destruction of the multilayer assembly. The strong influence of the ionic strength on the multilayer build-up is also indicative of the major role played by electrostatic interactions.

In the following, we will discuss to what extent theories on the adsorption of polyelectrolytes on solid substrates, based on electrostatic interactions, can provide insight into the mechanisms governing polyelectrolyte multilayer build-up. Indeed, the ensemble consisting of a substrate and the polyelectrolyte multilayer covering it may be seen as a substrate for another polyelectrolyte adsorption, although the complexation in the loose outer layer, with chains dangling around, must be somewhat different from the adsorption on a rigid surface. Accordingly, a part of this section is devoted to phenomena particular to the adsorption at this swollen interface, or occurring over several dipping cycles. Another reason to present the theories of the adsorption of polyelectrolytes on solid substrates is that this adsorption is the initial step for polyelectrolyte multilayer build-up. In the last part of this section, the effects of different parameters influencing the layer-by-layer deposition are examined.

The simplest adsorption theory of polyelectrolytes on oppositely charged surface is based on an ion-exchange model [127]: the polyelectrolyte competes with the counterions to pair with the charged sites on the surface. In the pure electrosorption model, repulsion between the charged segments of the polyelectrolyte opposes accumulation in the surface region, but the segment–surface interaction promotes adsorption. The polyelectrolyte looses entropy upon adsorption, but more entropy is gained by the liberated counterions, resulting in a total gain of entropy. Since surface counterions and polyelectrolyte counterions interactions are replaced by surface–polyelectrolyte and anion–cation interactions, the enthalpic change should be small, and polyelectrolyte adsorption is thus mainly an entropy-driven phenomenon. Long chains should thus replace short ones. The entropy loss experienced by a stiffer polyelectrolyte is smaller, and it can be anticipated that the adsorbed amount is more important for such polymers [128]. As a consequence of intrachain repulsion, highly charged polyelectrolytes adsorb with a flat conformation, forming rather thin adsorbed layers. By contrast, the balance between entropic and enthalpic energy will lead to the formation of much more tails and loops in the case of polyelectrolytes with moderate segment charge. Indeed, such conformations are entropically favored, and, due to the long-range character of electrostatic interactions, adsorption energy is also gained for segments not attached to the surface [129]. The behavior of weak polyelectrolytes is more complex: their degree of ionization depends on the local pH of the solutions, which, close to the surface, is different from that in the bulk of the solution [130]. Aside from the pure electrosorption, specific interactions between the polyelectrolyte and the surface may also play a role in the adsorption behavior of the polyelectrolytes. In this respect, the effect of ionic strength is instructive. On one side, at high ionic strength, the segment–segment interaction is screened, and the polyelectrolyte behaves like an uncharged polymer: it adopts conformations with loops and tails, and the adsorbed amount is expected to increase (screening-enhanced regime). On the other side,

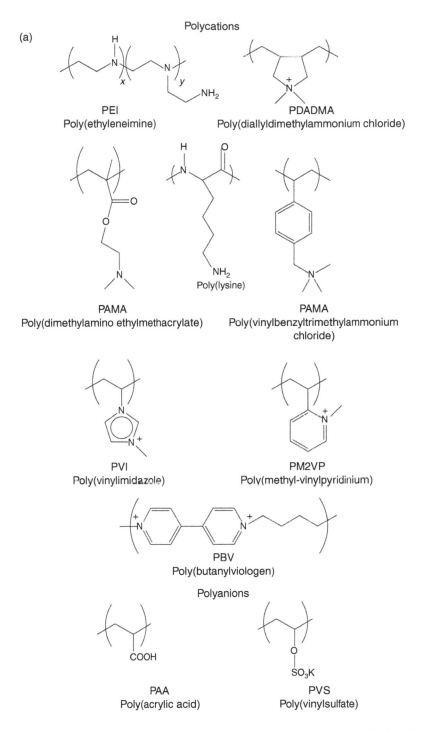

Figure 6 (a) Chemical structure of some polyelectrolytes that have been used for ESA. (b) Top: chemical structure of some conducting polyelectrolytes that have been used for ESA. Bottom: chemical structure of some precursors of conducting polyelectrolytes that have been used for ESA. (c) Top: chemical structure of some polysaccharides that have been used for ESA. Bottom: chemical structure of some polynucleotides that have been used for ESA.

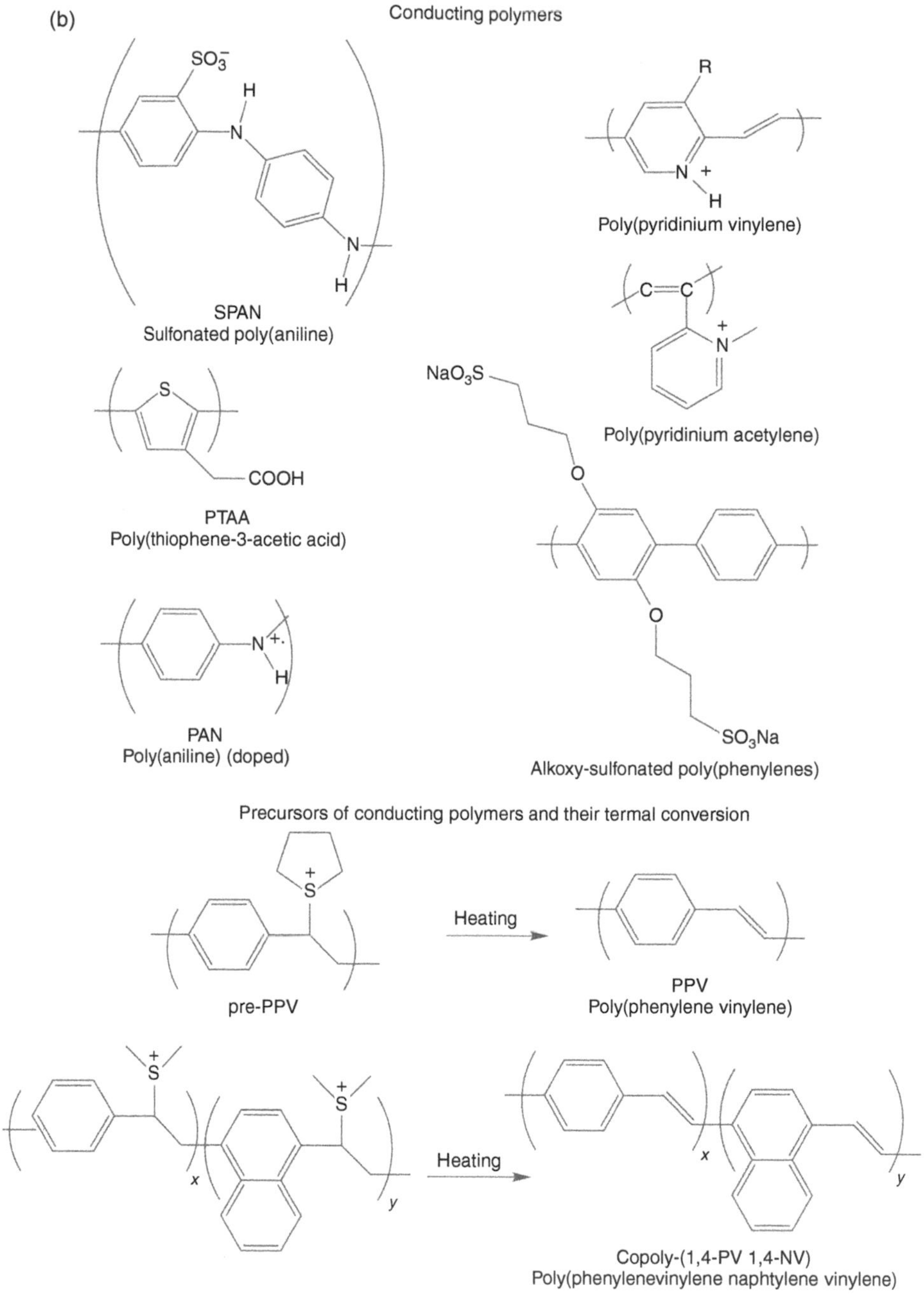

(b)   Conducting polymers

SPAN
Sulfonated poly(aniline)

PTAA
Poly(thiophene-3-acetic acid)

PAN
Poly(aniline) (doped)

Poly(pyridinium vinylene)

Poly(pyridinium acetylene)

Alkoxy-sulfonated poly(phenylenes)

Precursors of conducting polymers and their termal conversion

Heating

pre-PPV

PPV
Poly(phenylene vinylene)

Heating

Copoly-(1,4-PV 1,4-NV)
Poly(phenylenevinylene naphtylene vinylene)

Figure 6 *Continued*

the electrostatic attraction by the surface is also reduced, which can lead to a decreased adsorption (screening-reduced regime) or even to complete desorption of the polyelectrolyte chains: a fully screened polyelectrolyte can only adsorb if there is an attractive nonelectrostatic interaction with the surface. The balance between electrostatic and nonelectrostatic interactions determines the final behavior of the system. An increase of the adsorbed amount is usually observed for strong polyelectrolytes. In the screening-reduced regime, a decrease in adsorbed amount does not necessarily result

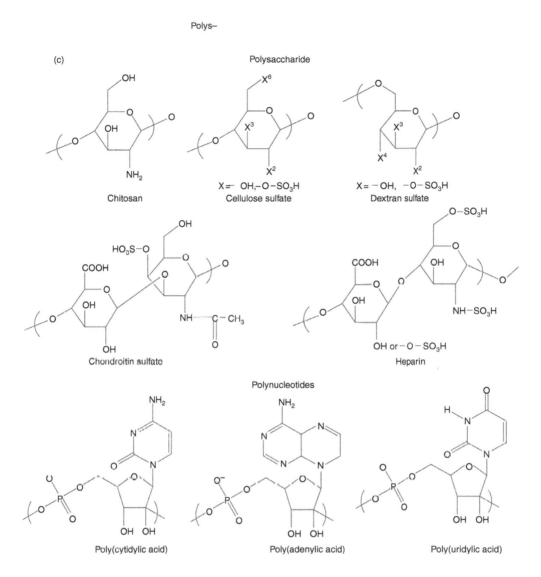

Figure 6 *Continued*

in a decrease in film thickness [131]. Another complication is the commonly encountered reduced solubility of screened polyelectrolytes, which may lead to the precipitation of the polyelectrolyte at very high ionic strengths. Counterions can also displace the adsorbed polyelectrolyte if they have a specific affinity for the surface, and if their concentration is high enough. Experimentally, the ion-exchange model has been qualitatively verified: the adsorbed amount versus segment charge, surface charge, or ionic strength follows the expected trends. Arguments coming from the ion-exchange theory are often used as a qualitative explanation of the effect of different parameters on the multilayer growth.

The ion-exchange model has nevertheless, few shortcomings. This theory based on a mean surface potential cannot explain why certain polycations adsorb on net positively charged TiO_2, where positive and negative charges coexist [132]. Theories taking into account the heterogeneity of surfaces have been developed [133–136]. Similarly, it has been observed that amphoteric molecules (i.e., bearing positive and negative charges) are able to adsorb on surfaces with a charge opposite to their net charge [137,138]. Also, correct matching of the average distance between the charges of the polyelectrolyte plays a role [139]. It is thus rather the topological possibility to form ion pairs than the net surface charge that determines adsorption.

A more fundamental shortcoming of the ion-exchange model is the assumption of a thermodynamic equilibrium. Even if only the strong and numerous electrostatic interactions with the charged surface are considered, adsorbed polyelectrolytes are expected to have a small mobility. Photobleaching experiments showed indeed that the diffusion coefficient for PAH adsorbed on mica is almost zero [140] (although movements of the whole molecule could be driven by capillarity); plastically deformed PSS/PAH bilayers did not anneal [140]. Polyelectrolyte adsorption may thus be expected to be an irreversible process. This problem has been tackled by different experiments; literature data indicate that polyelectrolyte adsorption on solid substrates is not a very reversible process [141]. Reversibility is enhanced in the presence of salt, and for low molar mass polyelectrolytes. As could be expected, irreversibility is also observed during the multilayer build-up by the layer-by-layer ESA. We have already seen that PAH/PSS multilayers remained unchanged when immersed in water (i.e., irreversibility upon dilution), and did not dissolve even at high ionic strength. Similarly, multilayers of poly-(L-lysine) (Figure 6(a)) and copper phthalocyanine tetrasulfonic acid resist 10 min immersion in 1 M NaCl [48]. A PSS/poly(vinylbenzyltrimethylammonium chloride) (PVBTA) (Figure 6(a)) bilayer immersed in pure water does not show desorption either [142]. The PVBTA layer is not displaced by the addition of 0.2 M YCl$_3$, though the highly charged Y^{3+} competes very strongly for the negatively charged sulfonate groups. Self-exchange experiments give similar results: the radiolabeled PSS situated in the outer layer of PSS/PVBTA bilayer is not exchanged by a large excess of unlabeled PSS of the same molar mass. Radiolabeled poly(N-methyl-2-vinylpyridinium) (PM2VP) was used to build-up multilayers with PSS. PM2VP-capped multilayers were not exchanged by poly(diallyldimethylammonium chloride) (PDADMA) (Figure 6(a)) for exposure times up to 2 h. Partial exchange of only about 20% was observed after several days [125]. On the contrary, the PAH outer layer of a PAH/PSS multilayer is at least partially exchanged by fluorescence-labeled PAH, but, in this later case, 1 M NaCl was added to the polyelectrolyte solutions, which probably increases adsorption reversibility, as just shown above. Experiments testing the reversibility upon pH changes have also been carried out [143]. In these experiments, multilayers of PSS and a weak polycation, poly(dimethylaminoethylmethacrylate) (PAMA) (Figure 6(a)) are built-up at pH 8. The pH is then decreased to 4, increasing the charge of PAMA by protonation. According to ion-exchange theories, PAMA desorption should occur. Nevertheless, no desorption occurs when PSS is the outer layer, proving that PAMA is effectively trapped inside the multilayer, and only limited desorption occurs when PAMA forms the outermost layer. The desorbed molecules are probably loosely attached PAMA molecules in the outer layer. Cassagneau et al. [144] mentioned that poly(pyrrole)/PSS and poly(pyrrole)/α-ZrP bilayers are so stable that even sonication in ethanol cannot destroy these assemblies. It is questionable whether an equilibrium theory can be used to study systems like polyelectrolyte multilayers exhibiting such pronounced adsorption hysteresis. Before concluding this discussion about the irreversibility, we want to stress that the aforementioned lack of mobility of the polyelectrolyte in the layers does not preclude local movements of parts of the polyelectrolyte chains, and especially of pendant group. Rearrangement of pendant chromophores trapped in the multilayers has indeed been revealed by second harmonic generation (SHG) measurements [145,146], and by the dependence of spectral shifts (metachromic effect) on the number of dipping cycles in UV measurements [139,147,148]. In this context, it is probably important to remind that water molecules are thought to occupy 40% of the volume in the PAH/PSS multilayers [100].

Another feature unexplained by the ion-exchange theory is the often-found charge overcompensation upon polyelectrolyte adsorption. In the case of pure physisorption, ion-exchange theory predicts only a minute overcompensation, due to a balance between entropic and electrostatic energy. Notwithstanding, nonelectrostatic attractive interactions may lead to some supplementary overcompensation. It is nevertheless hard to believe that nonelectrostatic interactions are present in all the observed cases of overcompensation, considering the huge number of different substrates and polyelectrolyte pairs concerned. Furthermore, it has been shown that quaternized poly(vinylpyridine) (namely, PM2VP)

(Figure 6(a)) has no specific interactions with TiO_2, by measuring the adsorbed amount as a function of ionic strength [132], but very stable multilayers could be built on this substrate by alternatively adsorbing PM2VP and PSS [143]. The successful multilayer build-up is an indirect evidence of overcompensation upon adsorption of PM2VP on TiO_2. An explanation of the overcompensation is best sought in kinetic phenomena. Adsorption is often observed as a two-step procedure: first, a rapid one taking place within minutes, and second, a slow one that can be as long as several hours or days, due to reconformation of the adsorbed layer [132]. If reconformation is slower than adsorption, charge overcompensation is expected: during adsorption, polyelectrolyte chains anchors by only some of their charged groups to the surface, and, before they have time to reconform in order to occupy the neighboring charged sites of the surface, these latter sites are occupied by other polyelectrolyte chains. The attached polymer molecules create progressively a surplus of charge which leads to an electrostatic barrier that prevents the attachment of another polyelectrolyte chains by repelling them; the phenomenon is self-regulating. Solution of higher ionic strength shields better the electrostatic interactions, leading to a lowering of the activation energy necessary for getting over the electrostatic barrier. Furthermore, increasing the ionic strength fastens the reconformation steps, by decreasing the strength of the ion pairs. The above-mentioned irreversibility corroborates this model of kinetically hindered equilibrium.

Charge overcompensation is theoretically a prerequisite for multilayer build-up. Every example of successful multilayer build-up is thus an indirect proof of charge reversal upon polyelectrolyte adsorption. By contrast, measurements of the ζ-potential, and of its sign reversal during multilayer build-up, are direct proofs of charge overcompensation. Aside from the already mentioned case of PAH/PSS multilayers [94–96,103,105], there have been only a few such measurements. Our group has presented ζ-potential values calculated from streaming potential measurements for planar substrates covered by multilayers made of poly(vinylsulfate) (PVS) (Figure 6(a)) and an ionene [67]. From streaming potential measurements again, the pH dependence of the ζ-potential of different multilayers assembled by ESA has been reported by Schwarz et al. [149]. Electrophoretic measurements have been used to determine the ζ-potential of multilayer-coated polystyrene latex particles [94,95]; alternatively positive and negative potentials were observed respectively when polycation and polyanion formed the outer layer, for the PSS/PDADMA and deoxyribonucleic acid (DNA)/PDADMA layer pairs. The observed reproducible values of the ζ-potential from one dipping cycle to the next are consistent with the usually observed linear increase in film thickness during polyelectrolyte multilayers build-up. Linear increase of the adsorbed amount also requires an approximately constant roughness at the multilayer surface. Hoogeveen et al. [143] studied ζ-potential of poly(vinylimidazole) (PVI)/PAA (Figure 6(a)) multilayers on silica particles as a function of d_q, the degree of quaternization of PVI. They showed that alternatively positive and negative potentials are obtained respectively when PVI and PAA are the outer layers if $d_q > {\sim}0.1$. The higher the d_q, the higher the absolute values of the measured ζ-potential are. Interestingly, no multilayer growth is observed when $d_q = 0$, and stable multilayers formation occurs only when $d_q > 0.18$. This demonstrates experimentally a direct link between polyelectrolyte charge, charge overcompensation, and multilayer build-up. Schlenoff et al. [125] have used radiolabeled counteranions to determine surface excess charge in PSS/PDADMA multilayers. They found 45 and 46% of uncompensated charges on the surface of multilayers capped respectively with the polycation and the polyanion.

The picture that emerges is that of a multilayer whose topmost adsorbed layer contains trains anchoring the polyelectrolyte to the multilayer, and loops or tails dangling around. These loops and tails are responsible for the charge overcompensation which, in turn, determine the amount adsorbed in the next step. The adsorbed amount depends thus on the nature of the underlying layer. For example, QCM monitoring of the deposition of PSS shows adsorbed amount depending on the nature of the underlying polycation [86] or protein layer [123]. Also, we observed that the thickness of xAyAxAy multilayers (where A denotes a polyanion, x and y denote polycations) was significantly different from the thickness of yAxAyAx multilayers although they both contain the same number of

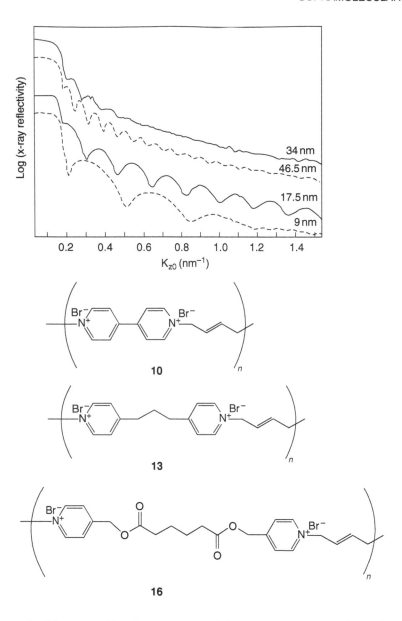

Figure 7 X-ray reflectivity versus K_{z0}, the component of the wavevector of the incident photon perpendicular to the interfaces, for different ESA multilayers containing aromatic ionenes (I0, I3, I6, below) and poly(vinylsulfate) (PVS). The curves have been shifted vertically for clarity. From top to bottom, the structure of the different samples is: Si/I3/PVS//I0/PVS/I3/PVS/I0, Si/I0/PVS/I3/PVS/I0/PVS/I3, Si/I0/PVS/I6/PVS/I0/PVS/I6, Si/I6/PVS/I0/PVS/I6/PVS/I0. The thickness of these multilayers in nm is reported in the figure. (Data taken from B. Laguitton, X. Arys, A.M. Jonas, and A. Laschewsky. Unpublished Results. With permission.)

layers of A, x, and y (see Figure 7) [150]. Similar observations have been made with other systems too [139,146,147,151,152], demonstrating that the thickness of the layers in multilayers built by ESA depends on the chosen polycation–polyanion pair. Frequently in the literature authors assume, however, that the polyelectrolyte layer thickness is an intrinsic property of the polyelectrolyte, and subtract the polyelectrolyte layer thickness measured on other systems from their own measured growth increment, in order to calculate the contribution of the other component of their multilayer to the overall growth step. According to the above discussion, the conclusion of such reasoning must be

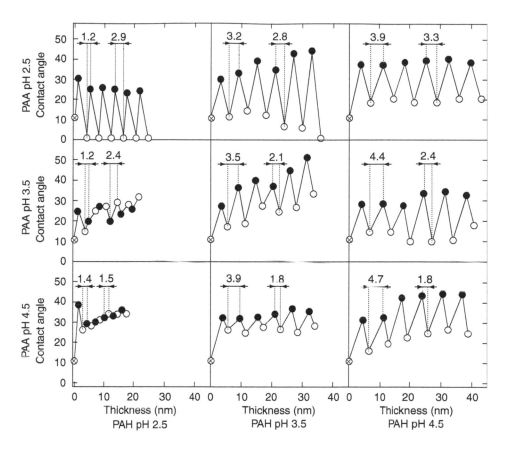

Figure 8 Contact angle versus thickness in nanometers for multilayers composed of 1 to 12 layers. The pH of the PAA and PAH dipping solutions are indicated, respectively to the left and under the graphs. Open circles: multilayers with PAA as outermost layer. Filled circles: multilayers with PAH as outermost layer. The average incremental contribution of PAH and PAA to the multilayer thickness are reported in nm in the figure; left: contribution of PAH; right: contribution of PAA. (Data taken from D. Yoo, S.S. Shiratori, and M.F. Rubner. *Macromolecules* 31.4309, 1008. With permission.)

made taking much care, as it may at most be only a crude approximation. In agreement with the ion-exchange theory, if the loops and tails determine the adsorbed amount, this amount should diminish with increasing polyelectrolyte charge density and decreasing surface charge density. This has been well illustrated in a study of the build-up of PAA/PAH multilayer as a function of the pH of the polyelectrolyte solutions used for the assembly [153] (Figure 8). In the pH range used in this study, PAH is fully protonated. Thus, with PAA as outer layer, the pH controls the surface charge, since PAA is a weak polyelectrolyte, without changing the linear charge density of the adsorbing PAH. It is found that the average PAH layer thickness increases with increasing surface charge density (increasing the pH of the PAH dipping solution). When PAH is the outer layer, the pH determines the linear charge density of the adsorbing PAA, without changing the surface charge density. It was found that the average PAA layer thickness increases with decreasing linear charge density (decreasing the pH of the PAA dipping solution). Absolute values of the linear charge density may however be difficult to determine accurately, and nonelectrostatic interactions may play a role when the linear charge density becomes weak.

We have just seen that the formation of loops and tails seems to determine the adsorbed amount by dipping cycle. The possibilities for loops and tails formation are different in the vicinity of a solid surface or in the loose outer layer of polyelectrolyte multilayers. An effect of the substrate on the first deposited layers is thus to be expected, causing a nonlinear growth of the multilayer thickness for the first layers versus number of dipping cycles. Such nonlinear growths have indeed been often observed, for example, in Refs. 75 and 154–157. In agreement with the above, Lösche et al. [100] pointed out that linear increase of the PAH/PSS multilayer thickness is observed to start for a thickness that coincides with the length scale of interdiffusion between these polyelectrolytes, suggesting a causal relationship. Interestingly, looking at the adsorption of different polyelectrolytes on the same surface, nonlinear growth of the initial layers may occur for only some of them [155,158,159]. In some cases, nonlinear growth may also be attributed to a lack of affinity between the polyelectrolyte and the substrate [157]. In this case, the multilayer growth begins by adsorption at imperfections, forming islands that finally merge together with increasing number of deposition cycles. Linear growth usually occurs after <5 dipping cycles, but in some instance up to 20 dipping cycles are needed [154]. Once the regime of linear growth has been reached, we may wonder whether the growth increment is the same for multilayers deposited on different substrates, if they are made of the same given polyelectrolyte pair. A growth increment independent of the substrate would strongly suggest that the structure and properties of the outer layer converge toward a state independent of the substrate. Unfortunately, this question is still open, and cases for [6,85,122,146,157,160–163] and against [88,92,93,164,165] this view have been reported, although there is a vast majority of "for" cases. As a result of the outpacing of "for" cases, polyelectrolyte multilayers have even been proposed as decoupling layers for the adsorption of proteins on a variety of substrates [166]. Finally, we want to stress that the influence of the substrate on polyelectrolyte multilayers build-up may also be detected by measuring other properties than the growth step per dipping cycle. In some instance, these methods are even more sensitive, revealing the effect of the solid surface although a linear growth of the film thickness versus number of dipping cycles is observed. Examples are the formation of aggregates resulting in a shift of the maximum absorbance in the UV-Vis measurements [139,147,167] or a decrease in the overall degree of orientation of NLO chromophores, has evidenced by SHG experiments [146,168–170]. An example from our group is presented in Figure 9. Shift in electroluminescence and photoluminescence with decreasing number of layers

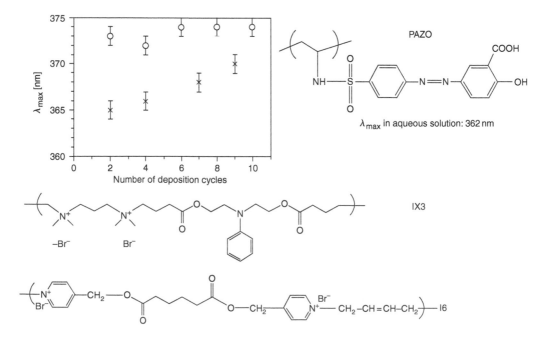

Figure 9   Shift of the absorbance maximum of the colored polyanion PAZO in growing multilayers made by ESA with different polycation partners (IX3 and I6). (Data taken from P. Fischer, A. Laschewsky, E. Wischerhoff, X. Arys, A. Jonas, and R. Legras. *Macromol. Symp.* 137:1, 1999. With permission.)

have also been reported, but they are rather due to a confinement of electron and holes than due to the presence of the substrate [171–173].

Since the multilayer build-up by the ESA does not proceed under thermodynamic equilibrium conditions, it is questionable whether the approximately 1:1 charge stoichiometry predicted by ion-exchange theory is found in the multilayers. Indeed, it is known from electrochemical manipulations after multilayer construction that a certain amount of extrinsic charge compensation does not lead to multilayer destruction [125]. Furthermore, it is also conceivable that some buried charges of the outer layer may not be accessible for polyelectrolyte complexation in the next adsorption step. In fact evidence for both stoichiometric [66,125,143,174] and nonstoichiometric [88,92,93,143] multilayers can be found in the literature, although only few data are available on this topic. It would be interesting to establish the minimum ion pair density required for the stability of the multilayer.

A related problem that may occur during multilayer build-up is the desorption of some of the polyelectrolyte from the outer layer caused by the adsorbing polyelectrolyte, in order to form a soluble complex. This problem is easily detected when using strongly colored weakly charged ionenes [175] (Figure 10), and has been observed for other systems too by *in situ* measurements, or simply from a smaller thickness of the film after an adsorption cycle [97,113,143,176]. Analysis by high-performance liquid chromatography of PSS solutions, which were used for PAH/PSS multilayer build-up, revealed also the presence of a small amount of PAH/PSS complex in the solution [94]. Serizawa et al. could circumvent the desorption problem by adsorbing from a solvent in which the complex was insoluble [176]. The problem of desorption is particularly acute in the case of multilayers containing small charged molecules, although it can be somewhat controlled by a proper setting of the ionic strength [177].

An interesting phenomenon arising from the use of polyelectrolytes for the build-up of multilayers is the observed self-healing capabilities of the layer-by-layer deposition process. An example of this is the restoration of a smooth film surface after the addition of a much rougher layer, like a virus [52] or globular protein layer [178]. The build-up of composite films made of montmorillonite (a clay platelet) and PDADMA leads to a surface presenting large pits, up to 700 nm diameter and 30 nm depth [179]. These pits also were smoothed during the layer-by-layer deposition; 188 nm diameter and 14 nm deep pits were covered after only one deposition cycle. These self-healing capabilities may be a drawback for patterning [180], but enable the build-up of polyelectrolyte multilayer on weakly charged surfaces [157,181]. In the latter case, adsorption occurs at isolated imperfections, forming islands that later grow vertically and laterally until they coalesce. The ability to bridge the gaps has been shown to be directly related to the molar mass of the polyelectrolyte [181]. We believe that the increase of the self-healing capabilities with increasing molar mass is a general trend.

Many parameters may influence the deposition process of polyelectrolyte multilayers: the linear charge density, flexibility, molar mass and polydispersity of the polyelectrolytes, the existence of specific interactions, the matching of the charge–charge distance, the substrate charge and its distribution on the surface, the ionic strength, pH, temperature and polyelectrolyte concentration of the dipping solutions, the nature of the solvent [176], the adsorption time, the presence or not of a drying step, or even the application of an external potential [179,182].

Concerning the linear charge density of the polyelectrolyte, we have already cited the work of Yoo et al. [153] and of Hoogeveen et al. [143]. The latter author showed the existence of a critical linear charge density for the successful build-up of polyelectrolyte multilayers, depending slightly on the ionic strength of the solutions. Our group has confirmed these results in the case of ionenes [156,183]; a critical charge density close to that found by Hoogeveen et al. was established. Also, liquid-crystalline (LC) polymers containing one charge per 80 carbon atoms could be used successfully for multilayer build-up, but not LC polymers with an ionic group content of one charge per 176 carbon atoms [184]. As the charge density is decreased, other interactions may come into play. This in turn may render the determination of a critical charge density difficult. Interestingly, decreasing the charge density may then even lead to better assemblies [139]. Further studies are clearly required to clarify this notion.

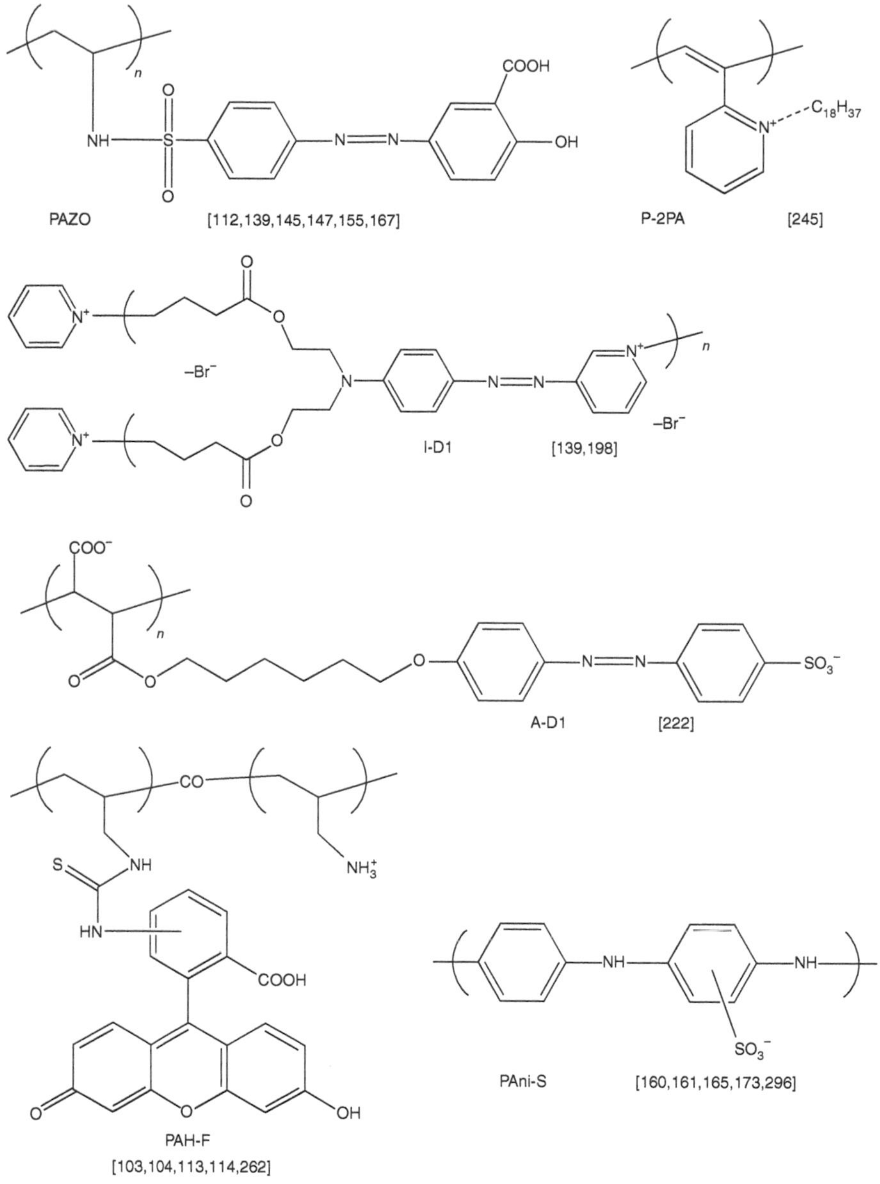

Figure 10 Examples for colored polyelectrolytes employed in ESA multilayers.

Aside from the already mentioned case of PAH/PSS multilayers, the effect of molar mass has also been studied in the case of polyaniline/PSS multilayers [123] (Figure 6(b)); no dependence on the molar mass of PSS was found when it was varied from 5,000 to 1,000,000. Very weak dependence on the molar mass of the polycation has been observed in the case of PAH/bolaamphiphiles multilayers [185]. Interestingly, even poly(thiophene-3-acetic acid) oligomers (Figure 6(b)) of only 8 to 12 repeat units may be useful to assemble multilayers [160].

As mentioned earlier, an increase in ionic strength generally leads to an increased thickness in the case of the adsorption of strong polyelectrolytes. Accordingly, an increased film thickness is generally observed in the case of polyelectrolyte multilayers prepared from solutions of high ionic strength [62,66,109,125,152,171,183,186–188] (see also Figure 5). The thickness can be changed

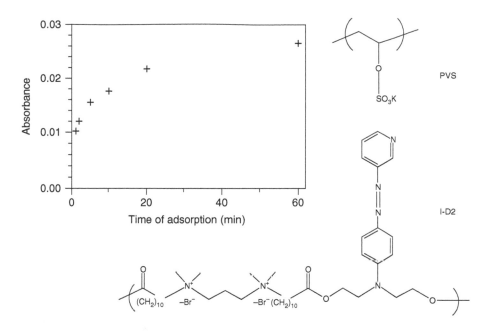

Figure 11 Kinetics of the adsorption of colored ionene I-D2 on a layer of polyvinylsulfate, as followed by Vis-spectroscopy. (Data taken from P. Fischer, A. Laschewsky, E. Wischerhoff, X. Arys, A. Jonas, and R. Legras. *Macromol. Symp.* 137:1, 1999. With permission.)

by as much as one order of magnitude. An increase in ionic strength may also increase [62,187,188] or reduce the film roughness [189].

Similarly to what is observed in the case of PAH/PSS multilayers, a slight increase in the adsorbed amount is observed for higher polyelectrolyte concentrations [122,142,160,161,190], leading to an increase in the growth step per dipping cycle [122,161,191].

Numerous kinetic investigations have been reported in the literature, aiming at a determination of an appropriate dipping time. Rapid adsorption usually takes place in the first few minutes [106,139,145,158,169,192–194], sometimes followed by a slower adsorption step as adsorption saturation sets in [131,155,161,186,195,196] (Figure 11). Nevertheless, adsorption may sometimes take more than 1 h [197–199]. Kinetics have been shown to depend on the substrate [162], the ionic strength of the solution [186], and the polyelectrolyte concentration [122,142,160,161]. Other parameters, like the molar mass, have also been shown to play a role in the case of the adsorption on a solid surface [132]. Since most of the physisorbed amount is deposited at the very beginning of the adsorption process, it is tempting to build multilayers with adsorption times shorter than the time required for saturation of the adsorption. Multilayers can indeed be successfully assembled for dipping times shorter than the time required for saturation of the adsorption [51,139,187], and, in some instance, dipping times shorter than 5 sec have been shown to be convenient [49,157,200]. Considering the amount of kinetic studies and the number of publications in the field of ESA, it must be recognized that dipping time is most of the time arbitrarily selected; this is not a very satisfactory situation, as the optimal dipping time is most certainly strongly system dependent.

As is the case for PAH/PSS multilayers, the presence of an apparently insignificant drying step may have different effects on the deposition process. It has indeed been mentioned that sample drying speeds the relaxation of the polyelectrolyte chains [200], increases the adsorbed amount [126,147,201], and, especially when performed by spinning the samples, results in a lower roughness of the films [152,202].

Studying the kinetics of adsorption of poly(*o*-methoxyaniline), it has been found that the adsorbed amount is notably increased when adsorption is interrupted several times for drying and measuring the sample [162]. Furthermore, the time required for reaching the saturation of the adsorption is enhanced. This is similar to what we mentioned previously in the case of the adsorption of PSS [86]. We believe this is a general phenomenon, and we must thus warn against kinetic measurements obtained by repeatedly dipping and drying the same sample: the time required for reaching saturation may be systematically overestimated.

IV. SUITABLE SUBSTRATES

Most of ESA multilayers have been built on glass, silicon, quartz, and ITO (indium-tin-oxide), since these substrates are particularly well suited for a number of experimental techniques. Another interesting point is that their surface physico-chemistry is well studied, and convenient procedures can thus be found in the literature to clean and charge their surface. When hydroxyl groups are present, a correct choice of the pH is sufficient to generate a useful surface charge. Other examples of procedures to generate charges on these substrates are: bolaform amphiphile adsorption, plasma deposition, or silanization (see introduction). A number of polyelectrolyte, like PEI or PDADMA [182], may even adsorb on the bare surface of these and other substrates, and thus be used as a precursor layer for further multilayer assembly.

A number of other substrates have also been used for the ESA. Among the metals, gold has been widely used. Its surface is easily charged by the chemisorption of alkanethiols (see introduction). Successful multilayer assembly on bare gold has also been reported [75]. Examples of other metals that have been successfully used are platinum [203], silver [123], and aluminum [204]. For the latter, the surface must be assumed to consist of aluminum oxide.

A whole range of polymer substrates has also been used. Aside from the polymers already mentioned in the PSS/PAH case study, poly(propylene) (PP), poly(methylmethacrylate) (PMMA) [197,205], poly(ethylene) (PE), polyurethane [206], and poly(vinyl chloride) (PVC) [207] have also been used, just to cite a few of them among the industrially most important. Successful multilayer assembly on polymeric substrates is obtained either by chemically modifying the polymer surface, or by functionalization of the polyelectrolyte so that nonelectrostatic interactions may occur.

Substrates covered by LB films have also been used [208]; as many different substrates are suitable for LB films, LB film deposition can be seen as a versatile way of bringing charges to these surfaces prior to ESA.

Substrates need not be planar for ESA, as exemplified by the already mentioned build-up of PAH/PSS multilayers on latex particles, and by the build-up of multilayers on optic fibers [209]. Substrates even need not be solid, as multilayers have been grown on alginate gel beads [210].

Although successful multilayer build-up has been demonstrated on many different substrates, very little is known on the adhesion of these multilayers to their supports (see below, the physical properties section). Especially when adsorption of the first layer is only due to specific interactions between the polyelectrolyte and the substrate, good adhesion is questionable [90,93,197].

V. APPLICATION RANGE

The adsorption of cationic bipolar amphiphiles has been shown to reverse the surface charge of a mica substrate, and to form well ordered monolayers [211,212]. Successful charge reversal is obtained only with a careful molecular design of the bolaamphiphiles, which must be long enough to avoid flat adsorption on the surface and favor lateral interactions, and short enough to avoid adsorption as a loop [44,213]. The problem of loop formation vanishes if a rigid core, like a biphenyl, azobenzene, or diacetylene unit, is inserted in the middle of the molecule, and consequently all the bolaamphiphiles presented in the following contain such rigid units. Nevertheless, successful self-assembly of a monolayer of a flexible amphiphile has also been reported [214]. Alternately adsorbing cationic and anionic bolaform amphiphiles leads to the formation of multilayers whose thickness depends linearly on the number of dipping cycles [44]. Multipolar molecules have also been used for the assembly of multilayers, offering the advantage of an increased tolerance against the formation of small defects [215,216]. It has been shown that composite multilayers made of bolaform amphiphiles and inorganic nanoparticles [208,217,218] or enzymes [219,220] could be successfully assembled. Nevertheless, so far most studies on multilayers containing bolaamphiphile and, more generally, multipolar dye molecules concerned composite assemblies with polyelectrolytes [48,148,154,163,177,196,221–224]. A number of studies have been driven by potential applications arising from the insertion of dye molecules in a multilayer. For example, electroluminescent [151,225–227], photoisomerizable [63,228,229], SHG-active [230,231], or pH sensitive dyes [148,232] have been assembled in multilayers with suitable polyelectrolytes. The possibility to manipulate several dyes in a multilayer opens the door to other applications, like the fabrication of tunable color filters [227]. The benefit of the association with polyelectrolytes appear clearly when looking at the case of composite assemblies with a pH-sensitive dye, congo red. Indeed, unlike dye monolayers on glass, the multilayer can be cycled from very low pH to very high pH many times without noticeable dye desorption [232]. A number of studies have also been devoted to the photopolymerization of acetylene and cinnamoyl-containing bolaamphiphiles layers [185,211,212,233–235], which may increase the chemical, thermal, and mechanical stability of the multilayers containing these molecules. The successful photopolymerization implies the existence of well-ordered amphiphilic domains in these layers, even when sandwiched between polyelectrolyte layers. Such well ordered layers might be thought to have interesting barrier properties [185]. We have already mentioned the importance of lateral interactions for the successful formation of bipolar amphiphiles monolayers. These interactions lead sometimes to aggregation in solutions. Amphiphiles with only one end charged may also lead to such aggregates in solution, forming charged bilayer membranes. Multilayers have been made by the ESA technique by adsorbing alternately such positively and negatively charged membranes, or by adsorbing such membranes alternatively with polyelectrolytes [236].

Polyelectrolyte/polyelectrolyte multilayers have even attracted more interest, partly due to their self-healing capabilities, leading to a reliable and easy build-up, and to the wide variety of useful polyelectrolytes, making it possible to tailor the properties of the multilayers. Many different polyelectrolytes have been used [58,139,237–241], among which some are commercially available, like PSS, PVS, PEI, PAH, PDADMA, and PAA, for example. LC polyelectrolytes [139,184,242], block copolymers [243–245], and dendrimers [51,246,247] are readily assembled in multilayer structures. Reproducible deposition of PEI/Cu^{2+} complex in alternation with PSS to produce multilayers has also been demonstrated [240]. A number of conducting polymers, such as poly(thiophene acetic acid), sulfonated poly(aniline), poly(pyridinium acetylene) (Figure 6(b)), bearing charged groups, have successfully been used for the electrostatic layer-by-layer self-assembly [161,245,248]. The partial doping of conjugated polymers that have no charged groups in their chemical structure, like polyaniline, leads to the presence of delocalized charges along the polymer backbone. This delocalized charge has been shown to be suitable for multilayer build-up by the ESA technique [122,191]. Alternate dipping of a substrate into a chemically active aqueous solution of an *in situ* polymerized polymer, such as poly(pyrrole) or poly(aniline), and a solution of a polyanion leads also to satisfactory multilayer assembly [249].

The biochemistry of living organisms relies heavily on macromolecules, not only for controlling the cell metabolism, but also as structural elements. Most of these macromolecules can be classified as polysaccharides, nucleic acids, and proteins. Polysaccharides are rather rigid long chains built from one or two monomeric units, and are mainly used by nature as building blocks or for food storage. Some polysaccharides are made of monomers bearing chemical groups like carboxylic acids or amines, making them suitable candidates for the layer-by-layer electrostatic multilayer build-up. Multilayers containing chitosan [126,176,186], chondroitin sulfate [186], dextran sulfate [124,207], cellulose sulfate [152], and heparin [174,206,207] (Figure 6(c)) have indeed been successfully assembled. Adsorption of some of these polysaccharides may enhance surface biocompatibility, a property needed for cell culture and human implants, for example. Since the long-term stability of these films in physiological solutions or in blood plasma is of importance, the presence of covalent bonds between the layers would be an advantage. Such bonds have been created by posttreatment of the multilayers, or from adsorption with a reactive polyelectrolyte [124,206,207].

Ribonucleic and desoxyribonucleic acids, which are used by the cell to store and transmit information, have covalently linked backbones made of alternating pentoses and highly negatively charged phosphates. They can thus be used as polyanions for the layer-by-layer deposition process. Whether DNA conserves its double helical structure or not is of importance for potential applications like the express-diagnostics of virus, or for the control of environmental pollution with carcinogenic molecules [250]. For example, the conservation of the double helix is of importance for biosensing of a number of drugs from alkaloid and antibiotic classes which intercalate in this double helix [251]. A conformation transition inside the multilayers may arise from temperature [47] or pH changes [252]. Homopolynucleotides like polycytidylic [252], polyuridylic, and polyadenylic acid (Figure 6(c)) [81] have also been deposited.

Proteins, which are used for structural or for metabolic purposes by the cell, are made of amino acids. Some of them bear a negatively charged lateral chain, like aspartic or glutamic acid, while others have a positively charged lateral chain, like lysine, arginine, or histidine. Most of the proteins are thus amphoteric, and may be globally negatively or positively charged when used respectively at a pH above or under their isoelectric point. Accordingly, composite protein-polyelectrolytes [80,113,123,124,158,178,253] and protein-DNA [195] multilayers have been prepared by the layer-by-layer deposition technique. Potential applications for such films range from nonthrombogenic surfaces [206] to biosensing [254,255], through the immobilization of enzymes for bio-catalysis [6]. It is of importance for these applications that the proteins are not denatured by the deposition process in order to remain biologically active. In this respect, ESA, which does not involve any enzyme modification or covalent bonding should preserve in most cases the functional characteristics of the proteins [80,123,158]. In biosensing, it may be hoped that multilayer build-up allows for a tunable sensitivity through a control of the number of layers, while the selectivity can be modified by an appropriate choice of the biospecific biomolecule [219,256,257]. Protein multicomponent films are also promising for sequential enzyme reactions through vectorial transfer of chemicals, electrons, or energy [5].

Supramolecular biological assemblies, like virus [52] or membrane fragments [258,259] containing bacteriorhodopsin have also been inserted in multilayers built by ESA. The insertion of inorganic nanoparticles or platelets with defined chemical and physical properties at precise locations in a multilayer is of interest for fundamental research, as well as for a number of potential applications. ESA may be useful to assemble such multilayers if suspensions of charged inorganics can be prepared [1,40]. For example, suspensions of delaminated clay platelets, which are intrinsically charged, have been successfully used to assemble multilayers. Both the positively charged hydrotalcite [242] and the negatively charged hectorite [49,260] and montmorillonite [80,123,146,179,192,242] clay platelets can be utilized for the build-up. Atomic force microscopy (AFM) and x-ray diffraction measurements have shown that the platelets lie flat on the substrate, as could be expected. Interestingly, it has been shown that, in some instance, more than one layer of platelets may be deposited [49,179]; charge balance considerations [125] seem to be able to explain this behavior. This is further corroborated by similar phenomenon observed during adsorption of purple membrane fragments [258]. Due in part to the high aspect ratio of the clay platelets, hybrid polymer-clay assemblies possess a set of unique mechanical, electrical, and gas permeation properties, and as a result have been used in gas permeation membranes [181] or as insulating layers [182]. Platelets may also be useful, simply to prevent interpenetration of adjacent layers. Exfoliated suspensions of other intrinsically charged inorganic sheets, like zirconium phosphate (α-ZrP) [193,261–264], MoS_2 [265], SnS_2 [265], or graphite oxide [266–268] have also been demonstrated for the ESA. It is noteworthy that after multilayer assembly graphite oxide can be reduced to graphite whose conductive and magnetoresistive properties are interesting.

The insertion of clusters and nanoparticles in multilayers may be useful too. Stable colloidal dispersions of negatively charged silica (at pH 10) [269], for example, have been successfully used for ESA. Various polyoxometalates [60,165,270–272], as well as gold [273–275] and platinum nanoparticles [276], were incorporated by this process in multilayer assemblies. A composite assembly containing polyelectrolytes and a membrane bilayer with a layer of gold particles on top were reported [277]. Many different semiconductor nanoparticles have been employed for the layer-by-layer deposition process, for example, cationic TiO_2 nanoparticles (isoelectric point corresponds to pH = 4.5 to 6.8) [203,278], but also PbI_2 [208,217], PbS [203], CdS [203,279], CdSe [144,280], and ZnS [281] and "coupled" TiO_2/PbS [282,283] nanoparticles. The considerable interest in these semiconducting systems arises in part from their special properties caused by quantum-size and surface effects, and the potential applications taking advantage of these effects. Preparation of multilayers containing these nanoparticles is not straightforward. Much attention has to be paid to the process parameters influencing the preparation of the colloidal suspension. Correct preparation must lead to a narrow size distribution of the particles and a correct surface charge, sufficiently important to allow for self-assembly but sufficiently small to prevent desorption during the rinsing step. Addition of stabilizers may be needed to prevent coagulation of the nanoparticles. As a last example of insertion of nanoscale particles, we cite multilayer magnetic thin films made by the consecutive adsorption of Fe_3O_4 particles stabilized by PDADMA and a polyamic acid salt [284].

A number of other systems have been shown to be suitable for multilayer build-up by the ESA, like charged latex nanospheres [50,199,285–287] or metallo-supramolecular complexes [288–290]. ESA has also been combined with Langmuir–Blodgett transfer [291–294], and even superlattice films have been produced.

Figure 12 presents a schematic of a virtual assembly in which all the kinds of molecules that have been shown to be usable for the ESA would be combined. It may seem from the preceding that any molecule or nanoobject bearing charges is readily deposited by ESA. Although this is probably close to the reality, we want to temper here this conclusion by a few remarks, and some examples to the contrary [47,123]. First, we have already mentioned the existence of a critical charge density for the ESA process, and this may explain some of the unsuccessful attempts with polyelectrolytes [181] and other charged entities [144,219,258]. Second, in the case of weak polyelectrolytes, the pK values of the polycation and polyanion (or the isoelectric point in the case of proteins) must fit in order for both polymers to be sufficiently charged and thus adsorb in a consecutive fashion; this also may explain some unsuccessful depositions [165]. The inhomogeneous repartition of the charges on the surface of proteins complicates their behavior, and has been claimed to be responsible for some multilayer deposition failure [113]. Finally, in the case of small molecules or nanoparticles, redissolution during rinsing or formation of a soluble complex [255] during adsorption may also be a major concern, as discussed above.

VI. Structure of the Multilayers

As shown in the case study on PAH/PSS multilayers, the interdiffusion between layers is so important in this system that no Bragg peaks are observed by XRR measurements, nor by NR measurements (when every PSS layer is deuterated) (Figure 3). However, it has been shown that these films are stratified. Is this combination of stratification and important interdiffusion a general feature of polyelectrolyte multilayers?

Stratification of PAA/PAH multilayers is suggested by contact angle measurements showing odd/even trends as a function of the number of deposited layers [153] (Figure 8). Also, stratification has been proved in the case of multilayers containing more than two polyelectrolytes: ${ABCB}_n$, ${(AB)_n(AC)}_m$, ${(AB)_n(AC)(AB)_m(AC)}_p$ films have been assembled which give rise to Bragg peaks detected by XRR [64,295] and NR [202,295]. By selective deuteration at varying

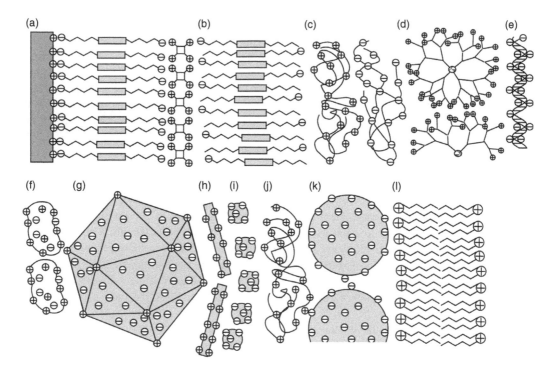

Figure 12 Scheme of the structure of a virtual multilayered heterostructure made from all kinds of molecules or particles that have been shown to be usable for ESA. Components: (a) boladications and other multipolar molecules; (b) lipid bilayers; (c) polyelectrolytes; (d) dendrimers; (e) DNA; (f) proteins; (g) viruses; (h) inorganic sheets; (i) inorganic nanoparticles; (j) latex nanospheres; (k) LB interlayers; (l) biotin/avidin complex. (Adapted from Y.M. Lvov and G. Decher. *Crystallogr. Rep.* 39:628, 1994. With permission.)

intervals along the film normal, the stratification of PAH/sulfonated polyaniline multilayers has been also demonstrated to be preserved at least for the deposition of 40 bilayers [296].

By contrast, neither x-ray nor neutron [297] reflectivity measurements on multilayers assembled from only one polyanion/polycation pair have shown Bragg peaks so far, except for one type of specific system, discussed later in this section. This may be due to either a lack of contrast between the polyelectrolytes constituting the multilayer or to a (too) large interfacial width. In our opinion, it is mainly the important interfacial width that is responsible for the absence of Bragg peaks. Indeed, as shown above, Bragg peaks have been observed in the case of multilayers made of more than two polyelectrolytes, showing that usually polyanions and polycations have sufficiently different electron densities to give rise to Bragg peaks. Indeed, even a drying step applied at periodic intervals during multilayer build-up has been claimed to induce sufficient contrast to give rise to Bragg peaks [64]. Finally, important interfacial widths have been demonstrated for a number of systems [66,153,202,295,296].

As discussed in the PAH/PSS case study, both the interfacial roughness and the polymer interdiffusion contribute to the observed interfacial width. The reported interfacial roughnesses represent thus an upper limit to the extent of the interdiffusion of polyelectrolytes in neighboring layers. Strong interdiffusion between polyelectrolyte layers has been shown in the case of multilayers containing poly(butanylviologen) (PBV) (Figure 6(a)) [66]: changing the distance between the redox active layers by interposing nonelectrochemically active layer pairs, it was determined that PBV is spread over a distance of at least 2.5 layer pairs. Covering a layer of partially biotinylated poly(L-lysine) by an increasing number of PSS and nonbiotinylated poly(L-lysine) layers, and looking at the amount of streptavidin that was still able to adsorb by avidin/biotin complexation, the interdiffusion of

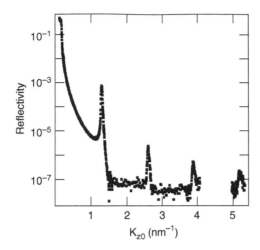

Figure 13 X-ray reflectivity versus K_{z0}, the component of the wavevector of the incident photon perpendicular to the interfaces, for an ESA multilayer containing 20 bilayers of I-D2 and poly(vinylsulfate) (PVS) deposited on silicon. (Data taken from P. Fischer, A. Laschewsky, E. Wischerhoff, X. Arys, A. Jonas, and R. Legras. *Macromol. Symp.* 137:1, 1999. With permission.)

the partially biotinylated poly(L-lysine) in at least four neighboring polyelectrolyte layers has been demonstrated [78]. Strong interdiffusion of the polyelectrolytes in the neighboring layers has also been shown in the case of PAA/PAH multilayers, by performing contact angle measurements as a function of polycation and polyanion layer thickness [153]. This has been corroborated by a simple surface dying technique, using methylene blue as dye [153]. Strong interpenetration of the layers in ESA multilayers is also supported by nonradiative energy transfer measurements [201].

Interdiffusion seems to be less important for the rigid conjugated polyelectrolytes, but large interfacial widths are nevertheless observed [202,295,296]. For example, PAH/sulfonated polyaniline (Figure 6(b)) multilayers exhibit interfacial width in the nanometer range, with an internal organization decaying monotonically away from the substrate. This suggests that the interfacial width is primarily due to the accumulation of defects as the multilayers are assembled, that is, primarily due to interfacial roughness. Neutron reflectivity measurements on multilayers containing poly(phenylenevinylene) (PPV) have shown that the interfacial width, although again in the nanometer range, was sufficiently small to maintain the PPV layers well separated from each other [202,295]. Confinement of electron–hole pairs in these layers is thus expected. PPV, which is prepared by thermal conversion of a charged precursor, is immiscible with PSS; phase separation is likely to occur during curing, resulting in a further decrease of the interdiffusion of the rigid PPV in the PSS layer [100]. By contrast, photoluminescent shifts roughly proportional to $1/d^2$, where d is the thickness of the assembly, were observed for multilayers containing another conjugated copolymer [171] and thin polyelectrolyte insulating layers (~ 7 Å), as expected for confined photogenerated electron–hole pair in an infinite square potential well (the whole film). Nevertheless, the photoluminescence of films with 40 Å thick insulating layers is independant of the multilayer thickness, suggesting a confinement of electron–hole pairs in the conducting layers, and thus an interdiffusion of the conducting polymer smaller than the insulating spacer layer thickness.

Measurements in our group are in sharp contrast with the preceding results showing high internal roughness and/or strong interdiffusion of adjacent layers. X-ray reflectivity measurements on multilayers containing some particular ionenes reveal indeed the presence of Bragg peaks [139,156,240,241,298]. In some instances, Bragg peaks up to the fourth order were observed [139] (Figure 13). This undoubtedly points to a small interdiffusion of the layers, coupled with a small waviness of the internal interfaces. The reason for this good ordering in the films is probably to be searched in the low linear charge density of the ionenes used, which let other nonelectrostatic interactions come into play. Interestingly, the same lamellar structure with a repetition distance of about 2.4 nm has been found both in these multilayers and in the polyelectrolyte complexes obtained by mixing the polyanion solution with the ionene solutions. However, such relationship between the structure of polyelectrolyte multilayers and the corresponding polyelectrolyte complexes does not seem to be an absolute rule [298].

Interdiffusion and, to a certain extent, interface waviness can be suppressed by using more rigid ionic blocks for the multilayer assembly. As a consequence, Bragg peaks may appear in x-ray measurements of some multilayers containing platelets [49,179,193,260,266], LB [292], or bolaamphiphile [219,222] interlayers. Bragg peaks may also be seen in the case of multilayers containing colloidal nanoparticles, but in this case, the spacer layer should be thick enough to avoid the overlap of the layers containing the nanoparticles [273]. Bragg peaks have also been claimed for samples made of bolaamphiphiles and colloidal nanoparticles [217]. The successful photocross-linking of some multilayers containing bolaamphiphiles implies good packing of these molecules in the multilayers, as seen above. An example of multilayer made of bolaamphiphiles and multipolar molecules sufficiently well ordered to give rise to Bragg peaks has also been reported [215]. Note however that even for systems discussed in this section, the appearance of Bragg peaks is exceptional.

The formation of aggregates [48,63,139,147,148,167,215,229,230] and the preferential orientation of molecules or molecular fragments has been investigated by many different techniques including SHG [61,145,146,168–170,258,299] (Figure 14), spectroscopic techniques [160,223,300–302],

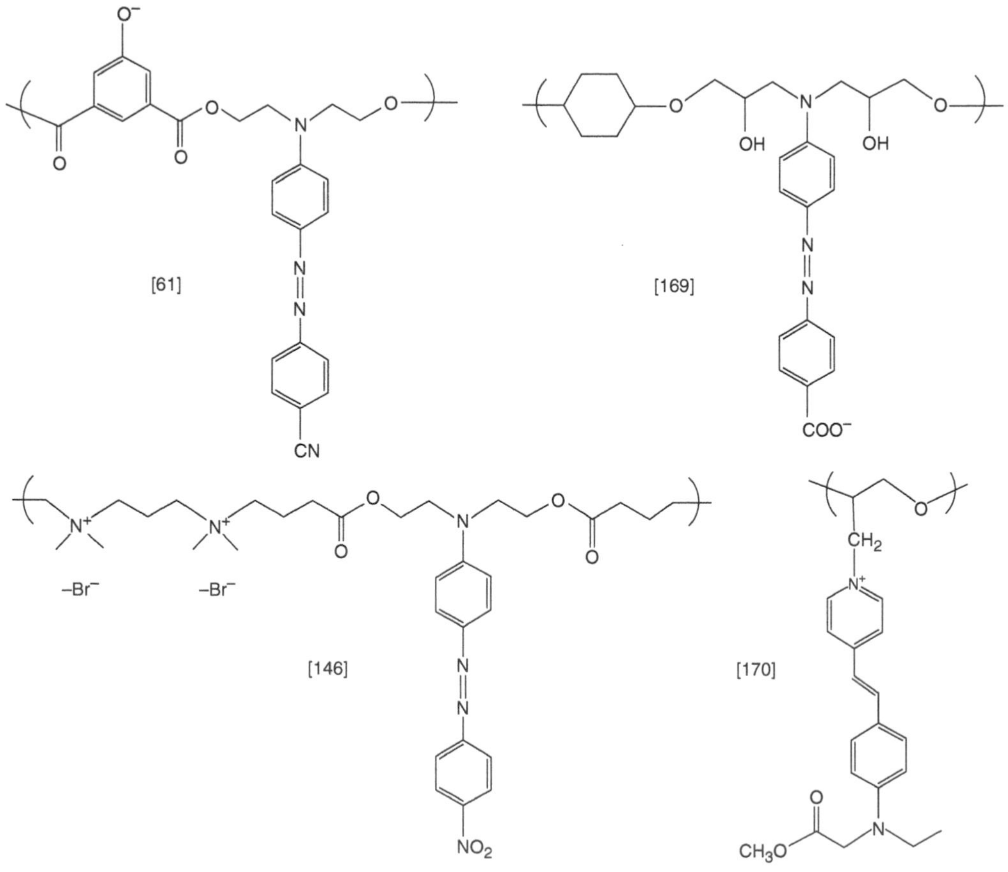

Figure 14 Examples of functional polyions employed in ESA multilayers to study second-order nonlinear optical effects.

and ellipsometry [98]. Naturally, the results depend heavily on the system under investigation. It is noteworthy that in one instance, orientation in the dipping direction has been reported [99].

The outer layer of the polyelectrolyte multilayers is expected to be significantly different from the bulk of the film, but not much is known about this particular layer. Though, XRR and NR, as well as AFM measurements have been used to measure the roughness of this layer, which range from molecularly flat [108,178] to tens of Ångstrøms [202], depending on the system under investigation. Diffusion experiments point to a looser molecular packing in the outer layer, at least for PAH/PSS multilayers [113,114]. As expected, more counterions are found in the outer layer than in the bulk of the film [125].

Of prime importance for potential applications in integrated optics or electronics is the possibility of patterning with micron resolution the polyelectrolyte multilayers. Both microcontact printing [180,302–305], ion beam lithography [306], and photostructuration [46] have been used to this end. The presence of salt in the polyion solutions affects both the selectivity of the deposition and the surface roughness of the film, and can even lead to a "negative" patterning when combined with a drying step [180,188,303,305].

VII. CHEMICAL MODIFICATIONS OF THE DEPOSITED FILMS

Chemical modifications of the layers after deposition open new opportunities to further tune the properties of the multilayers. The cross-linking of protein multilayers by different chemical reactions [124,206,207], and the photocross-linking of unsaturated bonds has already been cited [185,211,212,233–235]. Other examples of cross-linking reactions can be found in the literature devoted to ESA [116,197,307]. Cross-linking may be used for instance to improve the chemical [116,124,197,206,212,307], mechanical [197], and thermal [233] resistance of the films, reduce the interdiffusion between neighboring layers, or change the barrier properties of the multilayers [185].

Some interesting molecules cannot be incorporated in multilayers by ESA, simply because they are not charged. Chemistry is also very useful here, since it provides a way around this problem: a charged precursor of the molecule is used for the assembly, and is then converted into the desired molecule. Some examples are the reduction of graphite oxide in graphite, the thermal conversion of PPV-precursor [152,295,308–310] or of precursors of other conjugated polymers [72,171,172] repeat units (Figure 6(b)), and the use of polyimide precursors [62]. Remember however that, if no specific interactions come into play, the stability of the multilayers may be a concern during these chemical reactions, since these reactions suppress the charges that link together the neighboring layers [125].

Electrochemistry in multilayers [164,195,216,253,256,266], and particularly in multilayers containing viologen moieties [66,125,311,312], has been extensively studied, partly because of the potential applications in electrocatalysis, sensing, or electrochromism, but also because of the wealth of fundamental information that it may provide. Electrochemical methods provide also the opportunity to control the ionic content of the films [125,312].

Another benefit of chemistry results from its combination with ESA for building up thin organic films. Our group has indeed worked on a new route for the development of organic multilayers [126,156,241,313,314], consisting in the electrostatic adsorption of a polyelectrolyte, followed by a chemical reversal of its charge (Figure 15). The process can be cycled as many times as desired. It is noteworthy that in the case of one such activation reaction, a negative ζ-potential has been measured, although the total ratio of cationic to anionic groups could not be more than 2:1, even after 99% conversion [67]. This strongly suggests a noncentrosymmetric type of ordering, as confirmed by SHG measurements. The NLO properties of these films have prompted the extension of this type of reactions to obtain noncentrosymmetric ordering of functional dyes in normal ESA multilayers: in the classical ESA of a polyanion with a polycation, a step consisting of the chemical grafting of an uncharged NLO-active dye is inserted after every polycation adsorption step [239].

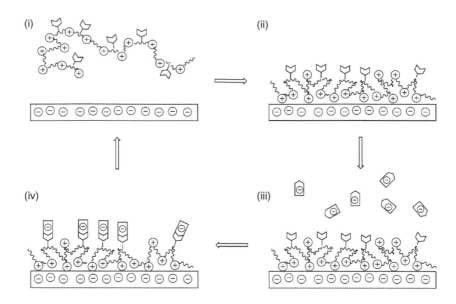

Figure 15 Schematic drawing of the multilayer preparation by combined ESA and chemical activation. (Adapted from A. Laschewsky, B. Mayer, E. Wischerhoff, X. Arys, A. Jonas, P. Bertrand, and A Delcorte. *Thin Solid Films* 284/285:334, 1996. With permission.)

VIII. PHYSICAL PROPERTIES

In the following, some physical properties of the polyelectrolyte multilayers will be presented. The number of studies focusing on these issues is limited, and often system dependent. Although we must thus warn against excessive generalization, these results may be useful as guidelines for further work in this field.

In sharp contrast to most of the LB films, the multilayers made by ESA usually show good ageing properties [113,202,297], although they are not in thermodynamical equilibrium, too. This is mostly due to the numerous ion pairs, which lead to a reduced mobility of the polyelectrolyte chains and thus to the observed irreversibility or reduced reversibility of the adsorption process.

Numerous ionic bonds are also present in the polyelectrolyte complexes obtained by mixing solutions of polycations and polyanions of high charge density. These bonds are responsible for the formation of compact precipitates that are usually insoluble even at high salt concentrations, but can be solubilized in ternary solvent mixtures [315]. Accordingly, good solvent resistance of multilayers made of strong polyelectrolytes is usually reported [62,166,313], even at high ionic strength [111], but dissolution in a ternary mixture may occur [307]. Multilayers made of weak polyelectrolytes may dissolve at a pH that reduces or inverts their charge [116,124,126,206]. As already mentioned, solvent resistance may be improved by cross-linking [116,124,197,206,207,212,307].

The ion pairing is also most probably responsible for the good thermal behavior reported [46,64,187], at least as long as no degradation or conformational transition occurs [47]. Cross-linking can also be used in an attempt to further increase the thermal stability [233]. Nevertheless, heating may lead to local mobility of the chains or pendant groups; rearrangement of chromophores trapped in the multilayers above a (system-dependent) critical temperature has been reported [145,167,170]. This local mobility is also probably responsible for the sometimes-mentioned increase of internal organization upon annealing of the multilayers [292,295].

So far, investigations of the mechanical properties of the multilayers assembled by ESA have been limited to tape peel tests [40,90,93,160,197,204,216]. Failure occurring in the layer, at the interface with the substrate, or in the tape have been reported, depending on the system under investigation. Again, cross-linking may be useful to improve the mechanical properties of the multilayers [197].

In contrast, much more studies have been devoted to the diffusion of small molecules in the multilayers, because of its importance in applications like permeation membranes or biosensing. Both IR spectroscopy [316] and fluorescence measurements [317] have shown the diffusion of protons in the multilayers, and thus an influence of the pH of the outer solution even far inside the films. The influence of water on the thickness of the multilayers is also well-documented [111,318]. Diffusion of radiolabeled salt ions has also been measured [125,312]. Voltamperometry showed that PAH/PAA films had little effect on the diffusion of $Fe(CN)_6^{3-}$, but that PAH/PSS films could hinder its transport [116]. 6-CF [96], acridine orange [81], daunomycin [251], $2'$-$3'$ cyclic adenosine monophosphate [252], bisulfite [313], and different diazonium salts [147,313] have been shown to permeate deeply in multilayers built by ESA. Immunoglobulin G (IgG) could permeate or not in a superlattice made of anti-IgG layers and PAH/PSS spacer layers, depending on the thickness of the spacer layer. The diffusion constants of rhodamine and of 2,2,6,6,-tetramethyl-4-piperidinol-1-oxide (TEMPOL) in PAH/PSS multilayers have been quantified [113,114].

Aside from the numerous studies over electrochemistry in the multilayers, and over multilayers containing conducting polymers, some more useful information concerning the electrical properties of multilayers may be found in the literature. Shinbo et al. have studied the electrical properties of PAH/PSS multilayers, and showed that the conductivity was due to hopping conduction, probably due to the migration of mobile ions [101]. As could be expected, humidity greatly influences the electrical properties. ESA has also been used to assemble a metal-insulator-gold nanocluster-insulator-metal heterostructure that allows the observation of single electron charging effects at ambient temperature [319].

IX. POTENTIAL APPLICATIONS

The huge number of organic and inorganic molecules that are suitable for the ESA, combined with the simplicity of this deposition technique paves the way to numerous potential applications. However, due to the relatively recent development of the technique, the number of patents in the field is still rather limited [320–331], and, in the best case, only feasibility studies have been performed up-to-date. The transfer of these technologies to industry will require the fulfillment of numerous other requirements, like cost effectiveness, stability of the devices, …. In the following, some of the potential applications proposed in the literature are reviewed.

To begin with, multilayers can be used to modify surfaces, and in particular their biocompatibility (e.g., for implants) [174,186,206,332] or their wettability (e.g., for antifogging mirrors or eyeglasses) [153]. Obviously, only a limited number of layers is needed for these applications; here, the fact that polyelectrolyte adsorbs readily on various substrates is a clear advantage.

Multilayers may also be used for their permeation properties. Accordingly, membranes covered by multilayers have been employed for gas permeability measurements and for pervaporation studies [88,181,185,333–335]. These measurements showed, for example, O_2/H_2O, CO_2/O_2, or toluene/heptane selectivity. The permeation properties of polyelectrolyte multilayers are also important when they are used for the encapsulation of enzymes [210] or living cells [336,337]. The deposition of polyelectrolyte multilayers or of hybrid polyelectrolyte/inorganic multilayers on latex particles, and the subsequent dissolution or calcination of the latex beads leads to the fabrication of hollow spheres [94–96,338–340]. Potential applications of such hollow spheres are numerous, for example, for the controlled release and targeting of drugs.

Polyelectrolyte multilayers are able to very effectively immobilize various molecules, like enzymes, for example. These multilayers have been shown to exhibit useful biocatalytic (i.e., enzymatic) [219,220] activities; hybrid multilayers containing polyoxometalate are promising for catalytic applications [164]. ESA multilayers have also been used for molecular recognition [198,341], or, more specifically, as biosensors [75,166,250–252,342,343]. Electrocatalytic [195,253,256] and electrosensing [254] capabilities of multilayers deposited on electrodes have

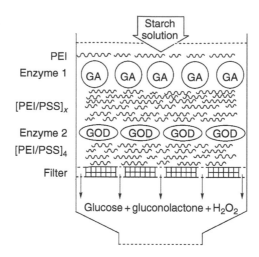

Figure 16 Scheme of the sequential enzymatic conversion of starch by glucoamylase (GA) and by glucose oxidase (GOD) in multilayers prepared by ESA. (Adapted from M. Onda, Y. Lvov, K. Ariga, and T. Kunitake. *J. Ferment. Bioeng.* 82:502, 1996. With permission.)

also been demonstrated. As already mentioned, the possibility to put different enzymes at specific locations in the multilayer has been employed to design systems for multi-step catalysis [6] (Figure 16).

Applications in the domain of data storage have also been envisioned. Multilayers with alternating layers of magnetic nanoparticles and polyimide have already been prepared, yielding an average magnetic flux per bilayer of 850 nT at a distance of 2 cm [284]. Optical data storage based on the Weigert effect, that consists in birefringence or dichroism induced by irradiation with polarized light, can be envisioned in the case of multilayers containing azobenzene functionalized polymers [344,345]. Indeed, these derivatives are known to undergo *cis–trans* and *trans–cis* photoisomerization upon adequate light irradiation. Multilayers containing azobenzene-functionalized polymers have also been shown to be useful for the photofabrication or photoerasure of surface relief gratings [346], and as command surfaces for LC display devices [63,229].

As stated in the application range section, electroactive, conducting, semiconducting, and insulating layers can be inserted in the multilayers assembled by ESA. Furthermore, we mentioned the possibilities of patterning the deposition of the multilayers. Various applications are thus expected in the electronic and optoelectronic fields. Thin conducting layers may find applications in transparent electrodes, antistatic coatings, or electromagnetic interference (EMI) shielding, for example [68]. Such thin conducting layers are also useful for the fabrication of organic LED. Potential advantages of organic LEDs are the possibility of fabrication of large area devices, and their relatively low cost [308]. Nevertheless, their long-term stability is a concern in some instances [280], and the processability of some electro-active polymers is challenging [173]. ESA enables an easy assembly of complex heterostructures containing very thin layers of these electroactive polymers, with a level of control higher than what would be possible by spin-coating, and more easily than by the LB technique [173]. As a consequence, organic LEDs have been prepared by the ESA; the quality of these films is of at least the same quality as spin-cast films [54]. Precursors of poly(phenylene vinylene) [308–310] and of copolymers (phenylenevinylene–naphtylenevinylene) [72,171,172], as well as poly(pyridinium vinylene) [165,173,347] and alkoxy-sulfonated poly(phenylenes) [159,348] (Figure 6(b)) are examples of electroactive polymers that have been used for the ESA of LEDs. Devices have also been prepared from an electroluminescent ruthenium complex, either incorporated in the main chain of a polyester, or used as such [151,225,227,232,349]. Multilayers made of conducting polymers and nanoparticles are also potentially interesting as electroluminescent devices

[280,350]. Indeed, their emission wavelength can be controlled by the size of the nanoparticles, while the transport properties of electrons and holes can be adjusted by a correct choice of the polymer. Further, CdSe/PPV devices showed longer lifetimes than devices prepared from only one of these components. LEDs based on a conjugated bipyridinium have also been reported in the literature [226]. The behavior and performances of the LEDs have been shown to depend on the nature of the counter-polyanion [309]. By combining chemical tuning with size effects, the emission can be finely tuned, and may cover the entire visible range [171]. The photocurrent generation has been measured in different systems containing colloidal nanoparticles [182,203,283]. In this respect, "coupled" type nanoparticles are particularly interesting, since charge injection from one semiconductor into another can lead to efficient and longer charge separation by minimizing the electron–hole recombination pathway [283]. Such systems are thus anticipated to have potential applications in solar energy conversion. Solar energy conversion by biomimetic light-harvesting systems would be very interesting, but requires the possibility to control the supramolecular arrangement of the assemblies [8]. The versatility of ESA, that enables an easy juxtaposition of electron donors and acceptors at controlled distances for the photoinduced electron transfer, would be an interesting way toward such systems [261,262]. Multilayers containing bacteriorhodopsin show also interesting photoelectric properties, in particular the generation of a differential photocurrent. This, together with the long-term stability and photochromic properties of bacteriorhodopsin, leads to potential applications ranging from imaging devices to light sensitive alarm devices through devices for motion and direction detection [258,259].

A number of other applications has been proposed, including SHG (see "Structure of the multilayers"), electrochromism (see "Chemical modifications of the multilayers"), high density rechargeable lithium-ion batteries [267,268], tunable color filters [227], or pH indicators [232]. A pH-sensitive optrode has also been proposed [209], but first attempts were unsuccessful due to the absence of response of the pH-sensitive dye used (phenol red) to the pH of the surrounding solution, when trapped in the multilayer.

Despite the wide range of possible applications of the ESA multilayers proposed, we want to somewhat temper here the widespread enthusiasm. It should be stressed for example, that these films are sensitive to atmospheric humidity and other environmental conditions [111,318]. The environment dependent swelling of the multilayers can be prejudicial for some applications. Among other potentially unwanted aspects of these multilayers, we can also cite the ionic conductivity due to small ions trapped in the films, although this has not been studied much so far [101]. Therefore, much work will be needed in the future to establish the practical scopes and limits of the ESA method.

X. EVOLUTION OF THE FIELD IN 2000 TO 2003

Since the first edition of "*Supramolecular Polymers*," the field of ESA films has evolved organically and continuously (Figure 17). The number of articles has been growing steadily and passed the bench mark of 1000 in 2002. The foundations laid in the 1990s have proven to be sound so that no change of paradigm has occurred. But our detailed knowledge has been much broadened, concerning mechanistic aspects, useful polymeric and colloidal systems, suitable supports, film structure, film properties, as well as potential applications. Though comprehensive overviews are scarce [351–353], several publications reviewing aspects of ESA films have appeared in recent years [354–360]. Therefore, this chapter will only highlight the most important general developments.

The past years have impressively demonstrated the broad applicability of the ESA process for charged polymers and colloids, including multiply charged colorants and their aggregates, independently whether they are of organic or inorganic nature, or of synthetic or natural origin. Characteristically, one cannot distinguish between "useful" or "useless" components as such, but the pair of oppositely charged partners must be always considered [139,361–365]. Although plentiful, most various examples for ESA films in the literature may be somewhat misleading — because

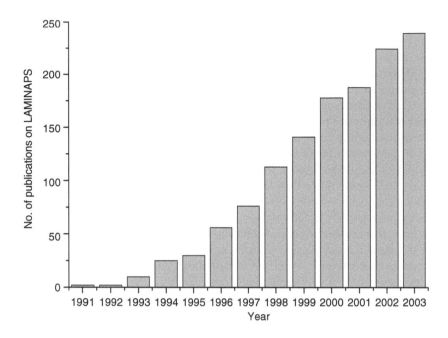

Figure 17 Evolution of publications concerning ESA multilayer films (number of papers in 2003 estimated on the basis of the first 9 months).

generally only successful systems are reported but not the failing ones — , it appears that for most charged systems suitable conditions (solvent, concentration, pH, ionic strength, temperature, drying periods, molar mass, etc., cf. Section II) can be elaborated to enable successful film growth by ESA. Recently, the successful co-adsorption of singly charged dyes with polyelectrolytes has opened the way to a simple functionalization of ESA films [366]. Despite the wealth of chemical structures successfully employed, simple standard polyelectrolytes dominate the field up to now. The system PAH/PSS continues to be the most widely used system, challenged only by PDADMA/PSS (cf. Figure 6(a) top line) that is increasingly employed by virtue of being made of two strong, that is, pH independent polyelectrolytes. Concerning the practical methodology, alternative methods for film build-up, namely spraying [367] and spin-coating [368–378] were introduced to speed up film deposition. Spin-coating may even provide particularly smooth surfaces [368,369,373], and sample spinning in the solution during deposition was claimed to provide more homogeneous films [379]. But layer-by-layer dipping continues to be the method of choice by virtue of its simplicity and of the regularity of the coatings obtained.

Despite many studies and much progress, the theory behind the ESA process [380–385] and the effective mechanism(s) [365,386–390] remain somewhat obscure. A complete theory of surface complexation of polyelectrolytes is still missing, and the nonequilibrium character of ESA contributes to the difficulties, as do nonelectrostatic interactions (see Section I and below). In any case, it is clear that subsequent adsorbed compounds may not be of the same type, that is, polycations do not adsorb on a positively charged surface, and polyanions generally do not adsorb on a negatively charged one [391–400]. Film growth by alternating adsorption apparently requires charge overcompensation, the formation of insoluble binary stoichiometric complexes (rare instances of pairs of polyelectrolytes are reported, able to form films while only forming soluble stoichiometric complexes in water [361]), and the immobilization of the complexes in the surface film by entanglement or vitrification [365,386,388]. At low ionic strength, 1:1 ion complexes with intrinsic charge compensation between the charged species are formed, and only the excess surface charge is balanced by low molar mass counterions [386,389,390]. A good charge density matching between the species employed seems to give best results [139,351,401,402]. Furthermore, particular good results were obtained for the combination of weak polycations with weak polyanions, or of strong polycations with strong polyanions [402]. Polyampholytes may be useful, too, if they are used together with polyelectrolyte partners [148,395,403–406]. Alternatively, even polyampholyte pairs may be useful if one type of the charged groups is at least partially neutralized [407].

Following the discussions in Section III, a general model has been widely accepted for the ESA process [365,386,388], but the discussion goes on [408,409]. For instance, adsorption seems to be not always irreversible, but depending on the ionic strength, stripping of the last adsorbed polymer layer in the consecutive adsorption step may occur for molar masses in the order of 10^4 D or smaller [410–413], to form presumably soluble nonstoichiometric complexes. A gel-to-liquid transition was put forward to explain the change from insoluble to soluble complexes [365,388]. The partially reversible binding of low to medium molar mass polyelectrolytes in ESA films seems also to be at the base of some spectacular cases of exponential film growth [414–417]. Here, local mobilization of excess polymers leads to its massive accumulation at the interface with correspondingly thick complexes formed with newly adsorbing partners. In agreement with the current main model, a critical linear charge density λ_c seems necessary for a successful film growth by ESA, when a charged polymer has to be deposited with a strong polyelectrolyte [418–424]. Studies on isomeric polyelectrolytes revealed that not only the absolute charge density, but also the relative position of the charged groups within the polymer plays a role [364]. The value of λ_c seems to increase linearly with the square root of the ionic strength [421]. Above λ_c, the variation in the increment of thickness versus charge density agrees well with the theories of polyelectrolyte solutions, passing successively through a charge-dependent "Debye–Hückel" regime, then through a charge-independent "strong-screening" regime where counterion condensation dominates the behavior [421]. However, as known for long, additional nonelectrostatic interactions between the oppositely charged partners may play an important role in ESA, too, such as hydrophobic forces [139,147,351,389,408,425–430] or H-bonding [427,431–434]. In fact, the strength of the latter is presumably responsible for the robustness of the system PAA/PAH at pH values lower than 10 [435–441], via $-NH_3^+ \div {}^-OOC-$ or $-NH_3^+ \div HOOC-$ sites at intermediate or low pH values, respectively (cf. Section II). It also explains easily the regularly recurring reports on successful alternating deposition of some noncharged polymer systems [421,442–455].

The value of λ_c can be adjusted by added salt via ion pairing and thus, the inclusion of mobile counterions and the films' thickness can be tuned (cf. Figure 5) [379,386,408,409,418,419,423,456–459]. But film thickness passes through a maximum with increasing ionic strength, and film growth is inhibited beyond a critical value [365,388,402,412,420,460–463]. The value at which the maximum is observed depends sensitively on both the polyelectrolyte pair used and on the low molar mass salt in the dipping solution [365,410,411,464,465].

The structural model of ESA films resulting from the combination of stratification with strong interdiffusion [53,352,353,386] (cf. Figure 3) presented in Section VI has been corroborated by many studies as a standard case. Accordingly, the local structure of binary ESA films is closer to the one of a scrambled-egg polyelectrolyte complex than to the one expected for a truly multilayered film. Though some internal order is obtained when employing polyelectrolytes with an oppositely charged rigid adsorbate [49,157,179,193,260,266,466–492], for instance, an exfoliated clay, such systems are best described by a simple model of multilayer growth, consisting in the random tiling of negatively charged platelets onto a dynamic cushion of an organic polyelectrolyte [179,489]. As a result of this random and dynamic process, platelets pack according to a distribution of vertical distances between nearest neighbors. The accumulation of such random fluctuations with ongoing film growth leads to the loss of layering of the platelets at a critical distance from the substrate. In this respect, most of such ESA films should be classified as nano-composites with preferential orientation of the platelets, rather than as true multilayers [488–492].

Nevertheless, recent studies on functional ionenes [139,361,364,387,492] demonstrate that ordered organic binary multilayers are possible, provided one of the polyelectrolytes bears groups which tend to induce structure in water, such as large hydrophobic segments or mesogens. In this respect, systems prone to form lyotropic mesophases are conducive to multilayers of higher order. But even under such circumstances, ordered multilayers emerge only when specific partners are employed. Further, the obtained substructure does not strictly coincide with the sequence of adsorption events, with more (or less) than one stratum being deposited per dipping cycle. Therefore, the order in such true multilayers results presumably from some degree of prestructuring of the water-swollen layer adsorbed during each step of deposition.

Different from the spontaneous structure formation in the vertical direction in special systems described above, lateral structure formation of ESA films has been only the result of patterning processes so far, either by selective adsorption on prepatterned substrates [303–305,493–514], or by patterning of the continuous films [264,514–517], for example, by lithographic techniques. Most reports on patterned multilayers were so far limited to the micron-scale, with patterning methods in the nanometer range only emerging [518].

In addition to the problem whether and how ESA films can be sub-structured, the possibility to obtain ordered arrangements of functional fragments within the films is of interest. Clearly, both types of internal order may occur independently. Particularly interesting are the cases where functional fragments are (preferentially) aligned, either in plane or normal to the film plane, or even oriented. Evidence for a preferential alignment of fragments parallel or normal to the substrate was given in several reports, mostly by optical studies [106,156,160,300,364,366,369,408,484,519–524] and occasionally by electrical conductivity studies [525–528]. This is understandable as in thin films, the size of functional fragments is comparable to the thickness increment per adsorbed layer so that interactions in plane and out of plane may be inherently different. In contrast, a preferential alignment of fragments in the film plane seems to be extremely rare [48,369,528,529] as even fast flow of the adsorbing solution over the substrate seems not to be effective to induce alignment [369], but may be favored by prepatterning the support [530,531].

Although a spontaneous orientation of functional fragments in the vertical direction is very rare, too, its realization has attracted substantial interest because of its importance for the function of certain proteins incorporated in the films [532–535], and in particular for the function of second-order nonlinear optical devices [536–548]. But the frequent hydration of the molecules poses several problems. On the one hand, some weak NLO effects are apparently due to oriented water molecules incorporated in the films [549,550]. On the other hand, the water molecules act as efficient plasticizer so that the resulting local mobility of the chromophores interferes with a stable orientation. Though a certain degree of orientation can be observed intermediately, reorientation to a more centrosymmetric arrangement takes place with ongoing film growth [146,538,542,549–554]. Still, special strategies based on orientation by surface effects via hydrophobic supports [541,542], chemical anchoring [314,315,548], intercalation [555,556], or inclusion complexation [537] have succeeded to provide stable oriented systems.

Concerning suitable substrates, new developments have mostly focused on nonplanar supports in the past 3 years. The initial studies using fiber substrates have been continued [557–559]. Colloidal supports with a size between 100 nm and a few microns found most interest [356–359,378,392,402,438,441,538,560–574]. In particular, such ESA systems have been used to prepare hollow polymer capsules [575–611] by removing the original support, applying selective solvents, chemical dissolution, or calcination. Besides serving for fundamental physical studies, such hollow capsules are investigated for encapsulation purposes, or as chemical microreactors. As for planar supports, hollow capsules can be made of polyelectrolyte pairs of opposite charge, as well as of a polyelectrolyte in combination with oppositely charged colloids (which of course have to be much smaller than the template used) or of colloid pairs of opposite charge. In the latter case, calcination of the sacrificial support offers an access to purely inorganic hollow capsules [462,612–621] or to hollow fibers, respectively [557]. Other nonplanar supports studied comprise nanorods [622–626], or the interior of capillaries [627–631] and even pores of bulk materials [558,624,632,633]. If the diameters of the cavities are larger than 0.5 μm, successful coating seems to proceed without problems. For smaller dimensions however, the situation is not clear. In fact due to their self-healing property, ESA films are able to span smaller pores thus forming coherent, dense films used, for example, as active layer for separation purposes [362,437,634–649]. Nevertheless most recently, nanorods have been prepared by growing ESA films inside small pores followed by removal of the external template [632,633]. Also, microchannels have been modified by ESA films [650,651]. Another interesting recent development are freestanding ESA films [484,624,652,653].

As pointed out in Section VII, chemical reactions in ESA films were focused on cross-linking for long time. Cross-linking seems to be of major interest in such reactions still, in order to stabilize the films [436,437,452,472,490,499,500,514,517,523,580,587,633,641,642,645,654–697] as well as for patterning [514,516,517,698]. For the latter purpose, noncross-linking alternatives have been reported, too [515,520,522,699–703] including surface relief gratings. Within a similar context, sol–gel chemistry has been studied within ESA films [466,618,625,704–707]. In contrast, the early interest in precursor strategies to obtain films loaded with functional polymers, particularly in the making of PPV [708–720], has dwindled. Instead, chemical reactions have found more interest for switching film properties photochemically [448,481,721–735] or electrochemically [736–746]. Further, the use of chemical reactions for obtaining noncentrosymmetrical arrangements of functional groups in the assemblies (often for nonlinear optical effects) has been pursued [313,314,536,548,747]. Another interesting development is the use of ESA films as matrix for chemical reactions [591,606,607,609,621,748–751], mostly for preparing colloidal metal deposits [593,752–761] or inorganic colloidal particles [574,762–764], sometimes combined with calcinations, etc. in order to prepare inorganic coatings [462,476–479,612–621,626,765–767].

Even without chemical cross-linking, the good inherent stability of many ESA films has been verified over the years, against ageing [639,646] as well as mechanical stress [425,525,581, 588,689,768,769]. Though measurements are difficult, some studies provided semiquantitative or even quantitative data on mechanical properties, moduli, hardness, adhesion, etc. [413,439,488,585, 587,602,605,678,690,770–772]. Also, the general high resistance to standard solvents has been corroborated in many studies [490,630,740,773,774]. The combination of two strong polyelectrolytes seems to provide more stable ESA films than the combination of a strong with a weak one [402,775,776]. Dissolution occurs typically only either by complex solvent mixtures [659–659,668], or by high ionic strengths [365,388,402,465,581,653,776–778], or by extreme pH-values when weak polyelectrolytes are employed [167,401,418,434,435,440,449,451,452,581,621,636,653,668,779,780]. Still, it is clear that most ESA films are at least somewhat hygroscopic and — as mentioned above — contain a substantial amount of bound water [408,457,570,781–784] that acts as plasticizer. In consequence under ambient conditions, films made of most standard polyelectrolytes are rather soft, internally mobile, relatively permeable to small and medium size molecules — but much less to polymers [362,415,458,581–583,587,588,594,595,600,601,604,611,621,635,644,665] and not dimensionally stable with changing humidity in air, or with ionic strength of contacting [457,683,775,783,785–789]. Nevertheless, integrity of such hydrogel coatings is guaranteed by the physical cross-linking via ion pairs. For instance, if ESA films are subjected to drying/swelling cycles, rearrangement of the polymer conformation may be so important that the films are smoothed [785,790,791]. Similarly, if they are exposed to aqueous media of differing ionic strength, rearrangement of the polymer conformation may smoothen the films [783,785,790,791], or may induce pore formation [434,436,449,453,589,792–796]. It is noteworthy that, ESA films tend to stabilize fragile molecules inside. For example, incorporation of proteins in ESA films tends to stabilize them against denaturation [523,799–801]. This is more remarkable as the secondary structure of polypeptides in ESA films is a complex function of the adsorption conditions and polyelectrolyte partner chosen [396,417,530,531,779,802,803].

Concerning possible applications, the proposed uses are still mainly based on the hydrogel character of ESA films, that is, on their tendency to swell efficiently in aqueous media without dissolution. As already discussed in Section IX up to the year 1999, such potential applications comprise biocompatible and nonfouling coatings [368,447,694,801,804–819], biodegradable coatings [774,814,820], enzyme immobilization or encapsulation [403,563,800,821–827], cell encapsulation

[770,828,829], controlled release/drug release [366,412,436,451,454,564,567,589,601–604,766, 774,780,792,793,818,830–832], separation membranes [362,437,634–649,683,685,691,833–836], (bio)sensors [406,499,571,692,693,773,837–865], or coatings for electrophoresis [411,627–631, 650,651]. Next to the hydrogel character, proposed applications exploit the controlled thinness of ESA films, for example, for protective coatings [439,684,686,819,866–868], light emitting and photovoltaic devices [426,472,502,695,711–719,767,869–883], optical coatings [436,619,689, 723,739,741,743,754,755,884–889], modified colloidal crystals [890–899], fluorescent marker particles [598,853,861,900,901], and modified electrodes [487,523,692,693,697,735,742,753,797, 808,838,848,854,857,860,880,902–923]. It is noteworthy that, though the layer-by-layer deposition and the versatility of useful materials allows the facile construction of highly complex ESA films [6,240,398,492,558,569,599,600,611,861,924–929], such films have been rarely suggested for uses yet.

XI. CONCLUSIONS

In comparison with other techniques for the preparation of thin organic multilayers, ESA presents a range of advantages, among which the most prominent is probably its versatility. Indeed, using this method, a broad range of organic and inorganic molecules can be assembled at a nanometer scale. Such control of the vertical stacking of molecules, complemented with patterning capabilities, will certainly lead to a broad range of applications in integrated optics and electronics as well as in biotechnologies. A disadvantage of this deposition technique is the often-encountered lack of internal organization of the multilayers.

Much progress in the understanding of the physical processes underlying the ESA has been gained over these last years, and the influence of the main processing parameters is now clearly established. Nevertheless, the optimization of the multilayer build-up remains essentially as an experimental trial-and-error process, and nonequilibrium theories taking into account all observed phenomena are still lacking. Consequently, the topic will remain an active field of research in the coming years.

We therefore think the future is bright for ESA, and many exciting developments are still to come.

ACKNOWLEDGMENTS

This work was supported by the Fonds National de la Recherche Scientifique F.N.R.S of Belgium, the DG Recherche Scientifique of the French Community of Belgium (conventions 94/99-173 and 00/05-261), and the Fraunhofer Society. The authors wish to thank the present and former members of their research teams working on ESA films, for their continuous collective efforts and stimulating interactions over the years.

REFERENCES

1. J.H. Fendler. *Chem. Mater.* 8:1616, 1996.
2. A. Dodabalapur. *Solid State Commun.* 102:259, 1997.
3. S. Mann. *Nature* 365:499, 1993.
4. H. Ringsdorf, B. Schlarb, and J. Venzmer. *Angew. Chem. Int. Ed. Engl.* 27:113, 1988.
5. T. Kunitake. *Thin Solid Films* 284:9, 1996.
6. M. Onda, Y. Lvov, K. Ariga, and T. Kunitake. *J. Ferment. Bioeng.* 82:502, 1996.
7. C. Nicolini. *Thin Solid Films* 284:1, 1996.
8. H. Byrd, E.P. Suponeva, A.B. Bocarsly, and M.E. Thompson. *Nature* 380:610, 1996.
9. A. Hagfeldt and M. Grätzel. *Chem. Rev.* 95:49, 1995.
10. V. Bach, D. Lupo, P. Comte, J.E. Moser, F. Weissörtel, J. Salbeck, H. Spreitzer, and M. Grätzel. *Nature* 395:583, 1998.

11. I. Langmuir. *Trans. Faraday Soc.* 15:62, 1920.
12. K.B. Blodgett. *J. Am. Chem. Soc.* 57:1007, 1935.
13. K.B. Blodgett and I. Langmuir. *Phys. Rev.* 51:964, 1937.
14. H. Kuhn, D. Möbius, and H. Bücher. In: A. Weissberger and B. Rossiter, Eds., *Physical Methods of Chemistry*. New York: Wiley, 1972, p. 577.
15. J.D. Swalen. *J. Mol. Electron* 2:155, 1986.
16. O. Albrecht and A. Laschewsky. *Macromolecules* 17:937, 1984.
17. V.V. Arslanov. *Adv. Colloid Interf. Sci.* 40:307, 1992.
18. T. Miyashita. *Prog. Polym. Sci.* 18:263, 1993.
19. F. Embs, D. Funhoff, A. Laschewsky, U. Licht, H. Ohst, W. Prass, H. Ringsdorf, G. Wegner, and R. Wehrmann. *Adv. Mater.* 3:25, 1991.
20. L. Netzer and J. Sagiv. *J. Am. Chem. Soc.* 105:675, 1983.
21. R. Maoz, L. Netzer, J. Gun, and J. Sagiv. *J. Chim. Phys.* 11/12:85, 1988.
22. A. Ulman. *Chem. Rev.* 96:1533, 1996.
23. R.J. Collins, I.T. Bae, D.A. Scherson, and C.N. Sukenik. *Langmuir* 12:5509, 1996.
24. T.J. Marks and M.A. Ratner. *Angew. Chem. Int. Ed. Engl.* 34:155, 1995.
25. R.G. Nuzzo and D.L. Allara. *J. Am. Chem. Soc.* 105:4481, 1983.
26. L.H. Dubois and R.G. Nuzzo. *Annu. Rev. Phys. Chem.* 43:437, 1992.
27. S.D. Evans, A. Ulman, K.E. Goppert-Beraduci, and L.J. Gerenser. *J. Am. Chem. Soc.* 113:5866, 1991.
28. T.L. Freeman, S.D. Evans, and A. Ulman. *Langmuir* 11:4413, 1995.
29. K. Hu, M. Brust, and J.A. Bard. *Chem. Mater.* 10:1160, 1998.
30. H. Lee, L.J. Kepley, H.-G. Hong, and T.E. Mallouk. *J. Am. Chem. Soc.* 110:618, 1988.
31. D.L. Feldheim and T.E. Mallouk. *Chem. Commun.* 2591, 1996.
32. A.C. Zeppenfeld, S.L. Fiddler, W.K. Ilam, B.J. Klopfenstein, and C.J. Page. *J. Am. Chem. Soc.* 116:9158, 1994.
33. M.A. Ansell, A.C. Zeppenfeld, K. Yoshimoto, E.B. Cogan, and C.J. Page. *Chem. Mater.* 8:591, 1996.
34. X.Q. Zhang, H.M. Wu, Z.H. Lu, and X.Z. You. *Thin Solid Films* 284/285:224, 1996.
35. H. Xiong, M. Cheng, Z. Zhou, X. Zhang, and J.C. Shen. *Adv. Mater.* 10:529, 1998.
36. I. Ichinose, T. Kawakami, and T. Kunitake. *Adv. Mater.* 10:535, 1998.
37. D.M. Sarno, D. Grosfeld, J. Snyder, B. Jiang, and W.E. Jones. *Polym. Prepr. Am. Chem. Soc. Polym. Chem. Div.* 39(2):1101, 1998.
38. D.L. Thomsen III and F. Papadimitrakopoulos. *Macromol. Symp.* 125:143, 1997.
39. H.E. Katz, G. Scheller, T.M. Putvinski, M.L. Schilling, W.L. Wilson, and C.E.D. Chidsey. *Science* 254:1485, 1991.
40. R.K. Iler. *J. Colloid Interface Sci.* 21:569, 1966.
41. C.-G. Gölander, H. Arwin, J.C. Eriksson, I. Lundstrom, and R. Larsson. *Colloids Surf.* 5:1, 1982.
42. P. Fromherz. In: A. Baumeister and W. Vogell, Eds., *Electron Microscopy at Molecular Dimensions*. Berlin: Springer, 1980, pp. 338–349.
43. R.H. Tredgold, C.S. Winter, and Z.I. El-Badawy. *Electron Lett.* 21:554, 1985.
44. G. Decher and J.D. Hong. *Makromol. Chem. Macromol. Symp.* 46:321, 1991.
45. G. Decher and J.D. Hong. *Ber. Bunsenges Phys. Chem.* 95:1430, 1991.
46. J.D. Hong, K. Lowack, J. Schmitt, and G. Decher. *Prog. Colloid. Polym. Sci.* 93:98, 1993.
47. Y. Lvov, G. Decher, and G. Sukhorukov. *Macromolecules* 26:5396, 1993.
48. T.M. Cooper, A.L. Campbell, and R.L. Crane. *Langmuir* 11:2713, 1995.
49. E.R. Kleinfeld and G.S. Ferguson. *Science* 265:370, 1994.
50. V.N. Bliznyuk and V.V. Tsukruk. *Polym. Prepr. Am. Chem. Soc. Div. Polym. Chem.* 38(1):963, 1997.
51. S. Watanabe and S.L. Regen. *J. Am. Chem. Soc.* 116:8855, 1994.
52. Y. Lvov, H. Haas, G. Decher, H. Möhwald, A. Mikhailov, B. Mtchedlishvily, E. Morgunova, and B. Vainshtein. *Langmuir* 10:4232, 1994.
53. G. Decher. *Science* 277:1232, 1997.
54. G. Decher, M. Eckle, J. Schmitt, and B. Struth. *Curr. Opin. Colloid Interface Sci.* 3:32, 1998.
55. A. Laschewsky. *Eur. Chem. Chronicle* 2:13, 1997.
56. W. Knoll. *Curr. Opin. Colloid Interface Sci.* 1:137, 1996.
57. G. Decher. In: J.C. Salamone, Ed., *The Polymeric Materials Encyclopedia: Synthesis, Properties, and Applications*. Boca Raton, FL: CRC Press Inc, 1996, pp. 4540–4546.

58. G. Decher. In: J.-P. Sauvage and M.W. Hosseini, Eds., *Comprehensive Supramolecular Chemistry.* Oxford: Pergamon Press, 1996, pp. 507–528.
59. M. Sano, Y. Lvov, and T. Kunitake. *Ann. Rev. Mater. Sci.* 26:153, 1996.
60. S.W. Keller, H.-N. Kim, and T.E. Mallouk. *J. Am. Chem. Soc.* 116:8817, 1994.
61. K.M. Lenahan, Y.X. Wang, Y.J. Liu, R.O. Claus, J.R. Heflin, D. Marciu, and C. Figura. *Adv. Mater.* 10:853, 1998.
62. J.W. Baur, P. Besson, S.A. O'Connor, and M.F. Rubner. *Mater. Res. Soc. Proc.* 413:583, 1996.
63. R.C. Advincula, D. Roitman, C. Frank, K. Wolfgang, A. Baba, and F. Kaneko. *Polym. Prepr.* 40(1):467, 1999.
64. G. Decher, Y. Lvov, and J. Schmitt. *Thin Solid Films* 244:772, 1994.
65. J. Schmitt, T. Grünewald, G. Decher, P.S. Pershan, K. Kjaer, and M. Lösche. *Macromolecules* 26:7058, 1993.
66. D. Laurent and J.B. Schlenoff. *Langmuir* 13:1552, 1997.
67. A. Laschewsky, B. Mayer, E. Wischerhoff, X. Arys, A. Jonas, P. Bertrand, and A. Delcorte. *Thin Solid Films* 284/285:334, 1996.
68. W.B. Stockton and M.F. Rubner. *Mater. Res. Soc. Proc.* 369:587, 1994.
69. L. Wang, Z. Wang, X. Zhang, and J. Shen. *Macromol. Rapid. Commun.* 18:509, 1997.
70. L. Wang, Y. Fu, Z. Wang, Y. Fan, and X. Zhang. *Langmuir* 15:1360, 1999.
71. W.B. Stockton and M.F. Rubner. *Macromolecules* 30:2717, 1997.
72. H. Hong, D. Davidov, M. Tarabia, H. Chayet, I. Benjamin, E.Z. Faraggi, Y. Avny, and R. Neumann. *Synth. Met.* 85:1265, 1997.
73. Y. Shimazaki, M. Mitsuishi, S. Ito, and M. Yamamoto. *Langmuir* 13:1385, 1997.
74. Y. Shimazaki, M. Mitsuishi, S. Ito, and M. Yamamoto. *Langmuir* 14:2768, 1998.
75. F. Caruso, K. Niikura, D.N. Furlong, and Y. Okahata. *Langmuir* 13:3422, 1997.
76. J.-I. Anzai, Y. Kobayashi, N. Nakamura, M. Nishimura, and T. Hoshi. *Langmuir* 15:221, 1999.
77. J.-I. Anzai and M. Nishimura. *J. Chem. Soc. Perkin Trans.* 2:1887, 1997.
78. T. Cassier, K. Lowack, and G. Decher. *Supramol. Sci.* 5:309, 1998.
79. C. Bourdillon, C. Demaille, J. Moiroux, and J.-M. Saveant. *J. Am. Chem. Soc.* 116:10328, 1994.
80. Y. Lvov, K. Ariga, I. Ichinose, and T. Kunitake. *Thin Solid Films* 285:797, 1996.
81. G.B. Sukhorukov, H. Möhwald, G. Decher, and Y.M. Lvov. *Thin Solid Films* 284/285:220, 1996.
82. Y.F. Nicolau. *Appl. Surf. Sci.* 22/23:1061, 1985.
83. S. Lindroos, T. Kanniainen, and M. Leskelä. *Appl. Surf. Sci.* 75:70, 1994.
84. Y. Lvov, G. Decher, H. Haas, H. Möhwald, and A. Kalachev. *Phys. B* 198:89, 1994.
85. Y. Lvov, H. Haas, G. Decher, H. Möhwald, and A. Kalachev. *J. Phys. Chem.* 97:12835, 1993.
86. Y.M. Lvov and G. Decher. *Crystallogr. Rep.* 39:628, 1994.
87. G. Decher, J.D. Hong, and J. Schmitt. *Thin Solid Films* 210/211:831, 1992.
88. J.-M. Leväsalmi and T.J. McCarthy. *Macromolecules* 30:1752, 1997.
89. V. Phuvanartnuruks and T.J. McCarthy. *Polym. Prepr. Am. Chem. Soc. Div. Polym. Chem.* 38:961, 1997.
90. V. Phuvanartnuruks and T.J. McCarthy. *Macromolecules* 31:1906, 1998.
91. M.C. Hsieh, R.J. Farris, and T.J. McCarthy. *Polym. Prepr. Am. Chem. Soc. Div. Polym. Chem.* 38:670, 1997.
92. M.C. Hsieh, R.J. Farris, and T.J. McCarthy. *Macromolecules* 30:8453, 1997.
93. W. Chen and T.J. McCarthy. *Macromolecules* 30:78, 1997.
94. G.B. Sukhorukov, E. Donath, H. Lichtenfeld, E. Knippel, M. Knippel, A. Budde, and H. Möhwald. *Colloids Surf. A* 137:253, 1998.
95. G.B. Sukhorukov, E. Donath, S. Davis, H. Lichtenfeld, F. Caruso, V.I. Popov, and H. Möhwald. *Polym. Adv. Technol.* 9:759, 1998.
96. F. Caruso, E. Donath, and H. Möhwald. *J. Phys. Chem. B* 102:2011, 1998.
97. A. Baba, F. Kaneko, and R.C. Advincula. *Polym. Prepr.* 40(1):488, 1999.
98. A. Tronin, Y. Lvov, and C. Nicolini. *Colloid. Polym. Sci.* 272:1317, 1994.
99. D. Korneev, Y. Lvov, G. Decher, J. Schmitt, and S. Yaradaikin. *Physica B* 213/214:954, 1995.
100. M. Lösche, J. Schmitt, G. Decher, W.G. Bouwman, and K. Kjaer. *Macromolecules* 31:8893, 1998.
101. K. Shinbo, K. Suzuki, K. Kato, F. Kaneko, and S. Kobayashi. *Thin Solid Films* 327/329:209, 1998.
102. P. Berndt, K. Kurihara, and T. Kunitake. *Langmuir* 8:2486, 1992.
103. K. Lowack and C.A. Helm. *Macromolecules* 31:823, 1998.
104. R. von Klitzing and H. Möhwald. *Langmuir* 11:3554, 1995.

105. E. Donath, D. Walther, V.N. Shilov, E. Knippel, A. Budde, K. Lowack, C.A. Helm, and H. Möhwald. *Langmuir* 13:5294, 1997.
106. J.J. Ramsden, Y.M. Lvov, and G. Decher. *Thin Solid Films* 254:246, 1995.
107. V.N. Bliznyuk, D.W. Visser, V.V. Tsukruk, A.L. Campbell, T. Bunning, and W.W. Adams. *Polym. Prepr.* 37(2):608, 1996.
108. V.V. Tsukruk, V.N. Bliznyuk, D. Visser, A.L. Campbell, T.J. Bunning, and W.W. Adams. *Macromolecules* 30:6615, 1997.
109. G. Decher and J. Schmitt. *Prog. Colloid. Polym. Sci.* 89:160, 1992.
110. V.V. Belyaev, A.L. Tolstikhina, N.D. Stepina, R.L. Kayushina. *Crystallogr. Rep.* 43:124, 1998.
111. G.B. Sukhorukov, J. Schmitt, and G. Decher. *Ber. Bunsenges. Phys. Chem.* 100:948, 1996.
112. G. Decher, Y. Lvov, and J. Schmitt. *Thin Solid Films* 244:772, 1994.
113. R. von Klitzing and H. Möhwald. *Thin Solid Films* 284/285:352, 1996.
114. R. von Klitzing and H. Möhwald. *Macromolecules* 29:6901, 1996.
115. R. Kayushina, Y. Lvov, N. Stepina, V. Belayev, and Y. Khurgin. *Thin Solid Films* 284/285:246, 1996.
116. M.L. Bruening, J.J. Harris, and P.M. DeRose. *Polym. Prepr.* 40(1):451, 1999.
117. T.P. Russell. *Mat. Sci. Rep.* 5:171, 1990.
118. J.S. Higgins and H.C. Benoît. *Polymers and Neutron Scattering*. Oxford: Clarendon Press, 1994.
119. V.W. Stone, A.M. Jonas, X. Arys, and R. Legras. *Macromolecules.* 33:3031, (2000).
120. E.M. Arnett. *Prog. Phys. Org. Chem.* 1:223, 1963.
121. H.C. Brown, D.H. McDaniel, and O. Hflinger. In: E.A. Braude and F.C. Nachod, Eds., *Determination of Organic Structures by Physical Methods*. New York: Academic Press, 1955, p. 567.
122. J.H. Cheung, W.B. Stockton, and M.F. Rubner. *Macromolecules* 30:2712, 1997.
123. Y. Lvov, K. Ariga, I. Ichinose, and T. Kunitake. *J. Am. Chem. Soc.* 117:6117, 1995.
124. E. Brynda and M. Houska. *Macromol. Rapid. Commun.* 19:173, 1998.
125. J.B. Schlenoff, H. Ly, and M. Li. *J. Am. Chem. Soc.* 120:7626, 1998.
126. M. Koetse, A. Laschewsky, and T. Verbiest. *Mat. Sci. Eng. C* 10:107, 1999.
127. G.J. Fleer, M.A. Cohen Stuart, J.M.H.M. Scheutjens, T. Cosgrove, and B. Vincent. *Polymers at Interfaces*. London: Chapman & Hall, 1993, pp. 343–373.
128. P. Linse. *Macromolecules* 29:326, 1996.
129. H.G.M. van de Stccg, M.A. Cohen Stuart, A. de Keizer, and B.H. Bijsterbosch. *Langmuir* 8:2538, 1992.
130. M.R. Böhmer, O.A. Evers, and J.M.H.M. Scheutjens. *Macromolecules* 23:2301, 1990.
131. O.J. Rojas, P.M. Claesson, D. Muller, and R.D. Neuman. *J. Colloid Interface Sci.* 205:77, 1998.
132. N.G. Hoogeveen, A. Martien, M.A. Cohen Stuart, and G.J. Fleer. *J. Colloid Interface Sci.* 182:133, 1996.
133. D. Andelman and J.-F. Joanny. *Macromolecules* 24:6041, 1991.
134. D. Andelman and J.-F. Joanny. *J. Phys. II* 3:121, 1993.
135. M. Muthukumar. *Curr. Opin. Colloid Interface Sci.* 3:48, 1998.
136. M. Muthukumar. *J. Chem. Phys.* 103:4723, 1995.
137. K.W. Mattison, P.L. Dubin, and I.J. Brittain. *J. Phys. Chem. B* 102:3830, 1998.
138. S. Neyret, L. Ouali, F. Candau, and E. Pefferkorn. *J. Colloid Interface Sci.* 176:86, 1995.
139. P. Fischer, A. Laschewsky, E. Wischerhoff, X. Arys, A. Jonas, and R. Legras. *Macromol. Symp.* 137:1, 1999.
140. K. Lowack and C.A. Helm. *Macromolecules* 28:2912, 1993.
141. N.G. Hoogeveen, A. Martien, M.A. Cohen Stuart, and G.J. Fleer. *J. Colloid Interface Sci.* 182:146, 1996.
142. J.B. Schlenoff and M. Li. *Ber. Bunsenges Phys. Chem.* 100:943, 1996.
143. N.G. Hoogeveen, A. Martien, M.A. Cohen Stuart, and G.J. Fleer. *Langmuir* 12:3675, 1996.
144. T. Cassagneau, T.E. Mallouk, and J.H. Fendler. *J. Am. Chem. Soc.* 120:7848, 1998.
145. Y. Lvov, S. Yamada, and T. Kunitake. *Thin Solid Films* 300:107, 1997.
146. A. Laschewsky, E. Wischerhoff, M. Kauranen, and A. Persoons. *Macromolecules* 30:8304, 1997.
147. D. Cochin and A. Laschewsky. *Macromol. Chem. Phys.* 200:609, 1999.
148. A. Laschewsky, B. Mayer, E. Wischerhoff, X. Arys, A. Jonas, P. Bertrand, and A. Delcorte. *Thin Solid Films* 284/285:334, 1996.
149. S. Schwarz, K.-J. Eichhorn, E. Wischerhoff, and A. Laschewsky. *Coll. Surf. A* 159:491, 1999.
150. B. Laguitton, X. Arys, A.M. Jonas, and A. Laschewsky. Unpublished Results.
151. A. Wu, J. Lee, and M.F. Rubner. *Thin Solid Films* 327/329:663, 1998.
152. B. Lehr, M. Seufert, G. Wenz, and G. Decher. *Supramol. Sci.* 2:199, 1995.
153. D. Yoo, S.S. Shiratori, and M.F. Rubner. *Macromolecules* 31:4309, 1998.

154. M. Lütt, M.R. Fitzsimmons, and D. Li. *J. Phys. Chem. B* 102:400, 1998.

155. R. Advincula, E. Aust, W. Meyer, and W. Knoll. *Langmuir* 12:3536, 1996.

156. X. Arys, A.M. Jonas, B. Laguitton, R. Legras, A. Laschewsky, and E. Wischerhoff. *Prog. Org. Coat* 34:108, 1998.

157. E.R. Kleinfeld and G.S. Ferguson. *Chem. Mater.* 8:1575, 1996.

158. Y. Lvov, K. Ariga, and T. Kunitake. *Chem. Lett.* 12:2323, 1994.

159. S. Kim, J. Jackiw, E. Robinson, K.S. Schanze, J.R. Reynolds, J. Baur, M.F. Rubner, and D. Boils. *Macromolecules* 31:964, 1998.

160. M. Ferreira and M.F. Rubner. *Macromolecules* 28:7107, 1995.

161. M. Ferreira, J.H. Cheung, and M.F. Rubner. *Thin Solid Films* 244:806, 1994.

162. M. Raposo, R.S. Pontes, L.H.C. Mattoso, and O.N. Oliveira. *Macromolecules* 30:6095, 1997.

163. F. Saremi, G. Lange, and B. Tieke. *Adv. Mater.* 8:923, 1996.

164. I. Moriguchi and J.H. Fendler. *Chem. Mater.* 10:2205, 1998.

165. J. Tian, C.C. Wu, M.E. Thompson, J.C. Sturm, and R.A. Register. *Chem. Mater.* 7:2190, 1995.

166. A. Diederich and M. Lösche. *Adv. Biophys.* 34:205, 1997.

167. S. Dante, R. Advincula, C.W. Frank, and P. Stroeve. *Langmuir* 15:193, 1999.

168. S. Yamada, A. Harada, T. Matsuo, S. Ohno, I. Ichinose, and T. Kunitake. *Jpn. J. Appl. Phys.* 36:L1110, 1997.

169. S. Balasubramanian, X.G. Wang, H.C. Wang, K. Yang, J. Kumar, S.K. Tripathy, and L. Li. *Chem. Mater.* 10:1554, 1998.

170. M.J. Roberts, G.A. Lindsay, W.N. Herman, and K.J. Wynne. *J. Am. Chem. Soc.* 120:11202, 1998.

171. H. Hong, M. Tarabia, H. Chayet, S. Davidov, E.Z. Faraggi, Y. Avny, R. Neumann, and S. Kirstein. *J. Appl. Phys.* 79:3082, 1996.

172. H. Hong, D. Davidov, Y. Avny, H. Chayet, E.Z. Faraggin, and R. Neumann. *Adv. Mater.* 7:846, 1995.

173. M. Onoda, H. Nakayama, T. Yamaue, K. Tada, and K. Yoshino. *Jpn. J. Appl. Phys. Part 1* 36:5322, 1997.

174. M. Houska and E. Brynda. *J. Colloid Interface Sci.* 188:243, 1997.

175. X. Arys, A.M. Jonas, A. Laschewsky, and R. Legras. Unpublished Results.

176. T. Serizawa, H. Goto, A. Kishida, T. Endo, and M. Akashi. *J. Polym. Sci. Part A. Polym. Chem.* 36:801, 1999.

177. M.R. Linford, M. Auch, and H. Möhwald. *J. Am. Chem. Soc.* 120:179, 1998.

178. M. Onda, Y. Lvov, K. Ariga, and T. Kunitake. *Jpn. J. Appl. Phys. Lett.* 36:L1608, 1997.

179. N.A. Kotov, T. Haraszti, L. Turi, G. Zavala, R.E. Geer, I. Dékány, and J.H. Fendler. *J. Am. Chem. Soc.* 119:6821, 1997.

180. S.L. Clark and P.T. Hammond. *Adv. Mater.* 10:1515, 1998.

181. N.A. Kotov, S. Magonov, and E. Tropsha. *Chem. Mater.* 10:886, 1998.

182. J.J. Fendler. *Stud. Surf. Sci. Catal.* 103:261, 1997.

183. X. Arys, A.M. Jonas, B. Laguitton, A. Laschewsky, R. Legras, and E. Wischerhoff. *Thin Solid Films* 329:734, 1998.

184. M. Paßmann, G. Wilbert, D. Cochin, and R. Zentel. *Macromol. Chem. Phys.* 199:179, 1998.

185. F. van Ackern, L. Krasemann, and B. Tieke. *Thin Solid Films* 327/329:762, 1998.

186. Y. Lvov, M. Onda, K. Ariga, and T. Kunitake. *J. Biomater. Sci. Polym. Ed.* 9:345, 1998.

187. Y. Lvov, G. Decher, and H. Möhwald. *Langmuir* 9:481, 1993.

188. S.L. Clark, M.F. Montague, and P.T. Hammond. *Polym. Prepr. Am. Chem. Soc. Div. Polym. Chem.* 38(1):967, 1997.

189. G. Mao, Y.-H. Tsao, M. Tirrell, H.T. Davis, V. Hessel, and H. Ringsdorf. *Langmuir* 11:942, 1995.

190. L.A. Godinez, R. Castro, and A.E. Kaifer. *Langmuir* 12:5087, 1996.

191. J.H. Cheung, A.F. Fou, and M.F. Rubner. *Thin Solid Films* 244:985, 1994.

192. Y. Lvov, K. Ariga, I. Ichinose, and T. Kunitake. *Langmuir* 12:3038, 1996.

193. H.N. Kim, S.W. Keller, T.E. Mallouk, J. Schmitt, and G. Decher. *Chem. Mater.* 9:1414, 1997.

194. W. Knoll. *Ann. Rev. Phys. Chem.* 569, 1998.

195. Y.M. Lvov, Z. Lu, J.B. Schenkman, X. Zu, and J.F. Rusling. *J. Am. Chem. Soc.* 120:4073, 1998.

196. K. Ariga, Y. Lvov, and T. Kunitake. *J. Am. Chem. Soc.* 119:2224, 1997.

197. A. Laschewsky, E. Wischerhoff, P. Bertrand, and A. Delcorte. *Macromol. Chem. Phys.* 198:3239, 1997.

198. A. Laschewsky, E. Wischerhoff, S. Denzinger, H. Ringsdorf, A. Delcorte, and P. Bertrand. *Chem. Eur. J.* 3:34, 1997.

199. T. Serizawa, H. Takeshita, and M. Akashi. *Langmuir* 14:4088, 1998.

200. Y.M. Lvov, J.F. Rusling, D.L. Thomsen, F. Papadimitrakopoulos, T. Kawakami, and T. Kunitake. *Chem. Commun.* 1229, 1998.
201. A. Laschewsky, F. Mallwitz, J.-F. Baussard, D. Cochin, P. Fischer, J.-L. Habib-Jiwan, E. Wischerhoff, *Macromol. Symp.* 211:135, 2004.
202. H. Hong, R. Steitz, S. Kirstein, and D. Davidov. *Adv. Mater.* 10:1104, 1998.
203. N.A. Kotov, I. Dekany, and J.H. Fendler. *J. Phys. Chem.* 99:13065, 1995.
204. G. Kim, R.J. Farris, and T.J. McCarthy. *Polym. Prepr. Am. Chem. Soc. Div. Polym. Chem.* 38(2):672, 1997.
205. A. Delcorte, P. Bertrand, E. Wischerhoff, and A. Laschewsky. *Langmuir* 13:5125, 1997.
206. E. Brynda and M. Houska. *J. Colloid Interface Sci.* 183:18, 1996.
207. H. Kim and M.W. Urban. *Langmuir* 14:7235, 1998.
208. M. Gao, X. Zhang, B. Yang, and J. Shen. *J. Chem. Soc. Chem. Commun.* 2229, 1994.
209. W. Fabianowski, M. Roszko, and W. Brodziñska. *Thin Solid Films* 327/329:743, 1998.
210. P. Rilling, T. Walter, R. Pommersheim, and W. Vogt. *J. Membr. Sci.* 129:283, 1997.
211. G. Mao, Y.-H. Tsao, M. Tirrell, H.T. Davis, V. Hessel, and H. Ringsdorf. *Langmuir* 9:3461, 1993.
212. G. Mao, Y.-H. Tsao, M. Tirrell, H.T. Davis, V. Hessel, and H. Ringsdorf. *Langmuir* 11:942, 1995.
213. G. Mao, Y.-H. Tsao, M. Tirrell, H.T. Davis, V. Hessel, J. van Esch, and H. Ringsdorf. *Langmuir* 10:4174, 1994.
214. B. Sellergren, A. Swietlow, T. Arnebrant, and K. Unger. *Anal. Chem.* 68:402, 1996.
215. X. Zhang, M. Gao, X. Kong, Y. Sun, and J. Shen. *J. Chem. Soc. Chem. Commun.* 1055, 1994.
216. K. Araki, M.J. Wagner, and M.S. Wrighton. *Langmuir* 12:5393, 1996.
217. J.S. Do, T.H. Ha, J.D. Hong, and K. Kim. *Bull. Korean Chem. Soc.* 19:257, 1998.
218. M. Gao, Mi Gao, X. Zhang, Y. Yang, B. Yang, and J. Shen. *J. Chem. Soc. Chem. Commun.* 2777, 1994.
219. W. Kong, X. Zhang, M.L. Gao, H. Zhou, W. Li, and J.C. Shen. *Macromol. Rapid. Commun.* 15:405, 1994.
220. W. Kong, L.P. Wang, M.L. Gao, H. Zhou, X. Zhang, W. Li, and J.C. Shen. *J. Chem. Soc. Chem. Commun.* 1297, 1994.
221. G. Decher and J.D. Hong. *Ber. Bunsenges Phys. Chem.* 95:1430, 1980.
222. M. Gao, X. Kong, X. Zhang, and J. Shen. *Thin Solid Films* 244:815, 1994.
223. T.M. Cooper, A.L. Campbell, and R.L. Crane. *Polym. Prepr.* 36(1):377, 1995.
224. D.Q. Li, M. Lutt, M.R. Fitzsimmons, R. Synowicki, M.E. Hawley, and G.W. Brown. *J. Am. Chem. Soc.* 120:8797, 1998.
225. J.K. Lee, D.S. Yoo, E.S. Handy, and M.F. Rubner. *Appl. Phys. Lett.* 69:1686, 1996.
226. H.P. Zheng, R.F. Zhang, Y. Wu, and J.C. Shen. *Chem. Lett.* 909, 1998.
227. D. Yoo, A. Wu, J. Lee, and M.F. Rubner. *Synth. Met.* 85:1425, 1996.
228. I. Ichinose, H. Tagawa, S. Mizuki, Y. Lvov, and T. Kunitake. *Langmuir* 14:187, 1998.
229. R. Advincula, E. Fells, N. Jones, J. Guzman, A. Baba, and F. Kaneko. *Polym. Prepr.* 40(1):443, 1999.
230. S. Yamada, A. Harada, T. Matsuo, S. Ohno, I. Ichinose, and T. Kunitake. *Jpn. J. Appl. Phys.* 36:L1110, 1997.
231. H. Fukumoto and Y. Yonezawa. *Thin Solid Films* 327/329:748, 1998.
232. D. Yoo, J.-K. Lee, and M.F. Rubner. *Mater. Res. Soc. Proc.* 413:395, 1996.
233. F. Saremi, E. Maassen, B. Tieke, G. Jordan, and W. Rammensee. *Langmuir* 11:1068, 1995.
234. F. Saremi, B. Tieke, G. Jordan, and W. Rammensee. *Supramol. Sci.* 4:471, 1997.
235. F. Saremi and B. Tieke. *Adv. Mater.* 7:379, 1995.
236. I. Ichinose, K. Fujiyoshi, S. Mizuki, Y. Lvov, and T. Kunitake. *Chem. Lett.* 257, 1996.
237. T.S. Lee, J. Kim, and J. Kumar. *Macromol. Chem. Phys.* 199:1445, 1998.
238. K.S. Alva, J. Kumar, K.A. Marx, and S.K. Tripathy. *Macromolecules* 30:4024, 1997.
239. V. Charlier, A. Laschewsky, B. Mayer, and E. Wischerhoff. *Macromol. Symp.* 126:105, 1997.
240. A. Laschewsky, B. Mayer, E. Wischerhoff, X. Arys, and A. Jonas. *Ber. Bunsenges Phys. Chem.* 100:1033, 1996.
241. A. Delcorte, P. Bertrand, X. Arys, A. Jonas, E. Wischerhoff, B. Mayer, and A. Laschewsky. *Surf. Sci.* 366:149, 1996.
242. D. Cochin, M. Paßmann, G. Wilbert, R. Zentel, E. Wischerhoff, and A. Laschewsky. *Macromolecules* 30:4775, 1997.
243. L. Balogh, L. Samuelson, K.S. Alva, and A. Blumstein. *J. Polym. Sci.: Part A: Polym. Chem.* 36:703, 1998.

244. L. Balogh, L. Samuelson, K.S. Alva, and A. Blumstein. *Macromolecules* 29:4180, 1996.
245. A. Blumstein and L. Samuelson. *Adv. Mater.* 10:173, 1998.
246. V.V. Tsukruk. *Adv. Mater.* 10:253, 1998.
247. V.V. Tsukruk, F. Rinderspacher, and V.N. Bliznyuk. *Langmuir* 13:2171, 1997.
248. J. Tian, M.E. Thompson, C.-C. Wu, J.C. Sturm, R.A. Register, M.J. Marsella, and T.M. Swager. *Polym. Prepr.* 35(2):761, 1994.
249. A.C. Fou and M.F. Rubner. *Macromolecules* 28:7115, 1995.
250. G.B. Sukhorukov, M.M. Montrel, A.I. Petrov, L.I. Shabarchina, and B.I. Sukhorukov. *Biosensors Bioelectronics* 11:913, 1996.
251. M.M. Montrel, G.B. Sukhorukov, A.I. Petrov, L.I. Shabarchina, and B.I. Sukhorukov. *Sensors Actuators B* 42:225, 1997.
252. M.M. Montrel, G.B. Sukhorukov, L.I. Shabarchina, N.V. Apolonnik, and B.I. Sukhorukov. *Mat. Sci. Eng. C* 5:275, 1998.
253. Y. Sun, J. Sun, X. Zhang, C. Sun, Y. Wang, and J. Shen. *Thin Solid Films* 327/329:730, 1998.
254. S.F. Hou, H.Q. Fang, and H.Y. Chen. *Anal. Lett.* 30:1631, 1997.
255. F. Caruso, D.N. Furlong, K. Ariga, I. Ichinose, and T. Kunitake. *Langmuir* 14:4559, 1998.
256. J. Hodak, R. Etchenique, E.J. Calvo, K. Singhal, and P.N. Bartlett. *Langmuir* 13:2708, 1997.
257. F. Caruso, K. Niikura, D.N. Furlong, and Y. Okahata. *Langmuir* 13:3427, 1997.
258. J.-A. He, L. Samuelson, L. Li, J. Kumar, and S.K. Tripathy. *Langmuir* 14:1674, 1998.
259. J.-A. He, L. Samuelson, L. Li, J. Kumar, and S.K. Tripathy. *J. Phys. Chem. B* 102:7067, 1998.
260. G.S. Ferguson and E.R. Kleinfeld. *Adv. Mater.* 7:414, 1995.
261. S.W. Keller, S.A. Johnson, E.S. Brigham, E.H. Yonemoto, and T.E. Mallouk. *J. Am. Chem. Soc.* 117:12879, 1995.
262. D.M. Kaschak and T.E. Mallouk. *J. Am. Chem. Soc.* 118:4222, 1996.
263. H.G. Hong. *Bull. Korean Chem. Soc.* 16:1145, 1995.
264. J. Kerimo, D.M. Adams, P.F. Barabara, D.M. Kaschak, and T.E. Mallouk. *J. Phys. Chem. B* 102:9451, 1998.
265. P.J. Ollivier, N.I. Kovtyukhova, S.W. Keller, and T.E. Mallouk. *Chem. Commun.* 15:1563, 1998.
266. N.A. Kotov, I. Dékany, and J.H. Fendler. *Adv. Mater.* 8:637, 1996.
267. T. Cassagneau and J.H. Fendler. *Adv. Mater.* 10:877, 1998.
268. J.H. Fendler. *Croatica. Chem. Acta* 71:1127, 1998.
269. K. Ariga, Y. Lvov, M. Onda, I. Ichinose, and T. Kunitake. *Chem. Lett.* 2:126, 1997.
270. F. Caruso, D.G. Kurth, D. Volkmer, M.J. Koop, and A. Muller. *Langmuir* 14:3462, 1998.
271. I. Ichinose, H. Tagawa, S. Mizuki, Y. Lvov, and T. Kunitake. *Langmuir* 14:187, 1998.
272. D. Ingersoll, P.J. Kulesza, and L.R. Faulkner. *J. Electrochem. Soc.* 141:140, 1994.
273. J. Schmitt, G. Decher, W.J. Dressick, S.L. Brandow, R.E. Geer, R. Shashidhar, and J.M. Calvert. *Adv. Mater.* 9:61, 1997.
274. W. Schrof, S. Rozouvan, Keuren, E. Van, E.D. Horn, J. Schmitt, and G. Decher. *Adv. Mater.* 3:338, 1998.
275. Y.J. Liu, Y.X. Wang, and R.O. Claus. *Chem. Phys. Lett.* 298:315, 1998.
276. Y. Liu and R.O. Claus. *J. Appl. Phys.* 96:419, 1991.
277. T. Yonezawa, S.-Y. Onoue, and T. Kunitake. *Adv. Mater.* 10:414, 1998.
278. Y. Liu, A. Wang, and R. Claus. *J. Phys. Chem. B* 101:1385, 1997.
279. M.Y. Gao, X. Zhang, B. Yang, F. Li, and J.C. Shen. *Thin Solid Films* 284:242, 1996.
280. M. Gao, B. Richter, S. Kirstein, and H. Möhwald. *J. Phys. Chem. B* 102:4096, 1998.
281. J. Sun, E. Hao, Y. Sun, X. Zhang, B. Yang, S. Zou, J. Shen, and S. Wang. *Thin Solid Films* 327/329:528, 1998.
282. Y. Sun, E. Hao, X. Zhang, B. Yang, M. Gao, and J. Shen. *Chem. Commun.* 2381, 1996.
283. Y. Sun, E. Hao, X. Zhang, B. Yang, J. Shen, L. Chi, and H. Fuchs. *Langmuir* 13:5168, 1997.
284. Y. Liu, A. Wang, and R.O. Claus. *Appl. Phys. Lett.* 71:2265, 1997.
285. V.N. Bliznyuk and V.V. Tsukruk. *Polym. Prepr. Am. Chem. Soc. Div. Polym. Chem.* 38(1):963, 1997.
286. K.-U. Fulda, D. Piecha, B. Tieke, and H. Yarmohammadipour. *Prog. Colloid. Polym. Sci.* 101:178, 1996.
287. K.-U. Fulda, A. Kampes, L. Krasemann, and B. Tieke. *Thin Solid Films* 327/329:752, 1998.
288. T. Salditt, Q. An, A. Plech, C. Eschbaumer, and U.S. Schubert. *Chem. Commun.* 2731, 1998.
289. U.S. Schubert, C. Eschbaumer, Q. An, and T. Salditt. *Polym. Prepr.* 40(1):414, 1999.
290. M. Schütte, D.G. Kurth, M.R. Linford, H. Cölfen, and H. Möhwald. *Angew. Chem. Int. End.* 37:2891, 1998.
291. B. Lindholm-Sethson. *Langmuir* 12:3305, 1996.

292. R. Advincula and W. Knoll. *Coll. Surf. A* 123/124:443, 1997.

293. Y. Lvov, F. Eßler, and G. Decher. *J. Phys. Chem.* 97:13773, 1993.

294. H. Bock, R.C. Advincula, E.F. Aust, J. Käshammer, W.H. Meyer, S. Mittler-Neher, C. Fiorini, J.-M. Nunzi, and W. Knoll. *J. Nonlinear Opt. Phys.* 7:385, 1998.

295. M. Tarabia, H. Hong, D. Davidov, S. Kirstein, R. Steitz, R. Neumann, and Y. Avny. *J. Appl. Phys.* 83:725, 1998.

296. G.J. Kellogg, A.M. Mayes, W.B. Stockton, M. Ferreira, M.F. Rubner, and S.K. Satija. *Langmuir* 12:5109, 1996.

297. R. Bijlsma, A.A. van Well, and M.A. Cohen Stuart. *Physica B* 234/236:254, 1997.

298. X. Arys. Understanding Ordering in Polyelectrolyte Multilayers: Effect of the Chemical Architecture of the Polycation. Ph.D. Dissertation, Université catholique de Louvain-La-Neuve, Louvain-La-Neuve, Belgium, 1999.

299. S. Balasubramanian, X. Wang, H.C. Wang, L. Li, D.J. Sandman, J. Kumar, S.K. Tripathy, and M.F. Rubner. *Polym. Prepr. Am. Chem. Soc. Div. Polym. Chem.* 38(2):502, 1997.

300. K. Yang, S. Balasubramanian, X.G. Wang, J. Kumar, and S. Tripathy. *Appl. Phys. Lett.* 73:3345, 1998.

301. X. Wang, S. Balasubramanian, L. Li, X. Jiang, D.J. Sandman, M.F. Rubner, J. Kumar, and S.K. Tripathy. *Macromol. Rapid Commun.* 18:451, 1997.

302. V.G. Gregoriou, R. Hapanowicz, S.L. Clark, and P.T. Hammond. *Appl. Spectrosc.* 51:470, 1997.

303. S.L. Clark, M. Montague, and P.T. Hammond. *Supramol. Sci.* 4:141, 1997.

304. P.T. Hammond and G.M. Whitesides. *Macromolecules* 28:7569, 1995.

305. S.L. Clark, M.F. Montague, and P.T. Hammond. *Macromolecules* 30:7237, 1997.

306. A. Delcorte. Static Secondary Ion Mass Spectrometry of Thin Organic Layers. Ph.D. Dissertation, Université catholique de Louvain-La-Neuve, Louvain-La-Neuve, Belgium, 1999.

307. J.Q. Sun, T. Wu, Y.P. Sun, Z.Q. Wang, X. Zhang, J.C. Shen, and W.X. Cao. *Chem. Commun.* 1853, 1998.

308. O. Onitsuka, A.C. Fou, M. Ferreira, B.R. Hsieh, and M.F. Rubner. *J. Appl. Phys.* 80:4067, 1996.

309. A.C. Fou, O. Onitsuka, M. Ferreira, M.F. Rubner, and B.R. Hsieh. *J. Appl. Phys.* 79:7501, 1996.

310. M. Onoda and K. Yoshino. *Jpn. J. Appl. Phys.* 34:260, 1995.

311. J. Stepp and J.B. Schlenoff. *J. Electrochem. Soc.* 144:L155, 1997.

312. J.B. Schlenoff, D. Laurent, H. Ly, and J. Stepp. *Adv. Mater.* 10:347, 1998.

313. M. Koetse, A. Laschewsky, B. Mayer, O. Rolland, and E. Wischerhoff. *Macromolecules* 31:9316, 1998.

314. A. Laschewsky, B. Mayer, E. Wischerhoff, X. Arys, A. Jonas, M. Kauranen, and A. Persoons. *Angew. Chem. Int. Ed. Engl.* 36:2788, 1997.

315. J. Smid and D. Fish. *Encyclopedia of Polymer Science & Engineering*, Vol. 11, New York: Wiley, 1988, p. 724.

316. M. Müller, T. Rieser, K. Lunkwitz, S. Berwald, J. Meier-Haack, and D. Jehnichen. *Macromol. Rapid. Commun.* 19:333, 1998.

317. R. von Klitzing and H. Möhwald. *Langmuir* 11:3554, 1995.

318. G.B. Sukhorukov, G. Decher, and J. Schmitt. *J. Am. Chem. Soc.* personal communication.

319. D.L. Feldheim, K.C. Grabar, M.J. Natan, and T.E. Mallouk. *J. Am. Chem. Soc.* 118:7640, 1996.

320. G. Decher and J.-D. Hong. European Patent 00 472 990 A2, 1992.

321. G. Decher and J.-D. Hong. U.S. Patent 00 5 208 111 A, 1993.

322. M.F. Rubner and J.H. Cheung. Int Patent 95/02251, 1995.

323. M.F. Rubner and J.H. Cheung. U.S. Patent 00 5 536 573 A, 1996.

324. M.F. Rubner and J.H. Cheung. U.S. Patent 00 5 518 767 A, 1996.

325. M.-S. Sheu and I.-H. Loh. U.S. Patent 00 5 700 559 A, 1997.

326. M.-S. Sheu and I.-H. Loh. U.S. Patent 00 5 837 377 A, 1998.

327. M.-S. Sheu and I.-H. Loh. U.S. Patent 00 5 807 636 A, 1998.

328. G.S. Ferguson and E.R. Kleinfeld. U.S. Patent 5 716 709 A, 1998.

329. H.-U. Siegmund, L. Heiliger, B. van Lent, and A. Becker. European Patent 00 561 239 A1, 1993.

330. H.-U. Siegmund, L. Heiliger, B. van Lent, and A. Becker. Int Patent 96/18498, 1996.

331. H.-U. Siegmund, L. Heiliger, B. van Lent, and A. Becker. U.S. Patent 00 5 711 915 A, 1998.

332. M. Houska and E. Brynda. *J. Colloid Interface Sci.* 188:243, 1997.

333. P. Stroeve, V. Vasquez, M.A.N. Coelho, and J.F. Rabolt. *Thin Solid Films* 284:708, 1996.

334. P. Zhou, L. Samuelson, K.S. Alva, C.-C. Chen, R.B. Blumstein, and A. Blumstein. *Macromolecules* 30:1577, 1997.

335. L. Krasemann and B. Tieke. *J. Membr. Sci.* 150:23, 1998.

336. B. Jacob, J. Schrezenmeir, R. Pommersheim, and W. Walter. *Diabetologia* 36:A189, 1993.

337. B. Jacob, A. Gaumann, R. Pommersheim, W. Vogt, and J. Schrezenmeir. *Diabetologia* 37:A216, 1994.
338. E. Donath, G.B. Sukhorukov, F. Caruso, S.A. Davis, and H. Möhwald. *Angew. Chem. Int. Ed. Engl.* 37:2202, 1998.
339. F. Caruso, H. Lichtenfeld, M. Giersig, and H. Möhwald. *J. Am. Chem. Soc.* 120:8523, 1998.
340. F. Caruso, R.A. Caruso, and H. Möhwald. *Science* 282:1111, 1998.
341. X. Yang, S. Johnson, J. Shi, T. Holesinger, and B. Swanson. *Sens. Actuators B* 45:87, 1997.
342. G. Decher, B. Lehr, K. Lowack, Y. Lvov, and J. Schmitt. *Biosensors Bioelectron.* 9:677, 1994.
343. G. Decher, F. Eßler, J.-D. Hong, K. Lowack, J. Schmitt, and Y. Lvov. *Polym. Prepr.* 34(1):745, 1993.
344. J.-D. Hong, E.-S. Park, and A.-L. Park. *Bull. Korean Chem. Soc.* 19:1156, 1998.
345. F. Saremi and B. Tieke. *Adv. Mater.* 10:388, 1998.
346. N.K. Viswanathan, S. Balasubramanian, L. Li, J. Kumar, and S.K. Tripathy. *J. Phys. Chem. B* 102:6064, 1998.
347. J. Tian, C.-C. Wu, M.E. Thompson, J.C. Sturm, R.A. Register, M.J. Marsella, and T.M. Swager. *Adv. Mater.* 7:395, 1995.
348. S. Kim, J. Jackiw, E. Robinson, K.S. Schanze, J.R. Reynolds, J. Baur, M.F. Rubner, and D. Boils. *Macromolecules* 31:964, 1998.
349. J.-K. Lee, D. Yoo, and M.F. Rubner. *Chem. Mater.* 9:1710, 1997.
350. H. Mattoussi, L.H. Radzilowski, B.O. Dabbousi, E.L. Thomas, and M.G. Bawendi. *Chem. Mater.* 9:1710, 1997.
351. P. Bertrand, A. Jonas, A. Laschewsky, and R. Legras. *Macromol. Rapid. Commun.* 21:319, 2000.
352. S.K. Tripathy, J. Kumar, and H.S. Nalwa, Eds., *Handbook of Polyelectrolytes and their Applications.* Vol 1, Los Angeles, CA: American Scientific Publishers, 2002.
353. G. Decher and J.B. Schlenoff, Eds., *Multilayer Thin Films — Sequential Assembly of Nanocomposite Materials.* Weinheim: Wiley-VCH, 2003.
354. P.T. Hammond. *Curr. Opin. Coll. Interface Sci.* 4:430, 2000.
355. J.C. Shen, J.Q. Sun, and X. Zhang. *Pure Appl. Chem.* 72:147, 2000.
356. F. Caruso. *Chem. Eur. J.* 3:413, 2000.
357. H. Möhwald, H. Lichtenfeld, S. Moya, A. Voigt, G. Sukhorukov, S. Leporatti, L. Dähne, A. Antipov, C.Y. Gao, and E. Donath. *Stud. Surf. Sci. Catal.* 132:485, 2001.
358. F. Caruso. *Adv. Mater.* 13:11, 2001.
359. G.B. Sukhorukov. In: D. Möbius and R. Miller, Eds., *Designed Nano-engineered Polymer Films on Colloidal Particles and Capsules.* Amsterdam: Elsevier, 2001, pp. 384–414.
360. M. Schönhoff. *Curr. Opin. Colloid Interface Sci.* 8:86, 2003.
361. X. Arys, P. Fischer, A.M. Jonas, M.M. Koetse, R. Legras, A. Laschewsky, and E. Wischerhoff. *J. Am. Chem. Soc.* 125:1859, 2003.
362. B. Tieke, L. Krasemann, and A. Toutianoush. *Macromol. Symp.* 163:97, 2002.
363. X.Y. Shi, T. Cassagneau, and F. Caruso. *Langmuir* 18:904, 2002.
364. M. Koetse, A. Laschewsky, A.M. Jonas, and W. Wagenknecht. *Langmuir* 18:1655, 2002.
365. D. Kovacevic, S. van der Burgh, A. de Keizer, and M.A. Cohen Stuart. *J. Phys. Chem. B* 107:7998, 2003.
366. E. Nicol, J.-L. Habib-Jiwan, and A.M. Jonas. *Langmuir* 19:6178, 2003.
367. J.B. Schlenoff, S.T. Dubas, and T. Farhat. *Langmuir* 16:9968, 2000.
368. J.H. Cho, K.H. Char, J.D. Hong, and K.B. Lee. *Adv. Mater.* 13:1076, 2001.
369. S.S. Lee, J.D. Hong, C.H. Kim, K. Kim, J.P. Koo, and K.B. Lee. *Macromolecules* 34:5358, 2001.
370. O. Mermut and C.J. Barrett. *Polym. Mat. Sci. Eng.* 85:348, 2001.
371. P. Chiarelli, M.S. Johal, J.L. Casson, J.B. Roberts, J.M. Robinson, and H.L. Wang, *Adv. Mater.* 13:1167, 2001.
372. P.A. Chiarelli, M.S. Johal, D.J. Holmes, J.L. Casson, J.M. Robinson, and H.L. Wang. *Langmuir* 18:168, 2002.
373. B.H. Sohn, T.H. Kim, and K. Char. *Langmuir* 18:7770, 2002.
374. C.J. Lefaux, J.A. Zimberlin, and P.T. Mather. *Polym. Prepr. Am. Chem. Soc. Div. Polym. Chem.* 43(2):356, 2002.
375. W. Li, D.E. Hooks, P. Chiarelli, Y. Jiang, H. Xu, and H.-L. Wang. *Langmuir* 19:4639, 2003.
376. S.S. Lee, K.B. Lee, and J.D. Hong. *Langmuir* 19:7592, 2003.
377. N. Song and Z.Y. Wang. *Macromolecules* 36:5885, 2003.
378. S.E. Burke and C.J. Barrett. *Biomacromolecules* 4:1773, 2003.
379. S.T. Dubas and J.B. Schlenoff. *Macromolecules* 32:8153, 1999.

380. J.F. Joanny. *Eur. Phys. J. B* 9:117, 1999.
381. R.R. Netz and J.F. Joanny. *Macromolecules* 32:9013, 1999.
382. M. Castelnovo and J.F. Joanny. *Langmuir* 16:7524, 2000.
383. A.V. Dobrynin, A. Deshkovski, and M. Rubinstein. *Macromolecules* 34:3421, 2001.
384. R. Messina, C. Holm, and K. Kremer. *Langmuir* 19:4473, 2003.
385. N. Dan. *Nano. Lett.* 3:823, 2003.
386. J.B. Schlenoff and S.T. Dubas. *Macromolecules* 34:592, 2001.
387. X. Arys, A. Laschewsky, and A.M. Jonas. *Macromolecules* 34:3318, 2001.
388. D. Kovacevic, S. van der Burgh, A. de Keizer, and M.A. Cohen Stuart. *Langmuir* 18:5607, 2002.
389. A.F. Xie and S. Granick. *J. Am. Chem. Soc.* 123:3175, 2001.
390. A.F. Xie and S. Granick. *Macromolecules* 35:1805, 2002.
391. Y.M. Lvov, G.N. Kamau, D.L. Zhou, and J.F. Rusling. *J. Colloid. Interface Sci.* 212:1570, 1999.
392. T. Okubo and M. Suda. *Colloid Polymer. Sci.* 27:380, 2000.
393. P. Fischer and A. Laschewsky. *Macromolecules* 33:1100, 2000.
394. D. Li, Y.D. Jiang, Z.M. Wu, X.D. Chen, and Y.R. Li. *Thin Solid Films* 360:24, 2000.
395. G. Ladam, C. Gergely, B. Senger, G. Decher, J.-C. Voegel, P. Schaaf, and F.J.G. Cuisinier. *Biomacromolecules* 1:674, 2000.
396. M. Müller, S. Heinen, U. Oertel, and K. Lunkwitz. *Macromol. Symp.* 164:197, 2001.
397. V. Pardo-Yissar, E. Katz, O. Lioubashevski, and I. Willner. *Langmuir* 17:1110, 2001.
398. Y. Lvov, R. Price, B. Gaber, and I. Ichinose. *Coll. Surf. A* 198/200:375, 2002.
399. K. Katagiri, R. Hamasaki, K. Ariga, and J.-I. Kikuchi. *Langmuir* 18:6709, 2002.
400. B. Kim and W.M. Sigmund. *Langmuir* 19:4848, 2003.
401. B. Schoeler, E. Poptoshev, and F. Caruso. *Macromolecules* 36:5258, 2003.
402. T. Okubo and M. Suda. *Colloid Polym. Sci.* 280:533, 2002; *Colloid Polym. Sci.* 281:782, 2003.
403. G. Ladam, P. Schaaf, F.J.G. Cuisinier, G. Decher, and J.C. Voegel. *Langmuir* 17:878, 2001.
404. P. Schwinté, J.-C. Voegel, C. Picart, Y. Haikel, P. Schaaf, and B. Szalontai. *J. Phys. Chem. B* 105:11906, 2001.
405. G. Ladam, P. Schaaf, G. Decher, J.C. Voegel, and F.J.G. Cuisinier. *Biomol. Eng.* 19:273, 2002.
406. H. Ai, M. Fang, S.A. Jones, and Y.M. Lvov. *Biomacromolecules* 3:560, 2002.
407. H.H. Rmaile, B.C. Bucur, and J.B. Schlenoff. *Polym. Prepr. Am. Chem. Soc. Polym. Chem. Div.* 44(1):540, 2003.
408. K. Büscher, K. Graf, H. Ahrens, and C.A. Helm. *Langmuir* 18:3585, 2002.
409. H.L. Tan, M.J. McMurdo, G. Pan, and P.G. Van Patten. *Langmuir* 19: 9311, 2003.
410. Z. Sui and J.B. Schlenoff. *Polym. Mater. Sci. Eng.* 84:682, 2001.
411. Z. Sui, D. Salloum, and J.B. Schlenoff. *Langmuir* 19:2491, 2003.
412. A.J. Khopade and F. Caruso. *Langmuir* 18:7669, 2002.
413. J. Lukkari, M. Salomäki, T. Ääritalo, K. Loikas, T. Laiho, and J. Kankare. *Langmuir* 18:8496, 2002.
414. C. Picart, P. Lavalle, P. Hubert, F.J.G. Cuisinier, G. Decher, P. Schaaf, and J.-C. Voegel. *Langmuir* 17:7414, 2001.
415. C. Picart, J. Mutterer, L. Richert, Y. Luo, G.D. Prestwich, P. Schaaf, J.-C. Voegel, and P. Lavalle. *Proc. Natl. Acad. Sci.* 99:12531, 2002.
416. P. Lavalle, C. Gergely, F.J.G. Cuisinier, G. Decher, P. Schaaf, J.-C. Voegel, and C. Picart. *Langmuir* 18:4458, 2002.
417. F. Boulmedais, V. Ball, P. Schwinté, B. Frisch, P. Schaaf, and J.-C. Voegel. *Langmuir* 19:440, 2003.
418. S.T. Dubas and J.B. Schlenoff. *Macromolecules* 34:3736, 2001.
419. R. Steitz, W. Jaeger, and R. von Klitzing. *Langmuir* 17:4471, 2001.
420. B. Schoeler, G. Kumaraswamy, and F. Caruso. *Macromolecules* 35:889, 2002.
421. K. Glinel, A. Moussa, A.M. Jonas, and A. Laschewsky. *Langmuir* 18:1408, 2002.
422. U. Voigt, V. Khrenov, K. Tauer, M. Hahn, W. Jaeger, and R. von Klitzing. *J. Phys. Condens. Matter* 15:S213, 2003.
423. U. Voigt, W. Jaeger, G.H. Findenegg, and R. von Klitzing. *J. Phys. Chem. B* 107:5273, 2003.
424. T. Serizawa, N. Kawanishi, and M. Akashi. *Macromolecules* 36:1957, 2003.
425. N.A. Kotov. *Nanostruct. Mater B* 12:789, 1999.
426. C. Luo, D.M. Guldi, M. Maggini, E. Menna, S. Mondini, N.A. Kotov, and M. Prato. *Angew. Chem. Int. Ed. Engl.* 39:3905, 2000.
427. S.L. Clark and P.T. Hammond. *Langmuir* 16:10206, 2000.

428. Z.F. Dai, A. Voigt, S. Leporatti, E. Donath, L. Dähne, and H. Möhwald. *Adv. Mater.* 13:1339, 2001.
429. M. Paßmann and R. Zentel. *Macromol. Chem. Phys.* 203:363, 2002.
430. F. Caruso, H. Lichtenfeld, E. Donath, and H. Möhwald. *Macromolecules* 32:2317, 1999.
431. M. Raposo and O.N. Oliveira Jr. *Langmuir* 16:2839, 2000.
432. L.G. Paterno and L.H.C. Mattoso. *Polymer* 42:5239, 2001.
433. M. Raposo and O.N. Oliveira Jr. *Langmuir* 18:6866, 2002.
434. V. Izumrudov and S.A. Sukhishvili. *Langmuir* 19:5188, 2003.
435. S.S. Shiratori and M.F. Rubner. *Macromolecules* 33:4213, 2000.
436. J.D. Mendelsohn, C.J. Barrett, V.V. Chan, A.J. Pal, A.M. Mayes, and M.F. Rubner. *Langmuir* 16:5017, 2000.
437. A.M. Balachandra, J.H. Dai, and M.L. Bruening. *Macromolecules* 35:3171, 2002.
438. N. Kato, P. Schuetz, A. Fery, and F. Caruso. *Macromolecules* 35:9780, 2002.
439. P.V. Pavoor, B.P. Gearing, A. Bellare, and R.E. Cohen. *Polym. Mat. Sci. Eng.* 87:473, 2002.
440. E. Kharlampieva and S.A. Sukhishvili. *Langmuir* 19:1235, 2003.
441. S.E. Burke and C.J. Barrett. *Langmuir* 19:3297, 2003.
442. T. Cassagneau, J.H. Fendler, and T.E. Mallouk. *Langmuir* 16:241, 2000.
443. L.Y. Wang, S.X. Cui, Z.Q. Wang, X. Zhang, M. Jiang, L.F. Chi, and H. Fuchs. *Langmuir* 16:10490, 2000.
444. E.C. Hao and T. Lian. *Chem. Mater.* 12:3392, 2000.
445. E.C. Hao and T.Q. Lian. *Langmuir* 16:7879, 2000; S.A. Sukhishvili, S. Granick. *J. Am. Chem. Soc.* 121:9550, 2000.
446. J. Choi and M.F. Rubner. *J. Macromol. Sci. Chem. A* 38:1179, 2001.
447. Y.X. Wang, W. Du, W.B. Spillman Jr., and R.O. Claus. *Proc. SPIE* 4265:142, 2001.
448. Y. Fu, H. Chen, D.L. Qiu, Z.Q. Wang, and X. Zhang. *Langmuir* 18:4985, 2002.
449. Y. Fu, S.L. Bai, S.X. Cui, D.L. Qiu, Z.Q. Wang, and X. Zhang. *Macromolecules* 18:9451, 2002.
450. T. Serizawa, S. Kamimura, N. Kawanishi, and M. Akashi. *Langmuir* 18:8381, 2002.
451. S.A. Sukhishvili and S. Granick. *Macromolecules* 35:301, 2002.
452. V. Kozlovskaya, S. Ok, A. Sousa, M. Libera, and S.A. Sukhishvili. *Macromolecules* 36:8590, 2003.
453. H.Y. Zhang, Y. Fu, D. Wang, L.Y. Wang, Z.Q. Wang, and X. Zhang. *Langmuir* 19:8497, 2003.
454. J. Cho and F. Caruso. *Macromolecules* 36:2845, 2003.
455. Y.J. Zhang, Y. Guan, S.G. Yang, J. Xu and C.C. Han. *Adv. Mater.* 15:832, 2003.
456. G. Ladam, P. Schaad, J.C. Voegel, P. Schaaf, G. Decher, and F. Cuisinier. *Langmuir* 16:1249, 2000.
457. R. Steitz, V. Leiner, R. Siebrecht, and R. von Klitzing. *Coll. Surf. A* 163:63, 2000.
458. T.R. Farhat and J.B. Schlenoff. *Langmuir* 17:1184, 2001.
459. H. Riegler and F. Essler. *Langmuir* 18:6694, 2002.
460. X.D. Wang, W.L. Yang, Y. Tang, Y.J. Wang, S.K. Fu, and Z. Gao. *J. Chem. Soc. Chem. Commun.* 2161, 2000.
461. I. Galeska, D. Chattopadhyay, and F. Papadimitrakopoulos. *J. Macromol. Sci. Chem. A* 39:1207, 2002.
462. T. Cassagneau and F. Caruso. *J. Am. Chem. Soc.* 124:8172, 2002.
463. D.M. DeLongchamp and P.T. Hammond. *Chem. Mater.* 15:1165, 2003.
464. O. Mermut and C.J. Barrett. *J. Phys. Chem. B* 107:2525, 2003.
465. T.R. Farhat and J.B. Schlenoff. *J. Am. Chem. Soc.* 125:4627, 2003.
466. J.H. Rouse, B.A. MacNeill, and G.S. Ferguson. *Chem. Mater.* 12:2502, 2000.
467. T. Cassagneau and J.H. Fendler. *J. Phys. Chem. B* 103:1789, 1999.
468. I. Moriguchi, Y. Teraoka, S. Kagawa, and J.H. Fendler. *Chem. Mater.* 11:1603, 1999.
469. T. Cassagneau, N. Kotov, and J.H. Fendler. *J. Disp. Sci. Technol.* 20:13, 1999; *J. Disp. Sci. Technol.* 20:1517, 1999.
470. T. Cassagneau, F. Guerin, and J.H. Fendler. *Langmuir* 16:7318, 2000.
471. G.S. Ferguson and E.R. Kleinfeld. U.S. Patent 6 022 590, 2000.
472. M. Eckle and G. Decher. *Nano. Lett.* 1:45, 2001.
473. B. Struth, M. Eckle, G. Decher, R. Oeser, P. Simon, D.W. Schubert, and J. Schmitt. *Eur. Phys. J. E* 6:351, 2001.
474. T. Sasaki, Y. Ebina, M. Watanabe, and G. Decher. *J. Chem. Soc. Chem. Commun.* 2163, 2000.
475. T. Sasaki, Y. Ebina, T. Tanaka, M. Harada, M. Watanabe, and G. Decher. *Chem. Mater.* 13:4661, 2001.
476. T. Sasaki, Y. Ebina, K. Fukuda, T. Tanaka, M. Harada, and M. Watanabe. *Chem. Mater.* 14:3524, 2002.
477. M. Harada, T. Sasaki, Y. Ebina, and M. Watanabe. *J. Photochem. Photobiol. A: Chem.* 148:273, 2002.

478. Z.S. Wang, T. Sasaki, M. Muramatsu, Y. Ebina, T. Tanaka, L. Wang, and M. Watanabe. *Chem. Mater.* 15:807, 2003.
479. L. Wang, Y. Omomo, N. Sakai, K. Fukuda, I. Nakai, Y. Ebina, K. Takada, M. Watanabe, and T. Sasaki. *Chem. Mater.* 15:2873, 2003.
480. Z.S. Wang, Y. Ebina, K. Takada, M. Watanabe, and T. Sasaki. *Langmuir* 19:9534, 2003.
481. Z.H. Chen, Y.A. Yang, J.B. Qiu, and J.N. Yao. *Langmuir* 16:722, 2000.
482. Z.H. Chen, Y. Ma, X.T. Zhang, B. Liu, and J.N. Yao. *J. Colloid Interface Sci.* 240:487, 2001.
483. Z.H. Chen, Y. Ma, and J.N. Yao. *Thin Solid Films* 384:160, 2001.
484. D.W. Kim, A. Blumstein, J. Kumar, L.A. Samuelson, B. Kang, and C. Sung. *Chem. Mater.* 14:3925, 2002.
485. X.W. Fan, J. Locklin, J.H. Youk, W. Blanton, C.J. Xia, and R. Advincula. *Chem. Mater.* 14:2184, 2002.
486. X. Fan, M.K. Park, C. Xia, and R. Advincula. *J. Mater. Res.* 17:1622, 2002.
487. Y. Zhou, Z. Li, N. Hu, Y. Zeng, and J.F. Rusling. *Langmuir* 18:8573, 2002.
488. K. Glinel, A. Laschewsky, and A.M. Jonas. *Macromolecules* 34:5267, 2001.
489. K. Glinel, A. Laschewsky, and A.M. Jonas. *J. Phys. Chem. B* 106:11246, 2002.
490. P.Y. Vuillaume, A.M. Jonas, and A. Laschewsky. *Macromolecules* 35:5004, 2002.
491. P.Y. Vuillaume, K. Glinel, A.M. Jonas, and A. Laschewsky. *Chem. Mater.* 15:3625, 2003.
492. K. Glinel, A.M. Jonas, A. Laschewsky, and P.Y. Vuillaume. In: *Multilayer Thin Films — Sequential Assembly of Nanocomposite Materials.* Weinheim: Wiley-VCH, 2003, G. Decher, J.B. Schlenoff eds., pp. 177–205.
493. D. Allard, S. Allard, M. Brehmer, L. Conrad, R. Zentel, C. Stromberg, and J.W. Schultze. *Electrochim. Acta* 48:3137, 2003.
494. S.L. Clark, E.S. Handy, M.F. Rubner, and P.T. Hammond. *Adv. Mater.* 11:1031, 1999.
495. B.P. Nelson, A.G. Frutos, J.M. Brockman, and R.M. Corn. *Anal. Chem.* 71:3928, 1999.
496. A. Delcorte, P. Bertrand, E. Wischerhoff, and A. Laschewsky. In: A. Benninghoven, P. Bertrand, and H.N. Migeon, Eds., *Secondary Ion Mass Spectroscopy*, SIMS XII. Amsterdam: Elsevier, 2000, p. 757.
497. X.P. Jiang and P.T. Hammond, *Langmuir* 16:8501, 2000.
498. K.M. Chen, X.P. Jiang, L.C. Kimerling, and P.T. Hammond. *Langmuir* 16:7825, 2000.
499. K. Sirkar, A. Revzin, and M.V. Pishko. *Anal. Chem.* 72:2930, 2000.
500. X.P. Jiang, S.L. Clark, and P.T. Hammond. *Adv. Mater.* 13:1669, 2001.
501. L.X. Shi, J.Q. Sun, J.Q. Liu, J.C. Shen, and M.Y. Gao. *Chem. Lett.* 1168, 2002.
502. M.Y. Gao, J.Q. Sun, E. Dulkeith, N. Gaponik, U. Lemmer, and J. Feldmann. *Langmuir* 18:4098, 2002.
503. L.X. Shi, Y.X. Lu, J. Sun, J. Zhang, C.Q. Sun, J.Q. Liu, and J.C. Shen. *Biomacromolecules* 4:1161, 2003.
504. M.C. Berg, J. Choi, P.T. Hammond, and M.F. Rubner. *Langmuir* 19:2231, 2003.
505. H.P. Zheng, M.F. Rubner, and P.T. Hammond. *Langmuir* 18:4505, 2002.
506. H.P. Zheng, I.S. Lee, M.F. Rubner, and P.T. Hammond. *Adv. Mater.* 14:569, 2002.
507. I.S. Lee, H.P. Zheng, M.F. Rubner, and P.T. Hammond. *Adv. Mater.* 14:572, 2002.
508. J.Q. Sun, M.Y. Gao, and J. Feldmann. *J. Nanosci. Nanotech.* 1:133, 2001.
509. J.Q. Sun, M.Y. Gao, M. Zhu, J. Feldmann, and H. Möhwald. *J. Mater. Chem.* 12:1775, 2002.
510. I. Lee, P.T. Hammond, and M.F. Rubner. *Chem. Mater.* 15:4583, 2003.
511. F. Hua, J. Shi, Y. Lvov, and T. Cui. *Nano. Lett.* 2:1219, 2002.
512. F. Hua, T.H. Cui, and Y. Lvov. *Langmuir* 18:6712, 2002.
513. M.A. Hempenius, M. Péter, N.S. Robins, E.S. Kooij, and G.J. Vancso. *Langmuir* 18:7629, 2002.
514. S.Y. Yang and M.F. Rubner. *J. Am. Chem. Soc.* 124:2100, 2002.
515. G.M. Credo, G.M. Lowman, J.A. DeAro, J.P. Carson, D.L. Winn, and S.K. Buratto. *J. Chem. Phys.* 112:7864, 2000.
516. J.Y. Chen, H. Luo, B.X. Yang, and W.X. Cao. *J. Appl. Polym. Sci.* 80:1983, 2001.
517. F. Shi, B. Dong, D.L. Qiu, J.Q. Sun, T. Wu, and X. Zhang. *Adv. Mater.* 14:805, 2002.
518. A. Pallandre, K. Glinel, A.M. Jonas, and B. Nysten. *Nano. Lett.* 4:365, 2004.
519. J.D. Hong, E.S. Park, and A.L. Park. *Langmuir* 15:6515, 1999.
520. J.A. He, S.P. Bian, L. Li, J. Kumar, S.K. Tripathy, and L.A. Samuelson. *J. Phys. Chem. B* 104:10513, 2000.
521. J. Locklin, J.H. Youk, C.J. Xia, M.K. Park, X.W. Fan, and R.C. Advincula. *Langmuir* 18:877, 2002.
522. A. Ziegler, J. Stumpe, A. Toutianoush, and B. Tieke. *Coll. Surf. A* 198/200:777, 2002.
523. V. Panchagnula, C.V. Kumar, and J.F. Rusling. *J. Am. Chem. Soc.* 124:12515, 2002.
524. J. Locklin, K. Shinbo, K. Onishi, F. Kaneko, Z. Bao, and R.C. Advincula. *Chem. Mater.* 15:1404, 2003.
525. G. Zotti, S. Zecchin, A. Berlin, G. Schiavon, and G. Giro. *Chem. Mater.* 13:43, 2001.

526. X. Wang, K. Naka, H. Itoh, T. Uemura, and Y. Chujo. *Macromolecules* 36:533, 2003.
527. V. Ruiz, P. Liljeroth, B.M. Quinn, and K. Kontturi. *Nano. Lett.* 3:1459, 2003.
528. T. Schneider and O.D. Lavrentovich. *Langmuir* 16:5227, 2000.
529. D.S. Santos Jr., A. Bassi, L. Misoguti, M.F. Ginani, O.N. Oliveira Jr., and C.R. Mendonça. *Macromol. Rapid Commun.* 23:975, 2002.
530. M. Müller. *Biomacromolecules* 2:262, 2001.
531. M. Müller, B. Kessler, and K. Lunkwitz. *J. Phys. Chem. B* 107:8189, 2003.
532. J.A. He, L. Samuelson, L. Li, J. Kumar, and S.K. Tripathy. *Adv. Mater.* 11:435, 1999.
533. M.L. Li, B.F. Li, L. Jiang, T. Jussila, N. Tkachenko, and H. Lemmetyinen. *Langmuir.* 16:5503, 2000.
534. M. Li, B. Li, L. Jiang, T. Jussila, N. Tkachenko, and H. Lemmetyinen. *Chem. Lett.* 266, 2000.
535. T. Jussila, M. Li, N.V. Tkachenko, S. Parkkinen, B. Li, L. Jiang, and H. Lemmetyinen. *Biosens. Bioelectr.* 17:509, 2002.
536. M. Koetse, A. Laschewsky, and T. Verbiest. *Mat. Sci. Eng. C* 10:107, 1999.
537. P. Fischer, M. Koetse, A. Laschewsky, E. Wischerhoff, L. Jullien, T. Verbiest, and A. Persoons. *Macromolecules* 33:9471, 2000.
538. M. Koetse, A. Laschewsky, A.M. Jonas, and T. Verbiest. *Coll. Surf. A.* 198/200:275, 2002.
539. J.R. Heflin, C. Figura, D. Marciu, Y. Liu, and R.O. Claus. *Appl. Phys. Lett.* 74:495, 1999.
540. K.M. Lenahan, Y.J. Liu, Y.X. Wang, and R.O. Claus. *Proc. SPIE* 3675:104, 1999.
541. G.A. Lindsay, M.J. Roberts, A.P. Chafin, R.A. Hollins, L.H. Merwin, J.D. Stenger-Smith, R.Y. Yee, and P. Zarras. *Chem. Mater.* 11:924, 1999.
542. M.J. Roberts, J.D. Stenger-Smith, P. Zarras, G.A. Lindsay, R.A. Hollins, A.P. Chafin, R.Y. Yee, and K.J. Wynne. *Nonlinear Opt.* 20:23, 1999.
543. M.J. Roberts. *Mat. Res. Soc. Symp. Proc.* 561:33, 1999.
544. M.J. Roberts, W.N. Herman, G.A. Lindsay, and K.J. Wynne. *Proc. SPIE* 3623:63, 1999.
545. G.A. Lindsay, M.J. Roberts, W.N. Herman, A.P. Chafin, R.A. Hollins, L.H. Merwin, J.D. Stenger-Smith, and P. Zarras. *Nonlinear Opt.* 22:3, 1999.
546. W.N. Herman and M.J. Roberts. *Adv. Mater.* 13:744, 2001.
547. J.B. Mecham, K.L. Cooper, K. Huie, and R.O. Claus. *Proc. SPIE* 4468:179, 2001.
548. K.E. Van Cott, M. Guzy, P. Neyman, C. Brands, J.R. Heflin, H.W. Gibson, and R.M. Davis. *Angew. Chem. Int. Ed. Eng.* 41:3236, 2002.
549. R.A. McAloney and M.C. Goh. *J. Phys. Chem. B* 103:10729, 1999.
550. J. Kim, G. Kim, and P.S. Cremer. *J. Am. Chem. Soc.* 124:8751, 2002.
551. Y. Shimazaki, S. Ito, and N. Tsutsumi. *Langmuir* 16:9478, 2000.
552. J.L. Casson, D.W. McBranch, J.M. Robinson, H.L. Wang, J.B. Roberts, P.A. Chiarelli, and M.S. Johal. *J. Chem. Phys. B* 104:11996, 2000.
553. J.L. Casson, H.L. Wang, J.B. Roberts, A.N. Parikh, J.M. Robinson, and M.S. Johal. *J. Phys. Chem. B* 106:1697, 2002.
554. Y. Niidome, S. Tagawa, and S. Yamada. *Coll. Surf. A* 198/200:467, 2002.
555. B. van Duffel, R.A. Schoonheydt, C.P.M. Grim, and F. DeSchryver. *Langmuir* 15:7520, 1999.
556. B. van Duffel, T. Verbiest, S. Vanelshocht, A. Persoons, F.C. DeSchryver, and R.A. Schoonheydt. *Langmuir* 17:1243, 2001.
557. Y.J. Wang, Y. Tang, X.D. Wang, W.L. Yang, and Z. Gao. *Chem. Lett.* 1344, 2000.
558. N.I. Kovtyukhova, B.R. Martin, J.K.N. Mbindyo, T.E. Mallouk, M. Cabassi, and T.S. Mayer. *Mat. Sci. Eng. C* 19:255, 2002.
559. L. Wågberg, S. Forsberg, A. Johansson, and P. Juntti. *J. Pulp Paper Sci.* 28:222, 2002.
560. T. Okubo and M. Suda. *Colloid Polym. Sci.* 277:813, 1999.
561. H. Möhwald, H. Lichtenfeld, S. Moya, A. Voigt, H. Bäumler, G. Sukhorukov, F. Caruso, and H. Donath. *Macromol. Symp.* 145:75, 1999.
562. H. Möhwald. *Coll. Surf. A* 171:25, 2000.
563. C. Schüler and F. Caruso. *Macromol. Rapid. Commun.* 21:750, 2000.
564. F. Caruso, W.J. Yang, D. Trau, and R. Renneberg. *Langmuir* 16:8932, 2000.
565. C.Y. Gao, S. Leporatti, E. Donath, and H. Möhwald. *J. Phys. Chem. B* 104:7144, 2000.
566. A.S. Susha, F. Caruso, A.L. Rogach, G.B. Sukhorukov, A. Kornowski, H. Möhwald, M. Giersig, A. Eychmuller, and H. Weller. *Coll. Surf. A* 163:39, 2000.
567. X.Y. Shi and F. Caruso. *Langmuir* 17:2036, 2001.
568. K. Furusawa and S. Satou. *Coll. Surf. A* 195:143, 2001.

569. M. Fang, P.S. Grant, M.J. McShane, G.B. Sukhorukov, V.O. Golub, and Y.M. Lvov. *Langmuir* 18:6338, 2002.

570. I. Estrela-Lopis, S. Leporatti, S. Moya, A. Brandt, E. Donath, and H. Möhwald. *Langmuir* 18:7861, 2002.

571. R.N. Smith, L. Reven, and C.J. Barrett. *Macromolecules* 36:1876, 2003.

572. K.S. Mayya, B. Schoeler, and F. Caruso. *Adv. Funct. Mat.* 13:183, 2003.

573. S. Moya, W. Richter, S. Leporatti, H. Bäumler, and E. Donath. *Biomacromolecules* 4:808, 2003.

574. G. Kaltenpoth, M. Himmelhaus, L. Slansky, F. Caruso, and M. Grunze. *Adv. Mater.* 15:1113, 2003.

575. G.B. Sukhorukov, M. Brumen, E. Donath, and H. Möhwald. *J. Phys. Chem. B* 103:6434, 1999.

576. S. Moya, G.B. Sukhorukov, M. Auch, E. Donath, and H. Möhwald. *J. Colloid Interface Sci.* 216:297, 1999.

577. F. Caruso, R. Caruso, and H. Möhwald. *Chem. Mater.* 11:3309, 1999.

578. F. Caruso, M. Schüler, and D.G. Kurth. *Chem. Mater.* 11:3394, 1999.

579. G.B. Sukhorukov, L. Dähne, J. Hartmann, E. Donath, and H. Möhwald. *Adv. Mater.* 12:112, 2000.

580. S. Leporatti, A. Voigt, R. Mitlohner, G. Sukhorukov, E. Donath, and H. Möhwald. *Langmuir* 16:4059, 2000.

581. I.L. Radtchenko, G.B. Sukhorukov, S. Leporatti, G.B. Khomutov, E. Donath, and H. Möhwald. *J. Colloid Interface Sci.* 230:272, 2000.

582. G.B. Sukhorukov, E. Donath, S. Moya, A.S. Susha, A. Voigt, J. Hartmann, and H. Möhwald. *J. Microencapsul.* 17:177, 2000.

583. S. Moya, G.B. Sukhorukov, M. Auch, E. Donath, and H. Möhwald. *Macromolecules* 33:4538, 2000.

584. R. Georgieva, S. Moya, S. Leporatti, B. Neu, H. Bäumler, C. Reichle, E. Donath, and H. Möhwald. *Langmuir* 16:7075, 2000.

585. H. Bäumler, G. Artmann, A. Voigt, R. Mitlohner, B. Neu, and H. Kiesewetter. *J. Microencapsul.* 17:651, 2000.

586. I. Gittins and F. Caruso. *Adv. Mater.* 12:1947, 2000.

587. I. Pastoriza-Santos, B. Schöler, and F. Caruso. *Adv. Funct. Mater.* 11:122, 2001.

588. C.Y. Gao, S. Moya, H. Lichtenfeld, A. Casoli, H. Fiedler, E. Donath, and H. Möhwald. *Macromol. Mater. Eng.* 286:355, 2001.

589. G. Ibarz, L. Dähne, E. Donath, and H. Möhwald. *Adv. Mater.* 13:1324, 2001.

590. B. Neu, A. Voigt, R. Mitlöhner, S. Leporatti, C.Y. Gao, E. Donath, H. Kiesewetter, H. Möhwald, H.J. Meiselman, and H. Bäumler. *J. Microencapsul.* 18:385, 2001.

591. L. Dähne, S. Leporatti, E. Donath, and H. Möhwald. *J. Am. Chem. Soc.* 123:5431, 2001.

592. B.D. Jung, J.D. Hong, A. Voigt, S. Leporatti, L. Dähne, E. Donath, and H. Möhwald. *Coll. Surf. A* 198/200:483, 2002.

593. A.A. Antipov, G.B. Sukhorukov, Y.A. Fedutik, J. Hartmann, M. Giersig, and H. Möhwald. *Langmuir* 18:6687, 2002.

594. A.A. Antipov, G.B. Sukhorukov, S. Leporatti, I.L. Radtchenko, E. Donath, and H. Möhwald. *Coll. Surf. A* 198/200:535, 2002.

595. R. Georgieva, S. Moya, M. Hin, R. Mitlöhner, E. Donath, H. Kiesewetter, H. Möhwald, and H. Bäumler. *Biomacromolecules* 3:517, 2002.

596. G. Berth, A. Voigt, H. Dautzenberg, E. Donath, and H. Möhwald. *Biomacromolecules* 3:579, 2002.

597. C.Y. Gao, S. Moya, E. Donath, and H. Möhwald. *Macromol. Chem. Phys.* 203:953, 2002.

598. N. Gaponik, I.L. Radtchenko, G.B. Sukhorukov, H. Weller, and A.L. Rogach. *Adv. Mater.* 14:879, 2002.

599. Z.F. Dai, L. Dähne, E. Donath, and H. Möhwald. *Langmuir* 18:4553, 2002.

600. Z. Dai, H. Möhwald, B. Tiersch, and L. Dähne. *Langmuir* 18:9533, 2002.

601. Z. Dai and H. Möhwald. *Chem. Eur. J.* 8:4751, 2002.

602. A.J. Khopade and F. Caruso. *Biomacromolecules* 3:1154, 2002.

603. C.G. Gao, E. Donath, H. Möhwald, and J.C. Shen. *Angew. Chem. Int. Ed.* 41:3789, 2002.

604. E. Donath, S. Moya, B. Neu, G.B. Sukhorukov, R. Georgieva, A. Voigt, H. Bäumler, H. Kiesewetter, and H. Möhwald. *Chem. Eur. J.* 8:5481, 2002.

605. V.V. Lulevich, I.L. Radtchenko, G.B. Sukhorukov, and O.I. Vinogradova. *J. Phys. Chem. B* 107:2735, 2003.

606. D. Shchukin, G. Sukhorukov, and H. Möhwald. *Chem. Mater.* 15:3947, 2003.

607. D.G. Shchukin and G.B. Sukhorukov. *Langmuir* 19:4427, 2003.

608. D.G. Shchukin, W. Dong, and G.B. Sukhorukov. *Macromol. Rapid Commun.* 24:462, 2003.

609. C.S. Peyratout, H. Möhwald, and L. Dähne. *Adv. Mater.* 15:1722, 2003.

610. M.K. Park, K. Onishi, J. Locklin, F. Caruso, and R.C. Advincula. *Langmuir* 19:8550, 2003.
611. K. Glinel, G.B. Sukhorukov, H. Möhwald, V. Khrenov, and K. Tauer. *Macromol. Chem. Phys.* 204:1784, 2003.
612. K.H. Rhodes, S.A. Davis, F. Caruso, B. Zhang, and S. Mann. *Chem. Mater.* 12:2832, 2000.
613. R.A. Caruso, A.S. Susha, and F. Caruso. *Chem. Mater.* 13:400, 2001.
614. F. Caruso, X.Y. Shi, R.A. Caruso, and A. Susha. *Adv. Mater.* 13:740, 2001.
615. F. Caruso, M. Spasova, V. Salgueiriño-Maceira, and L.M. Liz-Marzán. *Adv. Mater.* 13:1090, 2001.
616. V. Valtchev and S. Mintova. *Micropr. Mesopor. Mater.* 43:41, 2001.
617. V. Valtchev. *Chem. Mater.* 14:956, 2002.
618. D. Wang and F. Caruso. *Chem. Mater.* 14:1909, 2002.
619. T. Cassagneau and F. Caruso. *Adv. Mater.* 14:732, 2002.
620. L. Wang, T. Sasaki, Y. Ebina, K. Kurashima, and M. Watanabe. *Chem. Mater.* 14:4827, 2002.
621. D. Shchukin, Y. Fedutik, A.I. Petrov, G.B. Sukhorukov, and H. Möhwald. *Angew. Chem. Int. Ed. Eng.* 42:4472, 2003.
622. J.S. Yu, J.Y. Kim, S. Lee, J.K.N. Mbindyo, B.R. Martin, and T.E. Mallouk. *J. Chem. Soc. Chem. Commun.* 2445, 2000.
623. N.I. Kovtyukhova, A.D. Gorchinskiy, and C. Waraksa. *Mater. Sci. Eng. B* 69:424, 2000.
624. N.I. Kovtyukhova, B.R. Martin, J.K.N. Mbindyo, P.A. Smith, B. Razavi, T.S. Mayer, and T.E. Mallouk. *J. Phys. Chem. B* 105:8762, 2001.
625. K.S. Mayya, D.I. Gittins, A.M. Dibaj, and F. Caruso. *Nano. Lett.* 1:727, 2001.
626. Y.G. Guo, L.J. Wan, and C.L. Bai. *J. Phys. Chem. B* 107:5441, 2003.
627. T.W. Graul and J.B. Schlenoff. *Anal. Chem.* 71:4007, 1999.
628. Y. Liu, J.C. Fanguy, J.M. Bledsoe, and C.S. Henry. *Anal. Chem.* 72:5939, 2000.
629. L. Szyk, P. Schaaf, C. Gergely, J.C. Voegel, and B. Tinland. *Langmuir* 17:6248, 2001.
630. C.P. Kapnissi, C. Akbay, J.B. Schlenoff, and I.M. Warner. *Anal. Chem.* 74:2328, 2002.
631. Z. Sui and J.B. Schlenoff. *Langmuir* 19:7829, 2003.
632. S.F. Ai, G. Lu, Q. He, and J.B. Li. *J. Am. Chem. Soc.* 125:11140, 2003.
633. Z. Liang, A.S. Susha, A. Yu, F. Caruso, and M. Grunze. *Adv. Mater.* 15:1849, 2003.
634. L. Krasemann and B. Tieke. *Mat. Sci. Eng. C* 8/9:513, 1999.
635. L. Krasemann and B. Tieke. *Langmuir* 16:287, 2000.
636. J.J. Harris, J.L. Stair, and M.L. Bruening. *Chem. Mater.* 12:1941, 2000.
637. J. Meier-Haack, T. Rieser, W. Lenk, D. Lehmann, S. Berwald, and S. Schwarz. *Chem. Ing. Technol.* 23:114, 2000.
638. J. Meier-Haack, W. Lenk, D. Lehmann, and K. Lunkwitz. *J. Membr. Sci.* 184:233, 2001.
639. L. Krasemann, A. Toutianoush, and B. Tieke. *J. Membr. Sci.* 181:221, 2001.
640. B. Tieke, F. Vanackern, L. Krasemann, and A. Toutianoush. *Eur. Phys. J. E* 5:29, 2001.
641. J.L. Stair, J.J. Harris, and M.L. Bruening. *Chem. Mater.* 13:2641, 2001.
642. D.M. Sullivan and M.L. Bruening. *J. Am. Chem. Soc.* 123:11805, 2001.
643. K.P. Xiao, J.L. Harris, A. Park, C.M. Martin, V. Pradeep, and M.L. Bruening. *Langmuir* 17:8236, 2001.
644. J.H. Dai, A.M. Balachandra, J.I. Lee, and M.L. Bruening. *Langmuir* 35:3164, 2002.
645. M.L. Bruening and D.M. Sullivan. *Chem. Eur. J.* 8:3833, 2002.
646. A. Toutianoush, L. Krasemann, and B. Tieke. *Coll. Surf. A* 198/200:881, 2002.
647. A. Toutianoush and B. Tieke. *Mat. Sci. Eng. C* 22:135, 2002; *Mat. Sci. Eng. C* 22:459, 2002.
648. W. Jin, A. Toutianoush, and B. Tieke. *Langmuir* 19:2550, 2003.
649. B.W. Stanton, J.J. Harris, M.D. Miller, and M.L. Bruening. *Langmuir* 19:7038, 2003.
650. S.L.R. Barker, M.J. Tarlov, H. Canavan, J.J. Hickman, and L.E. Locascio. *Anal. Chem.* 72:4899, 2000.
651. S.L.R. Barker, D. Ross, M.J. Tarlov, M. Gaitan, and L.E. Locascio. *Anal. Chem.* 72:5925, 2000.
652. A.A. Mamedov and N.A. Kotov. *Langmuir* 16:5530, 2000.
653. S.T. Dubas, T.R. Farhat, and J.B. Schlenoff. *J. Am. Chem. Soc.* 123:5368, 2001.
654. E. Brynda, M. Houska, A. Brandenburg, A. Wikerstal, and J. Skvor. *Biosens. Bioelectr.* 14:363, 1999.
655. E. Brynda, M. Houska, A. Wikerstal, Z. Pientka, J.E. Dyr, and A. Brandenburg. *Langmuir* 16:4352, 2000.
656. J.Q. Sun, Z.Q. Wang, Y.P. Sun, X. Zhang, and J.C. Shen. *J. Chem. Soc. Chem. Commun.* 693, 1999.
657. J. Sun, L. Cheng, F. Liu, S. Dong, Z. Wang, X. Zhang, and J.C. Shen. *Coll. Surf. A* 169:209, 2000.
658. J.Q. Sun, T. Wu, F. Liu, Z.Q. Wang, X. Zhang, and J.C. Shen. *Langmuir* 16:4620, 2000.
659. J.Q. Sun, Z.Q. Wang, L.X. Wu, X. Zhang, J.C. Shen, S. Gao, L.F. Chi, and H. Fuchs. *Macromol. Chem. Phys.* 202:967, 2001.

660. X. Zhang, T. Wu, J.Q. Sun, and J.C. Shen. *Coll. Surf. A* 198/200:439, 2002.
661. Y. Fu, H. Xu, S.L. Bai, D.L. Qiu, J.Q. Su, Z.Q. Wang, and X. Zhang. *Macromol. Rapid Commun.* 23:256, 2002.
662. J. Chen, L. Huang, L. Ying, G. Lin, X. Zhao, and W. Cao. *Langmuir* 15:7208, 1999.
663. L. Huang, G.B. Luo, X.S. Zhao, J.Y. Chen, and W.X. Cao. *J. Appl. Polym. Sci.* 78:631, 2000.
664. H. Luo, J.Y. Chen, G.B. Luo, Y.N. Chen, and W.X. Cao. *J. Mater. Chem.* 11:419, 2001.
665. T.B. Cao, J.Y. Chen, C.H. Yang, and W.X. Cao. *Macromol. Rapid Commun.* 22:181, 2001.
666. J.Y. Chen, G.B. Luo, and W.X. Cao. *Macromol. Rapid Commun.* 22:311, 2001.
667. H. Zhong, J.F. Wang, X.R. Jia, Y. Li, Y. Qin, J.Y. Chen, X.S. Zhao, W.X. Cao, M.Q. Li, and Y. Wei. *Macromol. Rapid Commun.* 22:583, 2001.
668. B.X. Yang, J.Y. Chen, G.H. Yu, and W.X. Cao. *Macromol. Chem. Phys.* 202:2168, 2001.
669. T.B. Cao, J.Y. Chen, C.H. Yang, and W.X. Cao. *New J. Chem.* 25:305, 2001.
670. Y.J. Zhang and W.X. Cao. *New J. Chem.* 25:483, 2001.
671. Y.J. Zhang and W.X. Cao. *Langmuir* 17:5021, 2001.
672. T.B. Cao, S.M. Yang, J. Cao, M.F. Zhang, C.H. Huang, and W.X. Cao. *J. Phys. Chem. B* 105:11941, 2001.
673. Q. Li, J.H. Ouyang, J.Y. Chen, X.S. Zhao, and W.X. Cao. *J. Polymer. Sci. A. Polym. Chem. A* 40:222, 2002.
674. J.F. Wang, X.R. Jia, H. Zhong, Y.F. Luo, X.S. Zhao, W.X. Cao, M.Q. Li, and Y. Wei. *Chem. Mater.* 14:2854, 2002.
675. T.B. Cao, L.H. Wei, S.M. Yang, M.F. Zhang, C.H. Huang, and W.X. Cao. *Langmuir* 18:750, 2002.
676. Y.J. Zhang, W.X. Cao, and J. Xu. *J. Colloid Interface Sci.* 249:91, 2002.
677. Z.H. Yang, T.B. Cao, J.Y. Chen, and W.X. Cao. *Eur. Polym. J.* 38:2077, 2002.
678. L. Huang, X.R. Hou, Y.K. He, X.L. Zheng, F. Wei, X.S. Zhao, and W.X. Cao. *Chin. J. Polym. Sci.* 20:197, 2002.
679. Y.J. Zhang, S.G. Yang, Y. Guan, X.P. Miao, W.X. Cao, and J. Xu. *Thin Solid Films* 437:280, 2003.
680. Z.H. Yang and W.X. Cao. *J. Polym. Sci. Polym. Chem. A* 41:3103, 2003.
681. X.L. Hou, L.X. Wu, L. Sun, H. Zhang, B. Yang, and J.C. Shen. *Polym. Bull.* 47:445, 2002.
682. H. Zhang, B. Yang, R.B. Wang, G. Zhang, X.L. Hou, and L.X. Wu. *J. Colloid Interface Sci.* 247:361, 2002.
683. I. Ichinose, S. Muzuki, S. Ohno, H. Shiraishi, and T. Kunitake. *Polym. J. Jpn.* 31:1065, 1999.
684. J.H. Dai, D.M. Sullivan, and M.L. Bruening. *Ind. Eng. Chem. Res.* 39:3528, 2000.
685. J. Dai, A.W. Jensen, D.K. Mohanty, J. Erndt, and M.L. Bruening. *Langmuir* 17:931, 2001.
686. O. Inya-Agha, S. Stewart, T. Veriotti, M.L. Bruening, and M.D. Morris. *Appl. Spectrosc.* 56:574, 2002.
687. M.K. Park, C.J. Xia, R.C. Advincula, P. Schütz, and F. Caruso. *Langmuir* 17:7670, 2001.
688. M.R. Anderson, R.M. Davis, C.D. Taylor, M. Parker, S. Clark, D. Marciu, and M. Miller. *Langmuir* 17:8380, 2001.
689. J.A. Hiller, J.D. Mendelsohn, and M.F. Rubner. *Nat. Mater.* 1:59, 2002.
690. A.A. Mamedov, N.A. Kotov, M. Prato, D.M. Guldi, J.P. Wicksted, and A. Hirsch. *Nat. Mater.* 1:190, 2002.
691. D.M. Sullivan and M.L. Bruening. *Chem. Mater.* 15:281, 2003.
692. L. Zhou and J.F. Rusling. *Anal. Chem.* 73:4780, 2001.
693. L. Dennany, R.J. Forster, and J.F. Rusling. *J. Am. Chem. Soc.* 125:5213, 2003.
694. A.J. Khopade and F. Caruso. *Langmuir* 19:6219, 2003.
695. G.J. Yao, Y.P. Dong, T.B. Cao, S.M. Yang, J.W.Y. Lam, and B.Z. Tang. *J. Colloid Interface Sci.* 257:263, 2003.
696. S.G. Jiang, X.D. Chen, L. Zhang, and M.H. Liu. *Thin Solid Films* 425:117, 2003.
697. S.X. Zhang, Y.M. Niu, L. Zhang, L.X. Shi, X.F. Li, and C.Q. Sun. *Chem. Lett.* 32:960, 2003.
698. T.B. Cao, F. Wei, X.M. Jiao, J.Y. Chen, W. Liao, X.S. Zhao, and W.X. Cao. *Langmuir* 19:8127, 2003.
699. J.A. He, S.P. Bian, L. Li, J. Kumar, S.K. Tripathy, and L.A. Samuelson. *Appl. Phys. Lett.* 76:3233, 2000.
700. F. Kaneko, T. Kato, A. Baba, K. Shinbo, K. Kato, and R.C. Advincula. *Coll. Surf. A* 198/200:805, 2002.
701. K. Shinbo, A. Baba, F. Kaneko, T. Kato, K. Kato, R.C. Advincula, and W. Knoll. *Mat. Sci. Eng. C* 22:319, 2002.
702. V. Zucolotto, J.A. He, C.J.L. Constantino, N.M. Barbosa Neto, J.J. Rodrigues Jr., C.R. Mendonça, S.C. Zílio, L. Li, R.F. Aroca, O.N. Oliveira Jr., and J. Kumar. *Polymer* 44:6129, 2003.

703. C.S. Camilo, D.S. dos Santos Jr., J.J. Rodrigues Jr., M.L. Vega, S.P. Campana Filho, O.N. Oliveira Jr., and C.R. Mendonça. *Biomacromolecules* 4:1583, 2003.

704. T. Ito, Y. Okayama, and S. Shiratori. *Thin Solid Films* 393:138, 2001.

705. K.S. Mayya, D.I. Gittins, and F. Caruso. *Chem. Mater.* 13:3833, 2001.

706. S.S. Shiratori, T. Ito, and T. Yamada. *Coll. Surf. A* 198/200:415, 2002.

707. I. Ichinose, R. Takaki, K. Kuroiwa, and T. Kunitake. *Langmuir* 19:3883, 2003.

708. B. Richter and S. Kirstein. *J. Chem. Phys.* 111:5191, 1999.

709. S. Bourbon, M.G. Gao, and S. Kirstein. *Synt. Met.* 101:152, 1999.

710. J.W. Baur, M.F. Rubner, J.R. Reynolds, and S. Kim. *Langmuir* 15:6460, 1999.

711. P.K.H. Ho, J.S. Kim, J.H. Burroughes, H. Becker, S.F.Y. Li, T.M. Brown, F. Cacialli, and R.H. Friend. *Nature* 404:481, 2000.

712. S. Kirstein, S. Bourbon, M.Y. Gao, and U. Derossi. *Isr. J. Chem.* 40:129, 2000.

713. T. Sonoda, T. Fujisawa, A. Fujii, and K. Yoshino. *Appl. Phys. Lett.* 76:3227, 2000.

714. J. Cho, K. Char, S.Y. Kim, J.D. Hong, S.K. Lee, and D.Y. Kim. *Thin Solid Films* 379:188, 2000.

715. H. Mattoussi, M.F. Rubner, F. Zhou, J. Kumar, S.K. Tripathy, and L.Y. Chiang. *Appl. Phys. Lett.* 77:1540, 2000.

716. M.F. Durstock, B. Taylor, R.J. Spry, L. Chiang, S. Reulbach, K. Heitfeld, and J.W. Baur. *Synt. Met.* 121:373, 2001.

717. A. Marletta, F.A. Castro, D. Gonçalves, O.N. Oliveira Jr., R.M. Faria, and F.E.G. Guimarães. *Synt. Met.* 121:1447, 2001.

718. J.W. Baur, M.F. Durstock, B.E. Taylor, R.J. Spry, S. Reulbach, and L.Y. Chiang. *Synt. Met.* 121:1547, 2001.

719. T. Fujisawa, T. Sonoda, R. Ootake, A. Fujii, and K. Yoshino. *Synt. Met.* 121:1739, 2001.

720. A. Marletta, F.A. Castro, C.A.M. Borges, O.N. Oliveira Jr., R.M. Faria, and F.E.G. Guimarães. *Macromolecules* 35:9105, 2002.

721. A.L. Park and J.D. Hong. *Macromol. Symp.* 142:121, 1999.

722. J.D. Hong, E.S. Park, and B.D. Jung. *Mol. Cryst. Liq. Cryst.* 327:119, 1999.

723. S. Bian, J.A. He, L. Li, J. Kumar, and S.K. Tripathy. *Adv. Mater.* 12:1202, 2000.

724. X. Tuo, Z. Chen, L. Wu, X. Wang, and D. Liu. *Polym. Prepr. Am. Chem. Soc. Polym. Chem. Div.* 41(7):1405, 2000.

725. D.M. DeWitt and P.T. Hammond. *Polym. Prepr. Am. Chem. Soc. Polym. Chem. Div.* 41(7):815, 2000.

726. A. Baba, F. Kaneko, K. Shinbo, K. Kato, S. Kobayashi, and R.C. Advincula. *Mol. Cryst. Liq. Cryst.* 347:259, 2000.

727. R.C. Advincula, E. Fells, and M.K. Park. *Chem. Mater.* 13:2870, 2001.

728. L.F. Wu, X.L. Tuo, H. Chen, Z. Chen, and X.G. Wang. *Macromolecules* 34:8005, 2001.

729. M.K. Park and R.C. Advincula. *Langmuir* 18:4532, 2002.

730. J. Ishikawa, A. Baba, F. Kaneko, K. Shinbo, K. Kato, and R.C. Advincula. *Coll. Surf. A* 198/200:917, 2002.

731. I. Suzuki, T. Ishizaki, T. Hoshi, and J.I. Anzai. *Macromolecules* 35:577, 2002.

732. R. Advincula, M.K. Park, A. Baba, and F. Kaneko. *Langmuir* 19:654, 2003.

733. B.D. Jung, J. Stumpe, and J.D. Hong. *Thin Solid Films* 441:261, 2003.

734. D.S. dos Santos Jr., A. Bassi, J.J. Rodrigues Jr., L. Misoguti, O.N. Oliveira Jr., and C.R. Mendonça. *Biomacromolecules* 4:1502, 2003.

735. J.J. Kakkassery, D.J. Fermín, and H.H. Girault. *Chem. Commun.* 1240, 2002.

736. R.C. Millward, C.E. Madden, I. Sutherland, R.J. Mortimer, S. Fletcher, and F. Marken. *J. Chem. Soc. Chem. Commun.* 1994, 2001.

737. D. DeLongchamp and P.T. Hammond. *Adv. Mater.* 13:1455, 2001.

738. M. Pyrasch and B. Tieke. *Langmuir* 17:7706, 2001.

739. S.Q. Liu, D.G. Kurth, H. Möhwald, and D. Volkmer. *Adv. Mater.* 14:225, 2002.

740. S.Q. Liu, D.G. Kurth, B. Bredenkötter, and D. Volkmer. *J. Am. Chem. Soc.* 124:12279, 2002.

741. I. Moriguchi, H. Kamogawa, and Y. Teraoka. *Chem. Lett.* 310, 2002.

742. Z. Li and N. Hu. *J. Colloid Interface Sci.* 254:257, 2002.

743. C.R. Cutler, M. Bouguettaya, and J.R. Reynolds. *Adv. Mat.* 14:684, 2002.

744. A. Jaiswal, J. Colins, B. Agricole, P. Delhaes, and S. Ravaine. *J. Colloid Interface Sci.* 261:330, 2003.

745. A. Bandyopadhyay and A.J. Pal. *Adv. Mater.* 15:1949, 2003.

746. G. Zotti, S. Zecchin, G. Schiavon, B. Vercelli, and L.B. Groenendaal. *Chem. Mater.* 15:2222, 2003.

747. S.H. Lee, S. Balasubramanian, D.Y. Kim, N.K. Viswanathan, S. Bian, J. Kumar, and S.K. Tripathy. *Macromolecules* 33:6534, 2000.
748. D.G. Kurth, M. Schütte, and J. Wen. *Coll. Surf. A* 198/200:633, 2002.
749. H.S. Kim, B.H. Sohn, W. Lee, J.K. Lee, S.J. Choi, and S.J. Kwon. *Thin Solid Films* 419:173, 2002.
750. H. Krass, G. Papastavrou, and D.G. Kurth. *Chem. Mater.* 15:196, 2003.
751. M. Jiang, E.B. Wang, Z.H. Kang, S.Y. Lian, A.G. Wu, and Z. Li. *J. Mater. Chem.* 13:647, 2003.
752. S. Joly, R. Kane, L. Razilowski, T. Wang, A. Wu, R.E. Cohen, E.L. Thomas, and M.F. Rubner. *Langmuir* 16:1354, 2000.
753. J.Y. Liu, L. Cheng, Y.H. Song, B.F. Liu, and S.J. Dong. *Langmuir* 17:6747, 2001.
754. T.C. Wang, M.F. Rubner, and R.E. Cohen. *Langmuir* 18:3370, 2002.
755. T.C. Wang, R.E. Cohen, and M.F. Rubner. *Adv. Mater.* 14:1534, 2002.
756. J.L. Dai and M.L. Bruening. *Nano. Lett.* 2:497, 2002.
757. S. Minko, A. Kiriy, G. Gorodyska, and M. Stamm. *J. Am. Chem. Soc.* 124:10192, 2002.
758. A. Kiriy, S. Minko, G. Gorodyska, M. Stamm, and W. Jaeger. *Nano. Lett.* 2:881, 2002.
759. T.C. Wang, M.F. Rubner, and R.E. Cohen. *Chem. Mater.* 15:299, 2003.
760. J. Zhang, L. Bai, K. Zhang, Z. Cui, G. Zhang, and B. Yang. *J. Mater. Chem.* 13:514, 2003.
761. D.S. Shchukin and G.B. Sukhorukov. *Colloid Polym. Sci.* 281:1201, 2003.
762. A.K. Dutta, G. Jarero, L.Q. Zhang, and P. Stroeve. *Chem. Mater.* 12:176, 2000.
763. A.K. Dutta, T. Ho, L. Zhang, and P. Stroeve. *Chem. Mater.* 12:1042, 2000.
764. L. Zhang, A.K. Dutta, G. Jarero, and P. Stroeve. *Langmuir* 16:7095, 2000.
765. J.H. Rouse and G.S. Ferguson. *Adv. Mater.* 14:151, 2002.
766. W.L. Yang, X.D. Wang, Y. Tang, Y.J. Wang, C. Ke, and S.K. Fu. *J. Macromol. Sci. Chem. A* 39:509, 2002.
767. J.A. He, R. Mosurkal, L.A. Samuelson, L. Li, and J. Kumar. *Langmuir* 19:2169, 2003.
768. A. Gaumann, M. Laudes, B. Jacob, R. Pommersheim, C. Laue, W. Vogt, and J. Schrezenmeir. *Biomaterials* 21:1911, 2000.
769. T.C. Wang, B. Chen, M.F. Rubner, and R.E. Cohen. *Langmuir* 17:6610, 2001.
770. S. Schneider, P.J. Feilen, V. Slotty, D. Kampfner, S. Preuss, S. Berger, J. Beyer, and R. Pommersheim. *Biomaterials* 22:1961, 2001.
771. C.Y. Gao, S. Leporatti, S. Moya, E. Donath, and H. Möhwald. *Langmuir* 17:3491, 2001.
772. C. Gao, E. Donath, S. Moya, V. Dudnik, and H. Möhwald. *Eur. Phys. J. E* 5:21, 2001.
773. D. Trau, W. Yang, M. Seydack, F. Caruso, N.T. Yu, and R. Renneberg. *Anal. Chem.* 74:5480, 2002.
774. D.B. Shenoy, A.A. Antipov, G.B. Sukhorukov, and H. Möhwald. *Biomacromolecules* 4:265, 2003.
775. C. Gao, S. Leporatti, S. Moya, E. Donath, and H. Möhwald. *Chem. Eur. J.* 9:915, 2003.
776. T. Serizawa, T. Ohmori, and M. Akashi. *Polym. J. Jpn.* 35:810, 2003.
777. C. Schüler and F. Caruso. *Biomacromolecules* 2:921, 2001.
778. X.Y. Shi, R.J. Sanedrin, and F.M. Zhou. *J. Phys. Chem. B* 106:1173, 2002.
779. F. Boulmedais, M. Bozonnet, P. Schwinté, J.C. Voegel, and P. Schaaf. *Langmuir* 19:9873, 2003.
780. C. Gao, H. Möhwald, and J. Shen. *Adv. Mater.* 15:930, 2003.
781. T. Fahrhat, G. Yassin, S.T. Dubas, and J.B. Schlenoff. *Langmuir* 15:6621, 1999.
782. R. Steitz, V. Leiner, K. Tauer, V. Khrenov, and R. von Klitzing. *Appl. Phys. A* 74:S519, 2002.
783. R. Kügler, H. Schmitt, and W. Knoll. *Macromol. Chem. Phys.* 203:413, 2002.
784. M. McCormick, R.N. Smith, R. Graf, C.J. Barrett, L. Reven, and H.W. Spiess. *Macromolecules* 36:3616, 2003.
785. S.T. Dubas and J.B. Schlenoff. *Langmuir* 17:7725, 2001.
786. B. Schwarz and M. Schönhoff. *Langmuir* 18:2964, 2002.
787. E.S. Forzani, M. Otero, M.A. Pérez, M. López Teijelo, and E.J. Calvo. *Langmuir* 18:4020, 2002.
788. E.S. Forzani, M.A. Pérez, M. Lopez Teijelo, and E.J. Calvo. *Macromolecules* 35:9867, 2002.
789. J.L. Menchaca, B. Jachimska, F. Cuisinier, and E. Pérez. *Coll. Surf. A* 222:185, 2003.
790. R.A. McAloney, M. Sinyor, V. Dudnik, and M.C. Goh. *Langmuir* 17:6655, 2001.
791. R.A. McAloney, V. Dudnik, and M.C. Goh. *Langmuir* 19:3947, 2003.
792. G.B. Sukhorukov, A.A. Antipov, A. Voigt, E. Donath, and H. Möhwald. *Macromol. Rapid Comm.* 22:44, 2001.
793. A. Fery, B. Schöler, T. Cassagneau, and F. Caruso. *Langmuir* 17:3779, 2001.
794. G. Ibarz, L. Dähne, E. Donath, and H. Möhwald. *Macromol. Rapid Commun.* 23:474, 2002.
795. G. Ibarz, L. Dähne, E. Donath, and H. Möhwald. *Chem. Mater.* 14:4059, 2002.

796. B.Y. Kim and M.L. Bruening. *Langmuir* 19:94, 2003.

797. W.J. Li, Z. Wang, C.Q. Sun, M. Xian, and M.Y. Zhao. *Anal. Chim. Acta.* 418:225, 2000.

798. W. Jin, X. Shi, and F. Caruso. *J. Am. Chem. Soc.* 123:8121, 2001.

799. P. Schwinté, V. Ball, B. Szalontai, Y. Haikel, J.C. Voegel, and P. Schaaf. *Biomacromolecules* 3:1135, 2002.

800. L. Derbal, H. Lesot, J.C. Voegel, and V. Ball. *Biomacromolecules* 4:1255, 2003.

801. J. Chluba, J.C. Voegel, G. Decher, P. Erbacher, P. Schaaf, and J. Ogier. *Biomacromolecules* 2:800, 2001.

802. F. Boulmedais, P. Schwinté, C. Gergely, J.C. Voegel, and P. Schaaf. *Langmuir* 18:4523, 2002.

803. M. Debreczeny, V. Ball, F. Boulmedais, B. Szalontai, J.C. Voegel, and P. Schaaf. *J. Phys. Chem. B* 107:12734, 2003.

804. M. Müller, T. Rieser, M. Köthe, B. Keßler, M. Brissova, and K. Lunkwitz. *Macromol. Symp.* 145:149, 1999.

805. M. Müller, T. Rieser, K. Lunkwitz, and J. Meier-Haack. *Macromol. Rapid Commun.* 20:607, 1999.

806. M. Müller, T. Rieser, P.L. Dubin, and K. Lunkwitz. *Macromol. Rapid Commun.* 22:390, 2001.

807. I. Galeska, T. Hickey, F. Moussy, D. Kreutzer, and F. Papadimitrakopoulos. *Biomacromolecules* 2:1249, 2001.

808. D.S. Koktysh, X.R. Liang, B.G. Yun, I. Pastoriza-Santos, R.L. Matts, M. Giersig, C. Serra-Rodriguez, L. Liz-Marzan, and N.A. Kotov. *Adv. Funct. Mater.* 12:255, 2002.

809. T. Serizawa, M. Yamaguchi, and M. Akashi. *Biomacromolecules* 3:724, 2002.

810. S.Y. Yang, J.D. Mendelsohn, and M.F. Rubner. *Biomacromolecules* 4:987, 2003.

811. H.G. Zhu, J. Ji, Q.G. Tan, M.A. Barbosa, and J.C. Shen. *Biomacromolecules* 4:378, 2003.

812. D. Vautier, V. Karsten, C. Egles, J. Chluba, P. Schaaf, J.C. Voegel, and J. Ogier. *J. Biomater. Sci. Polym. Ed.* 13:713, 2002.

813. P. Tryoen-Toth, D. Vautier, Y. Haikel, J.C. Voegel, P. Schaaf, J. Chluba, and J. Ogier. *J. Biomed Mater. Res.* 60:657, 2002.

814. L. Richert, P. Lavalle, D. Vautier, B. Senger, J.F. Stoltz, P. Schaaf, J.C. Voegel, and C. Picart. *Biomacromolecules* 3:1170, 2002.

815. N. Jessel, F. Atalar, P. Lavalle, J. Mutterer, G. Decher, P. Schaaf, J.C. Voegel, and J. Ogier. *Adv. Mater.* 15:692, 2003.

816. J.D. Mendelsohn, S.Y. Yang, J.A. Hiller, A.I. Hochbaum, and M.F. Rubner. *Biomacromolecules* 4:96, 2003.

817. B. Thierry, F.M. Winnik, Y. Merhi, and M. Tabrizian. *J. Am. Chem. Soc.* 125:7494, 2003.

818. B. Thierry, F.M. Winnik, Y. Merhi, J. Silver, and M. Tabrizian. *Biomacromolecules* 4:1564, 2003.

819. V.A. Sinani, D.S. Koktysh, B.G. Yun, R.L. Matts, T.C. Pappas, M. Motamedi, S.N. Thomas, and N.A. Kotov. *Nano. Lett.* 3:1177, 2003.

820. W. Tachaboonyakiat, T. Serizawa, T. Endo, and M. Akashi. *Polym. J. Jpn.* 32:481, 2000.

821. F. Caruso, H. Fiedler, and K. Haage. *Colloids Surf. A* 169:287, 2000.

822. F. Caruso and C. Schüler. *Langmuir* 16:9595, 2000.

823. F. Caruso, D. Trau, H. Möhwald, and R. Renneberg. *Langmuir* 16:1485, 2000.

824. N.G. Balabushevitch, G.B. Sukhorukov, N.A. Moroz, D.V. Volodkin, N.I. Larionova, E. Donath, and H. Möhwald. *Biotech. Bioeng.* 76:207, 2001.

825. Y. Lvov and F. Caruso. *Anal. Chem.* 73:4212, 2001.

826. N.G. Balabushevich, O.P. Tiourina, D.V. Volodkin, N.I. Larionova, and G.B. Sukhorukov. *Biomacromolecules* 4:1191, 2003.

827. A. Antipov, D. Shchukin, Y. Fedutik, I. Zanaveskina, V. Klechkovskaya, G. Sukhorukov, and H. Möhwald. *Macromol. Rapid Commun.* 24:274, 2003.

828. A. Diaspro, D. Silvano, S. Krol, O. Cavalleri, and A. Gliozzi. *Langmuir* 18:5047, 2002.

829. H. Liu, K.M. Faucher, X.L. Sun, J. Feng, T.L. Johnson, J.M. Orban, R.P. Apkarian, R.A. Dluhy, and E.L. Chaikof. *Langmuir* 18:1332, 2002.

830. X.P. Qiu, S. Leporatti, E. Donath, and H. Möhwald. *Langmuir* 17:5375, 2001.

831. K. Sato, I. Suzuki, and J.I. Anzai. *Langmuir* 19:7406, 2003.

832. E. Vázquez, D.M. Dewitt, P.T. Hammond, and D.M. Lynn. *J. Am. Chem. Soc.* 124:13992, 2002.

833. T.R. Fahrhat and J.B. Schlenoff. *J. Am. Chem. Soc.* 125:4627, 2003.

834. H.H. Rmaile and J.B. Schlenoff. *J. Am. Chem. Soc.* 125:6602, 2003.

835. M. Pyrasch, A. Toutianoush, W. Jin, J. Schnepf, and B. Tieke. *Chem. Mater.* 15:245, 2003.

836. W. Jin, A. Toutianoush, M. Pyrasch, J. Schnepf, H. Gottschalk, W. Rammensee, and B. Tieke. *J. Phys. Chem. B* 107:12062, 2003.

837. F.J. Arregui, K.L. Cooper, Y.J. Liu, I.R. Matias, and R.O. Claus. *IEICE-Trans.-Electron.* E83C:360–365, 2000.

838. A.N. Shipway, M. Lahav, R. Blonder, and I. Willner. *Adv. Mater.* 12:993, 2000.

839. M. Lahav, A.N. Shipway, I. Willner, M.B. Nielsen, and J.F. Stoddart. *J. Electroanal. Chem.* 482:217, 2000.

840. A. Narvaez, G. Suarez, I.C. Popescu, I. Katakis, and E. Dominguez. *Biosens. Bioelectron.* 15:43, 2000.

841. J. Anzai and Y. Kobayashi. *Langmuir* 16:2851, 2000.

842. L. Cheng, J.Y. Liu, and S.J. Dong. *Anal. Chim. Acta.* 417:133, 2000.

843. M. Yamada and S.S. Shiratori. *Sensors Actuators B* 64:124, 2000.

844. M.K. Ram, M. Adami, S. Paddeu, and C. Nicolini. *Nanotechnology* 11:112, 2000.

845. L. Kumpumbu-Kalemab and M. Leclerc. *J. Chem. Soc. Chem. Commun.* 1847, 2000.

846. S.Q. Liu, Z.Y. Tang, Z.X. Wang, Z.Q. Peng, E.K. Wang, and S.J. Dong. *J. Mater. Chem.* 10:2727, 2000.

847. S.H. Lee, J. Kumar, and S.K. Tripathy. *Langmuir* 16:10482, 2000.

848. E.J. Calvo, F. Battaglini, C. Danilowicz, A. Wolosiuk, and M. Otero. *Faraday Disc* 116:47, 2000.

849. T. Chen, K.A. Friedman, I. Lei, and A. Heller. *Anal. Chem.* 72:3757, 2000.

850. J.F. Rusling, L.P. Zhou, B. Munge, J. Yang, C. Estavillo, and J.B. Schenkman. *Faraday Disc* 116:77, 2000.

851. E.S. Forzani, V.M. Solis, and E.J. Calvo. *Anal. Chem.* 72:5300, 2000.

852. X.C. Zhou, L.Q. Huang, and S.F.Y. Li. *Biosens. Bioelectr.* 16:85, 2001.

853. W.J. Yang, D. Trau, R. Renneberg, N.T. Yu, and F. Caruso. *J. Colloid Interface Sci.* 234:356, 2001.

854. E.J. Calvo, R. Etchenique, L. Pietrasanta, and A. Wolosiuk. *Anal. Chem.* 73:1161, 2001.

855. J. Sun, Y. Sun, Z. Wang, C. Sun, Y. Wang, X. Zhang, and J. Shen. *Macromol. Chem. Phys.* 202:111, 2001.

856. C. Pearson, J. Nagel, and M.C. Petty. *J. Phys. D Appl. Phys.* 34:285, 2001.

857. M.K. Ram, P. Bertoncello, H. Ding, S. Paddeu, and C. Nicolini. *Biosens. Bioelectr.* 16:849, 2001.

858. T. Hoshi, H. Saiki, S. Kuwazawa, C. Tsuchiya, Q. Chen, and J.I. Anzai. *Anal. Chem.* 73:5310, 2001.

859. A. Riul Jr., D.S. dos Santos Jr., K. Wohnrath, R. Di Tommazo, A.C.P.L.F. Carvalho, F.J. Fonseca, O.N. Oliveira Jr., D.M. Taylor, and L.H.C. Mattoso. *Langmuir* 18:239, 2002.

860. E.J. Calvo, C. Danilowicz, and A. Wolosiuk. *J. Am. Chem. Soc.* 124:2452, 2002.

861. D. Wang, A.L. Rogach, and F. Caruso. *Nano. Lett.* 2:857, 2002.

862. C.A. Constantine, S.V. Mello, A. Dupont, X. Cao, D. Santos Jr., O.N. Oliveira Jr., F.T. Strixino, E.C. Pereira, T.C. Cheng, J.J. Defrank, and R.M. Leblanc. *J. Am. Chem. Soc.* 125:1805, 2003; *J. Am. Chem. Soc.* 125:6595, 2003.

863. C.A. Constantine, K.M. Gattás-Asfura, S.V. Mello, G. Crespo, V. Rastogi, T.C. Cheng, J.J. DeFrank, and R.M. Leblanc. *J. Phys. Chem. B* 107:13762, 2003.

864. T. Serizawa, M. Yamaguchi, and M. Akashi. *Angew. Chem.* 115:1147, 2003.

865. A. Yu, Z. Liang, J. Cho, and F. Caruso. *Nano. Lett.* 3:1203, 2003.

866. J.J. Harris, P.M. DeRose, and M.L. Bruening. *J. Am. Chem. Soc.* 121:1978, 1999.

867. E.W. Taylor, K.L. Cooper, R.O. Claus, and L.R. Taylor. *Proc. SPIE* 4547:11, 2001.

868. T.R. Farhat and J.B. Schlenoff. *Electrochem. Solid-State Lett.* 5:B13, 2002.

869. G.S. Lee, Y.J. Lee, and K.B. Yoon. *J. Am. Chem. Soc.* 123:9769, 2001.

870. M. Gao, C. Lesser, S. Kirstein, H. Möhwald, A. Rogach, and H. Weller. *J. Appl. Phys.* 87:2297, 2000.

871. A. Rogach, D.S. Koktysh, M. Harrison, and N.A. Kotov. *Chem. Mater.* 12:1526, 2000.

872. J. Cho, K. Char, J.D. Hong, and D. Kim. *Mol. Cryst. Liq. Cryst.* 349:183, 2000.

873. H. Hong, R. Sfez, E. Vaganova, S. Yitzchaik, and D. Davidov. *Thin Solid Films* 366:260, 2000.

874. T. Piok, C. Brands, P.J. Neyman, A. Erlacher, C. Soman, M.A. Murray, R. Schroeder, W. Graupner, J.R. Heflin, D. Marciu, A. Drake, M.B. Miller, H. Wang, H. Gibson, H.C. Dorn, G. Leising, M. Guzy, and R.M. Davis. *Synt. Met.* 116:343, 2001.

875. A. Ikeda, T. Hatano, S. Shinkai, T. Akiyama, and S. Yamada. *J. Am. Chem. Soc.* 123:4855, 2001.

876. S. Roy and A.J. Pal. *Mater. Sci. Eng. C* 18:65, 2001.

877. H.L. Wang, D.W. McBranch, R.J. Donohoe, S. Xu, B. Kraabel, L.H. Chen, D. Whitten, R. Helgeson, and F. Wudl. *Synt. Met.* 121:1367, 2001.

878. L.S. Li and A.D.Q. Li. *J. Phys. Chem. B* 105:10022, 2001.

879. L.S. Li, Q.X. Jia, and A.D.Q. Li. *Chem. Mater.* 14:1159, 2002.

880. D.M. Guldi, C. Luo, D. Koktysh, N.A. Kotov, T. Da Ros, S. Bosi, and M. Prato. *Nano. Lett.* 2:775, 2002.
881. M.R. Pinto, B.M. Kristal, and K.S. Schanze. *Langmuir* 19:6523, 2003.
882. R. Mruk, S. Prehl, and R. Zentel. *Macromol. Rapid Commun.* 24:1014, 2003.
883. Y.G. Kim, J. Kim, H. Ahn, B. Kang, C. Sung, L.A. Samuelson, and J. Kumar. *J. Macromol. Sci. Chem.* A40:1307, 2003.
884. H.H. Yu, D.S. Jiang, C.W. Nan, Y.X. Wang, Y.J. Liu, and R.O. Claus. *Proc. SPIE* 4224:400, 2000.
885. F.J. Arregui, I.R. Matias, K.L. Cooper, and R.O. Claus. *Optics Lett.* 26:131, 2001.
886. F.J. Arregui, I.R. Matias, R.O. Claus, and K.L. Cooper. *Proc. SPIE* 4253:84, 2001.
887. H. Hattori. *Adv. Mater.* 13:51, 2001.
888. W.S. Sim and F.Y. Goh. *Surf. Rev. Lett.* 8:491, 2001.
889. V. Salgueiriño-Maceira, F. Caruso, and L.M. Liz-Marzán. *J. Phys. Chem. B* 107:10990, 2003.
890. A. Rogach, A. Susha, F. Caruso, G. Sukhorukov, A. Kornowski, S. Kershaw, H. Möhwald, A. Eychmüller, and H. Weller. *Adv. Mater.* 12:333, 2000.
891. D.Y. Wang and F. Caruso. *J. Chem. Soc. Chem. Commun.* 489, 2001.
892. D.Y. Wang, R.A. Caruso, and F. Caruso. *Chem. Mater.* 13:364, 2001.
893. Z. Liang, A.S. Susha, and F. Caruso. *Adv. Mater.* 14:1160, 2002.
894. G. Kumaraswamy, A.M. Dibaj, and F. Caruso. *Langmuir* 18:4150, 2002.
895. A.L. Rogach, N.A. Kotov, D.S. Koktysh, A.S. Susha, and F. Caruso. *Coll. Surf. A* 202:135, 2002.
896. D. Wang and F. Caruso. *Adv. Mat.* 15:205, 2003.
897. D. Wang, A.L. Rogach, and F. Caruso. *Chem. Mater.* 15:2724, 2003.
898. Z. Liang, A. Susha, and F. Caruso. *Chem. Mater.* 15:3176, 2003.
899. L.M. Goldenberg, B.D. Jung, J. Wagner, J. Stumpe, B.R. Paulke, and E. Görnitz. *Langmuir* 19:205, 2003.
900. P. Schuetz and F. Caruso. *Chem. Mater.* 14:4509, 2002.
901. N. Gaponik, I.L. Radtchenko, M.R. Gerstenberger, Y.A. Fedutik, G.B. Sukhorukov, and A.L. Rogach. *Nano. Lett* 3:369, 2003.
902. M. Lahav, A.N. Shipway, and I. Willner. *J. Chem. Soc. Perkin Trans.* 2:1925, 1999.
903. A.N. Shipway, M. Lahav, R. Blonder, and I. Willner. *Chem. Mater.* 11:13, 1999.
904. X. Zu, Z. Lu, J.B. Schenkman, and J.R. Rusling. *Langmuir* 15:7372, 1999.
905. G.M. Kloster and F.C. Anson. *Electrochim. Acta.* 44:2271, 1999.
906. H.Y. Ma, N.F. Hu, and J.F. Rusling. *Langmuir* 16:4969, 2000.
907. J.Y. Liu, L. Cheng, B.F. Liu, and S.J. Dong. *Langmuir* 16:7471, 2000.
908. L. Cheng and S. Dong. *J. Electroanal. Chem.* 481:168, 2000.
909. T. Hoshi, H. Saiki, S. Kuwazawa, Y. Kobayashi, and J. Anzai. *Anal. Sci.* 16:1009, 2000.
910. W.J. Li, M. Xian, Z.C. Wang, C.Q. Sun, and M.Y. Zhao. *Thin Solid Films* 386:121, 2001.
911. Z.L. Cheng, L. Cheng, S.J. Dong, and X.R. Yang. *J. Electrochem. Soc.* 148:E227, 2001.
912. L. Cheng and J.A. Cox. *Electrochem. Commun.* 3:285, 2001.
913. L.W. Wang and N.F. Hu. *Bioelectrochemistry* 53:205, 2001.
914. E.J. Calvo and A. Wolosiuk. *J. Am. Chem. Soc.* 124:8490, 2002.
915. P.L. He, N.F. Hu, and G. Zhou. *Biomacromolecules* 3:139, 2002.
916. D.M. Guldi, F. Pellarini, M. Prato, C. Granito, and L. Troisi. *Nano. Lett.* 2:965, 2002.
917. Y. Shen, J.Y. Liu, A.G. Wu, J.G. Jiang, L.H. Bi, B.F. Liu, Z. Li, and S.J. Dong. *Chem. Lett.* 550, 2002.
918. L. Cheng and J.A. Cox. *Chem. Mater.* 14:6, 2002.
919. Y.D. Jin, Y. Shao, and S.J. Dong. *Langmuir* 19:4771, 2003.
920. L.B. Shang, X.J. Liu, J. Zhong, C.H. Fan, I. Suzuki, and G.X. Li. *Chem. Lett.* 32:296, 2003.
921. N.F. Ferreyra, L. Coche-Guérente, P. Labbé, E.J. Calvo, and V.M. Solís. *Langmuir* 19:3864, 2003.
922. Y. Shen, J.Y. Liu, A.G. Wu, J.G. Jiang, L.H. Bi, B.F. Liu, Z. Li, and S.J. Dong. *Langmuir* 19:5397, 2003.
923. T. Cassagneau, J.H. Fendler, S.A. Johnson, and T.E. Mallouk. *Adv. Mater.* 12:1363, 2000.
924. J.H. Rouse and G.S. Ferguson. *Langmuir* 18:7635, 2002.
925. S. Das and A.J. Pal. *Langmuir* 18:458, 2002.
926. K. Esumi, S. Akiyama, and T. Yoshimura. *Langmuir* 19:7679, 2003.
927. J.F. Baussard, J.L. Habib-Jiwan, and A. Laschewsky. *Langmuir* 19:7963, 2003.
928. L.Q. Ge, H. Möhwald, and J.B. Li. *Chem. Eur. J.* 9:2589, 2003.
929. A. Carrillo, J.A. Swartz, J.M. Gamba, R.S. Kane, N. Chakrapani, B. Wei, and P.M. Ajayan. *Nano. Lett.* 3:1437, 2003.
930. I. Suzuki, T. Ishizaki, H. Inoue, and J.I. Anzai. *Macromolecules* 35:6470, 2002.

Chapter 20

Molecular Imprinting

Makoto Komiyama

Contents

I. Introduction

A number of elegant host molecules have already been synthesized and showed both high selectivity and binding activity toward the target guest compound. Through these studies, detailed and fundamental knowledge for molecular design of sophisticated receptors has been accumulated. Host–guest chemistry has been so fruitful and mature. From the viewpoint of the practical applications to industry and our daily lives, however, these synthetic receptors have several drawbacks. First, their synthesis usually requires many complicated reaction steps, and thus they are usually too expensive for common industrial use. Second, the design of receptors for large guest molecules is quite difficult, since the scaffolds available (e.g., cyclodextrin [CyD], crown ether, calixarene, and others) are in most cases not much greater than several angstroms. Under these conditions, it is hard to place two or more functional groups at notably remote sites (e.g., >10 Å) in the receptors. Third, precise guest recognition in water is difficult, since hydrogen bonding is easily broken due to the competition with the water.

Molecular imprinting is a newly developed method and is the most promising solution to these problems. By simple polymerization of appropriate functional monomers in the presence of template (the target molecule for the recognition or its analog), eminent receptors are cheaply prepared in a tailor-made fashion. Receptors for large guest molecules are also easily obtainable and no complicated organic synthesis is necessary. Furthermore, molecular recognition in water is possible by choosing appropriate functional monomers, cross-linking agents, and others and forming specific reaction fields at the recognition site. Also, molecular imprinting method is applicable to the formation of complicated supramolecules that are otherwise hard to form. Because of these features, this method has been attracting much interest. In this chapter, the roles of molecular imprinting method in supramolecular polymer chemistry are described. Although many papers have been published on molecular imprinting, only the typical example will be presented here to show the readers the trends of recent developments. More details should be referred to the recent book on this subject [1].

II. GENERAL PRINCIPLE OF MOLECULAR IMPRINTING METHOD

Molecular imprinting method is a technique in which the movements of molecules in solutions are frozen in polymeric structures (Figure 1). Thus, a template molecule (which is the target molecule for the recognition or its analog) and functional monomer are covalently bound and the conjugates are polymerized. Alternatively, functional monomers which show noncovalent interactions with a template molecule are polymerized in the presence of the template. During this polymerization processes, the structure of template is memorized in the resultant polymers, providing the receptors we need. Accordingly, molecular imprinting processes are composed of the following three steps: (1) preparation of covalent conjugates or noncovalent adducts between a functional monomer and a template molecule, (2) polymerization of this monomer–template conjugate (or adduct), and (3) removal of the template from the polymer.

In step (1), functional monomer and template are connected by covalent linkage (in "covalent imprinting"), or through noncovalent interactions (in "noncovalent imprinting"). In step (2), radical polymerization is often employed, although other polymerizations are also applicable. In this step, the structures of these conjugates (or adducts) are frozen in three-dimensional network of polymers, where the functional residues (derived from the functional monomers) are chemically and

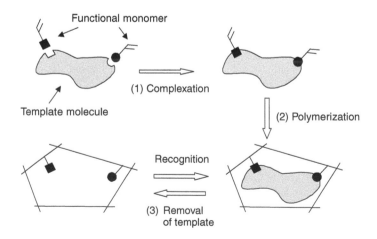

Figure 1 Schematic view of molecular imprinting.

topographically complementary to the template. In step (3), the template molecules are removed from the polymer. Here, the space in the polymer that is originally occupied by the template molecule is left as a cavity. Under appropriate conditions, these cavities satisfactorily remember the size, structure, and other physicochemical properties of template, and bind this molecule (or its analog) efficiently and selectively.

In most cases, cross-linking agents such as ethylene glycol dimethacrylate and divinylbenzene are used to fix the guest-binding sites firmly in the desired structure. They also make the imprinted polymers insoluble in solvents and facilitate their practical applications. By using different kinds of cross-linking agents, both the structure of the guest-binding sites and the chemical environments around them can be controlled.

III. COVALENT IMPRINTING AND NONCOVALENT IMPRINTING

Molecular imprinting method is divided into two categories depending on the nature of adducts between functional monomer and template (either covalent or noncovalent). In covalent imprinting, functional monomer and template are first bound to each other by covalent linkage (step 1). Then, this covalent conjugate is polymerized under the conditions where the covalent linkage is intact (step 2). After the polymerization, the covalent linkage is cleaved and the template is removed from the polymer (step 3). The first report on covalent molecular imprinting was made by Wulff and coworkers in 1977 (Figure 2) [2]. They synthesized 2:1 covalent conjugate of p-vinylbenzeneboronic acid with 4-nitrophenyl-α-D-mannopyranoside (the template), where the template was attached to the styrene

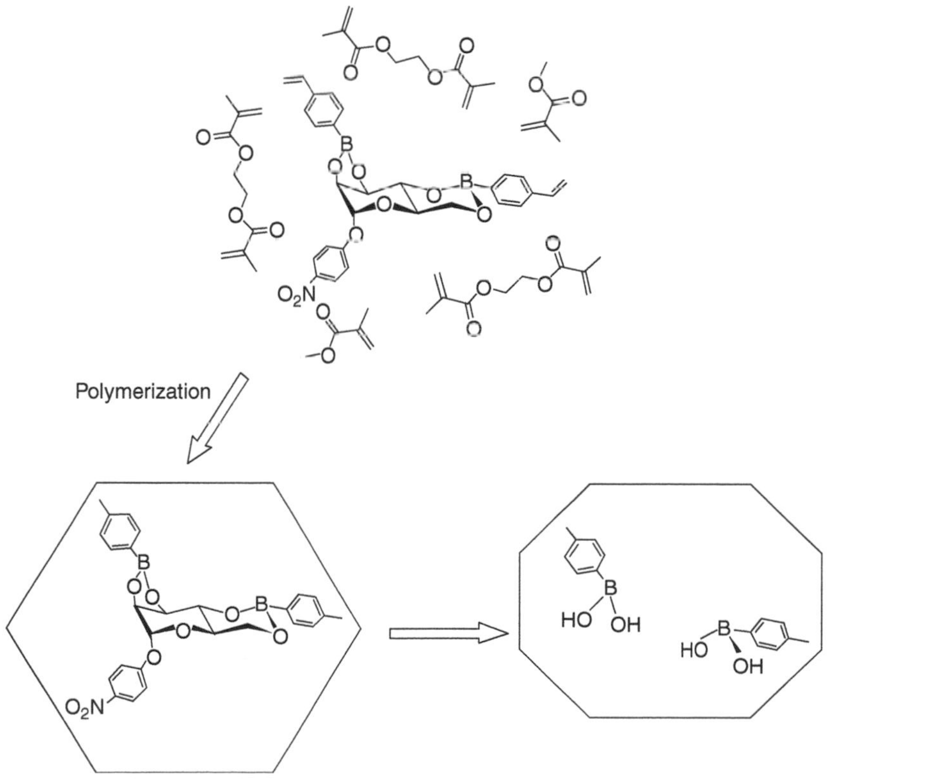

Figure 2 Covalent imprinting of mannopyranoside using its 4-vinylphenylboronic acid ester as a functional monomer.

derivative through the formation of boronic acid ester. Then, this vinyl monomer-conjugate was copolymerized with methyl methacrylate and ethylene dimethacrylate (a cross-linking monomer). After the polymerization, the boronic acid ester in the polymer was chemically cleaved, and the 4-nitrophenyl-α-D-mannopyranoside was removed. Exactly as designed, the resultant polymer strongly and selectively bound this sugar. The mutual conformation of the two boronic acid groups in the covalent conjugate was frozen in the polymer, and the structure of template was memorized in terms of the position and orientation of these groups.

On the other hand, noncovalent imprinting takes advantage of noncovalent interactions to connect functional monomer with template (step 1) [3]. Thus, the adducts are *in situ* formed in the reaction mixtures. Specific synthesis of covalent conjugate is not necessary. After the polymerization (step 2), the template is removed by extracting the polymer with appropriate solvents (step 3). In the molecular imprinting of methacrylic acid with theophylline (a drug), for example, noncovalent monomer-template adduct was formed through hydrogen bonding and electrostatic interaction (Figure 3). As is the case in covalent imprinting, both the selectivity and binding activity are greatly promoted by the imprinting. The same strategy was successful for various drugs, insecticides, and other practically important chemicals. This method is very easy to achieve and applicable to versatile templates.

IV. RECENT CHALLENGES AND PROGRESSES

On the basis of these pioneering works, further improvements of the binding activity and selectivity, as well as the applications of molecular imprinting method to still more versatile purposes, have been

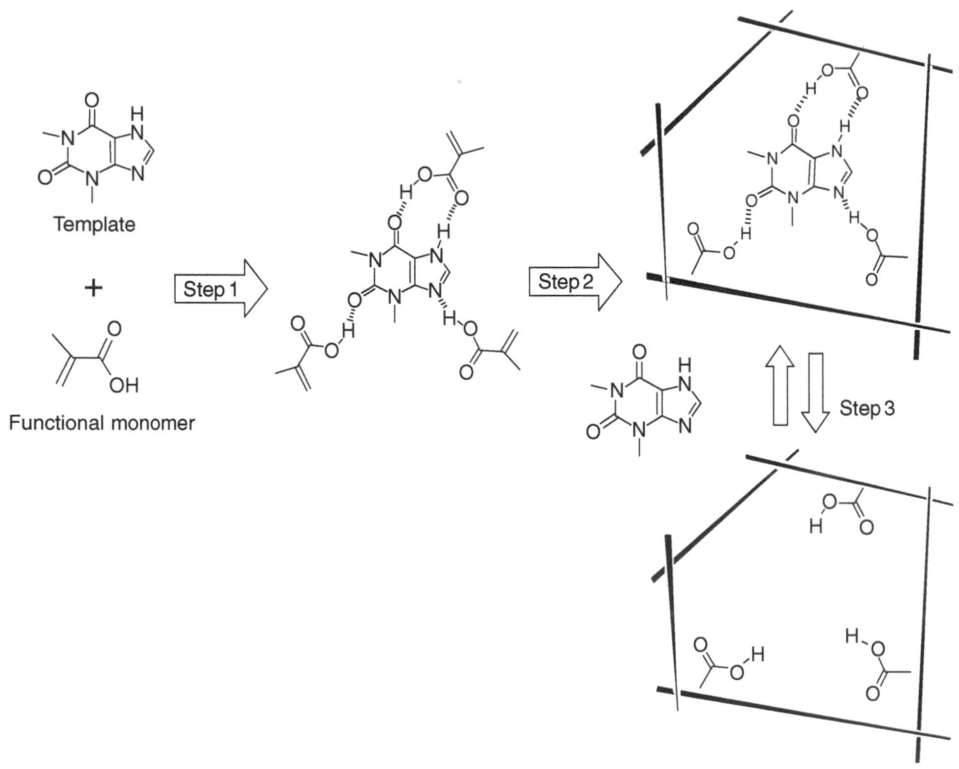

Figure 3   Noncovalent imprinting by using a drug (theophylline) as template. Step 1: Preorganization of functional monomers through noncovalent interactions. Step 2: Polymerization of preorganized functional monomers. Step 3: Removal of the template.

recently made. There are so many examples in which the aiming goals are notably different from each other. Among them, preparation of artificial antibodies is one of the most challenging targets. These supramolecules, if available, should be quite useful for industry, biotechnology, medicines, and many other applications. Molecular imprinting method is relevant to antibodies, in that the receptors can be synthesized in tailor-made fashion according to the target guest.

In order to mimic the functions of these naturally occurring receptors, large guest molecules must be precisely recognized in water. Accordingly, most of the attempts hitherto made involve the use of noncovalent interactions other than hydrogen bonding, since hydrogen bonds are easily broken in bulk water. It has also been attempted to provide specific reaction fields so that hydrogen bonds can be satisfactorily operative even in bulk water. Furthermore, the use of these artificial antibodies as highly active and selective catalysts is also a promising target. Compared with naturally occurring antibodies, synthetic polymers are far more stable and can be used under variety of reaction conditions. They are also much cheaper and obtainable in larger scale.

V.   MOLECULAR IMPRINTING OF HOST MOLECULES LEADING TO THEIR ORDERED ASSEMBLY

A.   Preparation of Ordered Assemblies of CyD as Receptors for Nanometer-Scaled Guests

Recent progresses in host–guest chemistry are so remarkable that selective hosts are rather easily obtainable as long as the target guest is small in size. However, design of artificial receptors for nanometer-scaled guests is far more difficult, and has not yet been sufficiently accomplished. These receptors are regarded as one of the keys for the future science and technology, where a wide spectrum of large molecules (e.g., proteins, nucleic acids, polysaccharides, and many other bioactive materials) must be strictly differentiated from each other and show the corresponding roles without too much cross talk.

The strategy we have developed for this purpose is to build up several host molecules so that each of them fits the designated portion of the target large guest [4–6]. Under these conditions, each of the host molecules in the ordered assembly binds a small portion of the target guest (as does the host in the monomeric form), but the assembly as a whole recognizes this large guest very exclusively. As the result, only the target nanometer-scaled guest should be selectively bound (Figure 4). The binding of the other guests is minimized due to the lack of cooperation of two (or more) host molecules and/or steric repulsion by some of the host molecules in the assemblies.

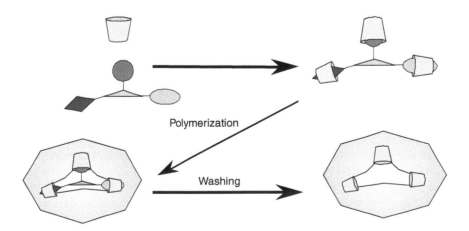

Figure 4   Molecular imprinting of CyD in the presence of a template in water.

These ordered assemblies can be easily prepared by using molecular imprinting method (Figure 4). First, host molecules (chemically modified when necessary) are bound to the target guest in solutions. Then, the host molecules in the host–guest adducts are cross-linked. By this procedure, the orientation of the host molecules in the adducts is immobilized in the polymeric structures. Since each of the host molecules accurately recognizes a small but predetermined portion of the target nanometer-scaled guest and they are orderly fixed in the receptor, the target guest is selectively bound. Apparently, this strategy is also applicable to the combination of various host molecules and variety of guest molecules.

For example, the molecular imprinting of CyD was achieved in either dimethyl sulfoxide (DMSO) or water, where CyD inclusion complexes are efficiently formed. In DMSO, CyD was reacted with diisocyanate compound (cross-linking agent) in the presence of a template (the guest molecule to be recognized by the receptor), as schematically depicted in Figure 4. In the resulting polymer, CyD molecules were connected to each other by urethane linkages. The typical cross-linking agents are toluene diisocyanate (TDI) and hexamethylene diisocyanate (HMDI), both of which are commercially obtainable. In DMSO, the inclusion complexes between CyD molecules and the template are successfully formed as in water, and thus the fundamental requirement of noncovalent molecular imprinting is fulfilled. Furthermore, DMSO is inert to the diisocyanate. Water cannot be used as the solvent, since it promptly reacts with the diisocyanate (the method for the imprinting in water is described below). After the polymerization, the CyD which did not react, the template, and the diisocyanate were removed by washing the polymer with various solvents. Although the receptors are generally obtained as white solids, water soluble receptors, if necessary, can be also synthesized by decreasing the amount of diisocyanate compound. In this method, artificial receptors can be prepared in a tailor-made fashion, even when the template (and the guest) is virtually insoluble in water.

The imprinted β-CyD polymers, prepared with the use of cholesterol as the template, bound cholesterol in water far more efficiently than did the polymer obtained in the absence of the template. The effect of molecular imprinting is evident. Cholesterol is too large to be accommodated in the cavity of one β-CyD molecule. Thus, two (or more) β-CyD molecules must simultaneously interact with one cholesterol molecule in order to bind cholesterol efficiently. Only when the molecular imprinting method is used and several β-CyD molecules are placed in the desired positions, eminent and practically useful polymeric receptors can be obtained.

The molecular imprinting of CyD can also be accomplished in water [7]. Since diisocyanates easily react with water and cannot be used as cross-linking agents (*vide ante*), receptors were synthesized by radical polymerization. Thus, a vinyl residue was first introduced to CyD (either the secondary hydroxyl side or the primary hydroxyl side). In the presence of the template in water, this vinyl monomer was copolymerized with N,N'-ethylene*bis*acrylamide (cross-linking agent). The imprinting polymer thus obtained also showed high binding activity and selectivity toward the template molecule in water.

B. Structure of Guest-Binding Sites and Analysis of the Molecular Imprinting Process [8]

Figure 5(a) shows the mass spectrum for the reaction mixture obtained with cholesterol as template in DMSO (the cross-linking agent = TDI). Under these imprinting conditions, dimers of β-CyD, as well as the β-CyD trimers, are efficiently formed. Monomeric β-CyDs are formed in only a small amount. Each of the signals in the dimer and the trimer regions (also in the monomer region) corresponds to different number of substitution by the cross-linking agent. For example, the signal at $m/z = 2444$ (the smallest molecule in the dimer region) is assigned to the β-CyD dimer in which two β-CyD molecules are bridged by one cross-linking agent. The signal at $m/z = 4223$ represents the trimer in which three β-CyDs are connected and two cross-linkers are tethered to these β-CyDs.

When the reactions are achieved in the absence of cholesterol, however, the mass spectra are dramatically changed (Figure 5(b)). Here, most of the products are monomeric β-CyDs. Dimers of

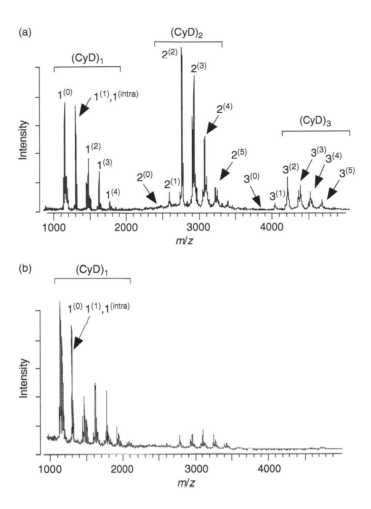

Figure 5 MALDI-TOF MS spectra of the products for cross-linking β-CyD with TDI in DMSO (a) in the presence of cholesterol and (b) in its absence.

β-CyD are only marginal, and trimers are not detectable. As described above, at least two β-CyD molecules are necessary to bind cholesterol efficiently. It is concluded that, only in the presence of an appropriate template, these ordered assemblies of β-CyD are efficiently formed and thus eminent receptors can be obtained. Direct spectroscopic evidence for the molecular imprinting has been obtained. These dimers and trimers of β-CyD cannot be easily formed by simple collision reactions in the absence of cholesterol template.

From these results, the mechanism of the present imprinting is proposed as depicted in Figure 6. First, one of the two isocyanate groups of cross-linking agent reacts with β-CyD. This reaction mainly occurs (either with or without the template) at the primary OH groups, which are more reactive. When the other isocyanate group reacts with another β-CyD molecule under the imprinting conditions, however, the reaction preferentially takes place at the secondary OH since this OH is placed near the second isocyanate group due to the orientation of inclusion complex. Because of thermodynamic stability, the cholesterol is included into the cavity from its secondary hydroxyl side (see Figure 6). As a result, these CyD dimers (or trimers) are efficiently formed in the imprinted polymers, and function as the cholesterol-binding sites. These phenomena in solutions give rise to the enormous molecular imprinting effect in the guest-binding properties.

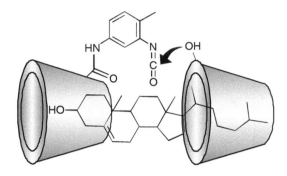

Figure 6   Proposed mechanism for the formation of cholesterol-binding site from two β-CyD molecules.

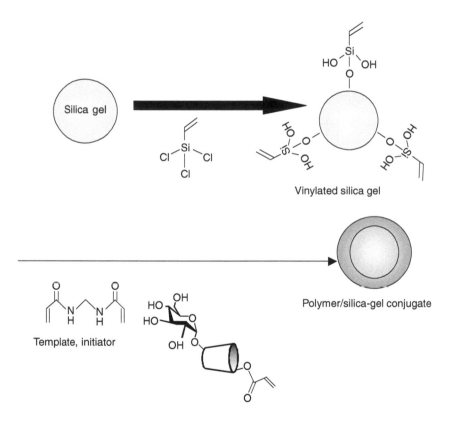

Figure 7   Immobilization of the imprinted CyD polymer on the silica-gel surface for the stationary phase of HPLC.

C.   Imprinting of CyD on the Surface of Silica-Gel Support [9]

As described earlier, imprinted CyD polymers are superb in molecular-recognizing activity. However, they are mechanically rather weak and inappropriate as the stationary phase of high performance liquid chromatography (HPLC). In order to compromise these two factors, CyDs were imprinted on the surface of silica-gel support. By this straightforward strategy, a stable stationary phase for HPLC can be obtained in a tailor-made fashion (Figure 7). First, vinyl groups were introduced to the surface of silica gel by treating the silica gel with trichlorovinylsilane. The mixture of this vinylated silica gel and vinylated β-CyD was reacted with cross-linking agent in the presence of template molecule. In the polymer/silica-gel composites thus obtained, the imprinted β-CyD polymer was

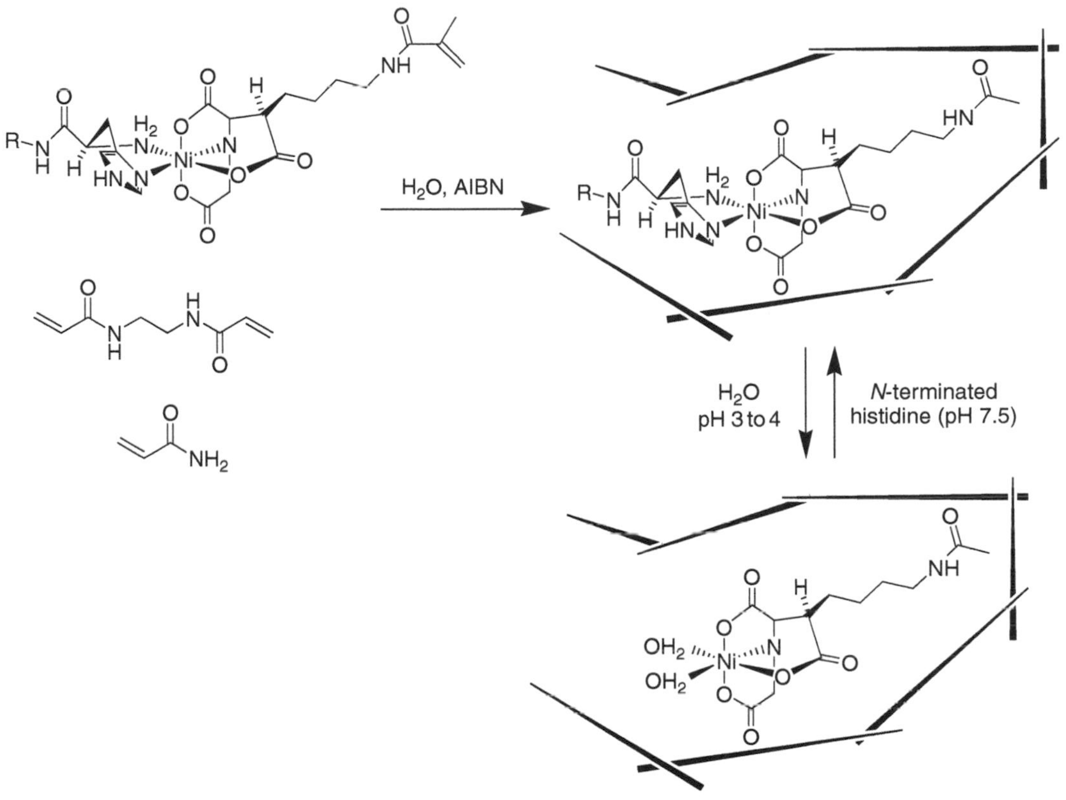

Figure 8 Molecular imprinting of peptides in water by use of metal complex formation.

covalently bound to the silica gel, and thus was never removed under wide varieties of operation conditions. The β-CyD-polymer/silica-gel composites were then packed into a stainless column, and used as a stationary phase of HPLC. When L-Phe-L-Phe was used as template, it was retained more strongly than D-Phe-D-Phe. With the conjugate prepared in the presence of D-Phe-D-Phe as the template, however, it was retained much strongly than L-Phe-L-Phe. The selectivity is absolutely ascribed to the imprinting effect, since the polymer/silica-gel conjugate prepared without template hardly discriminated these two isomers.

VI. USE OF METAL COMPLEXES AS GUEST-BINDING SITE IN WATER [10]

As the candidates for molecular recognition in water, polymer-bound metal ions are also promising. For example, polymeric receptors for peptides were prepared by using Ni(II) ion (see Figure 8). Two of the six coordination sites of Ni(II) ion were used to bind the template at its *N*-terminal amine and the imidazole of *N*-terminal histidine. An acrylate residue was attached to this Ni(II) ion through the coordination of the nitrilotriacetic acid bound to the acrylate. This complex was sufficiently stable in water at around pH 7 so that this preorganized complex could be directly copolymerized with acrylamide and *N*,*N*'-ethylene*bis*acrylamide. After the polymerization, the resultant polymer was treated with water at pH 3 to 4 to remove the template. The residual Ni(II)-nitrilotriacetate complexes on the polymer were the specific binding sites for the template in water. The imprinted polymer, obtained with the use of His–Ala as template, showed a significantly higher binding activity toward the template peptide over the other peptides such as His–Phe and His–Ala–Phe. Peptides containing no histidine group such as Ala–Phe had almost no affinity to this polymer. Apparently, the binding sites which are complementary to the template in both size and shape were formed in the imprinted polymer and served as the guest-binding sites.

There are many other attempts based on similar strategy, as presented in the recent book [1]. The use of metal ions, as well as host molecules, is highly promising for future applications.

VII. ARTIFICIAL ENZYMES BY MOLECULAR IMPRINTING

A. Use of Monomers Bearing Catalytically Active Groups

As well known, naturally occurring enzymes are basically composed of substrate-binding sites and catalytic sites. These sites are located close to each other and give rise to enormous catalysis for the specific reactions. Many artificial enzymes have been prepared on the basis of this concept. This strategy is also successful in molecular imprinting method where the corresponding catalytic residue is bound to the functional monomer. The catalytic group is placed near the selective guest-binding site formed through the molecular imprinting process.

For example, an artificial enzyme for the decomposition of atrazine (a herbicide) was synthesized by polymerizing 2-sulfoethyl methacrylate and methyl acrylate in chloroform in the presence of atrazine as template (see Figure 9) [11]. When the obtained artificial enzyme was added to the solution of atrazine together with methanol, the atrazine was efficiently decomposed to nontoxic atraton.

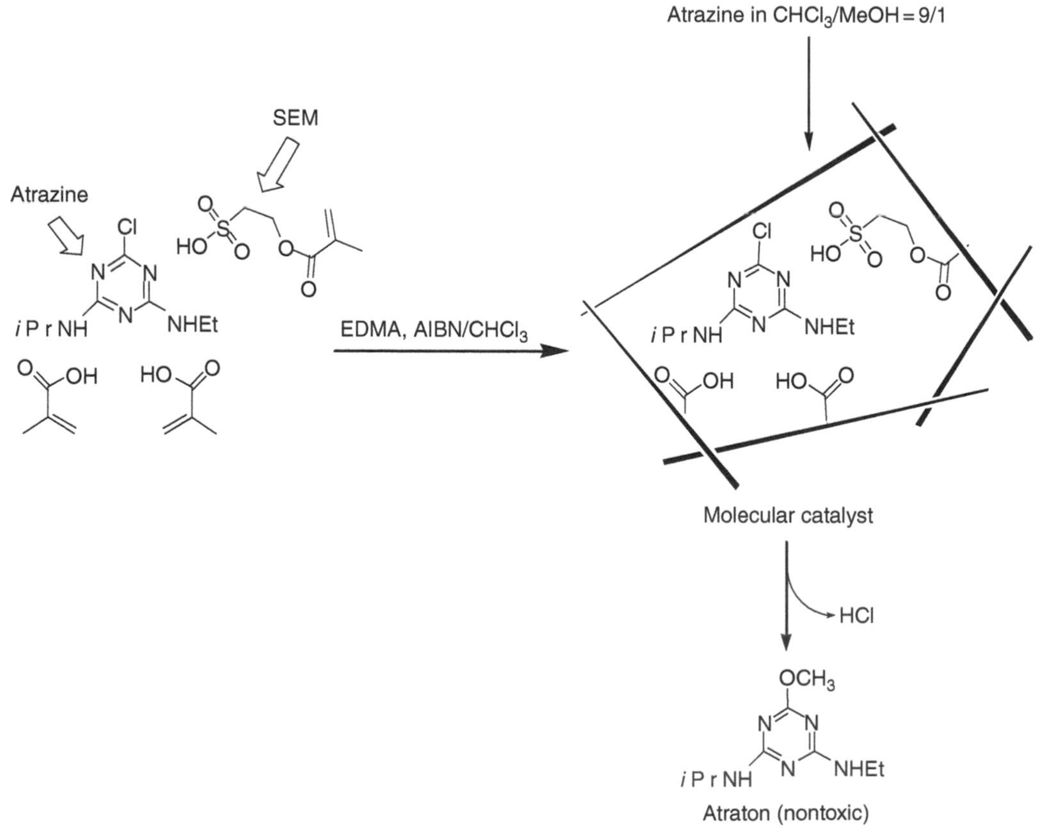

Figure 9 Preparation of catalyst for the decomposition of atrazine by using molecular imprinting technique.

Here, the sulfoethyl group as the catalytic site activates the methanol and promotes the nucleophilic substitution at the atrazine. Consistently, catalytic activities were far smaller when the polymerization of 2-sulfoethyl methacrylate and methyl acrylate was achieved in the absence of the template. Furthermore, upon polymerizing only methyl acrylate in the absence of 2-sulfoethyl methacrylate, the resultant polymer-bound atrazine but showed no catalytic activity for its decomposition. Apparently, the binding sites were sufficiently formed in the polymers but they are not sufficient to decompose atrazine efficiently.

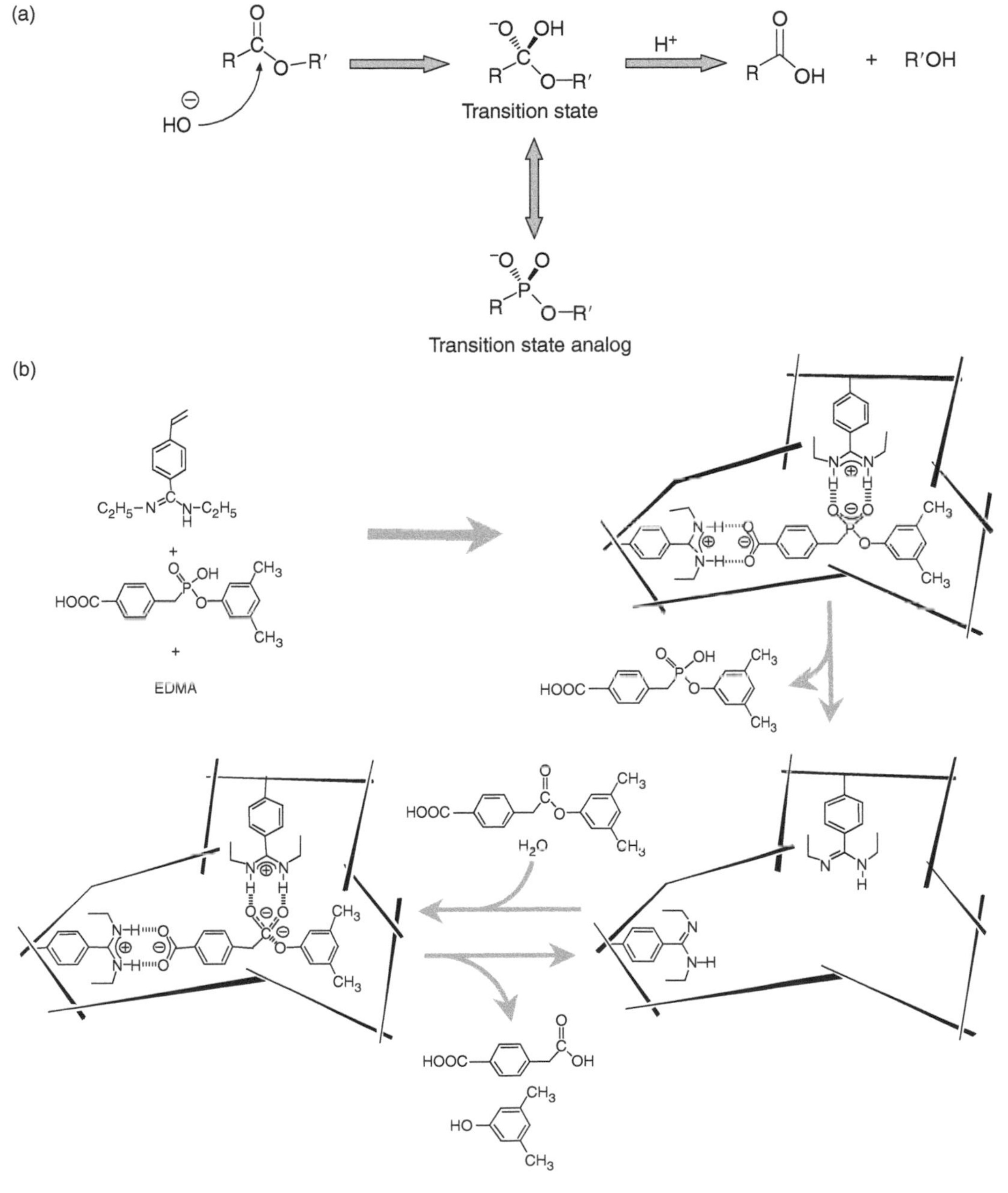

Figure 10 (a) Mechanism of ester hydrolysis and (b) molecular imprinting of transition-state analog (phosphonic acid).

B. Molecular Imprinting by Using Transition-State Analog as Template (Preparation of Synthetic Catalytic Antibody)

In addition to the conventional design of artificial enzymes described in Section VII.A, there exists an entirely new strategy based on the proposal made by Pauling about 50 years ago. As he pointed out, the difference between enzymes and antibodies is the fact that the former preferentially binds the transition-state of reaction whereas the latter binds the ground state. The enzymes enormously accelerate the reactions, since they bind the transition-state more strongly than the initial state and stabilize the transition-state to a greater extent. If an antibody selectively recognizes the transition-state of a reaction, it should catalyze the corresponding reaction. These antibodies are called "catalytic antibodies." For example, ester hydrolysis proceeds through the transition-state shown in Figure 10(a). Since this transition-state itself is too unstable to use, its analog is synthesized by replacing the central carbon with phosphorus. Exactly as designed, the monoclonal antibody, prepared with this transition-state analog as hapten (antigen), accelerated the hydrolysis of the corresponding carboxylic ester by a factor of 10^3 to 10^4 [12].

Similar approach is possible by using molecular imprinting method. In Figure 10(b), a phosphonic acid was used as the transition-state analog of the hydrolysis of carboxylic acid ester [13]. As the functional monomer, vinyl monomer of amidine was chosen because it readily forms stable complexes with both the carboxylic acid ester and the phosphonic acid monoester. From this monomer, the phosphonic acid (the template), and ethylene glycol dimethacrylate (cross-linking agent), imprinted polymers were synthesized. The obtained polymer efficiently catalyzed the hydrolysis of the target carboxylic acid ester, and the reaction followed typical Michaelis–Menten kinetics.

VIII. Conclusion

By using molecular imprinting method, various receptors have been prepared easily and economically. Tailor-made receptors thus obtained should be useful for various applications. Furthermore, supramolecules of host molecules have been synthesized by imprinting the host molecules toward the target guest compound. In these supramolecules, several host molecules are arranged exactly as dictated by the guest molecule and function, cooperatively. These approaches should open a new field in host–guest chemistry and supramolecular chemistry. Extension to artificial catalytic antibodies should also be promising.

Acknowledgments

This work was supported by a Grant-in-Aid for Scientific Research from the Ministry of Education, Science, Sports, Culture, and Technology, Japan. The support by the Bio-oriented Technology Research Advancement Institution is also acknowledged.

References

1. M. Komiyama, T. Takeuchi, T. Mukawa, and H. Asanuma, *Molecular Imprinting — From Fundamentals to Applications*, Wiley-VCH, Weinheim (2003).
2. G. Wulff, R. Grobe-Einsler, W. Vesper, and A. Sarhan, *Makromol. Chem.*, **178**, 2817 (1977).
3. R. Arshady and K. Mosbach, *Makromol. Chem.*, **182**, 687 (1981).
4. H. Asanuma, M. Kakazu, M. Shibata, T. Hishiya, and M. Komiyama, *Chem. Commun.*, **20**, 1971–1972 (1997).

5. T. Hishiya, M. Shibata, M. Kakazu, H. Asanuma, and M. Komiyama, *Macromolecules*, **32**, 2265–2269 (1999).
6. H. Asanuma, T. Hishiya, and M. Komiyama, *Adv. Mater.*, **12**, 1019–1030 (2000).
7. H. Asanuma, T. Akiyama, K. Kajiya, T. Hishiya, and M. Komiyama, *Anal. Chim. Acta*, **435**, 25–33 (2001).
8. T. Hishiya, H. Asanuma, and M. Komiyama, *J. Am. Chem. Soc.*, **124**, 570–575 (2002).
9. T. Hishiya, T. Akiyama, H. Asanuma, and M. Komiyama, *J. Incl. Phenom. Macrocyc. Chem.*, **44**, 365–367 (2002).
10. B.R. Hart and K.J. Shea, *J. Am. Chem. Soc.*, **123**, 2072–2073 (2001).
11. T. Takeuchi, D. Fukuma, J. Matsui, and T. Mukawa, *Chem. Lett.*, **30**, 530–531 (2001).
12. R.A. Lerner, S.J. Benkovic, and P.G. Schultz, *Science*, **252**, 659–667 (1991).
13. G. Wulff, T. Gross, and R. Schönfeld, *Angew. Chem. Int. Ed. Engl.*, **36**, 1962–1964 (1997).

Chapter 21

Protein Polymerization and Polymer Dynamics Approach to Functional Systems

Fumio Oosawa

Contents

I. Introduction

We presented, about 40 years ago, a theoretical framework for understanding the polymerization of protein molecules to helical polymers, based on the study of the G–F transformation of actin, a muscle protein [1–3].

In pure water, all actin molecules are in the G-actin (globular actin monomer) state and at the physiological concentration of salts, almost all are in the F-actin (fibrous actin polymer) state. In the intermediate condition of salts, at low concentrations of actin no F-actin exists. At a certain critical concentration, F-actin begins to be formed. With increasing concentration of actin, the amount of F-actin increases, coexisting with G-actin, which is kept at the critical concentration. In this condition, each monomer undergoes a cycle between two states, G- and F-actin. The macroscopic G–F balance is maintained by microscopic cycling. The critical concentration decreases with increasing salt concentration. Thus, the G–F transformation has similar features to gas–liquid condensation or crystallization. Actually, the transformation consists of two processes, nucleation and growth. Usually, nucleation is rate limiting.

A theory of helical polymerization was proposed to explain such crystallization-like features. F-actin was assumed to be a helical polymer of G-actin, where each actin monomer (G-actin) is

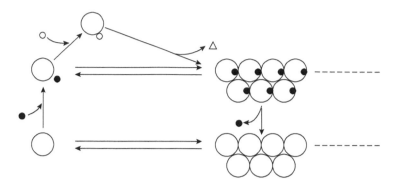

Figure 1   The G–F transformation of actin with ATP (open circle), ADP (closed circle), or no nucleotide; a classical scheme.

bound with four neighboring monomers through two kinds of bonds: one along the longitudinal strands and the other between two strands or along the genetic helix.

Soon after this proposal, electron micrographs showed that F-actin is a two-stranded helical polymer [4]. The bonding pattern predicted was finally confirmed by the structural analysis at atomic resolution. The three-dimensional structure of the actin molecule in a crystal of the complex with DNase I was determined by x-ray crystallography, and using this molecular structure, the structure of F-actin was built-up to give the best fit to the x-ray diffraction data from an oriented gel of F-actin [5,6]. Now, the amino acid residues taking parts in the two kinds of bonds in F-actin can be identified.

G-actin, when extracted from a muscle fiber into water, had bound ATP. During polymerization to F-actin, this ATP was hydrolyzed to ADP and inorganic phosphate [7]. The ADP was kept bound in F-actin. In the process of depolymerization of F-actin to G-actin, rephosphorylation of ADP did not happen. After depolymerization, bound ADP was replaced with ATP in solution. Later it was found that G-actin having ADP instead of ATP also polymerizes to F-actin, although the rate of polymerization is much slower than G-actin having ATP [8]. Even G-actin without ATP or ADP can polymerize, if denaturation of this nucleotide-free G-actin is inhibited by a high concentration of sucrose [9]. The G–F transformation of actin was described by the scheme shown in Figure 1.

Since then, the polymerization of various protein molecules was found to have similar features. There is a critical concentration for polymerization, and the polymerization consists of nucleation and growth. Bacterial flagella are helically curved tubular polymers of flagellin molecules. Flagella are formed from purified flagellin molecules in solution, although spontaneous nucleation of flagella can hardly occur [10]. The polymerization of tubulin molecules to microtubules is another example of tubular polymerization and in this case, the hydrolysis of GTP bound to monomers is associated with the polymerization process [11].

When we reached the idea of helical polymerization, we imagined immediately that the partial breaking or weakening of monomer–monomer bonds may produce a large conformational change of F-actin, keeping its filamentous continuity, as illustrated in Figure 2 [12,13]. Since then, we have been very interested in dynamic behaviors of helical polymers.

In this chapter, I briefly describe further development of the study on the polymerization of actin and other proteins and discuss regulation of the monomer–monomer bond and polymer conformation. Such regulation generates various dynamic functions of protein polymers. This chapter should help in understanding the design principle of protein polymers as a functional system.

II.   MONOMER–MONOMER BOND FORMATION

In the case of polymerization of actin having bound ADP, an equilibrium state between G- and F-actins is established. Under this condition the concentration of G-actin coexisting with F-actin,

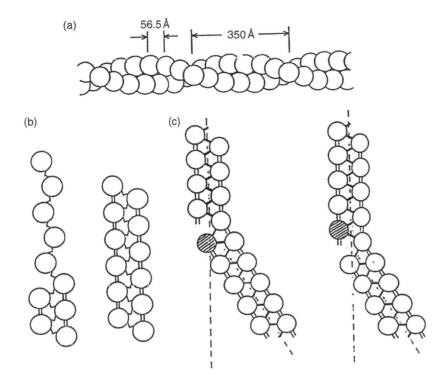

Figure 2   (a) A two-strand helical polymer structure of F-actin. (b) and (c) Illustration of hypothetical conformational changes of F-actin caused by weakening of the monomer–monomer bond possibly coupled with the ATP hydrolysis cycle.

the critical concentration, is related to the free energy of bonding of an actin monomer to the end of F-actin. It is estimated to be −6 to −13 kcal/mol [3]. (The bond free energy includes contributions of the two kinds of bonds.) The free energy depends on the concentration of salt ions, pH, and other environmental conditions. The polymerization is endothermic and the critical concentration decreases with rising temperature [14]. The addition of monomers to F-actin is driven by the entropy increase. Application of pressure promotes depolymerization [15]. The polymerization is associated with volume increase. The rearrangement of water molecules around monomers is probably involved in bonding. (Both too high temperature and too high pressure induce irreversible denaturation of actin molecules.)

F-actin is a semirigid filament. The flexural rigidity of F-actin was first estimated by quasielastic light scattering and later the thermal bending movement of F-actin was made directly visible by optical microscopy [16,17]. The average amplitude of bending movement was about 40 nm in F-actin of the length of 1 μm. Extensibility was examined by mechanically stretching the F-actin, both ends of which were bound to thin glass needles [18,19]. A force of a few hundred pN (picoNewton) stretched F-actin of the length of 1 μm by a few angstroms. The comparison of flexibility and extensibility suggests that the elasticity of the monomer–monomer bond and the intramonomer structure has no special anisotropy.

The rigidity decreased with rising temperature [20]. This means that the short-range recovery force of deformed F-actin comes from the enthalpy decrease, in contrast with the entropy increase for the monomer–monomer bonding. The monomer–monomer interaction free energy is a complex function of the distance and angle between monomers. The bond formation must be a multistep process. Structural analysis indicates that the interaction occurs between certain areas of the monomer surface. Bond formation is not simply attributable to a single pair of specific sites of monomers.

The polymerization of flagellin to flagellum and that of tubulin to microtubule are also endothermic. The monomer–monomer bond free energy is of the same order as that in F-actin, although the dependency on the salt concentration is different in each.

F-actin has a structural polarity, as demonstrated by the manner of binding of myosin fragments. The growth rate is much larger at the end named B-end (barbed end) than at the other end, named P-end (pointed end) [21]. The rate of depolymerization was also different at two ends. However, in the case of polymerization that is not accompanied by ATP hydrolysis, the ratio of the rates of association and dissociation of actin monomers must be equal at two ends, giving the critical concentration of G-actin in equilibrium with F-actin.

The growth of bacterial flagella is unidirectional [22]. In a living cell, the growth occurs only at the distal end; the *in vitro* growth also occurs at the same end. The binding of a flagellin monomer to the end of a flagellum was found to be associated with a large conformational change of the monomer, from a partly unfolded conformation to a folded one [23]. Such a conformational change is catalyzed by interaction of the monomer with monomers at the distal end. This makes spontaneous nucleation difficult. Accordingly, formation of free flagella from flagellin molecules inside the cell is inhibited. Correlation of the directionality of growth with the conformational change required was examined using flagellin monomers in which a part of the unfolded chain was chemically removed. The growth was not exclusively unidirectional.

Also, in the case of actin the polymerization is probably associated with a (small) conformational change of each monomer, which may be related to directionality of the growth. The conformational difference between free monomers and monomers in the polymer is one of the regulatory factors of the manner of polymerization.

III. BOND REGULATION COUPLED WITH THE CHEMICAL REACTION

Nucleotides bound to G-actin change the polymerization rate and the monomer–polymer balance. They have special characteristics as regulators of the monomer–monomer bond strength.

In the state of G-actin, the bound nucleotide molecule, ATP or ADP, is quickly exchangeable with the free molecule in solution. An equilibrium is established between bound and free molecules. On the other hand, in F-actin the bond nucleotide molecule is not easily exchangeable. According to the three-dimensional structural analysis, G-actin appears to be composed of four domains. ATP or ADP is found in a deep cleft in the middle of the molecule. Probably, in G-actin the cleft has some flexibility to expose bound ATP or ADP to the solvent, whereas in F-actin such flexibility is limited. Therefore, whether each actin monomer in F-actin has ATP or ADP depends on the history of the monomer.

Previously ATP hydrolysis was thought to be directly coupled with the polymerization of G-ATP-actin. However, it was later found that ATP hydrolysis occurs after polymerization [24]. At the end of F-actin, a newly incorporated actin monomer keeps ATP for a while. The actin monomer in F-actin has an ability to hydrolyze ATP bound to it. Near the growing end of F-actin, there are three kinds of actin monomers: those having ATP, those having ADP and inorganic phosphate, and those having ADP alone. The fraction of three kinds of monomers near the ends changes depending on the rate of growth, as shown in Figure 3 [25].

Figure 3 The monomers having ATP (T) or ADP (D) near two ends of F-actin during polymerization of G-actin having ATP; tread-milling may happen.

The hydrolysis of bound ATP weakens the monomer–monomer bond in F-actin. The bond free energy depends on whether the interacting monomers have ATP or ADP. The polymerization of G-actin without nucleotide can be performed in a sucrose solution. In this case, the critical concentration is very low; the monomer–monomer bond must be strong [26]. Various kinds of pairs of monomers having ATP, ADP, and inorganic phosphate, ADP alone, or no nucleotide can be formed in F-actin. (Bound inorganic phosphate is exchangeable.) The bond free energy and the bond elasticity are different in different pairs. Quantitative comparison has not yet been fully carried out.

In the case of polymerization of G-ATP-actin, the ratio of the rates of association and dissociation of monomers is different at two ends. The critical concentration of G-ATP-actin is defined by the condition that the sum of the growth rates at the two ends and the sum of the depolymerization rates are equal. At one of the two ends, the depolymerization may be faster than the growth and at the other end, the growth may be faster. Then a cycling of actin monomers takes place from one end of F-actin to the free monomer to the other end. This phenomenon was named "tread-milling" [27]. As a result, the polymer translates. Consumption of the free energy of ATP hydrolysis is required for such unidirectional translation of F-actin. The coupling of the polymerization with an irreversible chemical reaction produces a dynamic function of the polymer.

Tubulin monomers having GTP polymerize to microtubules having GDP. In this case also, the hydrolysis of GTP in microtubules occurs with a time lag after polymerization. The growing ends of microtubules have monomers keeping GTP. Then, those GTP are hydrolyzed. This hydrolysis makes the monomer–monomer bond weaker. At the end of the polymer, if monomers have GDP instead of GTP, they are more quickly dissociated from the end than those having GTP. If all monomers at the ends have GDP, quick depolymerization occurs. When new monomers having GTP are added to the end, depolymerization is stopped and polymerization is made favorable. In an apparent balancing state of tubulin monomers having GTP and microtubules having mainly GDP, each polymer may repeat growth to long ones and depolymerization to short ones, as shown in Figure 4 [28,29]. This phenomenon was actually observed and named "dynamic instability." This is another dynamic function of the polymer coupling with an irreversible chemical reaction. In both cases of tread-milling and dynamic instability, the coupling is indirect.

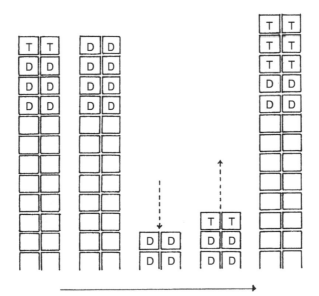

Figure 4 Dynamic instability of microtubules associated with the hydrolysis of bound GTP (T) to GDP (D) at the end; repetition of quick depolymerization and growth; a simplified model; microtubules are composed of more than 10 strands.

Divalent cations bound to actin molecules also change the polymerization rate and the monomer–monomer bond strength. Usually, G-actin tightly binds a magnesium ion or a calcium ion nearby ATP or ADP. The polymerization of G-actin having a magnesium ion is much faster than that of G-actin having a calcium ion. These cations are quickly exchangeable in G-actin but nonexchangeable in F-actin. Therefore, F-actin having magnesium ions and that having calcium ions can be formed separately. The bond free energy and the bond elasticity must be different. Actually, a structural difference was reported between these two kinds of F-actin and their structural dynamics was discussed [30].

IV. MONOMER AND POLYMER CONFORMATIONS

As described in Section III, the conformation of each actin monomer or each tubulin monomer in F-actin or microtubule is more or less different depending on the environmental condition and the species of bound nucleotides and bound cations. Does each monomer in the polymer assume the same conformation in the same environmental condition even when it has the same bound nucleotide and divalent cation? The monomer may take multiple conformations.

An interesting example was found in the case of bacterial flagella. Flagellin molecules form tubular polymers of a helical shape. The tube is composed of 11 longitudinal strands. In a straight tube, all monomers are in an equivalent position except those in the ends. However, in a helical tube, monomers in the innermost strand and those in the outermost strand are in different situations. They may take different conformations. To make the helical tube stable, it is likely that there are two free energy minima in the monomer conformation [3,31]. Inner and outer strands in the helical tube are composed of monomers in two different conformations, respectively. Helices of different shapes, handedness, pitch, and amplitude are constructed by different combinations of these two kinds of strands, as shown in Figure 5. All observed helical shapes of flagella can be explained by this idea. If all strands are composed of monomers in the same conformation, straight flagella are formed. In fact, two kinds of straight flagella exist. Flagellin monomers in these flagella have different bonding patterns [32,33].

The free energies of the monomer in two conformations depend on the environmental condition. A polymorphic transition of flagella between different helices occurs with changes in the salt concentration, pH, or temperature [34]. The transition propagates throughout the whole length of each flagellum in a definite direction, as shown in Figure 6. The transition of handedness can be caused also by a mechanical force [35]. Bacterial flagella give a typical case where polymorphism of polymers is generated from dimorphism of monomers.

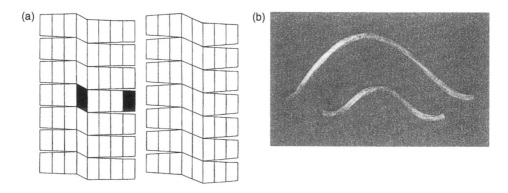

Figure 5 (a) A model showing the structure of helical tubes formed by combination of two kinds of strands, in which monomers are in two different conformations; different helical shapes are produced by changing the ratio of two kinds of strands. (b) Helical tubes made of the sheets of (a).

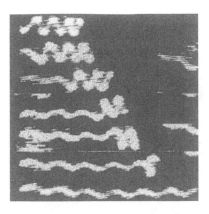

Figure 6 Unidirectional propagation of a polymorphic transition in a bacterial flagellum. (Observation by H. Hotani.)

Let us consider the case of actin. Actin molecules (from the same source) have the same amino acid sequence and take a specific three-dimensional structure in the crystal. Does each actin monomer in F-actin assume exactly the same conformation? Is it in the same state? The conformation of actin monomers in F-actin has been investigated by various methods. However, data obtained were only on an ensemble average of their conformation.

Recently, a new optical microscopic technique has been developed to investigate the conformation of a specified actin monomer in F-actin [36]. An actin molecule was labeled with two kinds of small fluorescent molecules at different amino acid residues, and their distance was measured by the method of fluorescence energy transfer. The result has shown that the distance is not kept constant, but changes with time, as in Figure 7. The change happens discontinuously and reversibly, indicating that actin monomer repeats transitions between two different conformations, compact and loose. The lifetime of each conformation is rather long. This experiment has given evidence that each monomer in F-actin does not always take the same conformation but possibly fluctuates among two or more conformations. It is very likely that the monomer conformation and the species of bound nucleotides or divalent cations have no one-to-one correspondence.

If monomers in F-actin make transitions between different conformations cooperatively, a remarkable change is expected to occur in the overall conformation of F-actin. Up to now, no such change has been observed. However, it was reported recently that the mobility of actin monomers in F-actin is remarkably changed by binding another protein molecule to its end [37]. The conformation of each monomer and the monomer–monomer bond seem to make a cooperative transition. The influence is spread over the whole F-actin.

Also, in the case of tubulin monomers in microtubules, the monomer conformation may not be uniquely determined by the species of bound nucleotides. Previous interpretation about dynamic instability seems to be too simple. A transition of conformation may propagate along the microtubule.

Block copolymers of bacterial flagellin molecules give another example. The helix of one block is influenced by that of the other block. Sometimes, flagellin molecules are forced to form a helix different from their own helix in the same environmental condition. Selection of the monomer conformation and the monomer–monomer bonding is cooperatively controlled in the polymer.

V. DYNAMICS OF POLYMERS INTERACTING WITH OTHER PROTEINS

Since single F-actin filaments were made directly visible under an optical microscope, various experiments were undertaken to observe dynamic behaviors of F-actin interacting with myosin and other actin-binding proteins.

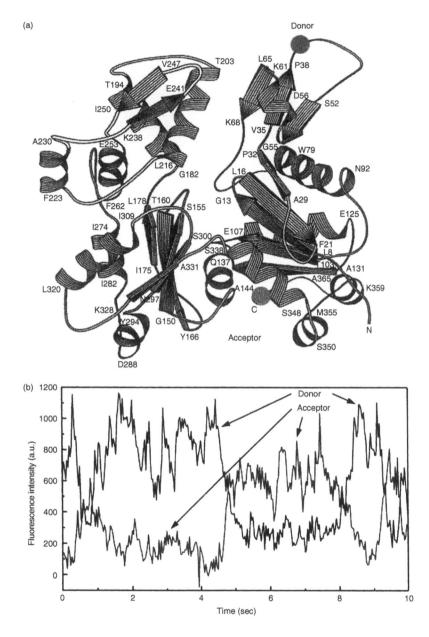

Figure 7   Transition of an actin monomer in F-actin between two conformations detected by fluorescence energy transfer: (a) donor and acceptor of fluorescence energy transfer in the three-dimensional structure of actin monomer; (b) record of fluorescence intensity from donor and acceptor; estimated distance is about 5 and 6 nm in two conformations, respectively. (Figures presented by Y. Ishii.)

When soluble myosin fragments and ATP were added to a solution of F-actin, F-actin showed large and fast bending and twisting movements, as shown in Figure 8. Without ATP such movements did not occur [17]. The free energy of ATP hydrolysis was transferred to the bending and twisting movements. The effective temperature of these degrees of freedom was estimated to be three or four times higher than room temperature. The free energy was injected into the monomer–monomer bonds. On the other hand, when F-actin was put on myosin fragments fixed on a plate, the addition of ATP induced sliding movements of F-actin on myosin fragments, as illustrated in Figure 9 [38].

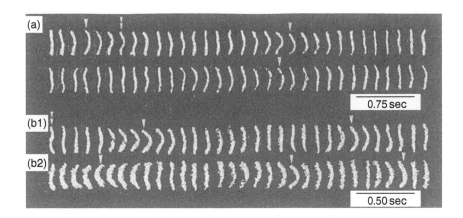

Figure 8 Bending movements of a thin filament (a complex of F-actin with tropomyosin and troponin) interacting with myosin fragments in the presence of ATP, where F-actin was labeled with fluorescent dye molecules to make it visible under a fluorescence microscope: (a) inactive state in the absence of calcium ions; (b) active state in the presence of calcium ions undergoing fast and large bending movements. The sequential micrographs were taken on 10 μm filaments at interval of 0.15 sec (a) and 0.10 sec (b1) and (b2). (From T. Yanagida, M. Nakase, N. Nishiyama, and F. Oosawa. *Nature* 307: 58, 1984. With permission.)

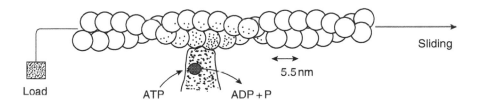

Figure 9 Illustration of sliding movements of F-actin on fixed myosin coupled with the ATP hydrolysis; the conformation of actin monomers during and after interaction with myosin may be changing.

Previously, concerning the mechanism of sliding movements between F-actin and myosin, we proposed the idea of loose coupling between the chemical reaction and the mechanical event. Namely, we assumed that the ATP hydrolysis and the sliding had no definite one-to-one correspondence [39–41]. Recent experiments have given evidence of such loose coupling. Long-distance multistep sliding of myosin on fixed F-actin or of F-actin on fixed myosin was produced during or after hydrolysis of one ATP molecule on myosin [42]. The sliding speed and distance are widely variable depending on the external force applied against sliding. There is often a time lag between start of the sliding and ending of the hydrolysis reaction [43]. Therefore, it is very likely that the free energy of ATP hydrolysis is transferred once and stored somewhere in myosin and/or F-actin and gradually released for sliding [44].

What kind of conformational change is induced in F-actin by interaction with myosin hydrolyzing ATP? How is the influence of myosin spread and maintained in F-actin? The helical polymer structure of F-actin may be useful to realize storage and gradual release of high free energy. The experiment described in Section IV showed that the monomer conformation in F-actin makes transitional changes and the lifetime of each conformation is rather long. Is a high free energy state stably formed in F-actin?

Combination of kinesin or dynein molecules and microtubules also generates sliding movements upon addition of ATP. How is the polymer structure of the microtubule influenced by interaction with kinesin or dynein hydrolyzing ATP? Recent experiments have suggested a long-range influence of binding of the kinesin molecule hydrolyzing ATP on a microtubule [45].

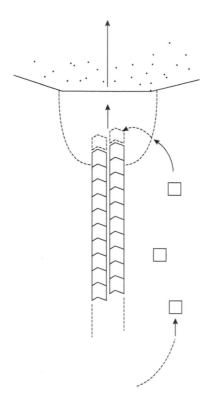

Figure 10   Illustration of translational movements produced by directional growth of F-actin; a simplified model, actually many kinds of proteins are involved to generate fast translation.

A large number of actin-binding proteins have been discovered from different cells. Their functions have been identified based on the scheme of helical polymerization. Some proteins bind to G-actin, inhibit the polymerization, and shift the G–F balance. Some bind to the end of F-actin and inhibit its growth. Some of them make nuclei for polymerization and break F-actin into fragments. Some proteins bind to the side of F-actin and change its interaction with other proteins. There are also proteins that make bundles or cross-links of F-actin and control the three-dimensional structure of the F-actin network.

In cooperation with many kinds of actin-binding proteins, the G–F transformation of actin generates translational movement of a bacterial cell in a host cell, as in Figure 10 [46]. Anchoring and nucleation proteins, depolymerizing proteins, and cross-linking proteins work to make possible a fast cycle of actin molecules from one end of F-actin to the other end, that is, tread-milling. This system has been artificially reconstructed *in vitro* [47].

The cytoplasmic streaming in a plant cell is due to interaction of F-actin bundles under the cell membrane with myosin in the streaming cytoplasm. The bundles, when squeezed out of the cell, often form circular rings. The ring shows active rotation interacting with myosin fixed on a plate [48]. Sometimes, polygons are formed and the positions of corners of the polygon move along the polygon, as shown in Figure 11 [49].

Besides sliding movements on myosin, we can suppose various kinds of active movements of F-actin: translation, rotation, undulation, and so on. It is valuable to investigate if these movements occur in living cells.

Similar questions must be asked in the case of microtubules. For example, active undulation of a bundle of microtubules is observed in some kinds of cells. Dynamics of microtubules, particularly depolymerization at the ends, has been supposed to be a mechanism of chromosome movements for cell division [50].

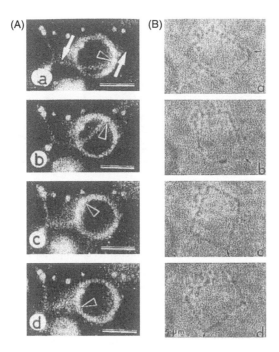

Figure 11 Active movements of F-actin bundles isolated from a plant cell: (A) rotational movements of a ring (presented by S.H. Fujime); (B) movements of corners of a polygon, where the polygon itself does not rotate (presented by K. Kuroda).

VI. DESIGN OF THE MONOMER STRUCTURE AND REGULATION OF INTERACTION

Regulation of the monomer conformation and the monomer–monomer bonding is a key process for dynamic behavior of protein polymers. How are protein monomers designed? Recent research has revealed that most globular protein molecules are constructed from several structural units called modules, like a building of bricks [51,52]. Each module has a compact three-dimensional structure which is formed by the folding of a short polypeptide chain, usually a sequence of 10 to 30 amino acid residues. The sequence is encoded in an exon in the gene DNA and the boundaries of a module correspond to the position of introns in DNA. It is likely that in the process of evolution, new protein molecules were produced by various combinations of modules, small ancient protein molecules.

A method to define modules in the three-dimensional structure of a protein molecule based on its distance map (a map giving distances between amino acid residues, alpha carbons) has been established. By application of this method to actin, its three-dimensional structure is expressed as an assembly of a number of modules. It appears to be composed of a central core and several modules attached to the core surface, as shown in Figure 12 [53]. The core itself is composed of many modules. However, in this figure, only modules involved in the monomer–monomer interaction to form F-actin are specified.

The three-dimensional structure of the core of actin molecule is very similar to that of hsp (heat-shock-protein)-70, another protein molecule, suggesting that these two proteins came from the same ancestor. Addition of new modules to the core gives a new function to the molecule, the ability of helical polymerization. Added modules have to work together.

The design principle is as follows. Protein molecules are constructed by additive assembly of modules. Each module has its own function. By combination of modules, a new function is generated. Regulation of the function greatly depends on communication and cooperation between modules.

In the case of actin, the monomer–monomer bond strength depends on ATP or ADP bound in the cleft. The conformation of modules in the central core has influences on the modules for

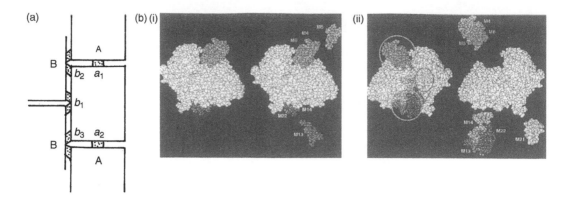

Figure 12 (a) The pattern of monomer–monomer bonding in F-actin. Each monomer is bound with four neighboring monomers by two kinds of bonds, A and B; a_1M5; a_2M13; b_1M21; b_2M4, 5, 6; b_3: M13, 14, 22. (b) The three-dimensional structure of actin monomer built-up from a number of modules: bond A, the contact of neighboring monomers in each longitudinal strand occurs between M5 of one monomer and M13 of the other monomer; (ii) bond B, the contact of monomers of different longitudinal strands occurs between M12 of one monomer and M4, 5, and 6 of the other monomer and also M13, 14, and 22 of the third monomer. (Figures presented by M. Go.)

the monomer–monomer bonding. The monomer–monomer bonding changes the conformation of modules in the core.

The expression of the three-dimensional structure of actin as an assembly of modules is useful for understanding the interaction of actin with many different kinds of actin-binding proteins. Probably, the whole surface of the actin molecule is utilized for such interaction. Overlapping or interference of interaction of actin with different proteins may be well expressed in terms of modules.

Including the monomer–monomer interaction in F-actin, the interaction of actin with many kinds of proteins has remarkable characteristics. In most cases, the mode of interaction is not unique but variable with some freedom. For example, the binding of myosin with F-actin has at least two modes, strong binding and weak binding; it probably has even more different modes. The mode of binding of tropomyosin to F-actin is also variable; it is changed by another protein, troponin, in the presence and absence of calcium ions. Such variability seems to be important for F-actin to exhibit its function. Its structural basis is not yet fully understood.

The structural analyses of tubulin monomers in microtubules or flagellin monomers in flagella at atomic resolution are now in progress. The design of these monomers will be made clear very soon.

VII. Concluding Remarks

As described in previous sections, protein polymers such as F-actin, microtubules, bacterial flagella, and the like show various dynamic behaviors. With development of new experimental techniques, I expect we will be able to observe further many different kinds of dynamic behaviors of protein polymers. Protein molecules and their polymers are a complex system having large degrees of freedom. The energy transfer among different degrees of freedom and the energy exchange with the environment are involved in their dynamics.

To end this chapter, I emphasize that in addition to chemistry and structural analysis, we have to accumulate information about energetics and thermodynamics or statistical mechanics of protein molecules and their polymers. We have to know what kind of state protein molecules and their polymers assume to exhibit dynamic behaviors, not only in structural terms but also in thermodynamic and statistical mechanical terms. This is important for finding a clue to designing functional systems similar to protein polymers.

VIII. ADDENDUM; RECENT PROGRESS

Recently, remarkable progress has been made in the research relating to the subject of this chapter. A few topics are described below.

i) Structural analysis and energetics

The 3-D structure of the straight bacterial flagella has been completely solved at atomic resolution (54). The structure of each monomer and the interacting sites between monomers in the tubular polymer have been determined. As previously mentioned, native flagella of living bacteria form helical tubular polymers, which are possibly constructed by regular arrangement of monomers of two different conformations (31). The energy calculation of the 3-D structure of the monomer suggested the presence of two slightly different conformations which give free energy minima (55). It is almost sure that different types of helices of flagella are composed by different combinations of two 3-D structures of the same monomers. The transition between different combinations results in the transition between different overall shapes of the polymer.

In the protein polymerization and polymer dynamics, water molecules must make important contributions. As already pointed out, in most cases the polymerization is endothermic. Very recently, it has been shown that a layer of hyper-mobile water molecules is formed around F-actin (56). In addition, most of the protein polymers are polyelectrolytes, so that counter ions are condensed and mobile along the polymer. As predicted earlier in the polyelectrolyte theory (57), it has been reported that the correlated wavy counter ion fluctuation produces an attractive force between F-actin filaments (58).

In F-actin and microtubules, many data have been accumulated to show that each monomer including surrounding water molecules and counter ions is not always in the same state. In general, protein molecules in the polymer may assume multiple states and the transitions between different states may occurs cooperatively in the polymer.

ii) Growth and branching of protein polymers

The growth process of single F-actin filaments has been made directly observable under an optical microscope, using actin molecules chemically labeled with fluorescent dyes which have no influences on the activity of monomers (59). After a steady state is established, association and dissociation at two ends apparently occur stepwise as if short fragments composed of several monomers behave as the structural units. The length of each F-actin filament shows quick and large fluctuation. It is not clear how the ATP hydrolysis is involved in this phenomenon. Since the polymerization of actin monomers which have no nucleotides has been performed (9) (60), it is interesting to examine if a similar phenomenon occurs in F-actin filament composed of nucleotide-free actin monomers. Reconstruction of bacterial flagella from flagellin molecules *in vitro* is not coupled with any chemical reaction. Nevertheless, direct observation revealed that in a solution of flagellin molecules the growth of each single flagellum is intermittently stopped for a while and then starts again. The growth seems to consist of two phases, growth and pause. Individual protein polymers do not always behave as expected in the statistically idealized theory.

When bacterial flagella are reconstructed from flagellin molecules *in vitro*, an equilibrium is established by balancing association and dissociation of monomers at the end (3). In living bacteria, the flagella grow unidirectionally at the distal end opposite to the bacterial cell. A regulatory protein named hap2 binds to the growing end, working to help addition of flagellin molecules transported through the central channel of tubular flagellum from inside of the cell and suppress dissociation of monomers from the end (61). This protein forms a circular pentamer, which rotates at the top of flagellum during its growth.

There are several proteins, which bind to the two ends of F-actin respectively. They inhibit association of actin monomers at the ends and regulate the unidirectional growth of F-actin. The inhibition is often released by association with other actin binding proteins (62). Then, keeping these proteins at the end, F-actin can continue the growth.

An actin binding protein named arp2/3 (a complex of two proteins) has been found to have the ability of branching of F-actin (63). When it binds to the side of F-actin, a new F-actin filament begins to grow from there and a tree-like structure is formed (64). The same protein can help nucleation of F-actin (65).

In different kinds of cells, actin monomers and polymers interact with a large variety of actin-binding proteins to produce active movements or morphological changes of their network (66). Microtubules interacting with various proteins are also multifunctional. The research in these fields is in extremely rapid progress.

iii) Functions of protein polymers in molecular machines and their switching

Experimental techniques, nanometry and nanomanipulation, have been extensively developed to analyze the behaviors of molecular machines. During hydrolysis of one ATP molecule, small forward steps of myosin or kinesin molecules, sometimes backward steps also, have been observed on F-actin or microtubules (42) (67) (68). It is becoming more probable that the input (chemical reaction) to output (mechanical event) coupling in the machine is not tight but loose. Sliding molecules of myosin or kinesin give long-range anisotropic influences on the state of actin or tubulin monomers along F-actin or microtubules. These polymers do not simply work as rigid rails but make more active contributions for sliding movements.

There are big families of myosin or kinesin molecules. Some of the family memebers of kinesin, for example, move on microtubules in the same (plus) direction, whereas other members move in the opposite (minus) direction. Each member binds a specified cargo to transport a specified molecule in the cargo along microtubules in the definite direction in the same cells (69). Most of myosin family members move an F-actin in the plus direction, whereas a few of them move in the minus direction. These members work in different cells for different purposes (70) (71). The mechanism for switching of direction of movements must be localized somewhere in the structure of myosin or kinesin molecule. In some of the kinesin members, the whole molecules moves in the plus direction but after a part of the molecule is removed, the movement is reversed, or made bidirectional, where the directional change seems to occur cooperatively on microtubule (72) (73) (74). It is interesting to design a sliding machine which has a switching mechanism in itself.

Living cells have also rotary motors such as the bacterial flagellar motor and the F1-ATPase or FoF1-ATPase. Protein molecules are arranged to form inner and outer cylindrical structures. Theoretical models on the molecular mechanism of rotation have been presented in few references; e.g., (40) (75). Rotation in a single particle of F1-ATPase during the ATP hydrolysis was directly observed (76). Very recently, the forced reversal of rotation has been demonstrated to promote the ATP synthesis (77). The flagellar motor is driven by the current of protons or sodium ions and rotates both clockwise and counterclockwise. It has an additional apparatus for switching the direction of rotation without changing the direction of current.

REFERENCES

1. F. Oosawa, S. Asakura, K. Hotta, N. Imai, and T. Ooi. *J. Polym. Sci.* 37: 323, 1959.
2. F. Oosawa and M. Kasai. *J. Mol. Biol.* 4: 10, 1962.
3. F. Oosawa and S. Asakura. *Thermodynamics of the Polymerization of Protein.* Academic Press, New York, 1975.
4. J. Hanson and J. Lowy. *J. Mol. Biol.* 6: 46, 1963.

5. W. Kabsch, H. Mannherz, D. Suck, P. Pai, and K. Holmes. *Nature* 347: 37, 1990.
6. K. Holmes, D. Popp, W. Gebhard, and W. Kabsch. *Nature* 347: 44, 1990.
7. F. Straub and G. Feuer. *Biochim. Biophys. Acta* 4: 455, 1950.
8. S. Higashi and F. Oosawa. *J. Mol. Biol.* 12: 843, 1965.
9. M. Kasai, E. Nakano, and F. Oosawa. *Biochim. Biophys. Acta* 94: 494, 1965.
10. S. Asakura, G. Eguchi, and T. Iino. *J. Mol. Biol.* 10: 42, 1964.
11. G. Borisy, J. Olmsted, and R. Klugman. *Proc. Natl. Acad. Sci.* 69: 2890, 1972.
12. F. Oosawa, S. Asakura, and T. Ooi. *Prog. Theor. Phys. Suppl.* 17: 14, 1960.
13. S. Asakura, M. Taniguchi, and F. Oosawa. *J. Mol. Biol.* 7: 55, 1963.
14. S. Asakura, M. Kasai, and F. Oosawa. *J. Polym. Sci.* 44: 35, 1960.
15. T. Ikkai and T. Ooi. *Biochemistry* 5: 1551, 1966.
16. S. Fujime. *J. Phys. Soc. Jpn* 29: 751, 1970.
17. T. Yanagida, M. Nakase, N. Nishiyama, and F. Oosawa. *Nature* 307: 58, 1984.
18. A. Kishino and T. Yanagida. *Nature* 334: 74, 1988.
19. H. Kojima, A. Ishijima, and T. Yanagida. *Proc. Natl. Acad. Sci.* 91: 12962, 1994.
20. T. Takebayashi, Y. Morita, and F. Oosawa. *Biochim. Biophys. Acta* 492: 35, 1977.
21. H. Kondo and S. Ishiwata. *J. Biochem.* 79: 159, 1976.
22. S. Asakura, G. Eguchi, and T. Iino. *J. Mol. Biol.* 35: 227, 1968.
23. Y. Uratani, S. Asakura, and K. Imahori. *J. Mol. Biol.* 67: 85, 1972.
24. M. Carlier, D. Pantaloni, and E. Korn. *J. Biol. Chem.* 259: 9983, 1984.
25. M. Carlier. *Adv. Biophys.* 26: 51, 1990.
26. E.M. De La Cruz and T. Pollard. Presented at the Actin Meeting, Maui, 1997.
27. A. Wegner. *J. Mol. Biol.* 108: 139, 1976.
28. T. Mitchison and M. Kirschner. *Nature* 312: 232, 1984.
29. T. Horio and H. Hotani. *Nature* 321: 605, 1986.
30. A. Orlova, E. Prochniewicz, and E. Egelman. *J. Mol. Biol.* 245: 598, 1995.
31. S. Asakura. *Adv. Biophys.* 1: 99, 1970.
32. K. Namba and F. Vonderviszt. *Quart. Rev. Biophys.* 30: 1, 1997.
33. I. Yamashita, K. Hasegawa, H. Suzuki, F. Vonderviszt, Y. Mimori-Kiyosue, and K. Namba. *Nat. Str. Biol.* 5: 125, 1998.
34. R. Kamiya and S. Asakura. *J. Mol. Biol.* 108: 513, 1976.
35. H. Hotani. *J. Mol. Biol.* 156: 791, 1982.
36. H. Yokota, Y. Ishii, T. Wazawa, T. Funatsu, and T. Yanagida. *Biophys. J.* 74: A46, 1998.
37. E. Prochiniewicz, Q. Zhang, P. Janmey, and D. Thomas. *J. Mol. Biol.* 260: 756, 1996.
38. S. Kron and J. Spudich. *Proc. Natl. Acad. Sci.* 83: 6272, 1986.
39. T. Yanagida, T. Arata, and F. Oosawa. *Nature* 316: 366, 1985.
40. F. Oosawa and S. Hayashi. *Adv. Biophys.* 22: 151, 1986.
41. R. Vale and F. Oosawa. *Adv. Biophys.* 26: 97, 1990.
42. K. Kitamura, M. Tokunaga, A. Iwane, and T. Yanagida. *Nature* 397: 129, 1999.
43. A. Ishijima, H. Kojima, T. Funatsu, M. Tokunaga, H. Higuchi, H. Tanaka, and T. Yanagida. *Cell* 92: 161, 1998.
44. F. Oosawa. *Genes Cells* 5, 2000.
45. E. Muto, T. Miyamoto, T. Funatsu, Y. Harada, A. Iwane, A. Ishijima, and T. Yanagida. To be submitted.
46. L. Tilney, D. DeRosier, and M. Tilney. *J. Cell Biol.* 118: 71, 1992.
47. T. Loisel, R. Boujemaa, D. Pantaloni, and M. Carlier. *Nature* 401: 613, 1999.
48. S. Higashi-Fujime. *J. Cell Biol.* 87: 569, 1980.
49. K. Kuroda. *Int. Rev. Cytol.* 121: 267, 1990.
50. S. Inoue and E. Salmon. *Mol. Biol. Cell* 6: 1619, 1995.
51. M. Go. *Nature* 291: 90, 1981.
52. K. Takahashi, T. Noguti, H. Hojo, K. Yamauchi, M. Kinoshita, S. Aimoto, T. Ohkubo, and M. Go. *Protein Eng.* 12: 673, 1999.
53. Y. Niimura and M. Go. To be submitted.
54. K. Yonekura, S. Maki-Yonekura, and K. Namba. *Nature* 424: 643, 2003.
55. F. Samatey, K. Namba, and Namba. *Nature* 410: 331, 2001.
56. S. Kabir, K. Yokoyama, K. Mihashi, T. Kodama, and M. Suzuki. *Biophys. J.* 85: 3154, 2003.
57. F. Oosawa. Polyelectrolytes, Dekker 1971.

58. T. Angellini, H. Liang, W. Wriggers, and G. Wong. *Proc. Natl. Acad. Sci.* 100: 8634, 2003.

59. I. Fujiwara, S. Takahashi, H. Tadakuma, T. Funatsu, and S. Ishiwata. *Nature Cell Biol.* 4: 666, 2002.

60. E. De La Cruz, A. Mandinova, M. Steinmetz, D. Stoffler, U. Aebi, and T. Pollard. *J. Mol. Biol.* 295: 517, 2000.

61. K. Yonekura, S. Maki, D. DeRosier, F. Vondervistz, K. Imada, and K. Namba. *Science* 290: 2148, 2000.

62. D. Kovar, J. Kuhn, A. Trichy, and T. pollard. *J. Cell Biol.* 161: 875, 2003.

63. L. Blanchoin, K. Amann, H. Higgs, J. Marchand, D. Kaiser, and T. Pollard. *Nature* 404: 1007, 2000.

64. K. Amann and T. Pollard. *Proc. Natl. Acad. Sci.* 98: 15009, 2001.

65. R. Boujemaa-Paterski, E. Goudin, G. Hansen, S. Samarin, C. Le Clainche, D. Didry, P. Dehux, P. Cossart, C. Kockss, M-F. Carlier, and D. Pantaloni. *Biochemistry* 40: 11390, 2001.

66. C. Schoeneberger, D. Steinmetz, D. Stoffler, A. Mandinova, and U. Aebi. *Microsc. Res. Tech.* 47: 38, 1999.

67. Y. Okada and N. Hirokawa. *Proc. Natl. Acad. Sci.* 97: 640, 2000.

68. M. Nishiyama, H. Higuchi, Y. Ishii, Y. Taniguchi, and T. Yanagida. *BioSystem* 71: 145, 2003.

69. N. Hirokawa. *Science* 279: 519, 1998.

70. J. Cramer. *J. Cell Biol.* 150: F121, 2000.

71. A. Inoue, J. Saito, R. Ikebe, and M. Ikebe. *Nature, Cell Biol.* 4: 302, 2002.

72. S. Endow and H. Higuchi. *Nature* 406: 913, 2000.

73. C. O'Connel and M. Mooseker. *Nature Cell Biol.* 5: 171, 2003.

74. M. Badooual, F. Julicher, and J. Prost. *Proc. Natl. Acad. Sci.* 99: 6521, 2002.

75. H, Wang and G. Oster. *Nature* 396: 279, 1998.

76. H. Noji, R. Yasuda, M. Yoshida, and K. Kinosita. *Nature* 386: 299, 1997.

77. H. Itoh, M. Yoshida, K. Kinosita. *Nature* 427: 465, 2004.

Chapter 22

Force Generation by Cellular Polymers

George Oster and Alex Mogilner

Contents

I. Introduction

Polymers perform many tasks in living cells. We discuss their roles in generating mechanical forces that drive important cellular motions. These can be classified into three general categories:

1. Actin and tubulin polymerization and depolymerization can generate mechanical forces using the free energy of monomer binding and/or nucleotide hydrolysis as their energy source.
2. Polymers can also store elastic energy during their polymerization that can be released later to generate mechanical forces that drive some of the most rapid of cellular motions.
3. Actin and tubulin are tracks for walking motors powered by nucleotide hydrolysis. These motors fall into three classes: myosins, kinesins, and dyneins. The plural applies because, while the force generating principle within each motor type is the same, there is substantial diversity in their dynamical behavior and the cargo they propel. However, the polymer tracks themselves probably play only a passive role as highways for intracellular transport.[1]

In this chapter we will focus on categories 1 and 2. Polymer motors do not operate in a cyclic fashion like other protein motors that cycle through a number of conformational steps and eventually reset themselves to their initial configuration. Rather they are "one-shot" engines that, after assembly, are disassembled. Nevertheless, these specialized motors play an important role in many cellular processes.

[1] There is some evidence that the polymer track may play a more active role in some protein motors [1,2].

II. POLYMERIZATION FORCES

Polymers can convert the binding free energy of their constituent monomers into mechanical force to push an axial load. Two mechanisms are generally referred to as *power strokes* and *Brownian ratchets* [3–5]. In a power stroke, the binding reaction is mechanically coupled to movement and generation of force. For example, if the chemical reaction of a monomer binding to a filament tip triggered a conformational change in the monomer that elongated it, then the conformational change would directly drive an object in front of the polymer tip. In a Brownian ratchet, the role of the monomer binding reaction is to prevent backward fluctuations of the load, rather than to apply a mechanical force directly to it. That is, the load is driven by its own Brownian fluctuations, and the binding energy of the polymer rectifies its diffusive motion [6,7].

To be concrete, suppose an object, say a small sphere with diffusion coefficient D, is aligned ahead of the growing polymer that is anchored at its left end (see Figure 1(a)). The polymer need not actually "push" the object, but it can "rectify" its Brownian motion. We can view the advancement of the polymerization as moving down on a free energy landscape, $\Delta G(x)$, as shown in Figure 1(b) and (c), where the load is moving up the load potential, $V_L \equiv F_L \cdot x$ (i.e., the load force, $F_L = dV_L/dx$). A ratchet potential energy has a "staircase" profile with step heights much larger than $k_B T$ as shown in Figure 1(b) ($k_B T \approx 4.1$ pN nm is the "unit" of thermal energy, where k_B is Boltzmann's constant, and T the absolute temperature [10]). A power stroke is an inclined path with barriers only a few $k_B T$ high, as shown in Figure 1(c). Of course, this nomenclature only characterizes the extreme situations; anything in between is possible.

The object, D, diffuses uphill against the load force until it reaches the vertical drop that represents the monomer binding step. If the free energy of binding is $\Delta G_b \gg k_B T$, a backward step is very unlikely, and so work is done against the load force. Thus chemical energy is expended to preferentially select forward steps (or prevent backward steps) and hence to favor forward motion of the load. The difference between ratchets and strokes is only a matter of degree: a power stroke

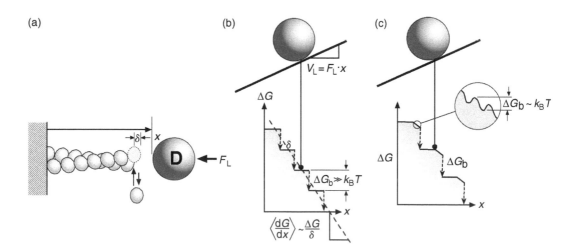

Figure 1   Power stroke and ratchet. (a) A helical polymer (e.g., actin) composed of monomers of size δ polymerizes against an object with diffusion coefficient D. A load force, F_L opposes the polymerization. (b) The Brownian ratchet [8]. The polymerization motor can be viewed as a point (black circle) moving on a staircase free energy surface whose step width is δ and whose height is equal to the free energy of polymerization, that is, the binding free energy of a monomer to the polymer tip is much larger than thermal energy: $\Delta G_b \gg k_B T$. (c) Power stroke. If, after binding, a monomer develops an internal stress (e.g., via nucleotide hydrolysis) that increases its axial length, then the load will be pushed to the right — corresponding to the inclined free energy segment. A closer inspection of this process would reveal that the power stroke comprises a sequence of thermally activated steps whose magnitude is of the order $\sim k_B T$ [9].

biases thermal motion by a sequence of small free energy drops while a ratchet *rectifies* diffusion by a sequence of large free energy changes [5,6,11]. We can make this intuitive picture quantitative as follows.

If the polymer assembly is unobstructed, its elongation rate is simply $V_p = \delta(k_{on}M - k_{off})$ where δ [nm] is the size of the monomer ($\sim$5.4 nm for actin), M [μM] the monomer concentration, and $k_{on}(1/[\mu M \sec]), k_{off}$ [1/sec] are the polymerization and depolymerization rate constants, respectively. In the case of a helical actin polymerization, as shown in Figure 1, the step size is half the monomer diameter $\frac{1}{2}5.4 = 2.7$ nm.

First, assume that the polymer is perfectly rigid, and that the diffusion coefficient, D, is large. In order for a monomer to bind to the end of the filament the object must open up a gap of size δ by diffusing away from the tip, and remaining there for a time $\sim 1/k_{on}M$ to allow polymerization to take place. In the limiting case when diffusion is much faster than polymerization, that is, $k_{on}M \ll D/\delta^2$, the elongation rate is given by the simple formula $v_p = \delta(k_{on}M \cdot p(\delta, F_L) - k_{off})$. That is, the polymerization rate is weighted by the probability $p(\delta, F)$ that the gap size is δ or larger [12,13]. This probability depends on the load force F_L pushing the object to the left; in this simple case $p = \exp(-F_L\delta/k_B I)$, where $F_L\delta$ is the work required for moving the object by a distance L. So the load–velocity relationship is given by

$$v = \delta(k_{on}M\, e^{-F_L\delta/k_BT} - k_{off}) \tag{1}$$

The stall load, F_{stall}, is reached when the work done in moving the object by a distance δ is just equal to the free energy of the binding reaction; that is, $v = 0$ when the load force is $F_{stall} = (k_BT/\delta)\ln(k_{on}M/k_{off})$, which corresponds to the equilibrium thermodynamic expression. For actin polymerization, M is usually in the micromolar range, the polymerization rate is $k_{on}M \sim 100/\sec$, and the depolymerization rate is $k_{off} \sim 1/\sec$, and each monomer added to the polymer tip increases its length by $\delta \approx 2.7$ nm. Therefore, without significant load, $v \approx 0.1$–1 μm/sec. When stalled, a filament generates a force of $\sim$5–7 pN, similar to that generated by myosin and kinesin [14,15]. These estimates apply in the limit when the object's diffusion is very fast, which is not always the case. However, actin filaments are not rigid, and their thermal bending undulations are very fast ($\sim 10^4/\sec$). The analysis in this case is similar, and it turns out that the above expression for the stall force of an "elastic ratchet" is still valid [12].

The filament length is an important factor in determining the amplitude of its thermal fluctuations. The effective elastic constant of an actin filament of length L tilted at angle σ to an obstacle is $k \approx 4\lambda k_BT/(L^3 \sin^2(\sigma))$ [12], where λ is the persistence length, which is in micron range [16]. This formula indicates that, if the filament is too short (less than $\sim$70 nm), or the angle α is too acute (less than $\sim$30°), the filament is effectively too stiff for the elastic ratchet to work because thermal fluctuations are insufficient to create a gap large enough for monomers to intercalate. On the other hand, if the filament is too long, it becomes too "soft," and it buckles under load forces of less than a picoNewton.

For microtubule polymerization, the mathematics is more involved [17–19]. If all 13 microtubule protofilaments are considered as independent force generators, "subsidizing" each other as described in the text, then the theoretically predicted stall force is $F_{stall} \approx 7$ pN (for relevant parameters, see [17,18]). More work is needed on treating the interdependence of the protofilament force generation [19].

Polymerization motors are simple and reliable, and in terms of energy consumption, they are moderately efficient. Indeed, the efficiency, η, can be estimated from the ratio of the work performed, $F_L \cdot \delta$, to the monomer binding free energy: $\Delta G_b = k_BT \ln(k_{on}M/k_{off}) = 4.1[pN\,nm] \cdot \ln(11.6[\mu M/\sec] \cdot 10[\mu M]/(1[1/\sec])) \approx 20[pN\,nm])$. Thus $\eta = 5[pN] \cdot 2.7[nm]/20[pN\,nm] \approx 0.68$. However, there is a large cost associated with controlling when and where polymerization occurs in the cell. This control depends on the enzymatic activity of actin: each monomer of actin binds and hydrolyzes one ATP molecule, whose free energy of hydrolysis is $\sim$80 pN nm. Comparing this to the work performed in a step gives a "control" efficiency of only $\eta_c \sim 15\%$. For microtubule, the energy of hydrolysis ~ 26 pN nm per dimer is used to generate ~ 7 pN $\cdot$ (8/13) nm of work, so the control efficiency is again $\sim 15\%$. When polymerization is fast, η_c drops below 10% far from stall.

III. DEPOLYMERIZATION FORCES

While polymerizing microtubules can generate a pushing force, depolymerizing microtubules can develop pulling forces, although the mechanism is less obvious. For example, depolymerizing microtubules can pull particles at $\sim 1 \mu m/sec$ against viscous drag forces of ~ 10 pN [20]. Also, plastic beads coated with plus-end-directed microtubule motors are carried toward the microtubule minus ends as the microtubule depolymerizes [21]. Depolymerizing microtuble pulling forces may be important in moving chromosomes to the cell pole.

Hill proposed the earliest model for depolymerization force generation, Figure 2(a). He assumed that the tip of the depolymerizing microtubule slides through the hole in a sleeve-like docking protein on the kinetochore that allows tubulin dimers to dissociate freely from the microtubule tip [22]. The interior of this sleeve has a high affinity for the microtubule lattice so that when subunits dissociate from the microtubule tip, the binding free energy gradient favors deeper insertion of the microtubule into the sleeve. Thus Brownian motion will drive the docking protein toward the microtubule minus end, producing a pulling force. Movement of the microtubule into the sleeve requires previous interactions to be broken and reformed. This creates a potential energy barrier to the movement of the sleeve that increases the deeper the microtubule penetrates the sleeve and slows further movement of a microtubule. However, subunit loss at the tip will allow the sleeve to follow the tip of the depolymerizing microtubule. According to this model the speed of depolymerization-coupled movement will remain constant over a wide range of load forces because the steady-state force generated by the sleeve adapts to an opposing load by adjusting the average length of the fiber inside the sleeve such that the speed of the load is equal to the depolymerizatio rate [24].

Another mechanism for generating pulling forces is the "conformation wave" model proposed by Mitchison [23] (Figure 2(b)). Depolymerizing microtubule ends consist of two-dimensional sheets that appear frayed and curved [25]. This is thought to arise from GTP binding to tubulin monomers inducing a conformational change that permits polymerization. When GTP is hydrolyzed the monomer tries to relax to its stress free state, but it cannot because it is trapped in the microtubule lattice. Thus the microtubule contains stored elastic energy that can only be released as the microtubule depolymerizes, thus the "banana peel" curvature of the frayed ends [26]. The outward curving protofilaments at the disassembling plus end drive the sliding collar toward the minus end as suggested in Figure 2(b). In this case, the force driving the movement of the sleeve is the release of mechanical strain stored in the lattice during microtubule polymerization. The bending of the protofilament sheets induced upon disassembly is analogous to a power stroke. This model has not been treated quantitatively, but the force it generates can be estimated knowing the strain energy stored in the microtubule lattice from the GTP hydrolysis: $\Delta G \sim 26$ pN nm/dimer [27]. Dividing by the fiber length increment after one act of unbinding gives 26 pN nm/(8 nm/13) ~ 45 pN.

Finally, Peskin et al. modeled experiments [21] in which a bead coated with high affinity tubulin-binding proteins undergoes rotational diffusion along the microtubule polymer lattice (Figure 2(c)) [8,28]. The binding energy gradient prevents the bead from detaching from the plus end of the microtubule. As it rolls, the bead binds to the microtubule via an immobile kinesin construct. This weakens the bonds between neighboring tubulin dimers and facilitates depolymerization at the tip. This mechanism is a ratchet: rotational diffusion of the bead is biased by the depolymerizing plus end of the polymer.

The ratchet model can be used to estimate the forces generated by the depolymerizing microtubule [8]. Each disassembly event allows the motor to rotate toward the minus end of the

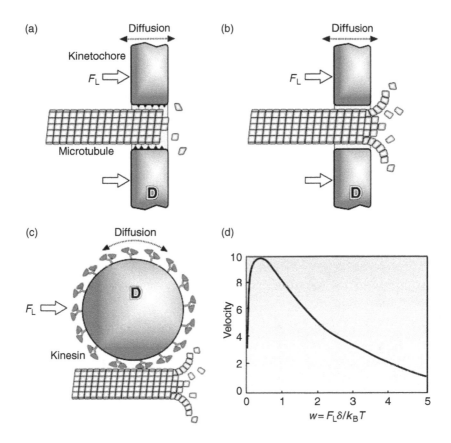

Figure 2 Depolymerization forces. (a) Hill's model for depolymerization driven transport of a kinetochore along a microtubule [22]. The free energy gradient due to the affinity of the sleeve for the microtubule keeps the microtubule within the sleeve as monomers dissociate from the plus (right) end. (b) The "conformation wave" model of pulling force generation by a depolymerizing microtubule [23]. The elasticity of the protofilaments that curve outward at the disassembling plus end drives a sliding collar on the kinetochore toward the minus end (power stroke). (c) A bead coated with tubulin-binding proteins undergoes rotational diffusion along the microtubule. The binding free energy gradient prevents the bead from detaching from the plus end of the microtubule and, as it rolls, the bead facilitates depolymerization at the plus end. Rotational diffusion of the bead is ratcheted by the depolymerizing plus end [8]. (d) The load–velocity curve passes through a maximum. This is because low load forces increase the residence time of the diffusing object at the microtubule end where it promotes depolymerization. Higher load forces obviate this effect and the curve decreases. w is the work (in units of $k_B T$) to move the load force F_L by a distance of one subunit δ.

microtubule with a step size, δ determined by the thermal energy required to step against the load force, $F_L : F_L \cdot \delta \sim k_B T$ or $\delta \approx k_B T / F_L$. Thus if dimers disassemble from the tip with a rate k_{off}, then the velocity of the bead is $V_p \approx k_{off} \delta = k_{off} k_B T / F_L$. Thus, the depolymerization velocity is inversely proportional to the load force. According to this formula the velocity can never reach zero. However, at a load force a few folds greater than $k_B T / \delta$, depolymerization would slow down significantly. Thus this depolymerization motor can develop force only in the pN range.

A peculiarity of the depolymerization motor is that the velocity rises at low loads, passes through a maximum, then decreases monotonically to zero at the stall force, as shown in Figure 2(d) [8]. This apparently paradoxical behavior has a simple explanation. Small pushing loads increase the residence time of the load at the microtubule end where it stimulates monomer dissociation, enhancing motion to the minus end (left in Figure 2(c)). At larger loads, this effect is overridden by the load and the velocity decreases monotonically.

IV. EXAMPLES OF CELLULAR POLYMERIZATION MOTORS

A number of intracellular pathogens propel themselves through their host's cytoplasm by hijacking the cell's actin machinery to its own ends. These organisms are studied as a simplified model system for eukaryotic cell motility. One particular bacterium, *Listeria monocytogenes*, propels itself by assembling the host cell actin into a comet-like tail of cross-linked filaments, with their polymerizing barbed ends oriented toward the bacterial surface [29]. *Listeria* moves through the host cytoplasm at velocities of $\sim$0.2 μm/sec [30]. Only one cell surface protein, ActA, is required for motility [31]. In addition to actin monomers and ActA, only ATP and a handful of cytoplasmic proteins are essential, including the nucleating and branching complex, Arp2/3, capping proteins, and the actin severing and depolymerizing factor, ADF–cofilin (see Figure 3(a)) [32].

The rigid polymerization ratchet theory was originally applied to *Listeria* propulsion by considering the cell itself as the thermally fluctuating object in front of the filament tips, as in Figure 1 [13].

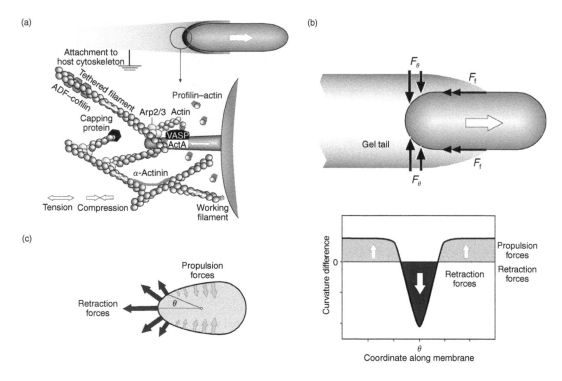

Figure 3   Models for the molecular propulsion machinery of *Listeria* [9,33]. (a) Microscopic model [9]. Actin polymerizes at the bacterial surface forming a "comet tail" of short filaments that is anchored to the host cytoskeleton. The posterior surface is coated with ActA that promotes actin polymerization at the surface. ActA in concert with VASP activates the Arp2/3 complex and forms a scaffold for actin assembly. This scaffold adds actin monomers to the growing tips by recruiting profilin–actin complexes. Activated Arp2/3 complex nucleates and branches actin filaments. Actin turnover is regulated by capping proteins that limit filament growth, while ADF–cofilin accelerates actin disassembly further from the bacterial surface. Profilin sequesters actin monomers and restricts polymerization to the region adjacent to the surface. α-Actinin cross-links actin filaments into a denser gel than the surrounding cellular cytoskeleton. The "mother" filaments are tethered transiently to the surface and resist the forward movement of the bacterium. The tethered filaments are in tension and eventually detach to become "working" filaments in compression whose thermal undulations exert the propulsive pressure on the bacterial wall [9]. (b) Continuum model. The actin gel polymerizes at the bacterial surface generating a circumferential stress. This sqeezing force produces an axial thrust that propels the bacterium forward [33]. (c) Vesicle propulsion by actin polymerization [34,35]. An actin tail growing from a lipid vesicle deforms the membrane. Measuring the difference in membrane curvature along the surface gives an estimate of the forces acting on the surface. The propulsive forces from the working filaments exceed the retraction forces from the attached filaments, so the vesicle is propelled forward.

This model predicted that the bacterial velocity should depend on its diffusion coefficient, and thereby on its size. However, experiments showed that the velocity did not depend on the cell size, so the model was modified to allow thermal fluctuations of the actin filament tips [36]. This resolved the size independence issue but the model ran afoul of another observation: the actin tail appeared to be attached to the surface of the cell [30,37,38]. This problem was resolved by a further generalization of the model. The "tethered ratchet" model assumed that the filaments are initiated while attached to the bacterial surface, but subsequently detach and become "working" filaments as in the elastic ratchet model (Figure 3(a), [19]). The attached fibers are in tension and resist the forward progress of the bacterium. At the same time, the dissociated fibers are in compression, and generate the force of propulsion, each filament developing a force of a few pN.

The effect of load forces on the velocity of *Listeria* can be measured by increasing the viscosity of the medium using methylcellulose. These experiments show that loads between 10 [39] and 100 pN [40] are required to slow the bacteria. The discrepancy between these measurements probably arises from differences in the biochemistry and experimental conditions that may alter the number of working filaments. Both are in agreement with the tethered ratchet theory, depending on the number of working/tethered filaments. Finally, there are suggestions that the actin tail generates not only axial propulsive forces, but torques as well (Robbins et al. observed that *Listeria* rotates about its long axis as it moves through the cytoplasm [41]). This torque could be generated by actin binding proteins that attach laterally and trap torsional fluctuations of the tethered filaments, analogous to the action of scruin in the acrosomal process of *Limulus* (see below) [42,43].

Other intracellular pathogens, such as *Shigella*, the spotted fever bacterium *Rickettsia*, and the *Vaccinia* virus all utilize the cellular actin assembly machinery for propulsion [44]. Although the molecular details vary, the physical mechanism is the same. For example, *Shigella* uses the surface protein IcsA instead of ActA to stimulate actin polymerization. *Vaccinia* utilizes tyrosine phosphorylation of a unique protein A36R (absent in *Listeria* and *Shigella*). *Rickettsia* does not employ Arp2/3 complex as an actin nucleating/branching center, and thus its tail consists of long actin filaments arranged in a parallel array, rather than short filaments cross-linked at acute angles as in *Listeria*, *Shigella*, or *Vaccinia*.

Plastic beads and lipid vesicles coated with either ActA, or WASP proteins, grow actin tails and move in the same way as the pathogens, and also deform the vesicles Figure 3(c) [31,34,35]. Measuring the curvatures of the vesicle gives an estimate of the balance of propulsive and retraction forces exerted on the surface by the working and tethered filaments, respectively. Experiments such as these promise to become useful tools in uncovering the secrets of cell motility.

V. OTHER MECHANISMS OF FORCE GENERATION

Although the ratchet model explains most of the experimental observations, there are alternative proposals for polymerization force generation. Generally, they fall into two categories: hypothetical protein motors and macroscopic phenomenological models. An example of the former is the molecular ratchet motor proposed by Laurent et al. that posits that frequent attachment and detachment of VASP on the cell surface to F-actin allows it to slide along a growing filament, driven by the free energy of monomer addition [45]. Another example of a hypothetical force generator is an affinity-modulated, clamped-filament elongation mechanism that exploits the intrinsic ATPase activity of actin monomers [46].

Kuo and McGrath observed that *Listeria* appeared to advance in discrete steps of 5.5 nm, similar to the size of an actin monomer [38]. These steps could suggest some intrinsic molecular scale mechanism at the interface between filaments and the surface. Finally, myosin is involved in some way during protrusion of filopodial-like actin bundles at the lamellipodial leading edge that are organized by inhibition of capping and subsequent cross-linking by fascin [47,48].

Gerbal et al. constructed a continuum model of *Listeria* propulsion based on the elastic shear stress developed by growth of the actin meshwork at the cell surface [33,49,50]. In this model, the polymerization of actin develops circumferential stresses in the actin meshwork of the tail surrounding the posterior portion of the cell. This developing stress "squeezes" the cell until a yield stress is reached whereupon the cell "squirts" forward, relieving the stress, and the cycle repeats (see Figure 3(b)). This stress-relaxation cycle produces step-like propulsion (with micronsized steps) similar to that observed in the movement of ActA-coated plastic beads [51]. The model predicts that the propulsive force should depend on the surface curvature; this can be tested experimentally. Being macroscopic, this model complements the microscopic elastic ratchet model; indeed, the latter provides a rationale for the polymerization induced stresses that develop in the continuum gel model.

Lipid vesicles coated with ActA also grow actin tails and move. The vesicles deform, and the stress distribution exerted by the actin can be computed from their shape [35]. These experiments confirm the existence of large ($\sim$nN) "squeezing" stresses on the vesicle, but "squirting" movement cycles were not observed. These experiments also indicate that a spatial separation between tethered and working filaments develops. Tethered filaments are swept to the very rear of the vesicle, while working filaments concentrate at the sides. A simple explanation for this separation phenomenon is that ActA attached to an immobile tethered filament drifts to the rear along the lipid surface as the vesicle is propelled forward by working filaments that keep up with the vesicle's sides.

Several mechanisms other than polymerization are probably involved in pushing out the cell's leading edge. In most crawling cells, the force of protrusion is generated locally [52]. Localized protrusive forces can be generated in actin gels because they are highly charged [53]. Because of the counterions to the actin fixed charges, the filaments of a cross-linked polyelectrolyte gel, such as the actin cytoskeletal network, are always in a state of elastic tension. At equilibrium, the elastic tension in the gel filaments is just balanced by the ion osmotic pressure. This is discussed in more detail in [54,55]. In transiently motile cells, the actin gel adjacent to the leading edge membrane may partially solate, for example, by the action of severing proteins, such as gelsolin triggered by calcium influx. This weakens elasticity of the gel so that the *local* osmotic pressure expands the gel boundary to a new equilibrium. Subsequently, the gel solidifies again stabilizing the protrusion. Some indirect evidence in favor of this scenario is the observations that raising external osmolarity inhibits protrusion, and that prior to protrusion, the lamellipodial leading edge of some cells swells and becomes softer [56].

A very simple and specialized cell, the nematode sperm of *Ascaris suum*, provides an important example of pushing out the front by a specialized form of gel swelling [57,58]. Nematode sperm lack the actin machinery associated with eukaryotic cell motility; instead, their movement is powered by a cytoskeleton built from *major sperm protein* (MSP) filaments [59]. This is a positively charged and partially hydrophobic protein that associates into symmetrical dimers that polymerize into helical filaments. Unlike actin, MSP polymerization and bundling does not require a broad spectrum of accessory proteins. The same hydrophobic and electrostatic interaction interfaces allow these filaments wind together in pairs to form larger bundles, and eventually congregate into higher order rope-like arrays [60]. MSP filaments are more flexible than actin, and so the polymerization ratchet mechanism may not be as effective in generating a protrusive force in nematode sperm. However, this assembly process forces the filaments within a higher order aggregate to assume an end-to-end distance that is larger than its persistence length in solution. In this fashion, bundles of MSP filaments are stiffer than, and contain the stored elastic energy of, their constituent filaments. These bundles of MSP form a thixotropic (i.e., shear thining) gel-like cytoskeleton within the lamellipod. This gel is a fibrous material, so that when filaments bundle laterally they generate a protrusive force longitudinally [58]).

Finally, osmotic gel swelling appears to propel certain bacterial locomotion, a phenomenon called "gliding motility" [61,62]. Cyanobacteria and myxobacteria glide on surfaces by hydration and extrusion of a gel from nozzle-like organelles [61,63,64].

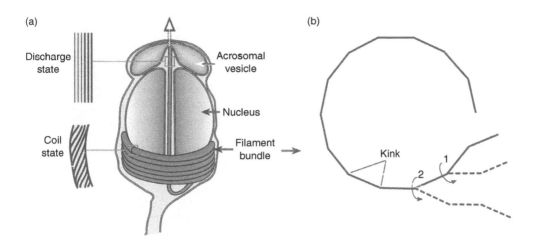

Figure 4 Energy storage in the acrosomes of *Limulus* acrosome (redrawn from Mahadevan, L., J. Shin, G. Waller, K. Langsetmo, and P. Matsudaira. *J. Cell Biol.* 2003, **162**:1183–1188. With permission). (a) The filaments are twisted in the coiled state but straight in the discharge state. (b) During the acrosomal reaction, the actin filaments untwist and unbend going from the coil to the discharge state. The actin bundle uncoils and projects rapidly from the top of the cell.

The smallest free-living organisms are the *Mollicutes* [65]. Some glide on surfaces, others swim, but all appear to generate the forces for propulsion by cyclically altering elastic properties of cytoplasmic filaments [66–69].

The largest and fastest reversible entropic motor is employed by spasmoneme in *Vorticellid* ciliates [70]. In this motor, a giant polymer chain is held in a distended configuration by the repulsion of its fixed charges. A rise in cytosolic calcium drastically reduces its rigidity by shielding the polymer-associated charges, triggering an entropic contractile force strong enough to retract at ~8 cm/sec [71]. Another dramatically fast "one-shot" polymerization engine is employed when the sperm of the sea cucumber *Thyone* encounters the egg jelly coat. An explosive actin polymerization reaction ensures pushing out the acrosomal process and enabling the sperm plasma membrane to penetrate the egg and fuse with the plasma membrane of the egg [72]. In this process, fast and transient actin polymerization is limited by actin delivery to the tip of the process, rather than by force. Water influx coupled with actin polymerization may contribute to force generation by a hydrostatic mechanism without treadmilling and/or nucleotide hydrolysis [73].

A particularly startling polymer motor is the acrosomal process of the *Limulus* sperm whose acrosomal process consists of a bundle of 30 to 50 actin filaments cross-linked by the protein scruin (Figure 4(a)). During polymerization of the acrosomal process, elastic energy is stored in the filament bundle by using scruin binding to trap torsional thermal fluctuations as elastic strain energy. Later, this strain energy is released to generate the force required to push the actin rod into the egg cortex [43,71]. Shin et al. have modeled this process and found that the protrusion is characterized by a stationary untwisting front that converts the coiled bundle into a straight bundle. Interestingly, while the torsional strain is distributed continuously along the filament bundle, the bending energy is concentrated in a series of "kinks." These kinks rotate sequentially so that the untwisting rotary motion is converted into axial extension, as shown in Figure 4(b) [43,74]. A mathematical model quantifying the mechanism leads to an equation for the balance of torques and an expression for the constant velocity of the front.

VI. Discussion

Force generation by polymerization of single filaments is fairly well understood. However, our knowledge is much less complete concerning how forces are generated during assembly and disassembly of cytoskeletal fibers and networks. Beyond the issue of force generation is the question of how these forces organize themselves, using complementary and antagonistic actions of microtubules and actin, coupled with protein motors, during cell locomotion and division. Moreover, force generation is intimately tied in with the processes of regulation [75], signal transduction [76], adhesion [77], and many other aspects of cell dynamics. Mechanochemical process generally involve many kinds of proteins, and are too complicated to understand without a mathematical model to expose the assumptions and to frame the qualitative picture in quantitative terms. The field needs new models to stimulate new experiments.

REFERENCES

1. Inoue, Y., A. Hikikoshi Iwane, T. Miyai, E. Muto, and T. Yanagida, Motility of single one-headed kinesin molecules along microtubules. *Biophys. J.*, 2001, **81**:2838–2850.
2. Ishii, Y. and T. Yanagida, A new view concerning an actomyosin motor. *Cell. Mol. Life Sci. (CMLS)*, 2002, **59**:1767–1770.
3. Oster, G., Darwin's motors. *Nature*, 2002, **417**:25.
4. Bustamante, C., D. Keller, and G. Oster, The physics of molecular motors. *Acc. Chem. Res.*, 2001, **34**:412–420.
5. Wang, H. and G. Oster, Ratchets, power strokes, and molecular motors. *Appl. Phys. A*, 2001, **75**:315–323.
6. Oster, G. and H. Wang, Rotary protein motors. *Trends Cell Biol.*, 2003, **13**:114–121.
7. Mogilner, A., T. Elston, H.-Y. Wang, and G. Oster, Molecular motors: theory and examples, in *Computational Cell Biology*, C.P. Fall, E. Marland, J. Tyson, and J. Wagner, Eds. 2002, Springer: New York. pp. 321–380.
8. Peskin, C.S. and G.F. Oster, Force production by depolymerizing microtubules: load–velocity curves and run-pause statistics. *Biophys. J.*, 1995, **69**:2268–2276.
9. Mogilner, A. and G. Oster, Force generation by actin polymerization II: the elastic ratchet and tethered filaments. *Biophys. J.*, 2003, **84**:1591–1605.
10. Howard, J., *Mechanics of Motor Proteins and the Cytoskeleton*. 2001, Sunderland, MA: Sinauer.
11. Oster, G. and H. Wang, Reverse engineering a protein: the mechanochemistry of ATP synthase. *Biochem. Biophys. Acta*, 2000, **1458**:482–510.
12. Mogilner, A. and G. Oster, The physics of lamellipodial protrusion. *Euro. Biophys. J.*, 1996, **25**:47–53.
13. Peskin, C.S., G.M. Odell, and G. Oster, Cellular motions and thermal fluctuations: the Brownian ratchet. *Biophys. J.*, 1993, **65**:316–324.
14. Visscher, K., M.J. Schnitzer, and S.M. Block, Single kinesin molecules studied with a molecular force clamp. *Nature*, 1999, **400**:184–189.
15. Ruegg, C., C. Veigel, J.E. Molloy, S. Schmitz, J.C. Sparrow, and R.H. Fink, Molecular motors: force and movement generated by single myosin II molecules. *News Physiol. Sci.*, 2002, **17**:213–218.
16. Janmey, P.A., S. Hvidt, J. Kas, D. Lerche, A. Maggs, E. Sackmann, M. Schliwa, and T.P. Stossel, The mechanical properties of actin gels. Elastic modulus and filament motions. *J. Biol. Chem.*, 1994, **269**:32503–32513.
17. Mogilner, A. and G. Oster, The polymerization ratchet model explains the force-velocity relation for growing microtubules. *Eur. J. Biophys.*, 1999, **28**:235–242.
18. van Doorn, G.S., C. Tanase, B.M. Mulder, and M. Dogterom, On the stall force for growing microtubules. *Eur. Biophys. J.*, 2000, **29**:2–6.
19. Stukalin, E. and A. Kolomeisky, Simple growth models of rigid multifilament biopolymers. *J. Chem. Phys.* **121**:1097–1104.
20. Coue, M., V.A. Lombillo, and J.R. McIntosh, Microtubule depolymerization promotes particle and chromosome movement *in vitro*. *J. Cell Biol.*, 1991, **112**:1165–1175.
21. Lombillo, V.A., R.J. Stewart, and J.R. McIntosh, Minus-end-directed motion of kinesin-coated microspheres driven by microtubule depolymerization. *Nature*, 1995, **373**:161–164.
22. Hill, T.L., Theoretical problems related to the attachment of microtubules to kinetochores. *Proc. Natl Acad. Sci. USA*, 1985, **82**:4404–4408.

23. Mitchison, T.J., Microtubule dynamics and kinetochore function in mitosis. *Annu. Rev. Cell Biol.*, 1988, **4**:527–549.

24. Joglekar, A.P. and A.J. Hunt, A simple, mechanistic model for directional instability during mitotic chromosome movements. *Biophys. J.*, 2002, **83**:42–58.

25. Arnal, I., E. Karsenti, and A.A. Hyman, Structural transitions at microtubule ends correlate with their dynamic properties in Xenopus egg extracts. *J. Cell Biol.*, 2000, **149**:767–774.

26. Hyman, A.A., D. Chretien, I. Arnal, and R.H. Wade, Structural changes accompanying GTP hydrolysis in microtubules: information from a slowly hydrolyzable analogue guanylyl-(alpha,beta)-methylene-diphosphonate. *J. Cell Biol.*, 1995, **128**:117–125.

27. Inoue, S. and E.D. Salmon, Force generation by microtubule assembly/disassembly in mitosis and related movements. *Mol. Biol. Cell*, 1995, **6**:1619–1640.

28. Tao, Y.C. and C.S. Peskin, Simulating the role of microtubules in depolymerization-driven transport: a Monte Carlo approach. *Biophys. J.*, 1998, **75**:1529–1540.

29. Tilney, L.G. and D.A. Portnoy, Actin filaments and the growth, movement, and spread of the intracellular bacterial parasite, *Listeria monocytogenes*. *J. Cell Biol.*, 1989, **109**:1597–1608.

30. Cameron, L.A., T.M. Svitkina, D. Vignjevic, J.A. Theriot, and G.G. Borisy, Dendritic organization of actin comet tails. *Curr. Biol.*, 2001, **11**:130–135.

31. Cameron, L.A., M.J. Footer, A. van Oudenaarden, and J.A. Theriot, Motility of ActA protein-coated microspheres driven by actin polymerization. *Proc. Natl. Acad. Sci. USA*, 1999, **96**:4908–4913.

32. Loisel, T.P., R. Boujemaa, D. Pantaloni, and M.F. Carlier, Reconstitution of actin-based motility of *Listeria* and *Shigella* using pure proteins. *Nature*, 1999, **401**:613–616.

33. Gerbal, F., P. Chaikin, Y. Rabin, and J. Prost, On the theory of the *Listeria* propulsion. *Biophys. J.*, 2000, **79**:2259–2275.

34. Upadhyaya, A., J. Chabot, A. Andreeva, A. Samadani, and A.V. Oudenaarden, Probing polymerization forces by using actin-propelled lipid vesicles. *Proc. Natl Acad. Sci. USA*, 2003, **100**:4521–4526.

35. Giardini, P.A., D.A. Fletcher, and J.A. Theriot, Compression forces generated by actin comet tails on lipid vesicles. *Proc. Natl Acad. Sci. USA*, 2003, **100**:6493–6498.

36. Mogilner, A. and G. Oster, Cell motility driven by actin polymerization. *Biophys. J.*, 1996, **71**:3030–3045.

37. Noireaux, V., R.M. Golsteyn, E. Friederich, J. Prost, C. Antony, D. Louvard, and C. Sykes, Growing an actin gel on spherical surfaces. *Biophys. J.*, 2000, **78**:1643–1654.

38. Kuo, S.C. and J.L. McGrath, Steps and fluctuations of *Listeria monocytogenes* during actin-based motility. *Nature*, 2000, **407**:1026–1029.

39. McGrath, J.L., N.J. Eungdamrong, C.I. Fisher, F. Peng, L. Mahadevan, T.J. Mitchison, and S.C. Kuo, The force velocity relationship for the actin-based motility of *Listeria monocytogenes*. *Curr. Biol.*, 2003, **13**:329–332.

40. Wiesner, S., E. Helfer, D. Didry, G. Ducouret, F. Lafuma, M.F. Carlier, and D. Pantaloni, A biomimetic motility assay provides insight into the mechanism of actin-based motility. *J. Cell Biol.*, 2003, **160**:387–398.

41. Robbins, J.R. and J.A. Theriot, *Listeria monocytogenes* rotates around its long axis during actin-based motility. *Curr. Biol.*, 2003, **13**:R754–R756.

42. Mahadevan, L. and P. Matsudaira, Motility powered by supramolecular springs and ratchets. *Science*, 2000, **288**:95–100.

43. Mahadevan, L., J. Shin, G. Waller, K. Langsetmo, and P. Matsudaira, Stored elastic energy powers the 60um extension of the *Limulus polyphemus* sperm actin bundle. *J. Cell Biol.*, 2003, **162**:1183–1188.

44. Goldberg, M.B., Actin-based motility of intracellular microbial pathogens. *Microbiol. Mol. Biol. Rev.*, 2001, **65**:595–626.

45. Laurent, V., T.P. Loisel, B. Harbeck, A. Wehman, L. Grobe, B.M. Jockusch, J. Wehland, F.B. Gertler, and M.F. Carlier, Role of proteins of the Ena/VASP family in actin-based motility of *Listeria monocytogenes*. *J. Cell Biol.*, 1999, **144**:1245–1258.

46. Dickinson, R.B. and D.L. Purich, Clamped-filament elongation model for actin-based motors. *Biophys. J.*, 2002, **82**:605–617.

47. Vignjevic, D., D. Yarar, M.D. Welch, J. Peloquin, T. Svitkina, and G.G. Borisy, Formation of filopodia-like bundles *in vitro* from a dendritic network. *J. Cell Biol.*, 2003, **160**:951–962.

48. Sheetz, M.P., D.B. Wayne, and A.L. Pearlman, Extension of filopodia by motordependent actin assembly. *Cell Motil. Cytoskeleton*, 1992, **22**:160–169.

49. Gerbal, F., P. Chaikin, Y. Rabin, and J. Prost, An elastic analysis of *Listeria monocytogenes* propulsion. *Biophys. J.*, 2000, **79**:2259–2275.

50. Gerbal, F., V. Laurent, A. Ott, M.-F. Carlier, P. Chaikin, and J. Prost, Elastic forces propel *Listeria monocytogenes*. *Euro. Biophys. J.*, 2000, **29**:134–140.

51. Bernheim-Groswasser, A., S. Wiesner, R.M. Golsteyn, M.F. Carlier, and C. Sykes, The dynamics of actin-based motility depend on surface parameters. *Nature*, 2002, **417**:308–311.

52. Grebecki, A., Membrane and cytoskeleton flow in motile cells with emphasis on the contribution of free-living amoebae. *Int. Rev. Cytol.*, 1994, **148**:37–80.

53. Oster, G.F., On the crawling of cells. *J. Embryol. Exp. Morphol.*, 1984, **83** (Suppl.):329–364.

54. Mogilner, A. and G. Oster, Shrinking gels pull cells. *Science*, 2003, **302**:1340–1341.

55. Mogilner, A. and G. Oster, Polymer motors: pushing out the front and pulling out the back. *Curr. Biol.*, 2003, **13**:R721–R733.

56. Rotsch, C., K. Jacobson, J. Condeelis, and M. Radmacher, EGF-stimulated lamellipod extension in adenocarcinoma cells. *Ultramicroscopy*, 2001, **86**:97–106.

57. Wolgemuth, C., A. Mogilner, and G. Oster, The hydration dynamics of polyelectrolyte gels with applications to cell motility and drug delivery. *Eur. Biophys. J.*, 2003, **33**:146–158.

58. Bottino, D., A. Mogilner, T. Roberts, M. Stewart, and G. Oster, How nematode sperm crawl. *J. Cell Sci.*, 2002, **115**:367–384.

59. Italiano, J.E., Jr., M. Stewart, and T.M. Roberts, How the assembly dynamics of the nematode major sperm protein generate amoeboid cell motility. *Int. Rev. Cytol.*, 2001, **202**:1–34.

60. Roberts, T.M., E.D. Salmon, and M. Stewart, Hydrostatic pressure shows that lamellipodial motility in Ascaris sperm requires membrane-associated major sperm protein filament nucleation and elongation. *J. Cell Biol.*, 1998, **140**:367–375.

61. McBride, M., Bacterial gliding motility: mechanisms and mysteries. *Am. Soc. Microbiol. News*, 2000, **66**:203210.

62. Hoiczyk, E., Gliding motility in cyanobacteria: observations and possible explanations. *Arch. Micro.*, 2000, **174**:11–17.

63. Hoiczyk, E. and W. Baumeister, The junctional pore complex, a prokaryotic secretion organelle, is the molecular motor underlying gliding motility in cyanobacteria. *Curr. Biol.*, 1998, **8**:1161–1168.

64. Wolgemuth, C., E. Hoiczyk, D. Kaiser, and G. Oster, How myxobacteria glide. *Curr. Biol.*, 2002, **12**:369–377.

65. Trachtenberg, S., Mollicutes-Wall-less bacteria with internal cytoskeletons. *J. Struct. Biol.*, 1998, **124**:244–256.

66. Wolgemuth, C., O. Igoshin, and G. Oster, The motility of Mollicutes. *Biophys. J.*, 2003, **85**:828–842.

67. Trachtenberg, S., R. Gilad, and N. Geffen, The bacterial linear motor of *Spiroplasma melliferum* BC3: from single molecules to swimming cells. *Mol. Microbiol.*, 2003, **47**:671–697.

68. Trachtenberg, S. and R. Gilad, A bacterial linear motor: cellular and molecular organization of the contractile cytoskeleton of the helical bacterium *Spiroplasma melliferum* BC3. *Mol. Microbiol.*, 2001, **41**:827–848.

69. Miyata, M. and A. Uenoyama, Movement on the cell surface of the gliding bacterium, mycoplasma mobile, is limited to its head-like structure. *FEMS Microbiol. Lett.*, 2002, **215**:285–289.

70. Moriyama, Y., H. Okamoto, and H. Asai, Rubber-like elasticity and volume changes in the isolated spasmoneme of giant *Zoothamnium* sp. under Ca^{2+}-induced contraction. *Biophys. J.*, 1999, **76**:993–1000.

71. Mahadevan, L. and P. Matsudaira, Motility powered by supramolecular springs and ratchets. *Science*, 2000, **288**:95–99.

72. Tilney, L.G. and S. Inoue, Acrosomal reaction of Thyone sperm. II. The kinetics and possible mechanism of acrosomal process elongation. *J. Cell Biol.*, 1982, **93**:820–827.

73. Oster, G., A. Perelson, and L. Tilney, A mechanical model for acrosomal extension in Thyone. *J. Math. Biol.*, 1982, **15**:259–265.

74. Winder, S.J., Structural insights into actin-binding, branching and bundling proteins. *Curr. Opin. Cell Biol.*, 2003, **15**:14–22.

75. Pollard, T.D. and G.G. Borisy, Cellular motility driven by assembly and disassembly of actin filaments. *Cell*, 2003, **112**:453–465.

76. Bourne, H. and O. Weiner, A chemical compass. *Nature*, 2002, **419**:21.

77. Webb, D.J., J.T. Parsons, and A.F. Horwitz, Adhesion assembly, disassembly and turnover in migrating cells — over and over and over again. *Nat. Cell Biol.*, 2002, **4**:E97–E100.

Index

9 780367 392956